# CUMULATIVE SUBJECT INDEX

VOLUMES 1–75 (including Revised Series Volumes 1–31)

PART I: A–E

# HANDBOOK OF CLINICAL NEUROLOGY

*Editors*

PIERRE J. VINKEN GEORGE W. BRUYN

*Executive Editor*

KENNETH ELLISON DAVIS



VOLUME 76

ELSEVIER

AMSTERDAM • LONDON • NEW YORK • OXFORD • PARIS •
SHANNON • TOKYO

# CUMULATIVE SUBJECT INDEX

## VOLUMES 1–75 (including Revised Series Volumes 1–31)

PART I: A–E

*Editors*

PIERRE J. VINKEN GEORGE W. BRUYN

*In collaboration with*

WILLEKE VAN OCKENBURG

*REVISED SERIES 32*

ELSEVIER

AMSTERDAM • LONDON • NEW YORK • OXFORD • PARIS •
SHANNON • TOKYO

ELSEVIER SCIENCE B.V.
Sara Burgerhartstraat 25
P.O. Box 211, 1000 AE Amsterdam, The Netherlands



First edition 2002

Library of Congress Cataloging in Publication Data
A catalog record from the Library of Congress has been applied for.

ISBN: 0-444-50919-4
ISSN: 0072-9752

∞ The paper used in this publication meets the requirements of ANSI/NISO Z39.48-1992 (Permanence of Paper).
Printed in The Netherlands.

# Epilogue

*As youngish professors at the University of Leyden, we once lost ourselves in a vain and self-serving phantasy, in which we saw the roots of clinical neurology emerge in the school of our illustrious predecessor at Leyden, the eighteenth-century physician, Herman Boerhaave. During the period 1730–1735, he delivered a total of 206 lectures on nervous diseases. The texts are preserved in* "Praelectiones Publice Habitae Morbus Nervorum", *which was published in Leyden in 1761, more than twenty years after Boerhaave's death. They cover the field on which today's neurosciences are based and they constitute the beginning of an imposing series of modern neurological textbooks and treatises.*

*Further in our phantasy (and with Fielding Garrison acting, often literally, as our quoting guide), the Faculty of Medicine at Edinburgh, shortly after its somewhat humble beginnings, achieved vigorous and fruitful growth, thanks to the work of a number of Boerhaave's pupils, including Alexander Monro and, of course, Robert Whytt, perhaps the foremost neurologist of the eighteenth century. His* Essay on the Vital and Other Involuntary Motions in Animals, *published in 1751, is considered to rank in importance with Harvey's* De Motu Cordis *and Sherrington's* Integrative Action. *Together with his extensive contributions to physiology, Whytt published a number of works in the field of clinical neurology. His book on "nervous hypochondriac or hysterical disorders" (1765) was the first important treatise in English on nervous diseases since Thomas Willis coined the word 'neurology' in 1664.*

*Of the three Alexander Monros to have occupied the Chair of Anatomy at Edinburgh over a period of 126 years, the first, following his return there from Leyden, published a number of neuroanatomical studies. These were followed, later on, by his son's numerous neurological publications. Some of the most eminent names in the history of neurology graduated from Edinburgh; Charles Bell, for example, who was the first to distinguish between the sensory and motor functions of the roots of the spinal nerves. In 1830, he published a number of his addresses to the Royal Society in book form under the title* The Nervous System of the Human Body. *The work of Marshall Hall, an Edinburgh graduate, on clinical neurological subjects published between 1837 and 1852, was also much quoted by colleagues in the nineteenth century.*

*The Edinburgh Medical School had a great influence on the development of clinical medicine in the United States, an influence far in excess of that produced by any university in continental Europe. In the first half of the nineteenth century, a number of neurological works found their way from the United Kingdom to the shelves of the medical library of the University of Pennsylvania in Philadelphia, the first great centre of U.S. medicine. Works by, for example, Charles Bell and Marshall Hall and also by John Cooke of London, another former student at the University of Leyden. Cooke's magnum opus,* A Treatise on Nervous Diseases, *appeared in the early 1820s. A generation or so later, U.S. doctors would undoubtedly have consulted Russell Reynold's* Diagnosis of Diseases of the Brain, Spinal Cord, Nerves and their Appendages, *which was published in London in 1855.*

*In the fullness of time, the Americans developed what was to become a particularly rich tradition of published works on clinical neurology. The first such appeared in 1871. It was a textbook entitled* Treatise on Disease of the Nervous System, *by William Hammond. It passed through nine editions and was translated into French, Italian and Spanish. Some 30 years were to elapse, however, before the U.S. saw the appointment of the first Professor of Neurology. The professor in question was Charles Mills, whose most celebrated work,* The Nervous System and its Diseases, *was published in 1898. It was received with wide acclaim in the United States. In the Preface, Mills observed that, "although excellent manuals of moderate size have been written", the great work of William Gowers was the only extensive treatise on nervous diseases then available in the English language. This was a reference to* A Manual of Diseases of the Nervous System, *which had been published in two volumes in London ten years earlier.*

*Meanwhile, in Europe, the first real advances in neurology were grounded on the vast clinical experience of Moritz Romberg, a professor of pathology and director of the University Hospital in Berlin. He had worked in Vienna under Johann Frank, the great pioneer of public health. The latter had himself been a pupil of Anton de Haen, head of the Vienna Clinical School, who, in turn, had had a thorough grounding in medicine by no other than, of course, Herman Boerhaave. He was called to Vienna by Gerard van Swieten, personal physician to the Empress Maria Theresia, and undoubtedly the most famous of all of Boerhaave's pupils. One of his publications was a neurological commentary on the latter's work.*

*In 1832, Romberg's German translation of Bell's* The Nervous System of the Human Body *appeared. His own* Lehrbuch der Nervenkrankheiten des Menschen *was published between 1840 and 1846; it was the first systemic treatise in neurology and was a milestone in the development of clinical neurology. If that was a milestone, the book published in 1894 under the title* Lehrbuch der Nervenkrankheiten *by Hermann Oppenheim may be designated a landmark. This classic work passed through seven German editions. In 1911, it was translated into English and thereafter into Spanish, Italian and Russian.*

*Although German neurology at the turn of the twentieth century was dominated by Oppenheim, important contributions were also made by his pupil Max Lewandowsky, who systematised the accrued knowledge of the period in his monumental* Handbuch der Neurologie, *published between 1910 and 1914. This was supplanted, some twenty years later, by a seventeen-volume "Meisterwerk", the* Handbuch der Neurologie, *published between 1935 and 1937 under the outstanding editorship of Oswald Bumke and Otfrid Förster.*

*When we first contemplated the publication of a new and comprehensive handbook in 1964, some thirty years after Bumke and Foerster's first volume had appeared, auxiliary disciplines and methodologies, such as neuroanatomy, neurophysiology, neurochemistry, electroencephalography, neurosurgery and neuroradiology had become more or less independent sciences. In the space of one generation, clinical neurology and its allied disciplines had advanced so rapidly that even those of us who were working in large neurological centres were hard put to keep up with the increases in knowledge.*

*After the conclusion of World War II, it was clear that the* "lingua franca" *of science had changed irrevocably from German to English. And although, as undergraduates in the 1950s, we ourselves had consulted, well nigh daily, the volumes of Bumke and Foerster, it was clear that any successor of their work would have to be in English: thus,* Handbuch *would have to become* Handbook. *This was not quite so easy as it might sound; and we had our critics. One eminent British neurologist, MacDonald Cirtchley, felt that the very word "handbook" was wanting in precision, since no human hand could possibly manage to hold 30 or 40 volumes. That is quite logical, but it obtained too for the German term* Handbuch, *which had already existed for centuries. And with two eminent German precursors bearing this title, we saw no objection to our declaring the English word "handbook" to be applicable to the multi-volume work that we had in mind. Meanwhile, no one could accuse us of a lack of respect for the English language, because in our daily sessions in the University library, we could see on a shelf not far from the German* Handbücher, *the* American Handbook of Psychiatry, *published under American editorship, whose three hefty volumes could scarcely be held in one human hand. Nevertheless, we sought unsuccessfully to find the word "handbook" in the* Encyclopaedia Britannica, *although the German "Handbuch" was covered.*

*It will be clear to the reader by this point that the two historic routes in the development of neurology started with Boerhaave in Leyden, the first running via Vienna to Germany and Russia and the other via Edinburgh to the United States, had, for one moment in time in the late 1960s, with our* Handbook of Clinical Neurology, *returned once again to Leyden.*

*For the sake of convenience, we have not considered here another of the roots of neurology which had already been formed in the 1770s in Edinburgh by the "neurological" lectures of William Cullen, subsequently published in his work* Of Neuroses and Nervous Disorders, *nor have we referred to John Cooke, physician at the London Hospital, who, in the early 1820s, published his comprehensive* Treatise on Nervous Disorders, *which contained the first historical survey of neurological thought from classical times to the turn of the nineteenth century. And we have also not considered neurology in France, where the third historic cornerstone underpinning current neurology was introduced, entirely autonomously so it seems, by Guillaume Duchenne and Jean Marie Charcot, alongside the English- and German-speaking ones.*

*Thus, there is also neurology whose provenance cannot be convincingly ascribed to Leyden. But we did warn the reader: this epilogue was merely phantasy, standing on the shoulders of giants. Yet it is not too far removed from reality!*

January 2002

Pierre J. Vinken
George W. Bruyn
Kenneth Ellison Davis

REFERENCES

McHENRY, LAWRENCE C.: Garrison's History of Neurology. Revised and enlarged with a bibliography of classical, original and standard works in neurology. Springfield, 1969.

SCHULTE, B.P.M.: Hermanni Boerhaave Praelectiones de Morbus Nervorum 1730–1735. Leyden, 1959.

*Handbook of Clinical Neurology, Vol. 76 (32):* Cumulative Subject Index, Part I: A–E


# Cumulative Subject Index to Volumes 1–75 (including Revised Series 1–31) Part I: A–E

*In collaboration with* W. van Ockenburg


A-alphalipoproteinemia
apolipoprotein AI, 51/396
apolipoprotein AII, 51/396
axonal degeneration, 51/397
cholesterol, 51/396
cholesterol ester, 51/396
corneal infiltration, 51/396
cranial neuropathy, 51/397
hemorrhage, 51/396
high density lipoprotein, 51/396
isoprotein abnormality, 51/396
metabolic polyneuropathy, 51/396
multifocal relapsing neuropathy, 51/397
nerve biopsy, 51/397
neuropathy, 51/396
pain, 51/397
Remak cell vacuole, 51/397
Schwann cell vacuole, 51/397
sensorimotor polyneuropathy, 51/397
syringomyelia like syndrome, 51/397
thromboplastin time, 51/396
yellow orange tonsil, 51/396
A band
muscle fiber, 62/26
normal muscle, 40/63
ultrastructure, 40/212
A-δ-fiber decrease
congenital pain insensitivity, 51/564
pain perception, 1/117-120
Aarskog-Scott syndrome, *see* Aarskog syndrome
Aarskog syndrome
cranial nerve defect, 30/100
dermatoglyphics, 43/313
growth retardation, 43/313
hypertelorism, 30/249, 43/313
mental deficiency, 43/314
ophthalmoplegia, 43/314
psychomotor retardation, 43/314
ptosis, 43/313
sex organ malformation, 30/249
shawl scrotum, 43/313
spinal cord compression, 43/314
swayback nose, 43/313
vertebral abnormality, 43/314
Abacterial encephalitis, *see* Viral encephalitis
Abadie sign
tabes dorsalis, 52/280
terminology, 1/97
Abdomen
wrinkled, *see* Wrinkled abdomen
Abdominal colic
hereditary sensory and autonomic neuropathy type III, 63/147
lead intoxication, 64/435
porphyria, 63/150
Abdominal injury
spinal injury, 26/198, 226
Abdominal migraine
*see also* Growing pain
cholecystectomy, 48/141, 159
concept, 48/158-160
Abdominal muscle absence
prune belly syndrome, 43/466
Abdominal muscle deficiency triad syndrome, *see*

Prune belly syndrome
Abdominal muscle hypoplasia
prune belly syndrome, 43/466, 50/515
Abdominal pain
acute, *see* Acute abdominal pain
Balaenoptera borealis intoxication, 37/94
couvade syndrome, 48/357
dissecting aorta aneurysm, 63/45
headache, 48/159
hereditary acute porphyria, 51/391
hyperlipoproteinemia type V, 10/277
hysteria, 48/357
ichthyohaemotix fish intoxication, 37/87
mackerel intoxication, 37/87
methanol intoxication, 36/353, 64/98
migraine, 48/3, 8, 38, 159
Minkowitz disease, 10/552
neophocaena phocaenoides intoxication, 37/95
ostrea gigas intoxication, 37/64
paralytic shellfish poisoning, 37/33
pellagra, 28/71
pheochromocytoma, 39/498
porpoise intoxication, 37/95
primary amyloidosis, 51/415
procainamide intoxication, 37/452
psychogenesis, 48/357
puffer fish intoxication, 37/80
subacute myelo-optic neuropathy, 51/296
urticaria pigmentosa, 14/790
venerupis semidecussata intoxication, 37/64
vincristine polyneuropathy, 51/299
vitamin A intoxication, 37/96
Abdominal reflex
diagnostic value, 1/245
instrumentation, 1/252
pelvic, *see* Pelvic abdominal reflex
Abdominal seizure
childhood, 15/141
Abducent nerve
clinical examination, 57/136
diabetes mellitus, 27/117
embryology, 30/396
Gradenigo syndrome, 48/302
head injury, 24/146
lesion localization, 2/310-319
neurinoma, 68/537, 539
nuclear lesion, 2/315, 24/78
pontine infarction, 12/16
root lesion, 2/314
subarachnoid hemorrhage, 55/18
superior cerebellar artery syndrome, 12/20
topographical anatomy, 2/312, 24/67
Abducent nerve abnormality
*see also* Congenital ptosis, Marcus Gunn phenomenon *and* Stilling-Türk-Duane syndrome
hypoplasia, 30/401
Marcus Gunn phenomenon, 30/403
Abducent nerve agenesis
*see also* Möbius syndrome
cranial nerve agenesis, 50/212
Möbius syndrome, 50/215
nuclear aplasia, 50/212
trigeminal nerve agenesis, 50/214
Abducent nerve injury
*see also* Ocular imbalance
anatomy, 24/67
bilateral, 24/68
brain injury, 24/67, 80
cavernous sinus thrombosis, 24/86
cerebellar tumor, 17/711
cervicomedullary injury, 24/146, 169, 171
fascicular lesion, 24/78
head injury, 24/68
incidence, 24/67, 77
intracranial hypertension, 24/69, 86
intracranial infection, 24/86
mechanism, 24/67, 84
nuclear lesion, 2/135, 24/78
orbital injury, 24/86
physiology, 24/67
prevalence, 2/310
prognosis, 24/69
root lesion, 2/314
sign, 24/67, 84
skull base fracture, 24/68, 82, 84
topical diagnosis, 24/80
treatment, 24/69
Abducent nerve paralysis
aqueduct stenosis, 50/312
Argentinian hemorrhagic fever, 56/366
benign intracranial hypertension, 42/740
brain lacunar infarction, 54/244
carotid dissecting aneurysm, 54/272
cavernous sinus thrombosis, 52/174
cerebellopontine angle syndrome, 2/318
Cestan-Chenais syndrome, 2/318
congenital, 42/293
Gradenigo syndrome, 2/91, 313, 16/210
Gruber petrosphenoid ligament, 2/91
hydrocephalus, 50/291
inferior petrosal sinus thrombosis, 54/398
Klippel-Feil syndrome, 42/37
Lyme disease, 51/204
mental deficiency, 43/299
Möbius syndrome, 42/324

multiple myeloma, 63/393
neuroborreliosis, 51/204
petrobasilar suture syndrome, 2/313
pure motor hemiplegia, 54/244
Stilling-Türk-Duane syndrome, 42/306
superior orbital fissure syndrome, 33/179
temporal arteritis, 48/313
traumatic carotid cavernous sinus fistula, 57/346, 350, 362
Abductor paralysis
laryngeal, *see* Laryngeal abductor paralysis
Abercrombie, J.
history, 1/6
neurologist, 1/6
Aberfeld syndrome, *see* Myotonic myopathy
Abetalipoproteinemia, *see* Bassen-Kornzweig syndrome
ABH secretor gene
myotonic dystrophy, 40/541
myotonic dystrophy linkage, 40/522
Abiotrophia retinae pigmentosa, *see* Primary pigmentary retinal degeneration *and* Secondary pigmentary retinal degeneration
Abiotrophic external ophthalmoplegia, *see* Progressive external ophthalmoplegia
Abiotrophy
cerebellar disease concept, 21/405
nosology, 21/419
pathogenesis, 21/436
primary pigmentary retinal degeneration, 13/194
Ablation
brain cortex, 6/852, 869
cerebellar, *see* Cerebellar ablation
cerebellar anterior lobe, 2/401
lobulus ansoparamedianus, 2/402
lobulus simplex, 2/402
opisthotonos, 2/402
pyramis, 2/402
sympathetic nervous system, *see* Sympathetic nervous system ablation
Abnormal acetylcholine-acetylcholine receptor interaction syndrome
congenital myasthenic syndrome, 62/437
EMG, 62/437
neuromuscular junction disease, 62/392
Abnormal karyotype
holoprosencephaly, 50/240
Abnormal myomuscular junction myopathy
congenital myopathy, 62/332
core like lesion, 62/353
facial weakness, 62/334
symptom, 62/353
Abnormal personality pattern
congenital pain insensitivity, 51/564
Abnormal visual clarity
occipital lobe tumor, 17/331
ABO incompatibility
kernicterus, 42/583
Abortion
anencephaly, 30/190
Brucella canis, 52/582
Edwards syndrome, 50/542
myotonic dystrophy, 62/228
spina bifida, 30/148
spontaneous, *see* Spontaneous abortion
trisomy 1, 50/541
trisomy 2, 50/542
trisomy 7, 50/542
trisomy 16, 50/542
Abortive migraine
migraine dissociée, 48/6, 157
Abscess
actinomycosis, 35/383
brain, *see* Brain abscess
brain stem, *see* Brain stem abscess
brain tumor, 67/218
cerebellar, *see* Cerebellar abscess
epidural, *see* Epidural empyema
extradural, *see* Epidural empyema
headache, 48/6
infectious striatopallidodentate calcification, 49/417
intracerebral, *see* Brain abscess
intracranial, *see* Intracranial abscess
intrasellar, *see* Intrasellar abscess
local anesthetic agent, 65/421
multiple brain, *see* Multiple brain abscess
nasal septum, 5/218
nocardial brain, *see* Nocardial brain abscess
opiate intoxication, 37/393
orbital, *see* Orbital abscess
paraspinal, *see* Paraspinal abscess
pituitary, *see* Pituitary abscess
posttraumatic brain, *see* Posttraumatic brain abscess
spinal cord, *see* Spinal cord abscess
spinal cord injury, 25/61
spinal epidural, *see* Spinal epidural empyema
spinal subdural, *see* Spinal subdural empyema
subarachnoid, *see* Subarachnoid abscess
subdural, *see* Subdural empyema
supratentorial brain, *see* Supratentorial brain abscess
temporal lobe, *see* Temporal lobe abscess
Absence
*see also* Petit mal epilepsy

abnormal theta wave, 42/667
akinetic, 15/138
astatic, 15/138
atypical, 15/134, 73/144
bowel motility, *see* Intestinal peristalsis absence
brain blood flow, 57/460
brain hemisphere, 42/12
centrencephalic EEG, 42/667
cerebellum, 50/83
claustrum, *see* Claustrum absence
definition, 73/143
drop attack, 15/138
external carotid artery, 12/291
falx cerebri, 42/18, 39
frontal sinus, 43/332
hereditary myoclonus ataxia syndrome, 21/511
history, 73/144
internal carotid artery, 30/664
lacrimal gland, 42/573
Lennox-Gastaut syndrome, 73/215
Lissauer tract, *see* Lissauer tract absence
muscle, *see* Muscle absence
myoclonic, 15/135
occipital delta rhythm, 42/672
organotin intoxication, 64/142
patella, *see* Patella absence
philtrum, *see* Philtrum absence
Probst bundle, 42/10
pyramidal tract, *see* Pyramidal tract absence
Rosenthal vein, 30/422
status epilepticus, 73/323
sweating, *see* Anhidrosis
tomacula, *see* Tomacula absence
triad, 15/142
tuberous sclerosis, 50/372
typical, 15/130, 132
zygomatic arch, *see* Zygomatic arch absence
Absence epilepsy
childhood, *see* Childhood absence epilepsy
juvenile, *see* Juvenile absence epilepsy
Absence seizure
juvenile myoclonic epilepsy, 73/162
Absence status
*see also* Petit mal status
EEG, 15/595
epilepsy, 73/474
status epilepticus, 73/474
twilight state, 15/595
Absinthol
neuropathy, 7/521
Absorption spectrometry
atomic, *see* Atomic absorption spectrometry
electrothermal atomic, *see* Electrothermal atomic absorption spectrometry
Abstinence syndrome, *see* Withdrawal syndrome
Abstract reasoning
Parkinson dementia, 46/375
Abt-Letterer-Siwe disease
*see also* Histiocytosis X
adenopathy, 42/441
anemia, 42/441
cerebral histiocytosis, 39/531
dermal manifestation, 42/441
encephalitis, 42/441
fever, 38/96, 42/441
hemorrhage, 42/441
hepatomegaly, 38/97, 42/441
histiocytosis X, 38/96
jaundice, 42/441
Langerhans cell histiocytosis, 70/56
leukocytosis, 42/441
otitis media, 38/96
space occupying lesion, 16/231
splenomegaly, 38/97, 42/441
teeth loss, 42/441
thrombocytopenia, 42/441
Abulia
anterior cerebral artery syndrome, 53/349
basofrontal syndrome, 53/349
brain lacunar infarction, 53/395
definition, 2/738
frontal lobe lesion, 2/741
frontal lobe syndrome, 53/349, 55/138
thalamic infarction, 53/395
vertebrobasilar system syndrome, 53/395
Acalculia
agnosia, 4/17
agnostic component, 3/30
agraphia, 45/480
alexia, 45/437, 480
anarithmetia, 45/474, 480
anatomic localization, 4/187-192
aphasia, 4/96, 45/476, 479
asemasia, 4/185
asymbolic, *see* Asymbolic acalculia
asyntactic, *see* Asyntactic acalculia
brain commissurotomy, 4/281
calculation, 4/182
cerebral dominance, 4/293, 465
cipher agraphia, 4/188, 191
cipher alexia, 4/191
constructive, *see* Constructive acalculia
counting, 4/191
definition, 4/181, 45/473
digit agraphia, 45/474
digit alexia, 45/474

frontal, *see* Frontal acalculia
frontal lobe lesion, 4/189
Gerstmann syndrome, 1/107, 2/686, 4/189, 222, 45/70, 474
Gestalt factor, 4/190
Henschen definition, 2/600
historic background, 45/473
historic survey, 4/184
information processing, 45/474-476
motor aphasia, 45/476
number notation system, 4/181
occipital, *see* Occipital acalculia
occipital lobe lesion, 4/188
parietal, *see* Parietal acalculia
parietal lobe lesion, 4/189
perseveration, 4/187
primary, *see* Primary acalculia
pure motor hemiplegia, 54/243
recent contribution, 45/476-479
secondary, *see* Secondary acalculia
semiotic system, 4/181
spatial, *see* Spatial acalculia
spatial orientation, 3/219
temporal lobe lesion, 4/189
Wernicke aphasia, 4/98, 45/476
Acanthamoeba
acquired immune deficiency syndrome, 71/304
culture, 52/318
granulomatous amebic encephalitis, 52/314, 316
isolation, 52/318
meningoencephalitis, 71/304
primary amebic meningoencephalitis, 35/25, 52/314
trophozoite, 52/311
ultrastructure, 52/311
Acanthamoeba castellani
Amoebae, 35/28, 32
granulomatous amebic encephalitis, 52/317
microscopy, 52/310
Acanthamoeba culbertsoni
acute amebic meningoencephalitis, 52/313
granulomatous amebic encephalitis, 52/317
Acanthamoeba Hartmannella
amebiasis, 35/25, 51
neuroamebiasis, 52/309
Acanthamoeba polyphaga
granulomatous amebic encephalitis, 52/317
Acanthamoeba rhysodes
acute amebic meningoencephalitis, 52/313
granulomatous amebic encephalitis, 52/317
Acanthaster planci intoxication
sea star, 37/66
Acanthocyte
anorexia nervosa, 49/332
Bassen-Kornzweig syndrome, 10/279, 549, 49/327
chorea-acanthocytosis, 49/327, 329
Critchley syndrome, 13/427
Eales disease, 49/332
Hallervorden-Spatz syndrome, 49/332
hereditary vitreoretinal degeneration, 49/332
hypobetalipoproteinemia, 49/327, 332
hypothyroidism, 49/332
Kearns-Sayre-Daroff-Shy syndrome, 49/332
liver cirrhosis, 49/332
McLeod phenotype, 49/330, 62/127
MELAS syndrome, 49/332
panhypopituitarism, 49/332
postsplenectomy state, 49/332
psoriasis, 49/332
pyruvate kinase deficiency, 49/332
reticulosarcoma, 49/332
Wolman disease, 49/332
Acanthocytosis
*see also* Neuroacanthocytosis
acquired type, 63/280
acute leukemia, 63/290
advanced malnutrition, 63/271
alcoholic cirrhosis, 63/289
anorexia nervosa, 63/271, 290
Bassen-Kornzweig syndrome, 13/427
burr cell, 63/289
carcinoid tumor, 63/290
celiac disease, 13/427
chromosome 19, 63/289
dietary hypobetalipoproteinemia, 63/280
Eales vitreous hemorrhage disease, 63/290
gait disturbance, 13/417
Hallervorden-Spatz syndrome, 63/271, 290
HARP syndrome, 63/290
hemoneurological disease, 63/271
hypothyroidism, 63/271, 290
idiopathic hemochromatosis, 63/290
In(Lu) red cell phenotype, 63/289
liver disease, 63/271
McLeod phenotype, 49/332, 63/271, 288
mitochondrial encephalomyopathy, 63/271
mitochondrial myopathy, 63/290
myxedema, 63/271
neuroacanthocytosis, 60/674
normotriglyceridemic abetalipoproteinemia, 63/271
panhypopituitarism, 63/271, 290
postsplenectomy, 63/271
red cell morphology, 63/271
red cell physiology, 63/271

reticulosarcoma, 63/290
spur cell, 63/289
spur cell anemia, 63/271
viper bite, 63/289
Acanthoma adenoides cysticum, *see* Multiple trichoepithelioma
Acanthosis nigricans
Alström-Hallgren syndrome, 42/391
associated disorder, 14/782
Bardet-Biedl syndrome, 13/294, 390, 462
Berardinelli-Seip syndrome, 42/591
hypertelorism, 30/248
mental deficiency, 30/248
Miescher syndrome type II, 14/119
neuropathy, 22/524
progressive sensory demyelinating neuropathy, 21/82
skin malformation, 30/248
vitamin PP intoxication, 65/573
Acanthuridae intoxication
family, 37/78
venomous spine, 37/78
Acanthurus intoxication
family, 37/78
venomous spine, 37/78
Acardiac monster
holocardiac twins, 30/51
various forms, 30/117
Acatalasemia, *see* Acatalasia
Acatalasia
peroxisomal disease, 51/383
peroxisomal disorder, 66/510
Acceleration injury, *see* Whiplash injury
Accessory facial anastomosis
facial nerve injury, 24/113
Accessory meningeal artery
anastomosis, 12/309
Accessory middle cerebral artery
radioanatomy, 11/84
Accessory nerve
abnormality, 30/410
congenital larynx stridor, 30/410
embryology, 30/398-400
injury, 24/179
Jackson syndrome, 1/189
lesion site symptom, 2/20
neurinoma, 68/541
petrobasilar suture syndrome, 2/313
spinal, *see* Spinal accessory nerve
topographical anatomy, 2/17
vocal cord paralysis, 30/411
Accessory nerve agenesis
cranial nerve agenesis, 50/219
Accessory neurenteric canal, *see* Neurenteric cyst
Accessory optic bundle
Ganser dorsal commissure, 2/529
Gudden-Meynert ventral commissure, 2/529
optic chiasm, 2/529
Accessory spleen
holoprosencephaly, 50/237
Accidental injury
glucagon, 74/324
hormone, 74/324
neuroendocrine response, 74/322
subdural effusion, 24/331
vasopressin, 74/324
Accommodation
characteristics, 74/422
demyelination, 47/37
disorder, 74/424
examination, 74/422
Krimsky-Prine rule, 74/423
ocular, *see* Ocular accommodation
physiology, 74/422
Accommodation disorder
cerebrum, 2/667
chloroquine intoxication, 65/484
convergence, 2/667
occipital lobe lesion, 2/667
postconcussional syndrome, 57/424
ACE inhibitor, *see* Dipeptidyl carboxypeptidase I inhibitor
Acebutolol
beta adrenergic receptor blocking agent, 65/434
migraine, 48/180
Acephalus spurius, *see* Anencephaly
Acephaly
*see also* Anencephaly
anencephaly, 30/174
semantics, 50/71
Acer pseudoplatanus
hypoglycin, 65/81
Acetaldehyde
alcohol intoxication, 64/109, 112
disulfiram polyneuropathy, 51/309
Acetaldehyde dehydrogenase, *see* Aldehyde dehydrogenase
Acetaldehyde intoxication
mushroom toxin, 65/39
Acetaldehyde production
Minamata disease, 64/414
Acetalphosphatide, *see* Plasmalogen
Acetaminophen, *see* Paracetamol
Acetazolamide
antiepileptic agent, 65/497, 73/350
brain edema, 63/417

brain infarction, 53/422
central sleep apnea, 63/463
convulsion, 37/418
epilepsy, 15/648, 73/350
hereditary periodic ataxia, 60/442
hydrocephalus, 50/296
hyperkalemic periodic paralysis, 62/470, 472
normokalemic periodic paralysis, 41/163, 43/172
periodic paralysis, 40/285, 556, 41/156, 161, 167
serum level, 15/691
side effect, 15/706
Acetazolamide intoxication
ataxia, 37/202
drowsiness, 37/202
epilepsy, 37/418
gastrointestinal disorder, 37/202
paresthesia, 37/202
Acetic acid
headache, 48/174
migraine, 48/174
Acetic thiokinase, *see* Acetyl coenzyme A synthetase
Acetoacetyl-coenzyme A carboxylase
pyruvate carboxylase deficiency, 66/417
Acetoacetyl-coenzyme A thiolase, *see* Acetyl-coenzyme A acetyltransferase
Acetonylacetone
chemical formula, 64/82
distal axonopathy, 64/12
neurotoxicity, 51/276
neurotoxin, 51/68, 276, 64/12, 81, 84
organic solvent, 64/82
sensorimotor polyneuropathy, 64/86
toxic encephalopathy, 64/12
toxic polyneuropathy, 51/68, 64/81, 86
Acetonylacetone intoxication
3-acetyl-2,5-hexanedione, 64/86
axon constriction, 64/90
axon diameter, 64/88
axonal flow, 64/89
biological half-time, 64/83
brain stem auditory evoked potential, 64/87
chemistry, 64/85
clinical symptom, 64/86
cranial neuropathy, 64/87
demyelination, 64/89
3,4-diethylhexanedione, 64/86
1,4-diketone, 64/85
3,4-dimethyl-2,5-hexanedione, 64/85
elimination time, 64/83
EMG, 64/87
giant axonal swelling, 64/85, 88
glue sniffing, 64/82
headache, 64/87
history, 64/81
Huffer neuropathy, 64/82
3,3'-iminodipropionitrile, 64/85
insomnia, 64/87
KSP repeats, 64/86
lysyl-ε-amino group, 64/84
mamillary body, 64/89
microtubule, 64/90
microtubule associated protein, 64/90
muscle spasm, 64/87
muscular atrophy, 64/86
myalgia, 64/87
nerve conduction velocity, 64/87
neurofilament, 64/89
neurofilament density, 64/90
neurofilament protein, 64/89
neuropathology, 64/88
optic pathway, 64/89
organic solvent, 64/81
paranodal myelin, 64/89
paresthesia, 64/86
pathogenesis, 64/85
pathomechanism, 64/86
polyneuropathy, 64/81
pyrrole oxidation, 64/86
Ranvier node, 64/89
recovery, 64/87
somatosensory evoked potential, 64/87
spastic paraparesis, 64/87
spasticity, 64/87
symptom, 64/12
threshold limit value, 64/84
treatment, 64/91
wallerian degeneration, 64/89
weakness, 64/86
Acetriozate, *see* Acetrizoic acid
Acetrizoic acid
iatrogenic neurological disease, 63/44
Acetylaminofluorene
brain tumor carcinogenesis, 17/10
*N*-Acetylaspartic acid
aspartoacylase, 66/663
brain, 9/16
spongy cerebral degeneration, 66/662
*N*-Acetylaspartic aciduria
organic acid metabolism, 66/639, 641
Acetylation
slow, *see* Slow acetylation
Acetylcholine
acetylcholinesterase, 40/129, 380
affective disorder, 29/451
akinesia, 45/167

Alzheimer disease, 46/266
antagonism, 40/143
autonomic nervous system, 74/92, 137, 75/85
basal ganglion, 49/33
biochemistry, 10/258
brain edema, 23/151
brain metabolism, 57/79
brain microcirculation, 53/79
CNS, 29/435-451
drug effect, 29/442-446
epilepsy, 15/64, 29/451, 72/89
experimental myasthenia gravis, 41/120
fever, 74/449
heart, 74/150
hemicholinium, 41/105, 111
hereditary sensory and autonomic neuropathy type III, 51/565, 60/35
Huntington chorea, 29/447-450
intestinal pseudo-obstruction, 51/493
malnutrition, 29/12
mesolimbic noradrenalin system, 49/51
migraine, 5/39
motor end plate, 7/86, 40/129
motor unit, 1/60
multiple neuronal system degeneration, 75/192
myasthenia gravis, 41/97, 105, 113, 126
myasthenic syndrome, 41/356
neglect syndrome, 45/155
neuromuscular transmission, 40/129
neurotransmitter, 74/92
Parkinson dementia, 49/178
Parkinson disease, 29/446, 42/246
pharmacology, 29/442-451, 74/142, 150
progressive dysautonomia, 59/147
regional distribution, 29/437-442
restoration of higher cortical function, 3/370
schizophrenia, 29/451
senile plaque, 46/268
supersensitivity, 40/130
synapse, 1/48, 53, 56
synaptic function, 29/435-437
temporal lobe, 46/266
vesicle, 40/129

Acetylcholine intoxication
organophosphate intoxication, 64/160

Acetylcholine receptor
α-bungarotoxin, 40/13, 41/97, 106, 224
α-bungarotoxin peroxidase stain, 40/5
denervation, 40/13, 142
Duchenne muscular dystrophy, 40/372, 384
experimental autoimmune myasthenia gravis, 41/97, 115-120
extrajunctional, 40/13, 142
Huntington chorea, 49/259
motor end plate, 40/5, 38
muscle reinnervation, 40/13, 142
muscle tissue culture, 40/186, 62/97
myasthenia gravis, 40/129, 41/106, 114, 125, 62/391
myasthenic syndrome, 62/391
myotubule, 41/114
neuromuscular transmission, 40/129
normal, 41/106
paraneoplastic syndrome, 69/338
Pick disease, 46/242

Acetylcholine receptor antibody, 41/114-136
experimental autoimmune myasthenia gravis, 41/115
myasthenia gravis, 41/114, 120, 129, 368, 62/408
neonatal, 41/120
plasmapheresis, 41/136

Acetylcholine receptor epitope
myasthenia gravis, 62/407

ε-Acetylcholine receptor subunit mutation syndrome
congenital, *see* Congenital ε-acetylcholine receptor subunit mutation syndrome

Acetylcholine transferase
regional distribution, 29/435-451

Acetylcholine transmission
botulism, 51/189
neuromuscular transmission, 7/85

Acetylcholinesterase
acetylcholine, 40/129, 380
Alzheimer disease, 46/266
anencephaly, 50/86
basal ganglion, 49/28
basal ganglion tier II, 49/21
biochemistry, 10/259
drug, 41/131
edrophonium test, 41/126
hereditary sensory and autonomic neuropathy type III, 21/114, 60/31
Huntington chorea, 49/257
motor end plate, 40/129, 380
muscle tissue culture, 62/97
myasthenia gravis, 41/107
organophosphate intoxication, 64/151
phenytoin, 40/558
Ranvier node, 9/27, 34
senile plaque, 46/255, 269
synaptic vesicle, 41/105

Acetylcholinesterase deficiency
decremental EMG, 41/130

Acetyl-coenzyme A
arsenic polyneuropathy, 51/267
pyruvate carboxylase deficiency, 42/586

pyruvate metabolism, 66/413
valproic acid induced hepatic necrosis, 65/132
Acetyl-coenzyme A acetyltransferase
acetyl-coenzyme A acyltransferase deficiency, 66/412
Acetyl-coenzyme A acetyltransferase deficiency
adrenocortical cell, 66/17
Acetyl-coenzyme A acyltransferase
inner mitochondrial membrane system, 66/412
mitochondrial matrix system, 66/412
Acetyl-coenzyme A acyltransferase deficiency
acetyl-coenzyme A acetyltransferase, 66/412
Acetyl-coenzyme A α-glucosaminide N-acetyltransferase
glycosaminoglycanosis type III, 66/65
glycosaminoglycanosis type IIIC, 66/282, 302
GNAT gene, 66/282
Acetyl coenzyme A:α-glycosaminedisulfate-*N*-acetyltransferase
glycosaminoglycanosis type III, 42/456
Acetyl coenzyme A synthetase
wallerian degeneration, 7/215
*N*-Acetylgalactosamine-4-sulfatase
ARSB gene, 66/282
biochemistry, 10/349
chemistry, 10/48
glycosaminoglycanosis type VI, 66/282, 309
metachromatic leukodystrophy, 10/28, 568, 29/359
*N*-Acetylgalactosamine-6-sulfatase
galactose-6-sulfurylase, 66/64
GALNS gene, 66/282
glycosaminoglycanosis type IV, 42/457, 66/303
glycosaminoglycanosis type IVA, 66/282
*N*-Acetylgalactosamine-6-sulfatase deficiency, *see* Glycosaminoglycanosis type IVA
α-*N*-Acetylgalactosaminidase
α-*N*-acetylgalactosaminidase deficiency, 66/62
gene, 66/347
genetics, 66/347
molecular defect, 66/348
type, 66/348
α-*N*-Acetylgalactosaminidase deficiency
α-*N*-acetylgalactosaminidase, 66/62
diagnosis, 66/345
enzyme defect, 66/345
epilepsy, 72/222
fibroblast, 66/346
genetics, 66/342
glycoprotein degradation, 66/62
metabolic defect, 66/345
neuroaxonal dystrophy, 66/16
pathology, 66/342
substrate, 66/346
type, 66/339
type I, 66/340
type II, 66/341
*N*-Acetyl-β-D-galactosaminidase
GM1 gangliosidosis, 10/559, 561
*N*-Acetylglucosamine kinase
metachromatic leukodystrophy, 10/48
*N*-Acetylglucosamine-6-sulfatase
glycosaminoglycanosis type III, 66/65
glycosaminoglycanosis type IIID, 66/64, 282, 302
G6S gene, 66/282
*N*-Acetylglucosamine sulfate sulfatase
glycosaminoglycanosis type VIII, 42/460
α-*N*-Acetylglucosaminidase
glycosaminoglycanosis type III, 66/65
glycosaminoglycanosis type IIIB, 66/282, 302
NAGLU gene, 66/282
*N*-Acetyl-α-D-glucosaminidase
glycosaminoglycanosis type III, 42/456
*N*-Acetyl-β-D-glucosaminidase
GM1 gangliosidosis, 10/477, 481, 561
*N*4-(β-*N*-Acetylglucosaminyl)asparaginase
deficiency, 66/63
degradation, 66/61
*N*-Acetylglucosaminylphosphotransferase
mannose residue, 66/83
*N*-Acetylglucosaminyltransferase
GM1 gangliosidosis, 66/83
*N*-Acetylglucosaminyltransferase II
carbohydrate deficient glycoprotein syndrome, 66/83
β-*N*-Acetylhexosaminidase
GM2 gangliosidosis, 66/257
β-*N*-Acetylhexosaminidase A
degradation, 66/59
gene localization, 42/434
GM2 gangliosidosis, 46/56, 66/58, 60, 64
GM2 gangliosidosis type I, 42/433
GM2 gangliosidosis type II, 27/156, 29/365, 42/435, 66/60, 64
GM2 gangliosidosis type III, 42/437
Hallervorden-Spatz syndrome, 66/714
pseudodeficiency, 66/58
β-*N*-Acetylhexosaminidase A deficiency
β-*N*-acetylhexosaminidase B deficiency, 59/402
adult Tay-Sachs disease, 29/365
areflexia, 60/660
choreoathetosis, 60/663
distal amyotrophy, 60/659
enzyme deficiency polyneuropathy, 51/369, 375
Ferguson-Critchley syndrome, 59/322
GM2 gangliosidosis type II, 59/402

hereditary spinal muscular atrophy, 59/25
juvenile amyotrophic lateral sclerosis, 59/402
juvenile Tay-Sachs disease, 29/365
neuropathology, 59/100
neuropathy, 60/660
Ramsay Hunt syndrome, 59/402
spinal muscular atrophy, 59/99-101, 401
vertical supranuclear ophthalmoplegia, 60/657
Wohlfart-Kugelberg-Welander disease, 59/100, 372
*N*-Acetyl-β-D-hexosaminidase
mucolipidosis type II, 42/448
mucolipidosis type III, 42/449
*N*-Acetyl-β-D-hexosaminidase A
*N*-acetyl-β-D-hexosaminidase B, 10/417
GM2 gangliosidosis, 10/297, 417
*N*-Acetyl-β-D-hexosaminidase A deficiency
GM2 gangliosidosis, 10/287
β-*N*-Acetylhexosaminidase B
gene localization, 42/435
GM2 gangliosidosis type I, 27/156, 42/434
GM2 gangliosidosis type II, 27/156, 29/365, 42/435, 66/60, 64
pseudodeficiency, 66/58
β-*N*-Acetylhexosaminidase B deficiency
β-*N*-acetylhexosaminidase A deficiency, 59/402
areflexia, 60/660
cherry red spot, 60/654
choreoathetosis, 49/385, 60/663
differential diagnosis, 59/402
enzyme deficiency polyneuropathy, 51/369, 375
GM2 gangliosidosis type II, 27/154, 59/402
juvenile amyotrophic lateral sclerosis, 59/402
Ramsay Hunt syndrome, 59/402
spinal muscular atrophy, 59/99
Wohlfart-Kugelberg-Welander disease, 59/372
β-*N*-Acetylhexosaminidase β subunit deficiency, *see* GM2 gangliosidosis type II
*N*-Acetyl-β-D-hexosaminidase B
*N*-acetyl-β-D-hexosaminidase A, 10/417
*N*-Acetyl-α-D-hexosaminidase deficiency
GM2 gangliosidosis, 27/156
Acetyl-β-methylcholine
primary pigmentary retinal degeneration, 13/198
*N*-Acetylneuraminic acid
biochemistry, 9/4, 10/252
CSF, *see* CSF *N*-acetylneuraminic acid
glycosphingolipid, 10/248
multiple sclerosis, 9/321
serum, 9/321
*N*-Acetylneuraminidase
mucolipidosis type I, 51/380
*N*-Acetylneuraminidase deficiency
enzyme deficiency polyneuropathy, 51/380
mucolipidosis type I, 46/55
*N*-Acetylpenicillamine
mercury intoxication, 64/394
Minamata disease, 64/416
3-Acetylpyridine
animal experiment, 22/31
Acetylsalicylic acid
anticoagulant, 63/306
brain infarction, 53/465
carotidynia, 48/339
cough headache, 48/371
cranial neuralgia, 5/378
headache, 48/39, 174
hemostasis, 63/306
internal carotid artery syndrome, 53/324
Kawasaki syndrome, 52/267
migraine, 5/98, 48/39, 174, 177
multiple sclerosis, 9/402
nociceptor, 45/229
prolonged transient ischemic attack, 53/274
Reye syndrome, 56/163, 238, 287, 63/438, 65/116, 122, 70/281
transient ischemic attack, 53/274, 324, 55/474
valproic acid induced hepatic necrosis, 65/127
Acetylsalicylic acid intoxication
coma, 65/460
confusion, 65/460
convulsion, 65/460
deafness, 65/460
epilepsy, 65/460
headache, 65/460
iatrogenic neurological disease, 65/460
self-medication, 37/417
tinnitus, 65/460
vertigo, 65/460
*N*-Acetylthreoceramide
sphingolipid, 9/9
Acetyltransferase
isoniazid neuropathy, 42/648
Acetylurea intoxication
anticonvulsant, 37/200
Achalasia
dysphagia, 1/504
esophageal, *see* Esophageal achalasia
esophageal motility disorder, 75/630
esophageal pressure, 74/634
esophageal sphincter, 75/628
esophagus, 75/627
Achalasia cardiae
hereditary motor and sensory neuropathy type I, 21/286
hereditary motor and sensory neuropathy type II,

21/286
Achilles reflex
Jendrassik maneuver, 1/246
lead polyneuropathy, 64/437
procainamide, 41/245
propranolol, 41/245
quinidine, 41/245
reserpine, 41/245
Achlorhydria
pellagra, 28/71
vitamin B12 deficiency, 51/339
Acholuric jaundice
hyperbilirubinemia, 6/492
kernicterus, 6/492
Achondrogenesis (Fraccaro-Parenti)
thanatophoric dwarfism, 31/272
Achondroplasia
basilar impression, 50/399
benign intracranial hypertension, 67/111
birth incidence, 43/317
CNS, 31/270-272
CSF, 43/316
dwarfism, 43/315-317, 46/88
epidemiology, 31/269
foramen magnum, 43/316, 50/395
genetics, 31/269
hydrocephalus, 50/289, 395
hypertelorism, 30/248
juvenile spinal cord injury, 61/233
juvenile vertebral column injury, 61/233
lumbar vertebral canal stenosis, 20/637, 777
macrencephaly, 31/271, 42/41, 43/316, 50/263
mental deficiency, 31/270, 46/88
mitochondrial abnormality, 41/214
mutation rate, 43/317
occipital dysplasia, 50/395
oxidative phosphorylation, 41/214
periventricular calcification, 50/258
prognosis, 31/272
progressive quadriparesis, 50/395
psychomotor retardation, 43/315
radiologic appearance, 31/270
scoliosis, 50/417
sensorineural deafness, 43/316
skeletal malformation, 30/248
spinal cord compression, 32/329-331
swayback nose, 43/315
vertebral abnormality, 30/100, 43/316
vertebral canal stenosis, 19/85, 43/316
Achromatic cortical neuron
Pick disease, 21/63
Achromatopsia, *see* Monochromatism
Achucarro, N.
history, 1/11
neurologist, 1/11
Aciclovir
adverse effect, 71/379
Behçet syndrome, 56/605
herpes simplex virus encephalitis, 56/219
herpes zoster ganglionitis, 51/181
neurotoxicity, 71/368
postherpetic neuralgia, 51/181, 56/234
Acid base balance, 28/507-513
brain blood flow, 55/208
brain injury, 23/116
CSF, 16/370, 28/509-513
general concept, 28/507-509
intracranial pressure, 55/208
metabolic acidosis, 28/512
metabolic alkalosis, 28/513
respiratory acidosis, 28/511
respiratory alkalosis, 28/512
salicylic acid intoxication, 37/418-420
Acid base metabolism
*see also* Acidosis
infantile lactic acidosis, 28/516
isovaleric acidemia, 28/517
lactic acidosis, 28/514-517
maple syrup urine disease, 28/517
β-methylcrotonylglycinuria, 28/518
α-methyl-β-hydroxybutyric aciduria, 28/519
methylmalonic aciduria, 28/521
neurologic aspect, 28/507-522
propionic acidemia, 28/520
spontaneous lactic acidosis, 28/515
subacute necrotizing encephalomyelopathy, 28/517
sweaty foot syndrome, 28/517
Acid ceramidase
Farber disease, 66/60
Acid ceramidase deficiency
enzyme deficiency polyneuropathy, 51/369, 379
Farber disease, 42/592-594, 51/379
Acid β-galactosidase
glycosaminoglycanosis type IVB, 66/247, 269
Acid β-glucosidase, *see* Glucosylceramidase
Acid glycolipid
globoid cell leukodystrophy, 10/79, 84
Acid glycosaminoglycan
heparin, 10/432
Acid lipase
Wolman disease, 21/50, 51/379
Acid lipase deficiency
enzyme deficiency polyneuropathy, 51/369, 379
Acid maltase
glycogen storage disease, 27/222

pH value, 27/226
Acid maltase deficiency, 41/175-181
adrenalin, 41/181
adult, *see* Adult acid maltase deficiency
allelic diversity, 62/481
autophagic vacuole, 40/153
biochemistry, 41/178, 62/481
cardiomyopathy, 43/178
chromosome 17, 62/481
clinical features, 40/280, 317, 41/59, 175, 62/480
creatine kinase, 62/480
differential diagnosis, 27/228, 40/438, 41/372
Duchenne muscular dystrophy, 27/228
ECG, 62/480
EMG, 62/480
genetic heterogeneity, 62/482
genetics, 41/180, 426
α-glucosidase, 41/180
α-1,4-glucosidase, 51/383, 62/481
α-1,6-glucosidase, 62/481
β-glycogen, 51/383
glycogen storage disease, 40/53, 99, 153, 190, 255, 280, 317, 438, 62/480
hereditary spinal muscular atrophy, 59/26
heterozygote detection, 43/178
histochemistry, 40/53
histopathology, 41/178
hydrolase, 41/179
infantile, *see* Infantile acid maltase deficiency
infantile spinal muscular atrophy, 22/99
inheritance, 62/481
ischemic exercise, 62/480
ischemic exercise test, 41/176
jaundice, 42/644
laboratory test, 41/177
late infantile, *see* Late infantile acid maltase deficiency
lysosomal storage disease, 41/178
lysosome, 40/153
metabolic myopathy, 41/175, 62/480
molecular genetics, 62/481
mRNA, 62/482
muscle biopsy, 62/481
muscle fiber feature, 62/17
muscle tissue culture, 40/190, 62/99
myotonia, 40/539, 41/176
neuronal storage dystrophy, 21/61
pathogenesis, 40/256
pathologic reaction, 40/255
pathology, 62/481
prenatal diagnosis, 43/178, 62/481
progesterone, 41/181
proximal muscle weakness, 43/178
rimmed muscle fiber vacuole, 41/178
rimmed vacuole distal myopathy, 62/481
treatment, 41/180, 62/482
ultrastructure, 40/99
vitamin A, 41/180
Wohlfart-Kugelberg-Welander disease, 59/372
Acid mucopolysaccharide
Bi-Col procedure, 9/25
cresyl violet, 40/3
experimental allergic encephalomyelitis, 9/38
Acid phosphatase
adult Gaucher disease, 10/528
demyelination, 9/38
Gaucher disease, 10/308, 42/437
globoid cell, 10/79, 303
globoid cell leukodystrophy, 10/303
glycosaminoglycanosis, 10/377
GM1 gangliosidosis, 10/410
GM2 gangliosidosis, 10/291
intervertebral disc, 20/538
juvenile Gaucher disease, 10/523
metachromasia, 10/53
metachromatic leukodystrophy, 10/48
multiple neuronal system degeneration, 42/75
multiple sclerosis plaque, 9/37
muscle cell, 40/5
Niemann-Pick disease, 10/487
Schwann cell, 51/10
senile plaque, 46/255
wallerian degeneration, 7/215, 9/36
Acid sialidase
glycoprotein, 66/377
Acid sphingomyelinase, *see* Sphingomyelin phosphodiesterase
Acidemia
hyperlactic, *see* Hyperlactic acidemia
hyperpipecolic, *see* Hyperpipecolatemia
hyperpyruvic, *see* Hyperpyruvic acidemia
isovaleric, *see* Isovaleric acidemia
lactic, *see* Lactic acidemia
methylmalonic, *see* Methylmalonic acidemia
propionic, *see* Propionic acidemia
pyruvate carboxylase deficiency, 42/516
pyruvate dehydrogenase deficiency, 42/516
respiratory encephalopathy, 63/414
Acidic lipid
phosphatidylethanolamine, 9/30
phosphatidylinositide, 9/30
phosphatidylserine, 9/30
Acidosis
*see also* Acid base metabolism
attack, 5/46
brain infarction, 53/128

brain ischemia, 53/128, 63/211
carnitine deficiency, 43/176
CSF, 28/511-513
delirium, 46/544
ethylene glycol intoxication, 64/125
headache, 48/422
hereditary renal tubular, *see* Hereditary renal tubular acidosis
hypoglycin intoxication, 65/80
intravascular consumption coagulopathy, 55/493
Jamaican vomiting sickness, 65/80
lactic, *see* Lactic acidosis
Lowe syndrome, 42/606
malignant hyperthermia, 38/551, 553, 43/117, 63/439
metabolic, *see* Metabolic acidosis
mucormycosis, 35/545
neuron death, 63/211
organic acid metabolism, 66/641
pyroglutamic aciduria, 29/230
recessive renal, *see* Recessive renal acidosis
renal, *see* Renal acidosis
renal tubular, *see* Renal tubular acidosis
respiratory, *see* Respiratory acidosis
sodium nitroprusside intoxication, 37/435
spontaneous lactic, *see* Spontaneous lactic acidosis
Aciduria
*N*-acetylaspartic, *see* *N*-Acetylaspartic aciduria
argininosuccinic, *see* Argininosuccinic aciduria
dicarboxylic, *see* Dicarboxylic aciduria
ethylmalonic, *see* Ethylmalonic aciduria
glutaconic, *see* Glutaconic aciduria
glutamic, *see* Glutamic aciduria
δ-glyceric, *see* δ-Glyceric aciduria
hereditary orotic, *see* Hereditary orotic aciduria
D-2-hydroxyglutaric, *see* D-2-Hydroxyglutaric aciduria
L-2-hydroxyglutaric, *see* L-2-Hydroxyglutaric aciduria
3-hydroxyisobutyric, *see* 3-Hydroxyisobutyric aciduria
3-hydroxy-3-methylglutaric, *see* 3-Hydroxy-3-methylglutaric aciduria
β-methylcrotonic, *see* β-Methylcrotonylglycinuria
3-methylglutaconic, *see* 3-Methylglutaconic aciduria
α-methyl-β-hydroxybutyric, *see* α-Methyl-β-hydroxybutyric aciduria
methylmalonic, *see* Methylmalonic aciduria
mevalonic, *see* Mevalonate kinase deficiency
orotic, *see* Orotic aciduria
pyridine dependent xanthurenic, *see* Vitamin B6 dependent xanthurenic aciduria
pyroglutamic, *see* Pyroglutamic aciduria
vitamin B6 dependent xanthurenic, *see* Vitamin B6 dependent xanthurenic aciduria
xanthurenic, *see* Xanthurenic aciduria
Acinetobacter
gram-negative bacillary meningitis, 52/110, 117
Acinetobacter calcoaceticus
gram-negative bacillary meningitis, 52/104
Acivicin intoxication
encephalopathy, 65/538
Ackee tree, *see* Blighia sapida
Acleiencephaly, *see* Anencephaly
ACNU, *see* Nimustine
Aconitase, *see* Aconitate hydratase
Aconitase deficiency, *see* Aconitate hydratase deficiency
Aconitate hydratase
aconitine, 27/16
defect, 66/396
myoglobinuria, 62/558
Aconitate hydratase deficiency
exercise intolerance, 62/503
mitochondrial disease, 62/503
myoglobinuria, 62/503
Aconitine
aconitate hydratase, 27/16
myotonia, 40/557
trigeminal neuralgia, 5/299
Acorn worm intoxication, *see* Hemichordata intoxication
Acoustic agnosia, *see* Auditory agnosia
Acoustic artery
primitive, *see* Primitive otic artery
Acoustic nerve
*see also* Statoacoustic nerve
chronic mercury intoxication, 64/371
Cogan syndrome type II, 51/454
hypothyroidism, 27/258
leprous neuritis, 51/220
malnutrition, 28/20
pneumococcal meningitis, 33/47, 52/51
radiation injury, 7/397
streptomycin polyneuropathy, 51/295
subarachnoid hemorrhage, 55/18
Acoustic nerve agenesis
associated disorder, 50/218
Bing-Siebenmann type, 50/217
congenital deafness, 50/217
cranial nerve agenesis, 50/216
genetics, 50/218
Klein-Waardenburg syndrome, 50/217
maternal infection, 50/219

Michel type, 50/216
Mondini-Alexander type, 50/216
Schiebe type, 50/217
sensorineural deafness, 50/217
Acoustic nerve compression
thalassemia, 63/257
Acoustic nerve tumor
*see also* Cerebellopontine angle syndrome *and* Cerebellopontine angle tumor
acoustic neuroma, 17/666, 42/756
alternate binaural loudness balance test, 17/676
audiogram, 17/671
audiometry, 2/74, 17/673
auditory fatigue, 17/672, 675
Békésy audiometry, 17/673-675
caloric test, 17/677
cerebellar symptom, 17/670
cholesteatoma, 17/684
CSF finding, 17/683, 685
differential diagnosis, 17/683
ENG, 17/677
epidermoid cyst, 17/684
headache, 17/670
histology, 17/668
history, 17/666
incidence, 17/669
iofendylate polytomography, 17/682
loudness balance test, 17/676
meningioma, 17/684
pathology, 17/667
posterior fossa, 17/670
pure tone audiometry, 17/671
radiology, 17/679, 681
short increment sensitivity index, 17/676
speech audiometry, 17/671
Stenvers projection, 17/680
surgery, 17/686-689
tinnitus, 17/669
tone decay test, 17/672, 675
Towne projection, 17/680
treatment, 17/684
tuning fork test, 17/676
vertigo, 17/670
vestibular function, 17/676
Acoustic neurilemoma, *see* Acoustic neuroma
Acoustic neurinoma, *see* Acoustic neuroma
Acoustic neuritis
Koch postulate, 56/105
viral infection, 56/105
Acoustic neurofibroma, *see* Acoustic neuroma
Acoustic neuroma, 42/756-758
acoustic nerve tumor, 17/666, 42/756
bilateral, *see* Bilateral acoustic neurinoma
caloric test, 16/318
cerebellopontine angle syndrome, 2/93-97
cochlear nerve, 2/361
diagnosis, 16/319
facial nerve injury, 24/109
hearing, 16/308, 42/756
imaging, 67/189
loudness recruitment, 16/308
multiple neurofibromatosis, 43/35
neurofibromatosis, 14/33, 151, 492, 18/402, 42/756
neurofibromatosis type I, 18/402, 42/756
nonrecurrent nonhereditary multiple cranial neuropathy, 51/571
nystagmus, 16/315
palisading, 16/19
Peutz-Jeghers syndrome, 14/602-604, 43/41
posterior fossa tumor, 18/402
Scarpa ganglion, 2/361
sensorineural deafness, 42/756
sex ratio, 42/756
spontaneous nystagmus, 2/361, 16/317
stapedius reflex, 16/315
statoacoustic nerve, 16/307
test, 16/315
tissue culture, 17/82
trigeminal nerve, 42/756
vestibular nerve, 2/361
Acoustic neuropathy
gentamicin, 64/9
kanamycin, 64/9
quinine, 64/9
salicylic acid, 64/9
streptomycin, 64/9
Acoustic system
brain cortex, 16/305
brain tumor, 16/301
pathway, 16/304, 306
subcortical pathway, 16/306
Acousticofacial artery, *see* Primitive otic artery
Acquired dysphasia
speech disorder in children, 4/425
Acquired epidermoid
exogenous origin, 32/453
Acquired epileptic aphasia, *see* Landau-Kleffner syndrome
Acquired generalized lipodystrophy, *see* Lawrence syndrome
Acquired hepatocerebral degeneration, 6/279-295
*see also* Bilirubin encephalopathy, Kernicterus *and* Posticteric encephalopathy
alcoholic dementia, 46/338
Alzheimer type II astrocyte, 46/338

astrocyte nucleus area, 6/287
bilirubin, 49/218
choreoathetosis, 46/338
chronic hepatocerebral degeneration, 49/219
clinical features, 46/338
Crigler-Najjar syndrome, 49/218
dementia, 49/213
differential diagnosis, 6/293
free fatty acid, 49/216
hepatic coma, 49/213
hyperammonemia, 6/282, 285
laboratory data, 6/284-286
liver disease, 27/352, 366
mental deficiency, 46/18
methionine, 49/216
neuropathology, 6/285-291, 27/352
neurotransmitter, 49/216
Opalski cell, 6/280, 46/338
pathology, 46/338
phenylalanine, 49/216
Reye syndrome, 49/217
symptom, 6/282-284, 27/366
treatment, 6/294, 49/216
tryptophan, 49/216
Acquired immune deficiency syndrome
*see also* Human immunodeficiency virus infection
Acanthamoeba, 71/304
acquired toxoplasmosis, 52/352, 356
acute inflammatory polyradiculoneuropathy, 56/493
acute meningoencephalitis, 56/492
adenovirus, 56/283, 287, 514
algae, 71/307
amyotrophic lateral sclerosis, 56/493
anhidrosis, 63/153
animal model, 56/453
antiretroviral therapy, 71/367
Aspergillus, 71/298
Aspergillus fumigatus, 56/515
autonomic nervous system, 74/335
autonomic neuropathy, 56/499
axonal degeneration, 56/518
axonal polyneuropathy, 56/517
B-lymphocyte lymphoma, 56/515
Bartonella, 71/291
blastomycosis, 71/296
brain abscess, 56/497
brain atrophy, 56/509
brain infarction, 56/498
Candida albicans, 56/515, 71/299
CD4 antigen, 56/511
central pontine myelinolysis, 56/512
Chagas disease, 71/305
chronic headache, 56/492
chronic lymphoid interstitial pneumonitis, 56/489
chronic meningitis, 56/644-646
cladosporiosis, 71/300
CNS invasion, 56/511
coccidioidomycosis, 52/411, 56/515, 71/296
computer assisted tomography, 56/509
congenital, 56/490
cranial nerve palsy, 56/492
creatine kinase, 56/498
cryptococcosis, 52/433
Cryptococcus neoformans, 56/494, 515, 71/292
CSF, 56/491
cytomegalovirus, 56/513, 516
cytomegalovirus encephalitis, 71/262
cytomegalovirus infection, 56/265
demyelination, 56/509, 518, 520, 583
diamorphine addiction, 55/517
diffuse leukoencephalopathy, 56/509
disease duration, 56/489
drug abuse, 55/517, 56/490
electron microscopy, 56/509
Epstein-Barr virus, 56/255
focal pontine leukoencephalopathy, 56/512
frequency, 56/489
Guillain-Barré syndrome, 56/493
headache, 56/492
hemophiliac, 56/490
herpes simplex, 56/495
herpes simplex virus, 56/478, 514
herpes simplex virus encephalitis, 56/478
herpes simplex virus myelitis, 56/221
herpes zoster, 56/498, 71/268
heterosexual contact, 56/490
histoplasmosis, 56/515, 71/297
homosexual man, 56/490
human immunodeficiency virus myopathy, 71/374
immunocompromise, 56/473
inflammatory demyelinating polyneuropathy, 56/517
inflammatory polyradiculoneuropathy, 56/518
intracranial mass lesion, 56/495
intrathecal IgG, 56/491
JC virus, 71/271
Kaposi sarcoma, 56/489, 515
Listeria, 71/291
Listeria monocytogenes, 56/495
Listeria monocytogenes meningitis, 56/515
listeriosis, 52/90
lymphoma, 56/515, 63/346
medication, 75/414
meningoencephalitis, 56/492, 494
microglia nodule, 56/508

mononeuritis, 56/517
mononeuritis multiplex, 56/493
mortality, 56/489
mucormycosis, 56/515
multinucleated giant cell, 56/507, 509
murine neurotropic retrovirus, 56/586
muscle fiber type II atrophy, 56/518
mycobacterium, 71/285
Mycobacterium avium intracellulare, 56/495
Mycobacterium tuberculosis, 56/497
myelomalacia, 63/55
myelopathy, 56/491, 493
myopathy, 56/518
necrotizing myelitis, 71/263
neuralgic amyotrophy, 56/493
neurocysticercosis, 71/307
neuropathology, 56/507
neuropathy, 56/490, 71/367
neurosyphilis, 71/289
Nocardia, 71/292
nocardiosis, 52/446, 56/515
obstructive hydrocephalus, 56/495
ophthalmologic infection, 56/499
opiate addict, 65/356
opportunistic infection, 56/490, 507, 513
orthostatic hypotension, 63/153
papovavirus JC, 56/18
pathogenesis, 56/511
Pneumocystis carinii pneumonia, 56/489
polymyositis, 56/202, 499, 518, 62/372
polyneuropathy, 56/493, 517
polyradiculomyelitis, 56/513, 71/264
posterior ganglionitis, 56/513
primary CNS lymphoma, 56/255, 498, 68/265
progressive multifocal leukoencephalopathy, 47/503, 506, 56/18, 495, 497, 514, 71/272
radiotherapy, 71/387
retinopathy, 56/491, 494
retrobulbar neuritis, 56/18
retrovirus, 56/18
sensorimotor polyneuropathy, 56/493, 517
in situ hybridization, 56/511
sleep, 74/554
slow virus disease, 56/544
sporotrichosis, 71/300
stress, 74/335
striatopallidodentate calcification, 49/424
Strongyloides stercoralis, 71/307
subacute combined spinal cord degeneration, 56/493, 512
subacute encephalitis, 56/507
sulfa agent, 71/383
syphilis, 56/494, 71/286
T-lymphocyte lymphoma, 56/489
Taenia solium, 71/307
temporal arteritis, 56/499
toxin, 75/414
toxoplasma encephalitis, 71/303
Toxoplasma gondii, 56/495, 514
toxoplasmosis, 52/356, 71/300
transfusion recipient, 56/490
treatment, 56/499, 71/367
tropical polymyositis, 62/373
Trypanosoma cruzi, 71/305
tuberculoma, 71/281
tuberculosis, 71/279, 281
tuberculous meningitis, 71/279
ultrastructure, 56/510
varicella zoster, 71/268
varicella zoster virus, 56/514
viral encephalitis, 56/490
viral meningitis, 56/490
viremia, 56/511
virus like structure, 47/215
vitamin B12 deficiency, 70/387
zidovudine, 56/492

Acquired immune deficiency syndrome dementia, 71/235
akinetic mutism, 56/491
animal model, 56/583
apathy, 56/491
child, 56/513
computer assisted tomography, 56/492
contact loss, 56/491
CSF, 56/491
epilepsy, 56/491
gait disturbance, 56/491
incontinence, 56/491
multinucleated giant cell, 56/507
myoclonus, 56/491
neuropsychologic defect, 56/491
opportunistic infection, 56/18
pseudodepression, 56/491
tremor, 56/491

Acquired immune deficiency syndrome dementia complex
antiretroviral therapy, 71/237
CSF, 71/245
definition, 71/239
early clinical sign, 71/241
early infetion, 71/252
epidemiology, 71/248
human immunodeficiency virus type 1 infection, 71/238
laboratory, 71/245
late clinical sign, 71/241

late infection, 71/252
neuroimaging, 71/245
neuropathology, 71/249
neuropsychologic test, 71/247
pathology, 71/250
stage, 71/240
survey, 71/235
terminology, 71/239
treatment, 71/251
Acquired immune deficiency syndrome treatment
*see also* Human immunodeficiency virus infection
neurotoxicity, 71/368
Acquired myoglobinuria
amphetamine, 62/573
bezafibrate, 62/576
Candida, 62/577
chronic myopathy, 62/577
ciclosporin, 62/576
clofibrate, 62/576
Clostridium, 62/577
cocaine, 62/577
cold, 62/574
Coxsackie virus, 62/577
crush, 62/574
diamorphine, 62/577
doxylamine, 62/573
drug, 62/575
drug overdose, 62/573
ECHO virus, 62/577
Epstein-Barr virus, 62/577
erythromycin, 62/577
Escherichia coli, 62/577
exercise, 62/572
gemfibrozil, 62/576
Haff disease, 62/577
heat stroke, 62/572
herpes simplex, 62/577
infection, 62/577
influenza, 62/577
ischemia, 62/574
isoniazid, 62/577
legionellosis, 62/577
leptospirosis, 62/577
loxapine, 62/573
marathon runner, 62/573
metabolic depression, 62/574
methadone, 62/577
mevinolin, 62/576
mycoplasma, 62/577
neuroleptic malignant syndrome, 62/572
osmolarity derangement, 62/575
oxprenolol, 62/577
phencyclidine, 62/573
phenmetrazine, 62/573
phentermine, 62/573
phenylenediamine, 62/573
polymyositis, 62/577
Salmonella, 62/577
salt depletion, 62/572
Shigella, 62/577
sickle cell anemia, 62/572
snake venom, 62/577
Staphylococcus, 62/577
status asthmaticus, 62/573
status epilepticus, 62/573
Streptococcus pneumoniae, 62/577
strychnine, 62/573
terbutaline, 62/577
theophylline, 62/573, 577
toxic agent, 62/575
toxic shock syndrome, 62/574, 577
toxin, 62/576
trauma, 62/572
tularemia, 62/577
typhoid fever, 62/577
viral infection, 62/572
vitamin PP, 62/577
Acquired porphyria, *see* Symptomatic porphyria
Acquired retraction syndrome
*see also* Stilling-Türk-Duane syndrome
mesencephalic lesion, 2/345
Acquired toxoplasmosis
acquired immune deficiency syndrome, 52/352, 356
associated disease, 52/353
ataxia, 35/126, 52/355
benign intracranial hypertension, 35/125, 52/354
brain biopsy, 52/356
brain miliary necrosis, 35/123, 52/357
cerebral toxoplasmosis, 35/123, 52/353
collagen disease, 52/353
coma, 35/125, 52/354
computer assisted tomography, 52/356
corticosteroid, 35/124, 52/353
CSF, 35/125, 52/354, 356
differential diagnosis, 52/353
encephalitis, 35/122, 52/352, 354, 357
headache, 52/354
histopathology, 52/357
Hodgkin disease, 35/124, 52/353
host defence, 52/356
immunocompromised host, 35/128, 52/353
intracranial hypertension, 52/354
meningoencephalitis, 35/122, 52/352, 354
microglia nodule, 35/123, 52/353, 357
multifocal necrosis, 52/357

multiple brain mass lesion, 52/354
muscle biopsy, 52/357
neuropathology, 52/352, 357
nuclear magnetic resonance, 52/356
organ transplantation, 52/353
polyneuropathy, 35/126, 52/355
psychiatric symptom, 35/125, 52/355
pyrimethamine, 35/128, 52/358
radiculitis, 52/355
relapsing encephalitis, 35/125, 52/355
reticulosarcoma, 35/124, 52/353
seizure, 35/125, 52/354
serology, 35/126, 52/356
somnolence, 35/125, 52/354
spiramycin, 35/128, 52/358
sulfadiazine, 35/128, 52/358
treatment, 35/128, 52/358

Acrania
*see also* Anencephaly
semantics, 50/71

Acraniata intoxication
*see also* Cephalochordata intoxication
chordata, 37/68

Acriflavinium chloride
myelin, 9/26

Acroatonia
congenital atonic sclerotic muscular dystrophy, 41/43

Acrobrachycephaly
acrocephalosyndactyly type II, 50/123
acrocephalosyndactyly type V, 46/87, 50/123
craniosynostosis, 50/113, 118
Saethre-Chotzen syndrome, 43/322

Acrocallosal syndrome
corpus callosum agenesis, 50/163

Acrocephalopolysyndactyly
Laurence-Moon-Bardet-Biedl syndrome, 46/87
Sakati-Nyhan-Tisdale syndrome, 43/326
type I, *see* Noack syndrome

Acrocephalosynanky
craniosynostosis, 46/87
mental deficiency, 46/87

Acrocephalosyndactyly, 43/317-327
incidence, 50/49
infant mortality, 50/49
prevalence, 50/49

Acrocephalosyndactyly type I
*see also* Craniosynostosis
acrocephalosyndactyly type II, 38/422, 46/87
acrocephalosyndactyly type V, 46/87
aqueduct stenosis, 43/320
birth incidence, 43/320
cavum septi pellucidi, 30/326
cheilopalatoschisis, 43/318
CNS, 31/232
corpus callosum agenesis, 43/320, 50/163, 168
craniosynostosis, 30/154, 222, 31/231, 38/422, 43/318-320, 50/115
differential diagnosis, 31/233
exophthalmos, 38/422, 43/318
face, 31/231
facial asymmetry, 43/318
fontanelle, 30/100
genetics, 31/231
glaucoma, 43/318
heart disease, 43/320
hydrocephalus, 43/320, 50/122
hydromyelia, 50/428
hypertelorism, 30/246, 43/318
intracranial pressure, 43/319
low set ears, 43/318
mental deficiency, 31/233, 43/320, 46/86
moyamoya disease, 42/751
mutation rate, 43/320
neurologic symptom, 46/87
optic atrophy, 13/81, 43/318
oral lesion, 31/232
oxycephaly, 13/401
polydactyly, 38/422
polysyndactyly, 50/122
ptosis, 43/318
septum pellucidum, 43/320
septum pellucidum agenesis, 30/315, 326, 43/320
skeletal abnormality, 31/231
skull dysplasia, 30/246
strabismus, 43/318
synaeretic change, 21/44
syndactyly, 38/422, 43/319

Acrocephalosyndactyly type II
*see also* Craniosynostosis
acrobrachycephaly, 50/123
acrocephalosyndactyly type I, 38/422, 46/87
antimongoloid eyes, 43/324
autosomal recessive inheritance, 50/123
brachydactyly, 38/422
brain dysplasia, 50/123
cloverleaf skull syndrome, 43/359, 50/123
congenital heart disease, 38/422
craniosynostosis, 30/154, 222, 31/235, 43/322-324
differential diagnosis, 31/236
dystopia canthorum, 43/324
ear, 31/236
epicanthus, 38/422
eye, 31/236
face, 31/235
fontanelle, 30/100

genetics, 31/235
genu valgum, 38/422
heart disease, 43/324
heart dysplasia, 50/123
hydrocephalus, 50/123
hypertelorism, 30/246, 38/422
hypogonadism, 38/422
Laurence-Moon-Bardet-Biedl syndrome, 38/422
low set ears, 38/422, 43/324
mental deficiency, 31/236, 38/422, 43/322, 324, 46/87, 50/123
micrognathia, 43/324
obesity, 38/422, 43/324
polydactyly, 43/322-324, 50/123
polysyndactyly, 38/422
ptosis, 43/324
skeletal abnormality, 31/235
skull, 31/235
skull dysplasia, 30/246
syndactyly, 38/422, 43/322-324, 50/123
Acrocephalosyndactyly type III
*see also* Craniosynostosis
autosomal dominant inheritance, 50/123
brachydactyly, 50/123
craniosynostosis, 50/123
plagiocephaly, 50/123
ptosis, 50/123
Saethre-Chotzen syndrome, 43/326
Acrocephalosyndactyly type V
*see also* Craniosynostosis
acrobrachycephaly, 46/87, 50/123
acrocephalosyndactyly type I, 46/87
antimongoloid eyes, 43/324, 50/123
autosomal dominant inheritance, 50/123
brachycephaly, 43/320
classification, 43/326
cloverleaf skull syndrome, 43/322
craniosynostosis, 43/320
exophthalmos, 43/320
fontanelle, 30/100
hypertelorism, 30/246, 43/320, 50/123
increased intracranial pressure, 50/123
mental deficiency, 46/87
orbital hypertelorism, 50/123
oxycephaly, 50/123
polydactyly, 43/321
Saethre-Chotzen syndrome, 43/322
skull dysplasia, 30/246
strabismus, 43/320
syndactyly, 43/320
vertebral abnormality, 43/320
Acrocephaly
craniosynostosis, 50/113
Gruber syndrome, 14/13
Klippel-Trénaunay syndrome, 14/524
Sakati-Nyhan-Tisdale syndrome, 43/325
Summitt syndrome, 43/325
Acrocyanosis
hereditary sensory and autonomic neuropathy type III, 43/58, 60
Raynaud disease, 8/322, 22/261, 43/70
reflex neurovascular syndrome, 8/341
Acrodermatitis atrophicans
Lyme disease, 52/260
Acrodermatitis enteropathica, *see* Danbolt disease
Acrodynia
*see also* Inorganic mercury intoxication
anorexia, 64/374
ascariasis, 64/374
ataxia, 64/374
autonomic polyneuropathy, 51/477
child, 64/373, 75/503
choreomyoclonus, 64/374
chronic mercury intoxication, 64/371
clinical features, 1/480, 36/155
dysautonomia, 64/371
Guillain-Barré syndrome, 64/374
history, 36/154
insomnia, 64/374
Kawasaki syndrome, 56/638
mercury intoxication, 1/480, 8/337, 51/272, 477
mercury polyneuropathy, 51/272
neuropathology, 64/374
pathology, 36/155
source, 36/154-156
symptom, 64/374
synonym, 1/480
teething powder, 64/374
tremor, 64/374
vitamin B6 deficiency, 7/581
Acrodysostosis
dermatoglyphics, 43/291
differential diagnosis, 31/230
genetics, 31/228
hydrocephalus, 30/100, 31/230, 43/291
hypogonadism, 31/230
mental deficiency, 31/230, 43/291
optic atrophy, 43/291
oral lesion, 31/230
seizure, 43/291
short stature, 43/291
skeletal abnormality, 31/229
skin lesion, 31/230
Acrodystrophic neuropathy, *see* Hereditary sensory and autonomic neuropathy type I
Acrogyria

mental deficiency, 14/791
Acromegalic myopathy
creatine kinase, 62/538
EMG, 62/538
muscle biopsy, 62/538
muscle ultrastructure, 62/538
pituitary gigantism myopathy, 41/249, 62/537
proximal limb muscle, 62/538
Acromegalic neuropathy
differential diagnosis, 7/556
metabolic neuropathy, 51/68
toxic neuropathy, 7/556
Acromegalic polyneuropathy
*see also* Endocrine polyneuropathy
muscle weakness, 41/249
type, 51/517
Acromegaly
adenohypophyseal syndrome, 2/454
Bardet-Biedl syndrome, 13/395, 401
bromocriptine, 68/353
carpal tunnel syndrome, 7/296, 41/249, 42/309
chronic axonal neuropathy, 51/531
concentric laminated body, 40/109
cutis verticis gyrata, 14/777, 43/244
diagnosis, 68/351
endocrine myopathy, 41/248, 250
fibrous dysplasia, 14/199
gigantism, 2/454
headache, 48/423
hypertrophic interstitial neuropathy, 21/145, 51/519
infantile optic glioma, 42/733
neurofibromatosis, 14/701, 740
obstructive sleep apnea, 63/454
octreotide, 68/353
pituitary, 68/350
pituitary adenoma, 17/393-395, 422
pituitary tumor, 68/351
polyneuropathy, 41/249
radiotherapy, 68/352
sleep apnea syndrome, 45/132
snoring, 63/451
Sturge-Weber syndrome, 14/512
surgery, 68/352
treatment, 68/352
Acromicria
congenital, *see* Down syndrome
mental deficiency, 42/341
polyneuritis, 42/341
Acro-osteolysis, *see* Neurogenic acro-osteolysis
Acroparesthesia
diphtheria, 65/240
ergot intoxication, 65/67
hyperventilation syndrome, 63/432
Melkersson-Rosenthal syndrome, 42/322
myxedema neuropathy, 7/169, 8/58
restless legs syndrome, 8/311
styrene intoxication, 64/54
Acropathie ulcéromutilante
familial, *see* Hereditary sensory and autonomic neuropathy type I
hereditary sensory and autonomic neuropathy type II, 8/180
Acrylamide
autonomic nervous system, 75/36
chemical formula, 64/65, 82
chemical structure, 64/65
chronic axonal neuropathy, 51/531
1,4-diketone neuropathy, 64/64
dimethylaminopropionitrile, 64/66
distal axonopathy, 64/12
element data, 64/65
grout, 64/66
hexacarbon neuropathy, 51/278
neurotoxicity, 51/276
neurotoxin, 7/522, 36/376, 46/393, 51/68, 276, 479, 531, 64/10, 12, 63
organic solvent, 64/82
polyacrylamide, 64/66
polymerization, 64/66
production, 64/65
synonym, 64/65
toxic encephalopathy, 64/12
toxic neuropathy, 64/10
toxic polyneuropathy, 51/68, 280
use, 64/66
Acrylamide intoxication
*N*-acetyl-*S*-(3-amino-3-oxypropyl)cysteine, 64/72
action site, 64/10
age, 64/75
annulospiral ending, 64/74
areflexia, 64/68
ataxia, 64/63, 67
ataxic gait, 64/63
autonomic nervous system, 75/504
axonal degeneration, 64/63
axonal flow, 64/75
axonal swelling, 64/74
axonopathy, 64/63
biomarker, 64/71
biotransformation, 64/72
cerebellar vermis, 64/74
dementia, 64/68
1,4-diketone neuropathy, 64/64
distal axonopathy, 64/64
distribution volume, 64/71

dorsal spinal column, 64/74
dysarthria, 64/68
dyssomnia, 64/67
EEG, 64/70
epidemic, 64/66
epilepsy, 64/63
excretion, 64/72
experimental intoxication, 64/72
exposure source, 64/66
fatigue, 64/68
glutathione, 64/72
glutathione transferase, 64/72
glycidamide, 64/72
H-reflex, 64/71
half-life, 64/71
hallucination, 64/68
hemoglobin adduct, 64/71
hexacarbon polyneuropathy, 64/63
hyperhidrosis, 64/67
kinetic, 64/71
large diameter fiber, 64/73
metabolism, 64/71
metabolite, 64/72
motor nerve terminal, 64/74
muscle action potential, 64/69
muscle spindle ending, 64/74
muscle weakness, 64/63
muscular atrophy, 64/67
nerve action potential, 64/70
nerve conduction velocity, 64/69
neuropathy, 75/504
nystagmus, 64/68
optic tract, 64/74
pacinian corpuscles, 64/74
pain sensitivity, 64/73
polyneuropathy, 64/63
polyphasic spike, 64/70
*S*-β-propionamide glutathione, 64/72
Purkinje cell, 64/74
recovery, 64/69
red blood cell, 64/71
regenerating fiber, 64/75
screening method, 64/70
sensorimotor polyneuropathy, 64/63, 67
skin lesion, 64/67
specie susceptibility, 64/63
sural nerve biopsy, 64/70
susceptibility, 64/63
symptom, 64/12, 67
tendon reflex latency, 64/71
threshold limit value, 64/67
tremor, 64/63, 68
ultrastructure, 64/74
urinary retention, 64/68
vibration sensitivity, 64/73
weight loss, 64/67
Acrylamide polyneuropathy
ataxia, 51/281
autonomic polyneuropathy, 51/281, 479
clinical features, 36/376
disorientation, 51/281
encephalopathy, 51/281
exposure source, 51/281
grouting process, 51/281
hallucination, 51/281
hyperhidrosis, 51/281
intention tremor, 51/281
laboratory finding, 36/377
neurofilament segregation, 51/282
pathogenesis, 36/377-380, 51/282
pathologic change, 51/282
psychosis, 51/281
sensorimotor neuropathy, 51/281
sensory polyneuropathy, 51/281
slow axon transport block, 51/282
symptom, 7/522, 51/280
ACTH, *see* Corticotropin
ACTH induced myopathy
EMG, 62/537
fatigue, 62/537
hyperpigmentation, 62/537
muscle biopsy, 62/537
muscle weakness, 62/537
Nelson syndrome, 62/537
ACTH secreting tumor
pituitary tumor, 68/344
ACTH test
adrenoleukodystrophy, 47/594
adrenomyeloneuropathy, 47/595
Actin
axon, 51/6
contractile protein, 40/212
nemaline myopathy, 41/10
nerve, 51/6
Actiniaria intoxication
local pain, 37/53
skin, 37/53
ulceration, 37/53
α-Actinin
dystrophin, 62/126
muscle fiber, 62/26, 35
nemaline myopathy, 41/10
Actinobacillus
bacterial endocarditis, 52/302
Actinomyces
actinomycosis, 35/383

bacterial meningitis, 52/4
brain abscess, 52/149
Actinomyces israelii
actinomycosis, 35/383
chronic meningitis, 56/645
spinal epidural empyema, 52/187
Actinomycetes, *see* Nocardia *and* Nocardiosis
Actinomycosis, 35/383-392
abscess, 35/383
Actinomyces, 35/383
Actinomyces israelii, 35/383
brain, 16/217
case report, 35/386
chronic meningitis, 56/644
CNS, 35/385
diagnosis, 35/387-390
epidemiology, 35/16
epidural abscess, 35/385
fibrosis, 35/383
focal neurologic deficit, 35/385
intracranial pressure, 35/385
mastoiditis, 35/386
meningitis, 35/385
Nocardia, 35/383
nonrecurrent nonhereditary multiple cranial neuropathy, 51/571
pathology, 35/390
sinusitis, 35/386
skull osteomyelitis, 35/386
Streptomyces, 35/383
subdural empyema, 35/385
third ventricle tumor, 17/445
treatment, 35/392
type, 35/384
Action dystonia
definition, 49/71
dystonia musculorum deformans, 6/523, 49/522
muscle tone, 6/523
Action myoclonus
anoxic encephalopathy, 63/217
brain embolism, 53/211
brain infarction, 53/211
cardiac arrest, 63/217
cerebrovascular disease, 53/211
definition, 49/612
dyssynergia cerebellaris myoclonica, 60/599
encephalitis lethargica, 6/767
intention tremor, 21/529
intracranial hemorrhage, 53/211
methyl bromide intoxication, 6/767
mucolipidosis type I, 60/662
mucolipidosis type II, 60/662
olivopontocerebellar atrophy (Wadia-Swami), 60/494
Ramsay Hunt syndrome, 6/767
restless legs syndrome, 51/356
transient ischemic attack, 53/211
Action potential
*see also* Nerve conduction
compound muscle, *see* Compound muscle action potential
dechronization effect, 7/139
evoked muscle, *see* Evoked muscle action potential
features, 7/63
motor unit, *see* Motor unit action potential
muscle, *see* Muscle action potential
nerve, *see* Nerve action potential
sensory, *see* Sensory action potential
Action tremor
*see also* Intention tremor
amiodarone intoxication, 65/458
hepatic coma, 49/213
hereditary tremor, 49/567
rate, 6/183
subgroup, 49/583
Sydenham chorea, 49/363
Activating system
reticular, *see* Reticular activating system
Acupuncture
anesthesia, 37/136
pain, 26/496, 45/231
spinal pain, 26/496
Acute abdominal pain
paraplegia, 26/211
Acute amaurosis
temporal arteritis, 55/342
Acute amebic meningoencephalitis
Acanthamoeba culbertsoni, 52/313
Acanthamoeba rhysodes, 52/313
ageusia, 52/315, 317
amebostome, 52/324
amphotericin B, 52/329
anosmia, 52/315, 317
bizarre behavior, 52/317
brain edema, 52/317
brain herniation, 52/317, 326
brain stem, 52/326
brain tissue, 52/316
clinical features, 52/316
CNS invasion, 52/326
cortex necrosis, 52/326
CSF, 52/316, 319
cytolysis, 52/323
diagnostic test, 52/316
differential diagnosis, 52/316

entry portal, 52/316
ependymitis, 52/326
epilepsy, 52/317
experimental infection, 52/323
headache, 52/317
hemorrhagic encephalitis, 52/326
histopathology, 52/316
immunology, 52/324
incidence, 52/313
invasion route, 52/316
laboratory diagnosis, 52/317
meningitis, 52/316, 326
Naegleria, 52/313
Naegleria fowleri, 52/313, 316
neuroamebiasis, 52/309
olfactory nerve, 52/323, 326
parosmia, 52/317
pathogenesis, 52/323, 326
pathology, 52/316
prognosis, 52/329
psychiatric symptom, 52/317
rifampicin, 52/329
season, 52/313
serologic test, 52/320
sex ratio, 52/316
treatment, 52/316, 329
trogocytosis, 52/323
Acute anterior poliomyelitis, *see* Poliomyelitis
Acute aseptic meningitis, *see* Meningitis serosa
Acute bacterial meningitis
epilepsy, 72/158
Acute benign ataxia
infectious disease, 21/563
Acute brachial neuritis, *see* Neuralgic amyotrophy
Acute brachial radiculitis
*see also* Acute paralytic brachial neuritis
neuralgic amyotrophy, 8/77
terminology, 8/77
Acute cerebellar ataxia, 34/619-634
bacterial meningitis, 34/624
brain stem encephalitis, 34/632
chickenpox, 34/162-170, 622, 627
Coxsackie virus, 56/329
Coxsackie virus A2, 56/326
Coxsackie virus A4, 56/326
Coxsackie virus A7, 56/326
Coxsackie virus A9, 56/326
Coxsackie virus B1-B6, 56/326
Coxsackie virus infection, 34/621
CSF, 56/253
differential diagnosis, 34/630-634
ECHO virus, 34/621, 56/329
ECHO virus 6, 56/326
ECHO virus 9, 56/326
encephalitis lethargica, 34/623
enterovirus 70, 56/329
enterovirus 71, 56/326
enterovirus infection, 34/621
Epstein-Barr virus, 56/253
Fisher syndrome, 34/629
fungal infection, 34/625
herpes simplex, 34/622
history, 34/619-621
immunocompromise, 56/477
immunocompromised host, 56/477
infectious mononucleosis, 34/623
legionellosis, 52/254
meningococcal meningitis, 33/23, 52/26
mumps, 34/628
Mycoplasma pneumoniae, 34/623
myoclonic encephalopathy, 34/629-632
opsoclonus, 34/613, 629
opsoclonus myoclonus ataxia syndrome, 34/629-632
poliomyelitis, 34/621, 56/329
poliovirus type 1, 56/326
poliovirus type 3, 56/326
postexanthemata, 34/626-628
rubella, 34/627
rubella encephalitis, 34/627
syphilis, 34/625
systemic bacterial infection, 34/624
tick paralysis, 34/627
toxoplasmosis, 34/626, 52/355
tuberculosis, 34/625
varicella zoster, 34/162-167, 622, 627
Acute childhood encephalopathy, *see* Reye syndrome
Acute confusional state, *see* Delirium
Acute demyelinating neuropathy, 47/605-612
*see also* Primary demyelinating neuropathy
diphtheria, 47/605, 608
Guillain-Barré syndrome, 47/605, 608-612
recurrent, *see* Recurrent acute demyelinating neuropathy
Acute diffuse lymphocytic meningoencephalitis
biphasic type, 34/557
clinical features, 34/553, 555-557
history, 34/553
monophasic type, 34/556
pathology, 34/553
treatment, 34/554
Acute disseminated encephalomyelitis
*see also* Postinfectious encephalomyelitis *and* Postvaccinal encephalitis
classification, 9/472

encephalitogenic myelin basic protein, 9/501
Guillain-Barré syndrome, 47/606
immunotherapy, 71/550
multiple sclerosis, 9/126, 135, 138, 199, 500, 47/72
pathogenesis, 71/549
pathology, 71/549
progressive bulbar palsy, 59/375
survey, 71/547
Sydenham chorea, 71/550
symptom, 71/548
treatment, 71/550
Acute dysautonomia
autonomic nervous system, 75/690
Guillain-Barré syndrome, 51/247
systemic lupus erythematosus, 51/447
test, 22/251
visceral function, 22/251
Acute encephalomeningitis, *see* Acute meningoencephalitis
Acute encephalopathy
Bordetella pertussis, 47/479
chickenpox, 47/479
eosinophilia myalgia syndrome, 64/257
influenza B, 47/479
influenza virus, 56/168
postinfectious complication, 47/483-489
Acute epiduritis, *see* Spinal epidural empyema
Acute experimental allergic encephalomyelitis
chronic multiple sclerosis, 47/430
features, 47/468
pathology, 47/437
Acute hemorrhagic conjunctivitis, 38/595-622
age, 56/350
age distribution, 38/597
age prevalence, 38/608
AHV virus, 38/600-606
clinical features, 38/596, 608, 615, 56/350
cranial nerve palsy, 56/351
CSF, 38/615, 56/351
EMG, 56/352
enterovirus, 56/307
enterovirus 70, 56/307, 349
epidemic, 56/315
epidemiology, 38/598
geographic distribution, 38/596, 606
history, 38/595
incidence, 38/606-608
multiple cranial neuropathy, 56/351
neurologic complication, 38/606, 56/349
neurologic complication incidence, 56/350
paralytic disease, 56/329
pathogenesis, 38/619
poliomyelitis, 56/349
prevention, 38/621, 56/353
serodiagnosis, 56/352
sex ratio, 38/608, 56/350
spinal disease, 56/350
treatment, 38/621
tropical myeloneuropathy, 56/526
virus isolation, 56/352
Acute hemorrhagic conjunctivitis virus, *see* Enterovirus 70
Acute hemorrhagic encephalitis, *see* Hyperacute disseminated encephalomyelitis
Acute hemorrhagic leukoencephalitis, 34/587-601
clinical features, 34/588, 47/472, 476
course, 34/589
CSF, 34/588
definition, 34/587
differential diagnosis, 34/600
epidemiology, 34/587
history, 34/587
incidence, 34/587
laboratory finding, 34/589
pathogenesis, 34/595-600
pathology, 34/590-595
premonitory illness, 34/588
prognosis, 34/590
treatment, 34/601
Acute hemorrhagic pancreatitis
Reye syndrome, 56/159
Acute hepatic coma, *see* Hepatic coma
Acute hepatic necrosis
*see also* Liver disease
carbon tetrachloride intoxication, 27/350
clinical consideration, 27/361-363
Acute humeroscapular paralysis, *see* Brachial neuralgia
Acute hypoxia
autonomic nervous system, 75/259
blood pressure, 75/260
cardiac output, 75/262
carotid body, 75/260
oxygen transport, 75/266
respiration, 75/260
vascular resistance, 75/263, 266
ventilatory acclimatization, 75/266
Acute hypertensive encephalopathy
headache, 5/154, 48/282
Acute immune mediated polyneuritis
animal model, 34/403
clinical syndrome, 34/393-395
epidemiology, 34/393
immune mediated syndrome related to virus infection, 34/392-403, 423-425

immunologic aspect, 34/401-403
pathology, 34/396-399
physiology, 34/400
prodromal factor, 34/400
Acute infantile hemiplegia
age, 12/341
arteritis, 12/345
artery occlusion, 12/342
brain hemorrhage, 31/169
brain infarction, 53/32
carotid artery thrombosis, 53/32
cause, 12/342
cerebrovascular disease, 53/32
course, 12/342
Coxsackie virus A9, 56/330
epilepsy, 12/342, 53/31
hemiathetosis, 53/31
hemorrhagic leukoencephalitis, 53/32
management, 12/347
neck injury, 12/345
onset, 12/342
phlebothrombosis, 53/32
2 presentation modes, 53/31
prodromal symptom, 12/341
sex ratio, 12/341
type, 12/342
Acute inflammatory demyelinating neuropathy, *see* Guillain-Barré syndrome
Acute inflammatory demyelinating polyneuropathy, *see* Guillain-Barré syndrome
Acute inflammatory polyradiculoneuropathy
*see also* Guillain-Barré syndrome
acquired immune deficiency syndrome, 56/493
Acute intermittent porphyria
5-aminolevulinate synthase, 51/390
aminolevulinic acid, 8/9, 366, 27/435
antiepileptic agent, 65/497
autonomic dysfunction, 63/150
autonomic nervous system, 27/434, 42/617
autonomic polyneuropathy, 51/479
basal ganglion degeneration, 42/618
biochemistry, 51/390
cranial nerve, 41/434, 42/618
delirium, 46/544
diagnosis, 27/435
differential diagnosis, 8/366
encephalopathy, 27/456
epilepsy, 72/222
exacerbation, 51/390
features, 27/434-437
flaccid paralysis, 42/618
hematologic function, 27/435
heme synthesis, 51/390
hyperbetalipoproteinemia, 42/618
hypercholesterolemia, 42/618
hypertension, 41/434, 42/617
metabolic neuropathy, 51/68
muscle weakness, 41/167
nerve conduction, 7/169
neurofibromatosis type I, 14/492
neuropathology, 63/150
neuropathy, 8/366, 27/434, 41/434, 60/120
orthostatic hypotension, 63/150
pain, 42/617
periodic ataxia, 21/570
polyneuropathy, 42/617
porphobilinogen, 8/9, 366, 41/334, 42/618
porphobilinogen deaminase, 51/390
porphyrin metabolism, 27/435-437
prevalence, 42/618
pyridoxal-5-phosphate, 28/130
recurrent nonhereditary multiple cranial neuropathy, 51/571
restless legs syndrome, 8/316, 51/546
sex ratio, 42/618
symptom, 27/434, 63/150
treatment, 27/436
uroporphyrinogen I synthetase, 42/618
vitamin B6, 28/130
Acute ischemic optic neuropathy
anemia, 38/25-27
optic disc, 48/312
temporal arteritis, 48/312
Acute ischemic papillopathy, *see* Acute ischemic optic neuropathy
Acute labyrinthitis, *see* Vestibular neuronitis
Acute leukemia
acanthocytosis, 63/290
immunocompromised host, 56/470
Acute lymphatic leukemia
age, 18/249
Acute lymphocytic leukemia
Down syndrome, 50/527
Acute meningoencephalitis
acquired immune deficiency syndrome, 56/492
meningococcal meningitis, 52/26
neurobrucellosis, 52/583
Acute mercury intoxication
coma, 64/375
diarrhea, 64/375
embolic mercury, 64/375
epigastric pain, 64/375
fatality rate, 64/375
metallic taste, 64/375
renal insufficiency, 64/375
stomatitis, 64/375

suicide, 64/375
symptom, 64/375
vomiting, 64/375
Acute metabolic encephalopathy
carnitine cycle, 66/402
carnitine deficiency, 66/406
fatty acid oxidation disorder, 66/406
mitochondrial matrix, 66/410
primary carnitine deficiency, 66/402
Acute mountain sickness
oxygen deficiency, 29/539-541
survey, 75/271
Acute multiple sclerosis
classification, 9/472
concentric sclerosis, 47/409, 416
histology, 47/222
remyelination, 47/245
Acute muscle contraction headache
incidence, 48/343
Acute muscle weakness, 41/287-292
*see also* Organophosphate intoxication *and* Tick paralysis
arsenic intoxication, 41/288
botulism, 41/288
classification, 41/287
diphtheritic neuropathy, 41/288
hyperkalemia, 41/287
onset weakness, 40/296
organophosphate intoxication, 41/288
saxitoxin intoxication, 41/291
succinylcholine, 41/291
tetrodotoxin intoxication, 41/291
tick paralysis, 41/290
Acute myeloid leukemia
Down syndrome, 50/274, 527
Acute myoclonic ataxia
differential diagnosis, 21/583
Acute necrotizing encephalitis, *see* Hyperacute disseminated encephalomyelitis
Acute necrotizing hemorrhagic leukoencephalitis
classification, 9/472
Acute necrotizing myopathy
paraneoplastic syndrome, 69/374, 386
Acute optic neuritis, *see* Anterior ischemic optic neuropathy
Acute panautonomic neuropathy, *see* Acute pandysautonomia
Acute pancreatitis
demyelination, 27/449-452
hyperosmolarity syndrome, 27/456
pancreatic encephalopathy, 27/449-452
Reye syndrome, 56/159
Acute pandysautonomia
anhidrosis, 51/475, 63/153
animal model, 75/123
autonomic nervous system, 75/13
autonomic polyneuropathy, 51/475
clinical features, 75/121
CSF protein, 51/475, 63/153
eye dryness, 51/475
fixed heart rate, 51/475
Guillain-Barré syndrome variant, 63/153
hypotonic bladder, 51/475
infectious disease, 74/336
intestinal peristalsis absence, 51/475
laboratory test, 75/122
mixed connective tissue disease, 51/450
nonreactive pupil, 51/475
orthostatic hypotension, 51/475, 63/152
Acute paralytic brachial neuritis
*see also* Acute brachial radiculitis
immune mediated syndrome related to virus infection, 34/395
Acute paroxysmal hypertension
headache, 48/282
Acute petrositis
headache, 5/210
Acute poliomyelitis
restless legs syndrome, 8/316, 51/546
Acute polymyositis
acute viral myositis, 56/199
adenovirus, 56/200
Coxsackie virus, 56/199
creatine kinase, 56/200
EMG, 62/71
Epstein-Barr virus, 56/200
herpes simplex virus, 56/200
influenza A virus, 56/199
influenza B virus, 56/199
parainfluenza virus, 56/199
trichinosis, 41/58, 373
Acute polyneuritis
sarcoid neuropathy, 51/197
Acute polyradiculoneuritis, *see* Guillain-Barré syndrome
Acute porphyria, *see* Acute intermittent porphyria
Acute postinfectious disseminated encephalomyelitis
*see also* Disseminated meningoencephalomyelitis
rubella virus, 56/407
varicella zoster virus, 56/229
varicella zoster virus meningoencephalitis, 56/229
Acute schistosomiasis
symptom, 52/536
Acute sensorimotor neuropathy
paraneoplastic syndrome, 69/374

Acute sensory neuropathy
secondary pigmentary retinal degeneration, 60/736
Acute serous encephalitis, *see* Reye syndrome
Acute subdural hematoma
angiography, 24/282-285
arterial rupture, 24/277
associated injury, 24/281
bleeding tendency, 24/277
boxing injury, 23/548
brain blood flow, 57/260
brain injury, 24/278, 280, 288
brain scintigraphy, 24/286
chronic subdural hematoma, 57/251
classification, 24/275
combined hematomas, 24/279
complication, 24/289
computer assisted tomography, 57/152
diagnosis, 24/279, 281
echoEG, 24/286
EEG, 24/286
epilepsy, 24/280
etiology, 24/276
head injury, 24/278
headache, 24/279
hemiparesis, 24/280
history, 24/275
incidence, 24/278
local effect, 57/260
lumbar puncture, 24/282
management, 24/645, 656
mechanism, 24/276
mortality, 24/290
newborn, 24/642
oculomotor nerve injury, 24/63
papilledema, 24/280
pathogenesis, 24/276
pathology, 24/278
postoperative care, 24/289
prognosis, 24/290
pupil, 24/280, 290
radiology, 24/282
recurrence, 24/289
surgery, 24/641, 645
symptom, 24/279, 645
treatment, 24/287, 289, 645, 656
twist drill, 24/287
unusual site, 24/280, 283, 288
ventriculography, 24/286
Acute symptomatic seizure, *see* Provoked seizure
Acute tension headache, *see* Acute muscle contraction headache
Acute toxic encephalitis, *see* Reye syndrome
Acute tubular necrosis
myoglobinuria, 40/338, 41/276
quail eater disease, 41/272
Acute urticaria
penicillamine intoxication, 49/235
Acute viral encephalitis, 56/131-144
arbovirus, 56/126, 132
arenavirus, 34/193, 294
Bunyamwera virus, 34/68
bunyavirus, 34/5
California encephalitis, 56/134
California encephalitis virus, 34/78
central European tick-borne encephalitis, 34/76, 56/134, 139
classification, 34/63
clinical features, 34/65, 56/131
clinical symptom, 56/142
Colorado tick fever, 34/77, 56/134
computer assisted tomography, 56/131
CSF, 34/66, 56/131
diagnosis, 56/141
diagnostic method, 56/143
diphasic meningoencephalitis, 34/76
eastern encephalitis, 56/134, 136
eastern equine encephalomyelitis, 34/72
EEG, 56/131
enterovirus, 56/126, 132
epidemiology, 34/64, 56/132
etiology, 34/63-65
Far Eastern tick-borne encephalitis, 56/134, 139
flavivirus, 56/137
herpes simplex virus, 56/126, 131
history, 56/142
Japanese B encephalitis, 34/75, 56/134, 138
Kyasanur Forest disease, 34/77
La Cross virus encephalitis, 56/134
laboratory diagnosis, 34/67
Leptospira, 56/126
louping ill, 34/77
louping ill encephalitis, 56/134
lymphocytic choriomeningitis virus, 56/126
mosquito, 56/133
mumps virus, 56/126
Murray Valley encephalitis, 34/76, 56/134, 139
myxovirus associated, 34/79
Omsk hemorrhagic fever, 34/77
pathology, 56/132
Powassan virus, 34/78, 56/134
prevention, 56/144
reovirus, 56/141
Reye syndrome, 34/80
sandfly, 56/133
season, 56/132

St. Louis encephalitis, 34/73-75, 56/134, 137
tick, 56/133
tick-borne encephalitis, 34/76
Togavirus, 34/68-78
treatment, 56/144
type, 56/125
Venezuelan equine encephalitis, 34/73, 56/134, 137
West Nile encephalitis, 34/76
western encephalitis, 56/134, 136
western equine encephalitis, 34/72
Acute viral meningitis, *see* Meningitis serosa
Acute viral myositis, 56/193-203
acute polymyositis, 56/199
adenovirus, 56/195
alphavirus, 56/193
animal model, 56/194-196
arbovirus, 56/195
benign acute childhood myositis, 56/198
Bornholm disease, 56/198
cowpox virus, 56/194
Coxsackie virus A, 56/194
Coxsackie virus B, 56/195
ECHO virus 9, 56/195
ECHO virus 11, 56/195
epidemic benign myalgia, 56/199
Epstein-Barr virus, 56/195
experimental, *see* Experimental acute viral myositis
hepatitis B virus, 56/195
herpes simplex virus, 56/195
herpes simplex virus type 1, 56/194
herpes zoster virus, 56/195
herpetovirus, 56/194
human immunodeficiency virus, 56/195
influenza A virus, 56/194
influenza B virus, 56/195
murine sarcoma virus, 56/194
muscle cell culture, 56/193
Newcastle disease virus, 56/193
orthomyxovirus, 56/193
papovavirus, 56/193
parainfluenza virus, 56/195
paramyxovirus, 56/193
Phlebotomus fever virus, 56/195
picornavirus, 56/193
polyomavirus, 56/194
retrovirus, 56/194
Ross river virus, 56/194
Rous sarcoma virus, 56/194
Semliki forest virus, 56/194
simian acquired immune deficiency virus type D, 56/194
simian virus 40, 56/193
Togavirus, 56/193
Acyclovir, *see* Aciclovir
Acyl-alkenyl-glycerophospholipid
biochemistry, 10/240
Acyl-alkyl-glycerophospholipid
biochemistry, 10/240
Acylcarnitine moiety
medium chain acyl-coenzyme A dehydrogenase, 66/411
unsaturated, 66/411
Acyl-coenzyme A
carnitine, 66/400
pyruvate carboxylase deficiency, 66/418
Acyl-coenzyme A dehydrogenase
hypoglycemia, 37/527
hypoglycin, 37/523-526, 65/130
valproic acid induced hepatic necrosis, 65/126
Acyl-coenzyme A oxidase
biochemistry, 66/519
Acyl-coenzyme A oxidase deficiency
peroxisomal disorder, 66/510, 517
Acylsphingosine deacylase
Farber disease, 66/219
Adamantinoma
melanotic, *see* Melanotic adamantinoma
Adamkiewicz artery
artificial embolization, 32/500, 503
ascending branch, 32/469
blood supply, 12/478
branch principle, 12/494
dissecting aorta aneurysm, 63/39
9th intercostal artery, 32/479
intramedullary angioma, 32/495
level, 55/96
origin, 32/481
pseudotetany, 32/507
retrospinal angioma, 32/483
schema, 20/502
spinal angioma, 32/474
spinal cord, 12/478
steal syndrome, 32/487
sulcocommissural artery, 32/476
6th thoracic artery, 32/490
Adamkiewicz artery syndrome
anterior spinal artery syndrome, 12/499
Adams, F.
history, 2/7
Adams-Oliver syndrome
scalp defect, 30/100
Adams, R.
diagnosis, 1/31
physical examination, 1/31

Adams-Richardson classification
diffuse sclerosis, 9/472
Adams-Stokes attack, *see* Adams-Stokes syndrome
Adams-Stokes syndrome
brain blood flow, 22/263, 53/47
cardiac dysrhythmia, 39/260-262
congenital heart disease, 63/5
Emery-Dreifuss muscular dystrophy, 40/389
Kearns-Sayre-Daroff-Shy syndrome, 43/143
myotonic dystrophy, 40/507, 522
pallidal necrosis, 49/466
Adaptation
rehabilitation, 55/240
visual, *see* Visual adaptation
Adaptation syndrome
general, *see* General adaptation syndrome
local, *see* Local adaptation syndrome
Adaptometer
nyctalopia, 13/159
ADD syndrome, *see* Attention deficit disorder syndrome
Addict, *see* Drug addiction
Addiction, *see* Drug addiction
Addison anemia, *see* Pernicious anemia
Addison disease
adrenal gland, 39/475-483
adrenal hypoplasia, 42/512
adrenal insufficiency, 70/210
adrenoleukodystrophy, 39/482, 42/513
anxiety, 46/431
behavior disorder, 42/512, 46/599
behavior disorder in children, 42/512
candidiasis, 42/512
contracture, 39/480
convulsion, 39/479
CSF, 39/481
delirium, 46/544
dementia, 42/512, 46/396
depression, 46/427
diffuse sclerosis, 39/482
EEG, 39/480
endocrine candidiasis syndrome, 42/642
flexion contracture, 39/480
headache, 48/6
hiccup, 39/479
hyponatremia, 28/499
infectious striatopallidodentate calcification, 49/417
laboratory finding, 39/476
leukodystrophy, 10/36, 129, 39/482
mental change, 39/477-479
muscle weakness, 41/250
myasthenia gravis, 41/240
myelinoclastic diffuse sclerosis, 9/481
myopathy, 39/477
neurofibromatosis, 14/732
neuropathy, 39/480
optic atrophy, 13/81
orthochromatic leukodystrophy, 10/114
papilledema, 39/479
pathology, 39/481
penis erection, 75/89
polyneuropathy, 42/512
pseudohypoparathyroidism, 6/706, 49/420
psychosis, 46/462
seizure, 42/512
spastic paraplegia, 42/512
syncope, 39/479
treatment, 39/476
tremor, 42/512
X-linked adrenoleukodystrophy, 66/457
Addison-Schilder disease, *see* Adrenoleukodystrophy
Addisonian crisis
corticosteroid, 63/359
Addisonian encephalopathy, *see* Adrenoleukodystrophy
Addisonian pernicious anemia, *see* Pernicious anemia
Adducted thumb syndrome, 43/328-331
antimongoloid eyes, 43/331
cheilopalatoschisis, 43/331
craniosynostosis, 43/331
dermatoglyphics, 43/331
MASA syndrome, 43/255
microcephaly, 43/331
myelin dysgenesis, 43/328
ophthalmoplegia, 43/331
talipes, 43/331
telecanthus, 43/331
Adductor reflex
tibial, *see* Tibial adductor reflex
Adenine arabinoside, *see* Vidarabine
Adenine nucleotides
migraine, 48/100
Adenocarcinoma
Huntington chorea, 49/291
Adenoglioma, *see* Ependymoma
Adenohypophyseal postpartum necrosis
Sheehan syndrome, 2/455
Adenohypophyseal syndrome
acromegaly, 2/454
Chiari-Frommel syndrome, 2/444, 456
gigantism, 2/454
hypophyseal tumor, 2/456-461
hypophysectomy, 2/461-463

pituitary dwarfism, 2/455
Sheehan syndrome, 2/455
Adenohypophysis
cell type, 2/442
corticotropin, 2/442
hormone, 2/442
human menopausal gonadotropin, 2/442
luteinizing hormone, 2/442
oxytocin, 2/443
prolactin, 2/442
reduced glutathione, 2/442
thyrotropic hormone, 2/442
Adenoid cystic carcinoma
skull base, 68/466
skull base tumor, 68/489
Adenoma
adrenal, *see* Adrenal adenoma
basophil, *see* Basophil adenoma
ceruminal gland, *see* Ceruminal gland adenoma
chromophilic, *see* Chromophilic adenoma
chromophobe, *see* Chromophobe adenoma
hypophyseal, *see* Hypophyseal adenoma
islet cell, *see* Islet cell adenoma
malignant, *see* Malignant adenoma
multiple endocrine, *see* Multiple endocrine adenoma
pancreas, *see* Pancreas adenoma
parathyroid, *see* Parathyroid adenoma
pituitary, *see* Pituitary adenoma
sebaceous, *see* Sebaceous adenoma
thyroid, *see* Thyroid adenoma
thyrotropic hormone secreting, *see* Thyrotropic hormone secreting adenoma
Adenoma sebaceum, *see* Sebaceous adenoma
Adenomatosis
multiple endocrine, *see* Multiple endocrine adenomatosis
Adenomatosis syndrome
multiple endocrine, *see* Multiple endocrine adenomatosis syndrome
Adenomorphous brown lipocytoblastoma, *see* Hibernoma
Adenopathy
Abt-Letterer-Siwe disease, 42/441
nasopharyngeal tumor, 17/205
Adenosine
brain blood flow, 53/54
brain infarction, 53/421
brain microcirculation, 53/80
smooth muscle function, 75/624
Adenosine monophosphate, *see* Adenosine phosphate
Adenosine monophosphate deaminase
Duchenne muscular dystrophy, 40/378
myoglobinuria, 62/558
Adenosine monophosphate deaminase deficiency
biochemistry, 62/512
chromosome 1p, 62/512
creatine kinase, 62/512
differential diagnosis, 62/511
histopathology, 62/512
ischemic exercise, 62/512
limb girdle syndrome, 62/190
muscle cramp, 62/511
myalgia, 62/511
myoglobinuria, 62/512
primary type, 62/511
secondary type, 62/511
Adenosine phosphate
brain infarction, 53/127
brain ischemia, 53/127
chronic migrainous neuralgia, 48/253
migraine, 48/92, 100
multiple sclerosis, 9/389
Adenosine triphosphatase
epileptic focus, 15/68
GM2 gangliosidosis, 10/291
multicore myopathy, 62/339
muscle fiber type, 62/4
myofibrillar, *see* Myofibrillar adenosine triphosphatase
NARP syndrome, 66/424
ouabain sensitive, *see* Ouabain sensitive adenosine triphosphatase
wallerian degeneration, 9/37
Adenosine triphosphatase 6 mutation syndrome
ataxia, 62/510
dementia, 62/510
epilepsy, 62/510
metabolic encephalopathy, 62/510
metabolic myopathy, 62/510
mitochondrial disease, 62/510
mitochondrial DNA point mutation, 62/509
multiple neuronal system degeneration, 62/510
muscle weakness, 62/510
proximal limb weakness, 62/510
secondary pigmentary retinal degeneration, 62/510
sensory neuropathy, 62/510
subacute necrotizing encephalomyelopathy, 62/510
Adenosine triphosphate
brain infarction, 53/127
brain ischemia, 53/127
Duchenne muscular dystrophy, 40/382, 385
hypoxia, 75/265

migraine, 48/100
muscle, 41/264
muscle contraction, 40/340
myophosphorylase deficiency, 41/188, 62/486
myotonic dystrophy, 40/497, 523
subacute necrotizing encephalomyelopathy, 42/588
synthesis, 66/390
wallerian degeneration, 7/215
Adenovirus
acquired immune deficiency syndrome, 56/283, 287, 514
acute polymyositis, 56/200
acute viral myositis, 56/195
adenovirus type 7, 18/133
astrocytoma, 56/291
benign acute childhood myositis, 56/198
cancer, 69/450
deafness, 56/107
DNA, 56/281
features, 56/8
gene therapy, 66/114
Guillain-Barré syndrome, 56/290
human, 56/2
immunocompromised host, 56/474
immunosuppression, 56/283
LLE 46 strain, 18/133
mast, *see* Mastadenovirus
mastadenovirus genus, 56/9
meningocerebral angiodysplasia, 56/290
morphology, 56/281
neurotropic virus, 56/8
nonneurologic syndrome, 56/282
organ transplantation, 56/283
protein, 56/281
retinoblastoma, 56/291
Reye syndrome, 9/549, 56/9, 150
species h1-h37, 56/9
structure, 56/281
taxonomic place, 56/2
vertigo, 56/107
viral meningitis, 56/283
virus-host relationship, 56/282
Adenovirus associated vector
gene therapy, 66/116
Adenovirus classification
aviadenovirus genus, 56/281
mastadenovirus genus, 56/281
6 subgenera (A-F), 56/281
Adenovirus encephalitis
brain infarction, 56/286
CSF, 56/286
encephalomalacia, 56/286
epilepsy, 56/283
immunosuppression, 56/286
Reye syndrome, 56/287-289
ultrastructure, 56/286
Adenovirus infection
Reye syndrome, 56/288
Adenovirus meningitis
prevalence, 56/283
Adenovirus meningoencephalitis
adenovirus type 1, 56/283
adenovirus type 3, 56/286
adenovirus type 5, 56/286
adenovirus type 7, 56/283-285
CSF, 56/284
EEG, 56/284
Adenovirus vector
brain tumor, 67/300
gene therapy, 66/115
Adenyl cyclase
classification, 74/149
Duchenne muscular dystrophy, 40/3, 382, 385
hemostasis, 63/303
Adenylate cyclase
neuroleptic agent, 65/279
pseudohypoparathyroidism, 42/578
Adenylate deaminase, *see* Adenosine monophosphate deaminase
Adenylate kinase
CSF, 54/199
Adenylosuccinate deficiency
epilepsy, 72/222
Jaeken syndrome, 60/673
ADH secretion
inappropriate, *see* Inappropriate antidiuretic hormone secretion
Adhesive arachnoiditis, *see* Spinal arachnoiditis
Adiadochokinesia
cerebellar ataxia, 1/323, 326
neocerebellar syndrome, 2/422
premotor cortex symptom, 2/731
Von Hippel-Lindau disease, 14/115
Adie syndrome, *see* Holmes-Adie syndrome
Adiposis dolorosa
alcoholism, 8/347
autonomic nervous system, 43/57
interstitial neuropathy, 43/57
Krabbe-Bartels disease, 14/406
menopause, 8/347
mental depression, 8/347
mental deterioration, 8/347
obesity, 43/57
pain, 8/347, 43/57
pituitary disorder, 8/347

subcutaneous lipoma, 8/347, 43/57
syphilis, 8/347
Adiposogenital dystrophy
Sturge-Weber syndrome, 14/512
Adiposogenital syndrome
Friedreich ataxia, 21/348
Adipsia
hypothalamic lesion, 1/500
Adolescent kyphosis, *see* Scheuermann disease
Adolescent spinal muscular atrophy
allelic nature, 59/41
Adolescent stigma scale
epilepsy, 72/434
Adoption
sexual function, 26/460
Adoptive transfer
brain tumor, 67/295
ADR syndrome
Babinski sign, 42/113
contracture, 42/113
mental deficiency, 42/112
nerve conduction velocity, 42/113
nystagmus, 42/112
osteoporosis, 42/113
paralysis, 42/112
Adrenal adenoma
hyperaldosteronism, 39/492
visceral lesion, 14/51
Adrenal calcification
*see also* Wolman disease
Wolman disease, 10/504, 546, 14/779, 60/165
Adrenal cortex atrophy, *see* Adrenoleukodystrophy
Adrenal cortex hypertrophy
general adaptation syndrome, 45/252
Adrenal disorder
adrenoleukodystrophy, 70/215
childhood, 70/216
congenital adrenal hyperplasia, 70/212
Cushing syndrome, 70/205
11-hydroxylase deficiency, 70/214
17-hydroxylase deficiency, 70/215
21-hydroxylase deficiency, 70/213
3β-hydroxysteroid dehydrogenase deficiency, 70/215
hypercortisolism, 70/209
hypocortisolism, 70/211
survey, 70/205
Adrenal gland
Addison disease, 39/475-483
congenital adrenocortical hyperplasia, 39/487
Conn syndrome, *see* Hyperaldosteronism
Cushing syndrome, 39/483-492
headache, 5/230
hyperaldosteronism, 39/492-496
multiple endocrine adenoma, 39/503
neurologic aspect, 39/469-504
pheochromocytoma, 39/496-504, 42/765
X-linked adrenoleukodystrophy, 66/460
Adrenal hemorrhage, *see* Waterhouse-Friderichsen syndrome
Adrenal hypoplasia
Addison disease, 42/512
anencephaly, 42/12
congenital ichthyosis, 43/406
pituitary gland, 42/11
Adrenal insufficiency
Addison disease, 70/210
adrenoleukodystrophy, 10/132, 47/597
Allgrove syndrome, 70/218
etiology, 70/210
hypoglycemia, 27/70
musculoskeletal abnormality, 70/211
orthostatic hypotension, 63/145
tuberculosis, 70/210
X-linked adrenoleukodystrophy, 66/457
Adrenal insufficiency myopathy
arthralgia, 62/536
corticosteroid withdrawal myopathy, 62/536
fatigue, 62/536
fever, 62/536
myalgia, 62/536
weight loss, 62/536
Adrenal medulla
autonomic nervous system, 74/35
general adaptation syndrome, 45/253
metadrenalin, 74/144
paraganglionic tissue, 8/473
Adrenal medullary neuroblastoma
neurofibromatosis type I, 50/370
Adrenal steroid
biosynthetic pathway, 70/212
Adrenal system
hypothalamic hypophyseal, *see* Hypothalamic hypophyseal adrenal system
Adrenalin
acid maltase deficiency, 41/181
adrenergic receptor, 74/163
allergic neuropathy, 5/234
anxiety tremor, 6/818
blood-brain barrier, 48/64
brain vasospasm, 11/126
caffeine intoxication, 65/260
headache, 48/91
hypertension, 75/471
hyperventilation syndrome, 63/434
migraine, 48/91

noradrenalin, 74/144
pharmacology, 74/144, 146
pial vessel, 11/515
postsynaptic, 74/146
progressive dysautonomia, 59/152
stress, 45/248, 74/313
subarachnoid hemorrhage, 12/112
synaptic transmission, 1/48
vascular response, 74/168
Adrenalin tremor
postural tremor, 6/818
Adrenergic neuron
malnutrition, 29/5
Adrenergic receptor
adrenalin, 74/163
alpha, *see* Alpha adrenergic receptor
catecholamine, 74/163
dopamine β-mono-oxygenase deficiency, 59/163
genetics, 75/242
noradrenalin, 74/163
pharmacology, 74/147
Adrenergic receptor blocking agent
pupil, 74/419
Adrenergic receptor stimulating agent
autonomic nervous system, 22/250
ephedrine intoxication, 65/260
orthostatic hypotension, 63/163
pupil, 74/419
subarachnoid hemorrhage, 12/112
vasomotor reflex response, 1/469
Adrenergic receptor subtype
human, 74/149
Adrenoceptor, *see* Adrenergic receptor
Adrenocortical carcinoma
Cushing syndrome, 22/550
SBLA syndrome, 42/769
Adrenocortical hyperplasia
congenital, *see* Congenital adrenocortical hyperplasia
hyperaldosteronism, 39/493
Adrenocortical insufficiency, *see* Addison disease
Adrenocortical lesion
neurofibromatosis, 14/732
Adrenocorticosteroids
brain fat embolism, 55/191
brain infarction, 53/427
chemotherapy, 39/120
neurologic toxicity, 39/120
polyarteritis nodosa, 8/126, 11/451, 39/300, 55/346, 357
Adrenocorticotropin, *see* Corticotropin
Adrenogenital syndrome
fever, 43/68
hemihypertrophy, 22/550
neurofibromatosis type I, 14/492
Adrenoleukodystrophy, 10/128-133
*see also* Adrenomyeloneuropathy
ACTH test, 47/594
Addison disease, 39/482, 42/513
adrenal disorder, 70/215
adrenal insufficiency, 10/132, 47/597
adrenomyeloneuropathy, 47/595
adult, *see* Adrenomyeloneuropathy
asymptomatic, 47/594
ataxia, 10/132, 51/384, 60/169
autosomal recessive inheritance, 51/383
behavior disorder, 42/483
behavior disorder in children, 42/483
biochemistry, 10/132
biopsy, 47/597
bone marrow transplantation, 66/87, 97
brain stem auditory evoked potential, 47/597
central pontine myelinolysis, 47/595
cerebrohepatorenal syndrome, 47/594, 596-598
child onset, 66/96
chorea, 60/169
chromosome Xq28, 60/169
classification, 66/5
clinical course, 60/169
clinical features, 10/131, 47/594
computer assisted tomography, 47/597, 60/169, 671
consciousness, 45/121
convulsive seizure, 60/169
cortical blindness, 60/169
cytoplasmic inclusion body, 47/595
deafness, 51/384, 60/169
dementia, 46/402, 60/169
demyelinating neuropathy, 47/622
diagnostic criteria, 47/595
differential diagnosis, 47/596, 59/402, 60/168
diffuse sclerosis, 47/419
dysarthria, 60/169
dysphagia, 42/483
encephalopathy, 45/119
endocrine disorder, 42/483
enzyme defect, 47/598
enzyme deficiency polyneuropathy, 51/383
epilepsy, 60/169, 72/222
fatty acid, 66/96
gene, 66/96
gene site, 60/169
genetics, 47/597
history, 60/168
hyperreflexia, 42/483
incontinence, 60/169

inheritance, 47/594
laboratory diagnosis, 47/596
lesion, 66/3
leukoencephalopathy, 47/593-598
linkage study, 47/597
melanoderma, 10/131, 42/483
minimal brain dysfunction, 47/595
multiple sclerosis, 47/595
myelopathy, 51/384
myoclonia, 60/169
neonatal, *see* Neonatal adrenoleukodystrophy
nerve biopsy, 51/384
nerve conduction velocity, 60/670
neuropathy, 51/384
nuclear magnetic resonance, 60/169
onset, 60/169
optic atrophy, 10/132, 42/483, 51/384, 60/655
orthochromatic leukodystrophy, 10/114
pathology, 10/129
peroxisomal disease, 51/383, 60/169
peroxisome, 47/598
phenotype, 47/594
pigmentation, 60/667
polyneuropathy, 60/168
prenatal diagnosis, 47/598
pseudobulbar paralysis, 60/658
psychiatric disorder, 60/665
psychosis, 60/169
Schwann cell, 47/593, 51/384
seizure, 10/132
skin fibroblast culture, 47/593
skin pigmentation, 60/169
spastic paraplegia, 60/169
spastic tetraplegia, 43/483
survey, 70/215
survival data, 66/98
testis, 47/595
treatment, 10/132, 47/598
type, 66/8
very long chain fatty acid, 47/597, 51/384
very long chain fatty acid β-oxidation, 60/169
vomiting bout, 10/131, 60/169
X-linked, *see* X-linked adrenoleukodystrophy
X-linked recessive ataxia, 60/638
Adrenoleukomyeloneuropathy, *see* Adrenomyeloneuropathy
Adrenomedullary hormonal system
sympathetic nervous system, 74/138
Adrenomyeloneuropathy
*see also* Adrenoleukodystrophy
ACTH test, 47/595
adrenocortical function, 59/354
adrenoleukodystrophy, 47/595
alopecia, 60/170
ataxia, 60/170
autonomic nervous system, 75/24
biochemical study, 60/169
bone marrow transplantation, 59/355
brain stem auditory evoked potential, 60/171
cerebrohepatorenal syndrome, 60/170
chronic inflammatory demyelinating polyradiculoneuropathy, 51/530
clinical features, 59/352, 60/170
demyelinating neuropathy, 47/622
diet, 59/354
differential diagnosis, 60/170, 660
EMG, 60/171
epidemiology, 60/169
familial spastic paraplegia, 59/308
frequency, 60/169
gene mapped to Xq28, 59/354
heterozygote, 59/355
history, 60/169
hyperpipecolatemia, 60/170
hypogonadism, 60/170
long chain fatty acid, 59/354
mimicking amyotrophic lateral sclerosis, 59/402
neuropathology, 60/171
onset, 66/96
polyneuropathy, 60/168, 170
psychosis, 60/170
review, 47/593-599
schizophrenic syndrome, 60/170
sexual impotence, 60/170
spastic paraparesis, 60/170
spastic paraplegia, 59/439
spasticity, 60/661
steroid treatment, 59/355
X-linked adrenoleukodystrophy, 66/453
Adriamycin
cardiomyopathy, 63/136
chemotherapy, 39/114
congestive heart failure, 63/136
neurologic toxicity, 39/114
neuropathy, 69/460, 471
rhabdomyosarcoma, 41/382
transient ischemic attack, 63/136
Adson, A.W., 1/5
Adson-Coffey maneuver
neurovascular compression syndrome, 2/150
thoracic outlet syndrome, 51/124
Adult acid maltase deficiency
clinical features, 27/222
limb girdle syndrome, 62/190
myopathy, 27/229, 62/480
pathology, 27/229

respiratory dysfunction, 62/480
tardive myopathy, 62/480
Adult celiac disease, *see* Celiac disease
Adult dominant myotubular myopathy
rimmed muscle fiber vacuole content, 62/18
rimmed muscle fiber vacuole type, 62/18
Adult Gaucher disease
acid phosphatase, 10/528
anemia, 10/564, 66/123
ataxia, 42/437
biochemistry, 29/351
bone fragility, 10/564, 66/123
bone marrow transplantation, 66/129
clinical features, 10/307, 60/154-156
enzyme replacement therapy, 66/127
genetic mutation, 66/126
glucosylceramidase, 66/126
hepatosplenomegaly, 66/123
leukopenia, 66/123
skin pigmentation, 10/564
thrombocytopenia, 10/564, 66/123
Adult GM1 gangliosidosis
corneal dystrophy, 60/652
facial dysmorphia, 60/666
mental deficiency, 60/664
myoclonic epilepsy, 60/663
myoclonus, 60/662
Niemann-Pick disease, 60/654
optic atrophy, 60/654
spasticity, 60/661
Adult leukodystrophy
spastic paraplegia, 59/439
Adult neuroaxonal dystrophy
primary neuroaxonal dystrophy, 49/408
progressive pallidal degeneration, 42/257
Adult neuronal ceroid lipofuscinosis, 42/465-467
autosomal recessive inheritance, 42/465
Batten disease, 66/671
cerebellar ataxia, 42/466
choreoathetoid movement, 42/466
Creutzfeldt-Jakob disease, 56/546
dementia, 42/466, 60/665
differential diagnosis, 66/683
EEG, 42/466
epilepsy, 60/665
familial amaurotic idiocy, 10/218, 225, 588, 622, 21/62
gene, 72/135
mental deterioration, 42/466
muscle fiber feature, 62/17
myoclonic epilepsy, 38/579, 60/662
myoclonus, 42/466
neurofibrillary tangle, 46/274
pathology, 66/684
rigidity, 42/466
Schaffer-Spielmeyer cell change, 10/218
seizure, 42/466
survey, 66/24
Adult onset spongiform leukodystrophy
classification, 66/5
type, 66/15
Adult parasomnia
non-REM sleep, 45/142
Adult polyglucosan body disease
axonal sensorimotor neuropathy, 62/484
brancher deficiency, 62/484
dementia, 62/484
lower motoneuron sign, 62/484
metabolic encephalopathy, 62/484
neurogenic bladder, 62/484
sensory loss, 62/484
upper motoneuron sign, 62/484
Adult Refsum disease
peroxisomal deficiency, 66/17
Adult T-cell leukemia
human T-lymphotropic virus type I, 56/525
Sézary syndrome, 18/235, 56/527
Adult Tay-Sachs disease
β-*N*-acetylhexosaminidase A deficiency, 29/365
Adverse Drug Events Profile
epilepsy, 72/431
Adverse drug reaction
epilepsy, 73/373
myasthenia gravis, 62/420
Adversive seizure
irritation, 2/731
premotor cortex, 2/731
Adynamia episodica, *see* Hyperkalemic periodic paralysis
Aedes sollicitans
eastern encephalitis, 56/136
Aedes triseriatus
California encephalitis, 56/140
Afebrile seizure
child, 72/27
prevalence, 72/27
Affect
emotion, 3/318, 333, 335-337
frontal lobe lesion, 45/31
Klüver-Bucy syndrome, 45/45
temporal lobe epilepsy, 45/45
temporal lobe syndrome, 45/44
Wernicke-Korsakoff syndrome, 45/199
Affect lability
pathologic laughing and crying, 45/220
Affect regulation

traumatic psychosyndrome, 24/555, 567
Affective disorder
acetylcholine, 29/451
bipolar, *see* Bipolar affective disorder
brain substrate, 46/425-434
catatonia, 46/423
delirium, 46/552
dopamine, 46/475
hypothalamus, 46/473
limbic system, 46/473
noradrenalin, 46/475
schizoidism, 43/209
serotonin, 46/475
unipolar, *see* Unipolar affective disorder
Affective psychosis
*see also* Manic depressive psychosis
classification, 46/444-446
computer assisted tomography, 46/452
Cushing syndrome, 46/462
EEG, 46/460
functional psychosis, 46/444-446
intelligence quotient, 46/457
lateralization, 46/503
neurochemistry, 46/448-450
psychopharmacology, 46/448-450
right temporal lobe, 46/504
Afferent
cerebellar, *see* Cerebellar afferent
flexor reflex, 1/57
hypothalamus, 2/437
nociceptor, *see* Nociceptor afferent
pupillary defect, 74/415
respiratory center, 63/430
striatal, *see* Striatal afferent
subthalamic nucleus, 49/10
Afibrinogenemia
brain hemorrhage, 38/72
coagulopathy, 63/312
congenital, 38/72
familial, 38/72
subdural hematoma, 38/72
Aflatoxin B
Reye syndrome, 49/217
toxic encephalopathy, 49/217
Aflatoxin intoxication
Reye syndrome, 29/338, 56/159
African histoplasmosis
*see also* Histoplasma duboisii
amphotericin B, 35/513
Histoplasma duboisii, 35/513
African tick-borne typhus
rickettsial infection, 34/657
African trypanosomiasis, 35/67-82
*see also* Trypanosoma brucei
arsenic encephalopathy, 52/343
arsenic intoxication, 52/343
athetosis, 52/341
benign intracranial hypertension, 52/342
brain edema, 52/343
bulimia, 52/341
cardiovascular, 52/340
chorea, 52/341
choroid plexus, 52/340
clinical features, 35/69-73, 52/340
clinical variant, 52/342
cold feeling, 52/340
cranial nerve palsy, 52/341
CSF, 35/75, 52/342
dementia, 52/341
demyelinating encephalitis, 35/72, 52/342
diagnosis, 35/79, 52/343
difluoromethylornithine, 52/343
dystonia, 52/341
EEG, 35/76-79, 52/341
encephalitis, 52/342
encephalopathy, 52/343
ependymitis, 52/340
epidemiology, 35/8-10
geographic distribution, 35/68
headache, 52/341
histopathology, 52/340
history, 35/67-69
hypersomnia, 52/341
incidence, 52/339
initial lesion, 35/69, 52/340
insomnia, 52/341
intracranial hypertension, 52/342
laboratory finding, 52/342
laboratory investigation, 35/74-79
lymphocytic meningitis, 52/342
lymphosanguine phase, 35/70, 52/340
melarsoprol, 52/343
meningoarachnoiditis, 52/340
meningoencephalitic phase, 35/71-73
meningoencephalitis, 52/341
Mott cell, 52/340
muscle cramp, 52/341
muscular hypertonia, 52/341
mutism, 52/341
myalgia, 52/341
neuropathology, 52/339
neuropathy, 52/341
papuloerythema, 52/340
parasitic disease, 35/8-10
paratonia, 52/341
pathogenesis, 52/339

pathology, 35/73
pentamidine, 52/343
plasmacytosis, 52/340
prophylaxis, 35/68
pruritus, 52/340
psychiatric symptom, 52/341
radiculitis, 52/341
radiculoneuropathy, 52/341
serologic test, 52/342
thalamic pain, 52/341
treatment, 35/80-82, 52/343
Trypanosoma, 35/67, 74, 52/339
tsetse fly, 35/67-69, 52/339
vaccine, 52/343
Afterdischarge
definition, 1/49
Afterimage, *see* Visual perseveration
After-nystagmus
diagnostic technique, 2/333
optokinetic nystagmus, 1/577
rotation vestibular nystagmus, 2/333
Afterpotential
features, 7/77
After-spasm
myotonia, 40/536
Agalagitakg intoxication, *see* Balaenoptera borealis intoxication
Agammaglobulinemia
herpes simplex virus, 56/211
polymyositis, 56/203, 62/373
viral encephalitis, 56/327
X-linked, *see* X-linked agammaglobulinemia
Aganglionic megacolon, *see* Congenital megacolon
Age at death
chorea-acanthocytosis, 63/281
Age at stroke
congenital heart disease, 63/3
Aged
antihypertensive treatment, 63/85
arterial hypertension, 63/85
autonomic nervous system, 75/9
cerebellar astrocytoma, 18/18
cerebrovascular disease, 53/42
drug induced dementia, 46/393
infectious endocarditis, 63/112
meningioma, 68/409
migraine, 48/19, 120
multiple sclerosis, 47/63
orthostatic hypotension, 1/473, 2/121, 63/154
paraplegia, 1/208
parasympathetic ganglion, 75/6
progressive dysautonomia, 59/158
reflex syncope, 15/824
senile purpura, 38/71
sympathetic nervous system, 75/5
Agenesis
abducent nerve, *see* Abducent nerve agenesis
accessory nerve, *see* Accessory nerve agenesis
acoustic nerve, *see* Acoustic nerve agenesis
caudal, *see* Caudal agenesis
cerebellar, *see* Cerebellar agenesis
cerebellar vermis, *see* Cerebellar vermis agenesis
coccyx, *see* Coccyx agenesis
corpus callosum, *see* Corpus callosum agenesis
cranial nerve, *see* Cranial nerve agenesis
definition, 30/4
dens, *see* Dens agenesis
facial nerve, *see* Facial nerve agenesis
glossopharyngeal nerve, *see* Glossopharyngeal nerve agenesis
hypoglossal nerve, *see* Hypoglossal nerve agenesis
internal carotid artery, 12/301
kidney, *see* Renal agenesis syndrome
lateral cerebellar, *see* Lateral cerebellar agenesis
oculomotor nerve, *see* Oculomotor nerve agenesis
olfactory bulb, *see* Olfactory bulb agenesis
optic chiasm, *see* Optic chiasm agenesis
optic nerve, *see* Optic nerve agenesis
pectoralis muscle, *see* Pectoralis muscle agenesis
pituitary gland, *see* Pituitary gland agenesis
renal, *see* Renal agenesis syndrome
sebaceous gland, *see* Sebaceous gland agenesis
septum pellucidum, *see* Septum pellucidum agenesis
sweat gland, *see* Sweat gland agenesis
thumb, *see* Thumb agenesis
tongue, *see* Tongue agenesis
trigeminal nerve, *see* Trigeminal nerve agenesis
trochlear nerve, *see* Trochlear nerve agenesis
urogenital, *see* Urogenital agenesis
vagus nerve, *see* Vagus nerve agenesis
vertebral, *see* Vertebral agenesis
Agent orange, *see* 2,4,5-Trichlorophenoxy acetic acid
Ageusia
*see also* Taste
acute amebic meningoencephalitis, 52/315, 317
anosmia, 24/17, 20
bilateral, 24/19
brain cortex, 24/15, 19
brain injury, 24/15
carbamazepine intoxication, 65/507
chorda tympani, 24/15
clinical test, 24/16
concomitant anosmia, 24/17

electrogustometry, 24/17
electrostimulation, 24/16
geniculate ganglion, 24/15
history, 24/1
hyponatremia, 63/545
incidence, 24/17
injury severity, 24/18
lesion site, 24/19
pathology, 24/19
pathway, 24/15, 20
posttraumatic amnesia, 24/18
psychogenic, 24/19
recovery, 24/18
specific, 42/422
taste center, 24/20
taste receptor, 24/16
Aggression
alcoholism, 46/178
brain command system, 45/274
central pontine myelinolysis, 2/263
cerebral gigantism, 31/333, 43/337
clinical features, 46/177
definition, 46/177
emotion, 3/337, 344
emotional disorder, 3/344-346
epidemiology, 46/177
epilepsy, 3/344, 73/479
etiology, 46/179
family background, 46/178
frontal lobe syndrome, 55/138
head injury, 46/626
hyperactivity, 46/177-180
hyperkinetic child syndrome, 4/341, 43/201
hypothalamus, 2/440
ictal, *see* Ictal aggression
intelligence, 46/178
interictal, *see* Interictal aggression
Klinefelter syndrome, 43/560
Klinefelter variant XXXXY, 43/560
Klinefelter variant XXXYY, 43/563
Klinefelter variant XXYY, 31/484, 43/562
Klüver-Bucy syndrome, 45/262
Lesch-Nyhan syndrome, 42/152, 63/256
lithium, 46/180
mental deficiency, 43/258, 268
migraine, 48/161
organolead intoxication, 64/131
organotin intoxication, 64/142
postictal, *see* Postictal aggression
prodromal, *see* Prodromal aggression
propranolol, 46/180
temporal lobe, 2/718
temporal lobe epilepsy, 46/433
traumatic psychosis, 24/527
treatment, 46/179
Aging
autonomic nervous system, 74/107, 225, 230, 75/5
baroreceptor, 74/227
bladder, 74/237
blood flow, 74/232
blood pressure, 74/228
brain blood flow, 54/150
cardiovascular function, 74/228
cell, *see* Cell ageing
CNS, *see* CNS aging
cognition, 46/211, 491
cold pressor test, 74/233
cutaneous blood flow, 74/232
dementia, 27/479, 46/199-211
fat metabolism, 74/237
heart rate, 74/228
hypernatremia, 74/235
hyponatremia, 74/235
hypotension, 74/229
hypoxia, 74/233
insulin, 74/233
intestinal function, 74/237
learning, 27/470
memory, 27/470, 46/210
muscarinic receptor, 74/226
muscle, *see* Muscle aging
muscle fiber, 40/199
myelin, 7/47
nerve cell, 74/227
neuroaxonal dystrophy, 49/396
noradrenalin, 74/228
organophosphate intoxication, 64/224
parasympathetic ganglion, 74/226
Parkinson disease, 49/127
perception, 46/491
performance, 46/210
postpoliomyelitic amyotrophy, 59/35
pulse interval, 74/230
pupillary response, 74/238
receptor, 74/226
renal blood flow, 74/232
salt regulation, 74/235
schizophrenia, 46/491
senile plaque, 42/288, 46/210
sleep, 74/542
space, 74/297
spheroid body, 6/626
stress, 74/232
sudomotor function, 74/235
sympathetic beta receptor, 74/226
sympathetic ganglion, 74/225

sympathetic nerve, 74/226, 657
thermoregulation, 74/233
Wernicke aphasia, 46/211
Agitation
*see also* Restlessness
amphetamine intoxication, 65/258
disulfiram intoxication, 37/321
hypernatremia, 63/554
lidocaine intoxication, 37/450, 65/454
migraine, 48/162, 165
OKT3, 63/536
phenylpropanolamine intoxication, 65/259
pontine infarction, 12/15
spasmodic torticollis, 6/799
Agnosia, 4/13-39
acalculia, 4/17
acoustic, *see* Auditory agnosia
acoustic music type, 2/607
agnosia concept, 4/29-32
ahylognosia, 4/34
Alzheimer disease, 42/275, 46/251-253
amorphognosia, 4/34
amorphosynthesis, 4/26
amusia, 4/38, 195-206
anatomic localization, 3/224, 4/20-22, 36-39
animal, 4/14, 17
aphasia, 4/28
apperceptive, *see* Apperceptive agnosia
apractognosia, 4/23
apraxia, 4/23
asomatognosia, 4/14, 37
associative, *see* Associative agnosia
astereognosia, *see* Tactile agnosia
asymbolia, 4/14
attention, 4/18, 30
attention disorder, 3/189, 4/38
auditory, *see* Auditory agnosia
Balint syndrome, 45/409
behavior characterization, 4/38
bilateral tactile agnosia, 4/35
body, *see* Body agnosia
body half, 1/107
bone membranous lipodystrophy, 42/279
brain commissurotomy, 4/275-287
carbon monoxide intoxication, 64/32
central dyschromatopsia, 4/17
cerebral asthenopia, 4/25
cerebral dominance, 4/298-301
classification, 4/15-18, 31, 38
clinical features, 4/16-20, 33-35, 38
clock hand, 3/30
color, *see* Color agnosia
color perception, 45/334-336
confabulation, 4/24, 28
congenital acoustic, *see* Congenital auditory agnosia
congenital central deafness, 4/38, 428
congenital tactile, *see* Congenital tactile agnosia
consciousness, 45/114
constructional apraxia, 4/23
conventional, 3/8
corpus callosum disconnection syndrome, 2/760
cortical ablation, 4/15
cortical blindness, 4/14, 20
craniocerebral trauma, 24/491
definition, 4/13, 32
dementia, 46/203
denial of illness, 4/23, 29
depth perception, 4/16
developmental, *see* Developmental agnosia
disconnection syndrome, 4/17, 21, 28, 37
dissolutive, 2/662
distance perception, 4/16
dyschromatopsia, 4/17
examination, 4/18, 22, 28, 35, 36, 38
experimental work, 4/14
eye movement, 4/18
eyelid, 4/18
finger, *see* Body scheme disorder *and* Finger agnosia
form, 4/15, 26
Geschwind definition, 2/599
Gestalt theory, 3/12, 4/27
gnosis, 4/26
gustatory, 4/14
gustatory hallucination, 4/14
hallucination, 4/14, 45/352
hemisphere dominance, 4/281
hidden figures test, 4/17-19
historic survey, 4/14, 33
hysteria, 4/23
image, 4/16
instability of function, *see* Funktionswandel
intellectual defect, 4/27
intermodal association, 4/280
Klüver-Bucy syndrome, 3/349-356
labiobuccal stereognosis, 4/35
letter recognition, 45/344
macrostereognosia, 4/35
migraine, 48/161
mind blindness, 4/14
monochromatism, 45/335
multisensorial, 4/13
musical, *see* Musical agnosia
neurofibrillary tangle, 46/260-263
neuropsychology, 4/13, 29

note blindness, 2/607
object recognition, 45/333-347
oculomotor disorder, 3/226
olfactory, *see* Olfactory agnosia
pain, *see* Pain agnosia
parietal lobe tumor, 17/301
pathopsychophysiogenesis, 4/24-29, 36
penetrating brain wound, 4/15
perception, 4/30
perception and behavior unity, 4/22
perception classification, 45/338-341
perceptual recognition, 4/14, 29, 33
physiognomics, 4/16, 380
picture, 2/607
Poppelreuter test, 4/16, 19
primary and secondary perception, 4/26, 30, 36
primary tactile, *see* Primary tactile agnosia
propositional, 4/26, 31
prosopagnosia, 4/16, 22, 31, 45/344
pseudo, *see* Pseudoagnosia
psychiatry, 4/23
psychic deafness, 4/38, 428
resulting from general disorder, 4/27
semantic classification, 45/341-343
semantics, 3/36
sensory defect in commissurotomy, 4/275-287
sensory extinction, 4/30
sensory impairment, 4/24-27
shape perception, 45/334
simultaneous, *see* Simultanagnosia
situation, *see* Simultanagnosia
situational set, 4/29
spatial, *see* Spatial agnosia
spatial memory, 45/416
stereoanesthesia, 4/34
subjective world, 4/23
symbolic, *see* Symbolic agnosia
tactile, *see* Tactile agnosia
temporal, *see* Temporal agnosia
temporal lobectomy, 45/336
terminology, 4/14, 33
theory, 3/224, 4/13, 24-29
thing, 4/15
topographical memory, 4/16
true entity, 4/27, 32, 36
tubular vision, 4/26
verbal, 4/38
visual, *see* Visual agnosia
visual acuity, 45/334-336
visual disorientation, 45/335
visual field defect, 45/334
visual memory, 45/344
visual object, *see* Visual object agnosia
visual sensory deficit, 45/333-336
visuospatial, *see* Visuospatial agnosia
Agnostic alexia
associated defect, 2/660-662
classification, 4/100, 116
definition, 2/660-662
form, 2/660-662
occipital lobe lesion, 4/95
occipital lobe tumor, 17/327
terminology, 4/122
visual agnosia, 2/656, 4/16, 95
Agony
brain death, 24/775
brain injury, 24/775
Agrammatism
agraphia, 45/459, 463
anterior alexia, 45/442
aphasia, 4/89-91, 45/306
frontal lobe lesion, 45/35
motor aphasia, 45/309-311
Agranulocytosis
gold intoxication, 64/356
immunocompromised host, 56/469
mephenytoin intoxication, 37/202
organic acid metabolism, 66/641
penicillamine intoxication, 27/405
procainamide intoxication, 37/451
Agraphesthesia
schizophrenia, 46/490
Agraphia, 4/141-174, 45/457-469
*see also* Dysgraphia
acalculia, 45/480
agrammatism, 45/459, 463
alexia, 2/608, 4/127-130, 45/437, 461
amusia, 45/487
anatomic localization, 4/152-155
anesthesia, 45/462
anomic aphasia, 45/461
anterior cerebral artery syndrome, 53/347
aphasia, 4/91, 95, 145, 167-169, 45/457-463
apraxia, 4/95, 146-149, 155
brain infarction, 53/347
cerebral dominance, 4/147, 293, 45/463
chain parasymbolia, 4/169
chorea, 45/466
cipher, *see* Cipher agraphia
classification, 4/143, 45/457
color anomia, 45/52
conduction, 4/149
confusional state, 45/462
constructional, 4/150, 155, 168
coprolalia, 45/468
corpus callosum syndrome, 53/347

deep, *see* Deep agraphia
deep dyslexia, 45/462
definition, 45/457
dementia, 45/469
developmental dysgraphia, 4/434-439
dictation writing, 4/169
digit, *see* Digit agraphia
direction disorder, 4/147, 168
disconnection syndrome, 2/608, 45/104, 463
drug abuse, 45/462
dysorthographia, 4/151, 435
dystonia, *see* Writers cramp
echographia, 4/151, 45/468
etiology, 4/169, 45/465-467
examination, 4/170-172, 45/465
extrapyramidal syndrome, 45/464
fluent, *see* Fluent agraphia
frontal lobe tumor, 17/256
Gerstmann syndrome, 1/107, 2/686, 4/95, 146, 222, 45/70, 461
Gilles de la Tourette syndrome, 45/466
global aphasia, 45/460
handedness, 4/150
historic survey, 4/143-152
Huntington chorea, 45/466
hypergraphia, 45/468
hyperkinetic, *see* Hyperkinetic agraphia
hysteria, 45/468
linguistic component, 4/145, 155
literal, 2/688
literal paraphasia, 4/169
localization, 45/462-464
metabolic encephalopathy, 45/462
micrographia, 4/168, 45/464
migraine, 48/166
mirror writing, 4/148, 168, 465
mixed transcortical aphasia, 45/460
motoneuron disease, 45/464
motor, 45/464
motor aphasia, 45/459, 461, 463
motor component, 4/100, 149-152
musical, *see* Musical agraphia
neuropathy, 45/464
nonaphasic, 45/457, 463-469
nonfluent, *see* Nonfluent agraphia
number, *see* Number agraphia
occipital lobe lesion, 4/5
paligraphia, 45/467
paragraphia, 4/145, 169
paraphasia, 4/169
paretic, 45/464
parietal lobe lesion, 4/154
parietal lobe syndrome, 45/75, 77, 80
perseveration, *see* Perseverative agraphia
Pick disease, 42/285
polyglot, 4/151, 169
praxic and space constructional component, 4/146-149
primary agraphia, 4/141-144
pure, 4/100, 150, 45/462
rehabilitation, 3/417-422, 4/102, 172-174
reiterative, *see* Reiterative agraphia
restoration of higher cortical function, 3/382, 384, 417
secondary, *see* Secondary agraphia
spatial, *see* Spatial agraphia
spelling disorder, 4/147
Sydenham chorea, 45/466
sympathetic dyspraxia, 45/463
syringomyelia, 45/464
tardive dyskinesia, 45/466
temporal lobe lesion, 4/3, 153
traumatic, 4/150
unilateral, *see* Unilateral agraphia
upper motoneuron disease, 45/464
verbal, 2/688
verbal paraphasia, 4/169
visual component, 4/149
visuospatial, 45/468
Wernicke aphasia, 45/313, 461
without alexia, 2/608, 4/127-130, 45/52, 70, 77, 104, 436-440
writers cramp, 45/467
writing, 4/141-143, 45/459
writing ability, 45/458
writing evolution, 4/142
Agyria
*see also* Lissencephaly
brain cortex disorganization, 50/249
complete, *see* Lissencephaly
epilepsy, 72/109
holoprosencephaly, 50/237
mental deficiency, 46/11, 48
Miller-Dieker syndrome, 50/592
polymicrogyria, 50/257
sincipital encephalocele, 50/103
Sturge-Weber syndrome, 14/66
Ahylognosia
agnosia, 4/34
tactile agnosia, 45/66
Aicardi syndrome
absent rib, 50/165
Arnold-Chiari malformation type I, 31/227
brain atrophy, 42/696
chorioretinal lacunae, 50/163
chorioretinopathy, 42/696

clinical features, 31/226
CNS, 31/226
corpus callosum agenesis, 30/100, 31/225, 42/8, 696, 50/151, 156, 159, 162
cortical heterotopia, 31/226
differential diagnosis, 31/228
EEG, 42/696, 50/163
epilepsy, 72/129
etiology, 31/226
eye, 31/227
genetics, 31/226
Goltz-Gorlin syndrome, 31/227
infantile spasm, 42/696
mental deficiency, 31/226, 42/696
microcephaly, 42/696
microphthalmia, 42/696, 50/165
neuronal heterotopia, 42/696
optic atrophy, 42/696
optic nerve coloboma, 50/165
prognosis, 31/227
scoliosis, 42/696
seizure, 31/226, 50/165
sex chromosomes, 50/165
sex ratio, 42/697
skin lesion, 31/227
spina bifida, 42/696
split brain EEG, 50/163
subependymal heterotopia, 50/168
vertebral abnormality, 42/696
X-linked gene defect, 72/136

AIDS, *see* Acquired immune deficiency syndrome

Air embolism
*see also* Decompression illness *and* Dysbarism
anesthesia, 23/614
arterial hypertension, 61/221
blood-brain barrier, 28/392
blood transfusion, 23/614
brain, *see* Brain air embolism
brain embolism, 53/160
brain infarction, 53/160, 63/419
brain injury, 23/609
brain ischemia, 53/160
brain microembolism, 9/588, 63/175
cardiac surgery, 23/614
decompression illness, 63/419
decompression myelopathy, 61/221
dysbarism, 12/671
experimental injury, 23/616
features, 11/400
hemiplegia, 61/216
hydrocephalus, 30/579
iatrogenic neurological disease, 55/192
prevalence, 11/444
pulmonary barotrauma, 61/221
systemic brain infarction, 11/437
treatment, 23/618

AJ 76, *see* 5-Methoxy-1-methyl-2-propylamino-tetralin

Akabane virus
animal viral disease, 34/301
hydranencephaly, 34/301, 50/340

Akathisia
anatomic classification, 1/279
calcium antagonist intoxication, 65/442
chlorpromazine, 1/290
drug induced extrapyramidal syndrome, 6/253
ethosuximide intoxication, 65/508
involuntary movement, 1/279, 290
neuroleptic, *see* Neuroleptic akathisia
neuroleptic agent, 49/191
neuroleptic dyskinesia, 65/274
nifedipine intoxication, 65/442
nimodipine intoxication, 65/442
Parkinson disease, 42/245, 49/67, 92
parkinsonism, 1/290, 6/186, 49/67
psychological factor, 1/290
restless legs syndrome, 1/290, 8/314, 51/546, 65/286
stereotypy, 65/286
tardive dyskinesia, 49/186
tremor, 65/286
vigabatrin intoxication, 65/511

Akee intoxication, *see* Hypoglycin A intoxication

Akinesia
acetylcholine, 45/167
alpha-gamma motoneuron, 6/100
anterior cerebral artery syndrome, 53/346
arcuate gyrus, 45/165
ballismus, 49/66
basal ganglion, 45/165
brain cortex, 45/162
brain infarction, 53/346
bromocriptine, 45/166
carbamate intoxication, 64/188
carbon disulfide intoxication, 64/188
carbon monoxide intoxication, 64/31
centromedian parafascicularis complex, 45/162
cerebellar, 2/398
contingent negative variation, 45/163
cyanide intoxication, 64/232
definition, 45/161
disulfiram, 64/188
dopamine, 49/67
drug induced extrapyramidal syndrome, 6/250
electrostimulation, 49/67
freezing, 49/67

frontal lobe, 45/162
frontal lobe syndrome, 55/138
globus pallidus, 49/68
homovanillic acid, 49/67
hypothalamus, 45/165
inferior parietal lobule, 45/162
limbic system, 45/162
mesencephalic reticular formation, 45/162, 166
mesocortical pathway, 45/165
mesolimbic pathway, 45/165
1-methyl-4-phenyl-1,2,3,6-tetrahydropyridine, 49/67
metirosine, 49/67
motor cortex, 45/166
motor extinction, 45/161
motor impersistence, 45/161
motor planning, 49/66
motor ritardando, 49/67
movement time, 49/66
neglect syndrome, 45/161-167
neuroleptic parkinsonism, 65/292
neuroleptic syndrome, 6/250
neuropharmacology, 49/58
nigrostriatal pathway, 45/165
OKT3, 63/536
oxidopamine, 49/67
pallidoluysionigral degeneration, 49/451
Parkinson disease, 45/165, 49/65, 91
pathophysiology, 45/161-167, 49/65
pellagra, 7/571
predicted movement, 49/66
pure, *see* Pure akinesia
pyramidal tract, 45/161
ramp movement, 49/66
readiness potential, 49/66
sensory neglect, 45/162
simultaneous performance, 49/66
striatum, 45/165
supplementary motor area syndrome, 53/346
thalamus, 45/163
ventral tegmental area, 49/67
Akinetic epilepsy
myoclonic, *see* Myoclonic akinetic epilepsy
Akinetic mutism
acquired immune deficiency syndrome dementia, 56/491
anterior cerebral artery syndrome, 53/342, 347
anterior cingulate gyrus, 45/33
brain death, 24/700, 703, 707
brain embolism, 53/209
brain infarction, 53/209, 347, 55/137
brain injury, 23/10, 24/589, 700
brain stem lesion, 3/56
carbon monoxide intoxication, 6/657, 49/474
centromedian parafascicularis complex, 45/163
cerebrovascular disease, 53/209, 55/137
cingulate gyrus, 2/766
coma, 3/66, 75, 113
consciousness, 45/118
corpus callosum syndrome, 2/762
diencephalic lesion, 2/450
EEG, 23/337
etiology, 46/424
frontal lobe syndrome, 55/138
history, 24/699
hydrocephalic dementia, 46/328
hypothalamus, 45/166
iatrogenic pallidus lesion, 6/634
insect sting, 6/676
intracranial hemorrhage, 53/209
Marchiafava-Bignami disease, 45/33
mesencephalic infarction, 53/209
neglect syndrome, 45/163
organic brain syndrome, 46/424
pineal tumor, 18/405
pontine infarction, 12/15, 20
postanesthetic encephalopathy, 49/479
psychogenic unresponsiveness, 46/424
reticular formation, 2/281
sign, 24/588, 700, 707-709, 714, 716
systemic lupus erythematosus, 55/376
thalamic dementia, 21/594
thalamic infarction, 53/209, 55/138
thalamic lesion, 2/483
transient ischemic attack, 53/209
tsukubaenolide intoxication, 65/558
venous sinus thrombosis, 54/432
Akinetic petit mal
*see also* Petit mal epilepsy
differential diagnosis, 7/453
Akinetic rigid syndrome
carbon monoxide intoxication, 49/475
Parkinson disease, 49/108
ALA, *see* Aminolevulinic acid
Alacrimia
hereditary sensory and autonomic neuropathy type III, 1/475, 43/58, 60/28
Alajouanine syndrome
Möbius syndrome, 2/325
Åland eye disease, *see* Forsius-Eriksson syndrome
Alanine
brain, 9/16
GM2 gangliosidosis, 10/295
hereditary periodic ataxia, 21/569
myophosphorylase deficiency, 41/189
β-Alanine

Friedreich ataxia, 42/144
Alanine aminotransferase
Reye syndrome, 56/151
Alanine glyoxylate aminotransferase
primary hyperoxaluria type I, 66/57
Alanine glyoxylate aminotransferase deficiency
peroxisomal disorder, 66/510
Alaninuria
diabetes mellitus, 42/515
dwarfism, 42/515
enamel hypoplasia, 42/515
lactic acidosis, 42/516
microcephaly, 42/515
optic atrophy, 42/516
Alarm clock syndrome, *see* Hypnic headache syndrome
Albers-Schönberg disease, 38/371-379, 43/455-457
*see also* Osteopetrosis
autosomal dominant inheritance, 43/456
clinical features, 38/371-374, 378
hereditary cranial nerve palsy, 60/43
history, 38/371
hypertelorism, 30/246
laboratory evaluation, 38/377
mental deficiency, 46/64
nonrecurrent nonhereditary multiple cranial neuropathy, 51/571
osteopetrosis, 70/15
pathology, 38/379
radiologic features, 38/374-376
skull dysplasia, 30/246
terminology, 43/445
treatment, 38/379
Albinism
albinoidism, 14/105
amblyopia, 14/106
ankyloblepharon, 14/105
cerebellar ataxia, 60/667
Chédiak-Higashi syndrome, 14/783, 60/667
circumscribed partial, *see* Leukism
congenital deafness, 14/781
deaf mutism, 14/781
differential diagnosis, 14/106
dominant, *see* Dominant albinism
Guérin-Stern syndrome, 14/105
hereditary, 14/106
hereditary nystagmus, 14/106
hypertelorism, 30/248
levodopa, 43/3
macular hypoplasia, 14/106
melanin, 14/106
mental deficiency, 14/781, 30/248
Nettleship-Falls, *see* Nettleship-Falls albinism
neurocutaneous syndrome, 14/101
nystagmus, 14/106
oculocutaneous, *see* Oculocutaneous albinism
partial, *see* Partial albinism
primary pigmentary retinal degeneration, 13/186
sea blue histiocyte, 60/674
skin malformation, 30/248
type, 14/105
tyrosine, 14/106
universal, 14/115
Albinoidism
albinism, 14/105
myopia, 42/404
sensorineural deafness, 43/3
Albright hereditary osteodystrophy, *see* Pseudopseudohypoparathyroidism
Albright syndrome
*see also* Fibrous dysplasia
café au lait spot, 14/12
CNS, 31/312
endocrine disorder, 31/312
female, 14/12
history, 14/164
hypermelanosis, 1/484
monostotic fibrous dysplasia, 14/163
pathology, 31/314
phakomatosis, 31/27, 50/377
polyostotic fibrocystic dysplasia, 14/12, 163
precocious puberty, 14/12
prognosis, 31/314
radiologic appearance, 31/313
skeletal abnormality, 31/312
skin lesion, 31/312
terminology, 14/163
treatment, 31/314
Albumin
brain fat embolism, 55/191
CSF protein, 9/343
multiple sclerosis, 47/86
temporal arteritis, 48/321
Albuminocytologic dissociation, *see* Proteinocytologic dissociation
Alcohol
absorption, 64/108
albumin leakage, 55/524
Anthopleura elegantissima, 37/54
arterial hypertension, 54/207
autonomic dysfunction, 75/493
biokinetic, 64/108
blood pressure, 75/170, 494
brain fat embolism, 55/191
brain hemorrhage, 54/288, 55/523
brain infarction, 55/523

brain injury, 57/130
brain vasospasm, 11/125
calcium channel blocking agent, 55/524
cardiomyopathy, 63/133
central alveolar hypoventilation, 63/464
cerebellar encephalopathy, 64/5
cerebrovascular disease, 53/17
chemical characteristic, 64/107
cluster headache, 48/122, 249
distribution, 64/108
disulfiram intoxication, 37/321
epilepsy, 72/234
ethylene glycol intoxication, 64/125
experimental myopathy, 40/162
fetal intoxication, *see* Fetal alcohol syndrome
headache, 48/6, 122, 445
hereditary tremor, 49/576-578
hypertension, 55/523
hyperthermia, 75/62, 495
hypomagnesemia, 63/562
hypothermia, 75/60
insomnia, 45/138
intrathecal chemotherapy, 61/152
late cortical cerebellar atrophy, 21/477-479, 483-489
leukoencephalopathy, 47/585
microcephaly, 30/511, 50/279
migraine, 48/85, 122, 146
motion sickness, 74/357
multiple neuronal system degeneration, 75/170
mushroom toxin, 65/39
myelinated fiber density, 75/495
myoglobinuria, 41/272
neuropathy, 75/494
neurotoxin, 4/331, 6/819, 7/166, 553, 613, 9/653, 12/113, 13/53, 61, 17/542, 21/478, 28/317, 331, 341, 40/54, 162, 317, 339, 438, 41/273, 43/197, 45/194, 46/335-348, 393, 540, 563, 47/557, 621, 48/420, 49/556, 590, 51/56, 315-319, 531, 53/18, 54/288, 55/523, 57/130, 259, 61/370, 62/601, 609, 63/151, 564, 64/2, 5, 107, 111-115, 125
obstructive sleep apnea, 63/454
optic atrophy, 13/61
pathobiochemistry, 64/108
physiologic effect, 64/107
platelet aggregation, 55/524
posttraumatic intracranial hematoma, 57/259
psychosis, 46/462
radial nerve palsy, 64/115
rebound thrombocytosis, 55/524
rhodactis howesi intoxication, 37/54
snoring, 63/451
spastic paraplegia, 61/369
spasticity, 26/483
spinal cord injury, 61/369
spinal spasticity, 61/369
subarachnoid hemorrhage, 12/113, 55/523
sweating, 75/495
teratogenic agent, 30/119-121
toxic encephalopathy, 64/2, 5
toxic myopathy, 62/601
transverse spinal cord lesion, 61/369
use, 64/107
vasodilatory effect, 75/493
vitamin B1 deficiency, 47/155, 51/331
wood, *see* Methanol
Alcohol abuse
chronic, *see* Alcoholism
epilepsy, 73/455
late cerebellar atrophy, 60/583
Alcohol dehydrogenase
disulfiram polyneuropathy, 51/309
ethambutol polyneuropathy, 51/295
metabolism, 43/198
methanol intoxication, 64/96
Alcohol flush reaction
survey, 70/342
Alcohol intoxication
abstinence, 64/116
acetaldehyde, 64/109, 112
acetaldehyde adduct, 64/112
acute insulin resistance, 64/111
acute type, 64/107
age, 64/110
alcohol dehydrogenase route, 64/109
alcohol neurotoxicity, 64/113
alcoholic blackout, 64/115
alcoholic headache, 64/115
aldehyde dehydrogenase isoenzyme I, 64/110
amnesia, 64/114
anion gap, 64/112, 125
argipressin, 64/114
ataxia, 51/317, 64/115
attention, 64/115
autonomic nervous system, 64/112
autonomic polyneuropathy, 51/316
autoscopia, 4/232
beta adrenergic receptor, 64/114
biomarker, 64/111
blood alcohol concentration, 64/111
blood lactate-pyruvate ratio, 64/111
blood level, 64/109
brain stem auditory evoked potential, 64/115
carbohydrate deficient transferrin, 64/112
cardiac dysrhythmia, 64/112

cardiac function, 64/112
chlorine ion flux, 64/114
chorea, 49/556
chronic type, 64/108
cytochrome p450 isoenzyme, 64/109
cytochrome Pek.450IIE1 pathway, 64/110
diarrhea, 64/112
dihydropyridine receptor, 64/114
diuresis, 64/113
dysarthria, 64/115
elimination rate, 64/109
endogenous ethanol, 64/109
flushing, 64/112
GABA receptor, 64/114
galactose intolerance, 64/110
general effect, 64/112
hallucination, 4/329, 331
hangover headache, 64/114-116
headache, 48/420, 64/112
hemodialysis, 64/116
hippocampus, 64/114
history, 64/107
holiday heart syndrome, 64/112
hyperlipidemia, 64/111
hypoparathyroidism, 64/113
inebriation state, 64/115
interindividual variation, 64/110
ion channel theory, 64/113
ketoacidosis, 64/112
lactic acidosis, 64/110
Lactobacillus acidophilus, 64/109
lipid theory, 64/113
locus ceruleus, 64/114
long chain, *see* Long chain alcohol intoxication
memory, 64/115
metabolism, 64/109
*N*-methyl-D-aspartic acid receptor, 64/114
Meyer-Overton rule, 64/113
microsomal ethanol oxidizing system, 64/109
muscular hypotonia, 64/115
myoglobinuria, 40/339
nausea, 64/112
neuropathology, 64/116
nucleus accumbens, 64/114
nystagmus, 64/115
opiate peptide, 64/114
osteocalcin, 64/113
osteoporosis, 64/113
pathobiochemistry, 64/111
peroxisomal catalase, 64/109
postural tremor, 51/317
race, 64/110
radial nerve palsy, 64/115
rhabdomyolysis, 40/339
Sarcina ventriculi, 64/109
Saturday night palsy, 64/115
serotonin, 64/114
serum osmolality, 64/111
sex difference, 64/110
spastic paraplegia, 59/438
substantia nigra, 64/114
symptom, 64/5
thrombocytopenia, 64/113
toxic metabolite, 64/112
treatment, 64/116
tremor, 64/115
uridine diphosphate glucose-4-epimerase, 64/110
vertigo, 64/112, 115
vomiting, 64/112
Alcohol neurotoxicity
alcohol intoxication, 64/113
Alcohol reaction
disulfiram, 37/321
Alcohol syndrome
fetal, *see* Fetal alcohol syndrome
Alcohol withdrawal delirium
chlordiazepoxide, 46/541
diazepam, 46/541
physostigmine, 46/541
vitamin $B_1$, 46/541
Wernicke encephalopathy, 46/541
Alcohol withdrawal syndrome
alcoholism, 70/346
Alcoholic amblyopia
clinical features, 28/332
etiology, 28/338-340
history, 28/332
malnutrition, 28/331
methanol, 13/62
pathology, 28/333-338
tobacco, *see* Tobacco alcoholic amblyopia
vitamin $B_{12}$, 28/174
Alcoholic blackout
alcohol intoxication, 64/115
epilepsy, 72/250
survey, 70/341
Alcoholic cirrhosis
acanthocytosis, 63/289
clinical features, 7/613
Alcoholic dementia, 46/335-348
acquired hepatocerebral degeneration, 46/338
alcoholism, 70/351
histology, 28/341-343
Marchiafava-Bignami disease, 17/542, 28/317, 46/337, 47/557
pellagra, 46/336

primary, *see* Primary alcoholic dementia
vitamin B3 deficiency, 46/336
vitamin B6 deficiency, 46/336
Wernicke-Korsakoff syndrome, 45/196, 46/335
Alcoholic deteriorated state, *see* Primary alcoholic dementia
Alcoholic encephalopathy
pellagra, 28/85, 46/336, 399
secondary neuroaxonal dystrophy, 49/408
Alcoholic hallucinosis
definition, 46/563-566
peduncular hallucinosis, 46/419
schizophrenia, 46/419
Alcoholic myelopathy
features, 28/343-345
Alcoholic myopathy, *see* Alcoholic polyneuropathy
Alcoholic neuropathy
alcoholism, 43/197, 70/351
demyelination, 47/621
diagnosis, 7/553
vitamin B1 deficiency, 7/166
Alcoholic polydipsia, *see* Dipsomania
Alcoholic polyneuropathy
acroparesthetic ulceromutilating form, 51/317
acute type, 62/609
alcoholism, 70/354
amyotrophy, 51/316
anemia, 51/317
apoenzyme-coenzyme dissociation, 51/315
ataxia, 51/316
autonomic abnormality, 75/695
autonomic neuropathy, 51/316
axonal degeneration, 51/318
beriberi, 51/318
burning feet syndrome, 51/316
chronic type, 62/610
clinical features, 40/162, 317
course, 51/317
demyelination, 47/621
differential diagnosis, 40/438
dysphagia, 51/316
EMG, 51/315, 318
epilepsy, 51/317
experimental, 40/162
histopathology, 62/609
hyperhidrosis, 51/316
hypophosphatemia, 63/565
incidence, 51/315
ischemic exercise test, 41/273
lancinating pain, 51/316
liver cirrhosis, 7/613
malnutrition, 51/316
myoglobinemia, 41/272
myoglobinuria, 40/162, 339
neuropathic arthropathy, 51/317
orthostatic hypotension, 63/151
pathogenesis, 51/318, 62/610
pathology, 40/54, 51/318
postural tremor, 51/317
proximal form, 51/317
rhabdomyolysis, 40/162, 339
rhinophyma, 51/317
segmental demyelination, 51/318
sensorimotor neuropathy, 51/316
sensory loss pattern, 51/56
sex ratio, 51/315
symptom, 62/609
transketolase, 51/315
transketolase assay, 51/319
treatment, 51/319
vocal cord paralysis, 51/316
Wernicke encephalopathy, 51/317
Alcoholic pseudoparesis, *see* Alcoholic dementia
Alcoholic tremor
postural tremor, 6/819, 49/590
Alcoholism, 43/197-199
acute response, 70/339
adiposis dolorosa, 8/347
aggression, 46/178
alcohol withdrawal syndrome, 70/346
alcoholic dementia, 70/351
alcoholic neuropathy, 43/197, 70/351
alcoholic polyneuropathy, 70/354
aldehyde dehydrogenase, 43/198
autonomic dysfunction, 63/148, 75/493
autonomic nervous system, 75/32
autonomic polyneuropathy, 51/476, 478
autoscopia, 4/232, 45/386
brain embolism, 63/133
brain hemorrhage, 63/313
cardiomyopathy, 63/133
central pontine myelinolysis, 9/636, 28/311, 47/585
cerebellar degeneration, 70/350
chronic, 70/347
chronic subdural hematoma, 24/299
craving, 70/342
deficiency neuropathy, 7/613, 51/322, 324
delirium, 46/532
delirium tremens, 3/99, 27/456, 43/197, 70/346
dementia, *see* Alcoholic dementia
demyelination, 9/634
dependence, 70/344
epidural hematoma, 63/313
epilepsy, 43/197
fetal alcohol syndrome, 30/119-121, 43/197,

46/81, 70/355
hallucination, 43/197
hepatic encephalopathy, 70/356
hypercortisolism, 70/209
hypomagnesemia, 63/562
hypophosphatemia, 63/564
insomnia, 45/138
listeriosis, 52/91
Marchiafava-Bignami disease, 9/653, 28/317, 46/337, 47/557, 70/353
neuroaxonal dystrophy, 6/626
neurologic soft sign, 46/491
nocardiosis, 52/446
orthostatic hypotension, 1/473, 63/148, 151
Parkinson disease, 49/96
pellagra, 7/573, 28/85, 46/399
polyneuritis, 43/197
polyneuropathy, 43/197, 51/68
positron emission tomography, 60/673
prevalence, 43/197
psychiatric disorder, 43/197
recurrent meningitis, 52/53
retrobulbar neuritis, 43/197
secondary neuroaxonal dystrophy, 49/398
sex ratio, 43/197
spinal epidural empyema, 52/73
striatal necrosis, 49/500, 504
subdural hematoma, 63/313
survey, 70/339
tolerance, 70/344
traumatic psychosis, 24/529
traumatic psychosyndrome, 24/568, 570
vitamin $B_c$ deficiency, 51/336
Wernicke encephalopathy, 70/348, 75/494
Wernicke-Korsakoff syndrome, 70/348
withdrawal seizure, 70/346
withdrawal syndrome, 70/347
Aldehyde dehydrogenase
alcoholism, 43/198
carbamate intoxication, 64/187
methanol intoxication, 36/354, 357, 64/96
migraine, 48/86
Aldehyde dehydrogenase isoenzyme I
alcohol intoxication, 64/110
Aldehyde reductase
diabetic polyneuropathy, 51/507
Alder abnormality
gyrate atrophy, 13/259
Alder-Reilly granule
GM1 gangliosidosis, 10/430
multiple sulfatase deficiency, 42/493
Aldolase, *see* Fructose-bisphosphate aldolase
Aldolase reductase, *see* Aldehyde reductase
Aldose reductase inhibitor
diabetes mellitus, 70/152
Aldosterone
anxiety, 1/502
headache, 5/231
pure autonomic failure, 22/234
salt appetite, 1/501
spinal cord injury, 26/320, 325
Aldosteronism
Bartter syndrome, 42/528
endocrine myopathy, 40/317
metabolic myopathy, 40/317
systemic brain infarction, 11/458
Aldrich-Mees line, *see* Mees line
Aldrin
chlorinated cyclodiene, 64/200
neurotoxin, 36/411, 64/197, 201
organochlorine, 36/393, 64/197
organochlorine insecticide intoxication, 36/411-417, 64/197
toxic encephalopathy, 64/201
Aldrin intoxication, 36/411-417
clinical features, 36/411
EEG, 36/414-416
laboratory finding, 36/414-416
pathology, 36/417
prognosis, 36/416
treatment, 36/417
Alertness
anterior cingulate gyrus, 2/765
attention, 3/138, 155, 168
orienting reaction, 3/156
Aleutian mink disease
animal viral disease, 56/585
blocking antibody, 56/61
multiple sclerosis, 9/126
viral mechanism, 56/20
Alexander disease, 10/94-102, 42/484-486, 66/701-708
adult type, 66/706
amaurotic idiocy, 10/101
chorea, 10/101
classification, 10/24, 36, 107, 114, 66/5
clinical features, 10/100
cystic leukoencephalopathy, 10/97, 100
differential diagnosis, 10/100
familial dysmyelination, 10/101
fever, 42/484
glial dystrophy, 21/67
histochemistry, 10/95
histology, 10/95, 42/484
history, 10/94
hyaline deposit, 10/99

hydrocephalus, 10/209
infantile type, 66/701
intracranial pressure, 42/484
juvenile type, 66/706
lipidosis, 10/351
macrencephaly, 10/101, 30/657, 42/485, 66/702
megalocephaly, 10/209, 42/485
mental deficiency, 10/209, 42/484, 46/49
pathogenesis, 66/707
Rosenthal fiber, 10/94, 100, 42/485, 66/701
spongy cerebral degeneration, 10/100, 209
treatment, 66/706
type, 66/14, 701
Alexander law
description, 2/366
ENG, 2/378
spontaneous nystagmus, 2/366, 378
Alexandrium genus
paralytic shellfish poisoning, 37/33, 65/151
saxitoxin, 37/33
Alexia, 4/112-137, 45/433-452
acalculia, 45/437, 480
acquired alexia in children, 4/135
agnostic, *see* Agnostic alexia
agraphia, 2/608, 4/127-130, 45/437, 461
amusia, 45/487
anatomic localization, 4/114, 124-126, 130
anomic aphasia, 45/461
anterior, *see* Anterior alexia
aphasia, 4/95, 116, 127, 129-133, 45/447
apraxia, 4/97
arteriovenous malformation, 45/438
associative, 45/438
asthenopic, *see* Asthenopic alexia
asyllabia, 4/119
auditory message, 4/96
auditory reading, 4/121
blindness, 4/135
Braille reading, 4/135
brain commissurotomy, 4/117, 126, 282
brain hematoma, 45/438
brain infarction, 45/438
brain tumor, 2/607, 45/438
calculation, 4/96
central, *see* Central alexia
cerebral dominance, 4/134, 293
cipher, *see* Cipher alexia
classification, 4/115-117, 45/435
clinical features, 4/117-134
color agnosia, 45/438
color anomia, 45/52, 438
color identification, 4/121, 45/438
comprehension, 45/433
computer assisted tomography, 45/12
conduction aphasia, 4/132, 45/447
confusional state, 4/134
constructional apraxia, 4/97
corpus callosum, 4/126, 45/12, 439
corpus callosum agenesis, 30/289
corpus callosum disconnection syndrome, 2/598
corpus callosum syndrome, 2/760
cortical, *see* Cortical alexia
deafness, 4/135
deep dyslexia, 45/445
defective Gestalt formation, 4/123, 45/439
definition, 4/112, 45/433
developmental dyslexia, 4/123, 135
digit, *see* Digit alexia
disconnection syndrome, 2/608, 4/126, 45/104
dysbarism, 2/607
dyslexia, 4/112, 115, 45/445
eye movement, 4/123
frontal, *see* Frontal alexia
gaze defect, 45/439
geometric form agnosia, 4/124
Gerstmann syndrome, 4/128, 45/70, 461
gesture, 4/96
global, 4/116, 119
hemi, *see* Hemialexia
historic survey, 4/113-115, 125, 45/433-435
homonymous hemianopia, 2/607, 609
intelligence, 4/97
Kana reading, 45/446
Kanji reading, 45/446
language cortex, 45/12
left half field, 2/597
lesion site, 2/587
letter naming difficulty, 4/120
literal, 2/610, 4/95, 112, 115, 45/433, 441
mental deficiency, 4/134
mimicry, 4/96
mind blindness, 4/112
mirror reading, 4/465
monochromatism, 45/438
motor aphasia, 4/99, 131, 45/436, 447
music reading, *see* Musical alexia
musical, *see* Musical alexia
neuropsychology, 45/11-13
neutral substrate, 45/449-452
nonverbal disorder, 4/96
number reading, 4/120, 45/437
occipital lobe lesion, 4/5, 125
occipital lobe tumor, 17/327, 45/436
occipital lobectomy, 45/436, 438
oriental language, 4/114, 45/446
pantomime, 4/96

paralexia, 4/129, 133, 45/433, 445, 448
paralexic substitution, 4/123
parietal lobe, 4/127
parietal lobe syndrome, 45/75, 77, 81
partial, 4/113
pathogenesis, 45/439
pathology, 45/438-440
posterior, *see* Posterior alexia
posterior cerebral artery syndrome, 53/411
primary and secondary, 4/113, 116
pseudo, *see* Pseudoalexia
psychogenic disorder, 4/134
pure word blindness, 4/17, 100, 112, 115, 118-127
pure word deafness, 4/131
reading aloud, 4/133, 136
reading speed, 4/136
reading test, 4/135-137
rehabilitation, 3/422-426, 4/137
semantic, *see* Semantic alexia
simultanagnosia, 4/123, 45/439
somatesthetic reading, 4/121, 45/437
spatial orientation disorder, 3/219, 4/124, 45/438
spelling, 4/123, 136
splenio-occipital, *see* Splenio-occipital alexia
subangular, *see* Subangular alexia
subcortical alexia, 4/115
surface dyslexia, 45/445
syndrome variation, 45/443
syntactic, *see* Syntactic alexia
temporal lobe lesion, 4/3, 45/13
terminology, 4/112, 115-117
transcortical alexia, 4/115
transcortical aphasia, 4/132
transcortical sensory aphasia, 45/447
unilateral paralexia, 4/133
verbal, 2/610, 4/95, 112, 115, 45/433
visual agnosia, 4/116, 122, 45/439
visual asymbolia, 4/116
visual cortex, 45/12
visual field defect, 45/438
visual message, 4/96
Wernicke aphasia, 4/130, 45/313, 447
without agraphia, 45/52, 70, 77, 104, 436-440
writing, 4/120
Alexia without agraphia
asthenopic alexia, 4/124
color agnosia, 4/121
left posterior cerebral artery, 53/412
migraine, 48/161
Alexia sine agraphia, *see* Alexia without agraphia
Alezzandrini syndrome
secondary pigmentary retinal degeneration, 13/186
Alfadolone acetate with alfaxalone intoxication
epilepsy, 37/409
Alföldi sign
diagnosis, 2/15
muscular atrophy, 2/15
nerve lesion, 2/15
Algae
acquired immune deficiency syndrome, 71/307
marine toxin, 37/97
Alganesthesia
terminology, 1/86
Algesia
terminology, 1/86
Algesthesia
terminology, 1/86
Algie meningée postérieure
cervico-occipital pain, 5/187
headache, 5/187
Algodiaphoria
terminology, 4/223
Alice in Wonderland syndrome
body scheme, 5/244
hallucination, 5/244
headache, 5/244
migraine, 5/244
Alien hand syndrome
anterior cerebral artery syndrome, 53/342, 346
brain infarction, 53/347
corpus callosum syndrome, 53/347
frontal lobe lesion, 45/32
split brain patient, 53/346
Aliphatic esterase
wallerian degeneration, 7/214, 9/37
Alkaline phosphatase
aminoaciduria, 43/126
congenital hyperphosphatasia, 31/258
craniodiaphyseal dysplasia, 43/356
enzyme, 26/403
experimental allergic encephalomyelitis, 9/38
fibrous dysplasia, 14/180
hypophosphatasia, 42/578
metachromatic leukodystrophy, 10/48
Morgagni-Stewart-Morel syndrome, 43/412
multiple sclerosis plaque, 9/37
myopathy, 43/126
osteitis deformans (Paget), 38/361, 43/451
Schwartz-Lelek syndrome, 31/257
sclerosteosis, 31/258, 43/475
Van Buchem disease, 31/258, 38/410, 43/411
wallerian degeneration, 7/215
Wolfson syndrome, 20/428
Alkalosis
Chvostek sign, 40/330

delirium, 46/544
metabolic, *see* Metabolic alkalosis
respiratory, *see* Respiratory alkalosis
Alkeran, *see* Melphalan
Alkylating agent
adverse effect, 69/488
antineoplastic agent, 65/528
chemotherapy, 39/93-98, 65/528
chlorambucil, 39/97
chlormethine, 39/93-97
cyclophosphamide, 39/93-98
neurologic toxicity, 39/93-98
neurotoxin, 67/354
oligodendroglioma, 68/132
ototoxicity, 67/354
thiotepa, 39/97
Alkyldihydroxyacetone phosphate synthase deficiency, *see* Alkylglycerone phosphate synthase deficiency
Alkylglycerone phosphate synthase deficiency
peroxisomal disorder, 66/510
Alkylmercury intoxication
central visual disturbance, 60/655
Alkyl tin intoxication
brain edema, 9/647, 23/151
Allachesthesia, *see* Allesthesia
Allaesthesia, *see* Allesthesia
Allbutt, Sir Clifford, 2/7
Allelic diversity
acid maltase deficiency, 62/481
Allelic expansion
myotonic dystrophy, 62/212
Allelic heterogeneity
Huntington chorea, 49/276
Allergic angiitis, *see* Allergic granulomatous angiitis
Allergic angiopathy
leukoencephalopathy, 9/609
polyarteritis nodosa, 9/609-612
Allergic encephalitis
arsenic, 9/650
consciousness, 45/119
rabies, 56/390
vaccination induced, 56/391
Allergic encephalomyelitis, 9/500-552
acute experimental, *see* Acute experimental allergic encephalomyelitis
autoimmune disease, 9/506
blood-brain barrier, 28/391
chronic relapsing experimental, *see* Chronic relapsing experimental allergic encephalomyelitis
definition, 9/501-504
experimental, *see* Experimental allergic encephalomyelitis
S-100 protein, 9/514, 520
terminology, 9/501-504
Allergic granulomatous angiitis
*see also* Polyarteritis nodosa
allergic rhinitis, 63/383
amaurosis fugax, 51/453
angiopathic polyneuropathy, 51/446
anorexia, 63/383
aortitis, 63/53
arteritis, 11/453
arthralgia, 51/453, 63/383
asthma, 11/134, 51/453, 63/383
brachial plexus, 63/383
brachial plexus neuritis, 63/383
brain amyloid angiopathy, 54/342
brain hemorrhage, 54/71, 63/383
brain infarction, 53/162
brain ischemia, 53/162
cerebrovascular disease, 11/133, 53/41
chronic bronchitis, 63/383
chronic meningitis, 56/644
classification, 71/4
coma, 63/383
computer assisted tomography, 54/63
confusion, 63/383
confusional state, 63/383
convulsion, 63/383
corticosteroid, 63/384
cranial neuropathy, 63/383
CSF, 54/198
cyclophosphamide, 63/384
differential diagnosis, 63/382
eosinophilia, 51/453, 63/371
eosinophilia myalgia syndrome, 64/261
epilepsy, 63/383
features, 63/382
fever, 63/383
granulomatous CNS vasculitis, 55/396
headache, 5/149
hypertension, 63/383
immune complex, 63/383
ischemic optic neuropathy, 51/453
kidney, 63/383
lung, 63/382
lymphoma, 63/350
malaise, 63/383
malignant lymphoma, 55/488, 490
meningococcal meningitis, 52/28
mononeuritis multiplex, 51/453, 63/383, 423
myalgia, 63/383
neurologic complication, 63/383
neurologic symptom, 63/423

optic neuropathy, 63/383
pathogenesis, 8/116, 63/383
polyarteritis nodosa, 51/453, 55/358, 63/382
polyneuropathy, 63/423
prognosis, 63/384
sensorimotor polyneuropathy, 51/453, 63/383
sex ratio, 51/453, 63/383
Sneddon syndrome, 55/408
subarachnoid hemorrhage, 63/383
survey, 71/156
systemic brain infarction, 11/453
temporal arteritis, 55/346
treatment, 63/384
trochlear nerve paralysis, 51/453
varicella zoster virus, 55/436
vascular dementia, 46/359
vasculitis, 71/8, 155
vessel type, 63/382
Wegener granulomatosis, 51/450
weight loss, 63/383
Allergic headache, 5/234-237
medical history, 5/236
migraine, 5/266
neuropathy, 5/234
treatment, 5/237
Allergic myositis
experimental, *see* Experimental allergic myositis
Allergic neuritis
autonomic nervous system, 75/16
experimental, *see* Experimental allergic neuritis
rabies, 9/501
tetanus toxoid, 52/240
Allergic neuroencephalopathy
definition, 9/501
terminology, 5/501
Allergic neuropathy
adrenalin, 5/234
corticosteroid, 7/555
definition, 9/501
neuralgic amyotrophy, 8/82
terminology, 9/501
Allergic purpura, *see* Henoch-Schönlein purpura
Allergic reaction
antidepressant intoxication, 37/319
crab intoxication, 37/56
crustacea intoxication, 37/56
malacostraca intoxication, 37/56
oyster, 37/64
pyrethroid intoxication, 36/435
subarachnoid hemorrhage, 12/114
tricyclic antidepressant intoxication, 37/319
Allergic vasculitis, *see* Allergic granulomatous angiitis
Allergy
arthropod envenomation, 65/198
brain thromboangiitis obliterans, 12/392
food, *see* Food allergy
Guillain-Barré syndrome, 8/87, 9/500
migraine, 5/266, 48/20
multiple sclerosis, 9/115
subarachnoid hemorrhage, 12/114
Allescheria boydii
allescheriosis, 35/558
fungal CNS disease, 35/558-560
Allescheriosis
Allescheria boydii, 35/558
brain abscess, 35/560
clinical features, 35/559
CSF, 35/559
history, 35/559
pathology, 35/560
Allesthesia
acoustic, 3/29
Anton-Babinski syndrome, 45/69
epilepsy, 45/171
hallucination classification, 45/56
optic, *see* Optic allesthesia
sensory extinction, 3/191
tactile, 3/29
terminology, 1/89
visual, *see* Visual allesthesia
visual hallucination, 45/356
Allethrin
neurotoxin, 64/212
pyrethroid, 64/212
pyrethroid intoxication, 64/212
Allgrove syndrome
adrenal insufficiency, 70/218
autonomic nervous system, 75/701
Allochiria
terminology, 1/89
Allodynia, *see* Hyperalgesia
Allogeneic mixed lymphocyte reaction
multiple sclerosis, 47/356
Allomethadione
side effect, 15/706
Allotype
classification, 47/363
IgG, *see* IgG allotype
Allotype G1m
multiple sclerosis, 47/363
Allotype G3m
multiple sclerosis, 47/363
Alloxan
diabetic polyneuropathy, 51/512
Alloxan Schiff staining

protein, 9/23
*N*-Allylmorphine, *see* Nalorphine
Almitrine
chronic obstructive pulmonary disease, 63/421
Almitrine dimesilate
drug induced polyneuropathy, 51/294
neurotoxicity, 51/294, 308
Almitrine dimesilate polyneuropathy
chronic obstructive respiratory tract disorder, 51/308
EMG, 51/308
sensory polyneuropathy, 51/308
Almitrine intoxication
polyneuropathy, 63/421
Alobar holoprosencephaly
definition, 30/438
genetic term, 50/230
gross anatomy, 50/227
terminology, 30/436
type, 30/433
Alogia
frontal lobe syndrome, 45/24
Alopecia
adrenomyeloneuropathy, 60/170
Alström-Hallgren syndrome, 42/391
Argentinian hemorrhagic fever, 56/364
dwarfism, 43/380
endocrine candidiasis syndrome, 42/642
eosinophilia myalgia syndrome, 63/375, 64/252
epilepsy, 42/684
Goltz-Gorlin syndrome, 43/19
hypogonadism, 42/374
ichthyohepatoxic fish intoxication, 37/87
incontinentia pigmenti, 14/214, 43/23
leprosy, 51/218
linear nevus sebaceous syndrome, 43/38
male pattern, *see* Baldness
Marinesco-Sjögren syndrome, 60/667
mental deficiency, 42/684
myotonic dystrophy, 62/238
β-naphthol, 13/224
orofaciodigital syndrome type I, 43/447
18p partial monosomy, 43/534
progeria, 43/465
progressive hemifacial atrophy, 14/776, 22/548, 31/253, 43/408, 59/479
pseudoprogeria Hallermann-Streiff syndrome, 43/405
Refsum disease, 60/231
Rothmund-Thomson syndrome, 43/460, 46/61
sensorineural deafness, 42/374, 395
striatopallidodentate calcification, 49/423
systemic lupus erythematosus, 43/420
thallium intoxication, 64/325
thallium polyneuropathy, 51/274
toxic oil syndrome, 63/380
trigeminal nerve agenesis, 50/214
Urbach-Wiethe disease, 42/595
Vogt-Koyanagi-Harada syndrome, 56/611, 622
Werner syndrome, 43/489
Alopecia areata
incontinentia pigmenti, 14/12
Alopecia circumscripta
anencephaly, 30/188
Alopecia universalis
Satoyoshi syndrome, 64/256
Alopex lagopus intoxication
vitamin A, 37/96
Alpers disease
basal ganglion degeneration, 42/486
blindness, 42/486
brain atrophy, 42/486
brain biopsy, 10/685
cerebellar atrophy, 42/486
CNS spongy degeneration, 42/486
EEG, 42/486
epilepsy, 42/486
glial dystrophy, 21/68
late cortical cerebellar atrophy, 21/477
mental deficiency, 42/486
microcephaly, 30/509, 514, 42/486
mitochondrial disease, 62/497
myoclonus, 42/486
psychomotor retardation, 42/486
sensorineural deafness, 42/486
sex ratio, 42/487
spinocerebellar degeneration, 10/210
spongy cerebral degeneration, 10/210
thalamus, 42/486
valproic acid induced hepatic necrosis, 65/128
Alpha adrenergic receptor
dihydroergotamine, 65/66
drug, 48/183
genetics, 75/242
Alpha adrenergic receptor blocking agent
ergot intoxication, 65/63
Alpha adrenergic receptor stimulating agent
autonomic nervous system, 22/251
brain infarction, 53/421
pain, 69/47
phenylpropanolamine intoxication, 65/259
Alpha blocking
electrical phenomenon, 2/543
Alpha cat
decerebration hypertonia, 1/262
Alpha coma

anoxic encephalopathy, 63/219
barbiturate, 55/220
cardiac arrest, 63/219
diazepam, 55/220
EEG, 57/187, 189
electrical injury, 61/192
Alpha driving
electrical phenomenon, 2/543
Alpha fucosidosis
enzyme defect, 66/330
Alpha mannosidosis
animal model, 66/330
clinical features, 66/330
enzyme defect, 66/330
survey, 66/329
type I, 66/330
type II, 66/331
Alpha motor system
decerebrate rigidity, 1/262
phasic type, 1/67
spastic paraplegia, 61/367
spinal cord injury, 61/367
spinal spasticity, 61/367
tonic type, 1/67
transverse spinal cord lesion, 61/367
Alpha oxidation
mitochondrial, *see* Mitochondrial alpha oxidation
phytanic acid, 27/524-526
Refsum disease, 27/519, 66/485, 492, 494, 497
Alpha rhythm
cortical blindness, 45/50
electrical phenomenon, 2/543
Alpha rigidity
nerve block, 41/307
Alpha system
basal ganglion control, 6/99
Alpha thalassemia, *see* Thalassemia major
Alpha wave
low voltage EEG, 42/675
Alphaprodine
chemistry, 37/366
opiate, 37/366, 65/350
Alphavirus
acute viral myositis, 56/193
Chikungunya virus, 56/12
eastern equine encephalitis, 56/12
Mayaro virus, 56/12
neurotropic virus, 56/12
O'nyong-nyong virus, 56/12
Ross river virus, 56/12
Venezuelan equine encephalitis, 56/12
Alport syndrome
cataract, 42/376, 60/770
cochlear hypoplasia, 42/376
Corti organ degeneration, 42/376
foam cell, 42/376
hereditary progressive cochleovestibular atrophy, 60/768-770
Lemieux-Neemeh syndrome, 22/522
lenticonus, 60/770
nephritis, 22/487, 42/376
retinal fleck, 60/770
retinal telangiectasia, 60/770
retrobulbar neuritis, 60/770
sex chromosome Xq, 60/770
spherophakia, 42/376
Ålport syndrome, *see* Forsius-Eriksson syndrome
Alprazolam
hereditary periodic ataxia, 60/442
Alprenolol
migraine, 48/180
Alprostadil
blood-brain barrier, 48/64
multiple sclerosis, 47/342, 355
ALS, *see* Amyotrophic lateral sclerosis
Alström-Hallgren syndrome
acanthosis nigricans, 42/391
alopecia, 42/391
Bardet-Biedl syndrome, 13/390, 22/503
cataract, 13/461, 42/391
deafness, 13/242, 460
diabetes mellitus, 13/460, 42/391
hereditary hearing loss, 22/502
hypogonadism, 22/502
Laurence-Moon-Bardet-Biedl syndrome, 13/462, 22/503
nystagmus, 13/461, 42/391
obesity, 13/460, 22/501
retinal aplasia, 13/460
secondary pigmentary retinal degeneration, 13/451, 460-463, 60/720
Tunbridge-Paley disease, 22/507
visual acuity, 13/461
Weiss-Alström syndrome, 13/463
Alström syndrome, *see* Alström-Hallgren syndrome
Alternaria
fungal CNS disease, 35/561, 52/480
Alternariasis
meningitis, 35/561, 52/481
Alternate binaural loudness balance
acoustic nerve tumor test, 17/676
Alternating hemiplegia
basilar artery aneurysm, 53/387
Alternating infantile hemiplegia
arteriography, 12/348
migraine, 48/146

Alternating ophthalmoplegia
Bielschowsky relapsing, *see* Bielschowsky relapsing alternating ophthalmoplegia
Alternating synergics
spinal cord injury, 26/250
Alternobaric vertigo
dysbarism, 23/624
Althesin intoxication, *see* Alfadolone acetate with alfaxalone intoxication
Altitude
autonomic nervous system, 74/586, 75/440
orthostatic intolerance, 74/326
respiratory function, 74/586
Altitude decompression
nitrogen, 12/665
Altitude illness
*see also* Mountain sickness
barogenic, 48/395-404
brain function, 48/396
carotid artery, 48/397
cocaine intoxication, 65/254
dysbarism, 12/665
EEG, 48/396
features, 48/10
headache, 48/395
hypoxia, 48/397
intracranial pressure, 48/6
migraine, 48/399
nervous system, 48/396
symptom, 48/400
toothache, 48/399
Altitude stress
sympathetic nervous system, 74/654
Altitudinal hemianopia
anterior ischemic optic neuropathy, 55/114
causation, 2/568
glaucoma, 2/563
ophthalmoscopy, 2/563
retinal lesion, 2/563
Altitudinal scotoma
glaucoma, 2/563
Altretamine
adverse effect, 69/493
antineoplastic agent, 65/528
chemotherapy, 39/117
encephalopathy, 65/528
neuropathy, 69/461, 471
neurotoxin, 39/95, 113, 117, 119, 46/394, 65/528, 534
polyneuropathy, 65/528
Altretamine intoxication
ataxia, 65/534
encephalopathy, 65/534
parkinsonism, 65/534
polyneuropathy, 65/534
Aluminum
Alzheimer disease, 46/265, 271
amyotrophic lateral sclerosis, 59/280
dementia, 63/525
epilepsy, 72/53
Guam amyotrophic lateral sclerosis, 59/264-266, 275
metabolism, 36/319
neurofibrillary tangle, 46/271
neurotoxin, 36/320, 46/392, 59/278, 63/525, 64/4, 10, 273
striatopallidodentate calcification, 6/716, 49/425
toxic encephalopathy, 64/4, 274
toxic neuropathy, 64/10
Aluminum encephalopathy
Alzheimer neurofibrillary change, 21/57
Aluminum intoxication
*see also* Dialysis encephalopathy
absorption factor, 64/275
action site, 64/10
acute form, 63/526
aluminum biokinetics, 64/274
aluminum clearance rate, 64/275
aluminum elimination, 64/275
aluminum plasma level, 64/279
Alzheimer disease, 36/320-323, 64/276, 278
amyotrophic lateral sclerosis-Parkinson dementia, 64/278
asterixis, 63/525, 64/279
axonal swelling, 64/276
body burden, 64/279
bone, 64/273
brain, 64/273
chronic, *see* Chronic aluminum intoxication
citric acid, 64/275
coma, 63/526
confusional state, 63/526
CSF aluminum, 63/534
deferoxamine, 63/525-527, 534, 64/279
delirium, 63/526
dementia, 46/392
diagnosis, 63/526, 64/279
dialysis, 63/523, 525
dialysis encephalopathy, 36/323, 64/276
diazepam, 63/527
distribution volume, 64/275
EEG, 63/526, 64/279
encephalopathy, 21/57
epilepsy, 36/323, 63/525
experimental encephalopathy, 64/276
exposure source, 64/274

Guam amyotrophic lateral sclerosis-Parkinson dementia complex, 64/276
intake threshold, 64/273
iron deficiency anemia, 64/274
kidney function, 64/276
lung, 64/273
metabolic encephalopathy, 63/525
microtubule associated protein, 64/276
mutism, 63/525, 64/279
myoclonic jerk, 63/525
neurofibrillary tangle, 36/320-324, 64/276
neurofilament, 64/276
neurotoxic effect, 64/276
neurotoxin mechanism, 64/276
oral citrate, 63/526
osteomalacia, 64/274
pH, 64/275
plasma half-life, 64/275
prevention, 64/279
prophylaxis, 63/527
pulmonary fibrosis, 64/274
slow axon transport, 64/276
symptom, 64/4
treatment, 64/279
vitamin C, 64/275
wallerian degeneration, 64/276

Aluminum neurotoxicity
asterixis, 63/525
dialysis, 63/525
epilepsy, 63/525
hallucination, 63/525
mutism, 63/525
myoclonus, 63/525
stuttering, 63/525

Alveolar/arteriolar J receptor
respiratory center, 63/431
respiratory control, 63/431

Alveolar hypoventilation
carotid body, 63/414
carotid endarterectomy, 63/414
central, *see* Central alveolar hypoventilation
chronic, *see* Chronic alveolar hypoventilation
congenital central, *see* Congenital central alveolar hypoventilation
multiple neuronal system degeneration, 63/146
myotonic dystrophy, 40/517, 62/238
obstructive sleep apnea, 63/455
orthostatic hypotension, 63/146
Parkinson disease, 49/130
parkinsonism, 49/130
pickwickian syndrome, 38/299
primary, *see* Primary alveolar hypoventilation

Alveolitis
fibrosing, *see* Fibrosing alveolitis

Alvodine intoxication, *see* Piminodine

Alzheimer, A., 1/8, 27

Alzheimer disease, 46/247-275
A4 protein, 59/287
acetylcholine, 46/266
acetylcholinesterase, 46/266
agnosia, 42/275, 46/251-253
aluminum, 46/265, 271
aluminum intoxication, 36/320-323, 64/276, 278
amnesia, 46/259, 622
amyloid angiopathy, 21/64
amyloid microangiopathy, 46/256-258
amyloidosis, 46/257
amyotrophic lateral sclerosis, 22/313, 59/284-287, 410
animal model, 46/269, 56/585
anomic aphasia, 4/98
aphasia, 42/275, 45/321, 46/203, 251, 253
apraxia, 42/275, 46/251, 253
arecoline, 46/267
autonomic nervous system, 75/9
autotopagnosia, 46/251
Balint syndrome, 46/252
biochemical aspect, 27/482-488
biological rhythm, 74/496
blood biochemistry, 46/265
blood-brain barrier, 46/265
brain atrophy, 46/247-249, 253, 364
brain biochemistry, 46/265
brain biopsy, 10/680, 685, 46/128
brain blood flow, 46/363, 54/144
bromocriptine, 46/266
bulbar paralysis, 22/142
C-reflex, 49/611
calcium, 46/265
cholecystokinin, 46/268
choline, 46/267
choline acetyltransferase, 46/266
chromatin, 46/273
chromosomal aberration, 42/276, 46/271, 273
circadian disorder, 74/496
circadian rhythm, 74/495
classification, 46/204-206
clinical features, 46/250-253
clinicopathologic correlation, 46/260-264
clock function, 74/468
computer assisted tomography, 46/264, 54/58
congenital pain insensitivity, 51/566
consciousness, 45/117
constructional apraxia, 46/251
cortical myoclonus, 49/612
corticosteroid, 46/265

corticotropin, 46/268
Creutzfeldt-Jakob disease, 6/746, 46/294, 56/546
L-cycloserine, 46/268
deanol, 46/267
definition, 27/482
delta sleep inducing peptide, 46/268
dementia, 27/482-488, 42/275-277, 46/199, 59/410
depression, 46/208
detection, 46/200
digital autotopagnosia, 46/251
disorientation, 46/250
dopamine, 46/266
Down syndrome, 46/20, 73, 273
dyshoric angiopathy, 46/256
EEG, 42/277, 46/264
endorphin, 46/268
enkephalin, 46/268
epidemiology, 46/272-274
epileptic seizure, 46/252
etiologic classification, 46/205
experimental model, 27/487, 46/269-272
finger agnosia, 46/251
form, 21/64
frontal dementia, 45/30
frontal lobe syndrome, 46/432
GABA, 46/267
genetics, 46/272-274
α-globulin, 46/265
glutamic acid, 46/267
granulovacuolar degeneration, 22/399, 46/259, 49/250
grasp reflex, 46/252
Guam Parkinson dementia, 27/488
Hallervorden-Spatz syndrome, 49/401
haptoglobin1 gene, 46/273
heterochromatism, 46/273
hippocampus, 46/249
historic note, 46/247-249
homogenous disintegration, 46/252
homovanillic acid, 46/266
Huntington chorea, 49/291
hydrocephalic dementia, 46/326, 329
5-hydroxyindoleacetic acid, 46/266
hyperdiploidy, 46/273
hypertonia, 46/252
ideational apraxia, 45/426
ideomotor apraxia, 46/251
Ig, 46/265
immature plaque, 59/287
immunochemistry, 46/265
incidence, 46/249
incontinence, 46/328
iron, 46/265
juvenile, *see* Juvenile Alzheimer disease
17-ketosteroid, 46/265
Klüver-Bucy syndrome, 45/260
language, 45/321
levodopa, 46/266
Lewy body, 42/276
light therapy, 74/496
limbic system, 46/268
lipoprotein, 46/265
malabsorption syndrome, 70/234
management, 46/209
MAO B, 46/266
melatonin, 74/495
memory, 46/250
mimicry, 46/252
monoamine, 46/266
multi-infarct dementia, 46/265
multiple neuronal system degeneration, 59/136
myoclonus, 46/252, 49/618
neurofibrillary tangle, 46/247, 258-260
neurologic symptom, 46/252
neuronal ceroid lipofuscinosis, 42/276
neuropathology, 27/483, 46/210
neuropsychologic assessment, 45/519
neurotensin, 46/268
neurotransmitter, 46/266-269
noradrenalin, 46/266
nuclear magnetic resonance, 54/96
oculomotor disorder, 46/252
olivopontocerebellar atrophy (Dejerine-Thomas), 60/516
pain reaction, 46/253
palilalia, 43/225
Papez circuit, 46/259
paraphasia, 46/251
parietal lobe syndrome, 46/201
Parkinson dementia, 46/376, 49/178, 59/284-287
Parkinson disease, 49/96, 126
pathogenesis, 46/274
pathology, 46/253-264
perseverative agraphia, 45/467
phosphatidylcholine, 46/267
physostigmine, 46/267
Pick disease, 42/285, 46/233
positron emission tomography, 46/264
presenile dementia, 15/343, 42/275-277, 46/248
prion, 56/570
progabide, 46/268
progressive dysautonomia, 59/136
reflexology, 46/252
scrapie, 46/270
semantic jargon, 45/315

senile dementia, 46/248, 287
senile plaque, 27/486, 42/275, 46/247, 253-258, 56/555, 569
serotonin, 46/266
Sjögren syndrome, 71/74
sleep, 74/496, 549
somatostatin, 46/268
spinocerebellar degeneration, 21/15
staging, 46/253
stereotypy, 46/252
striatonigral degeneration, 49/209
subcortical dementia, 46/312-314
substance P, 46/268
suprachiasmatic nucleus, 74/496
synkinesis, 46/252
temporal cortex, 46/249
temporal lobe, 46/266
thalamus degeneration, 2/491
therapeutic trial, 46/266-268
treatment principle, 46/275
tremor, 46/252
urine steroid excretion, 46/265
vasoactive intestinal peptide, 46/268
vasopressin, 46/268
ventricular dilatation, 42/277
visual agnosia, 46/251
Alzheimer neurofibrillary change
aluminum encephalopathy, 21/57
amyotrophic lateral sclerosis, 22/131, 313
argyrophilic dystrophy, 21/46
Chamorro people, 22/343
CNS degeneration, 21/44, 47
encephalopathy, 21/57
Hallervorden-Spatz syndrome, 6/623
neuronal argyrophilic dystrophy, 21/63
Parkinson dementia complex, 22/131
progressive supranuclear palsy, 22/222-224, 49/249
pseudoamyloid, 21/57
vincristine encephalopathy, 21/57
Alzheimer prelipid substance, *see* Prelipid substance (Alzheimer)
Alzheimer type 1 glia
Wilson disease, 6/268
Alzheimer type 2 glia
Wilson disease, 6/269, 292
Alzheimer type I astrocyte
hepatic coma, 9/635, 49/214
spongy cerebral degeneration, 10/207
Wilson disease, 9/635
Alzheimer type II astrocyte
acquired hepatocerebral degeneration, 46/338
argininosuccinic aciduria, 22/577
central pontine myelinolysis, 9/637
hepatic coma, 49/214
hyperammonemia, 22/577
leukoencephalopathy, 47/584
liver disease, 9/635, 47/584
Amalric syndrome
primary pigmentary retinal degeneration, 13/253
Amanita muscaria intoxication
ataxia, 36/534
confusion, 37/329
dizziness, 36/534
dysphoria, 37/329
GABA, 29/507, 36/534
hallucination, 37/329
prognosis, 36/534
Amanita pantherina intoxication
anxiety, 36/534
ataxia, 36/534
convulsion, 36/534
delirium, 36/534
hallucination, 36/534
mania, 36/534
muscle spasm, 36/534
mydriasis, 36/534
prognosis, 36/534
Amanita phalloides intoxication
amanitin, 36/539
clinical features, 36/539
pathology, 36/540
phallotoxin, 36/538
prognosis, 36/542
treatment, 36/541
Amanitin
*see also* Mushroom toxin
Amanita phalloides intoxication, 36/539
clinical features, 65/37
coagulopathy, 65/37
hepatorenal symptom, 65/37
lethality, 65/37
liver transplantation, 65/37
silibinin, 65/37
stage, 65/37
thioctic acid, 65/37
treatment, 65/37
Amanitin intoxication
brain edema, 65/38
encephalopathy, 65/38
Amantadine
behavior disorder, 46/592-594
Creutzfeldt-Jakob disease, 46/294
delirium, 46/539
hallucination, 46/594
influenza A2 virus, 22/24

insomnia, 46/594
neuroleptic akathisia, 65/290
neuroleptic parkinsonism, 65/296
postherpetic neuralgia, 51/181
Amatoxin, *see* Amanitin
Amaurosis
acute, *see* Acute amaurosis
bacterial endocarditis, 52/299
fumarate hydratase deficiency, 66/420
hereditary striatal necrosis, 49/495
mucolipidosis type III, 29/367
renal insufficiency, 27/234
temporal arteritis, 55/341
uremia, 27/334
Amaurosis congenita (Leber), *see* Congenital retinal blindness
Amaurosis fugax
allergic granulomatous angiitis, 51/453
anemia, 55/108
arterial hypertension, 55/108
arteritis, 55/108
brain infarction, 55/110
brain ischemia, 53/309
carotid artery, 48/338
carotid artery stenosis, 55/111
carotid dissecting aneurysm, 54/272
carotid system syndrome, 53/309, 315, 318
cause, 53/235
cerebrovascular disease, 55/107
clinical features, 55/107
coagulopathy, 55/108
cryoglobulinemia, 63/404
exposure to light, 53/318
Hollenhorst plaque, 55/108
internal carotid arteriosclerosis, 55/107
internal carotid artery syndrome, 53/315, 318
migraine, 55/108
mitral valve prolapse, 63/21
myopia, 55/108
optic disc dysplasia, 55/108
optic nerve hyaline body, 55/108
papilledema, 55/108
polyarteritis nodosa, 55/356
polycythemia, 55/108
retinal artery embolism, 55/107
retinal migraine, 55/111
thrombocythemia, 55/108, 473
transient ischemic attack, 53/265, 309, 315, 318
visual defect progression, 55/109
visual evoked response, 53/318
Amaurotic familial idiocy, *see* Familial amaurotic idiocy
Amaurotic idiocy
Alexander disease, 10/101
familial, *see* Familial amaurotic idiocy
ganglioside, 10/220
GM1 gangliosidosis, 10/390
juvenile, *see* Juvenile amaurotic idiocy
macrencephaly, 10/101, 50/263
metachromatic leukodystrophy, 10/53
prelipid substance (Alzheimer), 10/20
Amaurotic nerve cell change, *see* Schaffer-Spielmeyer cell change
Ambenonium chloride
myasthenia gravis, 41/131
Ambidexterity
handedness, 4/316, 348, 466
Ambient cistern
anatomy, 1/553
arachnoid cyst, 31/118
Ambient temperature
cryofibrinogenemia, 55/490
cryoglobulinemia, 55/490, 63/404
myotonia, 62/262
myotonia congenita, 62/264
myotonic hyperkalemic periodic paralysis, 62/267
myotonic syndrome, 62/262
paramyotonia congenita, 62/267
Amblyomma americanum
Lyme disease, 51/206
neuroborreliosis, 51/206
Amblyopia
*see also* Blindness
albinism, 14/106
alcoholic, *see* Alcoholic amblyopia
cinchonism, 65/452
congenital heart disease, 63/11
cyanogenic glycoside intoxication, 36/520-522
cyclic oculomotor paralysis, 42/332
diamorphine addiction, 55/518
ethchlorvynol intoxication, 37/354
malnutrition, 28/19-22
methanol intoxication, 13/62
nutritional, *see* Tobacco alcoholic amblyopia
opiate intoxication, 37/386
pellagra, 7/571
strabismus, 42/421
subacute combined spinal cord degeneration, 28/173-178
tobacco, *see* Tobacco amblyopia
tobacco alcoholic, *see* Tobacco alcoholic amblyopia
toxic, 37/386
vitamin B12, 28/173-178
vitamin PP intoxication, 65/573
Amebiasis, 35/25-60

Acanthamoeba Hartmannella, 35/25, 51
amebic meningoencephalitis, 35/25, 34, 38-41
cerebral, *see* Cerebral amebiasis
clinical features, 35/37-41
CSF, 35/44-47
direct observation, 35/41
Entamoeba histolytica, 35/25, 47
epidemiology, 35/11, 34-37
laboratory study, 35/41-47
Naegleria fowleri, 35/25, 48-52
nervous system, *see* Neuroamebiasis
parasitic disease, 35/11
pathogenesis, 35/47-52
pathology, 35/52-56
prognosis, 35/56-60
treatment, 35/56-60

Amebic meningoencephalitis
amebiasis, 35/25, 34, 38-41
primary, *see* Primary amebic meningoencephalitis

Amenorrhea
aqueduct stenosis, 50/311
pellagra, 28/71
POEMS syndrome, 47/619, 63/394
Richards-Rundle syndrome, 43/264
Turner syndrome, 50/543

American Spinal Injury Association
terminology, 25/148

American Spinal Injury Association Motor Scale
spinal cord injury, 61/424
transverse spinal cord lesion, 61/424

American trypanosomiasis, *see* Chagas disease

Amethopterin, *see* Methotrexate

Amfebutamone
neurotoxin, 65/311

Amidohydrolase
aspartylglycosaminuria, 42/528

Amiloride
antihypertensive agent, 63/87
diuretic agent, 65/438
prognosis, 63/105

Amimia
brain amyloid angiopathy, 54/337
Rett syndrome, 29/328, 63/437

Amine precursor uptake and decarboxylation cells, *see* APUD cell

Amino acid
blood-brain barrier, 56/81
brain amino acid metabolism, 29/22-24
citrullinemia, 29/96
CSF, *see* CSF amino acid
encephalitogenic myelin basic protein, 9/507-512
enteric nervous system, 75/617
epilepsy, 15/315, 72/56, 84
γ-glutamyl transpeptidase deficiency, 29/122
hepatic coma, 27/360, 49/215
malnutrition, 29/9
migraine, 48/73
myelin basic protein, 47/491
ornithine carbamoyltransferase deficiency, 29/91
pellagra, 7/574, 28/59, 46/336, 51/333
plasma, *see* Plasma aminoacid
progressive dysautonomia, 59/148
protein $P_2$, 47/453
saccharopinuria, 29/222
subacute combined spinal cord degeneration, 59/359
sulfite oxidase deficiency, 29/121
sulfur containing disorder, 29/111-124
synaptic transmission, 1/48, 53, 29/22
tuberculous meningitis, 33/220
vitamin $B_6$ deficiency, 70/423

Amino acid metabolism
brain, *see* Brain amino acid metabolism
branched chain, *see* Branched chain amino acid metabolism
enzymopathy, 21/49
Friedreich ataxia, 60/305
Huntington chorea, 6/348
hyperbeta-alaninemia, 29/230
hypoglycin, 37/521
Rett syndrome, 29/319, 323

Amino acid neurotransmitter, 29/485-516
aspartic acid, 29/512-514
GABA, 29/485-507
glutamic acid, 29/512-514
glycine, 29/508-511
taurine, 29/514-516

Amino acid sequence
dystrophin, 62/125

Aminoacetonitrile
osteolathyrism, 7/644

Aminoaciduria
alkaline phosphatase, 43/126
cataract, 60/652
cerebrohepatorenal syndrome, 43/339
complex III deficiency, 62/504
congenital ophthalmoplegia, 29/232, 43/134
creatine kinase, 43/126
diabetes mellitus, 42/411
dwarfism, 42/517
encephalopathy, 29/233
febrile seizure, 42/517
fructose intolerance, 42/548
galactosemia, 42/552
Hartnup disease, 7/577, 14/315, 29/148-163, 42/146

hereditary periodic ataxia, 21/564
hyperdibasic, *see* Hyperdibasic aminoaciduria
infantile hypotonia, 43/125-127
juvenile diabetes mellitus, 42/411
leprechaunism, 42/590
Lowe syndrome, 42/606
maple syrup urine disease, 10/138
mental deficiency, 42/517
metabolic ataxia, 21/579
microcephaly, 42/517
migraine, 5/42
mitochondrial myopathy, 43/188
mitochondrial respiratory chain defect, 42/587
muscle fiber size, 43/126
muscular dystrophy, 42/517
myopathy, 43/125-127
ophthalmoplegia, 43/125-127
osteoporosis, 42/517
Paine syndrome, 30/514, 43/433
Passwell syndrome, 21/216
phenylketonuria, 10/138
Rowley-Rosenberg syndrome, 43/473
sensorineural deafness, 43/126
spinal muscular atrophy, 59/371
vitamin B6, 28/117-123
Wilson disease, 27/387, 42/271, 49/229, 233
α-Amino-n-butyric acid
blood level, 13/339
CSF level, 13/339
epileptic seizure, 13/339
hearing loss, 13/339
mental deficiency, 13/339
myoclonus, 13/339
urine level, 13/339
visual disorder, 13/339
γ-Aminobutyric acid, *see* GABA
Aminocaproic acid intoxication
myoglobinuria, 40/339
rhabdomyolysis, 40/339
ε-Aminocaproic acid
brain aneurysm, 55/52
brain aneurysm neurosurgery, 55/52
subarachnoid hemorrhage, 12/177, 55/52
toxic myopathy, 62/598, 601
Aminoglutethimide
Cushing syndrome, 68/356
Aminoglycolipid
degradation, 10/302
Aminoglycoside
bacterial meningitis, 52/9
deafness, 65/482
drug induced polyneuropathy, 51/295
encephalopathy, 46/602
gram-negative bacillary meningitis, 52/105
hypomagnesemia, 63/562
myasthenic syndrome, 65/483
staphylococcal meningitis, 33/72-74
Aminoglycoside intoxication
deafness, 65/482
dysequilibrium, 65/482
hypoventilation, 65/483
myasthenic syndrome, 65/483
neuromuscular block, 65/483
ototoxicity, 65/482
predisposing factor, 65/482
respiratory depression, 65/483
Aminoguanidine
diabetes mellitus, 70/154
*p*-Aminohippuric acid
depigmentation, 22/548
α-Amino-3-hydroxy-5-methyl-4-isoxazolepropionic acid
antiepileptic agent, 73/349
excitotoxin, 64/20
α-Amino-3-hydroxy-5-methyl-4-isoxazolepropionic acid receptor
excitotoxic mechanism, 64/20
lathyrism, 65/14
β-Aminoisobutyric acid, *see* 3-Amino-2-methylproprionic acid
5-Aminolevulinate synthase
acute intermittent porphyria, 51/390
Aminolevulinic acid
acute intermittent porphyria, 8/9, 366, 27/435
erythrohepatic protoporphyria, 27/433
hereditary coproporphyria, 27/438
porphyria variegata, 8/9, 42/620
synthetase, 27/429, 431
Aminolevulinic acid dehydrase, *see* Porphobilinogen synthase
δ-Aminolevulinic acid, *see* Aminolevulinic acid
δ-Aminolevulinic acid dehydratase, *see* Porphobilinogen synthase
δ-Aminolevulinic acid dehydratase deficiency porphyria
neuropathy, 60/120
δ-Aminolevulinic acid synthetase, *see* 5-Aminolevulinate synthase
2-Amino-3-methylaminopropionic acid
*see also* Cycad intoxication
biokinetic, 65/22
excitotoxin, 64/20
Guam amyotrophic lateral sclerosis, 59/263, 274
Kii Peninsula amyotrophic lateral sclerosis, 59/274
Lytico Bodeg, 64/5

neurotoxin, 59/263, 274, 64/5, 65/22
source, 64/5
toxic encephalopathy, 64/5
2-Amino-3-methylaminopropionic acid intoxication
symptom, 64/5
*p*-Aminomethylbenzoic acid
subarachnoid hemorrhage, 12/178
2-Amino-4-methyl-5-phenyl-2-oxazoline, *see* 4-Methylaminorex
3-Amino-2-methylproprionic acid
metachromatic leukodystrophy, 10/45
Richards-Rundle syndrome, 22/516
Aminopeptidase
myelin, 9/34
Aminophylline
brain vasospasm, 11/125
delirium, 46/539
status epilepticus, 46/539
β-Aminoproprionitrile
osteolathyrism, 7/644
Aminopterin syndrome
anencephaly, 30/122, 50/80
hydrocephalus, 30/122
mental deficiency, 46/78
microcephaly, 30/508
vitamin Bc deficiency, 46/78
4-Aminopyridine
botulism, 65/212
Eaton-Lambert myasthenic syndrome, 41/360, 62/426
multiple sclerosis, 47/39-41
paraplegia, 61/401
spinal cord injury, 61/401
tetraplegia, 61/401
transverse spinal cord lesion, 61/401
Amiodarone
ataxia, 63/190
autonomic nervous system, 75/36
benign intracranial hypertension, 63/190, 67/111
cardiac pharmacotherapy, 63/190
cytoplasmic inclusion body, 65/484
drug induced demyelinating neuropathy, 47/620
dyskinesia, 63/190
Fabry corneal dystrophy, 66/228
iatrogenic neurological disease, 63/190, 65/457
neuropathy, 63/190
neurotoxicity, 51/294
optic neuropathy, 63/190
orofacial dyskinesia, 63/190
parkinsonism, 63/190
toxic myopathy, 62/606
toxic neuropathy, 51/302
tremor, 63/190
Amiodarone intoxication
action tremor, 65/458
ataxia, 65/458
benign intracranial hypertension, 65/459
corneal microdeposit, 65/459
corneal opacity, 60/172
dyskinesia, 65/458
hemiballismus, 65/458
insomnia, 65/458
jaw tremor, 65/458
lipidosis, 65/459
myoclonus, 65/458
nerve biopsy, 65/458
neurolipidosis, 65/458
neurologic adverse effect, 65/457
optic neuropathy, 65/459
sensorimotor neuropathy, 65/458
tremor, 65/458
Amiodarone myopathy
proximal, 65/458
sensorimotor polyneuropathy, 62/608
tremor, 63/190
Amiodarone polyneuropathy
ataxia, 51/302
chronic course, 51/303
chronic inflammatory demyelinating polyradiculoneuropathy, 51/530
drug induced polyneuropathy, 51/294, 303
EMG, 51/303
phospholipase A1, 51/303
phospholipase A2, 51/303
sensorimotor polyneuropathy, 65/458
tremor, 51/302, 63/190
Amitriptyline
anorexia nervosa, 46/588
chronic muscle contraction headache, 48/350
headache, 48/364
metabolism, 65/315
migraine, 5/101, 48/87, 179, 196
motoneuron disease, 59/467
neurotoxin, 51/68, 65/314
postherpetic neuralgia, 56/234
toxic polyneuropathy, 51/68
tricyclic antidepressant, 65/312
Ammon horn
epilepsy, 15/612
neuropathology, 21/45
Ammon horn sclerosis
cerebral hemiatrophy, 50/205
temporal lobe epilepsy, 2/718, 73/69
Ammonia
astrocytic response, 27/357
CSF, *see* CSF ammonia

hepatic coma, 27/355-360
hepatic encephalopathy, 27/355-360
ornithine carbamoyltransferase deficiency, 29/91
Reye syndrome, 9/550, 29/335, 338, 34/80, 168, 56/151, 238, 63/438
urea synthesis, 29/87
Ammonia intoxication
periodic, *see* Periodic ammonia intoxication
Ammonium chloride
hyperventilation syndrome, 63/429
periodic paralysis, 41/157
Amnesia
alcohol intoxication, 64/114
Alzheimer disease, 46/259, 622
amnesic shellfish poisoning, 65/166
anterior cerebral artery syndrome, 53/348
anterograde, *see* Anterograde amnesia
aphasia, *see* Memory disorder
basofrontal syndrome, 53/348
benzodiazepine intoxication, 65/337
bilateral occipital lobe infarction, 53/413
brain concussion, 23/420, 432, 24/477, 562
brain embolism, 53/206
brain infarction, 53/206, 55/139
brain injury, 23/12, 24/477
brain lacunar infarction, 53/395
cardiac arrest, 63/214, 217
cardiac catheterization, 63/176
cerebrovascular disease, 53/206, 55/139
chlormethine intoxication, 64/233
ciclosporin, 63/10, 537
color, *see* Color amnesia
confusion, 3/270
delirium, 46/528
dementia, 46/204, 210
domoic acid, 65/166
dysphrenia hemicranica, 5/79
epilepsy, 72/251
frontal lobe syndrome, 53/348
global, *see* Global amnesia
gunshot injury, 46/622
head injury, 3/293-295
hypereosinophilic syndrome, 63/372
hysteria, 43/207
intracranial hemorrhage, 53/206
Korsakoff, *see* Korsakoff amnesia
marine toxin intoxication, 65/142
mesencephalic infarction, 53/392
methanol intoxication, 36/353, 64/99
migraine, 48/8, 156, 161
migrainous transient global, *see* Migrainous transient global amnesia
musical, *see* Musical amnesia
neurofibrillary tangle, 46/260-263
obstructive sleep apnea, 63/454
Pick disease, 42/285, 46/234, 240
posterior cerebral artery syndrome, 53/412
posttraumatic, *see* Posttraumatic amnesia
presenile dementia, 46/253
primary alcoholic dementia, 46/342
prognosis, 46/611
progressive supranuclear palsy, 49/243
psychosis, *see* Wernicke-Korsakoff syndrome
retrograde, *see* Retrograde amnesia
temporal lobe epilepsy, 2/713
thalamic infarction, 53/392, 395
thalamic syndrome (Dejerine-Roussy), 45/92
topographical, *see* Topographical amnesia
toxic oil syndrome, 63/381
transient global, *see* Transient global amnesia
transient ischemic attack, 12/3, 53/206
traumatic, *see* Traumatic amnesia
vertebrobasilar system syndrome, 53/392, 395
visual, *see* Visual amnesia
visual agnosia, 4/16
vitamin $B_c$ deficiency, 63/255
Wernicke-Korsakoff syndrome, 28/251-254
Amnesia psychosis, *see* Wernicke-Korsakoff syndrome
Amnesic aphasia, *see* Memory disorder
Amnesic color anomia
occipital lobe syndrome, 45/52
Amnesic confabulatory psychosis, *see* Wernicke-Korsakoff syndrome
Amnesic shellfish poisoning
*see also* Marine toxin intoxication
amnesia, 65/166
anterior horn cell, 65/167
axonopathy, 65/167
coma, 65/167
computer assisted tomography, 65/167
confusion, 65/167
delirium, 65/167
dementia, 65/166
diagnosis, 65/168
diffuse axonopathy, 65/167
domoic acid, 65/166
EEG, 65/167
EMG, 65/167
epidemiology, 65/166
epilepsy, 65/167
headache, 65/167
hippocampus, 65/166
management, 65/168
marine toxin intoxication, 65/143
*N*-methyl-D-aspartic acid receptor, 65/167

mussels, 65/166
myoclonus, 65/167
neuronopathy, 65/167
neuropathology, 65/167
Nitzschia pungens, 65/166
nuclear magnetic resonance, 65/167
ophthalmoplegia, 65/167
pathomechanism, 65/166
treatment, 65/168
Amnesic state
ciclosporin intoxication, 63/537
Amnesic syndrome
basal forebrain stroke, 55/140
brain infarction, 55/139
diagnosis, 45/115
medial temporal lobe, 55/139
Amniocentesis
anencephaly, 30/178
Down syndrome, 31/444
Fabry disease, 29/363
Gaucher disease, 29/353
glycosaminoglycanosis type VII, 29/369
GM1 gangliosidosis, 29/369
GM2 gangliosidosis, 29/366
GM2 gangliosidosis type I, 29/366
Lesch-Nyhan syndrome, 29/273
mucolipidosis type III, 29/369
muscular dystrophy, 40/357
Niemann-Pick disease type A, 29/355
Niemann-Pick disease type B, 29/355
porencephaly, 50/359
prenatal diagnosis, 32/564-568
Amniotic band syndrome
amputation, 30/116
arthrogryposis multiplex congenita, 42/73
encephalocele, 30/116
limb deformity, 30/116
microcephaly, 30/116
scalp defect, 30/116
skull defect, 30/116
Amniotic fluid
α-fetoprotein, 32/563
glycoconjugate, 66/56
Amniotic fluid cell
prenatal diagnosis, 32/571
Amniotic fluid protein
prenatal diagnosis, 32/570
Amodiaquine
lupus erythematosus, 13/220
malaria, 13/220, 35/155, 52/370
rheumatoid arthritis, 13/220
Amoebae
Acanthamoeba castellani, 35/28, 32
classification, 35/26
Entamoeba histolytica, 35/28, 30
Hartmannella, 35/33
human pathogenetic specy, 35/28-33
Iodamoeba bütschlii, 35/30
Naegleria fowleri, 35/30-32
Amorphognosia
agnosia, 4/34
tactile agnosia, 45/66
Amorphosynthesis
agnosia, 4/26
apraxia, 4/57, 61
attention disorder, 3/189
body scheme disorder, 4/213, 215
constructional apraxia, 4/80
hemiasomatognosia, 45/375
neglect syndrome, 45/155
sensory extinction, 3/193
Amoxapine
heterocyclic antidepressant, 65/312
neurotoxin, 62/601, 65/311
toxic myopathy, 62/601
Amoxapine intoxication
chorea, 65/319
malignant neuroleptic syndrome, 65/319
tardive dyskinesia, 65/319
Amoxicillin
epilepsy, 65/472
neurotoxin, 65/472
Amoxycillin, *see* Amoxicillin
AMP, *see* Adenosine phosphate
AMPA, *see* α-Amino-3-hydroxy-5-methyl-4-isoxazolepropionic acid
Amphetamine
*see also* Hallucinogenic agent
acquired myoglobinuria, 62/573
autonomic nervous system, 75/501
behavior disorder, 46/592
brain hemorrhage, 55/519
brain infarction, 53/164
brain ischemia, 53/164
brain vasculitis, 55/520
cerebrovascular disease, 46/592
chemical structure, 13/232
depression, 46/474
drug abuse, 55/519
drug addiction, 65/256
epilepsy, 72/234
hallucinogenic agent, 65/41
hyperkinetic child syndrome, 3/196, 4/363, 46/177
hyperthermia, 75/64
ischemic myelopathy, 63/55
narcolepsy, 2/448, 3/94, 15/847

neurotoxin, 36/232, 48/420, 62/601, 65/251
pain, 69/48
related alkaloid, 65/50
spinal cord vasculopathy, 63/55
stereotypy, 46/592
thermoregulation, 75/64
toxic encephalopathy, 65/251
toxic myopathy, 62/601
turning behavior, 45/167
Amphetamine addiction
brain angiography, 55/520
brain edema, 55/519
brain hemorrhage, 55/519
brain infarction, 55/520
brain vasculitis, 55/520
coma, 55/519
computer assisted tomography, 55/520
drug induced angiitis, 55/520
headache, 48/420, 55/519
heat stroke, 55/519
hypertension, 55/519
polyarteritis nodosa, 55/360, 520
subarachnoid hemorrhage, 55/360, 520
Amphetamine intoxication
absorption, 65/256
abstinence, 65/258
acute symptom, 65/257
agitation, 65/258
anergia, 65/258
anhedonia, 65/258
appetite, 65/257
arterial hypertension, 65/257, 259
ataxia, 65/257
brain hemorrhage, 65/51, 258
brain infarction, 65/258
clinical effect, 65/51
CNS stimulant, 65/251
convulsion, 65/257
crash, 65/258
dependence, 65/258
dexamphetamine, 65/256
dopaminergic agent, 65/257
epilepsy, 65/51, 258
euphoria, 65/257
fenfluramine, 65/256
hallucination, 65/50
hedonia, 65/257
hypersomnia, 65/258
hyperthermia, 65/257
insomnia, 65/257
mental symptom, 65/257
methamphetamine, 65/256
mydriasis, 65/51
myoclonia, 65/257
necrotizing angiitis, 65/258
neurologic complication, 65/258
pharmacology, 65/51, 256
phentermine, 65/256
phenylisopropylamine, 65/256
schizophrenia, 65/257
schizophreniform psychosis, 65/257
stroke, 65/258
subarachnoid hemorrhage, 65/258
sudden death, 65/258
tachycardia, 65/257
tremor, 65/51, 257
vigilance, 65/257
Amphimixis
Down syndrome, 50/523
Amphiocularity
handedness, 4/258
Amphiphysin
paraneoplastic syndrome, 69/330, 336
Amphotericin B
acute amebic meningoencephalitis, 52/329
adverse effect, 69/504
African histoplasmosis, 35/513
aspergillosis, 35/399, 52/381
Candida meningitis, 52/404
candidiasis, 35/433, 52/403
cephalosporiosis, 52/482
cladosporiosis, 52/484
coccidioidomycosis, 35/450-452, 52/419
cryptococcosis, 35/487-490, 52/433
curvulariosis, 52/485
Histoplasma meningitis, 52/441
histoplasmosis, 35/511-513, 52/440
human immunodeficiency virus infection, 71/384
leukoencephalopathy, 47/570, 71/384
mucormycosis, 52/475
neuropathy, 7/516
neurotoxicity, 71/368
neurotoxin, 67/363
North American blastomycosis, 35/406, 52/390
paracoccidioidomycosis, 35/537, 52/460
Pseudoallescheria, 52/492
rhodotorulosis, 52/493
sporotrichosis, 52/495
toxic myopathy, 62/601
Amphotericin B intoxication
myoglobinuria, 40/339
rhabdomyolysis, 40/339
Amphotonia
diabetes mellitus, 22/251
AMPS, *see* Acid mucopolysaccharide
Ampulla venosa, *see* Galen vein

Amputation
amniotic band syndrome, 30/116
pain, 70/44
paresthesia, 70/44
phantom limb, 4/224, 400, 45/396
phantom limb pain, 45/400
phantom sensation, 45/396
traumatic neuroma, 70/44
Amsacrine
antineoplastic agent, 65/529
epilepsy, 65/529
neuropathy, 69/461, 472
neurotoxin, 65/529, 538
Amsacrine intoxication
epilepsy, 65/538
Amsterdam Neurological Society
foundation, 1/10
Amusia, 45/483-489
*see also* Musical agnosia
absolute pitch, 4/197, 45/487
agnosia, 4/38, 195-206
agraphia, 45/487
alexia, 45/487
amnesic amusia, 4/203
anatomic localization, 4/203
aphasia, 45/487
apraxia, 4/203
auditory agnosia, 4/37
cerebral dominance, 4/294
classification, 2/600
clinical features, 4/199-202, 45/485-487
clinicopathologic correlation, 45/488
cochlear, 4/203
emotion, 4/197
examination, 4/199-202
expressive, 4/197, 203
frontal lobe lesion, 45/31
history, 4/196
instrumental apraxia, 45/486
jargon aphasia, 45/487
language, 4/197, 45/488
memory disorder, 2/599, 691-693, 4/98, 203, 221
motor aphasia, 45/487
motor disturbance, 2/600
musical agraphia, 4/203, 45/486
musical alexia, 4/203, 45/486
musical amnesia, 45/486
musical deafness, 4/203, 45/485
musical function, 4/195-197
musical language, 4/197
musical memory, 45/485
musical perception, 45/485
musicianship, 45/484
musicogenic epilepsy, 4/204
paraphasia, 45/487
pitch perception, 45/487
prosody, 4/198, 45/488
receptive, *see* Receptive amusia
rhythmic sense, 4/196, 45/486
schizophrenia, 45/484
sense of sound, 4/196
sensory disturbance, 2/600
speech, 45/488
speech and music, 4/198
stroke, 45/487
temporal lobe lesion, 4/3, 204
temporal lobectomy, 45/46, 484
terminology, 4/196, 202
vocal, *see* Vocal amusia
vocal apraxia, 4/203
Wernicke aphasia, 45/487
Amygdala
anxiety, 74/309
autonomic nervous system, 74/58, 79
cardiac dysrhythmia, 63/239
emotion, 2/440, 719, 3/324, 329
epilepsy, 15/99
hippocampus, 73/61
seizure, 73/77
temporal lobe epilepsy, 73/61
Amygdala lesion
emotional disorder, 3/347
memory disorder, 2/713
Amygdalectomy
emotion, 45/276
schizophrenia, 46/497
sorrow, 45/276
Amygdalin
cyanide intoxication, 65/32
cyanogenic glycoside intoxication, 36/515, 65/26, 32
neurotoxin, 36/515, 65/26
Amygdalohippocampectomy
epilepsy, 73/390
Amygdaloid nucleus
anxiety, 45/275
basal ganglion, 49/1, 28
emotion, 45/273, 275
limbic encephalitis, 8/135
neurogenic arterial hypertension, 63/231
schizophrenia, 46/507
temporal lobe epilepsy, 73/55
ventral globus pallidus, 49/26
Amygdalostriate fiber
glutamic acid, 49/4
2-Amyl-methylphosphonofluoridate intoxication

chemical classification, 37/543
organophosphorus compound, 37/543
Amyl nitrate
scotoma, 48/65
Amyl nitrite
migraine, 5/379
Amyl nitrite intoxication
vasodilator agent, 37/453
Amylo-1,6-glucosidase
debrancher deficiency, 27/222, 230, 62/483
Amylo-1,6-glucosidase deficiency, *see* Debrancher deficiency
Amyloid
amyloid A protein, 51/414
carpal tunnel syndrome, 47/618
cerebral angiopathy, *see* Brain amyloid angiopathy
Creutzfeldt-Jakob disease, 46/295
Creutzfeldt-Jakob disease prion, 56/563
cystatin C, 51/414
demyelinating neuropathy, 47/618
Gerstmann-Sträussler-Scheinker disease, 60/629
kuru, 46/256
multiple myeloma, 39/132, 141, 156, 47/618
myeloma, 51/413
nucleotide base mutation, 51/414
P component, 51/414
pathogenesis, 46/255
prealbumin, 51/414
prion, 56/566, 568
pseudo, *see* Pseudoamyloid
psychochemical property, 51/413
senile dementia, 46/284
senile plaque, 46/253-258, 56/569
structure, 51/413
Virchow-Robin perivascular space, 46/257
Amyloid A protein
amyloid, 51/414
Amyloid angiopathy
Alzheimer disease, 21/64
argyrophilic dystrophy, 21/45
arterial hypertension, 63/74
brain, *see* Brain amyloid angiopathy
brain hemorrhage, 54/70
computer assisted tomography, 54/70
granulomatous CNS vasculitis, 55/388
headache, 48/278
hereditary cerebral, *see* Hereditary cerebral amyloid angiopathy
neuronal argyrophilic dystrophy, 21/64
various forms, 21/57
Amyloid autonomic polyneuropathy
orthostatic hypotension, 63/149
Amyloid deposit
dermatomyositis, 62/377
hereditary sensory and autonomic neuropathy type I, 60/9
inclusion body myositis, 62/377
multiple myeloma polyneuropathy, 51/431
polymyositis, 62/377
Amyloid disposition pattern
hereditary amyloid polyneuropathy, 60/104
Amyloid fibril, 51/414
amyloidosis, 71/504
protein, *see* Amyloid fibril protein
Amyloid fibril protein
Down syndrome, 50/530
Amyloid microangiopathy
Alzheimer disease, 46/256-258
Amyloid neuropathy, *see* Amyloid polyneuropathy
Amyloid nodule
amyloid polyneuropathy, 51/418
Amyloid plaque
Creutzfeldt-Jakob disease, 56/551
Gerstmann-Sträussler-Scheinker disease, 56/554
prion disease, 56/555
subacute spongiform encephalopathy, 42/655
Amyloid polyneuropathy
amyloid nodule, 51/418
analgesia, 51/416
anhidrosis, 63/149
ascending symmetrical neuropathy, 51/416
autonomic change, 75/691
autonomic dysfunction, 51/416, 63/148
autonomic nervous system, 75/31
autonomic neuropathy, 75/691
axonal degeneration, 51/417
baroreceptor sensitivity, 51/416
carpal tunnel syndrome, 51/416
CSF protein, 51/417
denervation, 51/417
dissociated sensibility, 51/416
dissociated sensory loss, 51/419, 63/393
dysphagia, 51/416
familial, *see* Hereditary amyloid polyneuropathy
hereditary, *see* Hereditary amyloid polyneuropathy
hypalgesia, 51/416
ischemia, 51/419
multiple myeloma polyneuropathy, 63/393
myelinated fiber loss, 51/418
nerve conduction velocity, 51/417
neuropathology, 51/417
orthostatic hypotension, 51/416
pain, 51/416
paranodal demyelination, 51/417
pathogenesis, 51/418

pupillary reflex abnormality, 51/416
sensorimotor polyneuropathy, 51/416
sex ratio, 51/416
sexual impotence, 51/416
small fiber damage, 51/419
sural nerve, 51/418
survival rate, 51/416
sweat test, 51/416
temperature sensory loss, 51/416
treatment, 51/419
ultrastructure, 51/417
Valsalva response, 51/416
Amyloid β-protein
dermatomyositis, 62/377
Guam amyotrophic lateral sclerosis, 59/259
inclusion body myositis, 62/377
polymyositis, 62/377
Amyloid tumor
dysarthria, 8/9
dysphagia, 8/9
Amyloidosis
*see also* Brain amyloid angiopathy
Alzheimer disease, 46/257
amyloid fibril, 71/504
apolipoprotein AI, 71/517
arteriolar fibrinohyalinoid degeneration, 11/581
autonomic dysfunction, 63/148
autonomic polyneuropathy, 51/476
back pain, 63/529
bowel, 75/650
brain embolism, 55/163, 63/137
brain hemorrhage, 11/611, 615, 42/727
brain infarction, 55/163
carpal tunnel syndrome, 8/7, 42/309
cerebrovascular disease, 53/28, 43, 55/163
Chédiak-Higashi syndrome, 42/536
classification, 51/414, 71/502
cranial neuropathy, 71/508
diagnosis, 71/509, 516
dyshoric angiopathy, 42/728
EMG, 71/505
heart disease, 71/508
hemiplegia, 63/137
hereditary, *see* Hereditary amyloidosis
hereditary primary, *see* Hereditary primary amyloidosis
hereditary sensory and autonomic neuropathy, 60/19
hereditary sensory and autonomic neuropathy type I, 60/19
histology, 11/140
hypertrophic interstitial neuropathy, 21/145
intestinal pseudo-obstruction, 51/493
ischemic neuropathy, 8/161
kidney, 71/508
β2-microglobulin, *see* β2-Microglobulin amyloidosis
Muckle-Wells syndrome, 22/489, 42/396
multinodular stenosing, *see* Multinodular stenosing amyloidosis
myopathy, 71/506
nerve ischemia, 12/658
neuropathy, 7/556, 8/4-8, 160, 71/517
nomenclature, 71/503
orthostatic hypotension, 1/472, 63/148
paraproteinemia, 69/290, 302
pathogenesis, 71/505
pathology, 71/506, 516
pathophysiology, 71/515
polyneuropathy, 71/504, 507
primary, *see* Primary amyloidosis
primary systemic, *see* Primary systemic amyloidosis
prognosis, 71/510
radiculopathy, 63/529
restless legs syndrome, 8/316, 51/546
restrictive infiltrative cardiomyopathy, 63/137
secondary, *see* Secondary amyloidosis
senile dementia, 46/285
sensorineural deafness, 42/396
survey, 71/501
symptom, 71/431
systemic, *see* Systemic amyloidosis
Takayasu disease, 12/413
tetraplegia, 63/529
transient ischemic attack, 63/137
Treacher-Collins, 71/511, 517
trigeminal neuralgia, 8/9, 42/353
vasa nervorum disease, 12/658
Amylo-1,4[1,6-transglucosidase
brancher deficiency, 27/222, 231
Amyoplasia congenita, *see* Arthrogryposis multiplex congenita
Amyotonia congenita
*see also* Hypotonia
classification, 40/286
clinical presentation, 40/336
controversy, 41/33
history, 41/2, 28
hypotonia, 1/271, 40/336, 43/79
motor retardation, 43/79
Amyotrophic cerebellar hypoplasia
cerebellar agenesis, 50/192
Amyotrophic dystonic paraplegia
Babinski sign, 42/199
brain atrophy, 42/200

EEG, 42/199
familial, *see* Familial amyotrophic dystonic paraplegia
hyperreflexia, 42/199
mental deficiency, 42/199
myalgia, 42/199
speech disorder, 42/199
ventricular dilatation, 42/199
Amyotrophic lateral sclerosis, 22/281-330, 59/169-204
*see also* Motoneuron disease, Parkinson dementia complex, Progressive muscular atrophy *and* Wetherbee ail
acquired immune deficiency syndrome, 56/493
age incidence, 59/179
age at onset, 22/290
altered receptor, 59/177
aluminum, 59/280
Alzheimer disease, 22/313, 59/284-287, 410
Alzheimer neurofibrillary change, 22/131, 313
annual incidence, 42/65
anterior horn syndrome, 42/65
anterior spinal artery, 22/19
arachnoiditis, 22/309
Aran-Duchenne disease, 22/305
Argyll Robertson pupil, 22/295
associated disease, 22/313
autonomic dysfunction, 59/196
Babinski sign, 42/65
ballismus, 59/235
basal ganglion cell death, 6/64
benign fasciculation, 59/204
blood vessel, 59/182
bulbar, *see* Progressive bulbar palsy
bulbar paralysis, 22/281, 42/65
Bunina body, 59/280
carcinoma, 59/202
carcinomatous encephalopathy, 22/144
carcinomatous form, 22/327
carcinomatous neuromyopathy, 22/327
cause, 22/125
cause of death, 22/115
celiac disease, 28/228
cell change, 59/181
cervical intervertebral disc prolapse, 22/309
cervical spondylarthrosis, 22/309
Chamorro people, 21/26, 34, 22/129, 283, 287, 339-347
chorea, 22/321, 59/235
chromatin, 59/280
clinical, 59/190-196
clinical features, 22/291, 309
communication, 22/126
conventional type, 22/305, 308
corneal lattice dystrophy, 22/139
cortical change, 59/234
corticobasal degeneration, 59/410
corticospinal tract degeneration, 42/65
cramp, 59/193
cranial nerve, 42/65, 59/193
Creutzfeldt-Jakob disease, 22/317, 59/203, 409
CSF, 22/296, 59/196
CSF homovanillic acid, 22/30
cutaneous collagen, 21/34
deafness, 22/138
death rate, 22/288
definition, 59/231
deglutition, 22/21
dementia, 22/313, 316, 318, 369, 42/67-69, 59/231, 409-412
dementia plus extrapyramidal disease, 22/316
dendritic damage, 59/177
depression, 59/195
deuteropathic motor system, 22/43
diabetes mellitus, 22/370
diabetic proximal neuropathy, 59/201
diagnosis, 22/112, 128, 294, 308, 310
differential diagnosis, 22/15, 59/200-204, 383-401
DNA, 59/177
dopamine, 21/34
drug medication, 59/197
duration of illness, 22/307, 59/180
dysarthria, 22/120, 42/65
dysphagia, 42/65
dystonia, 59/235
edrophonium test, 62/413
electrical injury, 22/329, 23/708, 710, 61/194, 196
electrophysiologic diagnostic criteria, 59/190
electrophysiology, 59/185-190
EMG, 1/639, 22/297-305, 59/185
epidemiology, 21/32, 22/287, 290, 339, 341, 349, 354, 59/179-181
esophagoscopy, 22/124
etiology, 59/174-179
excitatory aminoacid hypothesis, 59/178
extrapyramidal syndrome, 22/315, 59/203
familial, *see* Familial amyotrophic lateral sclerosis
familial incidence, 59/245
familial spastic paraplegia, 59/306
fasciculation, 22/292, 42/65, 59/185, 192
fatigability, 59/194
Friedreich ataxia, 21/354
gastrectomy, 22/29
gastrointestinal sign, 22/296
genetic abnormality, 59/175
geographic distribution, 59/179

geography, 22/287
glucose intolerance, 22/30
growth factor, 59/232
Guam, *see* Guam amyotrophic lateral sclerosis
Guam motoneuron disease, 65/21
H-reflex, 22/304
Hallervorden-Spatz syndrome, 6/623, 22/323, 383, 397, 59/410
hereditary cerebellar ataxia, 22/142
hereditary motor and sensory neuropathy type I, 22/326
hereditary motor and sensory neuropathy type II, 22/326
hereditary type, *see* Familial amyotrophic lateral sclerosis
Hirano body, 22/112, 131, 287, 325-327
Hirayama disease, 59/113
historic review, 59/231
historic survey, 59/231
history, 22/281-283, 59/169-171
HLA antigen, 42/66
homovanillic acid, 22/30
human T-lymphotropic virus type I associated myelopathy, 59/451-453
Huntington chorea, 22/138, 59/410
hyperlipoproteinemia type IIA, 22/46
hyperparathyroidism, 59/202
hyperreflexia, 42/65
hyperthyroidism, 59/202
idiopathic esophagus hypertrophy, 22/143
immunologic abnormality, 59/174
incidence, 21/32
inclusion body, 22/323
infantile, *see* Infantile amyotrophic lateral sclerosis
injury, 22/328
intention tremor, 59/235
Japanese, 22/353-416
juvenile, *see* Juvenile amyotrophic lateral sclerosis
juvenile Alzheimer disease, 59/285-287
Kii Peninsula, *see* Kii Peninsula amyotrophic lateral sclerosis
Klüver-Bucy syndrome, 59/234
laboratory investigation, 59/196
Lafora body, 42/69
late infantile neuroaxonal dystrophy, 49/405
laughing, 22/115
lead, 51/268
lead intoxication, 59/200
lead myelopathy, 64/436
lead polyneuropathy, 64/436
Lewy body disease, 59/410
lumbar type, 22/306, 308
lymphocyte infiltration, 59/183
Macaca irus, 59/291-294
Machado-Joseph disease, 59/410
Madras disease, 22/323
malignant, 22/31
manganese intoxication, 59/200, 64/307
Mariana Island type, 22/129
masseter muscle reflex, 22/293
mental change, 59/195
mercury intoxication, 59/200
methisoprinol, 22/24
Mills syndrome, 22/313
motor unit, 59/186-189
multiple neuronal system degeneration, 59/411
multiple sclerosis, 22/309, 59/202, 407
muscle, 59/184
muscular atrophy, 22/291, 42/65, 78-81
myasthenic syndrome, 22/295
myopathy, 59/201
nasopalpebral reflex, 22/293
natural history, 22/305
neoplasm, 22/31
nerve conduction, 59/189
nerve conduction block, 47/368
nerve conduction velocity, 22/303
neurofibrillary tangle, 22/399, 408, 42/70, 46/274, 49/249, 59/282-284
neurofibromatosis, 14/492
neurofilamentous swelling, 59/234
neuronal atrophic dystrophy, 21/62
neuropathology, 22/127, 59/232
neurosyphilis, 22/309, 52/278
New Guinea, 21/26, 35, 22/349
nosology, 22/128, 308, 310
nuclear volume, 59/280
nucleic acid metabolism, 59/280
nucleolar volume, 59/280
nystagmus, 22/114, 322
oculomotor paralysis, 22/295
olivopontocerebellar atrophy, 22/371, 59/411
olivopontocerebellar atrophy (Dejerine-Thomas), 60/531
ophthalmoplegia, 59/232, 411
optic atrophy, 13/97, 22/138
orthostatic hypotension, 63/147
osteitis deformans (Paget), 22/309
palatal myoclonus, 22/140
pallidal atrophy, 6/619
pallidoluysionigral degeneration, 49/447, 453, 456
pancreatic abnormality, 59/178
pancreatic dysfunction, 21/33, 22/30
parathyroid hormone, 59/177
Parkinson dementia, 49/167, 174, 401, 59/284-287

Parkinson dementia complex, 22/316, 403, 59/282-284, 409
parkinsonism, 59/231, 410
pathologic laughing and crying, 3/356, 22/115, 45/219
pathology, 22/283-288, 59/181-185
peroneal type, 22/306, 308
pharyngeal reflex, 22/293
Pick disease, 22/313, 59/410
poliomyelitis, 22/138, 34/114, 42/65
polyneuropathy, 62/413
posterior column degeneration, 22/325
postpoliomyelitic amyotrophy, 22/23, 59/35
presenting symptom, 22/112
prevalence, 42/65, 59/180
primary lateral sclerosis, 22/313
prognosis, 22/115, 307
progressive autonomic failure, 59/411
progressive bulbar palsy, 22/112, 293, 42/65, 59/220
progressive external ophthalmoplegia, 62/290
progressive muscular atrophy, 59/13, 18
progressive poliomyelitis, 22/329
progressive spinal muscular atrophy, 22/1, 42-44, 310-312
progressive supranuclear palsy, 22/226, 49/249, 64/5
pseudobulbar paralysis, 22/114, 294, 306
pseudobulbar sign, 59/194
pseudopoliomyelitic type, 22/310
pseudopolyneuritic type, 22/306, 308
psychiatric disorder, 22/296
pyramidal type, 22/292, 310
rehabilitation, 59/197
related syndrome, 59/180
respiration, 22/120, 296
retrovirus infection, 59/203
rigidity, 59/235
RNA, 59/177
sensory conduction, 59/190
sensory deficit, 59/412
sensory symptom, 22/114, 294, 59/195
sex ratio, 22/288, 341, 42/65, 59/179
sign, 59/193-196
skin lesion, 22/296, 341
spheroid body, 6/626, 22/17, 59/291
spinal cord electrotrauma, 61/194
spinal disease, 59/201
spinal muscular atrophy, 59/15, 113
spinal tumor, 19/87
spinocerebellar degeneration, 21/15, 22/371
spontaneous activity, 59/185
sporadic, *see* Sporadic amyotrophic lateral sclerosis
sporadic type, 22/129, 341
subacute anterior poliomyelitis, 22/311
subcortical change, 59/234
symptom, 59/191-193
symptomatic treatment, 59/459-471
thalamus degeneration, 22/141
therapeutic trial, 59/200
thyrotoxic myopathy, 62/529
torticollis, 59/235
toxin, 59/176
trauma, 22/309, 328, 59/178
treatment, 22/125, 59/196-200
tri-*o*-cresyl phosphate, 21/33, 22/310
trophic deficiency, 59/177
tropical spastic paraplegia, 59/451-453
underestimation, 59/245
unusual phenomena, 22/294
upper motoneuron, 59/194
viral infection, 59/172-174, 201
West New Guinea, 59/253
WFN classification, 59/3
white matter, 59/182
Wilson disease, 59/410
Amyotrophic lateral sclerosis-Parkinson dementia
aluminum intoxication, 64/278
Guam, *see* Guam amyotrophic lateral sclerosis-Parkinson dementia complex
Amyotrophic myelopathy
carcinoma, 38/671-675
malignancy, 38/671-675
Amyotrophic spondylotic myeloradiculopathy
progressive muscular atrophy, 59/17
Amyotrophy
alcoholic polyneuropathy, 51/316
antimony intoxication, 22/28
Aran-Duchenne, *see* Aran-Duchenne disease
autosomal dominant neuralgic, *see* Neuralgic amyotrophy
Bassen-Kornzweig syndrome, 21/580
blennorrhagic polyneuritis, 7/478
bulbar postpoliomyelitis, *see* Bulbar postpoliomyelitis amyotrophy
chorea-acanthocytosis, 51/398, 63/281
Cockayne-Neill-Dingwall disease, 51/399
Creutzfeldt-Jakob disease, 22/22, 46/292
deafness, 22/523
diabetes mellitus, 8/35, 19/86, 22/30, 27/121, 40/318, 326
distal, *see* Distal amyotrophy
familial spastic paraplegia, 21/16, 463, 22/427, 445-465, 469
Farber disease, 51/379

Friedreich ataxia, 21/320, 333, 355
gastrectomy, 22/29
Gerstmann-Sträussler-Scheinker disease, 21/553
gold polyneuropathy, 51/306
hereditary dystonic paraplegia, 22/460, 59/345
hereditary motor and sensory neuropathy type I, 21/274-279, 303
hereditary motor and sensory neuropathy type II, 21/274-279, 303
hereditary motor and sensory neuropathy variant, 60/244
hereditary neuralgic, *see* Hereditary neuralgic amyotrophy
hereditary olivopontocerebellar atrophy (Menzel), 21/443
hereditary sensory and autonomic neuropathy type I, 21/77
Huntington chorea, 49/291
hydromyelia, 50/429
hyperlipoproteinemia type IIA, 22/45
hyperlipoproteinemia type IV, 22/45
hypoglycemia, 22/29
jarretière, 21/276
Jeune-Tommasi disease, 22/516
Lemieux-Neemeh syndrome, 22/521
lithium, 41/333
lithium intoxication, 22/28
lower leg, 59/374
Marinesco-Sjögren syndrome, 21/556
mollet de coq, 21/276
multiple neuronal system degeneration, 59/138
muscle calcification, 43/258
myelomalacia, 22/19
myotonia, 1/269
neuralgia, *see* Brachial neuralgia
neuralgic, *see* Neuralgic amyotrophy
neurogenic, *see* Neurogenic muscular atrophy
oblique, *see* Oblique amyotrophy
oculopharyngeal, *see* Oculopharyngeal amyotrophy
olivopontocerebellar atrophy variant, 21/452, 455
paramyotonia, 22/32
paraplegia, 60/244
parietal, *see* Parietal amyotrophy
perhexiline maleate polyneuropathy, 51/302
peripheral, *see* Peripheral amyotrophy
peripheral neurolymphomatosis, 56/180
pes cavus, 59/374
postencephalitic, *see* Postencephalitic amyotrophy
postpoliomyelitic, *see* Postpoliomyelitic amyotrophy
posttraumatic type, 22/29
progressive, *see* Progressive amyotrophy
progressive dysautonomia, 59/138
Refsum disease, 21/198
Richards-Rundle syndrome, 22/516
Rosenberg-Bergström disease, 22/514
Roussy-Lévy syndrome, 21/176, 42/109
scapuloperoneal, *see* Scapuloperoneal spinal muscular atrophy
Small disease, 22/509
spastic paraplegia, 59/374
spinal cord tumor, 19/69
spinal muscular atrophy, 59/368, 374
spinocerebellar degeneration, 21/15
spinocerebellonigral, *see* Spinocerebellonigral amyotrophy
spinopontine degeneration, 21/446
Sylvester disease, 22/505
syringomyelia, 32/266
tick-borne encephalitis, 22/22
tri-*o*-cresyl phosphate intoxication, 22/28
tuberous sclerosis, 14/349
uremic polyneuropathy, 51/357
viral infection, 22/32
Amytal test
aphasia, 4/88, 45/303
body scheme disorder, 4/216
cerebral dominance, 3/55, 4/266
sensory extinction, 3/193
tactile, *see* Tactile amytal test
tactile agnosia, 4/36
temporal lobe epilepsy, 73/83
Anaerobic glycolysis
brain ischemia, 63/211
neuron death, 63/211
Anal atresia
axial mesodermal dysplasia spectrum, 50/516
Bardet-Biedl syndrome, 13/394
cardiofacial syndrome, 42/308
cat eye syndrome, 31/525, 43/548
caudal aplasia, 50/511
congenital megacolon, 42/3
Down syndrome, 50/527
holoprosencephaly, 50/237
renal agenesis syndrome, 50/512
trisomy 22, 43/548
VATER syndrome, 50/513
Anal disorder
fecal incontinence, 75/637
Anal fissure
autonomic nervous system, 75/639
Anal reflex
lower urinary tract, 61/295
paraplegia, 61/295
spinal cord injury, 61/295

thoracolumbar spine injury, 25/455
transverse spinal cord lesion, 61/295
urinary tract, 26/416
Anal sphincter
autonomic nervous system, 74/635
constipation, 74/635
fissure, 75/639
manometry, 74/635
Anal sphincter reflex
cutaneous reflex, 1/254
Analeptic agent, *see* CNS stimulant
Analgesia
amyloid polyneuropathy, 51/416
anterolateral myelopathy, 2/210
congenital, *see* Congenital analgesia
diamorphine intoxication, 65/352
hemi, *see* Hemianalgesia
hereditary perforating foot ulcer, 8/201
hereditary sensory and autonomic neuropathy type IV, 60/12
hypothalamus, 45/238-240
hysteria, 1/111
nalorphine, 37/366
nucleus raphe magnus, 45/238-240
periaqueductal gray matter, 45/238-240
terminology, 1/86
Analgesic agent
epilepsy, 72/229
headache, 48/6, 39
migraine, 48/39, 173
muscle contraction headache, 48/39
pain, 69/45
pain tolerance, 45/228
phantom limb syndrome, 45/401
spinal cord injury, 26/233
Analysis indication
chromosome, *see* Chromosome analysis indication
Anaphylactoid purpura, *see* Henoch-Schönlein purpura
Anaphylaxis
anemonia sulcata intoxication, 37/53
bee sting intoxication, 37/107
eosinophil, 63/370
pertussis vaccine, 52/236
Anaplastic astrocytoma, 18/3, 16
*see also* High grade glioma
age, 68/98
chromosome abnormality, 67/45
histology, 68/93
hypothalamic tumor, 68/71
survey, 68/87
thalamic tumor, 68/65
therapy, 68/99
Anarithmetia
acalculia, 45/474, 480
diagnosis, 45/479
Anarthria
aphasia, 4/89, 94, 99, 45/309
hereditary paroxysmal dystonic choreoathetotis, 49/349
motor aphasia, 45/309
operculum syndrome, 2/781
paralytic shellfish poisoning, 65/153
Parkinson dementia, 49/171
Anastomosis
accessory facial, *see* Accessory facial anastomosis
accessory meningeal artery, 12/309
ascending pharyngeal artery, 12/310
Blalock-Taussig, *see* Blalock-Taussig anastomosis
brain infarction, 53/109
brain ischemia, 53/109
carotid artery, *see* Carotid artery anastomosis
carotid basilar, *see* Carotid basilar anastomosis
carotid gasserian, *see* Carotid gasserian anastomosis
carotid rete mirabile, 12/308-311
carotid system syndrome, 53/298
carotid vertebral, *see* Carotid vertebral anastomosis
carotidocarotid, *see* Carotidocarotid anastomosis
congenital arterial malformation, 12/308-335
dura mater, 30/427
external maxillary artery, 12/308
facial nerve, 8/245
faciofacial, *see* Faciofacial anastomosis
hemangioendothelioma, 18/288
hypoglossal facial, *see* Hypoglossal facial anastomosis
internal carotid artery, 53/296, 298
internal maxillary artery, 12/308
ophthalmic artery, 53/296
stapedial artery, 12/309
Vidian artery, 12/308
Willis circle, 11/178
Anastomotic ansa of the conus
spinal artery, 12/494
Anastomotic artery
carotid rete mirabile, 12/309
Anastomotic vein
inferior, *see* Labbé vein
Anaxagoras
historic review, 2/5
Andermann syndrome
corpus callosum agenesis, 30/100, 50/163, 166
facial appearance, 50/166
mental deficiency, 50/166

sensorimotor neuropathy, 50/166
Andersen disease, *see* Brancher deficiency
Anderson-Fabry disease, *see* Fabry disease
André-Thomas extensibility test
handedness, 4/255
Anemia, 38/15-28
Abt-Letterer-Siwe disease, 42/441
acute ischemic optic neuropathy, 38/25-27
Addison, *see* Pernicious anemia
addisonian pernicious, *see* Pernicious anemia
adult Gaucher disease, 10/564, 66/123
alcoholic polyneuropathy, 51/317
amaurosis fugax, 55/108
arsenic intoxication, 41/288
atrial myxoma, 63/97
Bean syndrome, 43/11
blindness, 38/27
brain fat embolism, 55/179, 186
brain injury, 23/127
cerebrovascular disease, 55/463
Chédiak-Higashi syndrome, 42/536
classification, 38/17, 63/252
clinical features, 38/16
Cooley, *see* Thalassemia major
copper, 27/389
cytomegalic inclusion body disease, 56/267
dementia, 46/397
demyelination, 9/613
EEG dysrhythmia, 63/252
equine infectious, *see* Equine infectious anemia
erythropoiesis, 38/15
Fanconi, *see* Fanconi syndrome
fat embolism, 23/635
glucose-6-phosphate dehydrogenase deficiency, 42/643
hematologic features, 38/17
hemolytic, *see* Hemolytic anemia
hemoneurological disease, 63/249
hereditary hemorrhagic telangiectasia (Osler), 42/739
hereditary vitamin $B_{12}$ metabolism defect, 63/255
hypochromic microcytic, *see* Iron deficiency anemia
iron deficiency, *see* Iron deficiency anemia
Jervell-Lange-Nielsen syndrome, 43/388
juvenile Gaucher disease type A, 66/125
juvenile Gaucher disease type B, 66/125
lead intoxication, 9/642, 64/434
lead polyneuropathy, 51/269
leukoencephalopathy, 9/613
megaloblastic, *see* Megaloblastic anemia
mental deficiency, 43/278
metastatic cardiac tumor, 63/96
mevalonate kinase deficiency, 60/674
microcytic, *see* Iron deficiency anemia
multiple myeloma, 20/9, 39/138, 142, 63/392
myasthenia gravis, 62/402
neuropathy, 7/664
Niemann-Pick disease, 60/674
ocular manifestation, 38/16, 25-28
orotic aciduria, 42/607
10p partial trisomy, 43/520
pallidal necrosis, 49/466
pallidoluysionigral degeneration, 6/665
papilledema, 38/27, 63/252
pernicious, *see* Pernicious anemia
Peutz-Jeghers syndrome, 14/121, 43/41
phosphofructokinase deficiency, 43/183
polyarteritis nodosa, 8/125, 55/354
postgastrectomy syndrome, 28/234
pressure sore, 26/471
primary cardiac tumor, 63/96
restless legs syndrome, 8/315, 42/261, 51/545, 63/252
sickle cell, *see* Sickle cell anemia
sideroblastic, *see* Sideroblastic anemia
spinal cord injury, 26/380, 398
temporal arteritis, 55/342
transient ischemic attack, 55/463
transverse sinus thrombosis, 54/397
venous sinus thrombosis, 12/431
vitamin $B_6$ deficiency, 51/297
vitamin $B_{12}$ deficiency, *see* Vitamin $B_{12}$ deficiency anemia
vitamin $B_c$, 28/200
vitamin $B_c$ deficiency, *see* Vitamin $B_c$ deficiency anemia
Wolman disease, 10/546
Anemic anoxia
acute hemorrhage, 27/44
brain anoxia, 46/358
Anemonia sulcata intoxication
anaphylaxis, 37/53
Anencephalus, *see* Anencephaly
Anencephalus ichthyoides
history, 30/175
Anencephalus sequanensis
history, 30/175
Anencephaly, 30/173-201
*see also* Acephaly, Acrania *and* Meroanencephaly
abortion, 30/190
acephaly, 30/174
acetylcholinesterase, 50/86
adrenal hypoplasia, 42/12
adrenals, 30/180
alopecia circumscripta, 30/188

aminopterin syndrome, 30/122, 50/80
amniocentesis, 30/178
amniography, 30/178
animal, 32/179
anophthalmia, 50/83
anoxia, 30/184, 193
antenatal diagnosis, 30/176-178
autosomal recessive inheritance, 50/79
birth incidence, 42/13
brain hemisphere, 42/12
cerebellar agenesis, 42/12, 50/175
cerebellar hypoplasia, 42/12
cerebellum, 50/83
cheilopalatoschisis, 42/13
chemicals, 30/192, 196
choroid plexus, 42/12
chromosomal aberration, 50/79
classification, 30/174, 50/71
clinical features, 30/176-178
clomifene, 50/80
CNS, 50/83
CNS malformation, 50/57
coloboma, 50/83
corneal dermoid, 50/83
cranial dysraphia, 50/72
cranial nerve, 30/179
craniorachischisis, 30/175, 187, 50/72
cranioschisis, 50/72
CSF, 30/178, 181
cytoplasmic factor, 30/195
derencephalus, 50/73
dermal sinus tract, 30/188
diaphragm hernia, 50/84
diastematomyelia, 42/25
digestive system, 50/84
ear abnormality, 42/12
ectopia cordis, 50/84
Edwards syndrome, 43/537, 50/562
encephalomeningocele, 30/187
endocrine disorder, 50/85
environmental factor, 50/79
epidemiology, 30/150, 153, 188-191, 50/77
exencephaly, 30/173, 187, 32/179
experimental, 30/177, 181, 184, 188, 191
familial hydrocephalus, 50/289
familial tendency, 50/79
α-fetoprotein, 30/178, 42/13, 50/86
gastropleuroschisis, 50/84
genetics, 30/193
geographic variation, 50/78
growth retardation, 30/175
Gruber syndrome, 14/115, 505, 43/391
head abnormality, 50/83
heart disease, 42/13
heterotopic brain tissue, 50/84
history, 30/173-175
holoanencephaly, 30/175, 50/72
holoprosencephaly, 42/12, 50/235, 237
human embryo, 30/183
hydranencephaly, 50/347
hydrocephalus, 30/181, 42/36
hydronephrosis, 50/84
hydroureter, 50/84
hyperthermia, 50/80
hypogonadism, 42/13
hypoxia, 30/196
imperforate anus, 42/13
incidence, 50/58, 78
infection, 30/197
iniencephaly, 50/72, 131, 134
kyphoscoliosis, 42/12
limb, 50/84
lumbosacral myelocele, 50/77
lung growth, 50/84
malformation, 30/153, 176
management, 30/200
maternal influence, 30/194, 195
meningomyelocele, 50/71
mero, *see* Meroanencephaly
mesocephalic, *see* Mesocephalic anencephaly
metabolic abnormality, 30/178
morphology, 30/175
neurulation, 50/74
neurulation defect, 50/97
nonclosure hypothesis, 50/82
nosencephalus, 50/73
nutrition, 30/197, 200, 50/80
omphalocele, 50/84
optic atrophy, 42/12
optic nerve agenesis, 50/211
pathogenesis, 30/181-186, 50/81
pathology, 30/179-181, 50/83
pectoralis major muscle, 50/84
pes calcaneovalgus, 50/84
pituitary gland, 30/180, 42/12
polygenic inheritance, 50/79
polyhydramnios, 30/176, 178, 42/12
postneurulation defect, 50/97
prenatal diagnosis, 42/13, 50/85
prevention, 30/200
prune belly syndrome, 50/84
pteroylglutamic acid, 30/184
pulmonary hypoplasia, 42/12
pyramidal tract, 50/83
pyramidal tract absence, 50/83
13q partial monosomy, 43/526

r(13) syndrome, 50/585
rachischisis, 32/179-187, 42/12
radiation, 30/182
radiologic appearance, 30/178
recurrence risk, 42/13
reopening hypothesis, 50/82
respiratory virus infection, 31/219
rib abnormality, 50/84
rostral neuropore, 50/3
salicylic acid derivative, 50/80
schizencephaly, 50/72
seasonal variation, 50/81
seizure, 30/176
sex difference, 50/56, 79
skeletal deformity, 30/181
skull, 30/180, 50/83
socioeconomic class, 30/195, 42/13, 50/80
sphenoid bone defect, 42/12, 50/84
spina bifida, 30/163, 176, 188, 42/55, 50/75, 77, 79
survival, 30/176
swallowing reflex, 30/177
synonym, 30/173
talipes equinovarus, 50/84
teratogenic period, 30/184
terminology, 50/72
thymus, 30/181
thymus hyperplasia, 42/12
thyroid, 30/181
thyroid gland dysgenesis, 42/12
trypan blue, 30/184
twins, 30/185, 194, 50/79
ultrasound diagnosis, 50/86
valproic acid, 50/80
vertebral abnormality, 43/12
vertebral column, 50/83
vitamin A intoxication, 30/182, 185, 188, 192, 50/28
yawning, 63/493

Anergic syndrome
hebephrenic, *see* Hebephrenic anergic syndrome

Anesthesia
acupuncture, 37/136
agraphia, 45/462
air embolism, 23/614
blennorrhagic polyneuritis, 7/478
brain aneurysm, 12/211
brain aneurysm neurosurgery, 55/61
brain injury, 24/613
cervical vertebral column injury, 24/168
congenital facial, *see* Congenital facial anesthesia
epidural, *see* Epidural anesthesia
eye movement, 1/579
general, *see* General anesthesia
glycogen, 27/207
glycosaminoglycanosis type IH, 66/291, 297
hypokalemic periodic paralysis, 62/460
indifference to pain, 8/199
learning, 27/467
malignant hyperthermia, 38/550, 43/117
median nerve, 1/92
memory, 27/467
memory loss, 27/467
migraine, 48/143
missile injury, 23/510
nerve damage, 70/47
nerve lesion, 1/89
nitrous oxide, *see* Nitrous oxide anesthesia
pain indifference, 8/199
pain insensitivity, 8/199
pallidal necrosis, 6/658
patchy, *see* Patchy anesthesia
regional, *see* Regional anesthesia
sickle cell anemia, 55/466
spinal, *see* Spinal anesthesia
spinal nerve root avulsion, 1/89
ulnar nerve, 1/92

Anesthesia complication
myotonic dystrophy, 62/244
spinal epidural hematoma, 61/137

Anesthesia dolorosa
spinal meningioma, 20/198
thalamic syndrome (Dejerine-Roussy), 1/106
trigeminal neuralgia, 48/454

Anesthetic agent
epilepsy, 72/228
ether, 37/412
general, *see* General anesthetic agent
local, *see* Local anesthetic agent
memory disorder, 27/463, 467
pain, 69/46
vinyl ether, 37/412

Anesthetic agent intoxication, 37/401-413
general anesthetic agent, 37/408-413
local anesthetic agent, 37/401-407

Anetic syndrome
brain death, 24/701
brain injury, 24/702

Aneuploidy
*see also* Chromosome number abnormality
ataxia telangiectasia, 50/566
autosomal, 50/540-542
Bloom syndrome, 50/566
chromosomal, *see* Chromosomal aneuploidy
chromosome analysis indication, 50/566
chromosome number abnormality, 50/539

definition, 50/540
DNA marker, 50/567
double form, 50/556
double Y, *see* XYY syndrome
Down syndrome, 31/394, 50/566
hydronephrosis, 50/553
Klinefelter syndrome, 50/546
Lyon principle, 50/542
malignancy, 50/566
Patau syndrome, 31/504, 43/527
Prader-Labhart-Willi syndrome, 50/566
prevalence, 50/540
radioulnar synostosis, 50/553
sex chromosomes, 50/540-542
Turner syndrome, 50/543, 566
X-linked mental deficiency, 50/566
XXX syndrome, 50/553
XXXX syndrome, 50/554
XXXXX syndrome, 50/555
Aneurysm
anterior cavernous sinus syndrome, 12/89, 31/147
anterior communicating artery, *see* Anterior communicating artery aneurysm
anterior inferior cerebellar artery, 2/257
aorta arch, *see* Aorta arch aneurysm
aortic, *see* Aorta aneurysm
arteriosclerosis, 11/131
basilar artery, *see* Basilar artery aneurysm
brain, *see* Brain aneurysm
brain arteriovenous malformation, 12/237, 18/279
cardiac ventricular, *see* Cardiac ventricular aneurysm
carotid artery, *see* Carotid artery aneurysm
carotid artery bifurcation, *see* Carotid artery bifurcation aneurysm
carotid dissecting, *see* Carotid dissecting aneurysm
cerebellopontine angle syndrome, 12/95
cerebral, *see* Brain aneurysm
Charcot-Bouchard, *see* Charcot-Bouchard aneurysm
cirsoid, *see* Cirsoid aneurysm
coitus, 48/377
congenital, *see* Congenital aneurysm
coronary artery, *see* Coronary artery aneurysm
dissecting, *see* Dissecting aneurysm
dissecting aortic, *see* Dissecting aorta aneurysm
Galen vein, *see* Galen vein aneurysm
Garcin syndrome, 2/313
giant brain, *see* Giant brain aneurysm
headache, 48/6, 275
hypertensive encephalopathy, 11/562
intercostal artery, *see* Intercostal artery aneurysm
internal carotid artery, 12/221
intracerebral hematoma, 11/666
intracranial, *see* Intracranial aneurysm
intrasellar, *see* Intrasellar aneurysm
megadolichobasilar artery, 12/81
megadolichocarotid artery, 12/81
middle cavernous sinus syndrome, 12/89, 31/147
middle cerebral artery, *see* Middle cerebral artery aneurysm
mycotic, *see* Mycotic aneurysm
neoplastic, *see* Neoplastic aneurysm
optic chiasm compression, 68/75
pericallosal, *see* Pericallosal aneurysm
Pick supermiliary, *see* Pick supermiliary aneurysm
posterior communicating artery, *see* Posterior communicating artery aneurysm
posterior spinal artery, *see* Posterior spinal artery aneurysm
progressive bulbar palsy, 2/257
spinal angioma, 20/448
spinal cord arteriovenous malformation, 32/469
subarachnoid hemorrhage, 12/77, 55/2
superior cerebellar artery, *see* Superior cerebellar artery aneurysm
Takayasu disease, 12/414
Tolosa-Hunt syndrome, 31/147, 48/302
traumatic, *see* Traumatic aneurysm
trigeminal neuralgia, 48/450
vertebral artery, *see* Vertebral artery aneurysm
vertebral dissecting, *see* Vertebral dissecting aneurysm
vertebrobasilar, *see* Vertebrobasilar aneurysm
Aneurysm compression
cerebellopontine lesion, 2/257
lumbosacral plexus, 51/161
Aneurysma bilobata
subarachnoid hemorrhage, 12/81
Aneurysmal bone cyst, 20/1-6
age, 20/2, 4
cervical spine, 67/195
clinical features, 20/3
cranial vault tumor, 17/109
differential diagnosis, 19/300, 20/4
eggshell vertebra, 20/30
giant cell tumor, 19/301
history, 14/164
incidence, 68/530
orbital tumor, 17/202
papilledema, 20/3
paralysing sciatica, 20/30
pathogenesis, 20/1
pathology, 20/1-3
phantom limb, 3/29

pseudohypoparathyroidism, 42/622
radiation malignancy, 20/6
radiology, 20/3
sex ratio, 19/300, 20/2, 4
site, 20/2, 4
skull, 20/3
skull base tumor, 17/163
spinal cord, 67/194
spinal cord compression, 19/368
spinal tumor, 68/515
treatment, 20/5
tumeur à myéloplaxe, 19/170
vertebral column, 19/300, 20/29, 442
Angatea intoxication
suicide, 37/56
Angelman syndrome, *see* Happy puppet syndrome
Anger
dysphrenia hemicranica, 5/79
emotion, 3/336
migraine, 48/162
stress, 45/251
Angiitis
*see also* Vasculitis
brain infarction, 55/419
cerebrovascular disease, 53/40
granulomatous, *see* Allergic granulomatous angiitis
hypersensitivity, *see* Allergic granulomatous angiitis
isolated, 71/156
Angioblast
anatomy, 11/1, 33
collateral circulation, 11/168
embryology, 11/1
vascular system, 11/33
Angioblastic meningioma, *see* Hemangiopericytoma
Angioblastoma, *see* Hemangioblastoma
Angioblastosis
congenital polytopic, *see* Von Hippel-Lindau disease
Angiodysgenetic myelomalacia, *see* Foix-Alajouanine disease
Angiodysgenetic necrotizing encephalopathy, *see* Moyamoya disease
Angiodysgenetic necrotizing myelomalacia
*see also* Foix-Alajouanine disease
age, 20/482
arachnoiditis, 20/484, 486
associated malformation, 20/486
Brown-Séquard syndrome, 20/488
cardiac defect, 20/487
clinical features, 20/488
Cobb syndrome, 20/486
course, 20/487
CSF, 20/489
differential diagnosis, 20/488
extent, 20/484
hematomyelia, 20/487
histology, 20/484
history, 20/481
kyphoscoliosis, 20/489
location, 20/483
multiple lesions, 20/483
myelography, 20/489-491
pain, 20/487
pathogenesis, 20/485
precipitating cause, 20/486
prevalence, 20/482
radiology, 20/489
retinal angioma, 20/487
scoliosis, 20/487
sex ratio, 20/482
sphincter disturbance, 20/487
spinal angiography, 20/490-497
spinal bruit, 20/489
subarachnoid hemorrhage, 20/488
syringomyelia, 20/446
Angiodysgenetic necrotizing myelopathy, *see* Foix-Alajouanine disease
Angiodysgenetic radiculomyelomalacia, *see* Angiodysgenetic necrotizing myelomalacia
Angioedema
hereditary, *see* Hereditary angioedema
Angioendotheliomatosis
brain infarction, 53/165
brain ischemia, 53/165
iatrogenic brain infarction, 53/165
Angioendotheliosis
neoplastic, *see* Intravascular malignant lymphomatosis
Angiofibroma
facial, *see* Facial angiofibroma
juvenile, *see* Juvenile angiofibroma
Angiogenesis inhibitor
chemotherapy, 67/284
Angioglioma, *see* Cerebellar astrocytoma
Angiography
acute subdural hematoma, 24/282-285
anterior spinal artery, 32/48
axillary, *see* Axillary angiography
brachial, *see* Brachial angiography
brain, *see* Brain angiography
brain aneurysm, 12/205-208
brain arteriovenous malformation, 54/184
brain collateral blood flow, 11/176
brain death, 24/733, 55/264

brain glioma, 18/24-28
brain injury, 23/241, 248, 24/619
brain metastasis, 18/224
brain stab wound, 23/493
brain stem death, 57/460
brain thromboangiitis obliterans, 12/390, 39/203
brain tumor, 16/622
brain vasospasm, 11/521, 23/168, 175, 55/56
carotid artery, *see* Carotid angiography
carotid artery thrombosis, 26/45
carotid cavernous fistula, 24/418
carotid dissecting aneurysm, 54/272
cerebellar hemorrhage, 12/59
cerebellar infarction, 53/384
cerebral, *see* Brain angiography
cerebrovascular disease, 12/451
chronic subdural hematoma, 24/310
computer assisted tomography, 54/65
contraindication, 60/651
corpus callosum agenesis, 2/765, 50/154
Dandy-Walker syndrome, 32/59
death, 24/805, 819
digital subtraction, *see* Digital subtraction angiography
epidural hematoma, 24/271
femoral, *see* Femoral angiography
fibrillary astrocytoma, 18/25
fibromuscular dysplasia, 11/368, 55/283
fluorescein, *see* Fluorescein angiography
Foix-Alajouanine disease, 55/103
frontal lobe tumor, 17/265
granulomatous CNS vasculitis, 55/395
gunshot injury, 57/303
head injury, 57/165
headache, 48/287
iatrogenic neurological disease, 55/456
intermittent spinal cord ischemia, 12/520
intracerebral hematoma, 11/690, 24/360
intracranial pressure, 23/210
isotope, 57/462
Köhlmeier-Degos disease, 55/277
midbrain tumor, 17/639
migraine, 48/60, 73, 148-150
migraine coma, 60/651
moyamoya disease, 12/366, 42/749, 53/164, 55/297
neurocysticercosis, 35/307
penetrating head injury, 57/308
percutaneous transluminal, *see* Percutaneous transluminal angiography
pinealoma, 18/365, 42/767
pituitary adenoma, 17/302
posterior spinal artery, 32/48
repeating, 57/462
retinal fluorescence, *see* Retinal fluorescence angiography
retrograde brachial, *see* Retrograde brachial angiography
Rett syndrome, 29/317-319
Sjögren syndrome, 71/67
skull base tumor, 68/470
spinal, *see* Spinal angiography
spinal angioma, 20/455-458, 32/470-490
spinal arteriovenous malformation, 20/490-499
spinal artery injury, 26/65
spinal cord, 26/65
spinal cord injury, 61/515
spinal cord tumor, 19/229-243
spinal ependymoma, 20/374
spinal hemangioblastoma, 20/449
spinal tumor, 19/66, 229-243, 67/211
steal syndrome, 32/487
subclavian, *see* Subclavian angiography
subclavian steal syndrome, 53/374
subdural effusion, 24/337
subdural empyema, 57/323
Takayasu disease, 12/406, 71/202
third ventricle tumor, 17/462
thoracic intervertebral disc prolapse, 20/568
Tolosa-Hunt syndrome, 48/298
traumatic carotid cavernous sinus fistula, 57/347, 353
traumatic venous sinus thrombosis, 24/378
tuberculous meningitis, 33/222-224
venous sinus thrombosis, 54/405
vertebral artery, *see* Vertebral artery angiography
vertebral artery injury, 26/58
vertebral artery stenosis, 12/11
vertebral column injury, 61/515
vertebral column tumor, 20/23
vertebral dissecting aneurysm, 54/278
Angiohypertrophic spinal gliosis, *see* Foix-Alajouanine disease
Angioid streak
Bassen-Kornzweig syndrome, 63/273
Bruch membrane, 13/32
cardiovascular disease, 13/141
chorioretinal degeneration, 13/41
colloid body, 13/74
Ehlers-Danlos syndrome, 13/32, 141
Grönblad-Strandberg syndrome, 13/141, 14/114
optic atrophy, 13/41
optic disc, 13/74
osteitis deformans (Paget), 13/141, 38/364, 43/452
pseudoxanthoma elasticum, 11/465, 14/775, 43/45, 55/452

secondary pigmentary retinal degeneration, 13/237
senile cutaneous elastosis, 13/141
sickle cell anemia, 13/141, 55/452
Angioimmunoblastic lymphadenopathy
Hodgkin disease, 39/530
Angiokeratoma corporis diffusum, *see* Fabry disease
Angiokeratosis naeviformis
deafness, 14/775
Angiolipoma
renal, *see* Renal angiolipoma
spinal epidural lipoma, 20/405
spinal lipoma, 20/109
Angiolithic sarcoma, *see* Oligodendroglioma
Angioma
arteriovenous, *see* Arteriovenous angioma
barrel shaped vertebra, 20/27
brain, *see* Brain angioma
brain arteriovenous malformation, 14/440
brain hemorrhage, 54/288
capillary, *see* Capillary angioma
cavernous, *see* Cavernous angioma
cerebellar, *see* Cerebellar angioma
cerebellar hemorrhage, 54/291
cerebellar tumor, 17/708
cervical, *see* Cervical angioma
choroid, *see* Choroid angioma
cirsoid aneurysm, 12/236
classification, 14/51
facial, *see* Facial angioma
Garcin syndrome, 2/313
heterochromia iridis, 14/643
intracerebral hematoma, 11/668
lateral ventricle tumor, 17/605
Maffucci syndrome, 14/119
meningeal, *see* Meningeal angioma
neurofibromatosis, 14/8
orbital tumor, 17/202
racemose, *see* Racemose angioma
retinal, *see* Retinal angioma
skin, 14/392
spinal, *see* Spinal angioma
spinal epidural, *see* Spinal epidural angioma
spinal epidural tumor, 20/110
thoracic, *see* Thoracic angioma
transient infantile hemiplegia, 12/348
venocapillary type, 14/227
venous, *see* Venous angioma
vertebral column, *see* Spinal angioma
Angiomatosis
Bean syndrome, 43/11
differentiation, 14/51
diffuse corticomeningeal, *see* Divry-Van Bogaert syndrome
embryology, 14/87
leptomeningeal, *see* Leptomeningeal angiomatosis
leukodystrophy, 10/124
macrocephaly, *see* Riley-Smith syndrome
meningeal, *see* Meningeal angiomatosis
neurocutaneous, *see* Neurocutaneous angiomatosis
neurofibromatosis, 14/36
neuroretinal, *see* Bonnet-Dechaume-Blanc syndrome
systemic, *see* Ullmann syndrome
Ullmann syndrome, 14/72, 447
Angiomatosis osteohypertrophica, *see* Klippel-Trénaunay syndrome
Angiomatosis retinae, *see* Von Hippel-Lindau disease
Angiomatosis retinocerebellosa, *see* Von Hippel-Lindau disease
Angiomatous malformation
arteriovenous angioma, 14/52
congenital arteriovenous fistula, 31/200
moyamoya disease, 55/293
telangiectasia, 14/52
venous angioma, 14/52
Angiomatous meningioma
whorling, 18/297
Angiomyolipoma
epidural tumor, 20/110
renal, *see* Renal angiomyolipoma
renal lesion, 14/50
tuberous sclerosis, 50/371
Angionecrosis
brain hemorrhage, 54/203
pontine hemorrhage, 12/49
prehemorrhagic softening, 12/49
Angioneuromatosis
neurofibromatosis type I, 14/36
Angioneurotic edema
aphasia, 43/61
autonomic nervous system, 43/61
$C_1$ esterase inhibitor, 8/349, 43/61
dermatitis, 43/60
diamorphine addiction, 55/519
differential diagnosis, 8/228
diplopia, 43/61
eosinophilia, 63/371
hemiplegia, 43/61
hemiplegic migraine, 48/148
migraine, 48/148
preclinical detection, 43/61
prevalence, 43/61
speech disorder in children, 43/61

Angio-osteohypertrophy (Tobler), *see* Klippel-Trénaunay syndrome
Angiopathic polyneuropathy
allergic granulomatous angiitis, 51/446
arteriolar fibrinohyalinoid degeneration, 51/445
collagen vascular disease, 51/445
CREST syndrome, 51/446
dermatomyositis, 51/446
giant cell infiltration, 51/445
granuloma, 51/445
Henoch-Schönlein purpura, 51/446
immune complex, 51/445
inflammatory angiopathy, 51/445
ischemia, 51/445
Kawasaki syndrome, 51/446
Köhlmeier-Degos disease, 51/446
laboratory parameter, 51/445
lymphomatoid granulomatosis, 51/446
mixed connective tissue disease, 51/446
mixed cryoglobulinemia, 51/446
mononeuritis multiplex, 51/445
polyarteritis nodosa, 51/446, 452
polymyalgia rheumatica, 51/446
progressive systemic sclerosis, 51/446
rheumatoid arthritis, 51/446
sensorimotor polyneuropathy, 51/445
serogenetic, 51/446
Sneddon syndrome, 51/446
systemic lupus erythematosus, 51/446
Takayasu disease, 51/446
temporal arteritis, 51/446
vaccinogenetic, 51/446
Wegener granulomatosis, 51/446
Angiopathy
allergic, *see* Allergic angiopathy
amyloid, *see* Amyloid angiopathy
amyloid cerebral, *see* Brain amyloid angiopathy
brain amyloid, *see* Brain amyloid angiopathy
congophilic, *see* Amyloid angiopathy
diabetes mellitus, 42/544
diabetic polyneuropathy, 51/505
dyshoric, *see* Dyshoric angiopathy
immune induced, *see* Immune induced angiopathy
inflammatory, *see* Inflammatory angiopathy
Morel-Wildi dyshoric, *see* Brain amyloid angiopathy
Pantelakis meningocortical amyloid, *see* Brain amyloid angiopathy
reversible, *see* Reversible angiopathy
Scholz drüsige, *see* Brain amyloid angiopathy
subarachnoid hemorrhage, 12/110
Angioreticuloma
*see also* Cerebellar hemangioblastoma *and* Hemangioblastoma
vascular tumor, 20/449
Angiosarcoma
CNS, *see* CNS angiosarcoma
diagnostic features, 68/280
epidemiology, 68/278
hypervascular, 68/270
Maffucci syndrome, 14/14
malignant, *see* Malignant angiosarcoma
pathology, 68/279
primary cardiac tumor, 63/93
spinal cord compression, 19/368
survey, 68/270
treatment, 68/280
vertebral column, 20/35
vinyl chloride, 68/279
Angiosarcoma caroticum, *see* Carotid body tumor
Angiosarcoma (Del Rio Hortega), *see* Hemangiopericytoma
Angiostrongyliasis
Angiostrongylus cantonensis, 35/321, 327, 52/546
asymptomatic, 52/553
ataxia, 52/554
blindness, 52/555
brain microcavity, 52/550
Charcot-Leyden crystal, 35/334, 52/551
chorea, 52/554
clinical course, 35/327, 52/556
clinical features, 52/510
computer assisted tomography, 52/558
cranial nerve palsy, 35/323, 52/549, 553, 555
CSF, 35/323, 327, 330, 52/510, 553, 556-559
delirium, 52/554
diagnosis, 35/328, 52/558
differential diagnosis, 35/322, 329, 52/559
EEG, 35/327, 52/558
encephalitis, 52/548
eosinophilia, 35/328, 52/557
eosinophilic meningitis, 35/321, 327, 52/510, 553, 556
epidemiology, 35/322, 334, 52/547
eye involvement, 52/555
geography, 35/321, 52/545
gnathostomiasis, 35/329
headache, 35/323, 52/553, 556
helminthiasis, 52/510
incubation period, 35/322, 332, 52/551
intracranial hypertension, 52/553
lethargy, 52/554
meningitis, 35/329, 52/545, 548
meningoencephalitis, 35/329, 52/554, 556
mortality, 35/327, 52/556
myalgia, 52/548, 556

myelitis, 52/555
neuritis, 52/556
neurologic manifestation, 52/553
neuropathology, 52/549
nonneurologic clinical features, 52/551
optic atrophy, 52/555
optic neuritis, 52/549
prevention, 52/559
prognosis, 52/556
psychiatric symptom, 52/557
radiculitis, 52/549, 554
radiculomyelitis, 52/554, 556
radiculomyeloencephalitis, 52/551
recurrence, 52/557
seizure, 52/554
spastic paraplegia, 59/436
subarachnoid hemorrhage, 52/549
systemic symptom, 52/552
treatment, 35/337, 52/559
tropical myeloneuropathy, 56/526
Angiostrongylus cantonensis, 35/321-339
angiostrongyliasis, 35/321, 327, 52/546
chronic meningitis, 56/644
eosinophilia, 63/371
eosinophilic meningitis, 52/548
helminthiasis, 52/506, 545
life cycle, 35/322, 331, 52/546
nematode, 35/328, 331, 52/545
Angiotensin
Bartter syndrome, 42/528
blood-brain barrier, 48/64
blood pressure, 48/98
headache, 48/109
Angiotensin converting enzyme
antihypertensive agent, 63/84
neurosarcoidosis, 63/422
Angiotensin converting enzyme inhibitor, *see* Dipeptidyl carboxypeptidase I inhibitor
Angiotensin II
orthostatic hypotension, 63/145
spinal cord injury, 61/444
tetraplegia, 61/444
transverse spinal cord lesion, 61/444
Angiotome
terminology, 2/160
Angle
Bull, *see* Bull angle
cerebellopontine, *see* Cerebellopontine angle
Daubent, *see* Daubent angle
Decker-Breig, *see* Decker-Breig angle
Welcker, *see* Welcker angle
Angle syndrome
cerebellopontine, *see* Cerebellopontine angle syndrome
sphenocavernous, *see* Sphenocavernous angle syndrome
Angular artery
radioanatomy, 11/66, 88
topography, 11/66, 88
Angular kyphosis
cervical spine, 25/346
thoracolumbar spine, 25/459, 462
Angular vein
anatomy, 11/54
radioanatomy, 11/114
topography, 11/114
Angulation
definition, 25/243
spinal injury, 25/243
spinal nerve root, 25/399
Anhidrosis
acquired immune deficiency syndrome, 63/153
acute pandysautonomia, 51/475, 63/153
amyloid polyneuropathy, 63/149
autonomic dysfunction, 63/145
autonomic nervous system, 74/218, 407
behavior disorder in children, 42/348
belladonna alkaloid, 8/344
bulbar poliomyelitis, 8/344
chronic idiopathic, 75/66
cluster headache, 48/336
Cockayne-Neill-Dingwall disease, 13/432
congenital, 14/112
congenital pain insensitivity, *see* Hereditary sensory and autonomic neuropathy type IV
diabetes mellitus, 1/457
diabetic neuropathy, 1/457
diabetic polyneuropathy, 63/148
distal, 75/66
dominant, *see* Dominant anhidrosis
dorsal spinal nerve root ganglion degeneration, 42/346
dysautonomia, 22/260
Fabry disease, 60/172
forehead, *see* Forehead anhidrosis
global, 75/66
Guillain-Barré syndrome, 63/153, 75/685
heat injury, 8/342
hereditary amyloid polyneuropathy type 1, 42/518, 51/421
hereditary sensory and autonomic neuropathy type I, 60/6
hereditary sensory and autonomic neuropathy type II, 60/10
hereditary sensory and autonomic neuropathy type IV, 21/7, 78, 42/298, 346-348, 51/564, 60/12

hexane polyneuropathy, 51/277
2-hexanone polyneuropathy, 51/277
Holmes-Adie syndrome, 8/344
hypothermia, 42/346
leprous neuritis, 51/217, 219
Lissauer tract absence, 8/199, 42/346
mental deficiency, 42/346
multifocal, 75/66
orthostatic hypotension, 8/344, 63/145
paratrigeminal syndrome, 48/331
Parkinson disease, 59/138
poliomyelitis, 8/344
pure autonomic failure, 2/120, 22/232
Ross syndrome, 75/125
segmental, 75/66
Shy-Drager syndrome, 1/474
sweat test, 75/66
syringobulbia, 8/344
syringomyelia, 8/344
thermoregulation, 75/64
thermoregulatory sweat test, 75/108
Anhidrotic ectodermal dysplasia
Berlin syndrome, 14/112
congenital cataract, 46/91
diabetes insipidus, 14/788
differential diagnosis, 14/112
epilepsy, 14/788
Friedreich ataxia, 14/788
juvenile cataract, 46/91
linear nevus sebaceous syndrome, 14/788
mental deficiency, 14/788, 46/91
nystagmus, 14/788
schizoidism, 14/788
sensorineural deafness, 42/363
Anileridine
chemistry, 37/366
Aniline intoxication
headache, 48/420
Animal model
acquired immune deficiency syndrome, 56/453
acquired immune deficiency syndrome dementia, 56/583
acute immune mediated polyneuritis, 34/403
acute pandysautonomia, 75/123
acute viral myositis, 56/194-196
adenovirus type 12, 56/291
alpha mannosidosis, 66/330
Alzheimer disease, 46/269, 56/585
aorta surgery, 63/56
Argentinian hemorrhagic fever, 56/367
Arnold-Chiari malformation type II, 50/404
bone marrow transplantation, 66/88
brain abscess, 52/146
central pontine myelinolysis, 47/587
cerebellar medulloblastoma, 18/171
Creutzfeldt-Jakob disease, 56/585
critical illness polyneuropathy, 51/584
cytomegalovirus infection, 34/211
demyelination, 56/583, 587, 589
dystonia musculorum deformans, 49/525
epilepsy, 72/2, 52, 62
experimental demyelination, 47/32
genetic epilepsy, 72/63
Gerstmann-Sträussler-Scheinker disease, 56/585
globoid cell leukodystrophy, 47/599, 66/202-204
Guillain-Barré syndrome, 34/403, 51/243, 56/581
hereditary myotonic syndrome, 62/272
herpes simplex virus, 56/211
Huntington chorea, 49/261
hydrocephalus, 30/662, 50/289, 427
hydromyelia, 50/427
immune mediated encephalomyelitis, 34/421-423
JHM virus, 56/439
Krabbe-Bartels disease, 14/412
kuru, 56/558, 586
levodopa induced dyskinesia, 49/197
lipid storage disorder, 29/378
multiple sclerosis, 56/587
myotonia, 40/154, 495, 533, 548
myotonic syndrome, 62/263
neurotoxicity, 64/17
olivopontocerebellar atrophy (Dejerine-Thomas), 21/418
paraplegia, 56/455, 587
Parkinson disease, 49/100
porcine hemagglutinating encephalomyelitis virus, 56/439
progressive multifocal leukoencephalopathy, 47/514
progressive spinal muscular atrophy, 22/32
respiratory syncytial virus, 56/431
Reye syndrome, 56/167
slow virus disease, 56/439, 583
spongiform myelopathy, 56/455, 587
subacute sclerosing panencephalitis, 56/587
tardive dyskinesia, 49/190
triparanol, 40/548
varicella zoster virus infection, 56/240
viral infection, 34/43-54
viral neurological disease, 56/581
visna, 56/453
visna-maedi disease, 56/583
Wohlfart-Kugelberg-Welander disease, 59/91
Animal viral disease, 34/291-302, 56/581-590
Akabane virus, 34/301
Aleutian mink disease, 56/585

bluetongue virus, 34/297
border virus disease, 34/300-302
bovine viral diarrhea mucosal disease virus, 34/299
canine distemper encephalomyelitis, 9/669-675, 34/293
canine distemper virus, 56/587
congenital defect, 34/296-302, 379-384
demyelinating disease, 34/293-296
encephalitis, 56/588
encephalopathy, 34/297
enterovirus infection, 34/292
epidemiologic model, 34/292
equine encephalitis, 34/292
feline ataxia, 34/296
hog cholera virus, 34/298
human poliomyelitis, 34/292
hydranencephaly, 34/301
infectious leukoencephalomyelitis, 34/294
intrauterine infection, 34/296-302, 379-384
JHM virus encephalomyelitis, 34/293
metachromatic leukodystrophy of mink, 34/302
murine leukemia virus, 56/587
murine neurotropic retrovirus, 56/586, 59/447
old dog, 56/588
panleukopenia virus, 34/296
paralytic enterovirus infection, 34/292
respiratory syncytial virus, 56/431
retrovirus, 56/453
scrapie, 9/679-681, 34/275, 287, 302, 56/585
St. Louis encephalitis, 34/292
subacute degenerative disease, 34/302
teratogenic agent, 34/379-384
Theiler virus, 34/295, 302
visna, 9/681, 34/294
Anion
CSF, 28/436
Anion gap
alcohol intoxication, 64/112, 125
computation, 64/125
ethylene glycol intoxication, 64/122, 125
methanol, 64/96
methanol intoxication, 64/96, 125
Aniridia
ARG syndrome, 50/581
cerebellar ataxia, 42/123
congenital hemihypertrophy, 60/655
Gillespie syndrome, 60/655
mental deficiency, 42/123, 46/62
WARG syndrome, 50/581
Wilms tumor, 50/581, 60/655
Anisocoria
Behr disease, 42/416
chronic bromide intoxication, 36/300
cluster headache, 48/223
CNS degeneration, 42/97
flow chart, 74/407
headache, 75/283
Holmes-Adie syndrome, 74/411
metabolic encephalopathy, 69/400
middle cerebral artery syndrome, 53/358
miscellaneous, 74/413
oculomotor paralysis, 74/410
optic tract lesion, 2/577
polyneuropathy, 42/97
pupil, 74/406
Tournay phenomenon, 74/413
traumatic, 74/414
Ankle fracture
nerve damage, 70/42
peroneal nerve, 70/41
sural nerve, 70/41
Ankle injury
peroneal nerve, 70/40
Ankle jerk, *see* Achilles reflex
Ankyloblepharon
albinism, 14/105
Ankylosing spondylitis
aortitis, 63/52
atlas, 26/175
atrophy, 38/506
bamboo spine, 26/176
cauda equina syndrome, 38/508-513, 70/4
cervical subluxation, 38/514
cervical vertebral column injury, 26/175, 177-181
cervical vertebral fracture, 26/177, 268
classification, 71/4
complete cord lesion, 26/178-180
CSF, 38/505
dens, 26/175
dislocation, 70/3
fracture, 70/3
fusion, 26/182
history, 26/175
HLA loci, 41/77
iridocyclitis, 26/176
juvenile rheumatoid arthritis, 26/180
kyphosis, 50/419
multiple sclerosis, 38/517, 70/6
muscular atrophy, 38/506
myopathy, 70/7
neurologic complication, 38/505-519
paraplegia, 38/518
polymyositis, 62/373
radiculopathy, 38/506-508
radiotherapy, 26/182

railway spine, 26/176
rheumatoid arthritis, 19/330
single root lesion, 70/3
spastic paraplegia, 59/439
spinal cord compression, 19/375, 70/4
spinal epidural cyst, 20/163
spinal injury, 26/175, 38/515-517
spinal subluxation, 26/181
spondylotic kyphosis, 50/419
survey, 70/3
tetraplegia, 26/175-184
trauma to spine, 38/515-517
vertebral fusion, 26/182
vertebrobasilar insufficiency, 38/513

Ankylosis
multiple, *see* Multiple ankylosis
temporomandibular, *see* Temporomandibular ankylosis

Ankyrin
chorea-acanthocytosis, 63/285
chromosome 8, 63/263
dystrophin, 62/126

Annelida intoxication
lumbriconereis heteropoda, 37/55

Annual rhythm
*see also* Biological rhythm
sex hormone, 74/489
suprachiasmatic nucleus, 74/489
testosterone, 74/489
vasopressin, 74/490

Annulus calcification
mitral, *see* Mitral annulus calcification

Annulus fibrosus
anatomy, 25/124
injury, 25/481-511
intervertebral disc, 20/525

Anocutaneous reflex
lower urinary tract, 61/295
paraplegia, 61/295
spinal cord injury, 61/295
transverse spinal cord lesion, 61/295

Anodontia
familial ectodermal dysplasia, 14/112

Anoetic syndrome, *see* Anetic syndrome

Anomalad
definition, 30/8

Anomia
amnesic color, *see* Amnesic color anomia
anomic aphasia, 45/318
brain hemorrhage, 54/309
cancer, 67/390
color, *see* Color anomia
dialysis encephalopathy, 64/277
frontal lobe tumor, 17/255, 272
left temporal lobe lesion, 2/586
memory disorder, 2/599
quality of life, 67/390
semantic, *see* Semantic anomia
tactile, *see* Tactile anomia
thalamic aphasia, 54/309
unilateral tactile, *see* Unilateral tactile anomia
word production, *see* Word production anomia
word selection, *see* Word selection anomia

Anomic aphasia
agraphia, 45/461
alexia, 45/461
Alzheimer disease, 4/98
anomia, 45/318
definition, 45/318-320
head injury outcome, 57/404
memory disorder, 45/318
parietal lobe syndrome, 4/74
semantic anomia, 45/319
temporal lobe tumor, 4/98
tip of the tongue phenomenon, 45/319
type, 45/318-320
word production anomia, 45/318
word selection anomia, 45/319

Anomic dysphasia, *see* Anomic aphasia

Anophthalmia
anencephaly, 50/83
axial mesodermal dysplasia spectrum, 50/516
birth incidence, 42/14
cytomegalovirus, 56/268
Dandy-Walker syndrome, 50/333
definition, 30/469
embryology, 30/469
genetics, 30/470
holoprosencephaly, 30/470
Lenz syndrome, 43/419
microphthalmia, 42/404, 43/419
optic chiasm agenesis, 42/14
optic nerve agenesis, 42/14
optic tract, 2/528
Patau syndrome, 14/120, 31/504, 513, 43/527
pathogenesis, 30/470
pathologic anatomy, 30/470
polyploidy, 50/555
prenatal diagnosis, 42/14
tetraploidy, 43/567
trigeminal nerve agenesis, 50/214

Anorexia
acrodynia, 64/374
allergic granulomatous angiitis, 63/383
barotrauma, 63/416
brain edema, 63/416

brain hypoxia, 63/416
carbamate intoxication, 64/186
cat scratch disease, 52/127
cluster headache, 48/222
dysbarism, 63/416
headache, 48/32
jumpers of the Maine, 42/232
lead intoxication, 64/435
methanol intoxication, 36/353, 64/98
migraine, 48/125
mountain sickness, 63/416
organolead intoxication, 64/133
ostrea gigas intoxication, 37/64
pellagra, 28/86, 38/648
pheochromocytoma, 42/764
procainamide intoxication, 37/452
temporal arteritis, 48/309
toxic oil syndrome, 63/380
uremia, 63/504
valproic acid induced hepatic necrosis, 65/125
venerupis semidecussata intoxication, 37/64

Anorexia nervosa, 46/585-589
acanthocyte, 49/332
acanthocytosis, 63/271, 290
amitriptyline, 46/588
anorexigenic peptide, 46/587
Bartter syndrome, 46/586
basal metabolic rate, 46/586
behavior therapy, 46/589
brain cortex, 46/587
brain tumor, 46/588
bulimia, 46/589
cachexia, 7/634
chlorpromazine, 46/588
cholecystokinin, 46/587
corticotropin, 46/586
dexamethasone suppression test, 46/586
differential diagnosis, 46/588
EEG, 46/586
endocrinology, 46/587
etiology, 46/586-588
glucose tolerance test, 46/586
growth hormone, 46/586
human menopausal gonadotropin, 46/586
hydrocortisone half-life, 46/586
hyperalimentation, 46/589
hypothalamus, 46/587
hypothalamus tumor, 46/417
incidence, 46/585
laboratory investigation, 46/586
luteinizing hormone, 46/586
monoamine, 46/587
opiate, 46/587
potassium imbalance, 28/482
prognosis, 46/589
protirelin, 46/586
pseudo-Bartter syndrome, 46/586
psychotherapy, 46/589
schizophrenia, 46/588
Sheehan syndrome, 46/585, 588
sulpiride, 46/588
symptomatology, 46/585
thyroid disorder, 27/273
thyroid hormone, 46/586
treatment, 46/588

Anorexigenic agent
phenylpropanolamine addiction, 65/259

Anorexigenic peptide
anorexia nervosa, 46/587

Anosmia
*see also* Olfaction *and* Smell
acute amebic meningoencephalitis, 52/315, 317
ageusia, 24/17, 20
anatomy, 24/2, 8
anoxia, 24/11
ataxia, 60/652
biomechanics, 24/8, 10
brain concussion, 24/5
brain injury, 24/2, 8
cacosmia, 24/13
chordoma, 18/155
congenital, *see* Congenital anosmia
CSF fistula, 24/5
dysbarism, 24/11
electrical injury, 61/192
fibrous dysplasia, 14/190
frontal lobe tumor, 17/260
head injury, 57/135
hereditary motor and sensory neuropathy type I, 21/283, 308
hereditary motor and sensory neuropathy type II, 21/283, 308
hereditary sensory and autonomic neuropathy type II, 60/11
hypogonadism, 60/652
hysterical, *see* Hysterical anosmia
impact site, 24/6
incidence, 24/4, 8
Kallmann syndrome, 30/471, 43/418
law, 24/14
medicolegal aspect, 24/14
olfactory bulb, 24/8
olfactory nerve, 2/52
parosmia, 2/54, 24/13
pathology, 24/8
posttraumatic amnesia, 24/6, 12, 18

psychology, 24/2
recovery, 24/11
Refsum disease, 13/314, 21/197, 27/523, 42/148, 60/228, 231, 652, 728, 66/485, 490
rhinorrhea, 24/5
severe head injury, 24/5, 12
Sjögren syndrome, 71/74
skull fracture, 24/5
specific, *see* Specific anosmia
taste center, 24/20
treatment, 24/13
Anosodiaphoria
anterior cerebral artery syndrome, 53/349
Anton-Babinski syndrome, 45/69
basofrontal syndrome, 53/349
body scheme disorder, 2/478, 4/219, 45/379
frontal lobe syndrome, 53/349
Anosognosia
*see also* Cortical blindness
aphasia, 4/92, 98, 110
blindness, 3/284
body scheme disorder, 2/478, 4/23, 29, 92, 213, 215-219, 243-245, 297, 45/374, 379
cortical blindness, 2/593, 3/284
cortical lesion, 3/56
definition, 2/607
dementia, 46/200
denial of illness, 45/376
hemianopia, 4/216
hemiplegia, *see* Anton-Babinski syndrome
memory disorder, 3/284
middle cerebral artery syndrome, 53/363
neglect syndrome, 45/178
nondominant hemisphere, 55/138
nondominant parietal lobe, 2/695
occipital lobe syndrome, 45/51
painful, *see* Painful anosognosia
verbal, *see* Verbal anosognosia
Wernicke aphasia, 4/98
Anoxia
*see also* Hypoxia
anemic, *see* Anemic anoxia
anencephaly, 30/184, 193
anosmia, 24/11
brain, *see* Brain anoxia
brain metabolism, 24/602
cerebral, *see* Brain anoxia
demyelination, 9/573, 47/529
epilepsy, 72/362
glycogen, 27/206
histotoxic, *see* Histotoxic anoxia
late cortical cerebellar atrophy, 21/477
leukoencephalopathy, 9/572
Little disease, 1/209
metabolic encephalopathy, 69/400
organophosphate intoxication, 64/228
pallidal lesion, 6/652
pallidal necrosis, 6/654
prenatal, *see* Prenatal anoxia
rehabilitation, 46/611
stagnant, *see* Stagnant anoxia
traumatic psychosis, 24/519
Anoxia ischemia
Levine experimental, *see* Levine experimental anoxia ischemia
Anoxic encephalopathy
action myoclonus, 63/217
alpha coma, 63/219
antiepileptic agent, 63/222
barbiturate, 63/223
brain blood flow, 63/219
brain oxygen metabolic rate, 63/219
calcium channel blocking agent, 63/222
cardiac arrest, 63/214, 217
computer assisted tomography, 63/218
creatine kinase BB, 63/218
Creutzfeldt-Jakob disease, 56/546
EEG, 63/219
free radical scavenger, 63/223
isotope scan, 63/219
management, 63/221
*N*-methyl-D-aspartic acid receptor blocking agent, 63/223
neuropathology, 46/347
nuclear magnetic resonance, 63/218
symptomatic dystonia, 49/542
treatment, 63/221
Anoxic ischemic leukoencephalopathy, 47/525-543
anoxic leukomalacia, 47/526-528
antenatal period, 9/582
Binswanger disease, 9/589
brain edema, 9/583, 586, 47/531-535
brain hyperemia, 9/583, 586, 47/531
brain stem necrosis, 47/530
carbon monoxide intoxication, 9/578, 47/528, 530
cardiac arrest, 9/575
cardiac arrest encephalopathy, 47/530
cerebrovascular disease, 47/541
classification, 9/573
clinical features, 47/528
combined arrest, 9/576
delayed radionecrosis, 47/539
dexamethasone, 47/535
experimental, 47/530
experimental demyelination, 47/530
fat embolism, 47/538

glial-glial relationship, 9/593
glial-neuronal relationship, 9/592
head injury, 47/541
heart disease, 47/525
hypertension, 9/589
hypertensive encephalopathy, 47/536-538
hypotension, 9/576
lymphostatic encephalopathy, 47/533
Marchiafava-Bignami disease, 47/529
microangiopathy, 9/591, 593
microinfarction, 9/588
microthrombosis, 9/591
neonatal period, 9/579
neuronal oxygen need, 9/592
nonbacterial thrombotic endocarditis, 47/541
open heart surgery, 9/578
pancreatic encephalopathy, 47/538
Pelizaeus-Merzbacher disease, 47/529
postanesthetic encephalopathy, 9/577
synaptic function, 9/592
systemic hypotension, 9/576
thrombotic thrombocytopenic purpura, 47/541
trauma, 9/587
vasculitis, 47/541
vasogenic brain edema, 47/533
Waldenström macroglobulinemia, 47/541
Anoxic leukomalacia
anoxic ischemic leukoencephalopathy, 47/526-528
carbon monoxide intoxication, 47/527
gray matter necrosis, 47/526-528
pallidal necrosis, 47/527
white matter necrosis, 47/526-528
Anoxic seizure, *see* Syncope
Ansa insulare
radioanatomy, 11/86
topography, 11/86
Ansa lenticularis
course, 6/19-21
Ansa peduncularis
nucleus, *see* Nucleus ansa peduncularis
Antabuse, *see* Disulfiram
Antebrachial nerve
medial cutaneous, *see* Medial cutaneous antebrachial nerve
Antebrachial pain syndrome
brachial neuralgia, 32/406
brachialgia, 32/406
Anteflexion
*see also* Cervical hyperanteflexion injury
lower cervical spine, 25/234, 237
occipitoaxial connection, 25/230
Anterior alexia
agrammatism, 45/442
aphasia, 45/442
brain infarction, 45/442
cerebrovascular disease, 45/442
definition, 45/441-443
pathology, 45/442
reading, 45/442
symptomatology, 45/442
type, 45/441-443
writing, 45/442
Anterior cardinal vein
anatomy, 11/55
Anterior cavernous sinus syndrome
aneurysm, 12/89, 31/147
causative lesion, 2/321
cause, 2/90
symptom, 2/90
Anterior cerebellar lobe
decerebration, 2/393
Anterior cerebellar lobe syndrome
somatotopic localization, 2/407
Anterior cerebral artery
anatomy, 11/4, 7, 9, 53/339
aplasia, 11/74
brain aneurysm, 12/91
brain embolism, 53/340
brain infarction, 11/306, 53/339
collateral circulation, 53/341
congenital aneurysm, 31/152
Doppler sonography, 54/12
headache, 5/130
hypoplasia, 11/176
paracentral lobule syndrome, 53/345
postcommunical segment, 11/74, 76
precommunical segment, 11/74
radiologic segment, 11/68
supplementary motor area syndrome, 53/345
supply area, 53/339
thrombosis, 53/339
Willis circle, 11/176
Anterior cerebral artery syndrome, 53/339-349
*see also* Frontal lobe syndrome
abulia, 53/349
age, 53/342
agraphia, 53/347
akinesia, 53/346
akinetic mutism, 53/342, 347
alien hand sign, 53/346
alien hand syndrome, 53/342, 346
amnesia, 53/348
anosodiaphoria, 53/349
apathy, 53/349
apraxia, 53/342, 346
basofrontal syndrome, 53/348

behavior disorder, 53/348
bilateral infarction, 53/341
bilateral lesion, 53/346
brain aneurysm, 53/341
Caucasian patient, 53/340
clinical features, 53/342
collateral circulation, 53/341
computer assisted tomography, 53/341-343
confabulation, 53/349
consciousness, 53/343
corpus callosum syndrome, 53/347
death rate, 53/345
decorum loss, 53/349
diabetes insipidus, 53/349
dysphasia, 53/342, 346
dyspraxia, 53/346
euphoria, 53/349
fever, 53/349
frequency, 53/339
Huebner artery syndrome, 53/348
ideomotor apraxia, 53/346
inappropriate behavior, 53/349
initiative loss, 53/349
left hand agraphia, 53/347
left hand apraxia, 53/347
memory loss, 53/343
multiple embolism, 53/340
mutism, 53/346
oriental patients, 53/339
palilalia, 53/346
pathogenesis, 53/339
plaque site, 53/340
risk factor, 53/342
sex ratio, 53/342
side, 53/342
split brain syndrome, 53/347
subarachnoid hemorrhage, 53/341
supplementary motor area syndrome, 53/345
tactile anomia, 53/347
total infarction, 53/343
urinary incontinence, 53/342
vascular spasm, 53/341
Anterior cervical cord syndrome
hypalgesia, 25/270
hyperdeflexion, 25/359
hypoesthesia, 25/270
luxation, 25/335
paralysis, 25/270
spinal cord injury, 25/270, 283
Anterior choroidal artery
anatomy, 11/16, 53/297
brain aneurysm, 12/94
brain lacunar infarction, 54/236
course, 11/72
globus pallidus, 49/467
migraine, 5/85
sensorimotor stroke, 54/252
supply area, 11/16
Anterior choroidal artery occlusion
pallidal necrosis, 49/466, 469
Anterior cingulate gyrus
akinetic mutism, 45/33
alertness, 2/765
cardiac dysrhythmia, 63/239
Anterior cingulate syndrome
bilateral infarction, 53/346
hemineglect, 53/346
Anterior commissure, *see* Commissura anterior
Anterior communicating artery
brain aneurysm, 11/79, 12/92, 221
congenital aneurysm, 31/153
transcranial Doppler sonography, 54/12
Anterior communicating artery aneurysm
*see also* Brain aneurysm
brain aneurysm neurosurgery, 55/66
neurosurgery, 12/221
subarachnoid hemorrhage, 12/92
Anterior condylar canal syndrome
jugular foramen syndrome, 2/99
tumor, 2/99
Anterior corpus callosum syndrome, *see* Corpus callosum syndrome
Anterior cranial fossa fracture
panda eyes, 57/131
racoon sign, 57/131
subconjunctival hemorrhage, 57/132
Anterior ethmoidal artery
radioanatomy, 11/71
topography, 11/71
Anterior ethmoidal nerve syndrome
coryza, 5/215
headache, 5/215
Anterior fossa
CSF fistula, 24/186
epidural hematoma, 24/263, 266, 639
skull base fracture, 23/393
skull fracture, 23/393
Anterior horn
anterolateral myelopathy, 2/210
cell degeneration, *see* Anterior horn syndrome
central ischemic infarction, 12/586
familial amyotrophic lateral sclerosis, 21/30
Anterior horn cell
amnesic shellfish poisoning, 65/167
ataxia telangiectasia, 60/365
infantile spinal muscular atrophy, 59/64

lathyrism, 65/9
Pena-Shokeir syndrome type I, 59/369
Anterior horn cell degeneration
cyanide intoxication, 65/28
cyanogenic glycoside intoxication, 65/28
manganese intoxication, 64/307
Pena-Shokeir syndrome type I, 59/71, 369
Anterior horn syndrome
amyotrophic lateral sclerosis, 42/65
areflexia, 2/210
arteriosclerosis, 2/210
calf hypertrophy, 42/72
cerebellar hypoplasia, 42/19
cervical spinal canal stenosis, 42/58
ganglion cell, 2/210
hereditary motor and sensory neuropathy type I, 42/340
hereditary motor and sensory neuropathy type II, 42/340
hyporeflexia, 42/72
lower motoneuron dysfunction, 2/210
Marinesco-Sjögren syndrome, 42/184
multiple neuronal system degeneration, 42/75
muscular atrophy, 2/210
myelin dysgenesis, 42/498
oculopharyngeal muscular dystrophy, 43/101
orthostatic hypotension, 43/66
paralysis, 2/210
Pena-Shokeir syndrome type I, 43/438
poliomyelitis, 2/210
spastic paraplegia, 42/169, 177
spinocerebellar ataxia, 42/182
syphilis, 2/210
Anterior inferior cerebellar artery
anatomy, 11/92
aneurysm, 2/257
deafness, 55/132
embryology, 11/26, 38
Horner syndrome, 2/244
labyrinthine ischemia, 55/132
Meniere disease, 2/257
Anterior inferior cerebellar artery syndrome
ataxia, 53/396
brain stem infarction, 53/396
cerebellar infarction, 55/132
deafness, 53/396, 55/90
facial paralysis, 53/396, 55/90
Horner syndrome, 11/31, 53/396
symptom, 53/396
vertebrobasilar system syndrome, 53/396
vertigo, 55/132
Anterior intercavernous sinus
anatomy, 11/54
Anterior interosseous nerve syndrome
anatomy, 7/300
compression neuropathy, 51/92
median nerve lesion, 7/329
occupational neuropathy, 7/329
Anterior interosseous palsy, *see* Anterior interosseous nerve syndrome
Anterior ischemic optic neuropathy
altitudinal hemianopia, 55/114
cerebrovascular disease, 55/114
migraine, 55/112
papilledema, 55/114
retinal artery occlusion, 55/112
systemic lupus erythematosus, 55/377
temporal arteritis, 55/114
visual field defect, 55/114
Anterior language zone
motor aphasia, 45/18
Anterior lobe
cerebellar, *see* Cerebellar anterior lobe
Anterior longitudinal ligament
cervical vertebral column injury, 61/27
scoliosis, 50/415
Anterior middle cerebral artery
radioanatomy, 11/79
topography, 11/79
Anterior occipital condyle
skull base tumor, 17/182
Anterior poliomyelitis, *see* Poliomyelitis
Anterior pontomesencephalic vein
anatomy, 32/48
Anterior posterior asymmetry
computer assisted tomography, 46/485
schizophrenia, 46/485
Anterior posterior choroidal artery
radioanatomy, 11/95
topography, 11/95
Anterior sacral meningocele
arachnoid cyst, 31/93
associated malformation, 32/195-206
caudal aplasia, 32/206
clinical features, 32/195-205, 207
diagnosis, 32/207
epidemiology, 32/206
genetics, 32/206
location, 32/195-205
Marfan syndrome, 32/206
pathologic anatomy, 32/193
reported cases, 32/195-205
scimitar sacrum, 32/194
spina bifida, 50/500
treatment, 32/195-205, 208
vertebral abnormality, 32/193, 206

Anterior spinal artery
amyotrophic lateral sclerosis, 22/19
anatomy, 11/92, 26/6, 64
angiography, 32/48
Brown-Séquard syndrome, 12/499
cervical cord, 25/263
cervicomedullary injury, 24/156
compression, 26/100
congenital heart disease, 63/4
lesion, 12/600
spinal cord vascular disease, 12/499
spinal cord vascularization, 12/483
syndrome, 12/600
thrombosis, 61/111
topographical anatomy, 55/96
Anterior spinal artery infarction
diamorphine intoxication, 65/356
Anterior spinal artery syndrome
Adamkiewicz artery syndrome, 12/499
aorta thrombosis, 63/41
chronic traumatic thoracic aneurysm, 63/42
clinical features, 55/100
cocaine intoxication, 65/255
features, 63/42
lower type, 55/100
myelomalacia, 55/100
postpartum period, 63/42
pregnancy, 63/42
radiation myelopathy, 61/200
rheumatoid arthritis, 55/101
schistosomiasis, 52/538
spondylosis, 55/101
syphilis, 55/101
transverse myelopathy, 63/42
upper type, 55/100
Anterior spinal fusion
cervical, 25/297, 299-303
Anterior spinal meningocele
CSF, 42/43
dermoid cyst, 42/43
imperforate anus, 42/43
kyphoscoliosis, 42/44
neurofibromatosis, 42/44
neurofibromatosis type I, 42/44
sex ratio, 42/43
survey, 32/193-208
Anterior spinal nerve root, *see* Ventral spinal nerve root
Anterior spinal subluxation, *see* Spinal subluxation
Anterior spinal vein
medulla/cord border, 32/48
Anterior spinocerebellar tract, *see* Ventral spinocerebellar tract
Anterior tarsal tunnel syndrome
compression neuropathy, 51/105
extensor retinaculum, 51/105
Anterior temporal artery
radioanatomy, 11/86
topography, 11/86
Anterior temporal diploic vein
anatomy, 11/55
Anterior temporal lobectomy
epilepsy, 73/390
Anterior tephromalacia
carpal tunnel syndrome, 59/113
cervical rib, 59/113
Hirayama disease, 59/113
Anterior thalamic nucleus
connection, 2/472
Anterior thoracic nerve
origin lateral branch, 2/137
origin medial branch, 2/137
topographical diagnosis, 2/21
Anterior tibial phenomenon
synkinesis, 1/185
Anterior tibial syndrome
cauda equina intermittent claudication, 12/538, 20/798
edema, 8/153, 41/385
exercise, 12/538, 20/798, 41/385
ischemic neuropathy, 8/152
myalgia, 41/385
pain, 8/152, 12/538, 20/798
swelling, 12/538, 20/798
Anterior uveitis
headache, 5/206
Köhlmeier-Degos disease, 55/278
Anterior vestibular vein
cerebrovascular disease, 55/129
Anterocollis
dystonia musculorum deformans, 49/521
Anterograde amnesia
consciousness, 45/115
head injury outcome, 57/398
memory disorder, 3/269
Wernicke-Korsakoff syndrome, 45/197
Anterohyperflexion, *see* Cervical hyperanteflexion injury
Anterolateral cordotomy
pain, 26/497, 45/232
Anterolateral fiber
ascending, *see* Ascending anterolateral fiber
Anterolateral funiculus
cell origin, 45/232-234
Edinger tract, 45/232
pain, 45/231-234

reticular formation, 45/233
spinothalamic tract, 45/233
thalamus, 45/232
ventral spinocerebellar tract, 45/232
Anterolateral myelopathy
*see also* Changi Camp syndrome *and* Tropical ataxic neuropathy
analgesia, 2/210
anterior horn, 2/210
extramedullary tumor, 2/210
Jamaican neuropathy, 7/643
paralysis, 2/210
retrobulbar neuropathy, 7/643
spastic paraplegia, 7/643
spastic syndrome, 7/643
thrombosis, 2/210
Anterolateral thalamic syndrome
choreoathetoid movement, 2/485
intention tremor, 2/485
resting tremor, 2/485
thalamic hand, 2/485
Anterolateral tract syndrome, *see* Anterolateral myelopathy
Anterolisthesis
definition, 25/243
spinal injury, 25/243
Anthopleura elegantissima
alcohol, 37/54
intoxication, 37/54
neurotoxic, 37/54
pulmonary edema, 37/54
pupillary reflex abnormality, 37/54
sea anemone intoxication, 37/54
suicide, 37/54
treatment, 37/54
Anthozoa intoxication
coelenterata, 37/53
food, 37/54
sea anemone, 37/53
Anthracene-9-carboxylic acid, *see* 9-Anthroic acid
Anthranilic acid
headache, 48/174
migraine, 48/174, 176
Anthraquinone
intestinal pseudo-obstruction, 51/494
9-Anthroic acid
myotonic syndrome, 40/548
9-Anthroic acid intoxication
drug induced myotonia, 62/271
drug induced myotonic syndrome, 62/271
Antiarrhythmic agent
cardiac pharmacotherapy, 63/190
cardiovascular agent, 65/433
epilepsy, 72/229
iatrogenic neurological disease, 63/190
lidocaine, 37/449
procainamide, 37/451
quinidine, 37/447
Antiarrhythmic agent intoxication
lidocaine, 37/449-451
procainamide, 37/451
quinidine, 37/447-449
Antibiotic agent
acquired immune deficiency syndrome patient, 71/382
antineoplastic agent, 65/528
bacterial meningitis, 33/13, 52/9
behavior disorder, 46/601
brain abscess, 52/148, 155
candidiasis, 35/419
cavernous sinus thrombosis, 52/175
chemotherapy, 39/114, 65/528
cranial epidural empyema, 52/169
cranial subdural empyema, 52/171
CSF fistula, 24/188
CSF shunt infection, 52/74
deficiency neuropathy, 7/632
delirium, 46/538
gram-negative bacillary meningitis, 52/105
gunshot injury, 57/311
head injury, 57/234
Hemophilus influenzae meningitis, 52/64
hereditary deafness, 50/219
infectious endocarditis, 63/24, 118
intracranial infection, 24/224, 228
missile injury, 23/510, 57/311
mycotic aneurysm, 52/296
nerve stimulation, 7/105
neuropathy, 7/515
nocardiosis, 16/218, 52/449
North American blastomycosis, 35/407, 52/392
penetrating head injury, 57/311
pneumocephalus, 24/212
pneumococcal meningitis, 52/48
rhinorrhea, 24/188, 223
spinal cord abscess, 52/192
spinal epidural empyema, 52/188
staphylococcal meningitis, 52/72
streptococcal meningitis, 52/84
subdural empyema, 24/220
superior longitudinal sinus thrombosis, 52/179
vitamin B1, 7/632
Antibody
acetylcholine receptor, *see* Acetylcholine receptor antibody
antibrain, *see* Antibrain antibody

anticholinesterase receptor, *see* Anticholinesterase receptor antibody
anticytoplasmic, *see* Anticytoplasmic antibody
antinuclear, *see* Antinuclear antibody
antiphospholipid, *see* Antiphospholipid antibody
anti-Purkinje cell, *see* Anti-Purkinje cell antibody
autoimmune, *see* Autoimmune antibody
Borrelia burgdorferi, 52/263
brain, *see* Brain antibody
cerebellar ataxia, 69/356
CSF, *see* CSF antibody
CSF autoantigen, *see* CSF autoantigen antibody
CSF idiotypic, *see* CSF idiotypic antibody
CSF myelin basic protein, *see* CSF myelin basic protein antibody
CSF nonspecific, *see* CSF nonspecific antibody
CSF viral, *see* CSF virus antibody
cytomegalovirus, *see* Cytomegalovirus antibody
demyelination, 9/519, 47/445
Epstein-Barr virus, *see* Epstein-Barr virus antibody
experimental allergic encephalomyelitis, 9/136, 47/445
F protein, *see* F protein antibody
gliotoxicity, 9/519
GM1 ganglioside, *see* GM1 ganglioside antibody
H protein, *see* H protein antibody
helminthiasis, 52/509
herpes simplex virus, 56/210
Hu, *see* Hu antibody
idiotypic, *see* Idiotypic antibody
IgA, *see* IgA antibody
IgG, *see* IgG antibody
IgM, *see* IgM antibody
immune complex, 41/376
inflammatory myopathy, 41/75
kappa light chain, 9/502
lambda light chain, 9/502
lymphocyte, *see* Lymphocyte antibody
lymphocytic choriomeningitis, 9/544
lymphocytotoxic, *see* Lymphocytotoxic antibody
measles, *see* Measles antibody
molecular structure, 9/502
monoclonal, *see* Monoclonal antibody
monoclonal protein, *see* Monoclonal protein antibody
motor neuropathy, 71/442
multiple sclerosis, 9/141, 47/274, 339
myoglobin, 41/262
neuropathy, 71/442
NP protein, *see* NP protein antibody
oligodendrocyte, *see* Oligodendrocyte antibody
paraneoplastic syndrome, 69/354, 356, 359
paraproteinemia, 69/297
paraproteinemic neuropathy, 69/296
phosphorylase, 41/189
poliomyelitis, 34/101, 118-120
polyclonal, *see* Polyclonal antibody
sensory neuropathy, 71/443
serum, *see* Serum antibody
serum human T-lymphotropic virus type I, *see* Serum human T-lymphotropic virus type I antibody
serum nerve, *see* Serum nerve antibody
Sjögren syndrome, 71/69-73
synovial fluid, *see* Synovial fluid antibody
thymocyte, *see* Thymocyte antibody
thyroid autoimmune, *see* Thyroid autoimmune antibody
viral, *see* Viral antibody
visna, 56/61
Wegener granulomatosis, 71/183
Antibody absorption test
fluorescent treponemal, *see* Fluorescent treponemal antibody absorption test
Antibody dependent cytotoxicity
multiple sclerosis, 47/343, 352
viral infection, 56/60, 65
Antibody formation
*see also* Plasmapheresis
experimental allergic myositis, 40/165
experimental autoimmune myasthenia gravis, 40/163
inflammatory myopathy, 40/264
myasthenia gravis, 40/265
Antibody mediated demyelination
Guillain-Barré syndrome, 51/250
Antibody response
autonomic nervous system, 75/554
Hemophilus influenzae meningitis, 33/58
multiple sclerosis, 47/339
poliomyelitis, 34/118, 56/323
Antibrain antibody
multiple sclerosis, 9/141
Anticardiolipin
antiphospholipid antibody, 63/330
Anticholinergic agent, *see* Cholinergic receptor blocking agent
Anticholinergic agent intoxication, *see* Cholinergic receptor blocking agent intoxixation
Anticholinergic mydriasis, 74/412
Anticholinesterase
carbamate intoxication, 64/183
intoxication, 37/541
myasthenia gravis, 41/126, 62/412
neuromuscular transmission, 41/131

organophosphate intoxication, 64/224
Anticholinesterase deficiency
myasthenic syndrome, 41/130
Anticholinesterase receptor antibody
myasthenia gravis, 43/158
Anticoagulant
acetylsalicylic acid, 63/306
brain embolism, 12/456, 55/166, 169
brain hemorrhage, 54/71, 288, 291, 55/167, 476
brain hemorrhagic infarction, 63/19
brain infarction, 11/355, 53/425, 55/167, 169
cardiovascular agent, 65/433
cerebrovascular disease, 12/456-465, 55/157, 166, 169
completed stroke, 12/460
complication, 38/147
congestive heart failure, 63/133
embolism, 38/146
epidural hematoma, 26/18
femoral neuropathy, 8/13
fibrinolytic agent, 63/308
fibrinolytic agent risk, 63/320
head injury, 55/476
heparin action, 63/306
iatrogenic neurological disease, 55/476, 65/461
internal carotid artery syndrome, 53/325
International Normalised Ratio, 63/320
intracerebral hematoma, 24/365
intracerebral hemorrhage risk, 63/319
lumbosacral plexus, 51/164
myocardial infarction, 12/458
neuropathy, 7/520, 8/13
progressing stroke, 12/459
prolonged transient ischemic attack, 53/273
prothrombin time ratio, 63/320
risk, 63/19
spinal artery injury, 26/74
spinal cord injury, 26/316
spinal subdural hematoma, 12/566
streptokinase, 63/308
stroke, 12/459
subarachnoid hemorrhage, 12/113, 55/3
subdural hematoma, 26/32
tissue plasminogen activator, 63/308
transient ischemic attack, 12/11, 458, 53/273, 325
traumatic venous sinus thrombosis, 24/379
venous sinus thrombophlebitis, 33/181
vertebral artery stenosis, 12/11
vertebrobasilar insufficiency, 53/385
vertebrobasilar system, 53/385
vitamin E intoxication, 65/570
warfarin, 63/306
warfarin action, 63/307
Anticoagulant overdose
brain hemorrhage, 65/461
facial paralysis, 65/461
femoral nerve paralysis, 65/461
heparin, 65/460
ulnar nerve palsy, 65/461
warfarin, 65/460
Anticonvulsant
*see also* Antiepileptic agent
acetylurea intoxication, 37/200
barbiturate intoxication, 37/200
behavior disorder, 46/595, 600
benzodiazepine intoxication, 37/200
brain infarction, 53/432
brain metastasis, 71/626
brain tumor, 69/14
classic migraine, 48/159
clumsiness, 46/164
deoxybarbiturate intoxication, 37/200
dibenzazepine intoxication, 37/200
drug induced polyneuropathy, 51/307
glossary, 15/731
gunshot injury, 57/311
head injury, 57/230
hydantoin intoxication, 37/200
interaction, 15/729
metabolic effect, 15/729
migraine, 5/51, 102, 48/202
missile injury, 57/311
nystagmus, 2/361
osteoporosis, 70/19
oxazepam intoxication, 37/199
oxazolidinedione intoxication, 37/200
pain, 69/47
penetrating head injury, 57/311
potassium imbalance, 28/480
scorpion sting intoxication, 37/112
side effect, 15/727, 729
succinimide intoxication, 37/200
systemic lupus erythematosus, 55/381
tabes dorsalis, 45/230
teratogenic agent, 30/121
teratogenic effect, 15/726
teratogenicity, 37/211
trigeminal neuralgia, 5/300, 45/230
vitamin $B_c$, 28/203-205
vitamin $B_c$ deficiency, 28/203-205, 46/24
Anticonvulsant intoxication
downbeating nystagmus, 60/656
mental deficiency, 46/24
Anticytoplasmic antibody
Wegener granulomatosis, 51/450
Antidepressant

behavior disorder, 46/597
chronic muscle contraction headache, 48/350
cluster variant syndrome, 48/270
debrisoquine hydroxylation, 65/316
epilepsy, 72/229
heterocyclic, *see* Heterocyclic antidepressant
hyperventilation syndrome, 38/329, 45/254
lithium intoxication, 64/299
migraine, 48/97
multiple sclerosis, 47/178
stress, 45/254
tetracyclic, *see* Tetracyclic antidepressant
tricyclic, *see* Tricyclic antidepressant
Antidepressant intoxication
acute, 37/315
allergic reaction, 37/319
cardiovascular, 37/318
cardiovascular complication, 37/318
CNS, 37/317
hypersensitivity, 37/319
Antidiabetic agent
neuropathy, 7/518
Antidiuretic hormone, *see* Vasopressin
Antidiuretic hormone secretion
inappropriate, *see* Inappropriate antidiuretic hormone secretion
Antidromic occlusion
technique, 7/74
Antidromic response
nerve conduction, 7/148
Antiemetic agent
behavior disorder, 46/595
migraine, 48/197
pain, 69/47
Antiepileptic agent
*see also* Anticonvulsant
absorption, 65/496
acetazolamide, 65/497, 73/350
acute intermittent porphyria, 65/497
age, 65/497
α-amino-3-hydroxy-5-methyl-4-isoxazolepropionic acid, 73/349
anoxic encephalopathy, 63/222
arene oxide, 65/515
asterixis, 65/499
barbiturate, 65/497
basic mechanism, 73/347
bioavailability, 65/496
body mass, 65/497
brain ischemia, 63/222
carbamazepine, 37/200, 65/496
cerebellar atrophy, 65/499
cerebellar dysfunction, 65/499
choreoathetosis, 65/499
clobazam, 65/497
clomethiazole, 65/497
clonazepam, 65/497
congenital malformation, 65/513, 515
dermatomyositis, 65/499
diazepam, 37/199, 65/497
drug interaction, 65/498
dyskinesia, 65/499
EEG change, 65/499
encephalopathy, 65/498
epilepsy, 72/229
epoxide, 65/515
ethosuximide, 65/497
excipient effect, 65/496
extrapyramidal symptom, 65/499
felbamate, 65/496
fetal hydantoin syndrome, 65/514
fetal trimethadione syndrome, 65/514
free radical, 65/516
genetic variability, 65/497
hepatic enzyme induction, 65/497
juvenile myoclonic epilepsy, 73/168
lamotrigine, 65/496
lithium intoxication, 64/293
lorazepam, 37/358, 65/497
mesuximide, 37/200
metharbital, 37/200
midazolam, 65/497
myopathy, 65/499
nitrazepam, 65/497
nutrition, 65/497
orofacial dyskinesia, 65/499
oxcarbazepine, 65/496
paraldehyde, 65/497
perinatal development, 65/512
pharmacokinetics, 73/357
pharmacotoxicology, 65/495
phenobarbital, 65/497
phenytoin, 65/497
phenytoin parahydroxylation, 65/498
plasma concentration, 73/364
plasma level, 73/362
plasma protein, 65/498
polyneuropathy, 65/499
porphyria, 65/497
pregnancy, 65/512
primidone, 65/497
spina bifida, 65/516
sulpiride, 65/497
sultiame, 65/497
target range, 73/362
teratogenesis, 65/515

teratogenicity, 65/512
triazolam, 65/497
trimethadione, 37/200, 65/497
valproic acid, 65/497
vigabatrin, 65/496
vitamin Bc, 65/516
vitamin D3 turnover, 65/498
Antiepileptic agent intoxication, 37/199-216
acute, 37/201-203
ataxia, 60/662
chronic, 37/204
endocrine, 37/210
idiosyncrasy, 37/203
myoclonus, 60/662
skeletal, 37/210
vitamin Bc deficiency, 37/208-210
Antiepileptic agent treatment
action mechanism, 15/621
adverse event, 73/353
blood level, 15/624
cognitive side effect, 72/390, 394, 396
drug survey, 15/668, 693, 731
marketing data, 73/353
memory, 72/390
pancreatic insulinoma, 14/751
polytherapy, 72/392, 73/354
registration, 73/353
resistance, 15/670
serum level, 15/673, 73/355
side effect, 15/705, 720
survey, 73/351
trade name, 15/693
vitamin Bc, 15/729
Antigen
brain, *see* Brain antigen
carcinoembryonic, *see* Carcinoembryonic antigen
CD4, *see* CD4 antigen
CNS, 9/137, 514
CSF herpes simplex virus, *see* CSF herpes simplex virus antigen
CSF herpes simplex virus 1, *see* CSF herpes simplex virus 1 antigen
CSF HLA, *see* CSF HLA antigen
CSF rubella, *see* CSF rubella antigen
experimental allergic encephalomyelitis, 9/135-137
foreign, *see* Foreign antigen
Histoplasma capsulatum, 52/439
HLA, *see* HLA antigen
HLA-B7, *see* HLA-B7 antigen
HLA-B40, *see* HLA-B40 antigen
HLA-DR, *see* HLA-DR antigen
HLA-DR2, *see* HLA-DR2 antigen
HLA-DR4, *see* HLA-DR4 antigen
HLA-Dw2, *see* HLA-Dw2 antigen
HY, *see* HY antigen
K-1, *see* K-1 antigen
Kell blood group, *see* Kell blood group antigen
Kx blood group, *see* Kx blood group antigen
motor end plate, 69/338
multiple sclerosis, 47/250
Mycobacterium tuberculosis, 52/205
neural, *see* Neural antigen
paraneoplastic syndrome, 69/329, 357
polysaccharide, *see* Polysaccharide antigen
surface, *see* Surface antigen
viral, *see* Viral antigen
Yo, *see* Yo antigen
Antigen system
HLA histocompatibility, *see* HLA histocompatibility antigen system
Antihistaminic agent
behavior disorder, 46/595
epilepsy, 72/229
extrapyramidal disorder, 6/828
migraine, 48/201
toxic myopathy, 62/601
Anti-Hu syndrome
paraneoplastic syndrome, 71/684
Antihypertensive agent
adverse effect, 63/74, 82
amiloride, 63/87
angiotensin converting enzyme, 63/84
arterial flow, 63/82
arterial hypertension, 63/74, 77
arteriosclerosis, 63/82
atherosclerotic stroke risk, 63/83
beta adrenergic receptor blocking agent, 63/83, 87
blood velocity, 63/83
brain arteriosclerosis, 63/84
brain infarction, 63/74
calcium channel blocking agent, 63/84
cholesterol/high density lipoprotein ratio, 63/83
clonidine, 37/441-443, 63/87
colestyramine, 63/83
coma, 63/87
congestive heart failure risk, 63/83
diazoxide, 37/432
epilepsy, 63/87
fatigue, 63/87
fluoride absorption, 63/83
guanethidine, 37/435
hallucination, 63/87
high density lipoprotein, 63/83
hydralazine, 37/430, 63/83
hydrochlorothiazide, 63/87

intracerebral hemorrhage risk, 63/83
kaliuretic diuretic agent, 63/83
lacunar infarction risk, 63/83
lipoprotein, 63/82
lipoprotein pattern, 63/83
methyldopa, 37/438-440, 63/83, 87
mortality risk, 63/83
muscle cramp, 63/87
myocardial infarction risk, 63/83
myocardial ischemia, 63/74
myopathy, 63/87
neurologic adverse effect, 63/87
orthostatic hypotension, 63/152
parkinsonism, 63/87
polyneuropathy, 63/87
propranolol, 37/444-447, 63/83
rauwolfia alkaloid intoxication, 37/436-438
renal failure risk, 63/83
reserpine, 37/436, 63/87
sodium nitroprusside, 37/433
stroke risk, 63/83
triglyceride, 63/83
vitamin B6 deficiency polyneuropathy, 63/87

Antihypertensive agent intoxication
clonidine, 37/443
diazoxide, 37/433
guanethidine, 37/436
hydralazine, 37/431
methyldopa, 37/440
propranolol, 37/447
reserpine, 37/437
sodium nitroprusside, 37/435

Antihypertensive treatment
adverse effect, 63/74
aged, 63/85
arterial hypertension, 63/83
brain infarction, 53/463, 63/74
iatrogenic neurological disease, 63/87
myocardial ischemia, 63/74
neurologic complication, 63/86

Anti-infective agent
epilepsy, 72/229

Anti-inflammatory agent
behavior disorder, 46/602
epilepsy, 72/232
migraine, 48/173
vasculitis, 71/161

Antikinesin
neuroaxonal dystrophy, 49/395

Antilymphocyte globulin, *see* Lymphocyte antibody

Antimalarial agent
neurologic complication, 7/518, 38/496

Antimetabolite
antineoplastic agent, 65/528
azacitidine, 39/106
8-azaguanine, 39/106
chemotherapy, 39/98-106, 65/528
cytarabine, 39/105
demyelination, 9/627
fluorouracil, 39/104
immunocompromised host, 56/469
leukoencephalopathy, 9/627
mercaptopurine, 39/106
methotrexate, 39/98-104
neurologic toxicity, 39/98-106
pellagra, 9/627
tioguanine, 39/106
triazinate, 39/104
vitamin B1, 7/568
vitamin B6 deficiency, 7/581, 583

Antimongoloid eyes
acrocephalosyndactyly type II, 43/324
acrocephalosyndactyly type V, 43/324, 50/123
adducted thumb syndrome, 43/331
Armendares syndrome, 43/366
Coffin-Lowry syndrome, 43/238
craniofacial dysostosis (Crouzon), 43/359
Gorlin-Chaudry-Moss syndrome, 43/369
maxillofacial dysostosis, 43/425
mental deficiency, 43/257
monosomy 21, 43/540
9p partial trisomy, 43/515
10q partial trisomy, 43/517
r(21) syndrome, 43/543
Treacher-Collins syndrome, 43/421
trisomy 22, 43/548
Turner syndrome, 50/543
W syndrome, 43/271

Antimony intoxication
amyotrophy, 22/28

Antimycin A
mitochondrial respiration, 41/212

Antimyelin basic protein
experimental allergic encephalomyelitis, 47/446

Antineoplastic agent
*see also* Chemotherapy
acivicin, 65/529
alkylating agent, 65/528
altretamine, 65/528
amsacrine, 65/529
antibiotic agent, 65/528
antimetabolite, 65/528
asparaginase, 65/529
azacitidine, 65/528
behavior disorder, 46/600
bleomycin, 65/528

broxuridine, 65/529
carboplatin, 65/529
carmustine, 65/528
chlorambucil, 65/528
chlormethine, 65/528
cisplatin, 65/529
classification, 65/528
corticosteroid, 65/529
cyclophosphamide, 65/528
cyproterone acetate, 65/529
cytarabine, 65/528
doxorubicin, 65/528
elmustine, 65/528
encephalopathy, 46/600
etoposide, 65/528
fludarabine, 65/528
fluorouracil, 65/528
human immunodeficiency virus infection, 71/385
ifosfamide, 65/528
α-interferon intoxication, 65/529
β-interferon intoxication, 65/529
interleukin-2, 65/529
lomustine, 65/528
lonidamine, 65/529
melphalan, 65/528
methotrexate, 65/527
metodiclorofen, 65/528
misonidazole, 65/529
mitomycin C, 65/528
mitotane, 65/529
mitotic inhibitor, 65/528
mitoxantrone, 65/528
nimustine, 65/528
procarbazine, 65/528
side effect, 71/385
sparfosic acid, 65/529
spirogermanium, 65/529
spiromustine, 65/528
spongiform leukoencephalopathy, 47/568
suramin, 65/529
tamoxifen, 65/529
taxol, 65/529
taxotere, 65/529
teniposide, 65/528
thiotepa, 65/528
vidarabine, 65/528
vinblastine, 65/528
vincristine, 65/528
vindesine, 65/528
Antinuclear antibody
eosinophilic fasciitis, 63/385
Antiparkinsonism agent
pharmacodynamics, 6/222-224
Antiphospholipid antibody
anticardiolipin, 63/330
Behçet syndrome, 63/330
brain embolism, 63/122
brain infarction, 53/166, 63/122, 330
brain lacunar infarction, 54/240
cerebrovascular disease, 63/329
chorea, 63/330
eclampsia, 63/330
endocarditis, 63/122
Guillain-Barré syndrome, 63/330
hepatitis C, 63/330
human immunodeficiency virus, 63/330
hydralazine, 63/330
ischemic optic neuropathy, 63/330
laboratory finding, 63/331
livedo reticularis, 63/122, 330
malaria, 63/330
management, 63/331
migraine, 63/122, 330
phenothiazine, 63/330
phenytoin, 63/330
procainamide, 63/191
psoriatic arthropathy, 63/330
quinidine, 63/191
rheumatoid arthritis, 63/330
Sheldon syndrome, 63/122
Sjögren syndrome, 63/330
Sneddon syndrome, 55/403, 63/330
syphilis, 63/330
systemic lupus erythematosus, 55/372, 63/122, 330, 71/40
systemic sclerosis, 63/330
temporal arteritis, 63/330
thrombocytopenia, 63/122, 330
transient ischemic attack, 63/330
valvular heart disease, 63/330
Antiphospholipid antibody syndrome
clinical features, 71/41
Antiplatelet agent, *see* Antithrombocytic agent
Antipsychotic agent
lithium intoxication, 64/298
pain, 69/47
Antipsychotic agent intoxication
acute, 37/299-302
tardive dyskinesia, 37/307-312
tricyclic, 37/315-319
Anti-Purkinje cell antibody
late cerebellar atrophy, 60/585
Antipyresis
endogenous, 74/453
fever, 74/451
Antiretroviral therapy

acquired immune deficiency syndrome, 71/367
acquired immune deficiency syndrome dementia complex, 71/237
neuromuscular disorder, 71/367
Antithrombin
oral contraception, 55/525
Antithrombin III
intravascular consumption coagulopathy, 55/494
transverse spinal cord lesion, 26/394
Antithrombocytic agent
brain infarction, 53/464
cardiovascular agent, 65/433
vertebrobasilar system, 53/385
Antithymocyte globulin, *see* Thymocyte antibody
Antitremor agent
tremorine, 6/224
Antituberculosis agent
acquired immune deficiency syndrome patient, 71/382
toxicity, 71/382
Antitumor immune response
brain tumor, 67/292
Antivenene
cobra snake venom intoxication, 37/12
enhydrina schistosa intoxication, 37/94
hydrophis cyanocintus intoxication, 37/94
lapemis hardwicki intoxication, 37/94
lapemis semifasciata intoxication, 37/94
notechis scutatus intoxication, 37/94
sea snake intoxication, 37/11, 93
snake bite, 37/12
stonefish intoxication, 37/77
Anti-Yo syndrome
paraneoplastic syndrome, 71/689
Antley-Bixler syndrome, *see* Multisynostotic osteodysgenesis
Anton-Babinski syndrome
allesthesia, 45/69
anosodiaphoria, 45/69
aphasia, 45/378
body scheme disorder, 4/216-219, 45/376-378
disconnection syndrome, 45/378
hemisphere lesion, 3/29
nonconscious, 3/56
parietal lobe syndrome, 45/69
visuospatial agnosia, 3/29
Anton syndrome
*see also* Cortical blindness
bilateral occipital lobe syndrome, 53/413
cerebrovascular disease, 55/139, 144
cortical blindness, 2/697
occipital lobe syndrome, 45/50
posterior cerebral artery syndrome, 53/413
temporal arteritis, 48/316
Anton test
Listeria monocytogenes, 33/82
Antoni type A tumor
chromatin, 14/22
neurinoma, 14/22, 27, 33, 20/238
neurofibromatosis type I, 14/137
palisading, 14/22
Verocay nodule, 14/22, 27
Antoni type B tumor
arteriolar fibrinohyalinoid degeneration, 14/23
microcyst, 14/22
neurinoma, 14/22, 27, 33, 20/238
neurofibromatosis type I, 14/137
Antrum
maxillary, *see* Maxillary antrum
Anuria
aorta aneurysm, 63/50
marine toxin intoxication, 65/153
paralytic shellfish poisoning, 37/33, 65/153
Anus
imperforate, *see* Imperforate anus
Anxietas tibiarum, *see* Restless legs syndrome
Anxiety, 45/265-268
Addison disease, 46/431
aldosterone, 1/502
Amanita pantherina intoxication, 36/534
amygdala, 74/309
amygdaloid nucleus, 45/275
atlas syndrome, 5/373
attention, 3/138
brain circuit, 45/275
brain tumor, 46/430
caffeine intoxication, 65/260
carisoprodol, 5/168
cingulotomy, 1/110
cognition, 45/267
delirium, 46/430
dementia, 46/430
dysphrenia hemicranica, 5/79
EEG, 45/267
emotion, 3/336, 348
emotional disorder, 3/348
endocrine disorder, 46/431
epilepsy, 3/348, 15/95, 73/430, 454, 478
free floating, *see* Free floating anxiety
GABA, 45/267
head injury outcome, 57/407
headache, 5/254, 48/6, 20, 354, 359
hereditary acute porphyria, 51/391
hydralazine intoxication, 37/431
hyperthyroidism, 46/431
hyperventilation syndrome, 38/320, 63/436

hypoparathyroidism, 46/431
hypothalamus, 45/275
insomnia, 45/139
insulinoma, 46/431
lidocaine intoxication, 37/450
lobotomy, 1/110, 45/27
mephenesin, 37/353
meprobamate, 37/353
mercury polyneuropathy, 51/272
migraine, 5/381, 48/39, 79, 121, 161, 354
Morvan fibrillary chorea, 6/355
muscle contraction headache, 48/39, 355
noradrenergic nervous system, 74/309
occipital headache, 5/371
organolead intoxication, 64/133
pain, 1/110, 26/495
parietal lobe epilepsy, 73/102
pellagra, 28/86, 38/648, 46/336
personality, 45/266
phenylpropanolamine intoxication, 65/259
pheochromocytoma, 39/498, 500, 42/764, 46/431, 48/427
phobic, *see* Phobic anxiety
postpoliomyelitic amyotrophy, 59/35
prefrontal leukotomy, 1/110
presenile dementia, 42/286
propranolol, 37/447
psychiatric headache, 5/20
psychogenic headache, 5/254
rhizostomeae intoxication, 37/53
sea snake intoxication, 37/92
seizure, 72/420
sleep paralysis, 42/712
spasmodic torticollis, 6/799
stingray intoxication, 37/72
stress, 45/250, 253
temporal lobe, 45/275
transorbital leukotomy, 1/110
vitamin $B_c$ intoxication, 65/574

Anxiety neurosis
autonomic nervous system, 43/199
paresthesia, 43/199
sex ratio, 43/199

Anxiety tremor
adrenalin, 6/818
physiologic tremor, 6/818
postural tremor, 6/818
stress, 6/818
test, 6/818

Anxiolysis
benzodiazepine, 65/344
benzodiazepine intoxication, 65/335
buspirone, 65/344

Anxiolytic agent
behavior disorder, 37/348, 46/597
clorazepate, 37/355
doxepin, 37/360
headache, 48/203
inorganic salt intoxication, 37/348
pain, 69/47
tybamate, 37/354
ureides, 37/329

Aorta
arterial baroreflex, 75/110
arteritis, 48/316
dextroposed, *see* Dextroposed aorta
granulomatous arteritis, 48/316
headache, 48/316

Aorta aneurysm
anuria, 63/50
aorta disease, 63/37
arteriosclerosis, 39/250, 63/48
back pain, 63/48
brain embolism, 63/50
brain infarction, 63/50
dissecting, *see* Dissecting aorta aneurysm
eggshell calcification, 63/48
encephalomalacia, 63/50
extramedullary tumor, 20/444
femoral neuropathy, 63/49
genital ecchymosis, 63/50
hereditary, 63/48
hip pain, 63/49
Horner syndrome, 63/50
iliopsoas muscle hematoma, 63/49
location, 63/48
Marfan syndrome, 39/389, 63/48
multifocal encephalomalacia, 63/50
mycotic type, 63/51
neurologic complication, 63/48
nondissecting type, 63/48
obturator neuropathy, 63/49
pathogenesis, 63/48
prevalence, 63/48
recurrent laryngeal nerve palsy, 63/49
relapsing polychondritis, 63/48
rupture, 63/49
sciatic nerve palsy, 63/49
sciatic neuropathy, 63/49
sciatica, 63/49
sex ratio, 63/48
sexual dysfunction, 63/50
spinal cord hemorrhage, 63/37
spinal cord injury, 26/70, 74
swallow syncope, 63/58
testicular pain, 63/51

tuberous sclerosis, 63/48
Turner syndrome, 63/50
ureteric colic, 63/50
urologic symptom, 63/50
Aorta arch
right sided, *see* Right sided aortic arch
Aorta arch aneurysm
vascular dementia, 46/359
Aorta arch syndrome, *see* Takayasu disease
Aorta body tumor
diagnostic problem, 8/494
nomenclature, 8/494
pain, 8/494
respiratory distress, 8/494
Aorta branch arteritis, *see* Takayasu disease
Aorta coarctation
aorta surgery, 63/59
brain aneurysm, 12/101
brain blood flow, 53/48
clinical features, 63/4
congenital heart disease, 38/131-133, 63/4
CSF pressure, 63/5
dementia, 46/358
dissecting aorta aneurysm, 63/45
hypertensive encephalopathy, 54/212
migraine, 63/4
myelomalacia, 55/98, 61/112
neurologic complication, 38/131-133
paraparesis, 63/4
paraplegia, 61/115, 63/4
postoperative paraplegia, 63/4
spinal artery injury, 26/72
spinal cord injury, 26/72
subarachnoid hemorrhage, 55/2, 29, 63/4
tuberous sclerosis, 14/381, 50/373
Turner syndrome, 50/544
Aorta disease
aorta aneurysm, 63/37
aorta thrombosis, 63/41
aortitis, 63/37
aortoiliac atherosclerosis, 63/41
arteriosclerosis, 63/37
neurologic complication, 63/37
paraplegia, 61/111
poliomyelomalacia, 22/20
spinal cord intermittent claudication, 20/796, 32/338, 63/41
syphilitic aortitis, 63/37
Takayasu disease, 63/37
Aorta dissection
myelomalacia, 55/98
spinal cord injury, 61/76
thoracic spinal cord injury, 61/76
thoracic vertebral column injury, 61/76
Aorta lesion
spinal cord vascular disease, 12/592
Aorta rupture
flaccid paraplegia, 61/116
ischemic myelopathy, 63/43
mortality rate, 61/116
spinal artery injury, 26/72
spinal cord injury, 26/72
trauma, 61/112, 116
Aorta stenosis
aorta valve calcification, 63/25
brain embolism, 63/25
cardiac dysrhythmia, 63/25
cardiac valvular disease, 38/148, 63/24
exertional syncope, 63/24
infectious endocarditis, 63/112
neurologic complication, 38/148
retinal artery embolism, 63/25
rheumatic heart disease, 63/24
syncope, 15/823, 38/148, 63/24
Aorta surgery
animal model, 63/56
aorta coarctation, 63/59
brain embolism, 63/58
brain infarction, 63/58
bypass, 63/56
clamping, 63/56
femoral neuropathy, 63/58
iatrogenic neurological disease, 63/49, 55
myelomalacia, 55/99
neurologic complication, 39/251-253, 63/55
paraplegia, 61/116, 63/55
saphenous neuralgia, 63/58
segmental artery resection, 63/57
sexual dysfunction, 63/59
shunting, 63/57
somatosensory evoked potential, 63/57
spinal artery injury, 26/63, 69
swallow syncope, 63/58
traumatic aorta aneurysm, 63/58
traumatic aorta rupture, 63/58
Aorta thrombosis
abdominal aorta aneurysm, 63/41
anterior spinal artery syndrome, 63/41
aorta disease, 63/41
aorta injury, 63/41
injury induced, 61/115
oral contraception, 63/41
painful paraparesis, 63/41
paraparesis, 63/41
paraplegia, 61/115
spastic paraplegia, 59/437

spinal cord intermittent claudication, 63/41
Aortitis
allergic granulomatous angiitis, 63/53
ankylosing spondylitis, 63/52
aorta disease, 63/37
diamorphine, 63/54
drug induced, 63/54
giant cell arteritis, 63/52
large branch vasculitis, 63/51
myelomalacia, 63/51
Reiter disease, 63/55
rheumatoid disease, 63/55
small branch vasculitis, 63/51
syphilis, 63/52
systemic vasculitis, 63/51
Takayasu disease, 63/52
temporal arteritis, 11/452, 48/316, 63/54
Aortobrachiocephalic arteritis, *see* Takayasu disease
Aortography
iatrogenic neurological disease, 63/44
ischemic myelopathy, 63/43
spinal artery injury, 26/65
spinal cord lesion, 26/65
Aortoiliac disease
occlusive, *see* Occlusive aortoiliac disease
Apallic syndrome, *see* Akinetic mutism *and* Persistent vegetative state
Apamin
*see also* Bee venom
arthropod envenomation, 65/199
myotonic dystrophy, 62/222, 225
neurotoxin, 37/111, 62/222, 225, 65/194, 199
Apathy
acquired immune deficiency syndrome dementia, 56/491
anterior cerebral artery syndrome, 53/349
basofrontal syndrome, 53/349
brain infarction, 55/137
carbon monoxide intoxication, 64/32
cerebrovascular disease, 55/137
concentric sclerosis, 47/414
corpus callosum syndrome, 2/762
depressive pseudodementia, 46/222
dysphrenia hemicranica, 5/79
filariasis, 52/516
frontal lobe, 55/137
frontal lobe syndrome, 2/739, 53/349, 55/138
general paresis of the insane, 52/279
hereditary striatal necrosis, 49/495
Huntington chorea, 46/306
hyponatremia, 63/545
hypophosphatemia, 63/564
α-interferon intoxication, 65/562
γ-interferon intoxication, 65/562
migraine, 48/161
pellagra, 28/71
phenytoin intoxication, 65/500
presenile dementia, 42/287
progressive supranuclear palsy, 49/243
propionic acidemia, 27/72
thalamic dementia, 21/596, 46/357
thalamic syndrome (Dejerine-Roussy), 2/484
uremic encephalopathy, 63/504
uremic polyneuropathy, 51/356
vitamin B12 deficiency, 63/254
Wernicke-Korsakoff syndrome, 45/194
Whipple disease, 52/137
Apatite
hydroxyl, *see* Hydroxyl apatite
Aperistalsis
gastric, *see* Gastric aperistalsis
Apert-Crouzon syndrome, *see* Acrocephalosyndactyly type II
Apert disease, *see* Acrocephalosyndactyly type I
Apert syndrome, *see* Acrocephalosyndactyly type I
Apex syndrome
orbital, *see* Orbital apex syndrome
petrosal, *see* Gradenigo syndrome
Apgar score
congenital deafness, 42/364
mental deficiency, 46/4
Aphagia
*see also* Dysphagia *and* Feeding disorder
lateral hypothalamic syndrome, 1/500
Aphasia, 45/287-325
*see also* Dysphasia *and* Language
acalculia, 4/96, 45/476, 479
age, 45/304
agnosia, 4/28
agrammatism, 4/89-91, 45/306
agraphia, 4/91, 95, 145, 167-169, 45/457-463
alexia, 4/95, 116, 127, 129-133, 45/447
alternative communication, 46/615
Alzheimer disease, 42/275, 45/321, 46/203, 251, 253
amnesia, *see* Memory disorder
amnesic, *see* Memory disorder
amusia, 45/487
amytal test, 4/88, 45/303
anarthria, 4/89, 94, 99, 45/309
anatomic localization, 4/85-88, 97-100, 45/295-302
angioneurotic edema, 43/61
anomic, *see* Anomic aphasia
anosognosia, 4/92, 98, 110
anterior alexia, 45/442

Anton-Babinski syndrome, 45/378
apraxia, 4/108, 45/306, 429
assessment, 45/291-294
attention disorder, 3/189
auditory, *see* Auditory aphasia
auditory agnosia, 4/94
auditory phonatory system, 4/93
autopsy, 45/296
bilingualism, 45/305
Binswanger disease, 54/222
bone membranous lipodystrophy, 42/279
brain aneurysm, 42/723
brain atrophy, 45/297
brain commissurotomy, 4/280
brain cortex stimulation, 45/297
brain embolism, 53/214
brain hemorrhage, 45/17
brain infarction, 53/214
brain infection, 46/150
brain injury, 45/294, 46/153
brain surgery, 45/297
brain tumor, 16/731, 46/150
Broca, *see* Motor aphasia
buccofacial apraxia, 4/54
capsula interna, 45/321
catagorial thinking, 4/106
catastrophic reaction, 4/89
central, *see* Global aphasia
central alexia, 45/440
central word scheme, 4/93
cerebral dominance, 4/88, 262, 267, 292, 308, 316-320, 45/303, 46/148
cerebrovascular disease, 42/278, 53/214
child, 4/9, 109, 265, 361, 401, 425, 45/293, 46/147-155
ciclosporin, 63/10
cingulate gyrus, 2/767
classification, 3/35, 399, 4/97-100, 45/290
clinical features, 4/88-97, 46/150
color, *see* Color aphasia
complicated migraine, 5/63
computer assisted tomography, 45/299
conceptual thinking loss, 4/106
conduction, *see* Conduction aphasia
confused language, 57/404
congenital, 4/265
constructional apraxia, 4/78, 97, 108
cortical dysarthria, 4/89
cortical excision, 4/87
cortical motor aphasia, 4/99
cortical sensory aphasia, 4/98
Creutzfeldt-Jakob disease, 46/291
crossed, 4/5, 269, 291, 45/303
crossed aphasia, 46/149
deafness, 45/305
definition, 4/84
dementia, 45/296, 321, 46/203
depression, 46/426
developmental, *see* Articular defect
developmental dyslexia, 46/106, 110
diagnosis, 45/115
dysphasia, 2/598, 4/419
dysprosody, 4/89
EEG, 45/298, 46/150
electrical cortical stimulation, 4/87
epilepsy, 4/86, 15/80, 45/322, 46/150
etiology, 4/84, 46/150
evoked potential, 45/298
examination, 4/88
examination method, 3/5
expressive, *see* Motor aphasia
facial apraxia, 4/93
facilitation, 46/616
finger, *see* Finger aphasia
Flynn-Aird syndrome, 14/112, 42/327
frontal, *see* Frontal aphasia
frontal lobe tumor, 17/253, 271, 67/150
Gerstmann syndrome, 45/383
global, *see* Global aphasia
handedness, 4/6, 88, 264, 316-318, 45/303
head injury, 45/297, 46/150, 57/404
head injury outcome, 57/404
hemispherectomy, 46/152
historic note, 46/147-150
historic survey, 4/84-86
hypertensive encephalopathy, 54/212
ideational defect, 4/105-107
indifference to pain, 8/196
intelligence, 3/7, 4/97, 105-111, 422, 45/294, 46/154
intelligence test, 4/106
α-interferon intoxication, 65/562
γ-interferon intoxication, 65/562
internal speech, 4/28, 187
intracranial hemorrhage, 53/214
jargon, *see* Paraphasia
Köhlmeier-Degos disease, 55/277
lack of words, 4/91
Landau-Kleffner syndrome, 46/152, 73/282
language comprehension, 4/281-283
language recovery, 46/152-155
late infantile metachromatic leukodystrophy, 29/357
learning deficit, 4/109
linguistics, 45/306-309
literal paraphasia, 4/92

melodic intonation therapy, 46/617
memory disorder, 2/599, 691-693, 4/98, 221, 302
middle cerebral artery, 4/87
middle cerebral artery syndrome, 53/363
migraine, 4/86, 5/63, 45/322, 46/150, 48/70
minor hemisphere, 4/281-283
missile injury, 24/468
mixed transcortical, *see* Mixed transcortical aphasia
motor, *see* Motor aphasia
multiple sclerosis, 9/180
mutism, 45/322
neologistic jargon, 45/322
neurofibrillary tangle, 46/260-263
neurolinguistic analysis, 46/616
neuropsychologic test, 45/523
neuropsychology, 45/17-20
nominal, *see* Anomic aphasia
nonfluent, *see* Motor aphasia
nuclear magnetic resonance, 45/300-302
number, *see* Number aphasia
operant procedure, 46/616
optic, *see* Optic aphasia
pain insensitivity, 8/196
paragrammatism, 4/92, 45/306
paraphasia, 4/91-93, 45/17-20, 57/404
parietal lobe tumor, 17/305, 67/153
paroxysmal, *see* Paroxysmal aphasia
partial, *see* Paraphasia
pathogenesis, 4/85
perseveration, 4/91, 94
phonemic jargon, 4/92
phonetic disintegration, 4/89, 99
phonology, 45/307
Pick disease, 42/285, 46/203, 234
polyglot, 3/7, 45/305
positron emission tomography, 45/299
posttraumatic, 4/87, 268, 57/404
pragmatics, 45/309
presenile dementia, 46/201, 253
preverbitum, 4/28
prognosis, 4/100, 264, 267, 45/323, 46/150, 611
programmed procedure, 46/616
prosody, 45/309
pure agraphia, 4/100
pure alexia, 4/100
pure anarthria, 4/99
pure disorder, 4/99
pure motor, *see* Pure motor aphasia
pure motor hemiplegia, 54/243
pure word deafness, 4/100
pure word dumbness, 4/99
radioisotope diagnosis, 45/298
reception disorder, 4/94
receptive, *see* Sensory aphasia
regional brain blood flow, 45/297
rehabilitation, 3/373, 375, 378, 396-417, 4/101, 109, 24/689, 45/324, 46/614-617, 55/238
scale, 46/614
schizophrenia, 45/313
semantic defect, 4/106
semantic jargon, 4/92
semantics, 45/308
sensory, *see* Sensory aphasia
sensory extinction, 3/193
sex ratio, 45/304
sign language, 46/616
singing, 45/305
social capacity, 4/109
speech, 4/84, 45/306
speech deafness, 4/100
speech reduction, 4/89-91
speech therapy, 46/152-155, 616
split brain, 46/153
stereotypy, 4/89-91
stroke, 45/296, 46/150, 614, 629
stuttering, 46/171
subarachnoid hemorrhage, 12/132
subcortical auditory aphasia, 4/100
subcortical lesion, 45/320
subcortical motor aphasia, 4/99
subcortical visual aphasia, 4/100
syntaxis, 45/306
tactile, *see* Tactile aphasia
tactile agnosia, 4/34
telegram style, 4/90
temporal lobe tumor, 17/291, 67/152
terminology, 45/288
test for conceptual defect, 4/106
testamentary capacity, 4/105
thalamic, *see* Thalamic aphasia
thalamic hemorrhage, 45/320
thalamic infarction, 45/91
thalamic syndrome (Dejerine-Roussy), 45/90, 94
thalamic tumor, 45/320
thalamotomy, 45/320
total aphasia, 4/99
transcortical, *see* Transcortical aphasia
transcortical motor, *see* Transcortical motor aphasia
transcortical sensory, *see* Transcortical sensory aphasia
transient ischemic attack, 53/214
traumatic, *see* Traumatic aphasia
treatment, 46/152-155
uremic encephalopathy, 63/505

verbal, *see* Motor aphasia
visual, 4/96
visual action therapy, 46/615
visual communication therapy, 46/615
visuospatial agnosia, 45/498
Wada test, 4/88
Waldenström macroglobulinemia, 63/396
Wernicke, *see* Wernicke aphasia
zozymus aeneus intoxication, 37/56
Aphasia syndrome
parietal, *see* Parietal aphasia syndrome
Aphemia, *see* Motor aphasia
Aphthous ulcer
Behçet syndrome, 71/213
carotid artery, 48/338
carotidynia, 48/339
Aphthous uveitis
relapsing, *see* Behçet syndrome
Aphthovirus
picornavirus, 56/309
Apical periodontitis
dental headache, 48/391
headache, 5/224, 48/391
Apical petrositis
basal osteomyelitis, 33/180
Gradenigo syndrome, 33/181
Apistus intoxication
convulsion, 37/76
death, 37/76
delirium, 37/76
diarrhea, 37/76
hypotension, 37/76
hypothermia, 37/76
ischemia, 37/76
local pain, 37/76
nausea, 37/76
pallor, 37/76
radiation, 37/76
respiratory distress, 37/76
vomiting, 37/76
Aplasia
anterior cerebral artery, 11/74
bone marrow, *see* Bone marrow aplasia
caudal, *see* Caudal aplasia
cerebellar dysgenesis, 30/377
cerebellar vermis, *see* Cerebellar vermis aplasia
cranial nerve nucleus, 50/42
definition, 30/4
dens, 32/39
dentate fascia, *see* Dentate fascia aplasia
median, *see* Median aplasia
müllerian, *see* Müllerian aplasia
muscle, *see* Muscle aplasia
nasal bone, *see* Nasal bone aplasia
nuclear, *see* Nuclear aplasia
odontoid, *see* Odontoid aplasia
olfactory nerve, *see* Olfactory nerve aplasia
optic nerve, *see* Optic nerve aplasia
pectoralis muscle, *see* Pectoralis muscle aplasia
pituitary gland, *see* Pituitary gland aplasia
radial, *see* Radial aplasia
retinal, *see* Retinal aplasia
sphenoid body, *see* Sphenoid body aplasia
sphenoid wing, *see* Sphenoid wing aplasia
spinal cord, *see* Spinal cord aplasia
thymus, *see* Thymus aplasia
urogenital agenesis, 14/529
Aplasia axialis extracorticalis, *see* Pelizaeus-Merzbacher disease
Aplasia papillae N. opticorum
ophthalmoscopy, 22/537
Aploactidae intoxication
family, 37/78
venomous spine, 37/78
Aploactosoma intoxication
family, 37/78
venomous spine, 37/78
Aplysia intoxication
ingestion, 37/64
Apnea
brain cortex, 63/481
brain death, 55/257
brain stem death, 57/473, 476
central sleep, *see* Central sleep apnea
cingulate gyrus, 63/481
classification, 1/657
cold face test, 74/606, 622
Cushing reflex, 63/230
epilepsy, 63/481
episodic, *see* Episodic apnea
familial infantile myasthenia, 62/427
head injury, 57/65
hippocampus, 63/481
intracranial pressure, 1/659
Joubert syndrome, 60/667
local anesthetic agent, 65/420
multiple neuronal system degeneration, 59/159
obstructive sleep, *see* Obstructive sleep apnea
opiate intoxication, 65/353
orbital gyrus, 63/481
organophosphate intoxication, 64/174
17p partial monosomy, 50/592
Patau syndrome, 31/508, 50/557
posthyperventilation, 1/496, 63/482
progressive dysautonomia, 59/159
Rett syndrome, 60/636, 63/437

Reye syndrome, 29/333, 34/170, 56/151, 63/438
sleep, *see* Sleep apnea syndrome
Smith-Lemli-Opitz syndrome, 60/667
snake envenomation, 65/178
snoring, 63/450
spinal cord injury, 61/262
subarachnoid hemorrhage, 57/473
superior temporal gyrus cortex, 63/481
uncus, 63/481
Apneusis
brain stem infarction, 1/498, 63/485
definition, 1/498, 677, 63/485
demyelinating lesion, 63/485
nucleus parabrachialis medialis, 63/485
site, 63/478
trigeminal motor nucleus, 1/679, 63/485
Apneustic center
*N*-methyl-D-aspartic acid glutamate receptor, 63/479
pons, 63/479
respiratory center, 63/432, 478
respiratory control, 63/432
vagotomy, 63/479
Apneustic respiration, *see* Apneusis
Apocarboxylase
pyruvate carboxylase deficiency, 66/418
vitamin $B_W$, 66/418
β-Apo-8'-carotenal
primary pigmentary retinal degeneration, 13/193
Apoceruloplasmin
copper binding protein, 6/117
Apoceruloplasmin deficiency
secondary pigmentary retinal degeneration, 60/730
Apoenzyme-coenzyme dissociation
alcoholic polyneuropathy, 51/315
Apolipoprotein
classification, 66/536
Apolipoprotein A
hereditary amyloid polyneuropathy type 1, 60/108
Apolipoprotein A-I/C-III/A-IV gene complex
familial combined hyperlipidemia, 66/543
Apolipoprotein AI
a-alphalipoproteinemia, 51/396
amyloidosis, 71/517
neuropathy, 71/517
Tangier disease, 42/628
Apolipoprotein AII
a-alphalipoproteinemia, 51/396
Apolipoprotein B
Bassen-Kornzweig syndrome, 51/326
familial defective apolipoprotein B-100, 66/545
hypobetalipoproteinemia, 63/275
Apolipoprotein B-100 gene
Bassen-Kornzweig syndrome, 66/546
hypobetalipoproteinemia, 66/546
Apolipoprotein B containing lipoprotein disorder, 66/539
Apolipoprotein C-II gene
defect, 66/540
Apolipoprotein E gene
defect, 66/541
Apomorphine
chemical structure, 37/234
depression, 46/474
opiate, 65/350
puffer fish intoxication, 37/81
turning behavior, 45/167
Apoplectiform myelitis, *see* Spinal apoplexy
Apoplexy
capillary, *see* Capillary apoplexy
cerebral, *see* Brain hemorrhage
embolic striatum, *see* Embolic striatum apoplexy
miniature, *see* Miniature apoplexy
multinodular stenosing amyloidosis, 11/581, 613
pituitary, *see* Pituitary apoplexy
prehemorrhagic softening, 12/49
Spät, *see* Spät Apoplexie
spinal, *see* Spinal apoplexy
striatum, *see* Striatum apoplexy
subarachnoid hemorrhage, 12/165
traumatic delayed, *see* Traumatic delayed apoplexy
Apoptosis
brain tumor, 67/224
infantile spinal muscular atrophy, 59/60
Apparent death
brain injury, 24/788, 807, 809, 822
Apperceptive agnosia
associative agnosia, 4/25
classification, 4/25
classification history, 4/15
description, 45/336-338
perception classification, 45/338-341
temporal lobectomy, 45/336
Appetite
amphetamine intoxication, 65/257
elfin face syndrome, 31/317
hypothalamic lesion, 2/449
hypothalamus, 1/500, 2/441
hypothalamus tumor, 46/417
Prader-Labhart-Willi syndrome, 46/93
salt, *see* Salt appetite
Apractognosia
agnosia, 4/23
constructional apraxia, 45/492, 500

Apragmatism
  Pick disease, 46/233, 235, 240
Apraxia, 45/423-432
  agnosia, 4/23
  agraphia, 4/95, 146-149, 155
  alexia, 4/97
  Alzheimer disease, 42/275, 46/251, 253
  amorphosynthesis, 4/57, 61
  amusia, 4/203
  anatomic localization, 4/54-57
  anterior cerebral artery syndrome, 53/342, 346
  aphasia, 4/108, 45/306, 429
  asomatognosia, 4/61
  association cortex, 45/431
  ataxia, 1/310
  ataxic oculomotor, *see* Ataxic oculomotor apraxia
  attention disorder, 3/189
  avoidance, 4/51
  Binswanger disease, 54/222
  bladder, *see* Bladder apraxia
  body scheme disorder, 4/223
  bone membranous lipodystrophy, 42/279
  brain commissurotomy, 4/283-285, 45/427
  buccofacial, *see* Buccofacial apraxia
  buccolingual, *see* Buccolingual apraxia
  carbon monoxide intoxication, 64/32
  cerebral dominance, 4/55, 294-297
  classification, 4/50-54, 63, 45/425
  conduction aphasia, 45/429
  congenital oculomotor, *see* Cogan syndrome type I
  consciousness, 45/114
  constructional, *see* Constructional apraxia
  corpus callosum, 4/56, 45/426-428, 431
  corpus callosum agenesis, 45/100
  corpus callosum disconnection syndrome, 2/760
  corpus callosum infarction, 45/426-428
  corpus callosum lesion, 2/760
  corpus callosum tumor, 17/507
  deafness, 45/430
  definition, 4/48, 45/423
  dementia, 4/59, 62, 46/203
  developmental, *see* Developmental apraxia
  developmental dyspraxia, 4/362, 401, 443-464, 43/219
  diagnosis, 45/115, 424
  dialysis encephalopathy, 64/277
  disconnection symptom, 2/599
  disconnection syndrome, 4/58, 45/102
  dressing, *see* Dressing apraxia
  eye, *see* Oculomotor apraxia
  eyelid, *see* Eyelid apraxia
  facial, *see* Facial apraxia
  facial paralysis, 45/431
  finger, *see* Finger apraxia
  frontal lobe lesion, 4/54-57
  frontal lobe tumor, 17/252, 67/149
  Funktionswandel, 4/60
  gait, *see* Gait apraxia
  gaze, *see* Gaze apraxia
  genetic psychology, 4/61
  Gerstmann syndrome, 2/691, 4/53
  glioblastoma, 45/426
  handedness, 45/429
  historic survey, 4/48-50
  ideational, *see* Ideational apraxia
  ideomotor, *see* Ideomotor apraxia
  idiokinetic type, 2/601
  innervation, 4/50
  instrumental, *see* Instrumental apraxia
  intelligence disorder, 4/59
  kinesthetic, 2/731
  kinetic type, 2/600, 4/50
  left hemisphere, 45/428
  limb, *see* Limb apraxia
  magnetic, *see* Magnetic apraxia
  melokinetic, 4/50
  mental deficiency, 46/27
  metachromatic leukodystrophy, 66/168
  middle cerebral artery, 53/363
  middle cerebral artery syndrome, 53/362
  mimical, *see* Mimical apraxia
  motor, *see* Motor apraxia
  motor aphasia, 45/428
  neurofibrillary tangle, 46/260-263
  neuropsychologic test, 45/523
  object, *see* Object apraxia
  ocular, *see* Oculomotor apraxia
  oculomotor, *see* Oculomotor apraxia
  oral, 3/397
  pain, 4/223
  parietal lobe lesion, 5/50, 54-57
  parietal lobe syndrome, 45/74, 77
  parietal lobe tumor, 17/303, 67/153
  pathopsychophysiogenesis, 4/57-63
  Pick disease, 42/285, 46/234
  planotopokinesia, 4/54
  postural, *see* Postural apraxia
  praxic disintegration in dementia, 4/62
  pure motor hemiplegia, 54/243
  pyramidal tract, 45/431
  rehabilitation, 3/389, 46/620
  respiration, 63/481
  restoration of higher cortical function, 3/389
  Rett syndrome, 29/308, 60/636
  sensory aphasia, 45/429
  significance, 45/430-432

spatial thought, 4/60
split brain syndrome, 45/99
stroke, 45/430
symbolic activity, 4/58
sympathetic, *see* Sympathetic apraxia
tactile, *see* Tactile apraxia
tactile agnosia, 4/34
temporal lobe lesion, 4/54-57
tonic perseveration, 4/50
uremic encephalopathy, 63/505
verbal, *see* Verbal apraxia
visual disorder, 2/597
visual symptom, 2/760
vocal, 4/203
walling, 4/53
Wernicke aphasia, 45/429
Apraxia of gait, *see* Gait apraxia
Apraxia of gaze, *see* Gaze apraxia
Apresoline, *see* Hydralazine
Aprindine
delirium, 46/538
iatrogenic neurological disease, 65/455
Aprindine intoxication
ataxia, 65/455
delirium, 65/455
epilepsy, 65/455
hallucination, 65/455
headache, 65/455
tremor, 65/455
vertigo, 65/455
Aprosody
brain infarction, 55/143
cerebrovascular disease, 55/143
neuroleptic parkinsonism, 65/292
perisylvian region, 45/45
right hemisphere lesion, 55/143
right perisylvian region, 45/45
APUD cell
gastrointestinal disease, 39/454-456
multiple endocrine adenomatosis, 42/753
Aqueduct forking
aqueduct stenosis, 50/301
hydrocephalus, 30/538, 609, 613, 32/532, 540, 50/287
Aqueduct stenosis
*see also* Bickers-Adams syndrome
abducent nerve paralysis, 50/312
acrocephalosyndactyly type I, 43/320
amenorrhea, 50/311
aqueduct forking, 50/301
Argyll Robertson pupil, 50/311
arhinencephaly, 50/302
Arnold-Chiari malformation type I, 30/617, 50/301, 304, 309
Arnold-Chiari malformation type II, 30/620
astrocytoma, 18/8
astrocytosis, 30/614, 32/532, 540, 42/23, 35, 57, 50/302, 305
autosomal recessive inheritance, 50/62
bacterial encephalitis, 50/307
basilar artery aneurysm, 50/308
Bickers-Adams syndrome, 30/610
birth incidence, 42/57
blepharoptosis, 50/312
bobble head doll syndrome, 50/312
cerebellar ataxia, 50/311
clinical features, 30/618-620, 50/310
cloverleaf skull syndrome, 50/309
Collier sign, 2/279
computer assisted tomography, 50/312
corpus callosum agenesis, 50/302, 309
craniolacunia, 50/139
cuprizone intoxication, 50/303
cytomegalovirus infection, 50/308
Dandy-Walker syndrome, 42/21-23, 50/309
diagnosis, 30/620
embryology, 30/610
encephalomeningocele, 42/26
endocrine disorder, 50/311
epidemiology, 30/157, 50/305
epilepsy, 50/312
etiology, 50/306
experimental stenosis, 50/303, 307
forking, *see* Aqueduct forking
Fröhlich syndrome, 30/619
gait apraxia, 50/311
gait disturbance, 50/311
Galen vein aneurysm, 18/282, 50/308
genetics, 30/157, 609, 618, 50/306
glial septum, 50/302
head injury, 50/308
hemorrhage, 50/306
history, 30/609-611
hydrocephalus, 30/538, 609-622, 42/36, 56-58, 50/285-287, 306, 309
hyperreflexia, 42/57
incidence, 50/62
interventriculostomy, 50/315
intracranial pressure, 50/310
intraventricular hemorrhage, 50/309
karyotype, 30/613
Lyle syndrome, 2/302
lymphocytic choriomeningitis, 50/308
malformation, 30/618
mental deficiency, 42/57
metrizamide computer assisted tomography,

50/313
metrizamide ventriculography, 50/312
multiple neurofibromatosis, 43/34
mumps virus, 9/551, 50/308
neurofibromatosis, 14/37, 50/302, 309
neuropsychologic change, 50/311
nystagmus, 2/279, 42/57, 50/311
occipital cephalocele, 50/101
occipital encephalocele, 42/26
oculomotor paralysis, 50/311
ophthalmoplegia, 42/57
parasitic disease, 2/279, 50/308
Parinaud syndrome, 2/279, 50/311
pathogenesis, 30/609, 614-617
pathology, 50/301
pathophysiology, 50/309
periaqueductal syndrome, 2/279, 50/311
precocious puberty, 50/311
prenatal diagnosis, 42/58
pure stenosis, 50/301, 305
pyramidal syndrome, 50/311
radioisotope cisternography, 50/313
rubella, 50/308
septum, 30/614
sex linked recessive, *see* Sex linked recessive aqueduct stenosis
sex ratio, 42/57
spina bifida, 32/532, 42/55, 50/309
spontaneous CSF fistula, 50/310
superior cerebellar artery aneurysm, 50/308
surgical treatment, 50/314
survival, 30/619
syndrome, 30/610, 618, 42/57
syringomyelia, 50/302
tethered cord, 50/309
thalamic nucleus malformation, 50/309
third ventricle colloid cyst, 50/309
Toxoplasma gondii, 50/308
toxoplasmosis, 50/306
treatment, 30/620
treatment indication, 50/316
trigeminal nerve palsy, 50/312
trisomy 8 mosaicism, 43/511
ventriculoatrial shunt, 50/314
ventriculocisternostomy, 50/314
ventriculoperitoneal shunt, 50/314
ventriculostomy, 50/314
viral infection, 50/306
visual disorder, 50/310
Aqueduct of Sylvius, *see* Aqueduct stenosis
Aqueduct syndrome, *see* Aqueduct stenosis
Ara-A, *see* Vidarabine
Ara-C, *see* Cytarabine
Arabinoside
meningeal lymphoma, 63/348
Arachidonic acid
biochemistry, 66/35
brain edema, 67/80, 83
brain ischemia, 63/211
brain metabolism, 57/81
neuron death, 63/211
Arachnia propionica
chronic meningitis, 56/645
Arachnodactyly
Bardet-Biedl syndrome, 13/296
cataract, 60/652
Friedreich ataxia, 21/354
homocystinuria, 55/329
hypogonadal cerebellar ataxia, 21/467
Klippel-Trénaunay syndrome, 14/395, 524
Laurence-Moon-Bardet-Biedl syndrome, 13/303
mental deficiency, 43/293
9p partial monosomy, 50/579
Arachnoid abnormality
diastematomyelia, 50/437
diplomyelia, 50/437
Arachnoid cyst
ambient cistern, 31/118
anterior perforated substance, 31/102
anterior sacral meningocele, 31/93
brain base, 31/97-102
brain injury, 23/239, 307
brain scintigraphy, 23/307
brain ventricle diverticulum, 31/125-127
cavum vergae, 31/120
cerebellopontine angle, 31/114-116
cerebellopontine angle syndrome, 31/108
cerebral convexity, 31/102-107
cervicothoracic, 31/91
clinical features, 31/85, 87, 89-91
congenital, 31/75-128
corpus callosum agenesis, 31/119
CSF, 31/105
CSF fistula, 24/186
Dandy-Walker syndrome, 32/59
dermoid, 31/87
diagnosis, 68/311
differential diagnosis, 31/86
dilated cistern, 31/119
enterogenous, 31/88
ependymal, 31/88, 120
epidemiology, 31/85, 89
epidermoid, 31/87, 89
etiology, 31/90
extradural, *see* Epidural arachnoid cyst
genetics, 31/89

histology, 68/311
hydrocephalus, 30/541
intracranial, 31/96-128
intradural, 31/82-86, 89
lumbosacral, 31/92
Marfan syndrome, 31/103
meningitis serosa circumscripta spinalis, 20/86
myelographic finding, 31/86
myelography, 19/189
neural origin, 31/121
optic chiasm compression, 68/75
pathogenesis, 68/310
pathology, 31/90
perineural, 31/86
porencephaly, 31/119
posterior fossa, 31/107-127
posterior midline, 31/109-127
postoperative, *see* Postoperative meningeal diverticulum
posttraumatic leptomeningeal cyst, 31/127
sacral, *see* Sacral arachnoid cyst
sella turcica, 31/98
spinal, *see* Spinal arachnoid cyst
spinal intradural cyst, 20/72, 80, 82-87
subarachnoid hemorrhage, 12/184
subdural hygroma, 31/106
suprasellar, 31/99-102
symptomatology, 68/311
synonym, 31/75
temporal lobe agenesis syndrome, 31/105
tentorial notch, 31/117-127
treatment, 31/86, 68/312
ventricular diverticulum, 31/125
Arachnoid tear
posttraumatic headache, 5/182
Arachnoidal fibrosis
familial, *see* Familial spinal arachnoiditis
Arachnoidal sarcoma
cerebellar, *see* Cerebellar arachnoid sarcoma
Arachnoïdite cloisonnée, 20/86
Arachnoiditis
adhesive, *see* Spinal arachnoiditis
amyotrophic lateral sclerosis, 22/309
angiodysgenetic necrotizing myelomalacia, 20/484, 486
basal, *see* Basal arachnoiditis
cerebellar tumor, 17/716
cerebellomedullary, *see* Cerebellomedullary arachnoiditis
circumscripta, 20/86
coccidioidomycosis, 52/420, 422
cytarabine, 61/151, 63/357
familial spinal, *see* Familial spinal arachnoiditis
intrathecal chemotherapy, 61/151
local anesthetic agent, 65/421
lumbar puncture, 61/162
meningitis serosa, 20/86
methotrexate, 61/151
myelography, 19/189
neurobrucellosis, 33/312, 316, 52/587
neurocysticercosis, 52/530
opticochiasmatic, *see* Opticochiasmatic arachnoiditis
paratrigeminal syndrome, 48/337
posttraumatic syringomyelia, 61/395
serosa circumscripta spinalis, 32/393
spinal, *see* Spinal arachnoiditis
traumatic, 61/41
tuberculous meningitis, 52/216
Arachnopathy
delayed myelopathy, 12/612
Araeosoma intoxication
spinal injury, 37/67
Aran-Duchenne amyotrophy, *see* Aran-Duchenne disease
Aran-Duchenne disease
*see also* Progressive spinal muscular atrophy
amyotrophic lateral sclerosis, 22/305
bulbar paralysis, 22/134, 42/86
distal muscle weakness, 22/138, 42/86
epidemiology, 21/30
fasciculation, 42/86
hereditary motor and sensory neuropathy type I, 21/279
hereditary motor and sensory neuropathy type II, 21/279
history, 22/134, 282, 310
myalgia, 42/86
neuronal ceroid lipofuscinosis, 42/87
poliomyelitis, 42/86
scapulohumeral spinal muscular atrophy, 22/75
simian hand, 22/291
Aran, F.A.
biography, 22/2
Arbovirus
*see also* Arenavirus, Arthropod-borne virus, Bunyamwera virus *and* Togavirus
acute viral encephalitis, 56/126, 132
acute viral myositis, 56/195
definition, 34/5
tick bite intoxication, 37/111
viral meningitis, 56/126, 128
ARC 239, *see* 2-{2-[4-(2-Methoxyphenyl)-1-piperazinyl]ethyl}-4,4-dimethyl-1,3-(2h,4h)-isoquinolinedione
Arcade

Frohse, *see* Frohse arcade
Arch
C1 posterior, *see* C1 posterior arch
neural, *see* Neural arch
Arch abnormality
posterior, *see* Posterior arch abnormality
Arch hyperplasia
pedicular, *see* Pedicular arch hyperplasia
Arch hypertrophy
posterior, *see* Posterior arch hypertrophy
Archenteric cyst, *see* Intraspinal neurenteric cyst
Archeocerebellum
flocculonodular lobe, 1/329
Archery
rehabilitative sport, 26/525, 538
Archicerebellar syndrome
*see also* Cerebellar syndrome, Neocerebellar syndrome *and* Paleocerebellar syndrome
clinical symptomatology, 2/417
gait disturbance, 2/418
neocerebellar syndrome, 2/417
nystagmus, 2/418
paleocerebellar syndrome, 2/417
spasmodic torticollis, 2/418
terminology, 2/417
truncal ataxia, 2/418
Arcual kyphosis
cervical spine, 25/246
Arcuate gyrus
akinesia, 45/165
Arcuate nucleus
olivopontocerebellar atrophy (Dejerine-Thomas), 21/418
stress, 74/312
tuberoinfundibular system, 74/312
Area entorhinalis
connection, 73/64
Area medullovasculosa
neurenteric cyst, 20/57
spina bifida, 50/485
Areas
paranodal, *see* Paranodal area
perinodal, *see* Perinodal area
precentral, *see* Precentral area
premotor, *see* Premotor cortex
septal, *see* Septal area
ventral tegmental, *see* Ventral tegmental area
visual association, *see* Visual association area
Arecoline
Alzheimer disease, 46/267
Areflexia
*see also* Hyporeflexia
β-*N*-acetylhexosaminidase A deficiency, 60/660
β-*N*-acetylhexosaminidase B deficiency, 60/660
acrylamide intoxication, 64/68
anterior horn syndrome, 2/210
Argentinian hemorrhagic fever, 56/365
ataxia, 42/348
ataxia telangiectasia, 42/119, 60/365, 660
barium intoxication, 64/354
benign dominant myopathy, 43/111
Biemond syndrome, 21/379, 42/164, 299, 60/660
blennorrhagic polyneuritis, 7/478
cerebellar tumor, 17/710
cerebrohepatorenal syndrome, 43/338
Chédiak-Higashi syndrome, 42/536, 60/660
childhood encephalomyopathy, 60/661
chorea-acanthocytosis, 63/280
CNS degeneration, 42/97
congenital analgesia, 42/299
conidae intoxication, 37/60
corpus callosum agenesis, 42/100
Critchley syndrome, 13/427
critical illness polyneuropathy, 51/575
CSF paraprotein, 42/604
cytochrome related ataxia, 42/111
detrusor, *see* Detrusor areflexia
Ekbom syndrome type II, 60/660
Emery-Dreifuss muscular dystrophy, 62/149
eosinophilia myalgia syndrome, 63/376
facial spasm, 42/314
Friedreich ataxia, 42/143, 60/660
Gerstmann-Sträussler-Scheinker disease, 56/553, 60/660
glycogen storage disease, 43/180
Guillain-Barré syndrome, 42/465
Harvey syndrome, 42/222
hereditary ataxia, 60/660
hereditary hyperkalemic periodic paralysis, 43/151
hereditary multiple recurrent mononeuropathy, 42/316
hereditary neuropathy, 60/256
hereditary sensory and autonomic neuropathy type II, 60/10
Holmes-Adie syndrome, 43/72
hydromyelia, 50/429
hyperkalemia, 63/558
hyperkalemic periodic paralysis, 43/151
hypermagnesemia, 63/563
hypertrophia musculorum vera, 43/84
hypertrophic interstitial neuropathy, 42/317
hypokalemia, 63/557
hypophosphatemia, 63/564
inclusion body myositis, 62/373
infantile spinal muscular atrophy, 42/88

juvenile metachromatic leukodystrophy, 60/660
Machado-Joseph disease, 42/155
Matthews-Rundle syndrome, 60/660
mental deficiency, 43/258
Miller Fisher syndrome, 51/246
mitochondrial myopathy, 43/119
mitochondrial respiratory chain defect, 42/587
muscular atrophy, 42/82, 91
muscular dystrophy, 43/107, 109
nemaline myopathy, 43/122
neuroacanthocytosis, 42/209
neurogenic acro-osteolysis, 42/293
neuronal intranuclear hyaline inclusion disease, 60/660
oculopharyngeal muscular dystrophy, 43/100
olivopontocerebellar atrophy (Wadia-Swami), 60/660
organophosphate intoxication, 64/167
phenytoin intoxication, 42/641
polyneuropathy, 60/256
posterior column ataxia, 42/164
posttraumatic syringomyelia, 61/384
proximal muscle weakness, 43/107
pupillary change, 42/301
rabies, 56/388
Richards-Rundle syndrome, 43/264
Roussy-Lévy syndrome, 21/175, 42/108
Ryukyan spinal muscular atrophy, 42/93
scapulohumeroperoneal muscular atrophy, 42/343
scapuloilioperoneal muscular atrophy, 42/344
scapuloperoneal myopathy, 43/131
scapuloperoneal spinal muscular atrophy, 62/170
sensory neuropathy, 42/348
sensory radicular neuropathy, 42/351
serum paraprotein, 42/604
severe distal amyotrophy, 60/660
spina bifida, 42/54
spinocerebellar ataxia, 42/182
system degeneration, 42/97
thalidomide, 42/660
tremor, 60/660
X-linked ataxia, 60/660
X-linked optic atrophy, 42/417
xeroderma pigmentosum, 43/13, 60/336, 660
Areflexic dystaxia
hereditary, *see* Roussy-Lévy syndrome
Arenavirus
*see also* Arbovirus, Arthropod-borne virus, Bunyamwera virus *and* Togavirus
acute viral encephalitis, 34/193, 294
biology, 56/357
cerebellar hypoplasia, 50/179
complex, 56/355
features, 56/16
genus member, 56/355
geographic distribution, 56/356
host, 56/356
junin virus, 56/355
Lassa fever, 56/16
Lassa fever virus, 34/5, 56/16, 355
lymphocytic choriomeningitis, 9/193-196, 56/16
lymphocytic choriomeningitis lassa complex, 56/355
lymphocytic choriomeningitis virus, 34/5, 56/355
Machupo virus, 56/355
Tacaribe complex, 56/355
viral meningitis, 56/16
Arenavirus infection, 34/193-204
Argentinian hemorrhagic fever, 34/194-196, 200-203
Bolivian hemorrhagic fever, 34/194-196, 203
CSF, 34/195
laboratory diagnosis, 34/204
Lassa fever, 34/194-196, 203
lymphocytic choriomeningitis, 34/194-200
Areolar choroidal atrophy
central, *see* Central choroidal atrophy
Areolar choroidal sclerosis
central, *see* Central choroidal atrophy
Aretaeus
historic review, 2/7
ARG syndrome
aniridia, 50/581
cardiopathy, 50/582
cataract, 50/581
chromosome deletion syndrome, 50/581
cryptorchidism, 50/582
cytogenetics, 50/582
epilepsy, 50/582
facial dysmorphia, 50/582
glaucoma, 50/581
gonadoblastoma, 50/582
hypospadias, 50/582
megalocornea, 50/581
muscular hypertonia, 50/581
nystagmus, 50/581
11p partial monosomy, 50/581
pseudohermaphroditism, 50/582
strabismus, 50/581
Wilms tumor, 50/582
Argentina
neurology, 1/12
Argentinian hemorrhagic fever
abducent nerve paralysis, 56/366
alopecia, 56/364
animal model, 56/367

areflexia, 56/365
arenavirus infection, 34/194-196, 200-203
ataxia, 56/365
bleeding tendency, 56/364
cerebellar ataxia, 56/366
clinical features, 56/364
CSF, 34/195, 56/365
diagnosis, 56/368
diplopia, 56/366
encephalitis, 56/365
epidemiology, 56/367
epilepsy, 56/365
headache, 56/364
hypotension, 56/364
α-interferon, 56/367
intravascular consumption coagulopathy, 56/367
junin virus, 56/355, 364
late syndrome, 56/366
lethality, 56/366
leukopenia, 56/364
meningoencephalitis, 56/365
muscular hypotonia, 56/365
myalgia, 56/364
myocardial involvement, 56/364
palmomental reflex, 56/365
partial thromboplastin, 56/367
pathogenesis, 56/367
pathology, 56/366
relapse, 56/365
thrombocytopenia, 56/364
treatment, 56/368
tremor, 56/365
visual impairment, 56/366
Arginase
epilepsy, 72/222
Arginase deficiency, *see* Hyperargininemia
Arginine
autonomic nervous system, 74/137
Sakaguchi procedure, 9/24
Arginine deficiency
Reye syndrome, 56/167
Arginine vasopressin, *see* Argipressin
Argininemia
spastic paraplegia, 59/439
Argininosuccinase
epilepsy, 72/222
gene localization, 42/525
Argininosuccinase deficiency, *see* Argininosuccinic aciduria
Argininosuccinate lyase deficiency, *see* Argininosuccinic aciduria
Argininosuccinate synthetase
epilepsy, 72/222
Argininosuccinate synthetase deficiency
citrullinemia, 29/97, 42/538
Argininosuccinic aciduria
Alzheimer type II astrocyte, 22/577
argininosuccinate lyase, 29/99
ataxia, 42/525
basal ganglion degeneration, 42/525
biochemical finding, 29/99
birth incidence, 42/525
chronic variant, 29/89, 99
CSF, 42/525
dentate nucleus, 42/525
diet, 29/100, 42/526
dystonia, 60/664
feeding disorder, 42/524
genetics, 29/100
hair, 42/525
heterozygote detection, 42/525
hydrocephalus, 50/289
hyperammonemia, 22/577, 42/524-526
mental deficiency, 42/525
metabolic ataxia, 21/574, 577
myelin dysgenesis, 42/525
periodic ataxia, 21/577
prenatal diagnosis, 42/526
psychomotor retardation, 42/525
seizure, 42/524
subacute variant, 29/99
treatment, 29/100
trichorrhexis nodosa, 21/577
Arginosuccinate synthetase, *see* Argininosuccinate synthetase
Argipressin
alcohol intoxication, 64/114
basal ganglion, 49/33
brain edema, 67/80, 85
water intoxication, 64/240
Argyll Robertson phenomenon, *see* Argyll Robertson pupil
Argyll Robertson pupil
amyotrophic lateral sclerosis, 22/295
aqueduct stenosis, 50/311
Benedikt syndrome, 2/277
diabetes mellitus, 27/114, 124
dysautonomia, 22/258
facial spasm, 42/314
general paresis of the insane, 52/279
hereditary motor and sensory neuropathy type I, 21/285
hereditary motor and sensory neuropathy type II, 21/285
Holmes-Adie syndrome, 1/621
hypertrophic interstitial neuropathy, 21/151

light-near dissociation, 74/416
neurosyphilis, 1/621, 2/110, 280, 22/20, 33/356, 364, 52/280
pineal tumor, 17/656
polyneuropathy, 42/314
sarcoid neuropathy, 51/196
tabes dorsalis, 19/86, 33/364, 52/280
Argyrophil fiber
neurinoma, 14/24
Argyrophilic body
Pick disease, 21/54, 57, 27/489
Argyrophilic dystrophy
Alzheimer neurofibrillary change, 21/46
amyloid angiopathy, 21/45
brain amyloid angiopathy, 21/45
form, 21/56-58
neurofibrillary disease, 21/57
neuronal, *see* Neuronal argyrophilic dystrophy
subcortical, *see* Progressive supranuclear palsy
Argyrophilic inclusion
Pick disease, 21/54, 46/233, 236, 243
Argyrophobic cell
intestinal pseudo-obstruction, 51/493
Arhinencephaly
*see also* Holoprosencephaly
aqueduct stenosis, 50/302
cerebrohepatorenal syndrome, 50/242
diabetes mellitus, 9/582, 14/120
fetal alcohol syndrome, 50/242
Gruber syndrome, 50/242
incidence, 50/61
Kallmann syndrome, 30/471
Kundrat syndrome, 30/384
microgyria, 30/484
Patau syndrome, 14/120, 31/504, 508, 43/527, 50/556, 559
polyploidy, 50/555
13q partial monosomy, 43/526
r(13) syndrome, 50/585
Rubinstein-Taybi syndrome, 30/472, 50/242
syphilis, 9/582
trigonocephaly, 43/486
trisomy 22, 50/565
Ariëns Kappers, C.U., 1/10
Arieti-Gray syndrome
Sturge-Weber syndrome, 14/227
Aring, D., 1/17
Arithmetic
developmental dyslexia, 46/118
number aphasia, 2/600
Arm hypoplasia
Poland-Möbius syndrome, 43/461
Armadillo syndrome, *see* Isaacs syndrome
Armed forces
brain injury prognosis, 24/455
gainful employment, 24/462
head injury, 24/455
posttraumatic syndrome, 24/460
sequela missile injury, 24/455-475
traumatic epilepsy, 24/469
Armendares syndrome
antimongoloid eyes, 43/366
craniosynostosis, 43/366
dermatoglyphics, 43/366
dwarfism, 43/366
epicanthus, 43/366
microcephaly, 43/366
ptosis, 43/366
secondary pigmentary retinal degeneration, 43/366
Armstrong goggles
diplopia, 2/288
Arnold-Chiari malformation type I, 32/99-109
*see also* Spina bifida
Aicardi syndrome, 31/227
aqueduct stenosis, 30/617, 50/301, 304, 309
Arnold-Chiari malformation type II, 50/403
Arnold-Chiari malformation type III, 50/403
Arnold-Chiari malformation type IV, 50/403
atlanto-occipital synostosis, 50/393
atrophy, 32/104
basilar impression, 50/399
birth incidence, 42/16
cerebellar agenesis, 50/175
cerebellar ataxia, 42/16
cerebellar herniation, 32/104
cerebellar tumor, 17/716
clinical features, 32/105-107
CNS disorder, 31/122, 32/104
congenital laryngeal abductor paralysis, 42/319
corpus callosum agenesis, 50/163
cranial nerve, 42/15
craniolacunia, 32/187, 42/15, 50/139
CSF pressure, 19/92, 116
Dandy-Walker syndrome, 30/632, 32/59
definition, 32/100, 50/403
diagnostic procedure, 32/107
diastematomyelia, 32/247, 50/441
differential diagnosis, 32/107
diplomyelia, 50/441
downbeating nystagmus, 60/656
dysraphia, 14/111
Edwards syndrome, 50/562
encephalomeningocele, 42/26
epidemiology, 30/162, 32/103
epithelial neurenteric cyst, 32/437

foramen magnum, 42/15
genetics, 32/103
history, 32/100
hydrocephalus, 30/162, 538, 32/103, 42/15, 50/254
hydromyelia, 32/104, 231, 50/425, 428
incidence, 50/63
intracranial pressure, 42/16
intraspinal neurenteric cyst, 32/437
laryngeal abductor paralysis, 42/319
larynx stridor, 42/15
meningocele, 32/105, 42/15
mental deficiency, 46/8
microgyria, 30/484, 50/254
morphology, 32/100-102
neurologic syndrome, 32/106
nonrecurrent nonhereditary multiple cranial neuropathy, 51/571
obstructive hydrocephalus, 46/325
Patau syndrome, 31/513, 50/559
pathogenesis, 32/102
pathogenetic theory, 50/404
platybasia, 42/17
polyploidy, 50/555
posterior fossa tumor, 18/407
pseudotumor symptom, 32/106
scoliosis, 40/337
secondary factor, 32/99, 104
short dura mater syndrome, 5/372
spina bifida, 30/162, 32/521, 523-526, 42/54
spinal cord compression, 42/15
spinal cord traction, 32/105
spinal intradural cyst, 20/82
syncope, 75/218
syringomyelia, 32/104, 107, 300, 304, 42/16, 59, 50/444, 453, 458
treatment, 32/108
Arnold-Chiari malformation type II
*see also* Spina bifida
animal model, 50/404
aqueduct stenosis, 30/620
Arnold-Chiari malformation type I, 50/403
ataxia, 50/410
autonomic nervous system, 75/10
basilar impression, 50/399
C1 posterior arch, 50/407
cerebellar ataxia, 32/106
cerebellar herniation, 50/405
cerebellar vermis dysplasia, 50/405
cerebellar vermis heterotopia, 50/406
cervical root, 50/406
clinical features, 32/105
computer assisted tomography, 50/409
cranial nerve palsy, 50/410
craniolacunia, 32/187, 42/15, 50/408
deafness, 50/410
definition, 50/403
differential diagnosis, 50/410
downbeating nystagmus, 50/410
dysphagia, 50/410
embryology, 50/404
enlarged foramen magnum, 50/406
episodic apnea, 50/410
falx cerebri hypoplasia, 50/405, 408
foramen magnum, 50/408
fourth ventricle, 50/406
gray matter heterotopia, 50/407
heterotopia, 50/405
horizon XII, 50/405
hydrocephalus, 50/405, 407
hydromyelia, 50/407
interhemispheric fissure, 50/408
Klippel-Feil syndrome, 50/407
laryngeal abductor paralysis, 50/410
lumbar meningomyelocele, 50/403
meningeal malformation, 50/405
meningocele, 32/105
meningomyelocele, 50/403, 405, 407
mesencephalic tectum, 50/406
metrizamide, 50/409
microgyria, 50/405, 407
morphology, 32/100-102
myelography, 19/188
myelomeningocele, 50/484
myeloschisis, 50/405
neuroradiology, 50/408
occipital dysplasia, 50/404
opisthotonos, 50/410
persistence, 50/404
petrous pyramid, 50/408
polygyria, 50/407
polymicrogyria, 50/407
pontine plicature, 50/405
serrated ligament, 50/406
sign, 50/410
spina bifida, 32/523-526, 50/482, 490
stenogyria, 50/407
stridor, 50/410
symptom, 50/410
syringomyelia, 19/188
tentorial hypoplasia, 50/405, 408
tentorial incisure, 50/406
thalamus fusion, 50/405
treatment, 32/109, 50/410, 492
ultrasonography, 50/408
vitamin A, 50/404

Arnold-Chiari malformation type III
*see also* Spina bifida
Arnold-Chiari malformation type I, 50/403
craniolacunia, 32/187, 42/15
definition, 50/403
meningoencephalocele, 50/403, 410
morphology, 32/100
occipital cephalocele, 30/211, 50/101
occipital encephalocele, 42/26
spina bifida, 32/523
Arnold-Chiari malformation type IV
*see also* Spina bifida
Arnold-Chiari malformation type I, 50/403
cerebellar hypoplasia, 50/403
craniolacunia, 32/187, 42/15
Dandy-Walker syndrome, 50/403
definition, 50/403
morphology, 32/100
spina bifida, 32/523
Aromatic amino acid
metabolism, 75/235
tetrahydrobiopterin, 75/235
Aromatic levo amino acid decarboxylase
migraine, 48/86
Aromatic levo amino acid decarboxylase deficiency
defect, 75/236
Aromatic carboxylic acid intoxication
myotonic syndrome, 62/264
Arousal
cerebrovascular disease, 55/137
odor, 74/365
odorant, 74/364
olfaction, 74/364
relaxation, 74/365
right hemisphere, 45/173
sexual function, 26/455
Arousal disorder
brain cortex, 45/110
brain stem herniation, 45/112
cerebellar mass lesion, 45/113
consciousness, 45/107-113
diagnosis, 45/111-113
EEG, 45/108-110
encephalitis lethargica, 45/108
hippocampus, 45/109
hydrocephalus, 45/112
hypothalamus, 45/108, 111
subdural hematoma, 45/112
thalamus, 45/109
Wernicke encephalopathy, 45/108
Arousal reaction
attention, 3/138, 157, 168, 172
behavior change, 3/52
consciousness, 3/52
emotion, 3/321
subcortical, 3/63
terminology, 3/52
Arousal system
consciousness, 3/53, 171
localization, 45/111
progressive dysautonomia, 59/160
Arrested hydrocephalus
hydrocephalic dementia, 46/324
Arrhythmia
autonomic nervous system, 74/171
cardiac, *see* Cardiac dysrhythmia
diphtheria, 51/182
Guillain-Barré syndrome, 51/257
malignant hyperthermia, 38/551, 41/268, 62/567
stingray intoxication, 37/72
subarachnoid hemorrhage, 55/9
Arrhythmokinesis
neocerebellar syndrome, 2/422
Arsacetin
papilledema, 9/648
retrobulbar neuritis, 9/648
Arsanilic acid
arsenic intoxication, 9/648, 64/284
encephalomyelopathy, 9/648
toxic agent, 9/648, 13/204
ARSB gene
*N*-acetylgalactosamine-4-sulfatase, 66/282
glycosaminoglycanosis type VI, 66/282
Arsenate
arsenic intoxication, 64/285
arsenic polyneuropathy, 51/267
Arsenic
allergic encephalitis, 9/650
autonomic nervous system, 75/38
distal axonopathy, 64/12
element data, 64/284
neurotoxin, 5/299, 9/650, 37/199, 46/391, 48/420, 51/68, 264, 479, 531, 59/370, 62/284, 63/152, 64/10, 12
pentavalent, *see* Pentavalent arsenic
speciation, 64/284
spinal muscular atrophy, 59/370
toxic encephalopathy, 64/12
toxic neuropathy, 64/10
toxicity ranking, 64/284
trigeminal neuralgia, 5/299
trivalent, *see* Trivalent arsenic
Arsenic encephalopathy
African trypanosomiasis, 52/343
symptom, 51/266
Arsenic intoxication

action site, 64/10
acute muscle weakness, 41/288
African trypanosomiasis, 52/343
anemia, 41/288
arsanilic acid, 9/648, 64/284
arsenate, 64/285
arsenic trioxide, 64/284
arsenicals, 64/285
arsenobetaine, 64/284
arsine, 64/285
arsine gas, 64/284
arsphenamine, 9/648, 64/284
arsphenamine encephalopathy, 9/648
autonomic polyneuropathy, 51/479
biochemistry, 36/205
biological half-life, 64/285
blood-brain barrier, 64/285
bone marrow, 64/286
brain edema, 52/343
cancerogenic effect, 64/287
carbamazepine, 36/215
chronic exposure symptom, 64/287
clinical features, 36/200-204, 51/266
copper smelting, 64/284
CSF, 36/213
cupric acetoarsenite, 64/284
daily intake, 64/284
delayed effect, 64/287
dementia, 46/391, 64/288
demyelination, 9/647
deterioration, 64/288
differential diagnosis, 36/213
dimercaprol, 36/214, 64/289
distal axonopathy, 64/288
EMG, 36/204, 64/287
encephalomyelopathy, 9/648
enzyme system, 64/286
exposure source, 64/284
fourpenny beer, 64/284
Fowler solution, 64/284
fox intoxication, 37/96
gangrene, 64/287
gastrointestinal manifestation, 64/286
Guillain-Barré syndrome, 51/266
hair, 36/206
headache, 48/420, 64/284
history, 36/199, 64/283
hyperhidrosis, 64/287
hyperkeratosis, 64/287
individual sensitivity, 64/288
industry, 64/284
insecticide, 64/284
kinetic, 64/285
Korean herbal medicine, 64/284
laboratory diagnosis, 36/206-208, 64/288
leukoencephalopathy, 9/647, 650, 47/559
Mees line, 36/202, 64/286
melarsoprol, 52/343, 64/284
meningitis serosa, 9/648
metabolism, 64/285
metallurgic industry, 64/286
monitoring, 64/288
moonshine whiskey, 64/284
nail, 36/202, 206
neuralgic pain, 64/284
neurologic manifestation, 36/203
neuropathology, 47/559
neuropathy, 7/513, 36/203, 47/559, 64/286
neurotoxicity, 51/264
optic atrophy, 13/63
orthostatic hypotension, 63/152
oxidative phosphorylation, 64/286
paraparesis, 64/288
pathology, 36/208-213
pathomechanism, 64/286
paxillus involutus intoxication, 36/214
penicillamine, 36/214
polar bear intoxication, 37/96
polyneuropathy, 64/284, 75/504
potassium arsenite, 64/284
rash, 41/288
Raynaud phenomenon, 8/337
renal dysfunction, 41/288
risk evaluation, 64/288
rodenticide, 64/284
sensorimotor neuropathy, 41/288
skin lesion, 64/286
source, 36/199
succimer, 64/289
symptom, 36/200-202, 64/286
thiol binding, 64/285
treatment, 36/214, 64/289
tremor, 6/823
tryparsamide, 9/648, 64/284
unithiol, 64/289
urinary, 36/206-208
urinary concentration, 64/288
urinary excretion, 64/285
Wernicke-Korsakoff syndrome, 46/392
wolf intoxication, 37/96
Arsenic polyneuropathy
acetyl-coenzyme A, 51/267
arsenate, 51/267
arsenic trioxide, 51/267
atomic absorption spectrometry, 51/267
axonal degeneration, 51/267

bone marrow depression, 51/266, 64/286
chronic axonal neuropathy, 51/531
diagnostic test, 51/266
diarrhea, 51/266
dimercaprol, 51/266, 268
EMG, 51/267
encephalopathy, 51/266
Guillain-Barré syndrome, 51/266
mechanism, 51/267
nerve biopsy, 51/267
oxidative phosphorylation, 51/267
penicillamine, 51/266, 268
pentavalent arsenic, 51/267
psychosis, 51/266
pyruvic acid, 51/267
sensorimotor polyneuropathy, 51/266
stupor, 51/266
thioctic acid, 51/267
toxic polyneuropathy, 51/68, 266
treatment, 51/267
Arsenic trioxide
arsenic intoxication, 64/284
arsenic polyneuropathy, 51/267
Arsenite, *see* Arsenic trioxide
Arsphenamine
arsenic intoxication, 9/648, 64/284
encephalomyelopathy, 9/648
leukoencephalopathy, 9/647
papilledema, 9/648
polyneuritis, 7/529, 609, 9/648
retrobulbar neuritis, 9/648
Arsphenamine encephalopathy
arsenic intoxication, 9/648
pallidal necrosis, 49/479
Arsphenamine intoxication
pallidal necrosis, 6/652, 49/466
Arterial axis (Moniz)
radioanatomy, 11/86
topography, 11/86
Arterial baroreflex
aorta, 75/110
autonomic nervous system, 75/432
carotid sinus, 75/110
central site, 74/254
exercise, 74/251, 256
Oxford method, 75/432
space, 74/283
Arterial compression
thoracic outlet syndrome, 51/123
transient ischemic attack, 12/8
Arterial dilatation
baroreflex, 74/166
tetrodotoxin intoxication, 65/155
Arterial dissection
cerebrovascular disease, 53/34
headache, 48/286
pain, 48/287
Arterial hypertension
*see also* Vascular hypertension
aged, 63/85
air embolism, 61/221
alcohol, 54/207
amaurosis fugax, 55/108
amphetamine intoxication, 65/257, 259
amyloid angiopathy, 63/74
aneurysm rupture, 63/82
antihypertensive agent, 63/74, 77
antihypertensive treatment, 63/83
arterial consequence, 63/82
arterial shear rate, 63/77
arterial shear stress, 63/77
arteriolar adaptation, 63/74
arteriolar consequence, 63/82
arteriolar dilatation, 63/72
arteriolar fibrinohyalinoid degeneration, 63/75
arteriolar hypertrophy, 63/74
arteriolar microaneurysm rupture, 63/74
arteriolar wall thickness, 63/72
arteriosclerosis, 63/77
atheromatous debris, 63/77
atherosclerotic stroke, 63/71
autoregulation failure, 63/73
basal ganglion, 63/74
beta adrenergic receptor blocking agent, 54/208, 65/434
blood-brain barrier, 28/391
blood energy consequence, 63/82
blood kinetic energy, 63/78
blood lipid, 54/207
blood pressure energy, 63/78
brain blood flow, 11/126, 63/80
brain embolism, 54/203, 63/82
brain hemorrhage, 11/587, 591, 601, 54/203, 63/72, 82
brain infarction, 54/203, 63/82
brain lacunar infarction, 54/203, 63/72, 75
calcium antagonist, 54/209
calorie restriction, 54/207
central sleep apnea, 63/461
cerebellum, 63/74
Charcot-Bouchard aneurysm, 63/75
classification, 54/204
complication, 63/77
congenital heart disease, 63/10
congestive heart failure, 63/82
cortical blindness, 63/72

diabetes mellitus, 54/209
diet, 54/206
dipeptidyl carboxypeptidase I inhibitor, 65/444
dissecting aorta aneurysm, 63/82
diuretic agent, 54/206, 65/438
eclampsia, 63/72
economic burden, 63/71
Ekbom other syndrome, 63/85
endothelium, 63/78
epidemiology, 63/71
exercise, 54/206
Fabry disease, 60/172
fibrinoid necrosis, 63/72, 75
flow pattern, 63/82
forced vasodilation, 63/72
guanethidine monosulfate, 54/209
heart rate, 63/82
hemodynamics, 63/79
high density lipoprotein cholesterol, 54/208
hydralazine, 54/209, 63/74
hyperlipidemia, 63/77
hypertensive arteriolar disease, 63/72
hypertensive crisis, 63/72
hypertensive encephalopathy, 63/72
hypertensive stroke, 63/74
individual tolerance, 63/74
intracerebral hematoma, 54/203
intracerebral hemorrhage site, 63/74
intracranial pressure, 55/207
intraplaque hemorrhage, 63/77
ischemic heart disease, 54/209
kinetic energy, 63/82
lacunar stroke, 63/82
laminar blood flow, 63/77
large vessel lesion, 63/71
leukoaraiosis, 63/76
lipoprotein, 63/77
minoxidil, 54/209
neurogenic, *see* Neurogenic arterial hypertension
neurologic complication, 63/71
nifedipine, 63/74
normal pressure hydrocephalus, 63/231
obesity, 54/206
obstructive sleep apnea, 63/456
occipital cortex, 63/72
papilledema, 63/72
pathogenesis, 63/79
platelet thrombus, 63/77
pons, 63/74
pregnancy, 63/73
pressure energy, 63/82
prolonged transient ischemic attack, 53/259
race, 63/71
renal insufficiency, 63/71
shear rate, 63/82
small vessel lesion, 63/71
small vessel stroke, 63/71
smoking, 54/207
smooth muscle cell, 63/77
snoring, 63/451
sodium restriction, 54/207
stress reduction, 54/208
stroke type, 63/71
subarachnoid hemorrhage, 12/128
systemic brain infarction, 11/432
thalamus, 63/74
thiazide, 54/209
thrombosis, 11/130
transient ischemic attack, 53/259
treatment, 54/206
vascular centrencephalon, 63/74
vascular ectasia, 63/84
vasodilator agent, 54/209
vasospasm, 63/72
vessel geometry, 63/82
weight loss, 54/206
Arterial hypotension
*see also* Vascular hypotension
brain infarction, 63/74
clomethiazole intoxication, 65/330
hereditary sensory and autonomic neuropathy type III, 21/109, 60/7, 27
percutaneous transluminal valvuloplasty, 63/177
systemic brain infarction, 11/433
tetrodotoxin intoxication, 65/155
transient ischemic attack, 12/9
Arterial malformation
congenital, *see* Congenital arterial malformation
vertebral, *see* Vertebral arterial malformation
Arterial pressure
head injury, 57/227, 230
headache, 5/5, 48/278
vascular hypertension, 5/5
Arterial spasm
cervical, *see* Cervical arterial spasm
ergot intoxication, 65/69
migraine, 48/27
subarachnoid hemorrhage, 55/5
Arterial supply
basal ganglion, 49/468
cochlea, 55/130
labyrinth, 55/129
medulla oblongata, 2/226-228
mesencephalon, 2/275
multiple cranial neuropathy, 51/569
spinal cord, 26/6, 64, 55/95

visual pathway, 55/116
Arterial system
brain stem, 11/24-43
cerebellar, *see* Cerebellar arterial system
intramedullary, *see* Intramedullary arterial system
vertebrobasilar, *see* Vertebrobasilar arterial system
Arterial vascular injury
brain stab wound, 23/485, 499
history, 26/63
laser injury, 23/666
penetrating head injury, 57/305
spinal angiography, 19/232
spinal cord injury, 26/63-79, 61/527
Arteries
accessory meningeal, *see* Accessory meningeal artery
accessory middle cerebral, *see* Accessory middle cerebral artery
Adamkiewicz, *see* Adamkiewicz artery
anastomotic, *see* Anastomotic artery
angular, *see* Angular artery
anterior cerebral, *see* Anterior cerebral artery
anterior choroidal, *see* Anterior choroidal artery
anterior communicating, *see* Anterior communicating artery
anterior ethmoidal, *see* Anterior ethmoidal artery
anterior inferior cerebellar, *see* Anterior inferior cerebellar artery
anterior middle cerebral, *see* Anterior middle cerebral artery
anterior spinal, *see* Anterior spinal artery
ascending pharyngeal, *see* Ascending pharyngeal artery
basal ganglion, *see* Basal ganglion artery
basilar, *see* Basilar artery
Bernasconi-Cassinari, *see* Bernasconi-Cassinari artery
calcarine, *see* Calcarine artery
callosomarginal, *see* Callosomarginal artery
capsular, *see* Capsular artery
caroticotympanic, *see* Caroticotympanic artery
carotid, *see* Carotid artery
central retinal, *see* Central retinal artery
cerebellar, *see* Cerebellar artery
cerebral, *see* Brain artery
cervical intersegmental, *see* Cervical intersegmental artery
ciliary, *see* Ciliary artery
circumferential brain stem, *see* Circumferential brain stem artery
common carotid, *see* Common carotid artery
coronary, *see* Coronary artery
Desprogres-Gotteron, *see* Desprogres-Gotteron artery
diencephalic, *see* Diencephalic artery
dorsal callosal, *see* Dorsal callosal artery
dorsal meningeal, *see* Dorsal meningeal artery
Duret, *see* Duret artery
external carotid, *see* External carotid artery
external choroidal, *see* External choroidal artery
external maxillary, *see* External maxillary artery
femoral, *see* Femoral artery
floccular, *see* Floccular artery
frontobasilar, *see* Frontobasilar artery
frontopolar, *see* Frontopolar artery
Heubner, *see* Heubner artery
hypoglossal, *see* Hypoglossal artery
inferior cavernous sinus, *see* Inferior cavernous sinus artery
inferior frontal, *see* Inferior frontal artery
inferior hypophyseal, *see* Inferior hypophyseal artery
inferior paravermal, *see* Inferior paravermal artery
infraorbital, *see* Infraorbital artery
innominate, *see* Innominate artery
internal auditory, *see* Internal auditory artery
internal carotid, *see* Internal carotid artery
internal maxillary, *see* Internal maxillary artery
lacrimal, *see* Lacrimal artery
lateral occipital, *see* Lateral occipital artery
Lazorthes, *see* Lazorthes artery
lenticulostriate, *see* Lenticulostriate artery
long circumflex, *see* Long circumflex artery
marginal, *see* Marginal artery
maxillary, *see* Maxillary artery
megadolichobasilar, *see* Megadolichobasilar artery
megadolichocarotid, *see* Megadolichocarotid artery
meningeal, *see* Meningeal artery
meningohypophyseal, *see* Meningohypophyseal artery
mesencephalic, *see* Mesencephalic artery
mesenteric, *see* Mesenteric artery
middle cerebral, *see* Middle cerebral artery
middle meningeal, *see* Middle meningeal artery
middle peduncular, *see* Middle peduncular artery
middle posterior choroidal, *see* Middle posterior choroidal artery
middle temporal, *see* Middle temporal artery
occipitoparietal, *see* Occipitoparietal artery
ophthalmic, *see* Ophthalmic artery
otic, *see* Otic artery
paramedian pontine, *see* Paramedian pontine artery
parieto-occipital, *see* Parieto-occipital artery

pericallosal, *see* Pericallosal artery
perichiasmatic, *see* Perichiasmatic artery
perimedullary, *see* Perimedullary artery
persistent hyaloid, *see* Persistent hyaloid artery
posterior cerebral, *see* Posterior cerebral artery
posterior ethmoidal, *see* Posterior ethmoidal artery
posterior frontal, *see* Posterior frontal artery
posterior inferior cerebellar, *see* Posterior inferior cerebellar artery
posterior spinal, *see* Posterior spinal artery
posterior temporal, *see* Posterior temporal artery
prerolandic, *see* Prerolandic artery
primitive hypoglossal, *see* Primitive hypoglossal artery
primitive internal carotid, *see* Primitive internal carotid artery
primitive otic, *see* Primitive otic artery
primitive trigeminal, *see* Primitive trigeminal artery
quadrigeminal, *see* Quadrigeminal artery
radicular, *see* Radicular artery
radiculomedullary, *see* Radiculomedullary artery
radiculopial, *see* Radiculopial artery
retroauricular, *see* Retroauricular artery
Rolando, *see* Rolando artery
spinal, *see* Spinal artery
spinal nerve root, *see* Spinal nerve root artery
spinal sulcal, *see* Spinal sulcal artery
stapedial, *see* Stapedial artery
stylomastoid, *see* Stylomastoid artery
subclavian, *see* Subclavian artery
suboccipital intersegmental, *see* Suboccipital intersegmental artery
superficial temporal, *see* Superficial temporal artery
superior cerebellar, *see* Superior cerebellar artery
supratrochlear, *see* Supratrochlear artery
temporal, *see* Temporal artery
temporal occipital, *see* Temporal occipital artery
tentorial, *see* Tentorial artery
thalamic, *see* Thalamic artery
thalamogeniculate, *see* Thalamogeniculate artery
thalamolenticular, *see* Thalamolenticular artery
thalamoperforating, *see* Thalamoperforating artery
thalamostriate, *see* Thalamostriate artery
thalamotuberal, *see* Thalamotuberal artery
trigeminal, *see* Trigeminal artery
tuberal, *see* Tuberal artery
tuberothalamic, *see* Tuberothalamic artery
vertebral, *see* Vertebral artery
Vidian, *see* Vidian artery
Wallenberg, *see* Wallenberg artery
Arteriocirrhosis
Takayasu disease, 12/410
Arteriography
alternating infantile hemiplegia, 12/348
brain, *see* Brain angiography
brain abscess, 33/123
lateral medullary syndrome, 2/235
pericarotid syndrome, 48/335
pheochromocytoma, 39/500, 42/765
temporal, *see* Temporal arteriography
Arteriolar fibrinohyalinoid degeneration
amyloidosis, 11/581
angiopathic polyneuropathy, 51/445
Antoni type B tumor, 14/23
arterial hypertension, 63/75
Binswanger disease, 11/135, 157, 54/227
blood-CSF barrier, 19/132
brain hemorrhage, 54/203, 299
brain lacunar infarction, 53/364, 54/236
early stage, 11/129
heterogenous cause, 11/135-137
hypertension, 11/601
hypertensive encephalopathy, 11/560-562, 47/536, 54/213
malignant hypertension, 12/49
middle cerebral artery syndrome, 53/364
prehemorrhagic softening, 12/49
segmental, *see* Segmental fibrinoid necrosis
thrombus, 11/627
Arteriolar hyaline degeneration, *see* Arteriolar fibrinohyalinoid degeneration
Arteriolosclerosis
familial cortical, *see* Familial cortical arteriolosclerosis
intracerebral hematoma, 11/671
Arteriosclerosis
aneurysm, 11/131
anterior horn syndrome, 2/210
antihypertensive agent, 63/82
aorta aneurysm, 39/250, 63/48
aorta disease, 63/37
arterial hypertension, 63/77
autonomic nervous system, 74/120
basilar artery, *see* Basilar artery atherosclerosis
brain, *see* Brain arteriosclerosis
brain biopsy, 10/685
brain embolism, 11/389, 53/200, 304
brain hemorrhage, 11/603
brain infarction, 15/304, 53/159, 200
brain ischemia, 53/159
brain thromboangiitis obliterans, 12/392, 39/203
carotid artery, 11/327-365
carotid cavernous fistula, 24/407
carotid system syndrome, 53/304

cerebrovascular disease, 53/200
epilepsy, 15/304
fibromuscular dysplasia, 11/381, 55/288
hereditary hyperlipidemia, 11/462
histology, 11/129
hypertension, 75/479
internal carotid, *see* Internal carotid arteriosclerosis
intracerebral hematoma, 11/671
intracranial hemorrhage, 53/200
ischemic myelopathy, 63/38
ischemic neuropathy, 8/161
migraine, 48/156
multi-infarct, *see* Multi-infarct arteriosclerosis
multinodular stenosing amyloidosis, 11/613
myelomalacia, 22/20, 55/98, 63/38
myelopathy, 12/501, 532, 22/20
nerve ischemia, 12/654
neuropathy, 8/161
optic atrophy, 13/58
paraplegia, 1/208
Parkinson dementia, 46/376
Parkinson disease, 6/174, 230
plaque, 11/331, 437
progeria, 43/465
progressive spinal muscular atrophy, 22/19
radiation injury, 67/330
Refsum disease, 21/246
secondary pigmentary retinal degeneration, 13/218
spinal, 22/19
spinal apoplexy, 12/532
spinal artery injury, 26/66
spinal cord, *see* Spinal cord atherosclerosis
subclavian artery, *see* Subclavian artery atherosclerosis
Takayasu disease, 12/414
temporal arteritis, 48/322
transient ischemic attack, 53/200
traumatic psychosis, 24/529
vasa nervorum disease, 12/654
vascular dementia, 46/357
vertebral artery, *see* Vertebral artery arteriosclerosis
vertebrobasilar system syndrome, 53/371, 375
Werner syndrome, 43/489
Arteriosclerotic encephalopathy
subcortical, *see* Binswanger disease
Arteriosclerotic myelopathy
progressive muscular atrophy, 59/19
Arteriosclerotic papillitis, *see* Anterior ischemic optic neuropathy
Arteriospasm
brain, *see* Brain vasospasm
Arteriovenous aneurysm
brain, *see* Brain arteriovenous malformation
Arteriovenous angioma
angiomatous malformation, 14/52
Cobb syndrome, 14/440
subarachnoid hemorrhage, 12/106
Arteriovenous fistula
brain arteriovenous malformation, 54/364, 370
brain embolism, 55/450
cauda equina intermittent claudication, 20/796
congenital, *see* Congenital arteriovenous fistula
headache, 5/145, 148, 31/182
pulmonary, *see* Pulmonary arteriovenous fistula
vertebral artery injury, 53/377
Arteriovenous malformation
*see also* Brain hemorrhage
alexia, 45/438
brain, *see* Brain arteriovenous malformation
brain aneurysm, 42/723
brain scanning, 11/244
brain varix, 12/234
cerebellar hemorrhage, 55/89
cerebrovascular disease, 53/34
dura mater, 12/229, 236
epilepsy, 72/213
epileptic seizure, 31/181
headache, 48/273, 277
hemiballismus, 49/373
Klippel-Trénaunay syndrome, 43/25
lymphedema, 42/726
mental deficiency, 31/188
myelomalacia, 55/103
optic chiasm compression, 68/75
pulmonary hypertension, 42/726
spinal cord, *see* Spinal cord arteriovenous malformation
spinal cord tumor, 12/248
subarachnoid hemorrhage, 55/2
trigeminal neuralgia, 12/248
type I, *see* Arteriovenous malformation type I
type II, *see* Arteriovenous malformation type II
type III, *see* Arteriovenous malformation type III
type IV, *see* Arteriovenous malformation type IV
vascular dementia, 46/360
Arteriovenous malformation type I
spinal vascular tumor, 68/570
Arteriovenous malformation type II
spinal vascular tumor, 68/572
Arteriovenous malformation type III
spinal vascular tumor, 68/573
Arteriovenous malformation type IV
spinal vascular tumor, 68/573

Arteritis
acute infantile hemiplegia, 12/345
allergic granulomatous, *see* Allergic granulomatous angiitis
allergic granulomatous angiitis, 11/453
amaurosis fugax, 55/108
aorta, 48/316
bacterial, 11/132
bacterial endocarditis, 52/294
bacterial meningitis, 52/14
brain, *see* Brain arteritis
central retinal artery, 48/318
cerebrovascular disease, 53/28
ciliary artery, 48/319
coronary artery, 48/316
cranial, *see* Temporal arteritis
cryptococcal meningitis, 52/431
epilepsy, 72/216
giant cell, *see* Giant cell arteritis
granulomatous, *see* Granulomatous arteritis
headache, 5/119, 48/278
hepatic, *see* Hepatic arteritis
Heubner, *see* Heubner arteritis
HLA antigen, 48/310
hypertensive fibrinoid, *see* Hypertensive fibrinoid arteritis
intracerebral hematoma, 11/672
juvenile hypereosinophilic temporal, *see* Juvenile hypereosinophilic temporal arteritis
Kawasaki syndrome, 52/266
lymphocyte, 48/310
myelomalacia, 63/52
neurosyphilis, 52/278
Nissl-Alzheimer, *see* Nissl-Alzheimer arteritis
optic nerve, 48/319
orbital, *see* Orbital arteritis
primary granulomatous cerebral, *see* Primary granulomatous cerebral arteritis
recurrent nonhereditary multiple cranial neuropathy, 51/571
spinal cord blood flow, 12/498
subclavian artery, 48/316
Takayasu, *see* Takayasu disease
temporal, *see* Temporal arteritis
tuberculous meningitis, 55/429
venous sinus thrombosis, 54/411
vertebral, *see* Vertebral arteritis
Artery beading
brain vasospasm, 11/515, 517
Artery bifurcation
carotid, *see* Carotid artery bifurcation
Artery inflammation
renal, *see* Renal artery inflammation
Artery injury
spinal, *see* Spinal artery injury
vertebral, *see* Vertebral artery injury
Artery occlusion
acute infantile hemiplegia, 12/342
anterior choroidal, *see* Anterior choroidal artery occlusion
Behçet syndrome, 56/595
brain, *see* Brain artery occlusion
brain artery occlusion, 12/452
brain thromboangiitis obliterans, 55/308
carotid, *see* Carotid artery occlusion
central retinal, *see* Central retinal artery occlusion
cerebrovascular disease, 12/452
chronic superior mesenteric, *see* Chronic superior mesenteric artery occlusion
mesenteric, *see* Mesenteric artery occlusion
middle cerebral, *see* Middle cerebral artery occlusion
multiple intracranial, *see* Multiple intracranial arterial occlusion
myélomalacia en crayon, 12/498
pulmonary, *see* Pulmonary artery stenosis
retinal, *see* Retinal artery occlusion
subclavian, *see* Subclavian artery occlusion
superior cerebellar artery syndrome, 12/20
vertebral, *see* Vertebral artery occlusion
Artery stenosis
carotid, *see* Carotid artery stenosis
innominate, *see* Innominate artery stenosis
transient ischemic attack, 53/265
Artery syndrome
anterior spinal, *see* Anterior spinal artery syndrome
basilar, *see* Basilar artery syndrome
choroidal, *see* Choroidal artery syndrome
oculonuclear, *see* Oculonuclear artery syndrome
posterior spinal, *see* Posterior spinal artery syndrome
superior cerebellar, *see* Superior cerebellar artery syndrome
Arthralgia
adrenal insufficiency myopathy, 62/536
allergic granulomatous angiitis, 51/453, 63/383
ciguatoxin intoxication, 65/162
Colorado tick fever, 56/141
corticosteroid withdrawal myopathy, 62/536
cryoglobulinemia, 63/403
cryoglobulinemic polyneuropathy, 51/434
cubomedusae intoxication, 37/43
dermatomyositis, 62/374, 71/130
eosinophilia myalgia syndrome, 63/375, 64/252
hyperventilation syndrome, 38/319, 63/435

ichthyohepatoxic fish intoxication, 37/87
inclusion body myositis, 62/374
Kawasaki syndrome, 51/453
kuru, 56/556
lead intoxication, 64/435
Lyme disease, 51/203
lymphocytic choriomeningitis virus, 56/360
meningococcal meningitis, 33/23, 52/28
neuroborreliosis, 51/203
periodic, *see* Periodic arthralgia
polymyositis, 62/374, 71/130
rubella, 56/406
Takayasu disease, 55/337
Arthresthesia
terminology, 1/95
Arthritis
Bamberger-Marie syndrome, 1/488
Behçet syndrome, 56/594
classification, 71/15
familial spastic paraplegia, 22/425
indometacin responsive headache, 48/174
juvenile rheumatoid, *see* Juvenile rheumatoid arthritis
Kawasaki syndrome, 52/265
Lyme disease, 51/205, 52/260
meningococcal meningitis, 33/24, 52/26
multiple myeloma, 20/11
mummy, 74/187
neuroborreliosis, 51/205
rheumatoid, *see* Rheumatoid arthritis
rubella, 56/406
spinal rheumatoid, *see* Spinal rheumatoid arthritis
Arthritis myelitica
history, 19/399
terminology, 19/339
Arthrodento-osteodysplasia
basilar impression, 43/332
frontal sinus, 43/332
headache, 43/332
osteolysis, 43/331
osteoporosis, 43/331
pain, 43/332
Arthrogryposis
congenital acetylcholine receptor deficiency/synaptic cleft syndrome, 62/438
contracture, 40/337
creatine kinase, 43/78
Edwards syndrome, 50/561
hydranencephaly, 50/340, 345
joint fixation, 40/337
Kuskokwim, *see* Kuskokwim arthrogryposis
Möbius syndrome, 50/215
muscle fibrosis, 62/173
myopathy, 40/337
osteolysis, 43/85
Patau syndrome, 14/121
scoliosis, 50/417
weakness, 43/78
Arthrogryposis multiplex congenita, 32/511-517
amniotic band syndrome, 42/73
annual incidence, 43/78
associated malformation, 32/512
cerebellar vermis agenesis, 14/408
clinical features, 32/512, 41/29, 32, 46, 423
collagen synthesis, 32/512
congenital ophthalmoplegia, 43/134
corpus callosum agenesis, 14/408
counseling, 32/515
course, 32/513
Coxsackie virus, 56/200
differential diagnosis, 32/512
distal, 59/73
epidemiology, 30/165
fetal movement, 42/73
genetics, 30/165, 32/512
glycogen storage disease, 43/78
hypotonia, 41/32
infantile neuroaxonal dystrophy, 49/402
infantile spinal muscular atrophy, 59/71-74
intrauterine growth retardation, 46/89
jaundice, 59/73
joint deformity, 1/483
laboratory examination, 41/32
lower limb deformity, 32/517
main features, 21/24, 27
mental deficiency, 46/89
muscle contraction, 1/483
muscle fiber size, 43/78
myopathy, 32/512, 515, 40/337, 43/77
myosclerosis, 43/133
neurogenic, 59/369
neuropathy, 32/512, 515, 42/73-75
oligohydramnios, 42/74
orthopedic treatment, 32/516
pathogenesis, 32/511
pathology, 32/513-515, 41/33
Pena-Shokeir syndrome type I, 43/437, 59/369
polyneuropathy, 42/336
spinal deformity, 32/517
spinal muscular atrophy, 59/373
synonym, 32/511
talipes, 43/78
treatment, 32/516
tuberous sclerosis, 14/496
D-tubocurarine, 30/123
upper limb deformity, 32/516

X-linked recessive, 59/73
Arthrogrypotic like hand malformation
dermatoglyphics, 43/333
sensorineural deafness, 43/333
Arthro-ophthalmopathy
hereditary progressive, *see* Hereditary progressive arthro-ophthalmopathy
progressive, *see* Stickler syndrome
Arthropathy
central motor disorder, 1/183
hemiplegia, 1/183
human T-lymphotropic virus type I associated myelopathy, 59/450
neurogenic, *see* Neurogenic osteoarthropathy
neuropathic, *see* Neuropathic arthropathy
psoriatic, *see* Psoriatic arthropathy
Arthroplasty
hip, *see* Hip arthroplasty
shoulder, *see* Shoulder arthroplasty
Arthropod-borne virus
*see also* Arbovirus, Arenavirus, Bunyamwera virus *and* Togavirus
type, 34/5
Arthropod envenomation
acetylcholine release, 65/195
Agelenopsis aperta, 65/200
allergy, 65/198
Androctonus australis, 65/197
Androctonus australis hector, 65/196
Androctonus mauretanicus mauretanicus, 65/198
ant venom, 65/199
apamin, 65/199
apamin I 125, 65/198
Apis dorsata, 65/198
Apis mellifera, 65/198
Araneomorphae, 65/193
ascending paralysis, 65/195
ataxia, 65/195
Atrax, 65/195
Atrax robustus, 65/195
autonomic nervous system dysfunction, 65/196
bradykinin, 65/201
brain edema, 65/197
Buthus europeus, 65/197
Buthus martensii, 65/197
Buthus occitanus, 65/196
Buthus tamulus, 65/198
calcium channel, 65/194
calcium current, 65/200
Centruroides infamatous infamatous, 65/196
Centruroides noxius, 65/197
charybdotoxin, 65/198
convulsion, 65/196
dideazaphilanthotoxin 12, 65/200
dysphagia, 65/196
epilepsy, 65/197
facial paralysis, 65/195
Formicidae, 65/199
glutamatergic transmission, 65/200
glutamic acid receptor, 65/200
Hadronynche, 65/195
Hadronynche versutus, 65/195
hemiplegia, 65/197
hippocampus, 65/200
hyaluronidase, 65/198
hymenoptera venom, 65/193, 198
iberiotoxin, 65/198
Latrodectus mactans tredecimguttatus, 65/195
α-latroinsectotoxin, 65/196
α-latrotoxin, 65/196
Leiurus quinquestriatus, 65/197
Leiurus quinquestriatus hebraeus, 65/197
lethal dose, 65/198
lethality, 65/196
Loxosceles, 65/196
LqhαIT, 65/197
Megascolia flavifrons, 65/202
mortality rate, 65/197
motor end plate potential, 65/195
Mygalomorphae, 65/193
neurotoxic kinin, 65/201
noxiustoxin, 65/198
nystagmus, 65/196
paralysis, 65/194
Paraponera clavata, 65/199
philanthotoxin, 65/199
Philanthus triangulum, 65/199
Phoneutria nigriventer, 65/196
phospholipase A2, 65/198
polyamine blocking agent, 65/200
polyradiculoneuropathy, 65/198
poneratoxin, 65/199
potassium-calcium channel, 65/198
potassium channel, 65/194, 197
pseudoargiopin, 65/195
Purkinje cell, 65/200
robustotoxin, 65/195
scorpion venom, 65/193, 196
serrulatotoxin, 65/198
sodium channel, 65/194
sodium inactivation, 65/197
spider order, 65/193
spider venom, 65/193, 200
tarantula, 65/195
Theraphosidae, 65/195
tick paralysis, 65/194

tick venom, 65/193
Tityus serrulatus, 65/197
tremor, 65/199
venom, 65/198
Vespa affinis, 65/198
Vespa anulis, 65/198
Vespa tropica, 65/198
wasp kinin, 65/201
Arthropod venom
neurotoxin, 65/193
robustotoxin, 65/195
toxic encephalopathy, 65/193
versutoxin, 65/195
Arthropod venom origin
Agelenopsis aperta, 65/194
Apis mellifera, 65/194
Atrax robustus, 65/194
Buthus martensii, 65/194
Buthus tamulus, 65/194
Centruroides noxius, 65/194
Hadronynche versutus, 65/194
Ixodus holocyclus, 65/194
Latrodectus mactans tredecimguttatus, 65/194
Leiurus quinquestriatus hebraeus, 65/194
Megascolia specie, 65/194
Paraponera clavata, 65/194
Philanthus triangulum, 65/194
Phoneutria nigriventer, 65/194
Polistes, 65/194
Tityus serrulatus, 65/194
Vespa, 65/194
Vespula, 65/194
Arthropoda intoxication
horseshoe crab, 37/58
paralysis, 37/56
xiphosura, 37/56
Arthroscopy
nerve damage, 70/47
Arthrosis
arctic fox intoxication, 37/94
cervical, *see* Cervical arthrosis
pseudo, *see* Pseudoarthrosis
temporomandibular, *see* Temporomandibular arthrosis
Articular defect
developmental, 43/219
Articular process
cervical compression fracture, 25/258, 361
cervical tilting fracture, 25/365
vertebra, 25/258, 361, 365
Artificial brain tumor
intracranial hypertension, 17/1
Artificial insemination
sexual function, 26/459
Artificial respiration
cephalopoda intoxication, 37/62
conidae intoxication, 37/62
creatine kinase, 63/420
critical illness polyneuropathy, 63/420
cuirass respirator, 26/345
dinoflagellates intoxication, 37/35
hapalochlaena lunulata intoxication, 37/62
hapalochlaena maculosa intoxication, 37/62
myopathy, 63/420
octopoda intoxication, 37/62
puffer fish intoxication, 37/81
rhabdomyolysis, 63/420
spinal cord injury, 26/343
Artificial survival, *see* Brain death
Artificial ventilator
critical illness polyneuropathy, 51/575
Arylamidase
Duchenne muscular dystrophy, 40/380
Arylhexylamine
epilepsy, 72/234
Arylsulfatase A, *see* Cerebroside sulfatase
Arylsulfatase A deficiency, *see* Metachromatic leukodystrophy
Arylsulfatase B, *see* *N*-Acetylgalactosamine-4-sulfatase
Arylsulfatase B deficiency
glycosaminoglycanosis type IV, 42/458
glycosaminoglycanosis type VI, 27/152, 42/458
Arylsulfatase C
chemistry, 10/48
Arylsulfatase deficiency
corpus callosum agenesis, 50/151
Asari intoxication, *see* Venerupis semidecussata intoxication
Ascariasis
acrodynia, 64/374
clinical aspect, 35/358
delirium, 35/359
diagnosis, 35/360
eosinophil count, 21/569
epidemiology, 35/358
geographic distribution, 35/358
helminthiasis, 35/211
hereditary periodic ataxia, 21/569
lethargy, 35/359
meningeal sign, 35/359
opisthotonos, 35/359
parasite, 35/357
pathology, 35/360
prophylaxis, 35/361
restlessness, 35/359

stupor, 35/359
symptomatology, 35/358
treatment, 35/360
Ascending anterolateral fiber
pain, 45/234-236
Ascending cerebral vein
phlebography, 11/105
radioanatomy, 11/105
topography, 11/105
Ascending myelitis
herpes simplex virus, 56/478
immunocompromise, 56/477
immunocompromised host, 56/477
postinfectious complication, 47/481-489
Ascending neuritis
migrant neuritis, 8/165
terminology, 8/165
Ascending neuropathy
celiac disease, 28/229
Ascending occipital cerebral vein
radioanatomy, 11/108
topography, 11/108
vertebral phlebography, 11/108
Ascending paralysis
arthropod envenomation, 65/195
Ascending pharyngeal artery
anastomosis, 12/310
Ascending pharyngeal system
multiple cranial neuropathy, 51/569
Ascending polyneuropathy
rapid, *see* Rapid ascending polyneuropathy
Ascending reticular activating system
brain stem death, 57/459
coma, 3/64
consciousness, 3/53
wakefulness, 3/171
Ascending sensorimotor neuropathy
subacute myelo-optic neuropathy, 51/296
Ascending symmetrical neuropathy
amyloid polyneuropathy, 51/416
Ascending terminal
dorsal spinal nerve root ganglion, 49/395
Ascending transverse myelopathy, *see* Foix-Alajouanine disease
Ascites
glycosaminoglycanosis type VII, 66/311
liver cirrhosis, 27/351
POEMS syndrome, 47/619, 63/394
Ascorbic acid, *see* Vitamin C
Aseptic cavernous sinus thrombosis
Tolosa-Hunt syndrome, 48/303
Aseptic meningeal reaction, *see* Meningitis serosa
Aseptic meningitis, *see* Meningitis serosa
Aseptic meningoencephalomyelitis, *see* Meningitis serosa
Asexual ateleiotic dwarf, *see* Pituitary dwarfism type II
Ash leaf spot
tuberous sclerosis, 50/371
Ash-Upmark disease, *see* Takayasu disease
Asialia
Melkersson-Rosenthal syndrome, 8/221
parotid gland, 8/221
Asialoglycoprotein receptor
gene therapy, 66/114
Asimultagnosia
bilateral occipital lobe infarction, 53/413
posterior cerebral artery syndrome, 53/413
Asomatognosia
agnosia, 4/14, 37
apraxia, 4/61
cerebral dominance, 4/297
hemi, *see* Hemiasomatognosia
parietal lobe epilepsy, 73/101
Asparaginase
adverse effect, 69/494
antineoplastic agent, 65/529
brain hemorrhage, 69/426
brain infarction, 69/418
cerebrovascular disease, 69/418
chemotherapy, 39/117
empty delta sign, 69/495
encephalopathy, 63/359, 65/529
epilepsy, 65/529
neurologic toxicity, 39/117
neurotoxicity, 63/359
neurotoxin, 39/95, 113, 117, 119, 46/394, 63/345, 359, 65/529, 538
sagittal sinus thrombosis, 69/495
sensorimotor polyneuropathy, 63/359
venous sinus thrombosis, 63/345
Asparaginase intoxication
brain infarction, 65/538
encephalopathy, 65/538
Asparaginase linked protein
lysosomal degradation, 66/62
Aspartate aminotransferase
hypothyroid myopathy, 62/533
Reye syndrome, 9/549, 56/151
thyrotoxic myopathy, 62/529
Aspartic acid
amino acid neurotransmitter, 29/512-514
biochemical consideration, 29/512
brain ischemia, 63/211
clinical consideration, 29/513
epilepsy, 72/84

GM2 gangliosidosis, 10/296
Huntington chorea, 49/261
hyponatremia, 63/547
neuron death, 63/211
physiologic consideration, 29/512
respiratory encephalopathy, 63/416
spastic paraplegia, 61/367
spinal cord injury, 61/367
transverse spinal cord lesion, 61/367
white matter, 9/16
Aspartoacylase
*N*-acetylaspartic acid, 66/663
Aspartylglucosaminidase, *see* *N*4-(β-*N*-Acetylglucosaminyl)asparaginase
*N*-Aspartyl-β-glucosaminidase
enzyme deficiency polyneuropathy, 51/370
Aspartylglucosaminuria
biochemical diagnosis, 66/336
biochemical finding, 29/226-228
case, 66/330
clinical features, 29/225, 66/336
diagnosis, 29/228
differential diagnosis, 66/336
enzyme defect, 29/228, 66/330
enzyme deficiency polyneuropathy, 51/370
genetics, 29/228
infection, 29/225
mental deficiency, 29/225
speech disorder in children, 29/225
survey, 66/329
treatment, 29/228
Aspartylglycosaminuria
amidohydrolase, 42/528
aspartylglycosylaminase, 42/528
behavior disorder, 27/159, 42/527
birth incidence, 42/527
Finland, 66/62
glycoprotein degradation, 66/62
heart disease, 42/527
heterozygote detection, 42/528
mental deficiency, 27/159, 42/527
skeletal deformity, 27/159, 42/527
speech disorder in children, 27/159
vacuolated lymphocyte, 42/527
Aspartylglycosylaminase
aspartylglycosaminuria, 42/528
Aspergilloma
brain, *see* Brain aspergilloma
Aspergillosis, 35/395-399, 52/377-382
*see also* Aspergillus flavus, Aspergillus fumigatus, Aspergillus glaucus *and* Aspergillus niger
amphotericin B, 35/399, 52/381
biopsy, 52/381
blood culture, 52/380
brain abscess, 35/397, 52/148
brain aspergilloma, 52/380
brain granuloma, 16/231, 35/397
brain infarction, 35/398, 55/435
carotid cavernous thrombosis, 35/397, 52/380
cavernous sinus thrombosis, 52/173
cerebrovascular disease, 55/432
chronic meningitis, 56/644
computer assisted tomography, 52/380
consciousness, 52/378
convulsive seizure, 35/397
corticosteroid, 52/377
CSF, 35/398, 52/380
drug abuse, 35/396, 55/433
drug addict, 52/379
endocarditis, 52/379
eosinophilia, 63/371
flucytosine, 52/381
headache, 35/397, 52/378
hemorrhagic encephalitis, 52/379
histopathology, 52/381
host defence, 52/377
immunocompromised host, 52/378
infectious endocarditis, 52/302
ketoconazole, 52/381
lethargy, 52/378
leukemia, 35/396, 52/378, 63/345
leukopenia, 52/378
meningitis, 35/397, 52/378, 55/432
miconazole, 52/381
mycosis, 35/374, 395
mycotic aneurysm, 52/379
necrotizing encephalitis, 52/379
neurologic symptom, 52/378
neuropathology, 52/379
neurosurgery, 35/399, 52/382
nonrecurrent nonhereditary multiple cranial neuropathy, 51/571
nuclear magnetic resonance, 52/380
optic neuropathy, 52/380
orbital apex syndrome, 52/380
orbital infection, 52/379
predisposing factor, 52/377
prognosis, 35/399, 52/381
pulmonary infection, 35/396, 52/377
renal insufficiency, 27/343
seizure, 52/378
serum antigen detection, 52/381
sinusitis, 52/379
spinal cord compression, 35/398, 52/380
spinal epidural empyema, 52/187, 380
spinal subdural empyema, 52/189

sputum culture, 35/398, 52/380
subarachnoid hemorrhage, 35/397, 52/379, 55/435
systemic disease, 52/378
systemic infection, 52/377
toxin, 35/397, 52/377
treatment, 35/399, 52/381
Aspergillus
acquired immune deficiency syndrome, 71/298
CNS infection, 69/435
Aspergillus flavus
*see also* Aspergillosis
neurologic pathogen, 52/377
Aspergillus fumigatus
*see also* Aspergillosis
acquired immune deficiency syndrome, 56/515
brain infarction, 63/532
encephalitis, 63/532
epilepsy, 63/532
immunocompromised host, 65/549
meningitis, 63/532
neurologic pathogen, 52/377
renal transplantation, 63/531
Aspergillus glaucus
*see also* Aspergillosis
neurologic pathogen, 52/377
Aspergillus niger
*see also* Aspergillosis
neurologic pathogen, 52/377
Asphyxia
athetosis, 6/446
epilepsy, 72/344
neonatal, *see* Newborn asphyxia
perinatal, *see* Perinatal asphyxia
tetanus, 52/230
X-linked myotubular myopathy, 62/156
Asphyxiation, *see* Suffocation
Aspiny neuron
Huntington chorea, 49/256, 260
Aspiny neuron type I
neostriatum, 49/3
Aspiny neuron type II
neostriatum, 49/3
Aspiny neuron type III
neostriatum, 49/3
Aspiration pneumonia
dermatomyositis, 62/374
inclusion body myositis, 62/374
polymyositis, 62/374
Aspirin, *see* Acetylsalicylic acid
Assisted ventilation
Guillain-Barré syndrome, 51/256
Associated movement, *see* Synkinesis
Associated polyneuropathy
recurrent nonhereditary multiple cranial neuropathy, 51/570
Association cortex
apraxia, 45/431
Associative agnosia
apperceptive agnosia, 4/25
classification, 4/15
description, 45/341-343
semantic classification, 45/341-343
Associative areas
CNS, 3/13
visual, *see* Visual association area
Astaragazza
history, 6/300
Astasia
bismuth intoxication, 64/337
cerebellar ablation, 2/394-399
experimental animal, 2/394-399
giving way of legs, 1/347
opisthotonos, 2/394-399
Astasia abasia
*see also* Equilibrium, Förster syndrome *and* Posture
ataxia, 1/311, 314, 322, 325, 334
case report, 46/575
cerebellar atrophy, 1/345
eye closure, 1/311
Förster syndrome, 1/175
gait apraxia, 1/346
hypertonia, 1/346
hysterical, 1/345
incoordination, 1/345
labyrinthine, 1/345
Marchiafava-Bignami disease, 17/542, 28/324
movement visual control, 1/311
pseudobulbar paralysis, 1/197
syncope, 15/824
Von Romberg sign, 1/311, 345, 2/420
Astatic absence, *see* Drop attack
Astatic epilepsy
myoclonic, *see* Myoclonic astatic epilepsy
Astereognosia
agnosia, *see* Tactile agnosia
Astereopsis
neuropsychology, 45/16
occipitoparietal region, 45/16
Asterion
nonmidline cephalocele, 50/110
Asterixis
*see also* Tremor
aluminum intoxication, 63/525, 64/279
aluminum neurotoxicity, 63/525
antiepileptic agent, 65/499

barbiturate intoxication, 65/510
brain embolism, 53/211
brain infarction, 53/211
brain lacunar infarction, 54/247, 253
carbamazepine intoxication, 65/507
cerebrovascular disease, 53/211
Creutzfeldt-Jakob disease, 46/291
hepatic coma, 49/214
hypercalcemia, 63/561
intracranial hemorrhage, 53/211
lidocaine, 63/192
liver disease, 6/821
metrizamide, 61/166
mexiletine, 63/192
myoclonus, 49/622
OKT3, 63/536
Parkinson disease, 49/96
renal insufficiency, 27/324
respiratory encephalopathy, 63/414
Shapiro syndrome, 50/167
tocainide, 63/192
transient ischemic attack, 53/211
uremia, 63/504
uremic encephalopathy, 63/504
valproic acid intoxication, 65/504
Wilson disease, 6/274, 282, 284, 823, 49/225
Asteroidea intoxication
Acanthaster planci, 37/66
pain, 37/66
solaster papposus, 37/66
Asthenia
cerebellar, *see* Cerebellar asthenia
neocerebellar syndrome, 2/420
Asthenopia
cerebral, *see* Cerebral asthenopia
color, *see* Color asthenopia
cortical blindness, 45/51
Asthenopic alexia
alexia without agraphia, 4/124
Asthenopic scotoma
migraine, 48/162
Asthenosoma intoxication
spinal injury, 37/67
Asthma
allergic granulomatous angiitis, 11/134, 51/453, 63/383
granulomatous CNS vasculitis, 55/388
guanethidine intoxication, 37/436
muscular atrophy, 63/420
myopathy, 63/420
neuralgic amyotrophy, 51/174
poliomyelitis like syndrome, 63/420
progressive muscular atrophy, 59/22
rhizostomeae intoxication, 37/53
sleep, 74/553
Astroblastoma
astrocytoma grade II, 18/14
low grade glioma, 68/54
Astrocyte
Alzheimer type I, *see* Alzheimer type I astrocyte
Alzheimer type II, *see* Alzheimer type II astrocyte
astrocytoma, 16/10
Batten disease, 10/657
gliofibrillogenesis, 9/617, 625
globoid cell leukodystrophy, 10/76
lead, 64/437
lead intoxication, 64/437
1-methyl-4-phenyl-1,2,3,6-tetrahydropyridine, 65/382
1-methyl-4-phenyl-1,2,3,6-tetrahydropyridine intoxication, 65/382
nerve, 51/11
pernicious anemia, 9/617
plaque, 9/279-285
progressive multifocal leukoencephalopathy, 9/486, 39/524, 47/508, 56/514
protoplasmic type, 6/286-288
Ranvier node, 47/12, 51/16
subacute combined spinal cord degeneration, 9/614
viral infection, 56/66
Astrocyte death
cardiac arrest, 63/212
structure change, 63/212
Astrocytic response
ammonia, 27/357
interhemispheric, *see* Interhemispheric fissure
medullary, *see* Medullary fissure
transverse basilar, *see* Transverse basilar fissure
Astrocytic swelling
brain edema, 67/77
Astrocytoma, 42/731-733
adenovirus, 56/291
age, 18/4, 19/13
anaplastic, *see* Anaplastic astrocytoma
aqueduct stenosis, 18/8
astrocyte, 16/10
Au 198, 18/38
Bailey-Cushing classification, 18/1
Bergstrand classification, 18/3
biopsy, 67/226
brain, *see* Brain glioma
brain biopsy, 16/710, 713
brain stem, *see* Brain stem glioma
brain stem tumor, 18/503, 22/108
brain tumor, 18/1, 501

bromodeoxycytidine, 18/39
broxuridine, 18/39
brush cell, 18/2
calcification, 18/22
carmustine, 68/108
cerebellar medulloblastoma, 18/178, 42/742
cerebellar pilocytic, *see* Cerebellar pilocytic astrocytoma
cerebellar stellate cell, 18/2
cerebellar tumor, *see* Cerebellar astrocytoma
chemotherapy, 68/108
child, 18/17
chromosomal aberration, 42/732
chronic meningitis, 56/645
classification, 16/2, 4, 7, 18/1, 3, 5, 12, 68/1, 7
clinical malignancy, 18/12
Co 60, 18/38
congenital, 31/47
congenital brain tumor, 31/47-49
congenital spinal cord tumor, 32/355-386
corpus callosum tumor, 2/760, 17/498
CSF cytology, 16/395
cystic, *see* Cystic astrocytoma
cystic cerebellar, *see* Cystic cerebellar astrocytoma
cytostatic treatment, 18/39
differential diagnosis, 18/11
diffuse, *see* Diffuse glioblastosis
distribution, 67/130
EEG, 18/28
epidermal growth factor receptor, 67/297
epilepsy, 15/299, 18/19-22, 72/355
fibrillary, *see* Fibrillary astrocytoma
fluorouracil, 18/39
frequency, 19/13
frontal diffuse, 18/7
frontal lobe tumor, 17/238
frontodorsal, 18/6, 22
frontomedial, 18/7, 22
gamma encephalography, 18/29-33
ganglioside, 67/297
gemistocytic, *see* Gemistocytic astrocytoma
gene therapy, 67/291
giant cell, *see* Giant cell astrocytoma
gigantocellular, 18/2, 8-10
glomerate, 18/3
grade, 68/1, 7-12
grade I, *see* Astrocytoma grade I
grade II, *see* Astrocytoma grade II
grade III, *see* Astrocytoma grade III
grade IV, *see* Astrocytoma grade IV
head injury, 18/14
headache, 48/33
hemorrhage, 68/95
high grade, *see* High grade glioma
histogenesis, 67/11
histology, 20/330-332
hydroxyurea, 18/39
hypothalamic, *see* Hypothalamic glioma
hypothalamic tumor, 68/71
imaging, 67/169, 173
immune function, 67/294
immunohistochemistry, 67/11
immunotherapy, 67/291, 294
incidence, 67/132
infant, 42/733
intramedullary, 67/204
Ir 192, 18/39
isomorphic, 18/9
karyotype, 67/43
Kernohan classification, 18/1, 3, 11
local, 31/47
low grade, *see* Low grade glioma
low grade fibrillary, *see* Low grade fibrillary astrocytoma
lymphocytic cuff, 67/223
macrocystic, 18/13
macrocystic prognosis, 18/14
macroscope, 18/8
malignant, *see* Malignant astrocytoma
mesencephalon, 18/8
metastasis, 71/661
methotrexate, 18/39
microcyst, 16/10
midbrain tumor, 17/638
morbidity, 68/103
mortality, 68/103
multiple neurofibromatosis, 43/35
nasal, *see* Nasal glioma
natural history, 18/33
neoplastic satellitosis, 10/6
neurinoma, 68/565
neurofibromatosis, 20/330, 42/732, 68/289
neurofibromatosis type I, 14/140, 50/367
neuropathology, 18/2, 8-11
neuropsychology, 45/9
nonhereditary, 67/43
old age, 18/18
optic chiasm, 68/80
optic nerve, *see* Optic nerve glioma
optic nerve tumor, 17/350, 354, 356, 359
parietal lobe tumor, 17/297
parietolateral, 18/7
parietomedial, 18/8
PEG, 18/23
pilocytic, *see* Spongioblastoma

piloid, *see* Spongioblastoma
pituitary tumor, 17/425
plain roentgenogram, 18/22
plaque, 9/294
polymorphic, 18/9, 11
pons, *see* Pontine glioma
pontine tumor, 17/700
posterior fossa, *see* Posterior fossa astrocytoma
postoperative radiotherapy, 18/41
precocious puberty, 18/367
preferential site, 18/6
pregnancy, 18/19
prevalence, 18/4
protoplasmic, 18/2, 9, 13
pseudo, 14/12
radiation induced, 68/393
radioisotope diagnosis, 18/29-33
radionecrosis, 18/37
radiosensitivity, 18/12
radiotherapy, 18/36, 501, 67/269, 68/104
recurrence, 18/17, 35, 41
rehabilitation, 18/36
retinal, *see* Retinal astrocytoma
Ringertz classification, 18/6
Roussy-Oberling classification, 18/2
Russell-Rubinstein classification, 18/3
secondary structure, 10/7
seesaw nystagmus, 16/323
septum pellucidum tumor, 17/539
sex ratio, 18/4
site, 19/13
spider cell, 18/2
spinal cord, 19/13
spinal cord compression, 19/359
spinal cord tumor, 68/498
spinal glioma, 19/359, 20/329-331, 333, 336
subependymal, *see* Subependymal astrocytoma
subependymal giant cell, *see* Subependymal giant cell astrocytoma
subependymal glomerular, *see* Subependymal glomerular astrocytoma
surgery, 68/103
surgical treatment, 18/34
survival, 18/4, 33, 38, 68/5, 10-12
symptom, 18/19, 68/94
Ta 182, 18/38
tegmentum, 17/633
temporal, 18/7
terminology, 67/225
thalamic tumor, 18/8, 11, 408, 68/65
tissue culture, 16/38, 17/48, 50, 53, 60, 18/13
transferrin receptor, 67/297
treatment, 18/33
tuberous sclerosis, 14/46, 18/300, 42/732, 50/371
tumor infiltrating lymphocyte, 67/295
UICC classification, 18/3
Astrocytoma diffusum, *see* Diffuse glioblastosis
Astrocytoma grade I
supratentorial brain tumor, 18/310
Astrocytoma grade II
astroblastoma, 18/14
boron neutron capture, 18/39
chronic meningitis, 56/646
classification, 18/2
histology, 18/8-11
spinal ependymoma, 20/362
supratentorial brain tumor, 18/310
Astrocytoma grade III
boron neutron capture, 18/39
supratentorial brain tumor, 18/311
therapy, 67/296
Astrocytoma grade IV
*see also* Glioblastoma multiforme
anaplastic, 18/3, 16
boron neutron capture, 18/39
supratentorial brain tumor, 18/311
Astrocytosis
aqueduct stenosis, 30/614, 32/532, 540, 42/23, 35, 57, 50/302, 305
Binswanger disease, 54/227
blood-brain barrier, 53/87
brain arteriovenous malformation, 12/232
Creutzfeldt-Jakob disease, 56/551
GM2 gangliosidosis, 10/291, 398
Huntington chorea, 49/318
hydrocephalus, 50/290
late cortical cerebellar atrophy, 21/483-488
marginal, *see* Marginal gliosis
metachromatic leukodystrophy, 10/51
multiple sclerosis, 47/246
muscular atrophy, 42/83
pallidoluysionigral degeneration, 6/665
pencil, *see* Pencil tumor
Pick disease, 46/236, 238
piloid, *see* Piloid gliosis
plaque, 9/224
prion disease, 56/569
progressive spinal muscular atrophy, 22/17
progressive supranuclear palsy, 22/225, 46/302, 49/248, 250
radiation myelopathy, 26/90, 61/205
reactive, *see* Reactive astrocytosis
spinal, *see* Syringomyelia
tuberous sclerosis, 14/42
Astroglioma, *see* Astrocytoma
Astrogliosis, *see* Astrocytosis

Astroma
history, 18/2
terminology, 18/2
Asyllabia
alexia, 4/119
Asymbolia
agnosia, 4/14
body scheme disorder, 4/222
tactile, *see* Tactile asymbolia
tactile agnosia, 4/34
Asymbolia for pain, *see* Indifference to pain
Asymbolic acalculia
primary acalculia, 4/185, 192
Asymmetrical neuropathy
diabetic polyneuropathy, 51/499
lead polyneuropathy, 51/268
Asymmetry
anterior posterior, *see* Anterior posterior asymmetry
brain, *see* Brain asymmetry
brain venous system, 11/203
confluens sinuum, 11/112
cytoarchitectonic, *see* Cytoarchitectonic asymmetry
electrodermal, *see* Electrodermal asymmetry
facial, *see* Facial asymmetry
foramen magnum, 32/6
foramen magnum stenosis, 32/6
hemisphere, *see* Cerebral dominance
juvenile parkinsonism, 49/155
lateral, *see* Lateral asymmetry
limb, *see* Limb asymmetry
posterior cerebral artery, 11/98
posture, 2/405
Russell-Silver syndrome, 43/476
thalamic, *see* Thalamic asymmetry
Asymptomatic hyponatremia
parameter, 28/500
Asynergia
cerebellar ataxia, 1/323, 325
interpositus nucleus, 2/406
test, 1/325
Asyntactic acalculia
primary acalculia, 4/185, 192
Ataxia, 1/309-348
*see also* Incoordination
acetazolamide intoxication, 37/202
acquired toxoplasmosis, 35/126, 52/355
acrodynia, 64/374
acrylamide intoxication, 64/63, 67
acrylamide polyneuropathy, 51/281
acute, 1/348
acute benign, *see* Acute benign ataxia
acute cerebellar, *see* Acute cerebellar ataxia
acute myoclonic, *see* Acute myoclonic ataxia
adenosine triphosphatase 6 mutation syndrome, 62/510
adrenoleukodystrophy, 10/132, 51/384, 60/169
adrenomyeloneuropathy, 60/170
adult Gaucher disease, 42/437
alcohol intoxication, 51/317, 64/115
alcoholic polyneuropathy, 51/316
altretamine intoxication, 65/534
Amanita muscaria intoxication, 36/534
Amanita pantherina intoxication, 36/534
amiodarone, 63/190
amiodarone intoxication, 65/458
amiodarone polyneuropathy, 51/302
amphetamine intoxication, 65/257
angiostrongyliasis, 52/554
anosmia, 60/652
anterior inferior cerebellar artery syndrome, 53/396
antiepileptic agent intoxication, 60/662
apraxia, 1/310
aprindine intoxication, 65/455
areflexia, 42/348
Argentinian hemorrhagic fever, 56/365
argininosuccinic aciduria, 42/525
Arnold-Chiari malformation type II, 50/410
arthropod envenomation, 65/195
astasia abasia, 1/311, 314, 322, 325, 334
bacterial endocarditis, 52/292
barbiturate intoxication, 65/332, 509
Bardet-Biedl syndrome, 13/393
barotrauma, 63/416
basilar artery migraine, 48/136
basilar ischemia, 12/5
Bassen-Kornzweig syndrome, 10/548, 13/417, 21/574, 42/511, 51/327, 394, 63/272
Batten disease, 10/575
Bedouin spastic, *see* Bedouin spastic ataxia
Behçet syndrome, 56/597
Behr disease, 13/88, 42/109
benign intracranial hypertension, 42/741
benign late hereditary cerebellar, *see* Benign late hereditary cerebellar ataxia
benign peroxisomal dysgenesis, 60/670
benzodiazepine intoxication, 65/338, 510
benzodiazepine withdrawal, 65/340
Biemond syndrome, 21/377
Binswanger disease, 54/222
blood hyperviscosity, 55/484
bone membranous lipodystrophy, 42/280
brachium conjunctivum lesion, 21/514
brain abscess, 52/152

brain amyloid angiopathy, 54/337
brain edema, 63/416
brain embolism, 53/214
brain hypoxia, 63/416
brain infarction, 53/214
brain lacunar infarction, 54/246
brain microangiopathy, 55/381
brain stem arteriovenous malformation, 2/258
bromide intoxication, 36/299
bulbar lesion, 1/319
callosal, *see* Corpus callosum ataxia
carbamate intoxication, 64/190
carbamazepine intoxication, 37/202, 65/505
cerebellar, *see* Cerebellar ataxia
cerebellar ablation, 2/398
cerebellar agenesis, 50/193
cerebellar dysarthria, 42/137
cerebellar hemorrhage, 53/383, 54/314
cerebellar hypoplasia, 42/19
cerebellar infarction, 53/383
cerebellar vermis agenesis, 42/4
cerebello-olivary atrophy, 42/136
cerebral gigantism, 43/337
cerebral palsy, 46/14
cerebrotendinous xanthomatosis, 10/532, 546, 29/374, 51/399, 59/357, 60/167
cerebrovascular disease, 53/214
Charlevoix-Saguenay spastic, *see* Charlevoix-Saguenay spastic ataxia
Chédiak-Higashi syndrome, 21/583
child, 21/573
chlordecone intoxication, 64/204
chlormethine intoxication, 64/233
choreoathetosis, 60/656
chorioretinal degeneration, 13/39
chronic bromide intoxication, 36/299
chronic enterovirus infection, 56/330
chronic hepatocerebral degeneration, 49/219
chronic mercury intoxication, 64/370
ciclosporin, 63/10, 537
ciclosporin intoxication, 63/537, 65/552
cisplatin intoxication, 65/536
cisplatin polyneuropathy, 51/300
citrullinemia, 21/577, 42/538
citrullinuria, 42/538
clinical test, 1/324
Cockayne-Neill-Dingwall disease, 10/182, 13/433
colistin, 65/472
complex I deficiency, 62/504, 66/422
complex II deficiency, 66/422
complex III deficiency, 62/504, 66/423
complex IV deficiency, 66/423
complex V deficiency, 62/506
congenital cerebellar, *see* Congenital cerebellar ataxia
congenital pain insensitivity, 51/566
congenital retinal blindness, 13/298, 42/109
corpus callosum tumor, 17/508
Creutzfeldt-Jakob disease, 42/657, 56/545
cytarabine, 63/356
cytarabine intoxication, 65/533
cytochrome b deficiency, 42/111
cytochrome related, *see* Cytochrome related ataxia
Dandy-Walker syndrome, 50/328
deafness, 42/112, 60/657
deficiency neuropathy, 51/324, 326
Dejerine syndrome, 1/318
dementia, 42/114
dentate nucleus, 21/514
dentatorubropallidoluysian atrophy, 49/438
diabetes mellitus, 60/660, 674
dialysis, 63/528
dialysis encephalopathy, 63/528
diphtheria, 51/183
diplopia, 1/314
disequilibrium syndrome, 42/138-140
distal amyotrophy, 60/659
Divry-Van Bogaert syndrome, 55/319
dominant, *see* Dominant ataxia
dominant cerebello-olivary atrophy, 42/136
dysbarism, 63/416
early onset recessive stationary, *see* Early onset recessive stationary ataxia
elfin face syndrome, 43/259
endemic paresis, 9/684
eosinophil derived neurotoxin, 63/370
epilepsy, 42/114
Epstein-Barr virus meningoencephalitis, 56/252
equilibrium, 1/334
essential thrombocythemia, 63/331
ethylene glycol intoxication, 64/123
ethylene oxide, 51/275
examination, 1/313, 324
external ophthalmoplegia, 43/136
eye movement, 1/605
Fabry disease, 10/543
face, 1/314
familial hypobetalipoproteinemia, 13/427
familial late onset, *see* Familial late onset ataxia
familial paroxysmal, *see* Familial paroxysmal ataxia
familial spastic paraplegia, 22/425, 458, 59/305
familial striatopallidodentate calcification, 6/707, 49/420
felbamate intoxication, 65/512
feline, *see* Feline ataxia

Ferguson-Critchley syndrome, 42/142
flocculonodular lobe syndrome, 21/569
fluorouracil intoxication, 65/532
Flynn-Aird syndrome, 14/112, 42/327
Foix syndrome, 2/277
Förster syndrome, 42/202
Friedreich, *see* Friedreich ataxia
Friedreich ataxia, 60/311
frontal lobe, *see* Frontal lobe ataxia
gabapentin intoxication, 65/512
gait, *see* Locomotor ataxia
Gerstmann-Sträussler-Scheinker disease, 56/553
GM1 gangliosidosis type II, 42/432
gold polyneuropathy, 51/307
gonadal dysgenesis, 21/467
granulomatous amebic encephalitis, 52/317
granulomatous CNS vasculitis, 55/390
griseofulvin, 65/472
Guillain-Barré syndrome, 51/245
gynecomastia, 60/668
gyromitrine, 64/3
happy puppet syndrome, 30/515, 31/308
Hartnup disease, 7/577, 21/574, 579, 29/155
head, 1/314
hearing loss, 60/659
heat stroke, 21/583
hemolytic uremic syndrome, 63/324
hepatic polyneuropathy, 51/326
hereditary, *see* Hereditary ataxia
hereditary cerebellar, *see* Hereditary cerebellar ataxia
hereditary dystonic paraplegia, 22/458
hereditary hearing loss, 22/500-510, 513-517
hereditary motor and sensory neuropathy type I, 42/117
hereditary motor and sensory neuropathy type II, 42/117
hereditary motor and sensory neuropathy variant, 60/245
hereditary muscular atrophy, 13/306
hereditary periodic, *see* Hereditary periodic ataxia
hereditary sensory and autonomic neuropathy type III, 43/59, 60/27
hereditary spastic, *see* Hereditary spastic ataxia
hereditary spinocerebellar, *see* Hereditary spinocerebellar ataxia
heterocyclic antidepressant, 65/316
histoplasmosis, 52/438
history, 1/309
Holmes cerebellar, *see* Holmes cerebellar ataxia
horse, 9/686
Huntington chorea, 42/226, 49/280
hydrocephalic dementia, 46/327
hydromyelia, 50/429
hydrozoa intoxication, 37/40
hyperammonemia, 42/560
hyperargininemia, 42/562
hypereosinophilic syndrome, 63/372
hyperglycemia, 21/578
hyperparathyroidism, 27/292
hyperserotonemia, 29/225
hypertrophic interstitial neuropathy, 42/317
hyperuricemia, 42/571
hyperviscosity syndrome, 55/484
hypogonadal cerebellar, *see* Hypogonadal cerebellar ataxia
hypogonadism, *see* Matthews-Rundle syndrome
hypomagnesemia, 63/562
hypothyroid polyneuropathy, 51/515
hypothyroidism, 21/574, 27/260
idiopathic hypoglycemia, 42/576
IgA, 21/581
IgM paraproteinemic polyneuropathy, 51/437
incontinentia pigmenti, 14/12, 108
infantile neuronal ceroid lipofuscinosis, 42/461, 66/689
infantile sex linked cerebellar, *see* Infantile sex linked cerebellar ataxia
inherited spinocerebellar, *see* Inherited spinocerebellar ataxia
intention tremor, 42/230, 49/590
interictal, *see* Interictal ataxia
intracranial hemorrhage, 53/214
Joubert syndrome, 50/191, 63/437
juvenile Gaucher disease, 42/438
juvenile Gaucher disease type A, 66/125
Kearns-Sayre-Daroff-Shy syndrome, 62/508
kinesigenic, *see* Kinesigenic ataxia
kinetic, 1/313
Köhlmeier-Degos disease, 55/276
kuru, 42/657, 56/556
labyrinthine, 1/334-338
lamotrigine intoxication, 65/512
late cerebellar, *see* Late cerebellar ataxia
late cerebellar atrophy, 42/135
late onset, *see* Late onset ataxia
lateral medullary infarction, 53/382
lead encephalopathy, 64/435, 443
lead intoxication, 64/435
Leber hereditary optic neuropathy, 62/509
legionellosis, 52/254
leukemia, 63/342, 344
Leyden-Westphal, *see* Acute cerebellar ataxia
limb, 1/313
lipoprotein, 21/581
Listeria monocytogenes encephalitis, 52/93

Listeria monocytogenes meningitis, 52/93
lithium intoxication, 64/296
locomotor, *see* Locomotor ataxia
lower motoneuron disease, 60/657
lumbar vertebral canal stenosis, 20/705
Lundberg syndrome, 13/91
lysergide, 65/45
lysergide intoxication, 65/44
Machado-Joseph disease, 42/155
malignant hyperthermia, 63/439
malnutrition, 28/22
manganese deficiency, 64/305
manganese intoxication, 64/305
marine toxin intoxication, 37/33, 65/158
Marinesco-Sjögren syndrome, 42/184, 60/342
Matthews-Rundle syndrome, 60/578, 652
May-White syndrome, 22/511
mefloquine intoxication, 65/485
meningococcal meningitis, 33/23
mental deficiency, 42/112, 43/278, 46/14, 60/657, 659, 664, 722
mercury intoxication, 51/272
mercury polyneuropathy, 51/272
MERRF syndrome, 62/509
mesencephalon lesion, 2/277
metabolic, *see* Metabolic ataxia
metachromatic leukodystrophy, 8/16, 42/491, 493
methotrexate intoxication, 39/102, 65/530
methylmalonic aciduria, 55/329
metronidazole intoxication, 65/487
mexiletine intoxication, 65/455
micrographia, 45/464
migraine, 21/570, 48/147
Minamata disease, 7/642, 21/583, 36/74, 121, 64/413, 415, 419
minocycline, 65/472
mitochondrial myopathy, 60/649
mitotane, 65/529
mitotane intoxication, 65/538
mixed, 1/339
mountain sickness, 63/416
mucolipidosis type I, 42/475
multiple hamartoma syndrome, 42/755
multiple mitochondrial DNA deletion, 62/510
multiple myeloma, 55/487
multiple neuronal system degeneration, 75/162
multiple sclerosis, 9/179, 42/496
muscle fasciculation, 42/113
myatrophic, *see* Myatrophic ataxia
myelomalacia, 1/319
myoclonic epilepsy, 42/114
myoclonus, 21/510, 583
myopathy, 43/124
myxedema, 21/569, 574
myxedema neuropathy, 8/57, 59
nalorphine, 37/366
nerve conduction velocity, 60/670
neuritis, 1/318
neuroacanthocytosis, 21/581
neuroblastoma, 21/574
neurocysticercosis, 35/299, 52/530
neurogenic muscular atrophy, 60/674
neuronal ceroid lipofuscinosis, 10/550, 42/230
neurotoxic shellfish poisoning, 65/158
Niemann-Pick disease type C, 42/471
nigrospinodentate degeneration, 22/165, 168-171
nonprogressive cerebellar, *see* Nonprogressive cerebellar ataxia
nonprogressive cerebellar ataxia, 42/124
Northern Swedish recessive, *see* Northern Swedish recessive ataxia
nucleus ruber superior syndrome, 1/187
nystagmus, 5/312
ocular, *see* Ocular ataxia
OKT3, 63/536
olivopontocerebellar atrophy (Dejerine-Thomas), 21/425
opsoclonus myoclonus, *see* Opsoclonus myoclonus ataxia syndrome
optic, *see* Ocular ataxia
optic atrophy, 13/39, 42/117, 406, 412-417
optomotor, *see* Optomotor ataxia
organochlorine insecticide intoxication, 64/204
organolead intoxication, 64/131
organomercury intoxication, 36/74
organophosphate induced delayed polyneuropathy, 64/161
organophosphate intoxication, 37/555, 64/167
orthochromatic leukodystrophy, 42/500
20p partial trisomy, 43/539
pallidal degeneration, 49/450
pallidoluysionigral degeneration, 49/456
pallidonigral degeneration, 42/242
panleukopenia virus, 34/296
paralytic shellfish poisoning, 37/33, 65/158
paretic, 1/310
parietal, *see* Parietal ataxia
Pelizaeus-Merzbacher disease, 42/502, 505
pellagra, 28/90, 95, 46/399
pentaborane intoxication, 64/360
perhexiline maleate polyneuropathy, 51/302
periodic, *see* Periodic ataxia
periodic vestibulocerebellar, *see* Periodic vestibulocerebellar ataxia
phencyclidine intoxication, 65/368
phenytoin intoxication, 42/641, 65/500

photogenic epilepsy, 42/714
pneumococcal meningitis, 33/46, 52/44, 51
polyneuropathy, 42/117, 60/670
pontine infarction, 12/15, 53/390
pontine tuberculoma, 2/261
posterior column, *see* Posterior column ataxia
posture, 1/311, 314, 322
presenile dementia, 42/288
procainamide intoxication, 65/453
progressive, *see* Progressive ataxia
progressive cerebellar, *see* Progressive cerebellar ataxia
progressive external ophthalmoplegia, 22/193, 43/136
progressive multifocal leukoencephalopathy, 9/494
progressive myoclonus epilepsy, 42/114, 705
progressive rubella panencephalitis, 34/338, 56/409
pseudo, *see* Pseudoataxia
pure sensory stroke, 54/250
pushing test, 1/334
pyruvate dehydrogenase complex deficiency, 66/416
pyruvate dehydrogenase deficiency, 42/587, 63/439
pyruvic acid, 21/578
r(22) syndrome, 43/547
Refsum disease, 8/21, 21/9, 199, 36/348, 42/148, 51/384, 60/228, 230, 728, 66/490
rehabilitation, 3/389
restoration of higher cortical function, 3/389, 395
retinovestibular hereditary spastic, *see* Retinovestibular hereditary spastic ataxia
Rett syndrome, 60/636, 63/437
Reye syndrome, 63/438
rhodotorulosis, 52/493
Richards-Rundle syndrome, 22/516, 43/264
Roussy-Lévy syndrome, 42/108
Salla disease, 43/306, 59/360
sarcoid neuropathy, 51/196
sarcoidosis, 51/196
Schut family, *see* Schut family ataxia
scoliosis, 42/348
scrapie, 42/656, 56/585
secondary pigmentary retinal degeneration, 60/654, 660, 670, 674, 722
segmental, 1/313
selective serotonin reuptake inhibitor, 65/316
sensorineural deafness, 42/112, 386
sensory, *see* Sensory ataxia
sensory neuropathy, 42/348
Shapiro syndrome, 50/167
spastic, *see* Spastic ataxia
spastic paraplegia, 42/288
spinal, *see* Spinal ataxia
spinal cord lesion, 1/319
spinal lipoma, 20/400
spinal muscular atrophy, 59/374
spinal spongioblastoma, 19/360
spinocerebellar, *see* Spinocerebellar ataxia
spinopontine degeneration, 42/192
spirogermanium, 65/529
spirogermanium intoxication, 65/538
sporotrichosis, 52/495
Stewart-Holmes sign, 1/333
striatonigral degeneration, 49/206
striatopallidodentate calcification, 49/420
subacute myelo-optic neuropathy, 51/296
subacute necrotizing encephalomyelopathy, 42/588, 625, 51/387
subacute sclerosing panencephalitis, 56/419
subclavian steal syndrome, 53/373
sulfamethoxazole, 65/472
superior cerebellar artery syndrome, 53/397
systemic lupus erythematosus, 55/377
tabes dorsalis, 33/363, 52/280
tabetic, *see* Tabetic ataxia
Telfer syndrome, 22/513
temporal lobe, 1/343
temporal lobe tumor, 17/292
tetrahydrocannabinol intoxication, 65/47
tetrodotoxin intoxication, 65/156
thalamic infarction, 53/396
thalamic syndrome (Dejerine-Roussy), 1/319, 42/771
thallium intoxication, 64/325
thrombotic thrombocytopenic purpura, 55/472, 63/324
thyroid gland dysgenesis, 42/632
tick bite intoxication, 37/111
tick venom, 65/195
toluene, 64/3
toxoplasmosis, 52/355
transcobalamin II deficiency, 63/255
transient ischemic attack, 12/5, 53/214
trichinosis, 52/566
trichloroethanol, 65/329
trichloroethanol intoxication, 65/329
tri-*o*-cresyl phosphate polyneuropathy, 51/285
tricyclic antidepressant, 65/316-318
tricyclic antidepressant intoxication, 65/321
trigeminal nerve agenesis, 50/214
trimethoprim, 65/472
tropical ataxic neuropathy, 7/640, 51/322, 65/29
tropical spastic paraplegia, 56/536

Troyer syndrome, 42/194
truncal, *see* Truncal ataxia
tryptophanuria, 42/633
ubidecarenone, 62/504
undetermined significant monoclonal gammopathy, 63/399
Unverricht-Lundborg progressive myoclonus epilepsy, 42/705
upper motoneuron disease, 60/657
Usher syndrome, 42/392
valproic acid intoxication, 37/202, 65/504
vertebral artery arteriosclerosis, 53/376
vertebrobasilar insufficiency, 53/262
vertebrobasilar system syndrome, 53/373, 376, 390, 394, 397
vigabatrin intoxication, 65/511
vitamin $B_6$ abuse, 51/298
vitamin $B_6$ deficiency, 51/335
vitamin $B_{12}$ deficiency, 63/254, 70/377
vitamin $B_c$ transport defect, 43/278
vitamin $B_w$ dependent multiple carboxylase deficiency, 63/439
vocal cord, 1/314
Vogt-Koyanagi-Harada syndrome, 56/621
Von Stein test, 1/335
Waldenström macroglobulinemia, 63/396, 398
water intoxication, 64/241
Wernicke-Korsakoff syndrome, 28/248, 45/195
Whipple disease, 51/184
white hair, 42/230
Willvonseder syndrome, 42/194
X-linked, *see* X-linked ataxia
X-linked recessive, *see* X-linked recessive ataxia
xeroderma pigmentosum, 43/13, 51/399, 60/336
zinc intoxication, 36/338

Ataxia syndrome
erythrokeratodermia, *see* Erythrokeratodermia ataxia syndrome
hereditary myoclonus, *see* Hereditary myoclonus ataxia syndrome
opsoclonus myoclonus, *see* Opsoclonus myoclonus ataxia syndrome

Ataxia telangiectasia, 14/267-334, 60/347-392
age at onset, 60/367
aneuploidy, 50/566
anterior horn cell, 60/365
areflexia, 42/119, 60/365, 660
associated disease, 14/517
ataxic neuropathy, 14/316
ataxic oculomotor apraxia, 60/387
athetosis, 6/458, 21/581
autopsy, 14/307
autosomal recessive cerebellar hypoplasia, 60/387
bacterial pathogen, 60/371
basal cell carcinoma, 60/370
biochemistry, 14/274
Bloom syndrome, 14/315
bone marrow transplantation, 60/390
brain stem auditory evoked potential, 60/366
brain tumor, 46/58
bronchiectasis, 14/288, 60/371
cancer, 60/372
carcinoembryonic antigen, 60/375
case history, 14/318
cause, 14/280
cerebellar ataxia, 14/267
cerebellar medulloblastoma, 42/742
cerebellum, 14/73
chemotherapy, 60/390
childhood infectious disorder, 60/386
chorea, 21/581
choreoathetosis, 14/75, 267, 49/385, 60/361, 663
chromosomal aberration, 42/120, 60/350
chromosome 11q22-23, 60/427
chromosome translocation, 14/118
classification, 14/314
clinical features, 14/282, 302, 21/18, 31/24-27, 60/349, 360
clinical symptom, 14/118
Cockayne-Neill-Dingwall disease, 14/315
cognitive development arrest, 60/367
complementation group A, 60/425
complementation group C, 60/425
complementation group D, 60/425
complementation group E, 60/425
complementation group V1, 60/425
computer assisted tomography, 60/376, 671
course, 60/387
CSF, 14/296
cytogenetics, 60/348
dementia, 60/665
demyelination, 14/74
deoxyribosylphosphate modifying factor, 60/429
diabetes mellitus, 14/274, 293, 60/674
diagnosis, 14/302, 31/25-29
diarrhea, 60/668
differential diagnosis, 14/304, 315, 60/383-387
distal amyotrophy, 60/659
dwarfism, 14/517
dysarthria, 14/284
dyskeratosis congenita, 14/315
dystonia, 60/664
dystonia musculorum deformans, 49/524
ECG, 14/296
EEG, 14/296, 60/376
endocrine abnormality, 60/374

endocrinology, 14/274, 291
epidemiology, 21/21, 31/24
epilepsy, 15/325
epistaxis, 60/368
equable disposition, 60/361, 367
fasciculation, 60/659
fetal thymus transplantation, 60/389
α-fetoprotein, 60/353
flexor reflex, 60/364
forme fruste, 14/311
Friedreich ataxia, 14/75, 307-313, 316, 21/21, 60/383
gaze apraxia, 21/581
gaze paralysis, 14/652
genetics, 14/274, 515-517, 31/24, 60/358
gliovascular abnormality, 60/357
gliovascular nodule, 60/380
growth retardation, 14/290, 42/120, 60/361, 373
hair change, 60/370
Hartnup disease, 14/315
hereditary congenital cerebellar atrophy, 60/291
hereditary metabolic disease, 60/386
hereditary motor sensory neuropathy, 60/386
heterotopia, 14/74
heterozygote frequency, 42/121
historic review, 60/347
history, 14/267
Hodgkin disease, 46/58
hypogammaglobulinemia, 14/269, 42/120
hypogonadism, 14/274, 291, 517, 42/120, 60/374
hypotonia, 60/361
Ig turnover study, 60/351
IgA, 14/75, 269-273, 60/375
IgE, 14/273, 60/375
IgE deficiency, 60/351
IgG2, 60/375
IgM, 14/75
immunocompromised host, 56/468
immunodeficiency, 60/375
immunology, 14/269, 60/348
immunotherapy, 60/389
incontinentia pigmenti, 14/75
inferior olive, 14/75
insulin resistant diabetes mellitus, 60/361, 374
Joubert syndrome, 60/387
Klippel-Trénaunay syndrome, 14/252
laboratory finding, 14/296
laboratory study, 60/375
leukemia, 14/271, 46/58, 60/674
levamisole, 60/390
liver function, 14/297
lymph node, 31/24
lymphocyte, 14/270
lymphoid system, 14/75
lymphoid tissue abnormality, 60/349
lymphoma, 60/674
lymphopenia, 14/269, 60/375
lymphoreticular malignancy, 60/349
lymphoreticular tumor, 14/294
malignant tumor, 14/277, 294
management, 14/311
mental deficiency, 14/289, 31/3, 24, 42/120, 46/58
metabolic ataxia, 21/574, 581, 31/25
molecular genetics, 60/425
muscle biopsy, 14/321, 60/376
myoclonic jerk, 60/361
nerve conduction study, 60/365
nerve conduction velocity, 60/670
neuroaxonal degeneration, 14/279
neurocutaneous syndrome, 14/101
neurofibromatosis, 68/298
neurologic features, 60/361-367
neuropathology, 14/73, 275, 325, 60/353, 376-383
neuropathy, 1/318, 7/640, 14/316, 28/26, 36/522, 60/660
nonneurologic features, 60/367
nosology, 60/388
nuclear magnetic resonance, 60/376
nucleocytomegaly, 60/353, 382
Nijmegen breakage syndrome, 60/425
nystagmus, 42/119
ocular lesion, 14/75
oculomotor apraxia, 14/267, 283, 60/360, 387
ophthalmology, 14/651-653
optic atrophy, 60/654
pathogenesis, 14/281, 60/359
pathology, 14/276, 297-304, 31/24
peripheral nerve, 60/365
phakomatosis, 50/377
phosphoethanolamine, 60/383
phytohemagglutinin, 14/271
pigmentary change, 60/370
pituitary gland, 14/75, 293
plantar reflex, 60/364
polyneuropathy, 60/365
prevalence, 14/281, 42/120, 60/358
progeria, 14/267
progeric change, 60/353, 358, 361
prognosis, 14/310, 60/387
progressive cerebellar ataxia, 60/360
progressive oculocutaneous telangiectasia, 60/360
propranolol, 60/390
protoporphyria, 14/315
psychiatric disorder, 14/767
radiosensitivity, 60/352
radiotherapy, 60/390

Refsum disease, 60/235, 66/497
rehabilitation, 60/391
related syndrome, 14/101
roentgenology, 14/295
Rothmund-Thomson syndrome, 14/315
sarcoma, 46/58
seborrhea, 60/370
secondary pigmentary retinal degeneration, 60/654
secretory IgA, 60/350
sensation, 60/364
serum α-fetoprotein, 60/375
short description, 14/13
sinopulmonary infection, 14/268, 286, 60/360, 371, 647, 667
skeletal abnormality, 60/373
skeletal lesion, 14/291
skin change, 60/370
skin lesion, 14/75, 290
somatosensory evoked potential, 60/366
spinal cord, 14/74
spinal cord traction, 60/365
spinal muscular atrophy, 14/316
sural nerve biopsy, 60/376
symptomatic dystonia, 49/543
thymus, 14/76, 31/24
thymus aplasia, 60/361
torpedo, 14/75
total reported cases, 14/377
transfer factor, 60/390
treatment, 31/25-29, 60/389
tremor, 21/581
tuber, 14/74
tuberous sclerosis, 14/74, 279
tumor cell, 67/31
variant features, 60/367
virus, 60/371
Von Hippel-Lindau disease, 14/252
Von Romberg sign, 60/364
Werner syndrome, 14/315

Ataxic diplegia
cerebellar ataxia, 42/121
congenital, *see* Congenital ataxic diplegia
hyperreflexia, 42/121
mental deficiency, 42/121
prevalence, 42/122
strabismus, 42/121

Ataxic hemiparesis
brain lacunar infarction, 54/57, 247
capsular lacuna, 54/248
computer assisted tomography, 54/57
dysarthria, 54/248
hemisensory deficit, 54/248
nystagmus, 54/248
pontine hemorrhage, 54/313
pontine infarction, 53/394
pontine lacunar infarction, 54/248
thalamic infarction, 54/248
thalamic lacuna, 54/248
thalamic lacunar infarction, 54/248
transient ischemic attack, 54/247
vertebrobasilar system syndrome, 53/394
vertigo, 54/248

Ataxic myelopathy
tropical, *see* Tropical ataxic myelopathy

Ataxic neuropathy
ataxia telangiectasia, 14/316
cause, 65/30
clinical features, 1/318
cyanide intoxication, 7/642, 56/526, 65/30
cyanogenic glycoside intoxication, 65/30
differential diagnosis, 60/660
Furukara disease, 42/334
Jamaican polyneuropathy, 28/26-31
malnutrition, 28/21-26
motor nerve conduction velocity, 7/641
tri-*o*-cresyl phosphate, 56/526
tropical, *see* Tropical ataxic neuropathy
vitamin B2 deficiency, 28/206, 51/332

Ataxic oculomotor apraxia
ataxia telangiectasia, 60/387
differential diagnosis, 60/656

Ataxic respiration
brain injury, 57/129
brain stem encephalitis, 63/487
bulbar poliomyelitis, 1/498
cerebellar hemorrhage, 1/669, 63/487
encephalitis, 1/498

Atelectasis
Guillain-Barré syndrome, 51/255
respiratory dysfunction, 26/343
spinal cord injury, 26/343, 61/264

Ateleiotic dwarf
sexual, *see* Pituitary dwarfism type I

Ateleosis, *see* Levi-Lorain dwarfism

Atenolol
beta adrenergic receptor blocking agent, 65/434
carpal tunnel syndrome, 63/195
delirium, 63/195
diplopia, 63/195
migraine, 48/180
myasthenia, 63/195
paresthesia, 63/195
ptosis, 63/195
stupor, 63/195

Atergatis floridus intoxication

crustacea, 37/58
death, 37/58
Atherogenesis
lipid theory, 11/332
thrombogenic theory, 11/332
Atheroma
carotid artery, 11/331
cerebrovascular disease, 12/453
prevention, 12/453
Atheromatous plaque
arterial lesion, 11/331
atheromatous debris, 11/331
cholesterol, 11/331
embolism, 11/437
Atherosclerosis, *see* Arteriosclerosis
Atherothrombotic cerebrovascular disease
headache, 48/279
Athetoid movement, *see* Athetosis
Athétose double, *see* Double athetosis
Athetosis, 6/440-467
*see also* Chorea *and* Involuntary movement
African trypanosomiasis, 52/341
asphyxia, 6/446
ataxia telangiectasia, 6/458, 21/581
attitude, 2/481
axial dystonia, 49/382
barbiturate, 6/450
basal ganglion pathology, 6/59-61
Bassen-Kornzweig syndrome, 10/548, 21/580
birth incidence, 6/446
carbon disulfide intoxication, 6/450
carbon monoxide intoxication, 49/476
cerebellar stimulation, 49/387
cerebellar symptom, 6/444
chorea, 6/445
classification, 6/446
clinical definition, 6/153-157
clinical features, 6/440-445
congenital retinal blindness, 42/401
cortex, 6/459
cyanide intoxication, 64/232
definition, 49/381
dentatorubropallidoluysian atrophy, 21/520
differential diagnosis, 6/478
diffuse sclerosis, 6/457
distinctive feature, 1/287
double, *see* Double athetosis
drug treatment, 6/465, 49/387
dystonia, 6/445, 524, 526, 49/381, 521
dystonia musculorum deformans, 49/522
EEG, 6/446
EMG, 1/647
endogenous toxin, 6/450
Epstein-Barr virus, 56/252
Epstein-Barr virus meningoencephalitis, 56/252
familial paroxysmal choreo, *see* Familial paroxysmal choreoathetosis
familial spastic paraplegia, 22/453
fetal Minamata disease, 64/418, 421
focal, *see* Focal athetosis
glossal spasm, 49/382
GM2 gangliosidosis, 6/457
Hallervorden-Spatz syndrome, 6/61, 155, 454, 49/386
hand, 2/481
Harvey syndrome, 42/222
hemi, *see* Hemiathetosis
hereditary dystonic paraplegia, 22/453
hereditary olivopontocerebellar atrophy (Menzel), 21/441
hereditary striatal necrosis, 49/495
hollow hand sign, 1/184
hypomagnesemia, 63/562
hypoparathyroidism, 27/308
juvenile Gaucher disease, 10/307
kernicterus, 6/450, 503
Machado-Joseph disease, 42/262
micrographia, 45/464
Minamata disease, 64/414, 418, 421
mucolipidosis type IV, 51/380
muscular hypertonia, 6/445
oculocerebrocutaneous syndrome, 43/444
olivopontocerebellar atrophy variant, 21/452
organic acid metabolism, 66/641
orofacial dyskinesia, 49/382
pallidal necrosis, 2/502, 49/465
pallidoluysiodentate degeneration, 49/452
pallidoluysionigral degeneration, 49/446, 455
pallidonigral necrosis, 49/465
pallidostriatal necrosis, 49/465
pathogenesis, 6/482-485, 49/386
pathologic anatomy, 49/386
pathophysiology, 49/77
Pelizaeus-Merzbacher disease, 2/501, 6/457
pharyngeal spasm, 6/441
phenylketonuria, 6/456
posterior cervical spinal cord stimulation, 49/387
posthemiplegic, *see* Posthemiplegic athetosis
posthemiplegic chorea, 6/437
postural tremor, 49/591
posture, 6/445
progressive pallidal atrophy (Hunt-Van Bogaert), 6/455, 457, 554
pseudo, *see* Pseudoathetosis
rhinophonia aperta, 6/441
Salla disease, 43/306, 59/360

segmental, *see* Segmental athetosis
spastic dysphonia, 49/382
speech, 6/441
Spiegel-Baird review, 49/382
status marmoratus, 2/502, 6/16, 60
striatal necrosis, 49/499, 501
striatonigral degeneration, 42/262
striatopallidodentate calcification, 49/423
striatopallidum, 6/460-464
subacute necrotizing encephalomyelopathy, 51/387
subacute sclerosing panencephalitis, 56/419
surgical treatment, 6/463-467
systemic lupus erythematosus, 55/377
thalamotomy, 49/387
tonic avoidin reaction, 6/155
tonic grasp reflex, 6/155
treatment, 49/387
tuberous sclerosis, 6/457, 14/349
Vaquez disease, 6/458
Wilson disease, 6/459, 49/225
writers cramp, 49/382
Athetotic hemiplegia
brain atrophy, 42/201
cryptorchidism, 42/200
Dyke-Davidoff syndrome, 42/201
EEG, 42/201
hydrocephalus, 42/200
mental deficiency, 42/200
seizure, 42/200
speech disorder in children, 42/200
Athetotic membranous lipodystrophy
*see also* Nasu-Hakola disease
brain atrophy, 42/201
Athrepsia
diencephalic syndrome, 18/366
Athyrotic cretinism, *see* Thyroid gland dysgenesis
Atlantal ligament laxity
transverse, *see* Transverse atlantal ligament laxity
Atlantoaxial dislocation
*see also* Atlas *and* Dens
anterior, 25/131, 307
cervical myelopathy, 12/609
cervical vertebral column injury, 25/307
cervicomedullary injury, 24/143, 154, 157
clinical features, 32/83
definition, 25/240, 242
dialysis, 63/529
Down syndrome, 43/544
epidemiology, 32/83
football injury, 24/157
fracture odontoid process, 25/315-332
fusion, 25/298, 325
general, 25/30, 242
glycosaminoglycanosis type IV, 42/458
glycosaminoglycanosis type VI, 42/458
Grisel syndrome, 5/373, 32/83
head injury, 24/143
intermittent, 25/243
juvenile spinal cord injury, 61/233
juvenile vertebral column injury, 61/233
late cord lesion, 25/279
$\beta_2$-microglobulin amyloidosis, 63/529
myelopathy, 12/609
pain, 5/373
posterior, 25/132, 307
rheumatoid arthritis, 19/321, 323, 38/492-496
rotation, 25/308
self-reducing, 25/243
subluxation, 25/131
trauma, 32/83
treatment, 25/310, 32/83
Atlantoaxial joint
*see also* Atlas *and* Dens
anatomy, 19/323
compression, 26/289
head rotation, 19/323
movement, 61/5
Atlantoaxial luxation, *see* Atlantoaxial dislocation
Atlantoaxial offset
physiology, 25/247-249
Atlantoaxial subluxation
brain infarction, 53/165
brain ischemia, 53/165
cervicomedullary injury, 24/141
congenital spondyloepiphyseal dysplasia, 43/480
dens agenesis, 50/389
Down syndrome, 50/392
glycosaminoglycanosis type I, 50/392
glycosaminoglycanosis type IV, 50/392, 60/175
glycosaminoglycanosis type VI, 50/392, 66/308
pseudoachondroplastic dysplasia, 50/392
rheumatoid arthritis, 38/492-496, 50/392, 71/24
spondyloepiphyseal dysplasia, 50/392
Atlanto-occipital distance
normal value, 32/23
Atlanto-occipital instability
delayed cord lesion, 25/280
Atlanto-occipital junction
embryology, 50/381
Atlanto-occipital luxation
craniospinal injury, 25/310
Atlanto-occipital synostosis
Arnold-Chiari malformation type I, 50/393
Dandy-Walker syndrome, 50/393
foramen magnum tumor, 20/184

hydrocephalus, 50/393
Atlas
see also Atlantoaxial dislocation and Atlantoaxial joint
abnormal development, 32/5
ankylosing spondylitis, 26/175
anterior subluxation, 25/131
assimilation, 32/6, 25-28, 82
atlantoaxial compression, 26/289
basilar impression, 32/82
capitalization, 32/6
clinical features, 32/82, 93
defect, 32/37
embryology, 32/5, 50/381
epidemiology, 32/28, 82
fissure, 32/37
foramen arcuale, 32/38
lateral mass malformation, 32/40
lateral ponticulus, 32/39
normal development, 32/3
occipital dysplasia, 32/25
posterior luxation, 25/132
suboccipital dysplasia, 32/26
transverse process malformation, 32/40
Turner syndrome, 32/82
Atlas dysplasia
assimilation, 32/27
bone abnormality, 50/388
diagnosis, 50/392
foramen arcuale, 50/393
lateral ponticulus, 50/393
suboccipital dysplasia, 32/36-40
Atlas fracture
cervicomedullary injury, 24/152
mechanism, 25/130
spinal injury, 25/127, 314
surgery, 61/64
vertebral artery thrombosis, 25/263
Atlas syndrome
anxiety, 5/373
depression, 5/373
hypochondriasis, 5/373
muscular disorder, 5/373
pain, 5/373
restlessness, 5/373
Atmospheric hypoxidosis
pallidal necrosis, 49/466
Atomic absorption spectrometry
arsenic polyneuropathy, 51/267
electrothermal, see Electrothermal atomic absorption spectrometry
Minamata disease, 64/416
Atomic number
neurotoxicity, 51/265
Atomistic psychology
development, 3/12
Atonia astasia syndrome, see Förster syndrome
Atonic sclerotic muscular dystrophy
congenital, see Congenital atonic sclerotic muscular dystrophy
Atopic dermatitis
recessive, see Recessive atopic dermatitis
Atopomycterus nicthemerus intoxication, see Puffer fish intoxication
Atoxyl, see Arsanilic acid
ATP, see Adenosine triphosphate
ATP synthase, see Proton transporting adenosine triphosphate synthase
Atracurium
intensive care, 71/536
multiple organ failure, 71/536
Atresia
anal, see Anal atresia
biliary, see Biliary atresia
duodenal, see Duodenal atresia
esophagus, see Esophagus atresia
foramen of Luschka-Magendie, see Dandy-Walker syndrome
Luschka foramina, see Luschka foramina atresia
Magendie foramina, see Magendie foramina atresia
subclavian artery stenosis, 53/372
urethra, see Urethra atresia
vaginal, see Vaginal atresia
ventricular system, see Ventricular system atresia
Atretic cephalocele
parietal encephalocele, 50/110
Atrial arrhythmia, see Emery-Dreifuss muscular dystrophy
Atrial fibrillation
anticoagulant complication, 63/19
anticoagulation, 63/19
brain embolism, 53/157, 55/153, 168, 63/17
brain hemorrhage, 55/152
brain infarction, 53/157, 55/152
brain ischemia, 53/157
cardiac valvular disease, 38/144, 63/17
congestive heart failure, 63/133
hypertrophic cardiomyopathy, 63/136
mitral annulus calcification, 63/29
mitral stenosis, 63/18
mitral valve prolapse, 63/22
neurologic complication, 38/144
quinidine, 37/448
recurrent stroke, 63/19
rheumatic heart disease, 63/18

Atrial myxoma
age distribution, 63/96
anemia, 63/97
antimyolemmal antibody, 63/97
beaded artery, 63/99
brain aneurysm, 63/98
brain atrial myxoma metastasis, 63/99
brain embolism, 53/157, 55/163, 63/97
brain infarction, 53/157, 55/163, 63/97
brain ischemia, 53/157
cerebrovascular disease, 55/163
circulating immune complex, 63/97
dementia, 63/98
diagnosis, 63/97
embolic left/right ratio, 63/98
erythrocyte sedimentation rate, 63/97
fatigue, 63/97
fever, 63/97
frequency, 63/96
fusiform aneurysm, 63/99
generalized aching, 63/97
hyaluronic acid release, 63/97
IgG, 63/97
multi-infarct dementia, 63/98
multiple strokes, 63/98
myelomalacia, 63/41, 98
myelopathy, 63/41
natural history, 63/99
neurologic complication, 63/96
pathology, 63/102
prevalence, 63/93
prognosis, 63/96
recurrent laryngeal nerve compression, 63/99
retinal embolism, 63/98
saccular aneurysm, 63/99
spinal column atrial myxoma metastasis, 63/99
spinal cord embolism, 63/41, 98
systemic effect, 63/97
transient ischemic attack, 63/97
transverse myelopathy, 63/98
weight loss, 63/97
Atrial natriuretic peptide
brain edema, 67/80, 85
Atrial paralysis, *see* Emery-Dreifuss muscular dystrophy
Atrial septal defect
craniolacunia, 50/139
holoprosencephaly, 50/237
r(13) syndrome, 50/585
Atrioventricular block, *see* Cardiac dysrhythmia
Atromid-S, *see* Clofibrate
Atropa belladonna, *see* Belladonna alkaloid
Atrophic dystrophy
neuronal, *see* Neuronal atrophic dystrophy
Atrophic papulosis
malignant, *see* Köhlmeier-Degos disease
Atrophoderma idiopathica (Pasini-Pierini)
epilepsy, 14/791
neurocutaneous syndrome, 14/791
Atrophoderma vermiculatum
Down syndrome, 14/791
heart disease, 43/8
mental deficiency, 43/8
neurofibromatosis type I, 14/113
Atrophy
adolescent spinal muscular, *see* Adolescent spinal muscular atrophy
ankylosing spondylitis, 38/506
Arnold-Chiari malformation type I, 32/104
brain, *see* Brain atrophy
brain cortex, *see* Brain cortex atrophy
bulbospinal muscular, *see* Bulbospinal muscular atrophy
cardiac, *see* Cardiac atrophy
caudate nucleus, *see* Caudate nucleus atrophy
cerebellar, *see* Cerebellar atrophy
cerebellar granular layer, *see* Cerebellar granular layer atrophy
cerebellar vermis, *see* Cerebellar vermis atrophy
cerebellifugal, *see* Cerebellifugal atrophy
cerebello-olivary, *see* Cerebello-olivary atrophy
cerebellopallidoluysionigral, *see* Cerebellopallidoluysionigral atrophy
cerebral, *see* Brain atrophy
choroid, *see* Choroid atrophy
congenital cerebellar, *see* Congenital cerebellar atrophy
congenital cervical spinal, *see* Congenital cervical spinal atrophy
congenital hemifacial, *see* Congenital hemifacial atrophy
congenital progressive hemifacial, *see* Congenital progressive hemifacial atrophy
congenital recessive optic, *see* Congenital recessive optic atrophy
cortical cerebellar, *see* Cortical cerebellar atrophy
cortical muscle, *see* Cortical muscular atrophy
crossed cerebellar, *see* Crossed cerebellar atrophy
crossed cerebrocerebellar, *see* Crossed cerebrocerebellar atrophy
cutaneous, *see* Cutaneous atrophy
dentatorubral, *see* Dentatorubral atrophy
dentatorubropallidoluysian, *see* Dentatorubropallidoluysian atrophy
diabetic proximal neuropathy, 27/132
distal muscular, *see* Distal muscular atrophy

distal spinal muscular, *see* Spinal muscular atrophy
dominant cerebello-olivary, *see* Dominant cerebello-olivary atrophy
dominant infantile optic, *see* Dominant infantile optic atrophy
dominant olivopontocerebellar, *see* Dominant olivopontocerebellar atrophy
epilepsy, 73/399
facioscapulohumeral muscular, *see* Facioscapulohumeral muscular atrophy
facioscapulohumeral spinal muscular, *see* Facioscapulohumeral spinal muscular atrophy
frontal lobe, *see* Frontal lobe atrophy
frontotemporal, *see* Frontotemporal atrophy
Fuchs gyrate, *see* Gyrate atrophy (Fuchs)
granular brain cortex, *see* Granular cortical cerebral atrophy
granular layer, *see* Granular layer atrophy
gyrate, *see* Gyrate atrophy
hemicerebral, *see* Hemicerebral atrophy
hemifacial, *see* Hemifacial atrophy
hemispheric, *see* Hemispheric atrophy
hereditary congenital cerebellar, *see* Hereditary congenital cerebellar atrophy
hereditary congenital optic, *see* Hereditary congenital optic atrophy
hereditary muscular, *see* Hereditary muscular atrophy
hereditary progressive cochleovestibular, *see* Hereditary progressive cochleovestibular atrophy
hereditary progressive spinal muscular, *see* Hereditary progressive spinal muscular atrophy
infantile spinal muscular, *see* Infantile spinal muscular atrophy
juvenile progressive bulbar, *see* Juvenile progressive bulbar atrophy
Kjer optic, *see* Infantile optic atrophy
laminar brain, *see* Laminar brain atrophy
late cerebellar, *see* Late cerebellar atrophy
late cortical cerebellar, *see* Late cortical cerebellar atrophy
Leber hereditary optic, *see* Congenital retinal blindness
Leber optic, *see* Congenital retinal blindness
limb, *see* Limb atrophy
lobar, *see* Lobar atrophy
luysian, *see* Luysian atrophy
Menzel hereditary olivopontocerebellar, *see* Hereditary olivopontocerebellar atrophy (Menzel)
monomelic spinal muscular, *see* Monomelic spinal muscular atrophy
muscle fiber, 40/11, 142, 200, 62/13
muscle fiber type I, 40/14, 144
muscle fiber type II, *see* Muscle fiber type II atrophy
muscle filamentous body, 40/249
muscular, *see* Muscular atrophy
myopathy, 62/13
neurogenic muscular, *see* Neurogenic muscular atrophy
neuromuscular disease, 62/13
nonprogressive spinal muscular, *see* Nonprogressive spinal muscular atrophy
olivocerebellar, *see* Olivocerebellar atrophy
olivopontocerebellar, *see* Olivopontocerebellar atrophy
optic, *see* Optic atrophy
optic nerve, *see* Optic atrophy
opticocochleodentate, *see* Opticocochleodentate atrophy
pallidal, *see* Pallidal atrophy
pallidofugal, *see* Pallidofugal atrophy
pallidoluysian, *see* Pallidoluysian atrophy
ponto-olivary, *see* Ponto-olivary atrophy
primary granular cell layer, *see* Primary granular cell layer atrophy
progressive cochleovestibular, *see* Progressive cochleovestibular atrophy
progressive hemifacial, *see* Progressive hemifacial atrophy
progressive muscular, *see* Progressive muscular atrophy
progressive spinal muscular, *see* Progressive spinal muscular atrophy
recessive infantile optic, *see* Recessive infantile optic atrophy
recessive olivopontocerebellar, *see* Recessive olivopontocerebellar atrophy
recessive optic, *see* Recessive optic atrophy
reflex, *see* Reflex atrophy
Ryukyan spinal muscular, *see* Ryukyan spinal muscular atrophy
scapulohumeral muscular, *see* Scapulohumeral muscular atrophy
scapulohumeral spinal muscular, *see* Scapulohumeral spinal muscular atrophy
scapulohumeroperoneal muscular, *see* Scapulohumeroperoneal muscular atrophy
scapuloilioperoneal muscular, *see* Scapuloilioperoneal muscular atrophy
scapuloperoneal spinal muscular, *see* Scapuloperoneal spinal muscular atrophy
skin, *see* Cutaneous atrophy

spinal cord, *see* Spinal cord atrophy
spinal muscular, *see* Spinal muscular atrophy
tegmental, *see* Tegmental atrophy
temporal lobe, *see* Temporal lobe atrophy
testicular, *see* Testicular atrophy
tick bite intoxication, 37/111
tongue, *see* Tongue atrophy
tongue paralysis, 30/411
vestibular nerve, *see* Vestibular nerve atrophy
white matter, *see* White matter atrophy
Atropine
autonomic nervous system, 75/501
brain edema, 23/151
carbamate intoxication, 64/192
ciguatoxin intoxication, 37/84
delirium, 46/537
organophosphate intoxication, 37/555, 557, 64/170, 172, 230
parkinsonism, 6/833-838
pyrethroid intoxication, 64/216
synapse, 1/48
thermoregulation, 75/61
Atropine intoxication
headache, 48/420
Attachment disorder
infant, *see* Infantile attachment disorder
Attack
acidosis, 5/46
Adams-Stokes, *see* Adams-Stokes syndrome
bilious, *see* Bilious attack
brain edema, 67/80
carbamazepine, 5/381
catamenial migraine, 48/18, 85
classic migraine, 48/124
drop, *see* Drop attack
epileptic drop, *see* Drop attack
fructose intolerance, 42/548
headache, 48/44
hyperkalemic periodic paralysis, 62/461
hypokalemic periodic paralysis, 41/149, 62/457
influenza, *see* Influenza attack
migraine, 48/124
narcoleptic sleep, *see* Narcoleptic sleep attack
neuralgic, *see* Neuralgic attack
normokalemic periodic paralysis, 28/597, 41/162, 43/171, 62/463
pseudoepileptic, *see* Pseudoepileptic attack
syncopal, *see* Syncopal attack
thyrotoxic periodic paralysis, 62/461
transient ischemic, *see* Transient ischemic attack
traumatic vegetative syndrome, 24/593
vertebrobasilar transient ischemic, *see* Vertebrobasilar transient ischemic attack
vomiting, 5/46
Attention
activation, 3/138
afferent neuronal habituation, 3/167, 170
age, 3/145-147
agnosia, 4/18, 30
alcohol intoxication, 64/115
alertness, 3/138, 155, 168
anxiety, 3/138
apparent movement, 3/144
arousal reaction, 3/138, 157, 168, 172
attentive behavior, 3/137, 155
autoevoked, 3/179
catecholaminoceptive vigilance system, 3/172-174
cerebrovascular disease, 55/137
conditioned evoked potential, 3/156
conditioning, 3/156
conscious integration, 3/159-161
Continuance Performance Test, 3/143
Continuous Choice Reaction Test, 3/143
cortical excitability, 3/165
corticifugal projection, 3/172
delirium, 46/529
distraction, 3/157, 159-167
distraction in experimental animal, 3/159-166
electrophysiologic correlate, 3/155-171
evolutionary hypothesis, 3/174
exoevoked, 3/179
experimental work, 3/138, 156, 159
filter model, 3/137
focusing, 3/157
gaze, 3/23
habituation, 3/167-171
head injury outcome, 57/399
information processing, 3/137, 150, 176
information storage, 3/138
intersensory transfer, 3/142
introspection, 3/137
learning, 3/156
mental calculation, 3/158, 166
mental deficiency, 46/29
mental task, 3/166
microelectrode study, 3/157
motivation, 3/156
neural basis, 2/645
neurohumoral correlate, 3/173
neurophysiologic aspect, 3/155-181
neurophysiologic correlate, 3/167-171
neurophysiologic model, 3/179-181
neuropil habituation, 3/167
neuropsychologic defect, 45/524
organic solvent intoxication, 64/48
oscillation, 3/164

p-system, 3/138
parietal lobe, 45/172
perception span, 3/159, 161
phenytoin intoxication, 65/502
plastic inhibition, 3/167
psychological aspect, 3/137-151
reaction time, 3/143, 145
reinforcement, 3/157
reticular formation, 3/138, 160, 168, 171, 176-179
reticulofugal effect, 3/176-179
reticulofugal facilatory influence, 3/165
reticulofugal inhibitory influence, 3/165
s-system, 3/138
schizophrenia, 46/501
selective nature, 3/137, 155
senescence, 3/145-147
sensory evoked potential, 3/156-161, 166
sensory extinction, 3/160-167
sensory filtering rhythmicity, 3/164
sensory habituation, 3/159, 167-171
sensory inhibition, 3/159, 164, 166, 176
sensory transmission facilitation, 3/156, 159
sleep, 3/168
somatesthetic, 3/195
spiral aftereffect, 3/144
stimuli unawareness, 3/159
styrene intoxication, 64/54
theory, 3/137
variety, 3/179-181
vigilance, 3/142-144, 155, 168, 173
visual, *see* Visual attention
visual agnosia, 2/651
visual fixation, 2/651
voluntary, 3/157, 172
waking, 3/171-181
Attention deficit
head injury, 46/624
head injury outcome, 57/405
hemispatial, *see* Hemispatial attention deficit
rehabilitation, 46/624
Attention deficit disorder syndrome
developmental dysphasia, 46/141
Attention disorder, 3/137-151, 187-198
agnosia, 3/189, 4/38
amorphosynthesis, 3/189
anatomic localization, 3/189
aphasia, 3/189
apraxia, 3/189
brain injury, 3/140, 150, 188
brain injury in children, 4/363
cerebroasthenic syndrome, 3/142, 149
classification, 3/187-189
distractibility, 3/139, 352
distribution, 3/141
epilepsy, 3/150
examination, 3/187
figure ground syndrome, 3/139
forced responsiveness, 3/139
frontal lobe lesion, 3/148-150, 189
general disorder, 3/187
hemiasomatognosia, 3/194
hyperkinetic child syndrome, 3/195-198, 43/200-202
hyperosexia, 3/139
hysteria, 3/181
impaired visual attention, 4/18
incidence, 3/188
psychological study, 3/139-151
schizophrenia, 3/140, 188
senile dementia, 3/145-147
sensory disorder, 3/187, 195
sensory extinction, 3/189-195, 4/30
subcortical lesion, 3/150, 189
temporal lobe lesion, 3/147, 352
unconscious hemiasomatognosia, 4/215
unilateral auditory disorder, 4/38
visual agnosia, 4/18
visuospatial agnosia, 3/194, 4/215
Attitude en chien de fusil
spinal subarachnoid hemorrhage, 12/133
Atypical benign partial epilepsy
epilepsy with continuous spike wave during slow sleep, 73/275
seizure, 73/224
Atypical facial headache, *see* Sphenopalatine neuralgia
Atypical facial neuralgia, *see* Sphenopalatine neuralgia
Atypical giant cell, *see* Aneurysmal bone cyst
Atypical inclusion body
progressive myoclonus epilepsy, 15/414, 49/609
Atypical lipoma, *see* Hibernoma
Atypical psychosis
delirium, 46/552
Atypical somatoform disorder, *see* Hysteria
Atypical teratoid/rhabdoid tumor
survey, 68/384
Audiogenic startle syndrome
EMG, 49/620
Audiometry
acoustic nerve tumor, 2/74, 17/673
Békésy, *see* Békésy audiometry
brain stem, *see* Brain stem audiometry
brain stem tumor, 16/307, 321
cochlear nerve, 16/307
pure tone, *see* Pure tone audiometry

speech, *see* Speech audiometry
vestibular neurinoma, 68/426
Audition
disconnection syndrome, 45/102
handedness, 4/320-323
perception, 45/521
sensory extinction, 3/191
Auditory agnosia
amusia, 4/37
aphasia, 4/94
attention unilateral, 4/38
brain commissurotomy, 4/281
classification, 4/38
congenital, *see* Congenital auditory agnosia
examination, 4/38
language, 4/281
localization, 4/2, 38
music type, 2/607
pure word deafness, 4/100
sensory aphasia, 4/37
temporal lobe lesion, 4/38
verbal, *see* Verbal auditory agnosia
Auditory aphasia
Wernicke aphasia, 4/98
Auditory artery
internal, *see* Internal auditory artery
Auditory association area
auditory hallucination, 2/705
temporal lobe epilepsy, 2/705
Auditory aura
partial epilepsy, 72/133
Auditory evoked potential
brain stem, *see* Brain stem auditory evoked potential
Friedreich ataxia, 60/321
head injury, 57/194
hereditary multiple recurrent mononeuropathy, 51/556
hyperactivity, 46/184
language, 46/148
lateralization, 46/504
schizophrenia, 46/502
Auditory fatigue
acoustic nerve tumor, 17/672, 675
Auditory hallucination
*see also* Verbal hallucination
anatomoclinical correlation, 46/568
auditory association area, 2/705
delirium, 46/526
EMG, 45/356
phosphene, 45/368
schizophrenia, 45/357, 46/562
temporal lobe, 45/356
temporal lobe epilepsy, 2/706, 711
temporal lobe tumor, 46/461, 568
visual hallucination, 2/709
Auditory illusion
temporal lobe epilepsy, 73/72
Auditory imperception
congenital, *see* Congenital auditory agnosia
developmental dysphasia, 4/361
hearing loss, 4/428
Auditory nerve, *see* Acoustic nerve
Auditory ossicle hypoplasia
branchio-otorenal dysplasia, 42/361
Auditory perception
developmental dysphasia, 46/140
examination, 4/38
migraine, 45/510
Auditory recognition test
brain tumor, 16/336
Auditory stenosis
external, *see* External auditory stenosis
Auditory stimulation
epilepsy, 15/446
startle epilepsy, 15/441
Auditory system
stuttering, 46/170
Auditory threshold
schizophrenia, 46/505
Auditory vein
internal, *see* Internal auditory vein
Auerbach, L., 1/47
Auerbach myenteric plexus, 22/245, 32/438
Wolman disease, 29/373, 51/379
Aura
auditory, *see* Auditory aura
cheiro-oral paresthesia, 48/157
emotional, *see* Emotional aura
epileptic, *see* Epileptic aura
headache, 48/5
hereditary paroxysmal dystonic choreoathetotis, 49/351
hereditary paroxysmal kinesigenic choreoathetosis, 49/353
history, 2/6
migraine, 5/46, 48/4, 18, 35, 39, 59, 70, 74, 76, 117, 125, 144, 150, 156
ocular, 48/156
spreading depression, 48/59, 68, 113
tolazoline, 5/379
vitamin B3, 5/379
Aura status
migraine, 48/157
Auranofin
gold intoxication, 64/355

Auricular herpes
peripheral facial paralysis, 7/483
Auricular nerve
great, *see* Auricularis magnus nerve
Auricular neuralgia, *see* Intermedius neuralgia
Auricular vagus branch
intractable hiccup, 63/491
Auricularis magnus nerve
anatomy, 5/369
facial nerve, 5/286
leprous neuritis, 51/219
lesion localization, 2/134
neurectomy, 5/400
Auriculotemporal nerve
gustatory sweating, 8/216
Auriculotemporal syndrome
dysautonomia, 22/260
facial nerve injury, 1/457, 8/276
facial paralysis, 8/270
gustatory sweating, 1/457, 2/122, 8/276
intermedius neuralgia, 48/490
mastication, 2/103
sweating, 2/103
Aurothiomalate
gold intoxication, 64/355
Austin juvenile sulfatidosis
ichthyosis, 60/667
neuropathy, 60/660
Austin urinary test
metachromatic leukodystrophy, 10/25
Austria
neurology, 1/10
Autism
behavior disorder in children, 43/202
childhood schizophrenic syndrome, 43/203
developmental dysphasia, 46/141
disequilibrium syndrome, 42/139
epidemiology, 46/191
genetics, 46/191
hand flapping, 43/203
infantile, *see* Infantile autism
mental deficiency, 46/194
nonfamilial type, 60/673
nuclear magnetic resonance, 60/673
9p partial monosomy, 43/512
perceptual disturbance, 43/202
phenylketonuria, 29/30, 43/204
phosphoribosyl pyrophosphate synthetase, 42/573
Rett syndrome, 60/636, 63/437
rubella syndrome, 43/204
seizure, 43/204
speech disorder in children, 43/202
Autoallergy, *see* Autoimmune disease
Autoantigen
experimental allergic neuritis, 9/505
hypersensitivity, 9/504
myelin lipoprotein, 9/505
oligodendrocyte, 9/505
Schwann cell, 9/505
Autohallucination, *see* Autoscopia
Autohypnosis
headache, 48/39
migraine, 48/39
Autoimmune antibody
CSF, *see* CSF autoantigen antibody
dermatomyositis, 62/380
inclusion body myositis, 62/380
polymyositis, 62/380
thyroid, *see* Thyroid autoimmune antibody
Autoimmune disease
allergic encephalomyelitis, 9/506
autonomic nervous system, 75/554
brain containing vaccine, 9/538
experimental allergic encephalomyelitis, 9/506-536
Hashimoto thyroiditis, 41/77, 240
hemolytic anemia, 41/77
Huntington chorea, 6/382-384
hyperthyroidism, 41/235-240, 335
immunosuppressive therapy, 65/550
late cortical cerebellar atrophy, 21/495
mixed connective tissue disease, 41/78
multiple sclerosis, 9/107, 115, 135, 144-148, 539-543, 47/188, 338
myasthenia gravis, 41/95, 97, 62/391
myositis, 62/369
pernicious anemia, 41/73, 102, 240
polyarteritis nodosa, 41/89, 385
polymyalgia rheumatica, 41/60, 76, 83, 245
polymyositis, 41/54-83
progressive dysautonomia, 59/162
progressive multifocal leukoencephalopathy, 47/503
progressive spinal muscular atrophy, 22/24
progressive systemic sclerosis, 41/55, 59
rabies vaccine, 9/538
rheumatoid arthritis, 41/59, 76, 84, 240
Sjögren syndrome, 41/102, 240
systemic lupus erythematosus, 22/24, 41/55, 59, 102, 275, 55/370
temporal arteritis, 55/341
ulcerative colitis, 41/102
viral infection, 56/60
Waldenström macroglobulinemia polyneuropathy, 41/82
Wegener granulomatosis, 41/87

Autoimmune encephalomyelitis
  experimental, *see* Experimental autoimmune encephalomyelitis
Autoimmune myasthenia gravis
  experimental, *see* Experimental autoimmune myasthenia gravis
Autoimmune thyroiditis, *see* Hashimoto thyroiditis
Autoimmunity, *see* Autoimmune disease
Autokinesis
  migraine, 48/162
Autologous mixed lymphocyte reaction
  multiple sclerosis, 47/356, 361
Automatic movement
  basal ganglion control, 1/161
Automatism
  compulsive type, 2/737
  epilepsy, 73/474, 480
  migraine, 48/162
  narcolepsy, 45/148
  nonconvulsive seizure, 15/140
  petit mal epilepsy, 42/680
  reactive, 73/71
  reflex, *see* Reflex automatism
  spinal, *see* Spinal automatism
  temporal lobe epilepsy, 15/95, 42/685, 73/71
Automutilation, *see* Self-mutilation
Autoneuralization
  definition, 30/46
  Hensen node, 30/49
  neural antigen, 30/49
Autonomic cephalalgia, *see* Cluster headache
Autonomic crisis
  headache, 26/494
  hypothalamic lesion, 2/450
Autonomic dysfunction
  *see also* Autonomic nervous system
  acute intermittent porphyria, 63/150
  alcohol, 75/493
  alcoholism, 63/148, 75/493
  amyloid polyneuropathy, 51/416, 63/148
  amyloidosis, 63/148
  amyotrophic lateral sclerosis, 59/196
  anhidrosis, 63/145
  brain death, 24/721
  brain embolism, 53/207
  brain infarction, 53/207
  cardiac arrest, 63/215
  central, *see* Central autonomic dysfunction
  cerebellar ataxia, 69/356
  cerebrovascular disease, 53/207
  congenital, *see* Congenital autonomic dysfunction
  diabetic polyneuropathy, 63/148
  diffuse Lewy body disease, 75/188
  drug induced extrapyramidal syndrome, 6/254
  dyspnea, 63/482
  electrical injury, 23/717
  epilepsy, 15/140
  Fabry disease, 51/377
  familial, *see* Familial autonomic dysfunction
  frontal lobe tumor, 17/257
  genetics, 75/409
  Guillain-Barré syndrome, 51/256, 63/148
  head injury, 57/238
  human immunodeficiency virus infection, 75/412
  Huntington chorea, 49/280
  intracranial hemorrhage, 53/207
  leprous neuritis, 51/219
  multiple neuronal system degeneration, 63/146, 75/168
  olivopontocerebellar atrophy, 49/206, 63/146
  pain, 75/329
  paraneoplastic syndrome, 69/356
  Parkinson disease, 75/178
  porphyria, 63/148, 75/695
  posttraumatic syringomyelia, 61/383
  progressive supranuclear palsy, 75/189
  ptosis, 63/148
  respiration, 63/482
  rheumatoid arthritis, 71/20
  scleroderma, 71/113
  Shy-Drager syndrome, 63/146
  spinal cord injury, 26/253, 313-333
  striatonigral degeneration, 49/206, 63/146
  subacute sclerosing panencephalitis, 56/419
  syncope, 22/262
  syringomyelia, 32/268-270, 50/445
  test, 74/56
  tetanus, 65/217, 225
  thoracolumbar spine injury, 25/454
  transient ischemic attack, 53/207
  treatment, 26/327
  urinary tract, 75/670
  vibration neuropathy, 51/134
Autonomic faciocephalalgia, *see* Sphenopalatine neuralgia
Autonomic failure
  autonomic nervous system, 74/174, 75/34
  bending forward, 75/721
  blood pressure, 75/713
  chronic primary, 74/175
  dyssomnia, 74/560
  erythropoietin, 75/725
  head up tilt maneuver, 75/717
  multiple neuronal system degeneration, 75/644
  octreotide, 75/725
  orthostatic hypotension, 75/114, 713

pharmacology, 74/174
selective, 75/123
skeletal muscle pumping, 75/721
sleep disorder, 74/543
Autonomic ganglion
anatomy, 1/431
autonomic nervous system, 1/431
congenital disease, 75/230
transmission, 74/40
Autonomic hyperreflexia
definition, 75/571
history, 75/572
pharmacology, 75/574
pregnancy, 74/332
treatment, 75/573
Autonomic innervation
bladder, 1/368, 373
brain vasospasm, 11/142
iris, 74/419
larynx, 74/387
spinal cord, 22/246
Autonomic nerve
leprosy, 51/217
Autonomic nervous sign
Friedreich ataxia, 60/312
Autonomic nervous system
*see also* Autonomic dysfunction *and* Sympathetic nervous system
acetylcholine, 74/92, 137, 75/85
acid base homeostasis, 74/588
acquired immune deficiency syndrome, 74/335
acrylamide, 75/36
acrylamide intoxication, 75/504
activation, 74/9
acute dysautonomia, 75/690
acute hypoxia, 75/259
acute intermittent porphyria, 27/434, 42/617
acute pandysautonomia, 75/13
adiposis dolorosa, 43/57
adrenal medulla, 74/35
adrenergic innervation, 74/217
adrenergic receptor stimulating agent, 22/250
adrenomyeloneuropathy, 75/24
age, 74/61
aged, 75/9
aging, 74/107, 225, 230, 75/5
alcohol intoxication, 64/112
alcoholism, 75/32
allergic neuritis, 75/16
Allgrove syndrome, 75/701
alpha adrenergic receptor stimulating agent, 22/251
altitude, 74/586, 75/440
Alzheimer disease, 75/9
amiodarone, 75/36
amphetamine, 75/501
amygdala, 74/58, 79
amyloid polyneuropathy, 75/31
anal fissure, 75/639
anal sphincter, 74/635
anatomy, 1/428-432, 74/399
angioneurotic edema, 43/61
anhidrosis, 74/218, 407
anterior cingulate cortex, 74/58
antibody response, 75/554
anxiety neurosis, 43/199
area postrema, 74/62
arginine, 74/137
Arnold-Chiari malformation type II, 75/10
arrhythmia, 74/171
arsenic, 75/38
arterial baroreflex, 75/432
arteriosclerosis, 74/120
atropine, 75/501
autoimmune disease, 75/554
autonomic failure, 74/174, 75/34
autonomic ganglion, 1/431
axonal neuropathy, 75/15
Bamberger-Marie syndrome, 75/543
barbiturate intoxication, 75/501
baroreceptor disorder, 75/40
baroreceptor regulation, 74/213
baroreflex, 74/623
baroreflex pathway testing, 74/609
basal forebrain, 74/58
beta adrenergic receptor blocking agent, 22/251
beta adrenergic receptor stimulating agent, 22/251
biological rhythm, 74/467, 75/431
bladder, 74/116
blood flow, 74/73, 604, 75/429
blood pressure, 74/14, 248, 611, 613, 75/426, 429, 453
body fluid, 74/1
body function, 74/3
botulism, 75/38, 509
bradycardia, 75/428
brain artery, 74/216
brain blood flow, 75/361
brain cortex, 74/79
brain stem, 74/61, 71
brain stem dysfunction syndrome, 75/534
brain tumor, 75/527
breathlessness, 74/574
brown fat cell, 74/209
β-bungarotoxin intoxication, 75/512
calcitonin gene related peptide, 75/86

cancer, 75/527
Cannon view, 74/15
carbamate intoxication, 75/508
cardiac development, 74/213
cardiac dysrhythmia, 75/441
cardiac output, 74/248
cardiomyopathy, 75/447
cardiopulmonary receptor, 74/258
cardiopulmonary reflex, 75/111
cardiovascular control, 74/75
cardiovascular development, 74/210
cardiovascular function, 74/600
cardiovascular reflex, 74/72
cardiovascular rhythm, 74/71
cardiovascular system, 75/425
carotid artery, 75/434
carotid sinus massage, 74/623
causalgia, 75/36
cell death, 74/204
cell differentiation, 74/204
cell migration, 74/204
cell proliferation, 74/203
central command, 74/249
central pathway, 74/151
cerebrovascular development, 74/216
cerebrovascular disease, 75/361
Chagas disease, 74/335, 75/19, 700
chemoreceptor function, 74/214
chemosensitive area, 74/65
Chinese restaurant syndrome, 75/502
chlorine, 75/509
cholinergic dysautonomia, 75/13
cholinergic innervation, 74/217
cholinergic receptor blocking agent, 22/250
chronic fatigue, 75/126
chronic inflammatory demyelinating polyradiculoneuropathy, 75/16
chronic obstructive pulmonary disease, 75/12
ciclosporin, 75/501
ciguatoxin, 75/510
ciguatoxin intoxication, 75/510
circadian rhythm, 74/14, 75/410, 437
cisplatin, 75/38, 496
citrulline, 74/137
cluster headache, 1/513, 75/281
cocaine, 75/501
coital migraine, 75/301
cold face test, 74/606, 622
cold pressor test, 74/624
collagen vascular disease, 75/12
colocalization, 74/97
congenital chemosensitive hypoventilation syndrome, 74/583
congenital disease, 75/229
congenital megacolon, 75/231, 633
congenital pain insensitivity, 42/297
constipation, 74/116
cortical region, 74/57
cosmonaut, 74/273
cotransmission, 74/98
cough headache, 75/300
coughing, 74/584, 622
Coxsackie virus, 74/335
Crohn disease, 75/653
cutaneous vasoconstrictor neuron, 74/26
deficiency state, 75/491
descending outflow, 74/152
development, 74/106, 203, 207
diabetes mellitus, 75/26, 589
diabetic autonomic neuropathy, 75/29, 246, 648
diabetic neuropathy, 74/604, 75/29
diencephalic syndrome, 75/39
diencephalon, 74/59
digestion, 74/1
digestive system, 74/632
diphtheria, 75/20
disulfiram, 75/38
diving, 75/440
dopamine β-mono-oxygenase deficiency, 74/218, 75/21, 237
Down syndrome, 75/643
dysautonomia, 74/614
dyspnea, 74/575
Eaton-Lambert myasthenic syndrome, 75/16, 536
ECG, 74/621
Edinburgh Index, 74/621
Elapidae, 75/512
electrical injury, 7/378, 23/717
electrolyte balance, 1/499-501
electrolyte depletion, 75/492
emotion, 3/320, 74/79, 369, 75/439
enteric nervous system, 74/214
enteric neuronopathy, 75/535
epilepsy, 75/349, 353
episodic pulmonary edema, 43/62
Epstein-Barr virus, 74/335
ergoreceptor, 75/428
ergotropic reaction, 74/17
excretion, 74/1
exercise, 74/14, 81, 245
eye, 74/399
Fabry disease, 42/427, 75/21
facial nerve, 2/110
familial visceral neuropathy, 75/641
fatal familial insomnia, 75/21, 408
fetal heart, 74/211

fetal thermoregulation, 74/208
fetal transition, 74/214
fever, 43/68, 74/453, 75/543
fight, 74/1
final autonomic motor pathway, 74/21
final autonomic pathway, 74/20
flight, 74/1
forebrain, 74/152
forgotten respiration, 74/583
function, 74/1
functional anatomy, 74/4, 53
functional organization, 74/18
GABA, 74/56
gastrointestinal disorder, 75/571, 613
gastrointestinal tract, 74/14, 116
genetic influence, 74/212
genetics, 74/218
glossopharyngeal nerve, 2/110
glossopharyngeal neuralgia, 75/298
glutamic acid, 74/55
gravity, 74/274, 609
Groll-Hirschowitz syndrome, 75/642
guanethidine, 75/17
Guillain-Barré syndrome, 42/645, 51/245, 74/336, 75/14, 681
head paradoxical reflex, 74/574
head up tilt test, 74/615
headache, 75/281
heart, 74/14
heart disease, 75/425
heart failure, 74/173, 75/443
heart rate, 74/246, 581, 608, 611, 75/426
heart transplantation, 75/446
hemoglobin level, 74/589
hereditary, 75/229
hereditary amyloid polyneuropathy, 60/104
hereditary amyloid polyneuropathy type 1, 42/518, 522
hereditary motor and sensory neuropathy type I, 21/271-309, 75/24, 700
hereditary motor and sensory neuropathy type II, 75/24
hereditary sensory and autonomic neuropathy type I, 75/23
hereditary sensory and autonomic neuropathy type II, 75/23
hereditary sensory and autonomic neuropathy type III, 1/475-480, 8/345, 42/300, 43/59, 60/23, 26, 28, 35, 75/23, 154, 700
hereditary sensory and autonomic neuropathy type IV, 75/24
hereditary sensory and autonomic neuropathy type V, 75/24
Hering-Breuer deflation reflex, 74/572
Hering-Breuer inflation reflex, 74/572
herpes zoster, 75/21
Hess, 74/16
high altitude, 74/588, 75/259
Hirayama disease, 59/111
history, 1/427
Holmes-Adie syndrome, 43/72, 75/125
homeostasis, 74/15
Horner syndrome, 43/64, 74/406
human immunodeficiency virus, 74/335, 75/18, 412
human T-lymphotropic virus type I, 75/414
human T-lymphotropic virus type II, 75/415
hypertension, 74/118, 170, 75/246, 453-455
hyperventilation, 74/586, 75/261
hypocapnia, 75/261
hypotension, 74/229, 613
hypothalamic syndrome, 75/533
hypothalamus, 74/2, 14, 59, 76, 152, 156
hypoventilation, 1/427-515, 2/107-124, 74/583
hypoxia, 74/73, 108, 75/10, 259
idiopathic erythromelalgia, 75/36
imaging, 74/626
immune system, 75/551
immunology, 75/551
infantile development, 74/199
infantile neuroaxonal dystrophy, 49/402
infarction, 75/11
infectious disease, 74/334
inherited spinocerebellar ataxia, 75/35
inspiratory gasping, 74/625
insular cortex, 74/57
intoxication, 75/491
intrinsic nerve, 74/104
iris, 74/399
ischemic heart disease, 75/444
isometric exercise, 74/625
J receptor, 74/575
juvenile parkinsonism, 49/156
laryngology, 74/387
larynx, 74/387
lathyrism, 65/3
leprosy, 75/18
leukemia, 39/12
Lewis triple response, 1/485
limbic system, 2/720, 74/2
liver disease, 75/32
localization, 74/53
locus ceruleus, 74/155
longitudinal study, 74/595
Machado-Joseph disease, 75/35
macrophage, 75/556

MAO inhibitor intoxication, 37/320
medulla, 74/152
megacolon, 74/116
mental stress test, 74/624
mercury, 75/38
mercury intoxication, 75/503
metabolic hormone, 74/35
metabolism, 74/14, 77
microneurography, 74/608
micturition, 75/665
micturition reflex, 74/75
midbrain, 74/155
migratory pathway, 74/201
mineral deficiency, 75/492
miosis, 74/424
mitral valve prolapse, 75/40
MNGIE syndrome, 75/642
monoamine, 74/139
monoamine oxidase deficiency, 75/240
motion sickness, 74/351
motoneuron disease, 75/39
mountain medicine, 75/259
multiple endocrine neoplasia type 2b, 75/25
multiple neuronal system degeneration, 74/544, 75/32
multiple sclerosis, 47/68, 75/18
muscarinic acetylcholine, 75/553
muscle, 74/33
muscle vasoconstrictor neuron, 74/23
myocardial infarction, 74/171, 75/444
myocardial ischemia, 74/172
myotonic dystrophy, 40/518, 75/641
neck chamber, 75/434
neocortex, 74/2
neonatal development, 74/199
nervi nevorum, 75/703
neural crest, 74/201
neural induction, 74/203
neuroanatomy, 74/55
neuroblastoma, 42/758, 75/528
neurochemical coding, 74/10, 98
neurochemical organization, 74/87
neurocirculatory syndrome, 74/173
neuroeffector junction, 74/87
neuroendocrine function, 74/626
neurogenic orthostatic hypotension, 74/613
neurogenic osteoarthropathy, 1/487
neuroleptic malignant syndrome, 75/13, 497
neuromodulation, 74/98
neuron death, 74/206
neuropathy, 74/113, 75/681
neuropeptide, 74/10, 56
neurotransmission, 74/136
neurotransmitter, 74/38, 55, 89, 91
neurotrophin-3 deficiency, 74/218
nitric oxide, 74/137, 75/85
nitric oxide synthase, 74/137
noradrenalin, 74/92, 75/86
noradrenalin spillover, 75/458
notexin, 75/512
nucleus ambiguus, 74/63
oculomotor nerve, 2/107
oculosympathetic pathway, 74/409
odor, 74/367
odorant, 74/364, 366
Ogilvie syndrome, 75/633
old age, 74/3
olfaction, 74/363, 366
olivopontocerebellar atrophy, 22/265
Onufrowicz nucleus, 74/70
opioid analgesic agent, 75/501
organ regulation, 74/14
organ response, 22/248
organochlorine pesticide intoxication, 75/509
organophosphate intoxication, 64/162, 75/507
orthostatic hypotension, 22/253, 43/66, 75/715
Oxford method, 75/432
oxidopamine, 75/38
pain, 74/75, 75/309
pain insensitivity, 74/218
pain syndrome, 75/35
paleoneurobiology, 74/181
paleopathology, 74/183
palmar hyperhidrosis, 74/628
parabrachial region, 74/61
parainfluenza virus, 74/335
paralytic shellfish poisoning, 75/512
paraneoplastic dysautonomia, 75/15
paraneoplastic syndrome, 75/15, 536
parasympathetic activity, 74/213
parasympathetic nervous system, 74/201
parasympathetic neuron, 75/620
parasympathetic outflow, 74/152
Parkinson disease, 75/643
pathology, 75/1
patterning, 74/203
pellagra, 75/492
penis erection, 75/85
periaqueductal gray matter, 74/61
perinatal asphyxia, 74/214
periodic fever, 43/68
peripheral dysfunction, 75/105
peripheral reflex, 74/41
pesticide, 75/507
petit mal epilepsy, 42/680
petrosal nerve, 75/291

pharmacology, 74/135-137, 139
phenothiazine, 75/499
phenylketonuria, 75/234
pheromone, 74/373, 379
phosphate, 75/507
piloerection, 1/467
pilomotor neuron, 74/32
plasma catecholamine, 74/625
plasticity, 74/106
plethysmography, 1/466
poliomyelitis, 42/652, 75/20
POLIP syndrome, 75/643
pons, 74/152, 155
pontine micturition center, 74/62
porphyria, 75/31
postpoliomyelitic amyotrophy, 75/20, 416
postural reflex, 1/441
posture, 75/436
postviral fatigue syndrome, 75/127
pregnancy, 74/108, 328
prion disease, 75/407
proctalgia, 43/69
propionic acid, 74/367
puffer fish intoxication, 75/512
pulse interval, 74/230
pupillary response, 74/170
purine, 74/93
rabies, 75/418
rabies virus, 75/416
Raynaud disease, 43/70
referred pain, 22/253
reflex, 74/19, 25
reflex sympathetic dystrophy, 75/35, 309
Refsum disease, 21/244, 60/231
relative afferent pupillary defect, 74/413
REM sleep, 74/524
renal insufficiency, 27/338
reproductive behavior, 74/1, 14
respiration, 74/72, 581, 75/430, 568
respiratory function, 74/571, 75/425
retrovirus, 75/411
rheumatoid arthritis, 38/487
sacral parasympathetic outflow, 74/69
sacral reflex, 74/71
schizophrenia, 46/496-498
scorpion venom, 75/513
scrapie, 42/656
sensory receptor, 74/33
serotonin syndrome, 75/13, 499
sexual function, 74/540, 75/85
sexual impotence, 74/118
sexual organ, 74/14
Shapiro syndrome, 75/10
shivering, 1/451
skin flare, 74/607
skin response, 74/629
skin temperature, 75/430
sleep, 74/14, 523, 537, 75/437
sodium, 74/76
sodium glutamate, 75/502
somatomotor inhibition, 74/259
space, 74/273
specific nucleus, 74/59
spider bite intoxication, 75/512
spinal cord, 74/65, 70, 201, 75/567
spinal cord injury, 75/569
spinal efferent, *see* Spinal autonomous efferent
spinal meningioma, 20/198
spinal neurinoma, 20/252
spinal shock, 75/569
squatting, 74/619
squatting test, 74/619
standing, 74/609, 611
stiff-man syndrome, 75/18
stress, 74/79-81, 232, 307, 75/439
stroke volume, 74/247
structure, 74/1
styrene intoxication, 75/506
subacute combined degeneration, 75/30
subacute sensory neuronopathy, 75/534
subacute sensory neuropathy, 75/15
subarachnoid hemorrhage, 12/150, 55/14, 75/39
subnuclear organization, 74/62
sudden death, 74/171
sudomotor neuron, 74/31
sudomotor reflex, 74/629
SUNCT syndrome, 75/299
supraspinal control, 74/66
swallowing, 74/631
sweat disturbance, 74/660
sweat gland, 74/407
sweat test, 74/628
sweating, 1/452-459, 74/627
sympathectomy, 74/110, 75/17
sympathetic ganglion, 74/225
sympathetic nerve, 74/658
sympathetic neuron, 75/620
syphilis, 75/20
syringobulbia, 75/10
syringomyelia, 75/10
systemic sclerosis, 74/120
tachycardia, 74/614
taipoxin, 75/512
Tangier disease, 42/627, 75/32
target organ, 74/9
taxol, 75/38, 497

telencephalic structure, 74/57
temporal lobe, 2/720
test, 1/466
testing, 74/595
tetanus, 52/230
tetrahydrobiopterin metabolism, 75/234
tetrodotoxin intoxication, 75/511
thalamus, 74/61, 155
thallium, 75/38
thallium intoxication, 75/503
thermoregulation, 1/437-451, 74/1, 14, 76, 207
thyroid gland, 27/261-263
toluene intoxication, 75/506
transmitter substance, 74/38
trauma, 75/39
trichopoliodystrophy, 29/287, 75/22, 238
tricyclic antidepressant, 65/321
trigeminal neuralgia, 75/298
trophotropic reaction, 74/17
tyrosine hydroxylase, 75/233
ulcerative colitis, 75/653
uremia, 75/31
uremic neuropathy, 27/338
urinary bladder, 74/14
urinary tract, 75/571
urogenital tract, 74/116
vacor, 75/39
vacor intoxication, 75/509
vagal paraganglioma, 75/232
vagina, 2/110
vagus nerve, 2/110, 74/62, 619, 75/428, 529
Valsalva maneuver, 74/615
valvular heart disease, 75/447
vascular circulation, 74/213
vasculitis, 75/11
vasoactive intestinal polypeptide, 75/85
vasoconstrictor pathway, 74/23
vasodilatation, 75/428
vasodilator neuron, 74/32
venoarteriolar reflex, 74/607
venous return, 74/260
ventilation, 74/249
ventrolateral medulla, 74/63
ventromedial medulla, 74/65
ventromedial prefrontal cortex, 74/57
vestibular function, 74/351
vestibular system, 74/262
Vidian nerve, 75/291
vincristine, 75/39, 531
viral disease, 75/407, 411
visceral afferent neuron, 74/12
visceral vasoconstrictor neuron, 74/26
vision, 74/399
vital function, 74/13
vitamin $B_1$ deficiency, 75/491
vitamin $B_6$ overdose, 75/502
vitamin $B_{12}$ deficiency, 75/491
vitamin deficiency, 75/492
vitamin E deficiency, 75/492
vitamin overdose, 75/502
vocal fold, 74/388
voice, 74/388
vomiting, 74/74
water, 74/76
Wernicke encephalopathy, 75/30
yawning, 74/584
Autonomic neuron
recording, 74/22
Autonomic neuropathy
acquired immune deficiency syndrome, 56/499
alcoholic polyneuropathy, 51/316
amyloid polyneuropathy, 75/691
congenital pain insensitivity, 51/564
diabetes mellitus, 70/148, 156
diabetic, *see* Diabetic autonomic neuropathy
diabetic polyneuropathy, 51/499
dialysis, 63/523
dimethylaminopropionitrile polyneuropathy, 51/282
drug intoxication, 51/294
Epstein-Barr virus infection, 56/252
Guillain-Barré syndrome, 51/245
hexane polyneuropathy, 51/277
2-hexanone polyneuropathy, 51/277
Hodgkin disease, 75/535
human immunodeficiency virus neuropathy, 71/357
idiopathic, *see* Idiopathic autonomic neuropathy
infectious mononucleosis, 56/252
intestinal, *see* Intestinal autonomic neuropathy
neurotoxicity, 51/294
neurotoxicology, 64/14
organophosphate intoxication, 51/284
paraneoplastic syndrome, 69/374, 378
Pick disease, 46/235
Sjögren syndrome, 71/84
survey, 74/113
uremic polyneuropathy, 63/511
vincristine intoxication, 65/535
vincristine polyneuropathy, 51/299
vitamin $B_1$ deficiency, 28/5
vitamin $B_{12}$ deficiency, 51/340
Autonomic neurotropism
congenital muscle absence, 1/481
Autonomic paralysis
nerve injury, 7/249

Autonomic polyneuropathy
*see also* Visceral neuropathy
acrodynia, 51/477
acrylamide polyneuropathy, 51/281, 479
acute intermittent porphyria, 51/479
acute pandysautonomia, 51/475
alcohol intoxication, 51/316
alcoholism, 51/476, 478
amyloid, *see* Amyloid autonomic polyneuropathy
amyloidosis, 51/476
arsenic intoxication, 51/479
botulism, 51/477
caffeine, 51/487
Chagas disease, 52/346
chronic renal failure, 51/476
chronic sensory neuropathy, 51/482
diabetes mellitus, 51/476
diabetic, *see* Diabetic autonomic polyneuropathy
dihydroergotamine, 51/487
electrical injury, 61/192
Epstein-Barr virus, 56/255
ergotamine tartrate, 51/487
Fabry disease, 51/476
fludrocortisone, 51/487
Guillain-Barré syndrome, 51/478
heart rate control, 51/487
hereditary acute porphyria, 51/391
hereditary amyloid polyneuropathy type 1, 51/420, 479
hereditary motor and sensory neuropathy type I, 51/479
hereditary motor and sensory neuropathy type II, 51/479
Holmes-Adie syndrome, 51/475
hyperhidrosis, 51/486
lead polyneuropathy, 64/437
malignancy, 51/476, 479
mercury intoxication, 51/477
micturition syncope, 51/487
mixed connective tissue disease, 51/450, 479
multiple mucosal neuroma, 51/476
nosology, 22/251
organic solvent intoxication, 51/479
orthostatic hypotension, 51/486, 63/148
pain insensitivity, 51/476
perhexiline intoxication, 51/479
phenylmercury derivative intoxication, 51/478
porphyria variegata, 51/479
postpoliomyelitis, 51/482
primary amyloidosis, 51/416
pupillary abnormality, 51/487
reflex sympathetic dystrophy, 51/479, 482
rheumatoid arthritis, 51/446, 479
sexual impotence, 51/487
Shy-Drager syndrome, 51/479
systemic lupus erythematosus, 51/479
tabes dorsalis, 51/486
taxol intoxication, 65/537
taxotere intoxication, 65/537
thallium intoxication, 51/479
treatment, 51/487
trophic disorder, 51/481
uremia, 51/479
uremic polyneuropathy, 63/511
vincristine, 63/358
vincristine intoxication, 51/478
vitamin B6 intoxication, 51/479
Wolman disease, 51/379
Autonomic reflex
pathway lesion, 74/596
peripheral, 75/110
testing, 74/595
Autonomic symptom
epilepsy, 15/140
hereditary sensory and autonomic neuropathy type I, 60/7
hereditary sensory and autonomic neuropathy type II, 60/7
hereditary sensory and autonomic neuropathy type III, 60/7, 26
hereditary sensory and autonomic neuropathy type IV, 60/7, 12
hereditary sensory and autonomic neuropathy type V, 60/7, 15
multiple neuronal system degeneration, 75/168
multiple sclerosis, 9/180
Parkinson disease, 49/92, 75/179
spinal tumor, 19/59
Autonomic syncope
orthostatic hypotension, 22/262
Autonomic system
hereditary amyloid polyneuropathy type 1, 60/92
hereditary amyloid polyneuropathy type 2, 60/98
Autonomic test
pathway lesion, 74/596
pure autonomic failure, 75/119
result, 74/599
Autonomic visceral neuropathy
diabetic, *see* Diabetic autonomic visceral neuropathy
Autonomous bladder
*see also* Neurogenic bladder
symptom, 2/191
Autonomous micturition, *see* Neurogenic bladder
Autophagic vacuole
acid maltase deficiency, 40/153

histochemistry, 40/36
muscle fiber, 62/23
myelin like figure, 40/108
rimmed muscle fiber vacuole, 40/231
ultrastructure, 40/108
Autophonia
hallucinosis, 46/562
Autopsy incidence
brain aneurysm, 55/44
brain aneurysm neurosurgery, 55/44
paraganglioma, 42/763
subarachnoid hemorrhage, 55/44
Autoregulation
brain anoxia, 27/40
brain blood flow, *see* Brain blood flow autoregulation
brain death, 24/758
brain edema, 23/153
brain infarction, 53/112
brain microcirculation, 23/138, 142, 144
brain perfusion pressure, 54/120
head injury, 57/70
intracranial pressure, 55/208
spinal cord, *see* Spinal cord autoregulation
spinal cord blood flow, 63/39
transcranial Doppler sonography, 54/12
vascular, *see* Vascular autoregulation
Autoscopia
alcohol intoxication, 4/232
alcoholism, 4/232, 45/386
body scheme disorder, 3/31, 254, 4/210, 231-233, 330, 45/357, 386
brain hemorrhage, 45/387
brain tumor, 45/387
epilepsy, 4/231-233, 45/387, 509
hallucinosis, 2/714, 46/562
memory, 2/714
migraine, 4/231-233, 45/387, 510, 48/162
parietal lobe epilepsy, 73/102
posttraumatic, 24/492
psychosis, 4/232, 45/386
reduplication phenomenon, 3/254
sleep disorder, 4/232
spatial orientation disorder, 24/491
temporal lobe, 2/714, 45/357
temporal lobe epilepsy, 2/711
typhoid fever, 4/232, 45/386
Autosomal dominant compelling helio-ophthalmic outburst
sneezing, 74/431
Autosomal dominant early onset cerebellar vermis atrophy
hereditary congenital cerebellar atrophy, 60/288
Autosomal dominant hypoalphalipoproteinemia
brain amyloid angiopathy, 60/136
homocystinuria, 60/136
migraine, 60/136
mitral valve prolapse, 60/136
oral contraception, 60/136
Autosomal dominant infantile optic atrophy
dyschromatopsia, 13/91
Autosomal dominant inheritance
acrocephalosyndactyly type III, 50/123
acrocephalosyndactyly type V, 50/123
Albers-Schönberg disease, 43/456
basilar impression, 50/64
bilateral acoustic neurinoma, 42/757
congenital ataxic diplegia, 42/122
craniosynostosis, 42/21
epilepsy, 50/272
familial amyotrophic lateral sclerosis, 59/241
Finkel spinal muscular atrophy, 59/43
foramina parietalia permagna, 50/142
hereditary brachial plexus neuropathy, 60/71
hereditary motor and sensory neuropathy type II, 60/200
hereditary progressive arthro-ophthalmopathy, 13/41
hereditary vitreoretinal degeneration, 13/273
hyaloidoretinal degeneration (Wagner), 13/37, 273
ichthyosis, 13/476
Klippel-Feil syndrome, 50/64
Marfan syndrome, 39/393
microcephaly, 50/272-274
migraine, 48/146
multiple sclerosis, 47/291
Parry disease, 42/467
Pick disease, 46/243
pilonidal sinus, 50/65
primary pigmentary retinal degeneration, 13/26
pseudoinflammatory foveal dystrophy (Sorsby), 13/138
quadriceps myopathy, 43/128, 62/185
renal agenesis syndrome, 50/513
Rieger syndrome, 43/471
Wohlfart-Kugelberg-Welander disease, 59/84
Autosomal dominant myopathy
benign, *see* Benign autosomal dominant myopathy
Autosomal dominant nocturnal frontal lobe epilepsy
psychosis, 72/132
sleep disorder, 72/132
Autosomal dominant optic atrophy
chorioretinal degeneration, 13/38
Pelizaeus-Merzbacher disease, 13/38
spinopontocerebellar degeneration, 13/38
Autosomal dominant pain insensitivity

clinical features, 60/17
hyperplastic myelinopathy, 60/17
nerve biopsy, 60/17
tongue papilla, 60/17
Autosomal dominant primary dystonia
chromosome 9, 59/346
hereditary secondary dystonia, 59/346
Autosomal dominant pseudoglaucoma
optic atrophy, 13/120
Autosomal dominant spastic paraplegia
spinal muscular atrophy, 59/374
Autosomal recessive cerebellar hypoplasia
ataxia telangiectasia, 60/387
Autosomal recessive congenital optic atrophy
hereditary, *see* Hereditary autosomal recessive congenital optic atrophy
Autosomal recessive generalized myotonia
age at onset, 62/266
chloride conductance, 62/267
fall, 62/266
muscle hypertrophy, 62/267
muscle stretch reflex, 62/266
muscle weakness, 62/266
muscular atrophy, 62/267
myotonia distribution, 62/266
myotonic dystrophy, 62/240
myotonic syndrome, 62/263
prevalence, 62/267
startle, 62/266
Autosomal recessive inheritance
acrocephalosyndactyly type II, 50/123
adrenoleukodystrophy, 51/383
adult neuronal ceroid lipofuscinosis, 42/465
anencephaly, 50/79
aqueduct stenosis, 50/62
cebocephaly, 50/61
cerebrotendinous xanthomatosis, 29/375, 42/430, 59/357
congenital ataxic diplegia, 42/122
congenital erythropoietic porphyria, 27/433
congenital hyperphosphatasia, 31/258
congenital pain insensitivity, 51/565
craniosynostosis, 42/21, 50/60
cyclopia, 50/61
familial amyotrophic lateral sclerosis, 59/242
fructose-1,6-diphosphatase deficiency, 29/257
genetics, 13/25
Hallervorden-Spatz syndrome, 49/399
hereditary sensory and autonomic neuropathy type IV, 60/13, 17
hyaloidotapetoretinal degeneration (Goldmann-Favre), 13/37
infantile spinal muscular atrophy, 59/54
Klippel-Feil syndrome, 50/64
microcephaly, 50/262, 269-272, 274
Miller-Dieker syndrome, 50/260
multiple sclerosis, 47/290
multisynostotic osteodysgenesis, 50/123
renal agenesis syndrome, 50/513
Richner-Hanhart syndrome, 42/581
Rosenberg-Chutorian syndrome, 22/507
Smith-Lemli-Opitz syndrome, 50/276
Wohlfart-Kugelberg-Welander disease, 59/82
Autosomal recessive lissencephaly
inheritance, 72/133
Autosomal recessive optic atrophy
congenital retinal blindness, 13/38
ERG, 13/38
Autosomal recessive vitreoretinal degeneration, *see* Hyaloidotapetoretinal degeneration (Goldmann-Favre)
Autosome
mental deficiency, 46/68-73
Autotopagnosia
Alzheimer disease, 46/251
body orientation, 45/381-383
body scheme disorder, 1/107, 2/668, 4/210, 219-222, 297, 45/374, 381-383
brain infarction, 55/145
cerebrovascular disease, 55/145
congenital pain insensitivity, 4/222
description, 45/381-383
digital, *see* Digital autotopagnosia
finger agnosia, 45/381
Gerstmann syndrome, 4/222
hemineglect, 55/145
occipital lobe tumor, 17/329
parietal lobe lesion, 1/107
parietal lobe syndrome, 45/70
right left disorientation, 45/382
Wernicke aphasia, 4/98
Avellis syndrome
characteristic deficit, 2/99
hemianesthesia, 1/189
hemiplegia, 1/189
medulla oblongata syndrome, 2/217
nonrecurrent nonhereditary multiple cranial neuropathy, 51/571
nucleus ambiguus, 1/189
original description, 2/217
palate, 1/189
Avian herpes virus
Marek disease, 47/609, 51/250
Avidin-vitamin $B_W$ enzyme complex
neurotropic virus, 56/26
Avulsion

infraorbital nerve, 5/392
spinal nerve root, *see* Spinal nerve root avulsion
spinal nerve root injury, *see* Spinal nerve root avulsion
supraorbital neuralgia, 5/392
ventral spinal nerve root, *see* Ventral spinal nerve root avulsion
Avulsion lesion
brachial plexus, 51/144
Awake-sleep cycle
generalized tonic clonic seizure, 73/175
Awake state
sympathetic nerve, 74/652
Awakening
dreaming, 3/88
epilepsy with generalized tonic clonic seizure, 72/7
sleep, 3/84
Awakening disorder
narcolepsy, 45/149
Awakening epilepsy
barbiturate, 15/485
ependymoma, 18/126
familial disorder, 15/461
Awareness
brain tumor, 16/733
terminology, 3/52
Axial athetosis, *see* Axial dystonia
Axial dystonia
athetosis, 49/382
focal athetosis, 49/383
tardive dyskinesia, 49/186
Axial mesodermal dysplasia spectrum
*see also* Caudal aplasia
anal atresia, 50/516
anophthalmia, 50/516
cardiac defect, 50/516
caudal aplasia, 50/516
Goldenhar syndrome, 50/516
hydrocephalus, 50/516
hypospadias, 50/516
microcephaly, 50/516
microtia, 50/516
polycystic kidney, 50/516
Potter face, 50/516
Potter syndrome, 50/516
radial aplasia, 50/516
renal agenesis syndrome, 50/516
twins, 50/517
variant, 50/509
Axial mesodermal system
dysraphism pathogenesis, 32/162
Axial shift
brain herniation, 16/123
brain stem hemorrhage, 16/123
brain stem tumor, 16/123
Axial tomography
computerized, *see* Computer assisted tomography
Axillary angiography
brain tumor, 16/635
Axillary nerve
compression neuropathy, 51/102
innervation area, 2/137
occupational lesion, 7/331
shoulder arthroplasty, 70/31
topographical diagnosis, 2/21
Axis dysplasia
axis dysraphia, 50/393
suboccipital dysplasia, 32/40
Axis dysraphia
axis dysplasia, 50/393
Axolemma
axon, 51/4
internodal, *see* Internodal axolemma
nodal, *see* Nodal axolemma
paranodal, *see* Paranodal axolemma
potassium channel, 47/14-18
sodium channel, 47/14-18, 33-37, 51/47
structure, 47/14-19
Axon
actin, 51/6
axolemma, 51/4
axoplasm, 51/4
conduction property, 51/41
conduction velocity, 51/44
depolarization, 47/37
development, 7/11
excitability, 51/46
flow, 1/60
freeze fracture, 47/9-11, 15, 17, 51/47
glia regulation, 51/43
impulse conduction, 51/44
membrane reorganization, 51/52
mesaxon, 51/7
multiple sclerosis plaque, 9/38
myelination, 51/42
nonmyelinated region, 47/21
paranodal junction, 51/41
pathologic conduction, 51/51
potassium channel, 51/49
Ranvier node, 47/10, 51/5
saxitoxin, 51/47
size, 51/5
sodium channel, 51/47
spinal cord regeneration, 25/100, 103
voltage clamping, 51/48

volume, 51/4
Axon membrane, *see* Axolemma
Axon myelination ratio
nerve, 51/4
Axon-neuron ratio
nerve, 51/4
Axon organelle
axonal transport, 51/6
nerve, 51/6
Axon-perikaryon volume ratio, *see* Axon-neuron ratio
Axon reflex
Bonney test, 2/137, 173, 7/411
hereditary sensory and autonomic neuropathy type IV, 60/13
nerve injury, 7/252
sudomotor, *see* Sudomotor axon reflex
Axon reflex flare, *see* Skin flare
Axon regeneration
action, 61/483
myelin associated axon growth protein, 61/484
myelin sheath, 47/21
nerve growth factor, 61/484
nerve growth inhibitory protein, 61/482
paraplegia, 2/179, 61/407
spinal cord injury, 61/407
tetraplegia, 61/407
transverse spinal cord lesion, 61/407
tubulin, 61/483
Axon schollen, *see* Spheroid body
Axon terminal
neuroaxonal dystrophy, 49/393
Axonal cytoskeleton
nerve, 51/6
Axonal damage
Guillain-Barré syndrome, 51/244
migrant sensory neuritis, 51/462
pyrethroid intoxication, 64/215
Axonal degeneration
a-alphalipoproteinemia, 51/397
acquired immune deficiency syndrome, 56/518
acrylamide intoxication, 64/63
alcoholic polyneuropathy, 51/318
amyloid polyneuropathy, 51/417
arsenic polyneuropathy, 51/267
Bassen-Kornzweig syndrome, 51/394, 66/16
botulism, 51/188
carbamate intoxication, 64/186
cerebrotendinous xanthomatosis, 51/399
chloroquine, 65/484
chronic inflammatory demyelinating polyradiculoneuropathy, 51/532
cold neuropathy, 51/137
critical illness polyneuropathy, 51/584
cryoglobulinemia, 63/404
cryoglobulinemic polyneuropathy, 51/434
dapsone intoxication, 65/486
dialysis, 63/530
2,4-dichlorophenoxyacetic acid polyneuropathy, 51/286
dimethylaminopropionitrile polyneuropathy, 51/283
disulfiram polyneuropathy, 51/309
eosinophilia myalgia syndrome, 64/254
ethambutol polyneuropathy, 51/295
ethylene oxide polyneuropathy, 51/275
hereditary acute porphyria, 51/392
hereditary motor sensory neuropathy type I, 51/264
hereditary multiple recurrent mononeuropathy, 51/552
hereditary neuropathy, 60/258
hereditary sensory and autonomic neuropathy type I, 60/9
hydrocephalus, 50/290
hypereosinophilic syndrome, 63/373
isoniazid, 65/481
isoniazid polyneuropathy, 47/565
lathyrism, 65/4
lead, 51/263
local anesthetic agent, 65/423
methanol intoxication, 64/101
metronidazole intoxication, 65/487
misonidazole polyneuropathy, 51/301
mixed connective tissue disease, 51/450
multiple myeloma polyneuropathy, 51/431, 63/393
neuralgic amyotrophy, 51/173
neuroaxonal dystrophy, 49/393
neuropathology, 51/63
neurotoxicity, 51/294
organophosphate induced delayed neuropathy, 64/229
paraneoplastic polyneuropathy, 51/469
POEMS syndrome, 47/619, 51/431, 63/394
polyneuropathy, 60/258
predominant, *see* Predominant axonal degeneration
primary, *see* Primary axonal degeneration
primary hyperoxaluria, 51/398
renal insufficiency polyneuropathy, 51/358
rheumatoid arthritis, 51/446
Sjögren syndrome, 51/449
toxic oil syndrome, 63/381
tri-*o*-cresyl phosphate polyneuropathy, 51/285
tryptophan contaminant, 64/254
uremic neuropathy, 7/166, 8/2, 27/339

uremic polyneuropathy, 51/359, 63/511
vincristine polyneuropathy, 51/299
vitamin B6 intoxication, 51/298, 65/572
xeroderma pigmentosum, 51/399
Axonal demyelination
trichloroethylene polyneuropathy, 51/280
Axonal dystrophy
axonopathy, 49/393
hepatic polyneuropathy, 51/326
Axonal neuropathy
autonomic nervous system, 75/15
Bassen-Kornzweig syndrome, 51/394, 63/275, 278
carbon disulfide, 64/86
carbon disulfide intoxication, 64/25
chronic, *see* Chronic axonal neuropathy
ciclosporin intoxication, 65/554
dermatomyositis, 62/376
eosinophilia myalgia syndrome, 63/377
giant, *see* Giant axonal neuropathy
hereditary amyloid polyneuropathy type 3, 51/425
inclusion body myositis, 62/376
isoniazid intoxication, 65/481
lithium intoxication, 64/296
organophosphate intoxication, 64/160
polymyositis, 62/376
progressive spinal muscular atrophy, 59/404
undetermined significant monoclonal gammopathy, 63/402
Axonal polyneuropathy
acquired immune deficiency syndrome, 56/517
creatine kinase, 62/609
critical illness polyneuropathy, 51/575
distal, *see* Distal axonal polyneuropathy
muscle biopsy, 62/609
spheromembranous body, 62/609
taxol intoxication, 65/537
taxotere intoxication, 65/537
Axonal reaction
chemistry, 7/222
demyelinating neuropathy, 47/621
nerve cell change, 7/219
organotin intoxication, 64/140
progressive spinal muscular atrophy, 22/14
ultrastructure, 7/221
Axonal spheroid body, *see* Spheroid body
Axonal sprouting
muscle reinnervation, 40/140
Axonal swelling
acrylamide intoxication, 64/74
aluminum intoxication, 64/276
giant, *see* Giant axonal swelling
hexane polyneuropathy, 51/277
2-hexanone polyneuropathy, 51/277
Axonal transport
axon organelle, 51/6
diphtheria, 51/182
1-methyl-4-phenyl-1,2,3,6-tetrahydropyridine, 65/391
1-methyl-4-phenyl-1,2,3,6-tetrahydropyridine intoxication, 65/391
neurofilament polyneuropathy, 51/278
neurotropic virus, 56/33
poliomyelitis, 56/318
retrograde, *see* Retrograde axonal transport
vincristine intoxication, 65/535
Axonopathy
*see also* Neuroaxonal dystrophy
acrylamide intoxication, 64/63
amnesic shellfish poisoning, 65/167
axonal dystrophy, 49/393
chloroquine intoxication, 65/484
cisplatin polyneuropathy, 51/300
distal, *see* Distal axonopathy
metronidazole polyneuropathy, 51/301
neurofilament, *see* Neurofilament axonopathy
neurotoxicity, 51/263
vitamin B6, 51/264
Axonotmesis
brachial plexus, 51/145
definition, 7/244
histology, 51/77
nerve injury, 7/244, 51/75, 134
Axoplasm
axon, 51/4
nodal, *see* Nodal axoplasm
wallerian degeneration, 9/36
Axoplasmic flow
blockade, 40/143
cold neuropathy, 51/137
component, 7/81
compression neuropathy, 51/88
evidence, 7/78
mercury intoxication, 64/389
Azacitidine
antimetabolite, 39/106
antineoplastic agent, 65/528
chemotherapy, 39/106
encephalopathy, 65/528
myoglobinuria, 41/271
neurologic toxicity, 39/106
neuropathy, 69/461, 472
neurotoxin, 39/94, 99, 106, 119, 41/271, 65/528
polyneuropathy, 65/528
Azacitidine intoxication
myalgia, 65/534

Azacosterol
cataract, 40/514, 548
drug induced myotonia, 40/155, 328, 514, 548, 555, 62/271
experimental myotonia, 40/155
5-Azacytidine, *see* Azacitidine
8-Azaguanine
antimetabolite, 39/106
chemotherapy, 39/106
neurologic toxicity, 39/106
Azapropazone
headache, 48/174
migraine, 48/174
Azathioprine
dermatomyositis, 62/385
Eaton-Lambert myasthenic syndrome, 62/426
granulomatous CNS vasculitis, 55/397
immunocompromised host, 56/471
immunosuppressive therapy, 65/550
inclusion body myositis, 62/385
multiple sclerosis, 47/86, 196
myasthenia gravis, 41/96, 133, 136, 62/418
myoglobinuria, 41/271
neoplastic myopathy, 41/374
neurocysticercosis, 52/533
polyarteritis nodosa, 39/308, 55/346, 358
polymyositis, 41/73, 62/385
sarcoidosis, 71/493
systemic lupus erythematosus, 71/49
temporal arteritis, 48/324
vasculitis, 71/161
Azathioprine intoxication
meningitis, 65/557
Azethion intoxication
chemical classification, 37/546
organophosphorus compound, 37/546
Azidothymidine, *see* Zidovudine
Azobenzene intoxication
oxygen therapy, 36/434
Azocarmine
hypertensive fibrinoid arteritis, 11/136
Azoospermia
Down syndrome, 50/532
Klinefelter syndrome, 50/546
AZT, *see* Zidovudine
Aztreonam
bacterial meningitis, 52/12
gram-negative bacillary meningitis, 52/108

B-cell, *see* B-lymphocyte
B fiber
classification, 1/117-119
B-lymphocyte
CSF, *see* CSF B-lymphocyte
experimental allergic encephalomyelitis, 47/434, 442
function, 56/67
Guillain-Barré syndrome, 51/248
malignant lymphoma, 18/235
multiple sclerosis, 47/101, 112, 189, 309, 345, 360-363
myositis, 62/42
plasma cell dyscrasia, 63/391
viral infection, 56/66
B-lymphocyte lymphoma
acquired immune deficiency syndrome, 56/515
IgM, 18/235
B-lymphocyte neoplasm
demyelinating neuropathy, 47/615
B-mode sonography
*see also* Ultrasound diagnosis
angiographic correlation, 54/21
brain arteriovenous malformation, 54/43
brain infarction, 54/4
carotid artery, 54/21
carotid artery bifurcation, 54/21
carotid artery coiling, 54/21
carotid artery hemorrhage, 54/27
carotid artery kinking, 54/21, 26
carotid artery stenosis, 54/19
carotid artery thrombosis, 54/27
carotid body tumor, 54/25
child, 54/36, 40
continuous wave, 54/20
extracorporeal membrane oxygenation, 54/38
hydrocephalus, 54/36
infantile cerebrovascular disease, 54/35
method, 54/19
periventricular leukomalacia, 54/40
plaque size, 54/20
plaque ulcer, 54/23, 26
postinfarct encephalomalacia, 54/40
subarachnoid hemorrhage, 54/38
subependymal hemorrhage, 54/36
thyroid cyst, 54/21, 25
transducer, 54/20
vascular lumen diameter, 54/20
B wave
flicker fusion frequency, 13/19
head injury, 57/218
intracranial pressure, 23/202
intracranial pressure monitoring, 57/218
Babin cloth peg sign
vertebral canal stenosis, 50/466
Babin index
vertebral canal stenosis, 50/466

Babington disease, *see* Hereditary hemorrhagic
telangiectasia (Osler)
Babinski-Fröhlich syndrome
craniopharyngioma, 18/545
Babinski, J.F., 1/7
Babinski-Nageotte syndrome
Cestan-Chenais syndrome, 1/189, 2/221
Horner syndrome, 2/218
lateral medullary syndrome, 2/219
medial medullary syndrome, 2/219
medulla oblongata syndrome, 2/218, 330
Babinski sign
ADR syndrome, 42/113
amyotrophic dystonic paraplegia, 42/199
amyotrophic lateral sclerosis, 42/65
automatic pronation, 1/184
congenital cerebellar ataxia, 42/125
congenital retinal blindness, 42/401
cytochrome related ataxia, 42/111
dermatoleukodystrophy, 42/488
diagnostic value, 1/248-250
familial hypobetalipoproteinemia, 13/427
Förster syndrome, 42/202
Friedreich ataxia, 21/332, 42/143, 51/388
Gaucher disease, 42/438
Gilles de la Tourette syndrome, 49/631
Gordon reflex, 1/184, 250
Harvey syndrome, 42/222
hemiplegia, 1/190
hereditary motor and sensory neuropathy type I,
21/283, 42/339
hereditary motor and sensory neuropathy type II,
21/283, 42/339
hereditary olivopontocerebellar atrophy (Menzel),
21/442
hyperargininemia, 42/562
hyperosmolar hyperglycemic nonketotic diabetic
coma, 27/90
Kjellin syndrome, 42/174
laryngeal abductor paralysis, 42/319
Leber hereditary optic neuropathy, 66/428
Marinesco-Sjögren syndrome, 42/184
metachromatic leukodystrophy, 42/491-493
myelopathy, 42/156
myopathy, 43/123
oculocerebrocutaneous syndrome, 43/444
Oppenheim sign, 1/184, 250
opticocochleodentate degeneration, 42/241
pallidopyramidal degeneration, 42/244
Parkinson dementia complex, 42/249
pellagra, 46/336
peripheral lesion, 1/232
permanent, 6/641
pseudo, *see* Pseudo-Babinski sign
Reye syndrome, 31/168, 56/151
Richner-Hanhart syndrome, 42/581
Roussy-Lévy syndrome, 42/108
Salla disease, 43/306
Schäfer sign, 1/184, 250
Sjögren-Larsson syndrome, 43/307
spastic paraplegia, 42/180
striatonigral degeneration, 42/262, 49/206
thalidomide, 42/660
thigh and trunk flexion, 1/185
Troyer syndrome, 42/193
unilateral pontine hemorrhage, 12/40
Von Hippel-Lindau disease, 42/156
X-linked spinocerebellar degeneration, 42/191
Babinski-Weil test
labyrinthine, 1/337
Bacillary meningitis
infantile, *see* Infantile enteric bacillary meningitis
Bacillus Calmette-Guerin
adverse effect, 69/498
Back kneeing
limb girdle syndrome, 40/314
Back pain
amyloidosis, 63/529
aorta aneurysm, 63/48
bacterial endocarditis, 52/291, 299
cancer, 69/57
chrysaora intoxication, 37/51
dialysis, 63/529
diastematomyelia, 50/436
diplomyelia, 50/436
dissecting aorta aneurysm, 63/37
epidural lesion, 69/65
glucose-6-phosphate dehydrogenase deficiency,
42/644
ischialgia, 69/57
low, *see* Low back pain
Marie-Strümpell disease, 19/86
meningeal lymphoma, 63/347
β2-microglobulin, 63/529
multiple myeloma, 39/138, 141, 63/392
spinal anesthesia, 61/150
spinal cord abscess, 52/191
spinal epidural empyema, 52/187
spinal epidural lymphoma, 63/349
vertebral fracture, 19/86
Back response, *see* F wave
Backward masking
schizophrenia, 46/502
Baclofen
familial spastic paraplegia, 59/313
intrathecal chemotherapy, 61/152

motor unit hyperactivity, 41/297
multiple sclerosis, 47/157
nocturnal myoclonus, 45/140
restless legs syndrome, 51/548
spastic paraplegia, 61/368
spasticity, 26/482
spinal cord injury, 61/368
spinal spasticity, 61/368
tetanus, 65/224
transverse spinal cord lesion, 61/368
Bacterial encephalitis
aqueduct stenosis, 50/307
chorea, 49/553
dementia, 46/389
spinal cord, 33/187-193
Bacterial endocarditis
Actinobacillus, 52/302
amaurosis, 52/299
arteritis, 52/294
ataxia, 52/292
back pain, 52/291, 299
blindness, 52/299
blood culture, 52/301
brain abscess, 33/115, 52/290, 297
brain embolism, 52/292
brain hemorrhage, 52/291
brain infarction, 52/291
Brucella, 52/302
cardiac valve prosthesis, 52/289, 301
cerebrovascular disease, 52/290
computer assisted tomography, 52/292
congenital heart disease, 52/289
Coxiella burnetii, 52/302
CSF, 52/295, 300, 54/197
diagnosis, 52/289, 300
discitis, 52/299
drug addict, 52/289, 298, 301
echoCG, 52/302
Eikenella, 52/302
epilepsy, 52/291, 298
erythrocyte sedimentation rate, 52/291, 300
eye symptom, 52/299
facial paralysis, 52/300
headache, 52/291, 294, 296, 298
Hemophilus, 52/302
meningitis, 52/291, 297
middle cerebral artery, 52/293
mitral valve prolapse, 52/289
mononeuritis multiplex, 52/300
morbidity, 52/304
mortality, 52/304
mortality rate, 52/292
mycotic aneurysm, 52/291
myelitis, 52/300
myelomalacia, 52/291
Neisseria, 52/302
neurologic symptom, 52/292
neurologic symptom classification, 52/290
neurologic symptom incidence, 52/290
neuropathy, 52/291, 300
ophthalmoplegia, 52/299
ophthalmoscopy, 52/299
osteomyelitis, 52/299
panophthalmitis, 52/299
peroneal neuropathy, 52/300
posterior cerebral artery, 52/294
postoperative, *see* Postoperative bacterial endocarditis
predisposing factor, 52/289
prognosis, 52/304
Proteus mirabilis, 52/301
radiodiagnosis, 52/302
retinal hemorrhage, 52/299
rheumatic fever, 52/289
Roth spot, 52/299
Serratia marcescens, 52/298, 301
spinal cord abscess, 52/300
Staphylococcus aureus, 52/292, 295, 298, 301
Streptococcus, 52/292, 298
Streptococcus faecalis, 52/301
Streptococcus pneumoniae, 52/301
Streptococcus viridans, 52/300
subacute, *see* Subacute bacterial endocarditis
subarachnoid hemorrhage, 52/291, 293, 295, 55/3
toxic encephalopathy, 52/291, 296
transient ischemic attack, 52/291
treatment, 52/302
ulnar neuropathy, 52/300
ungual splinter hemorrhage, 52/300
Bacterial food poisoning
botulism, 37/88
crab intoxication, 37/56
crustacea intoxication, 37/56
Escherichia coli, 37/87
fish, 37/79
malacostraca intoxication, 37/56
Staphylococcus, 37/87
Bacterial meningitis, 33/1-15
acridine orange stain, 52/6
Actinomyces, 52/4
acute cerebellar ataxia, 34/624
age group, 52/11
aminoglycoside, 52/9
ampicillin, 52/10
antibiotic agent, 33/13, 52/9
arteritis, 52/14

aztreonam, 52/12
BMY-28142, 52/12
brain angiography, 52/14
brain edema, 52/13
brain herniation, 52/13
brain infarction, 52/14, 53/161, 55/419-421
brain ischemia, 53/161, 55/419
brain swelling, 52/8
brain vasculitis, 55/416
Brudzinski sign, 33/1
C-reactive protein, 52/7
carrier state, 52/5
cefotaxime, 52/10
ceftazidime, 52/10
ceftriaxone, 52/10
cefuroxime, 52/10
cell mediated immune response, 55/418
cephalosporin, 52/12
cerebrovascular disease, 55/416
chloramphenicol, 52/9
cilastatin, 52/12
clinical features, 33/1
clinical symptom, 55/418
CNS shunt, 52/4
coagulopathy, 52/7
complication, 33/14, 52/13
computer assisted tomography, 52/8, 55/419
confusion, 52/1
congenital dermal sinus, 33/3
co-trimoxazole, 52/10, 13
counterimmunoelectrophoresis, 52/7
CSF, 33/4-9, 52/2, 6
CSF anion gap, 52/8
CSF culture, 33/7
CSF endotoxin, 33/8
CSF enzyme, 52/8
CSF fistula, 52/5
CSF glucose, 52/6
CSF lactate dehydrogenase, 52/8
CSF lactic acid, 52/8
CSF leak, 52/3
CSF leukocytosis, 52/6
CSF lymphocytosis, 52/6
CSF lysozyme, 52/8
CSF protein, 33/8
CSF shunt, 52/2
CSF sugar, 33/7
deafness, 52/14
dementia, 46/387
diagnosis, 33/4-10, 52/6
diagnostic test, 52/7
differential diagnosis, 52/6
drug addiction, 33/2
epidemiology, 33/2-4, 52/3, 55/416
epilepsy, 52/14, 55/420
Escherichia coli, 52/2
fatality rate, 55/416
foreign body, 33/3
gentamicin, 52/10, 12
gram-negative bacillary meningitis, 52/3
gram-negative bacilli, 52/3
gram stain, 33/7
granulocytic dysfunction, 55/418
group B streptococci, 52/2, 55/416
Hemophilus influenzae, 52/2
Hemophilus influenzae type B, 52/3, 59, 55/416
history, 33/1
humoral immune mechanism, 55/418
hydranencephaly, 50/337
hydrocephalus, 52/13
hypoglycorrhachia, 33/7
identification, 33/7
imipenem, 52/12
immunocompromised host, 52/1
immunodeficiency, 52/3
immunosuppressed host, 52/4
immunosuppression, 52/5
infantile enteric, 33/63
intracranial hypertension, 52/13
intraventricular treatment, 52/12
Kernig sign, 33/1
Klebsiella, 52/3
Klebsiella pneumoniae, 52/7
latex particle agglutination, 52/7
lethargy, 52/1
limulus lysate assay, 52/7
Listeria monocytogenes meningitis, 52/2-4
lumbar puncture, 33/4-7, 61/162
management, 33/10-13
meningitis serosa, 52/6
meningococcal meningitis, 52/3
meningoencephalitis, 55/418
mental deficiency, 52/14
mental status, 52/1
metrizamide cisternography, 52/9
microcephaly, 52/14
moxalactam, 52/12
Mycobacterium tuberculosis, 52/4
Mycoplasma pneumoniae, 52/4
narcotic addiction, 33/2
Neisseria meningitidis, 55/416
neoplastic disease, 3/2
neurosurgery, 52/4
nitroblue tetrazolium test, 33/9
Nocardia, 52/4
nuchal rigidity, 52/1

nuclear magnetic resonance, 55/421
otitis media, 52/3
pathophysiology, 52/2
petechial rash, 52/2
pneumococcal meningitis, 33/36, 47, 52/3, 52
predisposing factor, 52/4
prevention, 52/15
prognosis, 33/15, 52/13
Pseudomonas, 52/3, 5
Pseudomonas aeruginosa, 52/7
radionuclide scan, 55/421
radionuclide scintigraphy, 52/9
recurrent, 33/3
recurrent meningitis, 52/5
respiratory disease, 52/1
reticuloendothelial system, 55/418
Reye syndrome, 56/159
rifampicin, 52/13
seasonal pattern, 52/3
sickle cell anemia, 63/258
skull fracture, 33/3, 52/6
specific bacterial antigen, 33/8
staphylococcal meningitis, 52/4
Staphylococcus aureus, 52/5
Staphylococcus epidermidis, 52/5
Streptococcus pneumoniae, 52/3, 55/416
subdural effusion, 33/15, 52/14
subdural hygroma, 52/8
sulfonamide, 52/9
symptom, 52/1
tetracycline, 52/9
Todd paralysis, 55/420
transverse myelitis, 52/14
treatment, 52/9-13
Treponema pallidum, 52/4
vaccine, 52/15
vancomycin, 52/11
venous occlusion, 52/14
venous sinus thrombosis, 55/420
ventricular enlargement, 52/8
ventriculitis, 52/12
Bacterial myositis
polymyositis, 62/372
Bacteroides
brain abscess, 52/149
gram-negative bacillary meningitis, 52/117, 121
Bacteroides fragilis
infantile enteric bacillary meningitis, 33/63
transverse sinus thrombosis, 52/177
Bacteroides melaninogenicus
brain abscess, 52/150
BAEP, *see* Brain stem auditory evoked potential
BAER, *see* Brain stem auditory evoked potential
Bailey, P., 1/23, 35
Bailey tumor, *see* Oligodendroglioma
Baillarger, J.G.F., 1/7
Bajonet posture
pallidal degeneration, 49/455
Baker, A.B., 1/35
BAL, *see* Dimercaprol
Balado, M., 1/17
Balaenoptera borealis intoxication
abdominal pain, 37/94
blood pressure, 37/94
diarrhea, 37/94
facial flushing, 37/94
headache, 37/94
histamine, 37/95
ingestion, 37/94
nausea, 37/94
vitamin A intoxication, 37/95
vomiting, 37/94
Balance
acid base, *see* Acid base balance
electrolyte, *see* Electrolyte balance
motion sickness, 74/354
sympathetic dystonia, 22/251
Balance test
equal loudness, *see* Equal loudness balance test
loudness, *see* Loudness balance test
Baldness
myotonic dystrophy, 43/152
Balint-Holmes syndrome, *see* Balint syndrome
Balint syndrome
agnosia, 45/409
Alzheimer disease, 46/252
bilateral occipital lobe infarction, 53/413
brain infarction, 55/144
brain metastasis, 45/16
Brodmann area 7, 55/144
Brodmann area 39, 55/144
cardiac arrest, 63/216
cerebrovascular disease, 45/16, 55/144
color naming defect, 2/651
definition, 1/311
diffuse lesion, 3/226
distance perception, 45/410
etiology, 45/407
eye movement disorder, 3/25
features, 2/667
fixation, 2/340
frontal lobe lesion, 45/410
gaze apraxia, 45/406, 55/144
hemianopia, 1/599
hemi-inattention, 45/410
hypereosinophilic syndrome, 41/65

neuropsychology, 45/15
occipital lobe lesion, 45/410
occipital lobe tumor, 17/329
occipitoparietal region, 45/15
ocular ataxia, 45/15, 408, 55/144
oculomotor apraxia, 45/15
parastriate cortex, 55/144
parietal lobe lesion, 45/410
parietal lobe syndrome, 45/73, 77
pathology, 45/410
posterior cerebral artery syndrome, 53/413
psychic gaze paralysis, 3/214
pursuit eye movement, 2/340
pursuit movement, 45/406
simultanagnosia, 2/657, 45/15, 409, 55/144
spatial orientation disorder, 45/406-411
visual agnosia, 2/602, 45/55
visual attention, 2/667, 45/409
visual fixation, 45/406
visual inattention, 45/410
Baller-Gerold syndrome
craniosynostosis, 43/370
heart disease, 43/370
polymicrogyria, 43/371
radial aplasia, 43/370
Ballismus, 6/476-488
absence during sleep, 49/370
akinesia, 49/66
amyotrophic lateral sclerosis, 59/235
anatomic classification, 1/279
anatomy, 49/371
bilateral, *see* Bilateral ballismus
contraction suppression, 49/370
cortex, 6/482
course, 6/477
dentatorubropallidoluysian atrophy, 21/520, 531
disease duration, 49/371
drug treatment, 6/485
EEG, 6/484
EMG, 6/484, 49/76
etiology, 6/479
familial paroxysmal choreoathetosis, 42/205
frequency, 49/369
GABA, 49/372
globus pallidus medialus, 49/372
glycine, 49/372
hemi, *see* Hemiballismus
hereditary olivopontocerebellar atrophy (Menzel), 21/441, 445
hereditary paroxysmal dystonic choreoathetotis, 49/349
incidence, 6/477
kinesigenic choreoathetosis, 42/207
luysian atrophy, 21/531
mesencephalic pedunculotomy, 6/855
mono, *see* Monoballism
neuropathology, 6/480-483
nosology, 6/477
para, *see* Paraballism
Parkinson disease, 49/66, 103
pathology, 1/288
pathophysiology, 6/484, 49/76, 371
physiology, 6/483
prognosis, 6/477, 49/371
reciprocal innervation, 49/76
striatum, 6/481
subacute sclerosing panencephalitis, 56/419
subthalamic body, 6/481
subthalamic nucleus, 6/59, 62
surgical treatment, 6/486-488
thalamus, 6/482
ventral lateral thalamotomy, 49/77
voluntary action interruption, 49/370
Balloon fish intoxication, *see* Puffer fish intoxication
Baló, J., *see* Von Baló, J.
Baltic myoclonus epilepsy
*see also* Dyssynergia cerebellaris myoclonica
dementia, 49/617
differential diagnosis, 60/662
dyssynergia cerebellaris myoclonica, 60/599
Lafora progressive myoclonus epilepsy, 49/617
progressive myoclonus epilepsy, 49/617
Unverricht-Lundborg progressive myoclonus epilepsy, 49/616
Bamberger-Marie syndrome
arthritis, 1/488
autonomic nervous system, 75/543
neurocutaneous melanosis, 1/488
neurotrophic influence, 1/488
periostitis, 1/488
Bamboo spine
ankylosing spondylitis, 26/176
cerebellar medulloblastoma, 18/187
Banding
oligoclonal, *see* Oligoclonal banding
Bandler syndrome
Bean syndrome, 14/106
Bang disease, *see* Brucellosis
Bannayan-Zonana syndrome
hemihypertrophy, 59/482
Bannwarth syndrome
*see also* Neuroborreliosis
chronic meningitis, 56/645
clinical features, 47/613
CSF, 51/200, 202

epidemiology, 51/200
erythema migrans, 51/200
facial paralysis, 51/200
Lyme disease, 47/612-614, 51/208
meningism, 51/202
multiple cranial neuropathy, 51/201
neuroborreliosis, 51/199, 52/260-264
pain, 51/201
sensory symptom, 51/201
Bantu paralysis, *see* Symptomatic porphyria
Baragnosis
parietal lesion, 3/191
terminology, 1/104
Bárány, R., 1/11
Bárány test, *see* Caloric nystagmus *and* Vestibular nystagmus
Barbiturate
alpha coma, 55/220
anoxic encephalopathy, 63/223
antiepileptic agent, 65/497
athetosis, 6/450
awakening epilepsy, 15/485
behavior disorder, 46/598
brain edema, 55/216
brain infarction, 53/116, 418
brain ischemia, 53/116, 55/522, 63/223
drug addiction, 65/330
epilepsy, 15/635, 670, 72/234
headache, 5/212
hyperkinetic child syndrome, 3/197
hypnotic agent, 37/349-353
intracranial hypertension, 55/215
intracranial pressure, 57/233
migraine, 5/98, 101, 48/202
multiple sclerosis, 47/71
myoglobinuria, 40/339, 41/270
neurologic intensive care, 55/209, 211, 215
neuropathy, 7/165, 518
neurotoxin, 6/450, 7/518, 15/713-715, 37/200, 352, 409, 410, 40/339, 41/270, 46/394, 598, 47/71, 49/478, 53/418, 55/216, 220, 522, 62/601, 65/332, 497, 508
phenobarbital, 40/561
porphyria, 37/410
postanesthetic encephalopathy, 9/577, 49/478
serum level, 15/678, 680, 683
side effect, 15/713-715
snoring, 63/451
status epilepticus, 73/329
thiopental intoxication, 37/410
toxic encephalopathy, 65/508
toxic myopathy, 62/601
Barbiturate coma
neuropathy, 7/165
Barbiturate intoxication
anticonvulsant, 37/200
asterixis, 65/510
ataxia, 65/332, 509
autonomic nervous system, 75/501
behavior, 65/509
body scheme disorder, 4/244
cerebellar sign, 65/332
Cheyne-Stokes respiration, 65/332
cognitive function, 65/509
confusion, 65/332
convulsion, 37/409
delirium, 65/332
dementia, 65/332, 509
dependence, 65/332
diplopia, 65/332
dysarthria, 65/332, 509
dyskinesia, 65/510
GABA chloride ionophore, 65/330
half-life, 65/330
hallucination, 65/332
hyperkinesia, 65/509
insomnia, 65/509
lethargy, 65/508
memory disorder, 65/509
mental depression, 65/509
muscular hypotonia, 65/332
nystagmus, 65/332, 509
pallidal necrosis, 6/652, 49/466
plasma level, 65/508
pyramidal syndrome, 65/332
respiratory depression, 65/331
torpor, 65/332
tremor, 65/332
vertigo, 65/332
withdrawal syndrome, 65/332
Bardet-Biedl syndrome, 13/380-407
acanthosis nigricans, 13/294, 390, 462
acromegaly, 13/395, 401
Alström-Hallgren syndrome, 13/390, 22/503
anal atresia, 13/394
arachnodactyly, 13/296
associated abnormality, 13/394
ataxia, 13/393
autopsy finding, 13/402-404
Biemond syndrome, 13/295, 381
biochemistry, 13/404
bleb disease, 13/400
brachycephaly, 13/385
brachydactyly, 13/385
bulimia, 13/395
cardinal symptom, 13/382

chorea, 13/392, 401
choreoathetosis, 13/392
congenital heart disease, 13/395
deafness, 13/390
diabetes insipidus, 13/395
diabetes mellitus, 13/463
differential diagnosis, 14/114, 22/536
dwarfism, 13/395
dystonia musculorum deformans, 13/401
EEG, 13/393, 43/233
endocrinology, 13/405
epilepsy, 13/392
ERG, 13/292, 383
extrapyramidal syndrome, 13/401, 43/233
familial spastic paraplegia, 59/330-332
geography, 13/396
histology, 13/293
hypogonadism, 13/389, 462, 22/508, 43/233
inheritance, 13/396-398
iris coloboma, 13/295
kidney malformation, 13/402
macrocephaly, 13/401
mental deficiency, 13/337, 387, 22/508, 43/233
myoclonia, 13/401
neurologic symptom, 13/391
nomenclature, 13/289, 380, 43/233, 254
nystagmus, 13/295
obesity, 13/383, 22/536, 43/233
ocular change, 13/291
ocular symptom, 13/382
oxycephaly, 13/386
parkinsonism, 13/392
pathogenesis, 13/399
polydactyly, 13/302, 385, 401, 22/508, 536, 43/233
polydipsia, 13/395
prenatal diagnosis, 43/233
prevalence, 13/395, 43/233
Refsum disease, 21/215
renal abnormality, 43/233
secondary pigmentary retinal degeneration, 13/242, 288-297, 337, 43/233, 60/719
skeletal abnormality, 13/384
spastic diplegia, 43/233
spastic paraplegia, 13/307
syndactyly, 13/302, 385
Turner syndrome, 13/400
uterus duplex, 13/394
vagina septa, 13/394
Baresthesia
terminology, 1/97
Barium
facial paralysis, 51/274
hypokalemic periodic paralysis, 64/16
muscle fiber, 51/274
neurotoxin, 36/325, 51/274, 64/16, 331, 337, 353
Barium carbonate intoxication
hypokalemia, 63/557
Barium intoxication
areflexia, 64/354
baritosis, 64/355
barium salt, 64/353
barium use, 64/353
Clarke column, 64/354
headache, 64/354
heart fibrillation, 64/354
hypokalemia, 64/354
lethal dose, 64/354
metal intoxication, 64/353
muscle weakness, 64/354
neuropathology, 64/354
organ absorption, 64/353
paralysis, 36/325
paraparesis, 64/354
pathology, 36/326
potassium channel, 64/354
prevention, 64/355
quadriparesis, 64/354
respiratory acidosis, 64/354
respiratory paralysis, 64/354
sensory neuropathy, 64/354
source, 36/324
sulfate sodium, 64/354
threshold, 64/354
treatment, 64/355
Barlow disease, *see* Mitral valve prolapse
Barnard-Scholz syndrome
ophthalmoplegia, 22/471
retinal degeneration, 22/471
spastic tetraplegia, 22/471
Barnes Akathisia Scale
neuroleptic akathisia, 65/286
Barognosis
sensory extinction, 3/191
terminology, 1/104
Baroreceptor
aging, 74/227
brain blood flow, 53/63
brain blood flow autoregulation, 53/63
carotid sinus syndrome, 11/533
disorder, 1/472
drug side effect, 1/470
pure autonomic failure, 22/231, 233, 63/147
water intoxication, 64/240
Baroreceptor-cardiac reflex, *see* Baroreflex
Baroreceptor disorder

autonomic nervous system, 75/40
Baroreceptor pathway
disorder, 75/123
Baroreceptor sensitivity
amyloid polyneuropathy, 51/416
Baroreflex
arterial, *see* Arterial baroreflex
arterial dilatation, 74/166
autonomic nervous system, 74/623
blood pressure, 75/467
cardiopulmonary, *see* Cardiopulmonary baroreflex
exercise, 75/439
heart failure, 75/444
heart rate, 75/467
hypertension, 75/476
low pressure, 75/468
orthostatic hypotension, 2/120, 63/143
pharmacology, 74/166
sensitivity, 74/623
solitary tract, 74/153
spinal cord injury, 61/444
syncope, 75/207
tetraplegia, 61/444
transverse spinal cord lesion, 61/444
Baroreflex pathway testing
autonomic nervous system, 74/609
Barotrauma
*see also* Decompression illness, Decompression myelopathy, Diving *and* Dysbarism
anorexia, 63/416
ataxia, 63/416
brain edema, 63/416
Cheyne-Stokes respiration, 63/482
consciousness, 63/416
course, 63/416
definition, 63/419
diving risk, 63/418
fatigue, 63/416
headache, 63/416
nausea, 63/416
nitrogen narcosis, 63/419
pathophysiology, 61/215
semantics, 61/215
sleep disorder, 63/416
Barr-Bertram body, *see* Sex chromatin
Barr body, *see* Sex chromatin
Barr-Hickmans disease, *see* Phosphatidylethanolamine lipidosis
Barraquer-Simons syndrome, *see* Lipodystrophia progressiva (Barraquer-Simons)
Barré-Liéou syndrome
*see also* Cervico-occipital pain
cervical arthrosis, 5/373
cervical migraine, 5/192, 7/449-451
cervical rib syndrome, 7/460
clinical symptom, 7/450
Barré sign
hamstring weakness, 1/185
Barré syndrome
disequilibrium, 2/219
fall, 2/219
lateropulsion, 2/219
medulla oblongata syndrome, 2/219
vertigo, 2/219
Barrel chest
glycosaminoglycanosis type IV, 66/304
Barrel shaped vertebra
angioma, 20/27
tumor, 20/30
Barrier
blood-brain, *see* Blood-brain barrier
blood-CSF, *see* Blood-CSF barrier
blood-nerve, *see* Blood-nerve barrier
Bart hemoglobin
thalassemia major, 42/630
Barth syndrome type I, *see* X-linked myotubular myopathy
Barth syndrome type II, *see* X-linked neutropenic cardioskeletal myopathy
Bartonella
acquired immune deficiency syndrome, 71/291
Bartonella infection
spastic paraplegia, 59/436
Bartonellosis
spastic paraplegia, 59/436
Bartter syndrome
aldosteronism, 42/528
angiotensin, 42/528
anorexia nervosa, 46/586
chloride reabsorption defect, 42/528
Chvostek sign, 42/529
convulsion, 42/529
EEG, 42/529
hypokalemia, 42/528
mental deficiency, 42/529
myalgia, 42/528
prostaglandin, 42/528
pseudo, *see* Pseudo-Bartter syndrome
short stature, 42/528
Basal arachnoiditis
syringomyelia, 32/281, 42/60, 50/455
Basal cell
congenital torticollis, 43/146
keloid, 43/146
Basal cell carcinoma
ataxia telangiectasia, 60/370

hereditary olivopontocerebellar atrophy, 60/647
xeroderma pigmentosum, 43/12
Basal cell carcinoma syndrome
multiple nevoid, *see* Multiple nevoid basal cell carcinoma syndrome
Basal cell nevus, 14/581
differentiation, 14/585
Basal cell nevus syndrome, *see* Multiple nevoid basal cell carcinoma syndrome
Basal encephalocele
*see also* Basal meningoencephalocele
age, 50/106
coloboma, 50/106
computer assisted tomography, 50/106
divergent strabismus, 50/106
endocrine disorder, 50/106
hypertelorism, 50/106
incidence, 50/105
intranasal mass position, 50/106
metrizamide cisternography, 50/106
microphthalmia, 50/106
nasal glioma, 50/106
optic disc pit, 50/106
optic nerve abnormality, 50/106
pathomechanism, 50/105-107
peripapillary staphyloma, 50/106
pulsation, 50/106
sphenoethmoidal type, 50/105
sphenolaryngeal type, 50/105
spheno-orbital type, 50/105
transethmoidal type, 50/105
transsphenoid type, 50/105
treatment, 50/106
Basal forebrain
autonomic nervous system, 74/58
brain infarction, 55/140
Basal forebrain stroke
amnesic syndrome, 55/140
Basal ganglion
acetylcholine, 49/33
acetylcholinesterase, 49/28
age related change, 49/38
akinesia, 45/165
amygdaloid nucleus, 49/1, 28
anatomy, 1/157-159, 6/1-55, 49/1, 101
argipressin, 49/33
arterial hypertension, 63/74
arterial supply, 49/468
behavior, 49/53
bilateral ballismus, 49/370
Binswanger disease, 54/224
biochemistry, 6/116-132
bombesin, 49/33
brain death, 24/738
brain hemorrhage, 54/68
calcification, *see* Striatopallidodentate calcification
calcium, 49/35
chemistry, 49/33
cholecystokinin, 49/33
circuit diaphragm, 49/101
CNS, 30/31
compartmentation, 49/25
computer assisted tomography, 46/486
copper, 49/35
cytology, 49/21
definition, 6/3
developmental anatomy, 30/31
dopamine, 49/33
dopamine receptor type, 49/47
dynorphin, 49/33
enkephalin, 49/28
flavivirus, 56/135
function, 6/90-115
GABA, 49/21, 33
glutamate decarboxylase, 49/21
glutamic acid, 49/33
glutaryl-coenzyme A dehydrogenase deficiency, 66/647
glycine, 49/33
Hallervorden-Spatz syndrome, 22/403
headache, 5/134, 138
hereditary motor and sensory neuropathy type I, 21/302
hereditary motor and sensory neuropathy type II, 21/302
hereditary olivopontocerebellar atrophy (Menzel), 21/438
histamine, 49/33
history, 6/56-89
hydranencephaly, 30/673, 50/347
hydrocephalus, 30/673
hypertensive fibrinoid arteritis, 11/136
hypoparathyroidism, 70/122
input zone, 49/20
intracerebral hematoma, 11/671
iron, 49/35
lateral habenular nucleus, 49/29
leucine enkephalin, 49/33
leukemia, 39/12
limbic system, 49/19, 25
loop circuit, 49/20
main transmitter pathway, 49/118
manganese, 49/36
manganese intoxication, 64/307
MAO, 6/120

mesolimbic dopamine system, 49/52
metabolism, 49/36
metenkephalin, 49/21, 33
neuropharmacology, 49/47-59
neurotensin, 49/33
neurotransmitter, 49/33
nigrostriatal dopamine system, 49/52
noradrenalin, 49/33
olivopontocerebellar atrophy (Dejerine-Thomas), 21/423
organotin intoxication, 64/142
output zone, 49/20
Parkinson disease, 49/471
pathology, 6/56-89
pathophysiology, 49/65
posture, 49/78
protirelin, 49/33
pyruvate dehydrogenase complex deficiency, 66/416
receptor, 49/34
schizophrenia, 46/486
serotonin, 49/28, 33
somatostatin, 49/28, 33
structure, 6/1-5
substance P, 49/33
symptomatic dystonia, 6/553
temporal lobe, 73/64
terminology, 6/1-5
tilting reaction, 49/78
tonic control center, 1/161
traumatic intracerebral hematoma, 24/356, 359
vascularization, 11/97
vasoactive intestinal peptide, 49/33
venous supply, 49/468
vestibular system, 16/333
Wilson disease, 6/62
Basal ganglion artery
radioanatomy, 11/82
topography, 11/82
Basal ganglion calcification, *see* Striatopallidodentate calcification
Basal ganglion degeneration
acute intermittent porphyria, 42/618
Alpers disease, 42/486
argininosuccinic aciduria, 42/525
cavernous angioma, 42/724
dyssynergia cerebellaris progressiva, 42/212
dystonia deafness syndrome, 42/372
dystonia musculorum deformans, 1/289, 42/217
Förster syndrome, 42/202
Huntington chorea, 42/226
kinesigenic choreoathetosis, 27/308, 42/208
myelin dysgenesis, 42/498
neuroacanthocytosis, 42/209-211
presenile dementia (Kraepelin), 42/286
progressive cerebellar dyssynergia, 42/212
progressive pallidal atrophy (Hunt-Van Bogaert), 6/164, 42/247
progressive supranuclear palsy, 6/557, 42/259
spastic paraplegia, 42/175, 177
subacute necrotizing encephalomyelopathy, 42/588, 625
Wilson disease, 42/270
Basal ganglion disease
cardiac arrest, 63/216
clinical symptomatology, 6/133-172
hemiplegic dystonia, 6/145-149
plastic rigidity, 6/139-142
Basal ganglion function
gain control, 6/92
motor control, 6/91-93
oscillation, 6/93
Basal ganglion hemorrhage
headache, 5/138
methanol intoxication, 64/100
Basal ganglion infarction
cyanide intoxication, 64/232
Galen vein thrombosis, 54/399
tuberculous meningitis, 52/208
Basal ganglion syndrome
neurotoxicology, 64/6
Basal ganglion tier I
neuroanatomy, 49/20
Basal ganglion tier II
*see also* Ventral globus pallidus
acetylcholinesterase, 49/21
connection, 49/23
cytoarchitecture, 49/21
histochemistry, 49/27
Huntington chorea, 49/24
neuroanatomy, 49/20
nucleus accumbens, 49/21
olfactory bulb, 49/21
substance P, 49/21
substantia innominata, 49/21
Basal ganglion tier III
*see also* Ventral neostriatum
connection, 49/23
enkephalin, 49/22
GABA, 49/22
Hallervorden-Spatz syndrome, 49/24
neuroanatomy, 49/20
substance P, 49/22
Basal ganglion tiers
connection, 49/26
cytoarchitecture, 49/21

cytology, 49/21
embryology, 49/24
pathology, 49/24
Basal ganglion tumor
thalamic tumor, 18/329
Basal meningoencephalocele
*see also* Basal encephalocele
age, 50/106
coloboma, 50/106
computer assisted tomography, 50/106
divergent strabismus, 50/106
endocrine disorder, 50/106
hypertelorism, 50/106
incidence, 50/105
intranasal mass position, 50/106
metrizamide cisternography, 50/106
microphthalmia, 50/106
nasal glioma, 50/106
optic disc pit, 50/106
optic nerve abnormality, 50/106
peripapillary staphyloma, 50/106
pulsation, 50/106
sphenoethmoidal type, 50/105
sphenolaryngeal type, 50/105
spheno-orbital type, 50/105
transethmoidal type, 50/105
transsphenoid type, 50/105
treatment, 50/106
Basal metabolic rate
anorexia nervosa, 46/586
Basal nucleus
embryo, 50/11
Meynert, *see* Meynert nucleus basalis
Basal osteomyelitis
apical petrositis, 33/180
Basal pachymeningitis
nonrecurrent nonhereditary multiple cranial neuropathy, 51/571
Basal vein, *see* Rosenthal vein
Basedow disease
*see also* Hyperthyroidism
external ophthalmoplegia, 22/179
Basedow paraplegia
hyperthyroid polyneuropathy, 51/516
Basement membrane, *see* Muscle cell membrane *and* Sarcolemma
Basic protein
antimyelin, *see* Antimyelin basic protein
CSF myelin, *see* CSF myelin basic protein
encephalitogenic myelin, *see* Encephalitogenic myelin basic protein
myelin, *see* Myelin basic protein
Basilar arterial intermittent insufficiency, *see* Transient ischemic attack
Basilar artery
anatomy, 32/45, 53/386
brain aneurysm, 12/95, 53/386
brain lacunar infarction, 54/236
branch, 12/20
cerebellopontine angle syndrome, 12/95
congenital aneurysm, 31/154-156
Darkschewitsch nucleus, 2/275
embolic occlusion, 53/388
embolism, 12/22
embryology, 11/34
fenestration, 32/47
headache, 5/134, 138
infarction, 12/20
migrainous acute confusional state, 48/166
migrainous transient global amnesia, 48/166
occlusion, 53/386
pathology, 53/386
rostral branches, 53/390
transcranial Doppler sonography, 54/12
Basilar artery aneurysm
alternating hemiplegia, 53/387
aqueduct stenosis, 50/308
dolichoectasis, 53/386
headache, 53/387
labyrinthine ischemia, 55/131
mortality, 53/387
site, 53/386
surgery, 12/222
thrombosis, 12/22
transient ischemic attack, 53/387
Basilar artery atherosclerosis
pattern, 53/386
plaque site, 53/386
Basilar artery insufficiency
visual hallucination, 46/567
Basilar artery migraine, 48/135-140
ataxia, 48/136
brain stem, 5/42
cerebellar migraine, 42/749, 48/157
classification, 48/6
complicated migraine, 48/166
confusional state, 48/138
consciousness, 48/137
differential diagnosis, 53/234
drop attack, 48/137
drowsiness, 48/137
EEG change, 48/137
EEG spike, 48/138
epilepsy, 48/137
hysteria, 46/575
nausea, 48/136

nystagmus, 48/138
pallor, 48/138
pathomechanism, 48/131, 166
photophobia, 48/138
sleep, 48/137
speech disorder, 48/136
tinnitus, 48/136
treatment, 48/139
vertigo, 48/136
vestibular test, 48/138
vomiting, 48/138
Basilar artery stenosis
coma, 53/386
craniovertebral abnormality, 32/53
death, 53/386
lesion site, 2/242-248
pontine infarction, 2/244, 12/13, 19-26
rostral site, 53/391
superior cerebellar artery, 53/386
Basilar artery syndrome
transient ischemic attack, 12/13
Basilar artery system, 11/24-42
Horner syndrome, 11/31
Basilar artery thrombosis
electrical injury, 61/192
headache, 48/274
local intra-arterial fibrinolysis, 54/180
streptokinase, 54/180
Basilar bone
iniencephaly, 50/134
Basilar dissecting aneurysm
coma, 54/280
headache, 54/280
subarachnoid hemorrhage, 54/280
Basilar fissure
transverse, *see* Transverse basilar fissure
Basilar impression
*see also* Platybasia
achondroplasia, 50/399
anterior, 32/21
Arnold-Chiari malformation type I, 50/399
Arnold-Chiari malformation type II, 50/399
arthrodento-osteodysplasia, 43/332
associated condition, 32/67
atlas, 32/82
autosomal dominant inheritance, 50/64
bilateral facial paresis, 32/76
Boogaard line, 32/18
Bull angle, 32/17
Chamberlain line, 32/15, 17, 20, 24, 26, 50/398
cleidocranial dysplasia, 50/399
clinical features, 32/67-69, 71-82, 50/398
CNS, 32/68-77
course, 32/67
CSF flow, 32/77
Daubent angle, 32/19
Decker-Breig angle, 32/25
digastric line, 50/398
diplopia, 42/17
disharmonic, 32/21
Down syndrome, 50/399
EEG, 32/78
epidemiology, 30/161, 32/67-70
exertional headache, 48/374
external appearance, 32/72
fibrous dysplasia, 50/399
foramen magnum, 32/18, 42/17
Francesconi index, 32/19
genetics, 30/161, 32/71
glycosaminoglycanosis type I, 50/399
harmonic, 32/21
Hultkranz line, 32/18
hydrocephalus, 32/63
hyperparathyroidism, 50/399
hypothyroidism, 50/399
incidence, 50/64
intracranial pressure, 42/17
Klaus index, 32/51
McGregor line, 32/17-19, 50/398
measurement, 32/17-19
medial, 32/21
occipital bone shortening, 32/14
occipital dysplasia, 32/16-20, 50/399
occipital encephalocele, 32/60
olivopontocerebellar atrophy (Wadia-Swami), 60/497
onset, 32/78
osteitis deformans (Paget), 32/67, 38/365, 50/399, 70/14
osteogenesis imperfecta, 32/67, 43/454, 50/399
osteomalacia, 50/399
papilledema, 42/17
pathogenesis, 32/79-82
periodic ataxia, 21/570
platybasia, 32/66, 42/17, 50/397
primary, 32/67-82
progressive bulbar palsy, 59/225
Prouzet line, 32/19
pseudo, *see* Pseudobasilar impression
radiographic criteria, 61/7
respiratory dysfunction, 32/74
rickets, 50/399
secondary, 32/67
sexual function, 32/72
speech disorder, 42/17
sphincter disturbance, 42/17

spina bifida, 32/73
spinal arachnoiditis, 32/70, 81
syringomyelia, 42/17, 59, 50/400, 446
tetraplegia, 42/17
Towne projection, 32/51
trauma, 32/68, 50/399
treatment, 32/82
Twining line, 32/19
type, 50/388
unilateral, 32/21
Basilar invagination, *see* Basilar impression
Basilar migraine, *see* Basilar artery migraine
Basilar process
bone abnormality, 50/388
features, 50/389
suboccipital dysplasia, 32/30
Basilar vein, *see* Rosenthal vein
Basilar venous plexus
radioanatomy, 11/113
topography, 11/113
Basion odontoid index
definition, 32/24
Basket cell
action, 2/416
cerebellar stimulation, 2/409, 411
Basketball
rehabilitative sport, 26/534
sport injury, 23/575
Basle, Switzerland
neurology, 1/10
Basofrontal syndrome
abulia, 53/349
amnesia, 53/348
anosodiaphoria, 53/349
anterior cerebral artery syndrome, 53/348
apathy, 53/349
behavior disorder, 53/348
brain infarction, 53/348
confabulation, 53/349
decorum loss, 53/349
diabetes insipidus, 53/349
euphoria, 53/349
fever, 53/349
inappropriate behavior, 53/349
initiative loss, 53/349
Basophil adenoma
corticotropin, 2/457
Cushing syndrome, 2/454, 457, 18/354
headache, 48/424
hypophyseal adenoma, 18/353
17-ketosteroid, 2/457
pituitary gland, 2/457
tomography, 2/457
Basophilic granular inclusion
dermatomyositis, 62/377
inclusion body myositis, 62/377
polymyositis, 62/377
Bass intoxication
classification, 37/78
Bassen-Kornzweig syndrome, 13/413-428, 29/401-424, 60/131-133
*see also* Neuroacanthocytosis
acanthocyte, 10/279, 549, 49/327
acanthocytosis, 13/427
age, 63/276
amyotrophy, 21/580
angioid streak, 63/273
apolipoprotein B, 51/326
apolipoprotein B-48, 63/275, 277
apolipoprotein B-50, 63/277
apolipoprotein B-100, 63/275, 277
apolipoprotein B-100 gene, 66/546
apolipoprotein B gene, 63/275
ataxia, 10/548, 13/417, 21/574, 42/511, 51/327, 394, 63/272
athetosis, 10/548, 21/580
autohemolysis, 63/279
autosomal dominant, 63/276
autosomal recessive, 63/275
axonal degeneration, 51/394, 66/16
axonal neuropathy, 51/394, 63/275, 278
biochemical features, 8/19, 60/131
bleeding tendency, 63/274
brain stem auditory evoked potential, 63/275
cardiac involvement, 51/327
cause, 51/394
celiac disease, 63/273
central disorder, 22/195
central scotoma, 51/394
chemical laboratory data, 63/273
chorea, 13/427
chorioretinal degeneration, 13/40
chylomicron, 42/512
classic, 63/276
clinical features, 10/548, 13/414, 29/406-419, 51/326, 60/131, 63/273
computer assisted tomography, 60/133
cuneate nucleus, 63/278
deficiency neuropathy, 7/596, 51/326
demyelinating neuropathy, 63/278
demyelination, 8/19, 10/548, 51/394
diagnosis, 29/423
differential diagnosis, 29/423
dorsal spinal nerve root, 63/278
dysarthria, 51/394
electrophysiologic study, 60/133

EMG, 51/394, 63/275
endocrine function, 63/277
endoneurial fibrosis, 51/394
ERG, 51/327, 394, 60/670, 63/275
erythrocyte change, 29/412-416
external ophthalmoplegia, 60/656
eye movement disorder, 40/303
familial hypobetalipoproteinemia, 29/391, 60/132
familial spastic paraplegia, 59/327
Friedreich ataxia, 10/548, 13/413, 418, 21/20, 357
gastrointestinal manifestation, 29/410-412
gene frequency, 21/580
genetics, 10/549, 13/425, 29/405, 410
gracile nucleus, 63/278
head tremor, 51/394
hematologic features, 51/394, 63/279
hematology, 8/19, 13/420
hereditary spinocerebellar degeneration, 21/14
history, 29/402-405, 63/272
hyperlordosis, 51/394
hypobetalipoproteinemia type, 63/275
hypocholesterolemia, 42/511
intention tremor, 21/580, 51/394, 63/272
internuclear ophthalmoplegia, 63/274
intestinal features, 13/421
kyphoscoliosis, 13/418, 51/327, 394
linoleic acid, 13/423
lipid study, 13/422
lipofuscin storage myopathy, 63/278
lipoprotein, 13/423-425, 29/419-421
β-lipoprotein, 8/19
low density lipoprotein, 42/511, 63/275
main features, 21/16
malabsorption, 7/596, 29/410-412
manifestation, 66/547-549
mental deficiency, 10/548, 42/511
metabolic ataxia, 21/574, 580
metabolic polyneuropathy, 51/393
mitral valve prolapse, 63/274
muscle biopsy, 63/278
muscle weakness, 13/417
myopathy, 60/132
nerve conduction velocity, 51/394
neuroacanthocytosis, 7/596, 8/19, 10/280, 549, 13/413, 21/20, 357, 580, 29/412-416, 51/327, 60/673, 63/271
neurologic features, 13/417, 29/417-419
neurologic sign, 51/395
neuropathologic aspect, 60/132
neuropathology, 60/133
neuropathy, 8/19, 42/511, 60/131
normotriglyceridemic abetalipoproteinemia, 63/275
nyctalopia, 51/327, 394
nystagmus, 63/273
onset, 29/405-410
ophthalmology, 8/19, 13/419
ophthalmoplegia, 22/191, 51/327, 394
optic atrophy, 13/40
pathogenesis, 13/427
pathophysiology, 29/421-423
periodic ataxia, 21/580
pes cavus, 10/548, 13/418, 51/394
pes equinovarus, 51/394
phosphatidylcholine sterol acyltransferase, 63/279
polyneuropathy, 21/52, 22/195, 60/165
posterior column, 63/275, 278
prevalence, 29/405
progressive external ophthalmoplegia, 22/179, 191, 204, 207, 60/656, 62/290, 304
prothrombin time, 51/327
pseudogenetic disorder, 63/277
ptosis, 63/273
Refsum disease, 13/316, 21/215, 60/233, 235, 66/497
related syndrome, 13/426
remyelination, 51/394
retinal lesion, 29/416
retinal pigmentary dystrophy, 63/272
retinitis punctata albescens, 13/317
scotoma, 10/549
secondary pigmentary retinal degeneration, 10/279, 13/316-318, 413-416, 21/580, 42/511, 51/326, 393-395, 60/653, 727
sensorimotor polyneuropathy, 51/394
somatosensory evoked potential, 51/327, 63/275
spheroid body, 63/278
spinocerebellar degeneration, 8/19, 10/279
spinocerebellar tract, 8/19, 63/278
steatorrhea, 13/421, 51/327
symptom frequency, 63/275
tongue atrophy, 60/132
treatment, 13/425, 29/424, 51/395, 63/278
triglyceride, 42/512
very low density lipoprotein, 63/275
visual evoked response, 51/327, 63/275
vitamin A deficiency, 51/326, 63/277
vitamin E, 63/277
vitamin E deficiency, 51/326, 394, 63/276, 70/443
vitamin K deficiency, 51/326
Bastian, C.C., 1/6
Bastian rule
incorrect rule, 2/180
myotatic areflexia, 2/180
transverse spinal cord lesion, 2/180
Bat ray intoxication

elasmobranch, 37/70
Bathing trunk hypalgesia
hereditary acute porphyria, 51/391
Bathyesthesia
terminology, 1/93
Batrachoides intoxication
family, 37/78
venomous spine, 37/78
Batrachoididae intoxication
family, 37/78
venomous spine, 37/78
Batrachotoxin
Phyllobates, 64/15
toxic neuromuscular junction disorder, 64/15
Batrachotoxin intoxication
action site, 64/15
pathomechanism, 64/15
symptom, 64/15
Batrachus intoxication
family, 37/78
venomous spine, 37/78
Battarismus
history, 43/220
terminology, 43/220
Batten disease
*see also* Neuronal ceroid lipofuscinosis
adult neuronal ceroid lipofuscinosis, 66/671
astrocyte, 10/657
ataxia, 10/575
biochemistry, 10/221, 598
brain atrophy, 42/461-464
cerebellar ataxia, 42/466
cerebromacular degeneration, 10/588
clinical course, 10/591
CSF, 10/617
dementia, 10/575, 42/466
differential diagnosis, 10/620
dystonia, 6/556
EEG, 10/617, 619
epilepsy, 10/575, 579, 608, 42/461, 463, 466, 60/666
eponymic classification, 10/217
familial amaurotic idiocy, 10/215
genetics, 10/576, 623
histopathology, 10/632
history, 10/589-591
laboratory finding, 10/613
lipid, 40/243
lymphocyte vacuolation, 10/567
lysosome, 40/37, 113
macular symptom, 10/602, 607
mental symptom, 10/609
metabolic ataxia, 21/575
microcephaly, 10/614, 42/461
morphology, 10/599
motor disorder, 10/611
muscle fiber feature, 62/17
neuronal, *see* Neuronal ceroid lipofuscinosis
neuronal ceroid lipofuscinosis, 66/671
neuropathology, 10/628, 631, 66/26
nomenclature, 10/588
ophthalmoplegia, 10/663
optic atrophy, 42/461, 463
pathogenesis, 10/670
pathology, 10/625, 627
phenotype variation, 10/224
proton transporting adenosine triphosphate synthase, 66/686
psychomotor retardation, 42/461-463
Purkinje cell loss, 42/461, 466, 468
retinal degeneration, 10/575, 42/461, 463
sea blue histiocyte, 60/674
secondary pigmentary retinal degeneration, 10/575, 13/327-330, 42/464
substantia nigra degeneration, 42/468
symptom, 10/601, 612
symptomatic dystonia, 6/556
treatment, 10/625
vacuolated lymphocyte, 10/576-578
visual disorder, 10/604, 606
white blood cell, 10/615
Batten, F.E., 10/227
Batten-Mayou disease, *see* Batten disease
Batten-Mayou-Spielmeyer-Vogt disease, *see* Batten disease
Batten, R.D., 10/227
Batten-Vogt disease, *see* Batten disease
Battered child syndrome, 23/603-607
brain injury, 23/603
head injury, 23/603
pediatric head injury, 57/328
prognosis, 23/607
subdural effusion, 24/331
subdural hematoma, 23/605
traumatic headache, 48/32
treatment, 23/606
Batwing tremor, *see* Asterixis
Bay rum
characteristics, 64/95
methanol, 64/95
use, 64/95
BBB syndrome, *see* Hypertelorism hypospadias syndrome
BCK, *see* Maple syrup urine disease
BCNU, *see* Carmustine
Beals syndrome

scoliosis, 30/101
Bean syndrome
anemia, 43/11
angiomatosis, 43/11
Bandler syndrome, 14/106
cerebellar medulloblastoma, 14/106
Cronckhite-Canada syndrome, 14/106
differential diagnosis, 14/106
dysraphia, 14/106
EEG, 43/11
Fabry disease, 14/106
fibrosarcoma, 14/106
glomus jugulare tumor, 14/106
Haferkamp syndrome, 14/106
hemangiomatosis, 14/106
hemangiosarcoma, 14/106
hereditary hemorrhagic telangiectasia (Osler), 14/106
Maffucci syndrome, 14/106
Mibelli angiokeratoma syndrome, 14/106
neurocutaneous syndrome, 14/101, 774
seizure, 43/11
speech disorder, 43/11
Bechterew disease, *see* Ankylosing spondylitis
Becker disease
cardiac decompensation, 38/213
fibrinoid necrosis, 38/213
hemoptysis, 38/213
Becker mild X-linked dystrophy, *see* Becker muscular dystrophy
Becker muscular dystrophy
age at onset, 62/135
carrier detection, 40/340
classification, 40/281, 389
clinical features, 40/306, 312, 317, 321, 323, 337, 342, 388, 442, 62/135
contracture, 43/86
creatine kinase, 40/281, 387, 43/86
differential diagnosis, 40/438, 62/135
drug trial, 62/136
Duchenne muscular dystrophy, 62/123
ECG, 43/86
genetics, 40/387, 41/411, 486
Haldane rule, 41/416
heart disease, 43/86
heterozygote detection, 43/87
histochemistry, 40/48
history, 40/434, 62/134
incidence, 40/387, 62/135
incomplete dystrophin deficiency, 62/135
Lyon principle, 41/408
malignant hyperthermia, 62/570
mental deficiency, 43/86
muscle biopsy, 62/123
muscle carnitine, 41/201
muscle necrosis, 40/389
myoglobinuria, 62/556, 558
necrosis, 40/389
neuromuscular disease, 41/411
pathology, 40/389
preclinical detection, 43/87
prevalence, 62/135
proximal muscle weakness, 43/86
pseudohypertrophy, 40/324, 388, 43/86
related phenotype, 62/135
ribosomal protein synthesis test, 41/411
treatment, 62/136
Wohlfart-Kugelberg-Welander disease, 59/372
X-linked muscular dystrophy, 40/387-389, 62/117
Becker generalized myotonia, *see* Generalized myotonia (Becker)
Beckwith-Wiedemann syndrome
birth incidence, 43/399
cardinal features, 31/329
CNS, 31/329
face, 31/330
genetics, 31/329
heart disease, 43/398
hepatoblastoma, 43/398
hypoglycemia, 43/398
mental deficiency, 43/398, 46/82
metabolic abnormality, 31/330
microcephaly, 31/329, 43/398
pathology, 31/331
prognosis, 31/332
radiologic appearance, 31/331
seizure, 31/330
treatment, 31/331
visceromegaly, 46/82
Wilms tumor, 43/398
Bedouin spastic ataxia
cataract, 60/653
computer assisted tomography, 60/671
corneal dystrophy, 60/652
tilted optic disc, 60/655
Bee sting
envenomation, 37/107-111
Bee sting intoxication
anaphylaxis, 37/107
circulatory disturbance, 37/107
consciousness, 37/107
edema, 37/107
erythema, 37/107
fatal, 37/107
histamine, 37/107
hypoxic ischemia, 37/107

neurologic complication, 37/107-111
pain, 37/107
respiratory tract disorder, 37/107
seizure, 37/107
swelling, 37/107
syncope, 37/107
urticaria, 37/107
Bee venom
*see also* Apamin
neurotoxin, 37/107, 111, 62/222, 225, 65/194, 199
polyradiculoneuropathy, 65/198
transient ischemic attack, 65/198
Beer potomania
water intoxication, 64/244
Beevor, C.E., 1/6
Behavior
barbiturate intoxication, 65/509
basal ganglion, 49/53
Binswanger disease, 46/319-321
bizarre, *see* Bizarre behavior
body odor, 74/377
brain lacunar infarction, 53/395
brain tumor, 16/728, 46/416
cluster headache, 48/223
consciousness, 3/51
cri du chat syndrome, 31/568
Down syndrome, 31/406, 43/544, 50/528
ectopallium, 2/720
effective, 2/720
emotion, 2/449
entopallium, 2/720
epilepsy, 72/267
epileptiform, *see* Epileptiform behavior
expressive, 2/720
feeding, *see* Feeding behavior
Gilles de la Tourette syndrome, 49/627
hypothalamus, 2/441
Klinefelter syndrome, 31/481, 50/549
Landau-Kleffner syndrome, 73/283
mesopallium, 2/720
motor unit, *see* Motor unit behavior
neuropsychology, 45/4, 515
newborn, *see* Newborn behavior
organophosphate intoxication, 64/226
pheromone, 74/377
prefrontal region, 2/736
psychomotor, *see* Psychomotor behavior
psychopathic, *see* Psychopathic behavior
purpose, *see* Purposive behavior
reflex, *see* Reflex behavior
reproductive, *see* Reproductive behavior
schizophrenia, 46/507-510
sebaceous secretion, 74/377
septal area, 2/771
sexual, *see* Sexual behavior
social, *see* Social behavior
stereotyped, *see* Stereotyped behavior
temporal lobe epilepsy, 45/45
temporal lobe lesion, 2/717
temporal lobectomy, 2/715
thalamic infarction, 53/395
thermoregulation, 75/54
total, 2/720
transient global amnesia, 45/209
Turner syndrome, 31/500, 50/545
turning, *see* Turning behavior
vertebrobasilar system syndrome, 53/395
water intoxication, 63/547
Wolf-Hirschhorn syndrome, 31/559
XYY syndrome, 31/487-489
Behavior development
abnormal, 4/360-365
adaptive behavior, 4/352-356
higher nervous activity disorder, 4/342-347
social behavior, 4/356-360
test, 4/353-356
tonic neck reflex, 4/346
Behavior disorder
Addison disease, 42/512, 46/599
adrenoleukodystrophy, 42/483
amantadine, 46/592-594
amphetamine, 46/592
anterior cerebral artery syndrome, 53/348
antibiotic agent, 46/601
anticonvulsant, 46/595, 600
antidepressant, 46/597
antiemetic agent, 46/595
antihistaminic agent, 46/595
anti-inflammatory agent, 46/602
antineoplastic agent, 46/600
anxiolytic agent, 37/348, 46/597
aspartylglycosaminuria, 27/159, 42/527
barbiturate, 46/598
basofrontal syndrome, 53/348
belladonna alkaloid, 46/595
benzodiazepine, 46/598
bleomycin, 46/600
body scheme disorder, 45/379
bone membranous lipodystrophy, 42/279
brain infarction, 53/348
brain injury, 4/360-363
bromocriptine, 46/594
butyrophenone, 46/597
cardiovascular agent, 46/601
cerebrovascular disease, 55/137
chloroquine, 46/602

cholinergic agent, 46/595
cholinergic receptor blocking agent, 46/595
chrysotherapy, 46/602
clonidine, 46/601
CNS stimulant, 46/591
corticosteroid, 46/599
corticotropin, 46/599
Cushing syndrome, 46/599
cyclophosphamide, 46/600
digitalis, 46/601
dopaminergic agent, 46/592-594
drug induced, *see* Drug induced behavior disorder
encephalitis lethargica, 34/455
frontal lobe syndrome, 53/348
Hallervorden-Spatz syndrome, 42/257
hallucinogenic agent, 46/594
head injury, 46/625-629
indometacin, 46/602
inorganic salt intoxication, 37/348
isoniazid, 46/601
juvenile Gaucher disease, 10/307, 523
juvenile metachromatic leukodystrophy, 66/168
Klinefelter syndrome, 43/557, 50/548
Lafora progressive myoclonus epilepsy, 15/383, 385, 27/172, 42/702
levodopa, 46/592-594
lisuride, 46/594
lithium, 46/597
MAO inhibitor, 46/597
methyldopa, 46/601
microcephaly, 30/513
multi-infarct dementia, 42/285
myoclonic epilepsy, 42/702
naproxen, 46/602
neuroleptic agent, 46/595-597
oral contraception, 46/599
orthochromatic leukodystrophy, 42/501
Parkinson dementia complex, 42/250
parkinsonism, 42/255
pergolide, 46/594
pernicious anemia, 42/609
phenothiazine, 46/597
phenytoin, 46/600
presenile dementia, 42/286
propranolol, 46/601
rehabilitation, 46/624-630
REM, *see* REM behavior disorder
REM sleep, 74/559
schizophrenia, 46/490
scopolamine, 46/596
serotoninergic agent, 46/596
spasmodic torticollis, 42/264
spreading depression, 48/67
stereotypy, 46/592
tetracyclic antidepressant, 46/597
tricyclic antidepressant, 46/595
Unverricht-Lundborg progressive myoclonus epilepsy, 42/705
Wilson disease, 42/271
Behavior disorder in children
acquired brain disease, 4/341, 360-365
Addison disease, 42/512
adrenoleukodystrophy, 42/483
anhidrosis, 42/348
autism, 43/202
brain maturation, 4/340, 468
developmental dyslexia, 4/385
Down syndrome, 43/544
Hallervorden-Spatz syndrome, 42/257
handedness, 4/465-469
hyperkinetic child syndrome, 4/363
infantile optic glioma, 42/733
Klinefelter syndrome, 43/557
mental deficiency, 46/26-30
motor disorganization, 4/455
myoclonic epilepsy, 42/702
organic nature, 4/340-342, 360-365
orthochromatic leukodystrophy, 42/500
progressive myoclonus epilepsy, 42/705-708
sensory neuropathy, 42/348
speech disorder in children, 4/423-425
verbal auditory agnosia, 43/228, 72/9
Wilson disease, 42/271
Behavior therapy
anorexia nervosa, 46/589
epilepsy, 73/442
infantile autism, 46/192
load, 45/253
stress, 45/253
traumatic psychosyndrome, 24/556
Behçet syndrome, 34/475-506, 56/593-605
aciclovir, 56/605
age distribution, 34/477
age at onset, 56/597
aneurysm formation, 56/595
antiphospholipid antibody, 63/330
aphthous ulcer, 71/213
artery occlusion, 56/595
arthritis, 56/594
articular manifestation, 71/217
ataxia, 56/597
bladder function, 56/598
blood vessel disorder, 34/481-487
brain stem, 56/595
brain stem syndrome, 56/597
bulbar paralysis, 56/598

cause, 56/603
chorioretinitis, 56/594
chronic meningitis, 56/645
classification, 71/4
clinical course, 56/597
clinical features, 34/478-487
CNS involvement, 71/217
CNS lesion distribution, 56/599
CNS lesion frequency, 56/596
colchicine, 56/605
computer assisted tomography, 56/600, 71/219
confusional state, 56/597
corticosteroid, 56/604
course, 34/491
cranial nerve palsy, 56/597
CSF, 34/487, 56/599, 71/222
CSF C-3 complement, 56/600
CSF IgA, 56/600
cutaneous hypersensitivity, 71/214
dementia, 46/390, 56/597
dermal manifestation, 34/479-481
diagnosis, 34/488-491, 56/593, 71/212
diagnostic criteria, 56/593, 71/213
differential diagnosis, 56/601
disseminated meningoencephalomyelitis, 56/598
dysarthria, 56/597
dysphagia, 56/597
EEG, 34/487, 56/600
encephalitis, 16/219
encephalomyelitis, 63/55
encephalopathy, 45/119
epidemiology, 34/476-478, 71/210
epididymitis, 56/594
erythema nodosum, 56/594, 71/214
etiology, 34/501-504
euphoria, 56/598
evoked potential, 56/601
familial cases, 34/478
fundus, 71/216
gastrointestinal disorder, 34/482
genital lesion, 34/480
genital ulcer, 56/594, 71/213
glomerulonephritis, 56/595
hallucination, 56/598
headache, 56/598
hemiballismus, 49/373
hemiparesis, 56/598
history, 34/475, 56/593, 71/209
hypopyon iritis, 71/215
immunology, 56/603
immunosuppression, 56/605
intention tremor, 56/597
interstitial nephritis, 56/595
intestinal ulcer, 56/594
intracranial hypertension, 71/224
involuntary movement, 56/598
iridocyclitis, 56/594
joint lesion, 34/482
laboratory finding, 34/487
locomotor ataxia, 56/598
meningeal irritation, 56/598
meningitis serosa, 71/225
meningomyelitis, 56/597
metal, 56/604
methisoprinol, 56/605
methylated vitamin B12, 56/605
microhematuria, 56/595
Mollaret meningitis, 34/549, 56/629
mortality, 56/598
mucous membrane, 34/479
multiple sclerosis, 9/206, 47/71, 56/597
multiple sclerosis like course, 56/597, 71/223
muscle involvement, 71/225
myopathy, 34/486
nervous system involvement, 34/483-487
neuro-Behçet, 71/217, 222
neurobrucellosis, 52/594
neuroimaging, 71/219
neuropathology, 34/493-501, 56/598, 71/218
neuropathy, 34/486
neuropsychiatric syndrome, 71/225
nodose lesion, 34/481
nodular lesion, 34/481
nonrecurrent nonhereditary multiple cranial neuropathy, 51/571
nuclear magnetic resonance, 56/600, 71/221
nystagmus, 56/598
ocular manifestation, 34/478, 71/215
onset, 71/211
oral lesion, 34/480
papulonodular lesion, 34/481
papulopustular lesion, 34/480
paraparesis, 56/598
pathogenesis, 71/212
pathologic laughing and crying, 56/597
pathology, 34/492-501
personality change, 56/597
polymyositis, 62/373
prognosis, 34/491, 56/597
proteinuria, 56/595
psychiatric symptom, 56/598
psychiatric syndrome, 34/485
pulmonary disorder, 34/483
pulmonary lesion, 56/594
recurrent hypopyon, 56/595
recurrent oral aphthous ulceration, 56/593

relapse, 56/597
rheumatism, 34/482
seizure, 71/225
sensory disturbance, 56/598
sex ratio, 56/596
skin lesion, 34/480, 71/213
spastic paraplegia, 59/436
subcutaneous thrombophlebitis, 56/594
survey, 71/209
symptom, 71/223
tetraparesis, 56/598
thrombophlebitis, 71/217
treatment, 34/504-506, 56/604, 71/226
urogenital lesion, 34/482
varix formation, 56/595
vascular disorder, 34/481-487
vasculitis, 71/8
venous occlusion, 56/595
venous sinus thrombosis, 71/224
Vogt-Koyanagi-Harada syndrome, 56/623

Behr disease, 13/88-91
anisocoria, 42/416
ataxia, 13/88, 42/109
autosomal recessive, 13/98
brain embolism, 53/213
brain infarction, 53/213
Bruyn-Went disease, 13/91
cerebrovascular disease, 53/213
clinical features, 13/88
color vision, 13/89
congenital retinal blindness, 13/67, 88, 91, 42/109
convergence paralysis, 60/657
differential diagnosis, 13/91
dyschromatopsia, 42/416
EOG, 13/90
epilepsy, 15/329, 331
ERG, 13/89
familial spastic paraplegia, 59/334
Friedreich ataxia, 13/91
hereditary, 13/90
hereditary adult foveal dystrophy (Behr), 13/88
hereditary cerebellar ataxia (Marie), 13/91
hereditary motor and sensory neuropathy type I, 21/308
hereditary motor and sensory neuropathy type II, 21/308
hereditary spastic ataxia, 21/372
histology, 13/90
hypertrophic polyneuropathy, 13/91
hypogonadism, 60/669
infantile optic atrophy, 42/408
intracranial hemorrhage, 53/213
late infantile neuroaxonal dystrophy, 49/405
lesion site, 53/213
mental deficiency, 13/88, 42/415
neuropathy, 60/660
nystagmus, 13/88, 42/416
olivopontocerebellar atrophy variant, 21/454
optic atrophy, 13/42, 81, 98
papilledema, 13/88, 60/655
pes cavus, 13/88
pyramidal symptom, 13/98
pyramidal syndrome, 13/88
strabismus, 42/416
transient ischemic attack, 53/213
visual acuity, 13/89
visual adaptation, 13/89
visual field, 13/89

Behr hereditary adult foveal dystrophy, *see* Hereditary adult foveal dystrophy (Behr)

Békésy audiometry
acoustic nerve tumor, 17/673-675
description, 2/74

Belgium
neurology, 1/10

Bell, C., 1/6, 8/242

Bell palsy, *see* Peripheral facial paralysis

Bell sign
operculum syndrome, 2/781
pineal tumor, 17/656
superior levator palpebrae, 2/318
vertical gaze palsy, 2/338
visual perseveration, 45/365

Belladonna alkaloid
anhidrosis, 8/344
behavior disorder, 46/595
delirium, 46/537
hyoscyamine, 65/48
Raynaud phenomenon, 8/337
scopolamine, 65/48

BEN intoxication, *see* Benzyl benzoate intoxication

Benactyzine
extrapyramidal disorder, 6/827

Bence Jones protein
multiple myeloma, 20/9, 11, 39/132, 134, 138

Bence Jones proteinuria
multiple myeloma, 18/255, 20/9-11, 13-16, 111, 39/133, 136, 138, 148
primary amyloidosis, 51/416

Bencze syndrome
hemifacial hypertrophy, 59/482

Bending forward
autonomic failure, 75/721

Bendrofluazide, *see* Bendroflumethiazide

Bendroflumethiazide
diuretic agent, 65/437

Bends, *see* Decompression illness
Benedikt syndrome
  Argyll Robertson pupil, 2/277
  cerebellar tremor, 49/588
  choreoathetosis, 1/187
  conjugate eye movement, 2/298
  extrapyramidal hyperkinesia, 2/298
  gaze paralysis, 2/298
  hemianesthesia, 2/298
  hemorrhage, 2/298
  involuntary movement, 2/276, 298
  medial lemniscus, 2/298
  midbrain syndrome, 2/298, 6/162
  nucleus ruber, 2/298
  oculomotor nerve, 2/276, 298
  oculomotor paralysis, 1/187, 6/162
  pineal tumor, 18/362
  ptosis, 2/298, 6/162
  pyramidal tract, 2/298
  rubral tremor, 21/529
  substantia nigra, 2/298
  terminology, 21/529
  tremor, 1/187, 6/162, 49/597
  tubercle, 2/298
  tumor, 2/298
  vascular disease, 2/298
Benign acute childhood myositis
  acute viral myositis, 56/198
  adenovirus, 56/198
  influenza A virus, 56/198
  influenza B virus, 56/198
Benign ataxia
  acute, *see* Acute benign ataxia
Benign autosomal dominant myopathy
  contracture, 43/111
  creatine kinase, 43/111
  subsarcolemmal myofibril, 43/112
Benign bone aneurysm, *see* Aneurysmal bone cyst
Benign bone cyst, *see* Aneurysmal bone cyst
Benign childhood epilepsy with centrotemporal spike, *see* Benign rolandic epilepsy
Benign childhood epilepsy with occipital paroxysm
  occipital lobe epilepsy, 73/109
Benign chondroblastoma
  spinal cord compression, 19/367
Benign choreoathetosis
  early onset, 42/204
  mirror writing, 42/204
  speech disorder, 42/205
  speech disorder in children, 42/205
Benign congenital hypotonia, *see* Amyotonia congenita
Benign cystic epithelioma
  multiple, *see* Multiple trichoepithelioma
Benign dominant myopathy
  areflexia, 43/111
  hyporeflexia, 43/111
  muscle fiber size, 43/112
  muscle fiber type I, 43/112
  proximal muscle weakness, 43/111
Benign essential tremor, *see* Hereditary tremor
Benign frontal lobe epilepsy, 73/28
Benign giant cell tumor, *see* Aneurysmal bone cyst
Benign hamartoma
  optic nerve glioma, 68/75
Benign infantile convulsion
  idiopathic partial epilepsy, 73/28
Benign inoculation lymphoreticulosis, *see* Cat scratch disease
Benign intracranial hypertension, 16/150-164
  abducent nerve paralysis, 42/740
  achondroplasia, 67/111
  acquired toxoplasmosis, 35/125, 52/354
  African trypanosomiasis, 52/342
  amiodarone, 63/190, 67/111
  amiodarone intoxication, 65/459
  ataxia, 42/741
  blood disease, 16/151, 159
  brain edema, 63/417
  brain hypoxia, 63/417
  brain phlebothrombosis, 12/425
  brain scanning, 16/153
  brain thrombosis, 16/157
  brain tumor, 16/157
  Castleman disease, 63/394
  cause, 16/151
  cerebellar granular cell hypertrophy, 67/111
  cerebral sinus thrombosis, 67/113
  cerebral toxoplasmosis, 52/354
  child, 48/34, 67/112
  chorionic gonadotropin beta subunit, 67/111
  ciclosporin, 67/111
  ciprofloxacin intoxication, 65/486
  clinical features, 33/171
  corticosteroid intoxication, 65/555
  corticosteroid withdrawal, 63/360
  CSF flow, 67/108
  cystic fibrosis, 67/111
  cytarabine, 67/111
  danazol, 67/111
  definition, 67/103
  diagnosis, 16/160, 33/171
  diplopia, 42/740
  drug reaction, 16/151
  endocrine disorder, 16/151, 157
  epidemiology, 67/104

etiology, 67/110
familial brucellosis, 67/111
galactosemia, 16/160, 42/551
glycosaminoglycanosis, 67/111
growth hormone, 67/111
headache, 5/155, 42/740, 48/6, 33
hexachlorocyclohexane, 67/111
hypercortisolism, 70/206
hypoparathyroidism, 67/111
hypothalamic pituitary adrenal axis disorder, 67/111
hypothyroidism, 67/111
indometacin, 67/111
intracranial pressure, 67/108
iron deficiency anemia, 67/111
isotretinoin, 67/111
leuprorelin, 67/111
levonorgestrel, 67/111
Leydig cell testicular tumor, 67/111
lindane intoxication, 65/479
lithium carbonate, 67/111
long-term assessment, 16/163
lumbar puncture risk, 61/164
maple syrup urine disease, 67/111
mastoiditis, 16/153
medroxyprogesterone, 67/111
Melkersson-Rosenthal syndrome, 8/277
minocycline, 65/472, 67/111
Monro-Kellie doctrine, 16/152, 67/108
mountain sickness, 63/417
nalidixic acid, 67/111
nalidixic acid intoxication, 65/486
nitrofurantoin, 67/111
obesity, 42/740
ofloxacin, 67/111
otitic hydrocephalus, 33/171
pathogenesis, 67/104
penicillin, 65/472
POEMS syndrome, 63/394
posterior fossa tumor, 18/405-407
quinolone intoxication, 65/486
recombinant somatomedin C, 67/111
renal insufficiency, 67/111
respiratory encephalopathy, 63/414
retinoic acid, 67/111
roseola infantum, 16/160
sarcoidosis, 16/160
sex ratio, 42/740
sinus obstruction, 16/153
skew deviation, 55/125
spinal ependymoma, 20/367
steroid therapy, 67/111
superior longitudinal sinus thrombophlebitis, 33/170-172
superior longitudinal sinus thrombosis, 33/170
survey, 67/103
Sydenham chorea, 16/160
symptom, 16/152
systemic disease, 16/157
systemic lupus erythematosus, 39/286
tetracycline, 65/472, 67/111
thyroxine, 67/111
treatment, 16/162, 33/172
uterine dysfunction, 42/740
venous drainage obstruction, 16/152
venous sinus thrombosis, 54/411
vitamin A, 16/159, 67/111
vitamin A deficiency, 28/43
vitamin A intoxication, 28/45, 37/27, 96, 65/568
Benign juvenile hereditary chorea, *see* Juvenile hereditary benign chorea
Benign late hereditary cerebellar ataxia
hypoesthesia, 60/661
Benign monoclonal gammopathy
immunocompromised host, 56/470
Benign monoclonal paraproteinemia
paraproteinemic polyneuropathy, 51/436
Benign myalgic encephalomyelitis, *see* Postviral fatigue syndrome
Benign neonatal convulsion
definition, 72/5
survey, 73/129
Benign neonatal familial convulsion
definition, 72/5
EEG, 73/133
epilepsy, 72/129, 131, 73/130, 253
gene locus, 72/129
survey, 73/129
treatment, 73/134
Benign neonatal idiopathic convulsion
epilepsy, 73/253
Benign neonatal sleep
myoclonus, 49/622
Benign osteoblastoma
age, 19/299
differential diagnosis, 19/300
pathology, 19/301
sex ratio, 19/299
symptom, 19/299
tabular survey, 19/300
treatment, 19/302
vertebral column, 20/442
Benign paraproteinemia
paraproteinemic polyneuropathy, 51/429
Benign paroxysmal positional nystagmus
cerebellar vermis, 2/371

differential diagnosis, 2/364
labyrinth, 2/370
nystagmus, 2/370
otitis media, 2/377
otolith lesion, 2/364, 370
reflex arc, 2/371
statoacoustic nerve, 2/377
test, 2/370
vertigo, 2/370
Benign paroxysmal torticollis
migraine, 48/38
Benign paroxysmal vertigo in children
migraine equivalent, 48/157
Benign partial epilepsy
atypical, *see* Atypical benign partial epilepsy
classification, 73/5
Benign partial epilepsy with centrotemporal spike
affective symptom, 73/18
EEG, 73/13
family, 73/8
imaging data, 73/5
interictal spike, 73/13
rolandic spike, 73/14
seizure, 73/10
sex ratio, 73/8
sleep, 73/12
survey, 73/6
treatment, 73/13
Benign partial epilepsy with occipital paroxysm, 73/18
age at onset, 73/20
brain mapping, 73/27
consciousness, 73/23
EEG, 73/24
family, 73/19
imaging data, 73/20
postictal deficit, 73/23
prevalence, 73/19
seizure, 73/20
treatment, 73/24
visual symptom, 73/21
Benign partial epilepsy with somatosensory evoked potential
EEG, 73/26
genetics, 73/26
seizure, 73/26
Benign peroxisomal dysgenesis
ataxia, 60/670
nerve conduction velocity, 60/670
neuropathy, 60/660, 670
Benign recurrent aseptic meningitis, *see* Mollaret meningitis
Benign rolandic epilepsy
definition, 72/5
epilepsy with continuous spike wave during slow sleep, 73/274
Benign spinal chondrogenic tumor, 19/293-309
classification, 19/294
Benign spinal osteogenic tumor, 19/293-309
classification, 19/294
Benign trigeminal paresis
chronic, *see* Chronic benign trigeminal paresis
Benorilate
headache, 48/174
migraine, 48/174
Benorylate, *see* Benorilate
Benzamide
chemical structure, 65/276
neuroleptic agent, 65/275
Benzathine penicillin
neurosyphilis, 33/373, 52/282
Benzatropine
motoneuron disease, 59/467
Benzatropine methanesulfonate
extrapyramidal disorder, 6/828
Benzedrine, *see* Amphetamine
Benzene hexachloride, *see* Hexachlorocyclohexane
γ-Benzene hexachloride, *see* Lindane
Benzhexol, *see* Trihexyphenidyl
Benzine, *see* Gasoline
Benzodiazepine
anxiolysis, 65/344
behavior disorder, 46/598
central sleep apnea, 63/463
drug addiction, 65/339
epilepsy, 15/645, 669, 72/234, 73/360
familial spastic paraplegia, 59/313
hydrocephalus, 50/297
hypnotic agent, 37/355-360
migraine, 5/101, 48/202
multiple sclerosis, 47/157
neonatal seizure, 73/257
organophosphate intoxication, 37/559, 64/172
pyrethroid intoxication, 64/216
restless legs syndrome, 51/548
snoring, 63/451
status epilepticus, 73/328
stress, 45/254
Benzodiazepine intoxication
abecarnil, 65/342
alpidem, 65/334, 342
amnesia, 65/337
anticonvulsant, 37/200
anxiolysis, 65/335
ataxia, 65/338, 510
behavior change, 65/336

benzodiazepine like compound, 65/342
benzodiazepine-GABAa-chloride, 65/333
bretazenil, 65/342
β-carboline, 65/334
childbirth, 65/338
chlordiazepoxide, 65/334
clorazepate, 65/334
dealkylflurazepam, 65/335
dementia, 65/337
dependence, 65/339, 341
diazepam, 65/334
driving ability, 65/338
drowsiness, 65/335
drug abuse, 65/339, 341
drug interaction, 65/339
dysarthria, 65/338
EEG, 65/338
elimination half-life, 65/334
flumazenil, 65/334
half-life, 65/331
hostility, 65/335
insomnia, 65/335
intellectual impairment, 65/337
inverse stimulating agent, 65/334
ionic gating receptor, 65/334
lactation, 65/338
lorazepam, 65/334
metabolism, 65/334
myoclonia, 65/339
nordazepam, 65/334
overdose, 65/339
oxazepam, 65/334
pharmacokinetics, 65/334
plasma level response ratio, 65/335
pregnancy, 65/338
psychomotor effect, 65/336
quazepam, 65/334
rebound insomnia, 65/340
receptor group, 65/334
rigidity, 65/339
suriclone, 65/343
temazepam, 65/335
torpor, 65/335
vertigo, 65/338, 343
withdrawal syndrome, 65/340
zolpidem, 65/334, 342
zopiclone, 65/343
Benzodiazepine receptor
Huntington chorea, 49/259
Benzodiazepine withdrawal
ataxia, 65/340
headache, 65/340
myalgia, 65/340
temperature reversal, 65/340
Benzomorphan intoxication
neurologic aspect, 37/366
Benzoylmethyl ecgonine, *see* Cocaine
Benzyl benzoate intoxication
skin manifestation, 36/435
Berant syndrome
radioulnar dysostosis, 43/371
Berardinelli-Seip syndrome
acanthosis nigricans, 42/591
brain atrophy, 42/592
corticospinal tract degeneration, 42/592
diabetes mellitus, 42/591
hepatomegaly, 42/591
Lawrence syndrome, 42/592
mental deficiency, 42/591
muscle hypertrophy, 42/591
neuronal heterotopia, 42/592
subcutaneous fat absence, 42/591
ventricular dilatation, 42/592
Bereitschaftspotential, *see* Readiness potential
Berenbruch-Cushing-Cobb syndrome, *see* Cobb syndrome
Bergeron chorea, *see* Chorea electrica
Bergeron-Henoch chorea, *see* Chorea electrica
Bergmann, E. Von, *see* Von Bergmann, E.
Bergmann fiber, 30/503
cerebellar cortical dysplasia, 30/496
Beriberi
*see also* Malnutrition, Vitamin B1 *and* Vitamin B1 deficiency
alcoholic polyneuropathy, 51/318
case report, 7/562
clinical features, 7/562, 28/6-9, 51
clinicopathologic features, 28/4
cocarboxylase, 28/3
demyelination, 7/563
diagnosis, 28/53
edema, 28/6
epidemiology, 7/636
etiology, 28/50
experiment, 7/566-568
historic aspect, 28/2, 49
laboratory study, 28/52
motor symptom, 28/8
muscular atrophy, 7/563-565
neuritis, 28/6
neuropathology, 7/564
neuropathy, 7/559, 28/4, 49-55
paresthesia, 28/8
pathology, 28/51
pellagra, 46/399
pregnancy, 8/14

prevalence, 7/563
prevention, 28/55
prognosis, 28/54
retrobulbar neuritis, 7/562
sensory involvement, 28/8
spastic ataxia, 7/644
treatment, 7/571, 28/54
vitamin B1 deficiency, 7/560, 636, 28/2-4, 6-9, 50, 51/331
weakness, 28/8
Wernicke encephalopathy, 28/4
Berland-Bergeron electric chorea, *see* Chorea electrica
Berlin, Germany
neurology, 1/8, 13, 19
Berlin syndrome
anhidrotic ectodermal dysplasia, 14/112
cutaneous pigmentation, 43/9
dental abnormality, 43/9
growth retardation, 43/9
hypotrichosis, 43/9
mental deficiency, 43/9
Berman disease, *see* Mucolipidosis type IV
Bernard, C., 1/7
Bernard-Horner syndrome, *see* Horner syndrome
Bernard-Soulier syndrome
hemostasis, 63/302
Bernasconi-Cassinari artery
brain angioma, 12/101
Bernfeld syndrome, *see* Stylohyoid syndrome
Bernhardt-Roth disease, *see* Meralgia paresthetica
Bernstein hypothesis
explanation, 7/63
Berry spot test
glycosaminoglycanosis type VII, 66/313
Bertolotti-Garcin syndrome, *see* Garcin syndrome
Bertrand classification
diffuse sclerosis, 10/18
Beryllium intoxication
Raynaud phenomenon, 8/337
Best disease, *see* Vitelliruptive macula degeneration
Beta adrenergic receptor
alcohol intoxication, 64/114
genetics, 75/243
Beta adrenergic receptor blocking
chronic hypoxia, 75/269
Beta adrenergic receptor blocking agent
acebutolol, 65/434
action mechanism, 37/444
antihypertensive agent, 63/83, 87
arterial hypertension, 54/208, 65/434
atenolol, 65/434
autonomic nervous system, 22/251
cardiac pharmacotherapy, 63/194
cardiovascular agent, 65/433
cluster headache, 48/240
dopamine β-mono-oxygenase deficiency, 59/163
hereditary tremor, 49/577
hyperthermia, 75/62
hyperventilation syndrome, 38/329, 63/442
iatrogenic neurological disease, 63/194, 65/435
indication, 65/434
metoprolol, 65/434
migraine, 48/151, 179-181, 65/434
myopathy, 63/87
nadolol, 65/434
orthostatic hypotension, 63/162
oxprenolol, 65/434
pharmacologic action, 65/434
pindolol, 65/434
propranolol, 37/444, 65/434
sotalol, 65/434
tetanus, 65/225
toxic myopathy, 62/604
Beta adrenergic receptor blocking agent intoxication
carpal tunnel syndrome, 65/437
CSF level, 65/435
delusion, 65/436
EMG, 65/437
hallucination, 65/436
insomnia, 65/436
mania, 65/436
memory disorder, 65/436
mental depression, 65/436
myalgia, 65/437
myasthenic syndrome, 65/437
myopathy, 65/437
myotonia, 65/437
neurologic adverse effect, 65/435
systemic adverse effect, 65/435
tremor, 65/437
Beta adrenergic receptor stimulating agent
autonomic nervous system, 22/251
Beta2 adrenergic receptor stimulation
physiologic tremor, 49/566
Beta blocker, *see* Beta adrenergic receptor blocking agent
Beta human chorionic gonadotropin, *see* Chorionic gonadotropin beta subunit
Beta mannosidosis
biochemical diagnosis, 66/333
clinical features, 66/332
enzyme defect, 66/330
genetics, 66/334
pathology, 66/333
survey, 66/329

Beta noradrenergic receptor blocking agent
neuroleptic akathisia, 65/290
Beta oxidation
defect, 66/396
fatty acid, 66/505
metabolic myopathy, 62/492
phytanic acid, 27/525
Beta thalassemia, *see* Thalassemia major
Betaine
homocystinuria, 55/325, 332
osmoregulation, 63/555
Betanidine
tetanus, 65/225
Bethanidine, *see* Betanidine
Bethe, A., 1/9
Bethlem-Van Wijngaarden syndrome
*see also* Limb girdle syndrome
autosomal dominant, 62/181
early onset, 62/181
Emery-Dreifuss muscular dystrophy, 62/155
features, 62/184
muscle contraction, 62/155, 181
Betz cell
Friedreich ataxia, 21/324
Betz cell degeneration
progressive spinal muscular atrophy, 22/18
Betz cell function
motor cortex, 1/58
Bi-Col procedure
acid mucopolysaccharide, 9/25
myelination, 9/40
wallerian degeneration, 9/35
Bibasal marginal atrophy (Jakob), *see* Focal cerebellar cortical panatrophy
Bibrachial paresis, *see* Brachial diplegia
Bicarbonate
CNS, 28/436
CSF, 28/436
diabetes mellitus, 70/201
methanol intoxication, 36/353, 357, 64/102
puffer fish intoxication, 37/81
renal tubular acidosis, 29/181
Biceps femoris reflex
diagnostic value, 1/247
Biceps reflex
diagnostic value, 1/242
Bickers-Adams syndrome
*see also* Aqueduct stenosis
aqueduct stenosis, 30/610
hydrocephalus, 30/610
Bickerstaff encephalitis, *see* Brain stem encephalitis
Biclonal gammopathy
paraproteinemia, 69/290
Bicornuate uterus, *see* Uterus bicornis
Bicuculline intoxication
excitotoxic change, 64/140
BIDS syndrome, *see* Hair brain syndrome
Bielschowsky body
pallidal degeneration, 49/459
pallidoluysionigral degeneration, 49/459
Bielschowsky-Henneberg classification
diffuse sclerosis, 10/12
Bielschowsky-Jansky disease, *see* Neuronal ceroid lipofuscinosis
Bielschowsky, M., 1/9, 27, 10/227
Bielschowsky relapsing alternating ophthalmoplegia
Tolosa-Hunt syndrome, 48/303
Bielschowsky test, *see* Doll head eye phenomenon
Biemond, A., 1/10
Biemond syndrome, 21/377-381, 60/445-449
ancillary study, 21/378
areflexia, 21/379, 42/164, 299, 60/660
ataxia, 21/377
Bardet-Biedl syndrome, 13/295, 381
cerebellum, 60/449
clinical features, 60/445-448
congenital analgesia, 42/299
cranial nerve, 42/164
differential diagnosis, 60/449
hypoesthesia, 60/661
inheritance, 60/448
kyphoscoliosis, 60/669
mental deficiency, 46/65
neuronal intranuclear hyaline inclusion disease, 60/449
neuropathology, 21/378, 60/446-448
nystagmus, 21/379
optic atrophy, 21/380, 42/164, 60/654
Purkinje cell loss, 42/164, 60/449
scoliosis, 42/164
sea blue histiocyte disease, 60/449
sensory deficit, 21/377
sensory loss, 42/164
spinothalamic tract degeneration, 42/299
trigeminal root, 21/378
trigeminal sensory deficit, 60/657
Von Romberg sign, 42/165
Biemond syndrome type II
congenital hip dislocation, 43/334
hypogonadism, 43/334
iris coloboma, 13/295, 43/334
mental deficiency, 43/334
polydactyly, 43/334
secondary pigmentary retinal degeneration, 60/720
sella turcica, 43/334

short stature, 43/334
syndactyly, 43/334
Biermer-Addison-Castle disease, *see* Pernicious anemia
Biermer syndrome, *see* Addison disease
Biernacki sign
tabes dorsalis, 52/280
terminology, 1/97
Bietti tapetoretinal degeneration with marginal corneal dystrophy, *see* Tapetoretinal degeneration with marginal corneal dystrophy (Bietti)
Bifid rib
multiple nevoid basal cell carcinoma syndrome, 14/459
Bifunctional enzyme deficiency
peroxisomal disorder, 66/510, 517
Bikeles test
upper extremity, 1/544
Bilateral acoustic neurinoma
autosomal dominant inheritance, 42/757
Gardner disease, 22/517
neurofibromatosis type I, 14/145, 151, 42/756
neurofibromatosis type II, 50/370
Bilateral acoustic neurofibromatosis, *see* Neurofibromatosis type II
Bilateral ballismus
basal ganglion, 49/370
frequency, 6/476, 49/369
Bilateral cataract
craniolacunia, 50/139
Bilateral cervical facet interlocking
cause, 25/329
clinical features, 25/136, 375
general, 25/136, 264
mechanism, 26/264
pathology, 25/331
radiodiagnosis, 26/265
radiology, 25/333, 336-338
treatment, 25/338
Bilateral facial paresis
basilar impression, 32/76
hereditary amyloid polyneuropathy type 3, 51/424
osteitis deformans (Paget), 32/76
Sjögren syndrome, 51/449
syringobulbia, 8/258
Bilateral massive myoclonus
classification, 15/107
myoclonic epilepsy, 15/126
Bilateral pallidal necrosis
brain anoxia, 6/658-661
cause, 27/42-44
Bilateral periorbital hematoma
clinical examination, 57/131
skull fracture, 57/131
Bilateral pontine hemorrhage
coma, 12/41
flaccid quadriplegia, 12/41
prodromal symptom, 12/41
Bilateral pontine syndrome
coma, 12/20
en chapelet malacia, 2/241
symptom, 12/13, 18-20, 23
Bilateral ptosis
Edwards syndrome, 50/561
Minkowitz disease, 10/552
Bilateral sensory trigeminal neuropathy
progressive systemic sclerosis, 51/447
Bilateral striatal necrosis
myoclonus, 21/514
Bile acid
cerebrotendinous xanthomatosis, 66/606
Bile salt disorder
deficiency neuropathy, 51/325
Biliary atresia
cytomegalic inclusion body disease, 56/267
secondary neuroaxonal dystrophy, 49/408
spheroid body, 6/626
Biliary cirrhosis
primary, *see* Primary biliary cirrhosis
Biliary cyst
Caroli disease, 75/640
Biliary system
ganglion, 75/615
Biliary tract
Chagas disease, 75/398
dysmotility, 75/640
Bilingualism
aphasia, 45/305
developmental dyslexia, 46/120
Bilious attack
classic migraine, 5/46
common migraine, 5/46
Bilirubin
acquired hepatocerebral degeneration, 49/218
CSF, 12/144, 54/196
excretion, 27/417
Friedreich ataxia, 42/144, 60/307
McLeod phenotype, 63/287
metabolism, 6/491, 27/415
transport, 27/416
vitamin $K_2$, 6/492
Bilirubin encephalopathy, 27/415-425
*see also* Acquired hepatocerebral degeneration, Kernicterus *and* Posticteric encephalopathy
biochemical aspect, 27/418

choreoathetosis, 27/415, 419, 421
cortical intoxication, 6/493, 513
Crigler-Najjar syndrome, 27/423
erythroblastosis fetalis, 27/422
exchange transfusion, 27/424
globus pallidus, 27/420
glucose-6-phosphate dehydrogenase deficiency, 27/423
glycogen storage disease type I, 27/423
hepatic encephalopathy, 49/218
hypotonia, 27/421
intellectual deficit, 27/421
kernicterus, 27/415, 419
newborn physiological jaundice, 27/422
opisthotonos, 27/421
pallidal necrosis, 49/465, 481
pallidonigral necrosis, 49/465, 482
pallidostriatal necrosis, 49/465
passive immunization, 27/425
pathogenesis, 27/420
pathology, 27/419
phototherapy, 27/422
symptomatic dystonia, 49/541
treatment, 27/423-425
Bilirubin glucuronoside glucuronosyltransferase
Crigler-Najjar syndrome, 42/540
Binasal hemianopia
optic nerve, 2/572-574
Binaural loudness balance
alternate, *see* Alternate binaural loudness balance
Bindearmchorea, *see* Rubral tremor
Bing-Neel syndrome, *see* Waldenström macroglobulinemia polyneuropathy
Bing, R., 1/10
Bing sign
diagnostic value, 1/250
Binswanger disease
*see also* Leukoaraiosis
age, 54/221
anoxic ischemic leukoencephalopathy, 9/589
aphasia, 54/222
apraxia, 54/222
arteriolar fibrinohyalinoid degeneration, 11/135, 157, 54/227
astrocytosis, 54/227
ataxia, 54/222
basal ganglion, 54/224
behavior, 46/319-321
blood pressure, 54/221
brain amyloid angiopathy, 54/231
brain arteriole, 46/318, 321
brain blood flow, 54/223
brain edema, 54/229
brain infarction, 53/158, 54/222
brain ischemia, 53/158
brain lacunar infarction, 54/224, 227, 229
brain white matter injury, 46/318
cerebrovascular disease, 46/318, 53/28
chronic ischemia, 54/227
clinical features, 46/318-321, 54/221
computer assisted tomography, 46/319, 321, 54/57, 59, 223
concomitant illness, 54/222
confusional state, 54/222
corpus callosum, 54/224
dementia, 42/278, 46/317-321, 54/221
demyelinating leukoencephalopathy, 10/9
demyelination, 9/613, 54/221, 227
differential diagnosis, 54/229
Divry-Van Bogaert syndrome, 10/125, 55/318
duration, 54/221
dysarthria, 54/222
dyslexia, 54/222
EEG, 54/223
emotion, 46/320
encephalopathy, 42/278, 49/471
epileptic seizure, 54/222
gait apraxia, 54/222
granular cortical cerebral atrophy, 54/224
hemiplegia, 54/222
historic note, 46/317
history, 54/221
hypertension, 46/318, 359
leukoencephalopathy, 9/589, 54/221
memory, 46/320
multi-infarct dementia, 54/229, 55/142
multiple white matter lesion, 54/224
neurochemistry, 27/478
neuropathology, 54/224
normal pressure hydrocephalus, 54/231
nuclear magnetic resonance, 54/94, 223
orthochromatic leukodystrophy, 10/125
parkinsonism, 46/319, 49/471
pathology, 46/318, 359
physiopathology, 54/227
progressive multifocal leukoencephalopathy, 54/231
pseudobulbar dysarthria, 54/222
pseudobulbar syndrome, 46/359
Rosenthal fiber, 54/227
seizure, 46/318
Sjögren syndrome, 71/74
speech, 46/319
stepwise evolution, 54/222
stroke, 46/318
transient global amnesia, 54/222

transient ischemic attack, 54/222
vascular dementia, 46/359
vertigo, 54/222
wallerian degeneration, 9/590
Wernicke-Korsakoff syndrome, 46/320
Binswanger encephalopathy, *see* Binswanger disease
Biocytin
pyruvate carboxylase deficiency, 66/418
Biofeedback
chronic muscle contraction headache, 48/346, 350
epilepsy, 73/445
headache, 48/39, 203, 350
hemiplegia, 46/621
hereditary tremor, 49/577
migraine, 48/39
rehabilitation, 55/242
Biogenic amine
biochemical anatomy, 29/464-467
biological inactivation, 29/462
biosynthesis, 29/459-461
brain function, 29/467-469
catabolism, 29/462
CNS, 29/459-478
cyclic adenosine phosphate, 29/464
glycogen, 27/208-210
Parkinson disease, 29/469-476
regional distribution, 29/464-467
release, 29/462-464
storage, 29/462
synthesis regulation, 29/463
tardive dyskinesia, 29/476-478
turnover, 29/463
Biological rhythm
*see also* Annual rhythm, Clock function *and* Seasonal rhythm
age, 74/495, 502
Alzheimer disease, 74/496
autonomic nervous system, 74/467, 75/431
blood pressure, 74/472
circadian rhythm, 74/501
circadian system, 74/467
circaseptan, 74/491
cluster headache, 74/507
corticotropin, 74/508
delivery, 74/492
depression, 74/504
disease, 74/491
disruption, 74/495
emotional disorder, 74/505
fetal, 74/492
gene, 74/467
hydrocortisone, 74/506
hypnic headache syndrome, 74/507
hypothalamus, 74/469, 505
irregular sleep-wake syndrome, 74/470
melatonin, 74/469, 495, 503
migraine, 74/507
oral contraception, 74/469
perinatal mortality, 74/491
pineal gland, 74/491, 499
premature baby, 74/492
prolactin, 74/508
retinohypothalamic tract, 74/476
Segawa dystonia, 74/473
sex, 74/502
sleep, 74/495
sleep-wake pattern, 74/495
subarachnoid hemorrhage, 74/472
sundowning, 74/495
suprachiasmatic nucleus, 74/476, 495
temperature, 74/492
testosterone, 74/508
time sense, 3/231-234
vasoactive intestinal polypeptide neuron, 74/495
vasopressin, 74/473, 495
vasopressin neuron, 74/471
Biomechanics
anosmia, 24/8, 10
brain injury, 23/68, 91, 24/160-168
cervical spine, 48/406
cervical vertebral column, *see* Cervical vertebral column biomechanics
cervical vertebral column injury, 25/227-239, 253-259, 61/23, 60
cervicomedullary injury, 24/160, 162, 168
child, 23/448
head injury, 23/67
intrauterine head injury, 23/471
ligamentum flavum hypertrophy, 20/810
pediatric head injury, 23/448
skull fracture, 23/69, 389
spinal injury, 25/123-140
stress wave, 23/94
thoracic vertebral column, *see* Thoracic vertebral column biomechanics
thoracolumbar vertebral column injury, 61/34
vertebral column, 61/3-19
Biopsies
brain, *see* Brain biopsy
cerebral, *see* Brain biopsy
conjunctiva, *see* Conjunctiva biopsy
muscle, *see* Muscle biopsy
nerve, *see* Nerve biopsy
sural nerve, *see* Sural nerve biopsy
temporal artery, *see* Temporal artery biopsy

Biopterin
dystonia musculorum deformans, 42/215
juvenile parkinsonism, 49/159
parkinsonism, 49/159
Bioresmethrin
neurotoxin, 64/213
pyrethroid, 64/213
pyrethroid intoxication, 64/213
Biot respiration
bilateral pneumotaxic center lesion, 63/487
definition, 63/485
lesion site, 63/487
periodicity, 1/650
sleep apnea syndrome, 63/487
Biotin, *see* Vitamin $B_W$
Biotin deficiency, *see* Vitamin $B_W$ deficiency
Biotin-serine complex, *see* Biocytin
Biotinidase
epilepsy, 72/222
Biotinidase deficiency
clinical finding, 66/650
organic acid metabolism, 66/641
Bipolar affective disorder
genetics, 43/209, 46/472
lithium intoxication, 64/293
Birbeck granule
histiocytosis X, 38/97-99
Bird headed dwarfism, *see* Seckel syndrome
Bird like face
Hallermann-Streiff syndrome, 43/403
hereditary sensory and autonomic neuropathy type III, 1/478
pseudoprogeria Hallermann-Streiff syndrome, 43/405
Rothmund-Thomson syndrome, 14/777
Birgus latro intoxication
chill, 37/58
clinical features, 37/58
death, 37/58
exhaustion, 37/58
gastrointestinal manifestation, 37/58
headache, 37/58
muscle weakness, 37/58
Birth asphyxia, *see* Newborn asphyxia
Birth incidence
achondroplasia, 43/317
acrocephalosyndactyly type I, 43/320
adult onset muscular atrophy, 42/84
anencephaly, 42/13
anophthalmia, 42/14
aqueduct stenosis, 42/57
argininosuccinic aciduria, 42/525
Arnold-Chiari malformation type I, 42/16
aspartylglycosaminuria, 42/527
athetosis, 6/446
autistic syndrome, 43/202
Beckwith-Wiedemann syndrome, 43/399
brachial paralysis, 2/141
brachial plexus, 51/143
brachial plexus injury, 7/416, 25/168, 395
cardiofacial syndrome, 42/308
caudal aplasia, 42/50
cerebral palsy, 46/14
cerebrocostomandibular syndrome, 43/237
cerebrohepatorenal syndrome, 43/340
cerebro-oculofacioskeletal syndrome, 43/241
cervical hyperdeflexion sprain, 25/163
congenital adrenal hyperplasia, 42/514
congenital cerebellar atrophy, 21/460
congenital deafness, 42/366
congenital megacolon, 42/3
corpus callosum agenesis, 42/103
cranial nerve, 26/55
craniosynostosis, 42/21
cystathioninuria, 42/541
Dandy-Walker syndrome, 42/23
De Lange syndrome, 43/246
Dejerine-Klumpke syndrome, 7/416
disequilibrium syndrome, 42/140
Down syndrome, 43/546
Duchenne-Erb syndrome, 7/416
Duchenne muscular dystrophy, 43/106
dystonia musculorum deformans, 42/217
Edwards syndrome, 43/537
Ellis-Van Creveld syndrome, 43/348
encephalomeningocele, 42/27
epidemiology, 25/169
epilepsy, 15/201, 544
Erb paralysis, 2/141
facial nerve injury, 24/109
facioscapulohumeral muscular dystrophy, 43/99
Fanconi syndrome, 43/17
galactosemia, 42/551
Gaucher disease, 42/437
globoid cell leukodystrophy, 42/490
glycosaminoglycanosis type I, 42/451
glycosaminoglycanosis type III, 42/456
glycosaminoglycanosis type IV, 42/457
$GM_2$ gangliosidosis, 42/434
Goldenhar syndrome, 43/443
Gruber syndrome, 43/392
hemophilia A, 42/738
hereditary sensory and autonomic neuropathy type III, 43/59, 60/24
histidinemia, 42/554
holoprosencephaly, 42/32

homocystinuria, 42/556
hydranencephaly, 42/34
hydrocephalus, 42/35
hypophosphatasia, 42/579
idiopathic scoliosis, 42/52
imperforate anus, 42/379
infantile spasm, 42/687
infantile spinal muscular atrophy, 42/89, 91
intermediate spinal muscular atrophy, 42/91
Jervell-Lange-Nielsen syndrome, 42/389
Klinefelter syndrome, 43/558
Klinefelter variant XXXY, 43/559
Klinefelter variant XXYY, 43/563
Klippel-Feil syndrome, 42/37
Lesch-Nyhan syndrome, 42/153
limb girdle syndrome, 43/104
maple syrup urine disease, 42/599
maternal, *see* Maternal birth injury
maxillofacial dysostosis, 43/426
meningocele, 42/27
metachromatic leukodystrophy, 42/494
microcephaly, 42/46
microphthalmia, 42/404
mortality, 25/171
multiple neurofibromatosis, 43/35
muscular atrophy, 42/84
nerve lesion, 25/167
neuroblastoma, 42/759
orofaciodigital syndrome type I, 43/447
osteogenesis imperfecta, 43/454
Patau syndrome, 43/527
Pena-Shokeir syndrome type I, 43/438
phenylketonuria, 42/611
pilonidal sinus, 42/47
Poland-Möbius syndrome, 43/461
porencephaly, 42/49
Potter syndrome, 43/469
progeria, 43/466
prune belly syndrome, 43/467
retinoblastoma, 18/339, 42/768
Seckel syndrome, 43/379
sensorimotor neuropathy, 42/103
sickle cell anemia, 42/624
sickle cell hemoglobin C disease, 42/624
Smith-Lemli-Opitz syndrome, 43/308
spastic cerebral palsy, 42/168
spina bifida, 42/55
spinal cord injury, 25/44, 94-96, 155-170
spinal shock, 25/157
Sprengel deformity, 43/483
Sturge-Weber syndrome, 43/49
subarachnoid hemorrhage, 12/158
subdural effusion, 24/330
thalassemia major, 42/624
thyroid agenesis, 42/632
traumatic venous sinus thrombosis, 24/371
trichopoliodystrophy, 42/584
triploidy, 43/564
trisomy 22, 43/549
tuberous sclerosis, 43/50
Turner syndrome, 43/551
vertebral artery, 25/168, 26/51
Werner syndrome, 43/489
Wilson disease, 42/271
Wohlfart-Kugelberg-Welander disease, 42/91
XX male, 43/556
XXX syndrome, 43/553
Birth order
hydrocephalus, 50/288
multiple sclerosis, 47/302
Birth weight
Down syndrome, 50/525
Edwards syndrome, 50/561
hereditary sensory and autonomic neuropathy type III, 21/109, 60/26
hexachlorophene intoxication, 37/499
r(13) syndrome, 50/585
Bis-cyclohexanone oxalyldihydrazone, *see* Cuprizone
Bis(diethoxyphosphoryl)ethylamine intoxication
chemical classification, 37/549
organophosphorus compound, 37/549
Bismuth
absorption, 64/332
absorption route, 64/334
analysis, 64/332
bioavailability, 64/333
brain scanning, 16/666
chemistry, 64/332
dementia, 64/4
distribution, 64/334
elimination, 64/332
insomnia, 64/4
myoclonus, 64/4
neurotoxin, 36/327, 46/392, 64/4, 10, 331
spastic paraplegia, 64/10
specification, 64/332
toxic encephalopathy, 64/4
toxic myelopathy, 64/10
Bismuth encephalopathy
reversible type, 51/265
Bismuth intoxication
action site, 64/10
astasia, 64/337
biokinetic, 64/331
blood level, 64/341

bone, 64/336
brain, 64/336
brain bismuth level, 64/342
clinical features, 64/337
computer assisted tomography, 64/343
confusion, 64/338
constipation, 64/331
CSF, 64/343
CSF level, 64/341
delirium, 64/338
dementia, 46/392, 64/337
diarrhea, 64/331
dimercaprol, 64/344
dysarthria, 64/337
dysbasia, 64/337
edetic acid, 64/344
EEG, 64/338, 343
elimination, 64/335
EMG, 64/338
encephalopathy, 36/327-329, 64/331
epidemiology, 64/338
epilepsy, 64/338
gait disturbance, 64/338
half-life, 64/336
hallucination, 64/337
history, 64/331
insomnia, 64/337
intestinal absorption, 64/333
kidney, 64/336
management, 64/344
mortality, 64/339
muscle cramp, 64/337
myoclonia, 64/337
nephrotoxicity, 64/331, 337
neuropathology, 64/340
neurotoxicity, 64/331, 337
organ level, 64/335
osteoarthropathy, 64/337
paraplegia, 36/327, 329
pathomechanism, 64/340
peptic ulcer, 64/331
plasma concentration, 64/333
risk, 64/331
sensory evoked potential, 64/343
stomatitis, 64/331
succimer, 64/344
symptom, 64/4
syphilis, 64/331
toxicity, 64/336
treatment, 64/344
tremor, 64/338
unithiol, 64/344
urinary incontinence, 64/338
wart, 64/331
Bis(*N*-oxopyridine-2-thianato) zinc
neuropathy, 40/141
Bite
insect, *see* Insect bite
sea snake, 37/92
viper, *see* Viper bite
Bitemporal hemianopia
brain injury, 24/43
brain tumor, 2/571
chiasmal lesion, 2/571
clinical examination, 57/135
false localization, 2/576
hemianopic pupillary phenomenon, 2/575, 577
optic chiasm injury, 24/43
optic chiasm syndrome, 18/354
trauma, 2/575
Bitter syndrome
dwarfism, 14/107
hemifacial hypertrophy, 14/107
mental deficiency, 14/107
microphthalmia, 14/107
neurocutaneous syndrome, 14/101
nevus sebaceous, 14/107
ptosis, 14/107
thalidomide, 14/107
Bivalves intoxication
ciguatoxin, 37/64
death, 37/64
experimental animal, 37/64
ingestion, 37/64
symptom, 37/64
tridacna maxima, 37/64
Bizarre behavior
acute amebic meningoencephalitis, 52/317
granulomatous amebic encephalitis, 52/317
BK virus
JC virus, 71/415
neural tumor, 71/418
progressive multifocal leukoencephalopathy, 34/313, 318-321, 47/504, 515, 518
Black widow spider venom intoxication
cramp, 40/547
motor unit hyperactivity, 41/310
muscle activity, 40/330
myokymia, 40/330
myotonia, 40/547
spinal muscular atrophy, 59/370
venom, 41/105
Blackwood classification
demyelinating disease, 10/22
diffuse sclerosis, 10/22
Bladder

*see also* Bladder function, Micturition *and* Urinary tract
afferent fiber, 75/667
aging, 74/237
anatomy, 1/357-359
autonomic innervation, 1/368, 373
autonomic nervous system, 74/116
biofeedback method, 75/676
cerebral influence, 1/386
condom drainage, 26/423
continence, 1/369, 373
degenerative disorder, 75/673
demyelinating disorder, 75/673
diabetes mellitus, 75/672
disorder, 1/356
electrical stimulation, 75/677
embryology, 1/356
emptying, 75/675
epileptic symptomatology, 15/140
genetic defect, 75/671
Guillain-Barré syndrome, 75/685
hereditary amyloid polyneuropathy, 75/672
histochemical anatomy, 75/667
hypotonic, *see* Hypotonic bladder
infection, *see* Bladder infection
innervation, 1/361, 75/666
intrinsic ganglion, 75/669
Kegel exercise, 75/677
manual expression, 26/409
motor nerve, 1/365
nerve, 74/112
neurogenic, *see* Neurogenic bladder
neurotransmitter, 75/669
old age, 75/669
pain, 1/363, 366, 379
pelvic plexus, 75/667
percussion, 26/322
progesterone, 75/670
reflex, 1/364, 375
rhythmic activity, 1/384
rupture, 25/470
sensation, 1/378, 75/672
spinal cord function, 1/370
stress incontinence, 75/670
tone, 1/381
training, *see* Bladder training
urinary, *see* Urinary bladder
urine storage, 75/675
vesicorelaxer center, 1/390
Bladder apraxia
corpus callosum tumor, 17/508
Bladder dysfunction
catheterization, 75/674
diabetes mellitus, 75/600
dimethylaminopropionitrile, 75/506
drug therapy, 75/674
electrical injury, 74/321
Friedreich ataxia, 60/326
management, 75/674
Parkinson disease, 75/178
treatment, 60/326
Bladder function, 1/356-398
*see also* Bladder
atonic bladder, 26/272
autonomous dysfunction, 26/320-322
autonomous influence, 26/412
Behçet syndrome, 56/598
caudal aplasia, 42/50
dorsal column stimulation, 47/166
drug treatment, 47/161-166
epilepsy, 15/95, 140
external urinary drainage, 26/423
female patient, 26/424
Ferguson-Critchley syndrome, 42/142
hereditary acute porphyria, 51/391
ice water test, 26/418, 424
infection, *see* Bladder infection
infranuclear bladder, 26/417, 424
intravesicular pressure, 26/420
lower motoneuron dysfunction, 26/272
management, 26/414, 425
multiple sclerosis, 47/160-166
new supply, 26/412, 417, 419
pharmacotherapy, 26/413, 421
radiodiagnosis, 26/271, 424
reflux, 26/273
residual urine, 26/422, 430
spina bifida, 32/550, 42/54
spinal arachnoiditis, 42/107
spinal cord injury, 26/412-426, 61/291
spinal shock, 2/188, 26/254
supranuclear bladder, 26/417, 424
surgery, 26/425, 47/166
training, *see* Bladder training
transverse spinal cord syndrome, 2/188-193
upper motoneuron dysfunction, 26/272
urodynamic measurement, 26/419
vincristine polyneuropathy, 51/299
Bladder infection
chemotherapy, 26/431
management, 26/431
mechanism, 26/428
micro-organism, 26/428-430
paraplegia, 26/208
prevention, 26/426
progressive dysautonomia, 59/140

residual urine, 26/426-432
Bladder neck
pelvic nerve, 1/370
Bladder neck resection
neurogenic bladder, 26/425
spinal cord injury, 61/304
urinary tract, 26/411
vesicourethral neuropathy, 61/304
Bladder percussion
autonomous dysfunction, 26/322
Bladder training
lower motoneuron bladder, 26/422, 424
neurogenic bladder, 26/210, 416
residual urine, 26/422, 430
upper motoneuron bladder, 26/420, 424
Bladder wall hypertrophy
prune belly syndrome, 43/467
Bladex, *see* Cyanazine
Blake pouch
persistence, 31/108-111
Blalock-Taussig anastomosis
congenital heart disease, 63/5
subclavian steal syndrome, 63/5
vertebrobasilar insufficiency, 63/5
Blast injury
brain fat embolism, 55/180
clinical features, 25/222
history, 25/221
pathology, 25/223
spinal cord injury, 25/221
Blastogenesis
measles, 47/357
multiple sclerosis, 47/357
Blastoma ependymale, *see* Ependymoma
Blastomatosis
genetics, 16/30
Blastomatous dysplasia
phakoma pathogenesis, 14/85
Blastomatous tuberous sclerosis, *see* Diffuse glioblastosis
Blastomyces, *see* Cryptococcus neoformans
Blastomyces brasiliensis
*see also* Paracoccidioidomycosis
CNS, 52/385
North American blastomycosis, 52/385
Blastomyces dermatitidis
*see also* North American blastomycosis
brain abscess, 52/151, 385
cerebrovascular disease, 55/432
hematoxylin-eosin stain, 52/390
meningitis, 52/385, 55/432
microbiology, 52/392
mycelial phase, 52/392
North American blastomycosis, 35/408, 52/392
tissue phase, 52/388
yeast phase, 52/392
Blastomycoides histolytica, *see* Cryptococcus neoformans
Blastomycoma
North American blastomycosis, 52/390
Blastomycosis
acquired immune deficiency syndrome, 71/296
brain granuloma, 16/230
chronic meningitis, 56/644
European, *see* Cryptococcosis
mycosis, 35/380
North American, *see* North American blastomycosis
South American, *see* Paracoccidioidomycosis
Bleb disease
Bardet-Biedl syndrome, 13/400
Bleeding tendency
acute subdural hematoma, 24/277
Argentinian hemorrhagic fever, 56/364
Bassen-Kornzweig syndrome, 63/274
chronic subdural hematoma, 24/298
coagulopathy, 38/80
CSF, 38/81
hemangioendothelioma, 18/288
Hermansky-Pudlak syndrome, 43/5
intracerebral hematoma, 11/673
neurologic complication, 38/53-83
sea blue histiocyte disease, 42/473
subdural effusion, 24/333
treatment, 38/83
trimethylaminuria, 29/231
vitamin C deficiency, 28/200, 208
Wolman disease, 10/546
Blennorrhagic polyneuritis
amyotrophy, 7/478
anesthesia, 7/478
antibiotic treatment, 7/478
areflexia, 7/478
hypoesthesia, 7/478
muscular atrophy, 7/478
paresthesia, 7/478
sphincter disturbance, 7/478
Bleomycin
antineoplastic agent, 65/528
behavior disorder, 46/600
chemotherapy, 39/114
cranial neuropathy, 65/528
neurologic toxicity, 39/114
neurotoxin, 39/111, 114, 119, 46/600, 65/528
Blepharitis granulomatosa
Melkersson-Rosenthal syndrome, 8/212

Blepharophimosis
chondrodystrophic myotonia, 40/284, 546
Edwards syndrome, 50/561
Marden-Walker syndrome, 43/424
10q partial trisomy, 43/517
vertebral fusion, 43/488
Blepharoplast
ependymoma, 18/105
spinal ependymoma, 20/362
Blepharoptosis
aqueduct stenosis, 50/312
dopamine β-mono-oxygenase deficiency, 59/163
Smith-Lemli-Opitz syndrome, 66/582
vertebral abnormality, 43/488
Blepharospasm
botulinum toxin A, 65/213
brain embolism, 53/211
brain infarction, 49/542, 53/211
brain lacunar infarction, 54/245
cerebrovascular disease, 53/211
dystonia musculorum deformans, 6/522, 42/218, 49/521, 523
epilepsy, 1/618
focal athetosis, 49/382
intracranial hemorrhage, 53/211
levodopa induced dyskinesia, 49/195
Melkersson-Rosenthal syndrome, 42/322
neuroleptic dystonia, 65/281
oculogyric crisis, 1/583
progressive supranuclear palsy, 22/219, 49/243
pseudoptosis, 2/318
sensorimotor stroke, 54/253
transient ischemic attack, 53/211
upper, *see* Upper eyelid spasm
Wilson disease, 49/228
Blighia sapida
fruit arillus, 37/516, 65/79
intoxications, *see* Hypoglycin A intoxication
Jamaican vomiting sickness, 37/511, 515, 65/79
Blighia sapida intoxication, *see* Hypoglycin A intoxication
Blind loop syndrome
deficiency neuropathy, 7/619, 51/325, 328
intestinal malabsorption, 7/619
neuropathology, 63/277
vitamin E deficiency, 63/277
Blind spot
enlargement, 2/566
papilledema, 2/567
pseudopapilledema, 2/567
Blind vertebra
radiologic finding, 20/22
spinal tumor, 20/22
Blindness
*see also* Amblyopia
acute, *see* Acute amaurosis
alexia, 4/135
Alpers disease, 42/486
anemia, 38/27
angiostrongyliasis, 52/555
anosognosia, 3/284
bacterial endocarditis, 52/299
Bonnet-Dechaume-Blanc syndrome, 43/36
brain embolism, 53/207
brain infarction, 53/207
cardiac arrest, 63/216
carnosinemia, 42/532
cavernous sinus thrombosis, 52/174
cerebral, *see* Cortical blindness
cerebral color, *see* Cerebral color blindness
cerebrovascular disease, 53/207
chloroquine, 65/484
chloroquine intoxication, 65/484
cisplatin intoxication, 64/358
cisplatin polyneuropathy, 51/300
clock function, 74/468
CNS spongy degeneration, 42/506
color, *see* Color blindness
congenital retinal, *see* Congenital retinal blindness
cortical, *see* Cortical blindness
cutis verticis gyrata, 43/244
deutan color, *see* Deutan color blindness
encephalomeningocele, 42/26
ethylhydrocupreine, 13/220
familial metaphyseal dysplasia, 46/88
fetal Minamata disease, 64/418
giant cell arteritis, 71/194
GM2 gangliosidosis type I, 22/537, 42/433
GM2 gangliosidosis type II, 42/435
headache, 48/280
hemangioendothelioma, 18/288
hereditary progressive arthro-ophthalmopathy, 13/41
hereditary striatal necrosis, 49/495
hexachlorophene intoxication, 37/486, 65/476
hypnotic, *see* Hypnotic blindness
hysteria, 24/51, 43/207
infantile neuroaxonal dystrophy, 49/402
infantile neuronal ceroid lipofuscinosis, 66/689
intracranial hemorrhage, 53/207
iron, 13/224
Laurence-Moon syndrome, 43/254
lead encephalopathy, 64/436
lead intoxication, 64/436
leprous neuritis, 51/217
Lowe syndrome, 42/606

meningeal leukemia, 63/341
metachromatic leukodystrophy, 10/568
methanol intoxication, 36/351, 356, 64/99
Minamata disease, 36/121, 64/418
mind, *see* Mind blindness
mitochondrial myopathy, 43/188
mucormycosis, 35/544, 52/471
multiple nevoid basal cell carcinoma syndrome, 43/31
myoclonic epilepsy, 42/702
neuromyelitis optica, 9/426, 428, 47/401
Niemann-Pick disease type A, 42/468, 51/373
night, *see* Nyctalopia
Norrie disease, 22/508, 31/291, 43/441, 46/66
nyctalopia, 13/224
occipital lobe tumor, 17/326
odor, *see* Specific anosmia
oligosaccharidosis, 66/329
olivopontocerebellar atrophy variant, 21/453
optic nerve injury, 24/27
opticocochleodentate degeneration, 21/538, 42/241, 60/752
organotin intoxication, 64/138
osteitis deformans (Paget), 43/451
pertussis, 52/234
pertussis encephalopathy, 33/289
pituitary adenoma, 17/385
polyarteritis nodosa, 51/452
porencephaly, 30/685, 50/359
postictal, *see* Postictal blindness
progressive myoclonus epilepsy, 42/702
psychic, *see* Psychic blindness
pure word, *see* Pure word blindness
Pyle disease, 46/88
retinal, *see* Retinal blindness
rhodotorulosis, 52/493
spinocerebellar ataxia, 42/181
streptococcal meningitis, 52/81
subacute necrotizing encephalomyelopathy, 42/625
superior orbital fissure syndrome, 33/180
syphilis, 52/281
taste, *see* Taste blindness
temporal arteritis, 11/133, 51/452
temporary, *see* Temporary blindness
thalamic, *see* Thalamic blindness (Störring)
transient cerebral, *see* Transient cerebral blindness
transient cortical, *see* Transient cortical blindness
transient ischemic attack, 53/207
transient monocular, 5/129
traumatic, *see* Traumatic blindness
Van Buchem disease, 43/410
visual evoked response, 2/545
word, *see* Word blindness
Blink reflex
corneal reflex, 2/56
diagnostic value, 1/239
facial paralysis, 2/65
Huntington chorea, 49/290
idiopathic facial paralysis, 8/263
multiple sclerosis, 47/131
Blinking
gaze apraxia, 45/406
Blitz-Nick-Salaam Krämpfe, *see* Infantile spasm
Blitzkrämpf
infantile spasm, 15/219
jumpers of the Maine, 42/232
Bloch-Sulzberger disease, *see* Incontinentia pigmenti
Bloch-Sulzberger syndrome, *see* Incontinentia pigmenti
Blood-brain barrier, 28/365-392, 53/83-87
*see also* Blood-CSF barrier
abnormal, 28/371-392
adrenalin, 48/64
air embolism, 28/392
allergic encephalomyelitis, 28/391
alprostadil, 48/64
Alzheimer disease, 46/265
amino acid, 56/81
anatomy, 28/366
angiotensin, 48/64
area postrema, 53/84, 56/80
arsenic intoxication, 64/285
arterial hypertension, 28/391
astrocytosis, 53/87
brain blood flow, 57/262
brain contusion, 57/262
brain edema, 23/148, 28/373-375, 53/135, 57/261, 67/80
brain edema mechanism, 53/137
brain hypoxia, 28/375-377
brain infarction, 28/377-380
brain injury, 23/148, 28/371-374
brain tumor, 28/381, 388, 67/71
brain tumor edema, 67/71
cancer, 69/432
capillary endothelium, 56/79
chemotherapy, 67/278
cholesterolemic xanthomatosis, 27/248
choroid plexus, 56/80
chronic mercury intoxication, 64/373
circumventricular organ, 56/80
CNS infection, 69/432
concentration gradient, 53/86
CSF, 30/530

development, 28/370
disseminated necrotizing leukoencephalopathy, 63/355
eminentia saccularis, 56/80
experimental allergic encephalomyelitis, 9/523, 47/438
glucose transport, 56/81
gram-negative bacillary meningitis, 52/105
headache, 48/64, 79
herpes simplex virus encephalitis, 56/85
herpes simplex virus myelitis, 56/85
historic aspect, 28/365
hyperosmosis, 53/85
hypertensive encephalopathy, 54/214
infection, 28/380
isoprenaline, 48/64
kernicterus, 28/388-391
lead encephalopathy, 28/392
manganese intoxication, 64/304
median eminence, 53/84, 56/80
meningitis, 28/380
mercury intoxication, 64/388
migraine, 48/61-63, 79, 93
mineralizing brain microangiopathy, 63/355
monoamine, 48/93
morphology, 56/79
multiple sclerosis, 47/79-121
nerve injury, 65/424
neuronoglial relationship, 9/624
neurotropic virus, 56/31
noradrenalin, 48/64
normal, 28/365-371
normal value, 53/107
optic recess, 56/80
paraphysis, 56/80
permeability, 53/86, 56/81-83
physiology, 28/367-370
pineal body, 53/84
pineal gland, 56/80
pinocytosis, 53/84
platinum intoxication, 64/357
poliomyelitis, 34/103, 42/652, 56/318
radiation, 28/392
radiation injury, 23/645
seizure, 28/391
serotonin, 48/64
Sindbis virus, 56/84
structural feature, 53/84
surface area, 56/80
tight junction, 53/84, 56/79
topographical distribution, 28/370
trauma, 28/371-374
tuber cinereum, 53/84, 56/80
tuberculous meningitis, 52/205
uremia, 28/391
uremic encephalopathy, 27/340, 28/391
vasogenic brain edema, 53/138
ventricle ependymoma, 56/80
viral infection, 56/79
Blood cell count
spinal cord injury, 26/379
Blood clotting disorder, *see* Coagulopathy
Blood clotting factor V deficiency
intraventricular hemorrhage, 63/312
Blood clotting factor VIII
Emery-Dreifuss muscular dystrophy, 62/157
hemophilia A, 42/737
Blood clotting factor VIII inhibitor
coagulopathy, 63/318
Blood clotting factor X deficiency
subdural hematoma, 63/312
Blood clotting factor XII deficiency
subarachnoid hemorrhage, 63/313
Blood clotting factor XIII deficiency
brain hemorrhage, 63/313
Blood clotting necrosis
brain infarction, 53/188
electrical injury, 61/191
Blood-CSF barrier
*see also* Blood-brain barrier
arteriolar fibrinohyalinoid degeneration, 19/132
Blood disease
benign intracranial hypertension, 16/151, 159
brain change, 12/109
demyelination, 9/593
leukoencephalopathy, 9/592
subarachnoid hemorrhage, 12/109
Blood dyscrasia
brain hemorrhage, 54/288
cerebellar hemorrhage, 55/89
epidural hematoma, 26/19
leukoencephalopathy, 9/572, 593
nonbacterial thrombotic endocarditis, 63/121
retinal artery occlusion, 55/112
spinal cord compression, 19/366
spinal subdural hematoma, 12/566
spontaneous cerebellar hemorrhage, 12/54
subarachnoid hemorrhage, 12/109, 55/2
Blood flow, 26/313-328
aging, 74/232
autonomic nervous system, 74/73, 604, 75/429
brain, *see* Brain blood flow
brain collateral, *see* Brain collateral blood flow
cerebral, *see* Brain blood flow
coronary, *see* Coronary blood flow
cough headache, 48/367-371, 75/300

cutaneous, *see* Cutaneous blood flow
epileptogenesis, 72/177
exercise, 75/438
extracranial, *see* Extracranial blood flow
headache, 75/285
hypothalamus, 1/459
intracranial pressure, 55/205
measurement, 74/601
migraine, 48/150, 75/295
muscle, 1/465
nervous control, 1/462
neurologic intensive care, 55/220
pial, *see* Pial blood flow
plethysmography, 1/463, 75/429
positron emission tomography, 72/371, 375
pressure sore, 26/463, 61/351, 355
regional brain, *see* Regional brain blood flow
regional cerebral, *see* Regional brain blood flow
renal, *see* Renal blood flow
single photon emission computer assisted tomography, 72/375
skeleton, 1/465
skin, *see* Cutaneous blood flow
skin temperature, 1/463
spinal cord, *see* Spinal cord blood flow
thermoregulation, 1/463, 26/371
vagus nerve, 75/620
vegetative nervous function, 1/466

Blood gas
Cheyne-Stokes respiration, 63/483
spinal cord injury, 26/385

Blood group ABO
tuberous sclerosis, 50/371

Blood hyperviscosity
*see also* Hyperviscosity syndrome
ataxia, 55/484
epilepsy, 55/484
headache, 55/484
vertigo, 55/484

Blood lead level
dementia, 64/438
diet, 64/433
erythrocyte, 64/433
hygiene, 64/432
lead intoxication, 36/23, 64/434, 443, 445
normal range, 64/434
occupation, 64/432
race, 64/432
relation with symptom, 64/435
sex and age, 64/432

Blood-nerve barrier
diabetic polyneuropathy, 51/505
nerve, 51/2
nerve injury, 65/424

Blood oxygenation level dependent nuclear magnetic resonance
brain tumor, 67/237

Blood patch
brain hemorrhage, 61/164
complication, 61/157

Blood platelet monoamine oxidase, *see* Platelet MAO

Blood pressure
*see also* Hypertension
acute hypoxia, 75/260
aging, 74/228
alcohol, 75/170, 494
angiotensin, 48/98
arterial, 5/5
autonomic failure, 75/713
autonomic nervous system, 74/14, 248, 611, 613, 75/426, 429, 453
Balaenoptera borealis intoxication, 37/94
baroreflex, 75/467
Binswanger disease, 54/221
biological rhythm, 74/472
brain blood flow, 11/123, 54/144, 55/208
brain collateral blood flow, 11/171
brain death, 24/758, 760, 821
brain embolism, 53/207
brain infarction, 53/109, 207
brain ischemia, 53/109
cerebrovascular disease, 53/207
chronic hypoxia, 75/270
ciclosporin, 75/476
classification, 54/205
cluster headache, 75/289
coital migraine, 75/301
desmopressin, 75/724
*threo*-3,4-dihydroxyphenylserine, 75/724
diurnal change, 75/715
dizziness, 75/715
electrical injury, 74/321
ergotamine, 65/64
exercise, 74/253, 75/716
glossopharyngeal neuralgia, 48/465, 467
Guillain-Barré syndrome, 51/245
head injury, 57/226
head up tilt test, 74/614
headache, 48/77, 282, 418, 75/282, 287
heart rate, 75/467
hypoxia, 75/264
intracranial hemorrhage, 53/207
intracranial pressure, 55/208
locus ceruleus, 74/155
measurement, 74/600

medulla, 74/154
midbrain syndrome, 24/585
midodrine, 75/724
migraine, 48/20, 73, 75/296
moxonidine, 74/171
multiple neuronal system degeneration, 75/170, 174
muscle metaboreflex, 74/255
neck chamber, 75/434
neurogenic, 75/373
neurologic intensive care, 55/217
obstructive sleep apnea, 63/456
pheochromocytoma, 75/476
pressure garment, 75/721
regulation, 1/466, 471, 74/260
response to standing, 74/597
rilmenidine, 74/171
secondary pigmentary retinal degeneration, 13/197
sleep, 3/82
space, 74/286
spinal cord injury, 26/205, 313-333, 61/439
squatting, 74/619
standing, 74/611
stress, 74/309
stroke, 53/459-466, 75/373
subclavian steal syndrome, 53/374
SUNCT syndrome, 75/299
syncope, 75/715
tetanus, 65/218
tetraplegia, 61/439
transient ischemic attack, 53/207
transverse spinal cord lesion, 61/439
traumatic vegetative syndrome, 24/578
trigeminal neurotomy, 48/108
Valsalva maneuver, 74/615

Blood residual N
subarachnoid hemorrhage, 12/151

Blood transfusion
air embolism, 23/614
Chagas disease, 52/345
human T-lymphotropic virus type I, 56/530
multiple sclerosis, 9/385
thrombocytopenia, 63/321

Blood vessel
amyotrophic lateral sclerosis, 59/182
cluster headache, 48/217-242, 75/291
Duchenne muscular dystrophy, 40/170, 251
globoid cell leukodystrophy, 10/77
glycosaminoglycanosis, 10/452
hereditary amyloid polyneuropathy, 60/102
inflammatory myopathy, 40/51, 164, 264
ischemic myopathy, 40/167-170
larynx, 74/390
migraine, 48/79
Niemann-Pick disease type A, 10/496
oligodendroglioma, 18/87
pathology, 40/43
retinal, *see* Retinal blood vessel
spina bifida, 32/541
vasa nervorum disease, 12/645

Blood viscosity
*see also* Hyperviscosity syndrome
blood cell, 55/483
brain blood flow, 11/124
brain collateral blood flow, 11/171
brain injury, 57/211
cell aggregation, 55/483
cerebrovascular disease, 55/483
congenital heart disease, 63/3
fibrinogen, 55/483
IgM, 55/483
malignancy, 55/483
plasma protein, 55/483
prolonged transient ischemic attack, 53/272
shear rate, 55/483
transient ischemic attack, 53/266, 272

Bloom syndrome
aneuploidy, 50/566
ataxia telangiectasia, 14/315
café au lait spot, 14/775, 43/10
carcinoma, 43/10
chromosomal aberration, 43/11
dwarfism, 13/436
growth retardation, 14/775, 43/10
hypopigmentation, 14/775, 43/10
Ig, 43/10
leukemia, 14/775, 43/10
lymphoma, 43/10
mental deficiency, 14/775, 43/10
microcephaly, 30/100, 508
micrognathia, 43/10
photosensitivity, 43/10
sex ratio, 43/10
telangiectasia, 14/775, 43/10

Blowfish intoxication, *see* Puffer fish intoxication

Blowout fracture
craniofacial injury, 23/378
orbital fracture, 24/99, 57/137
orbital injury, 24/35, 88, 99
radiodiagnosis, 24/100
treatment, 24/101

Blue diaper syndrome
Hartnup disease, 29/167

Blue nevus
cell, 14/595

dermatology, 14/599
Schwann cell, 14/599
Blue rubber bleb nevus syndrome, *see* Bean syndrome
Blue sclerae
Ekman-Lobstein disease, 22/495
incontinentia pigmenti, 14/108
osteogenesis imperfecta, 43/453
Russell-Silver syndrome, 43/477
sensorineural deafness, 42/381
Bluetongue virus
animal viral disease, 34/297
encephalopathy, 34/297
experimental acute viral myositis, 56/195
hydranencephaly, 50/340
teratogen, 30/662, 665
Blurred vision, *see* Visual impairment
BO syndrome, *see* Branchio-otodysplasia
Bobble head doll syndrome
aqueduct stenosis, 50/312
Boder-Sedgwick disease, *see* Ataxia telangiectasia
Body
argyrophilic, *see* Argyrophilic body
cabbage, *see* Cabbage body
carotid sinus, *see* Carotid sinus body
ciliary, *see* Ciliary body
colloid, *see* Colloid body
concentric laminated, *see* Concentric laminated body
curvilinear, *see* Curvilinear body
cytoplasmic inclusion, *see* Cytoplasmic inclusion body
Elzholz, *see* Elzholz body
eosinophilic, *see* Eosinophilic body
fingerprint, *see* Fingerprint body
Heinz, *see* Heinz body
hematoxylin, *see* Hematoxylin body
Herring, *see* Herring body
Hirano, *see* Hirano body
inclusion, *see* Inclusion body
intranuclear inclusion, *see* Intranuclear inclusion body
jugular, *see* Jugular body
Lewy, *see* Lewy body
lipid inclusion, *see* Lipid inclusion body
mamillary, *see* Mamillary body
Marinesco, *see* Marinesco body
membranous cytoplasmic inclusion, *see* Membranous cytoplasmic inclusion body
membranovesicular, *see* Membranovesicular body
muscle filamentous, *see* Muscle filamentous body
myxedema, *see* Myxedema body
Negri inclusion, *see* Negri inclusion body
paracrystalline, *see* Paracrystalline body
Pick, *see* Pick body
pineal, *see* Pineal body
Renaut, *see* Renaut body
rod, *see* Rod body
spheroid, *see* Spheroid body
temperature, *see* Thermoregulation
tuffstone, *see* Tuffstein granule
vagal, *see* Vagal body
zebra, *see* Zebra body
Body agnosia
body scheme, 2/603
Body damage
neuron cell, *see* Neuronopathy
Body experience
body scheme, 4/210, 392, 45/373
Body fluid
autonomic nervous system, 74/1
space, 74/293
sulfatide, 66/174
Body function
autonomic nervous system, 74/3
Body image, *see* Body scheme
Body length
short, *see* Short stature
tall, *see* Tall stature
Body mass
antiepileptic agent, 65/497
catamenial migraine, 5/231
Down syndrome, 50/527
headache, 5/231
malnutrition, 29/2
Body odor
*see also* Odor *and* Odorant
bacterium, 74/375
behavior, 74/377
menstrual cycle, 74/377
sexual activity, 74/377
sweat, 74/375
Body orientation
autotopagnosia, 45/381-383
body scheme, 4/219
body scheme disorder, 4/213
Body scheme
action schema, 4/210
Alice in Wonderland syndrome, 5/244
anatomic localization, 4/211
body agnosia, 2/603
body experience, 4/210, 392, 45/373
body orientation, 4/219
boundary, 4/219
brain commissurotomy, 4/278-280
cenesthesia, 4/207

constructional aspect, 4/209
definition, 4/210, 45/373
description, 4/209
development, 3/27, 4/392-397, 403
disorder, *see* Body scheme disorder
genesis, 4/207-209
hand, 4/221
hemi-inattention, 45/154
hemispherectomy, 4/211
historic survey, 4/207-209
linguistic aspect, 4/210, 221
migraine, 48/129, 161
narcolepsy, 45/511
nature, 4/207-209, 392, 400
normal, 4/207-209, 393, 400
peripheral conscious, 4/209
phantom limb, 4/210, 225, 227, 45/399
plasticity, 4/210, 231
postural schema, 4/207
right left discrimination, 4/221
schema, 4/208
sensory extinction, 3/195
somatesthetic cortex, 4/209
somatesthetic symptom, 4/278-280
somatognosia, 4/208
somatopsyche, 4/207
spatial orientation disorder, 3/23
structure, 4/210
terminology, 4/207, 211, 392
topesthesia, 4/279
visual body image, 4/219
Body scheme disorder, 4/207-233, 45/373-387
amnesic aphasia in autotopagnosia, 4/221
amorphosynthesis, 4/213, 215
amytal test, 4/216
anosodiaphoria, 2/478, 4/219, 45/379
anosognosia, 2/478, 4/23, 29, 92, 213, 215-219, 243-245, 297, 45/374, 379
anosognosia for hemianopia, 4/216
anosognosia for hemiplegia, 4/216
anosognosie douloureuse, 4/216
Anton-Babinski syndrome, 4/216-219, 45/376-378
apraxia, 4/223
aschematica, 4/207, 230
asymbolia, 4/222
autoscopia, 3/31, 254, 4/210, 231-233, 330, 45/357, 386
autotopagnosia, 1/107, 2/668, 4/210, 219-222, 297, 45/374, 381-383
barbiturate intoxication, 4/244
behavior disorder, 45/379
body boundary, 4/396
body half agnosia, 1/107
body inside, 4/397
body orientation, 4/213
cerebral dominance, 4/213, 297
cerebral palsy, 4/398
classification, 4/212
confabulation, 4/217
congenital pain insensitivity, 4/222
conscious hemiasomatognosia, 4/214, 45/374
deafferentation, 4/213, 218
denial of illness, 4/216, 219, 45/379
developmental disorder, 4/392-406
developmental dysphasia, 4/401
disconnection syndrome, 4/212, 217
disorientation, 3/208
draw-a-person test, 4/214, 231, 233
drawing, 4/396
dyslexia, 4/401
dysphasia, 4/401
dyspraxia, 4/401
emotion, 45/380
epilepsy, 4/214
examination, 4/213, 394-397
exosomatesthesia, 4/233, 242
figure assembly, 4/296
finger agnosia, 2/668, 4/219-222
finger aphasia, 4/220
finger apraxia, 4/220
genetic psychology, 4/392-394
Gerstmann syndrome, 1/107, 2/668, 4/222, 297, 398, 401, 45/383
hallucination, 3/31, 254
Head's hand-ear-eye test, 4/214, 220
hemiasomatognosia, 2/478, 4/22, 214, 297, 45/374-376
hemidepersonalization, 4/214, 45/375
hemiplegia denial, 4/218
historic survey, 4/212
hyperschematia, 2/478, 4/207, 230
hypochondriasis, 45/380
hyposchematica, 4/207, 230
illness denial, 4/216, 219, 402, 45/379
indifference to pain, 4/222, 45/383-385
interparietal syndrome, 4/233
kinesthetic hallucination, 4/218
localization, 4/213-217, 219, 221, 223, 230, 232
macrosomatognosia, 4/226, 230, 45/385
mental deficiency, 4/397
mental illness, 4/402
microsomatognosia, 4/230, 45/385
migraine, 4/214, 48/162
misoplegia, 4/219, 45/379
naming of body parts, 4/397

neglect syndrome, 45/380
normal development, 4/392-397
nosoagnosic overestimation, 4/219, 45/379
occipital lobe, 2/668
pain, 4/223
pain insensitivity, 4/222, 45/383-385
paralyzed limb personification, 4/219, 45/380
paraschematica, 4/207, 230
parietal lobe lesion, 2/668, 4/4, 207-240
parietal lobe syndrome, 45/69
peripheral nervous system lesion, 4/399
phantom limb, 2/478, 668
practosomatognosia, 4/394-396
psychophysiological method, 4/394
psychosis, 4/233, 241-247
right left discrimination, 4/219-222, 395
selective functional anosognosia, 4/216
sensory deficit, 4/400
sensory extinction, 4/213
somatesthetic deafferentation, 4/218
somatognosia, 2/668
somatoparaphrenia, 4/219, 45/380
spatial orientation disorder, 3/223
topesthesia, 4/394-396
type, 45/374
verbal anosognosia, 4/218, 45/378
visual agnosia, 2/668
visual controlled movement, 4/215
visual disorientation, 2/668
visuospatial agnosia, 4/215
Wernicke-Korsakoff syndrome, 4/217
withdrawal reaction, 45/380
Body temperature, *see* Thermoregulation
Body type
sympathotonia, 22/252
vagotonia, 22/252
Body weight, *see* Body mass
Boerhaave, H., 1/10
Bogaert classification, *see* Van Bogaert classification
Bogota, Columbia, 1/12
BOLD MRI, *see* Blood oxygenation level dependent nuclear magnetic resonance
Bolivian hemorrhagic fever
arenavirus infection, 34/194-196, 203
clinical features, 56/369
CSF, 34/195
delirium, 56/369
epidemiology, 56/369
epilepsy, 56/369
facial grimacing, 56/369
intention tremor, 56/369
Machupo virus, 56/355
mortality, 56/369
pathogenesis, 56/369
pathology, 56/369
tremor, 56/369
Bolk, L., 1/10
Bombesin
basal ganglion, 49/33
Parkinson disease, 49/121
Bone abnormality
atlas dysplasia, 50/388
basilar process, 50/388
craniovertebral region, 32/62-65
dens, 50/388
occipital dysplasia, 50/388
Rothmund-Thomson syndrome, 14/777, 39/405, 43/460
Saldino-Noonan syndrome, 31/243
third occipital condylus, 50/388
transverse basilar fissure, 50/388
Bone cyst
aneurysmal, *see* Aneurysmal bone cyst
tuberous sclerosis, 14/51
tuberous sclerosis complex, 68/292
Bone defect
craniolacunia, 30/270
posttraumatic epilepsy, 24/663
skull, *see* Skull defect
sphenoid, *see* Sphenoid bone defect
Bone dysplasia
congenital cerebellar hypoplasia, 60/671
generalized, 32/138
hereditary sensory and autonomic neuropathy type II, 60/17
hereditary sensory and autonomic neuropathy type III, 60/17
juvenile metachromatic leukodystrophy, 60/671
Marinesco-Sjögren syndrome, 60/669
methotrexate syndrome, 30/122
vertebral column, 32/138
Bone fragility
adult Gaucher disease, 10/564, 66/123
dysosteosclerosis, 43/393
Ekman-Lobstein disease, 22/494
hyperostosis corticalis deformans, 43/409
hypophosphatasia, 42/578
osteogenesis imperfecta, 43/453
osteopetrosis, 43/456
pyknodysostosis, 43/468
Bone hypertrophy
cranial, *see* Cranial bone hypertrophy
Bone marrow
arsenic intoxication, 64/286
cranial vault tumor, 17/121

sea blue histiocyte, 60/674
skull base tumor, 17/172
Bone marrow aplasia
leukoencephalopathy, 9/599
Bone marrow depression
arsenic polyneuropathy, 51/266, 64/286
procainamide intoxication, 37/452
toluene intoxication, 36/372
Bone marrow transplantation
adrenoleukodystrophy, 66/87, 97
adrenomyeloneuropathy, 59/355
adult Gaucher disease, 66/129
animal model, 66/88
ataxia telangiectasia, 60/390
autologous, 67/282
chemotherapy, 67/282
data, 66/96
galactosylceramidase deficiency, 66/89
Gaucher disease, 66/129
globoid cell leukodystrophy, 66/87, 91
glycosaminoglycanosis type IH, 66/87, 99
glycosaminoglycanosis type II, 66/298
glycosaminoglycanosis type IVB, 66/306
glycosaminoglycanosis type VI, 66/309
glycosaminoglycanosis type VII, 66/312
α-*L*-iduronidase deficiency, 66/89
intelligence quotient, 66/97
late infantile metachromatic leukodystrophy, 66/95
mannosidase deficiency, 66/89
metachromatic leukodystrophy, 66/87, 92-95, 176
microglia, 66/89
myasthenia gravis, 62/402
neurotoxicity, 67/364
nonbacterial thrombotic endocarditis, 63/28
oligosaccharidosis, 66/329
prevention, 66/88
quality of life, 67/399
treatment, 66/87
X-linked adrenoleukodystrophy, 66/470
Bone membranous lipodystrophy
agnosia, 42/279
aphasia, 42/279
apraxia, 42/279
ataxia, 42/280
behavior disorder, 42/279
brain atrophy, 42/279-281
EEG, 42/281
epileptic seizure, 42/280
myoclonus, 42/279
optic atrophy, 42/280
Bone pain
chronic renal failure myopathy, 62/541
congenital hyperphosphatasia, 31/258
juvenile Gaucher disease type B, 66/125
multiple myeloma, 20/11, 39/138, 146, 63/392
periodic, 42/301
Bonham-Carter syndrome
hypertelorism, 30/248
mental deficiency, 30/248
skin manifestation, 30/248
Bonhoeffer, K., 1/8, 13, 23
Bonnet-Bonnet syndrome, *see* Orbital apex syndrome
Bonnet-Dechaume-Blanc syndrome, 14/260-265
blindness, 43/36
brain angiography, 14/265
brain angioma, 12/101
brain stem aneurysm, 14/70
cirsoid aneurysm, 14/260, 265
clinical features, 14/109, 261
course, 14/263
cranial nerve, 14/109, 43/36
diagnosis, 14/260, 262, 43/36
embryology, 14/260
epilepsy, 14/109
epileptic seizure, 14/262
exophthalmos, 14/261
facial nevus, 14/71
general concept, 14/88
genetics, 14/514
headache, 43/36
heterochromia iridis, 14/263
hydrocephalus, 14/262, 43/36
mental deficiency, 14/109, 262, 43/36
Millard-Gubler syndrome, 14/109
neurocutaneous syndrome, 14/101
neurofibromatosis, 68/298
neuropathology, 14/69
nosology, 14/260
ophthalmology, 14/649-651
oxycephaly, 14/264
papilledema, 14/108
phakomatosis, 14/13, 50/377
psychiatry, 14/767
racemose angioma, 14/109
retinal angioma, 14/70, 31/200
retinal vascular tortuosity, 14/261
secondary pigmentary retinal degeneration, 13/335, 60/736
seizure, 43/36
syringomyelia, 31/200
Bonnet syndrome
lilliputian hallucination, 45/358
Bonnevie-Ullrich status, *see* Turner syndrome
Bonnevie-Ullrich syndrome, *see* Turner syndrome

Bonney test
  axon reflex, 2/137, 173, 7/411
  description, 2/137, 7/411
  plexus injury, 7/411
  vasomotor reflex response, 2/173
Bonnier syndrome
  lateral medullary infarction, 2/220
  medulla oblongata syndrome, 2/220
Bony fusion
  anterior spine, *see* Anterior spinal fusion
  atlantoaxial, 25/298, 325
  cervical hyperanteflexion sprain, 25/345
  cervical hyperdeflexion injury, 25/363
  cervical spine, 25/281, 297-303, 345, 353, 363, 43/428
  craniospinal, 25/294
  facet interlocking, 25/340
  lumbar intervertebral disc prolapse, 25/503
  posterior spine, *see* Posterior spinal fusion
  thoracolumbar spine injury, 25/462
Bony spur
  vertebral, *see* Vertebral bony spur
Bony tumor
  Maffucci-Kast syndrome, 43/29
Boogaard line
  basilar impression, 32/18
BOR syndrome, *see* Branchio-otorenal dysplasia
Borate intoxication
  headache, 48/420
Borax intoxication
  symptom, 64/360
Border virus disease
  animal viral disease, 34/300-302
Borderline leprosy
  classification, 51/216
Borderline polyneuropathy
  undetermined significant monoclonal gammopathy, 47/619
Borderline syndrome
  holoprosencephaly, 50/242
Borderzone infarction
  vascular dementia, 46/356
Bordetella pertussis
  acute encephalopathy, 47/479
  experimental allergic encephalomyelitis, 9/526, 47/433
  hyperacute disseminated encephalomyelitis, 9/526
  postinfectious encephalomyelitis, 47/479
  postvaccinial encephalomyelitis, 47/479
  serotype, 52/233
  synapse modulation, 52/232
  toxin, 52/232
  virulence, 52/232
Boric acid intoxication
  epilepsy, 64/360
Börjeson-Forssman-Lehmann syndrome
  brain atrophy, 42/530
  CNS, 31/311
  cranial hyperostosis, 42/530
  dwarfism, 42/530
  epilepsy, 31/311, 42/530
  genetics, 31/311
  hyperkinesia, 42/530
  hypogonadism, 42/530
  mental deficiency, 31/311, 42/530
  metabolic abnormality, 31/311
  microcephaly, 30/508
  neuronal heterotopia, 42/530
  pathology, 31/311
  prognosis, 31/312
  psychomotor retardation, 42/530
Bornholm disease
  *see also* Epidemic benign myalgia *and* Viral myositis
  acute viral myositis, 56/198
  Coxsackie virus A, 56/198
  Coxsackie virus B, 56/198, 332
  Coxsackie virus B6, 56/198
  Coxsackie virus infection, 34/137
  ECHO virus, 56/198
  headache, 56/198
  myalgia, 40/325
  polymyositis, 41/54
  viral myositis, 41/54
Boron hydride intoxication, *see* Decaborane intoxication, Diborane intoxication *and* Pentaborane intoxication
Boron intoxication
  symptom, 64/360
Boron neutron capture
  astrocytoma grade II, 18/39
  astrocytoma grade III, 18/39
  astrocytoma grade IV, 18/39
Boron nitride intoxication
  symptom, 64/361
Boron trifluoride intoxication
  symptom, 64/361
Borrelia burgdorferi
  *see also* Lyme disease *and* Neuroborreliosis
  antibody, 52/263
  chronic meningitis, 56/644
  Lyme disease, 52/261
  meningitis, 52/261
  neuroborreliosis, 51/199
  polymyositis, 62/373
Borrelia duttonii

Lyme disease, 51/206
neuroborreliosis, 51/206
Borreliosis
chronic meningitis, 56/645
Bottleneck tumor, *see* Hourglass tumor
Botulinum toxin
cell internalization, 65/247
chain, 65/245
domain organization, 65/245
flaccid paralysis, 65/245
intoxication mechanism, 65/246
membrane translocation, 65/247
myasthenic syndrome, 64/15
neuroexocytosis block, 65/247
neuromuscular block, 40/129, 143
neuromuscular transmission, 40/143
neurotoxin, 7/112, 37/88, 40/129, 143, 41/129, 288, 63/152, 64/2, 15, 65/209, 245, 247
peptide bond, 65/248
proteinase, 65/245
receptor binding, 65/245
synaptobrevin, 65/248
synaptosomal associated protein 25, 65/248
syntaxin, 65/248
toxic myopathy, 64/15
toxic neuromuscular junction disorder, 64/15
type A-G, 65/245
zinc binding, 65/245
zinc endopeptidase, 65/245
Botulinum toxin A
blepharospasm, 65/213
dystonia, 65/213
hemifacial spasm, 65/213
neurotoxin, 65/213
writers cramp, 65/213
Botulinum toxin intoxication
action site, 64/15
orthostatic hypotension, 63/152
pathomechanism, 64/15
symptom, 64/15
Botulism
acetylcholine transmission, 51/189
acute muscle weakness, 41/288
4-aminopyridine, 65/212
autonomic nervous system, 75/38, 509
autonomic polyneuropathy, 51/477
axonal degeneration, 51/188
bacterial food poisoning, 37/88
bulbar symptom, 65/210
cholinergic dysfunction, 75/701
classic type, 65/209
clinical features, 40/297, 304, 308
cranial neuropathy, 65/210
decremental EMG, 41/129
descending polyneuropathy, 65/210
diagnosis, 65/212
differential diagnosis, 41/129, 357, 71/558
diplopia, 65/210
dysarthria, 65/210
dysphagia, 65/210
Eaton-Lambert myasthenic syndrome, 62/425
EMG, 51/189, 65/212
epidemiology, 71/557
evoked muscle action potential, 41/290
extraocular muscle weakness, 41/289
food-borne, *see* Food-borne botulism
guanidine, 41/289
guanidine hydrochloride, 65/212
hidden type, 65/209, 211
infantile spinal muscular atrophy, 59/74
infantile type, 41/290, 51/188, 190, 65/209
miniature motor end plate potential, 41/290
muscle action potential, 65/212
muscular hypotonia, 65/210
myasthenia gravis, 41/129, 62/416
myasthenic syndrome, 51/189
nerve lesion, 7/476
nerve stimulation, 7/112
neuritis, 51/188
neuromuscular block, 41/290
pathogenesis, 71/557
polyneuropathy, 65/210
postexercise facilitation, 65/212
prevention, 71/558
prognosis, 71/558
progressive bulbar palsy, 59/225, 375
ptosis, 65/210
pupillary constriction, 51/189
sudden infant death, 51/190
survey, 71/555
symptom, 65/210
synaptic vesicle, 41/290
tick paralysis, 40/308
toxic myopathy, 62/612
toxin, 40/143, 65/209
toxin mediated syndrome, 71/555
toxin type, 51/188
treatment, 65/212
type A toxin, 65/209
type B toxin, 65/209
type C toxin, 65/209
unclassified, 51/191
wound, *see* Wound botulism
Boucher-Neuhäuser syndrome
choroid dystrophy, 60/654
Boulimia, *see* Bulimia

Bourneville-Pringle disease, *see* Tuberous sclerosis
Boustrophedon
developmental dyslexia, 46/116
Bovine rinderpest virus
morbillivirus genus, 56/587
Bovine viral diarrhea mucosal disease virus
animal viral disease, 34/299
Bowel function, *see* Intestinal function
Bowel motility absence, *see* Intestinal peristalsis absence
Bowen-Conradi syndrome
cerebellar atrophy, 43/335
cryptorchidism, 43/335
growth retardation, 43/335
microcephaly, 43/335
micrognathia, 43/335
renal abnormality, 43/335
Bowen syndrome
contracture, 43/335
corpus callosum agenesis, 43/335
heart disease, 43/335
pulmonary hypoplasia, 43/335
talipes, 43/335
Bowing spasm, *see* Infantile spasm
Bowling
rehabilitative sport, 26/531
Boxer dementia
cerebral lesion, 23/554
clinical features, 46/395
neurofibrillary tangle, 23/557, 46/274
Boxer encephalopathy
boxing injury, 12/121, 22/131, 23/541, 566
cavum septi pellucidi, 30/327
subarachnoid hemorrhage, 12/121
Boxfish intoxication
classification, 37/78
pahutoxic, 37/82
Boxing injury
*see also* Punch drunk syndrome
acute subdural hematoma, 23/548
animal experiment, 23/560
boxer encephalopathy, 12/121, 22/131, 23/541, 566
brain atrophy, 23/553
brain concussion, 23/562
brain contusion, 23/562
brain edema, 23/550
brain injury, 23/527
carotid artery thrombosis, 23/539, 550
cavum septi pellucidi, 23/544, 554
cerebellar syndrome, 23/541, 554
cerebellum, 23/560
chronic subdural hematoma, 23/547
clinical features, 23/538
commotio cerebri, 23/562
dysarthria, 23/545
EEG, 23/546
epidural hematoma, 23/547
epilepsy, 23/530
experimental injury, 23/536, 557-562
extrapyramidal disorder, 23/541, 554
head injury, 23/527
history, 23/531
hyperkinesia, 23/543
impact mechanic, 23/534
knock out, 23/540, 552, 566
late effect, 23/549, 551
mental deficiency, 23/544
pankration, 23/533
parkinsonism, 23/556
pathology, 23/547, 555
permanent damage, 23/551
prophylaxis, 23/528
punch drunk syndrome, 23/540
Purkinje cell, 23/556
rotation injury, 23/564, 566
statistic, 23/528, 537
subarachnoid hemorrhage, 23/548
subdural hematoma, 23/547
substantia nigra, 23/557
translational injury, 23/557
traumatic carotid thrombosis, 23/539, 550
traumatic epilepsy, 23/530
Boxjelly intoxication
fatty degeneration, 37/41
necrosis, 37/41
skin, 37/41
Brachial angiography
brain tumor, 16/637
retrograde, *see* Retrograde brachial angiography
Brachial diplegia
cardiac arrest, 63/218
Brachial monoplegia
spinal cord injury, 25/274
Brachial nerve
medial cutaneous, *see* Medial cutaneous brachial nerve
Brachial neuralgia
*see also* Brachialgia
antebrachial pain syndrome, 32/406
clinical presentation, 40/319, 326
Brachial neuritis, 42/303-305
*see also* Brachial plexus neuritis
acute paralytic, *see* Acute paralytic brachial neuritis
brachial plexus, 2/139

cheilopalatoschisis, 42/304
dysesthesia, 42/303
epicanthus, 42/304
hereditary recurrent, 21/88
Hodgkin disease, 39/56
human immunodeficiency virus neuropathy, 71/354
hypotelorism, 42/304
infection etiology, 2/139
malignant lymphoma, 39/56
muscular atrophy, 42/303
nerve compression, 7/430
neurobrucellosis, 52/590
neuropathy, 42/305
paraneoplastic syndrome, 69/374
rubella, 7/486
sensory loss, 42/303
short stature, 42/304
vascular compression, 7/430
vocal cord paralysis, 42/303
Brachial paralysis
birth incidence, 2/141
brachial plexus, 51/143
Horner syndrome, 43/64
Brachial plexitis, *see* Brachial plexus neuritis
Brachial plexopathy
cancer, 69/84
nerve sheath tumor, 69/79
paraneoplastic syndrome, 69/379
Brachial plexus
allergic granulomatous angiitis, 63/383
anatomic variation, 2/135
anatomy, 7/402
avulsion lesion, 51/144
axonotmesis, 51/145
birth incidence, 51/143
brachial neuritis, 2/139
brachial paralysis, 51/143
C5-C6 lesion, *see* Erb paralysis
C8-T1 lesion, *see* Brachial paralysis
cadet palsy, 51/144
computer assisted tomography, 51/148
cord, 2/135, 137
deafferentation pain, 51/153
diagram, 2/136
direct injury, 7/406, 51/144
direct pressure, 51/144
division, 2/135, 137
dorsal spinal nerve root entry zone lesion, 51/154
EMG, 51/144, 148
Erb paralysis, 51/143
gunshot injury, 51/144
heavy duty below elbow splint, 51/151
hereditary mononeuritis multiplex, 2/138
Horner syndrome, 43/64, 51/144
humeral fracture, 70/31
hyperabduction syndrome, 7/441
iatrogenic nerve injury, 51/144
knife wound, 51/144
latissimus dorsi replacement, 51/152
lipocalcinogranulomatosis, 43/185
median nerve lesion, 2/146
meningocele, 51/148
motor syndrome, 2/146
muscle strength test, 51/147
myelography, 51/148
nerve, 2/135, 137
neuroma sign, 51/148
neurosurgery, 51/147
pain, 51/152
pectoralis major muscle transfer, 51/152
physiotherapy, 51/150
postfixation, 25/406
prefixation, 25/406
prognosis, 51/147
reconstructive surgery, 51/152
rehabilitation, 51/149
root, 2/135, 137
Saturday night palsy, 51/145
sensory action potential, 51/148
spinal nerve root avulsion, 2/135, 137, 7/407, 51/145-148
spinal nerve root repair, 51/149
spinal nerve root rupture, 51/146
stab wound, 25/203
Steindler procedure, 51/152
surgery, 51/144
tendon transfer, 51/149
Tinel sign, 51/149
traction injury, 51/145
transcutaneous nerve stimulation, 51/154
triceps transfer, 51/152
trunk, 2/135, 137
tumor, 2/143
ulnar nerve, 2/146
upper lesion, 7/408, 416
vascularized ulnar nerve grafting, 51/149
Brachial plexus block
complication rate, 65/422
Brachial plexus injury
anatomy, 25/405
birth incidence, 7/416, 25/168, 395
closed injury, 7/404
Dejerine-Klumpke syndrome, 7/409
Duchenne-Erb syndrome, 7/408
electrical injury, 61/192

examination, 51/147
exploration, 25/424
Horner syndrome, 7/410
humeral dislocation, 7/406
ionizing radiation neuropathy, 51/137
lesion site, 2/136
lesion type, 2/138-143
lower lesion, 7/409, 416
mechanism, 25/405
middle lesion, 7/408
motor syndrome, 2/146
pack palsy, 2/139
policeman tip position, 2/140
reflex sympathetic dystrophy, 61/122
review, 51/143-154
surgery, 25/425
total, 7/409
total lesion, 7/409
traction injury, 7/405-407, 25/395, 409-435
upper lesion, 7/408, 416
Brachial plexus lesion
heart transplantation, 63/181
iatrogenic neurological disease, 63/181
Brachial plexus neuritis
*see also* Brachial neuritis
allergic granulomatous angiitis, 63/383
diamorphine intoxication, 65/356
diphtheria antitoxin, 2/139
Ehlers-Danlos syndrome, 55/456
Epstein-Barr virus, 56/254
hereditary, *see* Hereditary brachial plexus neuropathy
interleukin-2 intoxication, 65/539
nerve compression, 7/430
nerve conduction, 7/178
neuralgic amyotrophy, 8/76
opiate intoxication, 37/388, 65/356
serogenetic type, 2/139
tetanus toxoid, 2/139, 52/40
vascular compression, 7/430
Brachial plexus paralysis
Duchenne-Erb syndrome, 2/140
Ehlers-Danlos syndrome, 55/456
Horner syndrome, 2/136
Klumpke test, 2/141
severe total type, 2/141
traction injury, 2/137, 141
Brachial plexus syndrome, 2/128-153
neurofibromatosis, 2/143
Brachial radiculitis
acute, *see* Acute brachial radiculitis
Brachialgia
*see also* Brachial neuralgia
antebrachial pain syndrome, 32/406
historic background, 20/548
neck field, 20/554
Brachialgia paresthetica nocturna, *see* Carpal tunnel syndrome
Brachiocephalic panarteritis, *see* Takayasu disease
Brachiocrural hemiparesis
brain lacunar infarction, 54/243
pure motor hemiplegia, 54/243
Brachiofacial hemiparesis
brain lacunar infarction, 54/242
pure motor hemiplegia, 54/242
Brachiofacial hemiplegia
clinical examination, 57/135
Brachioradial reflex
reinforcement, 1/243
Brachium conjunctivum
chorea, 21/529
dyssynergia cerebellaris myoclonica, 21/513
Brachium conjunctivum lesion
ataxia, 21/514
bilateral, 21/515
chorea, 21/515
tremor, 21/514
Brachycephaly
*see also* Craniosynostosis
acrocephalosyndactyly type V, 43/320
Bardet-Biedl syndrome, 13/385
cerebrohepatorenal syndrome, 43/338
cleidocranial dysplasia, 43/348
cloverleaf skull syndrome, 30/245
convulsion, 30/226
corpus callosum agenesis, 42/101
craniodigital syndrome, 43/243
craniofacial dysostosis (Crouzon), 50/122
craniosynostosis, 30/154, 221, 42/20, 50/113
Goltz-Gorlin syndrome, 14/113
happy puppet syndrome, 31/309
Herrmann-Opitz syndrome, 43/325
Klein-Waardenburg syndrome, 14/116
mental deficiency, 30/226
9p partial monosomy, 50/578
Brachydactyly
acrocephalosyndactyly type II, 38/422
acrocephalosyndactyly type III, 50/123
Bardet-Biedl syndrome, 13/385
DOOR syndrome, 42/385
Herrmann-Opitz syndrome, 43/325
Möbius syndrome, 50/215
Brachymetacarpalism
multiple nevoid basal cell carcinoma syndrome, 14/462
pseudohypoparathyroidism, 14/466

Refsum disease, 21/206
Turner syndrome, 14/466
Brachymetatarsalism
Refsum disease, 21/206-208
Brachyphalangy
mental deficiency, 43/262
Bracing
ambulation, 41/468
Duchenne muscular dystrophy, 41/468
Milwaukee, *see* Milwaukee bracing
rehabilitation, 41/468
scoliosis, 41/477
upper extremity, 8/404
Bradbury-Eggleston syndrome, *see* Pure autonomic failure
Bradyarrhythmia
Guillain-Barré syndrome, 51/245
metastatic cardiac tumor, 63/94
primary cardiac tumor, 63/94
Bradycardia
autonomic nervous system, 75/428
breath holding syncope, 15/828
cerebellar tumor, 17/710
Cheyne-Stokes respiration, 63/483
ciguatoxin intoxication, 65/161
cold face test, 74/606, 622
Cushing reflex, 63/230
ergot intoxication, 65/63
glossopharyngeal neuralgia, 48/465, 467
guanethidine intoxication, 37/436
Guillain-Barré syndrome, 51/257
Huntington chorea, 49/284
hypothalamic syndrome, 2/450
insula, 63/205
migraine, 48/8
nereistoxin intoxication, 37/55
phenylpropanolamine intoxication, 65/259
pheochromocytoma, 39/498
reflex, *see* Reflex bradycardia
spinal cord injury, 61/436
tetraplegia, 61/436
transverse spinal cord lesion, 61/436
vagus nerve, 75/428
Bradykinesia
carbon monoxide intoxication, 64/31
extrapyramidal disorder, 2/344
gaze spasm, 2/344
manganese intoxication, 64/306, 308
Mast syndrome, 42/282
mental deficiency, 46/28
metabolism, 1/129
micrographia, 45/465
multiple neuronal system degeneration, 75/162
neuroleptic parkinsonism, 65/292
nociceptive, 1/142
ocular, *see* Ocular bradykinesia
pallidal necrosis, 49/465
pallidoluysionigral degeneration, 49/446
pallidonigral necrosis, 49/465
pallidostriatal necrosis, 49/465
Parkinson dementia, 46/375, 49/171
Parkinson disease, 49/103
parkinsonism, 46/376
progressive supranuclear palsy, 22/219, 221, 46/301, 49/240
pseudo-Stellwag sign, 2/344
Wilson disease, 49/225
Bradykinin
arthropod envenomation, 65/201
flushing, 48/97
headache, 48/97
intravenous infusion, 48/97
kallikrein, 48/97
kininogen, 48/97
migraine, 1/142, 5/39, 48/97
nociceptor, 45/229
pain, 1/129
vasodilatation, 48/97
Bradyphrenia
Lafora progressive myoclonus epilepsy, 15/385
lead encephalopathy, 64/435
mental deficiency, 46/28
migraine, 48/161
progressive supranuclear palsy, 22/221, 49/243, 246
Bradypnea
capnography, 1/660
case report, 1/660
classification, 1/657
hypopnea, 1/666
intracranial pressure, 1/660
sobbing syncope, 15/827
Bragard sign
lumbar intervertebral disc prolapse, 20/587
lumbar stenosis, 20/684
radiculopathy, 20/684
Braille reading
alexia, 4/135
Brain
*N*-acetylaspartic acid, 9/16
actinomycosis, 16/217
alanine, 9/16
aluminum intoxication, 64/273
arteriospasm, *see* Brain vasospasm
bismuth intoxication, 64/336
calcification, *see* Intracerebral calcification

calcinosis, *see* Striatopallidodentate calcification
calcium, 9/16, 28/433
carcinoma, 38/625-660
cauliflower, *see* Cauliflower brain
cavernous hemorrhage, 18/277
cerebroside sulfatide, 10/47
chloride, 9/16
coccidioidomycosis, 16/216
congenital aneurysm, *see* Brain aneurysm
dystrophin, 62/125
echinococcosis, 16/227
eosinophilic granuloma, 16/231
epilepsy prone, 72/70
extracellular space, 56/81
GABA, 9/16
gene transfer, 66/118
gigantism, *see* Cerebral gigantism
gliogenesis, 67/10
gluconeogenesis, 27/13
glucose, 27/19
glutamic acid, 9/16
glutamine, 9/16
glycine, 9/16, 29/508
glycogen storage disease, 62/490
glycolysis, 27/6-13
glycosaminoglycanosis, 10/47
GM2 gangliosidosis, 10/385
granulomatosis, 16/229, 233
hamartoma, 12/104
hemangioblastoma, 14/60, 244
hemangiopericytoma, 18/293-297
hemangiosarcoma, 18/288-292
hereditary amyloid polyneuropathy, 60/102
histiocytosis X, 16/231
Hodgkin disease, 18/262
hypermelanosis, 14/418
hyponatremia, 28/495-497
Langerhans cell histiocytosis, 70/58
lead intoxication, 64/445
leukemia, 18/246, 251, 321
lipid, 10/233
lipomatosis, 14/529
malignancy, 38/625-660
Melkersson-Rosenthal syndrome, 16/222
microgliomatosis, *see* Primary CNS lymphoma
multiple organ failure, 71/527
mummification process, 74/184
neoplasm, *see* Brain tumor
neurinoma, 18/321
neurocutaneous melanosis, 68/390
neuropsychology, 45/4
neutral lipid, 10/233
paraproteinemia, 16/233
peptide, 37/371
phlebostasis, 9/579
polycythemia vera, 16/235
polyradiculitis, 16/223
potassium, 9/16, 28/427, 464
primary amyloidosis, 51/415
pseudomycosis, 16/217
radiation injury, 67/325
rete mirabile, *see* Moyamoya disease
reticulosarcoma, 18/236-238, 246, 257, 321, 340
reticulosis, 18/246
Reye syndrome, 70/271
rheumatoid arthritis, 38/492-496
sarcoidosis, 16/229
schistosomiasis, 16/228
sodium, 9/16
split, *see* Split brain
Sturge-Weber syndrome, 12/234
taurine, 9/16
telangiectasia, 12/104, 232, 14/520, 54/365
temporal arteritis, 5/121
transamination reaction, 27/17
trauma, *see* Brain injury
tricarboxylic acid cycle, 27/14-17
tuberculosis, 16/230
vascular malformation, 18/270
venous angioma, 18/272
visceral, *see* Visceral brain
voltage, 72/284
Von Hippel-Lindau disease, 12/236
water, 56/81
Brain abscess, 33/107-141
*see also* Cerebellar abscess *and* Intracranial infection
acquired immune deficiency syndrome, 56/497
acquired immune deficiency syndrome coinfection, 56/497
Actinomyces, 52/149
allescheriosis, 35/560
animal model, 52/146
antibiotic agent, 52/148, 155
antibiotic agent choice, 52/156
antibiotic treatment, 33/135-137
arteriography, 33/123
aspergillosis, 35/397, 52/148
ataxia, 52/152
aural infection, 33/112
bacterial endocarditis, 33/115, 52/290, 297
bacteriology, 33/119-122
Bacteroides, 52/149
Bacteroides melaninogenicus, 52/150
Blastomyces dermatitidis, 52/151, 385
brain angiography, 52/155

brain edema, 52/148
brain herniation, 52/152
brain infarction, 11/314
brain stem, 33/139
brain tumor, 16/263, 67/218
Candida albicans, 52/151
Candida tropicalis, 52/151
candidiasis, 52/398
capsule formation, 52/148, 153
cephalosporin, 52/156
cerebellar abscess, 33/138, 52/155
cerebellar site, 52/144
cerebral amebiasis, 52/315, 325
cerebritis, 52/146, 153
cerebrovascular disease, 53/31, 55/418
child, 63/6
chloramphenicol, 52/156
chronic meningitis, 56/645
Citrobacter diversus, 52/150
cladosporiosis, 35/563, 52/483
Cladosporium trichoides, 52/151
classification, 68/310
clinical features, 33/116-118, 52/151, 57/139
clinical presentation, 57/322
Clostridium, 52/149
Coccidioides immitis, 52/151
coccidioidomycosis, 52/412, 415
coma, 52/152
computer assisted tomography, 33/126, 46/387, 52/153, 57/322
congenital heart disease, 33/115, 38/127-131, 52/145, 63/5
contiguous spread, 52/143
corticosteroid, 52/148, 158
co-trimoxazole, 52/156
cranial nerve palsy, 52/152
Cryptococcus neoformans, 52/151
CSF, 33/118, 52/152
Curvularia pallescens, 52/151
curvulariosis, 52/485
dementia, 46/387
dental disease, 33/115
depressed skull fracture, 24/225
diagnosis, 33/122-128, 52/152
diagnostic study, 57/322
differential diagnosis, 16/209, 212, 53/237
disability, 33/140
doughnut sign, 52/154
Echinococcus granulosus, 52/151
EEG, 23/354, 33/123, 52/155
encephalopathy, 52/400
endocarditis, 33/115, 473
Entamoeba histolytica, 52/149, 151
Enterobacter, 52/149
epilepsy, 15/307, 33/117, 139, 52/152, 400, 72/145
Escherichia coli, 52/149
esophageal stricture, 52/145
etiology, 33/112-117
Fallot tetralogy, 52/145, 63/5
fonsecaeasis, 35/566, 52/487
fusariosis, 52/488
Fusarium, 52/151
Fusobacterium, 52/149
gram-negative bacillary meningitis, 52/120
great artery transposition, 63/5
gunshot injury, 33/116
head injury, 52/145, 57/238
headache, 5/220, 52/151
heart transplantation, 63/179
hematogenous spread, 33/114, 52/144
hemiplegia, 1/194
Hemophilus aphrophilus, 52/150
Hemophilus influenzae, 52/150
Hemophilus parainfluenzae, 52/150
Hemophilus paraphrophilus, 52/150
hereditary hemorrhagic telangiectasia (Osler), 55/449
hereditary telangiectasia, 52/145
herpes simplex virus encephalitis, 56/218
histologic zone, 52/146
histology, 52/147
Histoplasma capsulatum, 52/151
history, 33/107
iatrogenic neurological disease, 52/144, 150
immunocompromise, 65/548
immunocompromised host, 52/143
immunosuppression, 52/148
increased intracranial pressure, 33/117
infant, 33/119
infectious endocarditis, 55/167, 63/23, 112, 115
intracavity contrast media, 33/127
intracranial, 16/209
intracranial gas bubble, 52/154
intractable hiccup, 63/490
intrasellar abscess, 52/152
Klebsiella, 52/149
laboratory data, 33/118
lethargy, 52/152
Listeria monocytogenes encephalitis, 33/86, 52/94, 150
macrophage defense defect, 65/548
madurelliasis, 35/567, 52/488
mastoiditis, 52/144
meningeal sign, 33/118
meningitis, 52/144
meningococcal meningitis, 33/23, 52/150

mental deficiency, 46/19
mental disability, 33/140
metastasis, 33/114-116, 52/144, 148
metronidazole, 52/156
microabscess, 52/298
microbiology, 57/322
missile injury, 23/524, 24/227
mucormycosis, 35/546, 52/151, 471
multiple, *see* Multiple brain abscess
multiple microabscess, 52/385, 387, 400
Mycobacterium tuberculosis, 52/150
nasal sinus infection, 33/113, 122
neuroamebiasis, 52/312
neurologic disability, 33/140
neurologic intensive care, 55/204
neurosurgery, 52/158
nocardial, *see* Nocardial brain abscess
nocardiosis, 16/217, 35/519, 52/150, 447-449
North American blastomycosis, 35/401, 52/385, 387, 389
nuclear magnetic resonance, 57/322
nystagmus, 52/152
odontogenic, 33/115
organism, 57/322
otitis media, 52/144
otogenic source, 33/112, 122, 52/143
Paragonimus, 52/151
Pasteurella, 52/149
pathogenesis, 33/107, 52/143
pathology, 33/107-122, 52/146
penetrating gunshot injury, 33/116
penicillin, 52/156
penicilliosis, 35/568, 52/490
Peptostreptococcus, 52/149
periodontal disease, 33/115
phagocyte defense defect, 65/548
pituitary, *see* Pituitary abscess
pituitary tumor, 17/412, 426
pneumococcal meningitis, 33/47, 52/43
poikilomycosis, 35/567, 52/489
polycythemia, 63/6
posttraumatic, *see* Posttraumatic brain abscess
predisposing factor, 52/145
prognosis, 52/160
Proteus vulgaris, 52/149
Pseudoallescheria, 52/492
Pseudoallescheria boydii, 52/151
Pseudomonas aeruginosa, 52/149
pulmonary arteriovenous malformation, 52/145
pulmonary infection, 33/115, 52/145
radionuclide scintigraphy, 52/154
Salmonella, 52/149
scanning, 33/124-126
Schistosoma japonicum, 52/151
Serratia marcescens, 52/150, 298
sinus infection, 33/113
sinusitis, 52/144
sporotrichosis, 52/494
stage, 52/146
Staphylococcus aureus, 52/145, 149
Staphylococcus epidermidis, 52/150
steroid, 33/137
Streptococcus milleri, 52/149
Streptococcus pyogenes, 52/150
Streptococcus viridans, 52/149
Strongyloides stercoralis, 52/151
supratentorial, *see* Supratentorial brain abscess
surgery, 24/658, 33/128-135, 57/322
T-lymphocyte, 65/548
temporal lobe, 52/144
thalamic, 33/139
thoragenic, 33/115
thrombophlebitis, 52/144
torulopsiasis, 35/571, 52/496
Toxoplasma gondii, 52/148, 151
trauma, 33/116
treatment, 33/128-138, 52/155, 57/322
tuberculosis, 18/421
vancomycin, 52/156
ventricular system, 33/118
ventriculography, 33/124
vertigo, 52/152
Brain air embolism
brain edema, 55/194
cardiac dysrhythmia, 55/193
cardiac mill wheel murmur, 55/193
diagnosis, 55/193
diving ascent risk, 63/419
epilepsy, 55/193
glossal pallor, 55/193
hyperbaric oxygen, 55/194
hypotension, 55/193
impaired consciousness, 55/193
incidence, 55/192
mortality, 55/193
neuropathology, 55/193
nuclear magnetic resonance, 55/193
100% oxygen, 55/194
pathogenesis, 55/193
retinal air embolus, 55/193
symptom, 55/193
tachycardia, 55/193
therapy, 55/194
Brain amino acid metabolism
amino acid, 29/22-24
amino acid distribution, 29/25

GABA, 29/19
glutamic acid, 29/17-22
glutamine, 29/19
Krebs-Henseleit urea cycle, 29/20
nutrition, 29/24-26
synaptic transmission, 29/22
Brain amyloid angiopathy, 54/333-342
*see also* Amyloidosis
allergic granulomatous angiitis, 54/342
amimia, 54/337
argyrophilic dystrophy, 21/45
asymptomatic form, 54/337
ataxia, 54/337
autosomal dominant hypoalphalipoproteinemia, 60/136
Binswanger disease, 54/231
biochemistry, 54/341
birefringency, 54/334
brain hemorrhage, 54/292, 337, 63/74
cerebellar hemorrhage, 54/292
clinical picture, 54/337
Congo red stain, 54/334
cortical hemorrhage, 54/334
cortical infarction, 54/334
dementia, 54/337
demyelinating disease, 54/341
diagnosis, 54/342
disease course, 54/337
gaze paralysis, 54/337
Gerstmann-Sträussler-Scheinker disease, 56/555
headache, 54/337
hemiparesis, 54/337
hereditary, *see* Hereditary cerebral amyloid angiopathy
intracerebral hematoma, 54/337-340
intracerebral hemorrhage site, 63/75
leukoencephalopathy, 54/341
lingual tremor, 54/337
multiple cranial neuropathy, 54/337
neurofibrillary tangle, 54/334
neuropathology, 54/333
nonfamilial type, 54/337
parietal lobe hemorrhage, 63/75
pathogenesis, 54/341
perivascular gliosis, 54/334
preferential site, 54/334
secondary pigmentary retinal degeneration, 54/337
senile plaque, 54/334
subcortical infarction, 54/334
thioflavine, 54/334
transient ischemic attack, 54/337
ultrastructure, 54/334
Brain aneurysm, 12/205-223
*see also* Anterior communicating artery aneurysm
age, 12/85, 55/44, 74
ε-aminocaproic acid, 55/52
anesthesia, 12/211
aneurysm distribution, 55/46
aneurysm site, 55/46
aneurysm size, 55/46
angiography, 12/205-208
anterior cerebral artery, 12/91
anterior cerebral artery syndrome, 53/341
anterior choroidal artery, 12/94
anterior communicating artery, 11/79, 12/92, 221
antifibrinolysis, 55/52
aorta coarctation, 12/101
aphasia, 42/723
arteriovenous malformation, 42/723
artery ligation, 12/216
artificial embolization, 12/222
associated disease, 14/519
atrial myxoma, 63/98
atrial myxoma course, 63/99, 101
autopsy incidence, 55/44
basilar artery, 12/95, 53/386
brain angiography, 12/205-208, 55/48
brain arteriovenous malformation, 12/237, 54/43, 55/73
brain embolism, 53/159
brain hemorrhage, 54/288
carotid artery ligation, 12/217, 220
carotid artery stenosis, 55/73
carotid artery thrombosis, 26/47
cause, 12/77
cerebrovascular disease, 53/41
Charcot-Bouchard aneurysm, 12/82
child, 12/348
classification, 12/90
clinical features, 12/86-88
clip ligation, 12/216
combined malformation, 12/100
computer assisted tomography, 55/48
congenital, *see* Congenital aneurysm
congenital heart disease, 63/4
controlled hypotension, 12/212
critical size, 55/47
deafness, 55/134
dehydrating agent, 12/215
diabetes insipidus, 55/53
diagnosis, 55/48
diagnosis confirmation, 55/48
dissecting aneurysm, 55/79
Ehlers-Danlos syndrome, 12/102, 55/454
electrolyte disorder, 55/53

epilepsy, 55/53
etiology, 12/78
familial aneurysm, 55/81
familial incidence, 12/86
fibromuscular dysplasia, 55/283
fluid derangement, 55/53
furosemide, 55/53
fusiform aneurysm, 55/78
Galen vein, 12/103
general pathology, 11/131
genetics, 14/517-520
giant, *see* Giant brain aneurysm
giant aneurysm, 55/47
giant aneurysm symptom, 55/71
head injury, 12/99
headache, 5/141, 143, 42/723
hereditary hemorrhagic telangiectasia (Osler), 42/739, 55/450
hypothermia, 12/211
inappropriate antidiuretic hormone secretion, 55/53
inflammatory aneurysm, 55/78
internal carotid artery, 12/89
intracerebral calcification, 49/417
intraoperative angiography, 55/74
intrasellar aneurysm, 17/409
α-methylcyclohexane carboxylic acid, 55/52
micro type, *see* Charcot-Bouchard aneurysm
middle cerebral artery, 12/91
migraine, 5/141
mortality, 55/44
moyamoya disease, 42/751
multiple aneurysm, 12/82, 55/47, 72
natural history, 55/44
negative angiography, 55/49
negative angiography cause, 55/49
neoplastic aneurysm, 55/80
neurobrucellosis, 52/585
neurogenic pulmonary edema, 55/53
neuroradiology, 12/96-100
ophthalmic artery, 12/91
optic atrophy, 13/69
optic nerve, 12/90
pathogenesis, 12/78
pericallosal, 12/222
polycystic renal disease, 55/81
polycystic visceral disease, 12/100, 102
posterior cerebral artery, 12/93
posterior communicating artery, 12/93
posterior cranial fossa, 12/95, 97
postoperative management, 55/74
pregnancy, 55/73
preoperative management, 55/47
prevalence, 12/86, 42/723
prognosis, 55/45
prophylactic neurosurgery, 55/70
proximal ligation, 12/216
pseudoxanthoma elasticum, 11/465, 43/45, 55/454
radiodiagnosis, 12/96
rebleed, 55/52
rebleed prevention strategy, 55/51
reinforcing aneurysm wall, 12/217
renal abnormality, 42/723
rupture, 12/83
ruptured aneurysm, 12/211
saccular aneurysm anatomy, 55/45
sex incidence, 55/44
sex ratio, 12/85, 42/723
site, 12/88-95, 221, 55/44
speech disorder, 42/723
subarachnoid hemorrhage, 12/77, 104, 42/723
surgery *vs* nonsurgery, 12/208
surgical treatment, 12/205-217
thrombosis, 12/84
trap ligation, 12/216
traumatic, 12/99
traumatic aneurysm, 55/78
type, 12/79-82
unruptured aneurysm management, 55/70
vascular dementia, 46/360
vascular spasm, 12/209
vascular striatopallidodentate calcification, 49/417
vertebral artery, 12/96
vertigo, 55/134
wall reinforcement, 12/217
warning leak, 55/48
Brain aneurysm neurosurgery
age, 55/44, 74
ε-aminocaproic acid, 55/52
anesthesia, 55/61
aneurysm dissection, 55/61
aneurysm distribution, 55/46
aneurysm exposure, 55/61
aneurysm neck occlusion, 55/62
aneurysm site, 55/46
aneurysm size, 55/46
anterior communicating artery aneurysm, 55/66
antifibrinolysis, 55/52
artificial embolization, 55/66
artificial hypothermia, 55/42
autopsy incidence, 55/44
basilar artery bifurcation aneurysm, 55/69
brain angiography, 55/48
brain arteriovenous malformation, 55/73
brain ischemia, 55/57
brain vasospasm, 12/209, 55/55-58

calcium channel blocking agent, 55/59
cardiac dysrhythmia, 55/54
carotid-anterior choroidal artery aneurysm, 55/69
carotid artery bifurcation aneurysm, 55/68
carotid artery stenosis, 55/73
carotid cavernous fistula, 55/68
carotid ligation, 55/41
carotid ophthalmic artery aneurysm, 55/68
clinical grading, 55/50
clip type, 55/63
clip variety, 55/42
collateral circulation, 55/65
computer assisted tomography, 55/48
contraindication, 12/209
critical size, 55/47
deep hypothermia, 55/42
deep vein thrombosis, 55/54
diabetes insipidus, 55/53
diagnosis, 55/48
diagnosis confirmation, 55/48
direct approach, 55/61
dissecting aneurysm, 55/79
Drake-Kees clip, 55/64
drug induced hypotension, 55/43
early literature, 55/41
early surgery, 55/59
ECG, 55/54
electrolyte disorder, 55/53
epilepsy, 55/53
external internal carotid bypass, 55/65
familial aneurysm, 55/81
fluid derangement, 55/53
furosemide, 55/53
fusiform aneurysm, 55/78
gastrointestinal bleeding, 55/54
general medical management, 55/50
giant aneurysm, 55/47, 71
grade result, 55/76
history, 55/41
horse hair, 55/42
hydrocephalus, 55/55
hypothermia, 55/61
inappropriate antidiuretic hormone secretion, 55/53
incompletely treated aneurysm, 55/75
indirect approach, 55/63
induced hypotension, 55/61
inflammatory aneurysm, 55/78
intracerebral hematoma, 55/54
intracranial hematoma, 55/54
intracranial hypertension, 55/55, 59
intraoperative angiography, 55/74
intraventricular hematoma, 55/54
iron filling, 55/42
α-methylcyclohexane carboxylic acid, 55/52
microsurgery, 55/43
middle cerebral artery aneurysm, 55/68
mortality, 55/44
multiple aneurysm, 55/47, 72
muslin gauze, 55/42
natural history, 55/44
negative angiography, 55/49
negative angiography cause, 55/49
neoplastic aneurysm, 55/80
neuroanesthesia, 55/43
neurogenic pulmonary edema, 55/53
nimodipine, 55/59
pericallosal aneurysm, 55/69
polycystic renal disease, 55/81
posterior communicating artery aneurysm, 12/221, 55/67
postoperative angiography, 55/75
postoperative management, 55/74
pregnancy, 55/73
preoperative condition, 55/77
preoperative management, 55/47
prognosis, 55/45
prophylactic neurosurgery, 55/70
pulmonary embolism, 55/54
rebleed, 55/52
rebleed prevention strategy, 55/51
retrograde brachial angiography, 12/207
saccular aneurysm anatomy, 55/45
sex incidence, 55/44
site, 55/44
site result, 55/75
size result, 55/77
streptokinase, 55/59
subarachnoid hemorrhage, 12/215
subdural hematoma, 55/54
surgery result, 55/75
surgery timing, 55/43, 59
surgical complication, 55/54
surgical innovation, 55/42
surgical technique, 55/60
tandem clip, 55/64
thrombogenic wire, 55/42
tissue plasminogen activator, 55/59
tourniquet technique, 55/66
traumatic aneurysm, 55/78
unruptured aneurysm management, 55/70
vasospasm cause, 55/55
vertebral artery angiography, 12/207
vertebrobasilar aneurysm, 55/69
warning leak, 55/48
Brain angiitis, *see* Brain vasculitis

Brain angiography, 54/169-190
see also Carotid angiography
amphetamine addiction, 55/520
bacterial meningitis, 52/14
Bonnet-Dechaume-Blanc syndrome, 14/265
brain abscess, 52/155
brain aneurysm, 12/205-208, 55/48
brain aneurysm neurosurgery, 55/48
brain arteriovenous malformation, 54/183, 187, 379
brain blood flow arrest, 57/461
brain collateral blood flow, 11/176, 54/176
brain death, 24/770, 805, 55/256, 259
brain embolism, 53/228, 54/177
brain hemorrhage, 54/315
brain infarction, 53/165, 228, 54/173
brain ischemia, 53/165
brain lacunar infarction, 54/254
brain metastasis, 18/224
brain vasculitis, 55/421
cavernous sinus thrombosis, 52/174
cerebellar hemangioblastoma, 54/186
cerebrovascular disease, 53/228
chronic paroxysmal hemicrania, 48/261
complication, 16/642, 647, 54/172
contrast media, 16/647
conventional type, 54/169
cortical blindness, 55/119
cranial epidural empyema, 52/169
craniopharyngioma, 18/549
diamorphine addiction, 55/518
Ehlers-Danlos syndrome, 55/455
fibromuscular dysplasia, 11/372, 381
glioblastoma multiforme, 18/59
granulomatous CNS vasculitis, 55/391
head injury, 57/212
hereditary hemorrhagic telangiectasia (Osler), 55/451
histoplasmosis, 52/439
iatrogenic brain infarction, 53/165
indication, 54/175
infectious endocarditis, 63/118
interventional, 54/169, 179, 187
intra-arterial digital subtraction angiography, 54/169
intracranial artery lesion, 54/177
intracranial hemorrhage, 53/228
intracranial vein, 54/179
intravenous, 57/462
intravenous digital subtraction angiography, 54/169
local intra-arterial fibrinolysis, 54/180, 182
mycotic aneurysm, 52/295
neuroamebiasis, 52/321
oligodendroglioma, 18/95
percutaneous transluminal angiography, 54/179
pseudoxanthoma elasticum, 55/453
Seldinger technique, 54/169
sickle cell anemia, 55/505-507
Sneddon syndrome, 55/402
Sturge-Weber syndrome, 55/446
subarachnoid hemorrhage, 12/124, 154, 55/24, 48
subdural empyema, 52/73
superior longitudinal sinus thrombosis, 52/179
supratentorial brain tumor, 18/336
syphilis, 55/413
systemic lupus erythematosus, 55/380
temporal arteritis, 55/344
transient global amnesia, 55/141
transient ischemic attack, 45/212, 53/228, 270
transverse sinus thrombosis, 52/177
traumatic intracranial hematoma, 57/267
tuberculoma, 18/418
tuberculous meningitis, 52/209, 55/424, 429
visual hallucination, 55/146
Brain angioma
see also Cerebellar angioma
Bernasconi-Cassinari artery, 12/101
Bonnet-Dechaume-Blanc syndrome, 12/101
cerebellar tumor, 17/708
headache, 5/145
hereditary hemorrhagic telangiectasia (Osler), 55/450
intracerebral calcification, 49/417
lateral ventricle tumor, 17/605
vascular striatopallidodentate calcification, 49/417
venous, 18/272
Brain anoxia
see also Brain hypoxia
anemic anoxia, 46/358
autoregulation, 27/40
bilateral pallidal necrosis, 6/658-661
brain blood flow, 27/40
brain edema, 23/146, 27/47
burst suppression, 57/189
carbon monoxide intoxication, 64/34
cerebral palsy, 46/14
cerebrovascular resistance, 27/40
clinical features, 27/39-48, 46/358
EEG, 27/42, 54/159
etiology, 46/358
histotoxic anoxia, 46/358
neuropsychology, 45/9
$O_2$ deprivation, 9/574
stagnant anoxia, 27/44, 46/358
striatal necrosis, 6/674

syncope, 15/815-834
treatment, 27/46-48
triphasic wave, 57/189
vascular dementia, 46/357
Brain antibody
CSF, *see* CSF brain antibody
multiple sclerosis, 47/108, 365
Brain antigen
cellular response, 47/357
multiple sclerosis, 47/234, 357
Brain arteriosclerosis, 53/91-100
*see also* Cerebrovascular disease
antihypertensive agent, 63/84
arterial anatomy, 53/91
artery remodelling, 63/77
atheromatous debris, 63/77
blood flow pattern, 12/25, 63/77
blood kinetic energy, 63/78
blood pressure energy, 63/78
blood vessel disorder, 11/129
brain hemorrhage, 11/603
calcification, 53/94
cause, 53/99
cerebrovascular disease, 53/91, 200
collagen, 53/97
definition, 53/92
diffuse form, 11/605
dystonia, 6/556
elastin, 53/97
elementary lesion, 53/93
endothelium, 63/78
epilepsy, 15/304
evolution, 53/95
focal form, 11/605
hemodynamics, 63/79, 82
hemorrhage, 53/95
hyperlipoproteinemia type II, 53/98
hyperlipoproteinemia type III, 53/98
hypertension, 63/77
intraplaque hemorrhage, 63/77
lipid, 63/77
lipid accumulation, 53/97
location, 53/94
macrophage, 53/98
Minamata disease, 36/101-105, 123
Mönckeberg sclerosis, 53/94
monocyte, 53/98
pallidal necrosis, 49/466
pathogenesis, 11/129, 53/96, 63/79, 82
platelet, 53/96
platelet thrombus, 63/77
prevention, 53/99
proteoglycan, 53/97
psychosis, 11/126
Reynolds number, 63/80
risk factor, 53/99
scalariform type, 11/605
senile dystonia, 6/556
shear stress, 63/77
smooth muscle cell, 53/96
spheroid body, 6/626
stenosing nonthrombosing type, 11/603
stenosis, 53/94
stress factor, 53/97
symptomatic dystonia, 6/556
thrombosing type, 11/603
thrombosis, 12/25, 53/95
ulceration, 53/94
uncomplicated plaque, 53/93
very low density lipoprotein, 53/97
β-very low density lipoprotein, 53/98
wall shear, 53/94
Brain arteriovenous malformation, 12/227-264, 54/361-387
*see also* Dural arteriovenous malformation *and* Venous malformation
acquired arteriovenous malformation, 54/363
age, 12/239
age range, 53/35
anatomic variation, 54/367
aneurysm, 12/237, 18/279
angiography, 54/184
angioma, 14/440
arteriovenous fistula, 54/364, 370
astrocytosis, 12/232
B-mode sonography, 54/43
bleeding, 12/232
blood supply, 12/230
brain aneurysm, 12/237, 54/43, 55/73
brain aneurysm neurosurgery, 55/73
brain angiography, 54/183, 187, 379
brain blood flow, 12/246, 54/150
brain cavernous angioma, 12/232
brain edema, 54/382
brain hemorrhage, 12/232, 241, 248, 53/34
brain injury, 23/138
brain ischemia, 12/232
brain scanning, 12/242, 31/182
brain stem, 2/257, 259, 54/371, 373
brain tumor, 12/248
brain white matter, 53/35
calvaria, 12/236
cardiomegaly, 12/232
carotid bruit, 31/187
cavernous angioma, 12/232, 54/367
cavernous malformation, 53/35

cerebral steal, 54/378
cerebrodural type, 54/369
cerebrovascular disease, 53/34
characteristic features, 54/364
circumscribed, 54/367
classification, 12/229
clinical features, 12/239, 31/179-189, 53/35, 54/374
coagulation, 12/254
computer assisted tomography, 54/62, 70, 184, 379
corpus callosum tumor, 2/257-259, 17/537
cortex vascularization, 54/363
cranial bruit, 31/186
cryosurgery, 12/254
cryptic type, 54/375
CSF, 31/191
cystic type, 54/373
deafness, 55/134
deep feeder, 54/368
deep site, 54/369
dementia, 54/378
diagnosis, 31/189-191
differential diagnosis, 12/246
Doppler sonography, 54/380
dural arteriovenous malformation, 54/372
dysembryogenesis, 54/361
dysgenesis, 12/232
ECG, 12/246
echoEG, 12/243, 31/191
EEG, 12/108, 242, 31/189
embolization, 12/254, 54/187
embolization risk, 54/382
embryology, 12/228, 31/175
epidemiology, 30/160, 31/177, 54/370
epilepsy, 12/241, 246, 15/303, 31/181, 54/377
etiology, 54/361
excision, 12/254
face, 12/229
familial, 54/361
fistula, 5/145-149
frequency, 53/34
Galen vein, 2/257, 12/103, 236, 18/179-284, 54/43
Galen vein aneurysm, 12/260
gamma radiation, 54/384
genetics, 30/160
headache, 5/145-149, 12/248, 31/182, 53/35, 54/376
heart failure, 31/183-185
hemianopia, 53/36
hemiataxia, 53/36
hemiathetosis, 53/36
hemiparesis, 53/36
hemorrhage, 54/91, 373
hemorrhage syndrome, 54/375
hereditary hemorrhagic telangiectasia (Osler), 55/450
histopathology, 54/364
history, 12/227
incidence, 12/237, 239, 54/372
interventional brain angiography, 54/187
intracerebral calcification, 54/93
intracerebral hematoma, 54/382
intracranial hemorrhage, 31/179-181
ischemia, 12/232
Klippel-Trénaunay syndrome, 31/200
laterality, 12/239
location, 31/178
lumbar puncture, 12/243
microscopy, 12/230
migraine, 53/35, 54/376
multiple sclerosis, 12/248
natural history, 54/373
nuclear magnetic resonance, 54/91, 93, 184, 379
orbital apex syndrome, 31/200
parenchymal hemorrhage, 54/375
*en* passage vessel, 54/364
pathology, 31/175
pathophysiology, 12/230, 31/176
posterior fossa, 54/369, 372
postoperative hemorrhage, 12/262
postoperative management, 12/261
postoperative neurological deficit, 54/378, 386
prevalence, 54/371
progressing neurological deficit, 54/378
proton radiation, 54/384
proton radiation therapy, 53/35
radiation, 12/244, 252
radiologic appearance, 31/189-191
radiotherapy, 54/384
recurrent hemorrhage, 54/373
regional brain blood flow, 54/380
rupture, 54/374
scalp, 12/229
seizure, 53/35
sex ratio, 12/239
sinus thrombosis, 54/364
site, 12/239
site frequency, 54/364
size, 31/178
spastic hemiplegia, 54/378
spinal cord arteriovenous malformation, 54/369
stereoencephalotomy, 12/254
steroid treatment, 12/261
stroke, 54/372
Sturge-Weber syndrome, 31/196-199

subarachnoid hemorrhage, 12/241, 247, 53/35, 54/372, 374, 376, 55/73
superficial type, 54/369
surgery, 12/253
surgical complication, 54/387
surgical technique, 12/255
surgical treatment, 54/385
survival, 54/374
symptomatic dystonia, 49/543
syringomyelia, 14/520
telangiectasia, 12/232, 54/364, 375
therapeutic embolization, 54/381
tracheostomy, 12/263
transcerebral, 54/363, 367
transcranial Doppler sonography, 53/226
transdural, 54/367
transient ischemic attack, 54/375
treatment, 12/250, 443, 31/191, 54/380
treatment choice, 12/251
tremor, 53/36
type, 54/367
ultrasound diagnosis, 54/35
varices, 12/234
venous angioma, 53/35
venous thrombosis, 54/364
ventricular hemorrhage, 54/374, 376
vertigo, 55/134
Von Hippel-Lindau disease, 31/199
Wadia sign, 31/188
wedge shape, 54/363
Brain arteritis
cat scratch disease, 52/129
headache, 5/119, 149
primary granulomatous, *see* Primary granulomatous cerebral arteritis
Takayasu disease, 12/411
temporal arteritis, 55/343
Brain artery
accessory middle, *see* Accessory middle cerebral artery
anterior, *see* Anterior cerebral artery
anterior middle, *see* Anterior middle cerebral artery
autonomic nervous system, 74/216
middle, *see* Middle cerebral artery
migraine, 48/61
posterior, *see* Posterior cerebral artery
subtraction method, 11/71
traumatic aneurysm, 24/389
Brain artery occlusion
*see also* Cerebrovascular disease *and* Cerebrovascular occlusive disease
artery occlusion, 12/452
corpus callosum tumor, 17/542
drop attack, 12/23
middle, *see* Middle cerebral artery occlusion
Brain artery system
*see also* Brain blood flow
anatomy, 11/1-23
capillary, 11/119
embryology, 11/1-23
headache, 5/125
radiologic anatomy, 11/65-101
topography, 11/65-101
Brain aspergilloma
aspergillosis, 52/380
Brain asymmetry
schizophrenia, 46/504-507
Brain atherosclerosis, *see* Brain arteriosclerosis
Brain atrial myxoma metastasis
antibiotic treatment, 63/101
atrial myxoma, 63/99
chemotherapy, 63/101
Brain atrophy
*see also* Cerebellar atrophy
acquired immune deficiency syndrome, 56/509
Aicardi syndrome, 42/696
Alpers disease, 42/486
Alzheimer disease, 46/247-249, 253, 364
amyotrophic dystonic paraplegia, 42/200
aphasia, 45/297
athetotic hemiplegia, 42/201
athetotic membranous lipodystrophy, 42/201
Batten disease, 42/461-464
Berardinelli-Seip syndrome, 42/592
bone membranous lipodystrophy, 42/279-281
Börjeson-Forssman-Lehmann syndrome, 42/530
boxing injury, 23/553
brain thromboangiitis obliterans, 12/385, 39/202
carbon disulfide intoxication, 64/25
carnosinemia, 42/532
cerebrohepatorenal syndrome, 43/340
citrullinuria, 42/538
computer assisted tomography, 46/344-346, 57/164
congenital heart disease, 63/8
congenital hyperammonemia, 29/305-328, 42/561
cri du chat syndrome, 43/501
De Lange syndrome, 43/246
dermatoleukodystrophy, 42/488
diabetes mellitus, 42/411
dwarfism, 43/380
echoEG, 23/279
epilepsy, 15/327, 543, 547
epileptogenic encephalopathy, 42/693
formiminotransferase deficiency, 42/546

Förster syndrome, 42/202
frontometaphyseal dysplasia, 38/410
Galloway-Mowat syndrome, 43/431
general paresis of the insane, 52/279
granular cortical, *see* Granular cortical cerebral atrophy
head injury, 57/164
hemi, *see* Hemicerebral atrophy
hemidysplasia, 43/416
hereditary cerebello-olivary atrophy (Holmes), 21/412
hereditary tremor, 42/269
heredopathia ophthalmo-otoencephalica, 42/150
homocarnosinase deficiency, 42/558
homocarnosinosis, 42/558
human immunodeficiency virus infection, 75/413
Huntington chorea, 6/331, 42/226, 46/305
hyperammonemia, *see* Rett syndrome
hyperargininemia, 42/562
hypereosinophilic syndrome, 63/372
hypernatremia, 63/554
ichthyosis, 43/416
infantile epileptogenic encephalopathy, 42/693
juvenile diabetes mellitus, 42/411
laminar, *see* Laminar brain atrophy
leptomeningeal angiomatosis, 18/274
linear nevus sebaceous syndrome, 43/38
Lowe syndrome, 42/607
maple syrup urine disease, 42/599
metabolic ataxia, 21/577
microcephaly, 43/427
Minamata disease, 64/415
multi-infarct dementia, 42/283-285
myoclonic epilepsy, 42/700
myoclonus, 42/700
Neu syndrome, 43/430
neuronal ceroid lipofuscinosis, 42/461-464
Niemann-Pick disease type A, 42/469
organic solvent intoxication, 64/43
orthochromatic leukodystrophy, 10/114
osteopetrosis, 70/16
Parkinson dementia complex, 42/250
Parkinson disease, 6/177, 42/246
parkinsonism, 6/177
Patau syndrome, 43/527
PEG, 46/343
Pena-Shokeir syndrome type I, 43/438
Pick disease, 27/489, 42/285, 46/236-239, 242
postopen heart surgery, 63/8
presenile dementia, 46/454
presenile dementia (Kraepelin), 42/286
primary alcoholic dementia, 46/343-346
progressive rubella panencephalitis, 56/410
pseudoxanthoma elasticum, 43/45
r(14) syndrome, 50/588
r(21) syndrome, 43/543
Rett syndrome, 29/305, 63/437
schizophrenia, 46/421, 450-452, 454
self-mutilation encephalopathy, 42/141
Sjögren-Larsson syndrome, 43/307
spastic paraplegia, 42/176
spinocerebellar ataxia, 42/182, 190
spinopontine degeneration, 42/192
Sturge-Weber syndrome, 14/65, 43/48, 55/445
traumatic psychosyndrome, 24/539
trichopoliodystrophy, 14/784, 42/584
uremic encephalopathy, 63/506
vascular dementia, 46/364
vertical gaze palsy, 42/188
vertical supranuclear ophthalmoplegia, 42/605
Brain biopsy, 10/681-686, 67/217
acquired toxoplasmosis, 52/356
Alpers disease, 10/685
Alzheimer disease, 10/680, 685, 46/128
arteriosclerosis, 10/685
astrocytoma, 16/710, 713
brain metastasis, 16/724, 726
brain tumor, 16/37, 708, 17/42, 67/219
cerebellar medulloblastoma, 16/719
chordoma, 16/721
chromophobe adenoma, 16/720
craniopharyngioma, 16/720
Creutzfeldt-Jakob disease, 10/685, 56/546
CSF, 67/219
dementia, 46/128
diagnosis, 10/860
diffuse sclerosis, 10/685
ependymoma, 16/718
familial amaurotic idiocy, 10/680, 685
germinoma, 16/723
glioblastoma multiforme, 16/713
globoid cell leukodystrophy, 10/685
glycosaminoglycanosis type IH, 10/685
granulomatous CNS vasculitis, 55/394
Hand-Schüller-Christian disease, 10/685
herpes simplex virus encephalitis, 10/685, 56/215, 217
Huntington chorea, 10/685
inclusion body encephalitis, 10/685
Lyme disease, 52/263
meningioma, 16/722
meningitis, 10/685
meningoencephalitis, 10/685
metachromatic leukodystrophy, 10/680, 685
microglioma, 16/724
multiple sclerosis, 10/685, 47/223

neurinoma, 16/723
Niemann-Pick disease, 10/681, 685
North American blastomycosis, 35/406, 52/390
oligodendroglioma, 16/713, 715-717
parkinsonism, 10/685
Pick disease, 10/685
pineal tumor, 16/723
progressive multifocal leukoencephalopathy, 34/310, 47/511, 515
progressive rubella panencephalitis, 34/338, 56/410
Reye syndrome, 70/272
sarcoidosis, 10/685
senile dementia, 10/685
spinocerebellar degeneration, 60/678
spongy degeneration, 10/685
striatopallidodentate calcification, 10/685
subacute sclerosing panencephalitis, 10/685, 56/427
subacute spongiform encephalopathy, 10/685
technique, 10/683, 18/480
toxoplasmosis, 52/356
tuberous sclerosis, 10/686
Unverricht-Lundborg progressive myoclonus epilepsy, 10/685
value, 10/680
xanthomatosis, 10/686
Brain blood flow, 53/47-66, 54/139-153
*see also* Brain artery system, Brain hemorrhage, Cardiac arrest *and* Hyperviscosity syndrome
absence, 57/460
acceleration, 53/65
acid base balance, 55/208
acute subdural hematoma, 57/260
Adams-Stokes syndrome, 22/263, 53/47
adenosine, 53/54
aging, 54/150
Alzheimer disease, 46/363, 54/144
anoxic encephalopathy, 63/219
aorta coarctation, 53/48
aorta nerve, 53/62
arterial hypertension, 11/126, 63/80
arterial injection method, 54/140
arterial reconstructive surgery, 54/146
arteriole contraction, 11/124
arteriovenous difference, 11/270
autonomic nervous system, 75/361
baroreceptor, 53/63
Bayliss effect, 53/52
Binswanger disease, 54/223
blood-brain barrier, 57/262
blood energy consequence, 63/82
blood pressure, 11/123, 54/144, 55/208
blood pressure limit, 53/50
blood velocity, 63/80
blood viscosity, 11/124
brain anoxia, 27/40
brain arteriovenous malformation, 12/246, 54/150
brain death, 24/727, 770, 55/260, 263
brain edema, 23/139, 141, 149, 154, 63/417
brain infarction, 11/348-355, 53/113, 54/147
brain injury, 23/10, 139, 145, 149, 201, 240, 24/607, 619, 57/211
brain ischemia, 54/148, 63/219
brain lacunar infarction, 54/256
brain metabolism, 57/69
brain microcirculation, 53/79
brain perfusion pressure, 54/118, 146, 55/206, 57/384
brain stem death, 57/460
brain vascular volume, 53/59
calcitonin gene related peptide, 53/63
calcium channel blocking agent, 57/77
carbohydrate metabolism, 27/2
carbon dioxide, 11/124
carbon dioxide inhalation, 54/146
carbon dioxide reactivity, 53/112
carbon dioxide tension, 53/54
carbon monoxide intoxication, 64/34
cardiac arrest, 63/209
cardiac decompensation, 53/47
cardiac function, 53/47
cardiopulmonary bypass, 63/175
carotid cavernous fistula, 24/411, 419
carotid rete mirabile, 11/119
carotid sinus hypersensitivity, 11/123
carotid sinus nerve, 53/62
carotid sinus reflex, 11/123
cell ischemia, 57/86
cerebrovascular resistance, 53/47
cervical sympathectomy, 53/50
cervical sympathetic nerve, 53/48
cervical sympathetic nerve stimulation, 53/59
chemical control, 53/50, 80
Cheyne-Stokes respiration, 1/671, 63/483
cholinergic system, 53/62
chronic paroxysmal hemicrania, 48/261
clearance rate, 54/149
cluster headache, 48/238, 54/149
collateral, *see* Brain collateral blood flow
compartiment analysis, 54/140
conducting vessel, 53/114
congenital heart disease, 63/9
control, 11/123
cortical electrical function, 57/70
dementia, 54/150

diastolic damping time, 63/81
dilution method, 11/270
dipeptidyl carboxypeptidase I inhibitor, 65/444
Doppler sonography, 54/14
EEG, 23/10
electrolyte shift, 53/119
electromagnetic meter, 11/270
endothelium, 53/64
endothelium derived relaxing factor, 53/64
enflurane, 63/175
ergotamine, 65/64
expired air curve, 54/143
extracerebral steal, 53/115
familial hemiplegic migraine, 48/27, 70
fast rotating gamma camera, 54/149
Fick principle, 11/120
flow pattern, 63/82
focal brain activation, 54/151
functional psychosis, 46/453
functional threshold, 53/117
fusaric acid, 53/62
glyceryl trinitrate, 37/454
glyceryl trinitrate intoxication, 65/443
gravity, 53/65
head injury, 54/150, 57/211
head injury outcome, 57/384
headache, 48/78
heart rate, 63/81
heart transplantation, 63/175
hemispheric flow, 11/270
hemorrhagic parameter, 63/79
hemorrheology, 53/110
herpes simplex virus encephalitis, 56/87
humoral control, 11/124
hydrocephalus, 54/147
hypertension, 54/144
hypertensive encephalopathy, 11/571, 53/51, 54/213
hyperventilation syndrome, 38/318, 63/435
hypokinesia, 73/477
hypothermia, 71/574
index, 54/116, 119
inhalation method, 54/142
initial slope index, 54/141, 143
intracerebral hematoma, 54/147
intracerebral steal, 53/114
intracranial hypertension, 16/138, 54/147
intracranial pressure, 11/124, 23/201, 205, 55/207, 57/211, 384
ischemic core, 53/119
ischemic penumbra, 53/119
Kety-Schmidt method, 11/120, 269, 54/139
kinetic energy, 63/78, 82
Kr 79, 11/121
Kr 85, 11/121
large variability, 54/123
look-through phenomenon, 54/149
lowest value, 55/207
luxury perfusion syndrome, 57/70
mean flow, 11/270
measurement, 11/119, 24/619, 54/139, 56/86
membrane failure, 53/119
metabolic control, 53/79
metabolic mechanism, 53/53
method, 54/139
microcirculation, *see* Brain microcirculation
migraine, 48/60, 63, 70, 73, 79, 54/149
Monro-Kellie doctrine, 11/118, 53/49
morphologic injury, 53/119
multi-infarct dementia, 46/362, 54/123, 144
multiple sclerosis, 9/180
myogenic mechanism, 53/52
neurogenic control, 11/124, 53/50, 80
neurogenic mechanism, 53/56-58
neurologic intensive care, 55/220
neuropeptide Y, 53/63
nimodipine, 57/77
nitrate intoxication, 65/443
nonlaminar flow, 63/80
noradrenergic nervous system, 53/58
normal value, 11/121
O2, 11/124
occipital vertebral steal, 53/115
open heart surgery, 63/175
ophthalmic steal, 53/115
organic solvent intoxication, 64/44
orthostatic hypotension, 11/123
6-oxoprostaglandin F1@a, 53/55
oxygen extraction, 53/114
oxygen tension, 53/54
pattern, 63/79
peak count rate, 54/149
penetrating vessel, 53/114
peptide histidine isoleucine, 53/63
peptidergic nervous system, 53/63
perfusion function, 53/47
perfusion pressure gradient, 53/107
pH, 11/125, 53/54
pheochromocytoma, 42/764
physiologic variation, 11/122
physiology, 11/118-127, 55/205
pial artery, 53/48
pial vessel diameter, 53/59
plasma skimming, 53/82
polycythemia vera, 55/468
pontine infarction, 53/61

postganglionic nerve denervation, 53/61
preganglionic nerve denervation, 53/61
pressure energy, 63/82
prognosis, 24/675
pulsatile flow, 63/81
regional, *see* Regional brain blood flow
regulation, 24/607, 53/47, 112
resting flow, 54/152
Reynolds number, 63/81
rheoencephalography, 11/270
schizophrenia, 46/486, 73/477
segmental vascular resistance, 53/114
serial assay, 54/145
serotonergic innervation, 53/63
shear rate, 63/82
Shy-Drager syndrome, 38/244, 247, 53/61
sickle cell anemia, 55/511
sleep, 3/82
spinal cord injury, 26/328
spreading depression, 48/68, 73
St. Louis encephalitis, 56/87
steal phenomenon, 24/608
Stewart principle, 11/120
stroke, 54/144
subarachnoid hemorrhage, 54/147
subclavian steal, 53/115
substance P, 53/63
syncope, 75/209
thermovelocity method, 11/270
tissue pressure hypothesis, 53/65
tobacco addiction, 55/524
transient ischemic attack, 54/146, 148
turbulence, 63/80
uremic encephalopathy, 63/506
vascular dementia, 46/362-364
vascular resistance, 2/117, 11/124, 53/107
vascularization rate, 11/41
vasoactive intestinal polypeptide, 53/63
vasospasm, 54/147, 57/76
venous pressure, 11/123, 53/55
vessel geometry, 63/82
viral infection, 56/79
viscosity, 53/82
$^{133}$xenon, 48/63, 72
xenon computer assisted tomography, 54/149
Brain blood flow autoregulation
acceleration, 53/65
anatomy, 11/118
baroreceptor, 53/63
blood pressure limit, 53/50-52, 55/208
brain injury, 57/70
carbon dioxide, 11/124
congenital heart disease, 63/9
definition, 53/48
failure, 63/73
gravity, 53/65
history, 53/49
intracranial pressure, 55/208
mechanism, 53/52
metabolic mechanism, 53/53-55, 81, 112
myogenic mechanism, 53/52, 81
neurogenic mechanism, 53/56-58
serial assay, 54/145
site, 53/49
Brain blood vessel
trigeminovascular system, 48/112
Brain blood volume
index, 54/116, 119
measurement, 54/109
Brain calcification, *see* Intracerebral calcification
Brain capillary
brain injury, 23/134, 145
density, 11/119
Brain cavernous angioma, 72/209
appearance, 54/366
brain arteriovenous malformation, 12/232
nuclear magnetic resonance, 54/93, 379
subarachnoid hemorrhage, 12/104
Brain cavitation
brain injury, 23/93
Brain chordoma
epilepsy, 72/181
Brain circulation, *see* Brain blood flow
Brain collateral blood flow, 11/168-181
*see also* Cerebrovascular occlusive disease
angiography, 11/176
Bayliss effect, 11/268
blood pressure, 11/171
blood viscosity, 11/171
blood volume, 11/171
brain angiography, 11/176, 54/176
cardiac output, 11/171
cardiovascular relationship, 11/170
carotid artery kinking, 11/173
carotid rete mirabile, 11/173, 178
cervical anastomosis, 11/177
embryology, 11/169
extracranial artery, 11/172
extraintracranial anastomosis, 11/177
intervascular relationship, 11/170
intracranial anastomosis, 11/178
intracranial artery, 11/174
mathematical basis, 11/169
mediastinal anastomosis, 11/176
occlusion, 11/170
physiology, 11/170

transcranial Doppler sonography, 54/12
vasomotor reflex response, 11/170
venous sinus thrombosis, 54/397
vertebral abnormality, 11/173
Willis circle abnormality, 11/174

Brain command system
aggression, 45/274
curiosity, 45/273
emotion, 45/273
expectancy, 45/273
interest, 45/273
rage, 45/274

Brain commissurotomy
acalculia, 4/281
agnosia, 4/275-287
alexia, 4/117, 126, 282
aphasia, 4/280
apraxia, 4/283-285, 45/427
auditory agnosia, 4/281
body scheme, 4/278-280
cerebral dominance, 4/275-289, 305
mutism, 45/104
sensory defect, 4/275-287
split brain syndrome, 45/99

Brain compression
cerebrovascular disease, 11/155

Brain concussion
*see also* Brain contusion, Brain injury, Cerebellar concussion, Commotio cerebri, Craniofacial injury, Head injury *and* Skull injury
alpha rhythm slowing, 57/184
amnesia, 23/420, 432, 24/477, 562
amnesic component, 57/25
anosmia, 24/5
atypical, 23/441
awareness return, 57/25
boxing injury, 23/562
brain edema, 23/438
brain herniation, 23/438
brain injury mechanism, 57/23
brain metabolism, 23/439
brain stem reflex, 23/430
cerebellum, 23/459
child, 23/445
classification, 23/74
coma, 3/69, 23/428
commotio, 57/101
contusio, 57/101
cortical blindness, 24/49
course, 23/421
CSF, 23/420, 434
depolarization, 57/182
diagnosis, 23/441
EEG, 23/76, 421, 434, 57/182
electrophysiologic data, 57/25
evoked potential, 57/182
examination, 23/433
experimental, 23/71, 57/375
football injury, 23/571
gliding contusion, 57/46, 257
head injury, 57/23
history, 23/3, 71, 417
hypothalamus, 23/465
ice hockey injury, 23/583
intracranial hematoma, 23/438
loss of consciousness, 23/420, 427, 432, 440
memory disorder, 23/420, 433
model, 23/73
oculovestibular reflex, 23/429
physiopathological bases, 57/183
pituitary gland, 23/465
postconcussional syndrome, 23/424, 454
posttraumatic amnesia, 23/430, 433, 24/477, 479, 488
psychogenic, 23/442
radiodiagnosis, 23/433
retrograde amnesia, 23/432
rugby football injury, 23/570
severity grades, 57/106
shock, 23/437
shock wave, 23/98
spectral analysis, 57/182
stress wave, 23/94, 98
syndrome, 23/74
terminology, 23/6, 419, 454, 57/101
traumatic amnesia, 23/427, 24/562
traumatic neurosis, 23/425
traumatic psychosis, 24/521
treatment, 23/422, 435, 439
uterine, 23/471
vibration, 23/95

Brain contusion
*see also* Brain concussion
blood-brain barrier, 57/262
boxing injury, 23/562
brain injury mechanism, 57/33
brain scintigraphy, 23/288, 292
classification, 23/428
clinical features, 57/104
compression, 23/84
computer assisted tomography, 23/262, 57/149, 381
contrecoup lesion, 23/39, 93, 57/46
contusion index, 57/46
coup contusion, 57/46
course, 23/421

CSF, 23/420, 434
diffuse, 23/3, 419
echoEG, 23/269
EEG, 23/421, 434
electron micrography, 57/263
focal brain damage, 57/44
fracture contusion, 57/46
gliding contusion, 57/46, 257
head injury, 57/33, 104
hemorrhagic contusion, 57/45
history, 23/3
ischemic neuronal necrosis, 57/261
local effect, 57/260
loss of consciousness, 23/427
mixed attenuation, 57/154
nuclear magnetic resonance, 57/169, 173, 176
parasagittal contusion, 57/46
pathogenesis, 57/46
pathology, 23/38, 85, 57/44
perivascular hemorrhage, 57/44
posttraumatic amnesia, 23/427, 430, 433, 45/185
radiodiagnosis, 57/149
retrograde amnesia, 23/432
scar, 57/45
spheroid body, 6/626
spinal injury, 26/188
surface contusion, 57/46
symptom, 23/427, 429
terminology, 23/6, 57/104
traumatic intracerebral hematoma, 24/353, 57/257
traumatic intracranial hematoma, 57/252, 280
Wernicke-Korsakoff syndrome, 23/427
white matter, 23/40
Brain cortex
*see also* Cerebellar cortex
ablation, 6/852, 869
acoustic system, 16/305
activation, 3/53, 64
ageusia, 24/15, 19
akinesia, 45/162
anorexia nervosa, 46/587
apnea, 63/481
arousal disorder, 45/110
autonomic nervous system, 74/79
brain death, 24/738, 763, 817
brain diaschisis, 54/126
brain edema, 16/169
cell migration, 50/246
CNS, 50/245
combined sensation, 1/104
consciousness, 3/49, 55, 122, 171, 45/114
consciousness disorder, 45/114
cortical field, 1/150
development, 50/245
Divry-Van Bogaert syndrome, 55/320
Down syndrome, 50/274
EEG, 72/290
embryo, 50/13
embryonic granular layer, 14/335
feeding behavior, 46/587
fever, 74/453
Guam amyotrophic lateral sclerosis, 22/287, 343
Hallervorden-Spatz syndrome, 49/396
hereditary olivopontocerebellar atrophy (Menzel), 21/438
Huntington chorea, 49/318
Lafora body, 15/394
lamination, 2/727
layer 3, 2/727
layer 4, 2/727
layer 5, 2/727
leukemia, 39/11
memory, 3/260
microcephaly, 50/272
neglect syndrome, 45/157
occipital ischemia, 12/3
olivopontocerebellar atrophy (Dejerine-Thomas), 21/423
pain, 45/237
Parkinson disease, 6/212, 49/110
respiration, 1/654, 63/481
restoration of function, *see* Restoration of higher cortical function
sensory extinction, 45/160
sensory representation, 1/85
subglial granular layer, 30/68
thalamic connection, 2/491
traumatic hydrocephalus, 24/242
undercutting, 6/853
visual, *see* Visual cortex
Wernicke-Korsakoff syndrome, 45/196
Whipple disease, 70/242
Brain cortex atrophy
ergot intoxication, 65/69
glycosaminoglycanosis type III, 66/301
schizophrenia, 46/481
Sturge-Weber syndrome, 14/63, 65, 43/48
Brain cortex calcification
Cockayne-Neill-Dingwall disease, 13/435
Sturge-Weber syndrome, 14/63-65
Brain cortex disorganization
agyria, 50/249
Edwards syndrome, 50/249
encephalocele, 50/249
exencephaly, 50/249
holoprosencephaly, 50/249

microcephaly, 50/249
pachygyria, 50/249
Patau syndrome, 50/249
Brain cortex dysplasia
epilepsy, 30/364
focal, *see* Brain cortex focal dysplasia
microgyria, 50/257
polymicrogyria, 50/257
status verrucosus deformis, 50/253
Brain cortex focal dysplasia
clinical features, 50/208
clinical finding, 30/364
definition, 30/362, 50/207
epilepsy, 50/207
inferior olivary hypertrophy, 50/207
macroscopy, 30/362
mental deficiency, 30/364, 50/208
microscopy, 30/363
pathology, 30/362, 50/207
Brain cortex heterotopia
*see also* Lissencephaly, Microgyria, Pachygyria, Polymicrogyria *and* Ulegyria
cerebrohepatorenal syndrome, 50/248
dystopic cortical myelinogenesis, 50/206
Edwards syndrome, 50/248, 275
fetal alcohol syndrome, 50/40, 279
lissencephaly, 50/248
macrencephaly, 50/263
meninges, 50/249
microcephaly, 50/248, 261
pachygyria, 50/248
reeler mutant mouse, 50/39
schizencephaly, 50/199
status verrucosus, 50/203
status verrucosus simplex, 50/253
tuberous sclerosis, 50/248
Brain cortex laminar necrosis
polymicrogyria, 50/257
Brain cortex stimulation
aphasia, 45/297
hallucination, 4/336
Brain cyst
classification, 68/310
CSF, 68/309
echinococcosis, 52/524
helminthiasis, 52/511
hydrocephalic dementia, 46/326
survey, 68/309
toxoplasmosis, 52/352
Brain damage, *see* Brain injury
Brain death
*see also* Brain stem death *and* Death
absent brain blood flow criteria, 55/270
accession criteria, 55/257
agony, 24/775
air study, 24/733
akinetic mutism, 24/700, 703, 707
anatomoclinical correlation, 3/76, 24/740
anetic syndrome, 24/701
angiographic criteria, 55/260
angiographic definition, 55/263
angiographic proof, 55/265, 271
angiography, 24/733, 55/264
apnea, 55/257
associated lesion, 24/722
autonomic dysfunction, 24/721
autoregulation, 24/758
balling-up, 24/714
basal ganglion, 24/738
blood pressure, 24/758, 760, 821
brain angiography, 24/770, 805, 55/256, 259
brain blood flow, 24/727, 770, 55/260, 263
brain blood flow arrest, 24/770
brain cortex, 24/738, 763, 817
brain injury, 24/699, 757-786, 821
brain ischemia, 55/256
brain scintigraphy, 24/771
brain stem, 24/739, 763, 817
brain stem death, 55/259, 57/259, 444, 448, 452
brain stem reflex, 55/258
brain white matter, 24/738
British system, 55/265
calorie, 24/721
carbon dioxide tension, 55/262
cardiac arrest, 24/770, 63/213
caveat, 55/264
cephalic reflex, 55/258
cerebellum, 24/739
child, 55/263
ciliospinal reflex, 24/719
classification, 24/700, 704
coma, 3/66, 70, 75, 77, 55/257
coma cause, 55/262
coma prognosis, 55/258
computer assisted tomography, 55/262
concept, 55/256
conflicting UK/USA view, 55/267
confusion, 55/268
conjugate deviation, 24/720
controversy, 55/268
death, 24/821
death evidence, 55/263
decerebration, 24/702, 706, 714
decision making, 55/268
decortication, 24/702, 706, 714
definition, 3/77, 24/757, 817, 821, 55/261

dementia, 24/702
demyelination, 9/587
determination, 55/255
development, 24/758, 762
diencephalon, 24/738
differential diagnosis, 24/710, 773, 788
drug intoxication, 55/263
dysconjugate eye movement, 24/720
echoEG, 24/771
EEG, 24/729-732, 765-769, 781, 55/263
EMG, 24/776, 778
encephalitis lethargica, 3/67
encephalopathy, 24/702
etiology, 24/757
European concept, 55/256
European criteria, 55/260
evoked potential, 55/218, 264
expense, 55/268
experiment, 24/741
eye movement, 24/720
false death, 24/775
family concurrence, 55/269
function cessation, 55/263
gamma camera imaging, 55/263
grasp reflex, 24/718
Harvard Criteria, 55/257
heart death, 55/257
historic review, 55/255
history, 24/699
hospital policy, 55/269
6-hr EEG criterion, 55/258
hydrocephalus, 24/744
hypersomnia, 24/701, 711, 713
hypothermia, 55/263
hypoxic ischemic encephalopathy, 55/264
intracranial hypertension, 24/742, 758, 760, 766
intracranial pressure, 24/758, 760
isoelectric EEG, 55/219, 257
isoelectric EEG criterion, 55/256
isotope angiography, 24/771
isotopic bolus, 24/772
Klüver-Bucy syndrome, 24/723, 745
lactic acidosis, 24/601, 604, 728
legal aspect, 24/787, 787-828, 797
level stage, 24/707, 717
levodopa, 24/743
locked-in syndrome, 24/702, 708
medical standard, 55/261
metabolic intoxication, 55/263
metabolism, 24/728, 773
midbrain syndrome, 24/587
motor dysfunction, 24/714
motor pattern, 24/714, 716
muscle stretch reflex, 24/718
neurogenic osteoarthropathy, 24/722
neurologic intensive care, 55/218
neuropathology, 9/587, 63/213
neuroradiology, 24/733
neurotransmitter, 24/744
news media, 55/266
no code policy, 55/269
nomenclature, 24/699, 57/443
not based on prognosis, 55/270
oculovestibular reflex, 24/719, 764
organ donation, 55/270
parasomnia, 24/701, 711, 713
partial death, 24/778
pathogenesis, 24/757
pathologic criteria, 55/258
pathology, 24/736, 738, 759, 762
pediatric guideline, 55/264
persistent vegetative state, 24/703
polemics, 55/271
policy, 55/268
postural reflex, 24/717
prognosis, 55/258
protocol, 55/257
pupil, 24/719, 764
radioisotope bolus angiography, 55/263
radiology, 23/241
radionuclide angiography, 55/264
radionuclide scintigraphy, 55/265
radionuclide study, 55/260
recovery, 63/218
recuperation stage, 24/722
rehabilitation, 24/745
respiratory brain, 55/257
reticular formation, 24/739-741
rooting reflex, 24/718
Scandinavian criteria, 55/259
shock, 55/263
sleep, 24/729, 733
social readjustment, 24/746
spinal cord, 24/763, 817
stage, 24/758, 762
stupor, 24/702
sucking reflex, 24/718
surgery, 24/744
terminology, 24/703
test cross validation, 55/264
transition state, 24/705, 713
treatment, 24/742
United Kingdom criteria, 55/259
urinary infection, 24/722
USA criteria, 55/256
USA current position, 55/261

USA modern quideline, 55/261
USA prospective study, 55/257
vegetative survival, 24/703
visual evoked response, 24/731
Wernicke-Korsakoff syndrome, 24/725
Brain degeneration
cystic, *see* Cystic brain degeneration
spongy, *see* Spongy cerebral degeneration
Brain development
cerebroside, 9/18
glycerophosphatide, 9/17
hypoglycemia, 27/65
lipid, 9/17
malnutrition, 29/1-13
phase II-IV, 9/17
sulfatide, 9/18
Brain diaschisis
brain cortex, 54/126
brain infarction, 54/125
cause, 54/126
crossed cerebrocerebellar atrophy, 54/126
oxygen, 54/125
positron emission tomography, 54/125
thalamus, 54/126
Brain dysfunction
minimal, *see* Minimal brain dysfunction
paroxysmal, *see* Paroxysmal brain dysfunction
respiratory encephalopathy, 63/416
Brain dysplasia
acrocephalosyndactyly type II, 50/123
verrucose, *see* Status verrucosus
Brain edema
*see also* Brain tumor edema *and* Brain water content
acetazolamide, 63/417
acetylcholine, 23/151
acute amebic meningoencephalitis, 52/317
African trypanosomiasis, 52/343
alkyl tin intoxication, 9/647, 23/151
amanitin intoxication, 65/38
amphetamine addiction, 55/519
anorexia, 63/416
anoxic ischemic leukoencephalopathy, 9/583, 586, 47/531-535
arachidonic acid, 67/80, 83
argipressin, 67/80, 85
arsenic intoxication, 52/343
arthropod envenomation, 65/197
astrocytic swelling, 67/77
ataxia, 63/416
atrial natriuretic peptide, 67/80, 85
atropine, 23/151
attack, 67/80
autoregulation, 23/153
bacterial meningitis, 52/13
barbiturate, 55/216
barotrauma, 63/416
benign intracranial hypertension, 63/417
Binswanger disease, 54/229
blood-brain barrier, 23/148, 28/373-375, 53/135, 57/261, 67/80
boxing injury, 23/550
brain abscess, 52/148
brain air embolism, 55/194
brain anoxia, 23/146, 27/47
brain arteriovenous malformation, 54/382
brain blood flow, 23/139, 141, 149, 154, 63/417
brain concussion, 23/438
brain cortex, 16/169
brain fat embolism, 55/187
brain hemorrhage, 54/319, 67/80
brain hypoxia, 16/180, 23/146, 63/416
brain infarction, 16/181, 53/135, 190, 54/87
brain injury, 16/170, 23/133, 143, 150, 360, 438, 550, 699, 718, 28/373
brain ischemia, 16/181, 53/135, 67/80
brain metabolism, 57/69
brain metastasis, 47/534
brain microcirculation, 23/134, 139
brain operation, 16/191, 193
brain perfusion pressure, 23/138, 67/76
brain tumor, 16/176, 194, 67/80
brain water content, 28/427
brain white matter, 16/169
calcium channel blocking agent, 67/89
cardiac arrest, 63/213
cellular, 67/75
cerebrovascular disease, 11/128
chemical, 16/170
cisplatin intoxication, 65/537
classification, 16/187
clinical features, 16/186, 188, 190, 67/85
clonus, 63/417
cognitive impairment, 63/417
cold injury, 23/151
colloid oncotic pressure, 67/87
coma, 3/70, 63/416
compression, 23/150
corticosteroid, 24/627-633, 57/214
corticotropin, 16/204
cranial nerve palsy, 63/417
cranial neuropathy, 63/417
craniotomy, 55/216
CSF, 16/189, 67/76
CSF drainage, 55/216
cytotoxic, *see* Cytotoxic brain edema

definition, 53/135, 57/105
dementia, 63/417
demyelination, 47/562, 564
dexamethasone, 63/417
diabetes mellitus, 70/199
diabetic coma, 70/199
diabetic ketoacidosis, 27/83-89
diagnosis, 16/190
dialysis disequilibrium syndrome, 63/523
diffuse brain swelling, 57/280
diuretic agent, 16/204, 67/88
dysbarism, 63/416
echoEG, 23/270
edema kinetics, 53/135
EEG, 16/189, 23/353
electrical injury, 23/699, 718
electricity, 7/360
emotion, 3/323
endothelial tight junction, 57/69
epilepsy, 42/690, 63/417
etacrynic acid, 16/204
ethylene glycol intoxication, 64/123
extracellular space, 56/81
extracorporeal membrane oxygenation, 54/40
fatigue, 63/416
features, 47/532-535
free radical, 67/81
freezing lesion, 16/171
fundamental aspect, 16/167
furosemide, 16/204
galactosemia, 42/551
glioblastoma multiforme, 18/53
glucocorticoid, 16/203, 67/88
glucosteroid, 67/74
glutamic acid, 67/80, 84
glycerol, 16/202, 57/215
granulomatous amebic encephalitis, 52/317
gyromitrine, 64/3
hallucination, 63/417
head injury, 57/69, 105, 213
headache, 48/6, 418, 63/416
hemodialysis, 16/179
Hemophilus influenzae meningitis, 52/61
hexachlorophene intoxication, 37/488
high altitude, *see* High altitude brain edema
histamine, 67/85
histology, 67/72
history, 67/71
hydrocephalus, 47/535
hydrochlorothiazide, 16/204
hydrostatic, 23/144, 67/75, 78
hyperbaric oxygenation, 16/204, 24/618
hypercapnia, 16/180
hyperosmolar therapy, 67/87
hypertensive arteriolar disease, 63/72
hypertensive encephalopathy, 11/564, 16/195, 54/213, 63/72
hypertonic therapy, 67/86
hypertonic urea, 16/198
hyponatremia, 63/546
hypo-osmotic, 57/281
hypoxanthine, 67/81
inflammatory, 28/374
intracranial hypertension, 23/134, 137, 145, 63/417
intracranial pressure, 16/189
irradiation encephalopathy, 63/355
kallikrein-kinin system, 67/80, 83
lead encephalopathy, 64/436
lead intoxication, 64/436
Levine experimental anoxia ischemia, 9/582
lipid, 23/152
mannitol, 16/198, 200, 202, 23/154
mannitol-fructose, 16/202
mediator, 67/80
meningococcal meningitis, 52/25
methanol intoxication, 36/354, 64/99, 101
methodology, 16/168
middle cerebral artery syndrome, 53/358
migraine coma, 60/651
mitochondria, 23/157
mountain sickness, 63/416
myoclonic epilepsy, 42/701
necrotic, 23/148
neurobrucellosis, 33/313, 52/585, 587
neurocysticercosis, 35/300, 52/529-531
neurologic intensive care, 55/212, 214
neurotransmitter, 23/151
noradrenalin, 23/152
nuclear magnetic resonance, 54/87, 92, 57/69, 262
organotin, 64/4
organotin intoxication, 64/135, 138
osmolality normalization, 63/545
osmotic, 67/75, 77
osmotic agent, 57/214
osmotic imbalance, 16/177
oxygen free radical, 67/80
papilledema, 63/417
paresthesia, 63/417
pathogenesis, 63/417, 67/75
pathology, 23/51
pathophysiology, 11/357
peritumoral, 28/374, 67/79
peroxidation, 23/152
pneumococcal meningitis, 33/38, 43, 52/49
postoperative brain swelling, 57/278

postoperative craniotomy, 57/280
posttraumatic, 57/213
prenatal onset epilepsy, 42/690
prevention, 70/200
radiation injury, 23/646
renal disease, 16/195
respiration, 23/156
respiratory encephalopathy, 16/196
retinal blood flow, 63/417
Reye syndrome, 9/549, 27/361, 364, 29/331, 337, 31/168, 33/292, 34/80, 169, 49/373, 56/15, 153, 238, 65/115
RIHSA content, 16/173
spirolactone, 16/204
steroid, 16/176, 203, 57/214, 67/73
stroke, 16/194
stupor, 63/416
subarachnoid hemorrhage, 55/6
surgery, 24/643
swelling, 16/169
terminology, 16/167, 187
thallium, 64/4
toxic shock syndrome, 52/260
traumatic, 16/170, 23/133, 143, 150, 360, 438, 550, 699, 718, 28/373, 67/80
traumatic psychosis, 24/519
treatment, 16/197, 57/214
triethyltin, 28/375
triethyltin intoxication, 16/179, 196
tuberculous meningitis, 52/198
tumor, 16/176
type, 16/195, 47/532-535, 53/135
ultrastructure, 16/174
uremia, 9/639
vascular autoregulation, 23/138, 142, 144
vascular permeability, 57/213
vasogenic, *see* Vasogenic brain edema
venous sinus thrombosis, 63/417
viral infection, 56/86
water intoxication, 63/546, 64/240
white matter, 16/169
xenon, 23/157
Brain embolism
*see also* Cerebrovascular disease
action myoclonus, 53/211
age, 53/200
age adjusted mortality rate, 53/8
air embolism, 53/160
akinetic mutism, 53/209
alcohol excess, 55/163
alcoholic cardiomyopathy, 63/133
alcoholism, 63/133
amnesia, 53/206
amyloidosis, 55/163, 63/137
anterior cerebral artery, 53/340
anticoagulant, 12/456, 55/166, 169
anticoagulant hemorrhage, 63/20
antiphospholipid antibody, 63/122
aorta aneurysm, 63/50
aorta stenosis, 63/25
aorta surgery, 63/58
aphasia, 53/214
arterial hypertension, 54/203, 63/82
arteriosclerosis, 11/389, 53/200, 304
arteriovenous fistula, 55/450
asterixis, 53/211
ataxia, 53/214
atrial fibrillation, 53/157, 55/153, 168, 63/17
atrial myxoma, 53/157, 55/163, 63/97
atrial myxoma course, 63/99
auditory symptom, 53/207
autonomic dysfunction, 53/207
bacterial endocarditis, 52/292
Behr disease, 53/213
bicuspid aortic valve, 63/26
blepharospasm, 53/211
blindness, 53/207
blood pressure, 53/207
brain aneurysm, 53/159
brain angiography, 53/228, 54/177
brain scintigraphy, 53/232
brain stem auditory evoked potential, 53/233
calcific embolus, 63/25
Canadian Neurologic Scale, 53/209
cardiac amyloidosis, 53/156
cardiac arrest, 53/157
cardiac catheterization, 55/161, 63/6, 177
cardiac disorder, 53/200
cardiac dysrhythmia, 53/157, 55/151
cardiac myxosarcoma, 63/103
cardiac surgery, 11/396
cardiac valve prosthesis, 53/156, 63/26
cardiac valve replacement, 63/26
cardiac valvular disease, 53/155, 55/164, 63/17
cardiac ventricular aneurysm, 55/162
cardiomyopathy, 53/156, 55/153, 162, 63/131
cardiopulmonary bypass, 63/175
carotid bruit, 53/216
carotid endarterectomy, 53/292
carotid sinus massage, 53/219
catecholamine, 55/157
causative heart disease, 55/161
causative myocardial infarction, 55/161
cause, 53/201
Chagas disease, 55/163, 63/135
Charcot-Bouchard aneurysm, 11/587

chronic vegetative state, 53/209
climate, 53/200
clinical course, 53/202
clinical evaluation, 53/199
clinical features, 11/403-407, 418
clinical parameter, 53/318
collagen vascular disease, 55/163
coma, 53/208
computer assisted tomography, 53/220, 54/58
conduction aphasia, 53/215
confusion, 53/206
congenital heart disease, 53/31, 157
congestive heart failure, 55/162, 63/131, 133
consciousness loss, 53/206
correct diagnosis, 53/199
creatine kinase MB, 55/156
crushed tablet, 55/518
CSF, 53/224, 54/197
CSF spectrophotometry, 53/224
decerebrate rigidity, 53/211
decompression illness, 53/160
decortication, 53/211
diagnostic error, 53/220
differential diagnosis, 53/233
diphtheria, 55/163
diplopia, 53/207
directional Doppler velocimetry, 53/225
distal hypertension, 11/402
drug abuse, 55/166, 518
duplex sonography, 53/225
duration, 53/203
dysphagia, 53/207
dysphonia, 53/207
dystonia, 53/211
ECG, 53/230, 55/151
echoCG, 53/230
echoEG, 53/233
EEG, 11/409, 53/232
Ehlers-Danlos syndrome, 55/455
emotional disorder, 53/206
endocarditis, 53/156
endomyocardial fibrosis, 63/138
enterococcus, 55/166
external internal carotid bypass, 53/316
eyelid nystagmus, 53/213
fat, 11/397
fat embolism, 53/161
fibromuscular dysplasia, 53/163
foreign body embolism, 53/161
Friedreich ataxia, 55/163
gaze deviation, 53/212
general pathology, 11/130
genetic predisposition, 53/199
Glasgow Coma Scale, 53/209
global aphasia, 53/215
glycogen storage disease, 55/163
glycosaminoglycanosis type I, 55/163
hallucination, 53/206
headache, 5/126
heart transplantation, 63/175
heart tumor, 53/157, 55/163
heart wall ostia, 55/163
hematotachygraphy, 53/225
hemianopia, 53/207
hemiballismus, 53/211
hemichorea, 53/211
hemiparesis, 53/206
hemochromatosis, 55/163
hemodialysis, 55/166
hemodynamic mechanism, 53/316
hemostasis, 11/417
heparin, 63/306
hereditary hemorrhagic telangiectasia (Osler), 55/449
history, 11/386, 53/199
Holter monitoring, 53/230
homocystinuria, 55/328
Horner syndrome, 53/213
hypereosinophilia, 63/138
hypereosinophilic syndrome, 63/372
hyperthermia, 53/207
hyperthyroidism, 70/92
hypertrophic cardiomyopathy, 63/136
iatrogenic neurological disease, 55/161, 168, 63/175
idiopathic hypereosinophilic syndrome, 63/138
idiopathic hypertrophic subaortic stenosis, 53/156
immunosuppression, 55/166
infected prosthetic valve, 55/166
infectious endocarditis, 55/166, 63/23, 114
internuclear ophthalmoplegia, 53/212
intravascular consumption coagulopathy, 53/156
intravenous digital subtraction angiography, 53/227
late life, 53/41
leukemia, 63/345
libido, 53/207
Libman-Sacks endocarditis, 53/34, 156, 63/29
macroembolism, 63/175
malignancy, 11/415-420
malignant lymphoma, 55/489
marantic endocarditis, 53/156
melanoma, 55/163
mental function, 53/214
metastatic cardiac tumor, 63/104
microembolism, *see* Brain microembolism

micturition, 53/207
middle cerebral artery, 53/355
misery perfusion syndrome, 53/314
mitral annulus calcification, 53/156
mitral stenosis, 55/164, 63/17
mitral valve prolapse, 53/155, 55/165, 63/20
monocular vision loss, 53/207
muscle tone, 53/211
mycotic aneurysm, 11/395, 63/23
myocardial infarction, 11/388, 53/155
myotonic dystrophy, 55/163
neurogenic pulmonary edema, 55/157
neuropathology, 11/401-403
nonbacterial thrombotic endocarditis, 11/392, 415, 417, 47/541, 55/165, 489, 63/120
noncardiac source, 55/177
nuchal rigidity, 53/207
nuclear magnetic resonance, 53/231
nystagmus, 53/213
ocular bobbing, 53/213
oculoplethysmography, 53/225
oculopneumoplethysmography, 53/225
onset, 53/202
open heart surgery, 55/168, 63/175
ophthalmodynamometry, 53/219
ophthalmoscopy, 53/218
oscillopsia, 53/207
palatal myoclonus, 53/211
paradoxical embolism, 11/391, 53/161, 55/163
paradoxical embolization, 55/450
Parinaud syndrome, 53/212
pathologic bruit, 53/217
pathology, 11/415
pathophysiology, 53/314
pentazocine addiction, 55/518
percutaneous transluminal coronary angioplasty, 63/177
peripartum period, 55/163
pethidine addiction, 55/518
pheochromocytoma, 63/136
phonoangiography, 53/218
physiologic bruit, 53/217
pontine hematoma, 53/207
positron emission tomography, 53/232
posterior cerebral artery, 53/409
precipitating factor, 53/202
prognosis, 11/408, 53/319
progressive muscular dystrophy, 55/163
prolapsing mitral valve, 55/153
prosthetic cardiac valve, 55/165
pruritus, 53/207
pseudoabducens nerve palsy, 53/213
pyribenzamine addiction, 55/518
race, 53/200
recovery, 53/203
respiration, 53/208
retinal artery embolism, 53/218, 319
retraction nystagmus, 53/212
rhabdomyoma, 55/163
rheumatic fever, 55/163
rheumatic heart disease, 11/387, 53/155, 63/17
rheumatic mitral valve, 55/153
risk factor, 53/200
sarcoidosis, 55/163, 63/137
seizure, 53/206
septic embolus, 53/31, 55/489
sex ratio, 53/200
sexual impotence, 53/207
sick sinus syndrome, 55/153
sickle cell anemia, 55/507
sinoatrial block, 55/153
sinus bradycardia, 55/153
skew deviation, 53/213
source, 11/387-393
speech disorder, 53/206
Staphylococcus, 55/166
status cribrosus, 53/316
Streptococcus viridans, 55/166
striatum apoplexy, 11/585
subacute bacterial endocarditis, 11/393
subarachnoid hemorrhage, 12/170, 55/167
subclavian bruit, 53/216
surgery, 11/410
systemic brain infarction, 11/434
systemic disease, 53/200
systemic lupus erythematosus, 53/156
systemic lupus erythematosus endocarditis, 63/122
talcum powder, 55/518
thermoregulation, 53/207
thrombotic endocarditis, 55/489
thrombotic endocarditis syndrome, 11/415-420
tinnitus, 53/207
toxin, 55/163
toxoplasmosis, 55/163
transcortical aphasia, 53/215
treatment, 11/409, 419, 12/450
tuberous sclerosis, 55/163
tumor, 16/6
tumor embolism, 11/391, 53/161
typhoid fever, 55/163
unilateral pulsating tinnitus, 53/207
vascular pulsation, 53/215
vertical gaze palsy, 53/212
vertical nystagmus, 53/213
vertical one and half syndrome, 53/212

vertigo, 53/207
viral infection, 55/163
visual evoked response, 53/232
visual hallucination, 53/207
visual manifestation, 53/207
vomiting, 53/206
watershed infarction, 53/314
xanthochromia, 53/224
Brain-face correlation
holoprosencephaly, 30/450, 50/235
Brain failure, *see* Dementia
Brain fat embolism
adrenocorticosteroids, 55/191
age pattern, 55/182
albumin, 55/191
alcohol, 55/191
anemia, 55/179, 186
biochemical theory, 55/179
blast injury, 55/180
brain edema, 55/187
coma, 55/177
computer assisted tomography, 55/182
coronary bypass surgery, 63/182
delirium, 55/177
dextran, 55/191
differential diagnosis, 55/182
dyspnea, 55/183
epilepsy, 55/183
fat stain, 55/188
features, 55/177
fever, 55/178
fibrinogen, 55/185
free fatty acid, 55/180
fulminant syndrome, 55/181
heart transplantation, 63/178
hemoglobin fall, 55/178
heparin, 55/191
historic landmark, 55/178
hypoalbuminemia, 55/179
hypoxia, 55/184
hypoxia correction, 55/191
iatrogenic neurological disease, 55/177, 63/182
impaired consciousness, 55/183
incidence, 55/178
injury, 55/177
intravascular consumption coagulopathy, 55/180
ischemic demyelination, 55/189
lucid interval, 55/178, 183, 185
major diagnostic criteria, 55/182
mechanical theory, 55/179
minor diagnostic criteria, 55/182
mortality, 55/182, 192
multiple cystic leukoencephalomalacia, 55/190
neurogenic theory, 55/180
neuropathology, 55/187
nuclear magnetic resonance, 55/182
onset of neurologic symptom, 55/183
pathogenesis, 55/179
pathomechanism, 23/634
petechia, 55/185
prognosis, 55/192
pulmonary edema, 55/184
pyrexia, 55/186
respiratory symptom, 55/183
retinal artery embolism, 55/178, 187
sex ratio, 55/182
shock, 55/179
sputum fat globule, 55/178
subacute syndrome, 55/181
subclinical syndrome, 55/181
surgery, 55/177
symptom, 55/177
systemic fat embolism, 55/181
tachycardia, 55/186
tachypnea, 55/183
thrombocytopenia, 55/186
treatment, 55/191
white matter hemorrhage, 55/188
white matter petechia, 55/187
Brain function
altitude illness, 48/396
biogenic amine, 29/467-469
hypothermia, 71/574
minimal, *see* Minimal brain dysfunction
Brain fungus
missile injury, 23/524
Brain glioma, 18/1-41, 72/178
angiography, 18/24-28
biopsy, 16/710, 713
brain scanning, 16/688
brain scintigraphy, 18/29-33
high grade, 72/180
injury, 18/14
Maffucci syndrome, 14/775
neurofibromatosis type I, 50/367
seizure, 72/179
symptom, 72/179
Brain glucose metabolism
*see also* Dopamine receptor
positron emission tomography, 49/36
Brain granuloma
aspergillosis, 16/231, 35/397
blastomycosis, 16/230
candidiasis, 16/230
coccidioidomycosis, 16/230
congenital, *see* Congenital brain granuloma

cryptococcosis, 16/230
histoplasmosis, 11/230, 52/437
hypothalamus, 16/230
Listeria monocytogenes meningitis, 52/93
paracoccidioidomycosis, 35/532, 52/457
schistosomiasis, 52/537
sporotrichosis, 16/230
trichinosis, 52/572
Brain growth
CNS, 30/28
craniosynostosis, 30/219
embryo, 50/14
Brain gumma, 18/427-434
*see also* Neurosyphilis
lymphocytic meningitis, 52/278
neurosyphilis, 52/278
penicillin, 52/279, 283
Brain-gut axis
organization, 75/626
Brain hemangioma
brain vascular tumor, 18/276
Brain hematoma
*see also* Cerebellar hematoma
alexia, 45/438
headache, 48/278
infectious endocarditis, 63/23
intracerebral, *see* Intracerebral hematoma
neurobrucellosis, 52/588
parasagittal operation, 24/639
rehabilitation, 55/233
Brain hemisphere
absence, 42/12
anencephaly, 42/12
development, 67/2
Friedreich ataxia, 60/303
hematocephalus externus, 12/159
hypoplasia, 42/12
sensory sequence stimulation, 15/78
verbal comprehension, 45/103
Brain hemisphere enlargement
diffuse glioblastosis, 18/73
Brain hemorrhage, 11/578-652
*see also* Arteriovenous malformation, Brain blood flow, Cerebellar hemorrhage, Cerebrovascular disease, Congenital aneurysm, Intracerebral hematoma, Intracranial hematoma, Intracranial hemorrhage *and* Stroke
acute infantile hemiplegia, 31/169
afibrinogenemia, 38/72
age, 54/290
age adjusted mortality rate, 53/8
alcohol, 54/288, 55/523
alcoholism, 63/313
allergic granulomatous angiitis, 54/71, 63/383
amphetamine, 55/519
amphetamine addiction, 55/519
amphetamine intoxication, 65/51, 258
amyloid angiopathy, 54/70
amyloidosis, 11/611, 615, 42/727
angioma, 54/288
angionecrosis, 54/203
annual incidence, 54/287, 55/2
anomia, 54/309
anticoagulant, 54/71, 288, 291, 55/167, 476
anticoagulant overdose, 65/461
anticoagulant risk, 63/319
aphasia, 45/17
area vascularization, 11/619
arrhythmia pathogenesis, 63/237
arterial hypertension, 11/587, 591, 601, 54/203, 63/72, 82
arteriolar fibrinohyalinoid degeneration, 54/203, 299
arteriosclerosis, 11/603
artery, 11/82
asparaginase, 69/426
atrial fibrillation, 55/152
autoscopia, 45/387
bacterial endocarditis, 52/291
basal ganglion, 54/68
blood clotting factor XIII deficiency, 63/313
blood dyscrasia, 54/288
blood patch, 61/164
brain amyloid angiopathy, 54/292, 337, 63/74
brain aneurysm, 54/288
brain angiography, 54/315
brain arteriosclerosis, 11/603
brain arteriovenous malformation, 12/232, 241, 248, 53/34
brain edema, 54/319, 67/80
brain infarction, 11/310
brain lacunar infarction, 46/355, 54/289
brain stem hemorrhage, 54/309
brain tumor, 31/166, 54/292, 69/420, 424
brain tumor headache, 5/141
cardiac catheterization, 63/177
cardiac dysrhythmia, 55/151, 63/237
cardiac valvular disease, 55/164
catecholamine, 55/157, 63/237
caudate nucleus, 54/306
causative cardiac arrhythmia, 55/152
causative heart disease, 55/161
causative myocardial infarction, 55/161
cause, 54/287
cerebellar hemorrhage, 54/314
cerebellum, 11/591

cerebral malaria, 35/146
cerebrovascular disease, 53/34
Charcot-Bouchard aneurysm, 11/578, 54/289, 299, 63/74
child, 31/165, 53/34
choriocarcinoma metastasis, 63/104
claustrum, 11/587
clinical features, 54/301
cocaine addiction, 55/521
computer assisted tomography, 54/67, 71, 295-297, 316
creatine kinase MB, 55/156
critical illness polyneuropathy, 51/575
CSF, 54/197, 315
CSF examination, 11/258
declining mortality, 54/287
diagnosis, 54/315
diamorphine, 55/517
diamorphine addiction, 55/517, 65/356
diamorphine nephropathy, 55/517
drug abuse, 55/517
ECG, 55/151
ECG pathogenesis, 63/234
ECG pathology, 63/233
ECG pathophysiology, 63/234
ECG QT interval, 63/233
ECG ST depression, 63/233
ECG T wave inversion, 63/233
eclampsia, 39/542
ectopic beat, 55/152
EEG, 11/280
electrical injury, 61/192
embolism, 11/585
encephalitis, 31/166
encephalomyelitis, 31/167
epidural, 31/164
epidural anesthesia, 61/164
epilepsy, 72/215
essential thrombocythemia, 54/294
Fabry disease, 55/457
fat embolism, 9/588
fibrinoid degeneration, 11/637
fibrinolytic agent overdose, 65/462
fibrinolytic treatment, 54/291
frequency, 54/287
general pathology, 11/154
Glanzmann thrombasthenia, 63/325
granulomatous CNS vasculitis, 55/390
headache, 5/138, 48/273, 277, 53/205, 54/302
heart transplantation, 63/179
hematoma, 54/89
hemodialysis, 54/291
hemophilia, 53/34, 54/71, 293, 63/310
heparin risk, 63/319
hereditary, *see* Hereditary brain hemorrhage
history, 11/582
hyperaldosteronism, 39/496
hypernatremia, 63/554
hypertension, 11/587-601, 54/288
hypertensive encephalopathy, 11/587
hypertensive fibrinoid arteritis, 11/139
iatrogenic neurological disease, 61/164
idiopathic thrombocytopenic purpura, 54/294
infant, 31/165
infectious endocarditis, 55/167, 63/23, 115
intravascular consumption coagulopathy, 38/72, 54/293, 55/497, 63/314, 344
intraventricular, 31/160-162
leukemia, 39/6-8, 53/34, 54/293, 63/343
liver disease, 55/476, 63/313
lobar type, 54/68
location, 54/289, 294
lumbar puncture, 61/164
lymphocytic leukemia, 63/343
lymphoma, 63/350
macroglobulinemia, 55/485
malignant hypertension, 55/517
malignant lymphoma, 55/487
management, 12/454, 54/318
meningitis, 31/166
methadone, 55/517
methanol intoxication, 64/99
3,4-methylenedioxymethamphetamine intoxication, 65/367
mitral stenosis, 55/164
mode of onset, 54/302
moyamoya disease, 42/749, 53/164, 55/294, 296
multiple, 54/303, 306
mutism, 54/309
myelography, 61/164
myeloid leukemia, 18/250, 63/343
neglect syndrome, 45/175
neonate, 31/165
neoplasm, 54/288
neurogenic pulmonary edema, 55/157
neurologic intensive care, 55/204, 212, 234
neuropathology, 54/294
neurosurgery, 54/321
nonbacterial thrombotic endocarditis, 11/419, 55/164
nonsteroid anti-inflammatory agent intoxication, 65/415
noradrenalin, 63/234
nuclear magnetic resonance, 54/88, 317
oral contraception, 55/524
pathogenesis, 11/616, 628-636, 54/288, 298

pathology, 54/287
pediatric head injury, 23/449
periventricular hemorrhage, 54/36
phencyclidine, 55/522
phencyclidine addiction, 55/522
phenethylamine intoxication, 65/51
phenylpropanolamine intoxication, 65/259
Pick supermiliary aneurysm, 11/582, 587
polyarteritis nodosa, 39/301, 48/284, 54/294
polycythemia, 63/251
postictal blood pressure, 54/206
postneurosurgery, 63/322
prognosis, 54/322
promyelocytic leukemia, 63/344
pseudoxanthoma elasticum, 55/453
pure motor hemiplegia, 54/242
putamen, 11/587
putaminal hemorrhage, 54/304
quinidine, 63/192
recurrent hemorrhage, 54/303, 304
rehabilitation, 55/233
Reye syndrome, 31/168
risk factor, 54/289
serum total cholesterol, 54/288
sick sinus syndrome, 55/152
sickle cell anemia, 38/37-40, 53/34, 55/465, 504, 510, 63/262
site frequency, 54/303
5 sites, 54/67
size, 54/289
smoking, 55/524
Spät Apoplexie, 24/354, 357, 365
spinal anesthesia, 61/164
spinal cord electrotrauma, 61/192
stage, 54/89
streptokinase, 65/461
striatum apoplexy, 11/585
Sturge-Weber syndrome, 55/444
subarachnoid hemorrhage, 12/165, 31/160, 55/20, 167
subcortical lobar hemorrhage, 54/307
subdural, 31/163
subependymal hemorrhage, 54/36
symptom, 54/302
systemic lupus erythematosus, 39/280, 53/34, 54/294, 55/371
tachycardia, 55/152
thalamic hemorrhage, 54/307
thalamus, 11/591, 54/68, 309
thalassemia, 55/466, 513
thrombocytopenia, 54/71, 55/472, 63/322
thrombotic thrombocytopenic purpura, 53/34, 63/322
tissue plasminogen activation, 65/461
tobacco addiction, 55/524
toxic shock syndrome, 52/260
trauma, 11/608
trichinosis, 52/565, 572
ultrasound diagnosis, 54/35
uremic polyneuropathy, 63/513
vasoparalysis, 11/626
ventricle, 11/591
vessel innervation, 11/620
Waldenström macroglobulinemia, 55/486, 63/396

Brain hemorrhagic infarction
anticoagulant, 63/19
anticoagulant effect, 63/20
candidiasis, 52/401
computer assisted tomography, 54/51
dissecting aorta aneurysm, 63/47
herpes simplex virus encephalitis, 54/72
Libman-Sacks endocarditis, 63/29
varicella zoster virus encephalitis, 54/72

Brain herniation, 1/550-571
*see also* Clivus ridge syndrome, Midline shift, Tonsillar herniation *and* Transtentorial herniation
acute amebic meningoencephalitis, 52/317, 326
axial direction, 16/97
axial shift, 16/123
bacterial meningitis, 52/13
brain abscess, 52/152
brain concussion, 23/438
brain injury, 23/235
brain tumor, 67/145
cerebellar tonsil, 16/122
cerebral amebiasis, 52/326
cervicomedullary injury, 24/162
characteristics, 16/114
cingulate gyrus, 16/108, 111
cranial nerve sign, 16/132
depressed skull fracture, 24/225
epidural hematoma, 24/264
granulomatous amebic encephalitis, 52/317
intracranial pressure, 23/204
lumbar puncture, 61/163
meningococcal meningitis, 52/26, 30
posterior fossa hematoma, 24/346
prevalence, 16/117
schema, 16/106
sphenoid body aplasia, 30/278
subarachnoid hemorrhage, 55/5
temporal pressure cone, 16/119
tentorial, 24/64
tentorial incisure, 1/565
vital sign, 16/135

Brain hyperemia
anoxic ischemic leukoencephalopathy, 9/583, 586, 47/531
head injury outcome, 57/384, 387
Brain hypoperfusion
*see also* Brain perfusion pressure
metastatic cardiac tumor, 63/94
pallidal necrosis, 49/465
pallidonigral necrosis, 49/465
pallidostriatal necrosis, 49/465
primary cardiac tumor, 63/94
Brain hypoxia
*see also* Brain anoxia
anorexia, 63/416
ataxia, 63/416
benign intracranial hypertension, 63/417
blood-brain barrier, 28/375-377
brain edema, 16/180, 23/146, 63/416
chronic obstructive pulmonary disease, 63/417
consciousness, 63/416
convulsive syncope, 72/259
course, 63/416
cranial nerve palsy, 63/417
cranial neuropathy, 63/417
dementia, 63/417
diabetes mellitus, 70/196
dysbarism, 63/416
EEG, 23/353
electrolyte, 15/62
energy metabolism, 27/23
epilepsy, 15/62, 612, 63/417, 72/214
fatigue, 63/416
hallucination, 63/417
head injury, 23/137, 141
headache, 63/416
intracranial hypertension, 63/417
ischemic edema, 23/146
nausea, 63/416
pallidal necrosis, 49/465
pallidonigral necrosis, 49/465
pallidostriatal necrosis, 49/465
papilledema, 63/417
persistent vegetative state, 57/473
respiratory encephalopathy, 63/416
sleep disorder, 63/416
traumatic intracranial hematoma, 57/262
vascular dementia, 46/357
Brain infarction, 53/107-139, 155-165, 175-194, 417-433, 441-455
*see also* Brain lacunar infarction, Brain phlebothrombosis, Brain venous infarction, Carotid artery thrombosis, Cerebrovascular disease, Multiple cystic encephalopathy *and* Stroke
acetazolamide, 53/422
acetylsalicylic acid, 53/465
acidosis, 53/128
acquired immune deficiency syndrome, 56/498
action myoclonus, 53/211
activation, 53/418
activation metabolism, 53/116
acute infantile hemiplegia, 53/32
adenosine, 53/421
adenosine phosphate, 53/127
adenosine triphosphate, 53/127
adenovirus encephalitis, 56/286
adrenocorticosteroids, 53/427
age, 11/297, 53/200
agraphia, 53/347
air embolism, 53/160, 63/419
akinesia, 53/346
akinetic mutism, 53/209, 347, 55/137
alcohol, 55/523
alcohol excess, 55/163
alexia, 45/438
alien hand sign, 53/346
alien hand syndrome, 53/347
allergic granulomatous angiitis, 53/162
alpha adrenergic receptor stimulating agent, 53/421
amaurosis fugax, 55/110
amino acid transmitter, 53/132
amnesia, 53/206, 55/139
amnesic syndrome, 55/139
amphetamine, 53/164
amphetamine addiction, 55/520
amphetamine intoxication, 65/258
amyloidosis, 55/163
anastomosis, 53/109
anastomotic circulation, 53/175
ancrod, 53/425
angiitis, 55/419
angioendotheliomatosis, 53/165
annual incidence, 55/2
anterior alexia, 45/442
anterior cerebral artery, 11/306, 53/339
anticoagulant, 11/355, 53/425, 55/167, 169
anticoagulant hemorrhage, 63/20
anticoagulant risk, 63/319
anticonvulsant, 53/432
antihypertensive agent, 63/74
antihypertensive treatment, 53/463, 63/74
antiphospholipid antibody, 53/166, 63/122, 330
antiserotonin antagonist, 53/421
antithrombocytic agent, 53/464
aorta aneurysm, 63/50

aorta surgery, 63/58
apathy, 55/137
aphasia, 53/214
aprosody, 55/143
arachidonic acid cascade, 53/130
arachnoid arterial ring system, 53/179
arrhythmia cause, 55/154
arterial hypertension, 54/203, 63/82
arterial hypotension, 63/74
arteriosclerosis, 15/304, 53/159, 200
asparaginase, 69/418
asparaginase intoxication, 65/538
aspergillosis, 35/398, 55/435
Aspergillus fumigatus, 63/532
asterixis, 53/211
asymptomatic carotid artery stenosis, 53/293, 447
ataxia, 53/214
atlantoaxial subluxation, 53/165
atrial fibrillation, 53/157, 55/152
atrial myxoma, 53/157, 55/163, 63/97
auditory symptom, 53/207
autonomic dysfunction, 53/207
autoregulation, 53/112
autotopagnosia, 55/145
B-mode sonography, 54/4
bacterial endocarditis, 52/291
bacterial meningitis, 52/14, 53/161, 55/419-421
Balint syndrome, 55/144
barbiturate, 53/116, 418
basal forebrain, 55/140
basofrontal syndrome, 53/348
Bayliss principle, 53/194
behavior disorder, 53/348
Behr disease, 53/213
bilateral infarction, 53/190
bilateral lesion, 53/346
Binswanger disease, 53/158, 54/222
biochemical aspect, 27/27-35
biochemistry, 53/127
blepharospasm, 49/542, 53/211
blindness, 53/207
blood-brain barrier, 28/377-380
blood clotting factor, 53/166
blood clotting necrosis, 53/188
blood flow restoration, 53/418
blood pressure, 53/109, 207
blood pressure regulation, 53/431
blood viscosity reduction, 53/423
brain abscess, 11/314
brain angiography, 53/165, 228, 54/173
brain blood flow, 11/348-355, 53/113, 54/147
brain blood flow index, 54/119
brain blood volume index, 54/119
brain diaschisis, 54/125
brain edema, 16/181, 53/135, 190, 54/87
brain edema mechanism, 53/137
brain energy consumption, 53/116
brain hemorrhage, 11/310
brain ischemia topography, 53/107
brain lacunar infarction, 46/355, 53/318, 54/235
brain metabolism, 11/348-355
brain oxygen metabolic rate, 53/116
brain perfusion pressure, 53/159
brain perfusion pressure increase, 53/423
brain scanning, 11/243
brain scintigraphy, 53/232
brain stem auditory evoked potential, 53/233
brain stem infarction, 2/249
brain stimulation, 53/423
brain swelling, 53/126
brain thromboangiitis obliterans, 53/163
brain tissue pressure, 53/136
brain tumor, 11/313, 69/414, 420
brain vasculitis, 55/421
brain volume, 53/135
brain water content, 53/135, 137
γ-butyrolactone, 53/419
calcium, 53/125
calcium antagonist, 53/428
calcium channel blocking agent, 63/193
Canadian Neurologic Scale, 53/209
candidiasis, 52/399, 401, 55/435
capillary anastomosis, 53/179
carboanhydrase inhibitor, 53/423
carbon dioxide inhalation, 53/422
carbon dioxide reactivity, 53/112
carbon dioxide tension, 53/127
carbon monoxide intoxication, 53/167
cardiac amyloidosis, 53/156
cardiac arrest, 53/157
cardiac disorder, 53/200
cardiac dysrhythmia, 53/157, 55/151
cardiac myocytolysis, 55/154
cardiac output, 53/177
cardiac surgery, 11/442-448
cardiac valve prosthesis, 53/156
cardiac valvular disease, 53/155, 55/164
cardiomyopathy, 53/156
carotid artery occlusion, 11/336
carotid bruit, 53/216
carotid dissecting aneurysm, 53/39, 164
carotid endarterectomy, 53/165, 292, 426, 441
carotid endarterectomy result, 53/446
carotid sinus compression, 53/165
carotid sinus massage, 53/219
cat scratch disease, 53/162

catatonia, 46/423
catecholamine, 55/157
causative cardiac arrhythmia, 55/152
causative heart disease, 55/161
causative myocardial infarction, 55/161
cause, 53/155, 201
cell membrane depolarization, 53/118
central, 12/586
cerebellar diaschisis, 53/121
cerebral malaria, 53/162
Chagas disease, 55/163
child, 54/40, 55/505
chiropractic manipulation, 53/165
cholinergic receptor blocking agent, 53/421
chronic meningitis, 56/645
chronic mercury intoxication, 64/372
chronic vegetative state, 53/209
circadian rhythm, 53/193
classification, 11/295
climate, 53/200
clinical course, 53/202
clinical evaluation, 53/199
clinical features, 11/292-326
clinical syndrome, 11/303
clinical trial methodology, 53/459-462
clofibrate, 53/425
clumsy hand dysarthria syndrome, 53/318
coagulopathy, 69/414
cobalt therapy, 53/33
cocaine addiction, 55/521
collagen vascular disease, 55/163
collateral circulation, 53/108, 177
coma, 53/208
complicated migraine, 11/465
computer assisted tomography, 53/220, 313, 54/47, 50, 52, 53, 84
computer simulation, 53/109
conducting vessel, 53/114
conduction aphasia, 53/215
confusion, 53/206
confusional state, 55/138
congenital heart disease, 53/157, 63/2
congestive heart failure, 63/132
consciousness loss, 53/206
coronary bypass surgery, 63/181
corpus callosum syndrome, 53/347
correct diagnosis, 53/199
cortical blindness, 55/119, 144
cortical steady potential, 53/124
creatine kinase MB, 55/156
critical ischemia, 54/123
crossed cerebrocerebellar atrophy, 54/127
cryoglobulinemia, 53/166, 63/404
Cryptococcus neoformans, 63/532
CSF, 53/224, 54/197
CSF examination, 11/238
CSF spectrophotometry, 53/224
cystic infarct, 53/185
decerebrate rigidity, 53/211
decompression illness, 53/160
decortication, 53/211
deep vein thrombosis, 53/432
definition, 53/176
delayed cerebral radiation necrosis, 53/164
delirium, 53/363, 55/138
depression, 55/143
determinant factor, 54/121
dextran, 11/356
diabetes mellitus, 53/17
diagnostic error, 53/220
diamorphine, 53/164, 55/517
diamorphine addiction, 55/518, 65/356
diamorphine intoxication, 65/355
diaschisis, 53/120
difference zone, 53/175
differential diagnosis, 11/310-315, 53/186, 233, 54/51
dimethyl sulfoxide, 53/430
diphtheria, 55/163
diplopia, 53/207
dipyridamole, 53/465
directional Doppler velocimetry, 53/225
disability factor, 11/323
Divry-Van Bogaert syndrome, 55/320
dopamine, 53/421
dorsomedial thalamic nucleus, 55/140
drug abuse, 55/517
drug induced hypotension, 63/74
duplex sonography, 53/225
duration, 53/203
dysphagia, 53/207
dysphasia, 53/346
dysphonia, 53/207
dyspraxia, 53/346
dystonia, 53/211
ECG, 53/230, 55/151
echoCG, 53/230
echoEG, 53/233
ectopic beat, 55/152
edema treatment, 53/426
EEG, 53/122, 232
electrolyte shift, 53/119
embolism, 53/30
emotional disorder, 53/206
encephalopathy, 11/314, 63/345
endarterectomy mortality, 53/442

endocarditis, 53/156
energy metabolism, 27/29-31
enhancement, 54/51
epilepsy, 15/304, 63/345, 72/214
ergotamine, 53/164
erythrocyte count, 53/432
erythrocyte flexibility, 53/424
essential thrombocythemia, 53/166
état vermoulu, 53/189
etomidate, 53/419
evoked potential, 53/123
experimental animal, 53/122
experimental infarction, 53/108
external internal carotid bypass, 53/316, 451, 54/128
extracellular shift, 53/125
extracerebral steal, 53/115
extracorporeal membrane oxygenation, 54/40
extracranial anastomosis, 53/178
eyelid nystagmus, 53/213
Fabry disease, 53/33, 165, 55/457
fat embolism, 53/161, 55/188
fibrinolysis, 53/426
fibrinolytic agent, 11/355
fibromuscular dysplasia, 11/369, 371, 53/163, 55/283
fluorocarbon, 53/422
foreign body embolism, 53/161
free radical, 53/130
free radical scavengers, 53/429
Friedreich ataxia, 55/163
frontal lobe syndrome, 55/138
functional threshold, 53/117
fungal vasculitis, 53/161
gaze deviation, 53/212, 55/122
gaze paretic nystagmus, 55/122
general management, 53/431
genetic predisposition, 53/199
geography, 11/296
gestation, 54/427
Glasgow Coma Scale, 53/209
glial swelling, 53/136
global aphasia, 53/215
glucose utilization, 53/129
glue sniffing, 55/523
glycerol, 53/427
glycogen storage disease, 55/163
glycolysis, 54/122
glycosaminoglycanosis type I, 55/163
granulomatous CNS vasculitis, 55/389
gross change, 11/144
gross pathology, 53/187
hallucination, 53/206
headache, 53/205
heart disease, 53/155
heart transplantation, 63/179
heart tumor, 53/157, 55/163
heart wall ostia, 55/163
hematotachygraphy, 53/225
hemianopia, 53/207
hemiballismus, 53/211
hemichorea, 53/211
hemi-inattention, 55/145
hemineglect, 55/145
hemiparesis, 53/206
hemiplegia, 1/192, 63/345
hemochromatosis, 55/163
hemodilution, 53/424
hemodynamic mechanism, 53/316
hemodynamics, 53/107
hemolytic uremic syndrome, 63/324
hemorrhage, 53/365
hemorrhagic infarction, 53/184
hemorrhagic necrosis, 54/438
hemorrheology, 53/110
heparin, 53/425, 63/306
heparin risk, 63/319
hereditary hemorrhagic telangiectasia (Osler), 55/449
herniation, 54/404
herpes zoster ophthalmicus, 53/161
heterogenous flow, 53/108
hexobendine, 53/422
histamine, 53/421
history, 11/292-294, 53/175, 199
Holter monitoring, 53/230
homocystinuria, 53/33, 165, 55/328
Horner syndrome, 53/213
Huebner artery, 53/343
Huebner artery syndrome, 53/348
hydralazine, 63/74
γ-hydroxybutyric acid, 53/419
hyperaldosteronism, 39/496
hyperbaric oxygen, 53/422
hypercapnia, 53/422
hyperglycemia, 53/128, 167
hypernatremia, 63/554
hypertension, 53/464
hypertensive encephalopathy, 11/559, 53/158, 54/215
hyperthermia, 53/207
hypertrophic cardiomyopathy, 63/137
hyperventilation, 53/422
hyperviscosity syndrome, 53/165, 55/489
hypoglycemia, 53/167
hypotension, 53/158

hypotensive crisis, 53/192
hypothermia, 53/417
hypoxia, 53/167
iatrogenic, *see* Iatrogenic brain infarction
ideomotor apraxia, 53/346
idiopathic hypertrophic subaortic stenosis, 53/156
incidence, 11/295
indifferent reaction, 55/143
indometacin, 53/429
infectious endocarditis, 55/167, 63/23, 112, 114, 116
injury potential, 53/111
insidious course, 55/427
insulin, 23/123
internal carotid artery, 11/307
internal carotid artery occlusion, 53/448
internal carotid artery stumpectomy, 53/448
internal carotid artery syndrome, 53/292
internuclear ophthalmoplegia, 53/212
intracerebral hematoma, 11/310-312
intracerebral steal, 53/114, 54/120
intracerebral steal syndrome, 53/159
intracranial anastomosis, 53/178
intracranial hypertension, 53/159
intravascular consumption coagulopathy, 53/156, 166, 55/497, 63/316, 345
intravenous digital subtraction angiography, 53/227
inverse steal, 54/120
ion homeostasis, 53/126
ion shift, 53/125
ischemia tolerance prolongation, 53/417
ischemic core, 53/119
ischemic penumbra, 53/119
isolated cerebral vasculitis, 53/162
jugular catheterization, 53/167
Kawasaki syndrome, 52/265
Köhlmeier-Degos disease, 14/789, 53/163, 55/276
lactacidosis reduction, 53/428
lactic acid, 53/128
lacuna, 53/41
laminar infarction, 53/189
last meadow concept, 53/180
late life, 53/41
left hand agraphia, 53/347
left hand apraxia, 53/347
lenticulostriate artery, 53/365
leukocytosis, 53/165
leukotriene, 53/131
libido, 53/207
Libman-Sacks endocarditis, 53/156, 63/29
lidocaine, 53/116
lipid metabolism, 53/130
liquefaction, 53/189
livedo racemosa, 55/401
luxury perfusion, 53/113
lymphoma, 63/350
lymphomatoid granulomatosis, 53/162
lysergide, 53/164
macroglobulinemia, 55/485
malignant lymphoma, 55/487, 489
mamillary body, 55/140
mannitol, 53/427
marantic endocarditis, 53/156
medical management, 53/417
melanoma, 55/163
MELAS syndrome, 62/509
membrane failure, 53/119
meningeal anastomosis, 53/175, 179
meningococcal meningitis, 52/23, 25
meningovascular neurosyphilis, 53/162
mental function, 53/214
metabolic activation, 53/423
methamphetamine intoxication, 65/366
microcirculation, 53/110
microembolism, 53/108
microglia, 11/148
microinfarction, 53/183
microscopy, 53/188
micturition, 53/207
midazolam, 53/419
middle cerebral artery, 11/304, 53/353, 355, 363
middle cerebral artery occlusion, 53/107
migraine, 53/37, 163
misoplegia, 55/145
mitral annulus calcification, 53/156, 63/29
mitral stenosis, 55/164
mitral valve prolapse, 53/155, 63/20
modern view, 11/294
monocular vision loss, 53/207
morbidity, 11/297
morphologic injury, 53/119
mortality, 54/427
moyamoya disease, 53/164, 55/294-296
multiple, *see* Multiple brain infarction
multiple brain microinfarction, 63/345
multiple occlusion, 12/346
muscle tone, 53/211
mutism, 53/346
mycotic aneurysm, 63/23
myocardial infarction, 53/155
myotonic dystrophy, 55/163
naloxone, 53/430
natural history, 11/143, 315, 53/319
neglect syndrome, 45/175
neonate, 53/30

neurochemical study, 27/29-33
neurocysticercosis, 52/530, 53/162
neurogenic pulmonary edema, 55/157
neurologic intensive care, 55/234
neuronal activity, 53/124
neuronal damage, 53/111
neuropathology, 11/142-154
neuropeptide, 53/133
neuropsychology, 45/9
neurotransmitter metabolism, 53/131
nicotine intoxication, 65/263
nifedipine, 63/74
nimodipine, 63/193
nonbacterial thrombotic endocarditis, 11/419, 55/165, 489, 63/28, 120
no-reflow phenomenon, 53/111
nuchal rigidity, 53/207
nuclear magnetic resonance, 53/231, 54/65, 83, 87
nystagmus, 53/213
occipital lobe, 54/404
occipital vertebral steal, 53/115
ocular bobbing, 53/213
oculoplethysmography, 53/225
oculopneumoplethysmography, 53/225
onset, 53/202
open heart surgery, 53/165
operculum syndrome, 53/359
ophthalmic steal, 53/115
ophthalmodynamometry, 53/219
ophthalmoscopy, 53/218
opiate, 55/517
opiate intoxication, 65/355
oral contraception, 53/36, 164, 55/524, 63/332
orthostatic hypotension, 63/156
oscillopsia, 53/207
osmotherapy, 53/427
ouabain, 53/116
oxidation reduction status, 53/127
oxygen, 54/121
oxygen consumption, 53/129
oxygen extraction, 53/114, 128
oxygen extraction rate, 54/119
oxygen extraction rate + brain metabolic rate for oxygen, 54/121
oxygen tension, 53/127
oxygenation, 53/431
palatal myoclonus, 53/211
palilalia, 53/346
papaverine, 37/455, 53/421
para-alexia, 55/145
paracentral lobule syndrome, 53/345
paradoxical embolism, 53/161, 55/163
paraproteinemia, 53/166
Parinaud syndrome, 53/212
paroxysmal nocturnal hemoglobinuria, 53/167, 55/467
pathogenesis, 53/176, 191
pathologic bruit, 53/217
pathologic laughing and crying, 55/142
pathophysiologic stage, 54/121
pathophysiology, 53/107, 314
pattern, 53/175
penetrating vessel, 53/114
penumbra, 53/117, 175, 54/125
peri-infarct area, 54/123
peripartum period, 55/163
phenytoin, 53/419
phonoangiography, 53/218
phosphate change, 54/83
physiologic bruit, 53/217
physiotherapy, 53/432
plaque fibromyélinique, 53/189
platelet, 55/474
poikilomycosis, 35/567, 52/489
polyarteritis nodosa, 39/302, 48/284, 53/162, 55/356
polycythemia, 53/165, 63/250
polycythemia vera, 55/468, 63/250
pontine, *see* Pontine infarction
pontine hematoma, 53/207
porphyria, 8/10
positron emission tomography, 53/232, 313, 54/65, 87, 120
posterior cerebral artery, 11/306, 53/409
postherpetic, 56/237
postictal blood pressure, 54/206
postoperative care, 53/445
potassium, 53/125
preceding transient ischemic attack rate, 53/312
precipitating factor, 53/202
pregnancy, 54/425
preoperative monitoring, 53/443
prevention, 53/459
prognosis, 11/315-320
progressive muscular dystrophy, 55/163
progressive systemic sclerosis, 53/163
propranolol, 37/446
prostacyclin, 53/130, 429
protein C deficiency, 53/167, 63/328
protein S deficiency, 63/328
protein synthesis, 53/133
proximal vertebral artery endarterectomy, 53/450
pruritus, 53/207
pseudoabducens nerve palsy, 53/213
pseudoxanthoma elasticum, 43/45, 53/165, 55/453
puerperium, 54/425

race, 53/200
recovery, 53/203
recurrent carotid stenosis, 53/448
reduced nicotinamide adenine dinucleotide, 53/127
regional brain blood flow, 54/118
regional oxygen extraction rate difference, 54/122
rehabilitation, 53/432, 55/233
reperfusion, 53/136
residual metabolism, 53/116
respiration, 53/208
retinal artery embolism, 53/218
retraction nystagmus, 53/212
rhabdomyoma, 55/163
rheumatic fever, 55/163
rheumatic heart disease, 53/155
rheumatoid arthritis, 53/163
risk factor, 53/200
saccadic palsy, 55/122
safety mantle, 53/175
sarcoidosis, 55/163
secondary damage prevention, 53/418, 426
sedation, 53/432
segmental vascular resistance, 53/114
seizure, 53/206
septal area, 55/140
sex ratio, 11/297, 53/200
sexual impotence, 53/207
sick sinus syndrome, 55/152
sickle cell anemia, 53/33, 166, 55/464, 504, 508, 63/259
sickle cell trait, 55/513
sign topography, 53/210
single photon emission computer assisted tomography, 53/313, 54/65
site frequency, 54/53
Sjögren syndrome, 53/163
skew deviation, 53/213
smoking, 55/524
Sneddon syndrome, 55/401
sodium, 53/125
sodium change, 54/83
sorbitol, 53/427
speech disorder, 53/206
split brain syndrome, 53/347
sporotrichosis, 35/570, 52/496
stab wound, *see* Brain stab wound
staging, 54/50
status cribrosus, 53/316
status marmoratus, 53/190
striatocapsular infarct, 53/364
Sturge-Weber syndrome, 55/444, 446
subangular alexia, 45/444
subarachnoid hemorrhage, 53/163, 55/5, 9, 20, 167
subclavian bruit, 53/216
subclavian steal, 53/115
subclavian steal syndrome, 53/159
subdural hematoma, 11/312
sulfinpyrazone, 53/465
superior longitudinal sinus occlusion, 54/397, 438
supplementary motor area syndrome, 53/345
surgical indication, 53/441
surgical management, 53/443
surgical result, 53/446
surgical risk, 53/445
surgical technique, 53/442
survival rate, 11/315
survivor condition, 11/318
syphilis, 55/412
systemic, *see* Systemic brain infarction
systemic lupus erythematosus, 53/156, 55/371, 374
systemic lupus erythematosus endocarditis, 63/122
systemic neoplastic disease, 53/167
tachycardia, 55/152
tactile anomia, 53/347
Takayasu disease, 53/162
temporal arteritis, 48/314, 319, 323, 53/162, 55/343
terminology, 53/176
theory, 11/150-154
therapeutic time window, 54/122
therapy principle, 53/417
thermoregulation, 53/207
three territory border, 53/181
three territory infarct, 53/180
threshold, 53/116
threshold regional cerebral metabolic rate for oxygen, 54/118
thromboangiitis obliterans, 53/179
thrombocythemia, 55/473
thrombocytosis, 53/166
thrombophilia, 63/328
thrombotic thrombocytopenic purpura, 53/34, 166, 63/323
tinnitus, 53/207
tissue acidosis, 53/123
tissue osmolality, 53/138
tissue Ph, 54/122
tissue plasminogen activator, 63/321
tissue survival time, 53/192
tissue vulnerability, 54/122
tobacco addiction, 55/524
toluene sniffing, 55/523

total brain volume, 53/126
toxic oil syndrome, 63/381
toxin, 55/163
toxoplasmosis, 55/163
transcortical aphasia, 53/215
transient global amnesia, 55/141
transient ischemic attack, 53/307
transmission computer assisted tomography, 54/117
trauma, 53/165
traumatic delayed apoplexy, 53/190
tremor, 63/345
trichinosis, 52/565, 572
tuberculous arteritis, 53/161
tuberculous meningitis, 52/200, 208, 55/424
tuberous sclerosis, 55/163
tumor embolism, 53/161
typhoid fever, 55/163
typhus, 53/162
ulcerative colitis, 53/163
ultrasound diagnosis, 54/35
unilateral pulsating tinnitus, 53/207
vascular pulsation, 53/215
vascular spasm, 53/163
vascular surgery, 53/441
vasoactive agent, 53/421
vasodilator agent, 11/356
vasogenic brain edema, 11/155, 53/138
venous infarction, 11/308, 53/160
venous obstruction, 53/166
vertebral artery transposition, 53/450
vertebral dissecting aneurysm, 53/39
vertical gaze palsy, 53/212
vertical nystagmus, 53/213
vertical one and half syndrome, 53/212
vertigo, 53/207
vincamine, 53/422
viral infection, 55/163
visual evoked response, 53/232
visual hallucination, 53/207, 55/145
visual manifestation, 53/207
vitamin B3, 53/421
vomiting, 53/206
Waldenström macroglobulinemia, 55/486, 63/396
water-electrolyte balance, 53/431
watershed infarction, 53/110, 159, 314
watershed zone, 53/183, 317
wedge shaped infarct, 53/184
Wegener granulomatosis, 53/162
Wernicke-Korsakoff syndrome, 55/140
white matter infarction, 53/318
xanthine, 53/421
xanthochromia, 53/224
xenon computer assisted tomography, 54/65
yawning, 63/493
yellow jacket intoxication, 37/107
Brain infection
*see also* Intracranial infection
aphasia, 46/150
brain stab wound, 23/483
electrical injury, 23/717
sickle cell anemia, 38/34
streptomycosis, 35/570
Brain injury
*see also* Brain concussion, Brain stem injury, Commotio cerebri, Craniofacial injury, Head injury *and* Skull injury
abducent nerve injury, 24/67, 80
acid base balance, 23/116
acute subdural hematoma, 24/278, 280, 288
ageusia, 24/15
agony, 24/775
air embolism, 23/609
airway, 24/611
akinetic mutism, 23/10, 24/589, 700
alcohol, 57/130
amnesia, 23/12, 24/477
anamnesis, 23/224
anemia, 23/127
anesthesia, 24/613
anetic syndrome, 24/702
angiography, 23/241, 248, 24/619
anosmia, 24/2, 8
aphasia, 45/294, 46/153
apparent death, 24/788, 807, 809, 822
arachnoid cyst, 23/239, 307
arteriovenous shunt, 23/138
ataxic respiration, 57/129
attention disorder, 3/140, 150, 188
autonomic function, 23/230
avoidable, 57/429
axonal disruption, 57/208
battered child syndrome, 23/603
behavior disorder, 4/360-363
biochemical cascade, 57/209
biomechanics, 23/68, 91, 24/160-168
bitemporal hemianopia, 24/43
blood-brain barrier, 23/148, 28/371-374
blood gas analysis, 57/129
blood viscosity, 57/211
boxing injury, 23/527
brain arteriovenous malformation, 23/138
brain blood flow, 23/10, 139, 145, 149, 201, 240, 24/607, 619, 57/211
brain blood flow autoregulation, 57/70
brain capillary, 23/134, 145

brain cavitation, 23/93
brain death, 24/699, 757-786, 821
brain edema, 16/170, 23/133, 143, 150, 360, 438, 550, 699, 718, 28/373
brain herniation, 23/235
brain ischemia, 57/209
brain microcirculation, 23/133, 139
brain perfusion pressure, 23/145
brain regeneration, 16/32
brain scintigraphy, 23/246, 287, 294
brain stab wound, 23/477
brain stem, *see* Brain stem injury
brain swelling, 57/124
brain vasospasm, 23/163, 180, 240
calcium, 23/115
calcium channel blocking agent, 57/212
calorie, 23/125
carbon dioxide, 23/117, 137
catecholamine, 23/124
central alexia, 45/440
cerebellum, 23/459
cerebral dominance, 45/495-500
cervicomedullary injury, 24/141
chemical, 16/170
Cheyne-Stokes respiration, 57/129, 63/483
child, 4/360-373, 455, 24/688
circulatory disturbance, 23/231
classification, 23/227, 229
clinical classification, 57/124
clinical features, 23/223
coding, 23/16
coma, 23/12, 428, 430
compression, 23/84
computer assisted tomography, 23/243, 255
confusion, 23/12
conjugate eye movement, 57/137
conjugate lateral gaze palsy, 57/137
consciousness, 23/226, 45/118, 57/129
constructional apraxia, 45/491-505
cortical blindness, 24/48
corticosteroid, 24/618
corticotropin, 23/121
cranial nerve, 23/235
CSF composition, 24/590
CSF fistula, 23/310
CSF flow, 23/310
CSF lactic acid, 54/198
decerebration, 24/702
decompression illness, 23/612
decortication, 24/702
delayed syndrome, 23/12
delirium, 46/532
dementia, 24/477, 702
developmental dysgraphia, 4/436
developmental dyslexia, 4/377, 436, 46/118
developmental dysphasia, 4/265, 361
diabetes mellitus, 23/123, 125
digestive disorder, 23/232
diphtheria pertussis tetanus vaccine, 52/241
disability, 24/671
Doppler ultrasonography, 57/212
drug treatment, 24/615
dysbarism, 23/609
dyskinesia, 6/104
dysphasia, 57/134
education difficulty, 4/362
EEG, 23/317, 335, 350, 24/620
EEG classification, 23/227, 229
electrical burn, 7/375
electrical injury, 11/464, 23/683
electrolyte balance, 23/109
electrothermal lesion, 23/716
endocrine disorder, 23/119, 123
energy failure, 57/210
epidemiology, 23/23
epidural hematoma, 24/264, 268
epilepsy, 15/561, 23/234, 24/445, 57/130, 134, 73/470
examination, 4/365-373, 23/226, 232
experimental, 23/67, 24/600
fat embolism, 23/631
fatty acid, 23/127
fluid balance, 23/109
focal, *see* Focal brain damage
freezing lesion, 16/171
frontobasal, *see* Frontobasal brain injury
Gaucher disease, 10/525
geniculocalcarine tract injury, 24/46
Glasgow Coma Scale, 57/127
Glasgow Database, 57/44
glucose, 23/123
glutamic acid antagonist, 57/210
glycogen, 27/206
grasp reflex, 24/718
gravity scale, 23/13, 74
growth hormone, 23/122
head injury, 57/4
headache, 5/178, 23/12
heart death, 24/811, 823
heat stroke, 23/669
hematologic finding, 23/127
hemiballismus, 49/374
hemoglobin, 23/127
hemorrhage, 23/49
higher nervous activity disorder, 4/360-365, 466
homonymous hemianopia, 24/45, 57/134

hospital organization, 23/15-17, 33
hydrocortisone, 23/120, 122
hyperactivity, 46/182
hypercapnia, 57/212
hyperglycemia, 23/122
hyperkinetic child syndrome, 3/195, 197, 46/177
hypersomnia, 24/593, 701
hyperventilation, 24/605, 611
hypocapnia, 57/212
hypoglycemia, 70/184
hypophyseal diencephalic disturbance, 23/11
hypotension, 57/210, 433
hypothalamus, 23/59, 119
hypoxia, 23/117, 137, 141, 147
impact, 23/98, 101
infection, 23/61
insulin, 23/123
internuclear ophthalmoplegia, 57/137
intracerebral hematoma, 24/351
intracranial hematoma, 57/124
intracranial hypertension, 23/43, 53, 145
ischemic, *see* Brain ischemia
Kohlu injury, 23/597
lactic acid, 23/116, 119
laser injury, 23/665
learning, 4/9
learning disability, 46/124
legal aspect, 24/833
libido, 24/593
locked-in syndrome, 24/702
lumbar puncture, 23/237
luteinizing hormone, 23/123
Maley syndrome, 23/12
mechanism, 23/67
memory disorder, 23/12
meningioma, 24/441
metabolic autoregulation, 57/211
metabolism, 23/109, 24/601, 605, 57/211
migraine, 23/12
missile injury, 23/505, 514, 520, 24/455
model, 23/70
monitoring, 24/610, 619
mortality, 24/671
motor disorder, 4/455
motor response, 57/128
multiple injury, 24/611
myoclonus, 38/580-582
narcolepsy, 24/593
neonatal encephalopathy, 9/578
neurogenic osteoarthropathy, 24/722
neuroglia relationship, 28/417
neuronal function alteration, 57/208
neurotransmitter, 57/210
nitrogen balance, 23/125
nutrition, 24/615
nystagmus, 1/596
optic chiasm, 24/30, 42
optic chiasm injury, 24/30, 42
optic nerve, 24/27
optic nerve injury, 24/27
optic tract injury, 24/30, 45
organ transplantation, 24/813
outcome, 24/679, 692, 57/66
outcome category, 24/672
oxygen, 57/211
pain reaction, 23/230
parasomnia, 24/701
parietal lesion, 23/233
parietal lobe syndrome, 57/134
Parinaud syndrome, 57/137
pathophysiology, 57/207
penetrating, *see* Penetrating brain injury
perception, 23/11, 229
perinatal, *see* Perinatal brain injury
phagocytosis, 23/128
physiopathology, 23/83
pituitary gland, 23/60, 119
pneumocephalus, 24/201
porencephaly, 23/307
postanoxic intention myoclonus, 38/580
postoperative management, 24/614
posttraumatic epilepsy, 24/834, 57/211
posttraumatic syndrome, 24/838, 48/386
potassium, 23/115
potency, 24/593
pressure autoregulation, 57/211
prevention, 24/835
primary, 57/124, 207
prognosis, *see* Brain injury prognosis
prolactin, 23/123
protein, 23/125
psychological test, 15/561
psychosocial study, 23/1, 14
pupil, 57/128
radiation injury, 23/639
radiodiagnosis, 23/239, 57/143
radiosensitivity, 18/481
radiotherapy, 67/327
recovery, 23/335
regional brain blood flow, 57/69
rehabilitation, 23/16, 24/683
renal insufficiency, 23/115
respiration, 23/230, 24/602, 57/129
respiratory arrest, 23/156
RIHSA cisternography, 23/301, 304
rotation, 23/77, 96

schistosomiasis, 52/537
secondary, 57/124, 208, 431, 433
shock wave, 23/98
skull fracture, 24/39
skull radiation, 57/143
sleep disorder, 24/592
socioeconomic aspect, 23/17
sodium, 23/109
speech, 23/233
speech disorder, 4/361, 23/233
sport injury, 23/527
stab wound, *see* Brain stab wound
status epilepticus, 15/180
stomach function, 23/126
stomach ulcer, 23/126
stress wave, 23/94
surgery, 24/613
survival, 23/16
systemic lupus erythematosus endocarditis, 55/372
testosterone, 23/123
tetany, 24/594
thermoregulation, 23/231
thyroxine, 23/123
tonsillar herniation, 57/134
tracheostomy, 24/611, 614
translation, 23/77, 96
transport, 24/610
transtentorial herniation, 23/55, 57/133
transverse axial tomography, 24/620
traumatic aphasia, 24/490
traumatic blindness, 24/27
traumatic psychosis, 24/520
traumatic psychosyndrome, 24/838
trochlear nerve injury, 24/65
tumor origin, 16/31, 78
vascular autoregulation, 23/138, 142, 144, 153
vascular permeability, 57/212
vascular spasm, 23/164, 180
vascular tone, 57/212
vasomotor response, 23/138
vasopressin, 23/114, 28/499
vasospasm, 57/212
venous sinus thrombosis, 24/369
ventilation, 24/612
vertigo, 23/11
volume pressure response, 23/140
vomiting, 23/12
Wernicke-Korsakoff syndrome, 23/11
whiplash injury, 23/96, 98
xenon, 23/157
Brain injury mechanism, 57/17-37
Abbreviated Injury Scale, 57/24
blunt impact, 57/27
bone failure, 57/21
brain concussion, 57/23
brain contusion, 57/33
brain stem, 57/26
centripetal hypothesis, 57/24
craniospinal junction, 57/27
critical parameter, 57/18
diffuse axonal injury, 57/35
diffuse brain injury, 57/32, 34
distributed loading, 57/30
epidural hematoma, 57/32
experimental brain concussion, 57/23
exposed intracranial content, 57/30
flexion bending, 57/27
fluid percussion model, 57/30, 66, 86
focal brain injury, 57/32
force and acceleration, 57/17
fracture, 57/31
glial cell failure, 57/65
gliding contusion, 57/34
Head Injury Criterion, 57/36
impulsive loading, 57/21
indirect head loading, 57/27
injury criterion, 57/35
injury severity level, 57/35
intracranial, 57/22
living isolated tissue loading, 57/31
load duration, 57/18
Mean Strain Criterion, 57/36
model, 57/27
natural frequency, 57/18
neuronal, 57/65
neuropathology, 57/24
Newton Second Law, 57/18
pathology, 57/24
pressure gradient, 57/25
primary, 57/31-35
rotational acceleration, 57/36
rotational head motion, 57/22
scalp, 57/20
sensitivity curve, 57/19
sharp impact, 57/27
shear stress, 57/26
skull, 57/20
skull deformation, 57/21
skull deformation-angular acceleration theory, 57/22
specific, 57/19-31
stress wave, 57/21
subdural hematoma, 57/34
tolerance level, 57/35
translation, 57/25

translocational acceleration, 57/35
translocational skull motion, 57/22
whiplash, 57/27
Brain injury prognosis
admission criteria, 24/674
armed forces, 24/455
disability, 24/671, 673
Glasgow Coma Scale, 24/677
intracranial pressure, 24/674
outcome category, 24/672
persistent vegetative state, 24/672
posttraumatic syndrome, 24/834
prediction, 24/669, 679
recovery, 24/673
vocational, 24/692
Brain ischemia, 53/155-165
absent alpha activity, 54/157
acidosis, 53/128, 63/211
activation metabolism, 53/116
adenosine phosphate, 53/127
adenosine triphosphate, 53/127
air embolism, 53/160
allergic granulomatous angiitis, 53/162
amaurosis fugax, 53/309
amino acid transmitter, 53/132
amphetamine, 53/164
anaerobic glycolysis, 63/211
anastomosis, 53/109
angioendotheliomatosis, 53/165
anterior choroidal territory events, 53/311
antiepileptic agent, 63/222
arachidonic acid, 63/211
arachidonic acid cascade, 53/130
arterial territory ischemia, 57/161
arteriosclerosis, 53/159
arteriovenous shunt, 12/232
aspartic acid, 63/211
atlantoaxial subluxation, 53/165
atrial fibrillation, 53/157
atrial myxoma, 53/157
bacterial meningitis, 53/161, 55/419
barbiturate, 53/116, 55/522, 63/223
Binswanger disease, 53/158
biochemistry, 53/127
blood clotting factor, 53/166
blood glucose level, 63/211
blood pressure, 53/109
boundary zone ischemia, 57/163
brain aneurysm neurosurgery, 55/57
brain angiography, 53/165
brain arteriovenous malformation, 12/232
brain blood flow, 54/148, 63/219
brain blood flow index, 54/116, 119
brain blood volume index, 54/116, 119
brain death, 55/256
brain edema, 16/181, 53/135, 67/80
brain edema mechanism, 53/137
brain energy consumption, 53/116
brain injury, 57/209
brain lacunar infarction, 53/158
brain microcirculation, 23/146
brain oxygen metabolic rate, 53/116, 63/210, 219
brain perfusion pressure, 53/159
brain swelling, 53/126, 57/126
brain thromboangiitis obliterans, 53/163
brain tissue pressure, 53/136
brain volume, 53/135
brain water content, 53/135, 137
calcium, 53/125, 63/211
calcium antagonist intoxication, 65/441
calcium channel blocking agent, 63/222
carbon dioxide tension, 53/127
carbon monoxide intoxication, 53/167
cardiac amyloidosis, 53/156
cardiac arrest, 53/157, 63/209
cardiac dysrhythmia, 53/157, 63/236
cardiac valve prosthesis, 53/156
cardiac valvular disease, 53/155
cardiomyopathy, 53/156
carotid dissecting aneurysm, 53/164
carotid endarterectomy, 53/165
carotid sinus compression, 53/165
cascade, 54/120
cat scratch disease, 53/162
cause, 53/155
cell membrane depolarization, 53/118
cerebral malaria, 53/162
chiropractic manipulation, 53/165
chronic, *see* Ischemic encephalopathy
circadian rhythm, 55/219
classification, 53/309
coma duration, 63/214
computer assisted tomography, 57/161, 63/218
computer simulation, 53/109
conducting vessel, 53/114
congenital heart disease, 53/157
cortical steady potential, 53/124
creatine kinase BB, 63/218
critical degree, 54/121
critical duration, 54/121
cryoglobulinemia, 53/166
decompression illness, 53/160
delayed cerebral radiation necrosis, 53/164
delta activity, 54/157
demyelination, 63/214
density, 53/107

diagnosis, 63/218
diamorphine, 53/164
diaschisis, 53/120
dissecting aorta aneurysm, 63/46
distribution, 53/107
drug induced hypotension, 65/433
EEG, 11/272-276, 54/159, 55/219, 63/219
electrolyte shift, 53/119
endocarditis, 53/156
energy metabolism, 27/29-31
ergotamine, 53/164
essential thrombocythemia, 53/166
evoked potential, 53/123
experimental model, 27/27-29
extracellular fluid, 53/125
extracerebral steal, 53/115
Fabry disease, 53/165
fat embolism, 53/161
fibromuscular dysplasia, 11/383, 53/163, 55/283
flow rate dependency, 54/116
foreign body embolism, 53/161
free radical, 53/130
free radical scavenger, 63/223
functional threshold, 53/117
fungal vasculitis, 53/161
general pathology, 11/128
glial swelling, 53/136
glucose utilization, 53/129
glutamic acid, 63/211
glyceryl trinitrate intoxication, 65/443
head injury, 23/43
head injury outcome, 57/370, 386
headache, 48/280
heart disease, 53/155
heart tumor, 53/157
hemispheric events, 53/311
herpes zoster ophthalmicus, 53/161
heterogenous flow, 53/108
homocystinuria, 53/165
hyperglycemia, 53/128, 167
hypertensive encephalopathy, 53/158
hyperviscosity syndrome, 53/165
hypoglycemia, 53/167
hypotension, 53/158, 57/210
hypoxia, 53/167
iatrogenic procedure, 53/165
idiopathic hypertrophic subaortic stenosis, 53/156
intracerebral steal, 53/114
intracerebral steal syndrome, 53/159
intracranial hypertension, 53/159, 55/206
intracranial pressure, 55/206, 57/68
intravascular consumption coagulopathy, 53/156, 166
ion homeostasis, 53/126
ion shift, 53/125
irreversible type, 11/349
ischemic core, 53/119
ischemic penumbra, 53/119
isoelectric EEG, 55/219
isolated cerebral vasculitis, 53/162
isotope scan, 63/219
jugular catheterization, 53/167
Köhlmeier-Degos disease, 53/163
lactic acid, 53/128
lactic acidosis, 63/211
leukocytosis, 53/165
leukotriene, 53/131
Libman-Sacks endocarditis, 53/156
lidocaine, 53/116
lipid metabolism, 53/130
lupus coagulant, 53/166
lymphomatoid granulomatosis, 53/162
lysergide, 53/164
management, 63/221
marantic endocarditis, 53/156
membrane failure, 53/119
meningovascular neurosyphilis, 53/162
*N*-methyl-D-aspartic acid receptor blocking agent, 63/223
migraine, 53/163
mitral annulus calcification, 53/156
mitral valve prolapse, 53/155
morphologic injury, 53/119
mortality, 63/210
moyamoya disease, 42/749, 53/164, 55/294
myocardial infarction, 53/155
neurochemical study, 27/29-33
neurocysticercosis, 53/162
neuron death, 63/212
neuronal activity, 53/124
neuropeptide, 53/133
neurotransmitter metabolism, 53/131
nitrate intoxication, 65/443
no-reflow phenomenon, 63/210
nuclear magnetic resonance, 57/177, 63/218
occipital vertebral steal, 53/115
open heart surgery, 53/165
ophthalmic steal, 53/115
oral contraception, 53/164
ouabain, 53/116
oxidation reduction status, 53/127
oxygen consumption, 53/129
oxygen deprivation, 9/574
oxygen extraction, 53/114, 128
oxygen extraction rate, 54/116, 119
oxygen tension, 53/127

pallidal necrosis, 49/465, 63/213
pallidonigral necrosis, 49/465
pallidostriatal necrosis, 49/465
paradoxical embolism, 53/161
paraproteinemia, 53/166
paroxysmal nocturnal hemoglobinuria, 53/167
pathology, 23/57
pathophysiologic stage, 54/121
pathophysiology, 11/268, 54/115, 57/210
penetrating vessel, 53/114
penumbra, 53/117
phencyclidine addiction, 55/522
pheochromocytoma, 39/501
polyarteritis nodosa, 39/302, 53/162
polycythemia, 53/165
positron emission tomography, 53/267
posterior communicating artery events, 53/310
postischemic hyperemia, 63/210
potassium, 53/125
prognosis, 63/214, 220
progressive systemic sclerosis, 53/163
prostacyclin, 53/130
protein C deficiency, 53/167
protein synthesis, 53/133
pseudoxanthoma elasticum, 53/165
radiodiagnosis, 57/161
recovery time, 63/214
reduced nicotinamide adenine dinucleotide, 53/127
regional brain blood flow, 54/118
regional threshold, 53/267
reperfusion, 53/136
residual rate, 53/116
retinal events, 53/309
reversible type, 11/349
rheumatic heart disease, 53/155
rheumatoid arthritis, 53/163
segmental vascular resistance, 53/114
sickle cell anemia, 53/166, 55/464
Sjögren syndrome, 53/163
Sneddon syndrome, 55/408
sodium, 53/125
sodium-potassium adenosine triphosphatase pump, 63/211
subarachnoid hemorrhage, 53/163, 55/57
subclavian steal, 53/115
subclavian steal syndrome, 53/159
symptom, 53/309
systemic lupus erythematosus, 53/156
systemic neoplastic disease, 53/167
Takayasu disease, 53/162
temporal arteritis, 48/314, 53/162
threshold, 53/116
threshold regional cerebral metabolic rate for oxygen, 54/118
thrombocytosis, 53/166
thrombotic thrombocytopenic purpura, 53/166
tissue osmolality, 53/138
total brain volume, 53/126
transient, *see* Transient ischemic attack
trauma, 53/165
traumatic, 23/57
treatment, 63/221
tuberculous arteritis, 53/161
tuberculous meningitis, 52/200
tumor embolism, 53/161
typhus, 53/162
ulcerative colitis, 53/163
vascular epilepsy, 15/302
vascular spasm, 53/163
vasogenic brain edema, 53/138
venous infarction, 53/160
venous obstruction, 53/166
watershed infarction, 53/159
Wegener granulomatosis, 53/162
Willis headache, 48/281
Brain lacunar infarction
*see also* Brain infarction
abducent nerve paralysis, 54/244
abulia, 53/395
action induced dystonia, 54/247
amnesia, 53/395
anterior choroidal artery, 54/236
anterior striatal artery, 54/236
antiphospholipid antibody, 54/240
arterial hypertension, 54/203, 63/72, 75
arteriolar fibrinohyalinoid degeneration, 53/364, 54/236
arteriolar lesion, 63/75
asterixis, 54/247, 253
ataxia, 54/246
ataxic hemiparesis, 54/57, 247
basilar artery, 54/236
behavior, 53/395
bilateral capsular, 54/241
bilateral thalamic, 54/239
Binswanger disease, 54/224, 227, 229
blepharospasm, 54/245
brachiocrural hemiparesis, 54/243
brachiofacial hemiparesis, 54/242
brain angiography, 54/254
brain blood flow, 54/256
brain hemorrhage, 46/355, 54/289
brain infarction, 46/355, 53/318, 54/235
brain ischemia, 53/158
cause, 54/236, 239

cerebrovascular disease, 53/28
Charcot-Bouchard aneurysm, 54/239, 63/74
Claude syndrome, 54/245
clinical features, 46/355, 54/253
clumsy hand dysarthria syndrome, 46/355, 54/57, 249
cognition, 53/395
computer assisted tomography, 54/56, 255
CSF, 54/254
dementia, 54/253
distribution, 54/237
Doppler sonography, 54/255
dysarthria, 54/243, 253
dysphagia, 54/253
dystonia, 54/247
EEG, 54/254
embolism, 54/239
evoked potential, 54/254
facial paralysis, 53/210
facial paresthesia, 53/395
flaccid hemiparesis, 54/243
focal dystonia, 54/247
frequency, 54/257
gait disturbance, 1/199
gaze paralysis, 54/245
headache, 5/137
headache absence, 5/137
hemianesthesia, 46/355
hemiballismus, 54/246
hemicerebellopyramidal syndrome, 46/355
hemichorea, 54/246
hemiplegia, 46/355
hemodynamics, 53/183
hiccup, 54/253
homolateral ataxia crural paresis syndrome, 54/57
horizontal gaze, 54/246
Huebner artery, 54/236
hyperkinesia, 54/247
hyperkinetic mutism, 54/247
hypersomnia, 53/395
hypertensive encephalopathy, 11/559, 47/537, 54/213
incontinence, 54/253
internuclear ophthalmoplegia, 54/246
lacuna definition, 54/239
lacuna histology, 54/239
lacuna site, 54/236, 241
lacuna size, 54/236
lenticulostriate artery, 54/236
limb ataxia, 54/247
lingual paresis, 54/243
malignant lymphoma, 55/488
memory, 46/355
microatheroma, 54/236
Millard-Gubler syndrome, 54/244
monoplegia, 54/243
motor hemiparesis variant, 54/243
multiple myeloma, 55/487
mutism, 54/243
neuropathology, 46/355, 54/235
nuclear magnetic resonance, 54/255
oculomotor paralysis, 54/244
one and one half syndrome, 54/246
operculum syndrome, 54/243
oral contraception, 54/240
parkinsonism, 54/247
pathologic laughing and crying, 54/253
patient history, 54/254
perforating artery, 54/237
physiopathology, 54/235
polycythemia, 63/251
positron emission tomography, 54/124, 256
predilection site, 53/187
pseudobulbar syndrome, 46/355, 54/243, 253
pure motor hemiplegia, 53/364, 54/57, 240
pure sensory stroke, 53/395, 54/57, 250
pure supranuclear facial palsy, 54/242
sensorimotor stroke, 54/251
short step gait, 54/253
single photon emission computer assisted tomography, 54/256
site, 54/56
small infarction, 63/75
striatal necrosis, 49/500, 502
striatum, 2/499, 6/673
subacute vaginalitis, 54/239
syphilis, 54/240
thalamic infarction, 54/239
thalamic syndrome (Dejerine-Roussy), 54/250
thalamogeniculate artery, 54/236
thalamoperforating artery, 54/236
thalamus, 2/486, 6/673, 54/58
topography, 54/237
transient ischemic attack, 53/264, 54/241, 253
tuberothalamic artery, 54/236
vascular dementia, 46/355
vasculitis, 54/240
verbal perseveration, 53/395
vertebrobasilar system syndrome, 53/394
vertical gaze palsy, 53/395, 54/245
Weber syndrome, 54/244
white matter, 55/488
Brain lipoma
epilepsy, 72/181
Brain localization, *see* Cerebral localization
Brain lymphoma

biopsy, 67/228
heart transplantation, 63/10
Brain maldevelopment
clumsiness, 46/162
Minamata disease, 64/419
Brain malignant reticulosis, *see* Primary CNS lymphoma
Brain matrix hemorrhage
*see also* Brain periventricular infarction
cerebrovascular disease, 53/30
periventricular infarction, 53/30
premature infant, 53/30
striothalamic vein, 53/30
venous rupture, 53/30
vitamin K deficiency, 53/30
Brain maturation
behavior disorder in children, 4/340, 468
glia cell, 30/71
handedness, 4/255
higher nervous activity disorder, 4/364
Brain metabolic rate for oxygen
regional, *see* Regional brain metabolic rate for oxygen
Brain metabolism
acetylcholine, 57/79
anion exchange, 57/74
anoxia, 24/602
arachidonic acid, 57/81
brain blood flow, 57/69
brain concussion, 23/439
brain edema, 57/69
brain infarction, 11/348-355
brain perfusion pressure, 57/66
brain scanning, 16/687
calcium, 57/76
calcium antagonist, 57/75
calcium channel, 57/77, 80
calcium channel blocking agent, 57/77
carbohydrate, 27/1-23
cholinergic system, 57/82
coma, 3/70
CSF lactic acid, 57/72
endorphin, 57/83
β-endorphin, 57/83
energy, 27/1-23
enzyme reaction, 57/75
excess lactate, 57/72
excitatory aminoacid, 57/79
excitatory aminoacid antagonist, 57/81
fluid percussion model, 57/66, 86
free fatty acid, 57/82
free radical, 57/81
GABA, 27/17
ganglioside, 57/84
glia promoting factor, 57/85
glucose, 54/110, 112, 57/71
glucose utilization, 57/71
glutamic acid, 57/80
glycolysis, 57/75
hypotension, 57/67
hypovolemic shock, 57/67
hypoxic, 57/65
ibuprofen, 57/82
intracellular acidosis, 57/72
intracellular pH, 57/72
intracranial pressure, 23/205, 57/67
ionic alteration, 57/73
isolated head perfusion, 27/28
kallikrein-kinin, 57/82
kynurenate, 57/81
lactacidosis, 24/601, 604
lactic acid, 57/70
lactic acid accumulation, 57/73
lipid metabolism, 57/81
magnesium, 57/74
*N*-methylaspartic acid receptor, 57/80
naloxone, 57/83
nerve growth factor, 57/84
neurite promoting activity, 57/85
neurotransmitter, 57/78
nimodipine, 57/77
nuclear magnetic resonance spectroscopy, 57/71
oxygen, 54/110, 121, 125, 57/70, 211, 463
phencyclidine, 57/81
physiology, 55/205
potassium, 57/73
protein induction, 57/85
protirelin, 57/84
shock, 57/66
sodium, 57/73
sodium-potassium adenosine triphosphatase pump, 57/73
steroid compound, 57/82
vasospasm, 57/76
viral infection, 56/79
Brain metastasis, 18/201-230
acute onset, 18/216
age, 18/202
angiography, 18/224
anticonvulsant, 71/626
Balint syndrome, 45/16
blood examination, 18/221
brain angiography, 18/224
brain biopsy, 16/724, 726
brain edema, 47/534
brain scanning, 18/222

breast cancer, 18/204, 69/191
bronchopulmonary cancer, 18/204
cadherin, 69/111
calcium binding protein, 69/109
calvarial, 69/123
cancer, 69/1
carcinoma, 17/413, 421, 426
CD44v, 69/111
cell adhesion, 69/108
cell adhesion molecule, 69/111
cell surface receptor, 69/110
chemotherapy, 18/229, 69/201, 71/629
child, 18/309, 322
choriocarcinoma, 18/205
clinical features, 18/215
colon cancer, 69/107
consciousness, 45/120
cortical nodular type, 18/207
corticotropin, 18/229
course, 18/218
cranial nerve, 18/207, 211
CSF, 18/220
CSF cytology, 16/404
curtain loop, 18/224
dementia, 46/386
development, 69/109
diagnosis, 18/219, 69/138
distribution, 69/115, 136
dural, 69/123, 71/612
EEG, 18/221
encephalopathy, 45/119
endocrine gland tumor, 18/205
Entamoeba histolytica, 69/115
epidemiology, 16/66, 71/611
epilepsy, 18/216, 72/182
evaluation, 71/623
Ewing sarcoma, 69/124
frequency, 18/201, 69/1, 135
genital tract tumor, 18/205
glucocorticoid, 69/139, 71/625
gradual onset, 18/216
growth factor, 69/109
growth factor receptor, 69/109
guanine nucleotide binding protein, 69/109
headache, 18/216
hemiballismus, 49/373
hemiplegia, 18/216
Hemophilus influenzae, 69/116
histology, 18/211
immunologic privilege, 18/215
incidence, 67/134
integrin receptor, 69/109, 111
intercellular adhesion molecule 1, 69/111
interstitial brachytherapy, 69/144
intestinal tract tumor, 18/204
invasion process, 69/106
jugular foramen syndrome, 71/612
kidney cancer, 18/204
latency period, 18/228
leptomeningeal, 71/612
leukocyte, 69/112
liver metastasis, 69/107
lung cancer, 69/221
lung carcinoma, 71/613
lymphocyte, 69/112
lytic enzyme, 69/111
macroscopic aspect, 18/207
malaria, 69/117
malignancy, 18/227
melanoma, 18/204, 69/112
meningocortical type, 18/207, 209
mental symptom, 18/217
methotrexate, 69/201
middle fossa syndrome, 71/612
miliary carcinosis type, 18/207, 209
mirror metastasis, 18/227
mode of onset, 18/215
molecular alteration, 69/108
molecular mechanism, 69/105
motility factor, 69/109
motility factor receptor, 69/109
multiplicity, 18/221
myofibroblast, 69/112
Neisseria meningitidis, 69/116
neuroradiology, 18/223
nodular stain, 18/224
nonreceptor kinase, 69/109
number of metastasis, 18/206
occipital condyle syndrome, 71/612
oncogene, 69/108
onset, 69/108
orbital syndrome, 71/612
organ specificity, 69/114
osteosarcoma, 18/205
parasellar syndrome, 71/612
parenchymal, 71/612
pathogenesis, 18/214
pathophysiology, 71/624
pituitary tumor, 17/413, 421, 426
Plasmodium falciparum, 69/116
pregnancy, 16/73
prevalence, 69/2
primary cardiac tumor, 63/93
primary site, 18/227
prognosis, 18/226, 228
prophylactic cranial irradiation, 69/222

protease, 69/109
pseudoglioblastic, 18/225
pseudomeningiomatous form, 18/207, 209, 224
pseudovascular course, 18/219
radiography, 71/622
radionuclide diagnosis, 18/222
radiosurgery, 69/143
radiotherapy, 18/229, 512, 69/140, 193, 71/626, 629
recurrent, 69/145
reflex, 18/218
route of metastasis, 18/215
sarcoma, 18/205
sex ratio, 18/203
single, 18/227
site, 18/206
skin cancer, 18/205
skull base, 69/123, 71/612
small cell lung cancer, 69/221
spinal cord metastasis, 69/191
steroid, 69/194
Streptococcus pneumoniae, 69/116
subarachnoid hemorrhage, 12/120, 18/216
surgery, 69/141, 196, 71/627
surgical mortality, 18/226
survey, 69/135, 191
survival, 18/227
symptom, 69/137, 72/182
thalamic dementia, 18/218
therapeutic indication, 18/226
tissue culture, 17/85
transcription factor, 69/109
treatment, 69/139, 195, 200, 222, 71/625
tumor suppressor gene, 69/108
vascular cell adhesion molecule 1, 69/111
vasculotropin, 69/113
very late antigen 4, 69/111
whole brain radiation therapy, 69/142
Brain microabscess
candidiasis, 52/398
coccidioidomycosis, 52/412
Serratia marcescens, 52/298
Brain microaneurysm, *see* Charcot-Bouchard aneurysm
Brain microangiopathy
ataxia, 55/381
dementia, 55/381
female, 55/381
mineralizing, *see* Mineralizing brain microangiopathy
purpura, 9/598
retinal infarction, 55/381
Brain microcirculation, 53/79-83
acetylcholine, 53/79
adenosine, 53/80
autoregulation, 23/138, 142, 144
brain blood flow, 53/79
brain edema, 23/134, 139
brain injury, 23/133, 139
brain ischemia, 23/146
brain perfusion pressure, 23/138, 145
carbon dioxide, 53/79
cerebral malaria, 52/368
chemical control, 53/80
cholecystokinin, 53/79
definition, 53/79
endothelial regulation, 53/81
endothelium, 53/80
free radical, 53/81
functional importance, 53/79
histamine, 53/80
innervation, 53/79
metabolic control, 53/79
neurogenic control, 53/80
neuropeptide Y, 53/79
noradrenalin, 53/79
oxygen, 53/79
pial arteriole, 53/79
plasma skimming, 53/82
prostacyclin, 53/80
segmental vascular resistance, 53/79
serotonin, 53/79
traumatic, 23/8, 133-135, 138, 142, 144
vascular spasm, 53/83
viscosity, 53/82
volume pressure response, 23/140, 142
Brain microembolism
air embolism, 9/588, 63/175
calcium aggregate, 9/588
cerebrovascular occlusive disease, 9/588
chylomicron, 9/588
fat embolism, 9/588
open heart surgery, 63/175
Brain microgliomatosis, *see* Primary CNS lymphoma
Brain microinfarction
leukemia, 63/345
toxic encephalopathy, 52/297
Brain midline cyst
dysgenetic, 68/315
Brain miliary necrosis
acquired toxoplasmosis, 35/123, 52/357
Brain mucormycosis
diamorphine intoxication, 65/355
mortality, 35/545, 52/475
Brain neuroblastoma

classic primitive neuroectodermal tumor, 68/221
Brain oxygen extraction rate
determinant factor, 54/117
Brain oxygen metabolic rate
anoxic encephalopathy, 63/219
brain infarction, 53/116
brain ischemia, 53/116, 63/210, 219
cardiac arrest, 63/210
congenital heart disease, 63/10
Brain peduncle
pathologic laughing and crying, 3/362, 45/223
Brain perfusion pressure
*see also* Brain hypoperfusion
autoregulation, 54/120
brain blood flow, 54/118, 146, 55/206, 57/384
brain edema, 23/138, 67/76
brain infarction, 53/159
brain injury, 23/145
brain ischemia, 53/159
brain metabolism, 57/66
brain microcirculation, 23/138, 145
cardiopulmonary bypass, 63/188
congenital heart disease, 63/9
coronary bypass surgery, 63/188
diazoxide intoxication, 37/433
head injury, 57/226
heart transplantation, 63/188
iatrogenic neurological disease, 63/177
intracranial hypertension, 53/159
intracranial pressure, 57/66, 130, 300, 384
open heart surgery, 63/188
penetrating head injury, 57/300
Reye syndrome, 56/161
subclavian steal syndrome, 53/159
Brain periventricular infarction
*see also* Brain matrix hemorrhage
premature infant, 53/30
Brain phlebothrombosis, 12/422-444
*see also* Brain infarction *and* Brain thrombophlebitis
adult, 12/434
benign intracranial hypertension, 12/425
central cerebral vein, 54/399
cerebrovascular disease, 53/37
computer assisted tomography, 54/65
diagnosis, 54/440
EEG, 11/278
Galen vein, 54/399
headache, 5/154
hemorrhagic necrosis, 54/438
history, 12/423
incidence, 12/422
infant, 12/429
internal cerebral vein, 54/399
Labbé vein, 54/399
nuclear magnetic resonance, 54/445
otitic hydrocephalus, 12/425-429
pathogenesis, 12/424
pathology, 12/429
pregnancy, 11/459
prognosis, 12/442
protein S deficiency, 54/410
radiology, 12/439
Rolando vein, 54/398
sickle cell trait, 55/466
space occupying lesion, 16/235
subarachnoid hemorrhage, 12/170, 55/3, 20
superior longitudinal sinus, 12/427
thalassemia, 55/466
transverse sinus thrombosis, 12/425
treatment, 12/443, 54/445
Brain regeneration
brain injury, 16/32
brain tumor origin, 16/32
Brain rete mirabile, *see* Moyamoya disease
Brain sarcoma
cerebellar, 16/20
CSF cytology, 16/400
fibroblastic type, 16/23
fibrosarcoma, 16/20
monstrocellular sarcoma, 16/20
polymorphic form, 16/23
Brain scanning
*see also* Brain scintigraphy
arteriovenous malformation, 11/244
benign intracranial hypertension, 16/153
bismuth, 16/666
brain arteriovenous malformation, 12/242, 31/182
brain glioma, 16/688
brain infarction, 11/243
brain metabolism, 16/687
brain metastasis, 18/222
brain tumor, 16/661
cerebral paragonimiasis, 35/257
cerebrovascular disease, 11/240-247
child, 16/692
choroid plexus papilloma, 17/579
clinical use, 11/242
di-iodofluorescein I 131, 16/663
fluorescein photography, 11/250-252
frontal lobe tumor, 17/264
gamma-emitting isotope, 16/669
glioblastoma multiforme, 16/685
instrumentation, 11/241
iodide, 16/664
lateral ventricle tumor, 17/600

meningioma, 16/684
mercury compound, 11/407, 16/670
neurinoma, 16/692
partition study, 11/245
pertechnetate, 16/673
phosphate, 16/663
pituitary adenoma, 17/405
pontine tumor, 17/699
position detection, 11/242
positron emitting isotope, 16/666
postoperative, 16/695
potassium, 16/664
progressive multifocal leukoencephalopathy, 9/495
radioisotope, 11/249, 269, 16/663
radiopharmaceuticals, 11/240
rheoencephalography, 11/252
RIHSA, 16/664, 669
scintillation, 11/247-249
sellar tumor, 16/690
spongy cerebral degeneration, 66/665
subarachnoid hemorrhage, 12/157
supratentorial brain abscess, 33/124
supratentorial brain tumor, 18/334
third ventricle tumor, 17/459
tuberculous meningitis, 33/224
vital dye, 16/661
Brain scintigraphy
*see also* Brain scanning
acute subdural hematoma, 24/286
arachnoid cyst, 23/307
brain contusion, 23/288, 292
brain death, 24/771
brain embolism, 53/232
brain glioma, 18/29-33
brain infarction, 53/232
brain injury, 23/246, 287, 294
carotid cavernous fistula, 23/293
cerebrovascular disease, 11/247-252, 53/232
chronic subdural hematoma, 24/309
craniopharyngioma, 18/547
CSF fistula, 23/310, 24/190
CSF flow, 23/294
ependymoma, 18/130
epidural hematoma, 24/272
glioblastoma multiforme, 18/55
hydrocephalus, 23/294
intracerebral hematoma, 24/362
intracranial hemorrhage, 53/232
normal pressure hydrocephalus, 23/300, 302
oligodendroglioma, 18/99
penetrating brain injury, 23/288, 294
porencephaly, 23/307
RIHSA cisternography, 23/301, 304
subdural effusion, 24/336
subdural hematoma, 23/291, 297, 299
transient ischemic attack, 53/232
traumatic hydrocephalus, 24/239, 248
Brain shape
Down syndrome, 50/531
Brain size
microcephaly, 42/46
Brain stab wound
angiography, 23/493
arterial vascular injury, 23/485, 499
brain infection, 23/483
brain injury, 23/477
causal instrument, 23/478
cranial nerve, 23/485
CSF fistula, 23/485
epilepsy, 23/484
intracranial hematoma, 23/482
pathology, 23/479
pneumocephalus, 23/492
site, 23/479
statistic, 23/479
surgery, 23/494
transorbital, 23/489, 496
traumatic aneurysm, 23/488
Brain stem
acute amebic meningoencephalitis, 52/326
arterial system, 11/24-43
autonomic nervous system, 74/61, 71
basilar artery migraine, 5/42
Behçet syndrome, 56/595
brain abscess, 33/139
brain arteriovenous malformation, 2/257, 259, 54/371, 373
brain death, 24/739, 763, 817
brain injury mechanism, 57/26
cat scratch disease, 52/130
cirsoid aneurysm, 14/70
coma, 3/56
consciousness, 3/171
diffuse axonal injury, 57/53
dorsal zone artery, 11/30
echinococcosis, 52/524
flavivirus, 56/135
Friedreich ataxia, 60/302
glioblastoma multiforme, 18/54, 397
hemiplegia, 1/186
hemorrhage, *see* Brain stem hemorrhage
hereditary cerebello-olivary atrophy (Holmes), 21/407
hereditary olivopontocerebellar atrophy (Menzel), 21/438

Huntington chorea, 49/320
hydrocephalus, 30/673, 32/531
iniencephaly, 50/131
intrinsic arterial pattern, 11/39-41
Japanese B encephalitis, 56/138
Klüver-Bucy syndrome, 45/260
Langley ganglion, 22/245
lateral zone artery, 11/30
lead, 64/434
leukemia, 39/12
Listeria monocytogenes meningitis, 33/85
malabsorption syndrome, 70/232
median zone artery, 11/30
migraine, 5/83
multiple sclerosis, 9/207
myoclonia, 6/775
nystagmus, 1/590, 595
olivopontocerebellar atrophy (Dejerine-Thomas), 21/420
organolead intoxication, 64/132
osteitis deformans (Paget), 38/361
paramedian zone artery, 11/30
phantom sensation, 45/400
posterior fossa tumor, 63/230
posttraumatic syringomyelia, 26/126
pseudobulbar paralysis, 2/250
radiotherapy, 18/503
schizophrenia, 46/509
sincipital encephalocele, 50/103
sleep, 3/81
spongy cerebral degeneration, 22/195
status verrucosus, 50/202
systemic lupus erythematosus, 39/281, 71/44
vascularization, 11/24-43, 95, 12/24
Wolman disease, 14/779
yellow jacket intoxication, 37/107
Brain stem abscess
autopsy finding, 33/139
excision, 33/139
middle fossa, 33/139
surgical drainage, 33/139
Brain stem artery
circumferential, *see* Circumferential brain stem artery
Brain stem audiometry
vestibular neurinoma, 68/427
Brain stem auditory evoked potential
acetonylacetone intoxication, 64/87
adrenoleukodystrophy, 47/597
adrenomyeloneuropathy, 60/171
alcohol intoxication, 64/115
ataxia telangiectasia, 60/366
Bassen-Kornzweig syndrome, 63/275
brain embolism, 53/233
brain infarction, 53/233
carbon disulfide intoxication, 64/25
central pontine myelinolysis, 47/588, 63/553
cerebellar infarction, 54/163
cerebrovascular disease, 53/233
Charlevoix-Saguenay spastic ataxia, 60/455
chorea-acanthocytosis, 63/282
Down syndrome, 50/529
EEG, 54/160
extra pontine myelinolysis, 63/553
head injury, 57/193
head injury outcome, 57/195
head injury prognosis, 57/195
human T-lymphotropic virus type I associated myelopathy, 56/538
Huntington chorea, 49/290
intracranial hemorrhage, 53/233
juvenile Gaucher disease type B, 66/128
labyrinthine ischemia, 55/131
lead polyneuropathy, 64/437
locked-in syndrome, 54/162
Meniere disease, 47/140
mesencephalic infarction, 53/393
middle cerebral artery infarction, 54/162
multiple sclerosis, 47/133, 137-140, 142
neurologic intensive care, 55/217-219
pontine infarction, 53/393
pontine ischemia, 54/162
prognostic value, 57/195
progressive dysautonomia, 59/139
transient ischemic attack, 53/233
tropical spastic paraplegia, 56/538
uremic encephalopathy, 63/508
vertebrobasilar transient ischemic attack, 54/162, 164
Brain stem compression
forgotten respiration, 63/490
Brain stem concussion
skull fracture, 23/459
Brain stem death
*see also* Brain death
acute intoxication, 57/471
American Collaborative Study, 57/454, 485
American President Commission, 57/456
angiography, 57/460
apnea, 57/473, 476
apnea definition, 57/455
apnea determination, 57/477
apnea testing, 57/454
ascending reticular activating system, 57/459
asystole, 57/483
basic mechanism, 57/441

brain blood flow, 57/460
brain death, 55/259, 57/259, 444, 448, 452
brain stem reflex, 57/452
bronchial stimulation reflex, 57/473
cardiac arrest, 57/441, 466
cardiorespiratory function, 57/446
cause, 57/455
cerebral pulsation, 57/464
child, 57/479
church tradition, 57/447
Circulaire Jeanneney, 57/486
clinical concept, 57/443
clinical diagnosis, 57/470
clinical judgment, 57/453
cognitive death, 57/444
computer assisted tomography, 57/464
concept, 57/445
corneal reflex, 57/458, 473
cranial nerve, 57/474
cultural argument, 57/468
death criteria, 57/442, 445
death declaration, 57/478
death diagnosis, 57/449
defining death, 57/457
diabetes insipidus, 57/483
diagnosis, 57/444, 460, 479
digital subtraction angiography, 57/464
drug intoxication, 57/454
early account, 57/446
EEG, 57/451, 463
eeg artefact, 57/466
electrocerebral silence, 57/464
endocrine disorder, 57/471
evoked potential, 57/469
exclusion, 57/453, 471
exogenous intoxication, 57/471
French contribution, 57/447
gag reflex, 57/473
Harvard Criteria, 57/450
historic aspect, 57/446
hypothermia, 57/454, 471
hypoxic prevention, 57/475
instrumental diagnosis, 57/449
intensive care unit, 57/472
international development, 57/485
irremediable structural brain damage, 57/470
isoelectric EEG, 57/451
isotope angiography, 57/462
Kansas Statute, 57/456
lawyer, 57/469
Lazarus sign, 57/477
life support, 57/445
litigation, 57/468
Locus of the Soul, 57/447
Lyon Group, 57/448
metabolic coma, 57/455
metabolic disturbance, 57/454, 471
Minnesota Criteria, 57/452
mort cérébrale, 57/449
motor response, 57/474
muscle stretch reflex, 57/451, 476
neonate, 57/481
oculocephalic reflex, 57/458
oculovestibular reflex, 57/458
organ support, 57/445
oropharyngeal reflex, 57/458
oxygen, 57/463
pathology, 57/458
patient fate, 57/483
pediatric head injury, 57/479
Pendelfluss, 57/464
perforating pontine branch, 57/458
persistent apneic coma, 57/472
persistent vegetative state, 57/445
personal identity loss, 57/444
pitfall, 57/475
Pope Pius XII, 57/447
posterior fossa hematoma, 57/461
practice code, 57/450
precondition, 57/453, 470
primary, 57/458
primary infratentorial lesion, 57/458
primary supratentorial lesion, 57/458
prognostic significance, 57/454, 481
prolonged somatic survival, 57/482
pseudopitfall, 57/475
pupillary reflex, 57/458, 473
putrefaction, 57/487
radionuclide angiography, 57/463
radionuclide cerebral perfusion scintigraphy, 57/462
repeated testing, 57/478
repeating angiography, 57/462
repeating EEG, 57/481
residual EEG activity, 57/467
reticular formation, 57/459
reversible isoelectric EEG, 57/466
secondary, 57/458
shock, 57/471
short latency evoked potential, 57/469
single photon emission computer assisted tomography, 57/464
social death, 57/444
somatosensory evoked potential, 57/469
soul, 57/442
spinal cord soul, 57/477

spontaneous movement, 57/477
subarachnoid hemorrhage, 57/473
survival, 57/466
Sydney Declaration, 57/450
terminology, 57/442
test, 57/453
timing, 57/472
ultrasound, 57/464
United Kingdom Code, 57/453, 478
vasopressin, 57/484
vegetative state, 57/473
ventilating to asystole, 57/481
ventilator, 57/472
ventilator disconnection, 57/451
vestibulo-ocular reflex, 57/473, 476
whole brain, 57/459
whole brain death, 57/443
withdrawing ventilation, 57/482
xenon computer assisted tomography, 57/464

Brain stem disorder
carbamazepine, 47/57
multiple sclerosis, 47/56, 170
sleep apnea syndrome, 45/133

Brain stem dysfunction
infantile Gaucher disease, 10/307
prognosis, 24/677

Brain stem dysfunction syndrome
autonomic nervous system, 75/534

Brain stem encephalitis, 34/605-609
acute cerebellar ataxia, 34/632
ataxic respiration, 63/487
enterovirus 70, 56/351
Fazio-Londe disease, 22/107
gasping respiration, 63/487
Guillain-Barré syndrome, 51/246
Listeria monocytogenes, 33/77, 52/93
Listeria monocytogenes encephalitis, 52/93
Miller Fisher syndrome, 51/246
paraneoplastic syndrome, 69/355, 71/684
tumor, 18/407

Brain stem encephalocele
misonidazole polyneuropathy, 51/301

Brain stem excitability
hyperexplexia, 42/228

Brain stem glioma
brain tumor, 18/503
Cheyne-Stokes respiration, 63/440
Fazio-Londe disease, 22/108
hyperventilation, 63/440
imaging, 67/179
syringomyelia, 50/453

Brain stem hematoma
computer assisted tomography, 54/69
posterior fossa hematoma, 24/346
traumatic intracranial hematoma, 57/152

Brain stem hemorrhage
axial shift, 16/123
brain hemorrhage, 54/309
cerebellopontine lesion, 2/263
EEG, 11/280
expanding lesion, 16/122
forgotten respiration, 1/654
symptom, 2/255

Brain stem herniation
arousal disorder, 45/112
computer assisted tomography, 45/113
consciousness, 45/112

Brain stem hypoplasia
Patau syndrome, 43/527
trigeminal nerve agenesis, 50/214

Brain stem infarction
anterior inferior cerebellar artery syndrome, 53/396
apneusis, 1/498, 63/485
brain infarction, 2/249
central sleep apnea, 63/462
coma, 12/15
forgotten respiration, 1/654, 63/441, 488
Foville syndrome, 12/16
hypoventilation, 63/413
iatrogenic neurological disease, 63/177
internuclear ophthalmoplegia, 55/125
neurogenic pulmonary edema, 63/496
palatal myoclonus, 38/584
peduncular hallucination, 53/392
percutaneous transluminal coronary angioplasty, 63/177
pupil, 55/125
temporal arteritis, 48/314
visual evoked response, 54/160
visual hallucination, 55/146

Brain stem injury
*see also* Brain injury
birth, 25/164
central neurogenic hyperventilation, 63/484
coma, 3/66, 68, 75
nystagmus, 1/590, 595, 57/137
pathology, 23/47
radiation injury, 23/656

Brain stem ischemia
EOG, 54/164

Brain stem lacunar infarction
medulla, 53/394
pons, 53/394

Brain stem lesion
akinetic mutism, 3/56

central motor disorder, 1/186-189
corneomandibular reflex, 1/173, 239
decompression myelopathy, 61/216
dissociated nystagmus, 2/367
Millard-Gubler syndrome, 1/187
Parkinson disease, 49/113
vertical nystagmus, 16/322
Brain stem neoplasm
progressive bulbar palsy, 59/375
Brain stem reflex
brain concussion, 23/430
brain death, 55/258
brain stem death, 57/452
head injury outcome, 57/383
hyperexplexia, 42/228
progressive bulbar palsy, 22/113
return, 57/484
Brain stem response
evoked, *see* Evoked brain stem response
Brain stem reticular formation
gray matter, 3/64
pain, 45/236
Brain stem spasm, *see* Decerebrate rigidity
Brain stem tegmentum
hydranencephaly, 30/673, 675, 50/347
Brain stem tumor
astrocytoma, 18/503, 22/108
audiometry, 16/307, 321
axial shift, 16/123
central neurogenic hyperventilation, 63/441
central sleep apnea, 63/462
classification, 68/367
clinical syndrome, 68/368
cranial nerve palsy, 16/320
cystic astrocytoma, 68/370
diffuse, 68/368
dissociated nystagmus, 16/322
dorsal exohyptic tumor, 68/370
epidemiology, 68/366
exhophytic, 68/368
focal cystic, 68/368
focal tectal, 68/370
hiccup, 63/440
histopathology, 68/367
hyperpnea, 63/440
intractable hiccup, 63/490
neurogenic pulmonary edema, 63/496
nystagmus, 16/322
pontine tumor, 68/369
posterior fossa tumor, 18/397
radiation injury, 23/657
radiotherapy, 18/503
survey, 68/365
symptom, 68/366
tectal, 68/368
tone decay, 16/321, 17/672, 675
treatment, 68/368
vestibular system, 16/321
Brain stem valve
spina bifida, 32/531
Brain stem vascular accident
central neurogenic hyperventilation, 63/441
Brain stem weight
Down syndrome, 50/530
Brain stone, *see* Striatopallidodentate calcification
Brain surgery
aphasia, 45/297
epilepsy, 45/9
neuropsychology, 45/9
Brain swelling
bacterial meningitis, 52/8
brain infarction, 53/126
brain injury, 57/124
brain ischemia, 53/126, 57/126
computer assisted tomography, 57/159, 381
definition, 57/105
diffuse, 57/380
diffuse axonal injury, 57/380
focal, 57/380
head injury, 57/58, 105, 159
intracranial pressure, 57/58
mass effect, 57/105
nuclear magnetic resonance, 57/177
radiodiagnosis, 57/159
slit like ventricle, 57/160
traumatic intracerebral hematoma, 57/159
Brain thromboangiitis obliterans, 39/201-209, 55/307-314
age, 12/389, 55/308, 312
allergy, 12/392
angiography, 12/390, 39/203
arterial arachnoid ring, 55/308
arteriosclerosis, 12/392, 39/203
artery occlusion, 55/308
brain atrophy, 12/385, 39/202
brain infarction, 53/163
brain ischemia, 53/163
brain thrombosis, 55/310
cause, 12/392
central type, 55/309
clinical features, 12/388, 39/202, 55/312
computer assisted tomography, 39/203
constitution, 12/392
diagnosis, 12/386
differential diagnosis, 12/391, 55/311
distinctive criterion, 55/308

EEG, 12/390, 39/203
embolism, 12/393, 55/313
epidemiology, 55/312
immune response, 55/313
incidence, 12/388, 39/201
involved vessel size, 55/308
leukoencephalopathy, 9/613
meningeal anastomosis, 55/309
microscopic change, 55/311
multiple brain infarction, 55/310
pathogenesis, 39/208, 55/312
pathologic anatomy, 12/385, 55/309
PEG, 39/203
peripheral type, 55/309
prevalence, 55/309, 312
prognosis, 12/394, 39/209, 55/313
race, 55/313
sex ratio, 12/389, 39/203, 55/308, 312
smoking, 39/201, 55/308
Sneddon syndrome, 55/403
spatial predilection, 55/308
systemic brain infarction, 11/454
terminology, 12/385, 55/309
thrombus formation, 39/204, 55/310
tissue antigen, 55/313
treatment, 12/394, 39/209, 55/313
type, 12/385
vascular occlusion, 12/393

Brain thrombophlebitis, 12/422-444
*see also* Brain phlebothrombosis *and* Venous sinus thrombophlebitis
adult, 12/434
clinical features, 12/436
dural sinus, 12/422
history, 12/423
incidence, 12/422
infant, 12/439
intracerebral hematoma, 11/671
pathogenesis, 12/424
prognosis, 12/442
radiology, 12/439
treatment, 12/443

Brain thrombosis
*see also* Cerebrovascular disease *and* Stroke
age adjusted mortality rate, 53/8
benign intracranial hypertension, 16/157
brain thromboangiitis obliterans, 55/310
dural sinus, 12/422
general pathology, 11/130
headache, 5/126
homocystinuria, 55/328
paraproteinemia, 11/461
protein C deficiency, 63/328
protein S deficiency, 63/328
sickle cell anemia, 38/34
subdural effusion, 24/332
thrombophilia, 63/328
venous, 26/315

Brain tissue
acute amebic meningoencephalitis, 52/316
dystonia musculorum deformans, 49/526
epilepsy, 72/66
exencephaly, 50/99
granulomatous amebic encephalitis, 52/316
heterotopic, *see* Heterotopic brain tissue
multiple sclerosis, 47/80-83, 86, 103-106, 111

Brain trauma, *see* Brain injury

Brain tumor
*see also* CNS tumor *and* Nervous system tumor
abscess, 67/218
acoustic nerve, *see* Acoustic nerve tumor
acoustic system, 16/301
adenovirus vector, 67/300
adjunctive treatment, 68/192
adoptive transfer, 67/295
age, 67/131
age distribution, 16/27, 67, 69, 71, 73
alexia, 2/607, 45/438
angiography, 16/622
animal, 16/39
anorexia nervosa, 46/588
anticonvulsant, 69/14
antitumor immune response, 67/292
anxiety, 46/430
aphasia, 16/731, 46/150
apoptosis, 67/224
artificial, *see* Artificial brain tumor
astrocytic neoplasma, 67/4
astrocytoma, 18/1, 501
ataxia telangiectasia, 46/58
atrial myxoma metastasis, 63/99
auditory recognition test, 16/336
autonomic nervous system, 75/527
autoscopia, 45/387
awareness, 16/733
axillary angiography, 16/635
B-mitten pattern, 46/496
behavior, 16/728, 46/416
benign intracranial hypertension, 16/157
benign malignant, 16/6
biochemical marker, 71/652
biochemistry, 27/503-514
biology, 16/1
bitemporal hemianopia, 2/571
blood-brain barrier, 28/381, 388, 67/71
blood group, 16/76

blood oxygenation level dependent nuclear magnetic resonance, 67/237
brachial angiography, 16/637
brain abscess, 16/263, 67/218
brain arteriovenous malformation, 12/248
brain biopsy, 16/37, 708, 17/42, 67/219
brain edema, 16/176, 194, 67/80
brain hemorrhage, 31/166, 54/292, 69/420, 424
brain herniation, 67/145
brain infarction, 11/313, 69/414, 420
brain scanning, 16/661
brain stem glioma, 18/503
brain vasospasm, 23/178
bromouridine, 18/491, 520
carcinogen, 17/5, 9, 11, 13
cardiac dysrhythmia, 63/236
carmustine, 18/520
catecholamine, 74/314
cellular growth control, 67/303
central alexia, 45/440
central herniation, 67/145
cerebellum, 67/9
cerebrovascular disease, 69/413
chemical carcinogenesis, 16/34
chemotherapeutic agent, 67/283
chemotherapy, 18/517-522, 67/277
child, 67/133, 173, 400
childhood epilepsy, 16/264
choriocarcinoma metastasis, 63/104
choroid plexus papilloma, 18/512
chromosome, 67/44
chromosome study, 16/29
classification, 16/1, 4, 6, 8, 24, 31/37, 67/1, 68/137
clinical features, 31/40
clinical trial, 67/311
cognition, 16/729
cognitive dysfunction, 67/371
comparative neuropathology, 16/39
congenital, *see* Congenital brain tumor
consciousness, 16/130
convulsion, 31/44, 47, 50
convulsive seizure, 31/41
corpus callosum, *see* Corpus callosum tumor
corticoid secretion, 16/354
cough headache, 69/25
cracked pot sign, 18/325, 329
craniopharyngioma, 18/508
CSF, 16/360
CSF cytology, 16/371, 375, 380, 389
CSF diagnosis, 16/363, 366
CSF enzyme, 16/367
CSF fistula, 24/185
CSF glucose, 16/367
CSF lactate dehydrogenase, 16/368
CSF lactic acid, 16/370
CSF lipid, 16/366
CSF protein, 16/363
Cushing syndrome, 16/355
cytology, 17/43
cytostatic treatment, 18/39
death, 67/320
definition, 31/35-37
delirium, 46/545
dementia, 46/385
demyelination, 9/588
depression, 46/425, 461
diagnosed under 1 year, 31/38
differential diagnosis, 16/209, 31/41, 67/218
direct tumor sign, 20/213, 230
dizziness, 16/129
DNA analysis, 17/100
droplet metastasis, 20/244, 285
dystonia, 6/560
echoEG, 16/455
EEG, 16/418
electron microscopy, 67/221
electrophoresis, 16/365
embryonal, *see* Embryonal brain tumor
embryopathology, 31/50-68
endocrine disorder, 16/341, 345
environmental factor, 16/33, 58, 77
environmental risk, 67/135
ependymoma, 18/502, 42/731
epidemiology, 16/50, 60, 63, 31/39, 67/129
epilepsy, 15/295, 16/254, 259, 733, 18/19-22, 326, 72/175
evaluation, 67/315
exertional headache, 48/374
experimental, *see* Experimental brain tumor
extracerebral, 16/6
familial, 67/59
familial occurrence, 16/74
femoral angiography, 16/641
fluorouridine, 18/491
focal lesion, 67/139
focal tectal, 68/370
foramen magnum, *see* Foramen magnum tumor
fox intoxication, 37/96
frequency, 16/33, 57
frontal lobe, *see* Frontal lobe tumor
functioning scale, 67/312
gene delivery method, 67/299
gene therapy, 67/291, 299
general management, 18/457-469
genetics, 14/500, 16/28-30
geographic factor, 16/58, 79

geography, 67/131
glucocorticoid, 67/315
glycolytic enzyme, 27/508
gradation, 17/46
hairy cell leukemia, 18/235
hemiplegia, 1/193
herpes simplex virus, 67/300
herpes simplex virus encephalitis, 56/218
histochemistry, 16/36
histologic classification, 17/47
histopathology, 67/217
history, 16/1
Hodgkin disease, 18/234
Hodgkin lymphoma, 63/348
Hope-Stone method, 18/497
hospital admission, 16/58
hydrocephalus, 31/40, 46/324-326
hydrocortisone, 16/344
hypophyseal adenoma, 18/506
hypopituitarism, 16/347
hypothalamus, 16/343
imaging, 67/167, 173, 238, 313
immunoelectrophoresis, 16/364
immunogene therapy, 67/301
immunohistochemistry, 67/221
immunotherapy, 67/291, 294
inappropriate antidiuretic hormone secretion, 28/499
incidence, 16/57, 62
intelligence, 16/729
α-interferon, 67/298
β-interferon, 67/298
interleukin-2, 67/298
intracranial hemorrhage, 69/422
intracranial hypertension, 67/145
intraoperative monitoring, 67/250
Karnofsky score, 67/313
karyotype, 67/43
language, 16/731, 67/396
large series, 16/61, 63
lateral ventricle, *see* Lateral ventricle tumor
Li-Fraumeni syndrome, 67/54
localization, 67/235
location, 31/40, 72/178
lymphomatoid granulomatosis, 63/424
lymphoreticular, 72/180
maldevelopment, 68/309
mania, 46/430
marital status, 16/80
maternal age, 16/73
Mayo Clinic statistics, 16/61
medicolegal aspect, 16/32
medulla oblongata, 17/693
memory disorder, 16/732, 46/623
meningioma, 18/506
mental disorder, 46/415
metabolism, 69/395
metastasis, *see* Brain metastasis
metastatic cardiac tumor, 63/93
methotrexate, 18/491, 520, 39/98
midbrain, *see* Midbrain tumor
midline, *see* Supratentorial midline tumor
mithramycin, 18/519
mitosis, 17/87, 90, 92
molecular genetics, 67/41
mortality, 16/59
Müller classification, 16/1
multinucleation, 17/94
multiple, *see* Multiple brain tumor
mummy, 74/187
mycosis fungoides, 18/235
nausea, 67/143
necropsy rate, 16/58
neglect syndrome, 45/175
neurobehavior, 67/395
neurocognitive effect, 67/395, 400
neurocysticercosis, 16/225
neuroendoscope, 67/249
neuroepidemiology, 16/56, 60
neurofibromatosis type I, 14/136
neuroimmunology, 67/293
neurologic examination, 67/312
neurophysiologic test, 67/312
neuropsychology, 45/9
neurosurgery, 67/235
neurovascular complication, 69/415
nomenclature, 16/1-44
nonhereditary, 67/43
non-Hodgkin lymphoma, 63/348
nuclear magnetic resonance, 67/235
nuclear size, 17/94
nucleus, 17/95
occipital lobe, *see* Occipital lobe tumor
oligodendroglioma, 18/502
oncogenic virus, 17/4
ontogeny, 67/1
optic nerve glioma, 18/503
optic pathway, *see* Optic pathway tumor
origin and development, 16/24
orthostatic hypotension, 63/155
pain, 69/31
papilledema, 67/143
parietal lobe, *see* Parietal lobe tumor
Parkinson disease, 6/240
Paterson-Farr scheme, 18/497
pathogenesis, 16/27

pathology, 67/222
PEG, 16/530
perception, 16/732
personality disorder, 16/735
pilot trial, 67/317
pineal gland, *see* Pineal tumor
pinealoma, 16/64, 18/508
pituitary gland, 16/342, 344
plasma cell dyscrasia, 71/438
polar bear intoxication, 37/96
pons, *see* Pontine tumor
posterior fossa, *see* Posterior fossa tumor
pregnancy, 16/76
prevalence, 16/58, 67
prognosis, 31/41, 67/319
psychiatry, 16/727
quality of life, 67/395, 397
race, 16/76, 67/131
radiation, 16/40, 76, 80
radiographic diagnosis, 68/95
radioisotope diagnosis, 45/298
radiologic appearance, 31/41
radiosensitivity, 18/485
radiosensitizer, 18/491
radiotherapy, 18/1, 39, 481-512, 67/273
randomization, 67/319
regenerative origin, 16/31
regional epidemiology, 16/64, 79
rehabilitation, 18/523
RES, *see* RES brain tumor
retrovirus vector, 67/300
RIHSA, 18/29
sample, 67/317
SBLA syndrome, 42/769
schizophrenia, 46/461
schizophreniform psychosis, 46/420
seizure, 67/143, 69/10
septum pellucidum, 17/539
serial EEG records, 16/429
sex incidence, 16/27, 70, 73
sex ratio, 67/131
skull base, *see* Skull base tumor
spheroid body, 6/626
spinal glioblastoma, 19/361
splenio-occipital alexia, 45/444
spontaneous, 16/39
statistic, 16/61, 67/319
status epilepticus, 16/261
stereotactic localization, 67/240
striatopallidodentate calcification, 49/423
subangular alexia, 45/444
subarachnoid hemorrhage, 12/118, 169
subclavian angiography, 16/633
subclavian herniation, 67/145
subdural hematoma, 69/424
suicide gene, 67/301
suppressor gene, 67/44
supratentorial, *see* Supratentorial brain tumor
surgery, 18/471-480, 69/15
survival distribution, 67/394
symptom, 67/140, 142
symptomatic dystonia, 49/543
syndrome type, 18/483
temporal lobe, *see* Temporal lobe tumor
terminology, 16/1-44
thalamic syndrome, *see* Thalamic syndrome (Dejerine-Roussy)
thalamus, *see* Thalamic tumor
third ventricle, *see* Third ventricle tumor
thymidine kinase, 67/301
time-dose effect plot, 18/483
tissue culture, 16/37, 17/42, 45-48
tissue diagnosis, 67/217
TNM system, 18/488
topistic aspect, 16/26
transgene system, 67/301
transient global amnesia, 45/211
transplantation, 17/2
traumatic origin, 16/31, 78
treatment, 31/41, 67/303, 316
tuberous sclerosis, 14/45, 380, 18/297, 299-303, 322
tumor necrosis factor-α, 67/298
tumor neovascularization, 67/302
tumor recognition, 67/292
twin study, 16/29
UICC classification, 16/9
UICC nomenclature, 16/9
ultrasound, 67/239
ultrastructure, 16/35
vascular, *see* Brain vascular tumor
vascular disease, 69/413
vascular tumor, 68/269
venous occlusion, 69/421
vertebral artery angiography, 16/632
vertigo, 16/129
vincristine, 18/519
viral gene delivery, 67/299
virus, 16/40
visual perception, 67/396
vomiting, 16/129, 67/143
wolf intoxication, 37/96
World Health Organization classification, 16/24, 68/166
Brain tumor carcinogenesis
*see also* Brain tumor origin

acetylaminofluorene, 17/10
adenovirus type 7, 18/133
carcinogen, 17/9, 11, 13
carcinoma, 38/625
cerebellar medulloblastoma, 18/170
chemical, 16/34, 41
dimethylnitrosourea, 17/18
ependymoma, 18/132
ethylnitrosourea, 17/19
experimental brain tumor, 16/40-43
hormone, 16/40
hydrocarbon, 17/6-8
methylnitrosourea, 17/11, 14, 16, 21
nutrition, 17/28
oncogenic virus, 17/4
pellet, 17/6, 9
phenyldimethyltriazene, 17/20
radiation, 16/40
rubber, 17/28
spontaneous, 16/39
trimethylnitrosourea, 17/16
virus, 16/34, 40

Brain tumor edema
*see also* Brain edema
blood-brain barrier, 67/71
peritumoral, 67/79

Brain tumor headache
brain hemorrhage, 5/141
cause, 69/23
chemotherapy, 69/23
child, 48/33
duration, 69/21
frequency, 5/30, 69/20
intracranial hypertension, 5/173, 16/129
intracranial pressure, 67/141
management, 69/26
metastasis, 69/23
pathophysiology, 69/20
radiotherapy, 69/23
severe, 69/22
site, 69/22
surgery, 69/23
survey, 69/19
timing, 69/21
type, 69/21
uncommon headache syndrome, 69/25
venous sinus thrombosis, 69/24

Brain tumor hemorrhage
bronchogenic carcinoma, 5/141
cerebrovascular disease, 5/141
computer assisted tomography, 54/72
morning headache, 5/141

Brain tumor origin
*see also* Brain tumor carcinogenesis
brain regeneration, 16/32
Cohnheim-Ribbert theory, 16/24
dysontogeny, 16/26
environment, 16/33, 58, 77
experimental, *see* Experimental brain tumor
radiation, 16/40
topistic aspect, 16/26
virus, 16/40

Brain varix
arteriovenous malformation, 12/234

Brain vascular tumor, 18/269-297
brain hemangioma, 18/276
idiopathic multiple hemorrhagic sarcoma, 18/288-292
radiotherapy, 18/510
telangiectasia, 18/269

Brain vasculitis
*see also* Cerebrovascular disease *and* Granulomatous CNS vasculitis
amphetamine, 55/520
amphetamine addiction, 55/520
angiographic characteristic, 55/421
bacterial meningitis, 55/416
brain angiography, 55/421
brain infarction, 55/421
cell mediated immune response, 55/418
clinical symptom, 55/418
cocaine addiction, 55/521
cochlea, 55/381
computer assisted tomography, 54/62, 71, 55/420
Creutzfeldt-Jakob disease, 56/546
cryoglobulinemia, 63/404
CSF, 55/422
diamorphine addiction, 55/518
differential diagnosis, 55/422
drug abuse, 55/522
drug induced type, 55/519
ephedrine, 55/521
epilepsy, 55/420
fungal meningitis, 55/416, 434
granulocytic dysfunction, 55/418
granulomatous, *see* Granulomatous CNS vasculitis
headache, 48/287
humoral immune mechanism, 55/418
idiopathic hypereosinophilic syndrome, 63/124
immunopathology, 55/416
insidious stroke, 55/437
Köhlmeier-Degos disease, 14/789, 55/277
lymphoma, 63/350
lysergide, 55/522
malignant lymphoma, 55/488, 490

meningoencephalitis, 55/418
methamphetamine, 55/520
methylphenidate, 55/520
methylphenidate addiction, 55/520
moyamoya leash, 55/422
neurologic complication, 55/419
nuclear magnetic resonance, 55/421
oral contraception, 55/525
polyarteritis nodosa, 39/302, 55/353, 396
radionuclide scan, 55/421
reticuloendothelial system, 55/418
retina, 55/381
systemic lupus erythematosus, 55/371
Todd paralysis, 55/420
treatment, 55/423
tuberculosis, 55/429
tuberculous meningitis, 55/424, 429
varicella zoster virus, 56/237
varicella zoster virus infection, 55/436
Brain vasospasm, 11/511-529
adrenalin, 11/126
alcohol, 11/125
aminophylline, 11/125
angiography, 11/521, 23/168, 175, 55/56
artery beading, 11/515, 517
autonomic innervation, 11/142
brain aneurysm neurosurgery, 12/209, 55/55-58
brain injury, 23/163, 180, 240
brain tumor, 23/178
calcium channel blocking agent, 53/83
cause, 55/56
chemistry, 23/172
clinical features, 11/520, 525, 23/166, 168, 170, 176
computer assisted tomography, 54/64
contraindication, 12/209
diazoxide intoxication, 37/433
differential diagnosis, 11/522
endothelial injury, 53/83
experimental, 11/511, 23/183
headache, 48/64, 287
histamine, 11/125
history, 55/56
hypercalcemia, 63/561
hypertensive arteriolar disease, 11/566-571
hypothermia, 11/515
incidence, 55/56
infection, 23/173
intracranial, 23/180
lysergide, 55/522
mechanism, 11/515, 527
meningeal infection, 11/520
migraine, 23/173
model, 55/56
mortality, 55/57
multiple thrombi, 11/124
papaverine, 11/125
pathogenesis, 23/167, 171, 183
pial blood flow, 11/511-520
postoperative, 23/174, 176
prevention, 11/519, 55/58
primary pontine hemorrhage, 12/49
prognosis, 11/527
sellar meningioma, 23/178
subarachnoid hemorrhage, 11/520, 12/155, 23/168, 174, 187, 55/10, 55-58
systemic hypotension, 11/124
topical agent, 11/512
traumatic, 8/327, 23/163, 177, 180, 182, 184, 187
treatment, 11/528, 23/167, 173, 176, 55/56, 58
vitamin B3, 11/125
xanthine, 11/125
Brain venous infarction
*see also* Brain infarction
hypernatremia, 63/554
intravascular consumption coagulopathy, 63/314
leukemia, 63/344
lymphoma, 63/350
polycythemia, 63/251
protein C deficiency, 63/329
protein S deficiency, 63/329
thrombosis, 54/438, 63/3
Brain venous system
anatomy, 11/45-64
asymmetry, 11/203
cerebral dominance, 11/48
diagram, 11/104
embryology, 11/45-64
radiologic anatomy, 11/65-114
superficial, *see* Superficial brain venous system
topography, 11/65-114
varix, 12/234
Brain venous thrombosis, *see* Brain phlebothrombosis
Brain ventricle
history, 2/4-7
tumor, *see* Lateral ventricle tumor
Brain ventricle diverticulum
arachnoid cyst, 31/125-127
Brain ventricle drainage
intracranial pressure, 23/213
Brain ventricle enlargement
hebephrenia, 46/482
paranoid schizophrenia, 46/482
schizophrenia, 46/481
Brain ventricle hemorrhage

Brodmann area 44, 2/727, 776
Brodmann area 45, 2/725, 776
Brodmann area 46, 2/725, 727, 735
Brodmann area 47, 2/776
Brodmann, K., 1/27
Brolamfetamine
  *see also* Hallucinogenic agent
  designer drug, 65/364
  hallucinogenic agent, 65/42
Brolamfetamine intoxication
  symptom, 65/46
Bromate intoxication
  headache, 48/420
Bromide
  CSF, *see* CSF bromide
  intoxications, *see* Bromide intoxication
  metabolism, 36/293
  pharmacology, 36/292
  physiology, 36/293
Bromide intoxication, 36/291-315
  absorption, 36/293
  acute, 36/296
  ataxia, 36/299
  bromide level measurement, 36/294-296
  chronic, *see* Chronic bromide intoxication
  CSF, 36/301
  distribution, 36/293
  dysphagia, 36/300
  EEG, 36/302
  excretion, 36/293
  headache, 36/300
  history, 36/291
  incidence, 36/297
  metabolism, 36/293
  methyl bromide, *see* Methyl bromide intoxication
  neonatal, 36/314
  nuchal rigidity, 36/300, 302
  organic, *see* Organobromide intoxication
  parkinsonism, 36/300
  pharmacology, 36/292
  physiology, 36/295
  teratogenic effect, 36/314
  treatment, 36/307-310
  tremor, 36/299
Bromism, *see* Bromide intoxication
Bromocriptine
  acromegaly, 68/353
  akinesia, 45/166
  Alzheimer disease, 46/266
  behavior disorder, 46/594
  classification, 74/149
  CNS stimulant, 37/236
  delirium, 46/539
  migraine, 48/92
  pituitary tumor, 68/349
  restless legs syndrome, 51/548
Bromodeoxycytidine
  astrocytoma, 18/39
Bromodeoxyuridine, *see* Broxuridine
4-Bromo-2,5-dimethoxyamphetamine, *see* Brolamfetamine
Bromoureide intoxication
  chronic syndrome, 36/305-307
  clinical features, 36/304
  history, 36/303
  optic neuropathy, 36/306
Bromouridine
  brain tumor, 18/491, 520
  glioblastoma, 18/491
  glioblastoma multiforme, 18/65
  radiosensitizer, 18/491
Bronchial asthma, *see* Asthma
Bronchial neuropathy
  hereditary, *see* Hereditary brachial neuropathy
Bronchiectasis
  ataxia telangiectasia, 14/288, 60/371
  neuropathy, 8/18
Bronchodilating agent
  epilepsy, 72/228
Bronchogenic carcinoma
  brain tumor hemorrhage, 5/141
  nerve conduction, 7/167
Bronchopulmonary carcinoma
  vertebral column tumor, 20/43
Brondgeest
  stretch reflex, 1/63
Bronze diabetes, *see* Hemochromatosis
Brooke epithelioma, *see* Multiple trichoepithelioma
Brooke syndrome, *see* Multiple trichoepithelioma
Brooke tumor, *see* Multiple trichoepithelioma
Brossard syndrome
  age at onset, 22/63
  hereditary motor and sensory neuropathy type I, 22/57
  hereditary motor and sensory neuropathy type II, 22/57
  muscular dystrophy, 22/57
  pes equinovarus, 22/57
  shoulder girdle, 22/57
Brouwer, B., 1/10, 13, 21
Brown fat lipoma, *see* Hibernoma
Brown glial pigment
  neuroaxonal dystrophy, 49/395
Brown hyperparathyroid tumor
  differential diagnosis, 19/301
  mandible, 19/301

pain, 19/301
sex ratio, 19/301
Brown metachromasia
alcohol sensitivity, 10/25
glycosaminoglycanosis, 10/52
metachromatic leukodystrophy, 10/22, 44, 47
phospholipid, 10/25
sulfatide, 10/25
Brown-Séquard, C.E., 1/6
Brown-Séquard syndrome
angiodysgenetic necrotizing myelomalacia, 20/488
anterior spinal artery, 12/499
bronchial artery embolization, 63/44
canal encroachment, 61/60
central motor disorder, 1/200
cervical, 25/241, 273, 284
cervical hyperanteflexion injury, 25/335
cervical hyperdeflexion injury, 25/359
cervical metastasis, 20/428
cervical spondylarthrotic myelopathy, 50/469
cervical spondylosis, 7/456
classic type, 2/211
extramedullary spinal tumor, 19/30, 59
form, 1/200-202
leukemia, 63/342
meningeal leukemia, 63/343
radiation myelopathy, 26/85, 91, 61/200
second compression stage, 20/205
Sjögren syndrome, 71/75
spinal angioma, 32/466
spinal apoplexy, 19/41
spinal cord experimental injury, 25/22
spinal cord injury, 25/273
spinal cord tumor, 19/24, 30
spinal cyst, 32/437
spinal ependymoma, 19/360, 20/368
spinal epidural hematoma, 61/138
spinal lesion classification, 25/269
spinal neurinoma, 20/256
stab wound, 25/199
symptom, 1/200-202
varicella zoster virus myelitis, 56/236
Brown spider venom
toxic myopathy, 62/601, 611
Brown syndrome
localization, 2/327
ocular pseudoparesis, 2/327
rheumatoid arthritis, 71/19
Brown tumor, *see* Giant cell tumor
Brown-Vialetto-Van Laere syndrome
deafness, 22/108
Fazio-Londe disease, 59/129
progressive bulbar palsy, 59/374
Broxuridine
antineoplastic agent, 65/529
astrocytoma, 18/39
neurotoxin, 65/529
polyneuropathy, 65/529
Broxuridine intoxication
polyneuropathy, 65/538
Brucella
bacterial endocarditis, 52/302
Guillain-Barré syndrome, 51/185
hypertrophic pachymeningitis, 51/185
meningitis, 51/185
multiple cranial neuropathy, 51/185
radiculitis, 51/185
retrobulbar neuritis, 51/185
spondylitis, 51/185, 52/585
Brucella abortus
cattle, 52/582
geography, 52/582
neuritis, *see* Brucella neuritis
transmission mode, 52/582
Brucella canis
abortion, 52/582
geography, 52/582
Brucella endocarditis
mycotic aneurysm, 52/588
Brucella melitensis
geography, 52/582
history, 52/581
neuritis, *see* Brucella neuritis
transmission mode, 52/582
Brucella neuritis
symptom, 51/185
Brucella suis
geography, 52/582
neuritis, *see* Brucella neuritis
placenta, 52/582
Brucellosis, 33/305-321
acute stage, 52/582
bacteriology, 33/306
child, 33/318
chronic meningitis, 56/644
chronic progressive stage, 52/583
clinical features, 33/309
CSF, 33/310-312
diagnosis, 33/319, 320
elderly man, 52/591
epidemiology, 52/581
eradication, 52/581
familial, *see* Familial brucellosis
geography, 52/581
granuloma, 52/585

headache, 48/31
history, 33/305, 306
immunology, 52/584
infection reservoir, 33/307
laboratory finding, 33/310
melitococcal sciatica, 7/481
meningovascular, *see* Meningovascular brucellosis
microbiology, 33/306, 52/584
neuro, *see* Neurobrucellosis
pathology, 33/307-309, 312
pregnancy, 33/318
spondylitis, 19/368, 52/592
subarachnoid hemorrhage, 12/111
synonym, 52/581
tick-borne typhus, 34/657
transmission, 33/307
treatment, 33/320
Bruch membrane
anatomy, 2/507
angioid streak, 13/32
degeneration, 13/32
drusen, 13/236
ERG, 2/511
pseudoinflammatory foveal dystrophy (Sorsby), 13/138
Stargardt disease, 13/126
Brudzinski sign
bacterial meningitis, 33/1
contralateral leg sign, 1/543
lymphocytic choriomeningitis, 56/359
neck sign, 1/544
pneumococcal meningitis, 33/38, 52/43
Brueghel syndrome, *see* Dystonia musculorum deformans
Brufen, *see* Ibuprofen
Bruit
carotid cavernous fistula, 24/414, 417
spinal, *see* Spinal bruit
subclavian, *see* Subclavian bruit
subclavian steal syndrome, 53/374
traumatic carotid cavernous sinus fistula, 57/352
Bruit de pot fêlé, *see* Cracked pot sign
Brun pontoneocerebellar atrophy, *see* Pontoneocerebellar atrophy (Brun)
Bruns-Garland syndrome, *see* Diabetic proximal neuropathy
Brush cell
astrocytoma, 18/2
history, 18/2
terminology, 18/2
Brushfield spot
cerebrohepatorenal syndrome, 43/339
congenital heart disease, 63/11
Down syndrome, 43/544, 50/526
Bruxism
Rett syndrome, 60/636
sleep disorder, 3/101
Bruyn-Went disease
Behr disease, 13/91
congenital retinal blindness, 13/99, 22/454
dystonia musculorum deformans, 49/524
optic atrophy, 13/42, 22/454
Bruyn-Went-Novotny disease, *see* Bruyn-Went disease
Bucca lobata
Melkersson-Rosenthal syndrome, 8/213
Buccal neuralgia, *see* Sphenopalatine neuralgia
Buccofacial apraxia
aphasia, 4/54
error type, 45/425
frontal lobe lesion, 4/53
motor aphasia, 45/428
Buccolingual apraxia
developmental dysphasia, 46/140
Buccopharyngeal herpes
cranial nerve palsy, 7/483
facial nerve, 7/483
glossopharyngeal nerve, 7/483
trigeminal nerve, 7/483
Buccopharyngeal spasm
progressive pallidal atrophy (Hunt-Van Bogaert), 21/531
spinopontine degeneration, 21/390
Buchem disease, *see* Van Buchem disease
Buckthorn toxin
action site, 64/10, 14
Karwinska humboldtiana, 64/14
neurotoxin, 64/10, 14
segmental demyelination, 64/14
toxic neuropathy, 64/10, 14
Buerger disease, *see* Thromboangiitis obliterans
Bufotenine
*see also* Hallucinogenic agent
hallucinogenic agent, 65/42
potency, 65/45
Bufotenine intoxication
hallucination, 65/45
Buhot cell
metachromatic leukodystrophy, 10/53
Bulb diggers palsy, *see* Compression neuropathy
Bulbar brain syndrome
acute, 24/586
secondary, 24/587, 705, 727
traumatic psychosyndrome, 24/583, 705, 727
Bulbar cervical dissociation pattern

spinal cord injury, 25/275
Bulbar encephalitis
Hodgkin disease, 39/54
malignant lymphoma, 39/54
Bulbar jugular syndrome
classification, 18/444
Bulbar nerve syndrome
*see also* Multiple cranial neuropathy
multiple cranial neuropathy, 51/569
Bulbar paralysis
Alzheimer disease, 22/142
amyotrophic lateral sclerosis, 22/281, 42/65
Aran-Duchenne disease, 22/134, 42/86
Behçet syndrome, 56/598
brain stem vascular process, 22/144
cerebellar heredodegeneration, 22/142
cerebellar hypoplasia, 60/658
Chédiak-Higashi syndrome, 60/658
child, 42/94
CNS degeneration, 42/97
corneal lattice dystrophy, 22/139
Creutzfeldt-Jakob disease, 22/143
deafness, 22/138, 523
dementia, 22/142
diphtheria, 52/229
enterovirus 71, 56/329
eunuchoidism, 60/658
external ophthalmoplegia, 22/140
familial spastic paraplegia, 22/139
Far Eastern tick-borne encephalitis, 56/140
giant axonal neuropathy, 60/76
gynecomastia, 22/137
headache, 65/178
hereditary cerebellar ataxia, 22/142
hereditary olivopontocerebellar atrophy, 60/658
hereditary spastic ataxia, 60/658
3,3'-iminodipropionitrile intoxication, 22/127
infantile spinal muscular atrophy, 22/135, 60/658
intermediate spinal muscular atrophy, 42/90
late cerebellar atrophy, 60/658
lymphocytic choriomeningitis, 56/361
Machado-Joseph disease, 60/658
Mast syndrome, 22/142
neuronal intranuclear hyaline inclusion disease, 60/658
Norman disease, 60/658
oculopharyngeal neuromuscular disease, 22/140
olivopontocerebellar atrophy, 60/658
olivopontocerebellar atrophy (Wadia-Swami), 60/494
optic atrophy, 22/138
organomercury intoxication, 22/144
pallidocerebello-olivary degeneration, 60/658
Parkinson dementia complex, 22/143
Pick disease, 22/142
polyneuropathy, 42/97
progressive, *see* Progressive bulbar palsy
progressive external ophthalmoplegia, 60/658
progressive spinal muscular atrophy, 22/134
progressive supranuclear palsy, 22/140, 219
pseudo, *see* Pseudobulbar paralysis
puffer fish intoxication, 37/80
Schut family ataxia, 60/658
sea snake intoxication, 37/92
Shy-Drager syndrome, 22/139
Sjögren syndrome, 71/74
snake envenomation, 65/178
subacute necrotizing encephalomyelopathy, 60/658
tetrodotoxin intoxication, 65/156
thallium intoxication, 64/325
trichloroethylene, 51/280
Wohlfart-Kugelberg-Welander disease, 22/135
Bulbar poliomyelitis
anhidrosis, 8/344
ataxic respiration, 1/498
congenital hemifacial atrophy, 22/548
cranial nerve palsy, 56/328
frequency, 34/109
hypoventilation, 63/413
progressive hemifacial atrophy, 22/549, 59/479
Bulbar postpoliomyelitis amyotrophy
progressive hemifacial atrophy, 59/479
Bulbar syncope
poliomyelitis, 15/824
polyradiculoneuritis, 15/824
porphyria, 15/824
rabies, 15/824
Bulbar syndrome
glomus jugulare tumor, 18/444
progressive bulbar palsy, 22/312
various forms, 1/188
Bulbar vesicorelaxer center
medial reticular formation, 1/390
Bulbocapnine
experimental catatonia, 21/527
Bulbocavernosus reflex
diagnostic value, 1/254
lower urinary tract, 61/295
paraplegia, 61/295
spinal cord injury, 61/295
thoracolumbar spine injury, 25/455
transverse spinal cord lesion, 61/295
urinary tract, 26/416
Bulbocavernous muscle
development, 1/361

function, 1/361
urethra, 1/361
Bulbopontine paralysis
cranial nerve, 42/96
deafness, 42/95
facial paralysis, 42/96
familial, *see* Familial bulbopontine paralysis
hyperreflexia, 42/96
sensorineural deafness, 42/95
tongue atrophy, 42/96
vocal cord paralysis, 42/96
Bulbospinal muscular atrophy
age at onset, 59/43
androgen receptor failure, 59/44
diabetes mellitus, 59/44
electrophysiologic study, 59/44
gynecomastia, 59/44
hypogonadism, 59/44
intention tremor, 59/44
neuropathology, 59/44
progressive, *see* Progressive bulbospinal muscular atrophy
weakness, 59/43
Wohlfart-Kugelberg-Welander disease, 59/91
X-linked recessive inheritance, 59/43
Bulbus retraction syndrome, *see* Stilling-Türk-Duane syndrome
Bulge cell
diffuse sclerosis, 10/73
Bulimia
African trypanosomiasis, 52/341
anorexia nervosa, 46/589
Bardet-Biedl syndrome, 13/395
meningeal leukemia, 63/341
oxidation-phosphorylation coupling defect, 62/506
Pick disease, 46/236, 240
Bull angle
basilar impression, 32/17
Bullous keratitis, *see* Keratitis bullosa
Bullous myringitis
cranial nerve lesion, 5/210
headache, 5/210
meningoencephalitis, 5/210
viral, 5/210
Bumke, O.C.E., 1/1, 20
α-Bungarotoxin
acetylcholine receptor, 40/13, 41/97, 106, 224
experimental myasthenia gravis, 41/115
myasthenia gravis, 41/97, 113, 125, 62/409
myasthenic syndrome, 64/16
myoglobinuria, 64/16
neuromuscular block, 40/129, 143
neuromuscular transmission, 40/129
neurotoxin, 40/129, 41/97, 106, 113, 125, 62/409, 64/16, 65/177, 181
peroxidase stain, 40/5
snake envenomation, 65/182
snake venom, 65/181
toxic encephalopathy, 65/177
toxic neuromuscular junction disorder, 64/15
α-Bungarotoxin intoxication
action site, 64/15
pathomechanism, 64/15
symptom, 64/15
β-Bungarotoxin
chemistry, 65/184
myasthenic syndrome, 64/16
myoglobinuria, 64/16
neuromuscular transmission, 40/143
neurotoxin, 37/17, 40/143, 64/15, 65/181
snake envenomation, 65/182
snake venom, 65/181
toxic neuromuscular junction disorder, 64/15
β-Bungarotoxin intoxication
action site, 64/15
autonomic nervous system, 75/512
pathomechanism, 64/15
symptom, 64/15
κ-Bungarotoxin
chemistry, 65/183
neurotoxin, 65/183
snake envenomation, 65/182
snake venom, 65/183
Bunge Institute, 1/10
Bunyamwera virus
*see also* Arbovirus, Arenavirus, Arthropod-borne virus, California encephalitis virus *and* Togavirus
acute viral encephalitis, 34/68
definition, 34/5
Bunyavirus
*see also* California encephalitis virus
acute viral encephalitis, 34/5
bunyavirus genus, 56/15
California encephalitis, 56/16, 140
California encephalitis virus, 56/16
human, 56/3
La Cross virus, 56/16
mosquito, 56/16
nairovirus genus, 56/15
neurovirulence, 56/39
phlebovirus genus, 56/15
replication, 56/15
RNA virus, 34/5, 56/15
taxonomic place, 56/3

viral classification, 34/68
viral meningitis, 56/16
Buphthalmos
congenital heart disease, 63/11
François syndrome, 14/635
glaucoma, 14/398
historic note, 14/224
neurofibromatosis type I, 14/635
Sachsalber syndrome, 14/398
Sturge-Weber syndrome, 14/224, 234
Bupivacaine intoxication
convulsion, 37/405
seizure, 37/405
Buprenorphine
opiate, 65/350
Bupropion, *see* Amfebutamone
Burkitt lymphoma
brain infiltration, 63/352
clinical features, 39/66-68
cranial nerve, 39/74
cranial neuropathy, 63/351
CSF, 39/76
CSF pleocytosis, 63/351
diagnosis, 39/76
encephalitis, 63/352
encephalopathy, 63/352
epidemiology, 63/350
epilepsy, 63/351, 72/180
Epstein-Barr virus, 56/8, 249, 255, 63/351
Epstein-Barr virus infection, 34/189, 39/63-65
etiology, 39/63-65
flaccid paraplegia, 63/352
geographic distribution, 39/65
Guillain-Barré syndrome, 63/352
history, 39/63
hydrocephalus, 63/351
hyperkalemia, 63/352
hyperphosphatemia, 63/352
hyperuricemia, 63/352
hypocalcemia, 63/352
hypoglycemia, 63/352
immunocompromised host, 56/470
incidence, 39/68
infection, 63/352
intracranial involvement, 39/77
lactic acidosis, 63/352
Listeria monocytogenes, 63/352
malignant lymphoma, 18/236
meningeal lymphoma, 63/346, 351
meningitis, 63/352
metabolic acidosis, 63/352
multiple cranial neuropathy, 63/351
neurologic involvement, 39/63-85
non-Hodgkin lymphoma, 63/345
Ommaya reservoir, 63/352
paraplegia, 39/75, 63/351
pathology, 39/65, 69-73
polyradiculopathy, 63/352
prevention, 39/84
prognosis, 39/84
radiculopathy, 39/75, 63/351
spinal cord, 39/75
spinal epidural lymphoma, 63/351
spinal epidural tumor, 20/128
tetany, 63/352
treatment, 39/76-84
ventriculoperitoneal shunt, 63/352
Burkitt tumor, *see* Burkitt lymphoma
Burn
delirium, 46/548
immunocompromised host, 56/469, 474
intravascular consumption coagulopathy, 55/495
leukoencephalopathy, 47/574-576
nerve injury, 51/136
neuropathy, 8/15
Burning feet syndrome
alcoholic polyneuropathy, 51/316
associated deficiency disease, 28/11
cardiovascular system, 28/10
clinical features, 28/9
etiology, 28/11-14
general sign, 28/9
hereditary sensory and autonomic neuropathy type I, 60/8
hypertension, 28/12
malnutrition, 28/9-14
neurologic sign, 28/10
pain production mechanism, 28/11
pathology, 28/11-14
pellagra, 7/573, 28/11, 71
peritoneal dialysis, 27/342
renal insufficiency, 27/338, 342
symptom, 28/9
treatment, 28/10
tropical ataxic neuropathy, 51/322
uremic neuropathy, 27/338
vitamin B complex, 28/14
vitamin $B_1$ deficiency, 28/13
vitamin $B_2$ deficiency, 28/13, 200, 51/332
vitamin $B_3$ deficiency, 28/13
vitamin $B_5$, 28/207
vitamin $B_5$ deficiency, 28/200, 51/341
Burr erythrocyte
thrombotic thrombocytopenic purpura, 9/600, 607
Burst fracture
cervical vertebral column injury, 25/67, 26/266,

291, 61/30
depressed skull fracture, 24/651
intervertebral disc hemorrhage, 61/36
thoracic vertebral column injury, 61/78
vertebral column injury, 61/519
Buspirone
anxiolysis, 65/344
Buspirone intoxication
fatigue, 65/344
half-life, 65/344
headache, 65/344
insomnia, 65/344
metabolite, 65/343
nausea, 65/344
serotonin 1A receptor, 65/343
vertigo, 65/344
vomiting, 65/344
2-Butanone
chemical formula, 64/82
organic solvent, 64/82
organic solvent intoxication, 64/44
2-Butanone potentiation
hexane polyneuropathy, 51/276
2-hexanone polyneuropathy, 51/276
Butaperazine
dystonia time effect, 65/284
neuroleptic agent, 65/284
neuroleptic dystonia, 65/284
Butazolidine, *see* Phenylbutazone
Buthotoxin
neurotoxin, 65/194
Butorphanol
opiate, 65/350
*O-N*-Butyl *O*-carbetoxymethyl
ethylphosphonothioate intoxication
chemical classification, 37/547
organophosphorus compound, 37/547
Butyrophenone
behavior disorder, 46/597
chemical structure, 65/276
designer drug, 65/274
neuroleptic agent, 65/274
Butyrophenone derivative
neuroleptic syndrome, 6/257
Butyrylcholinesterase
CSF, *see* CSF butyrylcholinesterase
oligodendrocyte, 9/14
wallerian degeneration, 7/215
Butyrylcholinesterase deficiency
epidemiology, 42/654
evoked muscle action potential, 41/291
Bypass
aorta surgery, 63/56
cardiopulmonary, *see* Cardiopulmonary bypass
gastric, *see* Gastric bypass
jejunoileal, *see* Jejunoileal bypass
Bypass surgery
coronary, *see* Coronary bypass surgery

C band
chromosome, 31/343, 351
C cell hyperplasia
Sipple syndrome, 42/752
C fiber
pain transmission, 1/117-119
C. immitis, *see* Coccidioides immitis
C-reactive protein
bacterial meningitis, 52/7
Guillain-Barré syndrome, 51/248
C-reflex
Alzheimer disease, 49/611
myoclonus, 49/611
C syndrome
corpus callosum agenesis, 50/163
C wave
CSF pressure, 19/113
intracranial pressure, 23/202, 57/219
intracranial pressure monitoring, 57/219
$C_1$ esterase inhibitor
angioneurotic edema, 8/349, 43/61
$C_1$ posterior arch
Arnold-Chiari malformation type II, 50/407
$C_1$-$C_2$ subluxation, *see* Atlantoaxial subluxation
$C_3$ receptor
multiple sclerosis, 47/250
C-1 complement
CSF, *see* CSF C-1 complement
C-3 complement
CSF, *see* CSF C-3 complement
multiple sclerosis, 47/115, 371
C-3a complement
critical illness polyneuropathy, 51/584
C-3b complement
herpes simplex virus, 56/208
C-4 complement
CSF, *see* CSF C-4 complement
multiple sclerosis, 47/115
C-5a complement
critical illness polyneuropathy, 51/584
C-5,C-9 complement deficiency
meningococcal meningitis, 52/29
C-6,C-8 complement deficiency
meningococcal meningitis, 52/29
C1 fracture, *see* Atlas fracture
Cabbage body
muscle fiber, 40/192

Cabergoline
pituitary tumor, 68/349
Cachectin, *see* Tumor necrosis factor
Cachetic myopathy
paraneoplastic syndrome, 69/374, 387
Cachexia
anorexia nervosa, 7/634
Huntington chorea, 49/284
metastatic cardiac tumor, 63/96
multiple myeloma, 20/9, 39/138
neuropathy, 8/15
optic atrophy, 13/64
pellagra, 7/574, 28/66, 92
Peutz-Jeghers syndrome, 14/121
Pick disease, 46/236
primary cardiac tumor, 63/96
Russell syndrome, 16/349
Simmonds, *see* Sheehan syndrome
starvation neuropathy, 8/15
Cacosmia
anosmia, 24/13
Cadet palsy
brachial plexus, 51/144
Cadherin
brain metastasis, 69/111
metastasis, 71/645
Cadmium
neurotoxin, 64/10
toxic neuropathy, 64/10
Cadmium intoxication
action site, 64/10
carbamate intoxication, 64/187
Caenorhabditis elegans
tumor cell, 67/29
Café au lait spot
Albright syndrome, 14/12
Bloom syndrome, 14/775, 43/10
Fanconi syndrome, 43/17
multiple endocrine adenomatosis type III, 42/753
multiple neurofibromatosis, 43/34
neurofibromatosis, 1/484, 50/365
neurofibromatosis type I, 14/564, 31/14, 50/366, 68/288
neurofibromatosis type II, 50/370
nevus unius lateris, 43/40
Peutz-Jeghers syndrome, 14/121
Russell-Silver syndrome, 43/476, 46/21, 82
tuberous sclerosis, 14/343
Caffeine
autonomic polyneuropathy, 51/487
contracture induction, 41/269
drug addiction, 65/259
headache, 48/6
malignant hyperthermia, 40/546, 41/269
migraine, 5/98
neurotoxin, 5/98, 40/546, 41/269, 48/6, 61/156, 65/259
orthostatic hypotension, 63/164
postlumbar puncture syndrome, 61/156
progressive dysautonomia, 59/140, 158
screening test, 40/546
Caffeine intoxication
adenosine receptor, 65/260
adrenalin, 65/260
anxiety, 65/260
calcium, 65/260
convulsion, 65/260
cyclic nucleotide, 65/260
half-life, 65/260
hyperesthesia, 65/260
insomnia, 65/260
noradrenalin, 65/260
tachycardia, 65/260
tremor, 65/260
Cage paralysis of primates
demyelinating disease, 9/671
Caisson disease, *see* Decompression illness *and* Dysbarism
Cajal interstitial nucleus
medial longitudinal fasciculus, 2/275
mesencephalic syndrome, 2/275
Cajal-Retzius cell
CNS, 50/245
Cajal, S. Ramón y
Ranvier node, 47/9
Calcarine artery
radioanatomy, 11/98
topography, 11/98
Calcarine cortex
dyshoric angiopathy, 42/728
Minamata disease, 36/84, 107, 121, 64/415
Calcavin
experimental chemical data, 40/157
experimental myopathy, 40/160
I band, 40/157
mitochondrial abnormality, 40/160
Z band, 40/157
Calcergy
definition, 6/703
Calcification
adrenal, *see* Adrenal calcification
astrocytoma, 18/22
basal ganglion, *see* Striatopallidodentate calcification
brain, *see* Intracerebral calcification
brain arteriosclerosis, 53/94

brain cortex, *see* Brain cortex calcification
cerebellar, *see* Cerebellar calcification
cerebral, *see* Intracerebral calcification
chordoma, 18/158
cortical, *see* Cortical calcification
cortical tuber, 14/356
corticostriatopallidodentate, *see* Striatopallidodentate calcification
dentate nucleus, *see* Dentate nucleus calcification
dura mater, *see* Dura mater calcification
falx cerebri, 14/114, 463
familial basal ganglion, *see* Striatopallidodentate calcification
familial striatopallidodentate, *see* Familial striatopallidodentate calcification
Geyelin-Penfield, *see* Geyelin-Penfield calcification
hippocampus, 14/777
infectious striatopallidodentate, *see* Infectious striatopallidodentate calcification
intervertebral disc, *see* Intervertebral disc calcification
intracerebral, *see* Intracerebral calcification
intracranial, *see* Intracranial calcification
intraocular, *see* Intraocular calcification
midline, *see* Midline calcification
mitral annulus, *see* Mitral annulus calcification
muscle, *see* Muscle calcification
muscle fiber, 62/18
neurocysticercosis, 16/225, 52/529
occipital lobe epilepsy, 73/110
oligodendroglioma, 18/86
periventricular, *see* Periventricular calcification
physiologic striatopallidodentate, *see* Physiologic striatopallidodentate calcification
pineal tumor, 17/659
pituitary adenoma, 17/397
posterior longitudinal ligament, *see* Posterior longitudinal ligament calcification
pseudo, *see* Pseudocalcium
spinal ependymoma, 19/360, 20/371
spinal tumor, 19/166
striatopallidocorticodentate, *see* Striatopallidodentate calcification
striatopallidodentate, *see* Striatopallidodentate calcification
subependymal nodule, 14/364
sulcus, *see* Sulcus calcification
toxic striatopallidodentate, *see* Striatopallidodentate calcification
tuberculoma, 18/418
tuberous sclerosis, 14/45-47
vascular striatopallidodentate, *see* Vascular striatopallidodentate calcification
white matter vein, *see* White matter vein calcification
Calcinosis
brain, *see* Striatopallidodentate calcification
cerebral, *see* Striatopallidodentate calcification
dermatomyositis, 62/386
inclusion body myositis, 62/386
polymyositis, 41/57, 62/386
tumoral, *see* Lipocalcinogranulomatosis
Calcinosis cutis universalis
myositis ossificans, 41/57
polymyositis, 41/57
Calcinosis nucleorum cerebri, *see* Striatopallidodentate calcification
Calcinosis Raynaud phenomenon Esophageal involvement Sclerodactyly and Telangiectasia, *see* CREST syndrome
Calcinosis Raynaud phenomenon Sclerodactyly and Telangiectasia, *see* CRST syndrome
Calciphylaxis
definition, 6/703
Calcitonin
hypercalcemia, 63/562
hyperparathyroidism, 27/290
osteitis deformans (Paget), 38/368, 43/452, 60/773
pain, 69/47
parathyroid gland, 27/273
parathyroid hormone, 27/284
Calcitonin gene related peptide
autonomic nervous system, 75/86
brain blood flow, 53/63
penis erection, 75/86
reflex sympathetic dystrophy, 61/126
Calcitriol, *see* 1,25-Dihydroxyvitamin D3
Calcium
*see also* Tetany
Alzheimer disease, 46/265
basal ganglion, 49/35
brain, 9/16, 28/433
brain infarction, 53/125
brain injury, 23/115
brain ischemia, 53/125, 63/211
brain metabolism, 57/76
caffeine intoxication, 65/260
CNS, 28/433
CSF, 28/434
Duchenne muscular dystrophy, 40/381, 387
Entamoeba histolytica, 52/322
epilepsy, 15/312, 72/54
excitation contraction coupling, 40/130
Guam amyotrophic lateral sclerosis, 59/264, 275
headache, 48/111

hemostasis, 63/303
histochemistry, 40/5
hypocalcemia, 28/434
influx with sarcolemmal change, 40/213, 553
leakage, 28/528
membrane surface charge, 28/529
motor end plate, 69/338
movement disorder, 70/120
myelopathy, 70/120
nerve cell, 72/50
nerve conduction, 47/39
neurogenic osteoarthropathy, 26/517
neurologic syndrome, 70/111
neuron death, 63/211
neurophysiologic aspect, 28/527-532
pain, 75/325
paraneoplastic syndrome, 69/338
pharmacology, 74/143
pseudo, *see* Pseudocalcium
secretion, 28/531
serum concentration, 63/558
spinal cord compression, 69/170
spinal cord injury, 26/382, 387, 390, 402
streptomycin polyneuropathy, 51/296
striatopallidodentate calcification, 6/715-719, 49/425
sympathetic nerve terminal, 74/143
vascular mineralization, 11/141
Calcium antagonist
arterial hypertension, 54/209
brain infarction, 53/428
brain metabolism, 57/75
cardiovascular agent, 65/433
dysgeusia, 65/443
dysosmia, 65/443
gallopamil, 65/440
iatrogenic neurological disease, 65/441
migraine, 48/200
nifedipine, 65/440
pharmacologic action, 65/440
verapamil, 65/440
Calcium antagonist intoxication
akathisia, 65/442
brain ischemia, 65/441
carbamazepine neurotoxicity, 65/442
drug interaction, 65/442
dysgeusia, 65/443
dysosmia, 65/443
dystonia, 65/442
dystonic posture, 65/442
hallucination, 65/442
headache, 65/441
indication, 65/441
mental depression, 65/441
migraine, 65/441
muscle cramp, 65/442
muscle spasm, 65/442
myoclonia, 65/442
myoclonic jerk, 65/442
neurologic adverse effect, 65/441
nimodipine, 65/441
parkinsonism, 65/442
phenytoin toxicity, 65/442
psychosis, 65/441
respiratory failure, 65/442
subarachnoid hemorrhage, 65/441
vertigo, 65/441
Calcium binding protein
brain metastasis, 69/109
Calcium carbimide
drug induced polyneuropathy, 51/309
neurotoxicity, 51/309
Calcium carbimide polyneuropathy
diethyldithiocarbamic acid, 51/309
Calcium channel
arthropod envenomation, 65/194
brain metabolism, 57/77, 80
Eaton-Lambert myasthenic syndrome, 62/392, 421, 423
electrophysiology, 28/528-532
head injury, 57/80
malignant hyperthermia, 62/567, 569
muscle tissue culture, 62/91
Calcium channel blocking agent
alcohol, 55/524
anoxic encephalopathy, 63/222
antihypertensive agent, 63/84
brain aneurysm neurosurgery, 55/59
brain blood flow, 57/77
brain edema, 67/89
brain infarction, 63/193
brain injury, 57/212
brain ischemia, 63/222
brain metabolism, 57/77
brain vasospasm, 53/83
cardiac pharmacotherapy, 63/193
cyanide intoxication, 64/231
head injury, 57/77
iatrogenic neurological disease, 63/193
migraine, 48/199
myotonia, 62/276
neurologic intensive care, 55/217
subarachnoid hemorrhage, 55/59
Calcium current
arthropod envenomation, 65/200
epilepsy, 72/45

nerve cell, 72/45
Calcium dietary deficiency
Guam amyotrophic lateral sclerosis, 59/266
Calcium disodium edetate, *see* Edetate calcium disodium
Calcium edetic acid
mercury intoxication, 64/394
Pick disease, 46/244
Calcium imbalance
*see also* Tetany
clinical syndrome, 28/532-539
hypocalcemia, 28/536-539
intracranial pressure, 28/536
mental change, 28/536
neurologic manifestation, 28/527-539
Calcium metabolism
intracerebral calcification, 42/534
malignant hyperthermia, 38/552, 43/118
poliomyelitis, 34/113
Calcospherite
spongioblastoma, 16/15
Calculation
acalculia, 4/182
alexia, 4/96
cerebral dominance, 4/281, 288
information processing, 45/474-476
perception, 45/475
recognition, 45/475
Calculation defect
Gerstmann syndrome, 45/70
Calf hypertrophy
anterior horn syndrome, 42/72
dominant ataxia, 60/659
Hartnup disease variant, 60/659
hereditary motor and sensory neuropathy type I, 60/247
hereditary motor and sensory neuropathy type II, 60/247
hereditary myokymia, 60/659
infantile distal myopathy, 62/201
late infantile acid maltase deficiency, 62/480
neurogenic weakness, 60/659
periodic ataxia, 60/659
phosphoglucomutase deficiency, 27/222, 236, 41/192
Wohlfart-Kugelberg-Welander disease, 59/372
California encephalitis
acute viral encephalitis, 56/134
Aedes triseriatus, 56/140
bunyavirus, 56/16, 140
epilepsy, 56/141
Jamestown Canyon virus, 56/140
La Cross virus, 56/140
snowshoe hare virus, 56/140
Tahyna virus, 56/140
California encephalitis virus
*see also* Bunyamwera virus *and* Bunyavirus
acute viral encephalitis, 34/78
bunyavirus, 56/16
Caliper
skull, *see* Skull traction
Calleja islet
anatomy, 6/32
Callosal dyspraxia, *see* Corpus callosum dyspraxia
Callosal section
epilepsy, 73/408
Callosectomy
schizophrenia, 46/504
Callosomarginal artery
radioanatomy, 11/79
topography, 11/79
Callosotomy
spatial thinking, 45/414
split brain syndrome, 45/99
Callosum agenesis
corpus, *see* Corpus callosum agenesis
Callosum defect
corpus, *see* Corpus callosum defect
Callosum infarction
corpus, *see* Corpus callosum infarction
Callosum syndrome
corpus, *see* Corpus callosum syndrome
Callosum tumor
corpus, *see* Corpus callosum tumor
Calmeil classification
epilepsy, 15/3
Calmodulin
neuroleptic agent, 65/279
organochlorine insecticide intoxication, 64/200
Caloric nystagmus, 2/94, 333, 372, 24/122, 126
abnormal type diagram, 2/376
canal paresis, 2/375
directional preponderance, 2/374
retinovestibular hereditary spastic ataxia, 21/387
test, 1/577
Caloric test
acoustic nerve tumor, 17/677
acoustic neuroma, 16/318
ENG, 24/133, 135
quantitation, 2/374
spasmodic torticollis, 6/575
vestibular function, 16/324
Calorie
brain death, 24/721
brain injury, 23/125
Calvarial metastasis

survey, 69/127
CAM, *see* Cell adhesion molecule
CAMAK syndrome
deafness, 60/658
secondary pigmentary retinal degeneration, 60/730
CAMFAK syndrome
deafness, 60/658
secondary pigmentary retinal degeneration, 60/730
cAMP, *see* Cyclic adenosine monophosphate
Camphor intoxication
headache, 48/420
Campimetry
occipital lobe tumor, 17/318-320
visual field, 13/13
Camptobrachydactyly
mental deficiency, 43/236
sensorineural deafness, 43/236
Camptocormia
gait disturbance, 1/291
psychiatric aspect, 1/291
valproic acid intoxication, 65/505
Camptodactyly
dialysis, 63/529
β2-microglobulin, 63/529
β2-microglobulin amyloidosis, 63/529
Pena-Shokeir syndrome type I, 43/437, 59/71
XXXXX syndrome, 50/555
Camptomelic syndrome
macrencephaly, 50/263
mental deficiency, 46/63, 88
Campus foreli, *see* Forel field H
Campylobacter fetus
cranial subdural empyema, 52/171
Camurati-Englemann disease, *see* Progressive hereditary craniodiaphyseal dysplasia
Canadian Neurologic Scale
brain embolism, 53/209
brain infarction, 53/209
cerebrovascular disease, 53/209
intracranial hemorrhage, 53/209
transient ischemic attack, 53/209
Canal
carotid, *see* Carotid canal
Cloquet, *see* Cloquet canal
fallopian, *see* Fallopian canal
Guyon, *see* Guyon canal
neurenteric, *see* Neurenteric canal
patent central, *see* Patent central canal
spinal, *see* Vertebral canal
Canal of Guyon syndrome, *see* Guyon canal
Canal paresis
caloric nystagmus, 2/375
Canalography
lumbar stenosis, 20/722
Canavan disease, *see* Spongy cerebral degeneration
Cancer
*see also* Carcinoma, Carcinomatous myopathy *and* Malignancy
adenovirus, 69/450
anomia, 67/390
ataxia telangiectasia, 60/372
autonomic nervous system, 75/527
back pain, 69/57
blood-brain barrier, 69/432
brachial plexopathy, 69/84
brain metastasis, 69/1
breast, *see* Breast cancer
cerebrovascular disease, 69/420
CNS infection, 69/431, 433
colon, *see* Colon cancer
deafferentation pain, 69/32, 48
dermatomyositis-polymyositis, 41/82, 347, 368-378
disability, 67/390
Eaton-Lambert myasthenic syndrome, 41/349-360
ectopic hormone production, 75/542
encephalopathy, 69/396
epidural space, 69/63
epilepsy, 69/9
Epstein-Barr virus, 69/450
fungus, 69/438
handicap, 67/390
history, 41/319
impairment, 67/390
Isaacs syndrome, 41/347
ischialgia, 69/57
laryngeal, *see* Laryngeal cancer
listeriosis, 52/91
localized muscle, 41/381-386
lung, *see* Lung cancer
measles, 69/450
metabolic encephalopathy, 69/396, 399, 401
metabolism, 69/395
metastasis to muscle, 41/378
motoneuron disease, 41/329-338
motor neuropathy, 41/338-347
multiple sclerosis, 47/202, 303
myasthenia gravis, 41/361-368
myopathy, 41/317-386
myotonia, 41/347
nervous system, 71/611
neurologic disease, 69/1, 3
neuromuscular disease, 41/317-386
neuromyopathy, 41/322

nonbacterial thrombotic endocarditis, 11/417, 47/541, 55/489, 63/28
pain, 69/31, 60
paraneoplastic syndrome, 69/325, 71/696, 75/533
peripheral nervous system, 69/71
plexopathy, 69/79
polymyositis, 41/81
polyneuropathy, 41/338
prevalence, 69/1
progressive multifocal leukoencephalopathy, 69/449
prostate, *see* Prostate cancer
quality of life, 67/390
radiculopathy, 69/74
retinopathy, 71/696
small cell lung, *see* Small cell lung cancer
speech, 67/390
spine metastasis, 67/200
thyroid, *see* Thyroid cancer
toxic encephalopathy, 69/396
uterus, *see* Uterus cancer
vertebral lesion, 69/57
visceral, *see* Visceral cancer
yeast, 69/438
Cancer family syndrome
high grade glioma, 68/88
Cancer gene
tumor cell, 67/18
Candida
*see also* Candidiasis
acquired myoglobinuria, 62/577
biology, 35/413-415
cerebrovascular disease, 55/432
CNS infection, 69/435, 439
CNS invasion, 52/398
epidemiology, 35/413-415
glossitis, 8/3229
guilliermondii, 35/417, 52/398
microbiology, 52/397
nitrate utilization, 52/397
parapsilosis, 35/417, 52/398
pathogenic species, 52/397
phenol oxidase lack, 52/397
pseudotropicalis, 35/416
stellatoidea, 35/417, 52/398
urease lack, 52/397
viswanathii, 35/417, 52/398
Candida albicans
acquired immune deficiency syndrome, 56/515, 71/299
brain abscess, 52/151
chronic meningitis, 56/645
disseminated infection, 52/397
infectious endocarditis, 52/301
Candida granuloma
spinal cord compression, 52/401
Candida krusei
endocarditis, 35/416
septicemia, 35/416
Candida meningitis
amphotericin B, 52/404
cranial nerve palsy, 52/400
CSF, 52/400
epilepsy, 52/400
flucytosine, 52/404
hydrocephalus, 52/400
mental deficiency, 52/400
neurologic symptom, 55/432
nystagmus, 52/400
Ommaya reservoir, 52/404
otitis media, 52/400
symptom, 52/399
Candida tropicalis
animal experiment, 35/416
brain abscess, 52/151
disseminated infection, 52/397
nonrecurrent nonhereditary multiple cranial neuropathy, 51/571
Candidiasis, 35/413-435
*see also* Candida
Addison disease, 42/512
amphotericin B, 35/433, 52/403
antibiotic agent, 35/419
associated disease, 52/399
brain abscess, 52/398
brain granuloma, 16/230
brain hemorrhagic infarction, 52/401
brain infarction, 52/399, 401, 55/435
brain microabscess, 52/398
candidemia, 52/398
cerebrovascular disease, 55/435
clinical features, 35/421-423
CNS, 35/420-423
computer assisted tomography, 52/403
corticosteroid, 35/418, 52/398
CSF, 35/428-431
CSF culture, 52/402
CSF shunt, 52/404
demyelinating encephalitis, 52/398
diabetes mellitus, 52/398
diagnosis, 35/431, 52/402
disseminated disease, 52/401
encephalopathy, 52/400
endocarditis, 52/401
endocrine, *see* Endocrine candidiasis syndrome
endophthalmitis, 52/401

epidemiology, 35/414
flucytosine, 52/403
histopathology, 52/398
history, 35/413
immunocompromised host, 52/398
immunologic deficiency, 35/418
ketoconazole, 52/405
meningitis, 52/398
meningoencephalitis, 55/435
miconazole, 52/405
multiple brain abscess, 52/398
mycosis, 35/374
mycotic aneurysm, 52/401
neutropenia, 52/398
pathogenesis, 35/415-420
pathology, 35/423-428
pathophysiology, 52/398
predisposing factor, 35/417-420, 52/399
prognosis, 35/434, 52/403
prophylaxis, 35/435
renal transplantation, 35/419
spinal, 52/401
spinal cord compression, 52/401
subarachnoid hemorrhage, 52/401
subdural granuloma, 52/401
treatment, 35/432-434, 52/403
vasculitis, 52/401
Candidiosis, *see* Candidiasis
Candle guttering, *see* Subependymal nodule
Cane crushing injury, *see* Kohlu injury
Cane fever, *see* Leptospirosis
Canicola fever, *see* Leptospirosis
Canine chorea
distemper virus, 6/680
Canine distemper
encephalitis, 34/293
multiple sclerosis, 9/668, 47/325
optic neuritis, 9/683
paramyxovirus, 34/293
slow virus disease, 56/544
subacute sclerosing panencephalitis, 9/673
Canine distemper encephalomyelitis
acute, 9/546
animal viral disease, 9/669-675, 34/293
dog, 9/669-675
multiple sclerosis, 47/430
Canine distemper virus
animal viral disease, 56/587
demyelination, 56/587
morbillivirus genus, 56/587
multiple sclerosis, 47/110, 364
paramyxovirus, 56/587
viral encephalitis, 56/587
Canis familiaris intoxication
vitamin A, 37/96
Canis latrans
rabies, 56/384
Cannabinoid
epilepsy, 72/234
Cannabis
*see also* Tetrahydrocannabinol
dronabinol, 65/46
epilepsy, 72/238
hallucination, 4/332
hallucinogenic agent intoxication, 37/329
11-hydroxydronabinol, 65/46
neurotoxin, 50/219
tetrahydrocannabinol, 65/46
tetrahydrocannabinolic acid, 65/46
Cannabis intoxication
headache, 48/420
hereditary deafness, 50/219
Cannabis sativa
tetrahydrocannabinol, 65/46
Cannibalism
kuru, 34/285, 56/556
Cap disease
congenital myopathy, 62/332
facial weakness, 62/334
inclusion body, 62/341
sarcoplasmic mass, 62/341
Capillary angioma
racemose, *see* Telangiectasia
spinal cord, 20/444
Capillary apoplexy
hemorrhage, 11/583
Capillary closure
diabetic polyneuropathy, 51/511
Capillary hemangioma, *see* Nevus flammeus
Capillary permeability
chronic subdural hematoma, 57/287
radiation myelopathy, 61/203
Caprine arthritis encephalitis
lentivirus, 56/460
Caps
muscle fiber, 62/44
Capsaicin
headache, 48/111
Capsid
definition, 56/1
herpes simplex virus, 56/208
Capsomer
definition, 56/2
Capsula interna
aphasia, 45/321
complicated migraine, 5/85

diffuse axonal injury, 57/55
motor representation, 1/156
motor syndrome, 1/186
pathologic laughing and crying, 3/362, 45/223
Capsula interna lesion
hemianopia, 2/585
Capsular artery
carotid cavernous fistula, 12/273
radioanatomy, 11/69
Capsular polysaccharide antigen, *see* K-1 antigen
Captopril
dipeptidyl carboxypeptidase I inhibitor, 65/444
toxic myopathy, 62/597
Captopril intoxication
epilepsy, 65/445
headache, 65/445
Huntington chorea, 65/445
parkinsonism, 65/445
polyneuropathy, 65/445
progressive supranuclear palsy, 65/445
sensorimotor polyneuropathy, 65/445
Caput medusae
venous sinus thrombosis, 54/397
Caput obstipum
wry neck, 6/567
Capute-Rimoin-Konigsmark syndrome
deafness, 14/110
lentigo, 14/110
neurocutaneous syndrome, 14/101
zygodactyly, 14/110
Caramiphen
extrapyramidal disorder, 6/827
parkinsonism, 6/219
Carbamate
hypnotic agent, 37/353
neurotoxin, 37/353, 64/15, 160, 183
organophosphate induced delayed polyneuropathy, 64/177
polyneuropathy, 64/160
toxic neuromuscular junction disorder, 64/15
toxic polyneuropathy, 64/160
Carbamate intoxication
action site, 64/15
akinesia, 64/188
aldehyde dehydrogenase, 64/187
aldicarb, 64/184
aminocarb, 64/185
anorexia, 64/186
anticholinesterase, 64/183
ataxia, 64/190
atropine, 64/192
autonomic nervous system, 75/508
axonal degeneration, 64/186
benomyl, 64/187
benthiocarb, 64/187
cadmium intoxication, 64/187
cancer risk, 64/184
carbaril, 64/185, 189
carbofuran, 64/185
carbon disulfide, 64/188
catatonia, 64/191
chemical structure, 64/183
chemotherapy, 64/187
cholinergic receptor stimulating agent, 64/183
cimetidine, 64/189
coma, 64/186
conjugation, 64/189
convulsion, 64/190
dermal exposure, 64/186
diabetes insipidus, 64/186
diagnosis, 64/192
diarrhea, 64/186
diethylcarbamazine, 64/187
diethyldithiocarbamic acid, 64/187
dimethylcarbamic acid 1-isopropyl-3-methylpyrazol-5-yl ester, 64/185
distribution, 64/188
disulfiram, 64/185, 187
dystonia, 64/188
elimination, 64/188
environment, 64/184
epoxyethyl carbamate, 64/187
exposure, 64/184
fasciculation, 64/186, 190
fatality, 64/191
filariasis, 64/187
genetic effect, 64/187
headache, 64/186
hexapropymate, 64/186
human immunodeficiency virus infection, 64/187
hydrolysis, 64/189
hypothermia, 64/186
insecticide, 64/183, 187
lacrimation, 64/190
maneb, 64/185
manganese intoxication, 64/190
memory, 64/191
mercury intoxication, 64/187
metabolism, 64/188
*N*-methyl-*D*-glucamine dithiocarbamate, 64/187
muscle spasm, 64/186
mutagenicity, 64/190
nabam, 64/185
neostriatal degeneration, 64/188
occurrence, 64/184
oxidation, 64/189

pallidal degeneration, 64/188
pathomechanism, 64/15
pirimicarb, 64/185
polyneuropathy, 64/186, 191
putaminal degeneration, 64/188
respiratory failure, 64/190
rhinorrhea, 64/190
salivation, 64/190
sensorimotor polyneuropathy, 64/186
sleep apnea, 64/186
sodium dihydroxyethyl dithiocarbamate, 64/187
spasm, 64/190
symptom, 64/15
*m*-tolylmethyl carbamate, 64/186
treatment, 64/192
tremor, 64/191
urethan, 64/184, 187
urinary incontinence, 64/190
vinyl carbamate, 64/187
vomiting, 64/186
weight loss, 64/186
Carbamazepine
antiepileptic agent, 37/200, 65/496
arsenic intoxication, 36/215
attack, 5/381
brain stem disorder, 47/57
cognitive improvement, 72/395
cranial neuralgia, 5/23
demyelination, 47/38
dyssynergia cerebellaris myoclonica, 60/603
epilepsy, 15/653, 669, 73/358
glossopharyngeal neuralgia, 5/357
intractable hiccup, 63/491
lithium intoxication, 64/299
migraine, 48/202
motor unit hyperactivity, 41/296
multiple sclerosis, 47/57
myokymia, 41/300
myotonic syndrome, 40/546, 561
neuromyelitis optica, 47/406
neurotoxin, 5/383, 15/653, 669, 707, 37/200, 47/38, 56/234, 64/177, 299, 65/496, 505
phantom limb pain, 45/401
polyneuropathy, 47/38
postherpetic neuralgia, 51/181, 56/234
serum level, 15/692
side effect, 15/707
superior laryngeal neuralgia, 48/498
temporal lobe epilepsy, 73/78
toxic encephalopathy, 65/505
trigeminal neuralgia, 5/383
water intoxication, 64/244
Carbamazepine intoxication
ageusia, 65/507
asterixis, 65/507
ataxia, 37/202, 65/505
carbamazepine-10,11-oxide, 37/200
chorea, 65/506
cognition, 65/506
diplopia, 37/202, 65/505
drowsiness, 37/202
dysarthria, 65/505
dysgeusia, 65/506
dystonia, 65/506
EEG, 65/506
extrapyramidal system, 37/203
fatality, 65/505
headache, 65/505
hyponatremia, 65/507
metabolite, 37/200
myoclonus, 65/506
nystagmus, 37/202, 65/507
oxcarbazepine, 65/507
personality change, 65/507
seizure exacerbation, 65/506
symptom, 65/506
vertigo, 65/505
vestibular system, 37/203
water intoxication, 65/506
Carbamazepine neurotoxicity
calcium antagonist intoxication, 65/442
nifedipine intoxication, 65/442
nimodipine intoxication, 65/442
Carbamoylphosphate synthase deficiency
valproic acid intoxication, 65/504
Carbamoylphosphate synthetase
epilepsy, 72/222
Carbamoylphosphate synthetase deficiency
diagnostic problem, 29/88
metabolic ataxia, 21/576
Carbamylcholine
myopathy, 40/150
Carbamylphosphate synthetase, *see* Carbamoylphosphate synthetase
Carbamylphosphate synthetase-1 deficiency, *see* Carbamoylphosphate synthase deficiency
Carbaril
carbamate intoxication, 64/185, 189
organophosphate induced delayed polyneuropathy, 64/177
Carbaryl, *see* Carbaril
Carbenoxolone
toxic myopathy, 62/601
Carbinol, *see* Methanol
Carbofenotion
organophosphate intoxication, 64/155

Carbohydrate deficient glycoprotein
type II, 66/634
type III, 66/634
type IV, 66/635
Carbohydrate deficient glycoprotein syndrome
*N*-acetylglucosaminyltransferase II, 66/83
adult stable disability stage, 66/627
asparagine linked oligosaccharide, 66/83
biochemistry, 66/631
childhood ataxia-mental deficiency stage, 66/625
clinical features, 66/625
diagnostic test, 66/633
genetics, 66/627
glycoprotein, 66/631
late infantile ataxia-mental deficiency stage, 66/625
metachromatic leukodystrophy, 66/83
stage, 66/624
symptom, 66/629-631
teenage leg atrophy stage, 66/627
transferrin, 66/633
type, 66/623
type I, 66/623
type II, 66/633
type III, 66/634
type IV, 66/634
Carbohydrate metabolism
brain blood flow, 27/2
Down syndrome, 31/433
Friedreich ataxia, 60/304
glycosaminoglycanosis, 27/162
$GM_1$ gangliosidosis, 27/163
$GM_2$ gangliosidosis, 27/163
$GM_2$ gangliosidosis type II, 27/163
hepatic coma, 27/353
hypothalamus, 2/440
inborn error, *see* Inborn error of carbohydrate metabolism
Marinesco-Sjögren syndrome, 42/185
multiple sclerosis, 9/133
periodic paralysis, 28/582, 597
progressive myoclonus epilepsy, 21/62
salicylic acid intoxication, 37/418
subacute necrotizing encephalomyelopathy, 42/626
vitamin $B_1$ deficiency, 28/3
β-Carboline
benzodiazepine intoxication, 65/334
Carbon dioxide
brain blood flow, 11/124
brain blood flow autoregulation, 11/124
brain injury, 23/117, 137
brain microcirculation, 53/79
Cheyne-Stokes respiration, 1/497, 670-673
memory disorder, 27/463
migraine, 48/73, 76, 75/295
neurotoxicity, 51/276
partial pressure, 11/349-353
Carbon dioxide intoxication
symptomatic dystonia, 49/543
Carbon dioxide narcosis
lateral medullary infarction, 63/441
poliomyelitis, 63/441
syringobulbia, 63/441
Carbon disulfide
axonal neuropathy, 64/86
carbamate intoxication, 64/188
chemical formula, 64/82
element data, 64/23
hexacarbon neuropathy, 51/278
industrial application, 64/23
neurotoxin, 36/382, 46/393, 48/420, 49/466, 51/68, 279, 531, 64/8, 10, 23, 25, 40, 82, 86, 188
optic neuropathy, 64/8
organic solvent, 64/82
organic solvent intoxication, 64/40
toxic neuropathy, 64/8, 10
toxic polyneuropathy, 51/68, 64/23
Carbon disulfide intoxication
action site, 64/10
acute type, 64/24
akinesia, 64/188
athetosis, 6/450
axonal neuropathy, 64/25
biomonitoring, 64/23
biotransformation, 64/24
brain atrophy, 64/25
brain stem auditory evoked potential, 64/25
clinical features, 36/383
CNS effect, 64/25
coma, 64/24
convulsion, 64/24
death, 64/24
delayed effect, 64/27
delirium, 64/24
dementia, 64/25
distal axonopathy, 64/26
dithiocarbamate, 64/24
dystonia, 64/188
EEG, 64/25
EMG, 64/25
encephalopathy, 64/25
epilepsy, 64/24
exposure limit, 64/23
fluorescein angiography, 64/26
headache, 48/420, 64/24

hypertension, 64/27
insomnia, 64/25
ischemic heart disease, 64/27
kinetic, 64/23
laboratory finding, 36/384
long-term exposure, 64/24
myasthenia-like syndrome, 64/25
myasthenic reaction, 64/25
myelopathy, 64/8
neostriatal degeneration, 64/188
nerve conduction, 7/174
neuropathology, 49/477
neuropathy, 7/522, 64/8
pallidal degeneration, 64/188
pallidal necrosis, 6/652, 49/466
Parkinson tremor, 6/243
parkinsonism, 64/25
pathogenesis, 36/384
polyneuropathy, 6/658, 64/25
pregnancy, 64/27
prevention, 36/385
prognosis, 36/385
putaminal degeneration, 64/188
retinal microaneurysm, 64/26
retinopathy, 64/26
secondary demyelination, 64/26
segmental demyelination, 64/26
sensorimotor polyneuropathy, 64/25
sexual impotence, 64/25
spastic paraparesis, 64/25
symptom, 6/658
2-thiothiazolidine-4-carboxylic acid, 64/24
toxic neuropathy, 64/25
toxic polyneuropathy, 64/25
tremor, 64/25
visual impairment, 64/8
vitamin B6 deficiency, 64/24
Carbon disulfide polyneuropathy
chronic axonal neuropathy, 51/531
clinical features, 49/477
dementia, 49/477
EMG, 51/279
giant axonal change, 51/279
histopathology, 51/279
parkinsonism, 49/477
psychosis, 51/279
Carbon monoxide
neurotoxin, 46/393, 48/420, 49/466, 556, 62/601, 64/7, 31
neurotransmitter, 74/94, 96
toxic encephalopathy, 64/7
toxic myopathy, 62/601
Carbon monoxide encephalopathy
acute type, 9/629
clinical features, 9/629, 49/476
focal demyelination, 9/472
globus pallidus, 49/472
late type, 9/629
myelinoclastic sclerosis, 9/481
neuropathology, 9/629
secondary polycythemia, 63/250
Carbon monoxide intoxication
agnosia, 64/32
akinesia, 64/31
akinetic mutism, 6/657, 49/474
akinetic rigid syndrome, 49/475
animal study, 64/35
anoxic ischemic leukoencephalopathy, 9/578, 47/528, 530
anoxic leukomalacia, 47/527
apathy, 64/32
apraxia, 64/32
athetosis, 49/476
biochemistry, 63/418
bradykinesia, 64/31
brain anoxia, 64/34
brain blood flow, 64/34
brain computer assisted tomography, 63/418
brain infarction, 53/167
brain ischemia, 53/167
brain nuclear magnetic resonance, 63/418
carboxyhemoglobin, 64/31
catatonia, 49/474
chorea, 49/476, 556, 64/31
choreoathetosis, 49/475
chronic, *see* Carbon monoxide encephalopathy
chronic type, 64/32
clinical features, 6/656, 49/473
coma, 49/473, 64/31
computer assisted tomography, 49/476, 64/33
confusion, 49/475, 64/32
congenital malformation, 64/32
cortical blindness, 55/119
course, 6/656, 64/35
death rate, 49/474
decerebrate rigidity, 49/474
delayed symptom, 64/32
dementia, 64/32
demyelination, 9/628-632
differential diagnosis, 49/476
dizziness, 64/31
dopaminergic therapy, 64/36
dystonia, 64/31
EEG, 64/33
epidemiology, 64/35
epilepsy, 63/418, 64/31

experimental study, 49/473
fatality, 63/417
globus pallidus, 64/33
Grinker myelinopathy, 64/35
headache, 48/417, 420, 64/31
hyperbaric oxygen, 64/36
intracerebral calcification, 49/417
latency period, 64/7
leukoencephalopathy, 9/578, 628, 47/553-555
mortality rate, 64/35
muscle spasm, 64/31
mutism, 49/475, 64/32
neuropathology, 29/545, 64/35
neuropathy, 7/521, 64/32
nuclear magnetic resonance, 64/33
oxygen deficiency, 29/541-547
pallidal necrosis, 6/652, 654, 656, 660, 47/527, 49/466, 471, 473, 63/418
pallidonigral necrosis, 49/466
pallidostriatal necrosis, 49/466
Parkinson disease, 49/108
parkinsonism, 6/242, 49/475, 64/7, 31
pathogenesis, 49/473, 64/34
pathology, 49/471
periodic fever, 64/32
permanent neuropsychiatric sequelae, 49/474
personality change, 49/475
polymicrogyria, 50/257
prognosis, 64/35
pseudobulbar paralysis, 1/198
psychosis, 49/475, 64/32
pyramidal sign, 49/474
single photon emission computer assisted tomography, 64/33
spasticity, 64/31
spheroid body, 6/626
striatal necrosis, 49/500, 504
striatopallidodentate calcification, 49/417
subarachnoid hemorrhage, 12/113
symptom, 64/7
tachycardia, 64/31
tachypnea, 64/31
tissue acidosis, 64/34
treatment, 29/546, 49/476, 63/418, 64/35
tremor, 6/823, 64/32
trismus, 49/474
urinary incontinence, 64/32
Carbon tetrachloride intoxication
acute hepatic necrosis, 27/350
headache, 48/420
leukoencephalopathy, 47/556
Carbonate dehydratase
glia, 9/14
McLeod phenotype, 63/287
nerve myelin, 51/25
Carbonate dehydratase II deficiency
osteopetrosis, 70/17
Carbonic anhydrase, *see* Carbonate dehydratase
Carbonic anhydrase II deficiency, *see* Carbonate dehydratase II deficiency
Carbonylcyanide-meta-chlorophenylhydrazone
experimental myopathy, 40/158
mitochondrial abnormality, 40/158
Carbophenothion, *see* Carbofenotion
Carboplatin
antineoplastic agent, 65/529
chemotherapy, 67/284
neuropathy, 69/460, 464, 467
neurotoxin, 65/529, 67/356
polyneuropathy, 65/529
Carboplatin intoxication
ototoxicity, 65/537
polyneuropathy, 65/537
Carboxylic acid
headache, 48/174
migraine, 48/174
Carcinoembryonic antigen
ataxia telangiectasia, 60/375
Carcinogen
brain tumor, 17/5, 9, 11, 13
brain tumor carcinogenesis, 17/9, 11, 13
DMNU, *see* Dimethylnitrosourea
ENU, *see* Ethylnitrosourea
isomorphic glioma, 17/28
MNU, *see* Methylnitrosourea
PDMT, *see* Phenyldimethyltriazene
polymorphic glioma, 17/28
teratogen, 31/37
TMNU, *see* Trimethylnitrosourea
Carcinogenesis
brain tumor, *see* Brain tumor carcinogenesis
ionizing radiation neuropathy, 51/137
Carcinoid myopathy
paraneoplastic syndrome, 69/374, 387
Carcinoid syndrome
choroidal, 70/315
embolic infarction, 70/315
epidural, 70/314
hormone production, 70/315
intracranial, 70/314
metastasis, 70/314
myopathy, 70/316
neurofibromatosis, 70/315
neuropathy, 70/316
paraneoplastic, 70/316
pellagra, 70/315

sinus thrombosis, 70/316
survey, 70/313
Carcinoid tumor
acanthocytosis, 63/290
ivory vertebra, 19/171
terminology, 18/438
Carcinoma
*see also* Cancer *and* Malignancy
adenoid cystic, *see* Adenoid cystic carcinoma
adrenocortical, *see* Adrenocortical carcinoma
amyotrophic lateral sclerosis, 59/202
amyotrophic myelopathy, 38/671-675
basal cell, *see* Basal cell carcinoma
Bloom syndrome, 43/10
brain, 38/625-660
brain metastasis, 17/413, 421, 426
brain tumor carcinogenesis, 38/625
breast, *see* Breast carcinoma
bronchogenic, *see* Bronchogenic carcinoma
bronchopulmonary, *see* Bronchopulmonary carcinoma
central pontine myelinolysis, 38/657
choroid plexus, *see* Choroid plexus carcinoma
chronic axonal neuropathy, 51/531
colon, *see* Colon carcinoma
cord, 38/669-676
cortical cerebellar degeneration, 38/625-631
dementia, 38/647-651, 657
dyskeratosis congenita, 43/392
ectopic hormone production, 38/651
embryonal cell, *see* Embryonal cell carcinoma
encephalomyelitis, 38/631-647
epidemiology, 68/173
Foix-Alajouanine disease, 38/669-671
history, 38/679-684
Hodgkin disease, 38/687
hypercalcemia, 38/652-654
kidney, *see* Renal carcinoma
limb girdle syndrome, 62/188
liver cell, *see* Liver cell carcinoma
lung, *see* Lung carcinoma
malignant lymphoma, 38/687
mammary, *see* Mammary carcinoma
medullary thyroid, *see* Medullary thyroid carcinoma
metabolic neuropathy, 51/68
middle ear, *see* Middle ear carcinoma
mononeuritis, 38/686
multiple myeloma, 38/687
multiple nevoid basal cell, *see* Multiple nevoid basal cell carcinoma syndrome
nasopharyngeal, *see* Nasopharyngeal carcinoma
neuropathy, 8/139
nonbacterial thrombotic endocarditis, 11/415, 417, 55/165, 63/111, 120
oat cell, *see* Oat cell carcinoma
opsoclonus, 34/614
optic neuritis, 38/658-660
ovarian, *see* Ovary carcinoma
pellagra, 38/648
pituitary, *see* Pituitary carcinoma
pituitary adenoma, 17/383
pituitary gland, 17/384
pituitary tumor, 17/384
polyneuropathy, 38/679-688
progressive multifocal leukoencephalopathy, 38/654-657
progressive spinal muscular atrophy, 22/31
prostatic, *see* Prostatic carcinoma
renal, *see* Renal carcinoma
restless legs syndrome, 8/315, 51/546
sensorimotor neuropathy, 38/685
sensory neuropathy, 8/369, 38/684-686
squamous, *see* Squamous carcinoma
tertiary myelopathy, 38/675
thyroid, *see* Thyroid carcinoma
vertebral metastasis, 20/415
vitamin B12 deficiency, 38/650
Wernicke-Korsakoff syndrome, 38/648
Carcinoma multiple neurinoma syndrome
pheochromocytoma thyroid, *see* Pheochromocytoma thyroid carcinoma multiple neurinoma syndrome
Carcinoma syndrome
multiple nevoid basal cell, *see* Multiple nevoid basal cell carcinoma syndrome
Carcinomatous cerebellar degeneration
*see also* Late cortical cerebellar atrophy
dentatorubropallidoluysian atrophy, 21/520
Hodgkin disease, 21/495
late cortical cerebellar atrophy, 21/477, 489-495
opsoclonus, 21/492
slow virus disease, 21/495
Carcinomatous encephalopathy
amyotrophic lateral sclerosis, 22/144
polycythemia, 9/596
progressive bulbar palsy, 22/143
Carcinomatous meningitis
chronic meningitis, 56/645
Carcinomatous myopathy, 41/317-386
*see also* Cancer *and* Malignancy
Eaton-Lambert myasthenic syndrome, 22/39
motoneuron degeneration, 22/39
Carcinomatous neuromyopathy
amyotrophic lateral sclerosis, 22/327
clinical features, 41/322

incidence, 41/322
paraneoplastic syndrome, 69/374, 386
progressive muscular atrophy, 59/21
ventral horn inflammation, 22/31
Carcinoscorpius rotundicauda intoxication
toxic, 37/58
Cardiac abnormality, *see* Heart disease
Cardiac arrest
*see also* Brain blood flow
action myoclonus, 63/217
alpha coma, 63/219
amnesia, 63/214, 217
anoxic encephalopathy, 63/214, 217
anoxic ischemic leukoencephalopathy, 9/575
astrocyte death, 63/212
autonomic dysfunction, 63/215
Balint syndrome, 63/216
basal ganglion disease, 63/216
behavior change, 63/217
blindness, 63/216
brachial diplegia, 63/218
brain blood flow, 63/209
brain death, 24/770, 63/213
brain edema, 63/213
brain embolism, 53/157
brain infarction, 53/157
brain ischemia, 53/157, 63/209
brain ischemic threshold, 63/210
brain oxygen metabolic rate, 63/210
brain stem death, 57/441, 466
brain stem necrosis, 63/213
brain stem syndrome, 63/215
brain watershed area, 63/213
cardiopulmonary resuscitation, 63/206
central cord necrosis, 61/114
clinical features, 38/156-159
coma duration, 63/214
cortical blindness, 63/215
cortical laminar necrosis, 63/214
delayed encephalopathy, 63/214
dementia, 63/215, 217
dextran, 55/210
electrical injury, 23/706, 61/192
electrocerebral silence, 57/466
epidemiology, 63/205
epilepsy, 63/217
free radical scavenger, 55/210
gaze paralysis, 63/215
heparin, 55/210
hereditary sensory and autonomic neuropathy type III, 21/109, 60/27
hypermagnesemia, 63/563
hypomania, 63/217
iatrogenic neurological disease, 61/165
irritability, 63/217
ischemic myelopathy, 63/40, 213
lumbar puncture, 61/165
mental depression, 63/217
metabolic encephalopathy, 63/214
mutism, 63/215
myelomalacia, 61/114, 63/40, 215
myocardial infarction, 63/207
myoclonic epilepsy, 63/214, 217
myoclonic epileptic status, 63/214
neurologic complication, 38/148-159, 63/205
neuron death, 63/210, 212
ocular bobbing, 63/215
oculomotor apraxia, 63/216
oligodendroglia death, 63/212
organophosphate intoxication, 64/169
orofacial dystonia, 63/216
outcome, 63/208
pallidal necrosis, 49/466, 63/213
pallidostriatal necrosis, 49/466
paraplegia, 63/215
parkinsonism, 63/216
pathogenesis, 38/142-155
permanent neurologic deficit, 63/214
persistent vegetative state, 38/157, 57/441
personality change, 63/217
postanoxic intention myoclonus, 38/580
prognosis, 63/220
pure autonomic failure, 2/120, 63/215
resuscitation, 55/209
resuscitation effect, 63/208
simultanagnosia, 63/216
spinal anesthesia, 61/165
spinal cord electrotrauma, 61/192
spinal cord injury, 61/439, 446
status epilepticus, 63/217
striatal necrosis, 49/500
survival, 63/206
survival predictor, 63/208
tetanus, 52/230
tetraplegia, 61/264, 439, 446, 63/215
thermoregulation, 75/61
transient neurologic deficit, 63/214
transverse spinal cord lesion, 61/439, 446
vegetative state, 63/218
violence, 63/217
Wernicke-Korsakoff syndrome, 63/214, 217
Cardiac arrest encephalopathy
anoxic ischemic leukoencephalopathy, 47/530
Cardiac arrhythmia, *see* Cardiac dysrhythmia
Cardiac atrophy
neuronal intranuclear hyaline inclusion disease,

60/668
Cardiac catheterization
amnesia, 63/176
brain embolism, 55/161, 63/6, 177
brain embolism frequency, 63/176
brain embolism incidence, 63/6
brain hemorrhage, 63/177
catheter flushing, 63/177
confusional state, 63/176
congenital heart disease, 63/6
diffuse encephalopathy, 63/177
guidewire, 63/177
iatrogenic neurological disease, 55/161, 63/175
migraine, 63/177
neurologic complication, 63/6, 175
neurologic deficit, 63/176
nonionic contrast material, 63/177
retinal infarction, 63/176
temporal lobe ischemia, 63/177
vertebrobasilar embolus, 63/176
visual perseveration, 63/176
Cardiac circulatory failure
pallidonigral necrosis, 49/466
Cardiac conduction
Leber hereditary optic neuropathy, 62/509
myotonic dystrophy, 62/217
Cardiac dysrhythmia
Adams-Stokes syndrome, 39/260-262
alcohol intoxication, 64/112
amygdala, 63/239
anterior cingulate gyrus, 63/239
aorta stenosis, 63/25
autonomic nervous system, 75/441
brain air embolism, 55/193
brain aneurysm neurosurgery, 55/54
brain embolism, 53/157, 55/151
brain hemorrhage, 55/151, 63/237
brain infarction, 53/157, 55/151
brain ischemia, 53/157, 63/236
brain tumor, 63/236
cardiac nerve, 63/239
cardiovascular agent intoxication, 37/426
central sleep apnea, 63/461
cerebral manifestation, 39/259-268
cerebrovascular disease, 75/370
cervical spondylosis, 7/455
chronic sinoatrial disorder, 39/262-264
congenital heart disease, 63/5
daytime hypersomnia, 45/135
diabetic autonomic polyneuropathy, 63/238
diabetic polyneuropathy, 63/238
epilepsy, 63/236
fibrillation, 38/144
glycoside intoxication, 37/426
Guillain-Barré syndrome, 51/255, 63/238
head injury, 63/236
hippocampus, 63/239
hypothalamus stimulation, 63/239
incidence, 55/151
insula, 63/205
insular cortex, 63/239
investigation, 39/266-268
long QT syndrome, 39/266
middle cervical ganglion, 63/238
myotonic dystrophy, 41/421, 486
neurogenic pulmonary edema, 55/157
organophosphate intoxication, 64/227
paroxysmal supraventricular tachycardia, 39/264
paroxysmal tachycardia, 39/264-266
paroxysmal ventricular fibrillation, 39/265
paroxysmal ventricular tachycardia, 39/264
petrobasilar suture syndrome, 2/313
polymyositis, 41/73
progressive external ophthalmoplegia, 41/215
propranolol, 37/444
REM sleep, 63/239
scapulohumeral distal muscular dystrophy, 43/130
stellate ganglion, 63/238
stress, 63/239
stroke, 63/238, 75/370
subarachnoid hemorrhage, 55/10, 15, 54, 154, 63/236
subiculum, 63/239
sudden death, 63/238
sympathetic activation, 63/239
tetanus, 52/230
transient ischemic attack, 12/10
transtentorial herniation, 63/237
Cardiac glycoside
cardiovascular agent, 65/433
Digitalis lanata, 37/426
Digitalis purpurea, 37/426
digoxin, 65/449
historic aspect, 37/425
hypernatremia, 28/447
squill, 37/425
squill intoxication, 37/425
toxic myopathy, 62/598
Cardiac interstitial fibrosis
Friedreich ataxia, 21/344, 346
Cardiac involvement
Bassen-Kornzweig syndrome, 51/327
childhood myoglobinuria, 62/561
congenital myopathy, 62/334
debrancher deficiency, 62/482
Duchenne muscular dystrophy, 62/119

Kearns-Sayre-Daroff-Shy syndrome, 38/229
limb girdle syndrome, 62/185
Lyme disease, 51/205
myotonic dystrophy, 40/509, 523
neuroborreliosis, 51/205
procainamide, 40/509, 523
quinidine, 40/509, 523
trichinosis, 35/276
Wohlfart-Kugelberg-Welander disease, 59/87, 372

Cardiac lesion
congenital rhabdomyoma, 14/51
Friedreich ataxia, 21/325, 342-345
hereditary motor and sensory neuropathy type I, 21/286
hereditary motor and sensory neuropathy type II, 21/286
Refsum disease, 21/200-202, 60/231
tuberous sclerosis, 14/50

Cardiac migraine
classification, 48/6
migraine equivalent, 48/157

Cardiac myxoma, 38/173-189
clinical features, 38/174-179
retinal artery occlusion, 55/112
surgery, 38/188

Cardiac myxosarcoma
age, 63/103
brain embolism, 63/103
spinal column metastasis, 63/103

Cardiac output
acute hypoxia, 75/262
autonomic nervous system, 74/248
brain collateral blood flow, 11/171
brain infarction, 53/177
chronic hypoxia, 75/269
obstructive sleep apnea, 63/457

Cardiac pacemaker
Guillain-Barré syndrome, 51/257
innervation, 74/42
progressive dysautonomia, 59/140

Cardiac papillary fibroelastoma
stroke, 63/103
transient ischemic attack, 63/103

Cardiac pharmacotherapy
amiodarone, 63/190
antiarrhythmic agent, 63/190
beta adrenergic receptor blocking agent, 63/194
bretylium, 63/192
calcium channel blocking agent, 63/193
digoxin, 63/195
disopyramide phosphate, 63/192
flecainide, 63/192
immunosuppressive agent, 63/189
lidocaine, 63/190, 192
mexiletine, 63/192
neurologic complication, 63/175, 189
osteoporosis, 63/189
procainamide, 63/190
quinidine, 63/190
steroid myopathy, 63/189
tocainide, 63/192

Cardiac resuscitation
method, 63/206

Cardiac standstill, *see* Cardiac arrest

Cardiac surgery
*see also* Open heart surgery
air embolism, 23/614
brain embolism, 11/396
brain infarction, 11/442-448
bypass equipment, 38/163
fat embolism, 53/161
hypothermia, 71/578
iatrogenic neurological disease, 55/168
neurologic complication, 38/160-166
open, *see* Open heart surgery
postoperative fungal endocarditis, 11/446
postoperative psychosis, 11/446
risk factor, 38/162
systemic brain infarction, 11/442, 444
valve replacement, 38/164-166

Cardiac valvular disease
anticoagulant complication, 63/19
anticoagulant risk, 63/19
anticoagulation, 63/18
anticoagulation timing, 63/19
aorta stenosis, 38/148, 63/24
atrial fibrillation, 38/144, 63/17
atrial septal aneurysm, 63/22
bicuspid aortic valve, 63/26
brain embolism, 53/155, 55/164, 63/17
brain hemorrhage, 55/164
brain infarction, 53/155, 55/164
brain ischemia, 53/155
embolism, 38/143-146
infectious endocarditis, 63/22
intravascular consumption coagulopathy, 53/156
ischemic heart disease, 63/18
Libman-Sacks endocarditis, 63/29
mitral annulus calcification, 63/29
mitral valve prolapse, 38/145, 63/20
multiple valve prolapse, 63/22
neurologic complication, 38/143-149, 63/17
paradoxical embolism, 63/22
prosthetic cardiac valve, 63/26
recurrent embolism, 63/19
retinal artery occlusion, 55/112

rheumatic fever, 63/17
rheumatic heart disease, 38/143
Sydenham chorea, 49/363
syncope, 38/148
thromboembolism rate, 63/27
Cardiac ventricular aneurysm
brain embolism, 55/162
Cardiac ventricular septum defect
craniolacunia, 50/139
holoprosencephaly, 50/237
infectious endocarditis, 63/112
Cardiac wave
ECG change, 63/235
Cardiazol, *see* Pentetrazole
Cardiofacial syndrome
anal atresia, 42/308
birth incidence, 42/308
depressor anguli oris muscle, 42/307
facial asymmetry, 42/307
facial paralysis, 42/307
heart disease, 42/307
quadratus labii inferioris muscle, 42/307
sex ratio, 42/308
VATER association, 42/308
vertebral abnormality, 42/308
Cardiolipin
biochemistry, 9/8
chemistry, 10/238
Cardiomegaly
brain arteriovenous malformation, 12/232
eosinophilia myalgia syndrome, 63/376
Fabry disease, 51/377
infantile acid maltase deficiency, 27/195, 227, 62/480
long chain acyl-coenzyme A dehydrogenase deficiency, 62/495
venous malformation, 12/232, 236
Cardiomotor fiber
test, 74/596
Cardiomyopathy
*see also* Heart disease
acid maltase deficiency, 43/178
adriamycin, 63/136
alcohol, 63/133
alcoholism, 63/133
autonomic nervous system, 75/447
brain embolism, 53/156, 55/153, 162, 63/131
brain infarction, 53/156
brain ischemia, 53/156
brancher deficiency, 62/484
carnitine deficiency, 43/175, 62/493
cerebrovascular disease, 53/43, 55/162
Chagas disease, 52/348, 63/135
chloroquine myopathy, 62/608
chorea-acanthocytosis, 63/281
classification, 63/131
cobalt intoxication, 36/330
combined complex I-V deficiency, 62/506
complex I deficiency, 66/422
complex III deficiency, 66/423
congenital heart disease, 63/2
congestive, *see* Congestive heart failure
dermatomyositis, 62/374
diphtheria toxin, 41/201
doxorubicin, 64/17
Duchenne muscular dystrophy, 43/106
Emery-Dreifuss muscular dystrophy, 62/145, 152
endomyocardial fibrosis, 63/138
frequency, 63/131
Friedreich ataxia, 21/342-346, 40/508, 51/388, 60/318, 667
glycogen storage disease, 62/490
glycosaminoglycanosis type VI, 66/307
hereditary amyloid polyneuropathy type 1, 60/92
hypertrophic, *see* Hypertrophic cardiomyopathy
idiopathic, *see* Idiopathic cardiomyopathy
idiopathic hypereosinophilic syndrome, 63/138
inclusion body myositis, 62/374
incomplete dystrophin deficiency, 62/135
infantile histiocytoid, *see* Infantile histiocytoid cardiomyopathy
Kearns-Sayre-Daroff-Shy syndrome, 22/211, 43/142, 60/667, 62/310
lipid metabolic disorder, 62/493
long chain 3-hydroxyacyl-coenzyme A dehydrogenase deficiency, 62/495
McLeod phenotype, 49/332, 63/287
mitochondrial DNA point mutation, 66/427
mitochondrial myopathy, 43/119, 188, 60/667
muscle carnitine deficiency, 66/403
myotonic dystrophy, 43/152
nemaline myopathy, 62/341
neurologic complication, 63/131
obstructive, *see* Hypertrophic cardiomyopathy
organic acid metabolism, 66/641
peripartum, 63/134
pheochromocytoma, 39/498, 42/764, 63/136
phosphorylase kinase deficiency, 62/487
polymyositis, 62/374
primary carnitine deficiency, 66/401
primary systemic carnitine deficiency, 62/493
progressive external ophthalmoplegia, 22/192, 62/310
Refsum disease, 8/22, 42/148, 60/228, 237, 66/491
restrictive infiltrative, *see* Restrictive infiltrative cardiomyopathy

Roussy-Lévy syndrome, 21/175
scapuloperoneal myopathy, 43/131
scapuloperoneal spinal muscular atrophy, 59/45
short chain 3-hydroxyacyl-coenzyme A dehydrogenase deficiency, 62/495
spinal muscular atrophy, 59/45
subacute necrotizing encephalomyelopathy, 51/388
Wohlfart-Kugelberg-Welander disease, 59/87
Cardiopathy
ARG syndrome, 50/582
complex I deficiency, 62/503
complex III deficiency, 62/505
congenital, *see* Congenital cardiopathy
MELAS syndrome, 62/510
Miller-Dieker syndrome, 50/592
muscular atrophy, 42/343-345
r(14) syndrome, 50/590
Sydenham chorea, 6/416-418
Cardiopulmonary baroreflex
space, 74/283
Cardiopulmonary bypass
brain arteriolar dilatation, 63/175
brain blood flow, 63/175
brain capillary dilatation, 63/175
brain embolism, 63/175
brain perfusion pressure, 63/188
neurologic complication, 63/175
neurologic complication prophylaxis, 63/188
postoperative dysrhythmia, 63/186
prior stroke, 63/186
prior stroke significance, 63/188
pump time *vs* stroke risk, 63/186
regional brain blood flow, 63/175
stroke risk, 63/186
Cardiopulmonary receptor
autonomic nervous system, 74/258
Cardiopulmonary reflex
autonomic nervous system, 75/111
Cardiorespiratory failure
complex I deficiency, 62/503, 66/422
Mallory body myopathy, 62/345
striatal necrosis, 49/500
Cardiorespiratory syndrome
obesity, *see* Pickwickian syndrome
Cardiotomy
delirium, 46/547
Cardiotonic steroid agent
hydrocephalus, 50/297
Cardiovascular abnormality
iniencephaly, 50/131
vagus nerve agenesis, 50/219
Cardiovascular agent
antiarrhythmic agent, 65/433
anticoagulant, 65/433
antithrombocytic agent, 65/433
behavior disorder, 46/601
beta adrenergic receptor blocking agent, 65/433
calcium antagonist, 65/433
cardiac glycoside, 65/433
dipeptidyl carboxypeptidase I inhibitor, 65/433
diuretic agent, 65/433
fibrinolytic agent, 65/433
neurologic side effect, 65/433
nitrate, 65/433
Cardiovascular agent intoxication, 37/425-430
cardiac dysrhythmia, 37/426
color vision, 37/426
fatigue, 37/426
gynecomastia, 37/426
headache, 37/426
nausea, 37/426
scotoma, 37/426
seizure, 37/426
syncope, 37/426
transient global amnesia, 37/426
trigeminal neuralgia, 37/429
visual acuity, 37/426
vomiting, 37/426
weakness, 37/426
Cardiovascular disturbance
poliomyelitis, 34/107, 110, 113
Cardiovascular dysautonomia
Chagas disease, 75/394
Cardiovascular function
aging, 74/228
autonomic nervous system, 74/600
CNS lesion, 63/229
gravity, 74/277
Guillain-Barré syndrome, 51/257
hypothalamus, 1/461
respiration, 74/597
space, 74/277
standing, 74/597
syncope, 75/208
test, 74/597
Valsalva maneuver, 74/597
Cardiovascular reflex
autonomic nervous system, 74/72
testing, 74/600
Cardiovascular response
epilepsy, 75/349
exercise, 74/245
larynx, 74/394
standing, 74/610
Cardiovascular system

autonomic nervous system, 75/425
burning feet syndrome, 28/10
carotid cavernous fistula, 24/412
cerebrotendinous xanthomatosis, 66/601
chironex intoxication, 37/49
chiropsalmus intoxication, 37/49
cluster headache, 75/288
headache, 75/288
hereditary amyloid polyneuropathy, 21/124
hereditary amyloid polyneuropathy type 2, 60/98
paraplegia, 61/435
rehabilitative sport, 26/525
spinal cord injury, 26/313-333, 61/435
tetraplegia, 61/435
transverse spinal cord lesion, 61/435
traumatic vegetative syndrome, 24/579

Cardiovirus
picornavirus, 56/309

Caretta caretta intoxication
ingestion, 37/89

Carisoprodol
anxiety, 5/168
insomnia, 5/168
muscle contraction headache, 5/168
muscle relaxation, 5/168

Carmustine
antineoplastic agent, 65/528
astrocytoma, 68/108
brain tumor, 18/520
chemotherapy, 39/115, 67/282, 284
encephalopathy, 65/528
epilepsy, 65/528
glioblastoma multiforme, 18/65
high grade glioma, 68/111
leukemia, 39/18
neurologic toxicity, 39/114
neurotoxin, 18/520, 39/115, 65/528, 67/356

Carnegie stage
CNS, 50/2
embryonic period, 30/18, 20, 25

Carney Complex
corticotropin, 70/210
neurinoma, 68/543

Carnitine
acyl-coenzyme A, 66/400
fatty acid, 66/400
lipid storage myopathy, 66/403
muscle, 66/403
myopathic phenotype, 66/402
valproic acid induced hepatic necrosis, 65/132
X-linked neutropenic cardioskeletal myopathy, 62/156

L-Carnitine, *see* Vitamin Bt

Carnitine acylcarnitine translocase deficiency
biochemistry, 66/405
heart dysfunction, 66/405
metabolic encephalopathy, 66/405

Carnitine cycle
acute metabolic encephalopathy, 66/402
component, 66/400
defect, 66/401, 404
fatty acid oxidation, 66/399, 402
genetic condition, 66/401
myoglobinuria, 66/402
myopathy, 66/401
primary carnitine deficiency, 66/401
renal dysplasia, 66/402
systemic carnitine deficiency, 66/401

Carnitine deficiency
acidosis, 43/176
acute metabolic encephalopathy, 66/406
biochemistry, 41/197, 200
cardiomyopathy, 43/175, 62/493
clinical features, 40/280, 308, 317, 41/196, 199
corticosteroid, 41/198
differential diagnosis, 40/422
ECG, 62/493
encephalopathy, 43/175
experimental myopathy, 41/200
Fanconi syndrome, 62/493
fatty acid oxidation disorder, 66/406
genetics, 41/198, 200, 427
hemodialysis, 62/494
hepatic encephalopathy, 62/493
hepatomegaly, 43/175
histopathology, 41/197, 199
hypoglycemia, 43/176
hypoglycemic encephalopathy, 62/493
infantile hypotonia, 62/493
inheritance, 62/493
kwashiorkor, 62/493
lipid accumulation, 40/95
lipid metabolic disorder, 41/196-201, 62/492
lipid storage myopathy, 40/95, 258, 280, 308, 317, 422, 43/175, 184, 62/493
metabolic myopathy, 41/196, 62/492
mitochondria lipid glycogen myopathy, 41/202
mitochondrial abnormality, 41/197
mitochondrial matrix, 66/410
muscle, *see* Muscle carnitine deficiency
neuropathy, 43/175, 66/406
pathologic reaction, 40/258
primary, *see* Primary carnitine deficiency
primary systemic carnitine deficiency, 62/493
retinopathy, 66/406
Reye like episode, 62/493

Reye syndrome, 49/217
rimmed muscle fiber vacuole, 41/197, 199
secondary systemic carnitine deficiency, 62/493
systemic carnitine deficiency, 62/493
toxic encephalopathy, 49/217
treatment, 41/198, 200
triglyceride, 43/175
valproic acid, 62/494
valproic acid intoxication, 65/504
Carnitine deficiency myopathy
age at onset, 62/492
fluctuating weakness, 62/492
Carnitine palmitoyltransferase
childhood myoglobinuria, 62/563
myoglobinuria, 62/558
Carnitine palmitoyltransferase II deficiency
biochemistry, 66/405
Carnitine palmitoyltransferase deficiency
carnitine palmitoyltransferase isoenzyme, 62/494
clinical features, 40/339, 342
cold exposure, 62/494
creatine kinase, 43/177
differential diagnosis, 40/438, 41/56, 60
exercise, 62/494
exercise intolerance, 43/176
fasting, 62/494
ischemic exercise, 62/494
ischemic exercise test, 41/205
lipid accumulation, 40/95
lipid metabolic disorder, 41/204-207, 62/494
lipid storage myopathy, 40/95, 258, 339, 342, 438, 43/177
liver, 62/494
metabolic myopathy, 41/204, 62/494
muscle biopsy, 62/494
muscle cramp, 62/494
muscle fiber type I, 43/177
myalgia, 43/176, 62/494
myoglobinuria, 40/339, 41/204, 265, 277, 428, 43/176, 62/494
myopathy, 62/494
palmitic acid, 41/206
pathologic reaction, 40/258
polymyositis, 41/56, 60
precipitating factor, 62/494
renal insufficiency, 62/494
rhabdomyolysis, 40/264
sex ratio, 62/494
triglyceride, 43/177
Carnosinase deficiency
biochemical finding, 29/198
carnosinemia, 42/532-534
clinical features, 29/197
diagnosis, 29/198
epilepsy, 72/222
genetics, 29/198
Carnosinemia
blindness, 42/532
brain atrophy, 42/532
carnosinase deficiency, 42/532-534
epileptic seizure, 42/532
mental deficiency, 29/197, 42/532
microcephaly, 42/532
Purkinje cell loss, 29/198, 42/532
pyramidal tract absence, 42/532
pyramidal tract degeneration, 42/532
spastic tetraplegia, 42/532
spinocerebellar degeneration, 42/532
Carol-Godfried-Prakken-Prick syndrome
mental deficiency, 43/8
Caroli disease
biliary cyst, 75/640
β-Carotene
primary pigmentary retinal degeneration, 13/193
Caroticotympanic artery
internal carotid artery syndrome, 11/308
Carotid angiography, 11/65-98, 102-115
*see also* Brain angiography
complicated migraine, 5/66
craniovertebral region, 32/44, 60
fibromuscular dysplasia, 11/372
headache, 5/66
hydrocephalus, 30/549-550
Matas test, 12/208
middle cerebral artery, 53/354
migraine, 5/19, 48/72
pure motor hemiplegia, 53/364
retinal artery occlusion, 55/112
stereoscopic view, 12/207
supratentorial brain abscess, 33/123
transient ischemic attack, 11/340
venous sinus occlusion, 54/442
Carotid artery
altitude illness, 48/397
amaurosis fugax, 48/338
aphthous ulcer, 48/338
arteriosclerosis, 11/327-365
atheroma, 11/331
auscultation, 11/216
autonomic nervous system, 75/434
B-mode sonography, 54/21
carotidynia, 5/375, 48/337
coiling, 11/173, 12/293
collateral, 12/308
common, *see* Common carotid artery
congenital arterial malformation, 12/289-308

congenital malformation, 12/289
external, *see* External carotid artery
fibrinolytic treatment, 54/182
fibromuscular dysplasia, 55/283
gasserian ganglion, 5/273, 335
headache, 5/128, 48/77
hypoplasia, 12/306
internal, *see* Internal carotid artery
malformation, 12/289-339
migraine, 48/77
murmur, 11/217, 220
murmur classification, 11/219
neck chamber, 75/434
pain, 5/375, 48/274
palpation, 11/215
pathologic murmur, 11/221-223
physiologic murmur, 11/220
primitive internal, *see* Primitive internal carotid artery
referred pain, 48/336
Reynolds number, 11/217
tortuosity, 12/293
transient infantile hemiplegia, 12/348
Carotid artery anastomosis
ophthalmic artery, 12/308
Carotid artery aneurysm
nonrecurrent nonhereditary multiple cranial neuropathy, 51/571
phentaramine addiction, 55/521
Carotid artery bifurcation
B-mode sonography, 54/21
carotid system syndrome, 53/303
Doppler sonography, 54/7
fibromuscular dysplasia, 55/284
flow velocity, 53/303
hemodynamics, 53/303
internal carotid artery syndrome, 53/303
level, 12/292
morphology, 53/302
swelling, 48/338
Carotid artery bifurcation aneurysm
brain aneurysm neurosurgery, 55/68
Carotid artery kinking
B-mode sonography, 54/21, 26
brain collateral blood flow, 11/173
frequency, 11/67
internal carotid artery, 11/173
significance, 12/293
Carotid artery ligation
brain aneurysm, 12/217, 220
carotid cavernous fistula, 24/423
complication, 12/220
contraindication, 12/218
indication, 12/218
Matas test, 12/219
result, 12/219
technique, 12/218
Carotid artery occlusion
*see also* Carotid artery thrombosis
brain infarction, 11/336
carotid cavernous fistula, 24/419
contralateral stenosis, 11/359
coronary bypass patient, 63/185
head injury, 26/43
hemiballismus, 49/373
histogenesis, 11/334
infarction site, 11/337
intermittent flow, 54/176
juvenile, 12/352
optic atrophy, 13/83
pallidal necrosis, 49/466, 469
pseudo-occlusion, 54/176
pseudoxanthoma elasticum, 43/45, 55/454
surgical treatment, 11/357
transient ischemic attack, 11/340
vascular dementia, 46/356
vascular spasm, 26/47
vertebral artery stenosis, 12/14, 23
Carotid artery patency
ophthalmodynamometry, 11/210
palpation, 11/208
Carotid artery stenosis
*see also* Carotid artery thrombosis
amaurosis fugax, 55/111
B-mode sonography, 54/19
brain aneurysm, 55/73
brain aneurysm neurosurgery, 55/73
computer assisted tomography, 54/58
coronary bypass patient, 63/185
Doppler sonography, 54/1
internal, *see* Internal carotid artery syndrome
ischemic optic neuropathy, 55/111
lysergide, 65/45
nicotine intoxication, 65/264
ocular pain, 55/111
ophthalmic artery, 54/2
ophthalmology, 55/111
orbital pain, 55/111
periorbital artery, 54/2
photopsia, 55/111
subarachnoid hemorrhage, 55/73
suction ophthalmodynamometry, 11/228
supratrochlear artery, 54/2, 4
Carotid artery thrombosis, 11/327-361
*see also* Brain infarction, Carotid artery occlusion, Carotid artery stenosis, Cerebrovascular disease

*and* Stroke
acute infantile hemiplegia, 53/32
age, 26/44
angiography, 26/45
associated disease, 11/342
B-mode sonography, 54/27
boxing injury, 23/539, 550
brain aneurysm, 26/47
cerebrovascular disease, 53/32, 37
child, 12/348
clinical features, 26/44
diagnosis, 26/44
football injury, 23/574
head injury, 57/374
management, 12/349
mortality, 26/48
mucormycosis, 35/545, 52/471, 473, 55/436
onset, 12/349
ophthalmodynamometry, 26/45
pathogenesis, 26/43
pathology, 12/349, 26/46
skiing injury, 23/579
skull base fracture, 26/43, 57/345
soccer injury, 23/569
spontaneous, 26/46
trauma, 12/348, 23/539, 550, 26/43, 45
treatment, 11/355, 26/48
Carotid basilar anastomosis
primitive hypoglossal artery, 12/311, 326-330
primitive otic artery, 12/311, 322-326
primitive trigeminal artery, 12/311-322
proatlantal intersegmental artery, 12/311, 330
various forms, 53/301
Carotid body
acute hypoxia, 75/260
alveolar hypoventilation, 63/414
cluster headache, 48/242
function, 1/495
paraganglioma, 42/762
synonym, 8/482
Carotid body carcinoma, *see* Carotid body tumor
Carotid body sarcoma, *see* Carotid body tumor
Carotid body tumor
B-mode sonography, 54/25
carotidynia, 48/339
epithelioid cell, 8/487
glomus jugulare tumor, 18/438
headache, 48/337
nerve tumor, 8/482-493
paraganglioma, 8/482
pericarotid syndrome, 48/337
synonym, 8/485
tissue culture, 17/69
Carotid canal
artery fixation, 53/293
congenital aneurysm, 31/146
Carotid cavernous fistula, 12/267-287
*see also* Traumatic carotid cavernous sinus fistula
anatomy, 24/401, 409, 434
angiography, 24/418
arteriosclerosis, 24/407
bilateral, 24/405, 410
brain aneurysm neurosurgery, 55/68
brain blood flow, 24/411, 419
brain scintigraphy, 23/293
bruit, 24/414, 417
capsular artery, 12/273
cardiovascular system, 24/412
carotid artery ligation, 24/423
carotid artery occlusion, 24/419
carotid rete mirabile, 11/69
cause, 12/268
chemosis, 24/415
complication, 24/435
compression carotid artery, 24/423
computer assisted tomography, 57/166
congenital, 24/406
congenital aneurysm, 31/150
course, 24/420
cranial nerve, 24/410, 416
craniotomy, 24/432
definition, 24/401
delayed, 24/408
diagnosis, 12/268, 24/417
differential diagnosis, 24/419
dorsal meningeal artery, 12/271
Ehlers-Danlos syndrome, 24/407, 55/455
electrothrombosis, 24/434
embolization, 24/426
etiology, 24/404
exophthalmos, 24/414
external carotid artery, 24/407, 420
extraocular muscle, 24/415
fibromuscular dysplasia, 55/288
glaucoma, 24/413, 416
Hamby method, 24/427
head injury, 24/399
headache, 5/149, 24/414, 48/374
hemorrhage, 24/420
hereditary hemorrhagic telangiectasia (Osler), 55/450
history, 24/399
immediate, 24/408
incidence, 24/405
inferior cavernous sinus artery, 12/272
mechanism, 24/404, 408

meningohypophyseal trunk, 12/271
muscle plugging, 24/427, 430
natural course, 24/420
nonrecurrent nonhereditary multiple cranial neuropathy, 51/571
normal anatomy, 12/270
ophthalmic artery, 24/403, 425
ophthalmoplegia, 24/415
optic atrophy, 24/416, 422
pathogenesis, 24/408
pathology, 12/274
pathophysiology, 24/411
penetrating head injury, 24/405, 57/309
pituitary adenoma, 17/413
postoperative, 24/405
predisposing factor, 24/405
pregnancy, 24/408
primitive trigeminal artery, 24/407
prognosis, 24/421
pulsation, 24/414
radiology, 24/418
retina, 24/415
sign, 24/413
skull base fracture, 24/406, 57/345
spontaneous, 24/405, 407
spontaneous cure, 24/421
subarachnoid hemorrhage, 12/102
surgery, 12/278
surgical procedure, 24/423
symptom, 31/150
thrill, 24/414
trapping procedure, 24/424
traumatic, 24/404
treatment, 24/422
trigeminal artery, 24/407
venous drainage, 24/401, 410, 434
venous ligation, 24/426
vision, 24/415, 422
Carotid cavernous thrombosis
aspergillosis, 35/397, 52/380
Carotid compression
syncope, 15/825
Carotid compression test
arteriographic correlation, 11/234
carotid sinus massage, 11/231
carotid sinus reflex, 11/230
cerebrovascular reserve, 11/230
complication, 11/234
contraindication, 11/229
indication, 11/229
interpretation, 11/231, 233
mechanical effect, 11/231, 233
parameter recording, 11/229
technique, 11/229
Carotid dissecting aneurysm
abducent nerve paralysis, 54/272
age, 54/271
amaurosis fugax, 54/272
angiography, 54/272
bilateral, 54/271
brain infarction, 53/39, 164
brain ischemia, 53/164
carotid artery coiling, 54/272
cerebrovascular disease, 53/39
chiropractic manipulation, 54/271
cluster headache, 48/225
computer assisted tomography, 54/274
coughing, 54/271
double barrel lumen, 54/273
fibromuscular dysplasia, 54/271, 273, 275, 55/283, 288
Fisher string sign, 53/40
glossopharyngeal nerve palsy, 54/272
headache, 53/40, 54/272
hemilingual paralysis, 54/272
histology, 54/276, 281
homocystinuria, 54/282
Horner syndrome, 54/272
hypoglossal paralysis, 54/272
intracranial, 54/275
Marfan syndrome, 39/390, 54/282
neck injury, 54/271
neck pain, 53/205
nuclear magnetic resonance, 54/275, 278
oculosympathetic paralysis, 53/40
paralysis, 54/272
physical stress, 54/271
polyarteritis nodosa, 54/282
sneezing, 54/271
spontaneous, 53/39, 54/271
string sign, 54/272
syphilitic arteritis, 54/282
thunderclap headache, 54/275
tinnitus, 54/272
transient ischemic attack, 54/272
trauma, 53/39, 54/271
treatment, 54/283
Carotid dissection
headache, 48/286
Carotid endarterectomy
alveolar hypoventilation, 63/414
brain embolism, 53/292
brain infarction, 53/165, 292, 426, 441
brain ischemia, 53/165
hypoventilation, 63/413
iatrogenic brain infarction, 53/165

internal carotid artery syndrome, 53/292, 322
methyldopa intoxication, 37/441
neurologic intensive care, 55/222
primary alveolar hypoventilation, 63/414
prolonged transient ischemic attack, 53/275, 277
retinal artery occlusion, 55/112
transient ischemic attack, 53/275-277, 292, 322
Carotid gasserian anastomosis
paratrigeminal syndrome, 5/335
Carotid gland, *see* Carotid body
Carotid glomus, *see* Carotid body
Carotid inferolateral trunk
multiple cranial neuropathy, 51/569
Carotid injury
bifurcation, 11/426
spasm, 11/426
Carotid murmur
endocarditis, 63/111
Carotid rete mirabile, 12/352-382
anastomosis, 12/308-311
anastomotic artery, 12/309
anatomy, 11/119
brain blood flow, 11/119
brain collateral blood flow, 11/173, 178
branch, 53/294
carotid cavernous fistula, 11/69
external carotid artery, 12/308-311
internal carotid artery, 11/173, 12/308-311
ophthalmic artery, 12/308-311
Carotid sheath
cluster headache, 48/338
fibrosis, 48/338
Carotid sinus
arterial baroreflex, 75/110
carotidynia, 48/338
Carotid sinus body
respiratory control, 63/431
Carotid sinus hypersensitivity
brain blood flow, 11/123
syncope, 75/214
Takayasu disease, 39/213-232, 217
temporal arteritis, 55/344
Carotid sinus massage
autonomic nervous system, 74/623
brain embolism, 53/219
brain infarction, 53/219
carotid compression test, 11/231
cerebrovascular disease, 53/219
intracranial hemorrhage, 53/219
transient ischemic attack, 53/219
Carotid sinus nerve
brain blood flow, 53/62
Cheyne-Stokes respiration, 63/484
Carotid sinus reflex
anatomy, 11/533
brain blood flow, 11/123
carotid compression test, 11/230
carotid sinus stimulation test, 11/536
epilepsy, 11/546
hyperactive, *see* Hyperactive carotid sinus reflex
innervation, 11/535
narcolepsy, 11/546
physiology, 11/533-535
Carotid sinus stimulation test
carotid sinus reflex, 11/536
Carotid sinus syncope, *see* Carotid sinus syndrome
Carotid sinus syndrome, 11/532-552
baroreceptor, 11/533
clinical features, 11/542
diagnosis, 11/543
differential diagnosis, 11/546
epilepsy, 11/546
incidence, 11/542
massage, 53/294
narcolepsy, 11/546
pathogenesis, 11/544
precipitating factor, 11/543
syncope, 7/452
treatment, 11/545
Carotid siphon
age, 53/294
intubation, 11/613
radioanatomy, 11/67
transcranial Doppler sonography, 54/12
variation, 11/75
Carotid system syndrome, 53/291-326
amaurosis fugax, 53/309, 315, 318
anastomosis, 53/298
anatomy, 53/293
angioscopy, 53/315
anterior choroidal territory events, 53/311
arteriosclerosis, 53/304
bilateral, 53/318
carotid artery bifurcation, 53/303
carotid artery disease, 53/308
cause, 53/308
classification, 53/309
collateral, 53/293, 300
computer assisted tomography, 53/315
dermatomyositis, 53/308
diagnosis, 53/312
Ehlers-Danlos syndrome, 53/309
episodic limb shaking, 53/318
external internal carotid bypass, 53/316
fibromuscular dysplasia, 53/308
glycosaminoglycanosis, 53/309

hemispheric events, 53/311
hemodynamic mechanism, 53/316
homocystinuria, 53/309
Marfan syndrome, 53/309
Melkersson-Rosenthal syndrome, 53/308
misery perfusion syndrome, 53/314
moyamoya disease, 53/308
natural history, 53/319
pathophysiology, 53/314
polyarteritis nodosa, 53/308
posterior communicating artery events, 53/310
Reiter disease, 53/308
retinal events, 53/309
Sjögren syndrome, 53/308
Sneddon syndrome, 53/308
status cribrosus, 53/316
symptom, 53/309
Takayasu disease, 53/308
temporal arteritis, 53/308
temporal profile, 53/312
thromboangiitis obliterans, 53/308
watershed infarction, 53/314
Wegener granulomatosis, 53/308
Carotid vertebral anastomosis
primitive hypoglossal artery, 12/311, 326-330
primitive otic artery, 12/311, 322-326
primitive trigeminal artery, 12/311-322
proatlantal intersegmental artery, 12/311, 330
scheme, 53/301
Carotidocarotid anastomosis
congenital, 12/308
Carotidynia, 48/329-339
acetylsalicylic acid, 48/339
aphthous ulcer, 48/339
carotid artery, 5/375, 48/337
carotid body tumor, 48/339
carotid sinus, 48/338
cluster headache, 48/226
cranial neuritis, 5/22
definition, 48/337
differential diagnosis, 5/376
ergotamine, 5/375, 48/339
etiology, 5/376, 48/337
headache, 5/375, 48/6
methysergide, 48/339
neck pain, 5/375, 48/338
pathogenesis, 48/337
propranolol, 48/339
sphenopalatine neuralgia, 48/226, 338
steroid, 48/339
temporal arteritis, 48/338
treatment, 48/339
viral, 5/375, 48/338
Carp mouth
4p partial monosomy, 43/497
11q partial monosomy, 43/523
6q partial trisomy, 43/503
10q partial trisomy, 43/517
19q partial trisomy, 43/538
tetraploidy, 43/567
Carpal tunnel syndrome, 42/309-311
*see also* Eosinophilia myalgia syndrome
acromegaly, 7/296, 41/249, 42/309
amyloid, 47/618
amyloid polyneuropathy, 51/416
amyloidosis, 8/7, 42/309
anatomy, 7/438
anterior tephromalacia, 59/113
atenolol, 63/195
beta adrenergic receptor blocking agent intoxication, 65/437
cases, 7/434
cause, 51/93, 98
clinical symptom, 7/288
compression neuropathy, 7/286-300, 437-440, 51/87, 93, 133
decompression result, 7/161
demyelinating neuropathy, 47/618
diabetes mellitus, 7/296
diabetic neuropathy, 7/296
diagnostic test, 7/290
dialysis, 63/529
differential diagnosis, 7/297
dysesthesia, 7/431
electrodiagnosis, 51/94
eosinophilia myalgia syndrome, 64/255
eosinophilic fasciitis, 63/385
etiology, 7/291
familial, 51/560
flick sign, 51/94
ganglion, 7/293
glycosaminoglycanosis type IH, 66/285, 291
glycosaminoglycanosis type IH-S, 66/292
glycosaminoglycanosis type IS, 10/441, 42/453, 66/292
glycosaminoglycanosis type II, 66/295
glycosaminoglycanosis type III, 66/302
glycosaminoglycanosis type VI, 42/458, 66/307
gout, 7/295, 42/309
hereditary amyloid polyneuropathy type 2, 42/520, 51/423
hereditary multiple recurrent mononeuropathy, 51/560
hereditary primary amyloidosis, 7/296
Hirayama disease, 59/114
hyperthyroid polyneuropathy, 51/517

hypothyroid polyneuropathy, 51/516
hypothyroidism, 41/244, 70/100
leprosy, 7/296
Léri pleonosteosis, 7/293
median nerve, 7/286-300, 329, 42/309
median nerve lesion, 7/329
β2-microglobulin amyloidosis, 63/528
mucolipidosis type II, 42/309
mucolipidosis type III, 42/309
multiple myeloma, 8/7, 39/135, 155, 42/309
multiple myeloma polyneuropathy, 63/393
myxedema, 7/296
myxedema neuropathy, 7/169, 8/57
nerve conduction, 7/157-160, 42/309
nerve ischemia, 12/660
neurapraxia, 51/77
occupational lesion, 7/329
occupational neuropathy, 7/291
pain, 42/309
paraproteinemia, 39/155-168
paresthesia, 42/309
pregnancy, 7/292, 8/14
prevalence, 42/309
progressive systemic sclerosis, 51/448
propranolol, 63/195
rheumatoid arthritis, 7/294, 38/482, 71/18
sarcoidosis, 7/295
sclerodactyly, 7/295
sensory loss, 42/309
sex ratio, 42/309
syringomyelia, 32/284
systemic lupus erythematosus, 7/296, 51/447
tardy neuritis, 7/260
Tinel sign, 51/94
toxic oil syndrome, 63/380
toxic shock syndrome, 52/259
trauma, 7/292
treatment, 7/297, 51/95
tuberculosis, 7/295
tumor, 7/294
uremic polyneuropathy, 63/511
vitamin B6 deficiency, 70/432
Carpenter syndrome, *see* Acrocephalosyndactyly type II
Carpilius convexus intoxication
eating, 37/56
Carpilius noxius intoxication
eating, 37/56
Carpopedal spasm
eosinophilia myalgia syndrome, 64/256
hyperventilation syndrome, 38/319, 326, 63/435
tetany, 41/301
uremic encephalopathy, 63/505
Carrier detection
Becker muscular dystrophy, 40/340
cerebrotendinous xanthomatosis, 66/608
creatine kinase, 41/413, 487
Duchenne muscular dystrophy, 40/340, 354, 41/415, 484, 489
genetic counseling, 41/415, 484, 489
glycosaminoglycanosis type VII, 66/312
GM1 gangliosidosis, 10/481
myotonic dystrophy, 40/520
testing method, 41/415, 484, 489
Carrión disease, *see* Verruga peruana
Carter-Sukavajana syndrome
dementia, 21/412
hereditary cerebello-olivary atrophy (Holmes), 21/411
hereditary olivopontocerebellar atrophy (Menzel), 21/412, 447
olivopontocerebellar atrophy, 21/411
rigidity, 21/412
Carukia barnesi intoxication
carukiism, 37/43
symptom, 37/43
Carukiism
Carukia barnesi intoxication, 37/43
Carybdea intoxication
local pain, 37/43
Carybdeidae intoxication
carybdea, 37/43
tamoya, 37/43
Caseous meningitis
tuberculous meningitis, 33/201
Cassava
cyanide intoxication, 65/30
cyanogenic glycoside, 64/8
cyanogenic glycoside intoxication, 36/515, 519, 65/27, 30
Cassava intake
deficiency neuropathy, 7/642, 51/321, 323
multiple sclerosis, 9/103
tropical ataxic neuropathy, 7/641, 28/26, 30, 64/12
tropical myeloneuropathy, 7/641, 56/526
tropical spastic paraplegia, 65/30
Cassava intoxication
familial spastic paraplegia, 59/306
symptom, 64/12, 65/6
Castleman disease
*see also* POEMS syndrome
benign intracranial hypertension, 63/394
chemotherapy, 63/396
confusional state, 63/395
corticosteroid, 63/396
cranial nerve palsy, 63/395

cranial neuropathy, 63/395
cyclophosphamide, 63/396
motor polyneuropathy, 63/394
paraproteinemia, 69/307
plasmapheresis, 63/395
POEMS syndrome, 47/619, 63/394
polyneuropathy pathogenesis, 63/395
radiotherapy, 63/395
treatment, 63/395
2 varieties, 63/394
vasculopathy, 63/395
Castro, F. De, *see* De Castro, F.
Cat eye reflex
retinoblastoma, 42/768
Cat eye syndrome
*see also* Chromosome 22
anal atresia, 31/525, 43/548
hypertelorism, 30/250
iris coloboma, 31/525, 43/548
microcephaly, 30/508
trisomy 22, 43/549
CAT scan, *see* Computer assisted tomography
Cat scratch disease, 34/459-472
age, 52/129
anorexia, 52/127
brain arteritis, 52/129
brain histology, 52/131
brain infarction, 53/162
brain ischemia, 53/162
brain stem, 52/130
cerebellar ataxia, 52/130
choreoathetoid movement, 52/130
choreoathetosis, 52/130
clinical features, 34/462
coma, 52/129
confusion, 52/129
confusional state, 52/130
convulsion, 52/129
CSF, 34/472, 52/130
CSF examination, 52/129
decortication, 52/130
delirium, 52/130
diagnosis, 34/463-470, 52/128
differential diagnosis, 34/470
disease course, 52/127
EEG, 52/130
encephalitis, 52/128
encephalopathy, 52/129
epidemiology, 34/460, 52/126
erythema nodosum, 52/128
etiology, 34/461, 52/126
exanthema, 52/128
facial paralysis, 52/130
fever, 52/127
granulomatous CNS vasculitis, 55/387
headache, 52/127, 130
hiccup, 52/130
histopathology, 52/128, 131
history, 34/459
incubation, 52/126
infantile cerebrovascular disease, 54/40
intradermal test, 52/128
lymphadenitis, 52/127
lymphnode histology *vs* Hodgkin disease, 52/128
mode of presentation, 34/462
myelitis, 52/128
neurologic complication, 34/464-472
neuropathology, 34/472
neuropathy, 52/128
nystagmus, 52/130
osteolytic lesion, 52/128
Parinaud syndrome, 34/462, 52/128
parotid swelling, 52/128
pathology, 34/463, 472
pneumonitis, 52/128
presentation mode, 52/127
purpura, 52/128
pyramidal sign, 52/130
radiculoneuritis, 52/130
radiculoneuropathy, 52/129
relapse, 52/127
season, 52/126
sex ratio, 52/126
skin test, 34/461
sphincter disturbance, 52/130
symptomatology, 52/127
thrombotic thrombocytopenic purpura, 52/18
transverse myelitis, 52/130
treatment, 34/472
vomiting, 52/127
Cat scratch fever, *see* Cat scratch disease
Catagenesis
hereditary cerebello-olivary atrophy (Holmes), 21/405
Catalase
free radical scavengers, 64/19
Catalepsy
catatonia, 49/57
cerebellar, *see* Cerebellar catalepsy (Babinski)
cortical cerebellar atrophy, 60/644
dentatorubropallidoluysian atrophy, 60/644
neuroleptic, *see* Neuroleptic catalepsy
neuroleptic difference, 65/280
neuroleptic dyskinesia, 65/277
Catamenial migraine
attack, 48/18, 85

biochemistry, 48/85
body mass, 5/231
diuretic agent, 5/381
endocrine status, 5/230, 48/119
estrogen, 5/232
flumedroxone, 48/434
hormone change, 48/425
incidence, 48/36
serotonin 2 receptor, 48/92
spiperone, 48/92
vasopressin, 5/232
water balance, 5/231
Cataplexy, 42/710-713
consciousness, 42/711
emotion, 42/711
flaccid paralysis, 42/711
giving way of legs, 1/347
hereditary, 42/712
idiopathic narcolepsy, 3/94
narcolepsy, 2/448, 3/94, 45/147
narcoleptic syndrome, 15/842, 42/711
neurobrucellosis, 33/314, 52/587
Niemann-Pick disease type C, 60/151
prazosin, 65/448
prazosin intoxication, 65/448
sex ratio, 42/713
weakness, 42/711
Catapres, *see* Clonidine
Cataract
*see also* Corneal opacity
Alport syndrome, 42/376, 60/770
Alström-Hallgren syndrome, 13/461, 42/391
aminoaciduria, 60/652
arachnodactyly, 60/652
ARG syndrome, 50/581
azacosterol, 40/514, 548
Bedouin spastic ataxia, 60/653
bilateral, *see* Bilateral cataract
cerebrohepatorenal syndrome, 43/339
cerebro-oculofacioskeletal syndrome, 43/341
cerebrotendinous xanthomatosis, 10/532, 547, 27/250, 29/374, 42/430, 51/399, 59/357, 60/167, 653
Cockayne-Neill-Dingwall disease, 10/182, 43/351, 60/653
congenital, *see* Congenital cataract
congenital heart disease, 63/11
congenital muscular dystrophy, 43/89
congenital retinal blindness, 13/272, 22/530, 532
congenital rubella, 50/278
Conradi-Hünermann syndrome, 43/346
corpus callosum agenesis, 50/162
Crome syndrome, 43/242
cutis verticis gyrata, 43/244
diazacholesterol, 40/514
Down syndrome, 50/529
early onset recessive stationary ataxia, 60/653
electrical injury, 61/192
Fabry disease, 51/377, 60/172
facio-oculoacousticorenal syndrome, 43/400
Flynn-Aird syndrome, 14/112, 22/519, 42/327, 60/653
Friedreich ataxia, 21/347, 60/310, 653
Hallermann-Streiff syndrome, 43/403
hereditary amyloid polyneuropathy type 1, 51/424
hereditary olivopontocerebellar atrophy (Menzel), 21/445
hereditary sensory and autonomic neuropathy type I, 60/8
hereditary vitreoretinal degeneration, 13/274
heredopathia ophthalmo-otoencephalica, 42/150, 60/652
hyaloidoretinal degeneration (Wagner), 13/274
hyaloidotapetoretinal degeneration (Goldmann-Favre), 13/37, 276
hypoparathyroidism, 42/577
juvenile, *see* Juvenile cataract
late cerebellar ataxia, 60/653
Laurence-Moon syndrome, 43/254
Lichtenstein-Knorr disease, 22/515
Lowe syndrome, 42/606
Marinesco-Sjögren syndrome, 21/555, 42/184, 60/342, 652
Marshall syndrome, 42/393
mental deficiency, 43/123-125, 268, 282, 293
metabolic ataxia, 21/575
mevalonate kinase deficiency, 60/653
microcephaly, 43/427, 50/278
mitochondrial myopathy, 43/188
mucolipidosis type I, 60/653
multiple mitochondrial DNA deletion, 62/510
muscular dystrophy, 43/89
myopathy, 43/123-125
myotonic dystrophy, 40/512, 41/486, 43/153, 62/217, 233
neurosensory hearing loss, 60/653
Norrie disease, 22/508, 31/292, 43/441, 46/66
Northern Swedish recessive ataxia, 60/652
oculodento-osseous dysplasia, 43/445
Oguchi disease, 13/262
olivopontocerebellar atrophy (Wadia-Swami), 60/653
optic atrophy, 42/406
Patau syndrome, 14/120, 31/513
primary pigmentary retinal degeneration, 13/178
progressive external ophthalmoplegia, 62/306

pseudohypoparathyroidism, 6/706, 42/621, 49/419
Refsum disease, 8/22, 13/314, 21/9, 193, 60/228, 653, 728, 66/489
Richner-Hanhart syndrome, 42/581
Rieger syndrome, 43/470
rigidity, 60/653
Rothmund-Thomson syndrome, 14/777, 39/405, 43/460, 46/61
rubella virus, 56/408
Rubinstein-Taybi syndrome, 13/83, 46/92
sensorineural deafness, 42/393
Smith-Lemli-Opitz syndrome, 66/582
spinal muscular atrophy, 59/373
Stickler syndrome, 43/485
striatopallidodentate calcification, 49/419, 423
sunflower, *see* Sunflower cataract
temporal arteritis, 48/319
triparanol, 40/514
Usher syndrome, 60/653
Werner syndrome, 43/489
Wilson disease, 27/386
X-linked ichthyosis, 60/653
XXXX syndrome, 43/554
XXXXX syndrome, 43/554

Cataract surgery
delirium, 46/535, 547

Catatonia
affective disorder, 46/423
brain infarction, 46/423
carbamate intoxication, 64/191
carbon monoxide intoxication, 49/474
catalepsy, 49/57
cerebellar catalepsy (Babinski), 21/527
cycloserine, 65/481
cycloserine intoxication, 65/481
echographia, 45/468
echolalia, 45/468
echopraxia, 45/468
encephalitis, 46/423
experimental, *see* Experimental catatonia
hysteria, 46/423
α-interferon intoxication, 65/538
mania, 46/423
neuroleptic agent intoxication, 46/423
neuropharmacology, 49/56
pallidoluysionigral degeneration, 49/456
Pick disease, 46/424
presenile dementia, 42/286
schizophrenia, 46/422-424
systemic lupus erythematosus, 55/376
Wilson disease, 49/228

Catecholamine
adrenergic receptor, 74/163
brain embolism, 55/157
brain hemorrhage, 55/157, 63/237
brain infarction, 55/157
brain injury, 23/124
brain tumor, 74/314
chemodectoma, 74/315
chromaffin cell, 74/142
depression, 46/448
distribution, 74/311
Duchenne muscular dystrophy, 40/381
epilepsy, 72/91
experimental ischemic myopathy, 40/167-170
Friedreich ataxia, 21/326
function, 74/160
headache, 48/437
hereditary sensory and autonomic neuropathy type III, 21/109, 115, 43/59, 60/27
Huntington chorea, 6/351-353
leukodystrophy, 10/48
manic depressive psychosis, 43/209
metabolic pathway, 75/241
migraine, 5/41, 48/91
narcolepsy, 3/96
narcoleptic syndrome, 42/713
nervous system tumor, 27/510
neural tube formation, 30/54
neuroastrocytoma, 74/316
neuroblastoma, 42/758, 74/315
neurulation, 30/54
noradrenalin, 74/162
noradrenalin transporter deficiency, 75/240
orthostatic hypotension, 43/66, 74/325
panic disorder, 74/325
paraganglioma, 42/762
pathway, 74/312
pharmacology, 74/142, 144, 147
pheochromocytoma, 9/496, 14/252, 42/764
plasma, *see* Plasma catecholamine
presynaptic, 74/144
progressive dysautonomia, 59/150
spinal cord injury, 26/317
stress, 74/310
sympathetic nerve, 74/162
sympathetic nervous system, 75/458
transtentorial herniation, 63/237

Catecholamine test
pheochromocytoma, 8/481

Catecholaminergic cell
central, 74/159
pharmacology, 74/159

Catecholaminergic pathway
central, 74/159
pharmacology, 74/159

Catechol methyltransferase
Parkinson disease, 49/120
Catechols
chemical structure, 37/232
CNS stimulant, 37/232
phenethylamine, 37/232
Cathepsin
myelin sheath, 9/34
myopathy, 40/380
Catheterization
bladder dysfunction, 75/674
cardiac, *see* Cardiac catheterization
child, 25/189
indwelling catheter, 26/416
intermittent, 26/410, 414
spinal cord injury, 25/290, 61/299
Cathine
khat leaf, 65/50
Cathinone
khat leaf, 65/50
Catlin mark, *see* Foramina parietalia permagna
Cauda equina
anatomy, 19/78
droplet metastasis, 20/285
ependymoma, 18/120
intermittent spinal cord ischemia, 12/507-547
osteitis deformans (Paget), 70/14
spinal cord metastasis, 69/176
spinal ependymoma, 19/359, 20/354-356
spinal epidural lesion, 19/86
spinal meningioma, 20/233
spinal nerve root injury, 25/450, 454, 470
spinal nerve root syndrome, 20/594
vascular supply, 12/525
Cauda equina compression
*see also* Vertebral canal stenosis
Hodgkin lymphoma, 63/346
lumbar stenosis, 20/627-637
lymphoma, 63/346
meningeal leukemia, 63/343
Cauda equina injury, 20/624-801
hangman fracture, 25/450
rare trauma lesion, 25/470
spinal cord injury, 61/47
Cauda equina intermittent claudication, 12/507-547
*see also* Spinal cord intermittent claudication *and* Vertebral canal stenosis
anterior tibial syndrome, 12/538, 20/798
arteriovenous fistula, 20/796
clinical features, 20/696, 699, 703
differential diagnosis, 20/795
history, 20/611-613
intervertebral foramen stenosis, 20/791
laterality, 20/701
lumbar disc protrusion, 12/520, 63/43
lumbar vertebral canal stenosis, 20/598, 749, 760, 32/337-339
metrizamide, 20/722
nature, 20/620-624
occlusive aortoiliac disease, 12/535
osteitis deformans (Paget), 38/367
pain, 20/756, 32/337-339
pathogenesis, 20/749
vertebral canal stenosis, 19/149, 20/617
Cauda equina meningioma
surgical treatment, 20/233
Cauda equina syndrome
ankylosing spondylitis, 38/508-513, 70/4
dimethylaminopropionitrile polyneuropathy, 51/282
iatrogenic neurological disease, 61/159
intrathecal chemotherapy, 61/159
leukemia, 63/343
lumbar intervertebral disc prolapse, 20/594
neurotoxicity, 61/159
pathogenetic mechanism, 20/577
polyneuropathy, 19/86
spinal anesthesia, 61/159
testicular atrophy, 20/199
Waldenström macroglobulinemia polyneuropathy, 51/432
Cauda equina tumor
cervical spondylotic myelopathy, 19/85
child, 19/87
CSF, 19/79
diastematomyelia, 19/87
differential diagnosis, 19/77-90
EMG, 19/80
Fincher syndrome, 20/367
hereditary motor and sensory neuropathy type I, 19/87
hereditary motor and sensory neuropathy type II, 19/87
muscular dystrophy, 19/87
neurogenic bladder, 19/87
pathology, 19/78
plain X-ray, 19/80
spina bifida, 19/87
symptomatology, 19/79
tethered cord syndrome, 19/87
Caudal agenesis
fecal incontinence, 42/50
Potter syndrome, 43/469
sensory loss, 42/50
Caudal aplasia, 32/347-354, 42/49-51
*see also* Axial mesodermal dysplasia spectrum,

MURCS syndrome, Potter syndrome, Prune belly syndrome, Renal agenesis syndrome, VACTERL syndrome *and* VATER syndrome
anal atresia, 50/511
anterior sacral meningocele, 32/206
autosomal dominant, 50/65
axial mesodermal dysplasia spectrum, 50/516
birth incidence, 42/50
bladder function, 42/50
child of diabetic mother, 31/299, 32/348
clinical features, 32/350
CNS, 32/351
coccyx agenesis, 50/511
congenital hip dislocation, 42/50
diabetes mellitus, 30/126
diaphragm hernia, 50/512
diastematomyelia, 50/436
diplomyelia, 50/436
domestic fowl, 32/347
environmental factor, 32/348
epidemiology, 30/164, 32/348
etiology, 32/347-349
fecal incontinence, 42/50
genetics, 32/348
genital malformation, 50/511
Goldenhar syndrome, 50/516
heart disease, 42/50
hypogonadism, 42/50
hypospadias, 50/512
imperforate anus, 42/50
incidence, 50/65, 511
insulin, 32/347
kidney malformation, 50/511
malformation, 32/350
maternal diabetes mellitus, 50/512
mechanism, 50/29
meningocele, 32/206
microcephaly, 50/512
myelodysplasia, 32/352-354
oligohydramnios, 50/513
omphalocele, 50/512
paralysis, 42/50
Potter face, 50/513
prevalence, 30/150, 164
pulmonary hypoplasia, 50/511
radiologic appearance, 32/350
renal agenesis syndrome, 50/511
review, 50/509-517
sex ratio, 50/65
spine malformation, 50/512
symmelia, 42/50
treatment, 32/351
urethra atresia, 50/512
vertebral abnormality, 42/50
Caudal neuropore
neural tube, 50/4
Caudal regression, *see* Caudal aplasia
Caudal tegmental syndrome
Foville syndrome, 12/20
gaze, 12/20
hemiplegia, 12/20
peripheral facial paralysis, 12/20
tegmental pontine syndrome, 12/20
Caudate loop
visual process, 2/534
Caudate nucleus
*see also* Putamen *and* Striatum
brain hemorrhage, 54/306
causalgia, 8/328
chorea-acanthocytosis, 63/283
diffuse axonal injury, 57/55
function, 6/110
Huntington chorea, 49/318
reflex sympathetic dystrophy, 8/328
Caudate nucleus atrophy
chorea-acanthocytosis, 49/331
Hallervorden-Spatz syndrome, 49/396
Huntington chorea, 60/671
neuronal intranuclear hyaline inclusion disease, 60/671
Caulerpa intoxication
anesthetic effect, 37/97
caulerpicin, 37/97
depression, 37/97
respiratory complication, 37/97
vertigo, 37/97
Caulerpa racemosa intoxication
pharmacology, 37/98
Caulerpicin intoxication
depression, 37/97
respiratory complication, 37/97
vertigo, 37/97
Cauliflower brain
status verrucosus, 50/202
Causalgia
autonomic nervous system, 75/36
caudate nucleus, 8/328
clinical features, 1/108
dental, *see* Dental causalgia
dental headache, 5/226
dialysis, 63/530
dysautonomia, 22/254
features, 51/480
gate control system, 22/255
gyrectomy, 1/110
H-substance, 8/328, 22/255

incomplete nerve lesion, 7/250
meningococcal meningitis, 52/26
missile injury, 1/512
nerve injury, 75/316
nerve lesion, 7/433, 8/328
pain, 26/491
pathogenesis, 1/102
reflex sympathetic dystrophy, 8/328, 61/124
survey, 75/309
sympathetic nervous system, 48/112
symptom, 1/512
thalamic syndrome (Dejerine-Roussy), 2/482
topectomy, 1/110
Cave foot, *see* Pes cavus
Cavernoma, *see* Cavernous angioma *and* Cavernous hemangioma
Cavernous angioma
basal ganglion degeneration, 42/724
brain, *see* Brain cavernous angioma
brain arteriovenous malformation, 12/232, 54/367
definition, 14/51
epilepsy, 42/724, 72/129, 213
hemiplegia, 42/724
intracerebral calcification, 42/724
intracranial pressure, 42/724
pontine syndrome, 2/259
seizure, 72/209
sensory loss, 42/725
status epilepticus, 42/724
subarachnoid hemorrhage, 12/104, 42/724, 55/2
thalamic tumor, 68/65
Ullmann syndrome, 14/450
Cavernous hemangioma
brain site, 18/277
optic chiasm compression, 68/75
seizure, 73/400
vertebral column, *see* Vertebral hemangioma
Cavernous hemorrhage
brain, 18/277
Cavernous malformation
brain arteriovenous malformation, 53/35
diagnosis, 68/574
spinal cord tumor, 68/574
spinal vascular tumor, 68/574
Cavernous sinus
anatomy, 11/48, 54
meningioma, 68/412
paratrigeminal syndrome, 48/332, 334
radioanatomy, 11/112
skull base tumor, 68/468, 474
topography, 11/112
traumatic aneurysm, 24/386
Cavernous sinus artery
inferior, *see* Inferior cavernous sinus artery
Cavernous sinus syndrome
anatomic factor, 2/89
anterior, *see* Anterior cavernous sinus syndrome
differential diagnosis, 48/301
middle, *see* Middle cavernous sinus syndrome
nonrecurrent nonhereditary multiple cranial neuropathy, 51/571
orbital apex syndrome, 48/301
posterior, *see* Posterior cavernous sinus syndrome
skull base tumor, 17/181
superior orbital fissure syndrome, 48/301
Cavernous sinus thrombophlebitis
anatomy, 33/175-177
bacteriology, 33/178
clinical features, 33/178
course, 33/179
diagnosis, 33/178
pathogenesis, 33/177
pathology, 33/178
prognosis, 33/179
Cavernous sinus thrombosis
*see also* Venous sinus thrombosis
abducent nerve injury, 24/86
abducent nerve paralysis, 52/174
acute *vs* chronic, 52/174
antibiotic agent, 52/175
aseptic, *see* Aseptic cavernous sinus thrombosis
aspergillosis, 52/173
blindness, 52/174
brain angiography, 52/174
chemosis, 52/173
cilastatin, 52/175
computer assisted tomography, 52/174
congenital heart disease, 52/173
consciousness, 52/173
corticosteroid, 52/175
co-trimoxazole, 52/175
dental infection, 52/173
differential diagnosis, 16/237
epilepsy, 52/174
ethmoidal sinusitis, 54/398
exophthalmos, 52/173
facial edema, 52/173
functional anatomy, 52/177
Garcin syndrome, 2/313
gaze paralysis, 52/173
headache, 5/154, 220, 48/283, 52/173, 54/398
heparin, 52/175
imipenem, 52/175
laboratory finding, 52/174
malignancy, 52/173
marasmus, 52/173

meningism, 52/173
metronidazole, 52/175
mortality, 52/176
mucormycosis, 35/545, 551, 55/436
nafcillin, 52/175
neurosurgery, 52/175
nuclear magnetic resonance, 52/174
ophthalmoplegia, 54/398
oral contraception, 52/173
orbital infection, 52/173
orbital phlebography, 52/174
orbital venography, 54/407
otitis media, 52/173
papilledema, 52/174
paroxysmal nocturnal hemoglobinuria, 52/173
polycythemia, 52/173
polycythemia vera, 54/410
postoperative period, 52/173
postpartum period, 52/173
pregnancy, 52/173
radiation, 52/174
sickle cell anemia, 52/173
sinusitis, 52/172
sphenoid sinusitis, 52/172
Staphylococcus aureus, 52/173
Streptococcus pneumoniae, 52/173
subarachnoid hemorrhage, 55/20
symptom, 54/398
treatment, 52/175
vancomycin, 52/175
Cavernous sinus tumor
surgery, 68/476
Cavitating leukoencephalopathy, *see* Cystic leukoencephalopathy
Cavity
central medullary, *see* Central medullary cavity
encephaloclastic, *see* Encephaloclastic cavity
intracerebral hematoma, 54/89
Cavum duncani, *see* Cavum septi pellucidi
Cavum psalterii, *see* Cavum vergae
Cavum septi pellucidi
acrocephalosyndactyly type I, 30/326
anatomy, 30/303
behavior change, 30/325
boxer encephalopathy, 30/327
boxing injury, 23/544, 554
cell lining, 30/302
clinical features, 30/325
communicating, 30/322, 326
corpus callosum agenesis, 50/158
Dandy-Walker syndrome, 42/23
embryology, 30/303
epidemiology, 30/156, 304, 322-324
epileptic seizure, 30/325
fetal life, 30/304, 322
histology, 30/303
location, 30/303
malformation, 30/325
mental deficiency, 30/325
noncommunicating, 30/322, 326
nonmalformative, 30/321
4p partial monosomy, 43/497
pathogenesis, 30/326
radiologic appearance, 30/322
spina bifida, 30/325, 50/487
toxoplasmosis, 30/326
trauma, 30/326
treatment, 30/326
twins, 30/326
Cavum veli interpositi
anatomy, 30/327, 329
cavum vergae, 30/325
clinical features, 30/328-330
CNS, 30/328
corpus callosum agenesis, 50/158
cyst, 31/120
epilepsy, 30/328
mental deficiency, 30/328
4p partial trisomy, 43/500
prevalence, 30/156, 323, 327
prognosis, 30/328
radiologic appearance, 30/329
syndrome, 30/328
Cavum vergae
arachnoid cyst, 31/120
cavum veli interpositi, 30/325
corpus callosum agenesis, 50/158
cyst, 2/767
epileptic seizure, 30/325
hydrocephalus, 2/767
mental deficiency, 30/325
septum pellucidum, 30/302-304, 321-324
septum pellucidum tumor, 17/537
spina bifida, 30/325
Caylor syndrome, *see* Cardiofacial syndrome
CBV, *see* Brain blood volume
CCK-8, *see* Cholecystokinin octapeptide
CCNU, *see* Lomustine
CD4 antigen
acquired immune deficiency syndrome, 56/511
human immunodeficiency virus, 56/490
CD44v
brain metastasis, 69/111
CD8+ cytotoxic T-lymphocyte
myositis, 62/42
Cebocephaly

*see also* Holoprosencephaly
autosomal recessive inheritance, 50/61
definition, 50/225
Edwards syndrome, 50/562
holoprosencephaly, 30/446, 42/32, 50/233
Patau syndrome, 31/510, 512, 50/559
sex ratio, 50/61
Cecocentral scotoma, *see* Centrocecal scotoma
Cefalea a grappolo, *see* Cluster headache
Cefaloridine
neurotoxin, 33/43, 65/472
pneumococcal meningitis, 33/43
Cefalotin
neurotoxin, 65/472
Ceftazidime
bacterial meningitis, 52/10
gram-negative bacillary meningitis, 52/108
neurotoxin, 65/472
superior longitudinal sinus thrombosis, 52/179
transverse sinus thrombosis, 52/177
Celiac disease
*see also* Malabsorption syndrome *and* Postgastrectomy syndrome
acanthocytosis, 13/427
amyotrophic lateral sclerosis, 28/228
ascending neuropathy, 28/229
Bassen-Kornzweig syndrome, 63/273
chronic axonal neuropathy, 51/531
complication, 28/227
deficiency, 51/331
deficiency neuropathy, 7/618-626, 51/329
definition, 28/225
dementia, 28/230
dyssynergia cerebellaris myoclonica, 60/599
Eaton-Lambert myasthenic syndrome, 62/421
etiology, 28/232
familial incidence, 28/222
incidence, 28/225
intestinal pseudo-obstruction, 51/493
Kearns-Sayre-Daroff-Shy syndrome, 62/313
manifestation, 28/226
mental change, 28/230
myopathy, 28/231
neuropathy, 28/227, 229, 232
pathology, 28/227
polymyositis, 62/373
progressive cerebellar ataxia, 60/599
progressive external ophthalmoplegia, 62/313
psychiatric change, 28/230
schizophrenia, 28/230
survey, 70/224
vitamin $B_1$ deficiency, 51/331
vitamin B6, 28/232
vitamin $B_6$ deficiency, 28/232
vitamin $B_{12}$ deficiency, 28/232
vitamin $B_c$, 28/232
vitamin $B_c$ deficiency, 28/232, 51/336
Celiac plexus
anatomy, 1/503
Celiac sprue, *see* Celiac disease
Cell adhesion molecule
brain metastasis, 69/111
Cell ageing
Parkinson disease, 49/124
Cell count
blood, *see* Blood cell count
CSF, *see* CSF cell count
CSF cytology, 11/236, 12/145, 16/372
Cell cycle
chemotherapy, 67/277
Cell death
autonomic nervous system, 74/204
CNS, 30/74-77, 50/7, 40
embryo, 50/7
morphogenesis, 50/40
neural plate formation, 30/41
normal development, 30/74-76
pathologic development, 30/76, 50/41
programmed, *see* Programmed cell death
Cell inclusion
Schwann, *see* Schwann cell inclusion
Cell layer
granular, *see* Granular cell layer
subependymal, *see* Subependymal cell layer
Cell mediated immune response
bacterial meningitis, 55/418
brain vasculitis, 55/418
Cell mediated immunity
*see also* Delayed hypersensitivity *and* Lymphocyte
Chagas disease, 52/345
Guillain-Barré syndrome, 51/249
inflammatory myopathy, 41/62, 74, 375
multiple sclerosis, 47/106
myasthenia gravis, 41/126
streptococcal meningitis, 52/80
Cell membrane
muscle, *see* Muscle cell membrane
Cell membrane potential
critical illness polyneuropathy, 51/584
Cell migration
abnormality, 30/479-506
autonomic nervous system, 74/204
brain cortex, 50/246
cerebellar cortex, 50/247
cerebellar cortical dysplasia, 30/496

cerebellar heterotopia, 30/496
CNS, 50/29
microgyria, 30/479, 481
neuroepithelium, 50/32-35
status verrucosus, 30/489
tissue culture, 17/86
Cell nucleus
muscle, *see* Muscle cell nucleus
Cell proliferation
autonomic nervous system, 74/203
malnutrition, 29/3
Schwann, *see* Schwann cell proliferation
Cells
amniotic fluid, *see* Amniotic fluid cell
anterior horn, *see* Anterior horn cell
APUD, *see* APUD cell
argyrophobic, *see* Argyrophobic cell
B, *see* B-lymphocyte
basal, *see* Basal cell
basket, *see* Basket cell
brush, *see* Brush cell
Buhot, *see* Buhot cell
bulge, *see* Bulge cell
Cajal-Retzius, *see* Cajal-Retzius cell
catecholaminergic, *see* Catecholaminergic cell
cerebellar stellate, *see* Cerebellar stellate cell
chromaffin, *see* Chromaffin cell
endoneurial mononuclear, *see* Endoneurial mononuclear cell
epithelioid, *see* Epithelioid cell
ethmoid, *see* Ethmoid cell
foam, *see* Foam cell
Gaucher, *see* Gaucher cell
germ, *see* Germ cell
glia, *see* Glia cell
globoid, *see* Globoid cell
granule, *see* Granule cell
HeLa, *see* HeLa cell
hematopoietic, *see* Hematopoietic cell
Ig binding lymphoid, *see* Ig binding lymphoid cell
inflammatory, *see* Inflammatory cell
killer, *see* Killer cell
Kryspin-Exner, *see* Opalski cell
Kupffer, *see* Kupffer cell
LE, *see* LE cell
mast, *see* Mast cell
matrix, *see* Matrix cell
Meynert giant pyramidal, *see* Meynert giant pyramidal cell
Meynert solitary, *see* Meynert solitary cell
microglia, *see* Microglia cell
mononuclear, *see* Mononuclear cell
Mott, *see* Mott cell
Müller, *see* Müller cell
multinucleated giant, *see* Multinucleated giant cell
muscarinic cholinergic nerve, *see* Muscarinic cholinergic nerve cell
natural killer, *see* Natural killer cell
nerve, *see* Nerve cell
neuroid nevus, *see* Neuroid nevus cell
Opalski, *see* Opalski cell
peripheral blood, *see* Peripheral blood cell
physaliphorous, *see* Physaliphorous cell
plaque forming, *see* Plaque forming cell
plasma, *see* Plasma cell
pseudoxanthoma, *see* Pseudoxanthoma cell
Purkinje, *see* Purkinje cell
Reed-Sternberg, *see* Reed-Sternberg cell
Renshaw, *see* Renshaw cell
retinal ganglion, *see* Retinal ganglion cell
satellite, *see* Satellite cell
Schwann, *see* Schwann cell
spider, *see* Spider cell
T, *see* T-lymphocyte
T-helper, *see* T-helper cell
T-helper/T-suppressor, *see* T-helper/T-suppressor cell
target, *see* Target cell
tumor, *see* Tumor cell
Cellulitis
congenital hemifacial atrophy, 22/548
orbital, *see* Orbital cellulitis
Celontin, *see* Mesuximide
Cenesthesia
body scheme, 4/207
Cenesthesiopathy
migraine, 48/162
Centaurea solstitialis intoxication, *see* Yellow star thistle intoxication
Central alexia
aphasia, 45/440
brain injury, 45/440
brain tumor, 45/440
cerebrovascular disease, 45/440
Gerstmann syndrome, 45/440
reading, 45/440
writing, 45/440
Central alveolar hypoventilation
acquired type, 63/464
alcohol, 63/464
central sleep apnea, 63/460
congenital, *see* Congenital central alveolar hypoventilation
congenital megacolon, 63/464
definition, 63/463
neuroastrocytoma, 63/464

neuroblastoma, 63/464
obesity, 63/464
pulmonary hypertension, 63/464
REM sleep, 63/464
respiratory encephalopathy, 63/415
sleep apnea syndrome, 63/464
sleep disorder, 63/449
sleep hypoventilation, 63/463
Central aphasia, *see* Global aphasia
Central areolar choroidal atrophy, *see* Central choroidal atrophy
Central areolar choroidal sclerosis, *see* Central choroidal atrophy
Central artery, *see* Rolando artery
Central autonomic dysfunction
dopamine, 75/190
eye movement, 75/169
multiple neuronal system degeneration, 75/161
neuropeptide, 75/190
Central autonomic pathway
test, 74/596
Central cerebral vein
brain phlebothrombosis, 54/399
diagram, 11/107
Central cervical cord syndrome
concomitant head injury, 24/167
spinal cord injury, 25/65, 271, 283, 335, 359
Central chorioretinitis
differential diagnosis, 13/142
Stargardt disease, 13/142
Central choroidal angiosclerosis, *see* Central choroidal atrophy
Central choroidal atrophy
classification, 13/33
differential diagnosis, 13/138
secondary pigmentary retinal degeneration, 13/279, 281
Stargardt disease, 13/138
Central choroidal sclerosis, *see* Central choroidal atrophy
Central choroidal vascular atrophy, *see* Central choroidal atrophy
Central chromatolysis, *see* Axonal reaction
Central command
autonomic nervous system, 74/249
exercise, 74/263
Central cord necrosis
*see also* Myelomalacia
cardiac arrest, 61/114
epidural anesthesia, 61/138
hemorrhage, 61/378
posttraumatic, 25/80
progressive hemorrhage, 25/84
radiation myelopathy, 61/202
spinal cord injury, 61/40
systemic hypotension, 61/114
Central core disease, *see* Central core myopathy
Central core myopathy
chromosome 19q, 62/333
classification, 40/288
clinical features, 40/306, 335, 337
clinical variety, 62/333
concentric laminated body, 40/109
congenital hip dislocation, 41/3, 43/80
congenital myopathy, 41/3-8, 62/332
contracture, 43/80
core morphology, 62/337
core myofiber, 62/338
cores in tenotomized muscle, 40/243
creatine kinase, 43/81
definition, 62/333
differential diagnosis, 22/99, 40/422
experimental core lesion, 62/334
facial paralysis, 43/80
facial weakness, 62/334
facioscapulohumeral syndrome, 62/168
genetics, 41/430
glycogen storage disease, 43/81
histochemistry, 40/45
histopathology, 62/336
kyphoscoliosis, 43/81
lactate dehydrogenase, 41/5
malignant hyperthermia, 41/3, 43/81, 62/333, 570, 63/440
mitochondria, 40/91, 43/81
multicore lesion, 62/338
multicore myopathy, 40/158
multicore pathology, 62/338
muscle biopsy, 62/338
muscle fiber type I, 43/81
muscle fiber type I predominance, 41/4, 62/338
myoglobinuria, 40/339, 62/558
myophosphorylase, 43/81
neurogenic basis, 40/142, 243
organophosphate intoxication, 62/333
pathologic reaction, 40/259
proximal muscle weakness, 43/80
split myofiber, 62/338
structured core, 40/75, 243
talipes, 43/80
tenotomy, 62/334
tetanus toxin, 62/338
ultrastructure, 40/75
unstructured core, 40/75, 243
Z band, 40/215, 43/81
Central cystoid dystrophy, *see* Vitelliruptive macula

degeneration
Central deafness
bitemporal lesion, 2/703
congenital, *see* Congenital central deafness
developmental dysphasia, 4/361
hearing loss, 4/428
Central diffuse schwannosis, *see* Diffuse glioblastosis
Central eastern tick-borne encephalitis, *see* Central European tick-borne encephalitis
Central European tick-borne encephalitis
acute viral encephalitis, 34/76, 56/134, 139
thalamus, 2/486
Central excitatory state
Sherrington study, 1/50
Central giant cell tumor, *see* Aneurysmal bone cyst
Central gray matter
dentatorubropallidoluysian atrophy, 49/442
Central hemorrhagic necrosis
gray matter, 12/582
hematomyelia, 25/272
myelomalacia, 12/582
Central hyperexcitable state
pain, 75/320
Central inhibition
Sherrington, 1/50
state, 1/50
Central ischemic infarction
anterior horn, 12/586
gray matter, 12/586
watershed zone, 12/586
Central liquefaction necrosis, *see* Central cord necrosis
Central medullary cavity
diastematomyelia, 50/438
diplomyelia, 50/438
Central monochromatism
neuropsychology, 45/14
Central motor disorder
arthropathy, 1/183
brain stem lesion, 1/186-189
Brown-Séquard syndrome, 1/200
corneomandibular reflex, 1/173
cutaneous reflex, 1/172
differential diagnosis, 1/190
hemiplegia, 1/178-185
hypotonicity, 1/174
infantile hemiplegia, 1/195
Little disease, 1/209
mass reflex, 1/173
Mills syndrome, 1/202
monoplegia, 1/176-178
motor diaschisis, 1/174
muscle stretch reflex, 1/171
muscle tone, 1/169
Opalski syndrome, 1/189
palmomental reflex, 1/174
paralysis, 1/175-185
paraplegia, 1/205-210
pathology, 1/171-173
pseudobulbar paralysis, 1/196
review, 1/169-210
rigidity, 1/170
spasticity, 1/169
synkinesis, 1/182
topographical diagnosis, 1/185-190
Central myelomalacia
posttraumatic, 25/80
progressive deficit, 25/84
Central necrosis, *see* Central cord necrosis
Central nervous system, *see* CNS
Central neuroblastoma
supratentorial brain tumor, 18/310
Central neurocytoma, 68/138
classification, 68/150
epilepsy, 72/181
nuclear magnetic resonance, 68/153
survey, 68/150
Central neurogenic hyperventilation
*see also* Respiration *and* Tachypnea
brain stem injury, 63/484
brain stem tumor, 63/441
brain stem vascular accident, 63/441
encephalitis, 63/441
hepatic coma, 63/441
hyperpnea, 63/436, 440
prognostic value, 63/485
respiration, 1/496, 675, 57/129
respiratory alkalosis, 63/484
site, 63/484
sleep, 63/484
thalamic hemorrhage, 63/441
Central peripheral distal axonopathy, *see* Distal axonopathy
Central pontine dystrophy
Creutzfeldt-Jakob disease, 9/638
leukoencephalopathy, 9/636
Marchiafava-Bignami disease, 9/638
Wilson disease, 9/638
Central pontine myelinolysis, 28/285-313
acquired immune deficiency syndrome, 56/512
adrenoleukodystrophy, 47/595
adverse diuretic effect, 65/439
aggression, 2/263
alcoholism, 9/636, 28/311, 47/585
Alzheimer type II astrocyte, 9/637

animal model, 47/587
biphasic course, 63/548
brain stem auditory evoked potential, 47/588, 63/553
carcinoma, 38/657
cerebellopontine lesion, 2/263
cirrhosis, 28/293
clinical features, 28/289-295
clinical picture, 2/263
coma, 2/263, 28/291
computer assisted tomography, 47/588, 63/552
CSF, 63/553
definition, 28/285
delirium tremens, 28/291
dementia, 46/400
demyelination, 2/263, 9/636-638
diagnosis, 47/588
disorientation, 2/263
diuretic agent, 65/439
dysarthria, 28/291
EEG, 28/293, 63/553
electrolyte balance, 47/588
epidemiology, 28/293
etiology, 28/311-313
focal demyelination, 9/472
history, 28/285-289, 47/585
Horner syndrome, 63/549
hyponatremia, 47/586, 63/548
hyponatremia correction, 63/548
hypotension, 28/293
iatrogenic neurological disease, 65/439
internuclear ophthalmoplegia, 63/549
intracranial hypertension, 28/310
intrathecal chemotherapy, 61/151
leukoencephalopathy, 9/636, 47/585-589
liver disease, 28/293, 312, 47/585
liver glia, 9/635
locked-in syndrome, 63/549
malignancy, 38/657
management, 63/553
Marchiafava-Bignami disease, 28/293, 308, 325
metabolic encephalopathy, 69/399
multiple sclerosis, 47/72
mutism, 28/291, 63/549
neuroaxonal dystrophy, 6/626
nuclear magnetic resonance, 47/588, 63/552
nutrition, 47/588
ocular bobbing, 47/588
pathogenesis, 28/307-310, 47/586, 63/553
pathology, 2/263, 28/295-307, 47/586
prevention, 47/588
prognosis, 47/588
pseudobulbar paralysis, 2/263, 63/549
recovery, 63/552
renal insufficiency, 27/343
rigidity, 2/263
silent lesion, 63/549
striatal necrosis, 49/507
subacute necrotizing encephalomyelopathy, 28/293, 308
symptom, 63/548
tetraplegia, 2/263, 63/549
thiazide derivative, 47/587
treatment, 28/295, 47/588
tremor, 2/263
urea infusion, 63/554
uremia, 47/583
ventral pons, 47/586
vitamin $B_1$ deficiency, 28/311
weakness, 28/292
Wernicke encephalopathy, 28/289, 293, 308, 47/586
Wilson disease, 28/293, 312
Central progressive areolar choroidal dystrophy, *see* Central choroidal atrophy
Central progressive areolar choroidal sclerosis, *see* Central choroidal atrophy
Central respiratory disturbance, 1/650-680
striatonigral degeneration, 49/208
Central retinal artery
arteritis, 48/318
temporal arteritis, 48/318
Central retinal artery occlusion
temporal arteritis, 48/313
Central retinal venous thrombosis
visual acuity, 16/275
Central scotoma
Bassen-Kornzweig syndrome, 51/394
bilateral, 2/572
cinchonism, 65/452
dazzling test, 9/174
ethambutol intoxication, 65/481
hereditary progressive cone dystrophy, 13/134
hyaloidotapetoretinal degeneration (Goldmann-Favre), 13/276
macula, 2/564
multiple sclerosis, 9/174
orbital apex syndrome, 2/322
pseudoinflammatory foveal dystrophy (Sorsby), 13/138
Central serous retinopathy
edema, 13/142
metamorphopsia, 13/142
micropsia, 13/142
scotoma, 13/142
swelling, 13/142

Central sleep apnea
acetazolamide, 63/463
arterial hypertension, 63/461
benzodiazepine, 63/463
brain stem infarction, 63/462
brain stem tumor, 63/462
carbon dioxide tension, 63/461
cardiac dysrhythmia, 63/461
central alveolar hypoventilation, 63/460
Cheyne-Stokes respiration, 63/462
continuous positive airway pressure, 63/463
daytime somnolence, 63/461
definition, 63/460
diabetes mellitus, 63/462
dysphoria, 63/461
dyspnea, 63/461
EMG, 63/460
encephalitis, 63/462
epidemiology, 63/461
forgotten respiration, 63/461
hereditary sensory and autonomic neuropathy type III, 63/462
high altitude, 63/463
hypercapnia, 63/461
hypercapnic ventilatory response, 63/460
hypoxemia, 63/461
insomnia, 63/461
multiple sclerosis, 63/463
muscular dystrophy, 63/463
myasthenia gravis, 63/463
nonhypercapnic patient, 63/461
obesity, 63/461
poliomyelitis, 63/462
REM sleep, 63/462
sex ratio, 63/460
Shy-Drager syndrome, 63/462
sleep apnea syndrome, 63/461
sleep disorder, 63/449
snoring, 63/461
symptom, 63/461
syringomyelia, 63/463
treatment, 63/463
Central spinal cord necrosis, *see* Central cord necrosis
Central spinal syndrome
hematomyelia, 2/211
syringomyelia, 2/211
Central stimulant, *see* CNS stimulant
Central tapetoretinal dystrophy, *see* Stargardt disease
Central tegmental tract
dentatorubropallidoluysian atrophy, 49/442
mesencephalic syndrome, 2/275
precise origin, 6/22-25
Central visual disturbance
alkylmercury intoxication, 60/655
leukodystrophy, 60/655
poliodystrophy, 60/655
Centrally started tapetoretinal dystrophy, *see* Centroperipheral tapetoretinal dystrophy
Centrencephalic epilepsy, 42/682-684
classification, 15/9, 42/682
EEG, 42/682
etiology, 42/683
Centrencephalic system
consciousness, 3/55
hypothalamus, 2/450
memory, 2/450
Centrocecal scotoma
bilateral lesion, 2/567
infantile optic atrophy, 42/408
optic nerve, 2/567
visual field, 13/113
Centrofacial lentiginosis
dwarfism, 43/26
dysraphia, 14/782
emotional disorder, 14/782
epilepsy, 14/782, 43/26
hemiplegia, 14/782
kyphoscoliosis, 43/26
mental deficiency, 14/782, 43/26
psychiatric disorder, 43/26
sella turcica, 43/26
spina bifida, 14/782, 43/26
vertebral abnormality, 43/26
Centrolobar sclerosis, *see* Pelizaeus-Merzbacher disease
Centromedian parafascicularis complex
akinesia, 45/162
akinetic mutism, 45/163
Centromedian thalamic nucleus
anatomy, 6/18
Centronuclear myopathy, *see* Myotubular myopathy
Centroperipheral tapetoretinal dystrophy
Stargardt disease, 13/127
Centropogon intoxication
cardiovascular posture response, 37/76
ischemia, 37/76
local pain, 37/76
Centrostephanus intoxication
pain, 37/67
spinal injury, 37/67
Cephalalgia, *see* Headache
Cephalgia, *see* Headache
Cephalic index
craniosynostosis, 50/119

Cephalic tetanus
hypoglossal paralysis, 7/476
multiple cranial neuropathy, 65/218
ophthalmoplegia, 7/476
prevalence, 52/230
Cephalin, *see* Phosphatidylethanolamine
Cephalin lipidosis, *see* Phosphatidylethanolamine lipidosis
Cephalocele
*see also* Encephalocele
atretic, *see* Atretic cephalocele
cranial, *see* Cranial cephalocele
definition, 50/97
genetic factor, 50/98
hydrocephalus, 30/212
mesodermal insufficiency, 50/98
nonmidline, *see* Nonmidline cephalocele
occipital, *see* Occipital cephalocele
parietal, *see* Parietal cephalocele
radiation, 50/98
sex ratio, 50/97
sincipital, *see* Sincipital cephalocele
sodium arsenate, 50/98
teratogenesis, 50/98
trypan blue, 50/98
vitamin A, 50/98
Cephalochordata intoxication
*see also* Acraniata intoxication
chordata, 37/68
Cephalopoda intoxication
artificial respiration, 37/62
cephalotoxin, 37/62
convulsion, 37/62
dopamine, 37/62
flaccid paralysis, 37/62
glycoprotein, 37/62
Hapalochlaena lunulata, 37/62
hapalochlaena maculosa, 37/62
histamine, 37/62
octopamine, 37/62
octopus, 37/62
serotonin, 37/62
treatment, 37/62
vomiting, 37/62
Cephaloridine, *see* Cefaloridine
Cephalosporin
bacterial meningitis, 52/12
brain abscess, 52/156
cranial subdural empyema, 52/171
gram-negative bacillary meningitis, 52/105, 107
Listeria monocytogenes meningitis, 52/96
meningococcal meningitis, 33/26, 52/30
neurosyphilis, 52/282
Cephalosporiosis
amphotericin B, 52/482
meningitis, 35/562, 52/481
mycetoma, 35/562, 52/481
Cephalosporium
fungal CNS disease, 35/562, 52/479, 481
Cephalosyndactyly
Vogt type, *see* Acrocephalosyndactyly type II
Cephalothin, *see* Cefalotin
Ceramidase, *see* Acylsphingosine deacylase
Ceramide
biochemistry, 9/3
catabolism, 66/60
chemistry, 10/136, 247
Farber disease, 66/211, 216
metabolism, 66/217
threo, *see* Threoceramide
uridine diphosphogalactose, 10/250
Ceramide dihexoside
chemistry, 10/472
Fabry disease, 10/300
Ceramide galactose
Gaucher disease, 10/509
juvenile Gaucher disease, 10/525
Ceramide galactose sulfate
late infantile metachromatic leukodystrophy, 10/311
Ceramide glucose
Gaucher disease, 10/307, 509, 512, 525, 564, 21/50
globoid cell, 10/87
infantile Gaucher disease, 10/307, 66/124
Niemann-Pick disease, 10/286
Ceramide hexose
hypertrophic interstitial neuropathy, 21/12
Ceramide hexoside
hypertrophic interstitial neuropathy, 21/148, 161
Ceramide lactoside lipidosis
enzyme, 29/350
Niemann-Pick disease type C, 29/370
sphingolipid, 29/350
Ceramide polyhexoside
chemistry, 10/250
Fabry disease, 10/300
Ceramide tetrasaccharide
chemistry, 10/472
Ceramide trihexosidase
Fabry disease, 10/544
Ceramide trihexoside
Fabry disease, 10/300, 344, 544, 579
$GM_2$ gangliosidosis type III, 10/298
Cerebellar ablation
animal experiment, 2/396-398

astasia, 2/394-399
ataxia, 2/398
compensation period, 2/394
corticopontocerebellar deficiency, 2/402
definition, 34/5
dentate nucleus, 2/406
excitatory release, 2/394
fastigial nucleus, 2/404
flexion dystonia, 21/521
flocculonodular lobe, 2/399
gamma efferent system, 2/396
gamma rigidity, 2/399
inhibitory release, 2/394
intermediate zone, 2/406
interpositus nucleus, 2/406
lateral zone, 2/406
magnet reaction, 2/396-398
medial zone, 2/404
placing reaction, 2/403
Schunkelreaktion, 2/397
spinocerebellar deficiency, 2/401
stabilized deficiency, 2/396
supporting reaction, 2/396, 402
symptomatology, 2/399
unstabilized deficiency, 2/394
vestibulocerebellar deficiency, 2/399

Cerebellar abnormality
Dandy-Walker syndrome, 50/323

Cerebellar abscess
*see also* Brain abscess
brain abscess, 33/138, 52/155
computer assisted tomography, 52/448
nocardiosis, 52/448
North American blastomycosis, 35/404, 52/388
transverse sinus thrombosis, 54/397

Cerebellar afferent
electrostimulation, 2/409
folial pattern, 2/400

Cerebellar agenesis
*see also* Cerebellar hypoplasia
amyotrophic cerebellar hypoplasia, 50/192
anencephaly, 42/12, 50/175
Arnold-Chiari malformation type I, 50/175
ataxia, 50/193
cerebellar hemiaplasia, 50/181, 184
cerebellar heterotopia, 50/175, 178
cerebellar vermis, 50/188
cerebellar vermis aplasia, 50/175, 188, 190
chromosomal aberration, 50/177
classification, 50/181
clinical features, 50/184, 187
concept development, 21/462
Dandy-Walker syndrome, 50/175
deafferentation, 50/181
Down syndrome, 50/183
Edwards syndrome, 50/177, 183
embryology, 50/175
encephalocele, 50/177
epilepsy, 30/383
etiology, 50/176
experimental ablation, 50/181
falx cerebri, 42/18
Gunn rat, 50/180
hemiaplasia, 50/181
hyperbilirubinemia, 50/180
hypoplasia, 50/181
inferior olivary nucleus, 50/188
Joubert syndrome, 50/190
lateral, *see* Lateral cerebellar agenesis
lateral hemiaplasia, 50/181, 184
mechanical cause, 50/177
median aplasia, 50/181
mental deficiency, 30/380
mesocephalic anencephaly, 50/175
mosaicism, 50/178
neurocutaneous syndrome, 50/191
occipital encephalocele, 50/177
parvovirus, 50/178
Patau syndrome, 50/177, 183
pathology, 50/184, 188
pontoneocerebellar hypoplasia, 50/192
radiation, 50/180
respiratory dysfunction, 50/190
rhombencephalosynapsis, 50/188
subtotal agenesis, 30/378-380, 50/181, 183
tectocerebellar dysraphia, 50/177
total agenesis, 30/378-380, 50/181, 183
toxic agent, 50/180
ventriculocele, 50/177
viral infection, 50/178

Cerebellar angioma
*see also* Brain angioma
Ullmann syndrome, 14/72

Cerebellar anterior lobe
ablation, 2/401
opisthotonos, 2/401
respiratory center, 63/431
somatotopic localization, 2/407

Cerebellar arachnoid sarcoma
architecture, 16/20
supratentorial, 18/321
terminology, 18/321

Cerebellar arterial system, 11/24-42
embryology, 11/33
infarction, 12/20

Cerebellar artery

congenital aneurysm, 31/156
inferior, *see* Posterior inferior cerebellar artery
superior, *see* Superior cerebellar artery
Cerebellar artery aneurysm
superior, *see* Superior cerebellar artery aneurysm
Cerebellar artery syndrome
anterior inferior, *see* Anterior inferior cerebellar artery syndrome
brain stem sign, 55/89
Horner syndrome, 55/90
superior, *see* Superior cerebellar artery syndrome
Cerebellar asthenia
diffuse defect, 60/644
Cerebellar astrocytoma
aged, 18/18
classification, 18/1
clinical features, 18/390
cystic, *see* Cystic cerebellar astrocytoma
histologic differentiation, 18/124
imaging, 67/180
incidence, 18/4
macrocystic, 18/13
posterior fossa tumor, 18/388
prominent venous involvement, 18/297
Rosenthal fiber, 18/2
tumor, 17/709, 715
Cerebellar ataxia
acute, *see* Acute cerebellar ataxia
adiadochokinesia, 1/323, 326
adult neuronal ceroid lipofuscinosis, 42/466
albinism, 60/667
aniridia, 42/123
antibody, 69/356
aqueduct stenosis, 50/311
Argentinian hemorrhagic fever, 56/366
Arnold-Chiari malformation type I, 42/16
Arnold-Chiari malformation type II, 32/106
asynergia, 1/323, 325
ataxia telangiectasia, 14/267
ataxic diplegia, 42/121
autonomic dysfunction, 69/356
Batten disease, 42/466
benign late hereditary, *see* Benign late hereditary cerebellar ataxia
cat scratch disease, 52/130
cerebrotendinous xanthomatosis, 10/537, 27/250, 42/430
choreoathetosis, 21/468
Cogan syndrome type I, 42/329
congenital, *see* Congenital cerebellar ataxia
Conradi-Hünermann syndrome, 43/345
Creutzfeldt-Jakob disease, 46/290
deafness, 60/657
dementia, 42/126, 129
differential diagnosis, 21/428
Dumon-Radermecker syndrome, 42/477
dyschronometria, 1/323
dyssynergia cerebellaris myoclonica, 60/597, 599
EEG, 42/124
epilepsy, 42/125
Epstein-Barr virus, 56/252
Epstein-Barr virus infection, 56/252
esophageal achalasia, 43/260
extrapyramidal system, 42/126
finger to ear test, 1/324, 333
finger to nose test, 1/324, 333
Gerstmann-Sträussler-Scheinker disease, 56/553
Hartnup disease, 14/315, 42/146
hereditary, *see* Hereditary cerebellar ataxia
hereditary olivopontocerebellar atrophy (Menzel), 21/444
hereditary tremor, 49/568
heredopathia ophthalmo-otoencephalica, 42/150
Holmes, *see* Holmes cerebellar ataxia
hypermetria, 1/323
hypogonadism, *see* Hypogonadal cerebellar ataxia
hypothyroidism, 70/95
infectious mononucleosis, 56/252
intermittent, *see* Hill-Sherman syndrome
juvenile dystonic lipidosis, 42/446
Kearns-Sayre-Daroff-Shy syndrome, 43/142
Klinefelter syndrome, 21/467
kuru, 56/556
late, *see* Late cerebellar ataxia
Loken-Senior syndrome, 60/735
Lyme disease, 51/205
Marinesco-Sjögren syndrome, 21/556
May-White syndrome, 42/698
mental deficiency, 21/468, 42/123, 130-132, 43/260
multiple sclerosis, 47/55
muscular atrophy, 42/132
myocardial sclerosis, 42/131
myoclonic epilepsy, 42/698
myoclonus, 21/510-512, 60/597
myxedema, 21/496
myxedema body, 21/496
neuroborreliosis, 51/205
neuronal ceroid lipofuscinosis, 42/466
neuropathy, 60/660, 69/356
neurosensory hearing loss, 60/657, 660
nomenclature, 2/394
nonprogressive, *see* Nonprogressive cerebellar ataxia
oculomotor apraxia, 42/329
olivopontocerebellar atrophy, 42/161-163

ophthalmoplegia, 42/132
paralysis, 42/129
paraneoplastic syndrome, 69/356
Parry disease, 42/467
pituitary deficiency, 42/129
poliomyelitis, 56/329
polyneuropathy, 60/653
posterior fossa tumor, 18/407
prevalence, 42/163
progressive, *see* Progressive cerebellar ataxia
progressive external ophthalmoplegia, 43/137, 60/657
progressive myoclonus epilepsy, 42/698
pyruvate dehydrogenase deficiency, 29/212, 42/516
Refsum disease, 21/9, 41/434, 42/148, 60/228, 231, 66/490
renal disease, 13/308
secondary pigmentary retinal degeneration, 13/304-307, 60/653, 735
sensorineural deafness, 42/131, 698
sex linked infantile, *see* Infantile sex linked cerebellar ataxia
sign, 1/320-323
spastic paraplegia, 13/307
spinal, *see* Spinocerebellar ataxia
symptom, 1/320-323
varicella zoster virus infection, 55/436
vertical supranuclear ophthalmoplegia, 42/605
X-linked, *see* X-linked cerebellar ataxia
Cerebellar ataxia (Marie)
*see also* Hereditary cerebellar ataxia (Marie)
brief survey, 42/127
old taxonomy, 21/3
Cerebellar atrophy
*see also* Brain atrophy
Alpers disease, 42/486
antiepileptic agent, 65/499
astasia abasia, 1/345
Bowen-Conradi syndrome, 43/335
cerebellar catalepsy (Babinski), 21/527
chronic mercury intoxication, 64/372
computer assisted tomography, 46/483
congenital, *see* Congenital cerebellar atrophy
congenital cerebellar ataxia, 42/124
congenital pain insensitivity, 51/566
cortical, *see* Cortical cerebellar atrophy
cri du chat syndrome, 43/501
crossed, *see* Crossed cerebellar atrophy
disequilibrium syndrome, 42/140
dyssynergia cerebellaris myoclonica, 42/211
dyssynergia cerebellaris progressiva, 42/212
Edwards syndrome, 43/537
GM2 gangliosidosis, 10/291, 394
hemiballismus, 42/224
hemisphere, 2/419
hereditary congenital, *see* Hereditary congenital cerebellar atrophy
hereditary motor and sensory neuropathy type I, 21/302
hereditary motor and sensory neuropathy type II, 21/302
infantile spinal muscular atrophy, 59/70
late, *see* Late cerebellar atrophy
late cortical, *see* Late cortical cerebellar atrophy
lissencephaly, 42/39
Listeria monocytogenes encephalitis, 52/95
Machado-Joseph disease, 42/155
Marinesco-Sjögren syndrome, 42/184
Minamata disease, 36/123, 64/415
nystagmus, 42/135
olivopontocerebellar atrophy, 42/161
organic acid metabolism, 66/641
Paine syndrome, 43/433
Patau syndrome, 43/527
phenytoin intoxication, 65/502
posterior lobe, 2/419
schizophrenia, 46/483
sensory loss, 42/135
spastic paraplegia, 42/177
speech disorder, 42/135
spinocerebellar ataxia, 42/182, 190
spinopontine degeneration, 42/192
thalamic lesion, 2/489
triploidy, 43/566
vertical gaze palsy, 42/188
xeroderma pigmentosum, 43/13
Cerebellar calcification
lead encephalopathy, 64/436
Cerebellar catalepsy (Babinski)
catatonia, 21/527
cerebellar atrophy, 21/527
clinical course, 21/521
critique, 21/527
hyperkinesia, 21/521
ideomotor apraxia, 21/526
rigidity, 21/521
symptom, 21/526
Cerebellar concussion
*see also* Brain concussion
closed head injury, 23/459
Cerebellar cone, *see* Tonsillar herniation
Cerebellar cortex
*see also* Brain cortex
arrangement, 2/410
cell migration, 50/247

cerebellar tumor, 17/716
CNS, 30/30
development, 50/247
dyssynergia cerebellaris myoclonica, 22/172
embryo, 50/10
état glacé, 21/496
evoked potential, 2/410
histogenesis, 50/176
infantile neuroaxonal dystrophy, 49/396
lamina dissecans, 50/176
opticocochleodentate degeneration, 21/546-548
Purkinje cell, 50/176
senile plaque, 21/64
Cerebellar cortex heterotopia
congenital rubella, 50/278
Down syndrome, 50/531
Cerebellar cortex hypertrophy
cerebellar tumor, 17/716
Cerebellar cortical atrophy sine ataxia, *see* Focal cerebellar cortical panatrophy
Cerebellar cortical dysplasia
Bergmann fiber, 30/496
cell migration, 30/496
microgyria, 30/496
Cerebellar cortical panatrophy
focal, *see* Focal cerebellar cortical panatrophy
Cerebellar Creutzfeldt-Jakob disease
computer assisted tomography, 60/671
dementia, 60/665
Cerebellar cyst
cerebellar tumor, 17/708
Von Hippel-Lindau disease, 14/53, 242-244, 16/31
Cerebellar degeneration
alcoholic, *see* Nutritional cerebellar degeneration
alcoholism, 70/350
carcinomatous, *see* Carcinomatous cerebellar degeneration
cortical, *see* Cortical cerebellar degeneration
nutritional, *see* Nutritional cerebellar degeneration
paraneoplastic, 69/358
paraneoplastic syndrome, 69/357, 71/688, 692
sleep, 74/547
subacute malignant, *see* Subacute malignant cerebellar degeneration
Wernicke-Korsakoff syndrome, 28/272
Cerebellar disorder
olivopontocerebellar atrophy (Wadia-Swami), 60/492
pineal tumor, 17/657
Cerebellar dysarthria
ataxia, 42/137
character, 2/423
mental deficiency, 42/137
progressive bulbar palsy, 59/409
psychomotor retardation, 42/137
rostral paravermal zone, 60/643
tremor, 42/137
ventricular dilatation, 42/137
Cerebellar dysfunction
antiepileptic agent, 65/499
classification, 2/417
multiple neuronal system degeneration, 75/162
Cerebellar dysgenesis
aplasia, 30/377
chromosomal aberration, 30/377
Dandy-Walker syndrome, 50/328
experimental, 30/377
frequency, 30/377
heterotopia, 30/377
pathology, 30/378
Cerebellar dysplasia
occipital cephalocele, 50/101
Cerebellar ectopia, *see* Cerebellar heterotopia
Cerebellar encephalopathy
alcohol, 64/5
cytarabine, 63/357
fluorouracil, 64/5
methylmercury, 64/5
toluene, 64/5
Cerebellar external granular layer
cerebellar medulloblastoma, 18/168-170
CNS, 50/35
embryology, 30/65
fetal, 18/168
Cerebellar fit, *see* Decerebrate rigidity
Cerebellar function
ablation experiment, 2/393-403
excitation, 2/413
facilitation, 2/411
history, 1/330, 2/392
inhibition, 2/408-412
summary, 2/417
Cerebellar granular cell hypertrophy
benign intracranial hypertension, 67/111
chromosome, 67/58
CNS tumor, 67/54
computer assisted tomography, 60/672
Cowden syndrome, 60/666
macrencephaly, 43/252
mental deficiency, 43/252
nuclear magnetic resonance, 68/146
Purkinje cell loss, 43/253
sex ratio, 43/253
survey, 68/144, 300
Cerebellar granular cell hypoplasia
congenital, *see* Congenital cerebellar granular cell

hypoplasia
congenital cerebellar ataxia, 42/125
Cerebellar granular cell loss
trichopoliodystrophy, 14/784, 42/584
Cerebellar granular layer atrophy
clinical features, 21/500
congenital cerebellar atrophy, 21/461
neuropathology, 21/500
Cerebellar granular layer hypoplasia
staggerer mouse, 50/39
weaver mutant mouse, 50/39
Cerebellar hemangioblastoma
*see also* Angioreticuloma *and* Hemangioendothelioma
angioma type, 18/269
brain angiography, 54/186
child, 18/319
familial prevalence, 18/285
neuropathology, 14/57
nonrecurrent nonhereditary multiple cranial neuropathy, 51/571
pheochromocytoma, 18/285
polycythemia, 14/249, 18/285
polycythemia vera, 14/505, 18/285
secondary polycythemia, 63/250
spinal cord, 18/286
true hemangioblastoma, 18/284-288
tumor, 17/715
Von Hippel-Lindau disease, 14/13, 242-244, 18/286, 401, 31/16, 50/375, 68/294
Cerebellar hematoma
*see also* Brain hematoma
clinical features, 12/60, 55/90
symptom, 11/684
Cerebellar hemiaplasia
cerebellar agenesis, 50/181, 184
Cerebellar hemiparesis, *see* Marie-Foix-Alajouanine disease
Cerebellar hemorrhage
*see also* Brain hemorrhage
age, 54/291
angiography, 12/59
angioma, 54/291
arteriovenous malformation, 55/89
ataxia, 53/383, 54/314
ataxic respiration, 1/669, 63/487
blood dyscrasia, 55/89
brain amyloid angiopathy, 54/292
brain hemorrhage, 54/314
Charcot-Bouchard aneurysm, 12/55, 54/292, 55/89
clinical features, 55/90
computer assisted tomography, 54/314, 55/91
diagnosis, 12/56
diamorphine addiction, 65/356
echoEG, 12/58
examination, 12/57
facial paralysis, 54/314
gait apraxia, 54/314
gasping respiration, 63/487
headache, 5/140, 48/273, 278, 54/314
horizontal gaze paralysis, 54/314
hydranencephaly, 50/340
hypernatremia, 63/554
hypertension, 54/291, 55/91
management, 55/92
nuchal rigidity, 53/383
onset type, 12/56
pathology, 12/54
pontine hemorrhage, 12/45
prognosis, 55/92
progression, 55/92
respiration, 12/57
skew deviation, 12/56
spontaneous, *see* Spontaneous cerebellar hemorrhage
subarachnoid hemorrhage, 12/167
surgical treatment, 12/64
survival, 12/66
symptom, 54/314
symptomatology, 12/56
telangiectasia, 12/55
treatment, 12/61
venous angioma, 12/55
ventriculography, 12/59
vertigo, 55/91
Cerebellar heredodegeneration
bulbar paralysis, 22/142
Cerebellar herniation
Arnold-Chiari malformation type I, 32/104
Arnold-Chiari malformation type II, 50/405
intracranial hypertension, 16/135
posterior fossa hematoma, 24/346
Cerebellar heterotopia
cell migration, 30/496
cell type, 30/494-496, 499-502
cerebellar agenesis, 50/175, 178
cerebellar medulloblastoma, 30/493
chromosomal aberration, 30/494
cortical dysplasia, 50/251
Edwards syndrome, 50/251
frequency, 30/493
histologic features, 30/494-499
holoprosencephaly, 50/251
hydromyelia, 50/425
lymphocytic choriomeningitis virus, 56/358

Marinesco-Sjögren syndrome, 21/560
Miller-Dieker syndrome, 50/592
Patau syndrome, 50/39, 251
polymicrogyria, 50/251
Tamiami virus, 56/358
trisomy, 50/38
Cerebellar hypertrophy
diffuse, *see* Diffuse cerebellar hypertrophy
Cerebellar hypogenesis syndrome
symmetrical, *see* Symmetrical cerebellar hypogenesis syndrome
Cerebellar hypoplasia
*see also* Cerebellar agenesis
amyotrophic, *see* Amyotrophic cerebellar hypoplasia
anencephaly, 42/12
animal, 30/387
anterior horn syndrome, 42/19
arenavirus, 50/179
Arnold-Chiari malformation type IV, 50/403
ataxia, 42/19
autosomal recessive, *see* Autosomal recessive cerebellar hypoplasia
bulbar paralysis, 60/658
chromosomal aberration, 42/18
clinical features, 30/387, 50/192
computer assisted tomography, 60/672
congenital cerebellar atrophy, 21/461
cri du chat syndrome, 50/275
cytomegalovirus, 56/271
Dandy-Walker syndrome, 42/18
diagnosis, 30/388
Down syndrome, 50/531
Edwards syndrome, 50/178, 562
frequency, 30/385
genetics, 30/387
Gunn rat, 50/180
hydrocephalus, 42/18
hyperventilation, 42/19
infantile spinal muscular atrophy, 30/386, 50/192
Joubert syndrome, 50/191
mental deficiency, 30/358, 42/19
mutation, 50/179
neuronal heterotopia, 42/18
pathology, 30/386, 50/191
picornavirus, 9/551
polyploidy, 50/555
Purkinje cell heterotopia, 50/179
reeler mutant mouse, 50/179
review, 50/175-193
seizure, 42/18
striatopallidodentate calcification, 49/421
stridor, 60/667
symmetrical cerebellar hypogenesis syndrome, 30/388
Togavirus, 50/179
trigeminal nerve agenesis, 50/214
Von Hippel-Lindau disease, 42/18
weaver mutant mouse, 50/179
Cerebellar hypotonia
rebound phenomenon, 1/329
Cerebellar infarction
angiography, 53/384
anterior inferior cerebellar artery syndrome, 55/132
ataxia, 53/383
bilateral, 53/382
brain stem auditory evoked potential, 54/163
cerebellar edema, 53/383
clinical picture, 55/90
computer assisted tomography, 53/383, 54/55
coronary bypass surgery, 63/187
cyanide intoxication, 64/232
diagnosis, 53/384
dizziness, 53/383
dysarthria, 53/383
facial paralysis, 53/383
headache, 53/383
iatrogenic neurological disease, 63/187
large infarct, 53/383
methanol intoxication, 64/100
nuchal rigidity, 53/383
nystagmus, 53/383
obstructive hydrocephalus, 53/383
open heart surgery, 63/8, 187
papilledema, 53/383
posterior inferior cerebellar artery territory, 53/382
small infarct, 53/382
stroke, 63/8
stupor, 53/383
symptom, 53/383
temporal arteritis, 55/344
tonsillar herniation, 16/122, 53/383
treatment, 53/384
upward transtentorial herniation, 53/383
vertebral artery occlusion, 53/380
vertebral dissecting aneurysm, 54/281
vertebrobasilar system syndrome, 53/380
Cerebellar lesion
postural reflex, 1/336
resting tremor, 49/585
supporting reaction, 1/264
tilting chair test, 1/336
Cerebellar mass lesion
arousal disorder, 45/113
Cerebellar medulloblastoma, 18/167-191,

42/742-744
age, 18/172
animal model, 18/171
astrocytoma, 18/178, 42/742
ataxia telangiectasia, 42/742
bamboo spine, 18/187
Bean syndrome, 14/106
brain biopsy, 16/719
brain tumor carcinogenesis, 18/170
cerebellar external granular layer, 18/168-170
cerebellar heterotopia, 30/493
cerebellar tumor, 17/715, 18/124
chemotherapy, 18/190
classification, 16/4, 7
clinical features, 18/172, 392, 31/49
CSF cytology, 16/390, 18/173
cytological differentiation, 18/177
definition, 16/22
desmoplastic, *see* Desmoplastic cerebellar medulloblastoma
diagnostic problem, 21/569
epidemiology, 31/50
ethylnitrosourea, 18/171
glioma polyposis syndrome, 42/742
hereditary, 18/172
histogenesis, 18/167
histology, 18/124, 174
Homer Wright rosette, 16/18, 18/174, 177
hydrocephalus, 42/742
location, 31/49
Macewen sign, 18/173
macroscopic feature, 18/173
medullomyoblastoma, 18/181
meninges, 18/187
metastasis, 18/172, 186-188, 394
methotrexate, 18/190
multiple nevoid basal cell carcinoma syndrome, 14/79, 464, 18/172, 42/742
neuroblast, 18/170
neuroblastoma, 18/171, 178
neuronal type, 17/716
oligodendroglioma, 18/178
palisading, 18/175
pathology, 17/715, 18/173
pigmented papillary, *see* Pigmented papillary medulloblastoma
posterior fossa tumor, 18/392
posterior medullary velum, 18/170
prevalence, 18/171
prognosis, 17/715, 18/188-190
radiation change, 18/180
radiotherapy, 18/189, 395, 492
research, 16/37
sex ratio, 18/392
site, 18/172
spinal epidural tumor, 20/129
survival, 18/189
symptom in children, 21/569
tissue culture, 17/71, 73-79
treatment, 18/188-190, 31/50
ultrastructure, 18/179-181
variant, 18/181
vincristine, 18/190
whorling, 18/177
Cerebellar middle lobe
Down syndrome, 50/531
Cerebellar migraine
basilar artery migraine, 42/749, 48/157
vertigo, 5/78, 48/157
Cerebellar peduncle
hereditary motor and sensory neuropathy type I, 21/302
hereditary motor and sensory neuropathy type II, 21/302
Marchiafava-Bignami disease, 9/653
superior, *see* Superior cerebellar peduncle
Cerebellar peduncle lesion
superior, *see* Superior cerebellar peduncle lesion
Cerebellar pilocytic astrocytoma
child, 68/54
Cerebellar plexus hemisphericus
radioanatomy, 11/93
topography, 11/93
Cerebellar pressure cone
transtentorial herniation, 16/119
Cerebellar speech
Sjögren syndrome, 71/75
Cerebellar stellate cell
action, 2/416
astrocytoma, 18/2
function, 2/411
Cerebellar stimulation
athetosis, 49/387
basket cell, 2/409, 411
climbing fiber, 2/409, 414, 416
evoked potential, 2/409
excitatory postsynaptic potential, 2/412
experiment, 2/408
Golgi cell, 2/409, 411
granule cell, 2/409, 411
inhibitory postsynaptic potential, 2/413
interneuron, 2/411
mossy fiber, 2/409
parallel fiber, 2/412
physiology, 2/408
Purkinje cell, 2/409

Cerebellar syndrome, 2/392-426
*see also* Archicerebellar syndrome *and* Neocerebellar syndrome
boxing injury, 23/541, 554
ependymoma, 18/128
fluorouracil intoxication, 65/532
Forssmann-Skoog syndrome, 2/345
glomus jugulare tumor, 18/445
Kearns-Sayre-Daroff-Shy syndrome, 62/308
neurotoxicology, 64/5
paleo, *see* Paleocerebellar syndrome
paraneoplastic syndrome, 69/364
progressive external ophthalmoplegia, 62/308
progressive multifocal leukoencephalopathy, 9/494
traumatic, 23/460
Cerebellar tremor
*see also* Rubral tremor
anterior lobe exstirpation, 2/401
Benedikt syndrome, 49/588
clinical features, 6/820
etiology, 45/466
experiment, 2/394
GABA, 49/588
hereditary tremor, 49/568
hyperkinetic agraphia, 45/466
multiple sclerosis, 49/588
nucleus ruber, 49/587
pathology, 49/597
postural tremor, 6/820, 49/587
resting tremor, 49/585
superior cerebellar peduncle, 49/587
thalamotomy, 49/588
type, 49/587
Cerebellar tumor
abducent nerve injury, 17/711
anatomy, 17/708
angioma, 17/708
arachnoiditis, 17/716
areflexia, 17/710
Arnold-Chiari malformation type I, 17/716
astrocytoma, *see* Cerebellar astrocytoma
benign tumor, 17/716
bradycardia, 17/710
brain angioma, 17/708
cerebellar cortex, 17/716
cerebellar cortex hypertrophy, 17/716
cerebellar cyst, 17/708
cerebellar medulloblastoma, 17/715, 18/124
cerebellar sign, 17/711
CSF, 17/713
dermoid cyst, 17/716
differential diagnosis, 17/716, 18/124
diffuse cerebellar hypertrophy, 17/716
EEG, 17/715
ependymoma, 17/716
ethylnitrosourea, 18/171
headache, 17/710
hemangioma, 17/715
histology, 18/124
history, 17/707
intracranial hypertension, 17/710
medially situated, 17/712, 716
molecular biology, 67/7
neurocysticercosis, 17/716
nystagmus, 16/325, 17/711
ocular dysmetria, 16/326
ocular flutter, 16/326
palatal myoclonus, 16/325
papilledema, 17/710
pathology, 17/708, 715
radiology, 17/713
symptom, 67/159
symptomatology, 17/710
treatment, 17/717
trigeminal nerve, 17/711
ventriculography, 17/713
vertigo, 17/710
vomiting, 17/710
Von Hippel-Lindau disease, 43/21
Cerebellar vein
great anterior, *see* Petrosal vein
inferior posterior, *see* Inferior posterior cerebellar vein
precentral, *see* Precentral cerebellar vein
superior posterior, *see* Superior posterior cerebellar vein
Cerebellar vermis
acrylamide intoxication, 64/74
autosomal dominant early onset, *see* Autosomal dominant early onset cerebellar vermis atrophy
benign paroxysmal positional nystagmus, 2/371
cerebellar agenesis, 50/188
coordination, 1/307
dystopic cortical myelinogenesis, 50/206
rhombencephaloschisis, 50/188
somatotopic localization, 2/407
Cerebellar vermis agenesis
arthrogryposis multiplex congenita, 14/408
ataxia, 42/4
Dandy-Walker syndrome, 30/624
dentate nucleus, 42/4
dysmorphogenesis, 30/100
encephalomeningocele, 42/4
hyperventilation, 42/4
Krabbe-Bartels disease, 14/410

Kundrat syndrome, 30/384
lateral cerebellar foramina atresia, 42/5
mental deficiency, 42/4, 43/257
myoclonus, 42/4
neuronal heterotopia, 42/5
nystagmus, 42/4
olivary nucleus, 42/5
seizure, 42/4
ventricular dilatation, 42/4
Cerebellar vermis aplasia
cerebellar agenesis, 50/175, 188, 190
clinical picture, 30/385
computer assisted tomography, 60/672
corpus callosum agenesis, 50/150
Dandy-Walker syndrome, 30/624, 32/59, 50/168
Joubert syndrome, 30/384, 50/190
neurocutaneous syndrome, 50/191
pathology, 30/383
renal retinal syndrome, 60/672
trisomy 9, 50/564
Cerebellar vermis atrophy
congenital ataxia, 60/664
hereditary periodic ataxia, 60/441
mental deficiency, 60/664
upbeating nystagmus, 60/656
Cerebellar vermis dysplasia
Arnold-Chiari malformation type II, 50/405
Cerebellar vermis heterotopia
Arnold-Chiari malformation type II, 50/406
Cerebellar vermis hypoplasia, *see* Cerebellar vermis aplasia
Cerebellifugal atrophy
dentatorubropallidoluysian atrophy, 21/530
Cerebellolental degeneration, *see* Marinesco-Sjögren syndrome
Cerebellomedullary arachnoiditis
differential diagnosis, 17/716
Cerebello-olivary atrophy
ataxia, 42/136
clinical features, 60/570
computer assisted tomography, 60/571, 671
dementia, 42/136
dentate nucleus, 42/136
differential diagnosis, 60/570
dominant, *see* Dominant cerebello-olivary atrophy
hereditary, *see* Hereditary cerebello-olivary atrophy (Holmes)
history, 60/569
late cortical cerebellar atrophy, 60/571
myoclonic epilepsy, 60/663
nonhereditary, *see* Marie-Foix-Alajouanine disease
nosology, 60/571
nucleus ruber, 42/136
olivopontocerebellar atrophy, 60/571
sporadic, *see* Marie-Foix-Alajouanine disease
substantia nigra degeneration, 42/136
tremor, 60/661
Cerebellopallidoluysionigral atrophy
Ferguson-Critchley syndrome, 59/322
Cerebellopontine angle
arachnoid cyst, 31/114-116
neurinoma, 8/417
trigeminal neuralgia, 48/450
Cerebellopontine angle syndrome
*see also* Acoustic nerve tumor
abducent nerve paralysis, 2/318
acoustic neuroma, 2/93-97
aneurysm, 12/95
arachnoid cyst, 31/108
basilar artery, 12/95
clinical features, 2/93
deafness, 2/93
etiology, 2/94-97
facial paralysis, 2/318
gait disturbance, 2/94
intracranial hypertension, 2/93-97
neurofibromatosis type I, 19/39
posterior fossa, 31/108
skull base tumor, 17/182
statoacoustic nerve, 2/318
tinnitus, 2/93
vertigo, 2/94
Cerebellopontine angle tumor
*see also* Acoustic nerve tumor
deafness, 2/377
ependymoma, 18/120
epidermoid cyst, 5/273
statoacoustic nerve, 2/377
symptom, 67/159
tinnitus, 2/377
trigeminal neuralgia, 5/273
vertigo, 2/377
Cerebellopontine lesion
aneurysm compression, 2/257
brain stem hemorrhage, 2/263
central pontine myelinolysis, 2/263
Cerebellopontine vein
ventrolateral, *see* Ventrolateral cerebellopontine vein
Cerebellum
absence, 50/83
anatomy, 2/400
anencephaly, 50/83
arterial hypertension, 63/74
ataxia telangiectasia, 14/73

Biemond syndrome, 60/449
blood supply, 55/89
boxing injury, 23/560
brain concussion, 23/459
brain death, 24/739
brain hemorrhage, 11/591
brain injury, 23/459
brain tumor, 67/9
cellular development, 67/7
cone, *see* Tonsillar herniation
coordination, 1/293, 295, 300, 305-308, 324
dysplastic gangliocytoma, 68/144
embryology, 30/368
epilepsy, 15/616
familial amyotrophic lateral sclerosis, 59/246
folial pattern, 2/400
Friedreich ataxia, 60/303
gaze, 2/345
glycosaminoglycanosis, 10/444
GM2 gangliosidosis, 10/399
hemiparesis, *see* Marie-Foix-Alajouanine disease
hereditary cerebello-olivary atrophy (Holmes), 21/406, 60/569
hereditary olivopontocerebellar atrophy (Menzel), 21/437
Huntington chorea, 6/340, 49/320
hydranencephaly, 30/671, 50/347
hydrocephalus, 30/671
iniencephaly, 50/131
intrinsic arterial pattern, 11/41
Langerhans cell histiocytosis, 70/59
lead, 64/434
leukemia, 39/11
malabsorption syndrome, 70/232
malformation, 30/367
medulloblast, 18/168
mercury intoxication, 64/381
Niemann-Pick disease type A, 10/491
nystagmus, 1/595
olivopontocerebellar atrophy (Dejerine-Thomas), 21/421
phylogenetic part, 1/329
porencephaly, 30/683, 50/357
Purkinje cell function, 1/58
respiration, 1/655
somatotopic localization, 2/407, 425
St. Louis encephalitis, 56/138
status verrucosus, 50/202
tonsillar herniation, 1/565-567
tuberous sclerosis, 14/47, 366
vascularization rate, 11/42
vestibular system, 16/302
vitamin E deficiency, 70/448
Wolman disease, 14/779
Cerebellum weight
Down syndrome, 50/530
Cerebral amebiasis
brain abscess, 52/315, 325
brain herniation, 52/326
chloroquine, 52/328
chronic, 52/326
clinical picture, 35/37
coma, 52/315
computer assisted tomography, 52/321
confusion, 52/315
cranial nerve palsy, 52/315
direct observation, 35/41
drainage, 52/327
EEG, 52/321
emetine, 52/328
encephalitis, 52/315
Entamoeba histolytica, 35/25
epidemiology, 35/34
epilepsy, 52/315
headache, 52/315
immunity, 52/323
meningitis, 52/315
metronidazole, 52/327
neuroamebiasis, 52/309
pathogenesis, 52/321
pathology, 52/325
predisposing factor, 52/322
primary amebic meningoencephalitis, 35/34-37
prognosis, 52/327
treatment, 52/327
Cerebral amyloid angiopathy, *see* Brain amyloid angiopathy
Cerebral aneurysm, *see* Brain aneurysm
Cerebral angiitis, *see* Brain vasculitis
Cerebral angiography, *see* Brain angiography
Cerebral angiopathy
amyloid, *see* Brain amyloid angiopathy
congophilic, *see* Brain amyloid angiopathy
Cerebral anoxia, *see* Brain anoxia
Cerebral arterial system, *see* Brain artery system
Cerebral arteriosclerosis, *see* Brain arteriosclerosis
Cerebral asthenopia
agnosia, 4/25
metamorphopsia, 2/662, 670
Cerebral atrophy, *see* Brain atrophy
Cerebral blindness, *see* Cortical blindness
Cerebral blood flow, *see* Brain blood flow
Cerebral blood volume, *see* Brain blood volume
Cerebral calcification, *see* Intracerebral calcification
Cerebral cholesterosis
late cortical cerebellar atrophy, 21/477

tendon xanthoma, 14/779
Cerebral color blindness
color agnosia, 45/53
Cerebral commissurotomy, *see* Brain commissurotomy
Cerebral concussion, *see* Brain concussion
Cerebral contusion, *see* Brain contusion
Cerebral cortex, *see* Brain cortex
Cerebral cortex ectopia, *see* Brain cortex heterotopia
Cerebral cortex heterotopia, *see* Brain cortex heterotopia
Cerebral cyst, *see* Brain cyst
Cerebral diplegia
double athetosis, 49/384
infantile, *see* Cerebral palsy
reflex, 4/349
Cerebral dominance, 4/248-270, 312-324
*see also* Handedness *and* Manual dexterity
acalculia, 4/293, 465
agnosia, 4/298-301
agraphia, 4/147, 293, 45/463
alexia, 4/134, 293
amusia, 4/294
amytal test, 3/55, 4/266
anatomy, 3/26, 36, 4/301-309
aphasia, 4/88, 262, 267, 292, 308, 316-320, 45/303, 46/148
apraxia, 4/55, 294-297
asomatognosia, 4/297
auditory function, 4/39, 288
body scheme disorder, 4/213, 297
brain commissure, 4/263, 273
brain commissurotomy, 4/275-289, 305
brain injury, 45/495-500
brain venous system, 11/48
calculation, 4/281, 288
chemical hemispherectomy, 4/266
child, 4/306, 465-469
consciousness, 3/55
constructional apraxia, 4/52, 56, 68, 74-76, 79, 295, 45/491, 493-500
corpus callosum, 4/263, 273-290
corpus callosum agenesis, 4/288, 30/289
corpus callosum section, 4/273
cortical stimulation, 4/265
crossed aphasia, 4/5, 269
dementia, 46/204
depression, 46/473
developmental dysarthria, 4/465
developmental dyscalculia, 4/465
developmental dysgraphia, 4/465-469
developmental dyslexia, 4/380, 465-469, 46/116
dichotic listening, 4/39
disconnection symptom, 4/275-287
disconnection symptom masking, 4/285
disconnection syndrome, 4/264, 266, 273, 275-289, 45/103
dominant hemisphere, 3/33-36, 4/262, 291
dominant hemisphere lesion, 4/263
double reaction time task, 4/287
dressing apraxia, 4/296
dyslexia, 46/132-134
EEG, 4/307, 466
experimental chemical data, 4/266
frontal lobe, 4/2
frontal lobe tumor, 17/235
function restoration, 3/379
functional aspect, 3/26, 55
gesture, 4/5
hemispherectomy, 4/265
hemispheric specialization, 4/262, 304-306
higher nervous activity, 3/27-33, 4/465-469
historic survey, 4/248, 291
hyperkinetic child syndrome, 4/465
intelligence disorder, 3/310
intermodal sensory disconnection system, 4/280
intermodal sensory function, 4/280
language, 45/302
language and minor hemisphere, 4/281-283
learning disability, 46/125, 132-134
left handedness, *see* Handedness
literacy, 45/304
memory, 3/148
mental deficiency, 4/465
mental function, 3/15
mental property, 4/286
minor hemisphere, 4/262, 286, 291
motor control, 4/283-285
motor function, 4/283-285
neglect syndrome, 45/172-175
neuropsychologic test, 4/267, 466
neuropsychology, 45/3
neurosurgical data, 4/265
occipital lobe, 4/5, 303
occipital lobe lesion, 4/303
olfactory function, 4/288
parietal lobe lesion, 4/4, 302
perception, 4/5
reading, 45/448
restoration of higher cortical function, 3/379
schizophrenia, 46/456-460
sensory feedback, 4/286
sensory representation, 4/303, 305, 307, 324
shifted dominance, 4/263
somatosensory disorder, 4/303, 305
somesthetic disconnection symptom, 4/278-280

somesthetic function, 4/278-280
spatial orientation disorder, 3/27
speech, 4/280, 292, 322, 43/223
speech disorder, 4/5
speech disorder in children, 4/465-469
speech and writing, 4/280
split brain animal study, 4/273
stereognosis, 4/278
stock brainedness, 4/262
stuttering, 46/170
temporal lobe, 4/3, 301
temporal lobe lesion, 4/301
thumb sucking, 4/468
topographical memory, 4/299
traumatic cerebral lesion, 4/267
variation, 4/5
vascular cerebral lesion, 4/267
visual agnosia, 4/20, 289-301
visual allesthesia, 3/29
visual disconnection symptom, 4/277
visual function, 4/277
visuospatial agnosia, 4/298, 45/174
Wada test, 4/2, 89, 266, 318-320
writing, 4/280
Cerebral dysplasia, *see* Brain dysplasia
Cerebral edema, *see* Brain edema
Cerebral fibrosarcoma, 42/729-731
glioblastoma multiforme, 42/730
prevalence, 42/730
radiation, 42/731
Cerebral gigantism
aggression, 31/333, 43/337
ataxia, 43/337
brain defect, 30/247
dermatoglyphics, 31/332, 43/336
EEG, 43/337
etiology, 31/332
hypertelorism, 30/247, 43/336
17-ketosteroid, 31/333
kyphoscoliosis, 43/337
macrencephaly, 31/332, 42/41, 50/263
macrocephaly, 43/336
mental deficiency, 31/333, 43/337, 46/83
neurologic disorder, 30/247
radiologic finding, 31/333
sign, 31/332
symptom, 31/332
syndactyly, 43/336
Cerebral granuloma, *see* Brain granuloma
Cerebral hemiatrophy
Ammon horn sclerosis, 50/205
clinical features, 50/206
definition, 50/204
epilepsy, 50/206
etiology, 50/206, 59/479
infantile hemiplegia, 50/206
mental deficiency, 30/356
pathology, 50/204
primary, *see* Primary cerebral hemiatrophy
progressive hemifacial atrophy, 59/479
status spongiosus, 50/206
ulegyria, 50/204, 206
Cerebral hemiplegia
infantile, *see* Cerebral palsy
Cerebral hemorrhage, *see* Brain hemorrhage
Cerebral herniation, *see* Brain herniation
Cerebral heterotopia
epilepsy, 30/493
mental deficiency, 30/492
Cerebral histiocytosis
Abt-Letterer-Siwe disease, 39/531
Hand-Schüller-Christian disease, 39/531
Cerebral hydatid cyst
echinococcosis, 35/183-193, 347
Cerebral hyperemia, *see* Brain hyperemia
Cerebral hypoxia, *see* Brain hypoxia
Cerebral infarction, *see* Brain infarction
Cerebral ischemia, *see* Brain ischemia
Cerebral lateralization, *see* Cerebral dominance
Cerebral lipidosis, *see* Neurolipidosis
Cerebral localization, 3/11-20
ability level variation, 4/5-9
ablation study, 3/14
anatomy, 3/309, 312
clinical features, 3/13-15, 309, 369
complicated migraine, 5/65
consciousness, 3/124
diaschisis, 3/369
experimental data, 3/13-15
functional localization, 3/15-18
hallucinosis, 46/566
higher nervous activity, 3/3
higher nervous activity disorder, 3/22-38
history, 3/11-13, 22
intelligence, 3/300-303, 312, 4/8
learning, 3/312
orientation, 3/207
pain, 4/223
restoration of higher cortical function, 3/381
variation, 4/5
visuospatial agnosia, 3/25
Cerebral lymphomatoid granulomatosis
clinical features, 39/517
course, 39/522
differential diagnosis, 39/526-535
disseminated necrotizing leukoencephalopathy,

39/524
hemorrhage, 39/524
Hodgkin disease, 39/522-524
laboratory finding, 39/517
pathology, 39/518-522
progressive multifocal leukoencephalopathy, 39/524
Cerebral malaria
artemisinin, 52/373
brain hemorrhage, 35/146
brain infarction, 53/162
brain ischemia, 53/162
brain microcirculation, 52/368
coma, 35/145, 52/368
consciousness, 35/149, 52/368
CSF, 52/369
diagnosis, 35/152, 52/369
differential diagnosis, 35/154, 52/368
epilepsy, 35/146, 52/369
headache, 35/145, 149
hemorrhagic leukoencephalopathy, 52/368
mental symptom, 35/147, 149, 52/368
mortality, 35/146, 52/368
pathogenesis, 35/149, 52/368
photophobia, 35/146
polyneuropathy, 35/148, 52/370
psychiatric symptom, 35/147, 149, 52/368
transverse myelitis, 35/149, 52/370
Cerebral maturation, *see* Brain maturation
Cerebral membrane
hydranencephaly, 30/673, 50/347
Cerebral metabolism, *see* Brain metabolism
Cerebral midline syndrome, 2/758-775
Cerebral mineralizing microangiopathy
intrathecal chemotherapy, 61/151
Cerebral motor cortex, *see* Motor cortex
Cerebral necrosis
radiation injury, 67/329
Cerebral oxygen extraction rate, *see* Brain oxygen extraction rate
Cerebral palsy
ataxia, 46/14
birth incidence, 46/14
body scheme disorder, 4/398
brain anoxia, 46/14
breech delivery, 46/14
category, 46/14
clinical features, 1/195, 49/482
clumsiness, 46/161-163
congenital heart disease, 63/8
developmental disorder, 4/398
disorder of higher nervous activity, 4/360-365
dyskinesia, 46/14
EEG, 46/493
familial spastic paraplegia, 59/306
fetal Minamata disease, 64/418
forceps delivery, 46/14
incontinentia pigmenti, 14/214
meningococcal meningitis, 52/28
mental deficiency, 46/13-16
Minamata disease, 36/135, 64/418
organic acid metabolism, 66/641
Paine syndrome, 60/293
pallidal necrosis, 49/481
pathology, 46/14-16
spastic, *see* Spastic cerebral palsy
spastic diplegia, 46/14
spastic hemiplegia, 46/14
spastic paraplegia, 42/167
spastic tetraplegia, 42/167, 46/14
spasticity, 60/661
status marmoratus, 6/678, 46/16
striatal necrosis, 49/506
tactile agnosia, 4/34
tremor, 49/604
ulegyria, 46/14
voice, 46/30
Cerebral paragonimiasis, 35/251-260
brain scanning, 35/257
clinical features, 35/251-253
CSF, 35/257
diagnosis, 35/258
EEG, 35/257
prognosis, 35/259
sign, 35/253
treatment, 35/259
Cerebral perfusion pressure, *see* Brain perfusion pressure
Cerebral sclerosis, *see* Diffuse sclerosis
Cerebral sclerosis (Schilder), *see* Myelinoclastic diffuse sclerosis
Cerebral swelling, *see* Brain swelling
Cerebral thromboangiitis obliterans, *see* Brain thromboangiitis obliterans
Cerebral thromboembolism
poikilomycosis, 35/567, 52/489
Cerebral thrombosis, *see* Brain thrombosis
Cerebral toxoplasmosis
acquired toxoplasmosis, 35/123, 52/353
benign intracranial hypertension, 52/354
demyelinating encephalitis, 52/357
headache, 52/354
multifocal necrotizing encephalopathy, 52/357
necrotizing encephalitis, 52/357
Cerebral tumefaction, *see* Brain edema
Cerebral tumor, *see* Brain tumor

Cerebral vascular disease, *see* Cerebrovascular disease
Cerebral vasculitis, *see* Brain vasculitis
Cerebral vasospasm, *see* Brain vasospasm
Cerebral vein
  ascending, *see* Ascending cerebral vein
  ascending occipital, *see* Ascending occipital cerebral vein
  central, *see* Central cerebral vein
  deep middle, *see* Deep middle cerebral vein
  descending, *see* Descending cerebral vein
  inferior, *see* Inferior cerebral vein
  internal, *see* Internal cerebral vein
  striatal necrosis, 49/500
  superficial middle, *see* Superficial middle cerebral vein
Cerebral venous system, *see* Brain venous system
Cerebral venous thrombosis, *see* Brain phlebothrombosis
Cerebroasthenic syndrome
  attention disorder, 3/142, 149
Cerebrocostomandibular syndrome
  birth incidence, 43/237
  cheilopalatoschisis, 43/237
  heterozygote frequency, 43/237
  mental deficiency, 43/237
  micrognathia, 43/237
  rib abnormality, 43/237
  vertebral abnormality, 43/237
Cerebrocuprein
  brain specific copper protein, 6/119
Cerebrohepatorenal syndrome
  adrenoleukodystrophy, 47/594, 596-598
  adrenomyeloneuropathy, 60/170
  aminoaciduria, 43/339
  areflexia, 43/338
  arhinencephaly, 50/242
  biochemistry, 41/213
  birth incidence, 43/340
  brachycephaly, 43/338
  brain atrophy, 43/340
  brain cortex heterotopia, 50/248
  brain defect, 30/247
  brain white matter degeneration, 43/340
  Brushfield spot, 43/339
  cataract, 43/339
  clinical comparison, 66/526
  CNS, 31/304
  convulsion, 31/306, 43/339
  corpus callosum agenesis, 31/308, 50/163
  cortical heterotopia, 31/307
  craniofacial features, 66/511
  cryptorchidism, 43/339
  dermatoglyphics, 43/339
  differential diagnosis, 31/243
  docosahexaenoic acid deficiency, 66/515
  epicanthus, 43/339
  epilepsy, 72/222
  genetics, 31/305
  glaucoma, 43/339
  Gruber syndrome, 31/243
  hepatomegaly, 43/339
  histochemistry, 41/213
  hypertelorism, 30/247, 43/338
  hyporeflexia, 43/338
  infantile hypotonia, 43/338
  inferior olivary nucleus heterotopia, 50/250
  iron, 43/339
  lissencephaly, 50/250
  low set ears, 43/339
  mental deficiency, 43/339, 46/64
  metabolic abnormality, 31/307
  micrognathia, 43/339
  neurologic disorder, 30/247
  neuronal heterotopia, 43/340
  neuronal migration, 43/340
  pachygyria, 43/340
  pathology, 31/304, 307
  peroxisomal disorder, 66/17, 510
  L-pipecolic acid, 66/515
  polymicrogyria, 43/340
  radiologic appearance, 31/306
  renal abnormality, 43/340
  rhizomelic chondrodysplasia punctata, 66/525
  Rubinstein-Taybi syndrome, 30/472
  seizure, 31/305
  survey, 66/505
  talipes, 43/339
  treatment, 31/308
  VACTERL syndrome, 50/514
Cerebromacular degeneration
  Batten disease, 10/588
Cerebrometacarpometatarsal dystrophy, *see* Pseudopseudohypoparathyroidism
Cerebromuscular dystrophy, *see* Fukuyama syndrome
Cerebron
  cerebroside sulfatide, 10/48
  chemistry, 10/249
  globoid cell leukodystrophy, 10/304
Cerebron sulfuric acid
  chemistry, 10/313
Cerebronic acid, *see* 2-Hydroxytetracosanoic acid
Cerebro-oculofacioskeletal dysplasia, *see* Cerebro-oculofacioskeletal syndrome
Cerebro-oculofacioskeletal syndrome

birth incidence, 43/241
brain white matter degeneration, 43/341
cataract, 43/341
contracture, 43/341
gene frequency, 43/342
growth retardation, 43/341
heterozygote frequency, 43/342
intracerebral calcification, 43/341
mental deficiency, 43/341
microcephaly, 43/341
micrognathia, 43/341
microphthalmia, 43/341
osteoporosis, 43/341
ventricular dilatation, 43/341
Cerebroside
biosynthesis, 10/250
brain development, 9/18
chemistry, 10/248
Gaucher disease, 10/308
glia, 9/6
globoid cell leukodystrophy, 10/71, 304, 327, 29/356
2-hydroxytetracosanoic acid, 9/6
metachromatic leukodystrophy, 10/28, 49
myelin, 7/44, 9/3, 29
myelin change, 10/136
Pelizaeus-Merzbacher disease, 10/33
synthesis, 9/9
15-tetracosanoic acid, 9/6
type, 10/249
Cerebroside ester
chemistry, 10/250
Cerebroside sulfatase
biochemistry, 10/349
enzyme deficiency polyneuropathy, 51/368
heat labile component, 10/47, 143
late infantile metachromatic leukodystrophy, 10/311
measurement, 66/174
metachromatic leukodystrophy, 10/27, 28, 47, 56, 568, 29/350, 357, 358, 42/491, 494, 47/589, 589-591, 51/368, 60/124, 66/92
mucolipidosis type II, 42/448
mucolipidosis type III, 42/449
myelin sulfatide storage, 21/65
pseudodeficiency, 66/58, 171
Tuffstein granule, 21/51
Cerebroside sulfatase deficiency, *see* Metachromatic leukodystrophy
Cerebroside sulfatase gene
structure, 66/172
Cerebroside sulfate ester
metachromatic leukodystrophy, 10/25, 44
Cerebroside sulfatidase
heat labile component, *see* Cerebroside sulfatase
metachromatic leukodystrophy, 10/27, 47
in vitro, 10/143
Cerebroside sulfatide
brain, 10/47
cerebron, 10/48
kerasin type, 10/48
metachromatic leukodystrophy, 8/16, 10/48
plaque, 9/314
sphingolipid, 10/327
Cerebroside sulfatidosis, *see* Metachromatic leukodystrophy
Cerebroside sulfotransferase
globoid cell leukodystrophy, 10/21, 87, 306, 29/356
Cerebrosidosis, 10/306
Cerebrospinal fever, *see* Meningococcal meningitis
Cerebrospinal fluid, *see* CSF
Cerebrospinal ganglion
development, 7/3
Cerebrospinal meningitis, *see* Meningococcal meningitis
Cerebrosulfate ester
metachromatic leukodystrophy, 21/51
Cerebrotendinous cholestanolosis, *see* Cerebrotendinous xanthomatosis
Cerebrotendinous cholesterinosis, *see* Cerebrotendinous xanthomatosis
Cerebrotendinous xanthomatosis, 10/532-540
ataxia, 10/532, 546, 29/374, 51/399, 59/357, 60/167
autosomal recessive inheritance, 29/375, 42/430, 59/357
axonal degeneration, 51/399
bile acid, 66/606
biochemistry, 66/605
cardiovascular system, 66/601
carrier detection, 66/608
case survey, 10/538
cataract, 10/532, 547, 27/250, 29/374, 42/430, 51/399, 59/357, 60/167, 653
cerebellar ataxia, 10/537, 27/250, 42/430
chemical pathology, 10/532-534
chenodeoxycholic acid, 51/399, 59/358, 60/168
cholestanol, 10/532, 547, 29/375, 51/399, 60/674, 66/16, 110
cholestanol storage, 59/357
cholesterol, 10/547, 27/250, 29/374, 66/606
cholesterol metabolism, 42/430
classification, 10/343
clinical features, 10/537, 59/357
CSF, 66/603

dementia, 10/532, 29/374, 42/430, 46/403, 51/399, 59/357, 60/166
demyelination, 10/547, 51/399
differential diagnosis, 10/539, 60/167
dioxin intoxication, 60/167
dysarthria, 10/546
endocrine abnormality, 66/601
enterohepatic system, 66/601
epilepsy, 60/167, 66/602
gene, 66/607
heterogeneity, 14/779
history, 66/599
hyperlipidemic neuropathy, 60/167
hyperlipoproteinemia type II, 10/537, 539
hyperlipoproteinemia type III, 10/537, 539
inheritance, 10/547, 29/375
mental deficiency, 10/532, 537, 547, 27/250, 60/166
metabolic ataxia, 21/575
metabolic manipulation therapy, 66/110
metabolic polyneuropathy, 51/399
molecular diagnosis, 66/606
muscle, 66/604
muscle biopsy, 66/605
muscular atrophy, 60/167
nephrolithiasis, 66/601
nerve conduction, 66/608
nerve conduction velocity, 60/670
neurologic sign, 66/602
neuropathology, 10/534-537, 60/167
neuropathy, 60/660
nuclear magnetic resonance, 59/358, 60/673, 66/602
ocular finding, 66/599
optic neuropathy, 60/167
palatal myoclonus, 51/399, 60/167
paraparesis, 59/357
pathologic anatomy, 10/534
peripheral nerve, 66/605
peripheral nervous system, 66/603
polyneuropathy, 59/357, 60/166
premature aging, 66/609
prevalence, 60/166
pseudobulbar paralysis, 10/532, 537, 51/399, 60/167
psychosis, 59/357
remyelination, 51/399
skeleton, 66/601
spasm, 29/374
spastic paraparesis, 60/167
spastic paraplegia, 59/439
spasticity, 51/399, 60/661
sural nerve biopsy, 60/167
symptom, 60/166
tendon xanthoma, 10/532, 547, 27/250, 29/375, 42/430, 59/357, 60/669
therapy, 66/607
treatment, 10/540, 59/358
xanthoma, 29/374, 66/600
Cerebrovascular accident, *see* Stroke
Cerebrovascular disease, 53/1-21, 27-44, 199-241, 55/411-437, 463-476
*see also* Brain arteriosclerosis, Brain artery occlusion, Brain embolism, Brain hemorrhage, Brain infarction, Brain thrombosis, Brain vasculitis, Carotid artery thrombosis, Cerebrovascular occlusive disease, Subarachnoid hemorrhage *and* Vertebral artery thrombosis
action myoclonus, 53/211
acute infantile hemiplegia, 53/32
age, 12/450, 53/200
age adjusted mortality rate, 53/5, 8
age group, 53/27
aged, 53/42
akinetic mutism, 53/209, 55/137
alcohol, 53/17
alcohol excess, 55/163
allergic granulomatous angiitis, 11/133, 53/41
amaurosis fugax, 55/107
amnesia, 53/206, 55/139
amphetamine, 46/592
amyloidosis, 53/28, 43, 55/163
analytic epidemiology, 53/16
anemia, 55/463
angiitis, 53/40
angiography, 12/451
angioma compression, 11/128
anoxic ischemic leukoencephalopathy, 47/541
anterior alexia, 45/442
anterior ischemic optic neuropathy, 55/114
anterior vestibular vein, 55/129
anticoagulant, 12/456-465, 55/157, 166, 169
antiphospholipid antibody, 63/329
Anton syndrome, 55/139, 144
aorta valve endocarditis, 53/34
apathy, 55/137
aphasia, 42/278, 53/214
aprosody, 55/143
arousal, 55/137
arterial dissection, 53/34
arteriosclerosis, 53/200
arteriovenous malformation, 53/34
arteritis, 53/28
artery occlusion, 12/452
asparaginase, 69/418

aspergillosis, 55/432
asterixis, 53/211
asymptomatic individual, 53/3
ataxia, 53/214
atheroma, 12/453
atherothrombotic, *see* Atherothrombotic cerebrovascular disease
atrial myxoma, 55/163
attention, 55/137
auditory symptom, 53/207
autonomic dysfunction, 53/207
autonomic nervous system, 75/361
autotopagnosia, 55/145
bacterial endocarditis, 52/290
bacterial meningitis, 55/416
Balint syndrome, 45/16, 55/144
ballistocardiography, 11/258
basilar artery bifurcation aneurysm, 55/124
behavior change, 46/416
behavior disorder, 55/137
Behr disease, 53/213
Binswanger disease, 46/318, 53/28
Blastomyces dermatitidis, 55/432
blepharospasm, 53/211
blindness, 53/207
blood lipid, 53/17
blood pressure, 53/207
blood viscosity, 55/483
brain abscess, 53/31, 55/418
brain aneurysm, 53/41
brain angiography, 53/228
brain arteriosclerosis, 53/91, 200
brain arteriovenous malformation, 53/34
brain compression, 11/155
brain edema, 11/128
brain hemorrhage, 53/34
brain lacunar infarction, 53/28
brain matrix hemorrhage, 53/30
brain phlebothrombosis, 53/37
brain scanning, 11/240-247
brain scintigraphy, 11/247-252, 53/232
brain stem auditory evoked potential, 53/233
brain tumor, 69/413
brain tumor hemorrhage, 5/141
Canadian Neurologic Scale, 53/209
cancer, 69/420
Candida, 55/432
candidiasis, 55/435
cardiac disorder, 53/200
cardiac dysrhythmia, 75/370
cardiomyopathy, 53/43, 55/162
cardiovascular state, 12/449
carotid arteritis, 53/32
carotid artery thrombosis, 53/32, 37
carotid bruit, 53/20, 216
carotid dissecting aneurysm, 53/39
carotid sinus massage, 53/219
case ascertainment, 53/10
causative heart disease, 55/161
causative myocardial infarction, 55/161
cause, 53/201
central alexia, 45/440
Chagas disease, 55/163
Charcot-Bouchard aneurysm, 53/41
chiasmal field defect, 55/115
child, 53/31
chronic ocular hypoxia, 55/114
chronic vegetative state, 53/209
classification, 53/2
climate, 53/200
clinical course, 53/202
clinical evaluation, 53/199
clinical trial, 53/4
coagulopathy, 55/474
cobalt therapy, 53/33
Coccidioides immitis, 55/432
cochlear artery, 55/129
cognition, 55/137
collagen vascular disease, 55/163
collateral circulation, *see* Brain collateral blood flow
coma, 53/208
computer assisted tomography, 53/221, 54/47, 53
conduction aphasia, 53/215
confusion, 53/206
confusional state, 55/138
congenital heart disease, 53/28, 31
congestive heart failure, 55/162
consciousness loss, 53/206
correct diagnosis, 53/199
cortical blindness, 55/139, 144
coumarin anticoagulant, 55/476
course, 12/448
creatine kinase MB, 55/156
Cryptococcus neoformans, 55/432
CSF, 11/235-240, 53/224
CSF spectrophotometry, 53/224
cystathionine synthetase deficiency, 53/33
cystic leukoencephalopathy, 53/29
death certificate, 53/4
death rate, 11/184, 188
decerebrate rigidity, 53/211
decortication, 53/211
delirium, 46/545, 55/138
dementia, 42/277-279, 53/42
depression, 55/143

diabetes mellitus, 27/126-130, 53/17, 75/366
diagnostic error, 53/5, 220
diagnostic test, 11/208-266
diagnostic value, 53/3
differential diagnosis, 53/233
diphtheria, 55/163
diplopia, 53/207
directional Doppler velocimetry, 53/225
disease magnitude, 53/1, 4
dorsal midbrain syndrome, 55/124
Down syndrome, 53/31
drug abuse, 55/166
drug induced, 69/482
Duchenne muscular dystrophy, 53/43
duplex sonography, 53/225
duration, 53/203
dysphagia, 53/207
dysphonia, 53/207
dystonia, 53/211
ECG, 53/230
echoCG, 53/230
echoEG, 11/254-257, 53/233
EEG, 11/267-291, 53/232, 54/157
Ehlers-Danlos syndrome, 53/43, 55/454
Eisenmenger complex, 53/31
embolic, *see* Embolic cerebrovascular disease
embolism, 12/450
emotional disorder, 53/206
enterococcus, 55/166
EOG, 54/157
epidemiologic index, 53/2
epidemiology, 11/183-203, 53/1
epilepsy, 53/36
evoked potential, 54/157
exogenous estrogen, 53/17
experimental study, 53/20
eyelid nystagmus, 53/213
Fabry disease, 53/28, 33, 55/456
Fallot tetralogy, 53/31
fibromuscular dysplasia, 11/371, 53/28
fragile X disease, 53/43
Friedreich ataxia, 55/163
frontomesencephalic pathway, 55/121
fungal infection, 55/432
gaze deviation, 53/212
gaze paralysis, 55/121
general pathology, 11/128
general treatment, 12/445
genetic predisposition, 53/199
geniculocalcarine tract, 55/117
geographic factor, 11/188
geographic incidence, 53/11
geographic mortality, 53/5
geographic variation, 53/5, 7, 13
geopathology, 53/1
Glasgow Coma Scale, 53/209
global aphasia, 53/215
glycogen storage disease, 55/163
glycosaminoglycanosis type I, 55/163
granuloma, 55/432
granulomatous arteritis, 53/41
hallucination, 53/206
headache, 5/124, 48/273, 280, 288
heart disease, 53/17
heart rhabdomyoma, 53/43
heart tumor, 55/163
heart wall ostia, 55/163
hematotachygraphy, 53/225
hemianopia, 53/207
hemiathetosis, 53/31
hemiballismus, 53/211
hemichorea, 53/211
hemi-inattention, 55/145
hemineglect, 55/145
hemiparesis, 53/206
hemochromatosis, 55/163
hemodialysis, 55/166
hemolytic uremic syndrome, 53/34
hemophilia, 55/475
hemorrhagic leukoencephalitis, 53/32
hereditary, 53/43
hereditary hemorrhagic telangiectasia (Osler), 53/31, 55/448
Histoplasma capsulatum, 55/432
history, 53/199
Holter monitoring, 53/230
homocystinuria, 53/28, 33, 43
homonymous hemianopia, 55/117
Horner syndrome, 53/213
human immunodeficiency virus infection, 71/335, 337
hyperlipidemia, 53/34
hypertension, 12/449, 453, 53/17, 75/368
hypertensive encephalopathy, 53/39, 41
hypertensive fibrinoid arteritis, 11/136-139
hyperthermia, 53/207
hyperviscosity syndrome, 55/470, 483
hypotension, 53/29
hypoxia, 53/29
ICD classification, 53/2
ICD classification inconsistency, 53/2
immunosuppression, 55/166
incidence, 53/1
incidence rate, 53/11
indifferent reaction, 55/143
infantile, *see* Infantile cerebrovascular disease

infantile hemiplegia, 53/31
infected prosthetic valve, 55/166
infection effect, 55/411
infectious endocarditis, 55/166
inherited coagulation factor deficiency, 55/474
internal auditory artery, 55/129
internuclear ophthalmoplegia, 53/212
intravenous digital subtraction angiography, 53/227
iron deficiency anemia, 55/463
ischemia, *see* Brain ischemia
juvenile, 12/340-351
lacunar stroke, 53/41
late adult, 53/41
lateral geniculate body, 55/115
latitude variation, 53/7
leptospiral arteritis, 55/415
leptospiral infection, 55/415
libido, 53/207
macular sparing, 55/119
Marfan syndrome, 53/33, 43
mechanism, 11/128
melanoma, 55/163
MELAS syndrome, 53/43
mental change, 55/137
mental deficiency, 53/31
mental function, 53/214
mental state, 12/450
micturition, 53/207
mid adult, 53/38
migraine, 45/510, 53/28
migrainous stroke, 53/37
migrant variation, 53/7
misoplegia, 55/145
mitral valve prolapse, 53/34, 43, 55/165
monocular vision loss, 53/207
morbidity data, 53/9
morbidity variation, 53/12
mortality, 53/1
mortality data, 53/4
moyamoya disease, 12/346, 374, 53/28, 31
mucormycosis, 35/545, 55/435
multi-infarct dementia, 53/14, 28, 42, 55/141
multinodular stenosing amyloidosis, 11/581, 613
multiple myeloma, 55/470
muscle tone, 53/211
myotonic dystrophy, 53/43, 55/163
natural history, 53/19
neuropsychology, 45/8
neurotransmitter, 75/362
newborn, 53/30
nonsurgical treatment, 12/447-455
normotension, 12/451
nuchal rigidity, 53/207
nuclear magnetic resonance, 53/231, 54/81
nystagmus, 53/213
obesity, 53/17
occipital lobe, 55/118
ocular bobbing, 53/213
oculomotor system, 55/120
oculoplethysmography, 53/225
oculopneumoplethysmography, 53/225
onset, 53/202
ophthalmic aspect, 55/107
ophthalmodynamometry, 53/219
ophthalmoscopy, 11/212, 53/218
optic atrophy, 55/114
optic tract, 55/115
oral contraception, 11/199, 53/18, 36
oscillopsia, 53/207
otology, 55/129
palatal myoclonus, 53/211
pallidal necrosis, 49/465
pallidonigral necrosis, 49/465
pallidostriatal necrosis, 49/465
para-alexia, 55/145
Paracoccidioides brasiliensis, 55/432
paradoxical embolus, 55/163
parasympathetic innervation, 75/363
parietal lobe, 55/117
parietal lobe epilepsy, 73/104
Parinaud syndrome, 53/212
paroxysmal nocturnal hemoglobinuria, 55/467
parturition disease, 53/28
pathologic bruit, 53/217
pathologic laughing and crying, 55/142
pathology, 12/449
peripartum period, 55/163
periventricular infarction, 53/28
phonoangiography, 53/218
physiologic bruit, 53/216
plethysmography, 11/253
polyarteritis nodosa, 11/139, 451, 39/301, 48/285, 53/41
polycythemia, 53/31, 43
pontine hematoma, 53/207
pontine paramedian reticular formation, 55/124
pontine syndrome, 55/124
positron emission tomography, 53/232, 54/105
posterior vestibular artery, 55/129
postnatal disease, 53/28
precipitating factor, 53/202
pregnancy, 53/28, 36
prenatal vascular disease, 53/27
prevalence, 53/1
prevalence rate, 53/12

progressive muscular dystrophy, 55/163
prospective study, 53/16
pruritus, 53/207
pseudoabducens nerve palsy, 53/213
pseudoxanthoma elasticum, 55/451, 453
psychiatry, 55/137
puerperal brain thrombophlebitis, 53/37
puerperium, 53/28
pulmonary stenosis, 53/31
pursuit eye movement, 55/121
race, 11/188, 53/200
radiation injury, 67/330
records linkage, 53/10
recovery, 53/203
recurrence, 53/20
rehabilitation, 12/468-477, 46/610
respiration, 53/208, 75/375
retinal artery embolism, 53/218
retinal fluorescence angiography, 11/250-252
retraction nystagmus, 53/212
rhabdomyoma, 55/163
rheoencephalography, 11/252
rheumatic fever, 55/163
risk factor, 53/15, 17, 200
saccade, 55/121
sarcoidosis, 53/41, 55/163
scleroderma, 71/114
segmental intubation, 11/581
seizure, 53/36, 206
selective nucleoperikaryon necrobiosis, 11/152
sex ratio, 53/14, 200
sexual impotence, 53/207
sickle cell anemia, 53/28, 31, 33, 43, 55/464
sickle cell trait, 55/466
single photon emission computer assisted tomography, 54/105
site, 12/448
skew deviation, 53/213
speech disorder, 53/206
Staphylococcus, 55/166
status marmoratus, 53/28
Streptococcus viridans, 55/166
Sturge-Weber syndrome, 55/443
subacute spongiform encephalopathy, 11/129
subarachnoid hemorrhage, 55/167
subclavian bruit, 53/216
subdural empyema, 55/418
sudomotor dysfunction, 75/376
sulfite oxidase deficiency, 53/43
survey, 75/361
suspected, *see* Suspected cerebrovascular disease
sympathetic innervation, 75/361
sympathetic nervous disorder, 2/121
symptomatic individual, 53/3
syphilis, 53/34
syphilitic meningovascular disease, 53/39
systemic disease, 53/200
systemic lupus erythematosus, 53/34
Takayasu disease, 53/41
temporal arteritis, 53/41
temporal crescent, 55/118
temporal decline, 53/7, 13
temporal lobe, 55/117
temporal profile, 54/53
thalassemia, 55/466
thalassemia major, 55/466
thermography, 11/253
thermoregulation, 53/207
thrombocythemia, 55/473
thrombocytopenia, 55/472
thrombophlebitis, 53/32
thrombotic thrombocytopenic purpura, 53/34, 55/472
tinnitus, 53/207
toxin, 55/163
toxoplasmosis, 55/163
transcortical aphasia, 53/215
transient global amnesia, 45/206, 216, 55/141
transient ischemic attack, 53/17
trigeminovascular system, 75/364
tuberculosis, 55/432
tuberous sclerosis, 53/43, 55/163
typhoid fever, 55/163
unilateral pulsating tinnitus, 53/207
varicella zoster virus, 55/436
vascular anatomy, 11/208
vascular pulsation, 53/215
venous sinus thrombosis, 53/31, 55/418
vertebral dissecting aneurysm, 53/39
vertebrobasilar insufficiency, 55/132
vertical gaze palsy, 53/212, 55/123
vertical nystagmus, 53/213
vertical one and half syndrome, 53/212
vertigo, 53/207
viral infection, 55/163
visual evoked response, 53/232
visual hallucination, 53/207, 55/145
visual manifestation, 53/207
vitamin K deficiency, 55/476
vomiting, 53/206
Von Willebrand disease, 53/43, 55/472, 475
Wadia test, 11/212
Wegener granulomatosis, 53/41, 71/180
World Health Organization classification, 11/184
xanthochromia, 53/224
young adult, 53/34, 38

young woman, 53/36
Zygomycetes, 55/432
Cerebrovascular dissection
headache, 48/286
Cerebrovascular insufficiency
perinatal, 12/341
procainamide intoxication, 37/452
Cerebrovascular occlusive disease, 11/327-361, 12/1-33
*see also* Brain artery occlusion, Brain collateral blood flow *and* Cerebrovascular disease
brain microembolism, 9/588
collateral circulation, 11/168-181
consciousness, 45/119
microinfarction, 9/588
occlusion time sequence, 11/172
ocular plethysmography, 11/228
ocular tonography, 11/228
oculosphygmography, 11/228
vasomotor reflex, 11/170
Cerebrovascular resistance
brain anoxia, 27/40
brain blood flow, 53/47
intracranial pressure, 55/207
Cerebrovascular sclerosis, *see* Brain arteriosclerosis
Cerebrum
accommodation disorder, 2/667
hereditary cerebello-olivary atrophy (Holmes), 21/407
subependymal cell layer, 30/65-68
Ceroid lipidosis, *see* Neuronal ceroid lipofuscinosis
Ceroid lipofuscin
Hallervorden-Spatz syndrome, 66/713
terminology, 10/589
Ceroid lipofuscinosis, *see* Neuronal ceroid lipofuscinosis
Ceroidosis, *see* Neuronal ceroid lipofuscinosis
Ceruloplasmin
copper, 49/229
copper bound protein, 6/118
free radical scavengers, 64/19
Hallervorden-Spatz syndrome, 66/714
normal metabolism, 27/391
trichopoliodystrophy, 42/584
Willvonseder syndrome, 42/194
Wilson disease, 21/52, 27/395-399, 42/270, 46/58, 49/231
Ceruloplasmin copper
Wilson disease, 49/229
Ceruminal gland adenoma
skull base tumor, 17/190
Cerveau isolé
*see also* Encéphale isolé
cat, 2/280, 3/49, 81
consciousness disorder, 2/280, 3/49
EEG, 2/280, 3/81
experimental animal, 2/280, 3/81
Cervical angioma
drainage, 32/486
Galen vein, 32/486
vascularization, 12/637
Cervical arterial spasm
craniocerebral trauma, 23/177
Cervical arthrosis
Barré-Liéou syndrome, 5/373
headache, 5/198
occipital neuralgia, 5/373
pain, 5/373
Cervical collar pattern
syringomyelia, 50/444
Cervical compression fracture
articular process, 25/258, 361
Pillar fracture, 61/30
Cervical cord injury, *see* Cervical spinal cord injury
Cervical cord lesion
cervical hyperanteflexion sprain, 25/354
cervical hyperdeflexion sprain, 25/354
prognosis, 35/281
respiratory dysfunction, 26/337
sweat test, 2/200
sweating, 2/200
Cervical cord syndrome
anterior, *see* Anterior cervical cord syndrome
central, *see* Central cervical cord syndrome
Cervical cordotomy
choreoathetosis, 6/854
coughing, 63/491
forgotten respiration, 63/489
hyperkinesia, 6/867
surgical procedure, 6/867
Cervical facet interlocking
bilateral, *see* Bilateral cervical facet interlocking
unilateral, *see* Unilateral cervical facet interlocking
Cervical facet subluxation
reduction, 61/62
traction, 61/62
Cervical fibrous band
cervical rib analogue, 51/120
thoracic outlet syndrome, 2/146
Cervical headache, 5/192-201
*see also* Cervicogenic headache
classification, 48/6
Schützenberger-Barré-Liéou syndrome, 5/192
Cervical hyperanteflexion injury
*see also* Anteflexion

Brown-Séquard syndrome, 25/335
cervical spine, 25/254-256, 329-354, 26/290
disruptive, 25/255
experimental, 25/3, 6
general, 25/62, 26/287, 292
spinal injury, 25/63
thoracolumbar spine, 25/447
Cervical hyperanteflexion sprain
bony fusion, 25/345
cervical cord lesion, 25/354
compression injury, 25/341-345
definition, 25/244
Cervical hyperdeflexion injury
*see also* Retroflexion
bony fusion, 25/363
Brown-Séquard syndrome, 25/359
cervical spine, 25/30, 62, 134, 254, 354-363, 26/285, 291, 293
experimental, 25/3, 6
spinal injury, 25/63, 26/285
thoracolumbar spine, 25/447
Cervical hyperdeflexion sprain
birth incidence, 25/163
cervical cord lesion, 25/354
Cervical injury
headache, 5/180, 200
Cervical intersegmental artery
anatomy, 11/36
malformation, 12/333
Cervical intervertebral disc prolapse, 20/547-561
amyotrophic lateral sclerosis, 22/309
dissociated motor loss, 20/555
history, 20/548
myelopathy, 7/456
neuralgia, 20/548, 552-555
posttraumatic, 25/353
prevalence, 25/494
radicular syndrome, 20/555-559
spinal nerve root pain, 20/548
surgical procedure, 20/561
treatment, 20/559
Cervical meningocele
vagus nerve agenesis, 50/219
Cervical metastasis
Brown-Séquard syndrome, 20/428
Cervical migraine
Barré-Liéou syndrome, 5/192, 7/449-451
cervicogenic headache, 48/408
complicated migraine, 5/200
vertebral artery, 5/196
Cervical myelography
risk, 61/149
Cervical myelopathy
atlantoaxial dislocation, 12/609
CSF pressure, 19/117
delayed, 12/609, 611
familial spastic paraplegia, 59/306
glycosaminoglycanosis type VI, 66/309
radiation myelopathy, 12/501, 26/84, 91, 61/200
rheumatoid arthritis, 71/21
spinal muscular atrophy, 59/113
spondylosis, 7/456
type, 12/609
Cervical neurotomy
glossopharyngeal neuralgia, 5/357
Cervical plexus
anatomy, 2/134, 7/419, 25/404
gustatory sweating, 2/121
injury, 7/420
phrenic nerve lesion, 2/135
Cervical plexus syndrome, 2/128-153
scheme, 5/287
syncope, 7/452
Cervical pterygium
Edwards syndrome, 50/561
Cervical radicular syndrome
*see also* Cervical syndrome
disc protrusion, 20/555
intervertebral disc prolapse, 20/555
spondylotic, *see* Spondylotic radiculopathy
Cervical rib
anterior tephromalacia, 59/113
Hirayama disease, 59/114
prolongation type, 51/120
Raynaud phenomenon, 43/71
spinal lipoma, 20/397
subclavian artery stenosis, 53/372
systemic brain infarction, 11/429
thoracic outlet syndrome, 51/120
thoracic rib syndrome, 8/332
Cervical rib syndrome
*see also* Cervicothoracic syndrome, Costoclavicular syndrome *and* First rib syndrome
anatomy, 20/548
Barré-Liéou syndrome, 7/460
clinical features, 20/553
features, 7/444
historic background, 20/548
neuralgia, 20/554
neurovascular compression, 2/146, 7/430-434
pain, 5/287
radicular syndrome, 20/556
syncopic, 7/452
thoracic outlet syndrome, 51/121
Cervical root

Arnold-Chiari malformation type II, 50/406
Cervical sclerotome
embryology, 50/381
segment, 50/384
Cervical segmental artery
primitive, *see* Proatlantal intersegmental artery
Cervical spinal canal
clinical features, 32/339-342
epidemiology, 32/340
history, 32/329
spinal deformity, 25/278-281
treatment, 32/342-344
Cervical spinal canal stenosis
anterior horn syndrome, 42/58
clinical features, 32/339-342
history, 32/329
sagittal diameter, 32/339, 42/58
sex ratio, 42/58
treatment, 32/342-344
Cervical spinal cord
child, 20/364
diamorphine myelopathy, 63/54
edema, 25/262
hemorrhagic necrosis, 25/266
overstretching, 25/261
pincers mechanism, 25/65, 260
subependymal astrocytoma, 20/362
syndrome, 1/189
tract, 1/85
Cervical spinal cord compression
*see also* Cervicomedullary injury
glycosaminoglycanosis type IH, 66/297
spinal cord injury, 61/37
spinal ependymal cyst, 32/440
Cervical spinal cord injury
animal experiment, 61/59
canal encroachment, 61/58
course, 61/58
delayed myelopathy, 12/608
diaphragmatic respiration, 61/56
dopamine, 61/61
GM1 ganglioside, 61/61
hospitalization time, 61/66
hypoxia, 61/58
level assay, 61/56
mechanism, 61/57
methylprednisolone, 61/61
missed diagnosis, 61/56
neurologic outcome, 61/66
outcome determining factor, 61/57
plantar reflex, 61/56
progressive injury, 61/58
progressive ischemia, 61/57
rehabilitation, 61/66
secondary lesion, 61/58
sepsis, 61/58
supraclavicular nerve, 61/56
sympathetic areflexia, 61/61
symptom, 61/56
systemic hypotension, 61/58
treatment axiom, 61/68
Cervical spinal cord stimulation
posterior, *see* Posterior cervical spinal cord stimulation
Cervical spinal root
injury, 25/274, 26/97-112
posterior ramus, 20/553
spondylotic radiculopathy, 26/97-112
Cervical spine
aneurysmal bone cyst, 67/195
angular kyphosis, 25/346
arcual kyphosis, 25/246
biomechanics, 48/406
bony fusion, 25/281, 297-303, 345, 353, 363, 43/428
cervical hyperanteflexion injury, 25/254-256, 329-354, 26/290
cervical hyperdeflexion injury, 25/30, 62, 134, 254, 354-363, 26/285, 291, 293
chiropractic manipulation, 7/464
congenital disorder, 25/249
head injury, 57/147
headache, 48/406, 411
hyperrotation injury, 25/259
instability, 25/293
laminectomy, 25/304
microcephaly, 43/428
orthopedic trauma, 70/27
osteoblastoma, 67/196
posttraumatic headache, 24/505, 507, 580
reduction, 25/292
rotation injury, 25/67, 125
spinal hyperflexion injury, 25/125, 259
spinal injury, 25/227-379
spinal subluxation, 25/131, 341, 26/181, 289
spine experimental injury, 25/3
stabilization, 25/293-302
Stryker frame, 25/293
unilateral cervical facet interlocking, 25/136
vertebral experimental injury, 25/3
Cervical spondylarthrosis, 20/547-561
amyotrophic lateral sclerosis, 22/309
occipital neuralgia, 5/373
pain, 5/373
vertebral canal stenosis, 42/58
Cervical spondylarthrotic myelopathy

Brown-Séquard syndrome, 50/469
cervical stenosis, 50/469
Miller syndrome, 50/470
syringomyelia, 50/452
vertebral canal stenosis, 50/469
Cervical spondylosis
*see also* Cervical spondylotic myelopathy
Brown-Séquard syndrome, 7/456
cardiac dysrhythmia, 7/455
differential diagnosis, 7/462
features, 25/250
Glisson sling, 7/464
headache, 7/450
Hirayama disease, 59/113
ischemic myelopathy, 63/43
octavus nerve, 7/454
pathology, 25/27-42
scotoma, 7/450
spinal cord compression, 7/456
spinal trauma, 25/252
survey, 7/447
syncope, 7/452
transient ischemic attack, 12/7
treatment, 7/462
vasovagal syndrome, 7/452
vertebral column, 26/98
vertebral ischemia, 12/7
Cervical spondylotic myelopathy
*see also* Cervical spondylosis
cauda equina tumor, 19/85
clinical features, 26/104
conus medullaris tumor, 19/85
differential diagnosis, 26/106
extramedullary tumor, 19/59
history, 26/97
laminectomy, 26/109
myelography, 26/104, 107
nerve root compression, 12/534
pathogenesis, 26/99
pathology, 26/98
pincers grip mechanism, 19/116, 26/100
postural manometry, 19/92, 106
quadriparesis, 12/534
radiology, 26/106
spinal cord injury, 26/97-112
spine manipulation, 7/464
spine traction, 7/464
surgery, 26/108
treatment, 26/108
vertebral fusion, 26/108
Cervical spondylotic radiculopathy, *see* Spondylotic radiculopathy
Cervical stenosis
cervical spondylarthrotic myelopathy, 50/469
pathomorphology, 50/470
symptom, 50/468
vertebral canal stenosis, 50/468, 470
Cervical subluxation
ankylosing spondylitis, 38/514
Cervical syndrome
*see also* Cervical radicular syndrome
headache, 5/194
pain, 5/287
posttraumatic headache, 24/509
syncopic, *see* Syncopic cervical syndrome
terminology, 20/547
Cervical tilting fracture
articular process, 25/365
Cervical vertebral column
coupled motion, 61/7, 9
facet joint plane, 61/8
instability, 61/10
instability checklist, 61/10
motion range, 61/8
Cervical vertebral column biomechanics
anatomy, 61/4
kinematics, 61/4
Cervical vertebral column injury
*see also* Spinal injury *and* Unilateral cervical facet interlocking
anesthesia, 24/168
angulation reduction, 61/62
ankylosing spondylitis, 26/175, 177-181
anterior element, 61/24
anterior longitudinal ligament, 61/27
associated lesion, 25/259, 288
atlantoaxial dislocation, 25/307
axial compression, 61/30
biomechanics, 25/227-239, 253-259, 61/23, 60
brace, 25/297
burst fracture, 25/67, 26/266, 291, 61/30
cause, 25/253
Circolectric bed, 25/293
classification, 26/289, 61/23, 25
closed reduction, 61/62
compression fracture, 25/291, 345
compression injury, 61/23
compressive hyperanteflexion, 25/256
compressive hyperdeflexion, 25/258
computer assisted tomography, 61/55
contusio cervicalis posterior, 25/273, 284
decompression, 61/65
delayed diagnosis, 61/56
diagnosis, 61/55
discography, 25/304
dislocation, 61/25

disruption, 25/340
Egerton-Stoke-Mandeville frame, 25/293
epidemiology, 23/31
erroneous diagnosis, 25/245-253
experimental, 25/3, 5
extension fracture, 61/28
extension injury, 61/23, 27
extension sprain, 61/27
external stabilisation, 61/61
facet fracture, 61/29
facet subluxation, 61/62
flexion injury, 61/23, 25
fracture dislocation, 61/26
frontal headache, 61/33
functional anatomy, 25/227, 233, 61/23
fusion, 25/281
Glisson sling, 25/291
halo-thoracic vest, 61/63
halter traction, 25/291
hangman fracture, 61/28
head up tilting, 61/66
hyperextension dislocation, 61/29
hyperrotation, 25/259
iatrogenic tetraplegia, 61/56
instability, 25/293, 61/60
instability definition, 61/60
internal carotid artery, 61/33
intervertebral disc, 61/27
intervertebral disc avulsion, 61/31
intervertebral disc lesion, 61/31
intervertebral disc prolapse, 61/27, 31
intubation, 24/168
laminectomy, 25/304
management, 61/55, 61
manipulation, 25/3
mechanism, 26/264, 61/26
metabolism, 25/288
Minerva POP, 61/63
missed diagnosis, 61/56
mobilization, 61/65
movent diagram, 25/228, 248
multilevel injury, 61/56
myelography, 25/268
nuclear magnetic resonance, 61/55
occipital headache, 61/33
odontoid fracture, 61/63
open reduction, 61/62
pathology specimen, 25/28
pitfall, 25/250
plaster cast, 25/297
posterior element, 61/24
prevertebral space, 25/252, 316
radiodiagnosis, 26/263
radiographic procedure, 61/57
radiology, 25/227, 241, 247, 267, 286
realignment, 61/61
reduction, 25/290
rehabilitation, 61/65
respiration, 25/288
resuscitation, 61/61
rim lesion, 61/27, 32
segmental frequency distribution, 61/23
shear strain, 25/254
skull traction, 25/277, 286, 291, 26/215-219, 301, 303
soft tissue damage, 61/30
spinal cord lesion, 61/27, 30
spine manipulation, 25/292, 26/302
sprain, 26/289
stabilization, 25/293
subluxation, 61/25
surgical immobilization, 61/64
surgical stabilization, 25/294-302
survey, 25/227-279
traction value, 61/62
transport, 25/277
vacuum cleft phenomenon, 61/28
vertebral angiography, 25/304
vertebral arch fracture, 61/23
vertebral artery, 25/263, 61/33
vertebral artery angiography, 25/304
vertebral body fracture, 61/23
vertebral end plate bowing, 61/30
vertebral end plate fracture, 61/30
vertebral fracture, 61/28
vertebral teardrop fracture, 25/346-349
wedge compression, 61/25
wiring, 25/297
wry neck, 25/311
Cervical vertebral facet joint, *see* Intervertebral facet joint
Cervical vertebral injury, *see* Cervical vertebral column injury
Cervicobrachial outlet syndrome, *see* Cervical rib syndrome
Cervicobrachialgia, *see* Cervical rib syndrome
Cervicocephalic syndrome, *see* Barré-Liéou syndrome
Cervicogenic headache, 48/405-411
*see also* Cervical headache
cervical migraine, 48/408
chronic paroxysmal hemicrania, 48/260
diagnosis, 48/409
frontal referred pain, 48/124
migraine, 48/405
neck injury, 48/124

osteoarthrosis, 48/405
school headache, 48/408
sleep, 48/408
subarachnoid hemorrhage, 55/19
treatment, 48/409
trigger point, 48/409
unilateral, 48/224
Cervicomedullary injury
*see also* Cervical spinal cord compression
abducent nerve injury, 24/146, 169, 171
anterior spinal artery, 24/156
ascending infarction, 24/170
atlantoaxial dislocation, 24/143, 154, 157
atlantoaxial subluxation, 24/141
atlas fracture, 24/152
biomechanics, 24/160, 162, 168
brain herniation, 24/162
brain injury, 24/141
complete, *see* Complete cord lesion
concomitant head injury, 24/141-178
cruciate paralysis, 24/149
dens hypoplasia, 24/143
football injury, 24/157-160, 174
Horner syndrome, 24/148
mandible fracture, 24/172
occipitoatlantal dislocation, 24/141
pathology, 24/150, 152, 162, 167, 173
spondylotic, *see* Cervical spondylosis
transtentorial herniation, 24/98
trigeminal descending spinal tract lesion, 24/163-167, 25/147
trigeminal nerve, 24/164
vertebral artery, 24/145, 156-159, 168
vertebral teardrop fracture, 24/163
Cervico-occipital pain
*see also* Barré-Liéou syndrome
algie meningée postérieure, 5/187
differential diagnosis, 7/465
Cervico-oculoacoustic dysplasia, *see* Wildervanck syndrome
Cervico-oculoacusticus syndrome
chromosome study, 32/129
clinical features, 32/129
component, 32/123
dermoid, 32/127
dysraphia, 32/128
ear malformation, 32/126
epidemiology, 30/162, 32/125
genetics, 32/127-129
history, 32/123
Klippel-Feil syndrome, 32/123-126
lipodermoid, 32/127
malformation, 32/127
management, 32/129
mental deficiency, 32/127
pathogenesis, 32/125-127
sensorineural deafness, 32/123, 125
sex ratio, 30/162, 32/123, 125
Sprengel deformity, 32/123-130
Stilling-Türk-Duane syndrome, 32/123, 127
Cervicothoracic somite dysplasia
MURCS syndrome, 50/515
Cervicothoracic syndrome
*see also* Cervical rib syndrome
case history, 7/455
heart disease, 7/455
Cervós-Navarro encephalitis
history, 18/235
reticuloendotheliosis, 18/241
CES, *see* Central excitatory state
Cestan-Chenais syndrome
*see also* Foville syndrome
abducent nerve paralysis, 2/318
Babinski-Nageotte syndrome, 1/189, 2/221
basilar artery occlusion, 2/336
eponymic classification, 12/13
Foville syndrome, 1/188, 2/239
gaze paralysis, 2/239
hemiparesis, 2/318
lateral gaze palsy, 2/261
lesion site, 2/301
medial lemniscus, 2/239
medulla oblongata syndrome, 2/221
pontine infarction, 12/13
rostral tegmental lesion, 2/265
spinothalamic tract, 2/239
superior cerebellar peduncle, 2/239
symptom, 1/188
tegmental lesion, 2/265, 301
tuberculoma, 2/261, 336
Cestode
Echinococcus, 52/523
epidemiology, 35/12
helminthiasis, 52/505
parasitic disease, 35/12
Taenia solium, 52/529
Cetacea intoxication
ingestion, 37/94-97
vitamin A intoxication, 37/94-97
CETP, *see* Cholesterol ester transfer protein
Cetylpyridiniumchloride-carbazole test
glycosaminoglycanosis type VII, 66/313
Chachalen
deficiency neuropathy, 7/636
Chaddock sign
diagnostic value, 1/250

Chagas disease, 35/85-109
*see also* Trypanosoma cruzi
acquired immune deficiency syndrome, 71/305
acute form, 35/102
amastigote pseudocyst, 75/389
antineuron antibody, 52/347
autonomic nervous system, 74/335, 75/19, 700
autonomic polyneuropathy, 52/346
biliary tract, 75/398
blood transfusion, 52/345
brain embolism, 55/163, 63/135
brain infarction, 55/163
cardiac form, 75/391
cardiomyopathy, 52/348, 63/135
cardiovascular dysautonomia, 75/394
cause, 52/345
cell mediated immunity, 52/345
cerebrovascular disease, 55/163
Chagas trypanosoma cruzi, 35/86-89, 52/345
chronic, 35/104-107, 52/347, 75/391
clinical features, 35/102, 52/347
colon disorder, 75/397
CSF, 35/107, 52/347
diagnosis, 52/345, 75/396, 399
embolic form, 35/104
encephalopathy, 52/347
epidemic area, 75/386
epidemiology, 35/10, 75/385
epilepsy, 52/347
etiology, 35/86-89
granulomatous encephalitis, 52/346
headache, 52/347
heart disease, 75/393
heart failure, 75/399
hemiplegia, 63/135
history, 35/85
immunodeficiency, 52/347, 75/390
infection, 75/389
inoculation chagoma, 75/390
intestinal pseudo-obstruction, 51/493
laboratory finding, 35/91
megacolon, 75/400
megaesophagus, 75/400
meningoencephalitis, 52/347
mental symptom, 52/347
mummy, 74/188
necrotizing encephalitis, 52/346
Panstrongylus megistus, 75/387
parasitic disease, 35/10
parasympathetic denervation, 52/347
partial epilepsy, 52/347
pathogenesis, 35/89-91, 75/392
pathology, 35/92-102
polyneuropathy, 52/346, 348
pregnancy, 52/345
prevalence, 35/85, 52/345
prophylaxis, 35/108, 52/346
Rhodnius prolixus, 75/387
Romana sign, 75/390
segmental demyelination, 52/346, 348
small intestine, 75/397
stomach, 75/398
survey, 75/385
transmission cycle, 75/388
treatment, 35/108, 52/346, 75/399
Triatoma, 52/345
Triatoma infestans, 75/387
Trypanosoma cruzi, 75/388
urinary bladder, 75/398
vector, 52/345, 75/387
Chagas trypanosoma cruzi
Chagas disease, 35/86-89, 52/345
Chain disease
heavy, *see* Heavy chain disease
mitochondrial respiratory, *see* Mitochondrial respiratory chain defect
Chain saw
vibration neuropathy, 51/135
Chalacoderma
Ehlers-Danlos syndrome, 14/111
Chalfont National Hospital Scale
epilepsy, 72/418
Chalfont Scale
epilepsy, 72/430
Chalfont Seizure Severity Scale
epilepsy, 72/430
Chamberlain line
basilar impression, 32/15, 17, 20, 24, 26, 50/398
definition, 61/7
Chance fracture
intact anterior column, 61/90
seat belt injury, 61/90
Changi Camp syndrome
*see also* Anterolateral myelopathy
vitamin B1 deficiency, 7/560, 568
Channel
calcium, *see* Calcium channel
chloride, *see* Chloride channel
potassium, *see* Potassium channel
sodium, *see* Sodium channel
sodium transport, *see* Sodium transport channel
Character change
thalamic syndrome (Dejerine-Roussy), 2/484
Charcot arthropathy, *see* Neurogenic osteoarthropathy
Charcot-Bouchard aneurysm

arterial hypertension, 63/75
brain aneurysm, 12/82
brain arterial hypertension, 54/300
brain embolism, 11/587
brain hemorrhage, 11/578, 54/289, 299, 63/74
brain lacunar infarction, 54/239, 63/74
cerebellar hemorrhage, 12/55, 54/292, 55/89
cerebrovascular disease, 53/41
Charcot illustration, 11/580
history, 11/582
hypertension, 12/447
pontine hemorrhage, 12/48-50
prehemorrhagic softening, 12/49
rupture, 63/74
spontaneous cerebellar hemorrhage, 12/54
type, 54/301
Charcot disease, *see* Amyotrophic lateral sclerosis
Charcot d'embléehe1miplégique
multiple sclerosis, 47/56
Charcot J.M., 1/7, 13
Charcot joint, *see* Neurogenic osteoarthropathy
Charcot-Marie disease, *see* Hereditary motor and sensory neuropathy type I *and* Hereditary motor and sensory neuropathy type II
Charcot-Marie-Tooth disease, *see* Hereditary motor and sensory neuropathy type I *and* Hereditary motor and sensory neuropathy type II
Charcot syndrome
history, 12/508
Charcot-Wilbrand syndrome
parietal lobe syndrome, 45/73
visual agnosia, 2/654
visual dreaming, 2/654
Charles-Bonnet syndrome
visual hallucination, 55/146
Charles-Ruppe disease
fibrous dysplasia, 14/196
Charles Symonds syndrome, *see* Multiple cranial neuropathy
Charlevoix-Saguenay spastic ataxia, 60/451-458
age at onset, 60/452
biochemical study, 60/457
brain stem auditory evoked potential, 60/455
clinical features, 60/452
computer assisted tomography, 60/455, 671
course, 60/452
CSF protein, 60/457
dementia, 60/665
diabetes mellitus, 60/457, 674
distal amyotrophy, 60/453, 659
EEG, 60/455
EMG, 60/454
ENG, 60/455
epidemiology, 60/451
epilepsy, 60/666
fecal incontinence, 60/454, 664
foot deformity, 60/453
genetic study, 60/452
hypoesthesia, 60/661
intelligence, 60/454
mitral valve prolapse, 60/454, 668
muscle biopsy, 60/456
myelinated retinal nerve fiber, 60/453
nerve biopsy, 60/456
nerve conduction study, 60/454
nerve conduction velocity, 60/670
neuropathology, 60/456
neuropathy, 60/660
nonprogressive ocular sign, 60/453
ocular pursuit, 60/657
pes cavus, 60/669
progressive sign, 60/452
secondary pigmentary retinal degeneration, 60/654
somatosensory evoked potential, 60/455
spasticity, 60/453
urinary incontinence, 60/454, 664
vibration response, 60/453
visual evoked response, 60/455
Charlin neuralgia, 48/483-486
associated sign, 48/485
characteristics, 5/217
ciliary ganglion, 48/483
clinical features, 48/484
cluster headache, 48/486
differential diagnosis, 5/316, 330
epiphora, 48/486
history, 48/483
intranasal, 48/486
nasal secretion, 48/486
nasociliary nerve, 48/483
nasofacial reflex, 5/217, 48/485
ocular lesion, 48/485
ocular pain, 48/484
sphenopalatine neuralgia, 48/478, 485
symptom, 5/217
Charlin syndrome, *see* Charlin neuralgia
Charybdotoxin
arthropod envenomation, 65/198
neurotoxin, 65/194, 198
Chastek paralysis
vitamin B1 deficiency, 7/561, 568
Chédiak-Higashi syndrome
albinism, 14/783, 60/667
amyloidosis, 42/536
anemia, 42/536

areflexia, 42/536, 60/660
ataxia, 21/583
bulbar paralysis, 60/658
convulsion, 42/536
cranial nerve, 14/783, 42/536
cytoplasmic inclusion body, 42/536
diagnosis, 14/783
experimental allergic encephalomyelitis, 9/526
hepatosplenomegaly, 14/783, 42/536, 60/668
heterozygote detection, 42/537
leukemia, 60/674
leukocyte inclusion, 60/674
lymphoma, 60/674
metabolic polyneuropathy, 14/783, 51/397
neurocutaneous syndrome, 14/783
neuropathy, 60/660
nystagmus, 42/536
oculocutaneous albinism, 14/783, 42/536
papilledema, 14/783, 60/655
partial albinism, 60/647
photophobia, 42/536
photosensitivity, 14/783
prenatal diagnosis, 42/537
progressive external ophthalmoplegia, 60/657
progressive neuropathy, 51/397
sensory loss, 42/536
spasticity, 60/661
strabismus, 42/536
tremor, 14/783, 60/661
Cheese reaction
MAO inhibitor, 37/320
subarachnoid hemorrhage, 12/112
Cheilitis
Melkersson-Rosenthal syndrome, 8/213
Cheilitis granulomatosa
Melkersson-Rosenthal syndrome, 8/205-209, 222
tuberculoid granuloma, 8/222
Cheilognathopalatoschisis
corpus callosum lipoma, 14/410
Krabbe-Bartels disease, 14/410, 529
Patau syndrome, 14/120
tuberous sclerosis, 14/102
Cheilopalatoschisis
acrocephalosyndactyly type I, 43/318
adducted thumb syndrome, 43/331
anencephaly, 42/13
brachial neuritis, 42/304
cerebrocostomandibular syndrome, 43/237
corpus callosum agenesis, 42/7
cryptophthalmos syndrome, 43/374
diastrophic dwarfism, 43/382
EEC syndrome, 43/395
Gruber syndrome, 30/247
hemivertebra, 42/31
hypertelorism, 30/247
hypertelorism hypospadias syndrome, 43/413
hypothalamic phakoma, 42/737
incontinentia pigmenti, 43/23
Juberg-Hayward syndrome, 30/247, 43/446
Klein-Waardenburg syndrome, 43/52
Klinefelter variant XXXXY, 43/560
Klippel-Feil syndrome, 42/37
Marden-Walker syndrome, 43/424
median facial cleft syndrome, 30/247
mental deficiency, 43/256, 269
Mohr syndrome, 43/449
multiple nevoid basal cell carcinoma syndrome, 43/32
orbital hypertelorism, 30/247
orofaciodigital syndrome type I, 30/247, 43/447
otopalatodigital syndrome, 43/457
4p partial monosomy, 43/497
10p partial trisomy, 43/525
11p partial trisomy, 43/525
Patau syndrome, 14/120, 31/504, 506, 510
polydactyly, 43/256
4q partial monosomy, 43/496
2q partial trisomy, 43/493
10q partial trisomy, 43/517
Roberts syndrome, 30/247
Robin syndrome, 43/471
Seckel syndrome, 43/379
Smith-Lemli-Opitz syndrome, 43/308
speech disorder in children, 4/415
spondyloepiphyseal dysplasia, 43/478
Treacher-Collins syndrome, 43/422
triploidy, 43/564-566
trisomy 22, 43/548
Cheiloschisis
median, *see* Median cleft lip
Mende syndrome, 14/119
mental deficiency, 43/256, 269
orofaciodigital syndrome type I, 50/167
Patau syndrome, 31/504, 506, 43/527, 50/556
Saldino-Noonan syndrome, 31/243
trigeminal nerve agenesis, 50/214
trisomy 9, 50/564
Cheiralgia paresthetica
terminology, 1/108
topographical diagnosis, 2/28
Cheiro-oral paresthesia
aura, 48/157
migraine, 48/157
Cheiro-oral syndrome
transient ischemic attack, 12/4
Cheirospasm, *see* Writers cramp

Chelidonichthys intoxication
family, 37/78
venomous spine, 37/78
Chelonia intoxication
coma, 37/91
death, 37/91
death rate, 37/89
diarrhea, 37/89
epigastric pain, 37/89
headache, 37/91
hepatomegaly, 37/91
ingestion, 37/89
lethargy, 37/91
nausea, 37/89
neurologic involvement, 37/91
somnolence, 37/91
stomatitis, 37/91
sweating, 37/89
tachycardia, 37/89
vertigo, 37/89
vomiting, 37/89
Chelonia mydas intoxication
ingestion, 37/89
Chemical meningitis
dermoid, 56/130
epidermoid, 56/130
ibuprofen, 56/130
isotope cisternography, 56/130
lumbar puncture, 61/162
metrizamide, 56/130
spinal dermoid, 19/364
spinal epidermoid, 19/364
sulindac, 56/130
tolmetin, 56/130
trimethoprim, 56/130
Chemical warfare agent, *see* Chlormethine intoxication, Cyanide intoxication *and* Organophosphate intoxication
Chemodectoma
*see also* Paraganglioma
catecholamine, 74/315
hypervascular, 68/270
juxtacarotid, *see* Juxtacarotid chemodectoma
stress, 74/315
survey, 68/270
terminology, 8/485, 18/438, 42/762
tissue culture, 17/69
Chemokine
collagen vascular disease, 71/9
Chemoreceptor reflex
innervation, 75/112
Chemotherapy
*see also* Antineoplastic agent
adrenocorticosteroids, 39/120
adriamycin, 39/114
alkylating agent, 39/93-98, 65/528
altretamine, 39/117
angiogenesis inhibitor, 67/284
antibiotic agent, 39/114, 65/528
antimetabolite, 39/98-106, 65/528
asparaginase, 39/117
astrocytoma, 68/108
ataxia telangiectasia, 60/390
azacitidine, 39/106
8-azaguanine, 39/106
bladder infection, 26/431
bleomycin, 39/114
blood-brain barrier, 67/278
bone marrow transplantation, 67/282
brain atrial myxoma metastasis, 63/101
brain metastasis, 18/229, 69/201, 71/629
brain tumor, 18/517-522, 67/277
brain tumor headache, 69/23
carbamate intoxication, 64/187
carboplatin, 67/284
carmustine, 39/115, 67/282, 284
Castleman disease, 63/396
cell cycle, 67/277
cell injury, 67/278
cerebellar medulloblastoma, 18/190
chlorambucil, 39/97
chlormethine, 39/93-97
choroid plexus tumor, 68/171
cisplatin, 39/115, 67/284
classification, 65/528
clonal population, 67/279
CNS leukemia, 69/243
cyclophosphamide, 39/97
cytarabine, 39/105
dacarbazine, 39/116
dementia, 67/378
DNA synthesis inhibitor, 67/284
drug induced seizure, 69/482
drug resistance, 67/279
encephalopathy, 67/355
ependymoma, 18/130, 68/171
etoposide, 67/284
fluorouracil, 39/104
germ cell tumor, 68/250
glioblastoma, 18/65
glioblastoma multiforme, 18/65
hydroxyurea, 39/117
intra-arterial, 67/281
intrathecal, *see* Intrathecal chemotherapy
intravenous, 67/280
leptomeningeal metastasis, 71/655

leukemia, 39/15, 63/356, 69/233-235
local, 67/280
lomustine, 39/115, 67/282, 284
low grade fibrillary astrocytoma, 68/48
lymphocytic leukemia, 63/356
medulloblastoma, 68/199
meningitis, 69/160
mercaptopurine, 39/106
metastasis, 71/655
methotrexate, 39/98-104, 65/527
methylhydrazine, 39/118-120
migraine, 67/355
mineralizing microangiopathy, 67/361
mitotane, 39/120
mitotic inhibitor, 65/528
multifocal pontine lesion, 67/365
multiple myeloma, 18/256, 20/14, 111, 63/396
myeloid leukemia, 63/356
myelopathy, 67/355
neurologic intoxication, 39/91-122
neuropathy, 69/459
neurotoxicity, 67/353, 359, 69/481
nimustine, 67/282, 284
nitrosourea, 39/114, 67/282, 284
nocardiosis, 35/526, 52/445
nongerminomatous germ cell tumor, 68/248
non-Hodgkin lymphoma, 69/266, 281
North American blastomycosis, 35/401, 406, 52/390
oligodendroglioma, 68/129
oral, 67/280
osteosclerotic myeloma, 63/396
ototoxicity, 67/359
pain, 69/35
pinealoblastoma, 68/248
pinealoma, 68/248
plant alkaloid, 39/106-114
platinum, 67/284
podophyllotoxin, 39/114
POEMS syndrome, 63/396
postherpetic neuralgia, 5/324
primary CNS lymphoma, 68/263
primitive neuroectodermal tumor, 68/216, 219-221
procarbazine, 39/118-120, 67/284
protease, 67/284
radiotherapy influence, 65/539
sensitivity testing, 67/279
sex hormone, 39/121
skull base tumor, 68/473
sparfosic acid, 39/120
spinal cord metastasis, 69/184
spinal ependymoma, 20/382
spinal meningioma, 20/233
spinal neurinoma, 20/285
spongioblastoma, 68/49
stroke, 67/355
supratentorial brain tumor, 18/338
taxol, 67/284
taxotere, 67/284
thiotepa, 39/97
thymidine, 39/120
tioguanine, 39/106
triazinate, 39/104
tuberculous meningitis, 71/284
vertebral metastasis, 20/421
vinblastine, 39/110, 67/284
vinca alkaloid, 67/284
vincristine, 39/106-110, 67/284
vindesine, 39/110-114
Chenodeoxycholic acid
cerebrotendinous xanthomatosis, 51/399, 59/358, 60/168
Cherry red spot
β-*N*-acetylhexosaminidase B deficiency, 60/654
diagnostic pitfalls, 10/219
familial amaurotic idiocy, 10/338
Farber disease, 60/166, 66/19
GM1 gangliosidosis, 10/528
GM2 gangliosidosis, 10/334, 385, 387, 558, 13/326, 42/433, 66/19
GM2 gangliosidosis type I, 22/537, 42/433
GM2 gangliosidosis type II, 42/435
Goldberg syndrome, 42/439
metachromatic leukodystrophy, 66/19
mucolipidosis type I, 27/162, 42/475, 51/380, 60/654, 66/19
mucolipidosis type II, 51/380
mucolipidosis type III, 29/367
myoclonus syndrome, 42/475
neuronal ceroid lipofuscinosis, 10/550
Niemann-Pick disease, 10/470, 503, 66/19
Niemann-Pick disease type A, 10/572, 42/469, 51/373
retinal artery occlusion, 55/112
Cherry red spot myoclonus syndrome, *see* Mucolipidosis type I
Cherubism
fibrous dysplasia, 14/170
pituitary dwarfism type III, 42/614
Chess and card epilepsy
reflex epilepsy, 2/622
Chest injury
spinal injury, 26/190, 204
Chest wall
mobility treatment, 26/346

respiratory system, 26/340, 346
Chester-Erdheim syndrome
Langerhans cell histiocytosis, 70/56
Chewing
amygdaloid stimulation, 1/509
Chewing disease
yellow star thistle intoxication, 6/665
Cheyne-Stokes respiration
barbiturate intoxication, 65/332
barotrauma, 63/482
blood gas, 63/483
bradycardia, 63/483
brain blood flow, 1/671, 63/483
brain injury, 57/129, 63/483
brain stem glioma, 63/440
carbon dioxide, 1/497, 670-673
cardiac insufficincy, 1/671
carotid sinus nerve, 63/484
central sleep apnea, 63/462
cerebral lesion, 1/497
congestive heart failure, 63/483
dysbarism, 63/482
heart dysfunction, 63/482
heart rate, 63/483
heat stroke, 23/674
high altitude, 63/482
hyperpnea, 1/497
hyperventilation, 1/497
intracranial pressure, 63/483
medulla oblongata, 1/671
middle cerebral artery syndrome, 53/358
organophosphate intoxication, 37/555, 64/167
periodicity, 1/650
peripheral chemoreceptor, 63/484
pontine glioma, 63/440
pontine hemorrhage, 2/253
pontine infarction, 12/15
prognosis, 24/675
pupil size, 63/483
respiratory alkalosis, 63/483
sleep, 63/482, 484
striatonigral degeneration, 49/208
stroke, 75/375
theophylline, 63/484
uremia, 1/497
Chiari-Frommel syndrome
adenohypophyseal syndrome, 2/444, 456
chlorpromazine, 2/444
reserpine, 2/444
Chiari malformation
type I, *see* Arnold-Chiari malformation type I
type II, *see* Arnold-Chiari malformation type II
type III, *see* Arnold-Chiari malformation type III
type IV, *see* Arnold-Chiari malformation type IV
Chick pea, *see* Cicer arietinum
Chickenpox
*see also* Herpes zoster
acute cerebellar ataxia, 34/162-170, 622, 627
acute encephalopathy, 47/479
adult, 34/162-170
child, 34/162-170
childhood myoglobinuria, 62/562
clinical features, 34/162-165
congenital, 34/162
congenital myopathy, 62/562
Guillain-Barré syndrome, 34/168
immune mediated encephalomyelitis, 34/416-418
meningoencephalitis, 34/162-167
multiple sclerosis, 34/437, 47/329
myelitis, 34/167
neonatal, 34/162
pathogenesis, 34/165-167
pathology, 34/165-167
polyneuritis, 34/168
polyradiculitis, 34/168
postinfectious encephalopathy, 47/479
Reye syndrome, 31/168, 34/168-170, 37/534, 56/149, 151
secondary pigmentary retinal degeneration, 13/214
teratogenic agent, 34/378
Chickling pea, *see* Lathyrus sativus
Chikungunya virus
alphavirus, 56/12
experimental acute viral myositis, 56/195
Childhood absence epilepsy
definition, 72/5, 73/145
EEG, 73/147
epidemiology, 73/145
etiology, 73/145
history, 73/144
onset, 73/146
pathology, 72/117
prognosis, 73/147
seizure, 73/146
survey, 73/143
treatment, 73/149
Childhood chronic multiple tic
Gilles de la Tourette syndrome, 49/627, 630, 635
Childhood encephalomyopathy
areflexia, 60/661
Childhood encephalopathy
acute, *see* Reye syndrome
Childhood epilepsy
brain tumor, 16/264
polymorphic, 73/264

Childhood epilepsy with occipital paroxysm
benign, *see* Benign childhood epilepsy with occipital paroxysm
definition, 72/5
migrainous headache, 72/5
Childhood hypertonia
infantile spinal muscular atrophy, 1/271
Childhood motor development, 4/347-350
autonomic features, 4/443-446
neck righting reflex, 4/348
parachute reflex, 4/348
postural control, 40/333
psychosocial aspect, 4/443-446
Childhood myoglobinuria
age at onset, 62/561
attack feature, 62/561
bulbar symptom, 62/561
cardiac involvement, 62/561
carnitine palmitoyltransferase, 62/563
chickenpox, 62/562
death, 62/561
dysarthria, 62/564
dysmorphic features, 62/564
dysphagia, 62/564
ECG, 62/561
EMG, 62/561
encephalopathy, 62/561
epilepsy, 62/564
Epstein-Barr virus, 62/562
Escherichia coli, 62/562
exertional type, 62/561
familial history, 62/561
fatty acid oxidation, 62/566
glycogen storage disease, 62/564
2 groups, 62/560
herpetic stomatitis, 62/562
hypoglycemic encephalopathy, 62/563, 566
idiopathic myoglobinuria, 62/562
incomplete syndrome, 62/567
infantile, 62/560
infection, 62/561
lipid metabolic disorder, 62/566
malignant hyperthermia, 62/567
metabolic encephalopathy, 62/561-563
mitochondrial disease, 62/567
muscle biopsy, 62/561
muscle cramp, 62/561
muscle hypertrophy, 62/561
muscular atrophy, 62/561
myophosphorylase deficiency, 27/233, 62/564
Neisseria, 62/564
pathophysiology, 62/565
permanent weakness, 62/561
phosphofructokinase deficiency, 40/264, 62/564
phosphoglycerate kinase deficiency, 62/565
phosphoglycerate mutase deficiency, 62/565
phosphorylase kinase deficiency, 62/565
precipitating factor, 62/561
psychomotor retardation, 62/564
renal insufficiency, 62/561
respiratory distress, 62/561
Reye like syndrome, 62/563
sex ratio, 62/561
Shigella, 62/561
short chain 3-hydroxyacyl-coenzyme A dehydrogenase, 62/563
streptococcal pharyngitis, 62/561
subacute necrotizing encephalomyelopathy, 62/563
toxic type, 62/561
Childhood psychosis
clinical features, 46/192
De Lange syndrome, 46/194
differential diagnosis, 46/194
elective mutism, 46/193
epidemiology, 46/193
etiology, 46/193
Lesch-Nyhan syndrome, 46/194
mental deficiency, 46/194
neuroleptic agent, 46/193
schizophrenia, 46/192
self-mutilation, 46/193
treatment, 46/193
vestibular system, 46/491
Childhood schizophrenia
mental deficiency, 46/25
Childhood schizophrenic syndrome
autism, 43/203
Chile
neurology, 1/12
Chimaera intoxication
family, 37/78
venomous spine, 37/78
Chimaeridae intoxication
family, 37/78
venomous spine, 37/78
Chin tremor
hereditary, *see* Hereditary chin tremor
Chinese restaurant syndrome
autonomic nervous system, 75/502
definition, 48/10
glutamic acid, 75/502
headache, 48/6, 419
migraine, 48/419
sodium glutamate, 75/502
Chinizarin, *see* Quinizarin

Chirodropus intoxication
cubomedusae, 37/45
fatal, 37/45
Chironex intoxication
cardiotoxin, 37/49
cardiovascular system, 37/49
cubomedusae, 37/45
death, 37/46, 48
hemolysin, 37/49
pharmacologic examination, 37/48
respiratory system, 37/49
Chiropractic manipulation
brain infarction, 53/165
brain ischemia, 53/165
carotid dissecting aneurysm, 54/271
cervical spine, 7/464
neck injury, 53/377
vertebral artery injury, 53/377
vertebral dissecting aneurysm, 53/377, 54/276
Chiropraxis, *see* Chiropractic manipulation
Chiropsalmus intoxication
cardiotoxin, 37/49
cardiovascular system, 37/49
cubomedusae, 37/45
death, 37/48
hemolysin, 37/49
pharmacologic examination, 37/48
respiratory system, 37/49
Chiropsoides intoxication
cubomedusae, 37/45
fatal, 37/45
Chitobiase
degradation, 66/61
Chlamydia
Guillain-Barré syndrome, 51/242
Chloral hydrate
status epilepticus, 73/331
Chloral hydrate intoxication
nausea, 37/348
vomiting, 37/348
Chloral metabolite, *see* Trichloroethanol intoxication
Chlorambucil
adverse effect, 69/491
alkylating agent, 39/97
antineoplastic agent, 65/528
chemotherapy, 39/97
demyelinating neuropathy, 47/618
epilepsy, 65/528
neurologic toxicity, 39/97
neurotoxin, 39/97, 46/601, 65/528
Chloramine-T, *see* Tosylchloramide sodium
Chloramphenicol
bacterial meningitis, 52/9
brain abscess, 52/156
drug induced polyneuropathy, 51/294
encephalopathy, 46/602
experimental myopathy, 40/158
gram-negative bacillary meningitis, 52/105, 108
Hemophilus influenzae meningitis, 33/56-58
iatrogenic neurological disease, 65/486
Luft syndrome, 41/210
meningococcal meningitis, 33/26, 52/30
Mimae meningitis, 33/103
mitochondrial abnormality, 40/158
neuropathy, 7/515, 530
neurosyphilis, 52/282
neurotoxicity, 51/294
neurotoxin, 7/516, 530, 41/210, 46/602, 51/68, 294, 65/486
phenobarbital, 40/561
retrobulbar neuritis, 51/294
staphylococcal meningitis, 33/72-74
toxic encephalopathy, 65/486
toxic polyneuropathy, 51/68
Chloramphenicol acetyltransferase
intrauterine gene therapy, 66/119
Chloramphenicol intoxication
chromatopsia, 65/486
optic atrophy, 13/63
optic neuritis, 65/487
optic neuropathy, 65/486
papillitis, 65/486
4-Chloramphetamine
designer drug, 65/364
Chlorate intoxication
headache, 48/420
Chlordane
chlorinated cyclodiene, 64/200
distal axonopathy, 64/13
neurotoxin, 36/408, 64/13, 201
organochlorine insecticide intoxication, 36/408-411
toxic encephalopathy, 64/202
toxic neuropathy, 64/13
Chlordane intoxication
acute, 36/408-410
chronic, 36/410
diagnosis, 36/411
prognosis, 36/411
symptom, 64/13
treatment, 36/411
Chlordecone
chemical formula, 64/204
organochlorine, 36/393, 64/197
organochlorine insecticide intoxication,

36/420-423, 64/197, 203
Chlordecone intoxication
ataxia, 64/204
chlordecone shake, 64/204
clinical features, 36/421-423
dysarthria, 64/204
epilepsy, 64/204
headache, 48/420
neurotoxicity, 51/286
opsoclonus, 64/204
tremor, 64/204
Chlordiazepoxide
alcohol withdrawal delirium, 46/541
benzodiazepine intoxication, 65/334
chemical structure, 37/356
description, 37/357
side effect, 15/708
Chlorfenvinphos, *see* Clofenvinfos
Chloride
blood biochemistry, 26/401
brain, 9/16
CNS, 28/436
CSF, 28/436
spinal cord injury, 26/381, 387
Chloride channel
myotonic dystrophy, 62/224
Chloride reabsorption defect
Bartter syndrome, 42/528
Chlorinated cyclodiene
aldrin, 64/200
chlordane, 64/200
CNS stimulation, 64/200
dieldrin, 64/200
endosulfan, 64/200
epilepsy, 64/200
GABA dependent chloride gate, 64/200
headache, 64/200
heptachlor, 64/200
LD50, 64/200
myoclonia, 64/200
organochlorine, 64/197, 200
organochlorine insecticide intoxication, 64/197, 200
picrotoxinin, 64/200
Chlorinated cyclodiene intoxication
epilepsy, 64/200, 202
fatality, 64/202
headache, 64/200
myoclonia, 64/200
Chlorine
autonomic nervous system, 75/509
Chlormethazanone, *see* Chlormezanone
Chlormethiazole, *see* Clomethiazole
Chlormethiazole intoxication, *see* Clomethiazole intoxication
Chlormethine
adverse effect, 69/491
alkylating agent, 39/93-97
antineoplastic agent, 65/528
chemotherapy, 39/93-97
encephalopathy, 46/601, 65/528
epilepsy, 65/528
Hodgkin lymphoma, 63/358
leukemia, 39/18, 63/358
lymphocytic leukemia, 63/358
myeloid leukemia, 63/358
neurologic toxicity, 39/93-97
neuropathy, 7/517, 537
neurotoxicity, 63/358
neurotoxin, 7/517, 537, 39/18, 94, 96, 119, 46/601, 63/358, 65/528
non-Hodgkin lymphoma, 63/358
Chlormethine intoxication
amnesia, 64/233
ataxia, 64/233
cholinomimetic effect, 64/233
confusion, 64/233
delayed effect, 64/233
epilepsy, 64/233, 65/534
hallucination, 65/534
headache, 64/233
hydrocephalus, 64/233
hyporeflexia, 64/233
organophosphate intoxication, 64/151
papilledema, 64/233
paralysis, 64/233
respiratory failure, 64/233
urinary incontinence, 64/233
Chlormezanone
muscle contraction headache, 5/168
muscle relaxation, 5/168
*p*-Chloroamphetamine, *see* 4-Chloramphetamine
2-Chlorodeoxyadenosine
neuropathy, 69/461
Chloroform intoxication
inhalation anesthetics, 37/412
Chloroma
definition, 63/343
epidural site, 63/343
epidural tumor, 63/343
exophthalmos, 18/252, 63/343
facial paralysis, 63/343
granulocytic sarcoma, 63/343
myeloblastoma, 63/343
myeloid leukemia, 18/247, 63/343
ophthalmoplegia, 63/343

protoporphyrin, 18/252
skull base tumor, 17/200, 18/252
spinal cord compression, 63/343
spinal site, 63/343
symptom, 69/237
vertebral column, 18/252
4-{2-{[2-(3-Chlorophenyl)-2-hydroxyethyl]amino}-propyl}phenoxyacetic acid
classification, 74/149
Chlorophyceae intoxication
anesthetic effect, 37/97
caulerpicin, 37/97
depression, 37/97
respiratory complication, 37/97
vertigo, 37/97
Chlorophyllum molybdites
mushroom intoxication, 36/542
Chloroprocaine intoxication
toxic dose, 37/405
Chloroquine
autophagic mechanism, 40/153
axonal degeneration, 65/484
behavior disorder, 46/602
blindness, 65/484
cerebral amebiasis, 52/328
cytoplasmic inclusion body, 65/484
experimental myopathy, 40/151
Fabry corneal dystrophy, 66/228
iatrogenic neurological disease, 65/483
lupus erythematosus, 13/220
malaria, 13/220, 35/155, 52/370
multiple sclerosis, 9/402
myasthenic syndrome, 65/484
neurolipidosis, 51/302
neurologic complication, 38/496
neuropathy, 7/518
neurotoxin, 7/518, 13/149, 220, 38/496, 40/151, 339, 438, 51/302, 62/372, 606, 65/484
nocturnal cramp, 41/310
nonprimary inflammatory myopathy, 62/372
optic atrophy, 65/484
optic neuropathy, 65/484
pigmentary retinopathy, 13/148
polyneuropathy, 65/484
primary pigmentary retinal degeneration, 13/149
rheumatoid arthritis, 13/220
rimmed muscle fiber vacuole, 40/151, 339, 438
rimmed vacuole distal myopathy, 40/151
sarcoplasmic reticulum, 40/151
segmental demyelination, 65/484
toxic myopathy, 62/606
toxic neuropathy, 51/302
transverse tubule, 40/151
Chloroquine intoxication
accommodation disorder, 65/484
axonopathy, 65/484
blindness, 65/484
corneal opacity, 60/172
deafness, 65/483
doughnut, 13/143
epilepsy, 65/483
headache, 65/483
hearing loss, 65/483
motor end plate, 65/484
myasthenic syndrome, 65/483
myoglobinuria, 40/339
neuromyopathy, 65/483
polyneuropathy, 65/484
psychosis, 65/483
retinopathy, 13/212, 65/483
rhabdomyolysis, 40/339
scotoma, 65/484
secondary pigmentary retinal degeneration, 60/736
segmental demyelination, 65/484
target, 13/143
Chloroquine myopathy
cardiomyopathy, 62/608
diplopia, 62/606
EMG, 62/608
histopathology, 62/608
myeloid body, 62/608
neuropathy, 62/606
symptom, 62/606
Chlorphenotane
epilepsy, 64/199
organochlorine, 36/393, 64/197
organochlorine insecticide intoxication, 36/398-405, 64/197
tremor, 64/199
Chlorphenoxamine
extrapyramidal disorder, 6/828
Chlorphenylacetylurea
side effect, 15/708
Chlorpromazine
akathisia, 1/290
anorexia nervosa, 46/588
chemical formula, 65/275
Chiari-Frommel syndrome, 2/444
cluster headache, 48/230
drug induced parkinsonism, 6/234
dyssynergia cerebellaris myoclonica, 60/603
migraine, 5/380
neuroleptic agent, 13/223, 65/275
phenothiazine, 65/275
retinopathy, 13/223

toxic myopathy, 62/601
Chlorpromazine intoxication
tardive dyskinesia, 37/307-312
Chlorpropamide
neuropathy, 7/518
Chlorprothixene
chemical formula, 65/276
neuroleptic agent, 65/276
neuropathy, 7/518
Chlorzoxazone
muscle contraction headache, 5/168
muscle relaxation, 5/168
Chocolate
*see also* Cocoa
migraine, 48/93
Choke, *see* Decompression illness
Choking attack, *see* Buccopharyngeal spasm
Cholecalciferol intoxication, *see* Vitamin D3 intoxication
Cholecystokinin
Alzheimer disease, 46/268
anorexia nervosa, 46/587
basal ganglion, 49/33
brain microcirculation, 53/79
gastrointestinal disease, 39/452
Huntington chorea, 49/258, 298
Parkinson disease, 49/123
Cholecystokinin octapeptide
headache, 48/109
mesolimbic dopamine system, 49/48
neostriatum, 49/47
Parkinson disease, 49/120
Cholera
hog cholera virus, 34/298
Cholestanol
cerebrotendinous xanthomatosis, 10/532, 547, 29/375, 51/399, 60/674, 66/16, 110
chemistry, 10/234
encephalitis lethargica, 66/16
Cholesteatoma
*see also* Epidermoid cyst
acoustic nerve tumor, 17/684
clinical features, 31/67
epidemiology, 31/67
middle ear, 5/211
parasellar, *see* Parasellar cholesteatoma
pathology, 31/67
skull base tumor, 17/196, 206
transverse sinus thrombosis, 52/176, 54/397
vertigo, 2/362
Cholesterinosis, *see* Cholesterosis
Cholesterol
a-alphalipoproteinemia, 51/396
atheromatous plaque, 11/331
cerebrotendinous xanthomatosis, 10/547, 27/250, 29/374, 66/606
chemistry, 10/233
CSF, *see* CSF cholesterol
Down syndrome, 31/433
esterified, 9/312
glia, 9/6
Hand-Schüller-Christian disease, 10/343, 38/96, 42/442
Liebermann-Burchardt reaction, 9/26
lipid metabolic disorder, 10/265
multiple sclerosis, 9/321
neuropathy, 8/20
perchloric acid naphthoquinone method, 9/26, 29
Refsum disease, 8/23, 21/202, 205, 27/521
Reye syndrome, 56/238, 65/118
Schultz procedure, 9/26, 29
serum, 9/321, 10/266
spinal cord injury, 26/382
synthesis, 66/45
unesterified, 9/5
white matter, 9/5
Wolman disease, 10/504, 546, 14/779
Cholesterol biosynthesis
disorder, 66/656
mevalonate kinase deficiency, 66/656
Smith-Lemli-Opitz syndrome, 66/587
Cholesterol ester
a-alphalipoproteinemia, 51/396
chemistry, 10/233
demyelination, 9/36
dystrophy, 66/3
hepatic insufficiency polyneuropathy, 51/363
multiple sclerosis plaque, 9/37
myelinoclastic diffuse sclerosis, 9/480-482
Tangier disease, 66/16
Wolman disease, 10/352, 546
Cholesterol ester storage disease
Wolman disease, 60/166
Cholesterol ester transfer protein
deficiency, 66/552
Cholesterol ester transfer protein gene
defect, 66/552
lipoprotein metabolism, 66/552
Cholesterol esterase
CSF, *see* CSF cholesterol esterase
Wolman disease, 10/546
Cholesterol homeostasis
Niemann-Pick disease type C, 60/153
Cholesterol metabolism
biochemistry, 66/46
cerebrotendinous xanthomatosis, 42/430

heredopathia ophthalmo-otoencephalica, 42/150
leukodystrophy, 66/3
orthochromatic leukodystrophy, 42/501
Tangier disease, 10/278, 42/627
Cholesterolemic xanthomatosis
blood-brain barrier, 27/248
Cholesterosis
cerebral, *see* Cerebral cholesterosis
enzyme defect, 21/50
spinal, *see* Spinal cholesterosis
Cholesteryl ester, *see* Cholesterol ester
Cholestyramine, *see* Colestyramine
Choline
Alzheimer disease, 46/267
cluster headache, 48/239
neptunea intoxication, 37/64
plasmalogen, 10/241
Choline acetylase
wallerian degeneration, 7/214
Choline acetyltransferase
Alzheimer disease, 46/266
Huntington chorea, 42/227, 49/75, 258
neurofibrillary tangle, 46/268
pallidosubthalamic projection, 49/7
Parkinson dementia, 49/178
Parkinson disease, 49/67, 111
progressive dysautonomia, 59/147
progressive supranuclear palsy, 49/251
Choline ester
neptunea intoxication, 37/64
Choline kinase
wallerian degeneration, 7/215
Choline level
red blood cell, *see* Erythrocyte choline level
Cholinephosphotransferase
wallerian degeneration, 7/215
Cholinergic agent
behavior disorder, 46/595
progressive external ophthalmoplegia, 22/183
Cholinergic crisis
myasthenia gravis, 41/131
organophosphate intoxication, 51/284, 64/170
Cholinergic deficiency
Down syndrome, 50/530
Cholinergic dysautonomia
autonomic nervous system, 75/13
Cholinergic dysfunction
botulism, 75/701
Cholinergic neuron
muscarinic, *see* Muscarinic cholinergic nerve cell
progressive dysautonomia, 59/147
Cholinergic receptor blocking agent
autonomic nervous system, 22/250
behavior disorder, 46/595
brain infarction, 53/421
chorea, 49/190
delirium, 46/537
eye, 74/418
learning, 27/467
memory, 27/467
mode of action, 6/829-832
Cholinergic receptor blocking agent intoxixation
delirium, 64/537
physostigmine, 37/541, 46/538, 596
Cholinergic receptor stimulating agent
carbamate intoxication, 64/183
pupil, 74/419
Cholinesterase
organophosphate, 37/541
pseudo, *see* Butyrylcholinesterase
restoration of higher cortical function, 3/371
wallerian degeneration, 7/214
Cholinesterase blocking agent
organophosphate intoxication, 37/554, 51/284
organophosphorus compound intoxication, 37/541-558
Chondroblastoma
benign, *see* Benign chondroblastoma
spinal cord compression, 19/367
Chondrocranium
cranioschisis, 32/180
Chondrodysplasia
metaphyseal, *see* Metaphyseal chrondrodysplasia
peroxisomal, *see* Peroxisomal chondrodysplasia
Chondrodysplasia hemangioma syndrome, *see* Maffucci-Kast syndrome
Chondrodysplasia punctata
rhizomelic, *see* Rhizomelic chondrodysplasia punctata
Chondrodysplastic myotonia, *see* Chondrodystrophic myotonia
Chondrodystrophia calcificans congenita, *see* Conradi-Hünermann syndrome
Chondrodystrophia fetalis
spinal cord compression, 19/365
Chondrodystrophia punctata, *see* Conradi-Hünermann syndrome
Chondrodystrophic myotonia
blepharophimosis, 40/284, 546
classification, 40/284
clinical symptom, 62/269
curare, 40/546
dwarfism, 40/284, 328, 546
electrical activity, 40/328, 546
genetics, 31/279, 41/424
histopathology, 31/279

hyporeflexia, 62/269
laboratory finding, 31/277, 279
malignant hyperthermia, 62/570
mental deficiency, 62/269
motor unit hyperactivity, 41/296
muscle biopsy, 62/269
muscle hypertrophy, 40/284, 62/269
myoglobinuria, 62/558
myopathy, 31/278
myotonic myopathy, 43/165
myotonic syndrome, 62/262
ocular abnormality, 40/328
prognosis, 31/279
radiologic finding, 31/277
skeletal abnormality, 31/278, 40/328, 546
succinylcholine, 41/299
D-tubocurarine, 41/299
Chondrodystrophy
mental deficiency, 46/88
Chondroectodermal dysplasia, *see* Ellis-Van Creveld syndrome
Chondrogenic tumor
benign spinal, *see* Benign spinal chondrogenic tumor
Chondroitin
glycosaminoglycanosis, 10/327
serine, 10/431
undetermined significant monoclonal gammopathy, 63/402
Chondroitin sulfate
glycosaminoglycanosis, 66/281
Chondroitin sulfate A
CNS myelin, 9/31
connective tissue, 10/432
glycosaminoglycanosis, 10/327, 432
intervertebral disc, 20/526, 535
Chondroitin sulfate B, *see* Dermatan sulfate
Chondroitin sulfate C
connective tissue, 10/432
glycosaminoglycanosis, 10/327, 432
intervertebral disc, 20/526, 535
Chondroma
age, 19/303
nasopharyngeal tumor, 17/205
pathology, 19/303
periosteal, *see* Periosteal chondroma
sex ratio, 19/303
skull base tumor, 17/147, 150, 154, 156, 158, 194
treatment, 19/306
vertebral column tumor, 20/25
Chondromyxofibroma
benign tumor, 19/367
spinal cord compression, 19/367
Chondrosarcoma
incidence, 68/530
Maffucci syndrome, 14/14
skull base, 68/466
skull base tumor, 17/168, 68/484
spinal cord, 67/197
spinal cord compression, 19/368
spinal cord tumor, 68/524
vertebral column tumor, 20/32
Chondrosis intervertebralis
terminology, 7/447
Chorda tympani
ageusia, 24/15
anatomy, 48/488
intermedius neuralgia, 48/487
section, 48/491
Chorda tympani neuralgia, *see* Intermedius neuralgia
Chorda tympani syndrome, *see* Intermedius neuralgia
Chordata intoxication, 37/68-87
Acraniata, 37/68
cephalochordata, 37/68
ciguatoxin, 37/82-85
clupeotoxin, 37/86
dolphin, 37/68
elasmobranch, 37/70, 85
fish, 37/68
hallucinatory fish, 37/68, 86
hemichordata, 37/68
ichthyohaemotix fish, 37/87
ichthyohepatoxic fish, 37/87
ichthyo-otoxism, 37/86
ingestion, 37/69
pisces, 37/68
porpoise, 37/68
protochordata, 37/68
ray, 37/70
scombroid, 37/85
scorpaena, 37/75
shark, 37/70
spinal injury, 37/69
stingray, 37/71-75
stonefish, 37/75-77
tetrodotoxin, 37/79-82
tunicata, 37/68
urochordata, 37/68
vertebrata, 37/68
whale, 37/68
zebrafish, 37/75
Chordoma
age, 19/288
anosmia, 18/155

biopsy, 18/159
brain, *see* Brain chordoma
brain biopsy, 16/721
calcification, 18/158
congenital, 32/355-386
dental, 18/151
differential diagnosis, 18/159
ecchordosis physaliphora, 18/152, 19/287
embryology, 19/287
exophthalmos, 18/155
incidence, 68/530
infratentorial, 18/154
intervertebral disc rupture, 18/152
intracranial, 18/151-164
location, 19/288
metastasis, 18/160
multiple cranial neuropathy, 18/153
myeloencephalography, 18/158
nonrecurrent nonhereditary multiple cranial neuropathy, 51/571
notochord, 18/151, 19/18
optic chiasm compression, 68/75
origin, 18/151
pain, 19/18
pathology, 18/160, 19/288
PEG, 18/158
physaliphorous cell, 18/160, 20/40
posterior fossa tumor, 18/404
prognosis, 18/160, 19/291
radiography, 19/290
radiotherapy, 18/160
sacrococcygeal, *see* Sacrococcygeal chordoma
Schmincke tumor, 18/159
sex ratio, 19/288
sexual impotence, 18/155
skull base, 68/466
skull base tumor, 18/404, 68/484
spheno-occipital group, 19/289
spinal, 32/355
spinal cord, 67/197
spinal cord compression, 19/369
spinal epidural tumor, 20/129
spinal tumor, 68/520
supratentorial, 18/155
symptomatology, 19/289
treatment, 18/160, 19/290
tuberous sclerosis, 50/373
ultrastructure, 18/161
vertebral column, 20/39
vertebral group, 19/289
Chorea
*see also* Athetosis, Choreoathetosis *and* Involuntary movement
adrenoleukodystrophy, 60/169
African trypanosomiasis, 52/341
agraphia, 45/466
alcohol intoxication, 49/556
Alexander disease, 10/101
amoxapine intoxication, 65/319
amyotrophic lateral sclerosis, 22/321, 59/235
angiostrongyliasis, 52/554
antiphospholipid antibody, 63/330
ataxia telangiectasia, 21/581
athetosis, 6/445
bacterial encephalitis, 49/553
Bardet-Biedl syndrome, 13/392, 401
Bassen-Kornzweig syndrome, 13/427
benign juvenile hereditary, *see* Juvenile hereditary benign chorea
biochemistry, 49/550
brachium conjunctivum, 21/529
brachium conjunctivum lesion, 21/515
brain stem syndrome, 2/277
canine, *see* Canine chorea
carbamazepine intoxication, 65/506
carbon monoxide intoxication, 49/476, 556, 64/31
cause, 6/354
cholinergic receptor blocking agent, 49/190
chorea-acanthocytosis, 49/328, 63/281
chorea electrica, 1/277
choreatic rigidity, 1/267
choreatic syndrome, 6/435-439
chronic hepatocerebral degeneration, 49/219
chronic meningitis, 56/645
chronic mercury intoxication, 64/370
clinical features, 2/500
Cockayne-Neill-Dingwall disease, 13/433
complex I deficiency, 62/504
contraceptive drug induced, *see* Contraceptive drug induced chorea
cyanide intoxication, 64/232
dentatorubropallidoluysian atrophy, 21/519-521, 49/440, 60/663
differential diagnosis, 49/549
digoxin intoxication, 65/450
diphtheria, 33/486, 65/240
dog distemper, 49/500
dopamine receptor stimulating agent, 49/190
drug induced, 49/557
drug treatment, 6/839
dystonia, 6/557, 42/219
dystonia musculorum deformans, 49/522
edentulous orodyskinesia, 49/558
EEG, 49/500
EMG, 1/647, 49/500
encephalitis, 6/438

Epstein-Barr virus, 56/252
Epstein-Barr virus meningoencephalitis, 56/252
estrogen, 49/76
familial spastic paraplegia, 22/453
Foix syndrome, 2/277
Friedreich ataxia, 6/435, 21/330
full list of causes, 49/551
fundus flavimaculatus, 13/263
gold intoxication, 49/559
hereditary benign, *see* Hereditary benign chorea
hereditary cerebellar ataxia, 21/369
hereditary disorder, 49/557
hereditary diurnal fluctuating juvenile dystonia, 49/42
hereditary dystonic paraplegia, 22/453
hereditary olivopontocerebellar atrophy (Menzel), 21/441, 444
hereditary paroxysmal dystonic choreoathetotis, 49/349
hereditary periodic ataxia, 21/569
hereditary spastic ataxia, 21/367, 383, 60/462, 663
heterocyclic antidepressant, 65/319
hollow hand sign, 1/184
Huntington, *see* Huntington chorea
hypernatremia, 63/554
hyperthyroid, *see* Hyperthyroid chorea
hypomagnesemia, 63/562
hypoparathyroidism, 27/308
immunologic mediated, 49/554
infantile spinal muscular atrophy, 59/70
infectious disease, 49/553
inorganic mercury intoxication, 49/556
interictal, *see* Interictal chorea
Isaacs syndrome, 49/559
juvenile hereditary benign, *see* Juvenile hereditary benign chorea
juvenile Huntington, *see* Juvenile Huntington chorea
kernicterus, 6/438
Köhlmeier-Degos disease, 55/276
late cerebellar atrophy, 60/663
late cortical cerebellar atrophy, 21/495
levodopa, 49/190
levodopa induced dyskinesia, 49/195, 197
lithium intoxication, 64/295
Lyme disease, 51/205
manganese intoxication, 49/556, 64/308
mercury intoxication, 49/559
mesencephalic pedunculotomy, 6/868
metabolic disorder, 49/554
micrographia, 45/464
migraine, 5/80
Minamata disease, 64/414
miscellaneous infection, 49/554
Morvan fibrillary, *see* Morvan fibrillary chorea
neoplasm, 49/550
neuroaxonal dystrophy, 60/663
neuroborreliosis, 51/205
neurobrucellosis, 33/314, 52/587
neuroleptic agent, 49/191
neuroleptic syndrome, 6/257
neurophysiology, 49/549
Nothnagel syndrome, 2/298, 302
nucleus ruber superior syndrome, 1/187
oculocerebrofacial syndrome, 43/280
olivopontocerebellar atrophy (Dejerine-Thomas), 60/663
olivopontocerebellar atrophy variant, 21/453
organomercury intoxication, 49/556
palilalia, 43/225
pallidal degeneration, 49/450
pallidoluysiodentate degeneration, 49/456
pallidoluysionigral degeneration, 49/455
Parkinson disease, 49/197
pathogenesis, 1/287
pathophysiology, 49/73
phenytoin induced, *see* Phenytoin induced chorea
physiology, 6/436
polycythemia, 63/250
polycythemia vera, 55/468, 63/250
positron emission tomography, 49/42
posthemiplegic, *see* Posthemiplegic chorea
previous Sydenham, *see* Previous Sydenham chorea
progressive bulbar palsy, 59/409
progressive multifocal leukoencephalopathy, 9/494
regressive generalized, *see* Regressive generalized chorea
renal insufficiency, 27/336
rigid Huntington, *see* Rigid Huntington chorea
Schimke syndrome, 62/304
senile, *see* Senile chorea
site, 1/279
spastic ataxia, 42/165
spinal muscular atrophy, 59/373
stationary generalized, *see* Stationary generalized chorea
striatal necrosis, 49/499-501
striatopallidodentate calcification, 49/422
subacute necrotizing encephalomyelopathy, 51/387
subacute sclerosing panencephalitis, 56/419
subsarcolemmal myofibril, 42/219
superior cerebellar artery, 55/90
superior cerebellar artery syndrome, 53/397

Sydenham, *see* Sydenham chorea
symptomatic chorea, 6/354
syphilis, 6/438
systemic lupus erythematosus, 55/374, 377
tardive dyskinesia, 49/186
tetanoid, *see* Tetanoid chorea
thalamic dementia, 21/590
thalamic infarction, 53/396
thalamic lesion, 6/59
thalamic syndrome (Dejerine-Roussy), 6/437
thallium intoxication, 49/556, 64/325
tic, 6/789
toxic encephalopathy, 49/556
trauma, 6/438, 49/550
tuber cinereum, 6/354
viral infection, 49/554
Wilson disease, 49/224
xeroderma pigmentosum, 14/788
Chorea-acanthocytosis, 49/327-332, 60/139-145
acanthocyte, 49/327, 329
age at death, 63/281
age at onset, 49/328, 63/281
amyotrophy, 51/398, 63/281
ankyrin, 63/285
areflexia, 63/280
autosomal dominant, 63/283
autosomal recessive, 63/283
biochemical study, 60/141
brain stem auditory evoked potential, 63/282
brain ventricle size, 63/282
cardiomyopathy, 63/281
caudate nucleus, 63/283
caudate nucleus atrophy, 49/331
chemical laboratory data, 63/281
chorea, 49/328, 63/281
clinical features, 49/328, 60/140
computer assisted tomography, 49/329, 63/281
creatine kinase, 63/281
creatine kinase MM, 49/330
CSF GABA, 49/330
CSF noradrenalin, 49/330
dementia, 49/328, 63/281
differential diagnosis, 49/332
distal amyotrophy, 63/280
dopamine, 63/283
dysphagia, 63/282
dystonia, 49/328
EEG, 49/329, 63/281
EMG, 49/329, 60/141, 63/281
epidemiology, 49/327
epilepsy, 49/329, 51/398, 63/281
erythrocyte band 2 protein, 63/285
erythrocyte band 3 protein, 63/285
erythrocyte band 4 protein, 63/285
erythrocyte enzyme, 49/331
erythrocyte membrane, 60/142
erythrocyte membrane protein, 49/330
fatty acid, 63/285
genetics, 49/327, 63/283
giant axonal swelling, 63/283
globus pallidus, 49/331
hematologic features, 63/285
hereditary motor and sensory neuropathy type I, 63/280
hereditary motor and sensory neuropathy type II, 63/280
histochemistry, 49/331
history, 60/139, 63/280
Huntington chorea, 49/292
involuntary vocalization, 63/280
Kell blood group, 49/330
β-lipoprotein, 49/328
management, 63/285
McLeod phenotype, 49/330, 63/284, 287
mental depression, 63/281
motoneuron disease, 60/140
motor neuropathy, 51/398
motor polyneuropathy, 49/329
multisystem disease, 60/144
muscle biopsy, 49/331, 60/141
mutilating choreiform movement, 60/140
nerve biopsy, 49/331
nerve conduction velocity, 51/398
nerve ultrastructure, 63/283
neuroacanthocytosis, 42/209, 63/271, 280
neurogenic muscular atrophy, 63/281
neuropathology, 60/141
neuropathy, 63/283
noradrenalin, 63/283
nuclear magnetic resonance, 63/281
oral self-mutilation, 63/281
orofacial dyskinesia, 63/281
orolingual facial dyskinesia, 63/280
palmitic acid, 63/286
parkinsonism, 63/281
personality change, 63/281
positron emission tomography, 63/281
presenting symptom, 60/140
progressive bulbar palsy, 59/225
putamen, 63/283
readiness potential, 49/74, 329, 63/282
review, 63/271-290
self-mutilation, 49/328, 63/280-282
sensory action potential, 63/282
sensory loss, 63/281
serum alanine aminotransferase, 49/330

serum aspartate aminotransferase, 49/330
serum creatine kinase, 49/329
serum lactate dehydrogenase isoenzyme 3, 49/330
serum lipoprotein, 60/144
somatosensory evoked potential, 63/282
spectrin, 63/286
Spencer variant, 63/283
stearic acid, 63/286
substance P, 63/283
substantia nigra, 63/283
sural nerve biopsy, 60/141
terminology, 63/284
urinary homoprotocatechuic acid, 49/330
visual evoked response, 63/282
vocal tic, 63/281
X-linked, 63/284
Chorea electrica
chorea, 1/277
Huntington chorea, 6/301, 49/293
myoclonia, 6/769
Chorea gravidarum, *see* Sydenham chorea
Chorea minor, *see* Sydenham chorea
Chorea mollis
Sydenham chorea, 6/427, 49/360
Chorea paralytica
Sydenham chorea, 6/427, 49/359
Chorea Sancti Viti, *see* Sydenham chorea
Choreatic movement, *see* Chorea
Chorée fibrillaire Morvan, *see* Morvan fibrillary chorea
Chorée variable des dégénérés
Brissaud, 6/435, 49/294
Huntington chorea, 49/294
tic, 6/301
Choreic dyskinesia, *see* Chorea
Choreiform dyskinesia
XYY syndrome, 50/551
Choreiform syndrome
clumsiness, 46/160
Choreoathetosis
*see also* Chorea
β-*N*-acetylhexosaminidase A deficiency, 60/663
β-*N*-acetylhexosaminidase B deficiency, 49/385, 60/663
acquired hepatocerebral degeneration, 46/338
antiepileptic agent, 65/499
ataxia, 60/656
ataxia telangiectasia, 14/75, 267, 49/385, 60/361, 663
Bardet-Biedl syndrome, 13/392
Benedikt syndrome, 1/187
benign, *see* Benign choreoathetosis
bilirubin encephalopathy, 27/415, 419, 421
carbon monoxide intoxication, 49/475
cat scratch disease, 52/130
cerebellar ataxia, 21/468
cervical cordotomy, 6/854
chronic hepatocerebral degeneration, 49/219
chronic psychotic, *see* Chronic psychotic choreoathetosis
Cockayne-Neill-Dingwall disease, 13/433
Creutzfeldt-Jakob disease, 46/291, 56/545
cytochrome reductase deficiency, 49/385
dentatorubropallidoluysian atrophy, 49/438, 60/663
differential diagnosis, 6/355
Divry-Van Bogaert syndrome, 14/514
dystonic, *see* Dystonic choreoathetosis
epilepsy, 42/699
eunuchoidism, 60/663
external pallidum, 49/457
Fabry disease, 10/543
familial inverted, *see* Familial inverted choreoathetosis
familial paroxysmal, *see* Hereditary paroxysmal kinesigenic choreoathetosis
filariasis, 52/516
full list of causes, 49/551
Gerstmann-Sträussler-Scheinker disease, 60/663
Hallervorden-Spatz syndrome, 42/257, 49/385, 399, 66/712, 714
hemiballismus, 42/224
hereditary cerebellar ataxia, 21/575
hereditary cerebellar palsy, 60/663
hereditary exertional paroxysmal dystonic, *see* Hereditary exertional paroxysmal dystonic choreoathetosis
hereditary paroxysmal dystonic, *see* Hereditary paroxysmal dystonic choreoathetotis
hereditary paroxysmal kinesigenic, *see* Hereditary paroxysmal kinesigenic choreoathetosis
hereditary periodic ataxia, 21/564
Huntington chorea, 49/385
hyperthyroidism, 49/385
kernicterus, 6/160, 42/583
kinesigenic, *see* Kinesigenic choreoathetosis
Köhlmeier-Degos disease, 55/276
late cerebellar atrophy, 60/663
legionellosis, 52/254
Lesch-Nyhan syndrome, 6/456, 22/514, 29/263, 42/153, 49/385, 59/360, 60/19
lithium intoxication, 49/385, 64/296
Machado-Joseph disease, 60/663
methyldopa, 37/440
methyldopa intoxication, 37/441
3-methylglutaconic aciduria, 49/385

myoclonic epilepsy, 42/699
nephrolithiasis, 49/385
neuroaxonal dystrophy, 60/663
neuropharmacology, 49/58
Niemann-Pick disease type C, 49/385
nonketotic hyperglycemia, 49/385
oculomotor apraxia, 60/656
olivopontocerebellar atrophy (Dejerine-Thomas), 21/426
olivopontocerebellar atrophy variant, 21/456
olivopontocerebellar atrophy (Wadia-Swami), 60/663
opticocochleodentate degeneration, 21/538, 60/754
orthochromatic leukodystrophy, 42/500
oxoglutarate dehydrogenase deficiency, 66/419
pallidal degeneration, 6/644, 49/454
pallidal syndrome, 2/502, 6/644
pallidocerebello-olivary degeneration, 60/663
pallidoluysian degeneration, 49/450
pallidoluysiodentate degeneration, 49/454
pallidoluysionigral degeneration, 49/454
pallidonigral degeneration, 49/454
pallidopyramidal degeneration, 49/454
paroxysmal dystonic, *see* Hereditary paroxysmal dystonic choreoathetotis
paroxysmal kinesigenic, *see* Paroxysmal kinesigenic choreoathetosis
phencyclidine intoxication, 49/385
phenytoin intoxication, 49/385
polyglucosan body disease, 49/385
portacaval shunt, 21/576
postanesthetic encephalopathy, 49/479
posterolateral thalamic syndrome, 2/485
postopen heart surgery, 63/7
progressive familial, *see* Progressive familial choreoathetosis
progressive glomerulopathy, 49/385
progressive myoclonus epilepsy, 42/699
progressive pallidal atrophy (Hunt-Van Bogaert), 21/531, 42/248
pure pallidal degeneration, 49/447
pyrethroid intoxication, 64/214
pyruvate dehydrogenase deficiency, 29/212, 42/516
reserpine, 37/437
Rosenberg-Bergström disease, 22/514
Salla disease, 49/385, 59/360
secondary pigmentary retinal degeneration, 49/385
spastic paraplegia, 42/175
subacute necrotizing encephalomyelopathy, 49/385
symptomatic paroxysmal kinesigenic, *see* Symptomatic paroxysmal kinesigenic choreoathetosis
systemic lupus erythematosus, 49/385
thalamic dementia, 6/355, 21/591
thalamic syndrome, 2/481
thalamic syndrome (Dejerine-Roussy), 42/771
Troyer syndrome, 42/194
viral infection, 49/385
xeroderma pigmentosum, 43/13, 49/385, 51/399, 60/336, 663

Choriocarcinoma
brain metastasis, 18/205
metastasis, 63/104
pathology, 68/236
spinal cord, 32/361

Choriomeningitis
lymphocytic, *see* Lymphocytic choriomeningitis

Chorionic gonadotropin beta subunit
benign intracranial hypertension, 67/111

Chorioretinal degeneration
angioid streak, 13/41
associated disease, 13/39-42
ataxia, 13/39
autosomal dominant optic atrophy, 13/38
Bassen-Kornzweig syndrome, 13/40
bone disease, 13/40
classification, 13/24-43
Cockayne-Neill-Dingwall disease, 13/40
deaf mutism, 13/41
deafness, 13/40
diabetes mellitus, 13/39
Ehlers-Danlos syndrome, 13/41
elliptocytosis, 13/40
external ophthalmoplegia, 13/40
familial amaurotic idiocy, 13/39
Friedreich ataxia, 13/39
Gaucher disease, 13/39
glycosaminoglycanosis, 13/40
Hallervorden-Spatz syndrome, 13/39
Hallgren syndrome, 13/40
hereditary ataxia, 13/39
hereditary motor and sensory neuropathy type I, 13/39, 21/285, 308
hereditary motor and sensory neuropathy type II, 13/39, 21/285, 308
hereditary muscular atrophy, 13/39
Hooft disease, 13/41
Klinefelter syndrome, 13/40
Laurence-Moon-Bardet-Biedl syndrome, 13/40
Marfan syndrome, 13/41
myoclonic epilepsy, 13/39
myotonic dystrophy, 13/40

parathyroid dysfunction, 13/41
Pelizaeus-Merzbacher disease, 13/39
Pierre-Robin syndrome, 13/41
primary pigmentary epithelium degeneration, 13/29
progressive external ophthalmoplegia, 13/39
progressive muscular dystrophy, 13/40
pseudoxanthoma elasticum, 13/41, 43/45
psoriasis, 13/40
receptor degeneration, 13/26
Refsum disease, 13/40
renal disease, 13/40
sickle cell anemia, 13/41
Sjögren-Larsson syndrome, 13/41
spastic paraplegia, 13/39
Stickler syndrome, 13/41
tapetal reflex, 13/27
tetraplegia, 13/39
Usher syndrome, 13/40
Van den Bosch syndrome, 13/41
Weiss-Alström syndrome, 13/40
Werner syndrome, 13/40
X-linked inheritance, 13/25

Chorioretinal lacunae
Aicardi syndrome, 50/163

Chorioretinitis
Behçet syndrome, 56/594
central, *see* Central chorioretinitis
congenital blindness, 22/537
congenital rubella, 50/278
cytomegalovirus infection, 50/277, 56/266
histoplasmosis, 35/510
lymphocytic choriomeningitis, 9/198, 50/308, 56/360
microcephaly, 30/510, 50/277
pigmented paravenous retinochoroidal atrophy, 13/268
subacute sclerosing panencephalitis, 56/419
toxoplasmosis, 50/277

Chorioretinitis juxtapapillaris (Jensen)
primary pigmentary retinal degeneration, 13/243

Chorioretinopathy
Aicardi syndrome, 42/696
congenital toxoplasmosis, 42/662
corpus callosum agenesis, *see* Aicardi syndrome
microcephaly, 30/512, 43/429

Choristoma
pituitary tumor, 17/425
survey, 68/333, 335

Choroid angioma
pseudoxanthoma elasticum, 43/45
Sturge-Weber syndrome, 14/68, 639-643, 31/18, 23

Choroid angiosclerosis
central, *see* Central choroidal atrophy

Choroid atrophy
central, *see* Central choroidal atrophy
central atrophy, 13/33
choroideremia, 13/35
generalized, 13/35
gyrate atrophy (Fuchs), 13/35
hereditary vitreoretinal degeneration, 13/274
hyaloidoretinal degeneration (Wagner), 13/274
malignant myopia, 13/32
pseudoinflammatory foveal dystrophy (Sorsby), 13/138
regional, 13/32
Sjögren-Larsson syndrome, 13/473

Choroid coloboma
congenital heart disease, 63/11

Choroid dystrophy
Boucher-Neuhäuser syndrome, 60/654

Choroid elastosis, *see* Angioid streak

Choroid epithelioma, *see* Choroid plexus papilloma

Choroid hamartoma
tuberous sclerosis, 50/372

Choroid plexus
African trypanosomiasis, 52/340
anatomy, 30/526-528
anencephaly, 42/12
blood-brain barrier, 56/80
Coxsackie virus, 56/318
CSF production, 30/529
Dandy-Walker syndrome, 42/21
ECHO virus, 56/318
embryology, 30/526, 32/530
ion active transport, 30/530
lead intoxication, 64/437
mercury intoxication, 64/381
Niemann-Pick disease type A, 10/496
physiology, 30/529-531
pressure change affecting, 30/530
Refsum disease, 8/23, 21/236
renal insufficiency, 27/340
tuberculous meningitis, 52/198
verruga peruana, 34/667
villous hypertrophy, 17/588
xanthomatosis, 27/244-246
xanthomatous transformation, 27/244-247

Choroid plexus carcinoma
age distribution, 17/587
clinical features, 17/582
diagnosis, 17/587
germinoma, 67/177
imaging, 67/177
pathology, 17/583, 585

terminology, 18/315
Choroid plexus embryogenesis
Dandy-Walker syndrome, 50/324
Choroid plexus endothelium
neurotropic virus, 56/31
Choroid plexus fibrosis
hydrocephalus, 50/290
Choroid plexus neoplasm
epilepsy, 72/181
Choroid plexus papilloma, 17/555-581, 42/760-762
brain scanning, 17/579
brain tumor, 18/512
child, 18/105
classification, 18/105
clinical diagnosis, 17/571
clinical features, 17/569, 31/44
congenital brain tumor, 31/44-46
CSF, 17/571, 42/761
CSF cytology, 16/399
CSF fistula, 24/186
epidemiology, 31/44, 68/173
histologic form, 18/120
history, 18/110
hydrocephalus, 17/570, 30/540, 50/286
hypervascular, 68/270
imaging, 67/177
incidence, 17/557
lateral ventricle tumor, 18/332
literature, 17/556
local effect, 17/567
microscopy, 17/562, 564, 567
papilledema, 42/761
pathology, 17/558
posterior fossa tumor, 18/396
prevalence, 17/556
radiology, 17/572, 575, 577, 31/45
radiotherapy, 18/512
sex ratio, 42/761
Sturge-Weber syndrome, 42/761
supratentorial brain tumor, 18/314
surgery, 17/581
survey, 68/270
third ventricle tumor, 17/442
treatment, 31/46
Choroid plexus tumor
chemotherapy, 68/171
child, 68/171
imaging, 68/169
infratentorial, 68/169
leptomeningeal metastasis, 71/663
metastasis, 71/663
supratentorial, 68/169
survey, 68/165
symptom, 68/167
Choroid plexus xanthoma
leukemia, 18/322
Choroid sclerosis
central, *see* Central choroidal atrophy
Friedreich ataxia, 21/348
generalized type, 13/282
hereditary vitreoretinal degeneration, 13/274
hyaloidoretinal degeneration (Wagner), 13/274
peripapillary type, 13/282
primary pigmentary retinal degeneration, 13/247
secondary pigmentary retinal degeneration, 13/278-288
Choroid vascular atrophy
central, *see* Central choroidal atrophy
Choroidal artery
anterior, *see* Anterior choroidal artery
anterior posterior, *see* Anterior posterior choroidal artery
external, *see* External choroidal artery
middle posterior, *see* Middle posterior choroidal artery
Choroidal artery occlusion
anterior, *see* Anterior choroidal artery occlusion
Choroidal artery syndrome
medial thalamic syndrome, 2/486
Choroidal vein
inferior, *see* Inferior choroidal vein
radioanatomy, 11/107, 109
topography, 11/107, 109
Choroideremia
case history, 13/253-256
choroid atrophy, 13/35
primary pigmentary retinal degeneration, 13/247
Choroiditis
disseminated, *see* Disseminated choroiditis
Choroidoependymal cyst
spinal, *see* Spinal ependymal cyst
Chotzen syndrome, *see* Acrocephalosyndactyly type III
Chouinard Scale
neuroleptic akathisia, 65/286
Christ-Siemens-Touraine syndrome
ectodermal dysplasia, 14/112
Christmas disease, *see* Hemophilia B
Chromaffin cell
catecholamine, 74/142
Chromaffinoma
tissue culture, 17/69
Chromakalim, *see* Cromakalim
Chromatin
Alzheimer disease, 46/273
amyotrophic lateral sclerosis, 59/280

Antoni type A tumor, 14/22
retotheliosarcoma, 18/234
sex, *see* Sex chromatin
Chromatism
hetero, *see* Heterochromatism
Chromatolysis, *see* Axonal reaction
Chromatopsia
chloramphenicol intoxication, 65/486
cinchonism, 65/452
digoxin intoxication, 65/450
ethambutol, 65/481
hallucination classification, 45/56
Chromoblastomycosis, *see* Cladosporiosis
Chromomycin A3
experimental, 13/203
vitreous body, 13/203
Chromophilic adenoma
pituitary gland, 2/457
Chromophobe adenoma
*see also* Pituitary adenoma
brain biopsy, 16/720
headache, 48/423
hypophyseal adenoma, 18/352
microscopy, 17/380-382
pituitary gland, 2/456
Refsum disease, 21/246
sella turcica, 17/389
Sheehan syndrome, 18/354
visual field, 17/385-387
Chromosomal aberration
Alzheimer disease, 42/276, 46/271, 273
anencephaly, 50/79
astrocytoma, 42/732
ataxia telangiectasia, 42/120, 60/350
autoimmune disease related factor, 30/103
Bloom syndrome, 43/11
cerebellar agenesis, 50/177
cerebellar dysgenesis, 30/377
cerebellar heterotopia, 30/494
cerebellar hypoplasia, 42/18
chromosomal aneuploidy, 30/101
CNS malformation, 30/104
congenital analgesia, 42/298
corpus callosum agenesis, 50/151
cri du chat syndrome, 30/249, 46/21-24
definition, 30/101
delayed fertilization, 30/102
deletion, 30/101
Down syndrome, 46/19-21
duplication, 30/101
environmental pollutant, 30/103
Fanconi syndrome, 43/17
incontinentia pigmenti, 43/23
inversion, 30/102
isochromosomes, 30/102
maternal age, 30/102
mental deficiency, 46/19-24, 68-76
microcephaly, 50/274
neuroblastoma, 42/758
numerical *vs* structural type, 31/354-364
Nijmegen breakage syndrome, 60/426
polyploidy, 30/101
Prader-Labhart-Willi syndrome, 43/464
radiation, 30/103
retinoblastoma, 42/768
Rubinstein-Taybi syndrome, 30/472, 31/284
Russell-Silver syndrome, 46/21
syringomyelia, 43/115
translocation, 30/102
trigonocephaly, 43/487
virus, 30/103
Chromosomal aneuploidy
*see also* Down syndrome
autosomal, 31/474
category, 31/471
chromosomal aberration, 30/101
incidence, 31/473
mental deficiency, 46/19-21
terminology, 31/472
X-linked, 31/474-477
Chromosomal disjunction syndrome
Down syndrome, 30/160, 31/378-395
meiotic, 30/101
Turner syndrome, 31/497
Chromosome, 31/341-364
aberration, *see* Chromosomal aberration
banding technique, 50/522
brain tumor, 67/44
C band, 31/343, 351
cerebellar granular cell hypertrophy, 67/58
Chicago Conference, 30/92, 31/343, 387
classification, 30/92, 31/343, 376, 50/520
deletion, 31/359
Denver Conference, *see* Denver Conference on Chromosomes
DNA replication, 30/92, 94
DNA and RNA, 30/91
ependymoma, 67/53
euchromatin, 30/94
female karyotype, 30/93
frequency, 31/363
G band, 31/344, 350
glioblastoma multiforme, 67/50
glioma polyposis syndrome, 67/55
globoid cell leukodystrophy, 10/71
heterochromatin, 30/94

history, 31/341-343, 374-376
identification, 31/351-354
Ig, 56/55
immunity genetic control, 56/55
insertion, 31/363
inversion, 31/360
isochromosomes, 31/363
juvenile myoclonic epilepsy, 72/137
London Conference, *see* London Conference on Chromosomes
male karyotype, 30/93
medulloblastoma, 67/51
meiosis, 30/94
meningioma, 67/52, 68/402
metachromatic leukodystrophy, 47/590
methodology, 31/348-351
mitosis, 30/92
morphology, 31/351-354
multiple nevoid basal cell carcinoma syndrome, 67/58
multiple sclerosis, 9/100, 47/303
neurofibromatosis type I, 67/55
neurofibromatosis type II, 67/56
nomenclature, 31/343, 353
numerical, 31/354-358
oligodendroglioma, 67/50
Paris Conference, *see* Paris Conference of Chromosomes
phakomatosis, 14/531
Philadelphia, *see* Philadelphia chromosome
pyruvate dehydrogenase complex, 66/413
Q band, 31/343, 350
R band, 31/344, 351
sex chromatin, 31/345-347
sex ratio, *see* Sex chromosomes
structure and function, 30/91-94
study, 30/105
tissue culture, 17/90
translocation, 31/360-363
Von Hippel-Lindau disease, 67/57
Wohlfart-Kugelberg-Welander disease, 22/71

Chromosome 1
Gaucher disease, 60/154
hereditary motor and sensory neuropathy type I, 60/200
hereditary neuropathy, 60/265
phosphofructokinase deficiency, 62/488
trisomy, 50/541

Chromosome 2
kappa light chain control, 56/55
partial trisomy of long arm, 43/493
trisomy, 50/542

Chromosome 3
partial monosomy of short arm, 43/495
partial trisomy of long arm, 43/495
recombinant chromosome 3 syndrome, 43/495

Chromosome 4
hereditary dystonic paraplegia, 59/346
partial monosomy of long arm, 43/496
partial monosomy of short arm, *see* Wolf-Hirschhorn syndrome
partial trisomy of long arm, 43/498
partial trisomy of short arm, 43/500

Chromosome 5
partial monosomy of short arm, *see* Cri du chat syndrome

Chromosome 6
late cerebellar atrophy, 60/583
partial trisomy of long arm, 43/503-506
Schut family ataxia, 60/488

Chromosome 7
partial trisomy, 30/250, 50/542
partial trisomy of long arm, 43/506
phosphorylase kinase deficiency, 62/487

Chromosome 8
ankyrin, 63/263
familial spherocytosis, 63/263
partial monosomy of short arm, 43/507
trisomy, 31/525-528, 46/69, 50/163, 563
trisomy mosaicism, 31/526, 43/509-511, 50/153, 167, 563

Chromosome 9
autosomal dominant primary dystonia, 59/346
Friedreich ataxia, 60/322
hereditary dystonic paraplegia, 59/346
partial monosomy of short arm, 43/512, 50/578
partial trisomy of long arm, 46/69
partial trisomy of short arm, 43/515, 46/69, 50/578
ring chromosome, 43/513, 50/580, 584
trisomy, 30/250, 43/514, 50/564

Chromosome 10
CNS tumor, 67/44
metachromatic leukodystrophy, 60/126
partial trisomy of long arm, 43/517-519
partial trisomy of short arm, 30/249, 43/519-522
phosphofructokinase deficiency, 62/488
phosphorylase isoenzyme, 62/486
Wolman disease, 60/159, 166

Chromosome 11
interstitial deletion, 50/581
lactate dehydrogenase deficiency, 62/490
partial monosomy of long arm, 43/522
partial monosomy of short arm, 50/581
partial trisomy of long arm, 43/523-525, 46/69
partial trisomy of short arm, 43/525, 46/69
phosphorylase kinase deficiency, 62/487

porphyria, 60/119
Chromosome 13
CNS tumor, 67/47
partial deletion, 50/583
partial monosomy of long arm, 43/526, 46/69
partial trisomy, 31/514, 46/69, 50/560
ring chromosome, 50/583-586
trisomy, *see* Patau syndrome
Chromosome 14
heavy chain control, 56/55
partial trisomy of long arm, 43/529
phosphorylase isoenzyme, 62/486
ring chromosome, 50/588-590
trisomy, 30/250
Chromosome 15
partial monosomy of long arm, 43/530, 50/591
partial trisomy of long arm, 43/531
ring chromosome, 43/530
Chromosome 16
trisomy, 50/542
Chromosome 17
acid maltase deficiency, 62/481
hereditary motor and sensory neuropathy type I, 60/202
hereditary neuropathy, 60/265
hyperkalemic periodic paralysis, 62/463
myotonic hyperkalemic periodic paralysis, 62/224, 267
partial monosomy of short arm, 50/591
Chromosome 18
hereditary amyloid polyneuropathy, 60/107
holoprosencephaly, 42/33
partial monosomy of long arm, 30/250, 508, 31/582-590, 43/532-534, 46/70
partial monosomy of short arm, 30/250, 508, 31/573-582, 43/534, 46/70
trisomy, *see* Edwards syndrome
Chromosome 19
acanthocytosis, 63/289
CNS tumor, 67/47
hereditary motor and sensory neuropathy variant, 60/247
partial trisomy of long arm, 43/538
Chromosome 20
partial trisomy of short arm, 43/539
phosphorylase isoenzyme, 62/486
Chromosome 21
*see also* Chromosome G
familial amyotrophic lateral sclerosis, 59/249
monosomy, 31/595-604, 43/540, 50/540
monosomy mosaicism, 31/598-603
partial monosomy, 31/604
partial monosomy of long arm, 43/541
phosphofructokinase deficiency, 62/488
ring chromosome, 31/592-597, 43/543
trisomy, *see* Down syndrome
Chromosome 22
*see also* Cat eye syndrome
lambda light chain control, 56/55
partial trisomy of long arm, 50/565
partial trisomy of short arm, 50/565
ring chromosome, 31/606-610, 43/547
trisomy, 30/508, 31/524-526, 43/548, 50/565
trisomy mosaicism, 50/564
Chromosome analysis
Down syndrome, 31/440, 445-447
indication, 31/529
Chromosome analysis indication
aneuploidy, 50/566
Chromosome C
trisomy, 31/525-528
Chromosome D
chromosome deletion syndrome, 50/583
ring chromosome, 30/249, 50/583
trisomy, *see* Patau syndrome
Chromosome deletion syndrome
ARG syndrome, 50/581
chromosome D, 50/583
lissencephaly, 50/591, 593
Miller-Dieker syndrome, 50/591
new type, 50/577
9p partial monosomy, 50/578
11p partial monosomy, 50/405
Prader-Labhart-Willi syndrome, 50/590
r(13) syndrome, 50/583, 585-587
review, 50/577
WARG syndrome, 50/581
Chromosome E
trisomy, 31/515-523
Chromosome fragility
sex ratio, *see* Sex chromosome fragility
Chromosome G
*see also* Chromosome 21
monosomy, 31/590-610
trisomy, 30/508, 31/524-526
Chromosome number abnormality
*see also* Aneuploidy, Polyploidy *and* Triploidy
aneuploidy, 50/539
definition, 50/539
fetal wastage, 50/541
general, 50/539
monosomy 21, 50/539
monosomy X, 50/541
polyploidy, 50/541
sex chromosomes, 50/541
spontaneous abortion, 50/541

tetrasomy, 50/539
trisomy, 50/539, 541
Chromosome translocation
ataxia telangiectasia, 14/118
Down syndrome, 31/386-391
Chromosome X, *see* Sex chromosomes
Chromosome Y, *see* Sex chromosomes
Chronaximetry
facial paralysis, 8/267
flaccid paralysis, 1/224
Chronic adrenal insufficiency
psychiatry, 70/211
symptom, 70/211
Chronic alcohol abuse, *see* Alcoholism
Chronic alcoholism, *see* Alcoholism
Chronic aluminum intoxication
dialysis, 63/525
Guam amyotrophic lateral sclerosis, 59/266
Chronic alveolar hypoventilation
sleep apnea syndrome, 45/133
Chronic anterior poliomyelitis, *see* Progressive spinal muscular atrophy
Chronic axonal neuropathy
acromegaly, 51/531
acrylamide, 51/531
arsenic polyneuropathy, 51/531
carbon disulfide polyneuropathy, 51/531
carcinoma, 51/531
celiac disease, 51/531
chronic obstructive respiratory tract disorder, 51/531
dapsone, 51/531
diabetes mellitus, 51/531
disulfiram, 51/531
fibrosing alveolitis, 51/531
giant axonal neuropathy, 51/531
gold, 51/531
hereditary ataxia, 51/531
hereditary motor sensory neuropathy type II, 51/531
hexacarbon, 51/531
hypoglycemia, 51/531
isoniazid, 51/531
lead, 51/531
lymphoma, 51/531
metronidazole, 51/531
monoclonal gammopathy, 51/531
multiple myeloma, 51/531
myxedema, 51/531
nitrofurantoin, 51/531
organophosphate, 51/531
platinum, 51/531
primary amyloidosis, 51/531
primary biliary cirrhosis, 51/531
rheumatoid arthritis, 51/531
sarcoidosis, 51/531
systemic lupus erythematosus, 51/531
thalidomide, 51/531
thallium, 51/531
uremia, 51/531
vasculitis, 51/531
vincristine, 51/531
vitamin $B_1$, 51/531
vitamin $B_{12}$, 51/531
vitamin $B_c$, 51/531
vitamin E, 51/531
Waldenström macroglobulinemia, 51/531
Chronic benign lymphocytic meningitis
chronic meningitis, 56/645
Chronic benign trigeminal paresis
trigeminal root, 2/59
Chronic brain syndrome, *see* Dementia *and* Organic psychosis
Chronic bromide intoxication
anisocoria, 36/300
ataxia, 36/299
autonomic disturbance, 36/299
CSF examination, 36/301
CSF pleocytosis, 36/301
diplopia, 36/300
dysphagia, 36/300
EEG change, 36/302
incidence, 36/297
language disorder, 36/300
mental change, 36/302
ptosis, 36/300
rigidity, 36/300
sign, 36/298
symptom, 36/298
tremor, 36/299
Chronic carbon monoxide intoxication, *see* Carbon monoxide encephalopathy
Chronic cluster headache, *see* Chronic migrainous neuralgia
Chronic demyelinating neuropathy
*see also* Primary demyelinating neuropathy
drug induced, 47/620
hereditary type, 47/621
paraproteinemia, 47/615-620
Chronic demyelinating polyneuropathy
systemic lupus erythematosus, 51/447
Chronic enterovirus infection
ataxia, 56/330
dementia, 56/330
dermatomyositis, 56/330
epilepsy, 56/330

headache, 56/330
tremor, 56/330
Chronic epidural hematoma
symptom, 54/357
Chronic fatigue
autonomic nervous system, 75/126
Chronic fatigue syndrome, *see* Postviral fatigue syndrome
Chronic granulomatosis
McLeod phenotype, 49/332
Chronic granulomatous disease
McLeod phenotype, 63/289
Chronic headache
acquired immune deficiency syndrome, 56/492
general features, 48/6, 33
Chronic hepatocerebral degeneration
*see also* Wilson disease
acquired hepatocerebral degeneration, 49/219
ataxia, 49/219
chorea, 49/219
choreoathetosis, 49/219
dementia, 49/219
memory, 49/219
past history, 49/219
tremor, 49/219
Chronic hereditary trophedema, *see* Milroy disease
Chronic hypertension
hypertensive encephalopathy, 47/538
Chronic hypoxia
acclimatization, 75/267
beta adrenergic receptor blocking, 75/269
blood pressure, 75/270
cardiac output, 75/269
heart rate, 75/268
overview, 75/265
regulation, 75/267
sympathetic hyperactivity, 75/269
Chronic idiopathic demyelinating polyradiculoneuropathy
clinical features, 47/614
steroid, 47/614
sural nerve, 47/614
systemic lupus erythematosus, 47/614
Chronic inflammatory demyelinating polyneuropathy
autonomic involvement, 75/699
Guillain-Barré syndrome, 51/247
multiple myeloma, 63/393
undetermined significant monoclonal gammopathy, 63/402
undetermined significant monoclonal gammopathy IgM, 63/402
Chronic inflammatory demyelinating polyradiculoneuropathy
*see also* Guillain-Barré syndrome
adrenomyeloneuropathy, 51/530
amiodarone polyneuropathy, 51/530
autonomic nervous system, 75/16
axonal degeneration, 51/532
clinical features, 51/535
Cockayne-Neill-Dingwall disease, 51/530
CSF Ig, 51/536
CSF protein, 51/534
diagnosis, 51/530
EMG, 51/536
etiology, 51/530
examination, 51/532
experimental allergic neuritis, 51/537
F wave, 51/536
globoid cell leukodystrophy, 51/530
Guillain-Barré syndrome, 51/537
hereditary motor sensory neuropathy type I, 51/530
hereditary motor sensory neuropathy type II, 51/530
HLA antigen, 51/538
IgA, 51/537
IgG, 51/537
IgM, 51/537
immunology, 51/537
incidence, 51/530
laboratory investigation, 51/533
metachromatic leukodystrophy, 51/530
monoclonal gammopathy, 51/530
motor nerve conduction, 51/532
multiple myeloma, 51/530
nerve biopsy, 51/533
onion bulb formation, 51/537
perhexiline, 51/530
predominant axonal degeneration, 51/534
progressive spinal muscular atrophy, 59/404-406
Refsum disease, 51/530
remyelination, 51/536
segmental demyelination, 51/532
spinal nerve root, 51/536
sural nerve biopsy, 51/536
symmetric deficit, 51/532
treatment, 51/538
Waldenström macroglobulinemia, 51/530
Chronic inflammatory polyradiculopathy
progressive muscular atrophy, 59/17
Chronic ischemic leukoencephalopathy
Sneddon syndrome, 55/408
Chronic lead intoxication, *see* Lead encephalopathy
Chronic leukoencephalopathy
methotrexate, 47/571, 63/358

Chronic lymphatic leukemia, *see* Chronic lymphocytic leukemia
Chronic lymphocytic leukemia
chronic meningitis, 56/644
cryoglobulinemia, 39/181
immunocompromised host, 56/470
polyarteritis nodosa, 55/359
survey, 71/436
symptom, 71/436
Chronic lymphoid interstitial pneumonitis
acquired immune deficiency syndrome, 56/489
Chronic manganese intoxication, *see* Manganese encephalopathy
Chronic meningitis, 56/643-648
acquired immune deficiency syndrome, 56/644-646
Actinomyces israelii, 56/645
actinomycosis, 56/644
allergic granulomatous angiitis, 56/644
Angiostrongylus cantonensis, 56/644
Arachnia propionica, 56/645
aspergillosis, 56/644
astrocytoma, 56/645
astrocytoma grade II, 56/646
Bannwarth syndrome, 56/645
Behçet syndrome, 56/645
blastomycosis, 56/644
Borrelia burgdorferi, 56/644
borreliosis, 56/645
brain abscess, 56/645
brain infarction, 56/645
brucellosis, 56/644
Candida albicans, 56/645
carcinomatous meningitis, 56/645
category, 56/644
cause, 56/645
chorea, 56/645
chronic benign lymphocytic meningitis, 56/645
chronic lymphocytic leukemia, 56/644
CINCA syndrome, 56/645
Cladosporium trichoides, 56/645
coccidioidomycosis, 56/644
Coenurus cerebralis, 56/645
complicated migraine, 56/644
congenital rubella, 56/645
Coxsackie virus, 56/201, 646
Coxsackie virus B1, 56/646
craniopharyngioma, 56/644
cryptococcosis, 56/644
Cryptococcus neoformans, 56/645
CSF pleocytosis, 56/643
definition, 56/643
dermal sinus, 56/645
diffuse leptomeningeal gliomatosis, 56/645
dura mater cyst, 56/645
duration, 56/643
echinococcosis, 56/645
ECHO virus, 56/645
ECHO virus B6-B9, 56/646
enterovirus, 56/646
ependymoma, 56/645
epidermoid cyst, 56/645
Fabry disease, 56/644
granulomatous CNS vasculitis, 55/391
herpes virus hominis, 56/646
Histoplasma capsulatum, 56/645
histoplasmosis, 56/645
Hodgkin lymphoma, 56/645
inappropriate antidiuretic hormone secretion, 56/644
incontinentia pigmenti, 56/644
leptospirosis, 33/402, 56/645
leukemia, 56/645
Lyme disease, 56/645
lymphocytic choriomeningitis, 56/645
lymphoma, 56/645
measles, 56/645
midline granuloma, 56/645
migraine associée, 56/645
multiple myeloma, 56/644
multiple sclerosis, 56/645
neurocysticercosis, 56/644
nocardiosis, 56/645
nonrecurrent, 56/644
North American blastomycosis, 35/403, 52/387
pachymeningitis, 56/644
physiochemical arachnoiditis, 56/645
Pseudoallescheria boydii, 56/645
reticulosis, 56/645
sarcoidosis, 56/644-647
Sjögren syndrome, 56/644
sphenoid sinusitis, 56/644
spinal arachnoiditis, 56/644
Sporothrix schenkii, 56/645
sporotrichosis, 56/645
Sydenham chorea, 56/645
syphilis, 56/645
systemic lupus erythematosus, 56/645
taeniasis, 56/645
toxoplasmosis, 56/645
treatment, 56/648
tuberculosis, 56/645
tuberculostatic agent, 56/648
uveomeningitis, 56/645
Vogt-Koyanagi-Harada syndrome, 56/645
Wegener granulomatosis, 56/644

Whipple disease, 56/644
Chronic mercury intoxication
acoustic nerve, 64/371
acrodynia, 64/371
ataxia, 64/370
blood-brain barrier, 64/373
brain infarction, 64/372
cerebellar atrophy, 64/372
cerebellar granular cell, 64/373
chorea, 64/370
cytoplasmic inclusion body, 64/372
dementia, 64/370
erethism, 64/370
extrapyramidal disorder, 64/370
goldsmith, 64/369
Guillain-Barré syndrome, 64/371
hatter, 64/369
hypersalivation, 64/371
individual susceptibility, 64/371
intention tremor, 64/370
ischemic encephalopathy, 64/372
mercuria lentis, 64/371
mercurial gum line, 64/371
miner, 64/369
mirror maker, 64/369
multifocal ischemic encephalopathy, 64/372
nerve conduction velocity, 64/372
neuropathology, 64/372
phenylmercury acetate, 64/373
postviral fatigue syndrome, 64/371
Purkinje cell, 64/372
rigidity, 64/370
sensorimotor polyneuropathy, 64/371
sensory impairment, 64/370
symptom, 64/369
tremor, 64/370, 394
visual field, 64/370
Chronic migrainous neuralgia, 48/247-254
adenosine phosphate, 48/253
age at onset, 48/249
chronic paroxysmal hemicrania, 48/250
classification, 48/9
definition, 48/248
differential diagnosis, 48/250
duration, 48/248
ergotamine, 48/251
erythrocyte choline, 48/253
glyceryl trinitrate, 48/251
history, 5/111
Horner syndrome, 48/251
hypothalamus, 48/253
incidence, 48/248
ketotifen, 48/252
lithium, 48/252
methysergide, 48/252
migraine, 5/265, 48/250
nomenclature, 5/115
oxygen, 48/251
paratrigeminal syndrome, 48/250
pizotifen, 48/251
prevalence, 48/248
primary, 48/248
prognosis, 48/250
secondary, 48/248
serotonin inhibitor, 48/251
sex ratio, 48/249
steroid, 48/252
superficial temporal artery, 48/254
surgery, 48/253
temporal arteritis, 48/250
terminology, 5/111
tonometry, 48/251
trigeminal neuralgia, 15/330, 48/250, 254
Chronic mountain sickness
headache, 48/397
Chronic multiple sclerosis
acute experimental allergic encephalomyelitis, 47/430
lesion, 47/230
olivopontocerebellar atrophy (Wadia-Swami), 60/497
remyelination, 47/241-244
Chronic multiple tic
childhood, *see* Childhood chronic multiple tic
Chronic muscle contraction headache, 48/343-351
age, 48/344
amitriptyline, 48/350
antidepressant, 48/350
biofeedback, 48/346, 350
cluster headache, 48/344
constitution, 48/345
differential diagnosis, 48/346
duration, 48/344
frequency, 48/344
hypnotic agent, 48/350
incidence, 48/343
location, 48/344
meditation, 48/350
migraine, 48/344
Minnesota Multiphasic Personality Inventory, 48/345
muscle relaxant, 48/350
muscle spasm, 48/349
myogenic mechanism, 48/349
neurotic measure, 48/345
neurotic personality, 48/349

pain, 48/344
personality, 48/345
posttraumatic, 48/346
psychological aspect, 48/346
psychopathology, 48/346
psychotherapy, 48/350
sex ratio, 48/344
stress, 48/345
survey, 48/348
symptom, 48/348
treatment, 48/350
vasodilator agent, 48/350
Chronic myelitis
*see also* Human T-lymphotropic virus type I associated myelopathy
human T-lymphotropic virus type I associated myelopathy, 56/529
tropical spastic paraplegia, 56/529
Chronic myelogenous leukemia
granulomatous CNS vasculitis, 55/388
juvenile, *see* Juvenile chronic myeloid leukemia
Chronic myositis
ECHO virus, 56/201
Chronic nerve compression
compression neuropathy, 51/88
Chronic neutropenia
mental deficiency, 43/289
Chronic nodular polioencephalitis
atypical category, 34/553
clinical features, 34/555-562
dementia, 34/560
dysarthria, 34/560
neuropathology, 34/558
Chronic obstructive pulmonary disease
almitrine, 63/421
autonomic nervous system, 75/12
brain hypoxia, 63/417
chronic respiratory insufficiency, 63/413
hypoventilation, 63/413
hypoventilation syndrome, 63/413
neurobehavior impairment, 63/417
neurologic manifestation, 63/414
neurologic symptom, 63/414
polyneuropathy, 63/421
respiratory disease, 63/413
sensory neuropathy, 63/421
sleep apnea syndrome, 45/133
snoring, 63/451
Chronic obstructive respiratory tract disorder
almitrine dimesilate polyneuropathy, 51/308
chronic axonal neuropathy, 51/531
Chronic paroxysmal hemicrania
age at onset, 48/258
audiological parameter, 48/262
autonomic involvement, 48/263
brain angiography, 48/261
brain blood flow, 48/261
cardiovascular alteration, 75/293
cervicogenic headache, 48/260
chronic migrainous neuralgia, 48/250
cimetidine, 48/265
classification, 48/257-265
cluster headache, 48/224, 257, 260
cluster headache syndrome, 48/257
corneal temperature, 48/261, 75/292
differential diagnosis, 48/260
duration, 48/258
dynamic tonometry, 48/261
eye tonometry, 75/292
features, 48/218
flowmetry, 48/261
forehead sweating, 48/262
frequency, 48/258
headache, 48/6
heart rhythm, 48/262
hemicrania continua, 48/260
histamine, 48/260, 265
ibuprofen, 48/264
ice pick headache, 48/260
imidazole acidic acid riboside, 48/261
indometacin, 48/263, 267
jabs and jolts syndrome, 48/260
ketoprofen, 48/264
lacrimation, 48/263, 75/293
mechanical precipitation, 48/259
miosis, 48/263
nasal secretion, 48/262, 75/293
nocturnal attack, 48/258
oxyphenbutazone, 48/264
pain localization, 48/258
pathogenesis, 48/262
phenylbutazone, 48/264
piroxicam, 48/264
prechronic stage, 48/265
prevalence, 48/260
prostaglandin, 48/263
pupil, 75/292
pupillometry, 48/261
REM sleep, 48/261
salivation, 48/262, 75/293
sex distribution, 48/258
sleep, 48/261
subarachnoid hemorrhage, 55/20
substance P, 48/263
sweating, 48/263, 75/292
symptom, 48/258

tolfenamic acid, 48/264
treatment, 48/264
Chronic polyneuritis
clinical form, 47/484
Chronic progressive demyelinating polyneuropathy
Refsum disease, 51/385
Chronic progressive external ophthalmoplegia
age at onset, 1/612, 2/323
EMG, 62/70
extraocular palsy, 2/323
ptosis, 2/323
secondary pigmentary retinal degeneration, 60/725
Chronic progressive panencephalitis
rubella virus, 56/405
Chronic progressive peripheral neuropathy
spinal muscular atrophy, 59/373
Chronic progressive polyneuritis
nerve conduction, 7/178
Chronic progressive subcortical encephalitis, *see* Binswanger disease
Chronic psychotic choreoathetosis
Huntington chorea, 6/355
Chronic recurring neuropathy
differential diagnosis, 8/366
Chronic regional pain syndrome
survey, 75/309
Chronic relapsing demyelinating polyneuropathy
undetermined significant monoclonal gammopathy IgM, 63/402
Chronic relapsing experimental allergic encephalomyelitis
multiple sclerosis, 47/430, 442, 470
pathology, 47/438-441
Chronic relapsing neuritis
tremor, 49/588
Chronic relapsing neuropathy
cramp, 34/396, 40/326
features, 34/396
peripheral nerve, 40/318
Chronic relapsing polyneuritis
immune mediated syndrome related to virus infection, 34/396
Chronic relapsing polyneuropathy
hepatitis B virus, 56/300
Chronic renal disease
onion bulb formation, 21/252
Chronic renal failure
autonomic polyneuropathy, 51/476
listeriosis, 52/91
restless legs syndrome, 51/356
Chronic renal failure myopathy
bone pain, 62/541
clinical features, 62/541
creatine kinase, 62/541
fatal gangrenous calcification, 62/541
proximal limb weakness, 62/541
Chronic respiratory insufficiency
chronic obstructive pulmonary disease, 63/413
clinical features, 38/288-290
CSF, 38/290
EEG change, 38/290
hypoventilation, 63/413
hypoventilation syndrome, 63/413
increased intracranial pressure, 38/290, 298
mental change, 38/288-290
neurologic complication, 63/413
obstructive apnea syndrome, 63/413
ocular sign, 38/290
pathogenesis, 38/291-298
pickwickian syndrome, 38/299, 63/413
respiratory encephalopathy, 63/413
treatment, 38/298
Chronic sensorimotor neuropathy
paraneoplastic syndrome, 69/374, 380
Chronic sensory neuropathy
autonomic polyneuropathy, 51/482
Chronic sinoatrial disorder
cardiac dysrhythmia, 39/262-264
cerebral manifestation, 39/262-264
clinical features, 39/263
embolism, 39/264
syncope, 39/262
Chronic spinal arachnoiditis, *see* Spinal arachnoiditis
Chronic subdural hematoma
accompanying disease, 54/353
acute subdural hematoma, 57/251
age, 24/303
age distribution, 54/354
alcoholism, 24/299
angiography, 24/310
bilateral, 24/303, 54/354
bleeding tendency, 24/298
boxing injury, 23/547
brain scintigraphy, 24/309
burr hole drainage, 24/314, 57/288
capillary permeability, 57/287
capsule resection, 24/316
cause, 54/354
child, 24/306, 316, 318
clinical features, 57/287
clinical grading, 54/355
complication rate, 57/288
computer assisted tomography, 54/354, 57/154, 287

conservative management, 54/357
craniotomy, 24/316
definition, 24/297
development, 24/300
diagnosis, 54/354
echoEG, 24/307
EEG, 24/308
epilepsy, 24/305
etiology, 24/297
external drainage, 24/315
fontanelle tap, 24/314
frequency, 24/302
head injury, 57/48, 154
hemisyndrome, 24/305
histology, 24/301
hydration, 24/317
incidence, 24/302
infant, 24/306, 314, 318
interhemisphere, 24/303
internal drainage, 24/315
intracranial hypotension, 24/299
lucid interval, 24/304
mannitol, 24/313
mental deterioration, 24/306
mortality, 24/315, 317
newborn, 24/306, 318
nontraumatic, 24/298
papilledema, 24/305
pathogenesis, 24/300, 57/287
pathology, 24/299, 301
pediatric head injury, 24/306, 316, 318, 57/336
PEG, 24/312
peritoneal drainage, 24/316
posterior fossa hematoma, 24/303
postoperative course, 54/355
precipitating factor, 54/353
prognosis, 24/313, 54/357
psychic disorder, 24/316
pupil, 24/305
radiodiagnosis, 57/154
radiology, 24/307
radiotherapy, 24/314
recurrence, 57/288
reduction cranial size, 24/317
result, 24/318
sex ratio, 24/303
site, 24/303
soccer injury, 23/568
spontaneous, 24/298
staged treatment, 24/317
steroid, 57/288
surgery, 24/314, 661, 57/288
surgical indicator, 54/355
symptom, 24/304
traumatic intracranial hematoma, 57/154, 251
treatment, 24/313, 317, 661, 57/287
twist drill craniotomy, 57/288
ventriculoatrial shunt, 24/318
ventriculography, 24/312
Chronic superior mesenteric artery occlusion
polyneuropathy, 7/631
Chronic tactile hallucinosis
lung cancer, 46/564
organic substrate, 46/564
Chronic tension headache, *see* Chronic muscle contraction headache
Chronic ulcerative colitis
myasthenia gravis, 62/402
Chronotaraxis
thalamic syndrome (Dejerine-Roussy), 2/486
Chrysaora intoxication
back pain, 37/51
cough, 37/51
cramp, 37/51
lacrimation, 37/51
mental depression, 37/51
morphine, 37/51
pain, 37/51
rash, 37/51
respiratory system, 37/51
restlessness, 37/51
suicide, 37/51
toxicology, 37/52
weakness, 37/51
Chrysotherapy
behavior disorder, 46/602
drug induced demyelinating neuropathy, 47/620
risk, 51/305
Churg-Strauss syndrome, *see* Allergic granulomatous angiitis
Churg-Strauss vasculitis, *see* Allergic granulomatous angiitis
Chvostek sign
alkalosis, 40/330
Bartter syndrome, 42/529
hyperkalemic periodic paralysis, 62/462
hyperventilation syndrome, 38/326, 63/432
hypocalcemia, 40/330
hypomagnesemia, 63/562
hypoparathyroidism, 27/299, 41/248
myotonic hyperkalemic periodic paralysis, 62/462
pseudohypoparathyroidism, 42/621, 63/560
renal insufficiency, 27/336
tetany, 40/330, 41/302, 63/560
Chylomicron
Bassen-Kornzweig syndrome, 42/512

brain microembolism, 9/588
hyperlipoproteinemia type V, 10/277
lipid metabolic disorder, 10/267
Chylomicron metabolism
disorder, 66/539
Chylomicronemia syndrome
disorder, 66/539
Cicer arietinum
nontoxic food, 65/2
Ciclosporin
acquired myoglobinuria, 62/576
adverse effect, 69/503
amnesia, 63/10, 537
aphasia, 63/10
ataxia, 63/10, 537
autonomic nervous system, 75/501
benign intracranial hypertension, 67/111
blood pressure, 75/476
confusional state, 63/537
cortical blindness, 63/537
dementia, 63/537
dermatomyositis, 62/386
Duchenne muscular dystrophy, 62/131
encephalopathy, 63/537
epilepsy, 63/10, 189, 537
headache, 63/179, 537
hypertension, 75/476
hypomagnesemia, 63/10
immunocompromised host, 56/469, 471
immunosuppressive therapy, 65/550
inclusion body myositis, 62/386
lymphoma, 63/533
migraine, 63/179
multiple sclerosis, 47/359
myalgia, 63/190
myasthenia gravis, 62/418
neurologic complication, 63/10
neuropathy, 69/461, 472
paraparesis, 63/10
polymyositis, 62/386, 71/142
primary CNS lymphoma, 63/533
renal transplantation, 63/534, 537
sarcoidosis, 71/493
tetraparesis, 63/10
tetraplegia, 63/537
tremor, 63/10, 189, 537
undetermined significant monoclonal
gammopathy, 63/403
vasculitis, 71/161
visual hallucination, 63/10
Ciclosporin intoxication
amnesic state, 63/537
ataxia, 63/537, 65/552
axonal neuropathy, 65/554
childhood, 65/554
confusional state, 63/537
consciousness, 63/537
cortical blindness, 63/537
CSF pleocytosis, 63/538
cyclophilin, 65/551
dementia, 63/537
EEG, 63/537, 65/552
EMG, 65/554
encephalopathy, 65/551
epilepsy, 63/537, 65/551
hallucination, 65/553
headache, 63/537, 65/552
leukoencephalopathy, 63/537
mechanism, 65/551
myalgia, 65/554
myelopathy, 65/552
myopathy, 65/552, 554
nephrotoxicity, 65/551
polyneuropathy, 65/552
reflex sympathetic dystrophy, 65/552
serum lipoprotein, 65/553
side effect, 65/551
status epilepticus, 65/552
tetraplegia, 63/537
tremor, 63/537, 65/551
urinary retention, 63/537
Ciguatera, *see* Ciguatoxin
Ciguatera intoxication, *see* Ciguatoxin intoxication
Ciguatoxin
autonomic nervous system, 75/510
bivalves intoxication, 37/64
chordata intoxication, 37/82-85
epidemiology, 65/159
fish intoxication, 37/82-84
Gambierdiscus, 65/160
Gambierdiscus toxicus, 64/15
geography, 65/159
history, 65/159
incidence, 65/159
lethal dose, 65/160
lyngbya majuscula intoxication, 37/97
maitotoxin, 65/159
marine toxin intoxication, 65/147, 159
neurotoxin, 37/4, 64, 82, 97, 64/15, 65/141, 147,
159
Plectonema tetebrans intoxication, 37/97
scaritoxin, 65/159
sodium channel, 65/160
structure, 65/148, 160
toxic dose, 65/141
toxic neuromuscular junction disorder, 64/15

treatment, 75/511
tridacna maxima intoxication, 37/64
Ciguatoxin intoxication
*see also* Marine toxin intoxication
action site, 64/15
arthralgia, 65/162
atropine, 37/84
autonomic nervous system, 75/510
bradycardia, 65/161
ciguatoxin 1, 65/161
coma, 65/163
convulsion, 65/163
dementia, 65/163
demyelination, 37/83, 65/163
diagnosis, 65/163
differential diagnosis, 65/163
diffuse pain, 65/162
EEG, 65/163
EMG, 65/163
epilepsy, 65/163
fatal case, 37/84
fatality, 65/162
Gambierdiscus toxicus, 65/160
hallucination, 65/163
headache, 65/162
history, 37/30
incidence, 65/160
maitotoxin, 65/160
management, 65/164
myalgia, 37/83, 65/162
nerve conduction, 65/161
nerve conduction velocity, 65/163
nerve induction block, 65/160
neuropathy, 65/163
origin, 37/82
orthostatic hypotension, 65/161
palytoxin, 65/161
paresthesia, 37/83, 65/162
pathomechanism, 64/15, 65/160
pharmacology, 37/83
polymyositis, 65/163
polyneuropathy, 65/163
prevalence, 65/160
prevention, 37/85
pruritis, 65/162
sequela, 65/162
symptom, 64/15, 65/161
temperature reversal, 65/162
treatment, 37/84, 65/164
Ciliary artery
arteritis, 48/319
temporal arteritis, 48/318
Ciliary body
anatomy, 74/420
tumor, *see* Ciliary body tumor
Ciliary body tumor
nerve tumor, 8/496
Ciliary ganglion
Charlin neuralgia, 48/483
congenital pain insensitivity, 51/565
Holmes-Adie syndrome, 43/72
oculomotor nerve injury, 24/61
orbit, 74/402
pupillomotor fiber, 2/107
sympathetic root, 5/335
tonic pupil, 74/410
Ciliary neuralgia, *see* Chronic migrainous neuralgia *and* Cluster headache
Ciliospinal reflex
brain death, 24/719
spinal cord injury, 25/386
Cimetidine
carbamate intoxication, 64/189
chronic paroxysmal hemicrania, 48/265
delirium, 46/539
eosinophilic fasciitis, 63/385
nonprimary inflammatory myopathy, 62/372
toxic myopathy, 62/597, 609
CINCA syndrome
*see also* Juvenile rheumatoid arthritis *and* Muckle-Wells syndrome
chronic meningitis, 56/645
Cinchonism
amblyopia, 65/452
central scotoma, 65/452
chromatopsia, 65/452
color perception, 65/485
color vision, 65/452
deafness, 65/452, 485
delirium, 65/485
dementia, 65/452
diplopia, 65/452
headache, 52/372, 65/452, 485
hearing loss, 65/452, 485
iatrogenic neurological disease, 65/451
nausea, 65/452
optic atrophy, 65/485
photophobia, 65/452
quinidine intoxication, 37/448, 65/451
quinine overload, 52/372
syncope, 65/485
systemic lupus erythematosus, 65/452
tinnitus, 65/452, 485
vertigo, 65/452
vomiting, 65/452
Cinematographic vision

hallucination classification, 45/56
Cingulate cortex
coma, 3/62-77
pharmacology, 74/159
temporal lobe epilepsy, 73/59
Cingulate gyrus
akinetic mutism, 2/766
anterior, *see* Anterior cingulate gyrus
aphasia, 2/767
apnea, 63/481
brain herniation, 16/108, 111
grasp reflex, 1/71
limbic encephalitis, 8/135
respiration, 63/481
tachycardia, 2/767
visceral function, 2/765
Cingulate gyrus lesion
memory disorder, 2/713, 3/286
urinary incontinence, 2/767
Cingulate herniation
neuroanatomy, 1/569
pathogenesis, 1/569
pathologic change, 1/569
radiologic finding, 1/570
sign, 1/570
symptom, 1/570
Cingulectomy
memory disorder, 2/713
memory loss, 2/713
monkey, 45/276
Cingulosynapsis, *see* Holoprosencephaly
Cingulotomy
anxiety, 1/110
pain, 1/110
Cipermethrin
neurotoxin, 64/212
pyrethroid, 64/212
pyrethroid intoxication, 64/212
Cipher agraphia
acalculia, 4/188, 191
Cipher alexia
acalculia, 4/191
Ciprofloxacin
human immunodeficiency virus infection, 71/385
iatrogenic neurological disease, 65/486
neurotoxicity, 71/368
neurotoxin, 65/486
quinolone intoxication, 65/486
side effect, 71/385
toxic encephalopathy, 65/486
Ciprofloxacin intoxication
benign intracranial hypertension, 65/486
epilepsy, 65/486
headache, 65/486
papilledema, 65/486
Circadian activity
CSF, 74/487
Circadian disorder
Alzheimer disease, 74/496
light therapy, 74/496
Circadian rhythm
Alzheimer disease, 74/495
autonomic nervous system, 74/14, 75/410, 437
biological rhythm, 74/501
brain infarction, 53/193
brain ischemia, 55/219
clock function, 74/468
cluster headache, 48/221
dentin, 74/492
intracranial hypertension, 55/219
inversion, 2/483
melatonin, 59/155, 74/501
migraine, 48/123
period, 1/510
primary pigmentary retinal degeneration, 13/196
progressive dysautonomia, 59/155
sleep, 74/536
sleep deprivation, 1/492
suprachiasmatic nucleus, 74/487, 492
vasopressin, 74/488
Circadian system
biological rhythm, 74/467
depression, 74/507
Circannual rhythm
suprachiasmatic nucleus, 74/488
Circle of Willis, *see* Willis circle
Circolectric bed
cervical vertebral column injury, 25/293
Circopharyngeal dysfunction
esophagus, 75/628
Circulating immune complex
atrial myxoma, 63/97
Guillain-Barré syndrome, 51/249
Circulation
blood, *see* Blood flow
CSF, *see* CSF flow
Circulatory arrest, *see* Cardiac arrest
Circumferential brain stem artery
anatomy, 11/28
Circumflex nerve, *see* Axillary nerve
Circumscribed arachnoidal cerebellar sarcoma, *see* Desmoplastic cerebellar medulloblastoma
Circumscribed lipomatosis
multiple, *see* Krabbe-Bartels disease
Circumventricular organ
blood-brain barrier, 56/80

pharmacology, 74/158
Cirrhosis
alcoholic, *see* Alcoholic cirrhosis
central pontine myelinolysis, 28/293
lead intoxication, 9/642
listeriosis, 52/91
liver, *see* Liver cirrhosis *and* Liver disease
primary biliary, *see* Primary biliary cirrhosis
Cirsoid aneurysm
angioma, 12/236
Bonnet-Dechaume-Blanc syndrome, 14/260, 265
brain stem, 14/70
orbit, 14/651
retina, 14/70, 650
Cisplatin
adverse effect, 69/488-490
antineoplastic agent, 65/529
autonomic nervous system, 75/38, 496
chemotherapy, 39/115, 67/284
drug induced polyneuropathy, 51/300
encephalopathy, 65/529
epilepsy, 65/529
hypomagnesemia, 63/562
neurologic toxicity, 39/115
neuropathy, 69/460, 464, 466
neurotoxin, 39/95, 112, 115, 119, 51/299, 64/358, 65/529, 67/354
optic neuropathy, 65/529
ototoxicity, 67/354
toxic neuropathy, 51/299
Cisplatin intoxication
*see also* Platinum intoxication
action, 65/536
ataxia, 65/536
blindness, 64/358
brain edema, 65/537
cranial nerve palsy, 64/359
cranial neuropathy, 64/359, 65/537
diuresis, 64/359
epilepsy, 64/358, 65/537
fatality, 65/537
hearing loss, 64/358
Lhermitte sign, 64/358
lumbar plexus neuropathy, 65/537
mechanism, 65/536
metal intoxication, 64/353
muscle spasm, 64/358
myelotoxicity, 64/358
nerve conduction velocity, 64/358
neurotoxicity, 64/358
optic nerve degeneration, 65/536
ototoxicity, 64/358
polyneuropathy, 64/358, 65/536
prevention, 64/359
retrobulbar neuritis, 64/359
risk factor, 65/536
sensorimotor polyneuropathy, 64/358
sensory polyneuropathy, 65/536
tinnitus, 64/358
treatment, 64/359, 65/536
vibration perception, 64/358
Cisplatin polyneuropathy
administration mode, 65/529
ataxia, 51/300
axonopathy, 51/300
blindness, 51/300
cytotoxic agent, 51/299
deafness, 51/300
EMG, 51/300
encephalopathy, 51/300
sensory neuropathy, 51/300
ultrastructure, 51/300
Cistern
ambient, *see* Ambient cistern
anatomy, 16/98
pneumography, 16/98
radiology, 16/98
terminology, 16/98
topography, 16/99
Cisterna interventricularis, *see* Cavum veli interpositi
Cisternography
isotope, *see* Isotope cisternography
metrizamide, *see* Metrizamide cisternography
radioisotope, *see* Radioisotope cisternography
RIHSA, *see* RIHSA cisternography
tuberculous meningitis, 33/224
Citelli syndrome
malignant tumor, 2/313
nasopharyngeal tumor, 2/313
nonrecurrent nonhereditary multiple cranial neuropathy, 51/571
Citrate synthase
tricarboxylic acid cycle, 27/14-16
Citric acid
aluminum intoxication, 64/275
CSF, 16/371
hepatic coma, 49/215
Citric acid cycle enzyme
white matter, 9/14
Citrobacter
gram-negative bacillary meningitis, 52/104, 117
Citrobacter diversus
brain abscess, 52/150
infantile enteric bacillary meningitis, 33/62
Citrovorum factor

glioblastoma, 18/65
Citrulline
autonomic nervous system, 74/137
Citrullinemia
amino acid, 29/96
ammonia level, 29/95, 42/538
argininosuccinate synthetase deficiency, 29/97, 42/538
ataxia, 21/577, 42/538
atypical case, 42/538
biochemical finding, 29/95-97
chronic variant, 29/95
clinical features, 29/94, 42/538
diagnosis, 29/97
enzyme defect, 29/97
genetics, 29/97
lethal neonatal variant, 29/94
metabolic ataxia, 21/574, 577
orotic aciduria, 29/96
third variant, 29/95
treatment, 29/98
urea production, 29/96
Citrullinuria
ataxia, 42/538
brain atrophy, 42/538
feeding disorder, 42/538
mental deficiency, 42/538
osteoporosis, 42/538
psychomotor retardation, 42/538
seizure, 42/538
sex ratio, 42/538
spastic paraplegia, 42/538
speech disorder, 42/538
tremor, 42/538
Cladosporiosis
acquired immune deficiency syndrome, 71/300
amphotericin B, 52/484
brain abscess, 35/563, 52/483
CNS, 35/563, 52/482
CSF, 35/563, 52/484
flucytosine, 52/484
meningitis, 35/563, 52/483
Cladosporium
fungal CNS disease, 35/563, 52/480, 482
Cladosporium trichoides
brain abscess, 52/151
chronic meningitis, 56/645
Cladribine, *see* 2-Chlorodeoxyadenosine
Clam intoxication, *see* Paralytic shellfish poisoning
Clarke column
barium intoxication, 64/354
familial amyotrophic lateral sclerosis, 21/30
Friedreich ataxia, 42/144
infantile spinal muscular atrophy, 59/62
spastic paraplegia, 42/169
spinopontine degeneration, 21/391, 42/192
Clasp knife phenomenon
medullar pyramid lesion, 6/137-139
Classic migraine
abdominal attack, 48/159
anticonvulsant, 48/159
attack, 48/124
bilious attack, 5/46
calcium entry blocker, 48/151
classification, 48/6
clinical features, 48/117
cluster headache, 48/227
common migraine, 5/16, 48, 48/117
confusion, 48/165
convulsion, 48/159
epidemiology, 48/18
epilepsy, 48/159
histamine, 5/17
ischemia, 48/118
motion sickness, 5/46
pathophysiology, 48/69
physiology, 5/48
psychodynamics, 5/251
regional brain blood flow, 48/68-76
scotoma, 48/156
symptom, 5/49, 48/124
vertigo, 48/157
visual phenomenon, 5/50
Classic primitive neuroectodermal tumor
brain neuroblastoma, 68/221
location, 68/212
neuroradiology, 68/214
pathology, 68/212
pinealoblastoma, 68/222
surgery, 68/215
symptom, 68/214
treatment, 68/222
Classic typhus, *see* Louse-borne typhus
Classification
acrocephalosyndactyly type V, 43/326
acute disseminated encephalomyelitis, 9/472
acute multiple sclerosis, 9/472
acute muscle weakness, 41/287
acute necrotizing hemorrhagic leukoencephalitis, 9/472
acute subdural hematoma, 24/275
acute viral encephalitis, 34/63
Adams-Richardson, *see* Adams-Richardson classification
adenovirus, *see* Adenovirus classification
adenyl cyclase, 74/149

adrenoleukodystrophy, 66/5
adult onset spongiform leukodystrophy, 66/5
affective psychosis, 46/444-446
agnosia, 4/15-18, 31, 38
agnostic alexia, 4/100, 116
agraphia, 4/143, 45/457
Alexander disease, 10/24, 36, 107, 114, 66/5
alexia, 4/115-117, 45/435
allergic granulomatous angiitis, 71/4
allotype, 47/363
Alzheimer disease, 46/204-206
Amoebae, 35/26
amusia, 2/600
amyloidosis, 51/414, 71/502
amyotonia congenita, 40/286
anemia, 38/17, 63/252
anencephaly, 30/174, 50/71
angioma, 14/51
ankylosing spondylitis, 71/4
anoxic ischemic leukoencephalopathy, 9/573
antineoplastic agent, 65/528
aphasia, 3/35, 399, 4/97-100, 45/290
apnea, 1/657
apolipoprotein, 66/536
apperceptive agnosia, 4/25
apraxia, 4/50-54, 63, 45/425
arterial hypertension, 54/204
arthritis, 71/15
associative agnosia, 4/15
astrocytoma, 16/2, 4, 7, 18/1, 3, 5, 12, 68/1, 7
astrocytoma grade II, 18/2
ataxia telangiectasia, 14/314
athetosis, 6/446
attention disorder, 3/187-189
auditory agnosia, 4/38
B fiber, 1/117-119
basilar artery migraine, 48/6
bass intoxication, 37/78
Becker muscular dystrophy, 40/281, 389
Behçet syndrome, 71/4
benign partial epilepsy, 73/5
benign spinal chondrogenic tumor, 19/294
benign spinal osteogenic tumor, 19/294
Bertrand, *see* Bertrand classification
Bielschowsky-Henneberg, *see* Bielschowsky-Henneberg classification
bilateral massive myoclonus, 15/107
Blackwood, *see* Blackwood classification
blood pressure, 54/205
body scheme disorder, 4/212
borderline leprosy, 51/216
boxfish intoxication, 37/78
bradypnea, 1/657
brain abscess, 68/310
brain aneurysm, 12/90
brain arteriovenous malformation, 12/229
brain concussion, 23/74
brain contusion, 23/428
brain cyst, 68/310
brain death, 24/700, 704
brain edema, 16/187
brain infarction, 11/295
brain injury, 23/227, 229
brain ischemia, 53/309
brain stem tumor, 68/367
brain tumor, 16/1, 4, 6, 8, 24, 31/37, 67/1, 68/137
Bravais, *see* Bravais classification
Bridge, *see* Bridge classification
Bright, *see* Bright classification
Brill-Symmers disease, 18/235
bromocriptine, 74/149
bulbar jugular syndrome, 18/444
Calmeil, *see* Calmeil classification
cardiac migraine, 48/6
cardiomyopathy, 63/131
carotid system syndrome, 53/309
central choroidal atrophy, 13/33
central core myopathy, 40/288
central neurocytoma, 68/150
centrencephalic epilepsy, 15/9, 42/682
cerebellar agenesis, 50/181
cerebellar astrocytoma, 18/1
cerebellar dysfunction, 2/417
cerebellar medulloblastoma, 16/4, 7
cerebrotendinous xanthomatosis, 10/343
cerebrovascular disease, 53/2
cervical headache, 48/6
cervical vertebral column injury, 26/289, 61/23, 25
chemotherapy, 65/528
4-{2-{[2-(3-chlorophenyl)-2-hydroxyethyl]-amino}propyl}phenoxyacetic acid, 74/149
chondrodystrophic myotonia, 40/284
chorioretinal degeneration, 13/24-43
choroid plexus papilloma, 18/105
chromosome, 30/92, 31/343, 376, 50/520
chronic migrainous neuralgia, 48/9
chronic paroxysmal hemicrania, 48/257-265
classic migraine, 48/6
clozapine, 74/149
cluster headache, 48/5, 218
cluster headache syndrome, 48/257
CNS malformation, 30/5-11
CNS spongy degeneration, 66/5
coagulopathy, 63/309
Cogan syndrome, 71/4
collagen vascular disease, 71/3

colloid cyst, 68/310
coma, 3/71, 75
common migraine, 48/6
complicated migraine, 48/6
concentric sclerosis, 9/472
congenital brain tumor, 31/35-37
congenital fiber type disproportion, 40/288
congenital muscular dystrophy, 40/280, 41/29
congenital myopathy, 40/286, 62/331
congenital Pelizaeus-Merzbacher disease, 10/10, 151
Conradi-Hünermann syndrome, 38/401-403
consciousness, 45/108
conversion headache, 48/5
convulsive epilepsy, 15/4
convulsive triad, 15/8
corbadrine, 74/149
cortical dementia, 46/204
Corynebacterium diphtheriae, 33/479
cough headache, 48/6
craniofacial injury, 23/371
craniolacunia, 30/270
craniometaphyseal dysplasia, 38/390
craniosynostosis, 30/221, 38/415, 42/20
Creutzfeldt-Jakob disease, 6/732-734, 741-745
cryoglobulin, 39/181
cryoglobulinemia, 63/403, 71/4
CSF pressure, 19/99
cyberomone, 74/374
cylindroma, 17/186
decompression illness, 63/419
Delasiauve, *see* Delasiauve classification
dementia, 27/481, 46/204-206
demyelinating disease, 10/22
dental headache, 48/5
Denver Conference on Chromosomes, 30/92, 31/343, 376, 50/520
depression, 46/444-446
dermatomyositis, 40/284, 71/4
dermoid cyst, 68/310
designer drug, 65/274, 364
desmoplastic infantile ganglioglioma, 67/4
developmental dysarthria, 4/408
developmental dyslexia, 46/118
developmental dysphasia, 46/139
developmental dyspraxia, 4/458
diabetic neuropathy, 70/134
diffuse glioblastosis, 10/6
diffuse myelitis, 10/22
diffuse sclerosis, 9/471-473, 10/1-37
diffuse simultaneous corticosubcortical epilepsy, 15/11
2-{[2-(2,6-dimethoxyphenoxy)ethyl]amino-methyl}-1,4-benzodioxan, 74/149
disability, *see* Disability classification
dissecting aorta aneurysm, 39/246, 63/45
distal myopathy, 62/198
dobutamine, 74/149
Dubowitz, *see* Dubowitz classification
Duchenne muscular dystrophy, 40/281
dura mater, 30/426
dysarthria, 3/396
dysembryoplastic neuroepithelial tumor, 67/5, 68/154
dysesthesia, 1/81
dysraphia, 32/163, 189
ECHO virus, 34/4
EEG, *see* EEG classification
Einarson, *see* Einarson classification
Einarson-Neel, *see* Einarson-Neel classification
encephalitis, 34/63
encephalopathy, 69/396
endocarditis, 63/111
enterovirus, 56/309
eosinophilia, 63/371
ependymoma, 16/4, 7, 22, 68/16
epidermoid cyst, 68/310
epilepsy, 15/1-8, 10-13, 23, 25, 90, 492, 42/682, 72/1-12, 18, 340
Epstein-Barr virus encephalitis, 34/188
Esquirol, *see* Esquirol classification
eupnea, 1/657
exertional headache, 48/6
facioscapulohumeral syndrome, 40/280, 416
familial amaurotic idiocy, 10/15, 225
familial hemiplegic migraine, 48/6
familial paroxysmal choreoathetosis, 49/349
familial visceral neuropathy, 75/642
febrile seizure, 73/309
fenoldopam, 74/149
fibrillary astrocytoma, 18/2
fibrous dysplasia, 14/169, 38/382
fingerprint body myopathy, 40/288
fish intoxication, 37/78
Fried-Emery, *see* Fried-Emery classification
frontal lobe epilepsy, 72/10
functional and result, 25/151
functional psychosis, 46/444-446
Galen, *see* Galen classification
gamma heavy chain disease, 69/290
gangliosidosis, 10/222
Garcin syndrome, 2/101
Gastaut, *see* Gastaut classification
Gaucher disease, 60/153
gemistocytic astrocytoma, 18/2
germ cell tumor, 68/230

giant cell arteritis, 71/4, 199
Gibbs, *see* Gibbs classification
glial dystrophy, 21/61, 65
glioblastoma multiforme, 18/3, 20/335
globoid cell leukodystrophy, 9/472, 10/18-21, 28, 37, 66/5
glycogen storage disease, 27/222
glycosaminoglycanosis, 60/176
GM2 gangliosidosis, 29/363-365
Gowers, *see* Gowers classification
Greenfield, *see* Greenfield classification
growth deficiency syndromes, 31/269-292
gunshot injury, 25/49
Hallervorden, *see* Hallervorden classification
Hallervorden-Spatz syndrome, 66/715
hallucination, *see* Hallucination classification
hallucinatory fish intoxication, 37/79
hallucinogenic agent, 65/42
haloperidol, 74/149
handedness, 4/257-259
Hausmanowa-Petrusewicz, *see* Hausmanowa-Petrusewicz classification
head injury, 23/1, 11, 24/684, 57/101-116
headache, 5/11-14, 48/1-10, 55
heat stroke, 23/670
heat syncope, 15/824
hemiatrophy, 59/475
hemihypertrophy, 59/475
hemiplegic migraine, 48/6
hemoglobinopathy, 63/256
hereditary amyloid polyneuropathy, 60/91
hereditary brachial neuropathy, 21/12
hereditary motor and sensory neuropathy type I, 60/185
hereditary motor and sensory neuropathy type II, 60/185
hereditary neuropathy, 21/5
hereditary progressive cochleovestibular atrophy, 60/762
hereditary recurrent polyneuropathy, 21/87
hereditary sensory and autonomic neuropathy type II, 42/349
hereditary tremor, 49/587
herpes zoster, 34/161
Hodgkin disease, 39/27, 69/263
holoprosencephaly, 30/438, 50/230
hot dog headache, 48/6
Huntington chorea, 46/205
hydatid cyst, 68/310
hydranencephaly, 50/345, 347
hydrocephalus, 30/536-538, 670, 46/323, 50/285
hyperactivity, 46/176-181
hypercortisolism, 70/207
hyperkalemic periodic paralysis, 40/537
hyperkinetic child syndrome, 3/197
hyperlipoproteinemia type I, 66/540
hyperpnea, 1/657
hypertelorism, 30/242, 246-251
hypertrophic interstitial neuropathy, 7/199
hypocortisolism, 70/211
hypoglycemia, 27/55, 70/175
hypopnea, 1/657
hypoventilation, 63/413
ICD, *see* ICD classification
ice cream headache, 48/6
ideomotor apraxia, 45/425
idiopathic convulsion, 15/5
idiopathic hypereosinophilic syndrome, 38/213-215
idiopathic hypoglycemia, 27/55
idiopathic Parkinson disease, 49/108
idiopathic partial epilepsy, 73/5
inborn error of metabolism, 46/17
inclusion body myositis, 40/284
indeterminate leprosy, 51/216
infantile Gaucher disease, 29/351, 42/437
infantile spasm, 15/107, 73/200
infantile spinal muscular atrophy, 59/51-53, 367
inflammatory diffuse sclerosis, 9/472, 10/6, 15, 24
inflammatory myopathy, 40/284
interferon, 47/348
intervertebral disc injury, 25/484
intestinal malabsorption, 7/618
intestinal pseudo-obstruction, 51/491, 70/320
intracerebral hematoma, 11/680
intraspinal neurenteric cyst, 32/424
iron deficiency anemia, 63/252
isoprenaline, 74/149
jackfish intoxication, 37/78
Jackson, *see* Jackson classification
Jasper-Kershman, *see* Jasper-Kershman classification
juvenile rheumatoid arthritis, 71/4
Kearns-Sayre-Daroff-Shy syndrome, 40/281
Kernohan, 18/105, 110
Koshevnikoff epilepsy, 73/118
Krabbe, *see* Krabbe classification
Krücke, *see* Krücke classification
Kurland-Hauser, *see* Kurland-Hauser classification
Langelüddeke, *see* Langelüddeke classification
late cerebellar atrophy, 60/583
late infantile metachromatic leukodystrophy, 10/17
lateral pontine syndrome, 12/22
lateral ventricle tumor, 17/596

Le Fort fracture, 23/371
Lennox, *see* Lennox classification
lentivirus, 56/460
lepromatous leprosy, 51/216
leprosy, 33/423-426, 51/216
leptospirosis, 33/396
leukocytoclastic vasculitis, 71/4
leukodystrophy, *see* Leukodystrophy classification
Lichtenstein, *see* Lichtenstein classification
limb apraxia, 45/425
limb girdle syndrome, 40/277, 279
lipid storage disorder, 29/346
lipidosis, 10/44
lipoprotein, 66/535
lipoprotein phenotype, 66/543
lobar holoprosencephaly, 30/438, 50/227
loss of consciousness, 23/7
low grade glioma, 68/34
lumbar vertebral canal stenosis, 20/613, 636
luysian atrophy, 21/531
lymphoma, 63/345
lymphomatoid granulomatosis, 71/4
macrencephaly, 30/648
malabsorption syndrome, 7/602
malignant astrocytoma, 18/3
malignant lymphoma, 39/27, 69/262
mania, 46/444-446
Marchiafava-Bignami disease, 9/472, 10/15
Marie-Foix, *see* Marie-Foix classification
McNaughton, *see* McNaughton classification
medulloblastoma, 67/3
medulloepithelioma, 16/12
megaloblastic anemia, 38/18
memory disorder, 3/269, 286-288
meningioma, 68/404
meningitis, *see* Meningitis classification
mental deficiency, 46/1
mesencephalic syndrome, 2/272-283
metabolic neuropathy, 7/552
metachromatic leukodystrophy, 66/5
metal intoxication, 37/79
5-methoxy-1-methyl-2-propylaminotetralin, 74/149
2-{2-[4-(2-methoxyphenyl)-1-piperazinyl]ethyl}-4,4-dimethyl-1,3-(2h,4h)-isoquinolinedione, 74/149
2-(1-methyl-2-isoindolinylmethyl)-2-imidazoline, 74/149
metoprolol, 74/149
microcephaly, 30/508
migraine, 5/45
migraine equivalent, 48/6
migraine sine hemicrania, 48/6, 162
migrainous transient global amnesia, 48/6
mitochondrial disease, 66/394
mitochondrial myopathy, 40/288
mixed connective tissue disease, 71/4
mixed neuronoglial tumor, 67/5
monoclonal gammopathy, 69/290
moray intoxication, 37/78
motor unit hyperactivity, 40/285
mucolipidosis, 66/378
mucolipidosis type I, 66/357
Müller, *see* Müller classification
multicore myopathy, 40/288
multi-infarct dementia, 46/205
multiple myeloma, 39/135, 69/290
multiple neuronal system degeneration, 75/164
multiple sclerosis, 9/327, 47/49-52
multiple white matter lesion, 54/96
muscle contraction headache, 48/6
myasthenia gravis, 41/99, 62/400
myelinoclastic diffuse sclerosis, 10/2, 6
myoclonic epilepsy syndrome, 73/262
myoclonus, 38/576, 49/611
myoglobinuria, 41/277, 62/558
myopathy, 40/275-289
myotonia, 40/538, 41/165, 62/263
myotonia congenita, 40/283, 557
myotonic dystrophy, 40/282, 537, 62/216
myotonic myopathy, 40/282
myotonic syndrome, 40/537, 62/263
myotubular myopathy, 40/288
nemaline myopathy, 40/287
neonatal seizure, 72/8
nerve, 7/199
nerve injury, 7/244, 51/75
nerve tumor, 8/415
Neubürger, *see* Neubürger classification
neuraminidase activity, 66/357
neuroastrocytoma, 67/5, 7
neuroblastoma, 67/4
neurocutaneous melanosis, 14/415
neurofibromatosis type I, 14/11, 136
neuroleptic agent, 65/274
neuroleptic parkinsonism, 6/234
neurolipidosis, 10/327, 353
neuromuscular junction disease, 62/392
neuromyelitis optica, 10/22
neuron disease, 7/201
neuronal ceroid lipofuscinosis, 10/224
neuronal dystrophy, 21/61
neuronal tumor, 68/137
neuropathology, 7/201
neuroretinal degeneration, 13/24-43
neurosyphilis, 33/353

neurotrophic osteoarthropathy, 38/432
neurotropic virus, 56/1
Niemann-Pick disease, 10/557, 60/147
Niemann-Pick disease type C, 42/470
Noack syndrome, 43/326
non-Hodgkin lymphoma, 63/345, 69/262
nonsteroid anti-inflammatory agent, 65/415
nonvasoconstrictor neuron, 74/30
Norman microcephalic familial leukodystrophy, 10/114
nystagmus, 1/585-587
occipital lobe epilepsy, 72/11
ocular myositis, 22/179
oligodendroglioma, 16/4, 7, 18/5, 68/12
oligosaccharidosis, 66/378
olivopontocerebellar atrophy, 21/20
olivopontocerebellar atrophy (Dejerine-Thomas), 21/415-417
ophthalmoplegic migraine, 48/6, 141
opioid analgesic agent, 69/40
optic atrophy, 13/45
orbital fracture, 24/92
orbital hypertelorism, 30/242, 246-251
orthochromatic leukodystrophy, 10/37
orthostatic hypotension, 75/714
osteosclerotic myeloma, 69/290
overlap syndrome, 71/4
oxymetazoline, 74/149
pahutoxic intoxication, 37/78
pallidal atrophy, 21/531
pallidal degeneration, 6/632
papovavirus, 34/7
paramyotonia congenita, 40/537
paraneoplastic polyneuropathy, 51/465
paraneoplastic syndrome, 69/320, 330, 350
paraproteinemia, 39/135
parasympathetic neuron, 74/37
pargos intoxication, 37/78
parietal cephalocele, 50/108
parietal lobe epilepsy, 72/11
parietal lobe tumor, 17/297
Peiffer, *see* Peiffer classification
Pelizaeus-Merzbacher disease, 10/189, 66/5, 572
pellagra, 28/67
perception, *see* Perception classification
periodic paralysis, 28/581, 40/285
peripheral nerve pathology, 7/199-201
pernicious anemia, 38/18
petit mal epilepsy, 15/3, 6, 22
phakomatosis, 14/11-14
phantom sensation, 26/490
phentolamine, 74/149
phenylephrine, 74/149
pheromone, 74/374
phopspholipase C-β, 74/149
pineal tumor, 68/230
pituitary adenoma, 17/381, 383
plasma cell dyscrasia, 69/290
pleomorphic xanthoastrocytoma, 67/5
pneumocephalus, 24/202
poliovirus, 34/4, 95
polyarteritis nodosa, 71/4
polymyalgia rheumatica, 71/4
polymyositis, 40/284, 41/55, 71/4
pontine syndrome, 2/264
porphyria, 27/430
Poser, *see* Poser classification
Poser-Van Bogaert, *see* Poser-Van Bogaert classification
posttraumatic encephalopathy, 23/9
posttraumatic headache, 48/5
posttraumatic syringomyelia, 61/375
prazosin, 74/149
premature closure syndrome, 38/415
presenile dementia, 46/204
pressure sore, 26/468, 61/354
Prichard, *see* Prichard classification
primary adrenal insufficiency, 70/211
primary cerebral reticulosarcoma, 18/321
primary meningeal sarcomatosis, 18/238
primary systemic amyloidosis, 69/290
primitive neuroectodermal tumor, 67/3
progressive dysautonomia, 59/133
progressive external ophthalmoplegia, 22/179, 40/281, 60/47, 62/289, 290
propranolol, 74/149
psychogenic headache, 48/6
puffer fish intoxication, 37/78
raclopride, 74/149
Rathke cleft cyst, 68/310
ray intoxication, 37/78
recurrent nonhereditary polyneuropathy, 21/88
rehabilitation, 24/683
Reiter disease, 71/4
relapsing polychondritis, 71/4
renal tubular acidosis, 29/181
RES tumor, 18/235
retinal migraine, 48/6
Reynolds, *see* Reynolds classification
rheumatoid arthritis, 71/4, 15
rickettsial infection, 34/642
Ridley-Jopling, *see* Ridley-Jopling classification
Roussy-Oberling, 18/110
rufous oculocutaneous albinism, 43/6
sarcotubular myopathy, 40/288
SCH 33390, 74/149

schizophrenia, 46/444
scleroderma, 71/4
scombroid intoxication, 37/78
Seckel dwarf, 30/508
secondary adrenal insufficiency, 70/211
secondary agraphia, 4/144
secondary subcortical epilepsy, 15/11
segmental dystonia, 49/521
seizure, 72/1-12, 28
semantic alexia, 4/116
semilobar holoprosencephaly, 30/438
semiochemical, 74/374
senile dementia, 46/204
shark intoxication, 37/78
Sherrington, *see* Sherrington classification
sickle cell anemia, 63/258
Sjögren syndrome, 71/4
skull base tumor, 17/137, 68/465, 467
sleep disorder, 45/129
spatial orientation disorder, 3/212
speech disorder in children, 4/408, 428
sphenopalatine neuralgia, 48/6
spinal cord cyst, 25/90
spinal cord injury, 25/46, 145, 147, 151
spinal cord missile injury, 25/49
spinal cord vascular disease, 12/497
spinal cord vascular malformation, 12/548, 68/569
spinal epidural empyema, 33/187
spinal hemangioblastoma, 19/362
spinal injury, 25/127, 145, 148, 26/289
spinal lesion, *see* Spinal lesion classification
spinal muscular atrophy, 59/367
spinal vascular tumor, 68/569
spinocerebellar degeneration, *see* Spinocerebellar degeneration classification
split notochord syndrome, 20/62
sport disability, 26/537
status epilepticus, 73/318
Streptococcus, 52/78
striatal degeneration, 6/632
striatonigral degeneration, 22/172
Suarez, *see* Suarez classification
subcortical dementia, 46/204
subependymal astrocytoma, 18/3, 110
subependymal giant cell astrocytoma, 67/5
sudanophilic leukodystrophy, 66/5
sunfish intoxication, 37/78
Symonds, *see* Symonds classification
sympathetic neuron, 74/24
symptomatic dystonia, 49/71
synergomone, 74/374
synkinesis, 1/407, 410
syphilis, 33/339
syringomyelia, 50/443
systemic lupus erythematosus, 71/4
systemic lymphoma, 69/262
tachypnea, 1/657
Takayasu disease, 55/335, 71/4
tamsulosin, 74/149
tardive dyskinesia, 49/186
tegmental pontine syndrome, 12/22
temporal lobe epilepsy, 15/89, 72/10
temporal lobe triad, 15/8
tension vascular headache, 48/6
terbutaline, 74/149
tetanus, 33/523
thalamic nucleus, 21/587
thalassemia, 38/44, 63/257
third ventricle tumor, 17/441
thoracic intervertebral disc prolapse, 20/567
thoracic outlet syndrome, 51/124
thoracolumbar spine injury, 25/440
thoracolumbar vertebral column injury, 61/34, 91
thyroid associated ophthalmopathy, 62/530
tic, 6/785
Tissot, *see* Tissot classification
tonic clonic epilepsy, 15/107
tonic seizure, 15/107
transitional sclerosis, 10/36
traumatic aneurysm, 24/381
traumatic carotid cavernous sinus fistula, 57/345
traumatic epilepsy, 15/281
traumatic intracranial hematoma, 57/251
tremor, 6/811
trilaminar myopathy, 40/288
trophic disorder, 1/481
tuberculoid leprosy, 51/216
tuberculous meningitis, 71/284
tuberous sclerosis, 14/11
tumeur royale, 8/445, 14/38
UICC, *see* UICC classification
unilateral epileptic seizure, 15/107
US Veterans Administration, *see* US Veterans Administration Classification
Van Bogaert, 10/13, 15
Van Bogaert-Bertrand, *see* Van Bogaert-Bertrand classification
varicella zoster, 34/161
vascular dementia, 46/353
vasculitis, 71/191
vasoconstrictor neuron, 74/24
ventral pontine syndrome, 12/22
vertebral column, 32/133-152
vertebral column tumor, 20/20
vesicourethral neuropathy, 61/291
virology, 34/4-9

visual agnosia, 4/15-18, 20
visual disorientation, 2/612
Vogt-Koyanagi-Harada syndrome, 56/612
Von Romberg, *see* Von Romberg classification
Von Willebrand disease, 63/311
Waldenström macroglobulinemia, 69/290
Waldenström macroglobulinemia polyneuropathy, 18/235
Wegener granulomatosis, 71/4
WFN, *see* WFN classification
WHO, *see* World Health Organization classification
Wicke, *see* Wicke classification
Wilson disease, 6/281, 27/199
Wohlfart-Kugelberg-Welander disease, 59/81, 367
Wolman disease, 10/546
World Federation of Neurology, *see* WFN classification
Classification history
apperceptive agnosia, 4/15
limb girdle syndrome, 40/433, 62/179
reticuloblastoma, 18/234
Claude Bernard-Horner syndrome, *see* Horner syndrome
Claude syndrome
brain lacunar infarction, 54/245
symptom, 2/277
Claudication
cauda equina intermittent, *see* Cauda equina intermittent claudication
forearm intermittent, *see* Forearm intermittent claudication
manibular intermittent, *see* Mandibular intermittent claudication
muscle intermittant, *see* Muscle intermittent claudication
nerve intermittent, *see* Nerve intermittent claudication
neurogenic intermittent, *see* Cauda equina intermittent claudication *and* Spinal cord intermittent claudication
spinal cord intermittent, *see* Spinal cord intermittent claudication
Claustrum
anatomy, 6/19
brain hemorrhage, 11/587
limbic encephalitis, 8/135
Claustrum absence
lissencephaly, 42/40
Clavicle hypoplasia
cleidocranial dysplasia, 43/348
Clavicle injury
nerve lesion, 70/31
Claw foot, *see* Pes cavus
Claw hand
flexion contracture, 20/827
Friedreich ataxia, 21/341
glycosaminoglycanosis type II, 42/454
hereditary motor and sensory neuropathy type I, 21/279
hereditary motor and sensory neuropathy type II, 21/279
Hurler-Hunter disease, 10/436
Jeanné sign, 2/36
leprosy, 51/218
leprous neuritis, 51/220
Richards-Rundle syndrome, 43/264
topographical diagnosis, 2/36
ulnar nerve, 2/36, 146
Clear fluid meningitis, *see* Meningitis serosa
Cleft
median facial, *see* Median facial cleft syndrome
palate, *see* Palatoschisis
Rathke, *see* Rathke cleft
Schmidt-Lantermann, *see* Schmidt-Lantermann incisure
Cleft brain, *see* Schizencephaly
Cleft cyst
Rathke, *see* Rathke cleft cyst
Cleft face syndrome
median, *see* Median facial cleft syndrome
Cleft jaw and palate, *see* Palatognathoschisis
Cleft lip, *see* Cheiloschisis
Cleft lip and palate, *see* Cheilopalatoschisis
Cleft palate, *see* Palatoschisis
Cleidocranial dysostosis
hypertelorism, 30/248
neurofibromatosis type I, 14/492
scoliosis, 50/418
skeletal malformation, 30/248, 32/41
Cleidocranial dysplasia, 43/348-350
basilar impression, 50/399
brachycephaly, 43/348
clavicle hypoplasia, 43/348
dental abnormality, 43/348
growth retardation, 43/348
swayback nose, 43/348
vertebral abnormality, 43/349
Cleidocranial hypoplasia
foramina parietalia permagna, 50/142
Cleland-Chiari malformation, *see* Arnold-Chiari malformation type II
Climbing eye movement
gaze paralysis, 2/334
test, 2/334, 338
vertical gaze palsy, 2/338

Climbing fiber
cerebellar stimulation, 2/409, 414, 416
excitatory action, 2/413
Clinical cervical stability
*see also* Clinical vertebral column stability
checklist, 61/12
radiography, 61/12
Clinical lumbar stability
*see also* Clinical vertebral column stability
disc, 61/16
ligament, 61/16
management, 61/18
radiography, 61/17
Clinical neurology, 1/1-38
coelenterata intoxication, 37/38
rehabilitation, 46/609-631
Clinical thoracic stability
*see also* Clinical vertebral column stability
checklist, 61/13
Clinical vertebral column stability
*see also* Clinical cervical stability, Clinical lumbar stability *and* Clinical thoracic stability
scoring system, 61/11
Clinodactyly
*see also* Digital abnormality
congenital torticollis, 43/146
DOOR syndrome, 42/385
Fukuyama syndrome, 43/92
keloid, 43/146
Miller-Dieker syndrome, 50/592
9p partial trisomy, 43/516
2q partial trisomy, 43/493
r(13) syndrome, 50/585
Rubinstein-Taybi syndrome, 31/283, 43/234
Russell-Silver syndrome, 43/476
Clinoparietal line
radioanatomy, 11/86
topography, 11/86
Clioquinol
distal axonopathy, 64/13
drug induced polyneuropathy, 51/296
neurotoxin, 37/115, 45/207, 51/68, 296, 56/526, 64/8, 10, 13
optic neuropathy, 64/8
polyneuropathy, *see* Subacute myelo-optic neuropathy
subacute myelo-optic neuropathy, 64/8
toxic neuropathy, 64/8, 10, 13
toxic polyneuropathy, 51/68
transient global amnesia, 45/207
tropical myeloneuropathy, 56/526
Clioquinol intoxication
action site, 64/10
myelopathy, 64/8
neuropathy, 64/8
subacute myelo-optic neuropathy, 37/115-194, 47/405
symptom, 64/13
visual impairment, 64/8
Clioquinol polyneuropathy, *see* Subacute myelo-optic neuropathy
Clitocybe
muscarine, 64/14
mushroom intoxication, 36/535
mushroom toxin, 65/39
Clitorimegaly
holoprosencephaly, 50/237
Clivus ridge syndrome
*see also* Brain herniation *and* Transtentorial herniation
acute, 2/298
herniation, 2/297
migraine, 2/298
progressive, 2/297
Clobazam
antiepileptic agent, 65/497
Clock function
*see also* Biological rhythm
Alzheimer disease, 74/468
blindness, 74/468
circadian rhythm, 74/468
depression, 74/468
disorder, 74/468
jet leg, 74/468
migraine, 74/468
Parkinson disease, 74/468
Shy-Drager syndrome, 74/468
sleep-wake syndrome, 74/468
subarachnoid hemorrhage, 74/468
Clofazimine
leprous neuritis, 51/232
Clofenvinfos
organophosphate intoxication, 64/155
Clofibrate
acquired myoglobinuria, 62/576
brain infarction, 53/425
fasciculation, 41/300
multiple sclerosis, 9/400
myopathy, 40/155, 509
myotonic dystrophy, 40/548
toxic myopathy, 62/597, 601
very low density lipoprotein, 41/272
Clofibrate intoxication
myotonic syndrome, 62/264
Clofibride
toxic myopathy, 62/597

Clomethiazole
antiepileptic agent, 65/497
pyrethroid intoxication, 64/216
status epilepticus, 73/330
Clomethiazole intoxication
arterial hypotension, 65/330
ethanol withdrawel, 65/330
Clomifene
anencephaly, 50/80
Clomipramine
myotonia, 62/276
tricyclic antidepressant, 65/312
Clonazepam
antiepileptic agent, 65/497
dyssynergia cerebellaris myoclonica, 60/603
epilepsy with continuous spike wave during slow sleep, 73/276
hereditary periodic ataxia, 60/442
neuroleptic akathisia, 65/290
nocturnal myoclonus, 45/140
restless legs syndrome, 51/356, 548
side effect, 15/708
spinal myoclonus, 41/308
Clone
antibody molecule, 9/503
Clonic seizure
EEG, 15/122
generalized epilepsy, 15/121
generalized tonic, *see* Generalized tonic clonic seizure
hereditary long QT syndrome, 42/718
pyrethroid intoxication, 64/214
Sturge-Weber syndrome, 31/22
Clonidine
action mechanism, 37/442
antihypertensive agent, 37/441-443, 63/87
antihypertensive agent intoxication, 37/443
behavior disorder, 46/601
chemical structure, 37/442
cough headache, 48/371
delirium, 46/538
glucagon, 74/167
hypertensive encephalopathy, 54/217
migraine, 37/443, 48/151, 180, 186
neuroleptic akathisia, 65/290
pharmacology, 74/165, 167
side effect, 37/443
Clonidine intoxication
convulsion, 37/444
CSF, 37/443
dementia, 37/444
depression, 37/443
drowsiness, 37/443
emotion, 37/443
levodopa adverse reaction, 37/444
nightmare, 37/443
sedation, 37/443
sexual impotence, 37/443
Clonopin, *see* Clonazepam
Clonus
brain edema, 63/417
clinical features, 1/250
eyelid, 2/319
facial, *see* Facial clonus
Cloquet canal
optic disc, 13/81
persistence, 13/81
Clorazepate
anxiolytic agent, 37/355
benzodiazepine intoxication, 65/334
half-life, 37/357
structural formula, 37/556
Closing-in phenomenon
constructional apraxia, 4/53, 45/502
Mayer-Gross, 2/692
Clostridium
acquired myoglobinuria, 62/577
brain abscess, 52/149
mackerel intoxication, 37/87
scombroid intoxication, 37/85
Clostridium botulinum, *see* Botulinum toxin
Clostridium difficile infection
bowel, 75/653
Clostridium perfringens, *see* Clostridium welchii
Clostridium tetani
*see also* Tetanus
bacteriology, 33/491-494
CNS, 41/303
microbiology, 65/216
tetanolysin, 52/229
tetanospasmin, 52/229, 64/10, 16
toxin, 65/215
Clostridium welchii
demyelination, 9/39
gangrene, 41/51
meningitis, 33/103
toxic myopathy, 62/612
toxin, 40/151
tropical polymyositis, 41/51
Closure failure
neural tube, *see* Neural tube closure failure
Closure syndrome
premature, *see* Premature closure syndrome
Clouding
corneal, *see* Corneal clouding
primary pontine hemorrhage, 12/39

Cloverleaf skull
craniosynostosis, 46/87
mental deficiency, 46/87
Cloverleaf skull syndrome
acrocephalosyndactyly type II, 43/359, 50/123
acrocephalosyndactyly type V, 43/322
aqueduct stenosis, 50/309
brachycephaly, 30/245
craniofacial dysostosis (Crouzon), 31/233, 43/359
craniosynostosis, 30/154, 38/422, 50/120
dolichocephaly, 30/245
epidemiology, 30/154
hydrocephalus, 38/422
hypertelorism, 30/245
orbital hypertelorism, 30/245
synostosis, 30/245
thanatophoric dwarfism, 31/272, 43/388
Clozapine
chemical formula, 65/276
classification, 74/149
neuroleptic agent, 65/275
specific action, 65/280
Club foot, *see* Talipes
Clumsiness
anatomic correlate, 46/162
anticonvulsant, 46/164
assessment, 46/161
brain maldevelopment, 46/162
cerebral palsy, 46/161-163
child, 4/362, 455, 46/159-166
choreiform syndrome, 46/160
clinical features, 46/160
computer assisted tomography, 46/163
definition, 46/159
developmental agnosia, 46/159
developmental apraxia, 46/159
developmental dyscalculia, 46/160
developmental dyslexia, 46/117
education, 46/164
EEG, 46/164
Gerstmann syndrome, 46/160
hereditary sensory and autonomic neuropathy type II, 60/11
incidence, 46/161
intelligence test, 46/163
Minamata disease, 36/105
minimal brain dysfunction, 46/160
motor performance, 46/163
neuropsychologic study, 46/163
physiotherapy, 46/165
prognosis, 46/165
Roussy-Lévy syndrome, 42/108
speech therapy, 46/164
treatment, 46/164
Clumsy hand dysarthria syndrome
brain infarction, 53/318
brain lacunar infarction, 46/355, 54/57, 249
computer assisted tomography, 54/57
pontine infarction, 53/394
pontine lacuna, 54/249
putaminal hemorrhage, 54/305
symptom, 54/249
white matter injury, 53/318
Clupeotoxin intoxication
fish, 37/79
Herring, 37/86
Cluster headache
age at onset, 48/218, 249
alcohol, 48/122, 249
anhidrosis, 48/336
anisocoria, 48/223
anorexia, 48/222
associated phenomena, 48/221
attack distribution, 48/259
attack provocation, 48/228
autonomic nervous system, 1/513, 75/281
behavior, 48/223
beta adrenergic receptor blocking agent, 48/240
biological rhythm, 74/507
blood pressure, 75/289
blood vessel, 48/217-242, 75/291
brain blood flow, 48/238, 54/149
cardiovascular system, 75/288
carotid body, 48/242
carotid dissecting aneurysm, 48/225
carotid sheath, 48/338
carotidynia, 48/226
carpal spasm, 48/223
cervicogenic, 48/224
Charlin neuralgia, 48/486
child, 5/241
chlorpromazine, 48/230
choline, 48/239
chronic, *see* Chronic migrainous neuralgia
chronic muscle contraction headache, 48/344
chronic paroxysmal hemicrania, 48/224, 257, 260
circadian rhythm, 48/221
classic migraine, 48/227
classification, 48/5, 218
clinical features, 48/247-254
cluster migraine syndrome, 48/268
cluster variant syndrome, 48/270
cluster vertigo, 48/226
computer assisted tomography, 48/237
conjunctival injection, 48/222, 241
corneal identation pulse, 48/234, 75/281

corneal sensitivity, 48/235
corneal temperature, 48/234
cortisone, 48/230
cranial neuralgia, 5/114, 48/267
Cushing reflex, 75/289
cycle, 48/219, 240
description, 48/225
diagnosis, 48/224
diarrhea, 48/222
drug treatment, 48/231
duration, 48/219, 221, 249
dysautonomia, 22/255
ECG, 75/289
edema, 48/222
EEG, 48/238
endocrine system, 48/437
β-endorphin, 48/233
episodic, 48/218, 229
epistaxis, 48/222
ergotamine, 5/117, 48/228, 231
ergotamine tartrate, 48/229
extracranial vessel, 48/223
facial flushing, 48/222
familial, 5/265
flushing, 48/9
forehead sweating, 48/236
forme fruste, 48/221
gastrin, 48/240
gastrointestinal ulcer, 48/224
glaucoma, 48/226
glossopharyngeal neuralgia, 48/468
glyceryl trinitrate, 37/454
glyceryl trinitrate test, 48/228
growth hormone, 74/508
heart rate change, 48/239
herpes simplex, 48/220
histamine, 5/4, 116, 48/232
histamine desensitization, 48/230
histamine H1 receptor antagonist, 48/230
histamine H2 receptor antagonist, 48/230
histamine skin test, 48/228, 233
history, 48/217
HLA histocompatibility antigen system, 48/233
Horner syndrome, 5/113-115, 48/9, 222
hydrocortisone, 48/240
hypercapnia, 48/242
hyperhidrosis, 48/9
hypermasculinism, 48/219
hyperventilation, 75/289
hypothalamus, 48/242
hypoxemia, 48/242
ice pick headache, 48/260
indometacin, 48/230
intermedius neuralgia, 5/114
intraocular pressure, 48/224, 234, 241
jabs and jolts syndrome, 48/260
kinin, 48/233
lacrimation, 48/9, 221, 237, 270
leonine-mouse syndrome, 48/219
β-lipotropin, 48/233
lisuride, 48/230
lithium, 48/229, 239
lower syndrome, 48/223
male, 48/241, 74/508
mandible, 48/223
mast cell, 48/232
melatonin, 48/233
menopause, 48/219
menstruation, 48/219
methysergide, 5/19, 117, 48/229
migraine, 5/265, 48/78, 227, 240, 269
muscle contraction headache, 5/115
nasal congestion, 48/9, 270
nasal secretion, 48/237
nausea, 48/222
nocturnal attack, 48/221, 224
nomenclature, 48/9
ocular pain, 48/241
oculosympathetic paralysis, 48/335
orbital pain, 48/241, 270
oxygen saturation, 48/239
oxygen therapy, 48/231, 242
pain, 48/225, 241
pain distribution, 48/223
parainfluenza virus, 74/335
paratrigeminal syndrome, 5/115, 48/226
paresthesia, 48/223
partial Horner syndrome, 48/222
PEG, 48/237
perivascular nerve fiber, 75/290
petrosal nerve, 5/116
petrosal neuralgia, 5/114
photophobia, 48/222
physiognomy, 48/219
physiologic factor, 48/220
pilocarpine, 48/236
pizotifen, 48/230
platelet, 48/234
platelet MAO, 48/94
posttraumatic dysautonomic cephalalgia, 48/227
pregnancy, 48/219
prevalence, 48/218
prostacyclin, 48/97
prostaglandin, 48/97, 233
psychogenic headache, 48/227
psychological test, 48/238

pupillometry, 48/235
REM sleep, 48/239
respiratory alteration, 75/292
retrobulbar neuritis, 48/270
rhinorrhea, 48/9, 222
salivation, 48/222, 237
seasonal prevalence, 48/221
sex distribution, 48/219, 249
single attack, 48/220, 231
skin, 48/219
sleep, 48/239
sleep apnea syndrome, 48/239
sphenopalatine neuralgia, 5/330, 48/217
subarachnoid hemorrhage, 55/20
substance P, 75/291
SUNCT syndrome, 75/299
suprachiasmatic nucleus, 74/507
supraorbital area, 48/251
supraorbital neuralgia, 5/115
sural nerve stimulation, 48/235
Symonds headache, 5/114
symptomatic, 48/227
syndrome de vasodilatation hemicéphalique d'origine sympathique, 48/217
synonym, 5/111
temporal artery, 48/78
testosterone, 48/239, 74/508
thermography, 48/235
Tolosa-Hunt syndrome, 48/226
treatment, 48/228
trigeminal neuralgia, 48/224
trigeminal pathway, 48/268
typical, 48/227
unilateral, 48/221, 240
upper syndrome, 48/223
vasoactive substance, 48/241
vasodilatation, 5/19
vertigo, 48/270
viral infection, 48/220
visual impairment, 48/222
volcano pattern, 48/240
vomiting, 48/222
Von Möllendorff Hemikranie, 48/217
Cluster headache syndrome
chronic paroxysmal hemicrania, 48/257
classification, 48/257
review, 48/218
Cluster migraine syndrome
cluster headache, 48/268
differential diagnosis, 48/226
migraine, 48/268
Cluster tic syndrome
clinical features, 48/267
surgery, 48/268
trigeminal nerve, 48/268
Cluster variant syndrome, 48/267-271
antidepressant, 48/270
cluster headache, 48/270
indometacin, 48/270
indometacin responsive headache, 48/227
lacrimation, 48/270
nasal congestion, 48/270
type, 48/227
vascular headache, 48/270
Cluster vertigo
cluster headache, 48/226
Cluttering
paraphrasia, 43/220
prevalence, 43/221
sex ratio, 43/221
stuttering, 43/221, 46/171
tachyphrasia, 43/220
Clutton joint
congenital syphilis, 33/370, 387
CMRO2, *see* Brain oxygen metabolic rate
Cnidarians intoxication, *see* Coelenterata intoxication
Cnidoglanis intoxication
family, 37/78
venomous spine, 37/78
CNS
acetylcholine, 29/435-451
achondroplasia, 31/270-272
acrocephalosyndactyly type I, 31/232
actinomycosis, 35/385
Aicardi syndrome, 31/226
Albright syndrome, 31/312
anencephaly, 50/83
antidepressant intoxication, 37/317
antigen, 9/137, 514
associative areas, 3/13
asymmetric, 30/28
basal ganglion, 30/31
basilar impression, 32/68-77
Beckwith-Wiedemann syndrome, 31/329
bicarbonate, 28/436
bilateral symmetry, 50/24
biogenic amine, 29/459-478
Blastomyces brasiliensis, 52/385
Börjeson-Forssman-Lehmann syndrome, 31/311
brain cortex, 50/245
brain growth, 30/28
Cajal-Retzius cell, 50/245
calcium, 28/433
candidiasis, 35/420-423
carbohydrate, 27/1-23

Carnegie stage, 50/2
caudal aplasia, 32/351
cavum veli interpositi, 30/328
cell death, 30/74-77, 50/7, 40
cell maturation, 30/70-72
cell migration, 50/29
cerebellar cortex, 30/30
cerebellar external granular layer, 50/35
cerebrohepatorenal syndrome, 31/304
chloride, 28/436
cladosporiosis, 35/563, 52/482
Clostridium tetani, 41/303
cobalt, 28/439
Cockayne-Neill-Dingwall disease, 31/238
coelenterata intoxication, 37/38
Coffin-Lowry syndrome, 31/245
Coffin-Siris syndrome, 31/240
Cogan syndrome type II, 51/454
commissure, 30/30
congenital muscular dystrophy, 41/35-43
Conradi-Hünermann syndrome, 31/275-277
copper, 27/393-399, 28/437
coronavirus infection, 56/440
corpus callosum agenesis, 30/287, 294
cranial nerve, 50/4
craniofacial dysostosis (Crouzon), 31/234
cri du chat syndrome, 31/566, 568
cryptic hemangioma, 12/104
cytoarchitectonic development, 50/245, 267
damsel fish intoxication, 37/86
De Lange syndrome, 31/281
dermal sinus and dermoid, 32/458
developmental, 30/15-40, 292
developmental anatomy, 30/30-32, 50/1
developmental disorder, *see* CNS developmental disorder
diabetes mellitus, 70/193
diastematomyelia, 32/244
differentiation error, 30/41-83, 50/19
Down syndrome, 31/367, 400, 407-412
drug induced parkinsonism, 6/234
drug induced symptom, 69/481
Duchenne muscular dystrophy, 62/121
echinococcosis, 52/523
Edwards syndrome, 31/521
electrical activity, 30/26
embryo, 30/18-20, 50/35
embryo EEG, 50/9
embryo movement, 50/4
embryo respiratory movement, 50/9
embryonal cell carcinoma, 68/235
endemic cretinism, 31/297
energy metabolism, 27/1-23
fetal alcohol syndrome, 30/121
fetal period, 30/26-28
filariasis, 35/163-168
Fukuyama syndrome, 41/38
function, 1/45-74
functional activity, 30/24
fungal, *see* Fungal CNS disease
glioependymal barrier, 50/250
glycine, 29/508
goatfish intoxication, 37/86
Goldenhar syndrome, 31/247
Goltz-Gorlin syndrome, 31/288
Gruber syndrome, 31/243
Hallermann-Streiff syndrome, 31/249
hallucinatory fish intoxication, 37/86
hexachlorophene intoxication, 37/488, 499
hexomonophosphate shunt, 27/13, 79
histology, 50/1
homocarnosinase deficiency, 42/558
hydrocephalus, 30/600
hypersensitivity, 9/504
hypoglycemia, 27/53-73, 70/171
hyponatremia, 28/430-432
imaging, 74/626
immunology, 9/501
incontinentia pigmenti, 31/242
infantile Gaucher disease, 42/438, 66/124
inorganic constituent, 28/423-439
iron, 28/438
isolated angiitis, *see* Granulomatous CNS vasculitis
isoniazid neuropathy, 7/582
Kawasaki syndrome, 51/453, 52/253
Klinefelter syndrome, 31/482
Laurence-Moon-Bardet-Biedl syndrome, 31/327
legionellosis, 52/253
lidocaine intoxication, 37/450
linear nevus sebaceous syndrome, 31/289, 43/38
Lowe syndrome, 21/49, 31/301
Lyme disease, 52/253
macroglia, 50/7
magnesium, 28/434-436
malaria, 31/221
maldevelopment, 50/28
malnutrition, 28/1-37
manganese, 28/439
mantle layer, 30/60-65, 50/32
MAO inhibitor intoxication, 37/320
metachromatic leukodystrophy, 66/166
methanol intoxication, 36/353
microglia, 50/7
monoamine neuron system, 30/33
multifocal, *see* Multifocal CNS disease

multiple organ failure, 71/527
multiple spinal ependymoma, 20/358
muscle tone, 1/64
myelin basic protein P1, 47/5
myelin lipid, 47/5
myelination, 7/31, 41
myotonic dystrophy, 40/517, 62/235
myxedema neuropathy, 8/57
neural connection, 30/73
neural crest, 30/18
neural induction, 50/24
neural tube formation, 30/19, 23, 25
neuroepithelium, 30/17, 50/29
neurogenic sarcoma, 14/30
neuroglia, 30/16
neuromery, 50/2
neuropore, 50/3, 27
neurotransmitter metabolism, 59/152-155
neurulation, 50/1, 19
nocardiosis, 35/518-523, 52/447
nongerminomatous germ cell tumor, 68/235
Norrie disease, 22/508, 31/292
nuclear migration, 30/16, 50/7
orofaciodigital syndrome type I, 31/251
overgrown brain, 50/30
phosphate, 28/437
poliomyelitis, 34/103
potassium, 28/427-431
Prader-Labhart-Willi syndrome, 31/322-324
primary neurulation, 50/4
progressive dysautonomia, 59/155
progressive hemifacial atrophy, 31/253, 43/408
proteolipid, 9/11
pyramidal tract, 30/30
Ranvier node, 7/22, 47/9, 51/15
rare tumor, 68/379
recessive infantile optic atrophy, 13/111
recurrent multifocal, *see* Recurrent multifocal CNS disease
remyelination, 47/41
rheumatoid arthritis, 38/487-496, 71/20
rhombencephalon, 50/2
Rubinstein-Taybi syndrome, 31/283
rudderfish intoxication, 37/86
S-G-M phase, 50/29
sarcoidosis, 38/529-531, 71/472, 477, 489
schistosomiasis, 52/537
schizencephaly, 30/346
scleroderma, 39/365-367
secondary proliferation layer, 30/65-70, 50/35
sensitization, 9/517
Sjögren syndrome, 39/429
slow virus disease, 34/275-288
Smith-Lemli-Opitz syndrome, 66/585
sodium, 28/431-433
spinal angioma, 32/466
spinal cord, 30/33
subependymal cell layer, 50/36
subpial granular layer, 50/37
teratogenesis, 30/34, 50/41
thalamus, 30/31
thallium intoxication, 64/323
thanatophoric dwarfism, 31/274
tick bite intoxication, 37/111
toxic shock syndrome, 52/253
tricyclic antidepressant intoxication, 37/317
triethyltin, 28/427, 51/274
trophic disorder, 1/481-488
Trypanosoma cruzi, 31/221
trypsin resistant protein residue, 9/32
Turner syndrome, 31/500
vascularization, 50/35
vasogenic brain edema, 28/426
ventricular system, 30/32
vertebral column, 32/141
vitamin E deficiency, 70/446
in vitro, 27/4-6
in vivo, 27/2-4
Wegener granulomatosis, 39/344
Whipple disease, 70/241
Wilson disease, 27/393-399, 28/438
Wolf-Hirschhorn syndrome, 31/551, 553, 557, 43/497
xeroderma pigmentosum, 31/290, 60/337
XYY syndrome, 31/489
zinc, 28/438
CNS aging
dementia, 27/479
memory, 27/470
CNS angiitis, *see* Brain vasculitis
CNS angiosarcoma
epidemiology, 68/278
CNS degeneration
Alzheimer neurofibrillary change, 21/44, 47
anisocoria, 42/97
areflexia, 42/97
bulbar paralysis, 42/97
polyneuropathy, 42/97
sensory loss, 42/97
CNS demyelination
demyelinating neuropathy, 47/606
neurobrucellosis, 52/586, 588
CNS developmental disorder
mental deficiency, 46/6-13
CNS disease
risk factor, 69/268

CNS granulomatous vasculitis
general features, 63/53
CNS hypersomnia
idiopathic, *see* Idiopathic CNS hypersomnia
CNS infection
Aspergillus, 69/435
bacterium, 69/434
blood-brain barrier, 69/432
cancer, 69/431, 433
Candida, 69/435, 439
coccidioidomycosis, 69/447
Cryptococcus neoformans, 69/446
fungus, 69/435, 438, 446
herpes simplex, 69/447
histoplasmosis, 69/447
Hodgkin disease, 39/41-45
host defence, 69/437
host reaction, 55/417
immunocompromised host, 56/467, 65/548
Listeria, 69/441
malignant lymphoma, 39/41-45
parasite, 69/444
renal transplantation, 63/530, 532
Strongyloides stercoralis, 69/446
Toxoplasma gondii, 69/444
treatment, 69/451
varicella zoster, 69/448
yeast, 69/438
CNS irradiation
disseminated necrotizing leukoencephalopathy, 63/355
meningeal leukemia, 63/340
CNS isolated angiitis, *see* Granulomatous CNS vasculitis
CNS lesion
cardiovascular function, 63/229
ECG change, 63/233
ECG pathology, 63/234
multiple nevoid basal cell carcinoma syndrome, 14/463
myocardial ischemia, 63/233
phantom limb, 4/224, 226
toluene intoxication, 36/373
CNS leukemia, 69/235
chemotherapy, 69/243
cytarabine, 69/250
early treatment, 69/243
intraCSF treatment, 69/250
methotrexate, 69/243, 250
symptom, 69/237
thiotepa, 69/251
CNS lymphoma
heart transplantation, 63/181
iatrogenic neurological disease, 63/181
metastasis, 69/264
primary, *see* Primary CNS lymphoma
CNS malformation
anencephaly, 50/57
ascertainment, 50/50
chromosomal aberration, 30/104
classification, 30/5-11
congenital heart disease, 63/11
craniovertebral region, 32/58-60
definition, 30/1-5, 11, 85, 87, 91-112
diagnosed at birth, 30/4
Down syndrome, 50/57
encephalocele, 50/57
environmental, 30/112-126
epidemiology, 30/139-171
ethnic variation, 30/148-151, 50/55
etiologic heterogeneity, 30/87-89
etiology, 30/85-138
exencephaly, 32/171, 50/57
gene, 30/126-129
genetics, 50/50
geographic variation, 30/148-151, 50/55
hydrocephalus, 50/57
incidence, 50/49, 52-54
live birth, 50/51
load, 50/51
major and minor, 30/5
microcephaly, 50/57
mortality, 30/151, 50/56
multifactorial causation, 30/87
multiple, 30/6-8
mutation, 30/99-101
pleiotropy, 30/88
prevalence, 50/49
proportion of all malformations, 50/51
proportional rate, 50/50
sex ratio, 30/151, 50/56
specific causation, 30/86
spina bifida, 50/57
spontaneous abortion, 50/55
stillbirth, 50/51
syndrome, 30/7
twins, 30/151
CNS metastasis, *see* Brain metastasis
CNS myelin
biochemistry, 9/11
chondroitin sulfate A, 9/31
histochemistry, 9/28
hyaluronic acid, 9/31
neurokeratin, 9/31, 33
peripheral nervous system, 47/5-9
proteolipid, 9/40

wallerian degeneration, 9/36
CNS-peripheral nervous system compound axon
nerve, 51/1
CNS reticulosis
peripheral neurolymphomatosis, 56/179
CNS spongy degeneration
Alpers disease, 42/486
blindness, 42/506
classification, 66/5
cranial nerve, 42/506
cytoplasmic inclusion body, 42/507
feeding disorder, 42/506
flaccid paralysis, 42/506
homocystinuria, 42/556
hyperreflexia, 42/506
macrencephaly, 42/506
myoclonus, 42/506
psychomotor retardation, 42/506
seizure, 15/426, 42/506
tetraplegia, 42/506
type, 66/15
CNS stimulant
amphetamine intoxication, 65/251
behavior disorder, 46/591
bromocriptine, 37/236
catechols, 37/232
cocaine intoxication, 65/251
epilepsy, 72/228
hypersomnia, 46/592
L-α-methylmetatyrosine, 37/234
narcolepsy, 46/592
nicotine intoxication, 65/251
psychosis, 46/591
sleep disorder, 46/592
CNS toxicity
leuprorelin, 69/500
nitroimidazole, 69/500
CNS transitional zone
nerve, 51/11
CNS tumor
*see also* Brain tumor
cerebellar granular cell hypertrophy, 67/54
chromosome 10, 67/44
chromosome 13, 67/47
chromosome 19, 67/47
cognitive dysfunction, 67/371
Cowden syndrome, 67/54
glioma polyposis syndrome, 30/101, 42/735, 67/54
hemangioendothelioma, 68/280
Li-Fraumeni syndrome, 67/54
microglia cell, 18/233
molecular genetics, 67/41
multiple endocrine adenomatosis syndrome, 30/101
multiple nevoid basal cell carcinoma syndrome, 30/101, 67/54
neurofibromatosis, 30/101
neurofibromatosis type I, 67/54
neurofibromatosis type II, 67/54
Rb gene, 67/47
syringomyelia, 32/286-293
vascular tumor, 68/269
Von Hippel-Lindau disease, 30/101, 67/54
CNS vasculitis, *see* Brain vasculitis
COACH syndrome
hereditary congenital cerebellar atrophy, 60/292
Coagulation
blood, *see* Hemostasis
brain arteriovenous malformation, 12/254
disorder, *see* Coagulopathy
Hageman factor, 1/129
intravascular consumption, *see* Intravascular consumption coagulopathy
Kirschner, *see* Kirschner coagulation
vasculitis, 71/153
Coagulopathy
acquired type, 63/313
afibrinogenemia, 63/312
amanitin, 65/37
amaurosis fugax, 55/108
bacterial meningitis, 52/7
bleeding tendency, 38/80
blood clotting factor VIII inhibitor, 63/318
brain infarction, 69/414
bruising, 63/308
cerebrovascular disease, 55/474
classification, 63/309
coagulation inhibitor, 63/318
congenital qualitative platelet disorder, 63/325
diagnosis, 63/308
dysbarism, 23/622
epistaxis, 63/308
essential thrombocythemia, 63/331
Glanzmann thrombasthenia, 63/325
head injury, 57/238
hemophilia A, 63/309
hemorrhage site, 63/309
intravascular consumption coagulopathy, 63/309
liver disease, 63/309, 313
menorrhagia, 63/308
oral contraception, 63/332
portal hypertension, 63/313
posttraumatic intracranial hematoma, 57/259
protein C deficiency, 63/328
protein S deficiency, 63/328

spinal cord injury, 26/391
subarachnoid hemorrhage, 55/3
symptom, 63/308
thrombocytopenia, 63/309, 321
thrombophilia, 63/328
vitamin K deficiency, 63/309
Von Willebrand disease, 63/309, 311
Von Willebrand inhibitor, 63/318

Coarctation
aorta, *see* Aorta coarctation

Coated pit
multiple sclerosis, 47/221

Coats disease
facioscapulohumeral muscular dystrophy, 62/167
facioscapulohumeral syndrome, 62/167
hearing loss, 62/167
primary pigmentary retinal degeneration, 13/249
Small disease, 22/509

Cobalamin, *see* Vitamin B12

Cobalt
CNS, 28/439
epilepsy, 72/53
mercury intoxication, 64/385
striatopallidodentate calcification, 6/716, 49/425

Cobalt edetate
cyanide intoxication, 64/233

Cobalt intoxication
animal study, 36/330
cardiomyopathy, 36/330
diarrhea, 36/330
polycythemia, 36/330
pulmonary edema, 36/330
rhabdomyosarcoma, 36/330

Cobalt therapy
brain infarction, 53/33
cerebrovascular disease, 53/33
child, 53/33
ionizing radiation neuropathy, 51/138

Cobb syndrome, 14/429-443
age, 14/439
angiodysgenetic necrotizing myelomalacia, 20/486
arteriovenous angioma, 14/440
definition, 14/429, 32/469
family, 14/438
hydrocephalus, 14/438
leptomeningeal angiomatosis, 14/14
lesion site, 14/440
neurologic symptom, 14/439
neuropathology, 14/440
nosology, 14/430, 442
paraplegia, 14/439
phakomatosis, 14/14, 72, 31/26
polycystic kidney, 14/438
prevalence, 14/438
reported cases, 14/430-437
sex ratio, 14/438
skin lesion, 14/441
spinal angioma, 14/429, 442
subarachnoid hemorrhage, 14/439
venous angioma, 14/441

Cobra snake venom
demyelination, 9/39
nerve growth factor, 37/12
snake envenomation, 65/181

Cobra snake venom intoxication
antivenene, 37/12
cardiotoxin, 37/12
cytotoxin, 37/12
myasthenia gravis, 41/122

Coca
Erythroxylum, 65/251

Cocaine
*see also* Hallucinogenic agent
acquired myoglobinuria, 62/577
anesthetic effect, 65/251
autonomic nervous system, 75/501
benzoylecgonine, 65/50
chemical structure, 37/236, 65/49
convulsion, 65/251
dopamine inhibition, 65/251
drug abuse, 55/521
ecgonine methyl ester, 65/50
epilepsy, 65/251, 72/236
half-life, 65/252
hallucinogenic agent, 65/41
metabolite, 65/50
neurotoxin, 37/236, 55/521, 62/577, 601, 65/41, 50, 251
noradrenalin, 74/145, 407
noradrenalin inhibition, 65/251
pharmacokinetics, 65/251
pharmacology, 65/49, 251, 74/145
serotonin inhibition, 65/251
street name, 65/49
toxic encephalopathy, 65/251
toxic myopathy, 62/601
tremor, 65/251

Cocaine addiction
brain hemorrhage, 55/521
brain infarction, 55/521
brain vasculitis, 55/521
headache, 55/521
myocardial infarction, 55/522
pharmacokinetics, 65/251
pharmacology, 65/251

platelet aggregation, 55/522
subarachnoid hemorrhage, 55/521
Cocaine intoxication
altitude illness, 65/254
anterior spinal artery syndrome, 65/255
CNS stimulant, 65/251
coca paste, 65/252
cocaine smoking, 65/252
color perception, 65/49
congenital malformation, 65/255
dopamine inhibition, 65/252
dystonia, 65/255
EEG, 65/253
epidemiology, 65/253
epilepsy, 65/251, 255
hallucination, 4/331, 65/49, 255
headache, 65/255
holoprosencephaly, 65/255
hyperprolactinemia, 65/253, 256
internuclear ophthalmoplegia, 65/50
intracranial hemorrhage, 65/255
irritability, 65/253
kindling, 65/253
lateral medullary infarction, 65/255
leaf chewing, 65/254
leukomalacia, 65/255
malignant hyperthermia, 65/252, 256
myalgia, 65/253
newborn, 65/255
nucleus accumbens, 65/252
panic attack, 65/256
paranoia, 65/253, 256
pathology, 65/254
pharmacology, 65/253
positron emission tomography, 65/254
pregnancy, 65/255
psychiatric disorder, 65/254
restlessness, 65/253
rhabdomyolysis, 65/252, 256
single photon emission computer assisted tomography, 65/254
subarachnoid hemorrhage, 65/50
sudden death, 65/256
transient ischemic attack, 65/255
tremor, 65/251, 253
withdrawal syndrome, 65/253
Cocarboxylase
beriberi, 28/3
deficiency neuropathy, 7/568, 51/330
multiple sclerosis, 9/401
necrotizing encephalomyelopathy, 28/355
subacute necrotizing encephalomyelopathy, 28/355, 42/588
Cocarboxylase loading test
vitamin B1 deficiency, 51/332
Coccidioides immitis
*see also* Coccidioidomycosis
bacteriology, 52/409
biological features, 35/444
brain abscess, 52/151
cerebrovascular disease, 55/432
estrogen, 52/410
hematogenous spread, 52/410
internal endospores, 52/410
meningitis, 55/432
methenamine silver stain, 52/411
neurotropism, 52/410
pregnancy, 52/410
race, 52/410
spherula, 52/410
spinal epidural empyema, 52/187
Coccidioidin skin test
coccidioidomycosis, 35/444, 449
Coccidioidomycosis, 35/443-454
*see also* Coccidioides immitis
acquired immune deficiency syndrome, 52/411, 56/515, 71/296
amphotericin B, 35/450-452, 52/419
arachnoiditis, 52/420, 422
basal meningitis, 52/411
brain, 16/216
brain abscess, 52/412, 415
brain granuloma, 16/230
brain microabscess, 52/412
chronic meningitis, 56/644
clinical features, 35/447
CNS infection, 69/447
coccidioidin skin test, 35/444, 449
computer assisted tomography, 52/418
confusion, 52/413
corticosteroid, 52/420
CSF, 35/449, 52/415
diagnosis, 52/414
EEG, 52/417
encephalopathy, 52/413
eosinophilia, 52/415
epidemiology, 35/16, 443-445
fluconazole, 52/422
flucytosine, 35/452, 52/419
geographic distribution, 35/16, 443
granuloma, 52/411
headache, 52/413
history, 35/443
host defence, 52/409
hydrocephalus, 52/412, 415
immunocompromised host, 52/411

intracranial hypertension, 52/413
itraconazole, 52/422
ivory vertebra, 19/171
ketoconazole, 52/421
laboratory finding, 35/448, 52/414
management, 35/453
meningitis, 52/412
mental symptom, 52/413
miconazole, 52/421
miconazole treatment, 35/452
microbiology, 52/409
mortality, 52/414
mycosis, 35/376-379
myelography, 52/417
neuropathology, 52/411, 414
neurosurgery, 52/415, 422
nuclear magnetic resonance, 52/419
Ommaya reservoir, 35/451, 52/422
paraspinal abscess, 52/416
pathogenesis, 35/445, 52/409
pathology, 35/446
PEG, 52/417
prognosis, 52/419
prophylaxis, 35/453
pulmonary infection, 52/409
radiation, 52/417
radionuclide scintigraphy, 52/418
serology, 52/416
skin test, 52/416
spinal cord compression, 52/414
spinal epidural empyema, 52/414
transfer factor, 52/421
transverse myelitis, 52/414, 420
treatment, 35/450-453, 52/419
ventriculitis, 52/419, 422
Coccygodynia
spinal ependymoma, 20/365
Coccygomedullary organ
ependymoma, 20/364
Coccyx agenesis
caudal aplasia, 50/511
Cochlea
arterial supply, 55/130
brain vasculitis, 55/381
Cochlear apparatus
radiation injury, 7/393
Cochlear hypoplasia
Alport syndrome, 42/376
branchio-otorenal dysplasia, 42/361-363
Cochlear nerve
acoustic neuroma, 2/361
audiometry, 16/307
function, 16/320
malnutrition, 28/21
vestibular neurinoma, 68/442
Cochlear spiral ganglion degeneration
Richards-Rundle syndrome, 21/359
Roussy-Lévy syndrome, 21/359
Cochleovestibular atrophy
hereditary progressive, *see* Hereditary progressive cochleovestibular atrophy
progressive, *see* Progressive cochleovestibular atrophy
Cockayne-Neill-Dingwall disease
amyotrophy, 51/399
anhidrosis, 13/432
ataxia, 10/182, 13/433
ataxia telangiectasia, 14/315
autosomal recessive, 14/315, 51/399
brain cortex calcification, 13/435
cataract, 10/182, 43/351, 60/653
chorea, 13/433
choreoathetosis, 13/433
chorioretinal degeneration, 13/40
chronic inflammatory demyelinating polyradiculoneuropathy, 51/530
clinical features, 13/431
CNS, 31/238
corneal dystrophy, 60/652
cortical calcification, 13/435
cranial hyperostosis, 43/350
deafness, 10/182, 13/433, 436, 60/658
demyelinating neuropathy, 47/622
dermatitis, 43/350
differential diagnosis, 13/236, 437
dwarfism, 10/182, 13/432, 43/350, 51/399
extrapyramidal sign, 13/433
eye, 31/238
face, 31/236
familial spastic paraplegia, 59/326, 332
Flynn-Aird syndrome, 14/113
genetics, 31/236
Gifford progeria, 13/323
hereditary hearing loss, 13/242, 22/503
incoordination, 13/433
intracerebral calcification, 43/350, 51/399
mental deficiency, 10/182, 13/433, 14/315, 31/236, 43/350, 46/62, 51/399, 60/664
metabolic polyneuropathy, 51/399
microcephaly, 13/432, 30/508, 31/236, 238, 43/350, 60/666
micrognathia, 43/350
muscle weakness, 13/433
muscular atrophy, 13/433
musculoskeletal abnormality, 31/236
myelin dysgenesis, 43/350

nerve conduction velocity, 51/399
neurofibromatosis type I, 14/492
neuropathology, 13/434
neuropathy, 60/660
nystagmus, 43/351
ocular lesion, 13/433
onion bulb formation, 51/399
optic atrophy, 10/182, 13/40, 43/350
Pelizaeus-Merzbacher disease, 10/115, 151, 182-184, 197, 13/436
photophobia, 43/351
photosensitive dermatitis, 51/399
photosensitivity, 13/432, 14/315
polyneuropathy, 43/351, 51/399
progeria, 13/436
prognathism, 10/182
Refsum disease, 21/217, 60/235, 66/497
retinal degeneration, 13/433
retinal pigmentation, 51/399
Rothmund-Thomson syndrome, 13/436, 31/238
Schwann cell inclusion, 51/399
secondary pigmentary retinal degeneration, 13/322-324, 43/350-352, 60/654, 729
seizure, 43/350
sensorineural deafness, 43/350
sex ratio, 43/352
skin lesion, 14/315, 31/237
speech, 13/433
striatopallidodentate calcification, 13/435
sudanophilic leukodystrophy, 13/436
ultraviolet light, 13/432
Usher syndrome, 13/438
ventricular dilatation, 43/350
white matter rarefaction, 10/194
Cocoa
*see also* Chocolate
migraine, 48/93
Codeine
migraine, 5/98
opiate, 65/350
Codeine intoxication
chemistry, 37/365
Co-dergocrine
multiple sclerosis, 9/390
Coding
brain injury, 23/16
Coelenterata intoxication
chironex, 37/38
clinical neurology, 37/38
CNS, 37/38
hydroids, 37/36
itch, 37/38
jellyfish, 37/36
local pain, 37/38
physalia, 37/38
sea anemone, 37/36
Coenesthopathias, *see* Cenesthesiopathy
Coenurus cerebralis
chronic meningitis, 56/645
spinal muscular atrophy, 59/370
Coenzyme A
deficiency neuropathy, 7/578
Coenzyme Q, *see* Ubiquinone
Coenzyme Q10, *see* Ubidecarenone
COER, *see* Brain oxygen extraction rate
Coffin-Lowry syndrome
antimongoloid eyes, 43/238
CNS, 31/245
cutis marmorata, 43/239
dental abnormality, 43/239
dermatoglyphics, 43/239
differential diagnosis, 31/245
EEG, 43/239
epicanthus, 43/238
face, 31/244
frontal bossing, 43/238
genetics, 31/244
hydrocephalus, 31/245, 43/239
hyperextensible skin, 43/239
hypertelorism, 30/248, 43/238
kyphoscoliosis, 43/238
low set ears, 43/238
mental deficiency, 31/245, 43/238, 46/46
microcephaly, 43/239
oral lesion, 31/245
pectus carinatum, 43/238
scoliosis, 30/101
seizure, 31/245
skeletal abnormality, 31/244
skeletal malformation, 30/248
skin lesion, 31/245
spina bifida occulta, 43/238
strabismus, 43/238
ventricular dilatation, 43/239
vertebral abnormality, 43/238
Coffin-Siris syndrome
CNS, 31/240
convulsion, 31/240
Dandy-Walker syndrome, 50/332
dental abnormality, 43/240
dermatoglyphics, 43/241
differential diagnosis, 31/240
digital abnormality, 43/240
Ellis-Van Creveld syndrome, 31/240
etiology, 31/239
face, 31/239

growth retardation, 43/240
hair, 43/240
mental deficiency, 31/240
microcephaly, 30/508, 43/241
musculoskeletal abnormality, 31/239
nail hypoplasia, 43/241
oral lesion, 31/240
patellar hypoplasia, 43/240
skin lesion, 31/239
swayback nose, 43/240
Coffin-Siris-Wegienka syndrome, *see* Coffin-Lowry syndrome
Cogan syndrome
classification, 71/4
Cogan syndrome type I, 42/329-331
cerebellar ataxia, 42/329
CSF, 51/454
eye movement disorder, 14/653
mental deficiency, 42/329
symptom, 1/605
Cogan syndrome type II
acoustic nerve, 51/454
CNS, 51/454
CSF, 51/454
deafness, 51/454
interstitial keratitis, 51/454
mononeuropathy, 51/454
multiple cranial neuropathy, 51/454
polyarteritis nodosa, 39/303, 51/446, 454
polyneuropathy, 51/454
retrobulbar neuritis, 51/454
vertigo, 2/360, 55/133
Cognition
aging, 46/211, 491
anxiety, 45/267
brain lacunar infarction, 53/395
brain tumor, 16/729
carbamazepine intoxication, 65/506
cerebrovascular disease, 55/137
computer assisted tomography, 46/482
delirium, 46/525-531
dementia, 46/124, 200
developmental dyslexia, 46/120
dyslexia, 46/127
hysteria, 46/579
intelligence, 3/298
learning disability, 46/123
mental deficiency, 46/4
migraine, 48/162
neuropsychologic defect, 45/517-521
neuropsychology, 45/516
Parkinson disease, 46/372
phenytoin intoxication, 65/501
schizophrenia, 46/482, 488, 490-492, 501
spatial, *see* Spatial cognition
stress, 45/246
test, 45/518
thalamic infarction, 53/395
vertebrobasilar system syndrome, 53/395
Wechsler Intelligence Scale, 45/521
Wechsler test, 45/521
Cognitive dysfunction
brain tumor, 67/371
CNS tumor, 67/371
cytarabine, 67/362
cytidine analog, 67/362
early delayed, 67/372
frontal lobe tumor, 67/148
irradiated patient, 67/374
radiation, 67/371
Cognitive function
barbiturate intoxication, 65/509
EEG, 72/289
epilepsy, 72/388
uremic encephalopathy, 63/509
Cogwheel phenomenon
Parkinson rigidity, 6/137
progressive supranuclear palsy, 22/221
tremor covariance, 6/161
Cogwheel rigidity
pellagra, 28/85, 46/336
tremor associated rigidity, 6/139, 180
Wilson disease, 49/225
Cogwheel saccade
progressive supranuclear palsy, 49/241
Cohnheim-Ribbert theory
brain tumor origin, 16/24
Coin-shaped vertebra
eosinophilic granuloma, 20/31
Coital cephalgia, *see* Exertional headache
Coital migraine
autonomic nervous system, 75/301
blood pressure, 75/301
muscle contraction headache, 48/379
Coitus
aneurysm, 48/377
exertional headache, 48/377
headache, *see* Exertional headache
migraine, 48/377
paraplegia, 61/315
sexual function, 26/448
spinal cord injury, 61/315
subarachnoid hemorrhage, 48/377
tetraplegia, 61/315
transverse spinal cord lesion, 61/315
vascular headache, 48/380

Colamine-cephalin, *see* Phosphatidylethanolamine
Colchicine
  Behçet syndrome, 56/605
  distal axonopathy, 64/13
  experimental myopathy, 40/154
  myotonia, 40/555
  neurofibrillary tangle, 46/271
  neurotoxin, 40/151, 154, 555, 46/271, 51/180, 62/597, 600, 606, 609, 64/13
  sarcoplasmic reticulum, 40/154
  toxic myopathy, 62/597, 600, 606
  toxic neuropathy, 64/13
  transverse tubule, 40/154
  varicella zoster, 51/180
Colchicine intoxication
  symptom, 64/13
Colchicine myopathy
  creatine kinase, 62/609
  muscle biopsy, 62/609
  spheromembranous body, 62/609
Cold
  acquired myoglobinuria, 62/574
  headache, 5/188
  migraine, 48/122
  nerve, 1/449
  nerve injury, 8/162, 51/136
  nerve lesion, 1/449
  Raynaud disease, 8/321, 22/261, 43/70
  sensitivity, 8/341
  stress, 74/318
Cold face test
  apnea, 74/606, 622
  autonomic nervous system, 74/606, 622
  bradycardia, 74/606, 622
  hereditary sensory and autonomic neuropathy type III, 75/152
  trigeminal nerve, 74/606, 622
Cold induced pain
  ice cream headache, 48/441
  postpoliomyelitic amyotrophy, 51/482
Cold injury
  brain edema, 23/151
  brain water content, 67/74
  ischemic neuropathy, 8/162
  paralysis periodica paramyotonica, 43/167
  paramyotonia congenita, 43/168
  Pernio syndrome, 8/341
  test, 8/341
  treatment, 8/341
Cold neuropathy
  axonal degeneration, 51/137
  axoplasmic flow, 51/137
  nerve conduction velocity, 51/136
Cold paresis
  Hirayama disease, 59/107
  spinal muscular atrophy, 59/107
Cold pressor test
  aging, 74/233
  autonomic nervous system, 74/624
  heart rate, 74/605
  hereditary sensory and autonomic neuropathy type III, 75/152
  orthostatic hypotension, 63/158
  pure autonomic failure, 22/233
Cold pressure test, *see* Cold pressor test
Cold vasodilatation test
  Bonney response, 2/174, 25/422
  spinal nerve root injury, 25/421
Colecalciferol intoxication, *see* Vitamin D3 intoxication
Colestyramine
  antihypertensive agent, 63/83
Collagen disease
  acquired toxoplasmosis, 52/353
  dermatomyositis, 8/122
  diastrophic dwarfism, 43/384
  encephalopathy, 45/118
  homocystinuria, 42/556
  migraine, 5/56
  muscular dystrophy, 43/103
  myosclerosis, 43/133
  neuropathy, 8/118
  Raynaud phenomenon, 43/71
  recurrent nonhereditary multiple cranial neuropathy, 51/570
  systemic brain infarction, 11/442
Collagen fiber
  hemangioblastoma, 14/55
  intervertebral disc, 20/526
  neurinoma, 14/25
Collagen like protein
  encephalitogenicity, 9/516
Collagen network
  nerve, 51/2
Collagen type hyaline, *see* Arteriolar fibrinohyalinoid degeneration
Collagen vascular disease
  *see also* Inflammatory myopathy
  angiopathic polyneuropathy, 51/445
  autonomic nervous system, 75/12
  brain embolism, 55/163
  brain infarction, 55/163
  cerebrovascular disease, 55/163
  chemokine, 71/9
  classification, 71/3
  clinical features, 40/317, 325

cytokine, 71/9
dissecting aorta aneurysm, 63/45
hereditary, *see* Hereditary connective tissue disease
hypertensive encephalopathy, 54/212
immunopathology, 71/3
inclusion body myositis, 62/373
mixed, *see* Mixed connective tissue disease
myasthenia gravis, 43/157
orthostatic hypotension, 63/152
polyarteritis nodosa, 8/124, 39/307, 51/446, 452, 54/294
progressive myositis ossificans, 43/191
retinal artery occlusion, 55/112
rheumatoid arthritis, 8/126
subarachnoid hemorrhage, 55/2
Collapsing cord sign
craniovertebral region, 32/65
Collateral neoformation
nerve regeneration, 7/226-230
neuritis, 7/226
Colle sign
diencephalic syndrome, 42/734
Collet-Bonnet syndrome
Villaret syndrome, 2/101
Collet-Sicard syndrome, *see* Sicard-Collet syndrome
Colliculi
inferior, *see* Inferior colliculi
superior, *see* Superior colliculi
Collier sign
aqueduct stenosis, 2/279
mesencephalic infarction, 53/391
Parinaud syndrome, 55/124
vertebrobasilar system syndrome, 53/391
Colloid body
angioid streak, 13/74
familial, *see* Hutchinson-Tay choroidopathy
fluorescein angiography, 13/75
retinal spot, 14/620
retinal tumor, 14/620
tuberous sclerosis, 13/75
Colloid cyst
*see also* Ependymal cyst
classification, 68/310
epilepsy, 72/181
hydrocephalus, 30/541, 543
third ventricle, *see* Third ventricle colloid cyst
third ventricle tumor, 17/440, 446-448, 450, 18/359
Colloid oncotic pressure
brain edema, 67/87
Coloboma
anencephaly, 50/83
basal encephalocele, 50/106
basal meningoencephalocele, 50/106
choroid, *see* Choroid coloboma
corpus callosum agenesis, 50/162
Edwards syndrome, 50/562
eyelid, *see* Eyelid coloboma
iris, *see* Iris coloboma
optic nerve, *see* Optic nerve coloboma
Patau syndrome, 14/120, 31/504, 513
Saethre-Chotzen syndrome, 43/322
trisomy 22, 50/564
Colon
innervation, 75/634
motility dysfunction, 75/633
neural regulation, 1/510
syndrome, 1/510
Colon cancer
brain metastasis, 69/107
Colon carcinoma
dural metastasis, 71/614
metastasis, 71/619
Colonic polyposis
Cowden syndrome, 42/754
Peutz-Jeghers syndrome, 14/121
Color agnosia
alexia, 45/438
alexia without agraphia, 4/121
anatomic localization, 4/22
cerebral color blindness, 45/53
color amnesia, 4/17
color aphasia, 4/17
color hemiagnosia, 4/16
definition, 2/607
hemisphere lesion, 4/300
occipital lobe lesion, 4/5, 300
occipital lobe syndrome, 45/52
type, 45/344
Color amnesia
color agnosia, 4/17
memory disorder, 4/17
Color anomia
agraphia, 45/52
alexia, 45/52, 438
amnesic, *see* Amnesic color anomia
occipital lobe tumor, 67/156
occipitoparietal tumor, 45/52
posterior cerebral artery syndrome, 53/411
Color aphasia
color agnosia, 4/17
occipital lobe syndrome, 45/52
Color asthenopia
features, 2/650
Color blindness

amnesic type, 2/649, 651
cerebral, *see* Cerebral color blindness
deutan, *see* Deutan color blindness
Friedreich ataxia, 21/347
lidocaine intoxication, 37/451
myopathy, 40/352, 388
occipital lesion, 2/648
occipital lobe tumor, 17/325
Color hemiagnosia
color agnosia, 4/16
Color misperception, *see* Color agnosia
Color naming defect
Balint syndrome, 2/651
neuropsychology, 45/13
occipital lobe syndrome, 45/52
Color perception
agnosia, 45/334-336
cinchonism, 65/485
cocaine intoxication, 65/49
infantile optic atrophy, 42/408
lysergide intoxication, 65/44
posterior cerebral artery syndrome, 53/411
Color perception disorder, *see* Color agnosia
Color sense
occipital lobe tumor, 17/324
Color vision
acquired, 13/13
Behr disease, 13/89
cardiovascular agent intoxication, 37/426
cinchonism, 65/452
digitoxin intoxication, 13/13
dominant infantile optic atrophy, 13/114-116
ethambutol intoxication, 65/481
glycoside intoxication, 37/426, 429
hereditary adult foveal dystrophy (Behr), 13/89
Stargardt disease, 13/129
testing, 13/13
Colorado tick fever
acute viral encephalitis, 34/77, 56/134
arthralgia, 56/141
Dermacentor andersoni, 56/141
headache, 56/141
myalgia, 56/141
orbivirus, 56/11
reovirus, 56/11
Colorado tick fever virus
erythrocyte, 56/30
Colpocephaly
corpus callosum agenesis, 50/155
definition, 50/149
Columbia, Bogota
neurology, 1/12
Columbia University
neurology, 1/5
Columbian spirits
characteristics, 64/95
methanol, 64/95
use, 64/95
Columbus, Christopher
syphilis, 33/338
Column
Clarke, *see* Clarke column
posterior, *see* Posterior column
vertebral, *see* Vertebral column
visual cortex, 2/539
Column degeneration
dorsal, *see* Dorsal column degeneration
Columnar cell glioma, *see* Ependymoma
Coma
*see also* Consciousness *and* Decortication
acetylsalicylic acid intoxication, 65/460
acquired toxoplasmosis, 35/125, 52/354
acute mercury intoxication, 64/375
agreement rate, 57/381
airway, 57/228
akinetic mutism, 3/66, 75, 113
allergic granulomatous angiitis, 63/383
alpha, *see* Alpha coma
aluminum intoxication, 63/526
amnesic shellfish poisoning, 65/167
amphetamine addiction, 55/519
amytal, 3/68
anoxic lesion, 3/70
antihypertensive agent, 63/87
apathic syndrome, *see* Decortication
apperceptivity areactive apathic and atonic syndrome, 3/66, 70, 75, 77
areactive apathic normotonic aperceptivity syndrome, 3/76
ascending reticular activating system, 3/64
autonomic reaction, 3/74
barbiturate, *see* Barbiturate coma
basilar artery stenosis, 53/386
basilar dissecting aneurysm, 54/280
bilateral pontine hemorrhage, 12/41
bilateral pontine syndrome, 12/20
brain abscess, 52/152
brain concussion, 3/69, 23/428
brain death, 3/66, 70, 75, 77, 55/257
brain edema, 3/70, 63/416
brain embolism, 53/208
brain fat embolism, 55/177
brain infarction, 53/208
brain injury, 23/12, 428, 430
brain metabolism, 3/70
brain stem, 3/56

brain stem infarction, 12/15
brain stem injury, 3/66, 68, 75
carbamate intoxication, 64/186
carbon disulfide intoxication, 64/24
carbon monoxide intoxication, 49/473, 64/31
cat scratch disease, 52/129
central pontine myelinolysis, 2/263, 28/291
cerebral amebiasis, 52/315
cerebral malaria, 35/145, 52/368
cerebrovascular disease, 53/208
chelonia intoxication, 37/91
Children Coma Score, 57/329
ciguatoxin intoxication, 65/163
cingulate cortex, 3/62-77
classification, 3/71, 75
cochleopalpebral reflex, 3/73
complex III deficiency, 62/504, 66/423
compressed spectral array, 57/190
conidae intoxication, 37/60
consciousness loss, 23/7, 207, 420, 427, 432, 440
coronary bypass surgery, 63/187
cortical lesion, 3/65, 75
cryoglobulinemia, 63/404
cyanide intoxication, 65/28
cyanogenic glycoside intoxication, 65/28
deafferentation, 3/63
decerebration syndrome, 3/65, 75
decortication syndrome, 3/65, 75
definition, 3/62
dépassé, *see* Brain death
diabetic, *see* Diabetic coma
diabetic ketoacidosis, 27/130
dialysis disequilibrium syndrome, 63/523
dialysis encephalopathy, 63/526
different scale, 57/381
dissecting aorta aneurysm, 63/47
duration, 57/109
EEG, 23/337, 340
EEG computer assisted analysis, 57/190-192
ergot intoxication, 65/68
ethylene glycol intoxication, 64/123
etiology, 3/63, 69-71
examination, 3/70-75, 77
filariasis, 35/168, 52/516
fish roe intoxication, 37/86
Galveston orientation and amnesia test, 57/383
Glasgow Coma Scale, 57/381
Glasgow Coma Sum Score, 57/330
Glasgow-Liège Coma Scale, 57/385
Glasgow Outcome Scale, 57/370
glutaric aciduria type II, 62/495
grade, 23/428
gram-negative bacillary meningitis, 52/119
gyromitrine intoxication, 65/36
head injury, 57/228
heat stroke, 71/588
hemoglobin SC disease, 55/465
hemolytic uremic syndrome, 63/324
hepatic, *see* Hepatic coma
herpes simplex virus encephalitis, 56/214
heterocyclic antidepressant intoxication, 65/323
hexachlorophene intoxication, 37/486, 65/476
hypernatremia, 63/554
hyperosmolar, *see* Hyperosmolar coma
hyperosmolar hyperglycemic nonketotic diabetic, *see* Hyperosmolar hyperglycemic nonketotic diabetic coma
hypertensive encephalopathy, 11/555, 54/212
hypocalcemia, 63/561
hypoglycemic, *see* Hypoglycemic coma
hypomagnesemia, 63/562
hyponatremia, 63/545
hypophosphatemia, 63/564
hypothyroid polyneuropathy, 51/515
hypothyroidism, 27/260, 70/93
iatrogenic neurological disease, 63/187
International Coma Data Bank, 57/370
intracranial hemorrhage, 53/208
intravascular consumption coagulopathy, 63/316
Jamaican vomiting sickness, 37/512, 533, 65/79
Kawasaki syndrome, 52/265, 56/639
ketotic, *see* Ketotic coma
ketotic hyperammonemia, 29/106
lead encephalopathy, 64/436
legal death, 3/77
lesion localization, 3/68
lidocaine intoxication, 65/454
lightning injury, 7/355, 360
lithium intoxication, 64/295
meningeal leukemia, 16/234, 63/341
meningococcal meningitis, 33/23
mesencephalic lesion, 2/280
metabolic encephalopathy, 45/123
methanol intoxication, 36/353, 356, 64/99
metharbital intoxication, 37/202
migraine, 48/144, 148, 156
mountain sickness, 63/416
multiple acyl-coenzyme A dehydrogenase deficiency, 62/495
mumps encephalitis, 34/628
muscle tone, 3/65, 75
mycotic aneurysm, 52/296
myoglobinuria, 40/338
neurologic intensive care, 55/204
nonketotic, *see* Nonketotic coma
open heart surgery, 63/187

opiate intoxication, 65/353
organic acid metabolism, 66/641
organophosphate intoxication, 37/555, 64/167
orienting reaction, 3/73
painful stimuli, 3/65, 73
paraproteinemic, *see* Paraproteinemic coma
parasomnia, 3/76
partial traumatic, 23/11
pathophysiology, 3/62-66
pellagra, 7/574, 38/648
pentaborane intoxication, 64/360
perception, 3/62, 71
pertussis, 52/234
pertussis encephalopathy, 9/551
phencyclidine intoxication, 65/54
phenobarbital intoxication, 37/202
phenytoin intoxication, 65/501
pontine hemorrhage, 12/44, 50
pontine infarction, 2/249, 12/15, 20
posttraumatic, *see* Posttraumatic coma
posttraumatic amnesia, 57/373, 382, 398
posture, 3/65
primary pontine hemorrhage, 12/37, 39
prognosis, 46/611
prolonged, *see* Brain death
prolonged transient ischemic attack, 53/262
propionic acidemia, 27/72
psychiatric aspect, 3/113
pyrethroid intoxication, 64/219
rabies, 34/254, 56/386, 388
reactive apathic hypoperceptive syndrome, *see* Brain death
reactive hyperpathic hypertonic aperceptivity syndrome, *see* Decortication
reactivity, 3/62, 71, 73
reflex, 3/75
resuscitation method, 3/71, 77
Reye syndrome, 27/363, 29/331, 333, 31/168, 34/169, 49/217, 56/150, 151, 238, 65/115
sickle cell anemia, 55/510
snake envenomation, 65/178
spasticity, 3/65
spinal cord electrotrauma, 61/192
spindle, 57/187-189
subarachnoid hemorrhage, 12/172, 55/21
superior longitudinal sinus thrombosis, 52/178
systemic lupus erythematosus, 55/376
thallium intoxication, 64/325
thrombotic thrombocytopenic purpura, 55/472, 63/324
toxic shock syndrome, 52/258
transient ischemic attack, 53/208, 262
Traumatic Coma Data Bank, 57/380
tricyclic antidepressant intoxication, 65/321
tuberculous meningitis, 52/203
uremia, 63/504
uremic encephalopathy, 63/504
uremic polyneuropathy, 51/355
valproic acid induced hepatic necrosis, 65/125
venous sinus thrombosis, 54/395
waking reaction, 3/73
Waldenström macroglobulinemia, 63/396, 398
water intoxication, 63/547, 64/241
zinc intoxication, 36/337
Coma dépassé, *see* Brain death
Coma Scale
Brussels Coma Scale, 57/107
Children Coma Score, 57/329
Glasgow, *see* Glasgow Coma Scale
Grady Coma Scale, 57/107
head injury, 57/107
Innsbruck Coma Scale, 57/107
mortality, 57/110
Ommaya Coma Scale, 57/107
Reaction Level Scale, 57/107
Coma vigil, *see* Akinetic mutism
Comb system
anatomy, 6/27
Edinger, *see* Edinger comb system
Combined complex I-V deficiency
cardiomyopathy, 62/506
encephalomyopathy, 62/506
lactic acidosis, 62/506
myopathy, 62/506
ragged red fiber, 62/506
Combined cord degeneration
subacute, *see* Subacute combined spinal cord degeneration
Combined myelopathies
deficiency neuropathy, 7/644
Combined peripheral and CNS inflammatory demyelinating syndrome
progressive spinal muscular atrophy, 59/406
Combined sensation
brain cortex, 1/104
Combined stenosis
vertebral canal stenosis, 50/466
Commissura anterior
embryo, 50/11
Marchiafava-Bignami disease, 9/653, 46/337
schizencephaly, 50/176
Commissure
CNS, 30/30
habenular, *see* Habenular commissure
hippocampus, *see* Hippocampus commissure
posterior, *see* Posterior commissure

Commissurotomy
  brain, *see* Brain commissurotomy
  cerebral, *see* Brain commissurotomy
Common carotid artery
  Doppler sonography, 54/5
  headache, 5/128
  origin variation, 12/294
  radioanatomy, 11/65
Common migraine
  bilateral, 5/19
  bilious attack, 5/46
  classic migraine, 5/16, 48, 48/117
  classification, 48/6
  clinical picture, 5/48
  clinical syndrome, 48/118
  ergotamine, 5/53
  hemiplegia, 48/144, 149
  history, 5/15
  motion sickness, 5/46
  natural history, 5/48
  occupation, 5/19
  personality, 5/51
  vertigo, 48/157
Common peroneal nerve
  entrapment, 7/315
  topographical diagnosis, 2/45
Commotio cerebri
  *see also* Brain concussion, Brain injury, Craniofacial injury, Head injury *and* Skull injury
  boxing injury, 23/562
  course, 23/421
  CSF, 23/420
  echoEG, 23/269
  EEG, 23/421
  history, 23/3
  loss of consciousness, 23/420
  pathology, 23/423
  posttraumatic amnesia, 23/420
  retrograde amnesia, 23/432
  terminology, 23/6, 419
  treatment, 23/422
Commotio medullae oblongata
  history, 23/3
Commotio spinalis, *see* Spinal cord concussion
Communicating artery
  anterior, *see* Anterior communicating artery
  posterior, *see* Posterior communicating artery
Communicating syringomyelia
  hydromyelia, 50/425
  *vs* noncommunicating type, 32/256, 287, 292, 295, 300-308, 313, 317
  type, 50/457
Comparative neurology
  demyelinating disease, 9/664-668
  striatal necrosis, 49/511
Compartment syndrome
  nerve damage, 70/43
Compelling helio-ophthalmic outburst
  autosomal dominant, *see* Autosomal dominant compelling helio-ophthalmic outburst
Compensation neurosis
  headache, 5/183
  minor head injury, 46/628
Complement
  C-3, *see* C-3 complement
  C-3a, *see* C-3a complement
  C-3b, *see* C-3b complement
  C-4, *see* C-4 complement
  C-5a, *see* C-5a complement
  inflammatory myopathy, 41/376
  multiple sclerosis, 47/366
  myasthenia gravis, 41/119
  senile dementia, 46/284
  viral infection, 56/58
Complement C1
  CSF, *see* CSF C-1 complement
Complement C3, *see* C-3 complement
Complement component
  multiple sclerosis, 47/303
Complement deficiency
  gram-negative bacillary meningitis, 52/105
  immunocompromised host, 56/468
  recurrent meningitis, 52/53
Complement factor, *see* Complement component
Complement mediated microangiopathy
  myositis, 62/369
Complement system
  intravascular consumption coagulopathy, 55/495
  multiple sclerosis, 47/309
Complete cord lesion
  ankylosing spondylitis, 26/178-180
  cervical, 25/269, 282, 335, 359
  child, 25/178
  definition, 25/146
  epidural hematoma, 26/21
  lumbodorsal, 25/443
  postural control, 26/524
  prognosis, 25/282, 26/307
  reflex pattern, 26/256
  respiration, 26/255-257
  sexual function, 26/448
  spinal cord injury prognosis, 26/307
  spinal deformity, 26/166, 172
  spinal epidural hematoma, 26/21
  thoracic, 25/442
  traumatic spine deformity, 26/166, 172

Completion phenomenon
homonymous hemianopia, 2/584
occipital lobe tumor, 17/342
Complex organic mercury
mercury polyneuropathy, 51/271
Complex I
mitochondrial DNA, 66/422
Complex I deficiency
ataxia, 62/504, 66/422
cardiomyopathy, 66/422
cardiopathy, 62/503
cardiorespiratory failure, 62/503, 66/422
chorea, 62/504
clinical syndrome, 66/422
dementia, 62/504, 66/422
epilepsy, 62/504
exercise intolerance, 66/422
fatal infantile multisystem disorder, 62/503, 66/422
hearing loss, 62/504, 66/422
hypotonia, 66/422
infantile hypotonia, 62/503, 66/422
lactic acidosis, 62/503
metabolic myopathy, 62/503
mitochondrial disease, 62/503
mitochondrial encephalomyopathy, 62/504, 66/422
multiple neuronal system degeneration, 62/503
myopathy, 62/503, 66/422
neuropathy, 66/422
ophthalmoplegia, 62/504, 66/422
pigmentary retinopathy, 62/504, 66/422
psychomotor retardation, 62/503, 66/422
reduced nicotinamide adenine dinucleotide dehydrogenase, 62/503
retinopathy, 66/422
sensory neuropathy, 62/504, 66/422
Complex II
myoglobinuria, 62/558
nuclear DNA, 66/422
respiratory chain complex, 66/422
Complex II deficiency
ataxia, 66/422
developmental delay, 66/422
encephalomyopathy, 62/504, 66/422
exercise intolerance, 62/504
hypotonia, 66/422
lethargy, 66/422
metabolic myopathy, 62/504
mitochondrial disease, 62/504
myoclonus, 66/422
myoglobinuria, 62/504
myopathy, 62/504
respiratory insufficiency, 66/422
succinate cytochrome-c reductase, 62/504
weakness, 66/422
Complex III
mitochondrial DNA, 66/423
Complex III deficiency
aminoaciduria, 62/504
ataxia, 62/504, 66/423
cardiomyopathy, 66/423
cardiopathy, 62/505
coma, 62/504, 66/423
dementia, 62/504, 66/423
dystonia, 62/504, 66/423
encephalomyopathy, 62/504
exercise intolerance, 62/504, 66/423
fatal infantile encephalomyopathy, 66/423
infantile histiocytoid cardiomyopathy, 62/505
infantile hypotonia, 62/504, 66/423
lactic acidosis, 62/504, 66/423
metabolic myopathy, 62/504
mitochondrial disease, 62/504
multiple neuronal system degeneration, 62/504
muscle weakness, 62/504, 66/423
myopathy, 62/504, 66/423
pigmentary retinopathy, 62/504, 66/423
pyramidal syndrome, 62/504, 66/423
seizure, 62/504, 66/423
sensorineural deafness, 62/504, 66/423
sensory neuropathy, 62/504
short stature, 62/504, 66/423
Complex IV
complex IV deficiency, 62/505
cyanide intoxication, 65/28
cyanogenic glycoside intoxication, 65/28
histotoxic hypoxia, 9/621
Kearns-Sayre-Daroff-Shy syndrome, 62/314
Levine experimental anoxia ischemia, 9/583
Marchiafava-Bignami disease, 9/653, 47/558
methanol intoxication, 64/96
mitochondrial disease, 40/31, 62/497
mitochondrial DNA, 66/423
mitochondrial respiratory chain defect, 42/587
progressive external ophthalmoplegia, 62/293, 314
senile plaque, 46/255
skeletal muscle, 40/4
wallerian degeneration, 9/37
Complex IV deficiency
ataxia, 66/423
combined complex deficiency, 62/509
complex IV, 62/505
DeToni-Fanconi-Debré syndrome, 62/505, 66/424
dystonia, 66/424

encephalomyopathy, 66/423
encephalopathy, 62/505
fatal infantile myopathy, 62/505, 66/424
hypotonia, 66/424
lactic acidosis, 62/505, 66/424
metabolic myopathy, 62/505
mitochondria, 41/211
mitochondrial disease, 62/505
mitochondrial DNA deletion, 62/508
myopathic syndrome, 66/424
myopathy, 62/505, 66/423
nystagmus, 66/424
ophthalmoplegia, 66/423
optic atrophy, 66/423
POLIP syndrome, 62/506
pyramidal tract sign, 66/424
pyruvate dehydrogenase complex deficiency, 66/425
respiratory insufficiency, 62/505, 66/424
subacute necrotizing encephalomyelopathy, 41/212, 62/505, 66/424
trichopoliodystrophy, 41/212
Complex V
mitochondrial DNA, 66/424
proton transporting adenosine triphosphate synthase, 66/424
Complex V deficiency
ataxia, 62/506
dementia, 62/506
metabolic encephalopathy, 62/506
metabolic myopathy, 62/506
mitochondrial disease, 62/506
multiple neuronal system degeneration, 62/506
myopathy, 62/506
neuropathy, 62/506
paracrystalline inclusion, 62/506
ragged red fiber, 62/506
retinopathy, 62/506
Complicated hereditary infantile optic atrophy, *see* Behr disease
Complicated hydromyelia
syringomyelia, 50/427
Complicated migraine, 42/747-749
aphasia, 5/63
basilar artery migraine, 48/166
brain infarction, 11/465
capsula interna, 5/85
carotid angiography, 5/66
cerebellar, 5/77, 42/749
cerebral localization, 5/65
cervical migraine, 5/200
child, 48/147
chronic meningitis, 56/644
classification, 48/6
clinical features, 5/60
definition, 5/59
dysphrenia hemicranica, 5/78-80
dysphrenic, 5/78
EEG, 5/64, 66
facioplegic, 5/75, 42/748
familial hemiplegic, 5/266, 42/748
hemiplegia, 5/63, 48/145
history, 5/59
involuntary movement, 5/80
localization, 5/84
morbid anatomy, 5/80
nomenclature, 48/155
ophthalmoplegia, 5/68, 42/747
paresthesia, 5/61
subgroup, 48/131
thalamic localization, 5/85
Components
P, *see* P component
serum M, *see* Serum M component
Compound muscle action potential
familial infantile myasthenia, 62/428
hereditary multiple recurrent mononeuropathy, 51/552
organophosphate intoxication, 64/163
Compound skull fracture
*see also* Depressed skull fracture
depressed skull fracture, 23/405, 410, 24/226, 57/146, 334
internal type, 24/650, 659
low velocity, 24/647
surgery, 24/650
type, 24/646
Compounds
grouting, *see* Grouting compound
Comprehension
alexia, 45/433
frontal lobe tumor, 67/150
verbal, *see* Verbal comprehension
Comprehension disturbance
traumatic psychosyndrome, 24/551
Compression
acoustic nerve, *see* Acoustic nerve compression
aneurysm, *see* Aneurysm compression
anterior spinal artery, 26/100
arterial, *see* Arterial compression
atlantoaxial joint, 26/289
brain, *see* Brain compression
brain contusion, 23/84
brain edema, 23/150
brain injury, 23/84
brain stem, *see* Brain stem compression

carotid, *see* Carotid compression
cauda equina, *see* Cauda equina compression
chronic nerve, *see* Chronic nerve compression
extramedullary, *see* Extramedullary compression
foramen magnum tumor, 2/208
gasserian ganglion, 5/397
headache, 48/6
intermittent nerve, *see* Intermittent nerve compression
ischemic neuropathy, 8/162
jugular vein, *see* Jugular vein compression
median nerve, *see* Median nerve compression
median nerve lesion, 7/329
nerve, *see* Nerve compression
odontoid peg, 71/23
optic atrophy, 13/69
optic chiasm, *see* Optic chiasm compression
recurrent laryngeal nerve, *see* Recurrent laryngeal nerve compression
spinal cord, *see* Spinal cord compression
spinal cord injury, 25/72, 260
spinal injury, 25/67
spinal nerve root, *see* Spinal nerve root compression
trigeminal root, 5/274
ulnar nerve, 7/162
vertebral, *see* Vertebral compression
vertebral artery, 12/7, 19/153
Compression fracture
cervical, *see* Cervical compression fracture
cervical vertebral column injury, 25/291, 345
mechanism, 25/134
spinal injury, 25/128, 345
thoracic vertebral column injury, 61/77
thoracolumbar vertebral column injury, 61/90
vertebral column, 25/125
Compression injury
cervical hyperanteflexion sprain, 25/341-345
cervical vertebral column injury, 61/23
spinal cord, 25/72, 260
thoracic spinal cord injury, 25/137
vertebral column, 25/67
Compression neuropathy, 7/276-283
*see also* Nerve compression
acute compression, 51/87
anterior interosseous nerve syndrome, 51/92
anterior tarsal tunnel syndrome, 51/105
autonomic change, 51/85
axillary nerve, 51/102
axoplasmic flow, 51/88
carpal tunnel syndrome, 7/286-300, 437-440, 51/87, 93, 133
chronic nerve compression, 51/88
clinical diagnosis, 51/85
computer assisted tomography, 51/86
concert musician, 51/87
conduction block, 51/91
crutch, 51/92
demyelination, 51/88
diabetic polyneuropathy, 51/501
diamorphine intoxication, 65/356
digital nerve, 51/102
drummer, 51/87
electroneurography, 51/86
EMG, 51/86
familial predisposition, 51/100
fascicle, 51/135
femoral nerve, 51/107
flutist, 51/87
genitofemoral nerve, 51/108
guitar player, 51/87
Guyon canal, 51/99
history, 51/85
honeymoon palsy, 51/92
hypothyroidism, 70/99
ilioinguinal nerve, 51/109
immobilization, 51/87
intercostal nerve, 51/109
intermittent nerve compression, 51/88
ischemia, 8/162, 51/88
lateral femoral cutaneous nerve, 51/108
Léri pleonosteosis, 7/293
long thoracic nerve, 51/102
median nerve, 51/91
microcephaly, 43/428
monomelic spinal muscular atrophy, 59/376
Morton neuralgia, 51/106
musculocutaneous nerve, 51/102
nerve, 51/2
nerve conduction, 7/157
nerve conduction velocity, 42/311
nerve injury, 51/134
nerve intermittent claudication, 20/792-795
neuralgic amyotrophy, 51/99, 102
obturator nerve, 7/314, 51/108
occupational circumstance, 51/87
opiate intoxication, 65/356
osteitis deformans (Paget), 38/361, 43/451
osteopetrosis, 43/456
paralysis, 21/89
pathophysiology, 51/87
peroneal nerve, 51/104
physical examination, 51/85
polyneuropathy, 51/86
predisposing factor, 51/86
radial nerve, 51/99

radial tunnel syndrome, 51/101
radiography, 51/86
rapid reversible conduction blocking, 51/88
recovery rate, 51/88
rectus abdominis muscle syndrome, 51/109
recurrent laryngeal nerve compression, 63/99
rheumatoid arthritis, 38/481-484, 71/17
rifle sling palsy, 51/92
saphenous nerve, 51/107
Saturday night palsy, 51/100
sciatic nerve, 51/102
sclerosteosis, 43/475
Sjögren syndrome, 71/85
spondyloepiphyseal dysplasia, 43/481
supracondylar ligament, 51/92
suprascapular nerve, 51/102
sural nerve, 51/106
symptom, 51/98, 106
tarsal tunnel syndrome, 51/105
terminology, 7/246
thoracic outlet syndrome, 51/87
tibial nerve, 51/105
ulnar nerve, 51/98
ulnar paralysis, 51/87
vibration neuropathy, 51/87
violinist, 51/87
wallerian degeneration, 51/90
Compression syndrome
costoclavicular syndrome, 7/430, 432, 442
differential diagnosis, 7/435
first rib syndrome, 7/430, 443
hyperabduction syndrome, 7/430, 432, 434, 441, 8/335
neurovascular, *see* Neurovascular compression syndrome
paresthesia nocturna, 7/437
shoulder girdle, 7/430
shoulder hand syndrome, 7/430
subclavian steal syndrome, 7/435
thoracic outlet syndrome, 7/430, 443
Compulsion
Pick disease, 46/234
Compulsive water drinking, *see* Water intoxication
Computer assisted tomography
acquired immune deficiency syndrome, 56/509
acquired immune deficiency syndrome dementia, 56/492
acquired toxoplasmosis, 52/356
acute contusion, 57/150
acute subdural hematoma, 57/152
acute viral encephalitis, 56/131
adrenoleukodystrophy, 47/597, 60/169, 671
affective psychosis, 46/452
alexia, 45/12
allergic granulomatous angiitis, 54/63
Alzheimer disease, 46/264, 54/58
amnesic shellfish poisoning, 65/167
amphetamine addiction, 55/520
amyloid angiopathy, 54/70
angiography, 54/65
angiostrongyliasis, 52/558
anoxic encephalopathy, 63/218
anterior cerebral artery syndrome, 53/341-343
anterior posterior asymmetry, 46/485
aphasia, 45/299
aqueduct stenosis, 50/312
Arnold-Chiari malformation type II, 50/409
aspergillosis, 52/380
ataxia telangiectasia, 60/376, 671
ataxic hemiparesis, 54/57
bacterial endocarditis, 52/292
bacterial meningitis, 52/8, 55/419
basal encephalocele, 50/106
basal ganglion, 46/486
basal ganglion hematoma, 57/162
basal meningoencephalocele, 50/106
Bassen-Kornzweig syndrome, 60/133
Bedouin spastic ataxia, 60/671
Behçet syndrome, 56/600, 71/219
bifrontal contusion, 57/150
Binswanger disease, 46/319, 321, 54/57, 59, 223
bismuth intoxication, 64/343
boundary zone ischemia, 57/163
brachial plexus, 51/148
brain abscess, 33/126, 46/387, 52/153, 57/322
brain abscess *vs* infarction, 54/51
brain aneurysm, 55/48
brain aneurysm neurosurgery, 55/48
brain arteriovenous malformation, 54/62, 70, 184, 379
brain atrophy, 46/344-346, 57/164
brain contusion, 23/262, 57/149, 381
brain death, 55/262
brain embolism, 53/220, 54/58
brain fat embolism, 55/182
brain glioma *vs* infarction, 54/51
brain hemorrhage, 54/67, 71, 295-297, 316
brain hemorrhagic infarction, 54/51
brain infarction, 53/220, 313, 54/47, 50, 52, 53, 84
brain injury, 23/243, 255
brain ischemia, 57/161, 63/218
brain lacunar infarction, 54/56, 255
brain phlebothrombosis, 54/65
brain stem death, 57/464
brain stem hematoma, 54/69
brain stem herniation, 45/113

brain swelling, 57/159, 381
brain thromboangiitis obliterans, 39/203
brain tumor hemorrhage, 54/72
brain vasculitis, 54/62, 71, 55/420
brain vasospasm, 54/64
Broca area, 46/486
bullet tract, 57/148
burst lobe, 57/149
candidiasis, 52/403
carbon monoxide intoxication, 49/476, 64/33
carotid artery stenosis, 54/58
carotid atherosclerosis, 54/58
carotid cavernous fistula, 57/166
carotid dissecting aneurysm, 54/274
carotid system syndrome, 53/315
cavernous sinus thrombosis, 52/174
central pontine myelinolysis, 47/588, 63/552
cerebellar abscess, 52/448
cerebellar atrophy, 46/483
cerebellar Creutzfeldt-Jakob disease, 60/671
cerebellar granular cell hypertrophy, 60/672
cerebellar hemorrhage, 54/314, 55/91
cerebellar hypoplasia, 60/672
cerebellar infarction, 53/383, 54/55
cerebellar vermis aplasia, 60/672
cerebello-olivary atrophy, 60/571, 671
cerebral amebiasis, 52/321
cerebromalacia, 57/164
cerebrovascular disease, 53/221, 54/47, 53
cervical vertebral column injury, 61/55
Charlevoix-Saguenay spastic ataxia, 60/455, 671
chorea-acanthocytosis, 49/329, 63/281
chronic subdural hematoma, 54/354, 57/154, 287
clumsiness, 46/163
clumsy hand dysarthria syndrome, 54/57
cluster headache, 48/237
coccidioidomycosis, 52/418
cognition, 46/482
compression neuropathy, 51/86
concentric sclerosis, 47/414
congenital olivopontocerebellar atrophy, 60/671
congenital retinal blindness, 60/672
congenital toxoplasmosis, 52/359
contrast enhancement, 54/51
contusional hematoma, 57/149
corpus callosum agenesis, 50/154
cranial epidural empyema, 52/169
craniopharyngioma, 18/547
Creutzfeldt-Jakob disease, 46/292
cytomegalic inclusion body disease, 56/268
Dandy-Walker syndrome, 50/329, 60/672
data presentation, 23/262
decompression myelopathy, 61/225
delayed hematoma, 57/156
dementia, 46/128, 202
density pattern, 54/47
dentatorubropallidoluysian atrophy, 60/671
diagnostic pitfalls, 54/51
dialysis disequilibrium syndrome, 63/524
diastematomyelia, 50/436, 439
diffuse axonal injury, 57/151, 153, 380
diffuse focal sign, 46/484
diplomyelia, 50/436, 439
dyslexia, 46/131
dyssynergia cerebellaris myoclonica, 60/601, 671
early onset recessive stationary ataxia, 60/671
echinococcosis, 52/525
eclampsia, 54/60
eosinophilia myalgia syndrome, 64/257
epidural hematoma, 23/259, 262, 54/347, 57/103, 152, 154, 256, 274
epilepsy, 72/337
ethylene glycol intoxication, 64/123
extra pontine myelinolysis, 63/552
familial ectodermal dysplasia, 60/671
fibromuscular dysplasia, 54/64
focal abnormality, 46/486
Friedreich ataxia, 60/320, 671
functional psychosis, 46/451-453
general paresis of the insane, 52/279
Gerstmann-Sträussler-Scheinker disease, 56/554, 60/671
Gillespie syndrome, 60/671
gliding contusion, 57/153
global aphasia, 45/312
glutaryl-coenzyme A dehydrogenase deficiency, 66/643
GM2 gangliosidosis type III, 60/671
Gruber syndrome, 60/672
guideline, 57/272
gunshot injury, 57/302
HARD syndrome, 60/672
Haw River syndrome, 60/671
head injury, 57/224, 272, 274, 381
head injury monitoring, 57/164
head injury outcome, 57/111
head injury prognosis, 57/164
headache, 48/277-279
Hemophilus influenzae meningitis, 52/61
hemorrhagic infarction course, 54/52
hemorrhagic infarction density, 54/51
hereditary hemorrhagic telangiectasia (Osler), 55/451
hereditary paroxysmal dystonic choreoathetotis, 49/351
hereditary paroxysmal kinesigenic

choreoathetosis, 49/353
hereditary striatal necrosis, 49/495
herpes simplex virus encephalitis, 54/72, 56/215
Hirayama disease, 59/111
histoplasmosis, 52/439
holoprosencephaly, 50/237
homolateral ataxia crural paresis syndrome, 54/57
Huntington chorea, 49/288
hydranencephaly, 50/351
hydrocephalic dementia, 46/328
hydrocephalus, 30/555, 50/293
hydromyelia, 50/430
hypereosinophilic syndrome, 63/372
hypertensive encephalopathy, 54/60, 212
infantile autism, 46/191
infantile spinal muscular atrophy, 59/57, 60/672
infarct edema, 54/47
infarct hypodensity, 54/47
infarct induced herniation, 54/48
infarct mass effect, 54/47
internal carotid artery syndrome, 53/315
intracerebellar hemorrhage, 54/68
intracerebral hematoma, 24/362, 54/72
intracranial hemorrhage, 53/220
intracranial pressure, 57/160, 276
intracranial vessel spasm, 57/166
intraventricular hemorrhage, 54/68, 57/148
Joubert syndrome, 60/672
juvenile spinal cord injury, 61/243
Kearns-Sayre-Daroff-Shy syndrome, 62/306, 311
lacunar stroke, 54/53
lateral asymmetry, 46/486
lead encephalopathy, 64/436
legionellosis, 52/256
lissencephaly, 50/268
Listeria monocytogenes encephalitis, 52/94
livedo reticularis, 54/64
lymphomatoid granulomatosis, 54/63
management guideline, 57/272
Marinesco-Sjögren syndrome, 60/671
Matthews-Rundle syndrome, 60/577, 671
meningitis, 54/63
mesencephalic infarction, 53/392
1-methyl-4-phenyl-1,2,3,6-tetrahydropyridine intoxication, 65/372
metrizamide, *see* Metrizamide computer assisted tomography
microcephaly, 50/268, 281, 60/672
middle cerebral artery infarction, 54/54
middle cerebral artery syndrome, 53/358
midline shift, 57/217
migraine, 48/145, 148, 151, 54/64
missile injury, 57/302
mitochondrial encephalomyopathy, 60/672
mixed attenuation, 57/154
motor aphasia, 45/310
moyamoya disease, 54/64, 55/296
multi-infarct dementia, 46/362, 54/58
multiple sclerosis, 47/66, 191
mycotic aneurysm, 52/295
myelinoclastic diffuse sclerosis, 47/424
neonatal adrenoleukodystrophy, 60/671
neuroamebiasis, 52/321
neuroaxonal dystrophy, 60/673
neurocysticercosis, 52/531
neuroleptic malignant syndrome, 71/592
neuromyelitis optica, 47/402
neuronal intranuclear hyaline inclusion disease, 60/671
neuropathology, 54/66
neuropsychology, 45/1, 7, 11
neurosarcoidosis, 63/422
neurosyphilis, 52/278
Niemann-Pick disease type C, 60/152
nocardiosis, 35/524, 52/447
Norman disease, 60/672
nuclear magnetic resonance, 54/65, 57/177, 381
obstructive hydrocephalus, 57/164
occipital encephalocele, 50/100
occipital reversal, 46/484
olivopontocerebellar atrophy (Dejerine-Thomas), 60/514-516, 519, 522-528, 671
olivopontocerebellar atrophy (Schut-Haymaker), 60/671
olivopontocerebellar atrophy (Wadia-Swami), 60/499, 671
orbital myositis, 62/299, 305
paracoccidioidomycosis, 52/459
parietal cephalocele, 50/108
parietal encephalocele, 50/108
Parkinson disease, 49/98
Patau syndrome, 50/560
pediatric head injury, 57/329
penetrating head injury, 57/308
pheochromocytoma, 42/765
Pick disease, 46/242
pneumococcal meningitis, 52/51
pontine hemorrhage, 54/68
pontocerebellar hypoplasia, 60/672
porencephaly, 42/49
positron emission tomography, 54/65
posterior cerebral artery infarction, 54/54
posterior fossa hematoma, 57/157, 253
posterior inferior cerebellar artery infarction, 54/55
posttraumatic amnesia, 45/187

posttraumatic syringomyelia, 50/454, 61/384
primary alcoholic dementia, 46/344-346
primary brain injury, 57/149
progressive dysautonomia, 59/139
progressive external ophthalmoplegia, 62/295, 306, 311
progressive multifocal leukoencephalopathy, 47/509, 511
progressive supranuclear palsy, 46/301, 49/245
prolonged transient ischemic attack, 53/258
pure motor hemiplegia, 53/364, 54/57, 241
pure sensory stroke, 54/57, 251
radiodiagnosis, 57/149
renal retinal syndrome, 60/672
Rett syndrome, 63/437
Reye syndrome, 56/153
Salla disease, 59/361
scanner, 23/255
schistosomiasis, 52/540
schizencephaly, 50/201
schizophrenia, 46/451, 481-487
Schut family ataxia, 60/484
senile dementia, 46/129, 286
sex chromosome fragility, 60/672
sickle cell anemia, 55/506
sincipital encephalocele, 50/104
single photon emission, *see* Single photon emission computer assisted tomography
slit like ventricle, 57/160
Sneddon syndrome, 54/64
sphenoid wing aplasia, 50/143
spinal cord abscess, 52/191
spinal cord injury, 61/515
spinal epidural empyema, 52/188
spinal tumor, 67/206
striatopallidodentate calcification, 49/428, 60/672
Sturge-Weber syndrome, 55/445
subarachnoid hemorrhage, 54/60, 70, 55/48, 57/157
subdural empyema, 52/73, 57/323
subdural hematoma, 23/260, 262, 54/350, 57/152, 274
succinate semialdehyde dehydrogenase deficiency, 60/671
superior longitudinal sinus thrombosis, 52/178
supratentorial brain abscess, 33/126
syphilis, 55/413
syringomyelia, 50/447
systemic lupus erythematosus, 54/63, 55/380, 71/47
thalamic hemorrhage, 54/310, 312
thalamic infarction, 53/392
thoracic vertebral column injury, 61/76
thoracolumbar vertebral column injury, 61/91
thyroid associated ophthalmopathy, 62/531
toxoplasmosis, 52/356
transient global amnesia, 45/213
transient ischemic attack, 53/220, 258, 268, 313, 315, 54/53
transmission, *see* Transmission computer assisted tomography
transverse sinus thrombosis, 52/176
traumatic carotid cavernous sinus fistula, 57/353
traumatic intracerebral hematoma, 57/149, 156
traumatic intracranial hematoma, 57/152, 267, 272-274
tuberculous meningitis, 52/207, 215, 55/429
tuberous sclerosis, 50/371
uremic encephalopathy, 63/506
vascular dementia, 46/361
venous sinus thrombophlebitis, 33/166
venous sinus thrombosis, 54/65, 400, 443
vertebral column injury, 61/515
vertebrobasilar system syndrome, 53/392
vertebrobasilar transient ischemic attack, 54/53
viral infection, 56/86
watershed infarction, 54/55, 64
Whipple disease, 52/138
Wilson disease, 49/233
X-linked adrenoleukodystrophy, 60/672
xenon, *see* Xenon computer assisted tomography
xeroderma pigmentosum, 60/337
Computer model
myotonia, 62/277
Concanavalin A
Entamoeba histolytica, 52/322
multiple sclerosis, 47/346-349, 374
Concentric cerebral sclerosis, *see* Concentric sclerosis
Concentric laminated body
acromegaly, 40/109
central core myopathy, 40/109
Dyggve-Melchior-Clausen syndrome, 40/109
histochemistry, 40/39
Marfan syndrome, 40/109
mitochondrial myopathy, 40/109
muscle fiber, 40/39, 109, 154, 249
nemaline myopathy, 40/109
periodic paralysis, 40/109
poliomyelitis, 40/109
polyneuropathy, 40/109
Concentric sclerosis, 9/437-450, 47/409-416
acute multiple sclerosis, 47/409, 416
age, 47/414
alternating demyelination, 9/449
apathy, 47/414

characteristic lesion, 9/446
classification, 9/472
clinical features, 9/448, 47/414
computer assisted tomography, 47/414
course, 47/414
CSF examination, 47/414
dementia, 46/404
diagnosis, 47/416
disseminated encephalomyelitis, 47/473
dysphagia, 47/414
encephalomalacia, 9/440
etiology, 9/439, 47/416
frequency, 47/411
gait disturbance, 47/414
headache, 47/414
hemihypesthesia, 47/416
hemiplegia, 47/416
histopathology, 9/448
history, 47/409-411
inflammatory diffuse sclerosis, 47/411, 416
initial symptom, 47/414
lecithinolytic ferment, 9/438
Liesegang ring, 9/438, 47/409
Mahoney technique, 9/441, 446
mode of onset, 47/414
multiple sclerosis, 9/440, 47/73, 409, 411, 414, 416
mutism, 47/414
original description, 10/10
pathologic characteristics, 47/412
pathology, 47/411-414
Philippines, 47/411
potassium cyanide intoxication, 9/440
seizure, 47/414
sex ratio, 47/414
spasticity, 47/414
storm center, 9/439, 447
stroke, 9/449
urinary incontinence, 47/414
vascular impairment, 9/441
Concomitant strabismus
differential diagnosis, 2/291
oculomotor paralysis, 2/291
paralytic strabismus, 2/291
Concordance
multiple sclerosis, 47/294
Concussio mercurialis, *see* Chronic mercury intoxication
Concussion
brain, *see* Brain concussion
brain stem, *see* Brain stem concussion
cerebellar, *see* Cerebellar concussion
labyrinth, *see* Labyrinth concussion
spinal cord, *see* Spinal cord concussion
spinal cord injury, 61/37
Conditional responsive extrapyramidal syndrome, *see* Hereditary paroxysmal kinesigenic choreoathetosis
Conditioned evoked potential
attention, 3/156
Conditioning
attention, 3/156
epilepsy, 15/550
prefrontal syndrome, 2/744
stress, 45/248
Condom drainage
bladder, 26/423
Conductance
muscle, *see* Muscle fiber conductance
muscle fiber, *see* Muscle fiber conductance
myelin sheath, 51/41
sarcolemma, 40/549
Conduction
agraphia, 4/149
cardiac, *see* Cardiac conduction
facial nerve, 7/165
femoral nerve, 7/164
impulse, *see* Impulse conduction
motor nerve, *see* Motor nerve conduction
nerve, *see* Nerve conduction
nerve decompression, 7/161
nerve regeneration, 7/153
pathologic, *see* Pathologic conduction
peroneal nerve, *see* Peroneal nerve conduction
radial nerve, 7/164
saltatory, *see* Saltatory conduction
sensory nerve, *see* Sensory nerve conduction
spinal cord, *see* Spinal cord conduction
spinal cord experimental injury, 25/11
Conduction aphasia
alexia, 4/132, 45/447
apraxia, 45/429
brain embolism, 53/215
brain infarction, 53/215
cerebrovascular disease, 53/215
intracranial hemorrhage, 53/215
parietal lobe, 45/18
parietal lobe syndrome, 45/75, 77
radioisotope diagnosis, 45/298
transient ischemic attack, 53/215
Wernicke aphasia, 4/98, 45/316
Conduction block
compression neuropathy, 51/91
demyelinated nerve fiber, 47/15
demyelination, 47/30, 38
experimental demyelination, 47/30

idiopathic facial paralysis, 2/68
nerve, *see* Nerve conduction block
nerve injury, 51/75
rapid reversible, *see* Rapid reversible conduction blocking
Conduction property
axon, 51/41
nerve, 51/41
Condylar hypoplasia
C1 evaluation, 50/396
Condyloid emissary foramen
anatomy, 11/55
Condylus
anterior occipital, *see* Anterior occipital condyle
occipital, *see* Occipital condylus
third occipital, *see* Third occipital condylus
Cone
cerebellar, *see* Tonsillar herniation
cerebellar pressure, *see* Cerebellar pressure cone
cerebellum, *see* Tonsillar herniation
temporal pressure, *see* Temporal pressure cone
Cone degeneration
primary pigmentary retinal degeneration, 13/28
Cone dysfunction, *see* Cone degeneration
Cone dystrophy
hereditary progressive, *see* Hereditary progressive cone dystrophy
Cone intoxication
envenomation, 37/60-62
Confabulation
agnosia, 4/24, 28
anterior cerebral artery syndrome, 53/349
basofrontal syndrome, 53/349
body scheme disorder, 4/217
delirium, 46/528
frontal lobe syndrome, 53/349
memory disorder, 3/275
Pick disease, 46/234
prefrontal syndrome, 2/749
primary alcoholic dementia, 46/342
thalamic dementia, 21/595
traumatic psychosis, 24/525
Wernicke-Korsakoff syndrome, 28/252
Confluens sinuum
asymmetry, 11/112
confluens venosum, 11/104
radioanatomy, 11/103
type, 11/53
Confusion
acetylsalicylic acid intoxication, 65/460
allergic granulomatous angiitis, 63/383
Amanita muscaria intoxication, 37/329
amnesia, 3/270
amnesic shellfish poisoning, 65/167
bacterial meningitis, 52/1
barbiturate intoxication, 65/332
bismuth intoxication, 64/338
brain death, 55/268
brain embolism, 53/206
brain infarction, 53/206
brain injury, 23/12
carbon monoxide intoxication, 49/475, 64/32
cat scratch disease, 52/129
cerebral amebiasis, 52/315
cerebrovascular disease, 53/206
chlormethine intoxication, 64/233
classic migraine, 48/165
coccidioidomycosis, 52/413
consciousness, 3/113, 45/117
consciousness content disorder, 45/114
cyanide intoxication, 64/232
cycloserine intoxication, 65/481
digoxin intoxication, 65/450
disopyramide intoxication, 65/453
eosinophilia myalgia syndrome, 64/256
epilepsy, 73/480
fluorouracil intoxication, 65/532
headache, 48/284
hypercalcemia, 63/561
hypereosinophilic syndrome, 63/372
hypernatremia, 63/554
hypocalcemia, 63/558, 561
hypomagnesemia, 63/562
hypophosphatemia, 63/564
intracranial hemorrhage, 53/206
lead encephalopathy, 64/435
lidocaine intoxication, 37/450, 65/454
lindane intoxication, 65/479
lymphocytic choriomeningitis, 9/197
mercury polyneuropathy, 51/272
methanol intoxication, 64/99
migraine, 5/79, 48/7, 144, 146, 162
North American blastomycosis, 35/406, 52/389
organic solvent intoxication, 64/42
organochlorine insecticide intoxication, 64/199
organophosphate intoxication, 64/226
organotin intoxication, 64/141
pellagra, 38/648, 46/336
petit mal status, 15/170
phenylpropanolamine intoxication, 65/259
posttraumatic amnesia, 24/479
procainamide intoxication, 37/451, 65/453
respiratory encephalopathy, 63/414
Shapiro syndrome, 50/167
superior longitudinal sinus thrombosis, 52/178
tabernathe iboga intoxication, 37/329

toxic oil syndrome, 63/381
transient ischemic attack, 53/206
uremia, 63/504
Waldenström macroglobulinemia, 63/396
Wernicke encephalopathy, 2/450
Confusional state
acute, *see* Delirium
agraphia, 45/462
alexia, 4/134
allergic granulomatous angiitis, 63/383
aluminum intoxication, 63/526
basilar artery migraine, 48/138
Behçet syndrome, 56/597
Binswanger disease, 54/222
brain infarction, 55/138
cardiac catheterization, 63/176
Castleman disease, 63/395
cat scratch disease, 52/130
cerebrovascular disease, 55/138
ciclosporin, 63/537
ciclosporin intoxication, 63/537
cytarabine, 63/357
dialysis, 63/528
dialysis encephalopathy, 63/526, 528
dysphrenia hemicranica, 5/78
hypercalcemia, 63/561
hypereosinophilic syndrome, 63/372
hyperviscosity syndrome, 55/470
hypocalcemia, 63/558, 561
hypomagnesemia, 63/562
hyponatremia, 63/545
intravascular consumption coagulopathy, 63/316
levodopa, 46/379
lidocaine, 63/192
memory disorder, 3/270
meningeal lymphoma, 63/347
methotrexate, 63/357
middle cerebral artery, 55/138
migraine, 48/6, 8
migrainous acute, *see* Migrainous acute confusional state
multiple myeloma, 63/395
nonsteroid anti-inflammatory agent intoxication, 65/416
osteosclerotic myeloma, 63/395
POEMS syndrome, 63/395
polyarteritis nodosa, 55/356
pontine infarction, 12/15, 20, 23, 30
posterior cerebral artery, 55/139
primary alcoholic dementia, 46/342
rabies, 56/386
rejection encephalopathy, 63/530
respiratory encephalopathy, 63/414
systemic lupus erythematosus, 55/370
temporal arteritis, 55/344
toxic oil syndrome, 63/381
transient ischemic attack, 12/3
uremic encephalopathy, 63/504
vincristine, 63/358
Waldenström macroglobulinemia, 63/396
Wernicke-Korsakoff syndrome, 46/341
Congenital abnormality
congenital megacolon, 75/635
hereditary sensory and autonomic neuropathy type I, 21/73, 60/9
neuropsychology, 45/4
progressive bulbar palsy, 59/375
Takayasu disease, 12/415
vertebral arterial malformation, 12/289-335
Congenital acetylcholine receptor deficiency/short channel opentime syndrome
congenital myasthenic syndrome, 62/437
EMG, 62/437
facial diplegia, 62/437
neuromuscular junction disease, 62/392
ophthalmoplegia, 62/437
Congenital acetylcholine receptor deficiency/synaptic cleft syndrome
arthrogryposis, 62/438
congenital myasthenic syndrome, 62/438
cyanotic episode, 62/438
exacerbation, 62/438
exercise intolerance, 62/438
neuromuscular junction disease, 62/392
ptosis, 62/438
respiratory distress, 62/438
Congenital ε-acetylcholine receptor subunit mutation syndrome
congenital myasthenic syndrome, 62/436
EMG, 62/436
neuromuscular junction disease, 62/392
symptom, 62/436
Congenital acromicria, *see* Down syndrome
Congenital adrenal hyperplasia
adrenal disorder, 70/212
birth incidence, 42/514
convulsion, 42/515
desmolase deficiency, 42/514
genetic linkage, 42/515
HLA antigen, 42/515
11-hydroxylase deficiency, 42/514
17-hydroxylase deficiency, 42/514
21-hydroxylase deficiency, 42/514, 70/213
3β-hydroxysteroid dehydrogenase deficiency, 42/514
hypospadias, 42/514

pseudohermaphroditism, 42/514
survey, 70/212
Congenital adrenocortical hyperplasia
adrenal gland, 39/487
Congenital amaurosis (Leber), *see* Congenital retinal blindness
Congenital amyelinating neuropathy
Pena-Shokeir syndrome type I, 59/369
Congenital analgesia, 42/295-300
*see also* Congenital pain insensitivity, Hereditary sensory and autonomic neuropathy type I *and* Hereditary sensory and autonomic neuropathy type III
areflexia, 42/299
Biemond syndrome, 42/299
chromosomal aberration, 42/298
congenital autonomic dysfunction, 60/18
hereditary sensory and autonomic neuropathy, 60/19
hereditary sensory and autonomic neuropathy type III, 60/19
hereditary sensory and autonomic neuropathy type IV, 60/19
hereditary sensory and autonomic neuropathy type V, 60/19
polyneuropathy, 42/296
sensorineural deafness, 42/299
spinothalamic tract degeneration, 42/299
Congenital aneurysm, 31/137-173
*see also* Brain hemorrhage
anterior cerebral artery, 31/152
anterior communicating artery, 31/153
associated abnormality, 31/143-145
basilar artery, 31/154-156
brain, *see* Brain aneurysm
carotid canal, 31/146
carotid cavernous fistula, 31/150
cerebellar artery, 31/156
clinical features, 31/145, 152
diagnosis, 31/157
epidemiology, 31/138
genetics, 31/139
infraclinoid, 31/147-150
internal carotid artery, 31/146
location, 31/142
middle cerebral artery, 31/152
multiple, 31/143
ophthalmic artery, 31/153
pathogenesis, 31/140
pathology, 31/138
posterior cerebral artery, 31/156
posterior communicating artery, 31/153
ruptured, 31/157-173
size, 31/142
supraclinoid, 31/151
unruptured, 31/145-157
vertebral artery, 31/156
Congenital anosmia
hypoparathyroidism, 42/357
trigeminal nerve, 42/357
vitamin A deficiency, 42/357
Congenital arterial malformation, 12/289-335
anastomosis, 12/308-335
carotid artery, 12/289-308
Congenital arteriovenous fistula
angiomatous malformation, 31/200
blunt trauma, 31/200
gunshot injury, 31/200
rare presentation, 31/202
stab wound, 31/200
Congenital arthrogryposis multiplex, *see* Arthrogryposis multiplex congenita
Congenital astereognosia, *see* Congenital tactile agnosia
Congenital ataxic diplegia, 42/121-123
acquired type, 42/122
autosomal dominant inheritance, 42/122
autosomal recessive inheritance, 42/122
diagnosis, 42/122
hyporeflexia, 42/121
hypotonia, 21/463, 42/121
intention tremor, 42/121
motor retardation, 42/121
pyramidal tract sign, 42/121
toe walker, 42/121
X-linked recessive inheritance, 42/122
Congenital atonic sclerotic muscular dystrophy
acroatonia, 41/43
myosclerosis, 41/45
taille de guèpe, 41/44
Congenital auditory agnosia
Osuntokun syndrome, 22/522
Congenital auditory imperception, *see* Congenital auditory agnosia *and* Verbal auditory agnosia
Congenital autonomic dysfunction
clinical features, 60/18
congenital analgesia, 60/18
deafness, 60/18
nerve biopsy, 60/18
optic atrophy, 60/18
Congenital bilateral perisylvian syndrome
gene, 72/138
Congenital brain granuloma
cytomegalic inclusion body disease, 42/736
fungal infection, 42/736
syphilis, 42/736

toxoplasmosis, 42/736
tuberculosis, 42/736
Congenital brain malformation
organic acid metabolism, 66/641
Congenital brain tumor, 31/35-68
astrocytoma, 31/47-49
cardinal sign, 31/40
cardinal symptom, 31/31-40
choroid plexus papilloma, 31/44-46
classification, 31/35-37
clinical diagnosis, 31/38-41
craniopharyngioma, 31/50
definition, 31/35-37
dermoid cyst, 31/61-67
differential diagnosis, 31/41
ependymoma, 31/46
epidemiology, 30/150, 160
epidermoid cyst, 31/61-68
incidence, 31/39
location, 31/40
medulloblastoma, 31/49
meningioma, 31/50
neuroradiology, 31/41
prognosis, 31/41
surgical procedure, 31/41
teratology, 31/37
teratoma, 31/41-43, 50-61
Congenital cardiopathy
tuberous sclerosis, 14/50
Congenital cataract
anhidrotic ectodermal dysplasia, 46/91
hereditary spastic ataxia, 60/464
incontinentia pigmenti, 14/12
β-naphthol, 13/204
Congenital central alveolar hypoventilation
autonomic nervous system dysfunction, 63/464
congenital megacolon, 63/415, 464
definition, 63/464
ganglioneuroblastoma, 63/415, 464
hypercapnia, 63/464
hypoxemia, 63/464
intestinal pseudo-obstruction, 63/415
neuroastrocytoma, 63/415, 464
neuroblastoma, 63/415, 464
REM sleep, 63/464
sleep disorder, 63/449
sudden death, 63/415
sudden infant death, 63/415
Congenital central deafness
agnosia, 4/38, 428
Congenital central hypoventilation syndrome
sleep, 74/560
Congenital cerebellar ataxia
Babinski sign, 42/125
cerebellar atrophy, 42/124
cerebellar granular cell hypoplasia, 42/125
nystagmus, 42/124
strabismus, 42/124
Congenital cerebellar atrophy, 21/459-465
birth incidence, 21/460
cause, 21/462
cerebellar cortical atrophy, 21/461
cerebellar granular layer atrophy, 21/461
cerebellar hypoplasia, 21/461
cerebellar white matter atrophy, 21/461
clinical features, 21/462
crossed cerebellar atrophy, 21/462
diagnosis, 21/464
differential diagnosis, 21/464
EEG, 21/463
hereditary, *see* Hereditary congenital cerebellar atrophy
neuropathology, 21/460
pathogenesis, 21/462
PEG, 21/464
treatment, 21/464
ulegyria, 21/460
Congenital cerebellar granular cell hypoplasia
dysmorphogenesis, 30/100
Congenital cervical spinal atrophy
spinal muscular atrophy, 59/370
Congenital chemosensitive hypoventilation syndrome
autonomic nervous system, 74/583
hypoventilation, 74/583
Congenital chondrodystrophy calcificans, *see* Conradi-Hünermann syndrome
Congenital cytomegalovirus infection, *see* Cytomegalic inclusion body disease
Congenital deafness
acoustic nerve agenesis, 50/217
albinism, 14/781
Alexander type, 42/366
Apgar score, 42/364
Bing-Siebenmann type, 42/366
birth incidence, 42/366
early onset, 42/364-367
gene frequency, 42/365
hyperbilirubinemia, 42/364
inner ear histopathology, 42/366
Jervell-Lange-Nielsen syndrome, 50/218
membranous labyrinth, 30/408
Michel type, 42/366
Mondini ear abnormality, 42/366
Patau syndrome, 14/120
Pendred syndrome, 42/366

prevalence, 42/366, 56/106
recurrence risk, 42/366
rubella virus, 56/107
Scheibe type, 42/366
sex ratio, 42/366
thyroid gland, 42/366
Usher syndrome, 50/218
viral infection, 56/106
Congenital dysgraphia
developmental dyslexia, 46/112
Congenital dyskeratosis
differential diagnosis, 43/392
Goltz-Gorlin syndrome, 14/113
hyperpigmentation, 43/392
mental deficiency, 43/392
Congenital ectodermal dysplasia, *see* Rothmund-Thomson syndrome
Congenital end plate acetylcholine receptor deficiency
congenital myopathy, 62/332
Congenital end plate acetylcholinesterase deficiency
congenital myasthenic syndrome, 62/430
congenital myopathy, 62/332
EMG, 62/430
muscle biopsy, 62/430
neuromuscular junction disease, 62/392
pathogenesis, 62/431
symptom, 62/430
Congenital erythropoietic porphyria
autosomal recessive inheritance, 27/433
hemolytic anemia, 27/433
photosensitivity, 27/433
Congenital facial anesthesia
trigeminal nerve agenesis, 50/214
Congenital facial diplegia, *see* Möbius syndrome
Congenital familial limb girdle myasthenia
congenital myasthenic syndrome, 62/439
EMG, 62/439
joint contracture, 62/439
neuromuscular junction disease, 62/392
symptom, 62/439
Congenital fiber type disproportion
classification, 40/288
clinical features, 40/306, 335, 337, 41/14-16
congenital hip dislocation, 41/15
congenital myopathy, 41/14-16, 62/332
etiology, 40/147
facial weakness, 62/334
fatality rate, 62/356
genetics, 41/431, 62/356
globoid cell leukodystrophy, 62/355
histochemistry, 40/19, 47
infantile spinal muscular atrophy, 22/99, 62/355
Lowe syndrome, 62/355
muscle fiber type I predominance, 41/15
respiratory dysfunction, 62/356
rigid spine syndrome, 62/355
scoliosis, 41/15
twins, 62/356
Congenital general fibromatosis
hemiatrophy, 59/480
Congenital glaucoma
neurofibromatosis type I, 50/367
Congenital heart disease
acrocephalosyndactyly type II, 38/422
actuarial survival, 63/7
Adams-Stokes syndrome, 63/5
age at stroke, 63/3
amblyopia, 63/11
anoxic cortical necrosis, 63/2
anoxic spell, 38/119-123
anterior spinal artery, 63/4
aorta coarctation, 38/131-133, 63/4
arterial hypertension, 63/10
arterial thrombosis, 63/4
artery transposition, 63/2
bacterial endocarditis, 52/289
Bardet-Biedl syndrome, 13/395
Blalock-Taussig anastomosis, 63/5
blood viscosity, 63/3
brain abscess, 33/115, 38/127-131, 52/145, 63/5
brain aneurysm, 63/4
brain atrophy, 63/8
brain blood flow, 63/9
brain blood flow autoregulation, 63/9
brain embolism, 53/31, 157
brain infarction, 53/157, 63/2
brain ischemia, 53/157
brain oxygen metabolic rate, 63/10
brain perfusion pressure, 63/9
brain protection, 63/10
Brushfield spot, 63/11
buphthalmos, 63/11
bypass time, 63/9
cardiac catheterization, 63/6
cardiac dysrhythmia, 63/5
cardiomyopathy, 63/2
cataract, 63/11
cavernous sinus thrombosis, 52/173
cerebral palsy, 63/8
cerebrovascular disease, 53/28, 31
choroid coloboma, 63/11
CNS malformation, 63/11
cognitive development, 63/1
congenital ptosis, 63/11
convulsion, 38/119-123

corneal clouding, 63/11
corpus callosum agenesis, 63/11
Coxsackie virus infection, 31/220
creatine kinase BB, 63/2
cyanosis, 63/1
cytomegalic inclusion body disease, 56/267
deafness, 63/11
deep hypothermia, 63/7
delayed choreoathetoid syndrome, 63/7
dendritic tree, 63/3
depressor anguli hypoplasia, 63/11
dexamethasone, 63/10
diffuse hypoxic ischemic encephalopathy, 63/7
Down syndrome, 50/274, 53/31
early surgical repair, 63/1
Edwards syndrome, 50/275
EEG, 38/131
emotional disorder, 63/2
epilepsy, 38/119-123, 63/3, 72/212
evaluation, 63/8
facial paralysis, 63/11
Fallot tetralogy, 63/2
glial scar, 63/2
Gruber syndrome, 14/505
headache, 38/133, 63/4
heart atrioventricular block, 63/5
heart transplantation, 63/10
hemidysplasia, 43/416
holoprosencephaly, 63/11
hypertelorism, 63/11
hypotelorism, 63/11
hypothermia, 63/9
hypoxemia, 63/3
hypoxemia duration, 63/1
hypoxia, 63/1
ichthyosis, 43/416
incidence, 63/1
infectious endocarditis, 63/5, 112
intellectual deficit, 63/2
intelligence quotient, 63/1
intercostal artery aneurysm, 63/4
iris coloboma, 63/11
iron deficiency, 63/3
ischemic myelopathy, 63/4
Klinefelter variant XXXXY, 31/483
leg pain, 63/4
LEOPARD syndrome, 14/117
long-term sequel, 63/7
low-flow bypass, 63/10
management, 63/8
mental deficiency, 38/133, 53/31, 63/3
mental function, 63/1
micrencephaly, 63/11
microcephaly, 30/510, 63/11
migraine, 63/4
muscle cramp, 63/4
mycotic aneurysm, 63/5
neurofibromatosis type I, 14/113
neurologic complication, 38/119-138
noninfectious endocarditis, 63/3
ocular abnormality, 63/11
optic disc hypoplasia, 63/11
9p partial monosomy, 50/579
paradoxical embolus, 63/3
paraplegia, 63/4
parietal lobe, 63/2
patent ductus arteriosus, 63/2
perceptual motor function, 63/1
polycythemia, 63/3
postperfusion syndrome, 63/10
postsurgical acute encephalopathy, 63/7
postsurgical cerebellar infarction, 63/8
postsurgical epilepsy, 63/7
postsurgical prolonged coma, 63/7
pseudopapilledema, 63/11
psychological consequence, 63/2
psychological deficit, 63/2
retinal pigment abnormality, 63/11
retinal vessel tortuosity, 63/11
school performance, 63/2
silent brain infarction, 63/3
spinal subarachnoid hemorrhage, 63/4
stellate iris, 63/11
strabismus, 63/11
stroke, 38/123-127, 63/3, 5
stroke incidence, 63/4
stroke mechanism, 63/4
subacute bacterial endocarditis, 38/136
subclavian steal, 63/5
subdural hematoma, 63/7
supranuclear ophthalmoplegia, 63/7
surgery, 38/137, 63/6
surgical mortality, 63/6
syncope, 38/119-123
synophthalmia cyclopia, 63/11
systemic brain infarction, 11/447
temporal lobe, 63/2
thromboembolism, 63/2
trisomy 22, 50/564
unusual stroke syndrome, 63/4
venous thrombosis, 63/3
ventriculomegaly, 63/11
white matter necrosis, 63/2
Congenital heart malformation
Down syndrome, 50/519, 526
foramina parietalia permagna, 50/142

microcephaly, 50/272
Patau syndrome, 14/120, 31/504, 43/528, 50/558
r(9) syndrome, 50/580
trisomy 9, 50/564
Congenital hemidysplasia
ichthyosis, *see* Unilateral hemimelic ichthyosiform erythromelia
Congenital hemifacial atrophy
bulbar poliomyelitis, 22/548
cellulitis, 22/548
scleroderma, 22/548
Congenital hemihypertrophy
aniridia, 60/655
type, 8/347
Congenital high conductance fast channel syndrome
congenital myasthenic syndrome, 62/438
EMG, 62/438
episodic apnea, 62/438
neuromuscular junction disease, 62/392
ptosis, 62/438
symptom, 62/438
Congenital hip dislocation
Biemond syndrome type II, 43/334
caudal aplasia, 42/50
central core myopathy, 41/3, 43/80
congenital fiber type disproportion, 41/15
Conradi-Hünermann syndrome, 43/346
diastematomyelia, 42/24
Dyggve-Melchior-Clausen syndrome, 43/266
Ehlers-Danlos syndrome, 14/111
idiopathic scoliosis, 42/52
mental deficiency, 42/75
multiple neuronal system degeneration, 42/75
myopathy, 40/337
porencephaly, 50/360
prune belly syndrome, 43/466
spina bifida, 42/54
Turner syndrome, 14/108
Congenital hyperammonemia
ammonia level, 27/368
brain atrophy, 29/305-328, 42/561
CSF glutamine, 27/368
Moro reflex, 42/560
neurologic manifestation, 27/367
pathogenesis, 42/560
pathology, 27/353
tachypnea, 42/560
treatment, 27/368, 42/561
Congenital hyperammonemia type I, *see* Carbamoylphosphate synthetase deficiency
Congenital hyperammonemia type II, *see* Ornithine carbamoyltransferase deficiency
Congenital hyperphosphatasia
alkaline phosphatase, 31/258
autosomal recessive inheritance, 31/258
bone pain, 31/258
dwarfism, 31/258
fever, 31/258
hearing loss, 31/258
hyperostosis, 31/258
retinal degeneration, 31/258
short neck, 31/258
uric acid, 31/259
Congenital hypothyroidism, *see* Cretinism
Congenital ichthyosiform erythroderma
corpus callosum agenesis, 42/7, 50/152
mental deficiency, 46/59
Congenital ichthyosis
adrenal hypoplasia, 43/406
edema, 43/406
hydrocephalus, 43/406
hyperkeratosis, 43/405
lipid storage myopathy, 41/203
mental deficiency, 43/276, 46/59
Passwell syndrome, 21/216
Sjögren-Larsson syndrome, 13/141, 66/615
thymus hypoplasia, 43/406
triglyceride storage disease, 62/496
ventricular cyst, 43/406
Congenital indifference to pain, *see* Congenital pain insensitivity
Congenital intestinal aganglionosis, *see* Congenital megacolon
Congenital kyphosis
cause, 50/419
myelography, 32/147
neurofibromatosis, 32/145
treatment, 32/145-147
Congenital laryngeal abductor paralysis
Arnold-Chiari malformation type I, 42/319
mental deficiency, 42/318
nucleus ambiguus, 42/319
stridor, 42/318
Congenital larynx stridor
accessory nerve, 30/410
etiology, 30/411
glossopharyngeal nerve, 30/410
mental deficiency, 30/411
syndrome, 30/411
vagus nerve, 30/410
Congenital limb agenesis
phantom limb, 4/224, 399
Congenital lordosis
vertebral column, 32/148
Congenital lysine intolerance
metabolic ataxia, 21/578

periodic ammonia intoxication, 29/104
periodic ataxia, 21/578
Congenital malformation
antiepileptic agent, 65/513, 515
carbon monoxide intoxication, 64/32
carotid artery, 12/289
cocaine intoxication, 65/255
fetal warfarin syndrome, 46/80
methotrexate, 46/79
respiratory virus infection, 31/219
tuberous sclerosis, 30/101
Congenital megacolon
anal atresia, 42/3
autonomic nervous system, 75/231, 633
birth incidence, 42/3
central alveolar hypoventilation, 63/464
congenital abnormality, 75/635
congenital central alveolar hypoventilation, 63/415, 464
dementia, 42/4
Down syndrome, 42/3
forgotten respiration, 63/489
intestinal pseudo-obstruction, 51/492
Klein-Waardenburg syndrome, 43/52, 75/635
Melkersson-Rosenthal syndrome, 8/229
metaphyseal chrondrodysplasia, 75/635
muscular atrophy, 42/3
neuroblastoma, 42/3, 758
neuronal migration, 42/3
pheochromocytoma, 42/3
respiratory distress, 42/3
Santos disease, 75/635
sensorineural deafness, 42/3
sex ratio, 42/3
Waardenburg-Shah syndrome, 75/635
Congenital miosis, *see* Microcoria
Congenital motoneuron disorder
WFN classification, 59/4
Congenital muscle absence
autonomic neurotropism, 1/481
myopathy, 40/323
pectoralis muscle, 43/416, 461
Congenital muscular dystrophy, 41/27-47
*see also* Fukuyama syndrome
atonic sclerotic, *see* Congenital atonic sclerotic muscular dystrophy
benign form, 41/33
cataract, 43/89
classification, 40/280, 41/29
CNS, 41/35-43
congenital myopathy, 41/27
contracture, 43/93, 95
differential diagnosis, 22/99
epilepsy, 43/91
facial paralysis, 43/93
feeding disorder, 43/93
genetics, 41/420
heart disease, 43/93
histochemistry, 40/50
history, 41/27
hypogonadism, 43/89
hypotonia, 41/36
infantile hypotonia, 43/92
mental deficiency, 43/91
miscellaneous congenital dystrophy, 41/46
muscle fiber size, 43/93-95
neuromuscular disease, 41/420
nonprogressive, 43/92
rapid progression, 43/93-95
severe form, 41/32
Congenital muscular hypoplasia
universal, 43/110
Congenital myasthenic syndrome
abnormal acetylcholine-acetylcholine receptor interaction syndrome, 62/437
congenital acetylcholine receptor deficiency/short channel opentime syndrome, 62/437
congenital acetylcholine receptor deficiency/synaptic cleft syndrome, 62/438
congenital ε-acetylcholine receptor subunit mutation syndrome, 62/436
congenital end plate acetylcholinesterase deficiency, 62/430
congenital familial limb girdle myasthenia, 62/439
congenital high conductance fast channel syndrome, 62/438
congenital myopathy, 62/332
congenital slow channel syndrome, 62/432
congenital synaptic vesical paucity syndrome, 62/429
Eaton-Lambert myasthenic syndrome, 62/425
edrophonium test, 62/413
facial weakness, 62/334
familial infantile myasthenia, 62/427
myasthenia gravis, 62/415
polyneuropathy, 62/413
progressive bulbar palsy, 59/375
Congenital myopathy, 41/1-22
*see also* Infantile hypotonia
abnormal myomuscular junction myopathy, 62/332
aspecific clinical criteria, 62/333
cap disease, 62/332
cardiac involvement, 62/334
central core myopathy, 41/3-8, 62/332
chickenpox, 62/562

classification, 40/286, 62/331
clinical features, 40/287
concept history, 62/331
congenital end plate acetylcholine receptor deficiency, 62/332
congenital end plate acetylcholinesterase deficiency, 62/332
congenital fiber type disproportion, 41/14-16, 62/332
congenital muscular dystrophy, 41/27
congenital myasthenic syndrome, 62/332
creatine kinase, 62/333, 350
cytoplasmic inclusion body myopathy, 62/332
definition, 62/331
differential diagnosis, 22/99
dysphagia, 62/335
EMG, 62/69, 333, 350
experimental model, 62/358
external ophthalmoplegia, 62/333
extraocular muscle weakness, 62/334
facial weakness, 62/334
familial infantile myasthenia, 62/332
familial lysis muscle fiber type I, 41/21
fingerprint body myopathy, 41/20, 62/332
granulofilamentous myopathy, 62/332
granulovacuolar lobular myopathy, 62/332
hereditary cylindric spirals myopathy, 62/332
history, 41/2
honeycomb myopathy, 62/332
hypotonia, 41/1
hypotrophy, 62/349
infantile spinal muscular atrophy, 41/1
malignant hyperthermia, 38/551, 62/335, 63/440
Mallory body myopathy, 62/332
mental deficiency, 62/335
microfiber myopathy, 62/332
minimal change myopathy, 62/332
mitochondria jagged Z line myopathy, 62/332
mitochondrial myopathy, 41/216
multicore myopathy, 41/16, 62/332
muscle cramp, 62/335
muscle fiber type I, 40/146, 62/349
muscle fiber type I hypotrophy, 62/332
muscle fiber type I predominance, 62/332
muscle fiber type II hypoplasia, 62/332
muscle hypertrophy, 62/335
myalgia, 62/335
myasthenia, 62/335
myofibrillar lysis myopathy, 62/332
myotonia, 62/335
myotubular myopathy, 41/11-14, 62/332
nemaline myopathy, 41/8-11, 62/332
neuromuscular disease, 41/3-15, 429
nucleodegenerative myopathy, 62/332
progressive bulbar palsy, 59/375
progressive external ophthalmoplegia, 62/297, 333
ptosis, 62/297, 333, 349
reducing body myopathy, 41/17-20, 62/332
respiration, 62/334
respiratory dysfunction, 62/334, 349
rigidity, 62/335
rimmed vacuole distal myopathy, 62/332
sarcoplasmic body myopathy, 62/332
sarcotubular myopathy, 41/21, 62/332
selective myosin degeneration myopathy, 62/332
skeletal deformity, 62/334
slow channel syndrome, 62/332
spheroid body myopathy, 62/332
structured type, 62/332
temporomandibular ankylosis, 62/335
tremor, 62/335
trilaminar myopathy, 41/22, 62/332
tubular aggregate myopathy, 62/332
tubulomembranous inclusion myopathy, 62/332
uniform muscle fiber type I myopathy, 62/332
unstructured type, 62/332
X-linked myotubular myopathy, 62/117
X-linked neutropenic cardioskeletal myopathy, 62/117
Z band plaque myopathy, 62/332
zebra body myopathy, 41/21, 62/332
Congenital negative variation, *see* Readiness potential
Congenital neuropathy
WFN classification, 60/2
Congenital nystagmus
clinical features, 1/597
congenital retinal blindness, 22/530
muscle fiber size, 43/93-95
optokinetic nystagmus, 42/158-161
oscillation, 2/367
prevalence, 42/158
primary pigmentary retinal degeneration, 42/159
spasmus nutans, 1/597, 16/323
strabismus, 42/159
vertigo, 2/361
Congenital obstruction
hydrocephalus, 50/285
Congenital oculomotor apraxia, *see* Cogan syndrome type I
Congenital olivopontocerebellar atrophy
computer assisted tomography, 60/671
contracture, 60/669
hereditary congenital cerebellar atrophy, 60/291
pericardial effusion, 60/668

recessive, 60/291
Congenital ophthalmoplegia
aminoaciduria, 29/232, 43/134
arthrogryposis multiplex congenita, 43/134
congenital ptosis, 22/189
floppy baby syndrome, 29/232
myopathy, 29/232
myopia, 43/134
nonprogressive, 43/134
Congenital optic atrophy, *see* Congenital retinal blindness
Congenital pain insensitivity
*see also* Congenital analgesia
A-δ-fiber decrease, 51/564
abnormal personality pattern, 51/564
Alzheimer disease, 51/566
anhidrosis, *see* Hereditary sensory and autonomic neuropathy type IV
ataxia, 51/566
autonomic nervous system, 42/297
autonomic neuropathy, 51/564
autosomal recessive inheritance, 51/565
autotopagnosia, 4/222
body scheme disorder, 4/222
cerebellar atrophy, 51/566
cerebral lesion, 4/222
ciliary ganglion, 51/565
classic type, 8/187
corneal reflex, 42/297
dementia, 51/566
differential diagnosis, 60/19
dizygotic twins, 51/565
dominant inheritance, 51/565
dopamine β-mono-oxygenase, 51/564
dorsal spinal nerve root ganglion, 51/564
β-endorphin, 51/566
epilepsy, 51/564
familial cases, 51/564
hereditary sensory and autonomic neuropathy, 60/19
hereditary sensory and autonomic neuropathy type II, 8/369, 21/5, 78, 51/564, 60/10, 14, 27
hereditary sensory and autonomic neuropathy type III, 16/33, 21/110, 41/435, 51/565, 60/7, 19, 27
hereditary sensory and autonomic neuropathy type IV, 21/5, 60/11, 17, 19
hereditary sensory and autonomic neuropathy type V, 60/19
lacrimation, 51/564
Lissauer tract, 51/564
mental deficiency, 51/564
multiple neuronal system degeneration, 51/566
naloxone, 51/564, 566
nerve growth factor, 51/566
neurogenic osteoarthropathy, 51/564
neurologic status, 51/563
neurotrophic osteoarthropathy, 38/467
nomenclature, 8/187
Osuntokun syndrome, 22/522
parasympathetic nervous system, 42/297
schizophrenia, 51/566
semantics, 51/563
subgroup, 51/564
sural nerve biopsy, 51/564
thalamic cyst, 51/565
Urbach-Wiethe disease, 14/777
Congenital Pelizaeus-Merzbacher disease
biochemistry, 10/160
classification, 10/10, 151
clinical finding, 10/158, 42/503
demyelination, 10/160
epilepsy, 15/376, 42/504
extrapyramidal hyperkinesia, 10/158, 42/503
genetics, 10/162
hyperkinesia, 42/503
neurochemistry, 10/167
nystagmus, 10/158, 42/503
review, 10/160
seizure, 15/426
Congenital photosensitive porphyria, *see* Congenital erythropoietic porphyria
Congenital pigmented epulis, *see* Pigmented papillary medulloblastoma
Congenital poikiloderma, *see* Rothmund-Thomson syndrome
Congenital polytopic angioblastosis, *see* Von Hippel-Lindau disease
Congenital progressive hemifacial atrophy
survey, 22/547
Congenital ptosis
*see also* Abducent nerve abnormality, Oculomotor nerve abnormality *and* Trochlear nerve abnormality
congenital heart disease, 63/11
congenital ophthalmoplegia, 22/189
genetics, 30/100, 401
Möbius syndrome, 30/401
nuclear agenesis of 3rd nerve, 30/401
oculomotor nerve abnormality, 30/401
oculomotor nerve agenesis, 50/212
oculopharyngeal muscular dystrophy, 50/212
ophthalmoplegia, 62/297
origin, 30/401
progressive external ophthalmoplegia, 22/189, 62/297
supranuclear defect, 30/401

trochlear nerve abnormality, 30/401
Congenital recessive optic atrophy
false diagnosis, 13/121
Congenital recurrent nerve palsy, *see* Vagus nerve agenesis
Congenital retinal amaurosis, *see* Congenital retinal blindness
Congenital retinal blindness, 13/94-108, 22/527-542
associated neural disease, 22/533
ataxia, 13/298, 42/109
athetosis, 42/401
autosomal recessive optic atrophy, 13/38
Babinski sign, 42/401
Behr disease, 13/67, 88, 91, 42/109
Bruyn-Went disease, 13/99, 22/454
case report, 59/339
cataract, 13/272, 22/530, 532
cause, 13/107, 22/537
clinical features, 22/530, 59/339
clinical picture, 13/139
computer assisted tomography, 60/672
congenital nystagmus, 22/530
deafness, 22/530, 60/724
diagnosis, 13/28
differential diagnosis, 13/97, 273, 22/535, 59/342
digito-ocular sign, 13/139, 272, 22/530
dystonia, 59/339
dystonia musculorum deformans, 49/524
EEG, 13/98, 272
Ehlers-Danlos syndrome, 22/528
embryogenesis, 22/537
epidemiology, 21/3
epilepsy, 13/272, 15/329, 22/529, 533
ERG, 13/98, 272, 22/528, 532
extrapyramidal syndrome, 42/400-402
eye fundus, 13/28
familial spastic paraplegia, 13/97, 22/454, 59/334
genetics, 22/539, 542, 59/342
Hallervorden-Spatz syndrome, 13/99
hereditary autosomal recessive, *see* Hereditary autosomal recessive congenital optic atrophy
hereditary cerebellar ataxia, 13/97
hereditary congenital optic atrophy, 13/96, 22/536
hereditary dystonic paraplegia, 59/346
hereditary motor and sensory neuropathy type I, 21/284
hereditary motor and sensory neuropathy type II, 21/284
hereditary motor and sensory neuropathy variant, 60/246
hereditary neurological disease, 13/98-102
hereditary secondary dystonia, 59/339-345
hereditary spastic ataxia, 21/372
histology, 22/528
history, 22/527
hyperreflexia, 42/401
hypertrophic polyneuropathy, 13/97
infantile dominant, 13/67
inheritance, 13/27
keratoconus, 13/272, 22/530
Lyon hypothesis, 13/96
mental deficiency, 13/272, 22/529, 533
microcephaly, 22/533
misdiagnosis, 13/102
mitochondrial disease, 59/344
mitochondrial DNA mutation, 59/342, 344
monochromatism, 13/121
multiple sclerosis, 9/103, 176, 13/102
neuropathology, 22/541
nystagmus, 13/272, 22/530, 42/407
olivopontocerebellar atrophy variant, 21/454
optic atrophy, 13/38, 83
pathology, 13/103, 273
photophobia, 13/272, 22/530
polycystic kidney, 22/529
prevalence, 13/94, 22/529, 42/399
primary pigmentary retinal degeneration, 13/27
psychosis, 13/272
pupillary reaction, 22/530
recessive, 13/67
renal dysplasia, 22/529
retinal aplasia, 22/537
secondary pigmentary retinal degeneration, 22/536, 60/724
septo-optic dysplasia, 42/407
sex ratio, 13/65, 42/399
speech disorder in children, 13/272, 42/401
symptomatology, 13/95, 98
thalamus degeneration, 21/598
tobacco alcoholic amblyopia, 13/67
treatment, 13/107, 22/537
vision, 22/530
vitamin B12, 28/175
Congenital retinal degeneration, *see* Congenital retinal blindness
Congenital rhabdomyoma
cardiac lesion, 14/51
Congenital rubella
cataract, 50/278
cerebellar cortex heterotopia, 50/278
chorioretinitis, 50/278
chronic meningitis, 56/645
deafness, 50/278
hydrocephalus, 50/278
microcephaly, 50/278
polymicrogyria, 50/278

septum pellucidum agenesis, 50/278
Congenital rubella syndrome
clinical features, 56/407
rubella virus, 56/405
Congenital seesaw nystagmus
secondary pigmentary retinal degeneration, 60/737
Congenital sensory neuropathy, *see* Hereditary sensory and autonomic neuropathy type II
Congenital slow channel syndrome
chemistry, 62/434
congenital myasthenic syndrome, 62/432
EMG, 62/432
muscle biopsy, 62/434
neuromuscular junction disease, 62/392
pathogenesis, 62/435
ptosis, 62/432
symptom, 62/432
Congenital spinal cord tumor
astrocytoma, 32/355-386
cases, 32/368-383
chlordoma, 32/355-386
clinical features, 32/358, 365-367
definition, 32/355
dermoid, 32/358-386
diagnosis, 32/364, 367-385
embryology, 32/360-362
epidemiology, 30/164, 32/355-357, 359, 362-364
epidermoid, 32/358-386
first year, 32/355-358
hemangioblastoma, 32/355-386
hemangioendothelioma, 32/355-386
histology, 32/361-363
lipoma, 32/358-386
location, 32/357, 364
lymphocytoma, 32/355-386
malformation, 32/357, 366
malignant astrocytoma, 32/355-386
melanoma, 32/355-386
meningioma, 32/355-386
meningitis, 32/366
motor disorder, 32/365
myelocytoma, 32/355-386
myeloma, 32/355-386
neuroastrocytoma, 32/355-386
neurofibroma, 32/355-386
onset of symptom, 32/364
operative result, 32/358
osteochondroma, 32/355-386
pain, 32/366
phakoma, 32/355-386
radiologic appearance, 32/366
sacrococcygeal teratoma, 32/355-386
sarcoma, 32/355-386
sensory disorder, 32/366
sphincter disturbance, 32/366
spina bifida, 32/357, 362
teratoma, 32/358-386
treatment, 32/358, 385
type, 32/358
Congenital spondyloepiphyseal dysplasia, 43/478-480
atlantoaxial subluxation, 43/480
autosomal dominant, 43/479
diagnosis, 43/478
genetic counseling, 43/480
genu valgum, 43/478
genu varum, 43/478
glycosaminoglycan, 43/480
hypotonia, 43/478
kyphosis, 43/478
lumbar lordosis, 43/478
myopia, 43/478
orthopedic correction, 43/480
palatoschisis, 43/478
pectus carinatum, 43/478
radiographic criteria, 43/478
retinal detachment, 43/478
retinal dysplasia, 43/478
scoliosis, 43/478
sex ratio, 43/479
short neck, 43/478
talipes, 43/478
Congenital stippled epiphyses, *see* Conradi-Hünermann syndrome
Congenital stridor, *see* Vagus nerve agenesis
Congenital suprabulbar paresis
dysphagia, 22/118
mental deficiency, 59/70
progressive bulbar palsy, 59/375
Congenital synaptic vesical paucity syndrome
congenital myasthenic syndrome, 62/429
EMG, 62/429
miniature motor end plate potential, 62/429
neuromuscular junction disease, 62/392
Congenital syphilis
Clutton joint, 33/370, 387
early, 33/340
Hutchinson teeth, 33/371
late, 33/340
mental deficiency, 46/76
microgyria, 30/483
symptom, 52/275
Congenital tactile agnosia
encephalopathy, 4/34
parietal lobe lesion, 4/34

porencephaly, 4/34
virgin hand, 4/34
Congenital tapetoretinal amaurosis, *see* Congenital retinal blindness
Congenital torticollis
basal cell, 43/146
clinodactyly, 43/146
facial asymmetry, 43/146
hyperpigmentation, 43/146
hypertension, 43/146
keloid, 43/146
progressive myositis ossificans, 43/191
renal abnormality, 43/146
rib abnormality, 43/146
Congenital toxoplasmosis
chorioretinopathy, 42/662
clinical features, 35/130-137
computer assisted tomography, 52/359
convulsion, 42/662, 52/358
CSF, 35/133, 42/662, 52/359
diagnosis, 35/137, 52/360
differential diagnosis, 52/360
echoEG, 52/359
EEG, 35/137
hydrocephalus, 35/130, 42/662, 52/358
intracerebral calcification, 52/359
intracranial calcification, 35/131-134
meningoencephalitis, 52/358
mental deficiency, 35/131, 46/77, 52/358
microcephaly, 35/130, 42/662
ocular involvement, 35/132
pathology, 35/128
pathophysiology, 35/129
pregnancy, 52/351
pyrimethamine, 52/360
radionuclide scintigraphy, 52/359
relapse, 52/359
retinochoroiditis, 52/358
spiramycin, 52/360
sulfadiazine, 52/360
syndrome, 35/134-153
transmission, 35/116
treatment, 35/137, 52/360
Congenital tremor
Siamese cat, 9/684
Congenital trigeminal anesthesia, *see* Trigeminal nerve agenesis
Congenital vascular vitreous veil, *see* Sex linked idiopathic retinoschisis
Congenital vitelliform macula cyst, *see* Vitelliruptive macula degeneration
Congenital vocal cord palsy, *see* Vagus nerve agenesis
Congestin intoxication
death, 37/37
diarrhea, 37/37
experimental animal, 37/37
gastrointestinal tract, 37/37
respiratory failure, 37/37
sea anemone, 37/37
vomiting, 37/37
Congestive heart failure
adriamycin, 63/136
alcoholic cardiomyopathy, 63/133
anticoagulant, 63/133
arterial hypertension, 63/82
atrial fibrillation, 63/133
beriberi heart disease, 63/135
brain embolism, 55/162, 63/131, 133
brain infarction, 63/132
cerebrovascular disease, 55/162
Cheyne-Stokes respiration, 63/483
dermatomyositis, 62/374
differential diagnosis, 63/135
eosinophilic polymyositis, 63/385
epilepsy, 63/132
glycosaminoglycanosis type VI, 66/309
hemiplegia, 63/132
hyponatremia, 28/499
inclusion body myositis, 62/374
myotonic dystrophy, 62/227
nuclear magnetic resonance, 63/132
peripartum cardiomyopathy, 63/134
polymyositis, 62/374
prolonged transient ischemic attack, 53/259
stroke, 63/134
sudden death, 63/131
symptom, 63/131
syncope, 63/132
transient ischemic attack, 53/259
Congophilic angiopathy, *see* Brain amyloid angiopathy
Congophilic cerebral angiopathy, *see* Brain amyloid angiopathy
Conidae intoxication
areflexia, 37/60
artificial respiration, 37/62
biochemistry, 37/61
burn toxin, 37/60
burning pain, 37/60
coma, 37/60
conus aulicus, 37/59
conus geographus, 37/59
conus marmoreus, 37/59
conus omaria, 37/59
conus striatus, 37/59

conus textile, 37/59
conus tulipa, 37/59
diplopia, 37/60
edema, 37/60
paralysed limb, 37/60
paralysis, 37/60
paresthesia, 37/60
pharmacology, 37/61
prevention, 37/62
respiratory failure, 37/60
suction, 37/62
taxonomy, 37/58
treatment, 37/62
visual impairment, 37/60
Conjugate deviation
brain death, 24/720
cortical gaze center, 2/652
diagram, 2/337
eye, 1/284, 2/586
Gasperini syndrome, 2/317
neuroanatomy, 2/335
operculum syndrome, 2/781
subarachnoid hemorrhage, 12/130
unilateral pontine hemorrhage, 12/40
Conjugate eye movement
Benedikt syndrome, 2/298
brain injury, 57/137
clinical examination, 57/137
higher nervous activity, 3/23-26
parietal lobe lesion, 2/588
posture, 3/24
tectoreticular system, 3/24
visual attention, 3/25
Conjugate lateral spasm
frontal gaze center, 2/336
Conjunctiva biopsy
spinocerebellar degeneration, 60/677
Conjunctival capillary biomicroscopy
diagnostic test, 11/215
Conjunctival tumor
nevus unius lateris, 43/40
Conjunctivitis
acute hemorrhagic, *see* Acute hemorrhagic conjunctivitis
kerato, *see* Keratoconjunctivitis
meningococcal meningitis, 33/24, 52/27
Conn syndrome, *see* Hyperaldosteronism
Connectin
limb girdle syndrome, 62/187
Connective tissue
chondroitin sulfate A, 10/432
chondroitin sulfate C, 10/432
dermatan sulfate, 10/432
disorder, 71/3
hyaluronic acid, 10/432
iduronic acid, 10/433
keratan, 10/432
motility disorder, 75/640
muscle, 62/40
myopathy, 40/40-43
neurogenic sarcoma, 8/446
persistent hyaloid artery, 13/81
Connective tissue disease
mixed, *see* Mixed connective tissue disease
Conradi disease, *see* Conradi-Hünermann syndrome
Conradi-Hünermann syndrome, 43/344-346
cataract, 43/346
cerebellar ataxia, 43/345
classification, 38/401-403
CNS, 31/275-277
congenital hip dislocation, 43/346
course, 38/406
Dandy-Walker syndrome, 46/50
epidemiology, 31/275
etiology, 38/405
general features, 38/402
genetics, 31/275, 38/401
history, 38/401
hypertelorism, 30/248
ichthyosis, 43/346
limb asymmetry, 43/344
mental deficiency, 43/346, 46/50
microcephaly, 43/346
neurologic manifestation, 38/403-405
optic atrophy, 43/345
prognosis, 38/406
radiologic features, 38/402
rhizomelic chondrodysplasia punctata, 66/525
rhizomelic type, 31/275, 43/346
skeletal malformation, 30/248
swayback nose, 43/345
treatment, 38/406
vertebral abnormality, 43/346
warfarin, 31/275
Consanguinity
Down syndrome, 50/525
Ferguson-Critchley syndrome, 22/435
Friedreich ataxia, 21/322
hereditary sensory and autonomic neuropathy type II, 21/80, 60/7
Marinesco-Sjögren syndrome, 21/21
multiple sclerosis, 47/298
primary pigmentary retinal degeneration, 13/149
progressive myoclonus epilepsy, 73/296
Conscious hemiasomatognosia
body scheme disorder, 4/214, 45/374

parietal lobe syndrome, 45/69
Consciousness, 3/48-58
*see also* Coma
adrenoleukodystrophy, 45/121
agnosia, 45/114
akinetic mutism, 45/118
allergic encephalitis, 45/119
Alzheimer disease, 45/117
amentia, 3/113
anatomic localization, 3/18, 54-57, 63, 68, 121, 171, 175, 45/111
anterior cerebral artery syndrome, 53/343
anterograde amnesia, 45/115
apraxia, 45/114
arousal disorder, 45/107-113
arousal reaction, 3/52
arousal system, 3/53, 171
ascending reticular activating system, 3/53
aspergillosis, 52/378
barotrauma, 63/416
basilar artery migraine, 48/137
bee sting intoxication, 37/107
behavior, 3/51
benign partial epilepsy with occipital paroxysm, 73/23
brain cortex, 3/49, 55, 122, 171, 45/114
brain hypoxia, 63/416
brain injury, 23/226, 45/118, 57/129
brain metastasis, 45/120
brain stem, 3/171
brain stem herniation, 45/112
brain tumor, 16/130
breath holding syncope, 15/828
cataplexy, 42/711
cavernous sinus thrombosis, 52/173
centrencephalic system, 3/55
cerebral dominance, 3/55
cerebral localization, 3/124
cerebral malaria, 35/149, 52/368
cerebrovascular occlusive disease, 45/119
ciclosporin intoxication, 63/537
classification, 45/108
clinical examination, 57/129
confusion, 3/113, 45/117
conscious experience, 3/50, 112, 118, 171, 174-176
content, 3/52, 56
definition, 24/789
delirium, 45/117, 46/530
dementia, 45/114, 117, 46/200
demyelination, 45/121
depersonalization, 3/116
disorder, 3/56, 62, 112, 45/107-124
disorder classification, 45/108
disturbance, 23/11
dreamy state, 3/123
drug overdose, 45/123
dysbarism, 63/416
dysphrenia hemicranica, 5/79
emotional focalization, 3/117
encephalitis, 45/119
encephalitis lethargica, 3/49
encephalopathy, 45/117
epilepsy, 3/123, 15/54, 130, 135, 594, 72/250
ergotropic reaction, 3/52
exogenous reaction, 3/114, 120
fainting attack, 3/113
Ganser syndrome, 3/126
general, 24/707
general disorder, 45/117-123
Glasgow Coma Scale, 57/132
hallucination, 3/116, 45/117
hallucinogenic agent, 3/125
hemidepersonalization, 3/56
herpes simplex virus encephalitis, 45/120
highest level, 3/54, 121
hypnosis, 3/117, 126
hypnotic blindness, 3/117
hyponatremia, 63/545
hysteria, 3/126
intention activity, 3/118
intracranial hypertension, 16/130, 136
introspection, 3/50
Lafora progressive myoclonus epilepsy, 15/384, 27/172
Lassa fever, 56/370
limbic system, 2/719
Listeria monocytogenes meningitis, 63/531
loss, *see* Loss of consciousness
manic depressive state, 3/116
memory, 3/48
meningitis, 45/120
metabolic encephalopathy, 45/121-123
metabolism, 3/53, 126
metachromatic leukodystrophy, 45/121
migraine, 48/162
mind-matter problem, 3/56, 175
mountain sickness, 63/416
moyamoya disease, 55/293
multifocal CNS disease, 45/118-121
multiple sclerosis, 45/119
neglect syndrome, 45/115
neurophysiologic basis, 3/52-54, 63, 171-181
neurosis, 3/119
obnubilation, 3/114
oneirism, 3/115

organic dynamic concept, 3/128
orienting reaction, 3/52
paranoia, 3/119
pellagra, 7/573, 28/85, 46/336
persistent vegetative state, 45/118
personality disorder, 3/119
pontine hemorrhage, 12/39
pontine infarction, 2/244, 12/15, 30
posttraumatic syndrome, 48/386
psychiatry, 3/112-128
psychodysleptics, 3/53, 126
psychopharmacology, 3/125
psychosis, 3/120
respiratory encephalopathy, 63/414
reticular activating system, 3/49, 53
reticular formation, 3/49, 53, 55, 171
Reye syndrome, 31/168, 49/218, 56/236
schizophrenia, 3/120
self-consciousness, 3/119
sensory deprivation, 3/49, 126
somnambulism, 3/117
split personality hallucination, 3/116
spongiform encephalopathy, 45/119, 121
state, 1/33, 3/52, 56
structure, 3/51, 118
stupor, 3/113
subacute sclerosing panencephalitis, 45/120
subarachnoid hemorrhage, 55/14
time sense, 3/118
toxicity, 3/125
traumatic psychosis, 24/527
twilight state, 3/115
visuospatial agnosia, 3/56
Von Hippel-Lindau disease, 14/115
Wernicke aphasia, 45/115
Wernicke-Korsakoff syndrome, 3/114, 28/246
Consciousness content disorder
confusion, 45/114
focal, 45/113-123
generalized, 45/113-123
Consciousness focal disorder
diagnosis, 45/114-117
pathophysiology, 45/114-117
Constipation
anal sphincter, 74/635
autonomic nervous system, 74/116
bismuth intoxication, 64/331
chronic idiopathic, 75/636
disulfiram intoxication, 37/321
elfin face syndrome, 31/317
gastrointestinal tract, 74/116
neurologic cause, 75/636
ostrea gigas intoxication, 37/64
pheochromocytoma, 39/498
progressive dysautonomia, 59/140
sacrococcygeal chordoma, 19/289
spasticity, 26/480
thallium intoxication, 64/324
venerupis semidecussata intoxication, 37/64
vincristine polyneuropathy, 51/299
Constriction
pupillary, *see* Pupillary constriction
Constructional apraxia, 4/67-81
agnosia, 4/23
alexia, 4/97
Alzheimer disease, 46/251
amorphosynthesis, 4/80
anatomic correlation, 45/491
anatomic localization, 4/56, 68, 72-74, 45/500-502
aphasia, 4/78, 97, 108
apractognosia, 45/492, 500
associated disability, 4/76-79
bilateral lesion, 2/698
block design, 4/69, 45/494, 498, 501
brain injury, 45/491-505
cerebral dominance, 4/52, 56, 68, 74-76, 79, 295, 45/491, 493-500
clinical features, 4/53, 72
closing-in phenomenon, 4/53, 45/502
constructional activity, 4/451-453
constructive task, 45/499
copying design, 45/494
copying geometrical figure, 4/69
defective 3-dimensional block construction, 3/221
definition, 4/67, 45/491
dementia, 2/698, 45/502
developmental dyspraxia, 4/458
disconnection syndrome, 4/80
dressing apraxia, 4/53
examination, 4/52, 69-72
experimental study, 45/498
extrapersonal space unilateral neglect, 4/77
form assembly test, 45/496
frontal lobe lesion, 45/500
Funktionswandel, 4/60
Gerstmann syndrome, 2/690-692, 4/78, 45/500
higher nervous activity disorder, 3/30
historic survey, 4/52, 67-69, 45/492-495
ideomotor apraxia, 45/492, 498
kinesthesia, 4/79
learning disability, 46/125
mental deterioration, 45/502
middle cerebral artery syndrome, 53/363
nondominant parietal lobe, 2/695
occipital lobe lesion, 45/500
oculomotor apraxia, 45/492

parietal lobe lesion, 4/4, 56, 45/500
parietal lobe syndrome, 45/74, 80
pathogenesis, 45/504
pathopsychophysiogenesis, 4/60, 68, 72, 80
perceptual task, 45/499
planning disturbance, 45/498-500
planotopokinesia, 4/68, 45/492
Raven Progressive Matrices, 45/495
rehabilitation, 3/428, 46/620
research, 4/72-79
restoration of higher cortical function, 3/428
Rey complex figure, 45/502
sensory aphasia, 45/501
shape perception, 45/413
spatial agnosia, 45/493
spatial analysis, 4/71
spatial orientation disorder, 4/52, 45/498
stick pattern, 4/71, 45/494
terminology, 4/67
test, 2/692
three dimensional construction, 4/71, 45/494
visual attention, 4/61
visual perception, 45/492-494
visuoconstructive disturbance, 45/495-500
visuospatial agnosia, 4/77, 45/492, 498, 504
Constructional dyspraxia
Huntington chorea, 46/308
Constructive acalculia
secondary acalculia, 4/190
Consumptive coagulopathy, *see* Intravascular consumption coagulopathy
Content disorder
consciousness, *see* Consciousness content disorder
Contingent negative variation
akinesia, 45/163
neglect syndrome, 45/163
psychosis, 46/502
schizophrenia, 46/499, 502
Continuous muscle fiber activity syndrome, *see* Isaacs syndrome
Continuous spike wave during slow sleep
epilepsy, *see* Epilepsy with continuous spike wave during slow sleep
Contraception
multiple sclerosis, 47/153
oral, *see* Oral contraception
Contraceptive drug induced chorea
hormone medicated dyskinesia, 49/365
predisposition, 49/364
previous Sydenham chorea, 49/365
Contractile protein
actin, 40/212
Duchenne muscular dystrophy, 40/380
myosin, 40/212
myotonic dystrophy, 40/523
troponin, 40/130
Contraction
flexion, *see* Flexion contracture
gastrointestinal tract, 75/622
idiomuscular, *see* Idiomuscular contraction
Morvan fibrillary chorea, 6/355
muscular, *see* Muscle contraction
paradoxical, *see* Paradoxical contraction (Westphal)
scalp muscle, *see* Scalp muscle contraction
Contraction band
Duchenne muscular dystrophy, 40/48, 208, 252, 369
Contracture
Addison disease, 39/480
ADR syndrome, 42/113
arthrogryposis, 40/337
Becker muscular dystrophy, 43/86
benign autosomal dominant myopathy, 43/111
Bowen syndrome, 43/335
caffeine induced, 41/269
central core myopathy, 43/80
cerebro-oculofacioskeletal syndrome, 43/341
clinical features, 41/187, 464
congenital muscular dystrophy, 43/93, 95
congenital olivopontocerebellar atrophy, 60/669
corpus callosum agenesis, 42/100
diaphyseal dysplasia, 43/376
dinitrofluorobenzene, 41/297
Duchenne muscular dystrophy, 40/337, 361, 41/465, 43/106, 62/119, 133
electrical phenomenon, 1/633
Emery-Dreifuss muscular dystrophy, 40/390, 43/88
facial, *see* Facial contracture
Farber disease, 42/593
FG syndrome, 43/250
flexion, *see* Flexion contracture
Freeman-Sheldon syndrome, 43/353
frontometaphyseal dysplasia, 43/401
Fukuyama syndrome, 43/91
glycosaminoglycanosis type IS, 42/453
glycosaminoglycanosis type III, 42/455
hemiballismus, 42/224
hemiplegia, 2/481
intermediate spinal muscular atrophy, 42/90
joint, 25/184
joint fixation, 40/327, 337
juvenile metachromatic leukodystrophy, 60/669
Lafora progressive myoclonus epilepsy, 60/669
limb girdle syndrome, 43/104

malignant hyperthermia, 41/269
Marden-Walker syndrome, 43/424
Mietens-Weber syndrome, 43/304
motor unit hyperactivity, 41/187, 269, 297, 464
multicore myopathy, 43/120
muscle, *see* Muscle contracture
muscle cramp, 40/330
muscular dystrophy, 43/109
myopathy, 40/337
myophosphorylase deficiency, 27/234, 40/327, 41/186-188, 297, 43/182, 62/485
nerve injury, 7/251
Neu syndrome, 43/430
neuronal intranuclear hyaline inclusion disease, 60/669
opticocochleodentate degeneration, 42/241
phosphofructokinase deficiency, 40/327, 547, 41/297, 43/183
physical rehabilitation, 61/465
6q partial trisomy, 43/503
rehabilitation, 41/466
scapulohumeral distal muscular dystrophy, 43/130
scapuloperoneal myopathy, 43/130
spastic cerebral palsy, 42/167
spastic paraplegia, 42/169
spasticity, 26/478-480, 60/461
spinal cord injury, 25/289
symmetrical spastic cerebral palsy, 42/167
Troyer syndrome, 42/193
vertebral fusion, 43/488
Wohlfart-Kugelberg-Welander disease, 22/70, 42/91, 59/86
xeroderma pigmentosum, 60/669
Contrast media
brain angiography, 16/647
epilepsy, 72/233
Contravoluntary movement
Huntington chorea, 49/279
Contusio cerebri, *see* Brain contusion
Contusio cervicalis posterior
cervical vertebral column injury, 25/273, 284
spinal cord injury, 25/273, 284
Contusion
nerve injury, 51/134
spinal cord, *see* Spinal cord contusion
spinal cord injury, 61/37
Conus
anastomotic ansa, *see* Anastomotic ansa of the conus
spinal tumor, *see* Conus medullaris tumor
Conus aulicus intoxication
animal experiment, 37/61
Gastropoda, 37/59
pharmacology, 37/61
picture, 37/59
Conus geographus intoxication
animal experiment, 37/61
Gastropoda, 37/59
pharmacology, 37/61
picture, 37/59
Conus marmoreus intoxication
animal experiment, 37/61
Gastropoda, 37/59
pharmacology, 37/61
picture, 37/59
Conus medullaris
anastomotic ansa, *see* Anastomotic ansa of the conus
anatomy, 19/78
artery, 12/494
lesion, 25/454
Conus medullaris central necrosis
thoracolumbar spinal cord injury, 25/443
Conus medullaris syndrome
differential diagnosis, 59/432
meningococcal meningitis, 52/26
Conus medullaris tumor
cervical spondylotic myelopathy, 19/85
CSF, 19/79
differential diagnosis, 19/77-90
EMG, 19/80
pathology, 19/78
symptomatology, 19/79
Conus omaria intoxication
Gastropoda, 37/59
picture, 37/59
Conus striatus intoxication
animal experiment, 37/61
$\gamma$-butyrobetaine, 37/61
Gastropoda, 37/59
homarine, 37/61
indole derivative, 37/61
*N*-methylpyridinium, 37/61
pharmacology, 37/61
picture, 37/59
quaternary ammonium compound, 37/61
Conus textile intoxication
animal experiment, 37/61
Gastropoda, 37/59
pharmacology, 37/61
picture, 37/59
Conus tulipa intoxication
animal experiment, 37/61
Gastropoda, 37/59
pharmacology, 37/61
picture, 37/59

Conventional tomography
vertebral column, 19/139
Convergence
accommodation disorder, 2/667
eye, *see* Eye convergence
Krimsky-Prine rule, 74/423
retinotopic organization, 2/546
Convergent strabismus
craniosynostosis, 50/125
Möbius syndrome, 50/215
Conversion disorder, *see* Hysteria
Conversion headache
child, 48/40
classification, 48/5
definition, 48/353
delusion, 5/249
features, 48/34
psychiatric headache, 5/20, 48/356
symptom, 48/356
Convulsion
*see also* Epilepsy *and* Seizure
acetazolamide, 37/418
acetylsalicylic acid intoxication, 65/460
Addison disease, 39/479
allergic granulomatous angiitis, 63/383
Amanita pantherina intoxication, 36/534
amphetamine intoxication, 65/257
apistus intoxication, 37/76
arthropod envenomation, 65/196
barbiturate intoxication, 37/409
Bartter syndrome, 42/529
benign infantile, *see* Benign infantile convulsion
benign neonatal, *see* Benign neonatal convulsion
benign neonatal familial, *see* Benign neonatal familial convulsion
benign neonatal idiopathic, *see* Benign neonatal idiopathic convulsion
brachycephaly, 30/226
brain tumor, 31/44, 47, 50
bupivacaine intoxication, 37/405
caffeine intoxication, 65/260
carbamate intoxication, 64/190
carbon disulfide intoxication, 64/24
cat scratch disease, 52/129
cephalopoda intoxication, 37/62
cerebrohepatorenal syndrome, 31/306, 43/339
Chédiak-Higashi syndrome, 42/536
ciguatoxin intoxication, 65/163
classic migraine, 48/159
clonidine intoxication, 37/444
cocaine, 65/251
Coffin-Siris syndrome, 31/240
congenital adrenal hyperplasia, 42/515
congenital heart disease, 38/119-123
congenital toxoplasmosis, 42/662, 52/358
corpus callosum agenesis, 50/152
craniodiaphyseal dysplasia, 43/356
craniometaphyseal dysplasia, 43/362
cri du chat syndrome, 31/568
cyanide intoxication, 65/28
cyanogenic glycoside intoxication, 65/28
cyclic oculomotor paralysis, 42/332
cytomegalovirus infection, 50/277
decerebration, 16/131, 134
diphtheria pertussis tetanus vaccine, 52/239
Down syndrome, 31/407, 43/544, 50/529
dwarfism, 43/380
eclampsia, 39/547
Edwards syndrome, 31/521, 43/535
enflurane intoxication, 37/412
ergot intoxication, 65/68
ethionamide intoxication, 65/481
febrile, *see* Febrile convulsion
fever, *see* Febrile convulsion
fish roe intoxication, 37/86
fluroxene intoxication, 37/412
Friedman-Roy syndrome, 43/251
gymnapistes intoxication, 37/76
halothane intoxication, 37/411
hapalochlaena lunulata intoxication, 37/62
hapalochlaena maculosa intoxication, 37/62
happy puppet syndrome, 31/309
heat stroke, 71/588
hereditary long QT syndrome, 42/718
hexachlorophene intoxication, 37/486
hydrocephalus, 30/555
hydrozoa intoxication, 37/40
hyperammonemia, 42/560
hypercalcemia, 63/561
hyperglycinemia, 42/565
hyperprolinemia type I, 29/134
hyperprolinemia type II, 29/138
hypomagnesemia, 63/562
hypoparathyroidism, 27/300, 303-305
hypophosphatasia, 42/578
idiopathic, *see* Idiopathic convulsion
incontinentia pigmenti, 14/85, 31/242, 43/23
infantile neuroaxonal dystrophy, 49/402
Jamaican vomiting sickness, 37/512, 65/79
Joseph syndrome, 29/231
ketamine intoxication, 37/408
Klinefelter syndrome, 50/549
Lassa fever, 56/370
lead intoxication, 9/642
lidocaine intoxication, 37/450
lissencephaly, 42/40

meningeal leukemia, 63/341
mental deficiency, 42/689, 691
mepivacaine intoxication, 37/404
methionine malabsorption syndrome, 29/127, 42/600
microcephaly, 30/510, 513, 50/272, 277
moyamoya disease, 12/360, 42/751
neonatal, *see* Neonatal convulsion
neurocutaneous melanosis, 14/422
neurocysticercosis, 35/300
nicotine intoxication, 36/425, 65/261
Niemann-Pick disease type C, 15/426
nothesthes intoxication, 37/76
octopoda intoxication, 37/62
organophosphate intoxication, 37/555, 64/174
organotin intoxication, 64/139
oxycephaly, 30/226
4p syndrome, 31/551
pallidoluysionigral degeneration, 6/665
palytoxin intoxication, 37/55
paracoccidioidomycosis, 35/531, 52/458
pediatric head injury, 57/329
pertussis encephalopathy, 9/551
pertussis vaccine, 52/239
pheochromocytoma, 39/501, 42/764
pituitary agenesis, 42/11
plagiocephaly, 30/226
propionic acidemia, 27/72
pseudohypoparathyroidism, 46/85
pterois volitans intoxication, 37/75
puffer fish intoxication, 37/80
pyridoxal-5-phosphate, 15/63
pyridoxine dependent, *see* Pyridoxine dependent convulsion
pyruvate carboxylase deficiency, 29/212
pyruvate dehydrogenase complex deficiency, 66/416
quinidine intoxication, 37/449
r(21) syndrome, 31/594, 597
renal insufficiency, 27/330-332
Reye syndrome, 9/549, 27/363, 29/331, 333, 31/168, 34/169
scorpaena intoxication, 37/76
scorpaenodes intoxication, 37/76
scorpaenopsis intoxication, 37/76
scorpion sting intoxication, 37/112
sebastapistes intoxication, 37/76
sensorineural deafness, 42/383
sickle cell anemia, 55/510
Sjögren-Larsson syndrome, 22/477
striatal, *see* Striatal convulsion
subacute necrotizing encephalomyelopathy, 42/588
subdural hematoma, 57/337
Taybi-Linder syndrome, 43/386
toxaphene intoxication, 36/419
transient neonatal hypoglycemia, 27/56
trichopoliodystrophy, 29/286
trisomy 8 mosaicism, 50/563
tuberous sclerosis, 14/108
uremic encephalopathy, 27/330-332
vitamin B1 intoxication, 65/572
vitamin B6 deficiency, 42/716, 51/297
vitamin B6 dependency, 42/690, 716
XYY syndrome, 31/489
Convulsive epilepsy
*see also* Epilepsy
classification, 15/4
history, 15/4
Convulsive seizure
adrenoleukodystrophy, 60/169
aspergillosis, 35/397
brain tumor, 31/41
cytoside lipidosis, 10/545
ethylene glycol intoxication, 64/122
lead encephalopathy, 64/435
Minamata disease, 36/105
moyamoya disease, 12/360, 374
pontine infarction, 12/15, 30, 51
primary pontine hemorrhage, 12/39
pseudopseudohypoparathyroidism, 6/707
subarachnoid hemorrhage, 12/132
tegmental pontine syndrome, 12/20
urticaria pigmentosa, 14/790
Convulsive syncope
brain hypoxia, 72/259
epilepsy, 72/259
Convulsive triad
classification, 15/8
Cooley anemia, *see* Thalassemia major
Coordinated associated movement, *see* Synkinesis
Coordination, 1/293-308
*see also* Incoordination
cerebellar vermis, 1/307
cerebellum, 1/293, 295, 300, 305-308, 324
clinical test, 1/324
control, 1/300
equilibrium, 1/306-308
gait, 1/308
hand-eye, *see* Hand-eye coordination
history, 1/293
motor function, 1/305-308
movement, 1/298
muscle, 1/297
oculomotor system, 2/274
posture, 1/306

reflectory, 1/295
Copolymer I
immunosuppression, 47/191-193
multiple sclerosis, 47/192
Copper
anemia, 27/389
basal ganglion, 49/35
biological function, 29/279-286
ceruloplasmin, 49/229
chemistry, 29/279-286
CNS, 27/393-399, 28/437
copper dependent lysyl oxidase, 42/584
deficiency, 28/438
demyelination, 9/645
homeostasis, 29/279-286
leukoencephalopathy, 9/645
metabolism, 6/117-123, 27/387-391
metallothionein, 49/229
multiple sclerosis, 9/112, 322
myelination, 9/634
serum, *see* Serum copper
striatopallidodentate calcification, 6/716, 49/425
swayback, 27/389
toxic encephalopathy, 49/230
toxicity, *see* Copper intoxication
trichopoliodystrophy, 29/279-286, 294-300, 42/584
urine, *see* Urine copper
Willvonseder syndrome, 22/227
Wilson disease, 27/393-401, 28/438, 42/270, 49/229
Copper dependent lysyl oxidase
copper, 42/584
Copper enzyme
tyrosinase, 49/229
Copper intoxication
dementia, 46/392
headache, 48/420
liver toxicity, 37/623
renal toxicity, 37/623
symptomatic dystonia, 49/543
Wilson disease, 27/392
Copper sulfate
toxic myopathy, 62/601
Coprine
*see also* Mushroom toxin
aldehyde dehydrogenase inhibition, 65/39
chemical structure, 65/39
Coprine intoxication
disulfiram like effect, 65/39
headache, 65/39
symptom, 65/39
Coprinus atramentarius
mushroom intoxication, 36/535-538
Coprographia, *see* Coprolalia
Coprolalia
agraphia, 45/468
Gilles de la Tourette syndrome, 1/291, 6/788, 42/221, 45/468, 49/627, 629
Coproporphyria
coproporphyrinogen oxidase, 51/390
hereditary, *see* Hereditary coproporphyria
sex ratio, 42/539
Coproporphyrin III
hereditary coproporphyria, 42/539
Coproporphyrinogen oxidase
coproporphyria, 51/390
hereditary coproporphyria, 42/539
lead, 64/434
Copropraxia
Gilles de la Tourette syndrome, 49/627
Coprostanol, *see* Coprosterol
Coprosterol
chemistry, 10/235
Cor de chasse
kernicterus, 6/494
Cor pulmonale
pickwickian syndrome, 38/299, 63/455
respiratory encephalopathy, 63/414
Coral snake intoxication
venom, 37/13
Corbadrine
classification, 74/149
Cord lesion
cervical, *see* Cervical cord lesion
complete, *see* Complete cord lesion
incomplete, *see* Incomplete cord lesion
late, *see* Late cord lesion
progressive, *see* Progressive cord lesion
spinal, *see* Spinal cord lesion
Cord necrosis
central, *see* Central cord necrosis
Cord sign
superior longitudinal sinus thrombosis, 52/178
venous sinus thrombosis, 54/402
Cordectomy
spinal, *see* Spinal cordectomy
Cordis-Hakim valve
hydrocephalus, 30/575-578
Cordotomy
anterolateral, *see* Anterolateral cordotomy
cervical, *see* Cervical cordotomy
dorsal, *see* Dorsal cordotomy
Core disease
central, *see* Central core myopathy
Core promotor element

gene transcription, 66/70
Cori disease
type I, *see* Glycogen storage disease type I
type II, *see* Infantile acid maltase deficiency
type III, *see* Debrancher deficiency
type IV, *see* Brancher deficiency
type V, *see* Myophosphorylase deficiency
type VI, *see* Hepatophosphorylase deficiency
type VII, *see* Glycogen synthetase deficiency
Cori-Forbes disease, *see* Debrancher deficiency
Cornea
glycosaminoglycanosis type VII, 66/310
keratitis bullosa, 8/350
reccurent erosion, 48/226
thyroid disease, 70/89
Cornea verticillata
Fabry disease, 60/172
Corneal clouding
congenital heart disease, 63/11
glycosaminoglycanosis type I, 10/435
glycosaminoglycanosis type IS, 10/437
glycosaminoglycanosis type II, 10/435, 42/454, 66/295
mucolipidosis type IV, 51/380
Corneal dermoid
anencephaly, 50/83
Corneal dystrophy
adult GM1 gangliosidosis, 60/652
Bedouin spastic ataxia, 60/652
Cockayne-Neill-Dingwall disease, 60/652
Fabry, *see* Fabry corneal dystrophy
glycosaminoglycanosis type VI, 66/309
Harboyan syndrome, 42/367
hereditary amyloid polyneuropathy type 3, 51/424
lattice type, *see* Corneal lattice dystrophy
macular, *see* Macular corneal dystrophy
multiple sulfatase deficiency, 60/652
palmoplantar keratosis, 42/581
Richner-Hanhart syndrome, 42/581
sensorineural deafness, 42/367
spinocerebellar degeneration, 60/652
X-linked ichthyosis, 60/652
Corneal identation pulse
cluster headache, 48/234, 75/281
Corneal infiltration
a-alphalipoproteinemia, 51/396
cyanea intoxication, 37/52
Corneal lattice dystrophy
amyotrophic lateral sclerosis, 22/139
bulbar paralysis, 22/139
hereditary amyloid polyneuropathy, *see* Hereditary amyloid polyneuropathy type 3
Corneal opacity
*see also* Cataract
amiodarone intoxication, 60/172
chloroquine intoxication, 60/172
differential diagnosis, 60/172
Edwards syndrome, 50/562
Fabry disease, 42/427, 51/376
glycosaminoglycanosis type I, 10/436, 42/450, 60/177
glycosaminoglycanosis type IH, 60/177
glycosaminoglycanosis type IS, 42/452, 60/177
glycosaminoglycanosis type II, 60/177
glycosaminoglycanosis type IV, 42/457, 60/177
glycosaminoglycanosis type VI, 42/458, 60/177, 66/308
Hurler-Hunter disease, 10/341, 436
mannosidosis, 42/597
Mietens-Weber syndrome, 43/304
mucolipidosis type I, 27/162, 42/475, 51/380
mucolipidosis type III, 29/367, 42/449
Corneal reflex
blink reflex, 2/56
brain stem death, 57/458, 473
congenital pain insensitivity, 42/297
hereditary sensory and autonomic neuropathy type II, 60/11
lateral medullary infarction, 53/381
trigeminal nerve, 2/56
trigeminal nerve lesion, 1/239
Corneal sensitivity
cluster headache, 48/235
Corneal temperature
chronic paroxysmal hemicrania, 48/261, 75/292
cluster headache, 48/234
Corneal transplantation
Creutzfeldt-Jakob disease, 56/549
glycosaminoglycanosis type IH, 66/291
glycosaminoglycanosis type IS, 66/292
Harboyan syndrome, 42/368
rabies, 56/385
Corneal ulceration
orbital cellulitis, 33/179
trigeminal nerve agenesis, 50/214
Cornelia De Lange syndrome, *see* De Lange syndrome
Corneomandibular reflex
brain stem lesion, 1/173, 239
central motor disorder, 1/173
trigeminal nerve, 2/56
Corneo-oculogyric reflex
painful stimuli, 2/57
Corona
migraine, 48/162
Coronal suture

methotrexate syndrome, 30/122
premature fusion, *see* Craniosynostosis
Coronal synostosis
craniosynostosis, 50/118
Coronary artery
arteritis, 48/316
Coronary artery aneurysm
Kawasaki syndrome, 56/637
Coronary artery bypass graft, *see* Coronary bypass surgery
Coronary artery disease, *see* Ischemic heart disease
Coronary blood flow
spinal cord injury, 26/206
Coronary bypass surgery
bilateral carotid occlusion, 63/185
brain fat embolism, 63/182
brain infarction, 63/181
brain infarction risk, 63/181
brain perfusion pressure, 63/188
calcific embolus, 63/187
carotid artery disease, 63/184
carotid artery trauma, 63/184
carotid bruit, 63/184
carotid stenosis frequency, 63/185
cerebellar infarction, 63/187
coma, 63/187
cortical blindness, 55/119
delirium, 46/547
encephalopathy, 63/187
heart transplantation, 63/181
iatrogenic neurological disease, 63/181
internal jugular vein cannulation, 63/184
neurologic complication prophylaxis, 63/188
prior stroke significance, 63/188
retinal artery embolism, 55/110
risk factor, 63/181, 184
risk variable, 63/184
stroke risk, 63/184
transient ischemic attack, 63/185
Coronary care unit
delirium, 46/535
Coronavirus
demyelination, 56/12
features, 56/12
human, 56/3
interrelationship, 56/440
JHM virus, 56/439
multiple sclerosis, 47/330, 56/13
neurotropic virus, 56/439
porcine hemagglutinating encephalomyelitis virus, 56/439
RNA virus, 56/12
serotype, 56/440
slow virus disease, 56/439
structure, 56/439
taxonomic place, 56/3
virus replication, 56/440
Coronavirus infection, 56/439-448
CNS, 56/440
Corpora amylacea
*see also* Glycogen
chemical analysis, 27/192
digestion, 27/191
distribution, 27/194
electron probe microanalysis, 27/193
genesis, 27/194
histochemical aspect, 27/189-191
infrared study, 27/192
significance, 27/194
tuberous sclerosis, 14/360
ultrastructure, 27/191
Corpus amygdaloideum
temporal lobe epilepsy, 73/58
Corpus callosum
alexia, 4/126, 45/12, 439
anatomy, 17/490
animal study, 2/758
apraxia, 4/56, 45/426-428, 431
Binswanger disease, 54/224
cerebral dominance, 4/263, 273-290
cyanide intoxication, 65/28
cyanogenic glycoside intoxication, 65/28
diffuse axonal injury, 57/53
discrimination learning, 2/759
embryo, 50/11
epilepsy, 15/43
experimental aspect, 2/758
focal brain damage, 57/51
function, 17/492
glioblastoma multiforme, 18/51
hemialexia, 4/117
history, 2/758-765
interhemispheric transfer, 2/758
laboratory finding, 2/758-760
Marchiafava-Bignami disease, 2/763, 28/317, 46/337, 47/558
necrosis, 17/542
olfactory bulb agenesis, 50/178
physiology, 3/36
schizophrenia, 46/454, 458
sensory extinction, 45/159
splenium, 2/763
tactile agnosia, 4/35
trauma, 17/542
Corpus callosum agenesis, 30/285-296, 42/6-9
acrocallosal syndrome, 50/163

acrocephalosyndactyly type I, 43/320, 50/163, 168
Aicardi syndrome, 30/100, 31/225, 42/8, 696, 50/151, 156, 159, 162
alexia, 30/289
anatomic aspect, 50/149
Andermann syndrome, 30/100, 50/163, 166
angiography, 2/765, 50/154
apraxia, 45/100
aqueduct stenosis, 50/302, 309
arachnoid cyst, 31/119
areflexia, 42/100
Arnold-Chiari malformation type I, 50/163
arthrogryposis multiplex congenita, 14/408
arylsulfatase deficiency, 50/151
associated brain abnormality, 50/150
associated malformation, 30/295
asymptomatic case, 50/159
behavior manifestation, 30/287-289
birth incidence, 42/103
Bowen syndrome, 43/335
brachycephaly, 42/101
C syndrome, 50/163
cataract, 50/162
cavum septi pellucidi, 50/158
cavum veli interpositi, 50/158
cavum vergae, 50/158
cerebellar vermis aplasia, 50/150
cerebral dominance, 4/288, 30/289
cerebrohepatorenal syndrome, 31/308, 50/163
cheilopalatoschisis, 42/7
chorioretinopathy, *see* Aicardi syndrome
chromosomal aberration, 50/151
clinical features, 30/287-289, 42/8, 50/159
CNS, 30/287, 294
coloboma, 50/162
colpocephaly, 50/155
computer assisted tomography, 50/154
congenital heart disease, 63/11
congenital ichthyosiform erythroderma, 42/7, 50/152
contracture, 42/100
convulsion, 50/152
corpus callosum lipoma, 30/285, 288, 50/149, 155, 157
cranial nerve, 42/7
craniofacial dysostosis (Crouzon), 50/168
cri du chat syndrome, 43/501
Dandy-Walker syndrome, 2/764, 42/23, 50/156, 163, 168, 333
dermoid cyst, 50/149
diagnosis, 2/765, 30/289-294, 50/153
dichotic listening, 50/161
differential diagnosis, 31/228, 50/157
disconnection syndrome, 50/160
dominant inheritance, 50/151
Down syndrome, 50/531
dwarfism, 43/385, 50/163
echoEG, 30/289, 292
Edwards syndrome, 31/522, 43/536, 50/152, 163, 167, 562
EEG, 30/289, 292, 42/10, 50/159
embryology, 30/292-294, 50/12, 150
encephalocele, 50/150
encephaloclastic cavity, 50/155
encephalomeningocele, 42/7, 26
epidemiology, 30/150, 155, 286
epilepsy, 42/7, 72/109
etiology, 50/151
eye abnormality, 50/150
face abnormality, 50/150
familial cases, 50/152
fetal alcohol syndrome, 50/153
FG syndrome, 30/100, 43/250, 50/163
genetics, 30/155, 285
glycogen storage disease, 42/8
glycosaminoglycanosis type I, 50/151
gray matter heterotopia, 50/165
Gruber syndrome, 50/163
hand-eye coordination, 50/161
heart disease, 42/8
hemiplegia, 50/162
heterozygote frequency, 42/103
hippocampus commissure, 50/150
history, 30/285
holoprosencephaly, 30/455, 50/155
hydranencephaly, 50/159
hydrocephalus, 2/764, 30/287, 42/7, 50/159, 161
hypertelorism, 42/101, 43/417, 50/162
hypothermia, 50/167
ichthyosiform erythroderma, 42/7
imperforate anus, 42/7
incidence, 50/153
infantile spasm, 42/8, 696, 50/162
interhemispheric cyst, 50/154, 156
interhemispheric lipoma, 2/764
Joubert syndrome, 63/437
Klinefelter syndrome, 50/163
Krabbe-Bartels disease, 14/408
kyphoscoliosis, 42/101
leprechaunism, 50/163
lipoma, 2/764, 14/410, 18/356, 30/285, 288
lissencephaly, 30/485, 42/39, 50/151, 153, 163, 168
macrencephaly, 50/263
macrocephaly, 50/152, 162
management, 30/296

median facial cleft syndrome, 30/243, 46/90, 50/150, 163
meningocele, 42/7
mental deficiency, 2/765, 30/287, 42/7-10, 100, 102, 46/12, 50/161
metabolic acidosis, 50/167
microcephaly, 30/287, 42/7, 43/427, 50/152, 162
micrognathia, 42/7
microgyria, 50/150
microphthalmia, 50/162
mirror movement, 42/233
monosomy 21, 31/604
multiple nevoid basal cell carcinoma syndrome, 14/79, 114, 50/163
myelin dysgenesis, 42/10, 498
Neu syndrome, 43/430, 50/163
neuronal heterotopia, 42/7, 10
neuropsychologic finding, 50/160
nuclear magnetic resonance, 50/156
occipital encephalocele, 42/26
optic atrophy, 42/8, 50/162
optic chiasm, 2/528
orofaciodigital syndrome, 50/163
orofaciodigital syndrome type I, 50/167
orthochromatic leukodystrophy, 10/114
4p partial monosomy, 43/497
17p partial monosomy, 50/592
18p partial monosomy, 43/534
4p partial trisomy, 43/500
4p syndrome, 31/557
18p syndrome, 31/579, 582
pachygyria, 50/150
parastriate cortex, 50/150
parietal cephalocele, 50/108
parietal encephalocele, 50/108
partial, 30/285, 288, 291, 295, 42/6, 9, 50/154
Patau syndrome, 14/120, 43/527, 50/152
pathogenesis, 30/286
pathology, 30/294
PEG, 2/765, 50/153
periventricular heterotopia, 50/151
polymicrogyria, 42/9, 50/165
polyploidy, 50/555
porencephaly, 2/764, 31/119, 42/7
prenatal diagnosis, 42/9
prevalence, 42/7, 50/61
Probst bundle, 30/294, 485, 42/10, 50/150
prosencephaly, 50/159
pyramidal tract absence, 42/7
pyramidal tract hypoplasia, 50/150
21q partial monosomy, 43/542
11q partial trisomy, 43/524
18q syndrome, 31/588
radiologic appearance, 30/289
renofacial dysplasia, 50/163
respiratory dysfunction, 50/167
Rubinstein-Taybi syndrome, 31/284, 50/163, 276
seizure, 2/765, 30/287, 42/7, 9, 101, 50/161
sensorimotor neuropathy, 42/100-103
septum pellucidum, 50/150
septum pellucidum tumor, 50/157
sex ratio, 50/61
Shapiro syndrome, 50/163, 167
spastic tetraplegia, 42/8
speech disorder in children, 42/100
spina bifida, 50/150, 487
status epilepticus, 42/7
status verrucosus, 50/203
strabismus, 42/101
subependymal heterotopia, 50/156
symptomatic case, 50/160
syndactyly, 42/101
talipes, 42/7
Taybi-Linder syndrome, 43/386
tetraplegia, 42/100, 50/152
thanatophoric dwarfism, 43/388
third ventricle dilatation, 50/149
total agenesis, 50/154
tremor, 42/100
trichopoliodystrophy, 50/61
triploidy, 43/566
trisomy 8, 31/528, 50/153, 163, 563
trisomy 8 mosaicism, 43/511, 50/167
trisomy 13-15, 14/774, 30/286
trisomy 22, 50/565
tuberous sclerosis, 30/286
ultrasonography, 50/154, 156
ventricular dilatation, 42/10
vertebral abnormality, 42/7
X-linked inheritance, 50/151
X-linked recessive inheritance, 50/151
Corpus callosum artery
dorsal, *see* Dorsal callosal artery
Corpus callosum ataxia
equilibrium, 1/343, 17/508
gait disturbance, 1/343, 17/508
tumor, 1/343, 17/508
Corpus callosum defect
hemialexia, 45/444
influenza, 9/582
Corpus callosum degeneration, *see* Marchiafava-Bignami disease
Corpus callosum disconnection syndrome
agnosia, 2/760
alexia, 2/598
apraxia, 2/760

schizophrenia, 46/501, 504, 508
Corpus callosum dyspraxia
Huntington chorea, 49/280
Corpus callosum hypoplasia
Edwards syndrome, 43/536, 50/275
homocystinuria, 55/330
Miller-Dieker syndrome, 50/592
XXXX syndrome, 50/555
Corpus callosum infarction
apraxia, 45/426-428
left hand agraphia, 2/763
left hand apraxia, 2/763
motor aphasia, 2/763
motor apraxia, 2/763
splenio-occipital alexia, 45/444
Corpus callosum lipoma
cheilognathopalatoschisis, 14/410
corpus callosum agenesis, 30/285, 288, 50/149, 155, 157
Krabbe-Bartels disease, 14/12, 409
Corpus callosum syndrome
agraphia, 53/347
akinetic mutism, 2/762
alexia, 2/760
alien hand syndrome, 53/347
anterior cerebral artery syndrome, 53/347
apathy, 2/762
brain infarction, 53/347
clinical features, 2/761
corpus callosum tumor, 2/760, 17/506
cranial nerve palsy, 2/761
emotional disorder, 2/761
headache, 2/761
imperviousness, 2/762
intracranial pressure, 2/761
memory, 2/762
mental deterioration, 2/761
mental symptom, 2/762
motor aphasia, 2/762
papilledema, 2/761
split brain syndrome, 45/100, 53/347
tactile anomia, 53/347
Corpus callosum transmission
schizophrenia, 46/500, 506
Corpus callosum tumor
age distribution, 17/497
anatomy, 17/490
apraxia, 17/507
astrocytoma, 2/760, 17/498
ataxia, 17/508
bladder apraxia, 17/508
brain arteriovenous malformation, 2/257-259, 17/537
brain artery occlusion, 17/542
clinical features, 17/503
corpus callosum syndrome, 2/760, 17/506
cranial nerve, 17/509
cranial nerve palsy, 17/509
CSF, 17/511
disconnection syndrome, 17/506
echoEG, 17/512
EEG, 17/512
extrapyramidal disorder, 17/511
function loss, 17/507
glioblastoma multiforme, 17/499
histology, 17/498
incidence, 17/497
intracranial hypertension, 17/509
lipoma, 14/409, 17/501, 18/355-357
localization, 17/500
Marchiafava-Bignami disease, 17/542
mental disorder, 17/504
psychiatric disorder, 17/504
pyramidal sign, 17/510
radiology, 17/513, 520, 524, 527, 530, 533
reflex behavior, 17/510
sex ratio, 17/497
speech disorder, 17/509
supratentorial midline tumor, 18/355
symptomatology, 17/511
treatment, 17/543
Corpus callosum vein
dorsal, *see* Dorsal callosal vein
Corpus caroticum, *see* Carotid body
Corpus paracaroticum, *see* Carotid body
Corpus pregeniculatum
neuroanatomy, 2/529
Corpus striatum
motor function, 1/160
Corpuscle rétrocarotidien, *see* Carotid body
Correct diagnosis
brain embolism, 53/199
brain infarction, 53/199
cerebrovascular disease, 53/199
intracranial hemorrhage, 53/199
transient ischemic attack, 53/199
Cortex
association, *see* Association cortex
athetosis, 6/459
ballismus, 6/482
brain, *see* Brain cortex
calcarine, *see* Calcarine cortex
cerebellar, *see* Cerebellar cortex
cerebral, *see* Brain cortex
cingulate, *see* Cingulate cortex
frontal, *see* Frontal cortex

inferior visual association, *see* Inferior visual association cortex
kernicterus, 6/503
language, *see* Language cortex
mesial, *see* Mesial cortex
migraine sine hemicrania, 48/165
motor, *see* Motor cortex
orbital frontal, *see* Orbital frontal cortex
peristriate, *see* Parastriate cortex
precentral, *see* Precentral cortex
prefrontal, *see* Prefrontal cortex
premotor, *see* Premotor cortex
temporal, *see* Temporal cortex
visual, *see* Visual cortex
Cortex focal dysplasia
brain, *see* Brain cortex focal dysplasia
Cortex function
motor, *see* Motor cortex function
Cortex hypertrophy
adrenal, *see* Adrenal cortex hypertrophy
cerebellar, *see* Cerebellar cortex hypertrophy
Cortex injury
visual, *see* Visual cortex injury
Corti organ
hereditary sensory and autonomic neuropathy type I, 60/9
Corti organ degeneration
Alport syndrome, 42/376
Muckle-Wells syndrome, 42/397
Usher syndrome, 43/392
Cortical alexia
angular gyrus syndrome, 4/127
reading disorder, 4/115
writing disorder, 4/115
Cortical arteriolosclerosis
familial, *see* Familial cortical arteriolosclerosis
Cortical atrophy, *see* Brain cortex atrophy
Cortical blindness
*see also* Anosognosia, Anton syndrome *and* Traumatic blindness
adrenoleukodystrophy, 60/169
agnosia, 4/14, 20
alpha rhythm, 45/50
anosognosia, 2/593, 3/284
Anton syndrome, 2/697
arterial hypertension, 63/72
asthenopia, 45/51
brain angiography, 55/119
brain concussion, 24/49
brain infarction, 55/119, 144
brain injury, 24/48
carbon monoxide intoxication, 55/119
cardiac arrest, 63/215
cerebrovascular disease, 55/139, 144
ciclosporin, 63/537
ciclosporin intoxication, 63/537
clinical features, 55/144
coronary bypass surgery, 55/119
Creutzfeldt-Jakob disease, 46/291
definition, 2/543, 22/536, 55/144
denial, 2/670, 24/49
diagnosis, 45/50
diagnostic symptom, 55/119
differential diagnosis, 24/51
EEG, 24/49, 45/50
ERG, 24/50
essential symptom, 2/592
form, 2/591-593
hallucination, 2/670
homonymous hemiachromatopsia, 2/650
hypoxia, 45/50
hysterical blindness, 24/51
late infantile metachromatic leukodystrophy, 66/167
leukemia, 63/342
MELAS syndrome, 62/510
memory disorder, 3/284
meningitis, 55/119
occipital lobe infarction, 55/119
occipital lobe syndrome, 45/49-51
occipital lobe tumor, 17/326
occipital symptom, 2/669
optokinetic nystagmus, 45/50
pathogenesis, 24/50, 45/50
penetrating head injury, 24/48, 50, 492
pheochromocytoma, 39/501
recovery, 45/51
temporal arteritis, 55/344
transient, *see* Transient cortical blindness
traumatic blindness, 24/48-52
venous sinus thrombosis, 54/432
visual evoked response, 45/50
Cortical calcification
Cockayne-Neill-Dingwall disease, 13/435
Cortical cerebellar atrophy
catalepsy, 60/644
hereditary type, 21/403
late, *see* Late cortical cerebellar atrophy
synkinesis, 60/644
Cortical cerebellar degeneration
carcinoma, 38/625-631
malignancy, 38/625-631
Cortical deafness
Wernicke aphasia, 45/315
Cortical dementia
classification, 46/204

Cortical dysplasia
cerebellar, *see* Cerebellar cortical dysplasia
cerebellar heterotopia, 50/251
neuronal migration disorder, 73/400
Cortical gaze center
conjugate deviation, 2/652
four sites, 2/652
Cortical heterotopia
Aicardi syndrome, 31/226
cerebrohepatorenal syndrome, 31/307
neurofibromatosis, 14/37
Sturge-Weber syndrome, 31/21
Cortical laminar necrosis
cardiac arrest, 63/214
Cortical laminar sclerosis
Marchiafava-Bignami disease, 28/317, 325, 342, 46/338, 47/558
Cortical muscular atrophy
parietal lesion, 2/684
Cortical myelinogenesis
dystopic, *see* Dystopic cortical myelinogenesis
Cortical myoclonus
Alzheimer disease, 49/612
epilepsy, 49/613
multifocal, 49/615
Cortical response
evoked, *see* Evoked cortical response
Cortical tuber
calcification, 14/356
number, 14/354
site, 14/354
umbilication, 14/355
Corticoid, *see* Corticosteroid
Corticomeningeal angiomatosis, *see* Divry-Van Bogaert syndrome
Corticopontocerebellar syndrome, *see* Neocerebellar syndrome
Corticospinal tract degeneration
amyotrophic lateral sclerosis, 42/65
Berardinelli-Seip syndrome, 42/592
spastic paraplegia, 42/180
Corticospinal tract disease
hyperthyroidism, 70/86
Corticosteroid
*see also* Corticotropin, Dexamethasone, 9α-Fluorohydrocortisone, Glucocorticoid, Steroid *and* Triamcinolone
acquired toxoplasmosis, 35/124, 52/353
addisonian crisis, 63/359
adverse effect, 69/501
allergic granulomatous angiitis, 63/384
allergic neuropathy, 7/555
Alzheimer disease, 46/265
antineoplastic agent, 65/529
aspergillosis, 52/377
behavior disorder, 46/599
Behçet syndrome, 56/604
brain abscess, 52/148, 158
brain edema, 24/627-633, 57/214
brain injury, 24/618
brancher deficiency, 41/185
candidiasis, 35/418, 52/398
carnitine deficiency, 41/198
Castleman disease, 63/396
cavernous sinus thrombosis, 52/175
clinical study, 24/627
coccidioidomycosis, 52/420
complication, 24/632
CSF, *see* CSF corticosteroid
demyelinating neuropathy, 41/346
dermatomyositis, 62/383, 71/141
Duchenne muscular dystrophy, 40/378, 62/129
encephalopathy, 63/359, 65/529
eosinophilic fasciitis, 63/385
epileptic treatment, 15/731
experiment, 24/629, 631, 634
facioscapulohumeral muscular dystrophy, 40/423
gastrointestinal bleeding, 24/632
glioblastoma multiforme, 18/65
granulomatous CNS vasculitis, 55/387, 397
gunshot injury, 57/311
headache, 48/325
hereditary motor and sensory neuropathy type I, 60/199
Herxheimer reaction, 52/285
Hodgkin lymphoma, 63/348
hypereosinophilic syndrome, 63/372, 374
immunocompromised host, 56/469
immunosuppressive therapy, 65/550
inclusion body myositis, 62/383
inflammatory myopathy, 41/73, 79, 374
intracranial hypertension, 24/618, 55/216
intrathecal chemotherapy, 61/151
Kawasaki syndrome, 52/267
leukemia, 39/15
lymphomatoid granulomatosis, 63/424
mania, 46/462, 599
missile injury, 57/311
mitochondrial abnormality, 41/251
multiple myeloma, 18/256, 20/11, 14, 63/396
multiple organ failure, 71/540
multiple sclerosis, 9/394-396, 411, 47/86, 101, 104
muscle fiber type II atrophy, 41/251
muscular atrophy, 40/160
myasthenia gravis, 41/93, 132, 62/417

myotonic syndrome, 40/561
neurologic intensive care, 55/216
neuromuscular transmission, 41/133
neuromyelitis optica, 47/406
neurotoxin, 63/359, 65/529, 555
non-Hodgkin lymphoma, 63/348
osteosclerotic myeloma, 63/396
paraplegia, 61/399
penetrating head injury, 57/311
periodic paralysis, 41/156
polymyositis, 41/59, 62/167, 187, 383, 71/141
postherpetic neuralgia, 51/181
progressive multifocal leukoencephalopathy, 47/506
proximal myopathy, 63/359
sarcoid neuropathy, 71/489
sarcoidosis, 71/492
schistosomiasis, 52/541
serum creatine kinase, 40/343
spinal cord edema, 24/631
spinal cord injury, 26/222, 61/399
steroid myopathy, 41/250, 62/384
systemic lupus erythematosus, 71/48
Takayasu disease, 55/339
temporal arteritis, 48/323
tetanus, 41/305
tetraplegia, 61/399
thyroid ophthalmopathy, 41/241
Tolosa-Hunt syndrome, 48/291, 299, 304
toxic myopathy, 62/605
transverse spinal cord lesion, 61/399
traumatic venous sinus thrombosis, 24/379
tremor, 63/360
trichinosis, 52/573
tuberculous meningitis, 52/213
vasculitis, 71/160
Corticosteroid intoxication
benign intracranial hypertension, 65/555
delirium, 65/555
hiccup, 65/555
myopathy, 65/555
spinal epidural compression, 65/555
spinal epidural lipoma, 65/555
Corticosteroid myopathy
clinical pattern, 65/556
incidence, 65/556
scleroderma, 71/110
Corticosteroid withdrawal myopathy
adrenal insufficiency myopathy, 62/536
arthralgia, 62/536
fatigue, 62/536
fever, 62/536
myalgia, 62/536
weight loss, 62/536
Corticosterone
Cushing syndrome, 70/206
Corticostriate encephalitis
monkey, 49/500
striatal necrosis, 6/679, 49/500, 512
Corticostriate fiber
glutamic acid, 49/4
Huntington chorea, 6/399, 403
terminal mosaic pattern, 49/3
Corticostriatocerebellar syndrome, *see* Gökay-Tükel syndrome
Corticostriatopallidodentate calcification, *see* Striatopallidodentate calcification
Corticostriatospinal degeneration, *see* Creutzfeldt-Jakob disease
Corticosubcortical epilepsy
diffuse simultaneous, *see* Diffuse simultaneous corticosubcortical epilepsy
Corticosubthalamic fiber
glutamic acid, 49/11
subthalamic nucleus, 49/11
Corticothalamic connection
neurosurgery, 2/491
Corticotropin
*see also* Corticosteroid
adenohypophysis, 2/442
Alzheimer disease, 46/268
anorexia nervosa, 46/586
basophil adenoma, 2/457
behavior disorder, 46/599
biological rhythm, 74/508
brain edema, 16/204
brain injury, 23/121
brain metastasis, 18/229
Carney Complex, 70/210
CSF cytology, 47/373
Cushing syndrome, 68/356, 70/205, 207
emotion, 45/279
happy puppet syndrome, 31/309
infantile spasm, 73/200
lentiginosis syndrome, 70/210
multiple myeloma, 18/256
multiple sclerosis, 9/393, 47/101, 104, 107, 116, 192, 356, 373
Parkinson disease, 49/123
pituitary adenoma, 17/393, 403
pituitary dwarfism type II, 42/613
polymyositis, 41/54, 74, 79
pro-opiomelanocortin, 65/350
stress, 45/248, 250, 74/308
Corticotropin releasing factor
brain center, 74/307

depression, 74/506
stress, 74/307
Corticotropin releasing hormone, *see* Corticotropin releasing factor
Cortinarius speciosissimus
mushroom intoxication, 36/542
mushroom toxin, 65/38
Cortisol, *see* Hydrocortisone
Cortisone
cluster headache, 48/230
migraine, 5/56
multiple myeloma, 18/256
multiple sclerosis, 9/393
neurotoxin, 50/219
Cortisone intoxication
hereditary deafness, 50/219
Corynebacterium diphtheriae
*see also* Diphtheritic neuritis
bacteriology, 52/227
classification, 33/479
features, 65/237
neuropathy, 47/608
tox-negative strain, 65/238
tox-positive strain, 65/238
toxin, 65/236
toxin fragment A, 51/182
toxin fragment B, 51/182
Coryza
anterior ethmoidal nerve syndrome, 5/215
sphenopalatine neuralgia, 48/477
Cosmonaut
autonomic nervous system, 74/273
Costen syndrome
*see also* Temporomandibular arthrosis
headache, 5/345, 48/6, 226
Costoclavicular maneuver
hyperabduction syndrome, 2/150
thoracic outlet syndrome, 51/124
Costoclavicular syndrome
*see also* Cervical rib syndrome
compression syndrome, 7/430, 432, 442
neurovascular compression syndrome, 2/146, 148
thoracic outlet syndrome, 51/121
Cotidine
nicotine intoxication, 65/262
Cotton seed intoxication
endemic periodic paralysis, 63/557
hypokalemia, 63/557
Cough
chrysaora intoxication, 37/51
CSF flow, 61/378
cyanea intoxication, 37/52
decompression illness, 63/419
eosinophilia myalgia syndrome, 64/257
esophageal pressure, 26/339
lateral medullary infarction, 53/382
posttraumatic syringomyelia, 61/378
respiratory encephalopathy, 63/414
syringomyelia, 61/378
toxic oil syndrome, 63/380
Cough headache
abdominal pressure, 48/370
acetylsalicylic acid, 48/371
associated lesion, 48/368
autonomic nervous system, 75/300
blood flow, 48/367-371, 75/300
brain tumor, 69/25
classification, 48/6
clonidine, 48/371
course, 48/368
CSF flow, 48/369
CSF pressure, 5/186, 48/370
differential diagnosis, 48/369
ergotamine, 48/371
etiology, 48/369
exertional headache, 48/375
expanding lesion, 48/367
indometacin, 48/371
intracranial tumor, 5/185
pathogenesis, 48/367
prognosis, 48/368
propranolol, 48/371
prostaglandin, 48/371
Queckenstedt test, 48/370
serotonin, 48/371
sex distribution, 5/185, 48/368
treatment, 48/370
vascular tone, 5/186
Cough syncope
dysautonomia, 22/263
laryngeal neuralgia, 48/497
vasomotor reflex, 1/499
Coughing
autonomic nervous system, 74/584, 622
carotid dissecting aneurysm, 54/271
cervical cordotomy, 63/491
medulla oblongata, 63/491
reflex syncope, 15/824
respiration reflex, 63/491
spinal cord injury, 61/262
Coughing reflex
physiology, 1/494
Coumafos
neurotoxicity, 37/546
organophosphate intoxication, 64/155
Coumafos intoxication

chemical classification, 37/546
organophosphorus compound, 37/546
Coumaphos, *see* Coumafos
Coumarin
relaxation, 74/365
Counter pressure phenomenon
spasmodic torticollis, 6/574, 42/264
Counterimmunoelectrophoresis
bacterial meningitis, 52/7
gram-negative bacillary meningitis, 52/104
infantile enteric bacillary meningitis, 33/64
meningococcal meningitis, 33/25, 52/25
pneumococcal meningitis, 33/40
Coup lesion
pathology, 23/38, 93
Coup de poignard rachidien, *see* Spinal subarachnoid hemorrhage
Coup de sabre deformity
progressive hemifacial atrophy, 22/548, 31/253, 59/479
Couplet periodic breathing
respiration, 1/674
Couvade syndrome
abdominal pain, 48/357
headache, 48/358
pain, 48/358
Cow-nose ray intoxication
elasmobranch, 37/70
Cowden syndrome
cerebellar granular cell hypertrophy, 60/666
CNS tumor, 67/54
colonic polyposis, 42/754
Flynn-Aird syndrome, 14/113
survey, 68/300
Cowdry A inclusion body
cytomegalovirus, 56/273
herpes simplex virus, 56/221
varicella zoster virus meningoencephalitis, 56/235
Cowpox encephalopathy
*see also* Reye syndrome
misdiagnosis, 56/149
Cowpox vaccine
Reye syndrome, 34/169, 56/150
Cowpox virus
acute viral myositis, 56/194
neurotropic virus, 56/5
neurotropism, 56/35
postinfectious encephalomyelitis, 47/326
Coxa valga
glycosaminoglycanosis type IV, 66/304
Coxiella burnetii
bacterial endocarditis, 52/302
Coxsackie meningitis
meningitis rash, 52/25
Coxsackie virus
acquired myoglobinuria, 62/577
acute cerebellar ataxia, 56/329
acute polymyositis, 56/199
arthrogryposis multiplex congenita, 56/200
autonomic nervous system, 74/335
choroid plexus, 56/318
chronic meningitis, 56/201, 646
deafness, 56/107
enterovirus infection, 34/134-139
epidemic, 56/315
group, 56/309
group A, 56/331
group B, 56/331
herpangina, 34/139
immunocompromised host, 56/476
myoglobinuria, 41/275, 56/199
myopathy, 56/332
myositis, 56/332
paralytic disease, 56/329
picornavirus, 34/5, 133-139, 56/307
polymyositis, 56/202
rhabdomyolysis, 56/200
spinal muscular atrophy, 59/370
transverse myelitis, 56/330
vertigo, 56/107
viral meningitis, 56/11, 128, 307
viral myositis, 41/20, 53, 73, 377
Coxsackie virus A
acute viral myositis, 56/194
Bornholm disease, 56/198
picornavirus, 56/195, 309
Reye syndrome, 29/332, 34/169, 49/217, 56/150
Coxsackie virus B
acute viral myositis, 56/195
Bornholm disease, 56/198, 332
dermatomyositis, 56/201
hydranencephaly, 50/340
picornavirus, 56/195, 309
Reye syndrome, 29/332, 34/169, 49/217, 56/150
Coxsackie virus infection
*see also* Enterovirus infection
acute cerebellar ataxia, 34/621
Bornholm disease, 34/137
clinical syndrome, 34/136-139
congenital, 34/138
congenital heart disease, 31/220
CSF, 34/136
encephalomyelitis, 34/136
epidemiology, 34/135
facial paralysis, 34/139
immunology, 34/134

meningitis serosa, 34/136
myopathy, 34/138
myositis, 41/20, 53, 73, 377
neonatal, 34/138
pathogenesis, 34/135
teratogenic agent, 34/378
Coyote, *see* Canis latrans
CPC-carbazole test, *see* Cetylpyridiniumchloride-carbazole test
CPT deficiency, *see* Carnitine palmitoyltransferase deficiency
Crab intoxication, 37/56-58
allergic reaction, 37/56
bacterial food poisoning, 37/56
dizziness, 37/58
eating, 37/56
headache, 37/58
horseshoe crab, 37/58
ingestion, 37/58
lassitude, 37/58
nausea, 37/58
neurotoxicity, 37/56
symptom, 37/56
vomiting, 37/58
Cracked pot sign
*see also* Macewen sign
brain tumor, 18/325, 329
hydrocephalus, 18/545, 50/292
intracranial pressure, 18/325, 329
Craft neurosis, *see* Occupational neurosis
Cramp
amyotrophic lateral sclerosis, 59/193
black widow spider venom intoxication, 40/547
chronic relapsing neuropathy, 34/396, 40/326
chrysaora intoxication, 37/51
cylindrical tubule, 40/39
disulfiram intoxication, 37/321
Finkel spinal muscular atrophy, 59/43
heat, *see* Heat cramp
hexachlorophene intoxication, 37/486
Isaacs syndrome, 40/330, 546
motor unit hyperactivity, 40/326
muscle, *see* Muscle cramp
myotonia, 40/327, 533
nocturnal, *see* Nocturnal cramp
pellagra, 7/573
progressive muscular atrophy, 59/13
stingray intoxication, 37/72
tetany, 40/330
writers, *see* Writers cramp
Cranial arteritis, *see* Temporal arteritis
Cranial base kyphosis
craniofacial dysostosis (Crouzon), 50/122
Cranial base synchondrosis, *see* Skull base synchondrosis
Cranial bone hypertrophy
thalassemia major, 42/629
Cranial cephalocele, 30/209-217
*see also* Cranium bifidum occultum
definition, 30/209
epidemiology, 30/154, 209
experimental induction, 30/209
frontal, 30/215-217
genetics, 30/209
nasal glioma, 30/209, 217
occipital, 30/210-213
parietal, 30/213-215
Cranial dura
*see also* Dura mater
anatomy, 33/149-153
Crowe test, 33/165
infection, 33/149-162
Cranial dysostosis (Franceschetti)
optic atrophy, 13/82
Cranial dysraphia
anencephaly, 50/72
Cranial epidural empyema
antibiotic agent, 52/169
brain angiography, 52/169
cause, 52/167
clinical features, 33/156, 52/168
computer assisted tomography, 52/169
co-trimoxazole, 52/169
course, 33/156
diagnosis, 33/156
Gradenigo syndrome, 52/168
head injury, 52/168
headache, 52/168
iatrogenic neurological disease, 52/168
imipenem, 52/169
laboratory finding, 52/168
mastoiditis, 52/170
metronidazole, 52/169
mucormycosis, 52/168
nafcillin, 52/169
neurosurgery, 52/169
otitis media, 52/167
pathology, 33/156
prognosis, 33/156
Propionibacterium acnes, 52/169
radiation, 52/168
sinusitis, 52/168
Staphylococcus aureus, 52/169
teichoplanin, 52/169
treatment, 33/157, 52/169
vancomycin, 52/169

Cranial half-base syndrome, *see* Garcin syndrome
Cranial hydatid disease
  echinococcosis, 35/202-207
Cranial hyperostosis
  Börjeson-Forssman-Lehmann syndrome, 42/530
  Cockayne-Neill-Dingwall disease, 43/350
  craniodiaphyseal dysplasia, 43/355
  craniometaphyseal dysplasia, 43/362-365
  hereditary cranial nerve palsy, 60/43
  myotonic dystrophy, 43/152
  osteitis deformans (Paget), 43/450
  Van Buchem disease, 38/407, 43/410
  xeroderma pigmentosum, 60/337
Cranial irradiation
  prophylactic, *see* Prophylactic cranial irradiation
Cranial meningocele
  terminology, 30/209
Cranial mononeuritis multiplex
  *see also* Multiple cranial neuropathy
  sarcoid neuropathy, 51/195
Cranial nerve
  abducent, *see* Abducent nerve
  accessory, *see* Accessory nerve
  acute intermittent porphyria, 41/434, 42/618
  amyotrophic lateral sclerosis, 42/65, 59/193
  anencephaly, 30/179
  Arnold-Chiari malformation type I, 42/15
  Biemond syndrome, 42/164
  birth incidence, 26/55
  Bonnet-Dechaume-Blanc syndrome, 14/109, 43/36
  brain injury, 23/235
  brain metastasis, 18/207, 211
  brain stab wound, 23/485
  brain stem death, 57/474
  bulbopontine paralysis, 42/96
  Burkitt lymphoma, 39/74
  carotid cavernous fistula, 24/410, 416
  Chédiak-Higashi syndrome, 14/783, 42/536
  CNS, 50/4
  CNS spongy degeneration, 42/506
  corpus callosum agenesis, 42/7
  corpus callosum tumor, 17/509
  craniodiaphyseal dysplasia, 43/356
  craniometaphyseal dysplasia, 43/364
  diabetes mellitus, 42/544
  diphtheria, 51/183
  embryology, 30/395-400
  facial, *see* Facial nerve
  familial spastic paraplegia, 59/305
  Fazio-Londe disease, 42/95
  focal brain damage, 57/51
  foramen magnum tumor, 17/725
  Friedreich ataxia, 60/302, 309
  frontal lobe tumor, 17/260
  Garcin syndrome, 2/313
  glomus jugulare tumor, 18/443
  glossopharyngeal, *see* Glossopharyngeal nerve
  glycosaminoglycanosis type I, 42/450
  Guillain-Barré syndrome, 7/497, 42/645
  headache, 48/6, 50
  hemidysplasia, 43/416
  hemophilia, 38/63
  hereditary amyloid polyneuropathy, 42/523
  hereditary motor and sensory neuropathy type I, 21/283
  hereditary motor and sensory neuropathy type II, 21/283
  heredopathia ophthalmo-otoencephalica, 42/150
  hyperlipoproteinemia type I, 42/443
  hypertrophic interstitial neuropathy, 21/151
  hypertrophic polyneuropathy, 56/183
  hypoglossal, *see* Hypoglossal nerve
  ichthyosis, 43/416
  ionizing radiation neuropathy, 51/137
  Kearns-Sayre-Daroff-Shy syndrome, 38/225
  leptomeningeal metastasis, 69/153
  leukemia, 39/9, 69/238
  lidocaine intoxication, 37/451
  lightning injury, 51/139
  linear nevus sebaceous syndrome, 43/38
  localization, 2/52-85
  malformation, 30/400-412
  malnutrition, 28/20
  Marcus Gunn phenomenon, 30/100
  Miller Fisher syndrome, 51/246
  multiple, *see* Multiple cranial nerve
  multiple lesions, 2/86-106
  multiple sclerosis, 9/167, 47/56
  muscular atrophy, 42/91
  neurinoma, 8/417, 68/535
  neurobrucellosis, 33/316
  occipital lobe tumor, 17/336
  occupational lesion, 7/334
  oculomotor, *see* Oculomotor nerve
  olfactory, *see* Olfactory nerve
  olivopontocerebellar atrophy (Schut-Haymaker), 42/162
  optic, *see* Optic nerve
  optic chiasm injury, 24/44
  osteitis deformans (Paget), 38/361, 70/12
  paratrigeminal syndrome, 48/336
  parietal lobe tumor, 17/301
  pertussis encephalopathy, 33/279
  pituitary adenoma, 17/390
  platybasia, 42/17

poliomyelitis, 42/652
polyarteritis nodosa, 39/303, 51/452
pontine hemorrhage, 12/22, 40
pontine infarction, 12/16, 22
posterior column ataxia, 42/164
posterior fossa hematoma, 24/347, 57/266
progressive external ophthalmoplegia, 43/136
radiation induced, 67/343
radiation injury, 67/342
sarcoid neuropathy, 38/525-529
sarcoidosis, 38/525-529
Sjögren syndrome, 39/426-429
skull fracture, 23/398
stab wound, 23/485
statoacoustic, *see* Statoacoustic nerve
subacute necrotizing encephalomyelopathy, 42/625
Tolosa-Hunt syndrome, 48/297
traumatic aneurysm, 24/386
trigeminal, *see* Trigeminal nerve
trochlear, *see* Trochlear nerve
vagus, *see* Vagus nerve
varicella zoster virus, 56/479
vertebral artery, 26/55
Wegener granulomatosis, 71/177
Wildervanck syndrome, 30/100
Wohlfart-Kugelberg-Welander disease, 59/86
Cranial nerve I, *see* Olfactory nerve
Cranial nerve II, *see* Optic nerve
Cranial nerve III, *see* Oculomotor nerve
Cranial nerve IV, *see* Trochlear nerve
Cranial nerve V, *see* Trigeminal nerve
Cranial nerve VI, *see* Abducent nerve
Cranial nerve VII, *see* Facial nerve
Cranial nerve VII palsy, *see* Facial paralysis
Cranial nerve VIII, *see* Statoacoustic nerve
Cranial nerve IX, *see* Glossopharyngeal nerve
Cranial nerve X, *see* Vagus nerve
Cranial nerve XI, *see* Accessory nerve
Cranial nerve XII, *see* Hypoglossal nerve
Cranial nerve agenesis
abducent nerve agenesis, 50/212
accessory nerve agenesis, 50/219
acoustic nerve agenesis, 50/216
clinical syndrome, 50/211
embryological development, 50/211
glossopharyngeal nerve agenesis, 50/219
hypoglossal nerve agenesis, 50/220
Möbius syndrome, 50/215
oculomotor nerve agenesis, 50/212
Poland-Möbius syndrome, 50/215
trigeminal nerve agenesis, 50/213
trochlear nerve agenesis, 50/212
vagus nerve agenesis, 50/219
Cranial nerve disorder
hypothyroidism, 70/96
Cranial nerve monitoring
surgery, 67/252
Cranial nerve nucleus
aplasia, 50/42
Möbius syndrome, 42/325, 50/42
Cranial nerve palsy
acquired immune deficiency syndrome, 56/492
acute hemorrhagic conjunctivitis, 56/351
African trypanosomiasis, 52/341
angiostrongyliasis, 35/323, 52/549, 553, 555
Arnold-Chiari malformation type II, 50/410
Behçet syndrome, 56/597
bilateral, 52/216
brain abscess, 52/152
brain edema, 63/417
brain hypoxia, 63/417
brain stem tumor, 16/320
buccopharyngeal herpes, 7/483
bulbar poliomyelitis, 56/328
Candida meningitis, 52/400
Castleman disease, 63/395
cerebral amebiasis, 52/315
cisplatin intoxication, 64/359
corpus callosum syndrome, 2/761
corpus callosum tumor, 17/509
Coxsackie virus A10, 56/326
Coxsackie virus B5, 56/326
cryptococcosis, 52/431
diphtheria, 52/229
ECHO virus 4, 56/326
enterovirus 70, 56/326, 329, 351
epidural anesthesia, 61/164
Epstein-Barr virus infection, 56/252
Guillain-Barré syndrome, 51/245
head injury, 57/135
hereditary, *see* Hereditary cranial nerve palsy
hereditary recurrent, *see* Hereditary recurrent cranial nerve palsy
herpes simplex virus encephalitis, 56/214
herpes zoster, 56/231
histoplasmosis, 52/438
hydromyelia, 50/429
hypertensive encephalopathy, 11/558
iatrogenic type, 61/163
infectious mononucleosis, 56/252
Kawasaki syndrome, 52/266, 56/639
legionellosis, 52/254
Listeria monocytogenes meningitis, 33/85, 52/93, 63/531
lumbar puncture, 61/163

Lyme disease, 52/261
meningeal lymphoma, 63/347
meningococcal meningitis, 33/23, 52/24, 26
mountain sickness, 63/417
multiple, *see* Multiple cranial neuropathy
multiple myeloma, 18/255, 39/142, 63/395
neurolymphomatosis, 56/182
organophosphate induced delayed neuropathy, 64/230
organophosphate intoxication, 64/230
osteitis deformans (Paget), 2/364
osteosclerotic myeloma, 63/395
pneumococcal meningitis, 33/47, 52/43, 51
POEMS syndrome, 63/395
poliomyelitis, 42/652, 56/328
poliovirus type 1-3, 56/326
postlumbar puncture syndrome, 61/163
progressive supranuclear palsy, 6/557
skull base fracture, 57/345
spinal anesthesia, 61/164
subarachnoid hemorrhage, 12/129
tetanus toxoid, 52/240
thrombotic thrombocytopenic purpura, 55/472
traumatic carotid cavernous sinus fistula, 57/346
tuberculous meningitis, 33/232, 52/216
venous sinus thrombosis, 54/431
vincristine polyneuropathy, 51/299
Vogt-Koyanagi-Harada syndrome, 56/621
Cranial neuralgia, 42/352-354
*see also* Facial pain *and* Trigeminal neuralgia
acetylsalicylic acid, 5/378
carbamazepine, 5/23
cluster headache, 5/114, 48/267
definition, 5/270
differential diagnosis, 5/316
drug treatment, 5/378
glossopharyngeal neuralgia, 5/312, 316
multiple sclerosis, 42/352, 47/56
neural pathway, 5/271
neurophysiology, 5/281
paracetamol, 5/378
pathologic anatomy, 5/270
pathophysiology, 5/270
posterior fossa tumor, 5/312
sex ratio, 42/353
surgery, 5/386
treatment, 5/378
trigeminal antidromic wave, 5/287, 289, 291
trigeminal neurinoma, 5/273
trigger, 1/108
Cranial neuritis
carotidynia, 5/22
herpes zoster ganglionitis, 51/181
neurobrucellosis, 33/312, 316, 52/585, 589
sphenopalatine neuralgia, 5/22
Cranial neuropathy
a-alphalipoproteinemia, 51/397
acetonylacetone intoxication, 64/87
allergic granulomatous angiitis, 63/383
amyloidosis, 71/508
bleomycin, 65/528
botulism, 65/210
brain edema, 63/417
brain hypoxia, 63/417
Burkitt lymphoma, 63/351
Castleman disease, 63/395
cisplatin intoxication, 64/359, 65/537
cryoglobulinemia, 63/404
diabetic, 70/138
eosinophilia myalgia syndrome, 64/254
Epstein-Barr virus, 56/253
fludarabine, 65/528
hemolytic uremic syndrome, 63/324
hereditary, *see* Hereditary cranial neuropathy
hereditary olivopontocerebellar atrophy (Menzel), 21/443
idiopathic hypereosinophilic syndrome, 63/124
ifosfamide, 65/528
Kawasaki syndrome, 56/639
Kearns-Sayre-Daroff-Shy syndrome, 62/313
leukemia, 63/354
lymphoma, 63/350
meningeal leukemia, 63/341
mountain sickness, 63/417
multiple, *see* Multiple cranial neuropathy
multiple myeloma, 55/470, 63/393, 395
neurosarcoidosis, 63/422
nonrecurrent nonhereditary multiple, *see* Nonrecurrent nonhereditary multiple cranial neuropathy
olivopontocerebellar atrophy variant, 21/455
osteosclerotic myeloma, 63/395
peripheral neurolymphomatosis, 56/181
POEMS syndrome, 63/395
polyarteritis nodosa, 51/452, 55/355
progressive external ophthalmoplegia, 62/313
recurrent multiple, *see* Recurrent multiple cranial neuropathy
recurrent nonhereditary multiple, *see* Recurrent nonhereditary multiple cranial neuropathy
sarcoidosis, 71/475
systemic lupus erythematosus, 51/447, 55/374, 376
Takayasu disease, 51/452, 63/53
tetanus, 65/218
thrombotic thrombocytopenic purpura, 63/324

uremic polyneuropathy, 63/512
vincristine intoxication, 65/528
Cranial polyneuritis, *see* Multiple cranial neuropathy
Cranial polyneuropathy, *see* Multiple cranial neuropathy
Cranial subdural empyema
antecedent factor, 52/170
antibiotic agent, 52/171
Campylobacter fetus, 52/171
cause, 52/167
cephalosporin, 52/171
clinical features, 33/158, 52/170
co-trimoxazole, 52/171
course, 33/160
diagnosis, 33/159
differential diagnosis, 33/160
epilepsy, 52/170, 172
Escherichia coli, 52/171
head injury, 52/170
headache, 52/170
Hemophilus influenzae, 52/170
iatrogenic neurological disease, 52/170
intracranial hypertension, 52/170
laboratory finding, 52/170
mastoiditis, 52/170
mortality, 52/171
neurosurgery, 52/171
otitis media, 52/167, 170
pathogenesis, 33/157
pathology, 33/157
prognosis, 33/160
radiation, 52/170
Salmonella typhi, 52/171
Salmonella typhimurium, 52/171
sinusitis, 52/170
skull fracture, 52/170
Streptococcus agalactiae, 52/171
treatment, 33/160
ventriculoperitoneal shunt, 52/170
Cranial vasculitis
headache, 48/273, 284
Cranial vault tumor
*see also* Subependymal astrocytoma
aneurysmal bone cyst, 17/109
blood diagnosis, 17/105
bone marrow, 17/121
Brill-Symmers disease, 17/123
dermoid cyst, 17/115
diagnosis, 17/105
eosinophilic granuloma, 17/116
epidermoid cyst, 17/113
Ewing sarcoma, 17/119
fibrosarcoma, 17/117
fibrous dysplasia, 17/118
giant cell tumor, 17/109
Hand-Schüller-Christian disease, 17/115
hemangioma, 17/110
history, 17/104
Hodgkin disease, 17/123
leukemia, 17/122
lipoma, 17/112
melanotic adamantinoma, 17/115
meningioma, 17/128
metastasis, 17/129, 131
neurofibromatosis, 17/109
ossifying fibroma, 17/109
osteitis deformans (Paget), 17/118, 170, 38/369
osteoma, 17/107
osteosarcoma, 17/117
radiation, 17/118
reticulosarcoma, 17/121
sarcoma, 17/117
soft tissue, 17/123, 125
spinal tumor, 68/515
thermography, 17/106
urine diagnosis, 17/105
Craniectomy
craniosynostosis, 42/21, 50/126
missile injury, 57/304
penetrating head injury, 57/304
short dura mater syndrome, 5/372
subdural effusion, 24/338
traumatic intracranial hematoma, 57/277
Craniocarpotarsal dysplasia, *see* Freeman-Sheldon syndrome
Craniocarpotarsal dystrophy, *see* Freeman-Sheldon syndrome
Craniocarpotarsal syndrome, *see* Freeman-Sheldon syndrome
Craniocephalic dysplasia
Patau syndrome, 14/120
Craniocervical bone abnormality
Stenvers projection, 32/10
submento-occipital projection, 32/10
survey, 50/381
Craniocervical junction, *see* Atlanto-occipital junction
Craniocervical region, 32/1-84
*see also* Craniovertebral region
endocrine disorder, 32/42
genetics, 32/7
history, 32/1
malformation, 32/5-84
radiologic appearance, 32/2, 5, 7-43
skeletal development, 32/3-5

tomography, 32/10-14
vertebral column, 32/41
Craniodiaphyseal dysplasia
alkaline phosphatase, 43/356
convulsion, 43/356
cranial hyperostosis, 43/355
cranial nerve, 43/356
foramen magnum, 43/356
headache, 43/356
hyperreflexia, 43/356
hypertelorism, 30/248
hyporeflexia, 43/356
optic atrophy, 43/354
progressive hereditary, *see* Progressive hereditary craniodiaphyseal dysplasia
sensorineural deafness, 43/355
skeletal malformation, 30/248
speech disorder in children, 43/355
Craniodigital syndrome
brachycephaly, 43/243
dermatoglyphics, 43/243
growth retardation, 43/243
hirsutism, 43/243
mental deficiency, 43/243
micrognathia, 43/243
syndactyly, 43/243
Cranioectodermal dysplasia
craniosynostosis, 43/357
neurofibromatosis type I, 43/35
Craniofacial dysostosis (Crouzon)
*see also* Craniosynostosis
antimongoloid eyes, 43/359
brachycephaly, 50/122
clinical features, 38/419-422
cloverleaf skull syndrome, 31/233, 43/359
CNS, 31/234
corpus callosum agenesis, 50/168
cranial base kyphosis, 50/122
craniosynostosis, 30/154, 222, 50/115, 121
cranium, 31/234
Dandy-Walker syndrome, 42/23
diaphyseal hyperplasia, 43/361
differential diagnosis, 31/234
dwarfism, 43/361
epilepsy, 31/234
exophthalmos, 50/122
eye finding, 31/233
face, 31/233
fontanelle, 30/100
foramina parietalia permagna, 50/142
genetics, 31/233
history, 38/419
hypertelorism, 30/246, 43/359-361
mental deficiency, 31/234, 43/359, 46/85
micrognathia, 43/361
optic atrophy, 13/70, 81, 43/359
oral lesion, 31/234
prevalence, 43/360
Saethre-Chotzen syndrome, 31/234
skull dysplasia, 30/246
strabismus, 43/359
Craniofacial injury
*see also* Brain concussion, Brain injury, Commotio cerebri, Head injury *and* Skull injury
blowout fracture, 23/378
classification, 23/371
diagnosis, 23/374
face, 57/309
Le Fort fracture, 23/371
maxillary fracture, 23/375-377
naso-orbital fracture, 23/379-381
penetrating head injury, 57/309
periorbital hematoma, 23/372
radiodiagnosis, 23/374
scalp, 57/309
treatment, 23/374
zygomaticomaxillary fracture, 23/375
Craniofenestria, *see* Craniolacunia
Craniolacunia
aqueduct stenosis, 50/139
Arnold-Chiari malformation type I, 32/187, 42/15, 50/139
Arnold-Chiari malformation type II, 32/187, 42/15, 50/408
Arnold-Chiari malformation type III, 32/187, 42/15
Arnold-Chiari malformation type IV, 32/187, 42/15
atrial septal defect, 50/139
bilateral cataract, 50/139
bone defect, 30/270
cardiac ventricular septum defect, 50/139
classification, 30/270
clinical features, 30/272, 50/139
course, 30/273, 50/140
cranium bifidum, 30/270
cryptorchidism, 50/139
definition, 50/137
dextrocardia, 50/139
diastematomyelia, 32/187, 50/139
embryopathogenesis, 50/137
embryopathology, 30/270, 50/137
encephalocele, 30/272, 50/137, 139
epidemiology, 30/272, 50/138
etiology, 30/271, 50/138
Floating Harbor syndrome, 50/139

hemivertebra, 32/187
hydrocephalus, 30/271, 50/139
inguinal hernia, 50/139
iniencephaly, 32/187
intelligence quotient, 50/139
Klippel-Feil syndrome, 32/187, 50/139
malformation, 30/273
meningomyelocele, 50/137, 139
mental deficiency, 30/273
phenylketonuria, 50/139
prognosis, 50/140
radiology, 30/270, 50/138
rib abnormality, 50/139
septum pellucidum agenesis, 50/139
sex ratio, 50/139
spina bifida, 32/521, 541, 50/487, 489
talipes, 50/139
Craniometaphyseal dysplasia
autosomal dominant, 43/362-364
autosomal recessive, 43/364
classification, 38/390
convulsion, 43/362
course, 38/394
cranial hyperostosis, 43/362-365
cranial nerve, 43/364
diagnosis, 38/394
dominant, *see* Pyle disease
etiology, 38/394
facial paralysis, 31/255, 43/362
foramen magnum, 43/362-365
general features, 38/390-392
genetics, 38/393
headache, 43/362
hearing loss, 43/364
history, 38/390
hyperreflexia, 43/362
hypertelorism, 30/246, 31/255, 43/362-364
laboratory finding, 38/394
neurologic features, 38/392
optic atrophy, 43/362-364
paresthesia, 43/362
Pyle disease, 13/83, 22/493, 31/255, 60/773
radiologic features, 38/393
sensorineural deafness, 31/255, 43/362
sensory loss, 43/362
tonsillar herniation, 43/365
treatment, 38/395, 396
Craniopharyngioma, 18/531-570
adamantoid structure, 18/537
adult, 18/565-570
age, 18/540
anatomic topography, 18/538
autonomic function, 18/544
Babinski-Fröhlich syndrome, 18/545
brachytherapy, 68/322
brain angiography, 18/549
brain biopsy, 16/720
brain scintigraphy, 18/547
brain tumor, 18/508
child, 18/540-565
chronic meningitis, 56/644
clinical features, 18/348, 545
clinical presentation, 68/318
computer assisted tomography, 18/547
congenital brain tumor, 31/50
congenital tumor, 31/50
CSF cytology, 16/402
diagnosis, 18/350, 546, 68/319
differential diagnosis, 18/134
distribution, 67/130
EEG, 18/547
embryology, 18/531
endocrine disorder, 16/347
endocrine function, 18/541
endocrinologic aspect, 68/323
headache, 18/541
histopathology, 18/537, 68/316
history, 68/315
hormonal test, 18/549
hypothalamic tumor, 68/71
hypothalamus tumor, 18/366
imaging, 67/188
infant, 18/540-565
Macewen sign, 18/545
management, 68/320
3-methylcholanthrene, 18/537
neuroradiology, 18/546
nonrecurrent nonhereditary multiple cranial neuropathy, 51/571
optic chiasm compression, 68/75
origin, 68/316
papillary structure, 18/537
papilledema, 18/541
pathology, 2/457-460, 18/347
Peter Pan syndrome, 18/545
pituitary adenoma, 17/397, 406
pituitary tumor, 17/428
prevalence, 18/540
prognosis, 18/469
radiology, 68/347
radiosurgery, 68/322
radiotherapy, 18/508, 68/322
Rathke pouch, 18/531
recurrence, 68/323
seesaw nystagmus, 16/323
sex ratio, 18/540

Sheehan syndrome, 2/460
skull base tumor, 17/194
spinal lipoma, 20/397
supratentorial brain tumor, 18/347-352
surgery, 68/320, 324
third ventricle tumor, 17/450
tissue culture, 17/82
topography, 18/533
treatment, 18/351, 68/322
tuberous sclerosis, 18/134
tumor type, 18/537
vascularization, 18/537
ventriculography, 18/551
visual field, 18/511
Cranioplasty
depressed skull fracture, 23/411
Craniorachischisis
*see also* Rachischisis *and* Spina bifida
anencephaly, 30/175, 187, 50/72
Gruber syndrome, 14/505
malformation, 32/181, 183
neurulation, 50/74
notochord alteration, 32/185
pathology, 32/186
radiologic appearance, 32/186
sphenoid wing aplasia, 50/144
Cranioschisis
anencephaly, 50/72
chondrocranium, 32/180
sphenoid wing aplasia, 50/144
type, 50/71
Craniospinal injury
*see also* Spinal injury
atlanto-occipital luxation, 25/310
delayed myelopathy, 25/93
functional anatomy, 25/231
spinal cord, 25/43, 66, 93
vertebral artery, 25/263
Craniospinal tumor, *see* Foramen magnum tumor
Craniostenosis, *see* Craniosynostosis
Craniosynostosis, 30/219-230, 50/113-127
*see also* Acrocephalosyndactyly type I, Acrocephalosyndactyly type II, Acrocephalosyndactyly type III, Acrocephalosyndactyly type V, Brachycephaly, Craniofacial dysostosis (Crouzon), Metopic suture, Oxycephaly, Plagiocephaly, Sagittal suture, Scaphocephaly *and* Trigonocephaly
acrobrachycephaly, 50/113, 118
acrocephalosynanky, 46/87
acrocephalosyndactyly type I, 30/154, 222, 31/231, 38/422, 43/318-320, 50/115
acrocephalosyndactyly type II, 30/154, 222, 31/235, 43/322-324
acrocephalosyndactyly type III, 50/123
acrocephalosyndactyly type V, 43/320
acrocephaly, 50/113
adducted thumb syndrome, 43/331
Armendares syndrome, 43/366
autosomal dominant inheritance, 42/21
autosomal recessive inheritance, 42/21, 50/60
Baller-Gerold syndrome, 43/370
birth incidence, 42/21
brachycephaly, 30/154, 221, 42/20, 50/113
brain growth, 30/219
brain weight, 30/219
cephalic index, 50/119
classification, 30/221, 38/415, 42/20
clinical features, 30/226
cloverleaf skull, 46/87
cloverleaf skull syndrome, 30/154, 38/422, 50/120
convergent strabismus, 50/125
coronal synostosis, 50/118
craniectomy, 42/21, 50/126
cranioectodermal dysplasia, 43/357
craniofacial dysostosis (Crouzon), 30/154, 222, 50/115, 121
craniovertebral abnormality, 32/42
cri du chat syndrome, 31/568
deafness, 50/125
dental malocclusion, 50/125
diagnosis, 30/228, 38/417, 42/21
divergent strabismus, 50/125
dolichocephaly, 50/113
dominant inheritance, 50/60
elfin face syndrome, 31/318
epidemiology, 30/154, 221, 225
etiology, 30/222-224, 50/115
exophthalmos, 50/125
external appearance, 30/226
facial growth, 50/124
features, 38/415
fibrous dysplasia, 14/199
fibular aplasia, *see* Lowry syndrome
genetic counseling, 42/21
genetics, 30/154, 221
Goltz-Gorlin syndrome, 14/788
Gorlin-Chaudry-Moss syndrome, 43/369
Herrmann-Opitz syndrome, 43/325
history, 30/219-221, 38/413
hydrocephalus, 30/228
hypertelorism, 50/125
hypophosphatasia, 42/578
incidence, 50/60, 116
lambdoid synostosis, 50/119
mental deficiency, 30/154, 227, 42/20, 43/262,

46/85, 50/126
metopic synostosis, 50/117
microcephaly, 50/267
optic atrophy, 50/126
orbitostenosis, 50/125
ossification, 42/20
oxycephaly, 30/154, 220, 42/20, 50/119
papilledema, 30/226
pathogenesis, 38/413, 50/115
plagiocephaly, 30/154, 222, 42/20, 50/113, 118
radial aplasia, 43/370
radiologic features, 38/416
radioulnar dysostosis, *see* Berant syndrome
respiratory dysfunction, 50/125
Saethre-Chotzen syndrome, 43/322
sagittal synostosis, 50/116
Sakati-Nyhan-Tisdale syndrome, 43/326
scaphocephaly, 30/154, 221, 42/20, 50/113, 116
Seckel syndrome, 43/379
secondary, 30/222
self-mutilation encephalopathy, 42/141
sex ratio, 42/21
sleep apnea syndrome, 50/125
speech disorder, 50/125
Summitt syndrome, 43/372-374
surgery indication, 30/229
synonym, 30/219
terminology, 30/221
treatment, 30/228, 230, 38/417-419
trigonocephaly, 42/20, 50/113, 117
Craniovertebral region
*see also* Craniocervical region
angiographic anatomy, 32/46-49, 61, 64
angiotomography, 32/45
bone abnormality, 32/62-65
carotid angiography, 32/44, 60
CNS malformation, 32/58-60
collapsing cord sign, 32/65
CSF space, 32/60-66
malformation, 32/66-84
occipital bone shortening, 32/14
vascular abnormality, 32/49-58
ventriculography, 32/66
vertebral angiography, 32/44, 50-60
vertebral artery occlusion, 32/55
Craniovertebral stenosis
vertebral canal stenosis, 50/468
Cranium bifidum
*see also* Iniencephaly
craniolacunia, 30/270
definition, 50/97
iniencephaly, 50/129
neurofibromatosis type I, 14/158
spina bifida, 50/477
terminology, 30/209
Cranium bifidum occultum
*see also* Cranial cephalocele
terminology, 30/209
Cranium bifidum occultum frontale, *see* Sincipital cephalocele
Creased hand sign
features, 1/184
significance, 1/184
Creatine
hyponatremia, 63/547
Creatine-creatinine
Duchenne muscular dystrophy, 40/379
Creatine kinase
acid maltase deficiency, 62/480
acquired immune deficiency syndrome, 56/498
acromegalic myopathy, 62/538
acute polymyositis, 56/200
adenosine monophosphate deaminase deficiency, 62/512
aminoaciduria, 43/126
arthrogryposis, 43/78
artificial respiration, 63/420
axonal polyneuropathy, 62/609
Becker muscular dystrophy, 40/281, 387, 43/86
benign autosomal dominant myopathy, 43/111
brancher deficiency, 62/484
carnitine palmitoyltransferase deficiency, 43/177
carrier detection, 41/413, 487
central core myopathy, 43/81
chorea-acanthocytosis, 63/281
chronic renal failure myopathy, 62/541
colchicine myopathy, 62/609
congenital myopathy, 62/333, 350
CSF, 54/199
debrancher deficiency, 62/482
dermatomyositis, 62/375
diethylstilbestrol, 40/343, 378
Duchenne carrier, 40/340, 356
Duchenne muscular dystrophy, 40/281, 376, 43/106, 62/118, 124
elevation, 40/341, 41/412
eosinophilia myalgia syndrome, 63/377
facioscapulohumeral muscular dystrophy, 40/420, 62/166
fatal infantile myopathy/cardiopathy, 62/510
Fukuyama syndrome, 43/91
generalized myotonia (Becker), 43/164
Guillain-Barré syndrome, 51/245
hyperaldosteronism myopathy, 62/536
hypertrophia musculorum vera, 43/84
hypoparathyroid myopathy, 62/541

hypothyroid myopathy, 62/533
hypothyroidism, 40/343
immunohistochemistry, 40/3
inclusion body myositis, 62/373, 375
infantile spinal muscular atrophy, 22/93
intensive care complication, 63/420
isoenzyme, 40/342, 41/410
Kawasaki syndrome, 56/639
limb girdle syndrome, 40/449, 62/185
low level medication, 40/343
malignant hyperthermia, 38/553, 40/339, 343, 43/117, 63/439
McLeod phenotype, 49/332, 63/287
muscle fiber, 43/116
muscle tissue culture, 62/87
muscular atrophy, 42/85, 91
muscular dystrophy, 43/88, 90, 96, 102
myofibrillar lysis, 43/116
myoglobinuria, 40/263, 339, 41/263, 271, 62/554, 556, 560
myopathy, 43/116, 126, 193
myophosphorylase deficiency, 27/233, 41/185, 187, 62/485
myotonic dystrophy, 43/153
myotonic myopathy, 43/166
normal value, 40/340
paralysis periodica paramyotonica, 43/167
pheochromocytoma myopathy, 62/537
phosphofructokinase deficiency, 62/487
phosphorylase kinase deficiency, 62/487
polymyositis, 22/25, 38/485, 41/60, 62/375
progressive external ophthalmoplegia, 22/183
quadriceps myopathy, 62/174
Reye syndrome, 29/335, 56/152, 159, 65/118
rhabdomyolysis, 40/339
rimmed vacuole distal myopathy, 62/204
Ryukyan spinal muscular atrophy, 42/93
sarcotubular myopathy, 43/129
scapuloilioperoneal muscular atrophy, 42/344
scapuloperoneal myopathy, 43/131
serum, *see* Serum creatine kinase
spinal cord injury, 26/384, 403
spinal muscular atrophy, 22/25
steroid myopathy, 62/535, 605
thyrotoxic myopathy, 62/529
toxic myopathy, 62/596
toxic shock syndrome, 52/259
ubidecarenone, 62/504
Wohlfart-Kugelberg-Welander disease, 22/71
xanthinuria, 43/193
Creatine kinase MM
chorea-acanthocytosis, 49/330
muscle tissue culture, 62/89
Creatine phosphokinase, *see* Creatine kinase
Creatinine
spinal cord injury, 26/389, 402
uremic encephalopathy, 63/514
uremic neuropathy, 7/554, 27/339
uremic polyneuropathy, 63/514
Creatinine clearance
renal insufficiency, 63/503
renal insufficiency polyneuropathy, 51/358
uremic polyneuropathy, 51/360
Creatinuria
steroid myopathy, 62/535
Cremasteric reflex
diagnostic value, 1/253
Creseis intoxication
ingestion, 37/64
CREST syndrome
angiopathic polyneuropathy, 51/446
progressive systemic sclerosis, 51/448
Crests
neural, *see* Neural crest
Cresyl violet
acid mucopolysaccharide, 40/3
Cretinism
athyrotic, *see* Thyroid gland dysgenesis
deafness, 27/256, 259
endemic, *see* Endemic cretinism
hypothyroidism, 70/92
Klinefelter syndrome, 50/549
mental deficiency, 31/297, 46/18, 83-85
neurologic manifestation, 27/256
Cretinismus furfuraceus, *see* Down syndrome
Creutzfeldt-Jakob disease, 6/726-755
adult neuronal ceroid lipofuscinosis, 56/546
air encephalography, 46/292
Alzheimer disease, 6/746, 46/294, 56/546
amantadine, 46/294
amaurotic type, 6/732, 741-745
amyloid, 46/295
amyloid plaque, 56/551
amyotrophic lateral sclerosis, 22/317, 59/203, 409
amyotrophic type, 6/733, 741-745, 46/292
amyotrophy, 22/22, 46/292
animal model, 56/585
anoxic encephalopathy, 56/546
aphasia, 46/291
asterixis, 46/291
astrocytosis, 56/551
ataxia, 42/657, 56/545
biochemical aspect, 27/491
biosafety level 2, 56/549
brain biopsy, 10/685, 56/546
brain vasculitis, 56/546

bulbar paralysis, 22/143
central pontine dystrophy, 9/638
cerebellar, *see* Cerebellar Creutzfeldt-Jakob disease
cerebellar ataxia, 46/290
cerebellar neuropathy, 21/501
choreoathetosis, 46/291, 56/545
classic type, 6/733, 741-745
classification, 6/732-734, 741-745
clinical course, 56/545
clinical features, 6/735-737, 46/290-292, 56/544
computer assisted tomography, 46/292
corneal transplantation, 56/549
cortical blindness, 46/291
course, 6/749
CSF, 56/545
CSF protein, 46/292
decontamination, 56/549
dementia, 27/491, 42/657, 46/128, 289-296, 389, 56/545, 59/409
diagnosis, 56/545
dialysis dementia, 56/546
differential diagnosis, 6/745-749, 41/333, 46/294, 56/546
diffuse sclerosis, 6/747
disease course, 46/291
disease duration, 56/552
dysarthria, 46/290
dyskinesia, 6/733, 741-745
dysphasia, 46/290
EEG, 6/732, 42/657, 46/292, 56/545
electrode implantation, 56/549
electron microscopy, 56/547
epidemiology, 56/546
epilepsy, 15/346, 46/291, 72/222
eye movement, 46/291
familial amaurotic idiocy, 6/748
Gerstmann-Sträussler-Scheinker disease, 46/295
glia cell, 6/750
glial dystrophy, 21/67
hepatic encephalopathy, 6/748
hereditary, 6/737-740
herpes simplex virus encephalitis, 56/546
histopathology, 56/551
historic note, 46/289
history, 6/726-731, 27/379
human growth hormone therapy, 56/547
Huntington chorea, 6/746, 49/293
immunology, 56/552
incidence, 6/735, 46/289
incubation period, 56/552
kuru, 34/285, 46/289, 292, 294
Lafora progressive myoclonus epilepsy, 56/546
late cerebellar atrophy, 21/495
late cortical cerebellar atrophy, 21/477
lithium intoxication, 64/297
motoneuron disease, 46/294
multiple sclerosis, 9/79, 120, 126
muscular atrophy, 42/657, 46/291
myoclonia, 6/731, 764
myoclonus, 6/166, 732, 741-745, 38/582, 42/657, 46/291, 49/618, 56/545, 60/662
neurologic symptom, 46/290
neuropathology, 6/749-754
neurosyphilis, 56/546
niger type, 6/745
nosology, 6/755
olivopontocerebellar atrophy (Dejerine-Thomas), 21/428, 60/534
optic atrophy, 56/545
pallidoluysionigral degeneration, 49/459
parkinsonism, 56/545
pathology, 46/292-294
pellagra, 6/748
Pick disease, 6/746
portosystemic anastomosis encephalopathy, 56/546
postencephalitic parkinsonism, 6/748
prion disease, 56/543
prodromal symptom, 46/290
progressive myoclonus epilepsy, 6/747
progressive rubella panencephalitis, 56/546
pseudodegeneration, 21/59
psychiatric symptom, 6/736
radiodiagnosis, 46/292
reflex disorder, 46/291
rigidity, 46/291
scrapie, 46/289
senile plaque, 56/551, 569
slow virus disease, 34/285
spastic paraplegia, 59/436
spasticity, 46/291
spinocerebellar degeneration, 21/15
spongiform degeneration, 56/551
spongiosis, 6/751
spongy cerebral degeneration, 6/744, 46/293
spongy dystrophy, 21/68
spongy type, 6/744
startle myoclonus, 56/545
startle reaction, 46/291
startle syndrome, 49/620
status spongiosus, 46/293
striatal lesion, 6/671, 699
striatal type, 6/744
striatonigral degeneration, 6/699
subacute sclerosing panencephalitis, 56/546

subacute spongiform encephalopathy, 6/735, 11/159, 42/655-659, 46/289, 291
surgical procedure, 56/549
synonym, 6/735
systemic neoplastic disease, 41/333
terminal stage, 46/291
thalamic dementia, 2/483, 6/355, 21/590
thalamus, 2/487, 489, 6/733, 741-745, 21/590-593
transmissible virus, 9/120
transmission, 46/295, 56/551
treatment, 46/294, 56/550
ultrastructure, 56/552
uremia, 56/546
vidarabine, 46/294
viral mechanism, 56/20
viral origin, 9/126
vision, 46/291
Wilson disease, 6/746
Creutzfeldt-Jakob disease prion
amyloid, 56/563
human prion protein gene, 56/566
Cri du chat syndrome, 31/561-573
associated malformation, 31/566-568
behavior, 31/568
brain atrophy, 43/501
brain weight, 50/275
cerebellar atrophy, 43/501
cerebellar hypoplasia, 50/275
chromosomal aberration, 30/249, 46/21-24
clinical features, 31/568, 46/21-24
CNS, 31/566, 568
convulsion, 31/568
corpus callosum agenesis, 43/501
craniosynostosis, 31/568
cytogenetics, 31/571-573
dermatoglyphics, 31/564-566, 43/501
diagnosis, 31/561
diphtheritic neuritis, 51/182
distinctive cry, 31/561, 563
EEG, 43/501
epidemiology, 30/161, 31/569-571
extensor plantar response, 50/275
external appearance, 31/563
face, 31/563
gene localization, 31/549, 568
Goldenhar syndrome, 31/566
history, 31/561
hydrocephalus, 31/566, 43/501
hypertelorism, 30/249, 43/501
hypertonia, 50/275
intelligence, 31/586
kyphoscoliosis, 43/501
laryngeal abnormality, 31/563
management, 31/568
mental deficiency, 31/568, 43/501, 46/21-24, 68, 50/275
microcephaly, 30/508, 31/563, 566, 43/501, 50/274
mosaicism, 31/572
parental karyotype, 31/570
recurrence risk, 43/502
ring chromosome, 31/573
seizure, 43/501
sex ratio, 43/502
talipes, 43/501
translocation, 31/572
twins, 31/569
ventricular dilatation, 50/275
Cribriform state, *see* Status cribrosus
Cricopharyngeal dysphagia
nemaline myopathy, 62/342
Cricopharyngeal myotomy
dermatomyositis, 62/386
dysphagia, 62/386
inclusion body myositis, 62/386
polymyositis, 62/386
Crigler-Najjar syndrome
acquired hepatocerebral degeneration, 49/218
bilirubin encephalopathy, 27/423
bilirubin glucuronoside glucuronosyltransferase, 42/540
glucuronosyltransferase, 6/492, 21/52
hepatogenic gliodystrophic encephalopathy, 21/52
jaundice, 42/540
kernicterus, 21/53, 27/420
uridine diphosphate glucuronosyltransferase, 42/540
Criminality, 43/204-206
epilepsy, 15/810
Klinefelter syndrome, 43/205, 46/75
sex ratio, 43/205
traumatic psychosyndrome, 24/567
XYY syndrome, 43/205, 46/75, 50/551
Crinoidea intoxication
Echinodermata, 37/66
Critchley, M., 1/4, 16, 24, 30, 35, 38
Critchley syndrome
acanthocyte, 13/427
areflexia, 13/427
involuntary movement, 13/427
mental deficiency, 13/427
Critical illness polyneuropathy
animal model, 51/584
areflexia, 51/575
artificial respiration, 63/420
artificial ventilator, 51/575

axonal degeneration, 51/584
axonal polyneuropathy, 51/575
brain hemorrhage, 51/575
C-3a complement, 51/584
C-5a complement, 51/584
cell membrane potential, 51/584
clinical features, 51/581
CSF, 51/576, 581
denervation, 51/581
distal muscular atrophy, 51/575
EMG, 51/575, 581
endothelium, 51/584
fever, 51/576, 578, 581, 584
Guillain-Barré syndrome, 51/576
H-reflex, 51/578
hemopathology, 51/575
hypotonic tetraparesis, 51/575, 581
infection, 51/575
intensive care, 63/420
intensive care complication, 63/420
interleukin-1, 51/584
interval, 51/584
laboratory examination, 51/581
lancinating pain, 51/575
leukocytosis, 51/576
major surgery, 51/575
motor conduction velocity, 51/578
motor unit action potential, 51/585
multiple organ failure, 51/575, 584, 63/420
multiple trauma, 51/575
muscle membrane dysfunction, 51/584
muscular atrophy, 51/575
myocardial infarction, 51/575
nerve conduction velocity, 51/575, 584
neuromuscular transmission blocking, 51/584
neuropathologic finding, 51/581
oxygen radical, 51/584
pathogenesis, 51/584
primary axonal degeneration, 51/581
primary muscle change, 51/581
primary muscle disease, 51/585
primary muscle membrane dysfunction, 51/584
prognosis, 51/575
respiratory insufficiency, 51/576, 581, 584
sepsis, 51/575, 584, 63/420
severe trauma, 51/581
spinal cord, 51/581
spinal nerve root, 51/581
spontaneous electrical muscle activity, 51/576, 578, 584
spontaneous respiration, 51/581
status asthmaticus, 51/575
sural nerve biopsy, 51/578, 581
tachycardia, 51/576, 581
tumor necrosis factor, 51/584
Crocodile tear, 74/429
Crocodile tear syndrome
*see also* Crying
dysautonomia, 22/259
facial nerve, 2/69, 123
facial paralysis, 2/69, 123, 8/270, 275
geniculate ganglion, 2/70, 123
lesion site, 2/70, 123
pathogenesis, 8/275
Stilling-Türk-Duane syndrome, 50/213
Vidian nerve, 48/480
Crohn disease
autonomic nervous system, 75/653
deficiency neuropathy, 7/620, 51/325, 329
hypomagnesemia, 63/563
malabsorption syndrome, 28/236, 70/234
neurologic manifestation, 28/236
polymyositis, 62/373
vitamin $B_{12}$ deficiency, 7/620
vitamin $B_c$, 7/620
vitamin $B_c$ deficiency, 51/336
Cromakalim
myotonia, 62/276
Crome syndrome
cataract, 43/242
CSF, 43/242
encephalopathy, 43/242
epilepsy, 43/242
mental deficiency, 43/242
microcephaly, 43/242
myelin dysgenesis, 43/242
nystagmus, 43/242
renal tubular dysfunction, 43/242
Crome-Zapella disease
Pelizaeus-Merzbacher disease, 10/197
sudanophilic leukodystrophy, 10/197
Cromoglicate disodium
migraine, 48/122
Cronckhite-Canada syndrome
Bean syndrome, 14/106
Cross innervation
histochemistry, 40/139
Crossed cerebellar atrophy
congenital cerebellar atrophy, 21/462
Crossed cerebrocerebellar atrophy
brain diaschisis, 54/126
brain infarction, 54/127
Crossed extensor reflex
experimental aspect, 2/408
stimulation, 2/408
Crossed hemiplegia

pontine syndrome, 1/187, 2/240-261
Crossed Lasègue test
intervertebral disc prolapse, 20/577, 587
Crossed legs
hereditary multiple recurrent mononeuropathy, 51/554
Crossed nucleus ruber syndrome
definition, 1/187
Crossed sensory paralysis, *see* Grenet syndrome
Crotamine
snake envenomation, 65/185
snake venom, 65/185
toxic myopathy, 62/610
Crotapotin, *see* Crotoxin A
Crotoxin
chemistry, 65/183
neurotoxin, 37/17, 65/182
snake envenomation, 65/182
snake venom, 65/183
toxic myopathy, 62/610
Crotoxin A
chemistry, 65/184
neurotoxin, 65/184
snake envenomation, 65/184
snake venom, 65/184
Crotoxin intoxication
venom, 37/17
Crouzon disease, *see* Craniofacial dysostosis (Crouzon)
Crouzon syndrome, *see* Craniofacial dysostosis (Crouzon)
Crow-Fukase syndrome, *see* POEMS syndrome
Crowe sign
transverse sinus thrombosis, 54/397
Crowe test
cranial dura, 33/165
superior petrosal sinus thrombophlebitis, 33/180
venous sinus thrombophlebitis, 33/165
CRSM syndrome, *see* Mucolipidosis type I
CRST syndrome
hereditary hemorrhagic telangiectasia (Osler), 14/120
Raynaud phenomenon, 43/71
Cruciate paralysis
cervicomedullary injury, 24/149
head injury, 24/149
hemiplegia, 1/189
oculomotor nerve injury, 24/63
pathology, 24/149, 152
spinal cord injury, 25/275, 391-393
symptom, 24/149, 151
Crush syndrome
opiate intoxication, 37/391
Crustacea intoxication
allergic reaction, 37/56
Atergatis floridus, 37/58
bacterial food poisoning, 37/56
birgus latro, 37/58
eating, 37/56
eriphia sebana, 37/58
neurotoxicity, 37/56
symptom, 37/56
Crutch
compression neuropathy, 51/92
Cruveilhier plexus, *see* Venous angioma
Crying
*see also* Crocodile tear syndrome
hydrozoa intoxication, 37/38
migraine, 48/161
neonatal myasthenia gravis, 41/100
pathologic laughing, *see* Pathologic laughing and crying
persistent, *see* Persistent crying
spasmodic, 1/196, 210
Crying spell
migraine, 48/162
Cryofibrinogen
hyperviscosity syndrome, 55/490
Cryofibrinogenemia
ambient temperature, 55/490
Cryoglobulin
classification, 39/181
Raynaud phenomenon, 43/71, 55/485
Cryoglobulinemia, 39/181-187
amaurosis fugax, 63/404
ambient temperature, 55/490, 63/404
arthralgia, 63/403
axonal degeneration, 63/404
brain infarction, 53/166, 63/404
brain ischemia, 53/166
brain vasculitis, 63/404
chronic lymphocytic leukemia, 39/181
classification, 63/403, 71/4
cold sensitivity, 63/403
coma, 63/404
course, 63/404
cranial neuropathy, 63/404
cyanosis, 63/403
distal muscle wasting, 63/404
EMG, 63/404
encephalopathy, 63/404
epilepsy, 63/404
hemolytic anemia, 39/181
hepatitis B virus, 63/403
hepatitis B virus infection, 56/302
human T-lymphotropic virus type I associated

myelopathy, 59/450
hyperviscosity syndrome, 55/484, 490, 71/439
idiopathic thrombocytopenic purpura, 39/181
IgM monoclonal gammopathy, 69/297
ischemic neuropathy, 63/404
lymphatic leukemia, 63/403
lymphoma, 39/181, 63/403
macroglobulinemia, 63/403
meningeal invasion, 71/439
mixed, *see* Mixed cryoglobulinemia
monoclonal IgA, 63/403
monoclonal IgG, 63/403
monoclonal IgM, 63/403
multiple myeloma, 39/135, 181, 63/403
muscular atrophy, 63/404
nerve biopsy, 63/404
nerve vasculitis, 63/404
neurologic complication, 63/403
neurologic manifestation, 39/182-185
neuropathy, 63/404, 71/441, 445
pain, 63/404
paraproteinemia, 63/403, 69/297, 308
paresthesia, 63/404
pathogenesis, 39/185
plasma cell dyscrasia, 63/391
plasmapheresis, 63/405
polyarteritis nodosa, 39/181, 51/434, 55/359
polyclonal, 63/403
primary type, 63/403
purpura, 63/403
Raynaud phenomenon, 51/434, 63/403
recurrent nonhereditary multiple cranial neuropathy, 51/570
rheumatoid arthritis, 39/181
secondary type, 63/403
Sjögren syndrome, 39/181
symptom, 71/431
syndrome, 71/437
systemic brain infarction, 11/461
systemic lupus erythematosus, 39/181
therapy, 39/186, 63/405
transient ischemic attack, 53/266, 63/403
type, 63/403, 71/432
vasculitis, 71/8
Waldenström macroglobulinemia, 39/181
Cryoglobulinemic polyneuropathy
arthralgia, 51/434
axonal degeneration, 51/434
demyelination, 51/434
EMG, 51/434
frequency, 51/434
laboratory study, 51/434
malignant lymphoma, 51/434
mixed type, 51/434
monoclonal gammopathy, 51/434
multiple myeloma, 51/434
nerve conduction velocity, 51/434
pathogenesis, 51/434
pathology, 51/434
polyarteritis nodosa, 51/434
primary condition, 51/434
purpura, 51/434
Raynaud phenomenon, 51/434
retinal hemorrhage, 51/434
secondary type, 51/434
treatment, 51/434
vasculitis, 51/434
Cryptic arteriovenous malformation, *see* Telangiectasia
Cryptic hemangioma
CNS, 12/104
hereditary hemorrhagic telangiectasia (Osler), 12/104
telangiectasia, 12/104
vascular malformation, 12/104
Cryptococcal meningitis
arteritis, 52/431
brain pseudoaneurysm, 55/433
cranial nerve deficit, 52/431
headache, 52/431
hydrocephalus, 52/432
intraventricular granuloma, 52/431
multi-infarct dementia, 55/435
neurologic symptom, 55/432
serologic study, 71/294
symptom, 52/431, 63/532
therapy, 71/294
Cryptococcal meningoencephalitis
*see also* Cryptococcosis
cause, 52/431
dementia, 52/432
epilepsy, 52/432
symptom, 52/431
Cryptococcoma
cryptococcosis, 52/432
CSF, 52/432
mature male, 52/432
multiplicity, 52/432
neurosurgery, 52/433
site, 52/432
Cryptococcosis, 35/459-491
*see also* Cryptococcal meningoencephalitis *and* Cryptococcus neoformans
acquired immune deficiency syndrome, 52/433
amphotericin B, 35/487-490, 52/433
associated disease, 35/460, 52/432

brain granuloma, 16/230
chronic meningitis, 56/644
clinical features, 35/461-479
course, 35/487
cranial nerve palsy, 52/431
cryptococcoma, 52/432
CSF, 35/479, 52/433
diabetes mellitus, 52/433
diagnosis, 52/433
epidemiology, 35/16
flucytosine, 35/490, 52/433
history, 35/459
immunocompromised host, 52/430
inappropriate antidiuretic hormone secretion, 52/431
intracranial calcification, 35/477
intracranial hypertension, 52/431
ketoconazole, 52/433
leukemia, 52/433
meningitis, 35/461, 52/431
meningoencephalitis, 35/461-477, 52/431
mental symptom, 52/431
miconazole, 52/433
myasthenia gravis, 52/433
mycosis, 35/375
neuropathology, 52/433
ophthalmoplegia, 52/431
pathology, 35/482-487
prevalence, 35/460, 52/430
psychiatric symptom, 52/431
renal insufficiency, 27/343
serology, 35/480-482
spinal granuloma, 52/432
surgical treatment, 35/491
systemic lupus erythematosus, 52/433
treatment, 35/487-491, 52/433
tuberculosis, 52/433
Cryptococcus
fungal CNS disease, 52/496
Hodgkin disease, 39/43
malignant lymphoma, 39/43
Cryptococcus histolyticus, *see* Cryptococcus neoformans
Cryptococcus meningitidis, *see* Cryptococcus neoformans
Cryptococcus neoformans
*see also* Cryptococcosis
acquired immune deficiency syndrome, 56/494, 515, 71/292
biology, 35/459
brain abscess, 52/151
brain infarction, 63/532
cerebrovascular disease, 55/432
chronic meningitis, 56/645
clinical features, 71/293
CNS infection, 69/446
CSF, 71/293
culture, 52/430
encephalitis, 63/532
epilepsy, 63/532
fungal CNS disease, 35/570, 52/480, 496
headache, 63/532
India ink stain, 52/430
meningoencephalitis, *see* Cryptococcal meningoencephalitis
microbiology, 52/429
nonencapsulated, 52/430
nonrecurrent nonhereditary multiple cranial neuropathy, 51/571
papilledema, 63/532
pigeon dropping, 52/429
radiology, 71/293
renal transplantation, 63/531
serotype, 52/430
sexual perfect form, 52/430
spastic paraplegia, 59/436
Cryptococcus neoformans infection
immunocompromised host, 65/549
Cryptophthalmos syndrome
cheilopalatoschisis, 43/374
encephalomeningocele, 43/374
external auditory stenosis, 43/374
hypogonadism, 43/374
mental deficiency, 46/61
renal abnormality, 43/374
syndactyly, 43/374
Cryptorchidism
ARG syndrome, 50/582
athetotic hemiplegia, 42/200
Bowen-Conradi syndrome, 43/335
cerebrohepatorenal syndrome, 43/339
craniolacunia, 50/139
Down syndrome, 50/274, 532
Edwards syndrome, 43/535, 50/561
FG syndrome, 43/250
holoprosencephaly, 50/237
hypertelorism hypospadias syndrome, 43/413
Klinefelter variant XXXXY, 31/484
Klinefelter variant XXYY, 31/484
LEOPARD syndrome, 43/28
Lowry syndrome, 43/368
Marinesco-Sjögren syndrome, 21/556
Miller-Dieker syndrome, 50/592
10p partial trisomy, 43/519
Patau syndrome, 31/508, 43/528, 50/558
Pena-Shokeir syndrome type I, 43/438

porencephaly, 50/360
prune belly syndrome, 43/466
Rubinstein-Taybi syndrome, 31/282, 284, 43/234, 50/276
Seckel syndrome, 43/379
Smith-Lemli-Opitz syndrome, 43/308, 50/276
snub-nosed dwarfism, 43/387
spasmodic torticollis, 43/146
Cryptorchism, *see* Cryptorchidism
Cryptotoxin intoxication
phanerotoxin intoxication, 37/29
CSF
A wave, 55/206
achondroplasia, 43/316
acid base balance, 16/370, 28/509-513
acidosis, 28/511-513
acquired immune deficiency syndrome, 56/491
acquired immune deficiency syndrome dementia, 56/491
acquired immune deficiency syndrome dementia complex, 71/245
acquired toxoplasmosis, 35/125, 52/354, 356
acute amebic meningoencephalitis, 52/316, 319
acute cerebellar ataxia, 56/253
acute hemorrhagic conjunctivitis, 38/615, 56/351
acute hemorrhagic leukoencephalitis, 34/588
acute viral encephalitis, 34/66, 56/131
Addison disease, 39/481
adenovirus encephalitis, 56/286
adenovirus meningoencephalitis, 56/284
adenylate kinase, 54/199
African trypanosomiasis, 35/75, 52/342
allergic granulomatous angiitis, 54/198
allescheriosis, 35/559
amebiasis, 35/44-47
amyotrophic lateral sclerosis, 22/296, 59/196
anencephaly, 30/178, 181
angiodysgenetic necrotizing myelomalacia, 20/489
angiostrongyliasis, 35/323, 327, 330, 52/510, 553, 556-559
anion, 28/436
ankylosing spondylitis, 38/505
anterior spinal meningocele, 42/43
arachnoid cyst, 31/105
arenavirus infection, 34/195
Argentinian hemorrhagic fever, 34/195, 56/365
argininosuccinic aciduria, 42/525
arsenic intoxication, 36/213
aspergillosis, 35/398, 52/380
ataxia telangiectasia, 14/296
bacterial endocarditis, 52/295, 300, 54/197
bacterial meningitis, 33/4-9, 52/2, 6
Bannwarth syndrome, 51/200, 202
Batten disease, 10/617
Behçet syndrome, 34/487, 56/599, 71/222
bicarbonate, 28/436
bilirubin, 12/144, 54/196
bismuth intoxication, 64/343
bleeding tendency, 38/81
blood-brain barrier, 30/530
Bolivian hemorrhagic fever, 34/195
brain abscess, 33/118, 52/152
brain arteriovenous malformation, 31/191
brain biopsy, 67/219
brain concussion, 23/420, 434
brain contusion, 23/420, 434
brain cyst, 68/309
brain edema, 16/189, 67/76
brain embolism, 53/224, 54/197
brain hemorrhage, 54/197, 315
brain infarction, 53/224, 54/197
brain lacunar infarction, 54/254
brain metastasis, 18/220
brain tumor, 16/360
brain vasculitis, 55/422
bromide intoxication, 36/301
brucellosis, 33/310-312
Burkitt lymphoma, 39/76
calcium, 28/434
Candida meningitis, 52/400
candidiasis, 35/428-431
cat scratch disease, 34/472, 52/130
cauda equina tumor, 19/79
cell turnover, 56/82
central pontine myelinolysis, 63/553
cerebellar tumor, 17/713
cerebral malaria, 52/369
cerebral paragonimiasis, 35/257
cerebrotendinous xanthomatosis, 66/603
cerebrovascular disease, 11/235-240, 53/224
Chagas disease, 35/107, 52/347
chloride, 28/436
choroid plexus papilloma, 17/571, 42/761
chronic respiratory insufficiency, 38/290
circadian activity, 74/487
citric acid, 16/371
cladosporiosis, 35/563, 52/484
clonidine intoxication, 37/443
coccidioidomycosis, 35/449, 52/415
Cogan syndrome type I, 51/454
Cogan syndrome type II, 51/454
commotio cerebri, 23/420
congenital toxoplasmosis, 35/133, 42/662, 52/359
conus medullaris tumor, 19/79
corpus callosum tumor, 17/511

Coxsackie virus infection, 34/136
creatine kinase, 54/199
Creutzfeldt-Jakob disease, 56/545
critical illness polyneuropathy, 51/576, 581
Crome syndrome, 43/242
cryptococcoma, 52/432
cryptococcosis, 35/479, 52/433
Cryptococcus neoformans, 71/293
CSF 3-methoxy-4-hydroxyphenylethylene glycol, 45/136
cytochrome related ataxia, 42/111
cytomegalic inclusion body disease, 56/268
dermoid cyst, 20/79
diabetic neuropathy, 8/43, 27/103
diabetic polyneuropathy, 27/103
dihydrophenylacetic acid, 45/136
dipeptidyl carboxypeptidase I inhibitor intoxication, 65/445
disseminated necrotizing leukoencephalopathy, 63/355
drechsleriasis, 52/485
dysraphia, 32/188
echinococcosis, 52/525
ECHO virus infection, 34/140
electrolyte, 19/127
embryo development, 32/529-531
empty sella syndrome, 17/409, 431
encephalitis lethargica, 34/456
eosinophilic meningitis, 35/323
ependymoma, 18/129
epidural hematoma, 26/21
Epstein-Barr virus infection, 56/257
Epstein-Barr virus meningoencephalitis, 56/252
ethylene glycol intoxication, 64/123
extra pontine myelinolysis, 63/553
Farber disease, 42/593
fatty acid, 16/371
α-fetoprotein, 32/563
filariasis, 35/167, 52/514, 516, 518
Foix-Alajouanine disease, 9/455
foramen magnum tumor, 17/728
Friedman-Roy syndrome, 43/251
fungal meningitis, 55/433
general paresis of the insane, 52/279
Gerstmann-Sträussler-Scheinker disease, 56/554
giant cell arteritis, 54/198
glioblastoma multiforme, 18/54
globoid cell leukodystrophy, 10/73, 350, 42/489
glycine, 6/348
gold polyneuropathy, 51/306
gram-negative bacillary meningitis, 33/103, 52/104, 119
granulomatous amebic encephalitis, 52/316, 319
granulomatous CNS vasculitis, 55/390
Guillain-Barré syndrome, 7/495, 498, 42/645, 51/247
heat stroke, 23/678
hemoglobin, 12/145
hemolysis, 12/144
Hemophilus influenzae meningitis, 33/55, 52/63
hepatitis B virus infection, 56/297, 302
hereditary hemorrhagic telangiectasia (Osler), 55/451
hereditary olivopontocerebellar atrophy (Menzel), 21/445
hereditary sensory and autonomic neuropathy type I, 42/348, 60/9
hereditary sensory and autonomic neuropathy type II, 42/350
hereditary sensory and autonomic neuropathy type IV, 60/13
herpes simplex virus encephalitis, 34/151
herpes simplex virus meningitis, 56/126
herpes zoster, 34/174
hippuran excretion test, 30/586
Hirayama disease, 59/110
histidinemia, 42/554
Histoplasma meningitis, 52/438
histoplasmosis, 52/438, 71/297
Hodgkin disease, 20/121
homocarnosinase deficiency, 42/558
homocarnosinosis, 42/557-559
homocystinuria, 55/328
hourglass tumor, 20/307
human T-lymphotropic virus type I associated myelopathy, 56/538
Huntington chorea, 6/348
hydrocephalus, 30/556, 32/528, 531, 42/35
5-hydroxyindoleacetic acid, 45/136
hyperaldosteronism, 39/496
hypereosinophilic syndrome, 63/372
hyperglycinemia, 42/565
hypernatremia, 28/452
hyperparathyroidism, 27/290
hypertensive encephalopathy, 11/557, 54/212
hypoxanthine, 54/199
infantile enteric bacillary meningitis, 33/64
infantile Gaucher disease, 60/157
infantile optic glioma, 42/734
infectious endocarditis, 63/117
infectious mononucleosis, 34/185
intervertebral disc prolapse, 25/499
intracerebral hematoma, 11/684
intracranial hemorrhage, 53/224
intravascular consumption coagulopathy, 38/73
Kawasaki syndrome, 51/453, 52/265, 56/639

Kearns-Sayre-Daroff-Shy syndrome, 38/225, 228, 43/142, 62/311
Köhlmeier-Degos disease, 55/277
lactic acid-pyruvic acid ratio, 54/196
lactic acidosis, 57/386
Lassa fever, 34/195
lead polyneuropathy, 51/270
legionellosis, 52/254-256
leptomeningeal metastasis, 69/154, 71/649
leukemia, 39/5, 69/239, 241
linear nevus sebaceous syndrome, 43/39
Listeria monocytogenes infection, 33/85
Listeria monocytogenes meningitis, 33/85, 52/93, 63/531
lumbar intervertebral disc prolapse, 20/591
lumbar puncture, 57/318
lumbar vertebral canal stenosis, 32/334
Lyme disease, 51/205
lymphocytic choriomeningitis, 9/544, 34/195, 197, 56/360
Machado-Joseph disease, 42/155
macrophage, 12/146
magnesium imbalance, 28/552
malignant cell, 71/651
mannitol, 57/214
meningeal leukemia, 16/234, 39/4-6, 63/341
meningism, 1/547
meningitis serosa, 34/86
meningococcal meningitis, 33/25, 52/24
metabolic acidosis, 28/512
metabolic alkalosis, 28/513
metachromatic leukodystrophy, 66/175
metastasis, 71/649
methemoglobin, 54/196
migraine, 48/143
mineralizing brain microangiopathy, 63/355
Mollaret meningitis, 34/547, 56/628
mucormycosis, 35/551, 52/473
multiple sclerosis, 9/289, 324-376, 16/369, 42/497, 47/79-82
mumps, 34/628
mumps encephalitis, 56/430
mumps meningitis, 34/411, 56/129, 430
muscular atrophy, 42/82, 545
mycotic aneurysm, 52/295
neural tube formation, 32/530
neuroamebiasis, 52/319
neuroborreliosis, 51/200, 202, 205
neurobrucellosis, 33/310, 52/593
neurocutaneous melanosis, 14/422, 43/33
neurocysticercosis, 16/226, 20/92, 35/305, 52/531
neurosarcoidosis, 38/533
neurotropic virus, 56/31
Nocardia meningitis, 52/449
nocardiosis, 35/524, 52/447, 449
Nonne-Froin syndrome, 7/498, 19/125, 20/459
North American blastomycosis, 35/404, 406, 52/390
olivopontocerebellar atrophy (Dejerine-Thomas), 21/427
organic acid, 66/646
oxyhemoglobin, 54/196
Paine syndrome, 30/514, 43/432, 60/293
paracoccidioidomycosis, 35/536, 52/460
paragonimiasis, 35/257, 261-263
Paragonimus meningitis, 35/261
paraneoplastic polyneuropathy, 51/467
parkinsonism, 42/255
pathway, 71/643
peripheral neurolymphomatosis, 56/180, 182
pertussis, 52/234
pertussis vaccination, 33/298
phosphate, 28/437
pinealoma, 42/767
pituitary adenoma, 17/405
plaque, 9/289
plateau wave, 55/206
pneumococcal meningitis, 33/39-42, 52/44-46
poliodystrophy, 42/588
poliomyelitis, 34/117, 56/333
pontine tumor, 17/697
porencephaly, 42/48
potassium, 28/427-429, 468-472
prealbumin, 9/340
presenile dementia, 42/288
primary amebic meningoencephalitis, 52/319
primary cerebral reticulosarcoma, 18/246
progressive external ophthalmoplegia, 62/311
progressive multifocal leukoencephalopathy, 9/495, 71/275
progressive rubella panencephalitis, 34/334, 56/409
proteinocytologic dissociation, 7/498
pyroracemic acid, 16/371
pyruvic acid, 6/349
rabies, 34/257, 56/389
radiation injury, 23/657
Refsum disease, 10/345, 42/148, 60/233, 66/491, 494
relapsing fever, 33/414
renal insufficiency, 27/340
respiration, 1/655
respiratory acidosis, 28/511
respiratory alkalosis, 28/512
respiratory encephalopathy, 38/290
Reye syndrome, 56/153

rubella encephalitis, 56/407
rubella encephalomyelitis, 56/407
rubella encephalopathy, 56/407
saccharopinuria, 29/222
sarcoid neuropathy, 51/196
sarcoidosis, 38/533, 71/483
schistosomiasis, 52/540
sepedoniumosis, 52/493
serogenetic neuropathy, 8/102
Sjögren syndrome, 71/65, 69
snake envenomation, 65/179
sodium, 28/431, 445
spina bifida, 32/528, 532, 42/55
spinal angioma, 20/459
spinal arachnoiditis, 42/107
spinal cord metastasis, 20/419
spinal ependymoma, 20/369
spinal epidural empyema, 52/73, 187
spinal epidural lipoma, 19/362, 20/407
spinal epidural tumor, 20/106
spinal glioma, 20/231
spinal lipoma, 20/401
spinal meningioma, 20/216-219, 231
spinal nerve root injury, 25/417
spinal neurinoma, 20/231, 270-274, 283
spinal paragonimiasis, 35/263
spinal subarachnoid block, 19/125-138
spinal subdural empyema, 52/189
spinal tumor, 19/61, 125-138
spinocerebellar degeneration, 60/676
sporotrichosis, 35/569, 52/495
staphylococcal meningitis, 33/71
streptococcal meningitis, 33/71, 52/81
striatonigral degeneration, 42/262
subacute necrotizing encephalomyelopathy, 28/352, 42/625
subacute sclerosing panencephalitis, 9/327, 349, 352, 34/354, 56/419
subarachnoid hemorrhage, 12/144, 54/195, 55/23
subdural hematoma, 26/34, 37
succinic acid, 16/371
superior longitudinal sinus thrombosis, 11/52, 52/178
suprachiasmatic nucleus, 74/487
supratentorial brain abscess, 33/118
supratentorial brain tumor, 18/337
syphilis, 33/345-347, 52/274, 276
syringomyelia, 32/277, 42/60, 50/446
systemic lupus erythematosus, 39/287, 43/420, 54/198, 55/381, 71/46
taurine, 42/255
temporal arteritis, 55/344
temporal lobe agenesis syndrome, 31/105
third ventricle tumor, 17/456
thoracic intervertebral disc prolapse, 20/567
threonine, 6/348
toluene intoxication, 64/50
toxic hydrocephalus, 16/236
toxic shock syndrome, 52/259
toxoplasmosis, 52/356
transient ischemic attack, 53/224
transverse sinus thrombosis, 52/176
traumatic hydrocephalus, 24/237
traumatic hypotension, 24/255
trichinosis, 35/277, 52/569
triethyltin intoxication, 36/282
tropical spastic paraplegia, 56/527, 538
tuberculosis, 71/282
tuberculous meningitis, 33/218-221, 52/203, 215, 55/428, 71/282
tumor cell, 67/220
undetermined significant monoclonal gammopathy, 63/399
uremic encephalopathy, 63/507
varicella zoster virus meningitis, 55/437
varicella zoster virus meningoencephalitis, 56/235
varicella zoster virus myelitis, 56/236
vasopressin, 74/487
venous sinus occlusion, 54/441
venous sinus thrombophlebitis, 33/165
venous sinus thrombosis, 54/399
ventricular diverticulum, 31/125
verruga peruana, 34/665
vertebral metastasis, 20/419
viral encephalitis, 34/66
vitamin $B_c$ deficiency, 51/336
Vogt-Koyanagi-Harada syndrome, 34/527, 56/621
Waldenström macroglobulinemia, 18/246
Wegener granulomatosis, 71/184
Wernicke-Korsakoff syndrome, 28/249
Whipple disease, 52/138
white blood cell count, 54/196
xanthochromia, 12/144, 54/196
zinc sulfate, 9/335

CSF *N*-acetylneuraminic acid
multiple sclerosis, 9/369

CSF amino acid
multiple sclerosis, 9/368, 16/369

CSF ammonia
hepatic coma, 27/357, 368

CSF antibody
multiple sclerosis, 9/364

CSF antibody specificity
multiple sclerosis, 47/107-112

CSF autoantigen antibody
multiple sclerosis, 47/108

CSF B-lymphocyte
multiple sclerosis, 47/373-375, 377
CSF brain antibody
multiple sclerosis, 47/378
CSF bromide
multiple sclerosis, 9/369
CSF butyrylcholinesterase
multiple sclerosis, 9/367
CSF C-1 complement
multiple sclerosis, 9/370
CSF C-3 complement
Behçet syndrome, 56/600
multiple sclerosis, 47/115
CSF C-4 complement
multiple sclerosis, 47/115
CSF cell antigen stimulation
multiple sclerosis, 47/375
CSF cell count
multiple sclerosis, 47/231, 372
CSF cerebral edema
hepatic coma, 27/350
CSF cholesterol, 19/127
multiple sclerosis, 9/368, 47/116
CSF cholesterol esterase
multiple sclerosis, 47/119
CSF circulation, *see* CSF flow
CSF composition
brain injury, 24/590
Dandy-Walker syndrome, 50/325
CSF corticosteroid
multiple sclerosis, 9/366
CSF 2',3'-cyclic nucleotide 3'-phosphodiesterase
multiple sclerosis, 47/118
CSF cytology
astrocytoma, 16/395
brain metastasis, 16/404
brain sarcoma, 16/400
brain tumor, 16/371, 375, 380, 389
cell count, 11/236, 12/145, 16/372
cell necrobiosis, 16/373
cerebellar medulloblastoma, 16/390, 18/173
choroid plexus papilloma, 16/399
corticotropin, 47/373
craniopharyngioma, 16/402
CSF pleocytosis, 9/546
ependymoma, 16/397
glioblastoma multiforme, 16/392
hypernephroma, 16/407
hypophyseal adenoma, 16/402
Kawasaki syndrome, 56/639
melanosarcoma, 16/407
meningeal lymphoma, 63/347
meningioma, 16/401
midbrain tumor, 17/635
mitogen, 47/374
mitogen stimulation, 47/374
multiple sclerosis, 9/337-340
myelin basic protein, 47/376
neurilemoma, 8/417, 16/401
neuroastrocytoma, 16/401
oligodendroglioma, 16/396
pinealoma, 16/399, 18/365, 42/767
spinal tumor, 19/61
spongioblastoma, 16/394
tissue culture, 17/43
tumor cell, 16/375, 378, 381, 383, 389
tumor cell cultivation, 16/409
viral meningitis, 56/126
CSF dynamics, *see* CSF flow
CSF electrophoresis
viral infection, 56/84
CSF endotoxin
bacterial meningitis, 33/8
gram-negative bacillary meningitis, 52/105
meningococcal meningitis, 52/25
CSF enzyme
bacterial meningitis, 52/8
brain tumor, 16/367
CSF esterase
multiple sclerosis, 9/368
CSF examination
brain hemorrhage, 11/258
brain infarction, 11/238
cat scratch disease, 52/129
chronic bromide intoxication, 36/301
colloid gold curve, 9/335
composition, 19/127, 30/556
concentric sclerosis, 47/414
contraindication, 11/235
daytime hypersomnia, 45/136
dementia, 46/127
electrophoresis, 16/364, 19/127-130, 133
enzyme, 11/237
epidural hemorrhage, 11/239
flocculation test, 9/355
Guillain-Barré syndrome, 47/610
Ig subfractionation, 9/361
indication, 11/235
mastic curve, 9/335
Mimae meningitis, 33/103
multiple myeloma, 18/255
multiple sclerosis, 9/325, 372-376, 47/79-121, 362, 372-379
myelinoclastic diffuse sclerosis, 9/479, 47/424
neuralgic amyotrophy, 51/172
neuromyelitis optica, 9/429, 47/401

noninfectious purulent meningitis, 33/5
Nonne-Froin syndrome, 20/217
Parkinson disease, 49/122
pathologic value, 19/130
phlebothrombosis, 11/239
progressive multifocal leukoencephalopathy, 47/510
subarachnoid hemorrhage, 11/238
subdural hematoma, 11/239
subdural hemorrhage, 11/239
syphilis, 11/239
systemic lupus erythematosus, 11/239
spectroscopy, 16/369
vasculitis, 11/239
xanthochromia, 11/236

CSF fistula
*see also* Otorrhea *and* Rhinorrhea
anosmia, 24/5
anterior fossa, 24/186
antibiotic agent, 24/188
arachnoid cyst, 24/186
bacterial meningitis, 52/5
brain injury, 23/310
brain scintigraphy, 23/310, 24/190
brain stab wound, 23/485
brain tumor, 24/185
choroid plexus papilloma, 24/186
clinical diagnosis, 24/189
contrast study, 24/191
craniopharyngeal canal, 24/186
delayed closure, 24/191
delayed traumatic, 24/187
dye test, 24/190
EEG, 23/352
empty sella syndrome, 17/431
ethmoid bone fracture, 24/184
etiology, 24/183
frontobasal, 24/184
head injury, 24/186, 57/311
history, 24/183
hydrocephalus, 24/186
immediate traumatic, 24/187
incidence, 24/184
intracranial bone fracture, 24/185
laboratory diagnosis, 24/189
lumbar puncture, 61/158
meningitis, 24/188, 57/311
missile injury, 23/523, 57/311
nonmissile injury, 57/311
nontraumatic, 24/185
normal pressure leak, 24/185
operation moment, 24/191
penetrating head injury, 57/311
pituitary tumor, 24/185
pneumocephalus, 24/211
posttraumatic meningitis, 57/318
prognosis, 24/191
radioactive scanning, 24/190
radiography, 24/190
recurrent meningitis, 52/53
rhinorrhea, 23/310, 485, 24/211
site, 24/186, 188
skull fracture, 24/185, 57/135, 139
space occupying lesion, 16/246
spontaneous, *see* Spontaneous CSF fistula
spontaneous closure, 24/191
surgery, 24/192, 660
temporal bone defect, 24/187
temporal bone fracture, 24/121, 125, 185, 57/139
traumatic, 24/184
tuberculoma, 18/417

CSF flow
basilar impression, 32/77
benign intracranial hypertension, 67/108
brain hypertension, 16/125
brain injury, 23/310
brain scintigraphy, 23/294
cough, 61/378
cough headache, 48/369
Dandy-Walker syndrome, 50/324
development, 50/324
escape, 32/531
headache, 48/32
intracranial hypertension, 16/125
intracranial pressure, 16/91-95
normal diagram, 46/324
osteitis deformans (Paget), 38/366
pinealoma, 18/362
posttraumatic syringomyelia, 61/379, 396
primary alcoholic dementia, 46/345
sink action, 9/370

CSF formation
Dandy-Walker syndrome, 50/325

CSF GABA
chorea-acanthocytosis, 49/330
pyrethroid intoxication, 64/219

CSF glial antibody
multiple sclerosis, 47/109, 119

CSF glial protein
multiple sclerosis, 47/119

CSF β-2-globulin
multiple sclerosis, 9/340

CSF glucose
bacterial meningitis, 52/6
brain tumor, 16/367
gram-negative bacillary meningitis, 52/104

posttraumatic meningitis, 57/318
subarachnoid hemorrhage, 54/196
CSF β-glucuronidase
multiple sclerosis, 9/368
CSF glutamic oxalacetic transaminase
multiple sclerosis, 9/367
CSF glutamine
congenital hyperammonemia, 27/368
hepatic encephalopathy, 27/349, 369
CSF glycolytic enzyme
multiple sclerosis, 9/368
CSF herpes simplex virus antigen
herpes simplex virus encephalitis, 56/218
CSF herpes simplex virus 1 antigen
multiple sclerosis, 47/114
CSF HLA antigen
multiple sclerosis, 47/375
CSF homocarnosine
familial spastic paraplegia, 59/312
CSF homovanillic acid
amyotrophic lateral sclerosis, 22/30
hyperthyroid chorea, 27/280
Machado-Joseph disease, 42/262
Rett syndrome, 63/437
sleep-wake cycle, 45/136
xeroderma pigmentosum, 60/337
CSF humoral immunity
multiple sclerosis, 47/376
CSF hypertension
*see also* CSF pressure, Intracranial hypertension *and* Intracranial pressure
headache, 5/155
subarachnoid hemorrhage, 12/177
CSF hypotension
hearing loss, 61/165
intracranial subdural hematoma, 61/164
postlumbar puncture syndrome, 61/154
subarachnoid hemorrhage, 12/172
traumatic, *see* Traumatic CSF hypotension
CSF idiotype
multiple sclerosis, 47/379
CSF idiotypic antibody
multiple sclerosis, 47/108, 362, 379
CSF Ig
chronic inflammatory demyelinating polyradiculoneuropathy, 51/536
multiple sclerosis, 9/356, 47/377
pleocytosis, 9/333
CSF Ig subclass
multiple sclerosis, 47/113
CSF IgA
Behçet syndrome, 56/600
multiple sclerosis, 9/336, 363, 47/103, 113
CSF IgD
multiple sclerosis, 9/363, 47/113
CSF IgE
multiple sclerosis, 47/113
CSF IgG
herpes simplex virus encephalitis, 56/215
multiple sclerosis, 9/140, 326, 328, 333, 335, 348-352, 360, 42/497, 47/65, 71, 80
Sjögren syndrome, 51/449
subacute sclerosing panencephalitis, 56/419
CSF IgG1
multiple sclerosis, 47/104, 377
CSF IgG2
multiple sclerosis, 47/104
CSF IgG3
multiple sclerosis, 47/104
CSF IgG kappa chain
multiple sclerosis, 47/105
CSF IgG lambda chain
multiple sclerosis, 47/105
CSF IgG synthesis
multiple sclerosis, 47/79-82, 87-107
CSF IgM
multiple sclerosis, 9/336, 363, 47/103, 112
CSF immune complex
multiple sclerosis, 47/114, 371
CSF intrablood-brain barrier IgG synthesis
multiple sclerosis, 47/79-82, 87-107
CSF isocitrate dehydrogenase
multiple sclerosis, 9/367
CSF keratin
epidermoid cyst, 20/79
CSF lactate dehydrogenase
bacterial meningitis, 52/8
brain tumor, 16/368
CSF lactic acid
bacterial meningitis, 52/8
brain injury, 54/198
brain metabolism, 57/72
brain tumor, 16/370
gram-negative bacillary meningitis, 52/119
CSF leukocyte
multiple sclerosis, 47/82-86
pleocytosis, 9/330
in vitro Ig synthesis, 47/84
CSF leukocyte count
intrablood-brain barrier IgG synthesis, 47/84
multiple sclerosis, 9/326, 329, 47/83, 100
CSF leukocytosis
bacterial meningitis, 52/6
CSF lipid
brain tumor, 16/366
multiple sclerosis, 9/368, 47/115

CSF listeria antigen
Listeria monocytogenes meningitis, 52/98
CSF lymphocyte subpopulation
multiple sclerosis, 47/85
CSF lymphocyte transformation
tuberculous meningitis, 52/207
CSF lymphocytosis
bacterial meningitis, 52/6
diphtheria, 51/183
herpes zoster ganglionitis, 51/181
CSF lysozyme
bacterial meningitis, 52/8
CSF magnesium
hypomagnesemia, 28/435, 552
CSF manometry, *see* CSF pressure
CSF marker
white matter disease, 47/115-119
CSF measles antibody
multiple sclerosis, 47/377
subacute sclerosing panencephalitis, 56/420
CSF 3-methoxy-4-hydroxyphenylethylene glycol
CSF, 45/136
daytime hypersomnia, 45/136
dystonia musculorum deformans, 49/524
Parkinson disease, 49/123
Rett syndrome, 63/437
CSF mixed lymphocyte reaction
multiple sclerosis, 47/375
CSF monoclonal antibody
multiple sclerosis, 47/85
CSF mononuclear cell
multiple sclerosis, 47/83, 231
CSF myelin basic protein
multiple sclerosis, 47/116
CSF myelin basic protein antibody
multiple sclerosis, 47/108, 117
CSF neutral protease
multiple sclerosis, 47/371
CSF nonspecific antibody
multiple sclerosis, 47/378
CSF noradrenalin
chorea-acanthocytosis, 49/330
CSF oligoclonal band
progressive rubella panencephalitis, 56/409
subacute sclerosing panencephalitis, 56/419
CSF oligoclonal IgG
infectious disease, 47/96
multiple sclerosis, 47/65, 94-96, 101, 362, 377
pattern shift visual evoked response, 47/137
CSF otorrhea, *see* CSF fistula
CSF oversecretion
hydrocephalus, 50/286
CSF 2-oxoglutaramic acid
hepatic coma, 27/349, 359, 364
hepatic encephalopathy, 27/349, 359, 364
Reye syndrome, 27/364
CSF paraprotein
areflexia, 42/604
hyporeflexia, 42/604
muscular atrophy, 42/603
paralysis, 42/603
polyneuropathy, 42/604
sensory loss, 42/603
CSF peptidase
multiple sclerosis, 9/368
CSF pH
respiratory encephalopathy, 63/414
CSF phosphatidylcholine
multiple sclerosis, 9/368
CSF phosphatidylethanolamine
multiple sclerosis, 9/368
CSF phosphohexose isomerase
multiple sclerosis, 9/367
CSF phospholipid
multiple sclerosis, 9/321, 368, 47/115
CSF pipecolic acid
hyperpipecolatemia, 29/222
CSF plasma cell
multiple sclerosis, 47/83, 112
CSF pleocytosis
Burkitt lymphoma, 63/351
chronic bromide intoxication, 36/301
chronic meningitis, 56/643
ciclosporin intoxication, 63/538
CSF cytology, 9/546
cytology, 9/546
eosinophilic granuloma, 63/422
histiocytosis X, 63/422
human T-lymphotropic virus type I associated myelopathy, 59/450
migraine coma, 60/651
mumps, 34/411, 56/129
OKT3, 63/536
persistence, 56/643
tick bite intoxication, 37/111
Vogt-Koyanagi-Harada syndrome, 56/611
CSF pressure, 19/91-120
*see also* CSF hypertension
abdominal compression, 19/107-110
amplitude frequency curve, 19/95
Antoni sign, 19/103
aorta coarctation, 63/5
Arnold-Chiari malformation type I, 19/92, 116
C wave, 19/113
cervical myelopathy, 19/117
classification, 19/99

cough headache, 5/186, 48/370
Dandy-Walker syndrome, 50/326
dialysis disequilibrium syndrome, 63/524
differential pressure, 19/92
dynamic physiologic resistance, 19/92
electromanometry, 19/94, 99, 118
epidural pressure, 19/113
examination, 11/237
frequency response, 19/94
fructose, 16/361
head posture, 19/100
hydrocephalus, 30/532-534
hysteresis, 19/96
indication, 19/117
intermittent spinal cord ischemia, 12/529
linearity, 19/96
myeloscintigraphy, 19/118
physiology, 30/532
postural manometry, 19/92, 106
pressure transducer, 19/93
Queckenstedt test, 19/91, 97, 107, 110
reliability, 19/118
Reye syndrome, 29/335, 31/168, 34/169, 49/218
sensitivity, 19/94
sorbitol, 16/361
spinal muscular atrophy, 59/110
spinal tumor, 19/114
stability, 19/94
striatopallidodentate calcification, 49/422
subarachnoid hemorrhage, 12/150, 172
syringomyelia, 19/116
Traube-Hering wave, 19/113
urea, 16/361
vagus stimulation, 19/113
Valsalva maneuver, 19/110
valve phenomenon, 19/107
venous pressure, 30/533
vitamin A intoxication, 37/96
CSF production
absorption, 32/529
choroid plexus, 30/529
ouabain intoxication, 37/426
physiology, 30/526-531
secretion, 32/530
CSF profile
multiple sclerosis, 9/326, 329, 47/119-121
CSF prostaglandin E
multiple sclerosis, 47/376
CSF protein
acute pandysautonomia, 51/475, 63/153
albumin, 9/343
amyloid polyneuropathy, 51/417
bacterial meningitis, 33/8
brain tumor, 16/363
Charlevoix-Saguenay spastic ataxia, 60/457
chronic inflammatory demyelinating polyradiculoneuropathy, 51/534
Creutzfeldt-Jakob disease, 46/292
dipeptidyl carboxypeptidase I inhibitor intoxication, 65/445
eosinophilic granuloma, 63/422
familial spastic paraplegia, 59/312
Friedreich ataxia, 21/350, 60/320
globoid cell leukodystrophy, 10/304, 59/355
hereditary acute porphyria, 51/392
hereditary amyloid polyneuropathy, 21/127
hereditary amyloid polyneuropathy type 1, 51/424, 60/92
hereditary cerebello-olivary atrophy (Holmes), 21/410
hereditary motor and sensory neuropathy type I, 60/199
hereditary motor and sensory neuropathy type II, 60/199
hereditary paroxysmal dystonic choreoathetotis, 49/351
herpes zoster ganglionitis, 51/181
histiocytosis X, 63/422
human T-lymphotropic virus type I associated myelopathy, 59/450
hydrodynamic radius, 56/80
hydromyelia, 50/428
hyperparathyroidism, 27/291
hypertrophic interstitial neuropathy, 21/161, 60/216
Joseph syndrome, 29/231
Kearns-Sayre-Daroff-Shy syndrome, 62/508
migraine, 48/145
multiple myeloma polyneuropathy, 51/430
myxedema neuropathy, 8/57, 59
Nonne-Froin syndrome, 19/125
normal value, 19/126
perhexiline maleate polyneuropathy, 51/302
primary amyloidosis, 51/417
progressive external ophthalmoplegia, 22/183, 43/137
progressive systemic sclerosis, 51/448
reflective type, 19/132
Refsum disease, 8/21, 10/345, 13/314, 21/9, 202, 210, 36/348, 41/434, 42/148, 51/384, 60/228, 233, 728, 66/494
spinal ependymoma, 19/360, 20/369
spinal meningioma, 19/68, 358, 20/216
spinal neurinoma, 20/271
spinal tumor, 19/61
subarachnoid hemorrhage, 12/149

syndrome, 42/545
trace, 9/367
transudative type, 19/132
uremic polyneuropathy, 51/355, 360
viral meningitis, 56/126
Whipple disease, 51/184
CSF protein S100
multiple sclerosis, 47/119
CSF proteinase
multiple sclerosis, 9/368, 47/119
CSF proteolipid
multiple sclerosis, 47/118
CSF reabsorption
Dandy-Walker syndrome, 50/326
physiology, 30/531
CSF retention
Dandy-Walker syndrome, 50/328
CSF rhinorrhea, *see* CSF fistula
CSF rubella antigen
progressive rubella panencephalitis, 34/334, 56/409
CSF shunt
bacterial meningitis, 52/2
candidiasis, 52/404
hydrocephalic dementia, 46/329
hydrocephalus, 30/573-575
prognosis, 46/330
seizure, 46/330
staphylococcal meningitis, 52/73
subdural hematoma, 46/329
CSF shunt infection
antibiotic agent, 52/74
risk factor, 52/73
Staphylococcus epidermidis, 52/73
treatment, 52/74
CSF space
craniovertebral region, 32/60-66
CSF sphingomyelin
multiple sclerosis, 9/368
CSF T-helper/T-suppressor cell
multiple sclerosis, 47/85
CSF T-lymphocyte
multiple sclerosis, 47/372
CSF thromboplastic activity
multiple sclerosis, 47/119
CSF total protein
multiple sclerosis, 9/340
CSF trace metal
multiple sclerosis, 9/369
CSF virus antibody
multiple sclerosis, 47/110, 377
viral infection, 56/100
CSF vitamin $B_{12}$
multiple sclerosis, 9/368
CSF vitamin $B_c$
multiple sclerosis, 9/367
CT scanning, *see* Computer assisted tomography
Ctenochaetis intoxication
family, 37/78
venomous spine, 37/78
Cubital canal syndrome
rheumatoid arthritis, 38/483
Cubitus valgus
Turner syndrome, 50/543
Cuboid reflex, *see* Mendel-Von Bechterew reflex
Cubomedusae intoxication
arthralgia, 37/43
headache, 37/43
local pain, 37/41
nausea, 37/43
taxonomy, 37/41
vomiting, 37/43
Cubomedusan intoxication
venom, 37/48-50
Cuirass respirator
artificial respiration, 26/345
respiratory dysfunction, 26/345
Culex annulirostris
Murray Valley encephalitis, 56/139
Culex mosquito
Venezuelan equine encephalitis, 56/137
Culex nigripalpus
St. Louis encephalitis, 56/137
Culex pipiens
St. Louis encephalitis, 56/137
Culex tarsalis
St. Louis encephalitis, 56/137
western encephalitis, 56/136
Culex tritaeniorhyncus
Japanese B encephalitis, 56/138
Culiseta melanura
eastern encephalitis, 56/136
Culmen
boundary, 2/407
Cupric acetoarsenite
arsenic intoxication, 64/284
Cuprizone
copper detector, 64/17
experimental demyelination, 47/563
neurotoxin, 47/563, 64/10, 17
oligodendrocyte inclusion body, 64/17
toxic neuropathy, 64/10
Cuprizone intoxication
action site, 64/10
aqueduct stenosis, 50/303
giant mitochondria, 64/17

leukoencephalopathy, 47/563
oligodendrocyte, 64/17
Curare
chondrodystrophic myotonia, 40/546
Isaacs syndrome, 40/546
motor end plate, 7/87, 40/129, 143
myasthenia gravis, 41/127, 62/413
myasthenic syndrome, 64/16
myoglobinuria, 64/16
myotonia, 41/299
neuromuscular block, 40/129, 143
neuromuscular transmission, 40/143
organophosphate intoxication, 64/165
progressive external ophthalmoplegia, 22/183
synapse, 1/48
toxic neuromuscular junction disorder, 64/15
Curare intoxication
action site, 64/15
pathomechanism, 64/15
symptom, 64/15
Curare sensitive ocular myopathy
progressive external ophthalmoplegia, 22/179, 188, 62/302
Curschmann-Steinert disease, *see* Myotonic dystrophy
Curtius syndrome type V
neurocutaneous syndrome, 14/101
Curvilinear body
membrane fragment, 21/52
metachromatic leukodystrophy, 60/126
neuronal ceroid lipofuscinosis, 10/226
Curvularia
fungal CNS disease, 52/479, 484
mycetoma, 52/484
Curvularia pallescens
brain abscess, 52/151
Curvulariosis
amphotericin B, 52/485
brain abscess, 52/485
keratitis, 52/484
miconazole, 52/485
mycetoma, 52/484
paraspinal abscess, 52/484
Cushing disease
exertional headache, 48/375
Cushing, H., 1/5, 19
Cushing reflex
apnea, 63/230
bradycardia, 63/230
cluster headache, 75/289
headache, 75/289
intracranial hypertension, 16/137, 63/230
intracranial pressure, 57/130
medulla oblongata distortion, 63/230
Cushing syndrome
*see also* Nelson syndrome
adrenal disorder, 70/205
adrenal gland, 39/483-492
adrenocortical carcinoma, 22/550
affective psychosis, 46/462
aminoglutethimide, 68/356
basophil adenoma, 2/454, 457, 18/354
behavior disorder, 46/599
blood sampling, 68/355
brain tumor, 16/355
corticosterone, 70/206
corticotropin, 68/356, 70/205, 207
delirium, 46/544
dementia, 46/396
depression, 46/427, 450
diagnosis, 68/354, 356
differential diagnosis, 41/202
etiology, 70/206
headache, 5/230
hydrocortisone, 70/206
ketoconazole, 68/355
metyrapone, 68/357
mifepristone, 68/357
muscle fiber type II atrophy, 41/251
myopathy, 41/250
obesity, 2/449
penis erection, 75/89
pituitary adenoma, 2/454, 17/422
proximal muscle weakness, 40/317
radiodiagnosis, 68/354
radiotherapy, 68/355
steroid myopathy, 40/162, 41/250-253, 62/535
treatment, 68/355
Wermer syndrome, 42/752
Cutanea tarda hereditaria, *see* Porphyria variegata
Cutanea tarda symptomatica, *see* Symptomatic porphyria
Cutaneocerebrovisceral angiomatosis, *see* Ullmann syndrome
Cutaneomandibular polyoncosis
hereditary, *see* Multiple nevoid basal cell carcinoma syndrome
Cutaneomeningomedullary angiomatosis, *see* Cobb syndrome
Cutaneomeningospinal angiomatosis, *see* Cobb syndrome
Cutaneotendinous xanthoma
hyperlipoproteinemia type II, 10/272
Cutaneous antebrachial nerve
medial, *see* Medial cutaneous antebrachial nerve
Cutaneous atrophy

Hallermann-Streiff syndrome, 43/403
hereditary amyloid polyneuropathy, 42/524
Cutaneous blood flow
aging, 74/232
nervous control, 1/464
odor, 74/367
Cutaneous brachial nerve
medial, *see* Medial cutaneous brachial nerve
Cutaneous dysplasia, 14/101-125
Divry-Van Bogaert syndrome, 14/101
dysraphia, 14/110
Ehlers-Danlos syndrome, 14/101
Grönblad-Strandberg syndrome, 14/101
hereditary hemorrhagic telangiectasia (Osler), 14/101
neurofibromatosis type I, 14/101
tuberous sclerosis, 14/101
Cutaneous ecchymosis
migraine, 48/160
Cutaneous femoral nerve
lateral, *see* Lateral femoral cutaneous nerve
meralgia paresthetica, 2/40, 19/86, 42/323
Cutaneous hemangiomata
multiple, *see* Multiple cutaneous hemangioma
Cutaneous hypopigmentation
tuberous sclerosis, 50/371
Cutaneous hypoplasia
Goltz-Gorlin syndrome, 43/19
Cutaneous innervation, *see* Skin innervation
Cutaneous leiomyoma
nerve tumor, 8/458
Cutaneous lesion, *see* Skin lesion
Cutaneous nerve
lateral, *see* Lateral cutaneous nerve
lateral femoral, *see* Lateral femoral cutaneous nerve
leprosy, 33/428
migrant sensory neuritis, 51/461
topographical diagnosis, 2/19
Cutaneous pigmentation
Berlin syndrome, 43/9
Gaucher disease, 10/307
Hallervorden-Spatz syndrome, 6/605, 610
hemochromatosis, 42/553
Jeune-Tommasi disease, 22/513, 516
nerve lesion, 7/266
Niemann-Pick disease type A, 42/468
porphyria variegata, 42/619
sensorineural deafness, *see* Margolis-Ziprkowski syndrome
tryptophanuria, 42/633
Cutaneous reflex
anal sphincter reflex, 1/254
central motor disorder, 1/172
Friedreich ataxia, 21/332
hereditary motor and sensory neuropathy type I, 21/282
hereditary motor and sensory neuropathy type II, 21/282
physiology, 1/252
Cutaneous scalp defect
Patau syndrome, 50/556
Cutaneous vasoconstrictor neuron
autonomic nervous system, 74/26
Cutis hyperelastica dermatorrhexis, *see* Ehlers-Danlos syndrome
Cutis laxa
definition, 14/32
mental deficiency, 46/60
neurofibromatosis type I, 14/32
penicillamine intoxication, 49/235
Wilson disease, 49/235
Cutis marmorata
Coffin-Lowry syndrome, 43/239
Divry-Van Bogaert syndrome, 42/725, 55/318
leukodystrophy, 10/120
microcephaly, 43/429
orthochromatic leukodystrophy, 10/113, 123
Rothmund-Thomson syndrome, 43/460
semantic confusion, 55/401
Sneddon syndrome, 55/401
Cutis verticis gyrata
acromegaly, 14/777, 43/244
blindness, 43/244
cataract, 43/244
epilepsy, 43/244
keratoconus, 43/244
mental deficiency, 14/777, 46/47
microcephaly, 30/512, 42/45, 43/244, 50/269
Miescher syndrome type II, 14/120
multiple dermal cylindroma, 14/591
nystagmus, 43/244
pachydermoperiostosis, 43/244
prevalence, 43/244
sella turcica, 43/244
sex ratio, 43/244
strabismus, 43/244
tuberous sclerosis, 14/496
Cyanazine
organophosphate, 64/153
organophosphate intoxication, 64/153
Cyanea intoxication
corneal infiltration, 37/52
cough, 37/52
eye injury, 37/52
morphine, 37/52

sweating, 37/52
swelling, 37/52
Cyanide
cyanogenic glycoside intoxication, 36/518, 65/27
neurotoxin, 36/518, 49/466, 65/25
Cyanide encephalopathy
cyanide intoxication, 64/231
Cyanide intoxication
*see also* Cyanogenic glycoside intoxication
acute intoxication, 64/232
acute lethal dose, 64/232
akinesia, 64/232
4-aminopropiophenone, 64/233
amygdalin, 65/32
anaerobic metabolism, 64/230
anterior horn cell degeneration, 65/28
ataxic neuropathy, 7/642, 56/526, 65/30
athetosis, 64/232
basal ganglion infarction, 64/232
calcium channel blocking agent, 64/231
cassava, 65/30
cerebellar infarction, 64/232
chorea, 64/232
clinical features, 49/477, 65/28
cobalt edetate, 64/233
coma, 65/28
complex IV, 65/28
confusion, 64/232
convulsion, 65/28
corpus callosum, 65/28
cyanide encephalopathy, 64/231
cyanogenesis, 65/29
cysteine, 65/32
cytochrome oxidation, 64/230
deficiency neuropathy, 51/323
delayed toxicity, 64/232
dementia, 49/477, 64/232, 65/29
demyelination, 9/39, 632, 64/231, 65/28
detoxication, 64/231, 65/28
4-dimethylaminophenol, 64/233
dominant inherited optic atrophy, 65/32
dysarthria, 64/232
epilepsy, 65/28
exposure time, 64/231
globus pallidus infarction, 64/232
headache, 65/28
hyperammonemia, 64/231
hyperventilation, 65/28
inhalation rate, 64/232
laetrile, 65/32
leukoencephalopathy, 9/632, 47/552
megaloblastic madness, 38/21
metabolism, 65/28
neuropathology, 49/477, 64/232, 65/28
optic atrophy, 65/30
organophosphate intoxication, 64/151
pallidal necrosis, 6/652, 49/466
pallidostriatal lesion, 6/657
parkinsonism, 6/242, 49/477, 64/232
pathophysiology, 64/230
plasma thiocyanate, 65/31
polioencephalomyelitic syndrome, 65/29
Pseudomonas pyocyanea, 65/29
putamen infarction, 64/232
retrobulbar neuropathy, 65/32
rhodanese, 65/28
rigidity, 64/232
sodium nitrite, 64/233
sodium thiosulfate, 64/233
striatal necrosis, 49/500, 504
sudanophilic leukodystrophy, 9/634
sulfurtransferase, 64/231
systemic acidosis, 64/231
thiocyanate, 65/28, 30
tobacco amblyopia, 7/643, 13/62, 65/30
tobacco smoke, 65/29
treatment, 64/233, 65/31
tropical ataxic myelopathy, 65/29
tropical ataxic neuropathy, 65/29
tropical myeloneuropathy, 56/525
tropical neuropathy, 65/30
tropical spastic paraplegia, 65/30
vincennite, 64/230
vitamin B2 deficiency, 51/332
vitamin B12, 65/32
vitamin C, 65/32
volatility, 64/230
zyklon B, 64/230
Cyanocobalamin, *see* Vitamin B12
Cyanocobalamin deficiency, *see* Vitamin B12 deficiency
Cyanogenic glycoside
cassava, 64/8
distal axonopathy, 64/12
neurotoxin, 36/515, 518, 520-524, 64/8, 10, 12, 65/25
optic neuropathy, 64/8
toxic encephalopathy, 64/12, 65/25
toxic neuropathy, 64/8, 10
tropical ataxic neuropathy, 64/12
Cyanogenic glycoside intoxication, 36/515-524
*see also* Cyanide intoxication
aglycone, 65/27
almond, 36/515, 65/27
amblyopia, 36/520-522
amygdalin, 36/515, 65/26, 32

anterior horn cell degeneration, 65/28
ataxic neuropathy, 65/30
bamboo, 36/515, 519, 65/26
cassava, 36/515, 519, 65/27, 30
clinical features, 65/28
coma, 65/28
complex IV, 65/28
convulsion, 65/28
corpus callosum, 65/28
cyanide, 36/518, 65/27
cyanogenesis, 65/29
cysteine, 65/32
dementia, 65/29
demyelination, 65/28
detoxication, 65/28
dhurrin, 36/515, 65/26
dominant inherited optic atrophy, 65/32
flax, 65/26
β-glucosidase, 65/27
headache, 65/28
hyperventilation, 65/28
laetrile, 65/32
lima bean, 36/519, 65/26
linamarase, 65/27
linamarin, 36/515, 65/27
lotaustralin, 36/515, 65/26
Manihot esculenta, 36/515, 65/27
metabolism, 65/27
myelopathy, 64/8
neuropathology, 65/28
neuropathy, 64/8
optic atrophy, 36/520-522
plant HCN content, 65/27
plasma thiocyanate, 65/31
polioencephalomyelitic syndrome, 65/29
Pseudomonas pyocyanea, 65/29
rhodanese, 65/28
symptom, 64/12
tapioca, 65/27
taxiphyllin, 65/26
thiocyanate, 65/28, 30
tobacco amblyopia, 36/521, 65/31
tobacco smoke, 65/29
treatment, 65/31
tropical ataxic myelopathy, 65/29
tropical ataxic neuropathy, 7/642, 36/522-524, 65/29
tropical neuropathy, 65/30
tropical spastic paraplegia, 65/30
visual impairment, 64/8
vitamin B12, 36/518, 521, 65/32
vitamin C, 65/32
wild cherry, 36/516, 65/26

Cyanosis
breath holding syncope, 15/828
congenital heart disease, 63/1
cryoglobulinemia, 63/403
hydrozoa intoxication, 37/40
neophocaena phocaenoides intoxication, 37/95
pickwickian syndrome, 3/96, 38/299
porpoise intoxication, 37/95
sea snake intoxication, 37/92
thalamic syndrome (Dejerine-Roussy), 2/482
Cyanotic heart disease
porencephaly, 50/359
Cybernetics
emotion, 3/325-332
Cyberomone
classification, 74/374
Cycad
Guam amyotrophic lateral sclerosis, 22/130, 344
Kii Peninsula amyotrophic lateral sclerosis, 22/130
Cycad intoxication
*see also* 2-Amino-3-methylaminopropionic acid *and* Guam motoneuron disease
botany, 65/21
epidemiology, 65/21
geography, 65/21
methylazoxymethanol glucoside, 65/23
neurotoxicology, 65/22
sotetsu, 65/21
Cycadism, *see* Cycad intoxication
Cycas circinalis
toxic neuronopathy, 22/27
Cycas circinalis neurotoxin
Guam amyotrophic lateral sclerosis, 59/274
Cycasin, *see* Methylazoxymethanol glucoside
Cyclazocine
chemistry, 37/366
opiate, 65/350
Cyclencephaly, *see* Holoprosencephaly
Cyclic adenosine monophosphate
dopamine receptor, 49/49
neuroleptic agent, 65/279
Parkinson disease, 49/123
pseudohypoparathyroidism, 42/622
Cyclic AMP, *see* Adenosine phosphate
Cyclic guanine monophosphate
Parkinson disease, 49/123
2',3'-Cyclic nucleotide 3'-phosphodiesterase
CSF, *see* CSF 2',3'-cyclic nucleotide 3'-phosphodiesterase
2',3'-Cyclic nucleotide-3'-phosphohydrolase
white matter, 9/11, 14
Cyclic oculomotor paralysis

amblyopia, 42/332
clinical features, 1/611
convulsion, 42/332
EEG, 42/332
miosis, 42/331
mydriasis, 42/331
oculomotor nucleus, 2/304
pontine glioma, 42/332
ptosis, 42/331
strabismus, 42/331
Cyclic vomiting
migraine, *see* Abdominal migraine
Cyclicism, *see* Biological rhythm
Cyclist palsy
nerve lesion, 7/334
occupational lesion, 7/334
Cyclizine
migraine, 5/99
Cycloleucine
spongiform leukoencephalopathy, 47/569
vitamin B12 deficiency, 51/339
Cyclophilin
ciclosporin intoxication, 65/551
Cyclophosphamide
alkylating agent, 39/93-98
allergic granulomatous angiitis, 63/384
antineoplastic agent, 65/528
behavior disorder, 46/600
Castleman disease, 63/396
chemotherapy, 39/97
dermatomyositis, 62/385
granulomatous CNS vasculitis, 55/387, 397
immunocompromised host, 56/471
inclusion body myositis, 62/385
lymphomatoid granulomatosis, 63/424
multiple myeloma, 20/14, 111, 63/396
multiple sclerosis, 47/86, 194-196
myasthenia gravis, 41/96, 134, 136, 62/418
neuroblastoma, 8/471
neurocysticercosis, 52/533
neurotoxin, 39/119, 46/600, 62/418, 64/244, 65/528, 67/357
osteosclerotic myeloma, 63/396
POEMS syndrome, 47/619, 63/396
polyarteritis nodosa, 39/308, 55/357
polymyositis, 62/385, 71/142
systemic lupus erythematosus, 55/381, 71/49
Takayasu disease, 55/339
vasculitis, 71/160
water intoxication, 64/244
Cyclophyllin, *see* Cyclophilin
Cyclopia
*see also* Holoprosencephaly
autosomal recessive inheritance, 50/61
Edwards syndrome, 50/562
face, 30/442
head, 30/422
holoprosencephaly, 42/32, 50/232
hydranencephaly, 50/347
iniencephaly, 50/131
mouth, 30/444
neural induction, 50/25
nose, 30/443
ocular finding, 30/443
optic nerve agenesis, 50/211
otocephaly, 30/442
Patau syndrome, 31/504, 510, 513, 50/559
polyploidy, 50/555
proboscis, 30/443
scalp, 30/442
sex ratio, 50/61
synophthalmia, *see* Synophthalmia cyclopia
teeth, 30/444
Cyclopic vision
vertical meridian, 3/31
Cyclopropane
malignant hyperthermia, 41/268
Cyclopropane intoxication
inhalation anesthetics, 37/412
Cycloserine
catatonia, 65/481
dysarthria, 65/481
epilepsy, 28/116, 65/481
headache, 65/481
iatrogenic neurological disease, 65/481
neurotoxin, 28/116, 33/229, 52/212, 65/481
tremor, 65/481
tuberculous meningitis, 33/225-229, 52/212
vertigo, 65/481
vitamin B6 metabolism, 70/430
L-Cycloserine
Alzheimer disease, 46/268
Cycloserine intoxication
catatonia, 65/481
confusion, 65/481
dysarthria, 65/481
epilepsy, 65/481
headache, 65/481
psychosis, 65/481
tremor, 65/481
vertigo, 65/481
Cyclosporin, *see* Ciclosporin
Cyclosporin A, *see* Ciclosporin
Cyclosporine, *see* Ciclosporin
Cyclothymia
manic depressive psychosis, 43/209

Cycrimine
parkinsonism, 6/219
Cyfluthrin
neurotoxin, 64/212
pyrethroid, 64/212
pyrethroid intoxication, 64/212
Cylindric spirals myopathy
hereditary, *see* Hereditary cylindric spirals myopathy
Cylindrical tubule
cramp, 40/39
electron micrography, 40/39
spinocerebellar degeneration, 40/39
Cylindroma
classification, 17/186
dermal, *see* Dermal cylindroma
history, 17/187
incidence, 17/187
multiple dermal, *see* Multiple dermal cylindroma
origin, 17/187
site, 17/187
skull base tumor, 17/186
symptom, 17/187
treatment, 17/187
Cypermethrin, *see* Cipermethrin
Cyphenothrin
neurotoxin, 64/215
pyrethroid, 64/215
pyrethroid intoxication, 64/215
Cyproheptadine
intermediate spinal muscular atrophy, 41/326
migraine, 5/101, 48/190, 192
pure autonomic failure, 22/240
serotonin 1 receptor, 48/88
systemic neoplastic disease, 41/326
Cyproterone acetate intoxication
optic neuropathy, 65/538
Cyst
aneurysmal bone, *see* Aneurysmal bone cyst
arachnoid, *see* Arachnoid cyst
biliary, *see* Biliary cyst
bone, *see* Bone cyst
brain, *see* Brain cyst
brain midline, *see* Brain midline cyst
cavum veli interpositi, 31/120
cavum vergae, 2/767
cerebellar, *see* Cerebellar cyst
cerebral hydatid, *see* Cerebral hydatid cyst
colloid, *see* Colloid cyst
cysticercotic, *see* Cysticercotic cyst
dermoid, *see* Dermoid cyst
desmoid, *see* Desmoid cyst
dura mater, *see* Dura mater cyst
Entamoeba histolytica, 52/309
enterogenous, *see* Neurenteric cyst
ependymal, *see* Ependymal cyst
ependymoma, 18/119
epidermoid, *see* Epidermoid cyst
epidural, *see* Epidural cyst
epidural arachnoid, *see* Epidural arachnoid cyst
epidural ganglion, *see* Epidural ganglion cyst
epithelial neurenteric, *see* Epithelial neurenteric cyst
fourth ventricle, *see* Fourth ventricle cyst
germ cell, *see* Germ cell cyst
hemiballismus, 49/373
hydatid, *see* Hydatid cyst
interhemispheric, *see* Interhemispheric cyst
intramedullary, *see* Intramedullary cyst
intrasellar, *see* Intrasellar cyst
intraspinal neurenteric, *see* Intraspinal neurenteric cyst
jaw, *see* Jaw cyst
lung, *see* Pulmonary cyst
macula lutea, 13/179
mandibular, *see* Mandibular cyst
meningeal, *see* Meningeal cyst
mesenteric, *see* Mesenteric cyst
multilocular pineal, *see* Multilocular pineal cyst
neurenteric, *see* Neurenteric cyst
neuroepithelial, *see* Neuroepithelial cyst
ovarian, *see* Ovarian cyst
pancreatic, *see* Pancreatic cyst
pineal, *see* Pineal cyst
pituitary adenoma, 17/427
pituitary tumor, 17/427
posterior fossa, *see* Posterior fossa cyst
posttraumatic syringomyelia, 12/604, 26/144
pseudo, *see* Pseudocyst
Rathke cleft, *see* Rathke cleft cyst
renal, *see* Renal cyst
sacral arachnoid, *see* Sacral arachnoid cyst
septum pellucidum, 30/322, 326
septum pellucidum tumor, 2/767, 17/537, 18/134
spastic paraplegia, 59/434
spina bifida occulta, 20/76
spinal, *see* Spinal cyst
spinal arachnoid, *see* Spinal arachnoid cyst
spinal cord, *see* Spinal cord cyst
spinal ependymal, *see* Spinal ependymal cyst
spinal epidural, *see* Spinal epidural cyst
spinal intradural, *see* Spinal intradural cyst
spinal perineural, *see* Spinal perineural cyst
spinal subdural, *see* Spinal subdural cyst
spinal subdural hematoma, 20/92
synovial, *see* Synovial cyst

terminal ventricle, *see* Terminal ventricle cyst
thalamic, *see* Thalamic cyst
third ventricle colloid, *see* Third ventricle colloid cyst
third ventricle tumor, *see* Ependymal cyst
thyroid, *see* Thyroid cyst
tumor, *see* Tumor cyst
ventricular, *see* Ventricular cyst
visceral, *see* Visceral cyst
Cystadenoma
epididymal, *see* Epididymal cystadenoma
Cystalgia
migraine, 48/160
Cystathionase deficiency
cystathioninuria, 29/113, 116, 42/541-543
Cystathionine
neuroblastoma, 8/471
Cystathionine synthetase deficiency
cerebrovascular disease, 53/33
homocystinuria, 42/555-557, 46/52
tyrosinemia, 42/635
Cystathionine β synthase deficiency, *see* Homocystinuria
Cystathioninuria
birth incidence, 42/541
classic, 29/116-118
cystathionase deficiency, 29/113, 116, 42/541-543
mental deficiency, 42/541
neuroblastoma, 8/471, 29/118
pyridoxal-5-phosphate, 28/119, 29/118
secondary, 29/118
treatment, 29/117
vitamin $B_6$, 28/117-120, 29/116-119
vitamin $B_6$ metabolism, 28/118
Cystatin C
amyloid, 51/414
Cysteine
cyanide intoxication, 65/32
cyanogenic glycoside intoxication, 65/32
Pelizaeus-Merzbacher disease, 66/561
Cysteine dioxygenase
Hallervorden-Spatz syndrome, 66/713
Cysteine peptiduria
γ-glutamyl transpeptidase deficiency, 29/123
Cystic astrocytoma, 18/13
brain stem tumor, 68/370
Cystic brain degeneration
porencephaly, 30/682, 50/356
Cystic cerebellar astrocytoma
striatopallidodentate calcification, 49/424
Cystic encephalomalacia
cytomegalic inclusion body disease, 56/271
Galen vein, 30/665
multiple, *see* Multiple cystic encephalopathy
Cystic fibrosis
benign intracranial hypertension, 67/111
neuropathology, 63/277
vitamin $B_{12}$ deficiency, 51/339
vitamin E deficiency, 63/277
Cystic kidney
holoprosencephaly, 50/237
Cystic leukoencephalopathy
Alexander disease, 10/97, 100
cerebrovascular disease, 53/29
$GM_2$ gangliosidosis, 10/394
perinatal hypoxia, 53/29
periventricular infarction, 53/29
Cystic medial necrosis, *see* Mönckeberg sclerosis
Cystic myelopathy, *see* Posttraumatic syringomyelia
Cystic teratoma, *see* Intraspinal neurenteric cyst
Cysticercosis, *see* Neurocysticercosis
Cysticercotic cyst
epilepsy, 72/149
Cysticercotic granuloma
epilepsy, 72/168
Cysticercus
inflammatory myopathy, 62/373
polymyositis, 62/373
Cysticercus cellulosae, *see* Taenia solium
Cystine
Hallervorden-Spatz syndrome, 49/402, 66/713
striatopallidodentate calcification, 6/715
Cystine metabolism
hereditary amyloid polyneuropathy type 3, 42/524
Cystinosis
adolescent, 29/177
benign, 29/177
biochemical aspect, 29/179
clinical features, 29/174-177
dissecting aorta aneurysm, 63/45
Fanconi syndrome, 29/173
genetics, 29/178
history, 29/173
intermediate late onset, 29/177
nephropathic, 29/174-177
penicillamine, 29/179
renal insufficiency, 29/179
vitamin C, 29/179
Cystinuria
dibasic amino acid malabsorption, 7/596
genetics, 29/188
history, 29/187
intestinal transport, 29/188
renal transport, 29/188
transport defect, 29/188
treatment, 29/189

Cystitis
interstitial, *see* Interstitial cystitis
Cystoid dystrophy
central, *see* Vitelliruptive macula degeneration
Cystometry
bladder tone, 1/381
human T-lymphotropic virus type I associated myelopathy, 56/537
spinal cord injury, 61/296
spinal cord lesion, 26/424
Cystopathy
diabetes mellitus, 70/141
Cystourethrography
bladder neck obstruction, 26/424
bladder tone, 1/381
pressure volume curve, 1/381
urethral sphincter obstruction, 26/424
Cytarabine
adverse effect, 69/487
antimetabolite, 39/105
antineoplastic agent, 65/528
arachnoiditis, 61/151, 63/357
ataxia, 63/356
benign intracranial hypertension, 67/111
cerebellar encephalopathy, 63/357
chemotherapy, 39/105
CNS leukemia, 69/250
cognitive dysfunction, 67/362
confusional state, 63/357
dysarthria, 63/357
encephalopathy, 63/356, 65/528, 67/362
herpes simplex, 34/154
herpes simplex virus encephalitis, 34/154
human immunodeficiency virus infection, 71/386
intrathecal chemotherapy, 61/151
late cerebellar atrophy, 60/586
leukemia, 39/17, 63/356
lymphocytic leukemia, 63/356
meningeal leukemia, 18/251, 63/342
meningitis, 61/151
myeloid leukemia, 63/356
myelopathy, 63/357, 65/528, 67/362
neurologic toxicity, 39/105
neuropathy, 69/461, 472
neurotoxicity, 63/357, 71/368, 386
neurotoxin, 39/94, 99, 105, 119, 60/586, 61/151, 63/356, 65/528, 532, 67/362
nystagmus, 63/356
progressive multifocal leukoencephalopathy, 34/324, 47/518, 56/497
Cytarabine intoxication
ataxia, 65/533
dysarthria, 65/533
encephalomyelopathy, 65/533
epilepsy, 65/533
locked-in syndrome, 65/533
meningeal reaction, 65/532
myelopathy, 65/532
nystagmus, 65/533
optic neuropathy, 65/533
polyneuropathy, 65/533
spastic paraparesis, 65/533
Cytidine analog
cognitive dysfunction, 67/362
encephalopathy, 67/362
myelopathy, 67/362
neurotoxin, 67/362
Cytidine diphosphate choline
biochemistry, 9/8
Cytidine diphosphate diglyceride
biochemistry, 9/7
Cytidine diphosphate diglyceride inositol phosphatide transferase
biochemistry, 9/16
Cytidine diphosphate ethanolamine
biochemistry, 9/8
Cytoarchitectonic asymmetry
parietal lobe, 45/78
Cytoarchitectonic development
CNS, 50/245, 267
Cytoarchitecture
basal ganglion tier II, 49/21
basal ganglion tiers, 49/21
frontal lobe, 45/24
Cytochrome
Wilson disease, 49/229
Cytochrome aa3, *see* Complex IV
Cytochrome b deficiency
ataxia, 42/111
mitochondrial respiratory chain defect, 42/587
Cytochrome c oxidase, *see* Complex IV
Cytochrome oxidase, *see* Complex IV
Cytochrome oxidase aa1
Huntington chorea, 49/262
Cytochrome oxidase deficiency, *see* Complex IV deficiency
Cytochrome P450
1-methyl-4-phenyl-1,2,3,6-tetrahydropyridine intoxication, 65/400
Parkinson disease, 49/133
tetrahydroisoquinoline, 65/400
Cytochrome PIID6
identified gene, 65/316
Cytochrome reductase deficiency
choreoathetosis, 49/385
Cytochrome related ataxia

areflexia, 42/111
Babinski sign, 42/111
CSF, 42/111
dementia, 42/111
Cytogenetics
ARG syndrome, 50/582
ataxia telangiectasia, 60/348
cri du chat syndrome, 31/571-573
Down syndrome, 31/376-395, 50/520, 522
Edwards syndrome, 31/515
Miller-Dieker syndrome, 50/593
9p partial monosomy, 50/579
r(9) syndrome, 50/581
r(13) syndrome, 50/586
r(14) syndrome, 50/590
Turner syndrome, 31/497
Wolf-Hirschhorn syndrome, 31/550, 552, 561
XYY syndrome, 31/490
Cytokine
collagen vascular disease, 71/9
eosinophil, 63/369
fever, 74/449
pyrogenic, *see* Pyrogenic cytokine
signal transduction, 67/24
sleep, 74/534
systemic lupus erythematosus, 71/41
vasculitis, 71/163
Cytokine mediator
multiple organ failure, 71/540
Cytolipin H, *see* Ceramide polyhexoside
Cytology
basal ganglion, 49/21
basal ganglion tiers, 49/21
brain tumor, 17/43
CSF, *see* CSF cytology
CSF pleocytosis, 9/546
globus pallidus, 49/5
neostriatum, 49/2
oligodendrocyte, 18/84
striatum, 49/2
substantia nigra, 49/7
subthalamic nucleus, 49/10
Cytomegalic inclusion body disease
anemia, 56/267
animal experiment, 56/112
asymptomatic, 56/270
biliary atresia, 56/267
clinical features, 34/214-217, 56/110
computer assisted tomography, 56/268
congenital brain granuloma, 42/736
congenital heart disease, 56/267
CSF, 56/268
cystic encephalomalacia, 56/271
diagnosis, 56/110
epidemiology, 56/264
hepatitis, 56/267
hydranencephaly, 56/268
hypospadias, 56/267
intracranial calcification, 56/110, 268
lymphocytic meningoencephalitis, 56/271
management, 56/110
maternal infection, 56/265
multicystic encephalomalacia, 56/268
necrotizing encephalitis, 56/271
neuropathology, 56/271
petechia, 56/267
progressive deafness, 56/110
prophylaxis, 34/225-227
purpuric rash, 56/267
sensorineural deafness, 56/109
splenomegaly, 56/267
temporal bone pathology, 56/111
thrombocytopenia, 56/267
treatment, 34/225-227
Cytomegaloviruria
neonatal asymptomatic, 34/217
Cytomegalovirus
acquired immune deficiency syndrome, 56/513, 516
anophthalmia, 56/268
biology, 56/263
cerebellar hypoplasia, 56/271
Cowdry A inclusion body, 56/273
deafness, 56/107, 266, 270
demyelination, 56/273
detection, 71/264
enhancer, 56/37
facial paralysis, 56/270
herpes virus, 56/8
hydranencephaly, 56/268
immunocompromise, 56/477
immunocompromised host, 56/476
intracranial calcification, 56/268
meningitis, 56/477
meningoencephalitis, 56/272
mental deficiency, 56/269
microcephaly, 56/268
microphthalmia, 56/268
multicystic encephalomalacia, 56/268
myelomalacia, 63/55
nervous system, 71/261
neuropathy, 71/358
optic atrophy, 56/268
polymicrogyria, 56/271
porencephaly, 56/271
proteinocytologic dissociation, 56/270

renal transplantation, 63/531
Reye syndrome, 56/150
structure, 56/263
vertigo, 56/107
virology, 56/263
Cytomegalovirus antibody
multiple sclerosis, 47/274
Cytomegalovirus encephalitis, 56/477
acquired immune deficiency syndrome, 71/262
heart transplantation, 63/180
Cytomegalovirus inclusion disease, *see* Cytomegalic inclusion body disease
Cytomegalovirus infection, 34/209-227, 56/263-275
acquired, 34/219-222, 56/270
acquired immune deficiency syndrome, 56/265
animal model, 34/211
aqueduct stenosis, 50/308
chorioretinitis, 50/277, 56/266
clinical features, 31/212, 215, 34/214-222, 56/266
cochleitis, 56/112
compromised host, 34/220-222
congenital, *see* Cytomegalic inclusion body disease
convulsion, 50/277
cytomegalovirus mononucleosis, 34/218
deafness, 50/277, 56/112
diagnosis, 56/273
epidemiology, 31/215, 34/212-214, 56/264
ganciclovir, 56/274
granulomatous CNS vasculitis, 55/387
Guillain-Barré syndrome, 51/240
heart transplantation, 63/180
hepatitis, 34/219
history, 34/209
hydrocephalus, 31/313, 315, 56/266
immunologic characteristics, 34/211
immunology, 56/266
incidence, 56/265
infectious striatopallidodentate calcification, 49/417
intestinal pseudo-obstruction, 51/493
laboratory diagnosis, 56/274
laboratory finding, 31/215
latency, 34/8
mental deficiency, 31/212, 215, 46/19, 77, 50/277
microcephaly, 30/508, 510, 31/212, 215, 50/277, 56/266
microgyria, 30/483, 50/277
multifocal neuritis, 71/264
multiple sclerosis, 47/330
natural history, 34/219-222
neonatal asymptomatic cytomegalovirus, 34/217
neuropathology, 56/271
organ transplantation, 56/263, 265
pathology, 34/222-224
polymicrogyria, 50/257
prophylaxis, 34/225-227
seizure, 31/212, 215, 56/268
sensorineural deafness, 56/266, 270
seroconversion, 56/264
socioeconomic factor, 31/215
spastic paraplegia, 50/277
teratogenic agent, 34/374-377
transfusion acquired, 56/265
transmission, 56/264
treatment, 31/216, 34/224-227, 227, 56/274
ultrastructure, 56/271
virologic characteristics, 34/209-211
Cytomegalovirus mononucleosis
cytomegalovirus infection, 34/218
Cytomegalovirus polyradiculopathy
human immunodeficiency virus neuropathy, 71/358
nuclear magnetic resonance, 71/359
Cytoplasmic inclusion body
adrenoleukodystrophy, 47/595
amiodarone, 65/484
Chédiak-Higashi syndrome, 42/536
chloroquine, 65/484
chronic mercury intoxication, 64/372
CNS spongy degeneration, 42/507
droplet metastasis, 20/285
Lafora body, 1/282
lead intoxication, 64/437
membranous, *see* Membranous cytoplasmic inclusion body
mitochondrial myopathy, 43/120
muscle fiber, 40/26, 107, 245
myelinoclastic diffuse sclerosis, 9/481
perhexiline maleate, 65/484
Salla disease, 43/306
spondyloepiphyseal dysplasia, 43/480
spondylometaphyseal dysplasia, 43/483
Cytoplasmic inclusion body myopathy
age, 62/344
congenital myopathy, 62/332
desmin intermediate filament, 62/344
emetine intoxication, 62/345
experimental injury, 62/345
facial weakness, 62/334
ipecac intoxication, 62/345
organophosphate intoxication, 62/345
Cytoside
Gaucher disease, 10/308, 544
globoid cell leukodystrophy, 10/304
Cytoside lipidosis

convulsive seizure, 10/545
foam cell, 10/545
jaundice, 10/545
mental deficiency, 10/545
spastic tetraplegia, 10/545
splenomegaly, 10/545
Cytosine arabinoside, *see* Cytarabine
Cytoskeleton
axonal, *see* Axonal cytoskeleton
muscle fiber, 62/19
muscle fiber type, 62/3
Cytosol
muscle fiber type, 62/4
Cytosol tyrosine aminotransferase
Richner-Hanhart syndrome, 42/581
tyrosinemia, 42/635
Cytosome
GM2 gangliosidosis, 10/365
lysosome, 40/37, 113
metachromatic leukodystrophy, 10/374
muscle fiber, 40/113
Cytostatic agent
delirium, 46/539
polyneuropathy, 69/459
Cytostatic treatment
astrocytoma, 18/39
brain tumor, 18/39
glioblastoma multiforme, 18/65
leukemia, 18/251
Cytotoxic agent
cisplatin polyneuropathy, 51/299
delirium, 46/539
epilepsy, 72/232
multiple myeloma, 18/256
multiple sclerosis, 9/146, 47/194-197
osteitis deformans (Paget), 43/452
toxic myopathy, 62/597
Cytotoxic brain edema
animal experiment, 11/155
characteristics, 67/76
definition, 47/531
intracellular water, 67/73
metabolic encephalopathy, 57/213
neuronal damage, 57/261
osmotic agent, 63/213
pathogenesis, 67/75
pathomechanism, 53/137
toxic encephalopathy, 57/213
triethyltin intoxication, 11/155
Cytotoxic T-lymphocyte
glioma antigen sensitized, *see* Glioma antigen sensitized cytotoxic T-lymphocyte
herpes simplex virus, 56/210
major histocompatibility complex, 56/68
multiple sclerosis, 47/189
response, 56/69
viral infection, 56/65, 68
Cytotoxicity
antibody dependent, *see* Antibody dependent cytotoxicity
Czechoslovakia
neurology, 1/11

D1 dopamine receptor, *see* Dopamine receptor type D1
Da Costa syndrome, *see* Hyperventilation syndrome
Dacarbazine
chemotherapy, 39/116
neurologic toxicity, 39/116
Dacrystic epilepsy
pathologic laughing and crying, 55/142
Dactinomycin
DNA metabolism, 64/17
neurotoxin, 47/568, 64/10, 17, 67/363
spongiform leukoencephalopathy, 47/568
toxic neuropathy, 64/10
Dactinomycin intoxication
action site, 64/10
Dalen-Fuchs nodule
Vogt-Koyanagi-Harada syndrome, 56/613
Dalrymple sign
hyperthyroidism, 2/318
Von Gräfe sign, 2/318
Damsel fish intoxication
CNS, 37/86
hallucination, 37/86
Danazol
benign intracranial hypertension, 67/111
toxic myopathy, 62/597
Danbolt disease
zinc deficiency, 36/336
Dance
St. Modestus, *see* Sydenham chorea
St. Vitus, *see* Sydenham chorea
Dancing eyes, *see* Opsoclonus
Dandy operation
history, 5/305
trigeminal neuralgia, 5/306, 396
Dandy syndrome, *see* Infraclinoid syndrome
Dandy vein, *see* Petrosal vein
Dandy-Walker like malformation
trisomy 9, 50/564
Dandy-Walker syndrome
age at diagnosis, 30/635
age at onset, 50/328
angiography, 32/59

anophthalmia, 50/333
aqueduct stenosis, 42/21-23, 50/309
arachnoid cyst, 32/59
Arnold-Chiari malformation type I, 30/632, 32/59
Arnold-Chiari malformation type IV, 50/403
ataxia, 50/328
atlanto-occipital synostosis, 50/393
autosomal recessive, 50/63
birth incidence, 42/23
cavum septi pellucidi, 42/23
cerebellar abnormality, 50/323
cerebellar agenesis, 50/175
cerebellar dysgenesis, 50/328
cerebellar hypoplasia, 42/18
cerebellar vermis agenesis, 30/624
cerebellar vermis aplasia, 30/624, 32/59, 50/168
choroid plexus, 42/21
choroid plexus embryogenesis, 50/324
clinical features, 30/623, 635
Coffin-Siris syndrome, 50/332
computer assisted tomography, 50/329, 60/672
Conradi-Hünermann syndrome, 46/50
corpus callosum agenesis, 2/764, 42/23, 50/156, 163, 168, 333
craniofacial dysostosis (Crouzon), 42/23
CSF composition, 50/325
CSF flow, 50/324
CSF formation, 50/325
CSF pressure, 50/326
CSF reabsorption, 50/326
CSF retention, 50/328
De Lange syndrome, 42/23
definition, 50/323
diagnosis, 30/642
diastematomyelia, 42/22
differential diagnosis, 32/59
dysraphia, 14/111
embryology, 30/628-630
encephalomeningocele, 42/26
epidemiology, 30/634, 50/328
etiology, 30/630
fetal alcohol syndrome, 50/333
fourth ventricle cyst, 30/624, 50/328
Galen vein, 30/638
genetics, 30/158, 630, 50/332
HARD syndrome, 50/593
HARDE syndrome, 50/593
history, 30/623, 628, 50/323
hydranencephaly, 50/344
hydrocephalus, 30/539, 623-646, 31/110-112, 32/59, 42/21, 36, 50/286, 324
hydromyelia, 42/23, 50/425
incidence, 50/62
iniencephaly, 42/22
intracranial pressure, 42/22
Klippel-Feil syndrome, 42/23
Luschka foramina atresia, 30/623, 627, 632, 50/323
Magendie foramina atresia, 30/623, 627, 632, 50/323
malformation, 30/628
mental deficiency, 30/636, 46/8, 49
micrencephaly, 42/23
Mohr syndrome, 50/332
murine, 50/324
myelomeningocele, 42/23
neuronal heterotopia, 42/23
neuroschisis, 30/632
nystagmus, 50/328
obstructive hydrocephalus, 46/325, 50/323
occipital cephalocele, 50/101
occipital encephalocele, 42/26
optic atrophy, 50/328
papilledema, 50/328
parietal cephalocele, 50/108
parietal encephalocele, 50/108
pathogenesis, 30/631-634, 50/331
pathology, 30/623-628
polycystic kidney, 50/332
polyploidy, 50/555
porencephaly, 42/23
posterior fossa cyst, 32/59, 42/21
posterior fossa enlargement, 30/627
posterior fossa tumor, 18/407
prognosis, 30/644, 50/330
radiologic appearance, 30/637-641
sex ratio, 50/63
sign, 50/328
skeletal malformation, 32/59, 50/332
spina bifida, 42/23, 50/487
symptom, 50/328
syringomyelia, 30/632, 32/59, 42/23
time of origin, 30/633
Torkildsen procedure, 30/643
treatment, 30/643, 50/330
trisomy 9, 43/514
ultrasonography, 50/329
warfarin, 50/333
Dandy, W.E., 1/5
Dantrolene
familial spastic paraplegia, 59/313
malignant hyperthermia, 38/554, 41/270, 298, 62/567
multiple sclerosis, 47/157
muscle relaxant, 41/297
spastic paraplegia, 61/368

spinal cord injury, 61/368
spinal spasticity, 61/368
transverse spinal cord lesion, 61/368
Danzig, Germany
neurology, 1/9
Dapsone
chronic axonal neuropathy, 51/531
human immunodeficiency virus infection, 71/384
iatrogenic neurological disease, 65/485
leprosy, 51/217
leprous neuritis, 51/232
neuropathy, 71/384
neurotoxicity, 71/368
neurotoxin, 51/68, 531, 65/485
toxic neuropathy, 65/485
toxic polyneuropathy, 51/68
vasculitis, 71/162
Dapsone intoxication
axonal degeneration, 65/486
polyneuropathy, 65/485
Darier disease
autosomal dominant transmission, 14/785
dwarfism, 43/380
Darier-Grönblad-Strandberg syndrome, *see* Grönblad-Strandberg syndrome
Darier lentiginosis profusa, *see* Lentiginosis profusa (Darier)
Dark adaptation, *see* Visual adaptation
Darkschewitsch nucleus
basilar artery, 2/275
stimulation, 2/343
vascular supply, 2/275
vertical gaze palsy, 2/338
Darvon, *see* Dextropropoxyphene
Dasyatidae intoxication
elasmobranch, 37/70
Daubent angle
basilar impression, 32/19
measurement, 32/19
Dauer Babinski sign, *see* Pseudo-Babinski sign
Davidenkow syndrome, *see* Scapuloperoneal spinal muscular atrophy
Davies disease
idiopathic hypereosinophilic syndrome, 38/212
Dawidenkow syndrome, *see* Scapuloperoneal spinal muscular atrophy
Dawson artery, *see* Perichiasmatic artery
Dawson encephalitis, *see* Subacute sclerosing panencephalitis
Dawson finger
multiple sclerosis, 9/237, 248, 251, 253, 258-260
plaque, 9/237, 248, 251, 253, 258-260
Day blindness, *see* Hemeralopia
Daytime hypersomnia
*see also* Narcolepsy
cardiac dysrhythmia, 45/135
CSF examination, 45/136
CSF 3-methoxy-4-hydroxyphenylethylene glycol, 45/136
dihydrophenylacetic acid, 45/136
EEG, 45/131
homovanillic acid, 45/136
5-hydroxyindoleacetic acid, 45/136
idiopathic CNS hypersomnia, 45/135
intermittent, *see* Intermittent daytime hypersomnia
menstrual associated syndrome, 45/137
multiple sleep latency test, 45/131, 135
obesity, 45/134
polysomnography, 45/135
REM sleep, 45/131, 134
sleep deprivation, 45/135
Dazzling test
central scotoma, 9/174
multiple sclerosis, 9/174
optic neuropathy, 9/174
DDAVP, *see* Desmopressin acetate
DDD, *see* 1,1-Dichloro-2,2-bis(4-chlorophenyl)ethane
DDD technique
protein, 9/24
DDT, *see* Chlorphenotane
DDT intoxication
acute, 36/398-402
chronic, 36/402
clinical features, 36/402
diagnosis, 36/404
laboratory finding, 36/302
late cortical cerebellar atrophy, 21/497
neuropathy, 7/522
pathology, 36/403
treatment, 36/405
DDVP intoxication, *see* Dichlorvos intoxication
De Castro, F., 1/11
De Kleijn, A., 1/10
De Lange syndrome
birth incidence, 43/246
brain atrophy, 43/246
childhood psychosis, 46/194
CNS, 31/281
Dandy-Walker syndrome, 42/23
dermatoglyphics, 31/281, 43/246
diagnostic criteria, 31/279
digital abnormality, 43/246
epidemiology, 31/279
etiology, 31/282
eye abnormality, 31/280

face, 31/280
growth retardation, 43/246
heart disease, 43/246
hirsutism, 43/246
hypertelorism, 30/248
hypertrichosis, 14/784
intrauterine growth retardation, 46/91
mental deficiency, 31/279-281, 43/246, 46/91
metabolic abnormality, 31/282
microcephaly, 30/508, 31/280, 50/276
muscle hypertrophy, 40/324
nystagmus, 43/246
oligodactyly, 50/276
optic atrophy, 43/246
phocomelia, 50/276
radiologic appearance, 31/281
recurrence risk, 43/247
seizure, 43/246, 50/276
skeletal abnormality, 31/280
skeletal malformation, 30/248
strabismus, 43/246
swayback nose, 43/246
syndactyly, 50/276
synophrys, 43/246, 50/276
De Lisi, L., 1/35
De Martell, T., 1/17
De Sanctis-Cacchione syndrome, *see* Xeroderma pigmentosum
Deaf mutism
albinism, 14/781
chorioretinal degeneration, 13/41
Klein-Waardenburg syndrome, 14/115
Laurence-Moon-Bardet-Biedl syndrome, 13/41, 390
Lindenov-Hallgren syndrome, 13/453
optic atrophy, 13/41
Roussy-Lévy syndrome, 21/175
secondary pigmentary retinal degeneration, 13/150, 239-242, 453
Usher syndrome, 13/241
Deafferentation pain
brachial plexus, 51/153
cancer, 69/32, 48
Deafferentation state
visual hallucination, 46/567
Deafness
*see also* Hearing loss *and* Sensorineural deafness
acetylsalicylic acid intoxication, 65/460
adenovirus, 56/107
adrenoleukodystrophy, 51/384, 60/169
alexia, 4/135
Alström-Hallgren syndrome, 13/242, 460
aminoglycoside, 65/482
aminoglycoside intoxication, 65/482
amyotrophic lateral sclerosis, 22/138
amyotrophy, 22/523
angiokeratosis naeviformis, 14/775
anterior inferior cerebellar artery, 55/132
anterior inferior cerebellar artery syndrome, 53/396, 55/90
aphasia, 45/305
apraxia, 45/430
Arnold-Chiari malformation type II, 50/410
ataxia, 42/112, 60/657
bacterial meningitis, 52/14
Bardet-Biedl syndrome, 13/390
brain aneurysm, 55/134
brain arteriovenous malformation, 55/134
Brown-Vialetto-Van Laere syndrome, 22/108
bulbar paralysis, 22/138, 523
bulbopontine paralysis, 42/95
CAMAK syndrome, 60/658
CAMFAK syndrome, 60/658
Capute-Rimoin-Konigsmark syndrome, 14/110
central, *see* Central deafness
cerebellar ataxia, 60/657
cerebellopontine angle syndrome, 2/93
cerebellopontine angle tumor, 2/377
chloroquine intoxication, 65/483
chorioretinal degeneration, 13/40
cinchonism, 65/452, 485
cisplatin polyneuropathy, 51/300
Cockayne-Neill-Dingwall disease, 10/182, 13/433, 436, 60/658
Cogan syndrome type II, 51/454
congenital, *see* Congenital deafness
congenital autonomic dysfunction, 60/18
congenital central, *see* Congenital central deafness
congenital heart disease, 63/11
congenital retinal blindness, 22/530, 60/724
congenital rubella, 50/278
cortical, *see* Cortical deafness
Coxsackie virus, 56/107
craniosynostosis, 50/125
cretinism, 27/256, 259
cytomegalovirus, 56/107, 266, 270
cytomegalovirus infection, 50/277, 56/112
deferoxamine, 63/534
dialysis, 63/523, 530
dipeptidyl carboxypeptidase I inhibitor intoxication, 65/446
diuretic agent intoxication, 65/440
dominant progressive sensorineural, *see* Dominant progressive sensorineural deafness
double athetosis, 49/384
dyssynergia cerebellaris myoclonica, 60/599

Ellis-Van Creveld syndrome, 14/788
enalapril intoxication, 65/446
encephalomyocarditis virus, 56/107
endemic cretinism, 46/84
epilepsy, 42/688
Epstein-Barr virus, 56/107, 252
Epstein-Barr virus infection, 56/252
erythromycin, 65/472
ethionamide, 65/481
ethionamide intoxication, 65/481
familial amyotrophic lateral sclerosis, 22/138
familial bulbopontine paralysis, *see* Brown-Vialetto-Van Laere syndrome
familial ectodermal dysplasia, 60/657
familial spastic paraplegia, 22/471
fat embolism, 55/132
fibrous dysplasia, 14/170
Flynn-Aird syndrome, 14/112, 60/657
foramina parietalia permagna, 50/142
Friedreich ataxia, 21/336, 347, 22/517, 51/388
Friedreich ataxia like condition, 60/658
furosemide, 65/440
Gasperini syndrome, 2/317
Gellé syndrome, 2/241
gene frequency, 43/43
gentamicin, 65/483
Gerstmann-Sträussler-Scheinker disease, 56/553
globoid cell leukodystrophy, 51/373
glycosaminoglycanosis, 60/176
glycosaminoglycanosis type I, 10/441
glycosaminoglycanosis type IH, 60/177
glycosaminoglycanosis type II, 10/436, 441, 42/454, 60/177
glycosaminoglycanosis type IV, 60/177
gram-negative bacillary meningitis, 52/123
Hallgren syndrome, 13/242, 338
Hemophilus influenzae meningitis, 52/63, 66
hepatitis virus, 56/107
hereditary, *see* Hereditary deafness
hereditary amyloid polyneuropathy type 1, 51/424
hereditary cerebellar ataxia, 13/298
hereditary interstitial nephritis, 27/335
hereditary motor and sensory neuropathy type I, 21/284, 308
hereditary motor and sensory neuropathy type II, 21/284, 308
hereditary perforating foot ulcer, 8/201
hereditary sensory and autonomic neuropathy type I, 21/73, 77
hereditary sensory and autonomic neuropathy type II, 60/11
heredopathia ophthalmo-otoencephalica, 60/657
herpes simplex virus, 56/107
Herrmann disease, 22/510
Huntington chorea, 49/291
Hurler-Hunter disease, 10/436
hyperparathyroidism, 27/293
hyperprolinemia type I, 29/133
hypoesthesia, 60/661
hypothyroid polyneuropathy, 51/515
hypothyroidism, 27/258
infantile neuroaxonal dystrophy, 49/402
infantile Refsum disease, 60/657
infectious mononucleosis, 56/252
influenza virus, 56/107
internal auditory artery occlusion, 55/131
Jaeken syndrome, 60/658
Jeune-Tommasi disease, 60/657
juvenile metachromatic leukodystrophy, 60/657
Kearns-Sayre-Daroff-Shy syndrome, 62/313
kernicterus, 6/507, 513, 49/384
Kjellin syndrome, 22/471
Klein-Waardenburg syndrome, 14/115
Klippel-Trénaunay syndrome, 14/523
labyrinthine hemorrhage, 55/133
Lassa fever, 56/370
Laurence-Moon syndrome, 13/302
LEOPARD syndrome, 14/117, 600
levodopa, 43/3
Lindenov-Hallgren syndrome, 13/452
lower motoneuron disease, 60/657
lymphocytic choriomeningitis, 56/361
lymphocytic choriomeningitis virus, 56/107
lymphomatoid granulomatosis, 51/451
macroglobulinemia, 55/132, 485
malnutrition, 28/20
measles, 55/131, 56/115
mefloquine intoxication, 65/485
Meniere disease, 2/377
meningococcal meningitis, 52/24, 26, 28
mental deficiency, 60/657, 664
mercury polyneuropathy, 51/272
microcephaly, 30/510, 50/277
migraine, 48/27
Minamata disease, 7/642, 36/74, 64/414, 419
mitochondrial disease, 66/394
Muckle-Wells syndrome, 14/113, 22/489
mucolipidosis type II, 51/380
mucormycosis, 52/471
mumps, 55/131, 56/113, 115, 129, 430
mumps virus, 56/107
musical, *see* Musical deafness
myoclonus, 60/657
myxedema neuropathy, 8/57
neonatal adrenoleukodystrophy, 60/657
neurobrucellosis, 33/317, 52/589

neuronal intranuclear hyaline inclusion disease, 60/657
neuropathy, 60/660
neurosensory hearing loss, 60/657
Norrie disease, 22/508, 31/292, 43/441
occipital lobe tumor, 17/337
optic atrophy, 13/40, 22/138, 60/660
opticocochleodentate degeneration, 21/538, 60/657, 752
organomercury intoxication, 36/74
osteitis deformans (Paget), 70/13
parainfluenza virus, 56/107
Patau syndrome, 14/120
perhexiline maleate polyneuropathy, 51/302
pertussis, 52/234
pneumococcal meningitis, 33/47, 52/43, 51
poliomyelitis, 56/107
polyneuropathy, 60/661
progressive bulbar palsy, 22/108, 114, 138, 516
progressive external ophthalmoplegia, 22/193, 60/658, 62/313
progressive sensory demyelinating neuropathy, 21/82
psychic, *see* Psychic deafness
pure word, *see* Pure word deafness
Pyle disease, 22/493, 46/88, 60/774
Refsum disease, 8/22, 10/345, 13/242, 314, 479, 21/196, 27/523, 60/228
renal insufficiency, 27/335
Richards-Rundle syndrome, 22/516, 43/264, 60/657
Rinne test, 2/74
Ron-Pearce syndrome, 60/658
Rosenberg-Chutorian syndrome, 60/658
rubella syndrome, 31/212
rubella virus, 56/12, 107, 408
Rud syndrome, 51/398, 60/721
Schwabach test, 2/73
secondary pigmentary retinal degeneration, 60/722, 724
sensorineural, *see* Sensorineural deafness
sensory neuropathy, 8/181
sex linked early onset neural, *see* Sex linked early onset neural deafness
spinocerebellar ataxia, 42/181
St. Louis encephalitis virus, 56/107
Stargardt disease, 13/133
streptococcal meningitis, 52/81
streptomycin, 51/295
striatopallidodentate calcification, 49/423
subarachnoid hemorrhage, 55/134
syphilis, 52/275
systemic lupus erythematosus, 55/377
tabes dorsalis, 52/280
Telfer syndrome, 60/658
thromboangiitis obliterans, 55/132
tick-borne encephalitis virus, 56/107
Townes-Brocks syndrome, 42/379
transient ischemic attack, 55/357
trichinosis, 52/566
tropical ataxic neuropathy, 7/640, 65/29
tropical spastic paraplegia, 56/536
tuberculosis, 51/186
tuberculous meningitis, 52/216
upper motoneuron disease, 60/657
uremic polyneuropathy, 63/512
Usher syndrome, 13/441, 60/657
varicella zoster virus, 56/107
variola virus, 56/107
viral infection, 56/105
Vogt-Koyanagi-Harada syndrome, 56/613, 622
Waldenström macroglobulinemia, 55/486, 63/398
Waldenström macroglobulinemia polyneuropathy, 51/432
Walsh syndrome, 13/326
Weber test, 2/74
western equine encephalitis virus, 56/107
word, *see* Word deafness
X-linked dystonia musculorum deformans, 49/521
xeroderma pigmentosum, 51/399, 60/336, 657
XXXX syndrome, 50/554
yellow fever virus, 56/107
Dealkylflurazepam
benzodiazepine intoxication, 65/335
Deanimation encephalopathy, *see* Brain death
Deanol
Alzheimer disease, 46/267
hemiballismus, 49/376
Death
*see also* Brain death *and* Law
abnormal case, 24/808
acceptable death criterion, 24/810, 819, 824
age, *see* Age at death
angiography, 24/805, 819
apistus intoxication, 37/76
apparent, *see* Apparent death
astrocyte, *see* Astrocyte death
Atergatis floridus intoxication, 37/58
basilar artery stenosis, 53/386
birgus latro intoxication, 37/58
bivalves intoxication, 37/64
brain, *see* Brain death
brain death, 24/821
brain stem, *see* Brain stem death
brain tumor, 67/320
California Code, 24/823

Capron-Kass proposal, 24/823
carbon disulfide intoxication, 64/24
cell, *see* Cell death
cerebral, *see* Brain death
chelonia intoxication, 37/91
childhood myoglobinuria, 62/561
chironex intoxication, 37/46, 48
chiropsalmus intoxication, 37/48
congestin intoxication, 37/37
cortical death, 24/819
death declaration, 24/799, 804, 809, 812, 821
definition, 24/788, 798, 809, 814
diagnosis, 24/799, 804, 808, 817, 819
eriphia sebana intoxication, 37/58
ethics, 24/800, 820
fluroxene intoxication, 37/412
German diagnosis, 24/804
Glasgow Outcome Scale, 57/132
gymnapistes intoxication, 37/76
head injury, 57/7, 433
head injury epidemiology, 57/7
head injury outcome, 57/368
heart, *see* Heart death
hydrozoa intoxication, 37/40
insect bite, 6/676
Jamaican vomiting sickness, 37/512
Kansas Statute, 24/822
law, 24/791
lead encephalopathy, 64/436
legal aspect, 24/787-828
legal model, 24/811
legal standard, 24/810-813
logic, 24/803
moment of death, 24/816
neophocaena phocaenoides intoxication, 37/95
nerve cell, *see* Neuron death
neuron, *see* Neuron death
nomenclature, 24/788, 807, 821
normal case, 24/808
nothesthes intoxication, 37/76
organ transplantation, 24/813
organophosphate intoxication, 37/554, 556, 64/167
organophosphorus compound intoxication, 37/541
palytoxin intoxication, 37/55
paralytic shellfish poisoning, 37/33, 65/151
pellagra, 28/87, 46/336, 399
phenytoin intoxication, 65/501
posttraumatic brain abscess, 57/323
pterois volitans intoxication, 37/75
pyrethroid intoxication, 64/214
quinidine intoxication, 37/449
safety requirement, 24/819
scorpaena intoxication, 37/76
scorpaenodes intoxication, 37/76
scorpaenopsis intoxication, 37/76
sebastapistes intoxication, 37/76
stingray intoxication, 37/70
striatal necrosis, 6/676
sudden, *see* Sudden death
sudden infant, *see* Sudden infant death
synanceja intoxication, 37/69
temporal arteritis, 48/326
traumatic carotid cavernous sinus fistula, 57/347
tridacna maxima intoxication, 37/64
water intoxication, 63/547
zozymus aeneus intoxication, 37/56
Débilité motrice
developmental dyspraxia, 4/454
Debrancher deficiency
adult type, 27/230, 232
amylo-1,6-glucosidase, 27/222, 230, 62/483
biochemical aspect, 27/230, 41/182
biochemistry, 62/483
cardiac involvement, 62/482
clinical course, 62/482
clinical features, 27/230, 41/181
creatine kinase, 62/482
distal muscle weakness, 43/179
ECG, 62/482
echoCG, 62/483
EMG, 62/482
epilepsy, 62/482
genetics, 41/183, 427
glucantransferase, 43/179
glycogen storage disease, 27/222, 40/256, 41/181, 62/482
hepatomegaly, 62/482
heterozygote detection, 43/179
histopathology, 62/483
hypoglycemia, 27/72, 43/179
infantile type, 27/230
inheritance, 62/483
ischemic exercise test, 41/182
liver disease, 62/482
metabolic myopathy, 41/181, 62/482
molecular genetics, 62/483
muscle biopsy, 62/483
muscle fiber feature, 62/17
muscle tissue culture, 62/99
muscular atrophy, 62/482
myophosphorylase, 43/179
oligo-1,4→1,4-glucantransferase, 62/483
pathology, 27/230
phosphorylase limit dextrin, 62/483
prenatal diagnosis, 43/179

rimmed muscle fiber vacuole, 41/182
rimmed vacuole distal myopathy, 62/483
type IIIA, 62/483
type IIIB, 62/483
type IIIC, 62/483
Debrisoquine
Parkinson disease, 49/133
parkinsonism, 49/133
perhexiline maleate polyneuropathy, 51/302
toxic neuropathy, 51/302
Debrisoquine hydroxylation
antidepressant, 65/316
desipramine, 65/316
nortriptyline, 65/316
Decaborane intoxication
dizziness, 64/360
drowsiness, 64/360
epilepsy, 64/360
headache, 64/360
metal intoxication, 64/353
opisthotonos, 64/360
prevention, 64/360
treatment, 64/360
vertigo, 64/360
Decamethonium
myasthenia gravis, 41/113
Decerebrate posture
Reye syndrome, 27/363, 56/150
Decerebrate rigidity
alpha motor system, 1/262
brain embolism, 53/211
brain infarction, 53/211
carbon monoxide intoxication, 49/474
cerebellar influence, 2/408
cerebrovascular disease, 53/211
clinical features, 1/267
dystonia, 6/152
false diagnosis, 6/149
gamma motor system, 1/66
hydranencephaly, 30/675, 50/349
intracranial hemorrhage, 53/211
lesion site, 53/211
mesencephalic lesion, 2/281-283
pathophysiology, 2/281
pontine infarction, 2/242, 12/15
pontine lesion, 2/242
porencephaly, 30/685, 50/359
primary pontine hemorrhage, 12/39
subacute sclerosing panencephalitis, 56/419
transient ischemic attack, 53/211
Decerebration
*see also* Gamma rigidity
anterior cerebellar lobe, 2/393
brain death, 24/702, 706, 714
brain injury, 24/702
convulsion, 16/131, 134
intracranial hypertension, 16/131, 134
lissencephaly, 30/486, 42/40
motor function, 24/714, 716
opisthotonos, 6/149
pontine infarction, 53/389
Reye syndrome, 56/238, 65/116
sign, 24/702, 706, 711
sulfite oxidase deficiency, 29/121
trichopoliodystrophy, 14/784
tuberculous meningitis, 52/202
Decerebration spasm, *see* Decerebrate rigidity
Decerebration syndrome
coma, 3/65, 75
Dechaume syndrome
temporomandibular joint dysfunction, 5/345
Dechronization effect
action potential, 7/139
Decker-Breig angle
basilar impression, 32/25
measurement, 32/25
technique, 32/25
Decomposition of movement
definition, 2/394
Decompression
cervical vertebral column injury, 61/65
dysbarism, 12/669, 23/619
foramen magnum, *see* Foramen magnum decompression
gasserian ganglion, 5/397
nerve, *see* Nerve decompression
surgical, *see* Surgical decompression
thoracic spinal cord injury, 61/83
trigeminal neuralgia, 48/449
Decompression illness
*see also* Air embolism, Barotrauma *and* Dysbarism
air embolism, 63/419
altitude illness, *see* Altitude decompression
brain embolism, 53/160
brain infarction, 53/160
brain injury, 23/612
brain ischemia, 53/160
classification, 63/419
clinical features, 23/615, 617, 61/219
cough, 63/419
dioxide breathing, 61/225
diving risk, 63/418
dysbarism, 23/612
dyspnea, 63/419
epilepsy, 63/420

free interval, 61/220
headache, 61/221
hearing loss, 61/219
helium/dioxide breathing, 61/225
hemoptysis, 63/419
lidocaine, 61/225
lung rupture, 63/419
management, 23/613
mediastinal emphysema, 63/419
migraine, 61/222
myelomalacia, 53/160
myoclonus, 49/618
pathology, 23/623
pneumothorax, 63/419
prevention, 23/615
spinal apoplexy, 12/532
symptom, 23/613, 61/218
treatment, 23/618
type I, 61/219
type II, 61/219
venous infarction, 53/160
vertigo, 63/419
Decompression injury, *see* Decompression illness
Decompression myelopathy
*see also* Barotrauma
age, 61/224
air embolism, 61/221
arterial air, 61/224
brain stem lesion, 61/216
cinephotomicrography, 61/223
clinical features, 61/219
complete factor activation, 61/224
computer assisted tomography, 61/225
differential diagnosis, 61/216
dioxide breathing, 61/225
epidemiology, 61/221
epidural vein air embolism, 61/224
epidural vein thrombosis, 61/223
epidural vertebral venous system, 61/223
Foix-Alajouanine disease, 61/223
foramen ovale patency, 61/224
free interval, 61/220
Haldane ratio, 61/217
helium/dioxide breathing, 61/225
helium-oxygen mixture, 61/217
hemiplegia, 61/217
Hill method, 61/217
history, 61/216
impredictability, 61/224
inert gas physiology, 61/218
lidocaine, 61/225
mental change, 61/221
multiple microemboli, 61/222
nuclear magnetic resonance, 61/221
obesity, 61/224
paraplegia, 61/216
pathology, 61/222
pathophysiology, 61/222
predilection site, 61/221
predisposing factor, 61/224
pulmonary artery pressure, 61/223
pulmonary venous air embolism, 61/224
recompression, 61/217, 224
spastic paraplegia, 61/220
spinal cord congestion, 61/223
spinal cord hemorrhage, 61/222
surface decompression, 61/217
symptom, 61/218
tetraplegia, 61/216
treatment, 61/224
type I, 61/219
type II, 61/219
Decompression sickness, *see* Decompression illness
Decorticate posture
striatal necrosis, 49/499
subacute sclerosing panencephalitis, 56/419
uremic encephalopathy, 63/505
Decortication
*see also* Coma
brain death, 24/702, 706, 714
brain embolism, 53/211
brain infarction, 53/211
brain injury, 24/702
cat scratch disease, 52/130
cerebrovascular disease, 53/211
intracranial hemorrhage, 53/211
motor function, 24/714, 716
Reye syndrome, 65/116
rigidity, 1/268
sign, 24/702, 706, 711
transient ischemic attack, 53/211
tuberculous meningitis, 52/203
uremic encephalopathy, 63/505
Decortication syndrome
coma, 3/65, 75
Decremental EMG
acetylcholinesterase deficiency, 41/130
botulism, 41/129
Eaton-Lambert myasthenic syndrome, 41/130
experimental autoimmune myasthenia gravis, 41/117
myasthenia gravis, 41/128
Decubitus ulcer, *see* Pressure sore
Decussation
Meynert, *see* Meynert decussation
pyramidal, *see* Pyramidal decussation

visual fiber, 2/550
Deep agraphia
dyslexia, 45/462
semantic paragraphia, 45/463
Deep dyslexia
agraphia, 45/462
alexia, 45/445
semantic paralexia, 45/463
Deep middle cerebral vein
anatomy, 11/47, 51
Deep pain
chemical elicitation, 1/133
clinical features, 1/97
location, 1/131
Deep sensibility
modality, 1/93
Deep vein thrombosis
brain aneurysm neurosurgery, 55/54
brain infarction, 53/432
intravascular consumption coagulopathy, 55/496
oral contraception, 55/524
paraplegic, *see* Paraplegic deep vein thrombosis
physical rehabilitation, 61/465
subarachnoid hemorrhage, 55/54
tetanus, 65/218
DEF, *see* Phosphorotrithioic acid tributyl ester
Defecation
frontal lobe tumor, 17/258
hydrozoa intoxication, 37/40
hypothalamus, 2/441
rectum, 1/508
Defective revisualization, *see* Charcot-Wilbrand syndrome
Deferoxamine
aluminum intoxication, 63/525-527, 534, 64/279
CSF aluminum, 63/534
deafness, 63/534
dementia, 63/534
dialysis encephalopathy, 63/527
encephalopathy, 63/534
increased risk patient, 63/534
neurotoxicity, 63/534
nyctalopia, 63/534
renal transplantation, 63/534
retinal pigmentary change, 63/534
retinal pigmentary dystrophy, 63/534
sensorineural deafness, 63/534
visual field defect, 63/534
Deferoxamine intoxication
secondary pigmentary retinal degeneration, 60/736
symptom, 63/534
Defibrination syndrome, *see* Intravascular consumption coagulopathy
Deficiency neuropathy, 7/558-645
*see also* Hepatic insufficiency polyneuropathy
alcoholism, 7/613, 51/322, 324
antibiotic agent, 7/632
ataxia, 51/324, 326
Bassen-Kornzweig syndrome, 7/596, 51/326
bile salt disorder, 51/325
blind loop syndrome, 7/619, 51/325, 328
cassava intake, 7/642, 51/321, 323
celiac disease, 7/618-626, 51/329
Chachalen, 7/636
cocarboxylase, 7/568, 51/330
coenzyme A, 7/578
combined myelopathies, 7/644
Crohn disease, 7/620, 51/325, 329
cyanide intoxication, 51/323
demyelination, 51/324
diabetes mellitus, 51/324
dibasic amino acid malabsorption, 7/596
diphyllobothrium latum myelopathy, 7/631
disaccharide deficiency, 7/596
EMG, 51/326
essential nutrient, 51/321
gastric bypass, 51/328
gastrocolic fistula, 7/618
glucose-galactose malabsorption, 7/596
Hartnup disease, 7/577
hepatic polyneuropathy, 51/324, 326
hepatogenic malabsorption, 7/608
intestinal lymphangiectasia, 51/326
intestinal malabsorption, 7/603, 618
iron deficiency, 51/327
isoniazid neuropathy, 7/582, 584, 588, 644
Jamaican neuropathy, 7/643, 51/323
jejunoileal bypass, 51/328
kwashiorkor, 7/635, 51/322
liver cirrhosis, 7/609, 613, 51/324
malabsorption, 51/321, 324, 328
malabsorption syndrome, 7/596-599
malnutrition, 51/321
mesenteric artery occlusion, 7/631
Minamata disease, 7/642
motor neuropathy, 51/324
nerve conduction, 7/166
optic atrophy, 51/326, 328
optic neuropathy, 51/321
osteoporosis, 51/327
2-oxoglutaric acid, 51/330
pancreatic insufficiency, 51/327
pancreatogenic malabsorption, 7/603
paresthetic causalgic syndrome, 7/636
pernicious anemia, 7/591

pigmentary retinopathy, 51/326
postgastrectomy syndrome, 51/327
postradiation syndrome, 51/326
protein deficiency, 7/633
protein energy malnutrition, 51/322
purupuru, 7/642
pyrimidine, 7/560
short bowel syndrome, 51/325, 328
snack food, 51/322
sprue, 7/618
starvation neuropathy, 7/644
steatorrhea, 51/324, 327
tapeworm, 7/631
transketolase reaction, 51/330
tropical ataxic neuropathy, 7/640, 51/322
tropical sprue, 7/628, 51/329
upward gaze paralysis, 51/326
vegetarian, 51/322
vitamin A deficiency, 51/326
vitamin $B_1$, 51/324
vitamin $B_1$ deficiency, 7/567, 51/330
vitamin $B_2$ deficiency, 7/577, 51/332
vitamin $B_3$, 7/572, 576, 51/325
vitamin $B_3$ deficiency, 7/576, 51/333
vitamin $B_4$ deficiency, 7/576
vitamin $B_5$, 7/578, 51/325
vitamin $B_6$, 51/325
vitamin $B_6$ deficiency, 7/580, 51/333
vitamin $B_{12}$, 51/325
vitamin $B_{12}$ deficiency, 7/587, 51/327, 335
vitamin $B_c$, 51/325
vitamin $B_c$ deficiency, 51/335
vitamin $B_w$ deficiency, 7/579
vitamin D deficiency, 51/326
vitamin E deficiency, 51/326
Whipple disease, 51/328
Zollinger-Ellison syndrome, 7/608
Deflazacort
Duchenne muscular dystrophy, 62/131
Deflexion, *see* Cervical hyperdeflexion injury *and* Retroflexion
Degenerations
acquired hepatocerebral, *see* Acquired hepatocerebral degeneration
anterior horn cell, *see* Anterior horn cell degeneration
arteriolar fibrinohyalinoid, *see* Arteriolar fibrinohyalinoid degeneration
axonal, *see* Axonal degeneration
basal ganglion, *see* Basal ganglion degeneration
Betz cell, *see* Betz cell degeneration
carcinomatous cerebellar, *see* Carcinomatous cerebellar degeneration
cerebellar, *see* Cerebellar degeneration
cerebromacular, *see* Cerebromacular degeneration
chorioretinal, *see* Chorioretinal degeneration
chronic hepatocerebral, *see* Chronic hepatocerebral degeneration
CNS, *see* CNS degeneration
cochlear spiral ganglion, *see* Cochlear spiral ganglion degeneration
cone, *see* Cone degeneration
Corti organ, *see* Corti organ degeneration
cortical cerebellar, *see* Cortical cerebellar degeneration
corticospinal tract, *see* Corticospinal tract degeneration
cystic brain, *see* Cystic brain degeneration
dorsal column, *see* Dorsal column degeneration
dorsal funiculi, *see* Dorsal funiculi degeneration
experimental pigmentary retinal, *see* Experimental pigmentary retinopathy
familial hepatocerebral, *see* Wilson disease
familial pallidoluysionigral, *see* Familial pallidoluysionigral degeneration
familial pallidonigral, *see* Familial pallidonigral degeneration
fibrinoid, *see* Fibrinoid degeneration
granulovacuolar, *see* Granulovacuolar degeneration
granulovacuolar neuronal, *see* Granulovacuolar neuronal degeneration
hepatolenticular, *see* Wilson disease
hereditary pallidoluysiodentate, *see* Hereditary pallidoluysiodentate degeneration
hereditary pallidoluysionigral, *see* Hereditary pallidoluysionigral degeneration
hereditary pallidopyramidal, *see* Hereditary pallidopyramidal degeneration
hereditary spinocerebellar, *see* Hereditary spinocerebellar degeneration
hereditary spinodentatonigro-oculomotor, *see* Hereditary spinodentatonigro-oculomotor degeneration
hereditary vitreoretinal, *see* Hereditary vitreoretinal degeneration
heterogeneous system, *see* Progressive supranuclear palsy
hyaline vessel wall, *see* Arteriolar fibrinohyalinoid degeneration
intervertebral disc, *see* Intervertebral disc degeneration
locus ceruleus, *see* Locus ceruleus degeneration
macular, *see* Macular degeneration
multiple neuronal system, *see* Multiple neuronal system degeneration

multisystem, *see* Multiple neuronal system degeneration
neostriatal, *see* Neostriatal degeneration
nerve cell, *see* Neuron degeneration
neuroaxonal, *see* Neuroaxonal degeneration
neuroretinal, *see* Neuroretinal degeneration
nigrospinodentate, *see* Nigrospinodentate degeneration
nigrostriatal, *see* Striatonigral degeneration
nutritional cerebellar, *see* Nutritional cerebellar degeneration
optic nerve, *see* Optic nerve degeneration
opticoacoustic nerve, *see* Opticoacoustic nerve degeneration
opticocochleodentate, *see* Opticocochleodentate degeneration
pallidal, *see* Pallidal degeneration
pallidocerebello-olivary, *see* Pallidocerebello-olivary degeneration
pallidodentate, *see* Pallidodentate degeneration
pallidoluysian, *see* Pallidoluysian degeneration
pallidoluysiodentate, *see* Pallidoluysiodentate degeneration
pallidoluysionigral, *see* Pallidoluysionigral degeneration
pallidoluysionigrothalamic, *see* Pallidoluysionigrothalamic degeneration
pallidonigral, *see* Pallidonigral degeneration
pallidopyramidal, *see* Pallidopyramidal degeneration
paraventricular nucleus, *see* Paraventricular nucleus degeneration
pigmentary, *see* Pigmentary degeneration
pigmentary macular, *see* Pigmentary macular degeneration
posterior column, *see* Posterior column degeneration
predominant axonal, *see* Predominant axonal degeneration
primary axonal, *see* Primary axonal degeneration
primary pigmentary epithelium, *see* Primary pigmentary epithelium degeneration
primary pigmentary retinal, *see* Primary pigmentary retinal degeneration
pure pallidal, *see* Pure pallidal degeneration
pure pallidoluysian, *see* Pure pallidoluysian degeneration
putaminal, *see* Putaminal degeneration
pyramidal tract, *see* Pyramidal tract degeneration
reticulocerebellar, *see* Reticulocerebellar degeneration
retinal, *see* Retinal degeneration
secondary pigmentary retinal, *see* Secondary pigmentary retinal degeneration
spinocerebellar, *see* Spinocerebellar degeneration
spinocerebellonigral, *see* Spinocerebellonigral degeneration
spino-olivocerebellar, *see* Spino-olivocerebellar degeneration
spinopontine, *see* Spinopontine degeneration
spinothalamic tract, *see* Spinothalamic tract degeneration
spongy, *see* Spongy degeneration
spongy cerebral, *see* Spongy cerebral degeneration
spontaneous striatal, *see* Spontaneous striatal degeneration
striatal, *see* Striatal degeneration
striatonigral, *see* Striatonigral degeneration
subacute cerebellar, *see* Subacute cerebellar degeneration
subacute combined, *see* Subacute combined degeneration
subacute combined spinal cord, *see* Subacute combined spinal cord degeneration
subacute malignant cerebellar, *see* Subacute malignant cerebellar degeneration
substantia nigra, *see* Substantia nigra degeneration
supraoptic nucleus, *see* Supraoptic nucleus degeneration
system, *see* System degeneration
tapetoretinal, *see* Secondary pigmentary retinal degeneration
thalamic, *see* Thalamic degeneration
thalamus, *see* Thalamus degeneration
vitelliruptive macula, *see* Vitelliruptive macula degeneration
vitreoretinal, *see* Vitreoretinal degeneration
wallerian, *see* Wallerian degeneration
white matter, *see* White matter degeneration
Degenerative diffuse sclerosis (Scholz), *see* Metachromatic leukodystrophy
Degenerative disease
*see also* Neurodegenerative disease
epilepsy, 15/325, 72/125
Degenerative subcortical encephalopathy
progressive, *see* Spongy cerebral degeneration
Dégénérescence systématisée opticocochléodentelée, *see* Opticocochleodentate degeneration
Deglutition
amyotrophic lateral sclerosis, 22/21
larynx, 74/392
progressive bulbar palsy, 22/121, 59/218
Deglutition spasm
rabies, 1/285
tetanus, 1/285

Degos disease, *see* Köhlmeier-Degos disease
Dehydrating agent
  brain aneurysm, 12/215
  intracranial pressure, 23/212, 215
  subarachnoid hemorrhage, 12/215
  transtentorial herniation, 1/564
Dehydration
  osmotic agent, 57/215
  pellagra, 7/574
  thermoregulation, 75/58
  venous sinus occlusion, 54/426
Dehydrogenase
  senile plaque, 46/255
Deiterospinal syndrome, *see* Barré syndrome
Déjà connu
  temporal lobe stimulation, 3/32
Déjà vécu
  migraine, 48/162
  temporal lobe epilepsy, 2/706
  temporal lobe stimulation, 3/32
Déjà vu
  memory imagery, 45/359
  migraine, 42/246
  temporal lobe epilepsy, 2/706, 46/433
  temporal lobe lesion, 4/3
  temporal lobe stimulation, 3/32
  temporal lobe tumor, 17/284
  visual hallucination, 45/356
Dejean syndrome, *see* Orbital floor syndrome
Dejection
  migraine, 48/162
Dejerine, J.J., 1/7
Dejerine-Klumpke syndrome
  birth incidence, 7/416
  brachial plexus injury, 7/409
  clinical features, 1/229
Dejerine onion skin pattern, *see* Trigeminal descending spinal tract lesion
Dejerine pattern, *see* Trigeminal descending spinal tract lesion
Dejerine-Roussy syndrome, *see* Thalamic syndrome (Dejerine-Roussy)
Dejerine-Sottas disease, *see* Hypertrophic interstitial neuropathy
Dejerine syndrome
  *see also* Intermittent spinal cord ischemia *and* Medial medullary syndrome
  ataxia, 1/318
  medial medullary syndrome, 2/221
  medulla oblongata syndrome, 2/221
Dejerine-Thomas olivopontocerebellar atrophy, *see* Olivopontocerebellar atrophy (Dejerine-Thomas)
Del Rio Hortega, P., 1/11
Delasiauve classification
  epilepsy, 15/4
Delayed hypersensitivity
  *see also* Cell mediated immunity
  chemomechanism, 9/503
  complement fixation, 9/520
  disseminated encephalomyelitis, 9/501, 47/476
  experimental allergic encephalomyelitis, 9/504, 47/429
  experimental allergic neuritis, 47/429
  multiple sclerosis, 9/147, 47/338
  mumps skin test, 47/339
  response degeneration, 56/70
  tetanus toxoid, 52/240
  viral infection, 56/65, 70
Delayed myelopathy
  arachnopathy, 12/612
  cervical, 25/93, 276
  cervical spinal cord injury, 12/608
  craniospinal injury, 25/93
  neck injury, 12/608
  pathogenesis, 25/94
Delayed neuropathy
  organophosphate induced, *see* Organophosphate induced delayed neuropathy
  organophosphate intoxication, 51/284, 64/229
Delayed polyneuropathy
  organophosphate induced, *see* Organophosphate induced delayed polyneuropathy
Delayed radionecrosis
  anoxic ischemic leukoencephalopathy, 47/539
Delayed sleep phase syndrome
  shift worker, 45/142
Deletion syndrome
  chromosome, *see* Chromosome deletion syndrome
Delirium
  acidosis, 46/544
  acute intermittent porphyria, 46/544
  Addison disease, 46/544
  affective disorder, 46/552
  age, 46/532
  alcohol withdrawal, *see* Alcohol withdrawal delirium
  alcoholism, 46/532
  alkalosis, 46/544
  aluminum intoxication, 63/526
  Amanita pantherina intoxication, 36/534
  amantadine, 46/539
  aminophylline, 46/539
  amnesia, 46/528
  amnesic shellfish poisoning, 65/167
  angiostrongyliasis, 52/554

antibiotic agent, 46/538
anxiety, 46/430
apistus intoxication, 37/76
aprindine, 46/538
aprindine intoxication, 65/455
ascariasis, 35/359
atenolol, 63/195
atropine, 46/537
attention, 46/529
atypical psychosis, 46/552
auditory hallucination, 46/526
barbiturate intoxication, 65/332
belladonna alkaloid, 46/537
bilateral occipital lobe infarction, 53/413
bismuth intoxication, 64/338
Bolivian hemorrhagic fever, 56/369
brain fat embolism, 55/177
brain infarction, 53/363, 55/138
brain injury, 46/532
brain tumor, 46/545
bromocriptine, 46/539
burn, 46/548
carbon disulfide intoxication, 64/24
cardiotomy, 46/547
cat scratch disease, 52/130
cataract surgery, 46/535, 547
cause, 53/239
cerebrovascular disease, 46/545, 55/138
cholinergic receptor blocking agent, 46/537
cholinergic receptor blocking agent intoxixation, 64/537
cimetidine, 46/539
cinchonism, 65/485
clinical diagnosis, 46/549
clinical features, 46/525-531
clonidine, 46/538
cognition, 46/525-531
confabulation, 46/528
consciousness, 45/117, 46/530
coronary bypass surgery, 46/547
coronary care unit, 46/535
corticosteroid intoxication, 65/555
course, 46/531
Cushing syndrome, 46/544
cytostatic agent, 46/539
cytotoxic agent, 46/539
definition, 46/523
delusion, 46/528
dementia, 46/124, 206-208, 524, 532, 551
dextropropoxyphene, 46/540
diabetes mellitus, 46/544
dialysis dementia, 46/542
dialysis encephalopathy, 46/542, 63/526
differential diagnosis, 46/551
digitalis intoxication, 46/538
digoxin, 63/195
digoxin intoxication, 65/450
disorientation, 46/528
drug intoxication, 46/537-541
dysgraphia, 46/531
dysphoria, 46/531
dysphrenia hemicranica, 5/79
EEG, 46/524, 550
electrolyte balance, 46/543
emotion, 46/531
endocrine disorder, 46/544
epilepsy, 15/596, 46/544
ergot intoxication, 65/62
etiologic diagnosis, 46/550
etiologic table, 46/535
etiology, 46/532
exogenous poison, 46/545
exogenous reaction type, 46/523
facilitating factor, 46/533-535
functional psychosis, 46/532
Ganser syndrome, 46/552
gymnapistes intoxication, 37/76
gyromitrine intoxication, 65/36
hallucination, 46/525-527
head injury, 46/545
heat stroke, 46/546
hepatic encephalopathy, 46/542
heterocyclic antidepressant, 65/316
hydrozoa intoxication, 37/40
hypercalcemia, 46/543, 63/561
hyperkalemia, 46/543
hypermagnesemia, 46/543
hypernatremia, 46/543, 63/554
hyperparathyroidism, 46/544
hyperthyroidism, 46/544
hypnotic agent, 46/539
hypocalcemia, 46/543
hypoglycemia, 46/542
hypokalemia, 46/543
hypomagnesemia, 46/543, 63/562
hyponatremia, 46/543, 63/545
hypoparathyroidism, 46/544
hypothermia, 46/546
hypothyroidism, 46/544
hypoxia, 46/541
iatrogenic factor, 61/165
iatrogenic neurological disease, 61/165
immobilization, 46/535
inappropriate antidiuretic hormone secretion, 46/543
incidence, 46/524

indometacin, 46/540
intensive care syndrome, 46/534
interleukin-2 intoxication, 65/539
intracranial infection, 46/544
intravascular consumption coagulopathy, 55/488
lead encephalopathy, 64/436
lead intoxication, 9/642
legionellosis, 46/544
lergotrile, 46/539
levodopa, 46/539
lidocaine intoxication, 65/454
lithium, 46/539
Lyme disease, 52/262
malignancy, 46/546
memory, 46/528
metabolic encephalopathy, 46/541
methyldopa, 46/538
metrizamide, 46/540, 61/165
middle cerebral artery, 53/363
migraine, 46/545, 48/37, 162, 166
muscimol intoxication, 65/40
nalorphine, 37/366
norpethidine intoxication, 65/357
nothesthes intoxication, 37/76
OKT3, 63/536
organic factor, 46/535-548
organolead intoxication, 64/133
orientation, 46/528
pancreatic encephalopathy, 46/543
panhypopituitarism, 46/544
paracetamol, 46/540
paraphasia, 46/531
pathogenesis, 46/548
pathophysiology, 46/548
pellagra, 6/748, 7/571, 573
pentazocine, 46/539
perception, 46/525-527
pethidine, 46/539
pethidine intoxication, 65/357
phenytoin intoxication, 65/500
phrenitis, 46/530
physical factor, 46/546
pneumonia, 46/544
postoperative, *see* Postoperative delirium
predisposing factor, 46/532
procainamide, 46/538
prognosis, 46/532
propranolol, 46/538, 63/195
propranolol intoxication, 37/447
psychomotor behavior, 46/530
psychosis, 46/552
pterois volitans intoxication, 37/75
quinidine, 46/538
reactive psychosis, 46/552
REM sleep, 46/533
respiratory encephalopathy, 63/414
respiratory failure, 46/542
Reye syndrome, 27/363, 29/333, 31/168, 34/80, 46/544, 56/151, 238, 63/438, 65/115
salicylic acid, 46/540
schizophrenia, 46/419, 552
scopolamine, 46/537
scorpaena intoxication, 37/76
scorpaenodes intoxication, 37/76
scorpaenopsis intoxication, 37/76
sebastapistes intoxication, 37/76
sedation, 46/553
sedative hypnotic withdrawal, *see* Hypnotic agent withdrawal delirium
selective serotonin reuptake inhibitor, 65/316
sensory deprivation, 46/534
sensory overload, 46/535
sleep deprivation, 46/533
sleep-wake cycle, 46/529, 533
sodium nitroprusside, 46/538
solvent abuse, 46/546
stress, 46/533
surgery, 46/546-548
systemic infection, 46/544
systemic lupus erythematosus, 55/376
thinking, 46/528
tocainide intoxication, 65/455
toxic shock syndrome, 46/544, 52/258
trazodone intoxication, 65/320
treatment, 46/552
trichinosis, 52/566
tricyclic antidepressant, 65/316
tricyclic antidepressant intoxication, 65/321
uremic encephalopathy, 46/542, 63/504, 506
visual hallucination, 46/526
vitamin $B_1$ deficiency, 46/543
vitamin $B_3$ deficiency, 46/543
vitamin $B_{12}$ deficiency, 46/543
vitamin $B_c$ deficiency, 46/543
wakefulness, 46/529
water intoxication, 64/241
Wernicke-Korsakoff syndrome, 46/341
Whipple disease, 52/139
Delirium tremens
alcohol withdrawal, 46/540
alcoholism, 3/99, 27/456, 43/197, 70/346
central pontine myelinolysis, 28/291
neuropathology, 46/346
sleep disorder, 3/99
visual perseveration, 45/365
Delivery

biological rhythm, 74/492
sexual function, 26/452
viral gene, *see* Viral gene delivery
Delphinapterus leucas intoxication
fatal, 37/95
ingestion, 37/95
Delta lesion
*see also* Muscle cell membrane *and* Sarcolemma
Duchenne muscular dystrophy, 40/370
muscle fiber, 62/15
nerve fiber, 41/410
Delta sleep inducing peptide
Alzheimer disease, 46/268
Deltamethrin
neurotoxin, 64/212
pyrethroid, 64/211
pyrethroid intoxication, 64/212
Delusion
beta adrenergic receptor blocking agent intoxication, 65/436
conversion headache, 5/249
delirium, 46/528
dentatorubropallidoluysian atrophy, 49/440
dysphrenia hemicranica, 5/79
headache, 5/249, 48/5, 361
Huntington chorea, 49/283
Lafora progressive myoclonus epilepsy, 27/172
manic depressive psychosis, 43/208
migraine, 48/162, 361
psychiatric headache, 5/20, 249, 48/361
schizophrenia, 43/212
striatopallidodentate calcification, 49/423
Wernicke-Korsakoff syndrome, 45/194
Démange disease, *see* Spinal cord atherosclerosis
Dementia
*see also* Mental deterioration *and* Parkinson dementia
acquired hepatocerebral degeneration, 49/213
acquired immune deficiency syndrome, *see* Acquired immune deficiency syndrome dementia
acrylamide intoxication, 64/68
Addison disease, 42/512, 46/396
adenosine triphosphatase 6 mutation syndrome, 62/510
adolescent, 46/130
adrenoleukodystrophy, 46/402, 60/169
adult neuronal ceroid lipofuscinosis, 42/466, 60/665
adult polyglucosan body disease, 62/484
African trypanosomiasis, 52/341
age, 3/310-312
aging, 27/479, 46/199-211
agnosia, 46/203
agraphia, 45/469
alcoholic, *see* Alcoholic dementia
aluminum, 63/525
aluminum intoxication, 46/392
Alzheimer disease, 27/482-488, 42/275-277, 46/199, 59/410
amnesia, 46/204, 210
amnesic shellfish poisoning, 65/166
amyotrophic lateral sclerosis, 22/313, 316, 318, 369, 42/67-69, 59/231, 409-412
anemia, 46/397
anosognosia, 46/200
anxiety, 46/430
aorta coarctation, 46/358
aphasia, 45/296, 321, 46/203
apraxia, 4/59, 62, 46/203
arsenic intoxication, 46/391, 64/288
assessment, 46/121-131
ataxia, 42/114
ataxia telangiectasia, 60/665
atrial myxoma, 63/98
bacterial encephalitis, 46/389
bacterial meningitis, 46/387
Baltic myoclonus epilepsy, 49/617
barbiturate intoxication, 65/332, 509
Batten disease, 10/575, 42/466
Behçet syndrome, 46/390, 56/597
benzodiazepine intoxication, 65/337
Binswanger disease, 42/278, 46/317-321, 54/221
bismuth, 64/4
bismuth intoxication, 46/392, 64/337
blood lead level, 64/438
boxer, *see* Boxer dementia
brain abscess, 46/387
brain amyloid angiopathy, 54/337
brain arteriovenous malformation, 54/378
brain biopsy, 46/128
brain blood flow, 54/150
brain death, 24/702
brain edema, 63/417
brain hypoxia, 63/417
brain injury, 24/477, 702
brain lacunar infarction, 54/253
brain metastasis, 46/386
brain microangiopathy, 55/381
brain tumor, 46/385
bulbar paralysis, 22/142
carbon disulfide intoxication, 64/25
carbon disulfide polyneuropathy, 49/477
carbon monoxide intoxication, 64/32
carcinoma, 38/647-651, 657
cardiac arrest, 63/215, 217

Carter-Sukavajana syndrome, 21/412
celiac disease, 28/230
central pontine myelinolysis, 46/400
cerebellar ataxia, 42/126, 129
cerebellar Creutzfeldt-Jakob disease, 60/665
cerebello-olivary atrophy, 42/136
cerebral dominance, 46/204
cerebral pontine myelinolysis, 46/400
cerebrotendinous xanthomatosis, 10/532, 29/374, 42/430, 46/403, 51/399, 59/357, 60/166
cerebrovascular disease, 42/277-279, 53/42
Charlevoix-Saguenay spastic ataxia, 60/665
chemotherapy, 67/378
child, 46/130
chorea-acanthocytosis, 49/328, 63/281
chronic enterovirus infection, 56/330
chronic hepatocerebral degeneration, 49/219
chronic mercury intoxication, 64/370
chronic nodular polioencephalitis, 34/560
ciclosporin, 63/537
ciclosporin intoxication, 63/537
ciguatoxin intoxication, 65/163
cinchonism, 65/452
classification, 27/481, 46/204-206
clinical features, 67/379
clonidine intoxication, 37/444
CNS aging, 27/479
cognition, 46/124, 200
complex I deficiency, 62/504, 66/422
complex III deficiency, 62/504, 66/423
complex V deficiency, 62/506
computer assisted tomography, 46/128, 202
concentric sclerosis, 46/404
congenital megacolon, 42/4
congenital pain insensitivity, 51/566
consciousness, 45/114, 117, 46/200
constructional apraxia, 2/698, 45/502
copper intoxication, 46/392
cortical, *see* Cortical dementia
corticobasal degeneration, 59/410
Creutzfeldt-Jakob disease, 27/491, 42/657, 46/128, 289-296, 389, 56/545, 59/409
cryptococcal meningoencephalitis, 52/432
CSF examination, 46/127
Cushing syndrome, 46/396
cyanide intoxication, 49/477, 64/232, 65/29
cyanogenic glycoside intoxication, 65/29
cytochrome related ataxia, 42/111
deferoxamine, 63/534
definition, 46/121, 199
delirium, 46/124, 206-208, 524, 532, 551
demyelinating disease, 46/403-405
dentatorubropallidoluysian atrophy, 21/520, 49/440, 60/665
depression, 46/122, 206-208, 427
detection, 46/200
diabetes mellitus, 46/397
diagnosis, 46/206-208
diagnostic method, 46/125-128
diagnostic problem, 46/128-130
dialysis, *see* Dialysis dementia
dialysis encephalopathy, 64/277
differential diagnosis, 46/206-208, 54/96
differentiation, 54/96
diffuse Lewy body disease, 59/410, 75/189
diffuse sclerosis, 46/404
Divry-Van Bogaert syndrome, 14/14, 42/725, 55/318
domoic acid, 65/141
Down syndrome, 27/488, 46/403
drug induced, *see* Drug induced dementia
dysphasia, 3/312
dyssynergia cerebellaris myoclonica, 21/510, 60/665
EEG, 46/127, 202
Ekbom other syndrome, 63/85
Ekbom syndrome type II, 60/665
encephalitis, 46/128, 130, 388
encephalitis lethargica, 46/388
encephalomyelitis, 46/389
eosinophilia myalgia syndrome, 63/376, 64/256
epilepsy, 15/344, 42/114, 46/128
etiology, 46/122, 204-206
Fabry disease, 55/457
familial, *see* Familial dementia
familial amaurotic, *see* Familial amaurotic idiocy
familial amyotrophic lateral sclerosis, 59/236, 245
familial progressive myoclonus epilepsy, 49/617
familial spastic paraplegia, 22/424, 426, 433, 469
familial striatopallidodentate calcification, 46/401
focal cerebral lesion, 3/309, 4/6-9
focal neuropsychological syndrome, 46/208
Friedreich ataxia, 46/401
frontal, *see* Frontal dementia
frontal lobe syndrome, 45/30, 46/124
fungal meningitis, 46/387
Ganser syndrome, 46/124
Gaucher disease, 46/402
general paresis of the insane, 52/279
Gerstmann-Sträussler-Scheinker disease, 21/553, 56/553, 60/622
gliomatosis cerebri, 46/386
global ischemia, 54/123
glycogen storage disease, 43/132
glycosaminoglycanosis, 60/176
glycosaminoglycanosis type I, 10/441, 60/175

glycosaminoglycanosis type IH, 60/175
glycosaminoglycanosis type II, 60/175
glycosaminoglycanosis type III, 60/175, 66/302
glycosaminoglycanosis type VII, 60/175
GM2 gangliosidosis variant, 60/665
gold intoxication, 46/392
granulomatous CNS vasculitis, 55/390
Guam amyotrophic lateral sclerosis-Parkinson dementia complex, 65/22
Guam motoneuron disease, 65/22
Guam Parkinson, *see* Guam Parkinson dementia
Hallervorden-Spatz syndrome, 42/257, 46/130, 401, 49/399, 59/410
Haw River syndrome, 60/665
head injury, 46/394
heart disease, 46/397
hemianopia, 42/278
hemodialysis, 46/398, 542
hepatic coma, 49/213
hepatic encephalopathy, 46/128, 398
hereditary ataxia, 46/401
hereditary cerebello-olivary atrophy (Holmes), 21/404, 409, 412
hereditary deafness, 22/516
hereditary dystonic paraplegia, 22/445-465
hereditary olivopontocerebellar atrophy (Menzel), 21/443, 445
hereditary spastic ataxia, 60/462
hereditary spinocerebellar ataxia, 21/553
heredopathia ophthalmo-otoencephalica, 42/150
Histoplasma meningitis, 46/388
history taking, 46/201
homocarnosinosis, 42/557, 59/359
human immunodeficiency virus, *see* Human immunodeficiency virus dementia
Huntington chorea, 27/492-496, 42/225, 46/130, 305-309, 49/281-283, 59/410
hydrocephalus, *see* Hydrocephalic dementia
hypereosinophilic syndrome, 63/372
hyperinsulinemia, 46/397
hyperparathyroidism, 46/396
hyperthyroidism, 46/396
hypertrophic interstitial neuropathy, 42/317
hyperviscosity syndrome, 55/487
hypocalcemia, 63/561
hypogonadal cerebellar ataxia, 21/469
hypoparathyroidism, 46/396
hypothyroidism, 46/205, 396
hysteria, 46/124
idiopathic Parkinson disease, 46/378
inborn error of metabolism, 46/130
incidence, 46/123
infantile neuroaxonal dystrophy, 42/229
infantile neuronal ceroid lipofuscinosis, 60/665
inherited disease, 46/400-403
α-interferon intoxication, 65/538
intrathecal chemotherapy, 61/151
irradiation, 67/358
isoniazid intoxication, 65/480
juvenile dystonic lipidosis, 42/446
juvenile metachromatic leukodystrophy, 60/665
Kearns-Sayre-Daroff-Shy syndrome, 62/308, 508
Kii Peninsula amyotrophic lateral sclerosis, 22/369
kuru, 46/389, 56/556
laboratory examination, 46/202
Lafora progressive myoclonus epilepsy, 15/385, 27/172, 41/194, 42/702, 49/617
language, 45/321
late cerebellar atrophy, 21/495, 60/583, 665
lead, 64/438
lead encephalopathy, 46/391, 64/443
lead intoxication, 9/642, 46/391, 64/437
leukodystrophy, 46/130
levodopa, 46/379
Lewy body disease, 59/410
limbic encephalitis, 46/387
lipomembranous polycystic osteodysplasia, 42/279-281
Listeria monocytogenes meningitis, 33/88
lithium intoxication, 64/296
lysosomal storage disease, 46/130
Machado-Joseph disease, 59/410
malignancy, 38/647-651, 657, 46/386
manganese intoxication, 46/392
mania, 46/123
Marchiafava-Bignami disease, 28/324, 46/399
Marinesco-Sjögren syndrome, 21/556
mass effect, 3/309, 313
Mast syndrome, 22/142, 42/282
medial thalamic syndrome, 2/486
MELAS syndrome, 62/509
memory disorder, 3/270-273
mercury intoxication, 46/391
MERRF syndrome, 62/509
metabolic ataxia, 21/575
metabolic disorder, 46/395-400
metachromatic leukodystrophy, 10/568, 42/493, 46/401, 59/356
methanol intoxication, 64/100
Minamata disease, 46/391, 64/414
mitochondrial encephalomyopathy, 60/665
mitochondrial myopathy, 43/188
Morgagni-Stewart-Morel syndrome, 43/411
motoneuron disease, 42/67
mountain sickness, 63/417

movement disorder, 46/130
multi-infarct, *see* Multi-infarct dementia
multiple myeloma, 55/487
multiple neuronal system degeneration, 59/411
multiple sclerosis, 46/403, 47/58
muscular dystrophy, 46/401
myopathy, 43/132
neuroacanthocytosis, 42/209
neuroaxonal dystrophy, 60/665
neuroaxonal leukodystrophy, 49/406
neurochemistry, 27/477-499
neurocysticercosis, 46/389, 52/530
neurodegenerative disease, 21/3
neurofibrillary tangle, 27/486, 489-491
neurologic examination, 46/201
neuronal ceroid lipofuscinosis, 46/402
neuronal intranuclear hyaline inclusion disease, 60/665
neuropathology, 67/384
neuropsychologic assessment, 46/202-204
neuropsychologic examination, 46/201
neuropsychologic test, 45/523, 46/127
neurosyphilis, 46/390, 52/278
neurotoxicity, 67/358
Niemann-Pick disease type C, 46/402, 60/665
normal pressure hydrocephalus, 27/492
nuclear magnetic resonance, 54/94, 96
olivopontocerebellar atrophy, 21/397, 59/411
olivopontocerebellar atrophy (Dejerine-Thomas), 21/426, 60/665
olivopontocerebellar atrophy variant, 21/452, 455
olivopontocerebellar atrophy (Wadia-Swami), 60/494, 665
ophthalmoplegia, 59/411
organic solvent intoxication, 64/41
organochlorine insecticide intoxication, 64/207
organotin intoxication, 64/142
orthochromatic leukodystrophy, 42/500
osteitis deformans (Paget), 38/366, 46/400, 70/14
pallidal degeneration, 49/450
pallidoluysionigral degeneration, 49/452, 456
pallidopyramidal degeneration, 49/456
papaverine, 37/455
paralysis, *see* General paresis of the insane
parietal lobe syndrome, 45/76, 46/124
Parkinson, *see* Parkinson dementia
Parkinson dementia complex, 59/409
Parkinson disease, 42/246, 46/371-381, 593, 49/112, 130
Parkinson tremor, 46/373
parkinsonism, 49/130, 59/410, 63/481
Pelizaeus-Merzbacher disease, 42/502-505
pellagra, 6/748, 7/571, 28/87, 46/399, 51/333, 70/413
pernicious anemia, 42/609
phenytoin intoxication, 65/500
photogenic epilepsy, 42/714
physical examination, 46/201
Pick disease, 46/200, 233, 59/410
Pick lobar atrophy, 27/489
polyarteritis nodosa, 39/300, 55/356
polychlorinated biphenyl intoxication, 64/207
polychlorinated dibenzodioxin intoxication, 64/205
polycythemia, 46/398
portacaval shunt, 21/576
postanesthetic encephalopathy, 49/479
prazosin intoxication, 65/447
precipitating event, 46/415
presenile, *see* Presenile dementia
prevalence, 42/275
primary alcoholic, *see* Primary alcoholic dementia
progressive, 42/279-281, 63/98
progressive autonomic failure, 59/411
progressive external ophthalmoplegia, 62/308
progressive multifocal leukoencephalopathy, 38/656, 39/19, 46/388, 47/504, 54/231
progressive myoclonus epilepsy, 15/334, 42/114, 704, 49/617
progressive rubella panencephalitis, 34/338, 56/409
progressive supranuclear palsy, 6/557, 22/218, 220, 42/259, 46/301, 49/239, 243
proximal muscle weakness, 43/132
pseudo, *see* Pseudodementia
pseudohypoparathyroidism, 46/396
pulmonary function, 46/397
radiation induced, 67/377
Refsum disease, 60/231
rehabilitation, 3/426-430
respiration, 63/481
Rett syndrome, 60/636, 665
Rickettsia encephalitis, 46/389
sarcoidosis, 46/390
schizophrenia, 46/123, 419
secondary, 46/385-405
secondary demyelination, 46/404
secondary pigmentary retinal degeneration, 60/654, 722
senile, *see* Senile dementia
senile chorea, 6/436
senile plaque, 27/486
sensory deficit, 59/412
Sheehan syndrome, 46/397
Sjögren syndrome, 71/74
slow eye movement, 42/187

Sneddon syndrome, 55/402, 406
spastic ataxia, 42/166
spastic paraparesis, 46/401
spastic paraplegia, 42/176
spinocerebellar ataxia, 42/182
spinocerebellar degeneration, 46/130
spino-olivocerebellar degeneration, 60/665
sporadic amyotrophic lateral sclerosis, 59/233
sport injury, 46/395
striatopallidodentate calcification, 46/130, 49/419, 422
subacute necrotizing encephalomyelopathy, 46/402
subacute sclerosing panencephalitis, 46/389, 56/418
subcortical, *see* Subcortical dementia
subdural hematoma, 46/395
supranuclear ophthalmoplegia, 42/605
temporal arteritis, 55/344
temporal lobe syndrome, 46/124
test, 3/309
2,3,7,8-tetrachlorodibenzo-*p*-dioxin, 64/205
thalamic, *see* Thalamic dementia
thalamic degeneration, 21/590
thalamic syndrome (Dejerine-Roussy), 45/91
thallium intoxication, 46/392
toluene intoxication, 64/42
toxoplasmosis, 46/389
traumatic, *see* Traumatic dementia *and* Traumatic intellectual defect
treatability, 46/206
treatment, 46/208
tropical ataxic neuropathy, 65/29
tuberculous meningitis, 46/387, 52/215
Turner syndrome, 14/108
Unverricht-Lundborg progressive myoclonus epilepsy, 15/334, 42/705-708, 46/400, 49/617
uremic encephalopathy, 46/398, 63/504
valproic acid intoxication, 65/504
vascular, *see* Vascular dementia
viral encephalitis, 46/388
vitamin $B_1$ deficiency, 46/399
vitamin $B_3$ deficiency, 51/333
vitamin $B_4$ deficiency, 28/201
vitamin $B_{12}$ deficiency, 46/399
vitamin $B_c$, 28/201
vitamin $B_c$ deficiency, 46/399, 63/255
Von Hippel-Lindau disease, 43/21
Wernicke-Korsakoff syndrome, 46/399
Whipple disease, 46/389, 52/137
white matter, 67/380
Willvonseder syndrome, 42/194
Wilson disease, 46/130, 398, 400, 59/410
Wolman disease, 60/165
X-linked ataxia, 60/665
xeroderma pigmentosum, 60/335
Dementia paralytica, *see* General paresis of the insane
Dementia pugilistica, *see* Boxer dementia *and* Boxer encephalopathy
Dementia syndrome
depression, 46/207
Demerol, *see* Pethidine
Demeton
neurotoxicity, 37/548
organophosphate, 64/152, 156
organophosphate intoxication, 64/152, 155
Demeton intoxication
chemical classification, 37/548
organophosphorus compound, 37/548
Demeton-S intoxication
chemical classification, 37/547
organophosphorus compound, 37/547
Demeton sulfoxide
organophosphate, 64/154, 169
organophosphate intoxication, 64/154, 169
Demyelinated nerve fiber
conduction block, 47/15
Demyelinating disease
animal, 9/664-687
animal *vs* human multiple sclerosis, 9/665
animal viral disease, 34/293-296
Blackwood classification, 10/22
brain amyloid angiopathy, 54/341
cage paralysis of primates, 9/671
classification, 10/22
comparative neurology, 9/664-668
dementia, 46/403-405
experimental allergic neuritis, 9/531-536
Lyme disease, 52/262
Minamata disease, 36/89, 64/416
myelin, *see* Demyelination
neuromyelitis optica, 9/428, 431, 47/398
pure motor hemiplegia, 54/242
zoöneurology, 9/666-668
Demyelinating disorder
bladder, 75/673
symptomatic dystonia, 49/542
Demyelinating encephalitis
African trypanosomiasis, 35/72, 52/342
candidiasis, 52/398
cerebral toxoplasmosis, 52/357
multifocal, 52/387, 398
multifocal lesion, 52/342, 357
North American blastomycosis, 52/387
Demyelinating encephalomyelitis

JHM virus, 56/443
Demyelinating leukoencephalopathy
Binswanger disease, 10/9
Demyelinating neuropathy, 47/605-623
acute, *see* Acute demyelinating neuropathy
acute inflammatory, *see* Guillain-Barré syndrome
adrenoleukodystrophy, 47/622
adrenomyeloneuropathy, 47/622
amyloid, 47/618
axonal reaction, 47/621
B-lymphocyte neoplasm, 47/615
Bassen-Kornzweig syndrome, 63/278
carpal tunnel syndrome, 47/618
chlorambucil, 47/618
chronic, *see* Chronic demyelinating neuropathy
CNS demyelination, 47/606
Cockayne-Neill-Dingwall disease, 47/622
corticosteroid, 41/346
diabetes mellitus, 47/621
dipeptidyl carboxypeptidase I inhibitor intoxication, 65/445
diphtheria, 47/608, 65/241
drug induced, *see* Drug induced demyelinating neuropathy
eosinophilia myalgia syndrome, 63/377
globoid cell leukodystrophy, 47/622
Guillain-Barré, *see* Guillain-Barré syndrome
hereditary motor sensory neuropathy, 47/621-623
hereditary motor sensory neuropathy type I, 47/622
hereditary motor sensory neuropathy type II, 47/622
hereditary motor sensory neuropathy type III, 47/622
IgA, 47/616
IgG, 47/615
IgM, 47/616-618
3,3'-iminodipropionitrile intoxication, 51/264
lymphoma, 41/343
melphalan, 47/618
metachromatic leukodystrophy, 47/622
monoclonal gammopathy, 47/615-618
multiple myeloma, 47/616, 618
myeloma, 41/343
paraproteinemia, 47/614-620
plasma exchange, 47/618
POEMS syndrome, 47/619, 51/432, 63/395
primary, *see* Primary demyelinating neuropathy
progressive spinal muscular atrophy, 59/404
recurrent acute, *see* Recurrent acute demyelinating neuropathy
Refsum disease, 47/622, 66/490
sarcoid neuropathy, 51/197
tetanus toxoid, 52/240
Waldenström macroglobulinemia, 47/616, 619, 63/397
Waldenström macroglobulinemia polyneuropathy, 51/433
Demyelinating polyneuropathy
chronic, *see* Chronic demyelinating polyneuropathy
chronic inflammatory, *see* Chronic inflammatory demyelinating polyneuropathy
chronic progressive, *see* Chronic progressive demyelinating polyneuropathy
chronic relapsing, *see* Chronic relapsing demyelinating polyneuropathy
hereditary motor and sensory neuropathy type I, 60/199
herpes zoster, 51/182
inflammatory, *see* Inflammatory demyelinating polyneuropathy
Marek disease, 51/250, 56/582
multiple myeloma, 55/470
Demyelinating polyradiculoneuropathy
chronic idiopathic, *see* Chronic idiopathic demyelinating polyradiculoneuropathy
chronic inflammatory, *see* Chronic inflammatory demyelinating polyradiculoneuropathy
Demyelination
*see also* Myelin dysgenesis *and* Remyelination
accommodation, 47/37
acetonylacetone intoxication, 64/89
acid phosphatase, 9/38
acquired immune deficiency syndrome, 56/509, 518, 520, 583
acute pancreatitis, 27/449-452
alcoholic neuropathy, 47/621
alcoholic polyneuropathy, 47/621
alcoholism, 9/634
anemia, 9/613
animal model, 56/583, 587, 589
anoxia, 9/573, 47/529
antibody, 9/519, 47/445
antibody mediated, *see* Antibody mediated demyelination
antimetabolite, 9/627
arsenic intoxication, 9/647
ataxia telangiectasia, 14/74
autoimmune, 47/447
autoimmune reaction, 56/447
autolysis, 9/39
axonal, *see* Axonal demyelination
Bassen-Kornzweig syndrome, 8/19, 10/548, 51/394
beriberi, 7/563

Binswanger disease, 9/613, 54/221, 227
blocking factor, 47/41
blood disease, 9/593
brain death, 9/587
brain edema, 47/562, 564
brain ischemia, 63/214
brain tumor, 9/588
canine distemper virus, 56/587
carbamazepine, 47/38
carbon monoxide intoxication, 9/628-632
cause, 47/29-43
central pontine myelinolysis, 2/263, 9/636-638
cerebrotendinous xanthomatosis, 10/547, 51/399
chemistry, 9/311
cholesterol ester, 9/36
ciguatoxin intoxication, 37/83, 65/163
Clostridium welchii, 9/39
CNS, *see* CNS demyelination
cobra snake venom, 9/39
compression neuropathy, 51/88
concentric, 9/438, 10/10
conduction block, 47/30, 38
conduction mechanism, 47/33-38
congenital Pelizaeus-Merzbacher disease, 10/160
consciousness, 45/121
copper, 9/645
coronavirus, 56/12
corpus callosum plaque, 9/39
cryoglobulinemic polyneuropathy, 51/434
cyanide intoxication, 9/39, 632, 64/231, 65/28
cyanogenic glycoside intoxication, 65/28
cytomegalovirus, 56/273
deficiency neuropathy, 51/324
diphtheria, 51/182, 52/229, 65/241
diphtheritic neuritis, 9/38
discontinuous type, 10/104
drug intoxication, 47/551
edema, 47/43
electron microscopy, 47/214-251
ergot derivative, 9/652
experimental, *see* Experimental demyelination
experimental allergic encephalomyelitis, 9/38, 684-687, 47/445
experimental allergic neuritis, 9/505
extracellular potassium, 47/36
fiber-fiber interaction, 47/36
Foix-Alajouanine disease, 10/15
fucosidosis, 42/550
galactocerebroside, 47/447
globoid cell leukodystrophy, 10/20, 303
GM2 gangliosidosis, 10/397
granulomatous amebic encephalitis, 52/325
Guillain-Barré syndrome, 9/543, 42/645
head injury, 9/587
hepatic coma, 9/634
hepatic encephalopathy, 9/634
hereditary hemorrhagic telangiectasia (Osler), 9/607
hereditary motor and sensory neuropathy type I, 59/369
hereditary neuropathy, 60/258
hereditary sensory and autonomic neuropathy type III, 8/345, 21/112, 60/30, 33
histochemistry, 9/23-44
histology, 47/214-223
hydrocephalus, 50/290
hyperpipecolatemia, 29/222
hyperthyroidism, 9/627
hypoglycemia, 9/626
IgM paraproteinemic polyneuropathy, 51/437
3,3'-iminodipropionitrile intoxication, 51/264
inflammation, 47/43
intoxication, 9/617
intracellular sodium, 47/38
ischemia, 9/573
JHM virus, 56/442, 444
kernicterus, 9/638
lead intoxication, 9/641
leptospirosis, 9/682
Lhermitte sign, 47/32
lipid change, 9/35
local anesthetic agent, 65/423
lymphomatoid granulomatosis, 51/451
macrophage oxidase, 9/38
manganese intoxication, 9/643
Marchiafava-Bignami disease, 9/653, 46/399, 47/558
Marek disease, 47/609, 56/582
mechanism, 47/29-43, 249
melanoderma, 10/36
mental deficiency, 43/303
metachromatic leukodystrophy, 8/16, 10/19, 50, 29/357, 51/367
metachromatic prelipid, 10/20
microangiopathy, 9/573
microscopy, 47/214
mucolipidosis type I, 51/380
multifocal, *see* Multifocal demyelination
multiple sclerosis, 47/214-223, 226
multiple sclerosis plaque, 9/37
murine model, 56/589
myotonic dystrophy, 62/237
natural, 47/29
necrotizing, *see* Necrotizing demyelination
nerve, 51/2, 52
nerve conduction velocity, 40/128

nerve myelin, 51/32
neuroaxonal, *see* Neuroaxonal demyelination
neurobrucellosis, 33/309, 52/584
neuroglia relationship, 28/410-412
neurologic sign, 47/30
Niemann-Pick disease type C, 51/375
Norman microcephalic familial leukodystrophy, 10/103, 182
oligodendrocyte, 47/551
oligosaccharidosis, 66/329
optic nerve, *see* Optic nerve demyelination
organomercury intoxication, 9/643
organophosphorus compound intoxication, 9/652
organotin intoxication, 64/136
orthochromatic leukodystrophy, 42/501
oxidase, 9/38
pallidal, *see* Pallidal demyelination
pallidoluysionigral degeneration, 6/665
paranodal, *see* Paranodal demyelination
pathologic change, 47/36
Pelizaeus-Merzbacher disease, 42/503-505
Pena-Shokeir syndrome type I, 59/369
peripheral, *see* Peripheral demyelination
peripheral nervous system, 47/606
peroxidase, 9/38
phenytoin, 47/38
phenytoin polyneuropathy, 47/38
photogenic epilepsy, 42/714
Pick disease, 27/489, 46/236, 239
polycythemia vera, 9/594
polyneuropathy, 41/339, 42/98, 47/38, 60/258
pontine, *see* Pontine demyelination
porphyria, 9/639
posterior column, *see* Posterior column demyelination
presenile dementia, 42/288
primary, *see* Primary demyelination
progressive multifocal leukoencephalopathy, 9/486, 34/307, 46/388, 47/503, 507-509, 511-513, 56/497, 514, 63/354
pyruvate dehydrogenase deficiency, 29/212
radiation myelopathy, 61/201, 63/43
rattle snake venom, 9/39
recovery, 47/29-43
Refsum disease, 21/238, 41/434
Renpenning mental deficiency, 43/303
scapuloperoneal spinal muscular atrophy, 62/170
secondary, 47/29
segmental, *see* Segmental demyelination
serum antibody, 47/446
snake venom, 9/39
spastic paraplegia, 59/434
spinal cord compression, 69/170
spinal cord injury, 61/43
spontaneous impulse generation, 47/32, 38
stage, 10/35
subacute combined spinal cord degeneration, 9/614-617
subacute sclerosing panencephalitis, 56/417, 422
symptomatic disorder, 49/542
systemic lupus erythematosus, 55/373
tellurium intoxication, 9/645
thallium intoxication, 9/646
Theiler virus, 34/295
thermoregulation, 47/38
thrombotic thrombocytopenic purpura, 9/599-607
type, 47/29, 489
undetermined significant monoclonal gammopathy, 63/399
uremia, 9/639, 47/621
varicella zoster virus meningoencephalitis, 56/236
virus induced, *see* Virus induced demyelination
visna, 9/681, 56/583
visna virus, 56/583
vitamin $B_6$ intoxication, 65/572
vitamin $B_{12}$ neuropathy, 7/593
Whipple disease, 52/139
white matter, *see* White matter demyelination
white matter lipid, 9/1
Wilson disease, 9/634
xanthomatosis, 27/247
xanthomatous transformation, 27/247

Demyelinative disease
major form, 66/5

Denervation
*see also* Neuropathy
acetylcholine receptor, 40/13, 142
amyloid polyneuropathy, 51/417
critical illness polyneuropathy, 51/581
experimental, 40/138-143
experimental myopathy, 40/140
histochemistry, 40/13, 43-45, 140
membrane resistance, 40/555
motor unit, 62/62
muscle cell membrane, 40/127
muscle contraction, 62/53
muscle fiber, 40/11-14
muscular atrophy, 22/160, 181, 40/200
myotonia, 40/555
necrosis, 40/23
organophosphate induced delayed polyneuropathy, 64/177
parasympathetic, *see* Parasympathetic denervation
preferential atrophy, 40/140
segmental denervation, 40/205
target fiber, 40/31, 75, 142, 243

trophic influence, 40/13
Denervation atrophy
muscle fiber, 62/13
myopathy, 62/13
neuromuscular disease, 62/13
pituitary gigantism myopathy, 62/538
Denervation hypersensitivity
epileptogenesis, 72/178
Dengue fever
flavivirus, 56/12
meningoencephalitis, 56/12
mononeuropathy, 56/12
nerve lesion, 7/486
polyneuritis, 56/12
Dengue virus
Reye syndrome, 56/150
Denial of illness
agnosia, 4/23, 29
anosognosia, 45/376
body scheme disorder, 4/216, 219, 45/379
phantom limb, 4/224
Denny Brown, D., 1/14, 26
Denny Brown syndrome, *see* Sensory radicular neuropathy
Dens
*see also* Atlantoaxial dislocation *and* Atlantoaxial joint
ankylosing spondylitis, 26/175
aplasia, 32/39
atlantoaxial, 32/41
atlantoaxial compression, 26/289
bone abnormality, 50/388
developmental, 32/3-5
dysplasia, 32/27, 36, 40
embryology, 50/381
epiphyseal line, 25/250
features, 50/389
fracture, *see* Odontoid fracture
occipitoaxial injury, 25/229
ossification center, 25/253
radiation, 50/391
synchondrosis, 25/250
Dens agenesis
atlantoaxial subluxation, 50/389
features, 50/389
Dens fracture, *see* Odontoid fracture
Dens hypoplasia
cervicomedullary injury, 24/143
head injury, 24/143-147
Dense delta sign
venous sinus thrombosis, 54/401
Dental abnormality
Berlin syndrome, 43/9
cleidocranial dysplasia, 43/348
Coffin-Lowry syndrome, 43/239
Coffin-Siris syndrome, 43/240
dysosteosclerosis, 43/393
EEC syndrome, 43/395
elfin face syndrome, 31/318, 43/259
Ellis-Van Creveld syndrome, 43/347
enamel defect, 42/691
epilepsy, 42/691
frontometaphyseal dysplasia, 31/257, 43/402
Fukuyama syndrome, 43/92
Gorlin-Chaudry-Moss syndrome, 43/369
Hallermann-Streiff syndrome, 43/403
hydrocephalus, 42/691
incontinentia pigmenti, 14/214, 43/23
mental deficiency, 42/691
Miescher syndrome type II, 14/120
ophthalmoplegia, 43/141
osteogenesis imperfecta, 43/453
otodental dysplasia, 42/419
20p partial trisomy, 43/539
psychomotor retardation, 42/691
Rieger syndrome, 43/470
sensorineural deafness, 42/419
Dental causalgia
headache, 5/226
Dental headache, 48/391-394
apical periodontitis, 48/391
causalgia, 5/226
classification, 48/5
dry socket, 5/226
mastication, 48/392
osteomyelitis, 5/227, 48/392
pulpitis, 5/222, 48/391
stomatitis, 5/226
Dental malocclusion
craniosynostosis, 50/125
Dental pit
tuberous sclerosis, 50/374
Dentate epilepsy
theory, 21/515
Dentate fascia aplasia
Norman leukodystrophy, 10/104
Dentate gyrus
temporal lobe epilepsy, 73/80
Dentate ligament
spinal cord injury, 25/271
spinal nerve root, 25/400
Dentate nucleus
argininosuccinic aciduria, 42/525
ataxia, 21/514
cerebellar ablation, 2/406
cerebellar vermis agenesis, 42/4

cerebello-olivary atrophy, 42/136
dentatorubropallidoluysian atrophy, 21/528, 49/440
dyssynergia cerebellaris myoclonica, 21/509, 513, 42/211
Friedreich ataxia, 42/144
inferior artery, *see* Inferior artery to dentate nucleus
kernicterus, 6/502
lissencephaly, 42/40
myoclonic epilepsy, 42/701
myoclonus, 21/510, 42/701
striatonigral degeneration, 42/262
subacute necrotizing encephalomyelopathy, 42/588
superior artery, *see* Superior artery to dentate nucleus
tremor, 21/514
Dentate nucleus calcification
striatopallidodentate calcification, 49/426
Dentatorubral atrophy
Friedreich ataxia, 21/530
intention tremor, 21/529
Dentatorubral degeneration, *see* Dyssynergia cerebellaris myoclonica
Dentatorubral system
dentatorubropallidoluysian atrophy, 49/440
Dentatorubral tremor
dentatorubropallidoluysian atrophy, 21/529
Dentatorubropallidoluysian atrophy, 21/519-532, 60/607-617
*see also* Progressive myoclonus epilepsy
age at onset, 21/519
anatomy, 21/521
ataxia, 49/438
ataxic choreoathetoid type, 49/437, 440, 60/607
athetosis, 21/520
ballismus, 21/520, 531
carcinomatous cerebellar degeneration, 21/520
case report, 60/610
case series, 49/437
catalepsy, 60/644
central gray matter, 49/442
central tegmental tract, 49/442
cerebellifugal atrophy, 21/530
chorea, 21/519-521, 49/440, 60/663
choreoathetosis, 49/438, 60/663
clinical features, 21/519, 522, 49/437
clinical form, 60/607
clinicopathologic correlation, 60/616
computer assisted tomography, 60/671
course, 21/521
delusion, 49/440
dementia, 21/520, 49/440, 60/665
dentate nucleus, 21/528, 49/440
dentatorubral system, 49/440
dentatorubral tremor, 21/529
dentatothalamic pathway, 49/441
differential diagnosis, 60/616
distal amyotrophy, 60/659
duration of illness, 21/520
dyssynergia cerebellaris myoclonica, 21/528
dystonia, 21/519
external ophthalmoplegia, 60/656
fastigiovestibular system, 49/442
fecal incontinence, 60/664
Ferguson-Critchley syndrome, 59/322
Friedreich ataxia, 49/442, 60/613
GABA, 49/443
gait, 21/519
gaze paralysis, 49/440
genetics, 49/440
globus pallidus lateralis, 49/440
grimacing, 21/519
Hallervorden-Spatz syndrome, 21/520
history, 60/607
Huntington chorea, 49/437, 60/614
hypoesthesia, 60/661
inheritance, 72/133
intention tremor, 21/528
juxtarestiform, 49/442
kernicterus, 21/520
myoclonia, 49/439
myoclonic epilepsy, 49/440, 60/608
myoclonus, 60/662
neuropathology, 21/522-525, 49/440, 60/610-614
nosology, 21/532
nuclear magnetic resonance, 60/609, 673
nucleus ruber, 49/440
olivopontocerebellar atrophy, 21/521, 526, 49/443
opsoclonus, 21/520, 528, 49/440
pallidofugal atrophy, 21/531
pallidoluysian system, 49/440
pallidoluysiodentate degeneration, 49/452
pallidoluysionigral degeneration, 49/459
parkinsonism, 21/521
personality change, 49/440
posterior column, 49/442
progressive external ophthalmoplegia, 60/656
progressive myoclonus epilepsy, 21/528
pseudo-Huntington type, 49/437, 440, 60/608
ptosis, 60/655
reticular formation, 49/442
review, 49/437
rigidity, 21/520
slow eye movement, 49/440

speech, 21/519
spinal cord lesion, 60/615
spinocerebellar tract, 49/442
striatopallidodentate calcification, 21/520
subthalamic nucleus, 49/440
superior colliculi, 49/442
tegmental atrophy, 49/442
uncinate fasciculus, 49/442
urinary incontinence, 60/664
without spinal cord lesion, 60/615
Dentatospinonigro-oculomotor degeneration, *see* Woods-Schaumburg syndrome
Dentatothalamic pathway
dentatorubropallidoluysian atrophy, 49/441
Dentin
circadian rhythm, 74/492
Dentofacial syndrome (Weyers-Fülling)
ectodermal dysplasia, 14/112
Denver Conference on Chromosomes
classification, 30/92, 31/343, 376, 50/520
Deoxybarbiturate intoxication
anticonvulsant, 37/200
2-Deoxy-2-fluoroglucose F 18
Parkinson disease, 49/98
2-Deoxyglucose C 14
Parkinson disease, 49/98
Deoxyribonucleic acid, *see* DNA
Depakene, *see* Valproic acid
Depersonalization
consciousness, 3/116
epilepsy, 15/585
metamorphopsia, 2/662
sensory integration, 2/711
temporal lobe epilepsy, 2/711
temporal lobe lesion, 2/709, 711
temporal lobe psychosis, 24/532
temporal lobe tumor, 17/284
visual perception disorder, 2/709
Depigmentation
*p*-aminohippuric acid, 22/548
kwashiorkor, 7/635, 51/322
oculodermal, *see* Oculodermal depigmentation
substantia nigra, *see* Substantia nigra depigmentation
Vogt-Koyanagi-Harada syndrome, 56/613
Depressed skull fracture
*see also* Compound skull fracture
adult, 24/651
age, 23/404, 57/332
bone fragment, 57/322
brain abscess, 24/225
brain debridement, 57/322
brain herniation, 24/225
burst fracture, 24/651
cause, 57/332
child, 24/650, 57/331
closed fracture, 23/405, 412
complication, 23/407, 57/334
compound skull fracture, 23/405, 410, 24/226, 57/146, 334
cranioplasty, 23/411
definition, 23/403
diagnosis, 23/405
epilepsy, 57/334
growing fracture, 57/334
head injury, 24/225-227
herniation, 24/225
incidence, 57/331
intracranial hematoma, 23/409
intracranial infection, 24/224, 227
linear fracture, 57/333
mechanism, 23/94
meningitis, 23/407, 411, 24/225
missile injury, 23/511, 519, 523
newborn, 24/650
pneumocephalus, 24/211
posttraumatic brain abscess, 57/321
posttraumatic epilepsy, 24/448, 451
replacement, 57/322
sex ratio, 23/404
site, 23/405, 57/331
statistic, 23/404
surgery, 24/226, 649-654, 660, 57/334
traumatic epilepsy, 23/410
treatment, 23/410, 24/226, 57/322, 334
unconsciousness, 24/226
Depression, 46/471-477
*see also* Manic depressive psychosis
Addison disease, 46/427
Alzheimer disease, 46/208
amphetamine, 46/474
amyotrophic lateral sclerosis, 59/195
aphasia, 46/426
apomorphine, 46/474
atlas syndrome, 5/373
biochemistry, 46/475
biological rhythm, 74/504
bone marrow, *see* Bone marrow depression
brain infarction, 55/143
brain tumor, 46/425, 461
catecholamine, 46/448
caulerpa intoxication, 37/97
caulerpicin intoxication, 37/97
cerebral dominance, 46/473
cerebrovascular disease, 55/143
chlorophyceae intoxication, 37/97

circadian system, 74/507
classification, 46/444-446
clinicopathologic correlation, 46/427-429
clock function, 74/468
clonidine intoxication, 37/443
corticotropin releasing factor, 74/506
course, 57/414
Cushing syndrome, 46/427, 450
dementia, 46/122, 206-208, 427
dementia syndrome, 46/207
dexamethasone suppression test, 46/474, 477
diffuse Lewy body disease, 75/189
diurnal rhythm, 74/506
dysphrenia hemicranica, 5/79
ECG ST, *see* ECG ST depression
emotion, 3/336, 348
emotional disorder, 3/348
endocrine disorder, 46/427
endocrinology, 46/473-475
epilepsy, 3/348, 15/597, 46/461, 73/430, 454, 478
frontal cortex, 46/428
general paresis, 46/427
genetics, 46/471
green seaweed intoxication, 37/97
growth hormone, 46/449
growth hormone secretion, 46/474
head injury, 46/461
head injury outcome, 57/407, 413
headache, 48/6, 34, 359
hereditary, 29/515
hydrocortisone, 46/449, 74/506
hypercortisolism, 70/207
hyperparathyroidism, 46/427
hypothalamic hypophyseal system, 46/449
hypothalamus, 74/504
hypothyroidism, 46/427
insomnia, 45/139
insulin, 46/474
juvenile parkinsonism, 49/156
lateralized dysfunction, 46/489
levodopa, 46/474, 593
levodopa intoxication, 46/593
lidocaine intoxication, 37/450
light, 74/507
light therapy, 74/507
limbic system, 46/427-429
malignancy, 46/427
manic, *see* Manic depression
MAO, 46/448
melatonin, 74/504
memory, 46/123
mental, *see* Mental depression
methyldopa, 65/448
methyldopa intoxication, 37/441
migraine, 5/247, 253, 381, 48/73, 162
Morvan fibrillary chorea, 6/355
multiple sclerosis, 46/427, 47/59, 178
muscle contraction headache, 48/355
narcolepsy, 45/149
neurochemistry, 46/448
neuroleptic parkinsonism, 65/294
neurologic soft sign, 46/491
neuropsychology, 46/580
neurotic, *see* Neurotic depression
olivopontocerebellar atrophy (Dejerine-Thomas), 60/665
olivopontocerebellar atrophy (Schut-Haymaker), 60/665
oral contraception, 28/131, 46/599
organic affective syndrome, 46/425-429
organic brain syndrome, 46/476
organic mental disorder, 46/476
Parkinson dementia, 46/376
Parkinson disease, 46/208, 379, 477, 49/90, 108
parkinsonism, 29/515, 49/130
pellagra, 7/571, 28/71, 86, 46/336
physiology, 46/472
plasma hydrocortisone, 46/474
positron emission tomography, 46/428
posttraumatic etiology, 57/414
posttraumatic headache, 24/511
potassium imbalance, 28/480
prazosin intoxication, 65/447
progressive supranuclear palsy, 49/239, 243
prolactin secretion, 46/474
propranolol intoxication, 37/447
protirelin, 46/449, 474
protirelin test, 46/474
pseudo, *see* Pseudodepression
psychiatric headache, 5/20
psychobiology, 46/471-476
psychogenic headache, 5/247, 253
psychopharmacology, 46/448
quinidine intoxication, 37/448
reflex, *see* Reflex depression
REM sleep, 46/473
reserpine, 46/601
rhizostomeae intoxication, 37/53
spreading, *see* Spreading depression
stress, 74/504
stroke, 46/426
suprachiasmatic nucleus, 74/506
systemic lupus erythematosus, 46/427
taurine deficiency, 29/515
thyrotropic hormone, 46/474
traumatic psychosyndrome, 24/838

tricyclic antidepressant, 46/448
vasopressin neuron, 74/506
vitamin B6, 28/131
Wernicke aphasia, 46/428
Depressive pseudodementia
apathy, 46/222
concentration disturbance, 46/207, 222
forgetfulness, 46/207
psychomotor symptom, 46/207
subcortical dementia, 46/312
Depressor anguli oris muscle
cardiofacial syndrome, 42/307
facial paralysis, 42/313
unilateral facial paresis, 42/313
Depropanex, *see* Kallikrein
Depth perception
agnosia, 4/16
dissociated visual perception, 2/584
space perception, 45/412
Dercum disease, *see* Adiposis dolorosa
Derealization
sensory integration, 2/711
temporal lobe epilepsy, 2/711
temporal lobe lesion, 2/709, 711
visual perception disorder, 2/709
Derencephalus
anencephaly, 50/73
Dermacentor andersoni
Colorado tick fever, 56/141
tick paralysis, 41/290, 65/195
Dermal cylindroma
multiple, *see* Multiple dermal cylindroma
multiple trichoepithelioma, 14/591
Dermal dysplasia
focal, *see* Focal dermal dysplasia
Dermal hypoplasia
focal, *see* Goltz-Gorlin syndrome
Dermal nerve
leprous neuritis, 51/229
Dermal sinus
chronic meningitis, 56/645
diastematomyelia, 13/79, 50/436
diplomyelia, 50/436
spinal cord abscess, 52/191
spinal subdural empyema, 52/189
Dermal sinus and dermoid, 32/449-462
*see also* Dermoid cyst
clinical classification, 32/456
clinical features, 32/456
CNS, 32/458
course, 32/460
definition, 32/449, 454
diagnosis, 32/457-460
differential diagnosis, 32/460
embryology, 32/450-452
epidemiology, 30/165, 32/455
exogenous origin, 32/452-454
genetics, 32/455
history, 32/450
malformation, 32/458
pathogenesis, 32/452
pathology, 32/454
prognosis, 32/462
radiologic finding, 32/459
theory, 32/452
treatment, 32/461
Dermal sinus tract
anencephaly, 30/188
spina bifida, 50/496
Dermatan sulfate
chemistry, 27/144
CNS disorder, 10/327
connective tissue, 10/432
degradation, 66/65
glycosaminoglycanosis, 66/281
glycosaminoglycanosis type I, 27/146, 46/54
glycosaminoglycanosis type IH, 10/429
glycosaminoglycanosis type IS, 10/429
glycosaminoglycanosis type II, 10/429, 42/454
glycosaminoglycanosis type VI, 10/429
iduronic acid, 10/433
intervertebral disc, 20/526, 535
Dermatitis
angioneurotic edema, 43/60
Cockayne-Neill-Dingwall disease, 43/350
Dubowitz syndrome, 43/247
Hartnup disease, 42/146
histiocytosis X, 42/441
hyperphenylalaninemia, 42/611
lyngbya majuscula intoxication, 37/97
Muckle-Wells syndrome, 42/396
pellagra, 6/748, 28/59, 71, 86, 46/399, 51/333
phenylketonuria, 42/610
photosensitive, *see* Photosensitive dermatitis
recessive atopic, *see* Recessive atopic dermatitis
Rothmund-Thomson syndrome, 14/777, 43/460
seborrhea, 60/370
thalidomide, 42/660
tropical ataxic neuropathy, 7/640, 51/322
vitamin B3 deficiency, 51/333
Dermatitis herpetiformis
polymyositis, 62/373
Stargardt disease, 13/133
Dermatoglyphics
Aarskog syndrome, 43/313
acrodysostosis, 43/291

adducted thumb syndrome, 43/331
Armendares syndrome, 43/366
arthrogrypotic like hand malformation, 43/333
cerebral gigantism, 31/332, 43/336
cerebrohepatorenal syndrome, 43/339
Coffin-Lowry syndrome, 43/239
Coffin-Siris syndrome, 43/241
craniodigital syndrome, 43/243
cri du chat syndrome, 31/564-566, 43/501
De Lange syndrome, 31/281, 43/246
Down syndrome, 31/420-422, 43/545
Dubowitz syndrome, 43/249
Edwards syndrome, 31/507, 43/536
foramina parietalia permagna, 30/278
Huntington chorea, 6/300, 49/284
hypertelorism hypospadias syndrome, 43/413
Klinefelter variant XXYY, 31/484
Lenz syndrome, 43/419
Miller-Dieker syndrome, 50/593
monosomy 21, 31/598, 603
neuronal ceroid lipofuscinosis, 10/226
9p partial monosomy, 50/578
4p partial trisomy, 43/500
8p partial trisomy, 43/511
4p syndrome, 31/553-555
18p syndrome, 31/579
Patau syndrome, 31/507, 43/528, 50/558
Pena-Shokeir syndrome type I, 43/438
phakomatosis, 14/530
Prader-Labhart-Willi syndrome, 43/463
13q partial monosomy, 43/526
21q partial monosomy, 43/541
2q partial trisomy, 43/493
4q partial trisomy, 43/499
6q partial trisomy, 43/503
10q partial trisomy, 43/517
18q syndrome, 31/586
r(9) syndrome, 43/513, 50/581
r(13) syndrome, 50/586
r(21) syndrome, 31/594
r(22) syndrome, 31/609, 43/547
Rubinstein-Taybi syndrome, 31/284
Russell-Silver syndrome, 43/477
Saethre-Chotzen syndrome, 43/322
Sjögren-Larsson syndrome, 13/475, 22/477
Smith-Lemli-Opitz syndrome, 43/308
tetraploidy, 43/567
triploidy, 43/566
trisomy 8 mosaicism, 43/511
Turner syndrome, 31/499, 43/550
Wolf-Hirschhorn syndrome, 31/553-555
XXXX syndrome, 43/554
XXXXX syndrome, 43/554

Dermatographism
spinal cord injury, 26/259
spinal meningioma, 20/199
spinal shock, 26/259
Dermatoleukodystrophy
Babinski sign, 42/488
brain atrophy, 42/488
hyporeflexia, 42/488
mental deficiency, 42/488
neuroaxonal spheroid, 42/487-489
psychomotor retardation, 42/488
tonic seizure, 42/488
ventricular dilatation, 42/488
Dermatolysis, *see* Cutis laxa
Dermatome
*see also* Segmental innervation
arrangement, 2/158, 166
border, 2/158
chart, 1/94-96
definition, 1/89-91, 2/159
Head's zone, 2/159
modern limit, 1/98
muscle innervation, 2/132
scheme, 2/129
sympathetic, 1/453
topographical diagnosis, 2/170
Dermatomyositis
*see also* Inflammatory myopathy *and* Polymyositis
acute, 71/132
adult, 71/130
amyloid deposit, 62/377
amyloid β-protein, 62/377
amyopathic dermatomyositis, 62/371
angiopathic polyneuropathy, 51/446
antiepileptic agent, 65/499
anti-Jo-1 antibody, 62/374
arthralgia, 62/374, 71/130
aspiration pneumonia, 62/374
autoimmune antibody, 62/380
autoimmune mechanism, 40/264
axonal neuropathy, 62/376
azathioprine, 62/385
basophilic granular inclusion, 62/377
calcinosis, 62/386
cardiac disorder, 62/374
cardiomyopathy, 62/374
carotid system syndrome, 53/308
childhood, 71/130
chronic, 71/133
chronic enterovirus infection, 56/330
ciclosporin, 62/386
classification, 40/284, 71/4

collagen disease, 8/122
congestive heart failure, 62/374
corticosteroid, 62/383, 71/141
Coxsackie virus B, 56/201
creatine kinase, 62/375
cricopharyngeal myotomy, 62/386
cyclophosphamide, 62/385
diagnosis, 62/375
diagnostic criteria, 62/378
dyspnea, 62/374, 71/130
electronmicroscopy, 62/377
EMG, 1/642, 62/375, 71/139
eosinophilia myalgia syndrome, 62/371
eosinophilic cytoplasmic inclusion, 62/377
epidemiology, 71/129
experimental myopathy, 40/164
fatigue, 62/371
fever, 62/373
gastrointestinal symptom, 62/374
Gottron rash, 62/371
heliotrope rash, 41/57, 62/371
hematemesis, 62/374
hepatitis B virus infection, 56/303
histochemistry, 40/51
histology, 40/51
Hodgkin disease, 39/56
hypoxemia, 62/374
immune complex myositis, 56/201
immunopathology, 62/379
immunosuppressive therapy, 62/385
inclusion body, 40/264
intestinal pseudo-obstruction, 51/493
intravenous Ig, 62/386
joint contracture, 62/374
lactate dehydrogenase, 62/375
leukapheresis, 62/386, 71/143
lymphoid irradiation, 62/386
malaise, 62/373
malignancy, 62/371, 374, 71/131
malignant lymphoma, 39/56
melena, 62/374
membranolytic attack complex, 62/379
methotrexate, 62/385
mixed connective tissue disease, 62/371, 374
muscle biopsy, 62/375
muscle weakness, 62/371
myalgia, 62/371
myocarditis, 62/374
myoglobinuria, 40/339, 41/277, 62/556
nuclear magnetic resonance, 62/379
overlap syndrome, 62/374, 71/131
paraneoplastic syndrome, 69/374, 384, 71/673
pathogenesis, 62/379
pathologic reaction, 40/264
physiotherapy, 62/386
picornavirus, 41/71, 62/381
plasmapheresis, 62/386
pneumatosis intestinalis, 62/374
pneumonitis, 62/374
polyarteritis nodosa, 39/307, 55/362
polymyositis, 71/129-131
practical guideline, 62/386
primary inflammatory myositis, 62/369
prognosis, 71/130
pulmonary dysfunction, 62/374
Raynaud phenomenon, 41/56, 62/374
rheumatoid arthritis, 62/374
rimmed muscle fiber vacuole, 62/376
rimmed muscle fiber vacuole content, 62/18
rimmed muscle fiber vacuole type, 62/18
scleroderma, 62/374, 71/106
serum alanine aminotransferase, 62/375
serum aspartate aminotransferase, 62/375
serum muscle enzyme, 62/375
shawl sign, 62/371
Sjögren syndrome, 62/374, 71/74
skin, 71/130
spastic paraplegia, 59/439
steroid myopathy, 62/384
systemic lupus erythematosus, 62/374
systemic sclerosis, 62/371, 374
tachyarrhythmia, 62/374
treatment, 62/383
tubulofilament, 62/377
viral myositis, 62/381
virus, 40/264
virus isolation, 56/203
weight loss, 62/374
Dermatomyositis-polymyositis
cancer, 41/82, 347, 368-378
Dermochelys coriacea intoxication
ingestion, 37/89
Dermoid, 32/449-462
*see also* Dermoid cyst
arachnoid cyst, 31/87
cervico-oculoacusticus syndrome, 32/127
chemical meningitis, 56/130
congenital spinal cord tumor, 32/358-386
corneal, *see* Corneal dermoid
epibulbar, *see* Epibulbar dermoid
Mollaret meningitis, 56/630
spinal, *see* Spinal dermoid
spinal cord compression, 19/363
Treacher-Collins syndrome, 32/127
Dermoid cyst
*see also* Dermal sinus and dermoid *and* Dermoid

age, 20/72, 77, 68/330
anterior spinal, 20/79
anterior spinal meningocele, 42/43
associated defect, 20/72, 77
cerebellar tumor, 17/716
classification, 68/310
clinical features, 20/78, 31/64
clinical presentation, 68/331
congenital, 32/355-386
congenital brain tumor, 31/61-67
corpus callosum agenesis, 50/149
cranial vault tumor, 17/115
CSF, 20/79
diagnosis, 68/331
differential diagnosis, 31/64
epidemiology, 31/61-63
frequency, 20/77
histopathology, 68/330
lateral ventricle tumor, 17/606
location, 20/72, 77, 31/63
lumbar puncture, 20/75, 78
myelomeningocele, 20/76
orbital tumor, 17/202
origin, 20/75
pathogenesis, 31/61, 68/329
pathology, 20/73, 31/65, 68/236
pineal tumor, 17/654
posterior fossa tumor, 18/403
radiology, 20/78
retrospinal, 20/79
sex ratio, 20/72, 78, 68/330
spina bifida, 50/494, 496
spinal, *see* Spinal dermoid
spinal cord, 19/16
spinal lipoma, 20/397
treatment, 20/79, 31/66, 68/333
Dermoid tumor
epilepsy, 72/181
*N*-Desalkyl flurazepam, *see* Dealkylflurazepam
Descartes, R., 2/11
Descemet membrane, *see* Kayser-Fleischer ring
Descending cerebral vein
radioanatomy, 11/105
topography, 11/105
Descending medial occipital vein
radioanatomy, 11/109
topography, 11/109
Descending spinal fiber
stress, 74/313
Desensitization
epilepsy, 73/444
histamine, *see* Histamine desensitization
mechanism, 74/150
Desferrioxamine intoxication, *see* Deferoxamine intoxication
Designer drug
*see also* Neurotoxin
aliphatics, 65/274
arylhexylamine class, 65/364, 367
benzylfentanyl, 65/364
brolamfetamine, 65/364
4-bromo-2,5-dimethoxyphenethylamine, 65/364
butyrophenone, 65/274
chemical structure, 65/274
4-chloramphetamine, 65/364
classification, 65/274, 364
definition, 65/365
dimetamfetamine, 65/364
2,5-dimethoxyamphetamine, 65/364
epilepsy, 72/236
4-ethoxyamphetamine, 65/364
eticyclidine, 65/364
fentanyl, 65/349
fentanyl class, 65/364
4-fluorofentanyl, 65/364
β-hydroxy-(3-methyl)-thienylfentanyl, 65/364
β-hydroxythienylfentanyl, 65/364
illicit market, 65/365
legislature, 65/365
mecloqualone, 65/364
mefentanyl, 65/364
methadone, 65/349
methamphetamine, 65/364
methaqualone, 37/355, 65/364
methcathinone, 65/364
4-methoxyamphetamine, 65/364
5-methoxy-3,4-methylenedioxyamphetamine, 65/364
α-methyl-acetylfentanyl, 65/364
α-methyl-acrylfentanyl, 65/364
2-methylaminopropiophenone, 65/364
4-methylaminorex, 65/364
4-methyl-2,5-dimethoxyamphetamine, 65/364
4-methyl-2,5-dimethoxyphenethylamine, 65/364
3,4-methylenedioxyamphetamine, 65/364
3,4-methylenedioxyethamphetamine, 65/364
3,4-methylenedioxymethamphetamine, 65/364
α-methylfentanyl, 65/364
*N*-methyl-4-phenyl-2-butanamine hydrochloride, 65/364
1-methyl-4-phenyl-4-propionoxypiperidine, 65/364
1-methyl-4-phenyl-1,2,3,6-tetrahydropyridine, 65/364
3-methyl-thienylfentanyl, 65/364
nitromethaqualone, 65/364

1-(3-oxo-3-phenylpropyl)-4-propionoxypiperidine, 65/364
pethidine, 65/349
pethidine class, 65/364
phencyclidine, 65/364
phenethylamine, 65/366
phenethylamine class, 65/364
1-(2-phenethyl)-4-phenyl-4-acetoxypiperidine, 65/364
1-(2-phenethyl)-4-phenyl-1,2,5,6-tetrahydropyridine, 65/364
1-(2-phenethyl)-4-propionoxypiperidine, 65/364
phenothiazine, 65/274
piperazine, 65/274
piperidine, 65/274
piperidinocyclohexanecarbonitrile, 65/364
pyridine class, 65/368
pyridine structure, 65/368
quinazolinone class, 65/364
rolicyclidine, 65/364
street name, 65/364
tenocyclidine, 65/364
tetrahydroisoquinoline, 65/399
1-[1-(2-thienyl)cyclohexyl]-pyrrolidine, 65/364
thienylfentanyl, 65/364
thioxanthene, 65/274
trichlorotetrahydroharmanes, 65/399
2,4,5-trimethoxyamphetamine, 65/364
tryptoline, 65/398
Designer drug intoxication
latency period, 64/7
1-methyl-4-phenyl-1,2,3,6-tetrahydropyridine, 64/6
Desipramine
debrisoquine hydroxylation, 65/316
metabolism, 65/316
noradrenalin, 74/145
tricyclic antidepressant, 65/312
*N*-Desmethyldiazepam, *see* Nordazepam
Desmin intermediate filament
cytoplasmic inclusion body myopathy, 62/344
sarcoplasmic body myopathy, 62/345
Desmoid cyst
supratentorial brain tumor, 18/317
Desmoid tumor
muscle tumor, 41/383
Desmolase deficiency
congenital adrenal hyperplasia, 42/514
Desmoplastic cerebellar medulloblastoma
meninges, 18/184-186
prognosis, 18/189
subarachnoid space, 18/184
Desmoplastic infantile ganglioglioma, 68/147
classification, 67/4
Desmopressin
blood pressure, 75/724
orthostatic hypotension, 63/164
Desmopressin acetate
postlumbar puncture syndrome, 61/156
Desmosterol
myelin, 9/18
Desprogres-Gotteron artery
vascularization, 20/502
Deterioration
arsenic intoxication, 64/288
epilepsy, 15/586
head injury outcome, 57/376
mental, *see* Mental deterioration
procainamide intoxication, 37/452
traumatic intracranial hematoma, 57/281
DeToni-Fanconi-Debré syndrome
complex IV deficiency, 62/505, 66/424
mitochondrial DNA depletion, 66/430
Detrusor areflexia
spinal cord injury, 75/576
Deutan color blindness
genetic linkage, 40/352, 388, 43/88, 131
Deuteropathic motor system
amyotrophic lateral sclerosis, 22/43
Developmental agnosia
clumsiness, 46/159
Developmental apraxia
clumsiness, 46/159
Developmental articulatory defect
sex ratio, 43/219
socioeconomic class, 43/220
speech disorder in children, 43/219
Developmental deprivation
handedness, 4/468
Developmental diagnosis, 4/342-347
higher nervous activity, 4/365-373
Developmental disorder
body scheme disorder, 4/392-406
cerebral palsy, 4/398
CNS, *see* CNS developmental disorder
dihydrolipoamide dehydrogenase, 41/210
disconnection syndrome, 4/467
epilepsy, 72/340
higher nervous activity, 4/340-469
neonatal encephalopathy, 9/578
reading disability, 4/435-439
tactile apraxia, 4/34
tonic neck reflex, 4/346
virgin hand, 4/467
Developmental disorder of the body scheme, *see* Body scheme disorder

Developmental disorder of higher nervous activity, *see* Higher nervous activity
Developmental disorder of speech, *see* Speech disorder in children
Developmental dysarthria, 4/407-442
cerebral dominance, 4/465
classification, 4/408
dysgraphia, 4/434-439
dyslexia, 4/434-439
dysphonia, 4/409-413
higher nervous activity disorder, 4/465
lip abnormality, 4/413-419
palate abnormality, 4/413-419
secondary speech disorder, 4/419-427
specific retardation of speech development, 4/427-434
speech dysrhythmia, 4/409-413
stuttering, 4/409-413
tongue abnormality, 4/413-419
Developmental dyscalculia
cerebral dominance, 4/465
clumsiness, 46/160
higher nervous activity disorder, 4/465
Developmental dysgraphia
agraphia, 4/434-439
brain injury, 4/436
cause, 4/434
cerebral dominance, 4/465-469
examination, 4/435, 437
management, 4/438
specific developmental dysgraphia, 4/435
Developmental dyslexia, 4/377-390, 46/105-121
*see also* Reading disability
alexia, 4/123, 135
aphasia, 46/106, 110
arithmetic, 46/118
behavior disorder in children, 4/385
bilingualism, 46/120
boustrophedon, 46/116
brain injury, 4/377, 436, 46/118
brain pathology, 46/119
cerebral dominance, 4/380, 465-469, 46/116
chronic system disease, 4/384
classification, 46/118
clinical features, 4/387, 46/107-116
clumsiness, 46/117
cognition, 46/120
cognitive patterning, 4/379
congenital dysgraphia, 46/112
critical period phenomenon, 4/383
definition, 4/377, 46/107
deprivation effect, 4/383-386
differential diagnosis, 4/388
directional constancy, 4/381
etiology, 4/377-386, 434, 46/119
examination, 4/387, 435, 437
eye movement, 46/114-116
factor analysis, 4/378
familial trait, 4/383
form constancy, 4/382
genetics, 46/118
Gerstmann syndrome, 4/401, 46/117-119
handedness, 4/380, 383, 46/117
hearing defect, 4/384
historic note, 46/105-107
hyperactivity, 46/117
idiopathic, *see* Idiopathic developmental dyslexia
immaturity, 46/114
incidence, 4/386
learning, 46/108-113
maturation lag, 4/377
maturity, 46/120
memory, 46/113, 117
mental deficiency, 4/377
mental illness, 4/385
minimal brain injury, 4/378
motor aphasia, 4/134
neurologic sign, 4/383, 46/113-116
object constancy, 4/381
oculography, 46/115
Orton theory, 4/380, 46/106
perception, 46/120
polyglot, 46/120
prevention, 4/388
prognosis, 4/389, 46/120
reading, 46/108-113
reading disability, 4/134
reading test, 46/109
regressive dyslexia, 4/388
rehabilitation, 4/388, 438
right left disorientation, 4/401
serialization, 46/113
sex ratio, 43/222
specific, 43/221
speech, 46/114
spelling dyslexia, 46/111-113
strephosymbolia, 46/107
subtype, 46/118
synkinesis, 46/114
transient dyslexia, 4/388
treatment, 4/388, 438, 46/120
Turner syndrome, 4/379
visual disorder, 4/384
writing, 46/108-113
Developmental dysphasia, 4/407-442, 46/139-144
attention deficit disorder syndrome, 46/141

auditory imperception, 4/361
auditory perception, 46/140
autism, 46/141
body scheme disorder, 4/401
brain injury, 4/265, 361
buccolingual apraxia, 46/140
central deafness, 4/361
classification, 46/139
dyslexia, 46/126
EEG, 46/141
emotional disorder, 46/141
environment, 46/141
hearing loss, 46/141
incidence, 46/139
intelligence, 46/141
Landau-Kleffner syndrome, 46/141
language, 46/141
mental deficiency, 4/361, 46/141
minimal brain dysfunction, 46/141
prognosis, 46/142
psychomotor disorder, 4/401
schizophrenia, 46/141
sensory aphasia, 46/140
spatial orientation, 4/401
stuttering, 46/140
theory, 46/142-144
treatment, 46/142
verbal auditory agnosia, 46/140
Developmental dyspraxia
apraxia, 4/362, 401, 443-464, 43/219
classification, 4/458
clinical features, 4/458
constructional activity, 4/451-453
constructional apraxia, 4/458
débilité motrice, 4/454
drawing, 4/450
dressing apraxia, 4/459
everyday activity, 4/449
examination, 4/446-453
facial apraxia, 4/458
genetic psychology, 4/445
handedness, 4/465-469
intelligence, 4/460
mental illness, 4/462
motor activity, 4/443-446
motor efficiency, 4/448-450
object apraxia, 4/459
oculomotor apraxia, 4/459
postural apraxia, 4/459
praxis, 4/450-453
psychopathology, 4/459-462
spatial dyskinesia, 4/458
verbal apraxia, 4/459
virgin hand, 4/467
Developmental expressive dysphasia
speech disorder in children, 4/428
Developmental receptive dysphasia
speech disorder in children, 4/428
Devic disease, *see* Neuromyelitis optica
Devil grip, *see* Bornholm disease
Devil ray intoxication
elasmobranch, 37/70
Devonshire colic
lead intoxication, 64/431
Dexamethasone
*see also* Corticosteroid
anoxic ischemic leukoencephalopathy, 47/535
brain edema, 63/417
congenital heart disease, 63/10
hyperaldosteronism, 39/493
neurobrucellosis, 33/321, 52/596
pyruvate decarboxylase deficiency, 41/428
steroid myopathy, 41/252
vasogenic brain edema, 47/535
Dexamethasone suppression test
anorexia nervosa, 46/586
depression, 46/474, 477
mania, 46/450
manic depressive psychosis, 43/208
Dexamphetamine
amphetamine intoxication, 65/256
Dexterity
manual, *see* Manual dexterity
Dextran
brain fat embolism, 55/191
brain infarction, 11/356
cardiac arrest, 55/210
neurologic intensive care, 55/210
restless legs syndrome, 8/318, 51/548
Dextroamphetamine, *see* Dexamphetamine
Dextrocardia
craniolacunia, 50/139
Dextromethorphan
chemistry, 37/366
Dextromoramide
chemistry, 37/366
Dextroposed aorta
holoprosencephaly, 50/237
Dextropropoxyphene
chemistry, 37/366
delirium, 46/540
opiate, 65/350
Dextropropoxyphene intoxication
epilepsy, 65/358
hallucination, 65/358
symptom, 65/358

synthetic opioid, 37/366
Dextrose
intracranial pressure, 23/212
DFP intoxication, *see* Dyflos intoxication
Dhurrin
cyanogenic glycoside intoxication, 36/515, 65/26
Di Ferrante syndrome, *see* Glycosaminoglycanosis type VIII
Diabetes insipidus
anhidrotic ectodermal dysplasia, 14/788
anterior cerebral artery syndrome, 53/349
autosomal dominant, 42/543
Bardet-Biedl syndrome, 13/395
basofrontal syndrome, 53/349
brain aneurysm, 55/53
brain aneurysm neurosurgery, 55/53
brain stem death, 57/483
carbamate intoxication, 64/186
cause, 2/445
eosinophilic granuloma, 63/422
frontal lobe syndrome, 53/349
Hand-Schüller-Christian disease, 38/96, 42/442
head injury, 46/395, 57/237
histiocytosis X, 42/442, 63/422
hypernatremia, 63/554
infantile neuroaxonal dystrophy, 59/75
late cerebellar ataxia, 60/668
leukemia, 39/10, 63/342
Möbius syndrome, 50/215
multiple myeloma, 18/255
neurosarcoidosis, 63/422
optic atrophy, 42/411
orthostatic hypotension, 63/145
pertussis encephalopathy, 33/290
pituitary adenoma, 17/391
pseudohypoparathyroidism, 6/706, 49/420
recurrent meningitis, 52/52
skeletal abnormality, 60/668
spinocerebellar degeneration, 60/650
subarachnoid hemorrhage, 55/9, 53
temporal arteritis, 55/344
Wegener granulomatosis, 71/182
Diabetes mellitus
*see also* Hyperglycemia
abducent nerve, 27/117
alaninuria, 42/515
aldose reductase inhibitor, 70/152
Alström-Hallgren syndrome, 13/460, 42/391
aminoaciduria, 42/411
aminoguanidine, 70/154
amphotonia, 22/251
amyotrophic lateral sclerosis, 22/370
amyotrophy, 8/35, 19/86, 22/30, 27/121, 40/318, 326
angiopathy, 42/544
anhidrosis, 1/457
Argyll Robertson pupil, 27/114, 124
arhinencephaly, 9/582, 14/120
arterial hypertension, 54/209
ataxia, 60/660, 674
ataxia telangiectasia, 14/274, 293, 60/674
autonomic nervous system, 75/26, 589
autonomic neuropathy, 70/148, 156
autonomic polyneuropathy, 51/476
Bardet-Biedl syndrome, 13/463
Berardinelli-Seip syndrome, 42/591
bicarbonate, 70/201
bladder, 75/672
bladder dysfunction, 75/600
bowel, 75/594
brain atrophy, 42/411
brain edema, 70/199
brain hypoxia, 70/196
brain infarction, 53/17
brain injury, 23/123, 125
bulbospinal muscular atrophy, 59/44
candidiasis, 52/398
cardiovascular dysfunction, 75/596
carpal tunnel syndrome, 7/296
caudal aplasia, 30/126
central sleep apnea, 63/462
cerebrovascular disease, 27/126-130, 53/17, 75/366
Charlevoix-Saguenay spastic ataxia, 60/457, 674
chorioretinal degeneration, 13/39
chronic axonal neuropathy, 51/531
circadian dysrhythmia, 75/597
CNS, 70/193
conduction failure, 70/142
cranial nerve, 42/544
cryptococcosis, 52/433
cystopathy, 70/141
deficiency neuropathy, 51/324
delirium, 46/544
dementia, 46/397
demyelinating neuropathy, 47/621
electrophysiology, 70/141
embryopathy syndrome, 31/298
encephalopathy, 27/123-136
epidemiology, 70/131
erection dysfunction, 75/601
esophagus, 42/543-545, 75/592
Fabry disease, 60/172
facial nerve, 27/118
facial paralysis, 8/269, 27/118
facioscapulohumeral muscular dystrophy, 43/99

fatty acid, 70/153
female sexual function, 75/602
fibrous dysplasia, 14/186
Friedreich ataxia, 21/349, 42/144, 51/388, 60/319, 325, 674
gallbladder, 75/594
ganglioside, 70/153
gastrointestinal disorder, 75/591
gastrointestinal neuropathy, 70/141
gastroparesis, 75/593
genitourinary dysfunction, 75/600
glycogen storage disease, 42/543
granulomatous CNS vasculitis, 55/388
headache, 5/230
hemochromatosis, 42/553
hereditary, *see* Hereditary diabetes mellitus
hereditary cranial nerve palsy, 60/41
hereditary muscular atrophy, 13/306
hereditary paroxysmal kinesigenic choreoathetosis, 49/355
Herrmann disease, 22/510
HLA antigen, 42/544
hyperosmolality, 70/196
hypoglycemia, 27/131
hypogonadal cerebellar ataxia, 21/476
immunocompromised host, 56/469, 473
inositol, 70/154
inositol deficiency, 70/150
insulin administration, 70/198
insulin resistant, *see* Insulin resistant diabetes mellitus
intestinal pseudo-obstruction, 51/493
intrauterine growth acceleration, 46/82
ischemia, 70/142
ischemic-hypoxic cycle, 70/151
ischemic neuropathy, 8/157-161
juvenile, *see* Juvenile diabetes mellitus
Kearns-Sayre-Daroff-Shy syndrome, 62/309, 508
ketonemia, 70/195
Kii Peninsula amyotrophic lateral sclerosis, 22/371
lactic acidosis, 70/200
listeriosis, 52/91
local anesthetic agent, 65/421
Machado-Joseph disease, 42/155
maternal, *see* Maternal diabetes mellitus
mental deficiency, 42/335, 411
metabolic acidosis, 70/195
metabolic neuropathy, 51/68
microangiopathy, 42/544
Miescher syndrome type II, 14/119
mitochondrial disease, 66/394
mitochondrial DNA point mutation, 66/427
moyamoya disease, 55/295
mucormycosis, 35/543, 52/469, 474
mummy, 74/194
myelomalacia, 63/42
myelopathy, 27/121-123
myotonic dystrophy, 40/511
nerve ischemia, 12/655
neurogenic muscular atrophy, 60/674
neurogenic osteoarthropathy, 8/31, 27/115
neurogenic ulcer, 27/114
neurologic manifestation, 27/99-133
neuropathy, 42/544, 60/660, 70/143, 74/118, 75/589
neurotrophic osteoarthropathy, 38/462-464
nonketotic coma, 70/196
oculomotor nerve, 27/117
oculomotor neuropathy, 70/143
olfactory nerve, 27/118
onion bulb formation, 21/252
ophthalmoplegia, 27/117
ophthalmoscopy, 11/214
optic atrophy, 13/39, 60, 64, 42/411
optic nerve, 27/118
orthostatic hypotension, 63/148, 70/139, 75/599
pain, 70/154
papilledema, 16/286
pathogenesis, 70/152
peripheral arterial dysfunction, 75/599
photogenic epilepsy, 42/714
polyneuropathy, 42/334, 70/133, 144
Prader-Labhart-Willi syndrome, 31/324, 43/464
prevalence, 42/544
primary pigmentary retinal degeneration, 13/306, 42/334, 411
progressive external ophthalmoplegia, 62/309
pseudohypoparathyroidism, 6/706, 49/420
psychiatric disorder, 42/411
pulmonary dysfunction, 75/603
pupillary dysfunction, 75/590
pupillary function, 70/140
radiculopathy, 27/113, 70/135
rectum, 75/595
restless legs syndrome, 42/261, 51/546
retinal artery occlusion, 55/112
retinopathy, 27/119
Rud syndrome, 13/321, 38/9, 43/284
secondary pigmentary retinal degeneration, 60/660, 674
seizure, 42/411
sensorineural deafness, 42/391
sexual impotence, 27/115, 70/141, 74/118
silent ischemia, 75/598
spheroid body, 6/626

spinal cord embolism, 63/42
spinal muscular atrophy, 59/374
spinocerebellar degeneration, 60/674
stomach, 75/592
streptococcal meningitis, 52/79
stroke, 75/366
subarachnoid hemorrhage, 55/9
sudomotor dysfunction, 75/602
sudomotor impairment, 70/139
survey, 70/131
systemic brain infarction, 11/457
tachycardia, 75/597
treatment, 60/325
tremor, 49/589
trigeminal nerve, 27/119
trochlear nerve, 27/117
Turner syndrome, 50/544
Urbach-Wiethe disease, 42/595
urinary incontinence, 27/115
urinary retention, 27/115
vasa nervorum disease, 12/655-658
vitamin, 70/153
Weiss-Alström syndrome, 13/463
Werner syndrome, 43/489
Diabetic amyotrophy, *see* Diabetic proximal neuropathy
Diabetic autonomic neuropathy
autonomic nervous system, 75/29, 246, 648
cardiovascular, 75/591
gastrointestinal, 75/591
genitourinary, 75/591
incidence, 75/589
manifestation, 75/591
neuroendocrine, 75/591
prevalence, 75/589
pupillary, 75/591
respiratory, 75/591
risk, 75/589
sudomotor, 75/591
survey, 70/138
Diabetic autonomic polyneuropathy
cardiac dysrhythmia, 63/238
orthostatic hypotension, 1/472, 22/263, 63/148
Diabetic autonomic visceral neuropathy
atonic bladder, 8/37
manifestation, 27/114-116
neurogenic diarrhea, 27/115
orthostatic hypotension, 27/115
pathology, 27/116
sexual impotence, 8/38, 27/115
urinary retention, 8/37
Diabetic coma
*see also* Nonketotic coma
brain edema, 70/199
Diabetic ketoacidosis
brain edema, 27/83-89
coma, 27/130
encephalopathy, 45/122
mucormycosis, 35/543, 52/470
survey, 70/194
Diabetic myopathy, *see* Diabetic proximal neuropathy
Diabetic neuropathy, 8/29-52, 27/100-119
*see also* Diabetic polyneuropathy
anhidrosis, 1/457
asymmetric, 27/112
autonomic nervous system, 74/604, 75/29
carpal tunnel syndrome, 7/296
classification, 70/134
cranial, 8/40, 27/119
cranial nerve deficit, 5/230
CSF, 8/43, 27/103
differential diagnosis, 7/553, 8/50, 367
EMG, 8/43
facial paralysis, 27/118
hyperalgesic, *see* Hyperalgesic diabetic neuropathy
hyperglycemia, 27/92-94
ischemic, 8/157
nerve conduction, 7/169
nerve ischemia, 12/655
ophthalmoplegia, 27/117
osteoporosis, 70/20
pathogenesis, 8/50
pathology, 8/45, 27/104-110
polyol pathway, 27/93, 110
radiculopathy, 27/113
recurrent nonhereditary multiple cranial neuropathy, 51/570
sensory polyneuropathy, 8/31
sorbitol, 27/93, 110
treatment, 8/51
Valsalva ratio, 74/604
vitamin B6 deficiency, 70/432
Diabetic ophthalmoplegia, *see* Diabetic neuropathy
Diabetic polyneuropathy, 70/135
*see also* Diabetic neuropathy *and* Endocrine polyneuropathy
aldehyde reductase, 51/507
alloxan, 51/512
angiopathy, 51/505
anhidrosis, 63/148
asymmetrical neuropathy, 51/499
autonomic dysfunction, 63/148
autonomic neuropathy, 51/499
blood-nerve barrier, 51/505

capillary closure, 51/511
cardiac dysrhythmia, 63/238
compression neuropathy, 51/501
CSF, 27/103
diabetic tabes, 51/499
diagnosis, 27/103
distal polyneuropathy, 51/499
distal sensory neuropathy, 51/499
electrophysiology, 51/502
endoneurial hypoxia, 51/506
focal neuropathy, 51/501
form, 51/499
fructose, 51/507
H-reflex, 51/504
hereditary amyloid polyneuropathy, 21/129
history, 51/497
H-M interval discrimination, 51/504
hypoxia, 51/506
incidence, 27/102, 51/498
inositol, 51/508
inositol hypothesis, 51/509
manifestation, 27/101
metabolic hypothesis, 51/506
microangiopathy, 51/505
mononeuropathy, 51/505
motor nerve conduction, 51/507
motor neuropathy, 51/499
multifocal neuropathy, 51/501
multiple mononeuropathy, 51/499
nerve conduction, 27/103
nerve conduction velocity, 51/503
nerve fiber loss, 51/511
nerve lesion topography, 51/510
nerve protein nonenzymatic glycosylation, 51/508
onion bulb formation, 21/163, 51/511
orthostatic hypotension, 1/472, 22/263, 63/148
osmosis hypothesis, 51/507
pain control, 51/514
paradoxical ischemia, 51/503
pathogenesis, 27/102, 107-110, 51/505
pathology, 27/104-110
polyol pathway, 51/506
prevalence, 51/498
prognosis, 27/112
proximal neuropathy, 51/501
rapid reversible neuropathy, 51/498
sensorimotor neuropathy, 51/500
sensory loss pattern, 51/57
sensory neuropathy, 8/30, 367, 51/500
sexual impotence, 51/500
sodium-potassium adenosine triphosphatase, 51/508
streptozocin, 51/510, 512
sural nerve biopsy, 51/510
symmetrical polyneuropathy, 51/499
sympathetic nervous disorder, 2/119
thermal discrimination threshold, 51/504
treatment, 27/110-112, 51/513
vacor intoxication, 51/512
vascular hypothesis, 51/505
Diabetic proximal neuropathy
amyotrophic lateral sclerosis, 59/201
atrophy, 27/132
muscle weakness, 40/318
muscular atrophy, 40/318
pain, 40/326
pathogenesis, 62/542
survey, 70/137
Diabetic tabes
diabetic polyneuropathy, 51/499
Diabetic thigh muscle infarction
clinical features, 62/542
histopathology, 62/542
recurrence, 62/542
Diacetate
myasthenia gravis, 41/132
Diacetyl vinblastine amide sulfate, *see* Vindesine
Diacyl-glycerophospholipid
chemistry, 10/240
Diadematidae intoxication
pain, 37/67
spinal injury, 37/67
Diagnoses
correct, *see* Correct diagnosis
developmental, *see* Developmental diagnosis
faulty, 2/2
introduction, 2/1-3
localization theory, 2/14
neurologic, *see* Neurologic diagnosis
prenatal, *see* Prenatal diagnosis
radioisotope, *see* Radioisotope diagnosis
Diagonal band
Broca, 2/770, 6/32
Diagonal band (Broca), *see* Broca diagonal band
Dialkylphosphoryl enzyme
organophosphate intoxication, 64/224
Dialysis
*see also* Hemodialysis
aluminum intoxication, 63/523, 525
aluminum neurotoxicity, 63/525
ataxia, 63/528
atlantoaxial dislocation, 63/529
autonomic neuropathy, 63/523
axonal degeneration, 63/530
back pain, 63/529
camptodactyly, 63/529

carpal tunnel syndrome, 63/529
causalgia, 63/530
chronic aluminum intoxication, 63/525
confusional state, 63/528
deafness, 63/523, 530
dialysate aluminum content, 63/525
discopathy, 63/529
epilepsy, 63/525
Fabry disease, 66/240
foot drop, 63/530
hearing loss, 63/530
history, 63/528
hypercalcemia, 63/562
ischemic neuropathy, 63/523, 529
listeriosis, 52/91
β2-microglobulin amyloidosis, 63/528
mononeuritis multiplex, 63/530
myoclonus, 63/525
neurologic complication, 63/523
ophthalmoplegia, 63/528
peritoneal, *see* Peritoneal dialysis
radiculopathy, 63/529
red blood cell transketolase, 63/528
renal, *see* Renal dialysis
stuttering, 63/525
subdural hematoma, 63/523, 525
tetraplegia, 63/529
uremia, 63/504, 523
uremic encephalopathy, 63/505
vitamin B1, 63/528
Wernicke encephalopathy, 63/523

Dialysis dementia
*see also* Dialysis encephalopathy
Creutzfeldt-Jakob disease, 56/546
delirium, 46/542
uremia, 46/398

Dialysis disequilibrium syndrome
brain edema, 63/523
coma, 63/523
computer assisted tomography, 63/524
CSF pressure, 63/524
CSF urea, 63/524
disorientation, 63/523
epilepsy, 63/523
headache, 63/523
idiogenic osmoles, 63/524
muscle cramp, 63/523
muscular twitching, 63/523
myoclonus, 63/523
plasma osmolality, 63/524
prevention, 63/524
reverse urea shift, 63/524
symptom, 63/523

Dialysis encephalopathy
*see also* Aluminum intoxication *and* Dialysis dementia
aluminum bioavailability, 64/274
aluminum intoxication, 36/323, 64/276
anomia, 64/277
apraxia, 64/277
ataxia, 63/528
coma, 63/526
confusional state, 63/526, 528
deferoxamine, 63/527
delirium, 46/542, 63/526
dementia, 64/277
diazepam, 63/527
EEG, 63/526, 64/277
epilepsy, 63/526, 64/277
external ophthalmoplegia, 63/528
hallucination, 64/277
iteration, 64/277
metabolic encephalopathy, 63/528
β2-microglobulin amyloidosis, 63/528
myoclonia, 64/277
ophthalmoplegia, 63/528
oral phosphate binder, 64/277
paraphasia, 64/277
patient at risk, 63/526
peritoneal dialysis, 27/341
prophylaxis, 63/527
red blood cell transketolase, 63/528
symptom, 27/323
transketolase, 63/528
vitamin B1, 63/528
Wernicke encephalopathy, 63/528

Diamidine, *see* Mepyramine

Diaminobenzidine
neurotropic virus, 56/26

Diaminophenoxyalkane
experimental animal, 13/208
retinal degeneration, 13/208

3,4-Diaminopyridine
Eaton-Lambert myasthenic syndrome, 62/426

*cis*-Diamminodichloroplatinum, *see* Cisplatin

Diamorphine
acquired myoglobinuria, 62/577
acute effect, 65/352
aortitis, 63/54
brain hemorrhage, 55/517
brain infarction, 53/164, 55/517
brain ischemia, 53/164
chemical structure, 65/352
derivation, 65/352
endocarditis, 37/391
hepatitis, 37/393

lumbar plexus, 37/388
malaria, 37/393
myoglobinuria, 41/271
myopathy, 37/390
opiate, 65/350
osteomyelitis, 37/391
overdose, 65/352
polyneuropathy, 37/389
spastic paraplegia, 59/438
spinal cord vasculitis, 63/55
spongiform leukoencephalopathy, 47/566-568
tetanus, 37/393
toxic myopathy, 62/601
tuberculosis, 37/392
Diamorphine addiction
acquired immune deficiency syndrome, 55/517
amblyopia, 55/518
angioneurotic edema, 55/519
brain angiography, 55/518
brain hemorrhage, 55/517, 65/356
brain infarction, 55/518, 65/356
brain vasculitis, 55/518
cerebellar hemorrhage, 65/356
hypergammaglobulinemia, 55/519
hypersensitivity, 55/519
hypertension, 55/517
immune complex, 55/519
myelopathy, 65/356
occlusive vasculopathy, 55/518
renal disease, 55/517
temporal arteritis, 55/518
toxic myopathy, 62/600
transverse myelitis, 37/387
uremia, 55/517
Diamorphine encephalopathy
drug re-exposure, 63/54
Diamorphine intoxication
analgesia, 65/352
anterior spinal artery infarction, 65/356
brachial plexus neuritis, 65/356
brain infarction, 65/355
brain mucormycosis, 65/355
complication, 65/355
compression neuropathy, 65/356
EEG, 65/352
endocarditis, 65/355
euphoria, 65/352
fatality rate, 65/355
fungal infection, 65/355
hepatotoxicity, 65/355
human immunodeficiency virus infection, 65/356
infection, 65/355
infectious hepatitis, 65/355
malaria, 65/355
meningitis, 65/355
1-methyl-4-phenyl-1,2,3,6-tetrahydropyridine, 65/357
miosis, 65/352
mycotic aneurysm, 65/355
myoglobinuria, 65/357
neuropathy, 65/356
nocardial brain abscess, 65/355
orthostatic hypotension, 65/352
pinpoint pupil, 65/352
polyneuropathy, 65/356
postanoxic encephalopathy, 65/356
pulmonary hypertension, 65/355
REM sleep, 65/352
renal disease, 65/355
respiratory depression, 65/352
spongiform encephalopathy, 65/357
subarachnoid hemorrhage, 65/355
thrombophlebitis, 65/355
vasculopathy, 65/356
vertebral osteomyelitis, 65/355
wound botulism, 65/355
Diamorphine myelopathy
cervical spinal cord, 63/54
drug abuse, 55/519
myelomalacia, 55/519
necrotic myelopathy, 65/356
paraplegia, 55/519
symptom, 55/519
thoracic spinal cord, 63/54
Diamorphine withdrawal syndrome
symptom, 65/354
Diaphoresis
pheochromocytoma, 42/764
Shapiro syndrome, 50/167
Diaphragm
function, 26/335, 337
infantile spinal muscular atrophy, 59/62
myotonic dystrophy, 40/517
paralysis, *see* Diaphragm paralysis
plexus lesion, 7/409
rupture, 26/192, 195-197, 226
spinal shock, 26/337
stab wound, 25/205
Diaphragm hernia
anencephaly, 50/84
caudal aplasia, 50/512
Edwards syndrome, 50/275
Ehlers-Danlos syndrome, 43/14
Galloway-Mowat syndrome, 43/431
iniencephaly, 50/131
renal agenesis syndrome, 50/512

Diaphragm paralysis
C3 innervation component, 2/161
congenital rapidly progressive muscular dystrophy, 43/94
diphtheria, 51/183
spinal cord injury, 61/266
spinal shock respiration, 26/254
Diaphragmatic respiration
C3 level lesion, 61/56
cervical spinal cord injury, 61/56
Diaphyseal aclasis, *see* Hereditary multiple exostosis
Diaphyseal dysplasia, 43/375-377
contracture, 43/376
cranio, *see* Craniodiaphyseal dysplasia
exophthalmos, 43/376
facial paralysis, 43/376
gait, 43/376
optic atrophy, 43/376
pain, 43/376
progressive, *see* Progressive diaphyseal dysplasia
sensorineural deafness, 43/376
Diaphyseal hyperplasia
craniofacial dysostosis (Crouzon), 43/361
Diarrhea
abnormal secretion, 75/652
acute mercury intoxication, 64/375
alcohol intoxication, 64/112
apistus intoxication, 37/76
arsenic polyneuropathy, 51/266
ataxia telangiectasia, 60/668
Balaenoptera borealis intoxication, 37/94
bismuth intoxication, 64/331
carbamate intoxication, 64/186
chelonia intoxication, 37/89
cluster headache, 48/222
cobalt intoxication, 36/330
congestin intoxication, 37/37
disulfiram intoxication, 37/321
echinoidea intoxication, 37/67
fish roe intoxication, 37/86
fluroxene intoxication, 37/412
fox intoxication, 37/96
glycosaminoglycanosis type III, 66/301
gold intoxication, 64/356
guanethidine intoxication, 37/436
gymnapistes intoxication, 37/76
Hartnup disease, 29/155
hereditary acute porphyria, 51/391
hexachlorophene intoxication, 37/486
ichthyohepatoxic fish intoxication, 37/87
mackerel intoxication, 37/87
meduso-congestin intoxication, 37/37
methyl salicylate intoxication, 37/417
migraine, 48/8, 125, 156
MNGIE syndrome, 62/505
neurogenic, *see* Neurogenic diarrhea
nothesthes intoxication, 37/76
palytoxin intoxication, 37/55
paralytic shellfish poisoning, 37/33
pellagra, 28/66, 71, 87, 38/648, 46/336, 399, 51/333
polar bear intoxication, 37/96
primary amyloidosis, 51/415
primary carnitine deficiency, 66/403
procainamide intoxication, 37/452
pterois volitans intoxication, 37/75
salicylic acid intoxication, 37/417
Satoyoshi syndrome, 64/256
scorpaena intoxication, 37/76
scorpaenodes intoxication, 37/76
scorpaenopsis intoxication, 37/76
sebastapistes intoxication, 37/76
sodium salicylate intoxication, 37/417
stingray intoxication, 37/72
subacute myelo-optic neuropathy, 51/296
toxic shock syndrome, 52/258
vagotomy, 75/653
vitamin A intoxication, 37/96
vitamin B3 deficiency, 51/333
wolf intoxication, 37/96
zinc intoxication, 36/337, 64/362
Diarthrodial joint
anatomy, 19/323
Diaschisis
*see also* Spinal shock
brain, *see* Brain diaschisis
brain infarction, 53/120
brain ischemia, 53/120
cerebral localization, 3/369
faulty localization, 2/2
mechanism, 53/121
motor, *see* Motor diaschisis
restoration of higher cortical function, 3/369
spinal cord, 2/181
topography, 53/120
Diasialotransferrin deficiency, *see* Jaeken syndrome
Diastasis recti
Goltz-Gorlin syndrome, 14/788
holoprosencephaly, 50/237
Diastematomyelia, 32/239-253
*see also* Diplomyelia
adult, 32/251-253
anencephaly, 42/25
arachnoid abnormality, 50/437
Arnold-Chiari malformation type I, 32/247,

50/441
back pain, 50/436
cauda equina tumor, 19/87
caudal aplasia, 50/436
central medullary cavity, 50/438
clinical features, 32/243, 50/435
CNS, 32/244
computer assisted tomography, 50/436, 439
congenital hip dislocation, 42/24
congenital scoliosis, 32/253
craniolacunia, 32/187, 50/139
cutaneous manifestation, 32/243
Dandy-Walker syndrome, 42/22
definition, 32/239, 50/435
dermal sinus, 13/79, 50/436
diagnosis, 50/436
differential diagnosis, 19/87
diplomyelia, 20/58, 61, 30/164, 32/242, 42/24
dura mater abnormality, 50/437
dura mater reduplication, 50/436
dura mater sheath reduplication, 50/436
embryology, 32/240
epidemiology, 50/439
family, 50/440
filum reduplication, 50/438
filum terminale, 50/438
first symptom, 32/243
flatfoot, 50/436
2 groups, 50/436
hair tuft, 50/436
hammer toe, 50/436
hemangiomatous nevus, 50/436
hemicord, 50/438
hemivertebra, 42/24, 30
history, 32/239-241
hydrocephalus, 50/441
hyperlordosis, 50/436
hypertrichosis, 50/498
incidence, 50/64
intraspinal lipoma, 50/439
intraspinal neurenteric cyst, 32/433
kyphosis, 50/436
leg abnormality, 50/436
leg shortening, 50/436
lipoma, 50/436, 438
malformation, 32/242
medullary fissure, 50/438
meningomyelocele, 50/437, 441
monoplegia, 50/436
muscle stretch reflex, 50/436
muscular atrophy, 50/436
myelography, 19/87, 187, 50/437, 439
myelomeningocele, 42/25
neurogenic bladder, 42/24
neurologic syndrome, 50/436
occipital cephalocele, 50/101
orthopedic syndrome, 50/436
pathophysiology, 32/241
pes cavovarus, 50/436
pigmented nevus, 50/436
pilonidal sinus, 50/436
posterior arch abnormality, 50/436
radiologic appearance, 32/244-248, 50/436
scoliosis, 32/253, 42/24, 50/416, 435
sensory trouble, 50/436
sex ratio, 50/64, 440
skin abnormality, 50/436
sphincter disturbance, 50/436
spina bifida, 50/482, 497
split notochord syndrome, 20/62, 32/241
surgical technique, 50/440
surgical treatment, 32/248-251
talipes, 42/24
terminal ventricle, 50/438
tethered cord, 50/438
tethered cord syndrome, 50/435
treatment, 50/440
vertebral abnormality, 42/24
vertebral bony spur, 50/435, 437, 439
vertebral fusion, 50/436
vertebral interpedicular widening, 50/439
vertebral segmentation, 50/436
Diastematomyelia entodermica, *see* Neurenteric cyst
Diastrophic dwarfism
cheilopalatoschisis, 43/382
collagen disease, 43/384
ear abnormality, 43/382
kyphoscoliosis, 43/382-384
spinal cord compression, 43/384
talipes, 43/382-384
vertebral abnormality, 43/384
Diazacholesterol
cataract, 40/514
myopathy, 40/155, 328
myotonia, 40/155, 514, 548, 555
myotonia acquisita, 40/328
myotonic dystrophy, 40/548
20,25-Diazacholesterol, *see* Azacosterol
Diazepam
alcohol withdrawal delirium, 46/541
alpha coma, 55/220
aluminum intoxication, 63/527
antiepileptic agent, 37/199, 65/497
benzodiazepine intoxication, 65/334
chemical structure, 37/356
description, 37/357

dialysis encephalopathy, 63/527
dyssynergia cerebellaris myoclonica, 60/603
epilepsy, 15/645
hemiballismus, 49/376
motor unit hyperactivity, 41/296
multiple sclerosis, 47/157
neuroleptic akathisia, 65/290
pyrethroid intoxication, 64/216
serum level, 15/692
side effect, 15/709
spastic paraplegia, 61/368
spasticity, 26/482
spinal cord injury, 61/368
spinal spasticity, 61/368
status epilepticus, 15/184
stiff-man syndrome, 40/546, 41/306
structural formula, 37/356
tetanus, 41/305, 65/223
toxic myopathy, 62/601
transverse spinal cord lesion, 61/368
Diazepam intoxication
drug interaction, 37/199
higher cortical function, 37/203
metabolite, 37/199
nordazepam, 37/199, 357
oxazepam, 37/199, 357
reticular activating system, 37/203
vestibular system, 37/203
Diazinon, *see* Dimpylate
Diazoxide
action mechanism, 37/432
antihypertensive agent, 37/432
antihypertensive agent intoxication, 37/433
chemical structure, 37/432
chemistry, 37/366
hypertensive encephalopathy, 54/216
iatrogenic neurological disease, 65/446
neurotoxin, 37/432, 65/446
Diazoxide intoxication
brain perfusion pressure, 37/433
brain vasospasm, 37/433
hyperglycemia, 37/433
hyperuricemia, 37/433
myocardial infarction, 37/433
oculogyric crisis, 37/433, 65/446
parkinsonism, 37/433, 65/446
rigidity, 37/433
tremor, 37/433, 65/446
trismus, 37/433
Dibasic amino acid malabsorption
cystinuria, 7/596
deficiency neuropathy, 7/596
Dibenzazepine intoxication
anticonvulsant, 37/200
Dibenzodiazepine
chemical structure, 65/276
Dibenzo-1,4-dioxin
neurotoxin, 64/197
toxic polyneuropathy, 64/197
Diborane intoxication
dizziness, 64/359
headache, 64/359
hiccup, 64/359
metal intoxication, 64/353
muscle spasm, 64/359
muscle weakness, 64/359
myalgia, 64/359
respiratory system, 64/359
tremor, 64/359
vertigo, 64/359
DIC, *see* Intravascular consumption coagulopathy
Dicarboxylic aciduria
lipid metabolic disorder, 62/494
long chain acyl-coenzyme A dehydrogenase deficiency, 62/495
1,1-Dichloro-2,2-bis(4-chlorophenyl)ethane
organochlorine, 36/393, 64/197
organochlorine insecticide intoxication, 64/197
Dichlorodiphenylethane
organochlorine, 64/197
organochlorine insecticide intoxication, 64/197
Dichlorodiphenyltrichloroethane, *see* Chlorphenotane
Dichlorodiphenyltrichloroethane intoxication, *see* DDT intoxication
1,2-Dichloroethylene
organic solvent intoxication, 64/40
2,4-Dichlorophenoxyacetic acid
experimental myopathy, 40/160
experimental myotonia, 40/154
Isaacs syndrome, 41/301
mitochondrial abnormality, 40/160
myotonia, 40/548, 555
myotonia acquisita, 40/328
neurotoxicity, 51/286
toxic neuropathy, 7/523, 51/286
2,4-Dichlorophenoxyacetic acid intoxication
drug induced myotonia, 62/271
drug induced myotonic syndrome, 62/271
2,4-Dichlorophenoxyacetic acid polyneuropathy
axonal degeneration, 51/286
pain, 7/523
Dichlorvos intoxication
chemical classification, 37/545
organophosphorus compound, 37/545
Dichotic listening

cerebral dominance, 4/39
corpus callosum agenesis, 50/161
handedness, 4/320-323
language, 46/149
schizophrenia, 46/503, 505
stuttering, 46/171
Diclofenac
headache, 48/174
migraine, 48/174
Dicobalt edetate, *see* Cobalt edetate
Dicodid, *see* Hydrocodone
Dicofol
organochlorine, 36/393, 64/197
organochlorine insecticide intoxication, 64/197
Dicoumarol
multiple sclerosis, 9/384
Dictyotene
Down syndrome, 50/523
Didanosine
neurotoxicity, 71/368
toxic effect, 71/372
Dideazaphilanthotoxin 12
arthropod envenomation, 65/200
DIDMOAD syndrome, *see* Wolfram syndrome
Dieldrin
chlorinated cyclodiene, 64/200
neurotoxin, 36/393, 411, 64/197, 201
organochlorine, 36/393, 64/197
organochlorine insecticide intoxication, 36/411-417, 64/197
toxic encephalopathy, 64/201
Dieldrin intoxication
clinical features, 36/411-414
EEG, 36/414-416
laboratory data, 36/414-416
pathology, 36/417
prognosis, 36/416
treatment, 36/417
Diencephalic artery
radioanatomy, 11/76
topography, 11/76
Diencephalic epilepsy
EEG, 2/452
hypothalamus, 2/452
Diencephalic syndrome, 2/432-468
athrepsia, 18/366
autonomic nervous system, 75/39
Colle sign, 42/734
infantile optic glioma, 42/733
literature, 18/366
mental deficiency, 46/24
neurofibromatosis, 14/741-746
nystagmus, 18/367, 42/733
precocious puberty, 68/71
traumatic, *see* Traumatic diencephalic syndrome
Diencephalic tumor
hydrocephalus, 67/156
symptom, 67/156
Diencephalic vein
ventral, *see* Ventral diencephalic vein
Diencephalon
autonomic nervous system, 74/59
brain death, 24/738
dysphrenia hemicranica, 5/80
optic pathway, 68/69
respiration, 1/653
vascularization, 11/96
2,4-Dienoyl-coenzyme A reductase
mitochondrial matrix system, 66/412
rare presentation, 66/412
Diet
*see also* Nutrition
adrenomyeloneuropathy, 59/354
argininosuccinic aciduria, 29/100, 42/526
arterial hypertension, 54/206
blood lead level, 64/433
fatty acid, 9/399
fructose intolerance, 42/549
galactosemia, 42/552
glycogen storage disease, 43/178, 182
histidinemia, 42/555
hyperammonemia, 42/561
hyperargininemia, 29/102, 42/563
hyperdibasic aminoaciduria, 42/564
hyperlysinemia, 42/567
hyperphenylalaninemia, 29/37
hypervalinemia, 42/574
hypophosphatasia, 42/579
isovaleric acidemia, 42/580
maple syrup urine disease, 29/76-79, 42/599
mercury intoxication, 64/383
metachromatic leukodystrophy, 10/58
methionine malabsorption syndrome, 29/127, 42/600
methylmalonic acidemia, 29/207, 42/603
multiple sclerosis, 9/103, 113, 399, 47/153-155
ornithine carbamoyltransferase deficiency, 29/90
pellagra, 28/66, 85, 46/336
phenylketonuria, 29/43, 42/611, 59/75
Refsum disease, 8/24, 21/190, 218-221, 27/520, 530-537, 36/347, 42/149, 60/229, 234, 236, 66/497
Richner-Hanhart syndrome, 42/582
saccharopinuria, 29/222
spina bifida, 50/479
tea and toast, *see* Tea and toast diet

Diethyl bisdimethyl pyrophosphordiamide intoxication
chemical classification, 37/549
organophosphorus compound, 37/549
Diethylcarbethoxydichloromethylphosphonate intoxication, *see* Forstenon intoxication
Diethyl 4-chlorophenyl phosphate intoxication
chemical classification, 37/545
organophosphorus compound, 37/545
Diethyl 2-chlorovinyl phosphate intoxication
chemical classification, 37/545
organophosphorus compound, 37/545
*O,O*-Diethyl *S*-2-diethylmethylammoniumethyl phosphonothiolate methylsulfate intoxication
chemical classification, 37/550
organophosphorus compound, 37/550
Diethyldithiocarbamic acid
calcium carbimide polyneuropathy, 51/309
carbamate intoxication, 64/187
disulfiram polyneuropathy, 51/309
thallium intoxication, 64/326
Diethyl ether, *see* Ether
*O,O*-Diethyl *S*-ethsulfonylmethyl phosphorothioate intoxication
chemical classification, 37/547
organophosphorus compound, 37/547
*O,O*-Diethyl *S*-eththiomethyl phosphorothioate intoxication
chemical classification, 37/547
organophosphorus compound, 37/547
*O,O*-Diethyl *S*-eththionylmethyl phosphorothioate intoxication
chemical classification, 37/547
organophosphorus compound, 37/547
*O,O*-Diethyl *O*-(4-methylumbelliferyl) phosphorothioate intoxication
chemical classification, 37/546
organophosphorus compound, 37/546
*O,O*-Diethyl *S*-(4-nitrophenyl) phosphorothioate intoxication
chemical classification, 37/548
organophosphorus compound, 37/548
*O,S*-Diethyl *O*-(4-nitrophenyl) phosphorothioate intoxication
chemical classification, 37/548
organophosphorus compound, 37/548
Diethyl phosphorocyanidate intoxication
chemical classification, 37/544
organophosphorus compound, 37/544
Diethyl phosphorofluoridate intoxication
chemical classification, 37/543
organophosphorus compound, 37/543
*O,O*-Diethyl *O*-(8-quinolyl) phosphorothioate intoxication
chemical classification, 37/547
organophosphorus compound, 37/547
Diethylstilbestrol
creatine kinase, 40/343, 378
muscular dystrophy, 40/343, 378
*N,N*-Diethyl-*m*-toluamide intoxication
animal study, 36/443
skin test, 36/443
Diethyl trichloromethylphosphonate intoxication
chemical classification, 37/545
organophosphorus compound, 37/545
Diezel, P.B., 10/24
Diffuse astrocytoma, *see* Diffuse glioblastosis
Diffuse axonal injury
axonal retraction ball, 57/53
brain injury mechanism, 57/35
brain stem, 57/53
brain swelling, 57/380
capsula interna, 57/55
caudate nucleus, 57/55
cerebellar folia, 57/55
computer assisted tomography, 57/151, 153, 380
corpus callosum, 57/53
external capsule, 57/55
head injury, 57/52, 151, 153, 380
nuclear magnetic resonance, 57/169
pathology, 57/52
prognosis, 57/380
radiodiagnosis, 57/151
scar, 57/53
traumatic intracerebral hematoma, 57/156
Diffuse cerebellar hypertrophy
cerebellar tumor, 17/716
Diffuse cerebral sclerosis, *see* Diffuse sclerosis *and* Leukodystrophy
Diffuse cerebral sclerosis (Krabbe), *see* Globoid cell leukodystrophy
Diffuse cerebral sclerosis (Schilder), *see* Diffuse sclerosis
Diffuse cerebral sclerosis (Scholz), *see* Metachromatic leukodystrophy
Diffuse corticomeningeal angiomatosis, *see* Divry-Van Bogaert syndrome
Diffuse cutaneous systemic sclerosis
scleroderma, 71/101
Diffuse disseminated sclerosis, *see* Transitional sclerosis
Diffuse dysrhythmia
epilepsy, 15/9
Diffuse glioblastosis, 67/5
age, 18/77
brain hemisphere enlargement, 18/73

classification, 10/6
diagnosis, 18/78
glial heterotopia, 14/145
histology, 18/74
macroscopy, 18/73
multiple sclerosis, 18/77
neurofibromatosis, 18/75
neurofibromatosis type I, 18/75-77
pathogenesis, 18/75
prevalence, 18/77
symptomatology, 18/77
tuberous sclerosis, 18/77
Diffuse glioma, *see* Diffuse glioblastosis
Diffuse lentiginosis
multiple endocrine adenomatosis type III, 42/753
Diffuse leptomeningeal gliomatosis
chronic meningitis, 56/645
Diffuse leukoencephalopathy
acquired immune deficiency syndrome, 56/509
Diffuse Lewy body disease
autonomic dysfunction, 75/188
dementia, 59/410, 75/189
depression, 75/189
differential diagnosis, 59/410
hallucination, 75/189
paranoia, 75/189
parkinsonism, 59/410
progressive external ophthalmoplegia, 62/304
Diffuse lymphocytic meningoencephalitis
acute, *see* Acute diffuse lymphocytic meningoencephalitis
Diffuse myelinoclastic sclerosis, *see* Myelinoclastic diffuse sclerosis
Diffuse myelitis
classification, 10/22
Diffuse plasmacytosis
multiple myeloma, 18/257
Diffuse schwannosis
central, *see* Diffuse glioblastosis
Diffuse sclerosis
*see also* Leukodystrophy
Adams-Richardson classification, 9/472
Addison disease, 39/482
adrenoleukodystrophy, 47/419
athetosis, 6/457
Bertrand classification, 10/18
Bielschowsky description, 47/419
Bielschowsky-Henneberg classification, 10/12
Blackwood classification, 10/22
brain biopsy, 10/685
bulge cell, 10/73
chemical classification, 10/24
classification, 9/471-473, 10/1-37
concentric, *see* Concentric sclerosis
Creutzfeldt-Jakob disease, 6/747
definition, 10/36
degenerative sporadic type, 10/15
degenerative type, 10/6, 10
dementia, 46/404
Einarson classification, 9/472
Einarson-Neel classification, 10/15
false concept, 60/168
familial heredodegenerative type, 9/472, 10/11, 14, 16
familial infantile, *see* Familial infantile cerebral sclerosis
familial 2 subtypes, 10/14
glioblastomatous type, 9/472, 10/15
gliomatosis type, 10/6
globoid cell, 10/68-70, 74
globoid cell type, *see* Globoid cell leukodystrophy
Greenfield classification, 9/472, 10/15-17
Hallervorden classification, 10/15, 24
history, 10/1-37, 43
infantile, *see* Infantile diffuse sclerosis
inflammatory, *see* Inflammatory diffuse sclerosis
Krabbe classification, 10/4
Krabbe type, *see* Globoid cell leukodystrophy
lead intoxication, 9/642
leukodystrophic type, 9/472, 10/19
metachromatic, *see* Metachromatic leukodystrophy
multifocal noninflammatory demyelinating type, 9/472
multifocal perivenous inflammatory demyelinating type, 9/472
multiple sclerosis, 9/138
myelinoclastic, *see* Myelinoclastic diffuse sclerosis
Neubürger classification, 9/472, 10/6
nosology, 10/1-37, 60/168
optic atrophy, 13/83
Pelizaeus-Merzbacher disease, 9/472
Poser classification, 9/472
Poser-Van Bogaert classification, 9/472, 10/18
progressive multifocal leukoencephalopathy, 9/472
Schilder type, 9/139, 10/18
Scholz-Bielschowsky-Henneberg type, *see* Metachromatic leukodystrophy
sporadic degenerative type, 9/472
Strümpell description, 47/419
subacute sclerosing panencephalitis, 9/470, 472, 10/20, 47/421
sudanophilic type, 9/472
three types, 10/8

true type, 9/472
Van Bogaert-Bertrand classification, 10/13
Van Bogaert classification, 10/13, 15
venous sinus thrombosis, 12/433
Wicke classification, 10/14
Diffuse sclerosis with meningeal angiomatosis, *see* Divry-Van Bogaert syndrome
Diffuse simultaneous corticosubcortical epilepsy classification, 15/11
Diffuse tapetoretinal dystrophy, *see* Centroperipheral tapetoretinal dystrophy
Di-2-fluoroethyl phosphorofluoridate intoxication
chemical classification, 37/543
organophosphorus compound, 37/543
Digalactosyl ceramide
Fabry disease, 42/428
Digastric line
basilar impression, 50/398
DiGeorge syndrome
hypoparathyroidism, 70/121
Digestion
autonomic nervous system, 74/1
corpora amylacea, 27/191
Digestive neural reflex
vagovagal reflex, 75/624
Digestive system
*see also* Intestine
anencephaly, 50/84
autonomic nervous system, 74/632
ethosuximide intoxication, 37/203
neural control, 1/503
primidone intoxication, 37/203
valproic acid intoxication, 37/203
Digestive system function disorder
cause, 70/224
Digit agraphia
acalculia, 45/474
Digit alexia
acalculia, 45/474
Digital abnormality
*see also* Clinodactyly
Coffin-Siris syndrome, 43/240
De Lange syndrome, 43/246
Down syndrome, 43/545
Dubowitz syndrome, 43/249
Edwards syndrome, 43/535
FG syndrome, 43/250
Freeman-Sheldon syndrome, 38/413, 43/353
Juberg-Hayward syndrome, 43/446
Lenz syndrome, 43/419
mental deficiency, 43/262, 269
orofaciodigital syndrome type I, 43/447
progressive myositis ossificans, 43/190
13q partial monosomy, 43/526
18q partial monosomy, 43/533
21q partial monosomy, 43/541
10q partial trisomy, 43/517
19q partial trisomy, 43/538
r(22) syndrome, 43/547
sensorineural deafness, 42/378
tetraploidy, 43/467
Turner syndrome, 43/550
XXXX syndrome, 43/554
XXXXX syndrome, 43/554
Digital autotopagnosia
Alzheimer disease, 46/251
Digital nerve
compression neuropathy, 51/102
occupational lesion, 7/32
Digital subtraction angiography
advantage intra-arterial over intravenous, 54/170
brain stem death, 57/464
complication, 54/172
contraindication, 54/170
indication, 54/170
principle, 54/169
transient ischemic attack, 53/270
Digitalgia paresthetica
terminology, 1/108
Digitalis
behavior disorder, 46/601
Digitalis intoxication
delirium, 46/538
gynecomastia, 37/427
headache, 48/420
historic aspect, 37/425
lithium, 46/538
Digito-ocular sign
congenital retinal blindness, 13/139, 272, 22/530
Digitoxin
neurotoxin, 37/426, 65/449
Digitoxin intoxication
color vision, 13/13
physiologic variation, 37/426
D-α,β-Diglyceride
biochemistry, 9/8
Digoxin
cardiac glycoside, 65/449
cardiac pharmacotherapy, 63/195
delirium, 63/195
iatrogenic neurological disease, 63/195, 65/449
indication, 65/449
irritability, 63/195
neurologic adverse effect, 65/449
neurotoxin, 37/426, 63/195, 65/449
restlessness, 63/195

stupor, 63/195
trigeminal neuralgia, 63/195
Digoxin intoxication
chorea, 65/450
chromatopsia, 65/450
color discrimination, 65/450
confusion, 65/450
delirium, 65/450
hallucination, 65/450
lethargy, 65/450
nausea, 65/450
physiologic variation, 37/426
transient global amnesia, 65/450
treatment, 65/451
visual disorder, 65/450
vomiting, 65/450
Dihexoside sulfate
late infantile metachromatic leukodystrophy, 10/141
Dihydrocodeine
opiate, 65/350
toxic myopathy, 62/601
Dihydrocodeinone, *see* Hydrocodone
Dihydroergotamine
alpha adrenergic receptor, 65/66
autonomic polyneuropathy, 51/487
chemistry, 65/63
orthostatic hypotension, 63/164
pharmacokinetics, 65/63
pharmacology, 65/63, 66
serotonin receptor, 65/66
side effect, 5/98, 380
Dihydroergotoxine, *see* Co-dergocrine
Dihydrofolate
vitamin Bc deficiency, 51/336
Dihydrofolate reductase
vitamin Bc deficiency, 51/336
10,11-Dihydro-5H-dibenz[*b,f*]azepine
extrapyramidal disorder, 6/829
Dihydroindolone
chemical structure, 65/276
Dihydrolipoamide acetyltransferase
monomer, 66/415
pyruvate dehydrogenase complex, 66/413, 415
Dihydrolipoamide dehydrogenase
developmental disorder, 41/210
homodimer, 66/415, 420
α-hydroxybutyric acid, 66/417
α-hydroxyisovaleric acid, 66/417
lactic acidosis, 41/210, 66/419
microcephaly, 41/210
2-oxoglutaric acid, 66/417
psychomotor retardation, 41/210
pyruvate dehydrogenase complex, 66/413, 415
subacute necrotizing encephalomyelopathy, 66/419
Dihydrolipoamide dehydrogenase deficiency
maple syrup urine disease, 66/651
Dihydrolipoyl dehydrogenase, *see* Dihydrolipoamide dehydrogenase
Dihydrolipoyl transacetylase, *see* Dihydrolipoamide acetyltransferase
Dihydromorphinone, *see* Hydromorphone
Dihydrophenylacetic acid
CSF, 45/136
daytime hypersomnia, 45/136
Dihydropteridine reductase deficiency
biochemistry, 29/39
epilepsy, 72/222
phenylalanine, 29/38-42
phenylketonuria, 29/38-42, 42/611
Dihydropyrimidine dehydrogenase
epilepsy, 72/222
Dihydrosphingosine
biochemistry, 9/8
chemistry, 10/246
1,25-Dihydroxycholecalciferol, *see* 1,25-Dihydroxyvitamin D3
Dihydroxycholestanemia
facial dysmorphia, 60/666
3,4-Dihydroxyphenylacetic acid, *see* Homoprotocatechuic acid
*levo-threo*-3,4-Dihydroxyphenylserine, *see* *threo*-3,4-Dihydroxyphenylserine
*threo*-3,4-Dihydroxyphenylserine
blood pressure, 75/724
freezing gait, 49/69
Parkinson disease, 49/69
progressive dysautonomia, 59/140
3,4-Dihydroxyphenylserine therapy
dopamine, 75/239
1,25-Dihydroxyvitamin D3
vitamin D3 intoxication, 65/569
Diisopropyl phosphorofluoridate, *see* Dyflos
Diisopropyl phosphorofluoridate intoxication, *see* Dyflos intoxication
Diisopropyl phosphoroiodidate intoxication
chemical classification, 37/544
organophosphorus compound, 37/544
1,4-Diketone
acetonylacetone intoxication, 64/85
diketone polyneuropathy, 64/84
1,4-Diketone neuropathy
acrylamide, 64/64
acrylamide intoxication, 64/64
Diketone polyneuropathy

carbonyl spacing, 64/84
1,4-diketone, 64/84
Dilantin, *see* Phenytoin
Dilated cardiomyopathy, *see* Congestive heart failure
Dilaudid, *see* Hydromorphone
Diltiazem
migraine, 48/200
myotonia, 62/276
parkinsonism, 63/194
DiMauro disease, *see* Carnitine palmitoyltransferase deficiency
Dimaval, *see* Unithiol
Dimefox intoxication
chemical classification, 37/543
organophosphorus compound, 37/543
Dimercaprol
arsenic intoxication, 36/214, 64/289
arsenic polyneuropathy, 51/266, 268
bismuth intoxication, 64/344
lead encephalopathy, 64/443
lead intoxication, 36/25-28, 64/439
lead polyneuropathy, 51/270
mercury intoxication, 36/181, 64/394
organolead intoxication, 64/134
thallium intoxication, 36/272
Wilson disease, 27/380, 402, 404
Dimercaprol intoxication
headache, 48/420
2,3-Dimercaptopropane sulfonate, *see* Unithiol
2,3-Dimercapto-1-propanesulfonic acid, *see* Unithiol
2,3-Dimercaptopropanol, *see* Dimercaprol
2,3-Dimercaptosuccinic acid, *see* Succimer
Meso-Dimercaptosuccinic acid, *see* Succimer
Dimetamfetamine
designer drug, 65/364
2-{[2-(2,6-Dimethoxyphenoxy)ethyl]amino-methyl}-1,4-benzodioxan
classification, 74/149
3,4-Dimethoxyphenylethylamine, *see* Homoveratrylamine
Dimethylaminoethanol, *see* Deanol
2-Dimethylaminoethyl dimethylphosphinate intoxication
chemical classification, 37/545
organophosphorus compound, 37/545
Dimethylaminopropionitrile
acrylamide, 64/66
bladder dysfunction, 75/506
chemical formula, 64/82
neuropathy, 75/506
organic solvent, 64/82
Dimethylaminopropionitrile polyneuropathy
autonomic neuropathy, 51/282
axonal degeneration, 51/283
cauda equina syndrome, 51/282
sensory neuropathy, 51/282
symptom, 51/280
urinary hesitancy, 51/282
*N,N*-Dimethylamphetamine, *see* Dimetamfetamine
Dimethylcarbamic acid 1-isopropyl-3-methylpyrazol-5-yl ester
carbamate intoxication, 64/185
3,4-Dimethyl-2,5-hexanedione
acetonylacetone intoxication, 64/85
chemical formula, 64/82
neurotoxicity, 51/278
organic solvent, 64/82
*N,N*-Dimethylhydrazine
chemical formula, 64/82
organic solvent, 64/82
*O,O*-Dimethyl *S*-(4-nitrophenyl) phosphorothioate intoxication
chemical classification, 37/548
organophosphorus compound, 37/548
Dimethylnitrosourea
brain tumor carcinogenesis, 17/18
Dimethyl-*p*-phenylenediamine
experimental chemical data, 40/157
experimental myopathy, 40/160
I band, 40/157
mitochondrial abnormality, 40/160
Z band, 40/157
Dimethyl sulfoxide
brain infarction, 53/430
nerve conduction, 7/174
Dimethyl 1,2,2,2-tetrachloroethyl phosphate intoxication
chemical classification, 37/545
organophosphorus compound, 37/545
Dimethylthiambutene
chemistry, 37/366
5-(3,3-Dimethyl-1-triazeno)imidazole-4-carbox-imide, *see* Dacarbazine
Dimethyltryptamine
psychosis, 46/594
*N,N*-Dimethyltryptamine
*see also* Hallucinogenic agent
hallucinogenic agent, 65/42
*N,N*-Dimethyltryptamine intoxication
visual hallucination, 65/45
Dimitri-Biemond disease, *see* Biemond syndrome
Dimpylate
neurotoxicity, 37/546
organophosphate intoxication, 64/154

Dimpylate intoxication
chemical classification, 37/546
organophosphorus compound, 37/546
Dimyelia, *see* Diplomyelia
Dincman disease, *see* Chronic psychotic choreoathetosis
Dinitrofluorobenzene
contracture, 41/297
creatine kinase inhibition, 41/264
2,4-Dinitrophenol
experimental myopathy, 40/158
Luft syndrome, 41/210
muscle tissue culture, 40/187
Dinitrophenol intoxication
action mechanism, 36/442
animal study, 36/440
treatment, 36/443
2,4-Dinitrophenylhydrazine test
maple syrup urine disease, 42/599
Dinoflagellates intoxication
*see also* Saxitoxin intoxication
artificial respiration, 37/35
Dinophysiotoxin
neurotoxin, 65/149
okadaic acid, 65/149
shellfish poisoning, 65/165
Dinoprostone
multiple sclerosis, 47/351, 355
spinal cord compression, 69/170
Diodone
iatrogenic neurological disease, 63/44
Dioxathion, *see* Dioxation
Dioxation
organophosphate intoxication, 64/155
Dioxin
neurotoxicity, 51/286
Dioxin intoxication
cerebrotendinous xanthomatosis, 60/167
differential diagnosis, 60/167
polyneuropathy, 60/167
Dipeptidyl carboxypeptidase I inhibitor
arterial hypertension, 65/444
brain blood flow, 65/444
captopril, 65/444
cardiovascular agent, 65/433
enalapril, 65/444
epilepsy, 65/445
headache, 65/445
Huntington chorea, 65/445
iatrogenic neurological disease, 65/445
indication, 65/444
lisinopril, 65/444
neuropharmacology, 65/444
parkinsonism, 65/445
pharmacologic action, 65/444
polyneuropathy, 65/445
progressive supranuclear palsy, 65/445
sensorimotor polyneuropathy, 65/445
Dipeptidyl carboxypeptidase I inhibitor intoxication
CSF, 65/445
CSF protein, 65/445
deafness, 65/446
demyelinating neuropathy, 65/445
dysgeusia, 65/446
epilepsy, 65/445
fatigue, 65/445
headache, 65/445
Huntington chorea, 65/445
hyponatremia, 65/445
motor nerve conduction, 65/445
myasthenic syndrome, 65/446
neurologic adverse effect, 65/445
orthostatic hypotension, 65/445
parkinsonism, 65/445
progressive supranuclear palsy, 65/445
psychosis, 65/445
sensorimotor neuropathy, 65/445
sensory neuropathy, 65/445
systemic adverse effect, 65/445
vertigo, 65/445
Diphasic meningoencephalitis
acute viral encephalitis, 34/76
Diphenhydramine
dyskinesia, 6/828
dystonia, 6/828
parkinsonism, 6/219, 828
Diphenylbutylpiperidine
chemical structure, 65/276
Diphenylhydantoin, *see* Phenytoin
Diphenylhydantoin intoxication, *see* Phenytoin intoxication
Diphenyl phosphorofluoridate intoxication
chemical classification, 37/543
organophosphorus compound, 37/543
Diphenyl-2-trimethylammoniumethyl phosphate bromide intoxication
chemical classification, 37/550
organophosphorus compound, 37/550
Diphosphoinositide
biochemistry, 9/10
Diphosphonate
hypercalcemia, 63/562
muscle cell membrane, 40/387
osteitis deformans (Paget), 43/452
pain, 69/46
Diphosphopyridine nucleotide diaphorase

GM2 gangliosidosis, 10/291
Diphtheria, 33/479-487
acroparesthesia, 65/240
acute demyelinating neuropathy, 47/605, 608
anatomic classification, 52/228
antitoxin, 65/237
arrhythmia, 51/182
asymptomatic carrier, 52/228
ataxia, 51/183
autonomic function, 75/700
autonomic nervous system, 75/20
axonal transport, 51/182
bacteriology, 33/479, 65/238
bacterium, 52/227
brain embolism, 55/163
brain infarction, 55/163
bulbar dysfunction, 40/297, 304, 308
bulbar paralysis, 52/229
bull neck, 65/237
cerebrovascular disease, 55/163
chorea, 33/486, 65/240
clinical aspect, 65/239
clinical features, 33/480
clinical picture, 71/561
Corynebacterium, *see* Corynebacterium diphtheriae
cranial nerve, 51/183
cranial nerve palsy, 52/229
CSF abnormality, 65/240
CSF lymphocytosis, 51/183
cutaneous, 33/481, 484
demyelinating neuropathy, 47/608, 65/241
demyelination, 51/182, 52/229, 65/241
diaphragm paralysis, 51/183
diphtheritic neuropathy, 9/38
dysautonomia, 65/240
dysphagia, 65/240
dysphonia, 65/240
EMG, 65/240
encephalitis, 51/182, 65/240
encephalopathy, 33/485
epidemiology, 52/227, 65/237
exotoxin, 52/227
extraocular palsy, 51/183
Guillain-Barré syndrome, 7/503, 51/183
hemichorea, 33/486
hemiparesis, 33/485
history, 33/479, 65/239
hyperhidrosis, 51/183
immunity, 52/228
immunization, 52/228, 235
larynx paresis, 51/183
lethal dose, 65/239
meningitis, 33/486, 65/240
mortality rate, 65/237
multiple cranial neuropathy, 65/240
myocarditis, 33/487, 52/227
myocardium, 51/182
natural infection, 52/227
nerve conduction velocity, 65/240
neurologic complication, 33/482-486
neurologic sequela, 52/227
neuropathology, 65/241
neurotoxicity, 65/239
ocular accommodation, 51/183
ocular manifestation, 33/483
odor, 65/237
oropharyngeal, 33/481-484
palatal weakness, 33/482, 51/183
pharyngeal hypoesthesia, 51/183
pharyngeal membrane, 65/237
pharynx paresis, 51/183
postdiphtheric paralysis, 65/239
prevention, 52/237
progressive bulbar palsy, 59/225
sensorimotor polyneuropathy, 51/183
survey, 71/560
symptom, 65/237
tachycardia, 51/183, 65/241
tetraplegia, 65/240
toxin fragment A, 65/238
toxin fragment B, 65/238
treatment, 65/241, 71/561
vaccination, 52/228
vaccine, 52/235
Diphtheria antitoxin
brachial plexus neuritis, 2/139
onion bulb formation, 21/252
Diphtheria pertussis tetanus immunization
postinfectious encephalomyelitis, 47/479
postvaccinial encephalomyelitis, 47/479
Diphtheria pertussis tetanus vaccine
anaphylactoid reaction, 52/238
Arthus reaction, 52/238
brain injury, 52/241
collapse, 52/245
composition, 52/235
convulsion, 52/239
encephalopathy, 52/243, 245
febrile convulsion, 52/243
infantile spasm, 52/239
neurologic complication, 52/238
persistent crying, 52/245
reaction prognosis, 52/244
Diphtheria toxin
cardiomyopathy, 41/201

chemical pathology, 33/480
chemistry, 65/238
leukoencephalopathy, 47/569
neurotoxin, 33/480, 51/68, 64/10, 14, 65/236
onion bulb formation, 21/252
segmental demyelination, 64/14
toxic neuropathy, 64/10, 14
toxic polyneuropathy, 51/68
Diphtheria toxin intoxication
action site, 64/10, 14
Diphtheria toxoid
composition, 52/235
efficacy, 52/237
neurotoxicity, 52/241
side effect, 52/235, 240
transverse myelitis, 52/241
Diphtheritic neuritis
*see also* Corynebacterium diphtheriae
clinical features, 52/227
cri du chat syndrome, 51/182
demyelination, 9/38
toxin, 51/182
Diphtheritic neuropathy
acute muscle weakness, 41/288
diphtheria, 9/38
nerve conduction, 7/155
nerve lesion, 7/475
segmental demyelination, 51/70
symptom, 52/229
vaccination, 8/91
Diphtheritic polyneuropathy
mousey odor, 41/288
ocular manifestation, 33/483
progressive bulbar palsy, 59/375
Diphyllobothrium latum
vitamin $B_{12}$ deficiency, 7/631, 51/339
Diphyllobothrium latum myelopathy
deficiency neuropathy, 7/631
malabsorption syndrome, 7/631
pernicious anemia, 7/631
subacute combined spinal cord degeneration, 7/631
vitamin $B_{12}$ deficiency, 7/631
Diplegia
ataxic, *see* Ataxic diplegia
brachial, *see* Brachial diplegia
cerebral, *see* Cerebral diplegia
congenital ataxia, *see* Congenital ataxic diplegia
congenital ataxic, *see* Congenital ataxic diplegia
congenital facial, *see* Möbius syndrome
facial, *see* Facial diplegia
infantile spastic, *see* Symmetrical spastic cerebral palsy
pontine infarction, 53/389
spastic, *see* Spastic diplegia
Diplegia spinalis brachialis traumatica, *see* Myelomalacia
Diplococcus pneumoniae, *see* Streptococcus pneumoniae
Diploic vein
anatomy, 11/52
anterior temporal, *see* Anterior temporal diploic vein
frontal, *see* Frontal diploic vein
occipital, *see* Occipital diploic vein
transverse sinus, 11/55
Diploidy
definition, 50/540
Diplomyelia
*see also* Diastematomyelia
arachnoid abnormality, 50/437
Arnold-Chiari malformation type I, 50/441
back pain, 50/436
caudal aplasia, 50/436
central medullary cavity, 50/438
computer assisted tomography, 50/436, 439
dermal sinus, 50/436
diagnosis, 50/436
diastematomyelia, 20/58, 61, 30/164, 32/242, 42/24
dura mater abnormality, 50/437
dura mater reduplication, 50/436
dura mater sheath reduplication, 50/436
epidemiology, 50/439
family, 50/440
filum reduplication, 50/438
filum terminale, 50/438
flatfoot, 50/436
2 groups, 50/436
hair tuft, 50/436
hammer toe, 50/436
hemangiomatous nevus, 50/436
hemicord, 50/438
hydrocephalus, 50/441
hyperlordosis, 50/436
intraspinal lipoma, 50/439
kyphosis, 50/436
leg abnormality, 50/436
leg shortening, 50/436
lipoma, 50/436, 439
medullary fissure, 50/438
meningomyelocele, 50/437, 441
monoplegia, 50/436
muscle stretch reflex, 50/436
muscular atrophy, 50/436
myelography, 50/437, 439

neurologic syndrome, 50/436
orthopedic syndrome, 50/436
pes cavovarus, 50/436
pigmented nevus, 50/436
pilonidal sinus, 50/436
posterior arch abnormality, 50/436
radiologic appearance, 50/436
review, 50/435-441
sensory trouble, 50/436
sex ratio, 50/440
skin abnormality, 50/436
sphincter disturbance, 50/436
surgical technique, 50/440
terminal ventricle, 50/438
tethered cord, 50/438
treatment, 50/440
vertebral bony spur, 50/437, 439
vertebral fusion, 50/436
vertebral interpedicular widening, 50/439
vertebral segmentation, 50/436
Diplopia
angioneurotic edema, 43/61
Argentinian hemorrhagic fever, 56/366
Armstrong goggles, 2/288
ataxia, 1/314
atenolol, 63/195
barbiturate intoxication, 65/332
basilar impression, 42/17
basilar ischemia, 12/4
benign intracranial hypertension, 42/740
botulism, 65/210
brain embolism, 53/207
brain infarction, 53/207
carbamazepine intoxication, 37/202, 65/505
cerebrovascular disease, 53/207
chloroquine myopathy, 62/606
chronic bromide intoxication, 36/300
cinchonism, 65/452
conidae intoxication, 37/60
facial spasm, 42/314
felbamate intoxication, 65/512
Ferguson-Critchley syndrome, 42/142
Foville syndrome, 2/278
fox intoxication, 37/96
frontometaphyseal dysplasia, 38/410
gabapentin intoxication, 65/512
gyromitrine intoxication, 65/36
Hartnup disease, 42/146
heterophoria, 2/290
idiopathic narcolepsy, 3/94
intracranial hemorrhage, 53/207
lateral medullary infarction, 53/381
lidocaine intoxication, 37/451
median longitudinal fasciculus syndrome, 42/321
mefloquine intoxication, 65/485
meningeal leukemia, 63/341
mental, 73/74
multiple sclerosis, 42/496, 47/55, 172
myasthenia gravis, 43/156
myotonic hyperkalemic periodic paralysis, 62/462
narcolepsy, 3/94
ophthalmoplegic migraine, 48/132
orbital apex syndrome, 2/322
orbital fracture, 24/93
paralysis periodica paramyotonica, 43/167
paralytic strabismus, 2/291
periodic vestibulocerebellar ataxia, 42/116
phenytoin intoxication, 42/641, 65/500
polar bear intoxication, 37/96
pontine infarction, 12/22
postconcussional syndrome, 57/424
propranolol, 63/195
quinidine intoxication, 37/448
sleep, 3/90
spinopontine degeneration, 42/192
subclavian steal syndrome, 53/373
temporal arteritis, 48/313
test, 2/288-291
thyroid associated ophthalmopathy, 62/531
tick bite intoxication, 37/111
transient ischemic attack, 12/4, 53/207
vertebrobasilar system syndrome, 53/373
vigabatrin intoxication, 65/511
vitamin A intoxication, 37/96
Waldenström macroglobulinemia, 63/397
wolf intoxication, 37/96
Diplorhinotrichum
fungal CNS disease, 35/565
Dipropylacetate
epilepsy, 15/669
side effect, 15/709
Di-*N*-propyl phosphorofluoridate intoxication
chemical classification, 37/543
organophosphorus compound, 37/543
Dipsomania
filariasis, 52/516
Dipyridamole
brain infarction, 53/465
migraine, 48/199
prolonged transient ischemic attack, 53/275
transient ischemic attack, 53/275
Direct muscle response
F wave, 7/142
H-reflex, 7/142
intermediate latency response, 7/142
Direct tumor sign

brain tumor, 20/213, 230
spinal neurinoma, 20/258
Directional preponderance
caloric nystagmus, 2/374
nystagmus, 2/707
Disabilities
rating, 8/531
reading, *see* Reading disability
Disability
brain abscess, 33/140
brain injury, 24/671
brain injury prognosis, 24/671, 673
cancer, 67/390
language, *see* Language disability
learning, *see* Learning disability
pain, 8/533
psychogenic headache, 48/40
quality of life, 67/390
Disability classification
rehabilitative sport, 26/537
Disability Status Scale
multiple sclerosis, 9/162, 409, 421, 47/147
Disabling neuropathy
subacute, *see* Subacute disabling neuropathy
Disaccharide deficiency
deficiency neuropathy, 7/596
Disc
clinical lumbar stability, 61/16
concealed, 20/576
Hatfield-Turner, *see* Hatfield-Turner disk
intervertebral, *see* Intervertebral disc
Merkel, *see* Merkel disc
optic, *see* Optic disc
temporal pallor, *see* Temporal optic disc pallor
tilted optic, *see* Tilted optic disc
Disc hernia, *see* Intervertebral disc prolapse
Disc protrusion
cervical radicular syndrome, 20/555
lumbar, *see* Lumbar disc protrusion
Discitis
bacterial endocarditis, 52/299
discography, 19/208
infectious endocarditis, 63/117
intervertebral disc, 19/208
rheumatic disease, 19/329
Discography, 19/205-225
assessment, 19/222
cervical, 7/457, 19/223
cervical vertebral column injury, 25/304
clinical value, 19/213
complication, 61/147
contraindication, 19/210
diagnostic error, 19/220
disc degeneration pattern, 19/207
discitis, 19/208
history, 19/205
indication, 19/206, 210
injection pressure, 19/222
intermittent spinal cord ischemia, 12/520
intervertebral disc prolapse, 25/490, 500
intracranial subdural hematoma, 61/164
lateral disc hernia, 19/211
lumbar disc protrusion, 12/520
normal appearance, 19/217
pain, 19/221
pathologic disc type, 19/219
principle, 19/205
recurrent disc hernia, 19/211
review, 19/209
side effect, 19/207
technique, 19/214
Discomplete spinal cord injury
definition, 61/43
EMG, 61/45
frequency, 61/45
hyperacute stage, 61/45
Disconnection syndrome
agnosia, 4/17, 21, 28, 37
agraphia, 2/608, 45/104, 463
alexia, 2/608, 4/126, 45/104
Anton-Babinski syndrome, 45/378
apraxia, 4/58, 45/102
audition, 45/102
body scheme disorder, 4/212, 217
cerebral dominance, 4/264, 266, 273, 275-289, 45/103
constructional apraxia, 4/80
corpus callosum, *see* Corpus callosum disconnection syndrome
corpus callosum agenesis, 50/160
corpus callosum tumor, 17/506
developmental disorder, 4/467
Klüver-Bucy syndrome, 2/603
language, 45/103
motor aphasia, 45/463
somatesthesia, 45/102
speech disorder, 4/17, 21, 28, 37
splenium corporis callosi, 2/597
split brain syndrome, 45/101-104
unilateral agraphia, 45/103
unilateral tactile anomia, 45/103
visual hemifield, 45/101
visual perception disorder, 2/597, 4/17
Discoordination
eye-hand, *see* Ocular ataxia
Discopathy

dialysis, 63/529
$\beta_2$-microglobulin, 63/529
Discrete lipomatosis, *see* Krabbe-Bartels disease
Discrimination
H-M interval, *see* H-M interval discrimination
right left, *see* Right left discrimination
spatial, *see* Spatial discrimination
speech, *see* Speech discrimination
tactile, *see* Tactile discrimination
two point, *see* Two point discrimination
visual, *see* Visual discrimination
Discriminative sensibility
parietal lobe syndrome, 2/682, 45/64-67
tactile, 1/88
two point, 1/88, 105
Discus throwing
rehabilitative sport, 26/530
Disequilibrium syndrome
ape, 2/400
ataxia, 42/138-140
autism, 42/139
birth incidence, 42/140
cerebellar atrophy, 42/140
dialysis, *see* Dialysis disequilibrium syndrome
dopamine β-mono-oxygenase, 42/410
experimental ablation, 2/400
hemodialysis, 28/501
hyponatremia, 28/501
hyporeflexia, 42/139
mental deficiency, 42/140
psychomotor retardation, 42/138-140
speech disorder in children, 42/139
tremor, 42/139
Disialosyl ganglioside
sensory neuropathy, 71/443
Disjunctive movement, *see* Synkinesis
Dislocation
ankylosing spondylitis, 70/3
atlantoaxial, *see* Atlantoaxial dislocation
cervical vertebral column injury, 61/25
congenital hip, *see* Congenital hip dislocation
humeral, *see* Humeral dislocation
sacroiliac fracture, *see* Sacroiliac fracture dislocation
Disodium etidronate, *see* Etidronic acid
Disopyramide intoxication
confusion, 65/453
epilepsy, 65/453
psychosis, 65/453
visual impairment, 65/453
Disopyramide phosphate
cardiac pharmacotherapy, 63/192
iatrogenic neurological disease, 63/192
Disorganization
brain cortex, *see* Brain cortex disorganization
disulfiram intoxication, 37/321
Disorientation, 3/251-256
acrylamide polyneuropathy, 51/281
Alzheimer disease, 46/250
body part, 3/253
body scheme disorder, 3/208
central pontine myelinolysis, 2/263
delirium, 46/528
dialysis disequilibrium syndrome, 63/523
dysphrenia hemicranica, 5/79
hexachlorophene intoxication, 37/486
hydrozoa intoxication, 37/40
memory disorder, 3/276
migraine, 48/165
neuropsychologic defect, 45/524
object, 3/254
pellagra, 7/573, 38/648
person, 2/668, 3/252
place, 2/668, 3/251
posttraumatic amnesia, 23/430, 45/188, 57/398
respiratory encephalopathy, 63/414
right left, *see* Right left disorientation
self, 3/254
spatial, *see* Spatial orientation disorder
temporal, *see* Temporal disorientation
time, 3/30, 32, 253, 276
traumatic psychosis, 24/525
uremia, 63/504
visual, *see* Visual disorientation
Wernicke-Korsakoff syndrome, 45/194
Disruptive hyperanteflexion
mechanics, 25/255
Disruptive hyperdeflexion
mechanics, 25/257
Dissecting aneurysm
basilar artery, *see* Basilar dissecting aneurysm
brain aneurysm, 55/79
brain aneurysm neurosurgery, 55/79
carotid artery, *see* Carotid dissecting aneurysm
pericarotid syndrome, 48/337
spinal artery injury, 26/71
spinal cord injury, 26/71, 74
subarachnoid hemorrhage, 55/79
traumatic aneurysm, 24/389
vertebral artery, *see* Vertebral dissecting aneurysm
Dissecting aorta aneurysm
abdominal pain, 63/45
Adamkiewicz artery, 63/39
aorta coarctation, 63/45
arterial hypertension, 63/82
back pain, 63/37

brain complication, 63/47
brain hemorrhagic infarction, 63/47
brain ischemia, 63/46
cardiac tamponade, 61/114
childhood, 63/45
classification, 39/246, 63/45
collagen vascular disease, 63/45
coma, 63/47
cystinosis, 63/45
definition, 39/246
description, 39/244
dysarthria, 63/44
headache, 63/47
history, 39/244, 63/44
hypertension, 63/46
inflammation, 63/45
intravascular consumption coagulopathy, 63/47
ischemic myelopathy, 61/115
ischemic necrosis, 39/248
leg pain, 63/46
limb weakness, 63/46
Marfan syndrome, 39/389, 63/45
Mönckeberg sclerosis, 63/45
myelomalacia, 61/112
necropsy finding, 61/114
neurologic complication, 39/247, 63/46
ocular involvement, 39/249
oliguria, 63/46
pain, 63/46
paraplegia, 61/114, 63/37, 39, 46
pathogenesis, 39/246
peripheral nerve dysfunction, 63/47
peripheral nerve ischemia, 63/46
predisposing factor, 39/247
prevalence, 63/45
prognosis, 39/249, 63/47
recurrent paraplegia, 61/115
risk factor, 61/114
superior vena cava obstruction, 63/47
survival, 61/114
syncope, 63/45
tetraplegia, 63/37, 44
transesophageal echoCG, 63/45
transient global amnesia, 63/47
trauma, 63/45
treatment, 39/249
two-dimensional echoCG, 63/45
Dissecting aortic hematoma, *see* Dissecting aorta aneurysm
Dissecting hematoma
carotid artery, *see* Carotid dissecting aneurysm
vertebral artery, *see* Vertebral dissecting aneurysm
Disseminated choroiditis
spastic paraplegia, 42/178
Disseminated disease
candidiasis, 52/401
Disseminated encephalomyelitis, 47/467-495
*see also* Postinfectious encephalomyelitis
acute, *see* Acute disseminated encephalomyelitis
concentric sclerosis, 47/473
delayed hypersensitivity, 9/501, 47/476
experimental allergic encephalomyelitis, 47/467-470
hyperacute, *see* Hyperacute disseminated encephalomyelitis
incubation period, 47/476
neuromyelitis optica, 9/426, 431, 47/473
nuclear magnetic resonance, 47/481
postinfectious, *see* Acute postinfectious disseminated encephalomyelitis
protein P2, 47/467
Reye syndrome, 47/479
viral infection, 47/474
Disseminated intravascular clotting, *see* Intravascular consumption coagulopathy
Disseminated meningoencephalomyelitis
*see also* Acute postinfectious disseminated encephalomyelitis
Behçet syndrome, 56/598
Disseminated necrotizing leukoencephalopathy
blood-brain barrier, 63/355
cerebral lymphomatoid granulomatosis, 39/524
CNS irradiation, 63/355
combined treatment, 63/355
CSF, 63/355
delayed irradiation, 63/355
intrathecal chemotherapy, 63/355
irradiation encephalopathy, 63/355
leukemia, 63/355
radiotherapy, 63/355
Disseminated sclerosis, *see* Multiple sclerosis
Dissociated nystagmus
bilateral, 2/367, 16/322
brain stem lesion, 2/367
brain stem tumor, 16/322
definition, 1/592
internuclear ophthalmoplegia, 2/367, 16/322
multiple sclerosis, 2/367, 16/322
test, 2/369
unilateral, 2/367, 16/322
Dissociated sensibility
amyloid polyneuropathy, 51/416
Dissociated sensory loss
amyloid polyneuropathy, 51/419, 63/393
dorsal horn pain, 20/428
hereditary amyloid polyneuropathy type 1, 51/420

hereditary sensory and autonomic neuropathy type I, 60/8
hydromyelia, 50/429
multiple myeloma polyneuropathy, 63/393
posttraumatic syringomyelia, 61/382
progressive systemic sclerosis, 51/448
spinal ependymoma, 20/368
spinal lipoma, 20/400
spinal tumor, 19/59
syringomyelia, 50/444
Dissociated visual perception
depth perception, 2/584
occipital lesion, 2/584
occipital lobe lesion, 2/584
Dissociations
apoenzyme-coenzyme, *see* Apoenzyme-coenzyme dissociation
light-near, *see* Light-near dissociation
motor, *see* Motor dissociation
protein cell, *see* Proteinocytologic dissociation
Distal amyotrophy
β-*N*-acetylhexosaminidase A deficiency, 60/659
ataxia, 60/659
ataxia telangiectasia, 60/659
Charlevoix-Saguenay spastic ataxia, 60/453, 659
chorea-acanthocytosis, 63/280
dentatorubropallidoluysian atrophy, 60/659
Ferguson-Critchley syndrome, 59/321
Friedreich ataxia, 60/658
hearing loss, 60/659
hereditary motor and sensory neuropathy type I, 21/303
hereditary motor and sensory neuropathy type II, 21/303
hereditary motor and sensory neuropathy variant, 60/244
hereditary spastic ataxia, 60/659
late cerebellar atrophy, 60/659
Machado-Joseph disease, 60/659
mental deficiency, 60/659
multiple neuronal system degeneration, 60/659
neonatal adrenoleukodystrophy, 60/659
neuronal intranuclear hyaline inclusion disease, 60/659
olivopontocerebellar atrophy (Dejerine-Thomas), 60/659
olivopontocerebellar atrophy (Wadia-Swami), 60/494, 659
spino-olivocerebellar degeneration, 60/659
Sylvester disease, 60/659
Wieacker-Wolff syndrome, 60/659
X-linked ataxia, 60/659
xeroderma pigmentosum, 60/659
Distal axonal polyneuropathy
scleroderma, 71/111
Distal axonopathy
acetonylacetone, 64/12
acrylamide, 64/12
acrylamide intoxication, 64/64
arsenic, 64/12
arsenic intoxication, 64/288
carbon disulfide intoxication, 64/26
chlordane, 64/13
clioquinol, 64/13
colchicine, 64/13
cyanogenic glycoside, 64/12
freon, 64/13
gold, 64/13
hexacarbon, 64/11
hexane, 64/12
2-hexanone, 64/12
lead, 64/13
lupin seed, 64/13
organophosphate, 64/11
perhexiline maleate polyneuropathy, 51/302
permethrin, 64/13
thallium, 64/12
tri-*o*-cresyl phosphate, 64/11
vinblastine, 64/13
vincristine, 64/13
Distal hypertension
brain embolism, 11/402
Distal motor polyneuropathy
lead polyneuropathy, 51/268
Distal muscle weakness
Aran-Duchenne disease, 22/138, 42/86
debrancher deficiency, 43/179
facioscapulohumeral muscular dystrophy, 40/481
hereditary amyloid polyneuropathy, 42/518-521
hereditary motor and sensory neuropathy type I, 42/338
hereditary motor and sensory neuropathy type II, 42/338
infantile myopathy, 43/114
Isaacs syndrome, 43/160
Lemieux-Neemeh syndrome, 42/370
neuropathy, 40/318, 320
paralysis periodica paramyotonica, 43/167
Roussy-Lévy syndrome, 42/108
Welander muscular dystrophy, 43/96
Distal muscular atrophy
critical illness polyneuropathy, 51/575
progressive, *see* Progressive distal muscular atrophy
Distal muscular dystrophy
distal myopathy, 62/199, 201

genetics, 41/421
hyporeflexia, 43/96
muscle fiber type I, 43/96
muscle weakness, 43/96
neuromuscular disease, 41/421
neuropathy, 43/97
scapulohumeral, *see* Scapulohumeral distal muscular dystrophy
sex ratio, 43/96
Distal myopathy
classification, 62/198
diagnosis, 62/197
diagnostic criteria, 62/198
differential diagnosis, 62/198
distal muscular dystrophy, 62/199, 201
EMG, 40/472
Gowers, *see* Gowers distal myopathy
hereditary motor and sensory neuropathy type I, 22/59
hereditary motor and sensory neuropathy type II, 22/59
heterogeneity, 62/197
infantile, *see* Infantile distal myopathy
juvenile, 62/199, 201
kyphoscoliosis, 43/114
myotonic dystrophy, 62/240
rimmed muscle fiber vacuole, 40/475, 62/204
rimmed vacuole distal myopathy, 62/199
spinal muscular atrophy, 40/478, 59/373
talipes, 43/114
treatment, 62/206
Wohlfart-Kugelberg-Welander disease, 62/199
Distal polyneuropathy
diabetic polyneuropathy, 51/499
undetermined significant monoclonal gammopathy, 63/402
Distal sensory neuropathy
diabetic polyneuropathy, 51/499
Distal symmetric polyneuropathy
human immunodeficiency virus neuropathy, 71/356
Distance perception
agnosia, 4/16
Balint syndrome, 45/410
parietal lobe, 45/362
Distemper
canine, *see* Canine distemper
Distemper encephalitis
canine, *see* Canine distemper encephalomyelitis
Distemper virus
canine chorea, 6/680
respiratory syncytial virus, 34/344
Distichia
spinal epidural cyst, 20/141
Distichiasis lymphedema syndrome
photophobia, 43/377
pterygium colli, 43/377
spinal epidural cyst, 43/377
Distractibility
attention disorder, 3/139, 352
juvenile Gaucher disease, 10/307
Klüver-Bucy syndrome, 3/352
pellagra, 28/86
Disulfiram
akinesia, 64/188
alcohol reaction, 37/321
autonomic nervous system, 75/38
carbamate intoxication, 64/185, 187
chronic axonal neuropathy, 51/531
drug induced polyneuropathy, 51/309
dystonia, 64/188
neostriatal degeneration, 64/188
neuropathy, 7/531
neurotoxicity, 51/308
neurotoxin, 7/531, 37/321, 46/394, 48/420, 51/68, 308, 531, 64/8, 188
optic neuropathy, 64/8
pallidal degeneration, 64/188
phenytoin, 40/561
putaminal degeneration, 64/188
toxic neuropathy, 64/8
toxic polyneuropathy, 51/68
Disulfiram intoxication
agitation, 37/321
alcohol, 37/321
constipation, 37/321
cramp, 37/321
diarrhea, 37/321
disorganization, 37/321
dizziness, 37/321
drowsiness, 37/321
drug dosage, 37/321
fatigue, 37/321
gastrointestinal disorder, 37/321
hallucination, 37/321
headache, 37/321, 48/419
myelopathy, 64/8
neuropathy, 64/8
odor, 37/321
paranoia, 37/321
rash, 37/321
sexual impotence, 37/321
taste, 37/321
visual impairment, 64/8
Disulfiram polyneuropathy
acetaldehyde, 51/309

alcohol dehydrogenase, 51/309
axonal degeneration, 51/309
diethyldithiocarbamic acid, 51/309
EMG, 51/309
Guillain-Barré syndrome, 51/309
mechanism, 51/309
neurofilament axonopathy, 51/309
optic neuritis, 51/309
pathomechanism, 51/308
trichloroethanol intoxication, 51/309
zinc, 51/309
Disuse
experimental myopathy, 40/143, 200
muscle fiber type II, 40/146, 200
muscular atrophy, 40/11, 200, 322
nutritional change, 7/273
Disystox intoxication
chemical classification, 37/547
organophosphorus compound, 37/547
Dithiocarbamate intoxication, *see* Carbamate intoxication
Dithiosystox intoxication, *see* Disystox intoxication
Diuretic agent
amiloride, 65/438
arterial hypertension, 54/206, 65/438
bendroflumethiazide, 65/437
brain edema, 16/204, 67/88
cardiovascular agent, 65/433
catamenial migraine, 5/381
central pontine myelinolysis, 65/439
EEG, 65/439
furosemide, 65/437
hypertensive encephalopathy, 54/217
hypokalemia, 65/438
hypomagnesemia, 63/562
hyponatremia, 65/438
iatrogenic neurological disease, 65/440
indication, 65/438
intracranial pressure, 23/215
lithium intoxication, 64/299
migraine, 5/102
myelinolysis, 65/439
neurologic adverse effect, 65/439
neurologic intensive care, 55/209, 212
potassium sparing diuretic agent, 65/438
spironolactone, 65/438
squill, 37/425
systemic adverse effect, 65/438
thiazide, 65/437
toxic myopathy, 62/597, 601
triamterene, 65/438
Diuretic agent intoxication
deafness, 65/440
hypokalemia, 65/440
muscle weakness, 65/440
myalgia, 65/440
pseudobulbar paralysis, 65/440
sodium correction rate, 65/440
spastic paraparesis, 65/440
tetany, 65/440
tinnitus, 65/440
vertigo, 65/440
Diurnal fluctuating juvenile dystonia
hereditary, *see* Hereditary diurnal fluctuating juvenile dystonia
Diurnal rhythm
depression, 74/506
sleep, 3/85
Dive bomber discharge, *see* Myotonic discharge
Diver disease, *see* Decompression illness
Divergence paralysis
differential diagnosis, 2/342
ocular imbalance, 24/75
Divergent strabismus
basal encephalocele, 50/106
basal meningoencephalocele, 50/106
craniosynostosis, 50/125
hyaloidoretinal degeneration (Wagner), 13/274
pontine infarction, 53/389
Diverticulitis
granulomatous CNS vasculitis, 55/388
hereditary deafness, 22/524
jejunoileal, *see* Jejunoileal diverticulitis
small intestine, *see* Small intestine diverticulitis
Diverticulosis
intestinal malabsorption, 7/619
intestinal pseudo-obstruction, 70/320
jejunal, 75/632
malabsorption syndrome, 28/235
small intestine, *see* Small intestine diverticulosis
Diverticulum
brain ventricle, *see* Brain ventricle diverticulum
ependymal, *see* Ependymal diverticulum
Meckel, *see* Meckel diverticulum
meningeal, *see* Meningeal diverticulum
postoperative meningeal, *see* Postoperative meningeal diverticulum
spinal nerve root injury, 25/419
ventricular, *see* Ventricular diverticulum
Diving
*see also* Barotrauma *and* Dysbarism
autonomic nervous system, 75/440
headache, 48/401
migraine, 48/402
syncope, 75/217
Diving response, *see* Cold face test

Divry-Van Bogaert syndrome, 10/120-127, 55/317-322
*see also* Sneddon syndrome
adult type, 10/123, 55/317
ataxia, 55/319
Binswanger disease, 10/125, 55/318
brain cortex, 55/320
brain infarction, 55/320
brain white matter degeneration, 42/726
capillary venous network, 55/320
choreoathetosis, 14/514
clinical picture, 10/121, 123
congenital type, 10/120
cutaneous dysplasia, 14/101
cutis marmorata, 42/725, 55/318
dementia, 14/14, 42/725, 55/318
differential diagnosis, 10/127, 55/320
differentiation, 14/108
disease duration, 55/322
epilepsy, 42/725, 55/318, 321
gait disturbance, 55/319
genetics, 14/512-514
granular cortical cerebral atrophy, 55/318
hemianopia, 14/108
hereditary, 55/321
Herman syndrome, 10/127
history, 55/317
hydrocephalus, 14/513
hypsarrhythmia, 10/121
infantile type, 55/317, 321
juvenile type, 10/123
leptomeningeal angiomatosis, 14/14, 71, 512, 55/320
livedo, 14/14
livedo reticularis, 55/317-319
mortality, 55/319
multiple brain infarction, 55/318
multiple cortical infarction, 55/320
multiple thrombi, 55/320
myoclonia, 55/322
myoclonic epilepsy, 55/322
Naegeli syndrome, 14/108
neurocutaneous syndrome, 14/101
neuropathology, 10/124, 14/71, 55/320, 322
optic atrophy, 14/514
orthochromatic leukodystrophy, 14/72
osteogenesis imperfecta, 14/513
prominent venous involvement, 55/320
pseudobulbar paralysis, 55/319
psychomotor impairment, 55/321
reticulodentatocerebellar pigmentation, 14/514
rigidity, 55/319, 321
Sneddon syndrome, 55/403, 408
Sturge-Weber syndrome, 14/227
symptom, 14/14
system degeneration, 10/124
visual disorder, 42/725
white matter demyelination, 55/320
Dizygotic twins
congenital pain insensitivity, 51/565
Dizziness
*see also* Vertigo
Amanita muscaria intoxication, 36/534
blood pressure, 75/715
brain tumor, 16/129
carbon monoxide intoxication, 64/31
cerebellar infarction, 53/383
crab intoxication, 37/58
decaborane intoxication, 64/360
diborane intoxication, 64/359
disulfiram intoxication, 37/321
echinoidea intoxication, 37/68
ethchlorvynol intoxication, 37/354
Gellé syndrome, 2/241
glossopharyngeal neuralgia, 48/466
head injury, 24/121, 126
hydralazine intoxication, 37/431
hyponatremia, 63/545
intracranial hypertension, 16/129
labyrinth concussion, 24/125, 136
lidocaine intoxication, 37/450
mackerel intoxication, 37/87
merostomata intoxication, 37/58
methanol intoxication, 36/353
neptunea intoxication, 37/64
pellagra, 7/573
pentaborane intoxication, 64/359
pheochromocytoma, 39/498
polycythemia, 63/250
polycythemia vera, 63/250
pontine infarction, 12/15, 23
postconcussional syndrome, 23/454, 57/412, 424
posttraumatic syndrome, 24/686
primary pontine hemorrhage, 12/39
procainamide intoxication, 37/452
puffer fish intoxication, 37/80
temporal bone fracture, 24/121
Toxopneustes pileolus intoxication, 37/68
traumatic vegetative syndrome, 24/577
urticaria pigmentosa, 14/790
Von Hippel-Lindau disease, 14/115
Waldenström macroglobulinemia, 63/398
DMAB nitrous acid
tryptophan, 9/24
DMNU, *see* Dimethylnitrosourea
DNA

adenovirus, 56/281
amyotrophic lateral sclerosis, 59/177
cytoplasmic, 30/110
gene transcription, 66/70
herpes simplex virus, 56/212
human chromosomes, 30/91
human T-lymphotropic virus type I, *see* Human T-lymphotropic virus type I DNA
memory, 3/262
mitochondrial, *see* Mitochondrial DNA
myotubular myopathy, 43/113
neurotoxicity, 64/17
nuclear, *see* Nuclear DNA
nucleotide sequence, 30/94
plasmid, *see* Plasmid DNA
repair capacity, *see* DNA repair mechanism
replication, 30/92, 94
RNA, 30/95
structure, 30/94
DNA marker
aneuploidy, 50/567
DNA mutation
mitochondrial, *see* Mitochondrial DNA mutation
spongy cerebral degeneration, 66/664
DNA repair mechanism
adverse factor, 30/98
xeroderma pigmentosum, 43/13, 60/335, 339
DNA synthesis
Werner syndrome, 43/480
DNA synthesis inhibitor
chemotherapy, 67/284
DNA virus
genome, 56/2
hepatitis B virus, 56/295
human, 56/2
neurotropic virus, 56/5
taxonomy, 56/2
transcription, 56/6
varicella zoster virus, 56/229
virology, 34/7-9
virus taxonomy, 56/2
DNP, *see* 2,4-Dinitrophenol
Dobutamine
classification, 74/149
Docetaxel, *see* Taxotere
Docosahexaenoic acid deficiency
cerebrohepatorenal syndrome, 66/515
Dogfish intoxication, *see* Squaliformes intoxication
Dolichocephalic skull
Edwards syndrome, 43/535, 50/561
Dolichocephaly
cloverleaf skull syndrome, 30/245
craniosynostosis, 50/113
Hurler-Hunter disease, 10/436
Dolichol
neuronal ceroid lipofuscinosis, 51/382, 66/685
Dolichol phosphate
nerve myelin, 51/30
Dolichol diphosphate oligosaccharide
neuronal ceroid lipofuscinosis, 66/685
Dolichostenomelia, *see* Arachnodactyly
Dolichylpyrophosphoryloligosaccharide, *see* Dolichol diphosphate oligosaccharide
Doll head eye phenomenon
anatomy, 2/275
ocular bradykinesia, 1/599
oculomotor disorder, 2/306
progressive supranuclear palsy, 22/209, 219, 221, 49/240, 243
trochlear nerve, 2/306
Dolorimeter
thermal, *see* Thermal dolorimeter
Dolorosa
adiposis, *see* Adiposis dolorosa
anesthesia, *see* Anesthesia dolorosa
paraplegia, *see* Paraplegia dolorosa
Doloxene, *see* Dextropropoxyphene
Dolphin intoxication
ingestion, 37/94
Dominance
cerebral, *see* Cerebral dominance
ocular, *see* Ocular dominance column
Dominant albinism
hereditary deafness, 50/218
Dominant anhidrosis
Helweg-Larsen-Ludvigsen syndrome, 22/491
Dominant ataxia
calf hypertrophy, 60/659
fasciculation, 60/659
Machado-Joseph disease, 60/476
myoclonus, 60/660
neurogenic weakness, 60/659
neuropathy, 60/660
polyneuropathy, 60/660
sensorineural deafness, 60/660
Dominant cerebello-olivary atrophy
age at onset, 42/136
ataxia, 42/136
dysarthria, 42/136
late cerebellar atrophy, 42/136
mental deterioration, 42/136
tremor, 42/136
Dominant hemisphere
*see also* Handedness
cerebral dominance, 3/33-36, 4/262, 291
hypochondriasis, 46/581

middle cerebral artery syndrome, 53/361, 363
migraine, 48/148
split brain syndrome, 45/103
Dominant ichthyosis
Goyer syndrome, 22/490
Dominant infantile optic atrophy, 13/111-121
associated abnormality, 13/117
color vision, 13/114-116
differential diagnosis, 13/120
EEG, 13/118
ERG, 13/117
eye fundus, 13/112
hereditary, 13/118
history, 13/111
sieve scotoma, 13/114
visual acuity, 13/112
visual adaptation, 13/116
visual field, 13/113
Dominant inheritance
autosomal, *see* Autosomal dominant inheritance
congenital pain insensitivity, 51/565
corpus callosum agenesis, 50/151
craniosynostosis, 50/60
hereditary progressive cone dystrophy, 13/131, 134
inclusion body myositis, 62/373
Mendel law, 30/95
mental deficiency, 46/42-46
myotonia fluctuans, 62/266
myotonic syndrome, 62/263
vitelliruptive macula degeneration, 13/266
Dominant low frequency hearing loss
hereditary progressive cochleovestibular atrophy, 22/482, 60/766
Dominant Meniere disease
hereditary progressive cochleovestibular atrophy, 22/486
Dominant midfrequency hearing loss
hereditary progressive cochleovestibular atrophy, 22/482, 60/765
Dominant olivopontocerebellar atrophy
myoclonus, 60/662
neuropathy, 60/660
rigidity, 60/661
secondary pigmentary retinal degeneration, 60/654
supranuclear pseudo-ophthalmoplegia, 60/656
Dominant progressive sensorineural deafness
hereditary progressive cochleovestibular atrophy, 22/481
Dominant urticaria
Muckle-Wells syndrome, 22/489, 490
Domoic acid
amnesia, 65/166
amnesic shellfish poisoning, 65/166
dementia, 65/141
glutamic acid receptor, 65/150
hippocampus necrosis, 65/151
kainic acid receptor, 65/151
marine toxin intoxication, 65/150
neurotoxin, 64/3, 65/141, 151, 166
source, 64/4
structure, 65/151
symptom, 64/4
toxic dose, 65/166
toxic encephalopathy, 64/3
Domperidone
migraine, 48/198
Doom
migraine, 48/162
DOOR syndrome
brachydactyly, 42/385
clinodactyly, 42/385
EEG, 42/385
yellow teeth, 42/385
Doose syndrome, *see* Epilepsy with myoclonic astatic seizure
Dopa, *see* Levodopa
L-DOPA, *see* Levodopa
DOPAC, *see* Homoprotocatechuic acid
Dopamine
affective disorder, 46/475
akinesia, 49/67
Alzheimer disease, 46/266
amyotrophic lateral sclerosis, 21/34
basal ganglion, 49/33
biochemical anatomy, 29/464
biological inactivation, 29/462
biosynthesis, 29/459
brain infarction, 53/421
catabolism, 29/460
central autonomic dysfunction, 75/190
cephalopoda intoxication, 37/62
cervical spinal cord injury, 61/61
chorea-acanthocytosis, 63/283
3,4-dihydroxyphenylserine therapy, 75/239
dystonia musculorum deformans, 6/532
epilepsy, 72/92
Huntington chorea, 49/75, 257
hyperthyroid chorea, 27/280
manganese intoxication, 64/306
mesolimbic pathway, 29/464
migraine, 48/92
neuroleptic agent, 46/446
neuronal ageing, 21/53
noradrenalin, 74/145

octopoda intoxication, 37/62
Parkinson disease, 6/180, 29/469-475, 49/118, 123
pharmacology, 74/145
postsynaptic, 74/146
progressive dysautonomia, 59/153, 160
progressive supranuclear palsy, 49/251
regional distribution, 29/464
release, 29/462
Reye syndrome, 56/152
rigidity, 49/70
schizophrenia, 46/422, 446
storage, 29/462
striatonigral fiber, 29/464, 49/4
suprachiasmatic nucleus, 74/489
tardive dyskinesia, 29/476-478
tuberoinfundibular pathway, 29/464
turning behavior, 45/167
turnover, 29/463
vitamin C, 74/145
Dopamine decreased turnover
tardive dyskinesia, 49/189
Dopamine β-hydroxylase, *see* Dopamine β-mono-oxygenase
Dopamine hypersensitivity
tardive dyskinesia, 49/191
Dopamine β-mono-oxygenase
congenital pain insensitivity, 51/564
disequilibrium syndrome, 42/410
dystonia musculorum deformans, 42/215, 49/524
Gilles de la Tourette syndrome, 42/222
kinesigenic choreoathetosis, 42/208
migraine, 48/91, 94
muscle contraction headache, 48/95
noradrenalin, 74/141
Parkinson disease, 42/246
parkinsonism, 49/69
reflex sympathetic dystrophy, 61/128
schizophrenia, 43/214
spinal cord injury, 61/448
test, 74/596
tetraplegia, 61/448
Dopamine β-mono-oxygenase deficiency
adrenergic receptor, 59/163
autonomic nervous system, 74/218, 75/21, 237
beta adrenergic receptor blocking agent, 59/163
blepharoptosis, 59/163
clinical features, 59/162
EEG, 59/163
hyponatremia, 59/163
noradrenergic dysfunction, 59/163
orthostatic hypotension, 59/163, 63/148
phenylephrine, 59/163
progressive dysautonomia, 59/162
REM sleep, 59/163
slow wave sleep, 59/163
tonic clonic seizure, 59/163
tyramine, 59/163
Dopamine receptor
*see also* Brain glucose metabolism
cyclic adenosine monophosphate, 49/49
distribution, 75/191
ergot intoxication, 65/68
genetics, 75/244
Huntington chorea, 49/259, 295
nigrostriatal, *see* Nigrostriatal dopamine receptor
schizophrenia, 46/422
Dopamine receptor antagonist, *see* Dopamine receptor blocking agent
Dopamine receptor blocking agent
orthostatic hypotension, 63/164
Dopamine receptor stimulating agent
chorea, 49/190
lysergide, 65/44
Dopamine receptor type
basal ganglion, 49/47
neuroleptic agent, 65/277
Dopamine receptor type D1
agonist, 49/49
antagonist, 49/49
cyclic adenosine monophosphate binding, 49/47
Huntington chorea, 49/260
Dopamine receptor type D2
agonist, 49/49
antagonist, 49/49
cyclic adenosine monophosphate binding, 49/47
Huntington chorea, 49/260
Parkinson disease, 49/99
Dopamine receptor type D3
cyclic adenosine monophosphate binding, 49/47
Dopamine system
mesolimbic, *see* Mesolimbic dopamine system
nigrostriatal, *see* Nigrostriatal dopamine system
Dopaminergic agent
amphetamine intoxication, 65/257
behavior disorder, 46/592-594
thyrotropic hormone secreting adenoma, 68/357
Dopaminergic intoxication
parkinsonism, 46/592-594
Dopaminergic nerve cell
age, 49/49
estrogen, 49/49
feedback regulation, 49/48
oral contraception, 49/49
presynaptic mechanism, 49/48
receptor modulation, 49/48
sprouting, 49/48

Doppelgänger, *see* Autoscopia
Doppler flowmetry
  skin, 75/109
  thoracic outlet syndrome, 51/127
Doppler sonography
  *see also* Ultrasound diagnosis
  anterior cerebral artery, 54/12
  autonomic frequency analysis, 54/1
  brachial artery, 54/9
  brain arteriovenous malformation, 54/380
  brain blood flow, 54/14
  brain blood flow value, 54/14
  brain lacunar infarction, 54/255
  carotid artery bifurcation, 54/7
  carotid artery stenosis, 54/1
  carotid flow volume, 54/14
  circulatory resistance, 54/5
  collateral compression, 54/4
  common carotid artery, 54/5
  diagnosis, 54/7
  diagnostic criteria, 54/8
  diastolic flow, 54/9
  duplex sonography, 54/2
  dural arteriovenous malformation, 54/9
  equipment, 54/1
  examiner skill, 54/7
  external carotid artery, 54/6
  flow reversal, 54/4
  flow velocity, 54/1
  frequency analysis, 54/10
  image, 54/13
  incident angle, 54/1
  internal carotid artery, 54/6
  internal jugular flow, 54/15
  peak frequency, 54/11
  phase shift, 54/2
  physical principle, 54/1
  pulsed wave, *see* Pulsed wave Doppler sonography
  resistance profile, 54/6
  second systolic peak, 54/10
  sensitivity, 54/7
  spectral broadening, 54/10
  stenosis grading, 54/11
  subclavian artery, 54/9
  superficial temporal artery, 54/9
  supratrochlear artery, 54/3
  transcranial, *see* Transcranial Doppler sonography
  transient ischemic attack, 53/268
  vertebral artery stenosis, 54/1
  vertebrobasilar artery, 54/8
*threo*-DOPS, *see threo*-3,4-Dihydroxyphenylserine
Dorsal cordotomy
  phantom limb pain, 45/400
Dorsal callosal artery
  radioanatomy, 11/97
  topography, 11/97
Dorsal callosal vein
  inferior, *see* Inferior posterior cerebellar vein
  radioanatomy, 11/109
  topography, 11/109
Dorsal column
  vitamin $B_{12}$ deficiency, 70/384
Dorsal column degeneration
  multiple myeloma polyneuropathy, 51/431
Dorsal column stimulation
  bladder function, 47/166
  multiple sclerosis, 47/159, 166
  pain, 26/497
  spinal pain, 26/497
Dorsal cuboid reflex, *see* Mendel-Von Bechterew reflex
Dorsal funiculi degeneration
  paraneoplastic polyneuropathy, 51/468
Dorsal globus pallidus
  functional anatomy, 49/25
Dorsal intestinal fistula, *see* Neurenteric cyst
Dorsal medullary syndrome
  clinical finding, 2/228
Dorsal meningeal artery
  carotid cavernous fistula, 12/271
  radioanatomy, 11/69
Dorsal midbrain syndrome
  cerebrovascular disease, 55/124
Dorsal neostriatum
  functional anatomy, 49/25
Dorsal 10th nucleus
  Parkinson disease, 6/191
Dorsal pallidum, *see* Dorsal globus pallidus
Dorsal paramedian nucleus
  Huntington chorea, 6/331
Dorsal rhizotomy
  *see also* Spinal pain
  pain surgery, 45/229
Dorsal root ganglion, *see* Dorsal spinal nerve root ganglion
Dorsal scapular nerve
  lesion localization, 2/137
  topographical diagnosis, 2/20
Dorsal spinal nerve root
  Bassen-Kornzweig syndrome, 63/278
  sensory change, 2/165-167
  spinal autonomous efferent, 2/167
Dorsal spinal nerve root entry zone
  pain, 45/229-231
Dorsal spinal nerve root entry zone lesion

brachial plexus, 51/154
Dorsal spinal nerve root ganglion
anti-Hu, 69/376
ascending terminal, 49/395
cell function, 1/58
congenital pain insensitivity, 51/564
doxorubicin, 51/300
hereditary amyloid polyneuropathy, 60/102
hereditary dystonic paraplegia, 22/452
hereditary sensory and autonomic neuropathy, 60/5
hereditary sensory and autonomic neuropathy type I, 60/5, 9
mercury polyneuropathy, 51/271, 273
misonidazole polyneuropathy, 51/301
nerve injury, 75/321
neuroaxonal dystrophy, 49/393
pain, 75/321
paraneoplastic polyneuropathy, 51/468
sensory neuropathy, 8/201
Dorsal spinal nerve root ganglion degeneration
anhidrosis, 42/346
episodic, 8/369
hereditary motor and sensory neuropathy type I, 42/340
hereditary perforating foot ulcer, 8/201
progressive, 8/369
sensory neuropathy, 42/346
sensory radicular neuropathy, 42/351
Dorsal spinal nerve root ganglionitis
Sjögren syndrome, 51/449
Dorsal striatum, *see* Dorsal neostriatum
Dorsal vagus nucleus, *see* Vagus nerve dorsal nucleus
Dorsolateral tract, *see* Lissauer tract
Dorsomedial hypothalamic nucleus
topography, 2/434-436
Dorsomedial thalamic nucleus
brain infarction, 55/140
ventral globus pallidus, 49/26
Wernicke-Korsakoff syndrome, 30/330
Dortmund, Germany
neurology, 1/9
Dosulepin
tricyclic antidepressant, 65/312
Dothiepin, *see* Dosulepin
Double athetosis
age at onset, 49/384
cerebral diplegia, 49/384
clinical features, 49/507
deafness, 49/384
dystonia musculorum deformans, 6/534
erythroblastosis fetalis, 49/384
frontopontine bundle, 6/448
genetics, 49/384
Hallervorden-Spatz syndrome, 6/155, 49/386
hyperbilirubinemia, 49/384
hypoxia, 49/384
kernicterus, 49/384
pallidoluysionigral degeneration, 49/459
perinatal hypoxia, 49/384
status marmoratus, 6/155, 447, 534, 49/384, 386, 507
striatal necrosis, 49/506
tonic postural reflex (Foix-Thévenard), 6/141
vertical gaze palsy, 49/384
Wilson disease, 6/155
Double barrel lumen
carotid dissecting aneurysm, 54/273
Double myelomeningocele, *see* Spina bifida
Double ureter
holoprosencephaly, 50/237
Double vagina
holoprosencephaly, 50/237
Double Y aneuploidy, *see* XYY syndrome
Double Y syndrome, *see* XYY syndrome
Doughnut sign
brain abscess, 52/154
glioblastoma, 18/55
Down syndrome, 31/367-451
*see also* Chromosomal aneuploidy
acute lymphocytic leukemia, 50/527
acute myeloid leukemia, 50/274, 527
Alzheimer disease, 46/20, 73, 273
amniocentesis, 31/444
amphimixis, 50/523
amyloid fibril protein, 50/530
anal atresia, 50/527
aneuploidy, 31/394, 50/566
arm-span, 50/527
art, 31/367-370
associated disorder, 46/20
atlantoaxial dislocation, 43/544
atlantoaxial instability, 50/529
atlantoaxial subluxation, 50/392
atrophoderma vermiculatum, 14/791
autonomic nervous system, 75/643
azoospermia, 50/532
basilar impression, 50/399
behavior, 31/406, 43/544, 50/528
behavior disorder in children, 43/544
biochemical aspect, 27/488
biochemistry, 50/531
birth incidence, 43/546
birth weight, 50/525
body height, 50/527

body mass, 50/527
brain cortex, 50/274
brain shape, 50/531
brain stem auditory evoked potential, 50/529
brain stem weight, 50/530
brain weight, 50/274, 530
Brushfield spot, 43/544, 50/526
carbohydrate metabolism, 31/433
cardinal sign, 31/400
cataract, 50/529
cause, 50/519
cerebellar agenesis, 50/183
cerebellar cortex heterotopia, 50/531
cerebellar hypoplasia, 50/531
cerebellar middle lobe, 50/531
cerebellum weight, 50/530
cerebrovascular disease, 53/31
cholesterol, 31/433
cholinergic deficiency, 50/530
chromosomal aberration, 46/19-21
chromosomal disjunction syndrome, 30/160, 31/378-395
chromosome analysis, 31/440, 445-447
chromosome translocation, 31/386-391
clinical features, 31/367, 399-409
CNS, 31/367, 400, 407-412
CNS malformation, 50/57
congenital heart disease, 50/274, 53/31
congenital heart malformation, 50/519, 526
congenital megacolon, 42/3
consanguinity, 50/525
convulsion, 31/407, 43/544, 50/529
corpus callosum agenesis, 50/531
counseling, 31/440-442
course, 31/438, 50/525
cryptorchidism, 50/274, 532
cytogenetics, 31/376-395, 50/520, 522
death rate, 50/525
dementia, 27/488, 46/403
dermatoglyphics, 31/420-422, 43/545
diagnosis, 31/438-440
dictyotene, 50/523
digital abnormality, 43/545
duodenal atresia, 50/274
duodenal obstruction, 50/527
dysplastic ear, 50/526
dysplastic pelvis, 50/526
early aging, 50/530
EEG, 31/401, 405, 408, 50/529
electrolyte, 31/433
Ellis-Van Creveld syndrome, 31/427
endocrine abnormality, 31/429-433
endocrinology, 50/531
epicanthus, 43/544
epidemic hepatitis, 31/398
epidemiology, 30/149-151, 160
epilepsy, 31/407
esophagus atresia, 50/527
extra chromosome 21, 50/520
extrapyramidal hyperkinesia, 50/526
eye finding, 31/415-417
face, 31/413
familial, 30/161, 50/524
α-fetoprotein, 50/531
fissure syndrome, 50/526
frontal lobe, 50/531
gait, 50/528
genetics, 31/372, 391-393
glycinamide phosphonucleotide, 50/531
granulovacuolar neuronal degeneration, 50/530
growth retardation, 50/519
Hashimoto thyroiditis, 50/532
head circumference, 50/525
hearing loss, 50/526
heart disease, 43/545
history, 31/370-374, 50/520
holoprosencephaly, 50/531
hyperuricemia, 50/531
hypogonadism, 43/545, 50/532
hypoplastic maxilla, 50/526
hypothyroidism, 50/532
hypozoospermia, 50/532
hypsarrhythmia, 50/529
ichthyosis, 13/476
immunocompromised host, 56/469, 473
immunodeficiency, 50/530
imperforate anus, 50/274
incidence, 31/367, 395, 46/2, 50/274, 519
infantile spasm, 50/529
institutionalization, 31/404, 50/533
intelligence quotient, 50/528
iris mesenchymal hypoplasia, 50/526
joint hyperlaxity, 50/519, 529
keratoconus, 50/526
Kuwait, 50/525
language, 31/404
late cortical cerebellar atrophy, 21/477
late infantile neuroaxonal dystrophy, 49/405
leukemia, 31/424-426, 43/545, 50/519
leukemoid reaction, 31/424-426
lipid, 31/433
management, 31/442-444
maternal age, 30/161, 31/373, 379-382, 43/545, 50/523, 525
maternal polysomy X, 50/524
mechanism, 31/381

megacolon, 50/527
meiosis type I, 50/522
meiosis type II, 50/522
meiotic, 31/378-383
mental deficiency, 31/401-406, 43/544, 46/19-21, 70-73, 50/519, 528-531
Meynert nucleus basalis, 50/530
microcephaly, 31/399, 410, 413, 43/544, 50/274
microdontia, 50/526
mitotic, 31/383-386
molecular genetics, 50/531
mosaicism, 31/383-386, 46/73, 50/521, 523
mosaicism parental, 31/382
moyamoya disease, 53/31
muscular hypotonia, 50/519, 526, 529
myopia, 50/526, 529
neurofibrillary tangle, 49/249, 50/530
neurologic symptom, 50/529
neuropathology, 21/45, 50/530
newborn period, 31/399-401
nystagmus, 50/526, 529
operculum hypoplasia, 50/531
otitis media, 50/526
9p partial monosomy, 50/578
paternal XYY syndrome, 50/524
pathology, 31/409-412
6-phosphofructokinase, 50/531
phosphoribosylamine glycine ligase, 50/531
physical characteristics, 31/412-420
pregnancy period, 31/399-401
prevalence, 31/367, 396, 43/546, 50/519
prevention, 31/444-447
prognosis, 31/438
protein metabolism, 31/437
psychological adjustment, 46/31
psychomotor development, 50/527
pyridoxal-5-phosphate, 28/129
r(9) syndrome, 50/580
radiation exposure, 31/398
radiologic features, 31/426-429
recurrence risk, 43/546
red cell enzyme, 31/435
risk, 31/381
Rubinstein-Taybi syndrome, 31/427
seasonal variation, 31/399
secondary pigmentary retinal degeneration, 13/338
seizure, 31/406-408, 50/529
senile dementia, 31/405
senile plaque, 50/530
serotonin, 31/433-435
sex difference, 31/396
sexual development, 31/432
short neck, 50/526
short stature, 43/544
simian crease, 50/526
skeletal abnormality, 31/400
slanting palpebral, 50/525
social adjustment, 46/31
somatosensory evoked potential, 50/529
spastic paraplegia, 50/529
speech, 31/404
spinal muscular atrophy, 59/371
strabismus, 50/526
striatopallidodentate calcification, 49/424, 50/531
superior temporal gyrus, 50/274
superior temporal lobe hypoplasia, 50/531
superoxide dismutase 1, 50/521, 530
symptomatology, 50/525-529
synonym, 31/367
thyroid autoimmune antibody, 50/532
thyroid function, 31/397, 429-431
thyroid gland, 50/532
thyrotropic hormone, 50/532
tongue protrusion, 50/526
tracheoesophageal fistula, 50/527
translocation, 30/160, 50/525
transverse atlantal ligament laxity, 50/392
treatment, 31/442-444, 50/532
tryptophan, 31/433-435
tuberous sclerosis, 14/350
tubulin, 50/530
twins, 30/161, 31/372, 393
vertebral abnormality, 43/544
visual evoked response, 50/529
vitamin B6, 28/129
white cell enzyme, 31/435
Downbeating nystagmus
anticonvulsant intoxication, 60/656
Arnold-Chiari malformation type I, 60/656
Arnold-Chiari malformation type II, 50/410
late cerebellar ataxia, 60/656
lithium intoxication, 64/296
multiple sclerosis, 60/656
neoplasm, 60/656
periodic ataxia, 60/656
syphilis, 60/656
Doxepin
anxiolytic agent, 37/360
tricyclic antidepressant, 65/312
Doxorubicin
antineoplastic agent, 65/528
cardiomyopathy, 64/17
DNA metabolism, 64/17
dorsal spinal nerve root ganglion, 51/300
drug induced polyneuropathy, 51/300

encephalopathy, 65/528
neuropathy, 69/471
neurotoxin, 51/300, 64/10, 17, 65/528
toxic myelopathy, 64/10
toxic neuropathy, 51/300, 64/10
Doxorubicin intoxication
action site, 64/10
Doxylamine
acquired myoglobinuria, 62/573
Doyne disease, *see* Hutchinson-Tay choroidopathy
Doyne-Grönblad-Strandberg syndrome, *see* Grönblad-Strandberg syndrome
Doyne honeycomb retinal degeneration, *see* Hutchinson-Tay choroidopathy
DPDA intoxication, *see* Iso-Ompa intoxication
DPNH, *see* Reduced nicotinamide adenine dinucleotide
DPT immunization, *see* Diphtheria pertussis tetanus immunization
Dracunculiasis
filariasis, 35/161, 52/514
spinal cord compression, 52/514
Dracunculosis, *see* Dracunculiasis
Drainages
brain ventricle, *see* Brain ventricle drainage
condom, *see* Condom drainage
urinary, *see* Urinary drainage
Drawing
body scheme disorder, 4/396
child, 4/353, 396, 450
developmental dyspraxia, 4/450
Dreaming
auditory discrimination, 3/89
awakening, 3/88
duration, 3/88
external stimuli, 3/88
history, 3/87
idiopathic narcolepsy, 3/94
ketamine intoxication, 37/408
limbic system, 3/89
memory, 3/87, 175
mental disease, 3/127
narcolepsy, 3/94
nightmare, 3/99
paradoxical sleep, 3/88, 103
penis erection, 3/89
plasma free fatty acid, 3/82
pontine lesion, 3/99
REM, 1/489, 3/88
seminal emission, 3/89
sexual activity, 3/89
symbolic representation, 3/89
visual, *see* Visual dreaming
Dreamy state
consciousness, 3/123
epileptic, 15/95
hypnagogic hallucination, 45/355
migraine, 48/162
musical hallucination, 2/705
olfactory hallucination, 2/702
visual hallucination, 2/616, 45/356
Drechslera
fungal CNS disease, 35/565, 52/485
Drechsleriasis
CSF, 52/485
encephalitis, 52/486
flucytosine, 52/486
keratitis, 52/486
ketoconazole, 52/486
lymphocytic meningitis, 35/565, 52/485
meningitis, 35/565, 52/485
Dressing apraxia
cerebral dominance, 4/296
constructional apraxia, 4/53
developmental dyspraxia, 4/459
hemiasomatognosia, 4/53, 215
parietal lobe syndrome, 2/692, 45/74
Drift
genetic, *see* Genetic drift
pronator, *see* Pronator drift
Driftwood cortex, *see* Dystopic cortical myelinogenesis
Drivers licence
epilepsy, 15/808
Driving
epilepsy, 72/421
head injury, 46/619
rehabilitation, 46/619
stroke, 46/619
transient global amnesia, 45/209
Dromedary gait
dystonia musculorum deformans, 49/522
Dromoran, *see* Levorphanol
Dronabinol
cannabis, 65/46
tetrahydrocannabinol, 65/46
Drooling
late cerebellar atrophy, 42/135
motoneuron disease, 59/466
Wilson disease, 49/225
Drop attack
absence, 15/138
basilar artery migraine, 48/137
basilar ischemia, 12/5
brain artery occlusion, 12/23
cause, 53/235

epilepsy, 72/250
giving way of legs, 1/347
head rotation, 12/23
progressive supranuclear palsy, 49/239
transient ischemic attack, 12/5, 53/262
vertebral artery insufficiency, 7/453
vertebrobasilar insufficiency, 2/281, 360, 53/262
vertebrobasilar stenosis, 12/23
Droplet metastasis
brain tumor, 20/244, 285
cauda equina, 20/285
cytoplasmic inclusion body, 20/285
laminectomy, 20/285
meningioma, 20/244
neurinoma, 20/244, 285
neurofibroma, 20/285
neurofibromatosis, 20/244
neurofibromatosis type I, 20/244, 285
spinal nerve root, 20/285
surgery, 20/285
Drowsiness
acetazolamide intoxication, 37/202
basilar artery migraine, 48/137
benzodiazepine intoxication, 65/335
carbamazepine intoxication, 37/202
clonidine intoxication, 37/443
decaborane intoxication, 64/360
disulfiram intoxication, 37/321
ethchlorvynol intoxication, 37/354
fox intoxication, 37/96
high-pressure nervous syndrome, 48/402
idiopathic narcolepsy, 3/94
lymphocytic choriomeningitis, 56/359
methyl salicylate intoxication, 37/417
migraine, 48/144, 161
narcolepsy, 45/147
pentaborane intoxication, 64/359
polar bear intoxication, 37/96
primidone intoxication, 37/202
reserpine intoxication, 37/438
salicylic acid intoxication, 37/417
sleep, 3/86, 90, 114
sodium salicylate intoxication, 37/417
valproic acid intoxication, 37/202
vitamin A intoxication, 37/96
wolf intoxication, 37/96
Drug abuse
acquired immune deficiency syndrome, 55/517, 56/490
agraphia, 45/462
amphetamine, 55/519
aspergillosis, 35/396, 55/433
benzodiazepine intoxication, 65/339, 341
brain embolism, 55/166, 518
brain hemorrhage, 55/517
brain infarction, 55/517
brain vasculitis, 55/522
cerebrovascular disease, 55/166
cocaine, 55/521
diamorphine myelopathy, 55/519
diamorphine nephropathy, 55/517
ephedrine, 55/521
granulomatous CNS vasculitis, 55/396
hepatitis, 37/393, 55/517
infectious endocarditis, 55/517, 63/112
juvenile cerebrovascular disease, 55/517
lysergide, 55/522
phencyclidine, 55/522
polyarteritis nodosa, 39/307, 55/359, 520
pyridoxine, *see* Vitamin B6 abuse
Staphylococcus aureus, 55/517
subarachnoid hemorrhage, 55/3
wound botulism, 65/209
Drug addiction, 37/365-394
amphetamine, 65/256
bacterial meningitis, 33/2
barbiturate, 65/330
benzodiazepine, 65/339
caffeine, 65/259
cocaine, *see* Cocaine addiction
diamorphine, *see* Diamorphine addiction
ephedrine, 65/259
ergotamine, 65/61
fenfluramine, 65/256
fetal, 37/377
methamphetamine, *see* Methamphetamine addiction
methylphenidate, *see* Methylphenidate addiction
nicotine, 65/251
nicotine intoxication, 65/263
opiate, *see* Opiate addiction
pentazocine, *see* Pentazocine addiction
pethidine, *see* Pethidine addiction
phencyclidine, *see* Phencyclidine addiction
phentaramine, *see* Phentaramine addiction
phentermine, 65/256
phenylpropanolamine, *see* Phenylpropanolamine addiction
pyribenzamine, *see* Pyribenzamine addiction
Drug holiday
headache, 48/6
levodopa induced dyskinesia, 49/199
Drug induced aseptic meningitis, *see* Chemical meningitis
Drug induced behavior disorder
drug list, 46/591-602

review, 46/591-602
Drug induced dementia
aged, 46/393
Drug induced demyelinating neuropathy
amiodarone, 47/620
chrysotherapy, 47/620
perhexiline maleate, 47/620
Drug induced dystonia
symptomatic dystonia, 49/543
Drug induced extrapyramidal syndrome, 6/248-264
akathisia, 6/253
akinesia, 6/250
autonomic dysfunction, 6/254
character, 6/254
clinical features, 6/249
dysarthria, 6/262
history, 6/249
hyperkinesia, 6/251
hypertonia, 6/251
hypokinesia, 6/250
incidence, 6/254
MAO inhibitor, 6/262
mental change, 6/254
metabolic disorder, 6/254
neuroleptic malignant syndrome, 6/258-260
neuropathology, 6/259-261
pathophysiology, 6/259-261
severe, 6/254
sex ratio, 6/255
sleep disorder, 6/254
treatment, 6/255, 261
tremor, 6/262
tricyclic derivative, 6/263
Drug induced hypotension
brain aneurysm neurosurgery, 55/43
brain infarction, 63/74
brain ischemia, 65/433
Drug induced myotonia
9-anthroic acid intoxication, 62/271
azacosterol, 40/155, 328, 514, 548, 555, 62/271
2,4-dichlorophenoxyacetic acid intoxication, 62/271
Drug induced myotonic syndrome
9-anthroic acid intoxication, 62/271
2,4-dichlorophenoxyacetic acid intoxication, 62/271
Drug induced parkinsonism
animal experiment, 6/102-104
chlorpromazine, 6/234
CNS, 6/234
motor disorder, 6/234
reserpine, 6/234
Drug induced polyneuropathy
*see also* Drug intoxication
almitrine dimesilate, 51/294
aminoglycoside, 51/295
amiodarone polyneuropathy, 51/294, 303
anticonvulsant, 51/307
calcium carbimide, 51/309
chloramphenicol, 51/294
cisplatin, 51/300
clioquinol, 51/296
disulfiram, 51/309
doxorubicin, 51/300
ethambutol, 51/294
gentamicin, 51/295
isoniazid, 51/294
lithium, 51/304
metronidazole, 51/301
misonidazole, 51/301
nitrofurantoin, 51/296
perhexiline, 51/294
perhexiline maleate, 51/301
phenytoin, 51/307
podophyllin, 51/294
salazosulfapyridine, 51/309
streptomycin, 51/295
thalidomide, 51/304
tricyclic antidepressant, 51/303
vincristine, 51/294, 299
zimeldine, 51/304
Drug induced seizure
chemotherapy, 69/482
vitamin B6, 28/114-117
Drug induced systemic lupus erythematosus
ethosuximide intoxication, 65/508
hydralazine, 16/222
isoniazid, 16/222
mephenytoin intoxication, 37/202
oral contraception, 16/222
phenytoin, 16/222
procainamide, 16/222
procainamide intoxication, 37/452
valproic acid intoxication, 65/505
Drug intoxication
*see also* Drug induced polyneuropathy
autonomic neuropathy, 51/294
brain death, 55/263
brain stem death, 57/454
delirium, 46/537-541
demyelination, 47/551
iatrogenic neuropathy, 7/527-546
neurologic intensive care, 55/220
platinum polyneuropathy, 51/274
visual hallucination, 55/146
Drug therapy

bladder dysfunction, 75/674
intrathecal, *see* Intrathecal chemotherapy
osteoporosis, 70/19
Drug trial
Becker muscular dystrophy, 62/136
Drug withdrawal, *see* Drug holiday
Drummer palsy
occupational lesion, 7/332
thumb, 7/332
Drusen
associated disease, 13/251
Bruch membrane, 13/236
dominant, *see* Hutchinson-Tay choroidopathy
familial, *see* Hutchinson-Tay choroidopathy
Grönblad-Strandberg syndrome, 13/251
primary pigmentary retinal degeneration, 13/249-251
tuberous sclerosis, 14/623
Urbach-Wiethe disease, 13/251
Drüsige angiopathy (Scholz), *see* Brain amyloid angiopathy
Dry socket
dental headache, 5/226
headache, 5/226
Dryness
eye, *see* Eye dryness
DTIC, *see* Dacarbazine
Duane retraction syndrome, *see* Acquired retraction syndrome *and* Stilling-Türk-Duane syndrome
Duane syndrome, *see* Stilling-Türk-Duane syndrome
Dubini chorea electrica, *see* Chorea electrica
Dubowitz classification
infantile spinal muscular atrophy, 59/51
Dubowitz syndrome
dermatitis, 43/247
dermatoglyphics, 43/249
digital abnormality, 43/249
ear abnormality, 43/248
epicanthus, 43/247
growth retardation, 43/247
high pitched voice, 43/249
hyperkinesia, 43/247
hypertelorism, 30/248
mental deficiency, 30/248, 43/247, 46/59
microcephaly, 30/100, 508, 43/247
micrognathia, 43/248
ptosis, 43/247
skin malformation, 30/248
swayback nose, 43/248
Duchenne de Boulogne, *see* Duchenne, G.B.A.
Duchenne disease, *see* Progressive bulbar palsy
Duchenne-Erb syndrome
birth incidence, 7/416
brachial plexus injury, 7/408
brachial plexus paralysis, 2/140
clinical features, 1/229
Duchenne erythrocyte
scanning electron microscopy, 40/386
Duchenne, G.B.A., 1/5, 7, 22/3-7, 9
Duchenne like muscular dystrophy
pseudohypertrophy, 43/102
recessive, 43/102
Duchenne muscular dystrophy
acetylcholine receptor, 40/372, 384
acid maltase deficiency, 27/228
adenosine monophosphate deaminase, 40/378
adenosine triphosphate, 40/382, 385
adenyl cyclase, 40/3, 382, 385
age at onset, 62/118
arylamidase, 40/380
atypical form, 62/122
Becker muscular dystrophy, 62/123
biochemistry, 40/376-382, 62/124
birth incidence, 43/106
blood vessel, 40/170, 251
bracing, 41/468
calcium, 40/381, 387
cardiac involvement, 62/119
cardiomyopathy, 43/106
carrier assessment, 41/415
carrier detection, 40/340, 354, 41/415, 484, 489
catecholamine, 40/381
cerebrovascular disease, 53/43
chromosome Xp21, 62/125
chronic ventilatory failure, 62/134
ciclosporin, 62/131
classification, 40/281
clinical features, 40/306, 308, 312, 321, 323, 335, 337, 342, 358-363, 41/408, 461
CNS, 62/121
contractile protein, 40/380
contraction band, 40/48, 208, 252, 369
contracture, 40/337, 361, 41/465, 43/106, 62/119, 133
corticosteroid, 40/378, 62/129
course, 62/119
creatine-creatinine, 40/379
creatine kinase, 40/281, 376, 43/106, 62/118, 124
deflazacort, 62/131
delta lesion, 40/370
differential diagnosis, 41/176, 62/122
dystrophin, 62/121, 125
ECG, 43/106
echinocytes, 41/410
electrophysiology, 40/373-375

Emery-Dreifuss muscular dystrophy, 40/389
EMG, 40/373
erythrocyte, 40/382, 385
facioscapulohumeral muscular dystrophy, 40/419
fasciotomy, 41/470
female, 40/352
fiber size variation, 40/367
fracture, 41/474
fructose-bisphosphate aldolase, 62/118
functional evaluation, 41/494
functional grading, 62/120
gastrointestinal features, 62/134
gastrointestinal manifestation, 62/120
genetic diagnosis, 62/128
genetics, 40/352-357, 41/487, 489, 62/125
genotype, 62/126
glycerol kinase deficiency, 62/127
Gowers sign, 40/314, 359, 43/105, 62/117
Haldane rule, 40/353, 41/416
heart disease, 40/362, 43/106
heart failure, 62/134
heterozygote detection, 43/106
histochemistry, 40/47
histology, 62/123
histopathology, 62/118
history, 40/350, 434, 62/117
hyporeflexia, 43/106
incidence, 62/118
incidence and prevalence, 40/351
infantile spinal muscular atrophy, 22/77
intellectual function, 41/478
intestinal pseudo-obstruction, 62/120
ischemic myopathy, 40/167-170
joint fixation, 40/361
lactate dehydrogenase, 40/376, 41/488
lactate dehydrogenase isoenzyme 5, 40/356
lipid, 40/380
lithium, 40/378
Lyon principle, 41/408, 50/542
malignant hyperthermia, 41/268, 62/570
management, 40/363-366
manifesting carrier, 41/415, 489
McLeod phenotype, 62/127, 63/289
mdx mouse model, 62/127
membrane theory, 40/252, 370, 384-387, 41/410
mental deficiency, 40/363, 41/479, 46/66, 62/121
3-methyl histidine, 40/380
microvasculature, 40/170, 251
misdirected muscle regeneration theory, 41/409
mitochondria, 40/382
motor unit, 40/374
muscle biopsy, 62/123
muscle carnitine, 41/201
muscle cell membrane, 40/74, 151, 252, 370, 384, 41/410
muscle fiber size, 43/106
muscle hypertrophy, 40/358, 41/408, 461
muscle ischemia, 41/409
muscle necrosis, 40/368
muscle regeneration, 40/371
muscle tissue culture, 40/191, 375, 62/100
mutation rate, 43/106
myoblast transfer, 62/131
myoglobinemia, 41/275
myoglobinuria, 62/555, 558
myotonic dystrophy, 40/507, 522
natural history, 62/119
necrosis, 40/251, 368
neurogenic theory, 40/383, 41/409
neuromuscular disease, 41/408
orthopedic aspect, 40/361
orthopedic management, 41/468
ouabain, 40/386
oxidative phosphorylation, 40/379
pathogenesis, 40/382-387
pathology, 40/250-252, 366-373
phenotype, 62/126
phosphorylcholine, 40/380
physiotherapy, 62/132
plasma membrane, 40/48, 151, 213, 252, 370, 43/106
polymyositis, 40/52, 253, 41/58
preclinical detection, 43/106
prenatal diagnosis, 41/417
prevalence, 43/106, 62/118
progressive, 43/105-107
prolylhydroxyproline, 40/380
protein metabolism, 40/379
pseudohypertrophy, 40/250, 366, 43/106
pulmonary disease, 40/360
pulmonary function, 41/494, 496
pyruvate kinase, 40/357, 43/106
regeneration, 40/371
respiratory dysfunction, 40/360
respiratory failure, 62/122, 133
ribosomal protein synthesis test, 41/411
sarcolemma, 40/74, 252, 370, 384
satellite cell, 40/371
scoliosis, 40/362, 62/119
secondary pigmentary retinal degeneration, 60/723
serine protease, 40/380
serotonin, 40/168, 383
serum creatine kinase, 41/80
sick motoneuron, 40/383
single fiber electromyography, 40/374

spinal fusion, 41/478
surgical treatment, 62/132
symptom, 62/118
tenotomy, 40/364
terminal innervation ratio, 40/384
toe walker, 40/321, 361
treatment, 62/129
treatment trial, 62/129
Turner syndrome, 40/352
ultrastructure, 40/68, 70
urinary carnosine, 62/124
urinary creatinine, 62/124
urinary 3-methylhistidine, 62/124
urinary putrescine, 62/124
urinary spermidine, 62/124
urinary spermine, 62/124
vascular theory, 40/167, 251, 382, 41/409
weakness, 40/358-360
X-linked muscular dystrophy, 40/277, 349-391, 41/408, 415, 62/117
Duchenne sign
hip, 2/43
topographical diagnosis, 2/43
Duct
frontonasal, *see* Frontonasal duct
müllerian, *see* Müllerian duct
Duhring-Brocq disease, *see* Dermatitis herpetiformis
Duhring disease, *see* Dermatitis herpetiformis
Dumbbell tumor, *see* Hourglass tumor
Dumon-Radermecker syndrome
cerebellar ataxia, 42/477
epilepsy, 42/477-479
mental deficiency, 42/477
optic atrophy, 42/477
speech disorder, 42/477
tremor, 42/477
Dumping syndrome, *see* Postgastrectomy syndrome
Duncan cave, *see* Cavum septi pellucidi
Dundee, United Kingdom
neurology, 1/7
Duodenal atresia
Down syndrome, 50/274
Duodenal obstruction
Down syndrome, 50/527
Dura
cranial, *see* Cranial dura
Dura mater
*see also* Cranial dura, Meninges, Meninx *and* Pachymeningitis
age, 30/419
anastomosis, 30/427
anatomy, 1/552, 30/417, 33/149-153
arteriovenous malformation, 12/229, 236
artery, 30/426
attachment, 30/420
bridge vein, 30/421
classification, 30/426
duplication, 30/419
embryology, 30/415
endothelioma, 20/187
fibrosarcoma, 16/20
Galen vein, 30/421
glioblastoma multiforme, 18/51
hemodynamic factor, 30/426
hypertrophic pachymeningitis, 33/190
inferior sagittal sinus, 30/423
innervation, 30/418
jugular foramen, 30/424
missile injury, 23/513, 520
occipital sinus, 30/424
short, 5/187, 372
sinus, 30/416-418
sinus pattern, 30/424
spinal, *see* Spinal dura mater
spinal nerve root injury, 25/397
splitting, 30/417
superior longitudinal sinus, 30/424
tissue culture, 17/48
transverse sinus, 30/424, 426
tuberculoma en plaque, 18/421
vascular bridge, 30/419, 420
venous malformation, 12/236
Dura mater abnormality
diastematomyelia, 50/437
diplomyelia, 50/437
Dura mater calcification
multiple nevoid basal cell carcinoma syndrome, 14/80, 114, 457
Dura mater cyst
chronic meningitis, 56/645
Dura mater reduplication
diastematomyelia, 50/436
diplomyelia, 50/436
Dura mater rupture
vertebral column injury, 61/96
Dura mater sheath reduplication
diastematomyelia, 50/436
diplomyelia, 50/436
Dural arteriovenous malformation
*see also* Brain arteriovenous malformation
brain arteriovenous malformation, 54/372
Doppler sonography, 54/9
Dural metastasis
colon carcinoma, 71/614
Ewing sarcoma, 69/124

survey, 69/123
Dural sac
dilatation, 12/519
Dural sinus occlusion, *see* Venous sinus occlusion
Dural sinus thrombophlebitis, *see* Venous sinus thrombophlebitis
Dural sinus thrombosis, *see* Venous sinus thrombosis
Dural venous plexus
anterior plexus, 11/57, 61
middle plexus, 11/60
posterior plexus, 11/60
Dural venous sinus, *see* Venous sinus
Duret artery
migraine, 5/85
Dutch Quality of Care
epilepsy, 72/433
Duvenhage virus
human pathogen, 56/384
rhabdovirus, 56/384
Dwarf
Harper, *see* Harper dwarf
Russell-Silver, *see* Russell-Silver syndrome
Seckel, *see* Seckel dwarf
Dwarfism
*see also* Short stature
achondroplasia, 43/315-317, 46/88
alaninuria, 42/515
alopecia, 43/380
aminoaciduria, 42/517
Armendares syndrome, 43/366
ataxia telangiectasia, 14/517
Bardet-Biedl syndrome, 13/395
bird headed, *see* Seckel syndrome
Bitter syndrome, 14/107
Bloom syndrome, 13/436
Börjeson-Forssman-Lehmann syndrome, 42/530
brain atrophy, 43/380
camptomelic, *see* Camptomelic syndrome
centrofacial lentiginosis, 43/26
chondrodystrophic myotonia, 40/284, 328, 546
Cockayne-Neill-Dingwall disease, 10/182, 13/432, 43/350, 51/399
congenital hyperphosphatasia, 31/258
convulsion, 43/380
corpus callosum agenesis, 43/385, 50/163
craniofacial dysostosis (Crouzon), 43/361
Darier disease, 43/380
diastrophic, *see* Diastrophic dwarfism
Dyggve-Melchior-Clausen syndrome, 43/265
EEG, 43/380
familial ectodermal dysplasia, 60/666
François syndrome, 13/436
glycosaminoglycanosis type I, 10/435
glycosaminoglycanosis type II, 10/435, 42/454
glycosaminoglycanosis type VI, 42/458
glycosaminoglycanosis type VII, 42/459
Goldberg syndrome, 42/439
Goltz-Gorlin syndrome, 14/113
Hallervorden-Spatz syndrome, 6/604, 607
Hartnup disease, 60/666
heart disease, 43/382, 435
Hurler-Hunter disease, 10/341, 436
hydrocephalus, 43/385
Kearns-Sayre-Daroff-Shy syndrome, 60/666
Laron, *see* Pituitary dwarfism type III
leukemoid reaction, 43/385
Levi-Lorain, *see* Levi-Lorain dwarfism
Marinesco-Sjögren syndrome, 60/666
MELAS syndrome, 60/666
mental deficiency, 43/265, 268, 276, 294, 381, 46/88
MERRF syndrome, 60/666
mesomelic, *see* Mesomelic dwarfism
microcephaly, 43/384
micrognathia, 43/380
mucolipidosis type III, 42/449
Mulibrey, *see* Mulibrey dwarfism
myotonic myopathy, 43/165
nanocephalic, *see* Seckel syndrome
neuralgic amyotrophy, 51/175
otopalatodigital syndrome, 43/457
Passwell syndrome, 21/216
polydystrophic, *see* Glycosaminoglycanosis type VI
ptosis, 43/382
Rothmund-Thomson syndrome, 13/436
Russell-Silver syndrome, 30/248
secondary pigmentary retinal degeneration, 13/152, 324-326
seizure, 43/385
snub-nosed, *see* Snub-nosed dwarfism
spondyloepiphyseal dysplasia, 43/478
Taybi-Linder syndrome, 43/385
thanatophoric, *see* Thanatophoric dwarfism
tryptophanuria, 42/633
vertebral abnormality, 43/381, 385
Werner syndrome, 13/436
xeroderma pigmentosum, 14/13, 51/399
Dyflos
chemical structure, 64/225
neurotoxin, 37/543, 64/152, 223
organophosphate, 64/152
organophosphate intoxication, 64/152, 223
Dyflos intoxication
chemical classification, 37/543

organophosphate, 64/152
organophosphorus compound, 37/543
Dyggve-Melchior-Clausen syndrome
concentric laminated body, 40/109
congenital hip dislocation, 43/266
dwarfism, 43/265
iliac crest abnormality, 43/266
kyphoscoliosis, 43/266
mental deficiency, 43/265
Smith-McCort syndrome, 43/266
Dying-back neuropathy, *see* Axonal degeneration
Dyke-Davidoff syndrome
athetotic hemiplegia, 42/201
Dynorphin
basal ganglion, 49/33
endogenous opioid, 57/83, 65/349
prodynorphin, 65/350
reflex sympathetic dystrophy, 61/124
Dysacusis
thalamic lesion, 2/480
Vogt-Koyanagi-Harada syndrome, 56/611
Dysarthria
*see also* Speech disorder *and* Stuttering
acrylamide intoxication, 64/68
adrenoleukodystrophy, 60/169
alcohol intoxication, 64/115
amyloid tumor, 8/9
amyotrophic lateral sclerosis, 22/120, 42/65
ataxia telangiectasia, 14/284
ataxic hemiparesis, 54/248
barbiturate intoxication, 65/332, 509
basilar ischemia, 12/5
Bassen-Kornzweig syndrome, 51/394
Behçet syndrome, 56/597
benzodiazepine intoxication, 65/338
Binswanger disease, 54/222
bismuth intoxication, 64/337
botulism, 65/210
boxing injury, 23/545
brain lacunar infarction, 54/243, 253
carbamazepine intoxication, 65/505
central pontine myelinolysis, 28/291
cerebellar, *see* Cerebellar dysarthria
cerebellar infarction, 53/383
cerebrotendinous xanthomatosis, 10/546
childhood myoglobinuria, 62/564
chlordecone intoxication, 64/204
chronic nodular polioencephalitis, 34/560
classification, 3/396
Creutzfeldt-Jakob disease, 46/290
cyanide intoxication, 64/232
cycloserine, 65/481
cycloserine intoxication, 65/481
cytarabine, 63/357
cytarabine intoxication, 65/533
developmental, *see* Developmental dysarthria
dissecting aorta aneurysm, 63/44
dominant cerebello-olivary atrophy, 42/136
drug induced extrapyramidal syndrome, 6/262
dystonia musculorum deformans, 6/523, 49/521
erythrokeratodermia ataxia syndrome, 21/216
ethylene glycol intoxication, 64/123
familial polyradiculopathy syndrome, 42/97
familial spastic paraplegia, 22/435
Fazio-Londe disease, 42/94
fluorouracil intoxication, 65/532
Friedreich ataxia, 21/337, 51/388
Gerstmann-Sträussler-Scheinker disease, 56/553
hereditary cerebello-olivary atrophy (Holmes), 21/403, 409, 60/569
hereditary diurnal fluctuating juvenile dystonia, 49/530
hereditary olivopontocerebellar atrophy (Menzel), 21/441
hereditary spastic ataxia, 21/367
hereditary striatal necrosis, 49/495
Huntington chorea, 49/277
hypophosphatemia, 63/564
juvenile hereditary benign chorea, 49/345
Köhlmeier-Degos disease, 55/276
kuru, 56/556
Lafora progressive myoclonus epilepsy, 15/385, 27/173
late infantile metachromatic leukodystrophy, 10/346, 29/357
lidocaine intoxication, 37/450, 65/454
lindane intoxication, 65/479
lip abnormality, 4/413-419
local anesthetic agent, 65/420
manganese intoxication, 64/308
Marchiafava-Bignami disease, 17/542, 28/317, 324, 46/337
marine toxin intoxication, 65/153
Marinesco-Sjögren syndrome, 42/184
Mast syndrome, 42/282
mercury polyneuropathy, 51/272
methanol intoxication, 64/99
migraine, 48/143
Minamata disease, 7/642, 36/74, 105, 64/413, 420
multiple neuronal system degeneration, 59/138
multiple sclerosis, 9/167, 179
muscle weakness, 40/306
myopathy, 40/306
myxedema neuropathy, 8/57, 59
neocerebellar syndrome, 2/423
neuroleptic dystonia, 65/282

nigrospinodentate degeneration, 22/170
olivopontocerebellar atrophy (Dejerine-Thomas), 21/425
organochlorine insecticide intoxication, 64/204
organomercury intoxication, 36/74
organophosphate intoxication, 64/167
palatal myoclonus, 38/584
palate abnormality, 4/413-419
pallidopyramidal degeneration, 49/456
paralytic shellfish poisoning, 65/153
Parkinson dementia, 49/171
paroxysmal, *see* Paroxysmal dysarthria
periodic, *see* Periodic dysarthria
phenytoin intoxication, 65/500
Pick disease, 42/285, 46/235, 240
polyradiculopathy syndrome, 42/97
pontine infarction, 2/242, 261, 12/15, 15/23, 53/389, 394
primary pontine hemorrhage, 12/39
progressive bulbar palsy, 22/115-121, 59/218
progressive external ophthalmoplegia, 22/187
pseudobulbar, *see* Pseudobulbar dysarthria
pseudobulbar paralysis, 2/241, 12/31, 40/307, 55/142
pure motor hemiplegia, 53/364
pyramidal syndrome, 46/240
rehabilitation, 3/396-399, 4/101
restoration of higher cortical function, 3/396-399
retinovestibular hereditary spastic ataxia, 21/384
Sjögren syndrome, 71/75
snake envenomation, 65/178
spinopontine degeneration, 21/390
striatopallidodentate calcification, 6/709, 49/419, 423
superior cerebellar artery syndrome, 12/20
Sydenham chorea, 49/360
tetrahydrocannabinol intoxication, 65/47
thalamic syndrome (Dejerine-Roussy), 2/482
thallium intoxication, 64/325
tongue abnormality, 4/413-419
transient ischemic attack, 12/5
traumatic psychosyndrome, 24/547
type, 22/119
uremic encephalopathy, 63/504
vertebrobasilar system syndrome, 53/394
Wilson disease, 6/272, 274, 49/225
Wohlfart-Kugelberg-Welander disease, 22/70

Dysautonomia, 22/243-272
acquired, 22/251
acrodynia, 64/371
acute, *see* Acute dysautonomia
anatomy, 22/243
anhidrosis, 22/260
Argyll Robertson pupil, 22/258
auriculotemporal syndrome, 22/260
autonomic nervous system, 74/614
cardiovascular, *see* Cardiovascular dysautonomia
causalgia, 22/254
cholinergic, *see* Cholinergic dysautonomia
cluster headache, 22/255
constitutional type, 22/252
cough syncope, 22/263
crocodile tear syndrome, 22/259
diphtheria, 65/240
erythromelalgia, 22/256
familial, *see* Hereditary sensory and autonomic neuropathy type III
glossopharyngeal neuralgia, 22/254
Holmes-Adie syndrome, 22/259
homeostasis, 22/245
Horner syndrome, 22/256
hyperhidrosis, 22/260
idiopathic small fiber neuropathy, 75/702
intestinal pseudo-obstruction, 51/493
lateral medullary infarction, 53/382
mercury intoxication, 75/503
micturition syncope, 22/264
nosology, 59/135
orthostatic hypotension, 22/262
pain, 22/253
paraneoplastic, *see* Paraneoplastic dysautonomia
Parkinson disease, 75/178, 182
parkinsonism, 22/253
pharmacology, 22/249, 74/170
physiology, 22/247
primary familial, *see* Primary familial dysautonomia
progressive, *see* Progressive dysautonomia
rabies, 56/386
reflex sympathetic dystrophy, 22/254
scoliosis, 50/418
secondary, 22/253
shoulder hand syndrome, 22/254
Shy-Drager syndrome, 22/253, 264
subacute, 22/251
swallow syncope, 22/264
symptom, 75/169
syncope, 22/262, 264
tetanus, 65/217
tympanic plexus neuralgia, 22/254
vagal neuralgia, 22/253
vasovagal syndrome, 22/263
vomiting syncope, 22/264

Dysbarism, 12/665-672
*see also* Air embolism, Barotrauma, Decompression illness *and* Diving

air embolism, 12/671
alexia, 2/607
alternobaric vertigo, 23/624
altitude illness, 12/665
alveolar rupture, 12/671
anorexia, 63/416
anosmia, 24/11
ataxia, 63/416
brain edema, 63/416
brain hypoxia, 63/416
brain injury, 23/609
Cheyne-Stokes respiration, 63/482
clinical features, 23/615, 617, 622
coagulopathy, 23/622
consciousness, 63/416
course, 63/416
decompression, 12/669, 23/619
decompression illness, 23/612
depth, 12/668
diving risk, 63/418
experimental injury, 23/616
fatigue, 63/416
headache, 63/416
high altitude, 63/416
high-pressure nervous syndrome, 23/623
history, 23/610
hyperbaric exposure, 23/610, 615
hypobaric exposure, 23/612
inert gas narcosis, 23/623
mountain sickness, 63/416
nausea, 63/416
nitrogen narcosis, 12/668, 63/419
oxygen intoxication, 12/668
pathology, 23/623
recompression, 23/619
sleep disorder, 63/416
spinal cord injury, 23/611
treatment, 23/618-623
treatment complication, 23/621
type, 23/611, 616
Dysbasia lordotica progressiva (Oppenheim-Fraenkel), *see* Dystonia musculorum deformans
Dysbasia spinalis intermittens, *see* Intermittent spinal cord ischemia
Dysbetalipoproteinemia, *see* Hyperlipoproteinemia type III
Dyscalculia
developmental, *see* Developmental dyscalculia
learning disability, 46/125
spatial orientation disorder, 24/491
Dyscephalia splanchnocystica, *see* Gruber syndrome
Dyschondromatosis
Maffucci-Kast syndrome, 43/29
Maffucci syndrome, 14/118
Dyschondroplasia hemangioma syndrome, *see* Maffucci syndrome
Dyschondrosteosis
Madelung deformity, 43/389-391
pain, 43/389, 391
short stature, 43/389-391
Dyschromatopsia
agnosia, 4/17
autosomal dominant infantile optic atrophy, 13/91
Behr disease, 42/416
ethchlorvynol intoxication, 37/354
Forsius-Eriksson syndrome, 42/397
juvenile optic atrophy, 42/410
Nettleship-Falls albinism, 42/405
pseudoinflammatory foveal dystrophy (Sorsby), 13/138
Dyschronometria
cerebellar ataxia, 1/323
definition, 1/328
Dysconjugate eye movement
brain death, 24/720
Dyscrasia
blood, *see* Blood dyscrasia
Ig, *see* Ig dyscrasia
plasma cell, *see* Plasma cell dyscrasia
Dysdiadochokinesis
Wilson disease, 49/226
Dysembryoma
third ventricle tumor, 17/450
Dysembryoplastic neuroepithelial tumor
classification, 67/5, 68/154
epilepsy, 72/181
nuclear magnetic resonance, 68/156
oligodendrocyte like cell, 67/6
survey, 68/154
Dysencephalia splanchnocystica, *see* Gruber syndrome
Dysentery
tropical ataxic neuropathy, 51/323
Dysequilibrium
aminoglycoside intoxication, 65/482
vestibular neurinoma, 68/446
Dysequilibrium syndrome, *see* Disequilibrium syndrome
Dysesthesia
brachial neuritis, 42/303
carpal tunnel syndrome, 7/431
classification, 1/81
familial hypobetalipoproteinemia, 13/427
hereditary sensory and autonomic neuropathy type III, 43/59

hypereosinophilic syndrome, 63/373
malignant radiculopathy, 69/73
pain fiber regeneration, 1/512
pellagra, 28/86
pernicious anemia, 70/376
postherpetic, *see* Postherpetic neuralgia
terminology, 1/89
Dysgenesia
frontal cortex nodular, *see* Frontal cortex nodular dysgenesia
nodular, *see* Nodular dysgenesia
toxic shock syndrome, 52/259
Dysgenesia nodularis disseminata, *see* Status verrucosus
Dysgénésie nodulaire disséminée, *see* Status verrucosus
Dysgenesis
benign peroxisomal, *see* Benign peroxisomal dysgenesis
brain arteriovenous malformation, 12/232
cerebellar, *see* Cerebellar dysgenesis
gonadal, *see* Gonadal dysgenesis
myelin, *see* Myelin dysgenesis
nucleus ambiguus, *see* Nucleus ambiguus dysgenesia
poliomyelencephalic, *see* Poliomyelencephalic dysgenesia
Dysgenesis neuroblastica gliomatosis, *see* Multiple nevoid basal cell carcinoma syndrome
Dysgenesis neuroepithelialis retinae, *see* Congenital retinal blindness
Dysgenesis neuroepithelialis (Waardenburg), *see* Congenital retinal blindness
Dysgeusia
calcium antagonist, 65/443
calcium antagonist intoxication, 65/443
carbamazepine intoxication, 65/506
dipeptidyl carboxypeptidase I inhibitor intoxication, 65/446
enalapril intoxication, 65/446
Sjögren syndrome, 71/74
Dysglobulinemia
hereditary cranial nerve palsy, 60/42
Dysgraphia
*see also* Agraphia
congenital, *see* Congenital dysgraphia
delirium, 46/531
developmental, *see* Developmental dysgraphia
developmental dysarthria, 4/434-439
Friedreich ataxia, 21/329
head injury outcome, 57/404
migraine, 48/162
parietal lobe syndrome, 45/75
rehabilitation, 4/102, 172-174, 438
Dyshoric angiopathy
Alzheimer disease, 46/256
amyloidosis, 42/728
calcarine cortex, 42/728
senile plaque, 42/728, 46/257
Dyshoric angiopathy (Morel-Wildi), *see* Brain amyloid angiopathy
Dyskeratosis congenita
ataxia telangiectasia, 14/315
blepharitis, 14/782, 43/392
carcinoma, 43/392
growth retardation, 43/392
hyperhidrosis, 14/784, 43/392
leukoplakia, 14/784, 43/392
mental deficiency, 14/784, 43/392
pancytopenia, 43/392
pigmentation, 14/784, 43/392
skin change, 14/784, 43/392
telangiectasia, 14/784, 43/392
testicular atrophy, 43/392
X-linked recessive inheritance, 14/784, 43/392
Dyskinesia
amiodarone, 63/190
amiodarone intoxication, 65/458
antiepileptic agent, 65/499
barbiturate intoxication, 65/510
brain injury, 6/104
cerebral palsy, 46/14
choreic, *see* Chorea
choreiform, *see* Choreiform dyskinesia
Creutzfeldt-Jakob disease, 6/733, 741-745
diphenhydramine, 6/828
ethosuximide intoxication, 65/508
glutaryl-coenzyme A dehydrogenase deficiency, 66/647
levodopa induced, *see* Levodopa induced dyskinesia
linguobuccofacial, *see* Linguobuccofacial dyskinesia
lithium, 51/304
mental deficiency, 46/14
mesencephalon, 2/277
methyldopa intoxication, 37/441
neuroleptic, *see* Neuroleptic dyskinesia
neuroleptic akathisia, 65/286
ocular, *see* Ocular dyskinesia
orofacial, *see* Orofacial dyskinesia
oromandibular, *see* Oromandibular dyskinesia
pallidal necrosis, 49/465
pallidonigral necrosis, 49/465
pallidostriatal necrosis, 49/465
Parkinson disease, 49/195

spatial, *see* Spatial dyskinesia
Sydenham chorea, 49/363
tardive, *see* Tardive dyskinesia
Dyslalia
speech disorder, 4/428
Dyslexia
alexia, 4/112, 115, 45/445
Binswanger disease, 54/222
body scheme disorder, 4/401
cerebral dominance, 46/132-134
cognition, 46/127
computer assisted tomography, 46/131
continuity theory, 46/126
deep, *see* Deep dyslexia
deep agraphia, 45/462
definition, 45/433-452
developmental, *see* Developmental dyslexia
developmental dysarthria, 4/434-439
developmental dysphasia, 46/126
discontinuity theory, 46/126
EEG, 46/131
etiology, 46/126
genetics, 46/130
Gerstmann syndrome, 2/691, 694, 46/127-129, 131
handedness, 46/129, 131
idiopathic developmental, *see* Idiopathic developmental dyslexia
language, 46/128, 132
learning disability, 46/125
model, 46/126-130
parietal aphasia syndrome, 45/75
parietal lobe syndrome, 45/75
rehabilitation, 46/617
selective activation model, 46/132
spatial orientation disorder, 24/491
spelling, *see* Spelling dyslexia
structural model, 46/132
subtype, 46/126-130
surface, *see* Surface dyslexia
type, 45/433-452
Dysmasesia
operculum syndrome, 2/779
Dysmegalopsia
higher nervous activity disorder, 3/29
parietal lobe, 45/362
Dysmetria
*see also* Hypermetria
amplitude error, 1/323
definition, 2/394
finger to ear test, 1/324, 333
finger to nose test, 1/324, 333
intention tremor, 49/590
juvenile Gaucher disease, 10/307
neocerebellar syndrome, 2/421
ocular, *see* Ocular dysmetria
prehension test, 1/324
serial, *see* Serial dysmetria
Dysmetropsia
central, 2/613
peripheral, 2/613
transcortical, 2/613
Dysmorphesthesia
migraine, 48/163
Dysmorphia
facial, *see* Facial dysmorphia
Rubinstein-Taybi syndrome, 13/83
Dysmorphic sign
Klinefelter syndrome, 50/550
Dysmorphopsia
hallucination classification, 45/56
Dysmyelination
familial, *see* Familial dysmyelination
neuroglia relationship, 28/410
proteolipid protein gene, 66/564
Dysmyelinative disease
major form, 66/5
Dysmyelinogenic leukodystrophy, *see* Alexander disease
Dysmyelogenic leukodystrophy, *see* Alexander disease
Dysorthographia
agraphia, 4/151, 435
Dysosmia
calcium antagonist, 65/443
calcium antagonist intoxication, 65/443
migraine, 48/162
Dysosteosclerosis
bone fragility, 43/393
dental abnormality, 43/393
facial paralysis, 43/394
micrognathia, 43/393
nystagmus, 43/393
optic atrophy, 43/393
sex ratio, 43/394
short stature, 43/393
strabismus, 43/393
Dysostosis
cleidocranial, *see* Cleidocranial dysostosis
cranial, *see* Cranial dysostosis (Franceschetti)
mandibulofacial, *see* Treacher-Collins syndrome
maxillofacial, *see* Maxillofacial dysostosis
mental deficiency, *see* Acrodysostosis
nasal hypoplasia, *see* Acrodysostosis
orocraniodigital, *see* Juberg-Hayward syndrome
peripheral, *see* Acrodysostosis

radioulnar, *see* Radioulnar dysostosis
Dysostosis craniofacialis (Crouzon), *see* Craniofacial dysostosis (Crouzon)
Dysostosis multiplex, *see* Glycosaminoglycanosis type IH
Dysostotic stenosis
vertebral canal stenosis, 50/465
Dyspallic syndrome, *see* Akinetic mutism
Dyspepsia
nonulcer, 75/631
Dysphagia
*see also* Aphagia *and* Feeding disorder
achalasia, 1/504
adrenoleukodystrophy, 42/483
alcoholic polyneuropathy, 51/316
amyloid polyneuropathy, 51/416
amyloid tumor, 8/9
amyotrophic lateral sclerosis, 42/65
Arnold-Chiari malformation type II, 50/410
arthropod envenomation, 65/196
basilar ischemia, 12/5
Behçet syndrome, 56/597
botulism, 65/210
brain embolism, 53/207
brain infarction, 53/207
brain lacunar infarction, 54/253
bromide intoxication, 36/300
cerebrovascular disease, 53/207
childhood myoglobinuria, 62/564
chorea-acanthocytosis, 63/282
chronic bromide intoxication, 36/300
concentric sclerosis, 47/414
congenital myopathy, 62/335
congenital suprabulbar paresis, 22/118
cricopharyngeal, *see* Cricopharyngeal dysphagia
cricopharyngeal myotomy, 62/386
differential diagnosis, 40/306
diphtheria, 65/240
dystonia musculorum deformans, 49/521
familial polyradiculopathy syndrome, 42/97
Friedreich ataxia, 21/348
Gerstmann-Sträussler-Scheinker disease, 56/553
glossopharyngeal nerve agenesis, 50/219
Guillain-Barré syndrome, 51/245
hereditary sensory and autonomic neuropathy type III, 21/109, 60/26
hereditary spastic ataxia, 21/367, 60/462
hereditary striatal necrosis, 49/495
Huntington chorea, 49/280
hyperosmolar hyperglycemic nonketotic diabetic coma, 27/90
hyperventilation syndrome, 38/320, 63/436
hypomagnesemia, 63/562
inclusion body myositis, 62/206, 373
infantile Gaucher disease, 42/438
intracranial hemorrhage, 53/207
lateral medullary infarction, 53/381
muscle weakness, 40/306
myasthenia gravis, 43/157
myotonic dystrophy, 40/494, 514
neuroleptic dystonia, 65/282
nigrospinodentate degeneration, 22/170
oculopharyngeal muscular dystrophy, 43/100
operculum syndrome, 2/779
ophthalmoplegia, 43/138
oropharyngeal, *see* Oropharyngeal dysphagia
palatal myoclonus, 38/584
paralysis periodica paramyotonica, 43/167
Parkinson dementia, 49/171
pellagra, 28/71, 38/648
petrobasilar suture syndrome, 2/313
polyradiculopathy syndrome, 42/97
pontine infarction, 12/15, 53/389
porencephaly, 30/685, 50/359
primary pontine hemorrhage, 12/39
progressive external ophthalmoplegia, 22/185, 187, 206
pseudobulbar paralysis, 2/241, 12/31, 40/307, 55/142
pyruvate dehydrogenase complex deficiency, 66/416
scorpion venom, 65/196
snake envenomation, 65/178
spastic ataxia, 42/165
spastic paraplegia, 42/175
syringobulbia, 50/445
tetanus, 52/230
thallium intoxication, 64/325
thyrotoxic myopathy, 62/528
transient ischemic attack, 12/5, 53/207
vagus nerve agenesis, 50/219
vertebral column tumor, 20/30
Wohlfart-Kugelberg-Welander disease, 22/70
Dysphasia
*see also* Aphasia
acquired, *see* Acquired dysphasia
anomic, *see* Anomic aphasia
anterior cerebral artery syndrome, 53/342, 346
aphasia, 2/598, 4/419
body scheme disorder, 4/401
brain infarction, 53/346
brain injury, 57/134
clinical examination, 57/134
Creutzfeldt-Jakob disease, 46/290
dementia, 3/312
developmental, *see* Developmental dysphasia

developmental expressive, *see* Developmental expressive dysphasia
developmental receptive, *see* Developmental receptive dysphasia
head injury, 46/614
learning disability, 46/125
migraine, 48/143, 148, 161
paroxysmal, *see* Paroxysmal dysphasia
progressive supranuclear palsy, 49/243
pure motor hemiplegia, 54/243
rehabilitation, 46/614-617
speech disorder in children, 4/419
subarachnoid hemorrhage, 55/18
supplementary motor area syndrome, 53/346
traumatic, *see* Traumatic aphasia
traumatic psychosis, 24/531
traumatic psychosyndrome, 24/546
Dysphonia
brain embolism, 53/207
brain infarction, 53/207
cerebrovascular disease, 53/207
developmental dysarthria, 4/409-413
diphtheria, 65/240
dystonia musculorum deformans, 49/521
hereditary whispering, *see* Focal dystonia musculorum deformans
intracranial hemorrhage, 53/207
muscle weakness, 40/306
myopathy, 40/306
progressive bulbar palsy, 22/116, 59/218
progressive external ophthalmoplegia, 22/206
pseudobulbar paralysis, 12/31, 40/307
spastic, *see* Spastic dysphonia
speech disorder in children, 4/409
thyrotoxic myopathy, 62/528
transient ischemic attack, 53/207
Dysphoria
Amanita muscaria intoxication, 37/329
central sleep apnea, 63/461
delirium, 46/531
dysphrenia hemicranica, 5/79
epilepsy, 5/79, 15/596
migraine, 5/79, 48/162
nalorphine, 37/366
neuroleptic akathisia, 65/287
obstructive sleep apnea, 63/453
tabernathe iboga intoxication, 37/329
Dysphrenia hemicranica
amnesia, 5/79
anger, 5/79
anxiety, 5/79
apathy, 5/79
complicated migraine, 5/78-80
confusional state, 5/78
consciousness, 5/79
delirium, 5/79
delusion, 5/79
depression, 5/79
diencephalon, 5/80
disorientation, 5/79
dysphoria, 5/79
emotional disorder, 5/79
excitation, 5/79
fatigue, 5/79
hallucination, 5/79
hallucinosis, 5/79
history, 5/78
hysteria, 5/79
illusion, 5/79
kleptomania, 5/79
logorrhea, 5/79
memory disorder, 5/78
migraine, 5/78-80
migraine psychosis, 5/79
paranoia, 5/79
pathologic laughing and crying, 5/79
phobia, 5/79
phylogenetic part, 5/80
psychosis, 5/79
restlessness, 5/79
terror, 5/78
twilight state, 5/79
vegetative disorder, 5/80
Dysphrenic migraine, *see* Migraine sine hemicrania
Dysplasia spectrum
axial mesodermal, *see* Axial mesodermal dysplasia spectrum
Dysplasia tarda
spondyloepiphyseal, *see* Spondyloepiphyseal dysplasia tarda
Dysplasias
anhidrotic ectodermal, *see* Anhidrotic ectodermal dysplasia
arthrodento-osteo, *see* Arthrodento-osteodysplasia
atlas, *see* Atlas dysplasia
axis, *see* Axis dysplasia
blastomatous, *see* Blastomatous dysplasia
bone, *see* Bone dysplasia
brain, *see* Brain dysplasia
brain cortex, *see* Brain cortex dysplasia
brain cortex focal, *see* Brain cortex focal dysplasia
brain verrucose, *see* Status verrucosus
branchio-oto, *see* Branchio-otodysplasia
branchio-otorenal, *see* Branchio-otorenal dysplasia
cardiac, *see* Heart dysplasia

cerebellar, *see* Cerebellar dysplasia
cerebellar cortical, *see* Cerebellar cortical dysplasia
cerebellar vermis, *see* Cerebellar vermis dysplasia
cerebral, *see* Brain dysplasia
cervico-oculoacoustic, *see* Wildervanck syndrome
cervicothoracic somite, *see* Cervicothoracic somite dysplasia
chondroectodermal, *see* Ellis-Van Creveld syndrome
cleidocranial, *see* Cleidocranial dysplasia
congenital ectodermal, *see* Rothmund-Thomson syndrome
congenital spondyloepiphyseal, *see* Congenital spondyloepiphyseal dysplasia
cortical, *see* Cortical dysplasia
craniodiaphyseal, *see* Craniodiaphyseal dysplasia
cranioectodermal, *see* Cranioectodermal dysplasia
craniometaphyseal, *see* Craniometaphyseal dysplasia
cutaneous, *see* Cutaneous dysplasia
diaphyseal, *see* Diaphyseal dysplasia
ectodermal, *see* Ectodermal dysplasia
epiphyseal, *see* Epiphyseal dysplasia
epiphysealis punctata, *see* Conradi-Hünermann syndrome
familial metaphyseal, *see* Familial metaphyseal dysplasia
fibromuscular, *see* Fibromuscular dysplasia
fibrous, *see* Fibrous dysplasia
focal, *see* Focal dysplasia
focal brain cortex, *see* Brain cortex focal dysplasia
focal dermal, *see* Focal dermal dysplasia
frontometaphyseal, *see* Frontometaphyseal dysplasia
hidrotic ectodermal, *see* Hidrotic ectodermal dysplasia
hypohidrotic ectodermal, *see* Hypohidrotic ectodermal dysplasia
iris, *see* Iris dysplasia
macula, *see* Macula dysplasia
metaphyseal, *see* Pyle disease
monostotic fibrous, *see* Monostotic fibrous dysplasia
neuroectodermal, *see* Neuroectodermal dysplasia
occipital, *see* Occipital dysplasia
oculodentodigital, *see* Oculodento-osseous dysplasia
olfactogenital, *see* Kallmann syndrome
optic disc, *see* Optic disc dysplasia
otodental, *see* Otodental dysplasia
polyostotic fibrocystic, *see* Polyostotic fibrocystic dysplasia
progressive diaphyseal, *see* Progressive diaphyseal dysplasia
progressive hereditary craniodiaphyseal, *see* Progressive hereditary craniodiaphyseal dysplasia
pseudoachondroplastic, *see* Pseudoachondroplastic dysplasia
renal, *see* Renal dysplasia
renofacial, *see* Renofacial dysplasia
retinal, *see* Retinal dysplasia
septo-optic, *see* Septo-optic dysplasia
sphenoid wing, *see* Sphenoid wing dysplasia
spondyloepiphyseal, *see* Spondyloepiphyseal dysplasia
spondylometaphyseal, *see* Spondylometaphyseal dysplasia
suboccipital, *see* Suboccipital dysplasia
thanatophoric, *see* Thanatophoric dysplasia
verrucose brain, *see* Status verrucosus
verrucose retinal, *see* Verrucose retinal dysplasia
Dysplastic gangliocytoma
cerebellum, 68/144
Dysplastic nail
Patau syndrome, 14/121
Dysplastic stenosis
vertebral canal stenosis, 50/465
Dyspnea
autonomic dysfunction, 63/482
autonomic nervous system, 74/575
brain fat embolism, 55/183
breathlessness, 1/495
central sleep apnea, 63/461
decompression illness, 63/419
dermatomyositis, 62/374, 71/130
diagnosis, 26/342
echinoidea intoxication, 37/68
eosinophilia myalgia syndrome, 63/375, 64/257
fish roe intoxication, 37/86
glyceryl trinitrate, 37/454
hydrozoa intoxication, 37/38, 40
hypothesis, 74/575
inclusion body myositis, 62/374
meduso-congestin intoxication, 37/37
metastatic cardiac tumor, 63/94
palytoxin intoxication, 37/55
pathophysiology, 1/495
pheochromocytoma, 39/498
pickwickian syndrome, 38/300
polymyositis, 62/374, 71/130
primary cardiac tumor, 63/94
progressive bulbar palsy, 22/114, 59/219
respiratory encephalopathy, 63/414
rhizostomeae intoxication, 37/53

tardive dyskinesia, 63/482
toxic oil syndrome, 63/380
Toxopneustes pileolus intoxication, 37/68
tremor, 63/482
zinc intoxication, 36/337
Dyspneumia
pseudobulbar paralysis, 22/119
Dyspraxia
anterior cerebral artery syndrome, 53/346
body scheme disorder, 4/401
brain infarction, 53/346
constructional, *see* Constructional dyspraxia
corpus callosum, *see* Corpus callosum dyspraxia
developmental, *see* Developmental dyspraxia
epilepsy, 72/133
middle cerebral artery syndrome, 53/362
Salla disease, 51/382
supplementary motor area syndrome, 53/346
sympathetic, *see* Sympathetic dyspraxia
Dyspropterin synthase deficiency
epilepsy, 72/222
Dysprosody
aphasia, 4/89
middle cerebral artery syndrome, 53/362
Dysproteinemia
subarachnoid hemorrhage, 12/113
Dysproteinemic neuropathy
macroglobulinemia, 8/72-76
melphalan, 8/74-78
Mott cell, 8/75
Dysraphia
*see also* Spina bifida
Arnold-Chiari malformation type I, 14/111
axis, *see* Axis dysraphia
Bean syndrome, 14/106
centrofacial lentiginosis, 14/782
cervico-oculoacusticus syndrome, 32/128
characteristics, 32/160
classification, 32/163, 189
cranial, *see* Cranial dysraphia
CSF, 32/188
cutaneous dysplasia, 14/110
Dandy-Walker syndrome, 14/111
developmental stage, 32/160
embryogenesis, 32/187
embryology, 32/161-165
epilepsy, 72/110
familial gliomatosis-glioblastomatosis, 14/498
Friedreich ataxia, 2/326
heterochromia iridis, 2/325
Horner syndrome, 2/318, 325
Krabbe-Bartels disease, 14/12, 407
malformation, 32/161
mechanics, 32/188
mental deficiency, 46/7
microgyria, 30/484
myelodysplasia (Fuchs), 14/111
neurocutaneous syndrome, 14/101
neurofibromatosis type I, 14/146
Nielsen syndrome, 14/109
occult spinal, 19/159
Ostrum-Furst syndrome, 14/111
posterior, *see* Posterior dysraphia
primary, 32/161
rachischisis, 32/159
secondary, 32/161
spinal, *see* Spina bifida
tectocerebellar, *see* Tectocerebellar dysraphia
tuberous sclerosis, 14/398
without schisis, 32/187
Dysrhythmia
cardiac, *see* Cardiac dysrhythmia
diffuse, *see* Diffuse dysrhythmia
gastric, *see* Gastric dysrhythmia
Dyssomnia
acrylamide intoxication, 64/67
autonomic failure, 74/560
glycosaminoglycanosis type III, 66/299
lead encephalopathy, 64/435
lead intoxication, 64/435
methyldopa intoxication, 37/441
vitamin Bc intoxication, 28/214
Dyssynergia, 21/509-516
progressive cerebellar, *see* Progressive cerebellar dyssynergia
Dyssynergia cerebellaris myoclonica, 21/509-516, 60/593-603
*see also* Baltic myoclonus epilepsy, Progressive myoclonus epilepsy *and* Ramsay Hunt syndrome
action myoclonus, 60/599
autosomal dominant, 21/510, 42/212
autosomal recessive, 21/510
Baltic myoclonus epilepsy, 60/599
brachium conjunctivum, 21/513
carbamazepine, 60/603
celiac disease, 60/599
cerebellar ataxia, 60/597, 599
cerebellar atrophy, 42/211
cerebellar cortex, 22/172
chlorpromazine, 60/603
clonazepam, 60/603
computer assisted tomography, 60/601, 671
deafness, 60/599
dementia, 21/510, 60/665
dentate nucleus, 21/509, 513, 42/211

dentatorubropallidoluysian atrophy, 21/528
diazepam, 60/603
differential diagnosis, 60/599
EEG, 21/514, 60/600
encephalopathy, 60/600
epilepsy, 15/332, 60/599
experimental study, 60/601
Friedreich ataxia, 21/20, 351, 354, 530, 42/211
hereditary myoclonus, 60/599
hereditary myoclonus ataxia syndrome, 21/510
history, 60/593
intention tremor, 21/509, 513, 515, 60/597
jejunal biopsy, 60/601
lactic acidemia, 60/601
lipofuscin, 10/550
literature, 60/594-596
main features, 21/17
MAO inhibitor, 60/603
maternal factor, 60/601
MERRF syndrome, 60/599
mucolipidosis type I, 60/599
muscle biopsy, 60/601
myoclonia, 6/766, 776
myoclonic epilepsy, 42/211
myoclonus, 21/511, 38/578, 49/616, 60/597, 599, 662
neuronal ceroid lipofuscinosis, 60/599
neuropathology, 60/599
neuropathy, 60/660
olivopontocerebellar atrophy, 60/599
olivopontocerebellar atrophy (Dejerine-Thomas), 60/532
olivopontocerebellar atrophy (Wadia-Swami), 60/665
oxitriptan, 60/603
pathophysiology, 21/512, 60/599
piracetam, 60/603
postural myoclonus, 6/767, 21/513
prognosis, 21/515, 60/602
progressive cerebellar ataxia, 60/599
progressive myoclonus epilepsy, 42/211, 60/599
Ramsay Hunt syndrome, 21/509, 42/211, 60/593
related syndrome, 60/599
REM sleep, 21/514
sensorineural deafness, 60/599
serotonin, 60/603
somatosensory evoked potential, 60/600
spinal muscular atrophy, 60/599
thalamotomy, 60/603
treatment, 21/516, 60/603
tremor, 42/211
twins, 21/509
Unverricht-Lundborg progressive myoclonus
epilepsy, 49/616
valproic acid, 60/603
Dyssynergia cerebellaris progressiva
*see also* Ramsay Hunt syndrome
basal ganglion degeneration, 42/212
cerebellar atrophy, 42/212
Friedreich ataxia, 42/213
hydrocephalus, 42/212
mental deficiency, 42/213
speech disorder, 42/212
tremor, 42/212
Dysthyroid ophthalmoplegia
exophthalmos, 22/184
progressive external ophthalmoplegia, 22/204
Dysthyroid orbitopathy, *see* Thyroid associated ophthalmopathy
Dystonia
*see also* Dystonia musculorum deformans *and* Involuntary movement
action, *see* Action dystonia
African trypanosomiasis, 52/341
agraphia, *see* Writers cramp
amyotrophic lateral sclerosis, 59/235
argininosuccinic aciduria, 60/664
ataxia telangiectasia, 60/664
athetosis, 6/445, 524, 526, 49/381, 521
autosomal dominant cerebellar ataxia, 60/664
axial, *see* Axial dystonia
basal ganglion angioma, 6/558
Batten disease, 6/556
botulinum toxin A, 65/213
brain arteriosclerosis, 6/556
brain embolism, 53/211
brain infarction, 53/211
brain lacunar infarction, 54/247
brain tumor, 6/560
calcium antagonist intoxication, 65/442
carbamate intoxication, 64/188
carbamazepine intoxication, 65/506
carbon disulfide intoxication, 64/188
carbon monoxide intoxication, 64/31
cerebrovascular disease, 53/211
chorea, 6/557, 42/219
chorea-acanthocytosis, 49/328
cocaine intoxication, 65/255
complex III deficiency, 62/504, 66/423
complex IV deficiency, 66/424
congenital retinal blindness, 59/339
craniocerebral trauma, 6/559
decerebrate rigidity, 6/152
definition, 49/521
dentatorubropallidoluysian atrophy, 21/519
diphenhydramine, 6/828

disulfiram, 64/188
drug induced, *see* Drug induced dystonia
dystonia musculorum deformans, 49/71
eating, 63/281
EEG, 42/219
EMG, 49/72
encephalitis, 6/559
encephalitis lethargica, 6/558
episodic, *see* Episodic dystonia
facial, *see* Facial dystonia
familial late onset ataxia, 60/664
familial spastic paraplegia, 22/433, 456
filariasis, 52/517
fixed, *see* Fixed dystonia
flexion, *see* Flexion dystonia
focal, *see* Focal dystonia
focal athetosis, 49/382
Friedreich ataxia, 6/555
fumarate hydratase deficiency, 66/420
GABA, 29/506
glutaryl-coenzyme A dehydrogenase deficiency, 66/647
GM2 gangliosidosis, 60/664
GM2 gangliosidosis type II, 60/664
Hallervorden-Spatz syndrome, 6/523, 551-554, 49/399, 66/712, 714
Hartnup disease, 60/664
hemiplegic, *see* Hemiplegic dystonia
hereditary diurnal fluctuating juvenile, *see* Hereditary diurnal fluctuating juvenile dystonia
hereditary dystonic paraplegia, 22/456
hereditary myoclonic, *see* Hereditary myoclonic dystonia
hereditary neurological disease, 49/541
hereditary paroxysmal dystonic choreoathetotis, 49/349
hereditary paroxysmal kinesigenic choreoathetosis, 42/205, 207, 49/352
hereditary secondary, *see* Hereditary secondary dystonia
hereditary torsion, *see* Dystonia musculorum deformans
heterocyclic antidepressant, 65/316
histology, 49/72
Huntington chorea, 6/533, 555, 49/280
hyperreflexia, 42/219
hypnogenic paroxysmal, *see* Hypnogenic paroxysmal dystonia
hypocalcemia, 63/561
hypoparathyroidism, 27/308
hyporeflexia, 42/253
hysteria, 6/561
idiopathic, *see* Idiopathic dystonia musculorum deformans
intracranial hemorrhage, 53/211
Japanese B encephalitis, 56/138
juvenile dystonic lipidosis, 60/664
juvenile parkinsonism, 49/163
kernicterus, 6/552
kinesigenic choreoathetosis, 42/207
lenticularis, 6/544, 553
levodopa, 42/213, 253
levodopa induced dyskinesia, 49/196
limb, *see* Limb dystonia
lingual, 63/281
lordosis, 42/253
Machado-Joseph disease, 42/262, 60/664
manganese intoxication, 64/308
metachromatic leukodystrophy, 66/168
methanol intoxication, 64/100
micrographia, 45/464
mitochondrial disease, 60/664
movement, 6/522
multiple sclerosis, 47/57
myoclonic torsion, *see* Myoclonic dystonia musculorum deformans
myoclonus, 42/240
neuroacanthocytosis, 60/664
neuroleptic, *see* Neuroleptic dystonia
neuroleptic agent, 49/191
neuroleptic dyskinesia, 65/274
neuronal ceroid lipofuscinosis, 60/664
neuropharmacology, 49/58
neurovegetative, *see* Neurovegetative dystonia
Niemann-Pick disease type C, 60/664
nifedipine, 63/194
nifedipine intoxication, 65/442
3-nitropropionic acid, 64/7
nuchal, *see* Nuchal dystonia
nystagmus, 42/240
oculofacial, *see* Progressive supranuclear palsy
oculofaciocervical, *see* Progressive supranuclear palsy
olivopontocerebellar atrophy (Dejerine-Thomas), 60/664
olivopontocerebellar atrophy (Wadia-Swami), 60/494
organic acid metabolism, 66/641
oromandibular, *see* Oromandibular dystonia
oxoglutarate dehydrogenase deficiency, 66/419
pallidal degeneration, 49/454
pallidal necrosis, 49/465
pallidoluysian atrophy, 6/642
pallidoluysiodentate degeneration, 49/454
pallidoluysionigral degeneration, 49/446, 451, 454, 456

pallidonigral necrosis, 49/465
pallidopyramidal degeneration, 42/244, 49/454, 456
pallidostriatal necrosis, 49/465
paraplegia, 6/520, 556
Parkinson disease, 6/555, 49/197
parkinsonism, 42/253
paroxysmal, *see* Paroxysmal dystonia
pathophysiology, 6/137, 140-144, 49/71
Pelizaeus-Merzbacher disease, 60/664
perinatal brain injury, 6/545, 553
periodic, *see* Hereditary paroxysmal kinesigenic choreoathetosis
phenothiazine, 6/560
positron emission tomography, 49/41
posthemiplegic, *see* Posthemiplegic dystonia
posture, 6/520
primary, *see* Primary dystonia
progressive bulbar palsy, 59/409
progressive external ophthalmoplegia, 43/136, 62/304
progressive multifocal leukoencephalopathy, 9/495
progressive pallidal atrophy (Hunt-Van Bogaert), 21/531, 42/247
progressive pallidal degeneration, 6/554
progressive supranuclear palsy, 22/217, 219, 42/259, 46/301, 49/240, 246
pure pallidal degeneration, 49/447
pyruvate dehydrogenase deficiency, 60/664
recessive olivopontocerebellar atrophy, 60/664
reflex sympathetic dystrophy, 61/122, 129
secondary, *see* Secondary dystonia
Segawa, *see* Segawa dystonia
segmental, *see* Segmental dystonia
segmental athetosis, 49/382
senile, *see* Senile dystonia
sensorineural deafness, 42/372
spasmodic torticollis, 42/263
spastic, *see* Spastic dystonia
spastic paraplegia, 42/175
speech disorder in children, 42/219
status marmoratus, 6/533, 546-551
stereoencephalotomy, 6/562
striatal hemiplegia, 2/504
striatal necrosis, 49/499, 501
striatal syndrome, 2/499
striatonigral degeneration, 42/262
striatopallidodentate calcification, 6/557, 60/664
subacute necrotizing encephalomyelopathy, 60/664
subsarcolemmal myofibril, 42/219
subthalamic nucleus, 49/457
sympathetic, *see* Sympathetic dystonia
symptomatic, *see* Symptomatic dystonia
syndrome, 6/310
syphilis, 6/559
tardive dyskinesia, 49/186
thalamic infarction, 53/396
torsion, *see* Dystonia musculorum deformans
transient ischemic attack, 53/211
treatment, 49/72
tremor, 42/253, 49/586
tricyclic antidepressant, 65/316
tuberculous meningitis, 6/559
vanilmandelic acid, 42/214
verapamil, 63/194
Wilson disease, 6/552, 27/384, 42/270, 49/224, 227
writers cramp, 45/467
xeroderma pigmentosum, 60/664
Dystonia deafness syndrome
basal ganglion degeneration, 42/372
levodopa, 42/372
mental deficiency, 42/372
self-mutilation, 42/372
speech disorder in children, 42/372
Dystonia musculorum deformans, 6/517-541, 49/519-526
*see also* Dystonia *and* Meige syndrome
action dystonia, 6/523, 49/522
age at onset, 6/526, 49/522
animal model, 49/525
anterocollis, 49/521
ataxia telangiectasia, 49/524
athetosis, 49/522
autosomal dominant, 42/215, 49/521
autosomal dominant transmission, 49/520
autosomal recessive, 42/216, 49/521
autosomal recessive transmission, 49/520
Bardet-Biedl syndrome, 13/401
basal ganglion degeneration, 1/289, 42/217
biopterin, 42/215
birth incidence, 42/217
blepharospasm, 6/522, 42/218, 49/521, 523
brain tissue, 49/526
Bruyn-Went disease, 49/524
characteristics, 6/149-151
chemistry, 49/523
chorea, 49/522
clinical features, 6/520-530
congenital retinal blindness, 49/524
course, 6/526, 528, 49/522
critique, 2/500
CSF 3-methoxy-4-hydroxyphenylethylene glycol, 49/524

diagnosis, 6/532
differential diagnosis, 6/532-534, 42/215, 49/524
dopamine, 6/532
dopamine β-mono-oxygenase, 42/215, 49/524
double athetosis, 6/534
dromedary gait, 49/522
dysarthria, 6/523, 49/521
dysphagia, 49/521
dysphonia, 49/521
dystonia, 49/71
dystonic choreoathetosis, 49/525
dystonic paraplegia, 49/525
EMG, 6/532, 49/523
epidemiology, 6/519, 49/520
fixed dystonia, 49/522
focal form, *see* Focal dystonia musculorum deformans
forme fruste, 6/530, 49/520
gene penetrance, 6/519
genetics, 49/520
H-reflex, 6/537
Hallervorden-Spatz syndrome, 49/524
hemidysplasia, 6/528
hereditary olivopontocerebellar atrophy (Menzel), 21/442
hereditary paroxysmal kinesigenic choreoathetosis, 49/525
hereditary tremor, 49/522
heterozygote frequency, 42/217
history, 49/519
homovanillic acid, 49/524, 526
Huntington chorea, 42/215, 49/524
5-hydroxyindoleacetic acid, 49/524
idiopathic, *see* Idiopathic dystonia musculorum deformans
incidence, 6/519
inheritance, 6/519
Jewish child onset type, 42/216, 49/521
juvenile parkinsonism, 49/163
laboratory diagnosis, 49/523
Lesch-Nyhan syndrome, 49/524
levodopa, 42/215-218
lipid accumulation, 6/535
lipofuscin, 6/535
Machado-Joseph disease, 49/524
Meige syndrome, 49/521
mutation rate, 6/519
myoclonic, *see* Myoclonic dystonia musculorum deformans
nasal deformity, 49/525
neuroleptic dystonia, 65/281
neurolipidosis, 49/524
neuropathology, 6/534-536
neurotransmitter, 49/526
non-Jewish adult onset type, 49/521
noradrenalin, 49/526
nosologic development, 6/517-519
pallidoluysionigral degeneration, 49/458
paradoxical kinesia, 6/522, 49/522
parkinsonism, 49/524
pathophysiology, 6/536
peacock gait, 49/521
PEG, 6/532
Pelizaeus-Merzbacher disease, 49/524
postural tremor, 49/522
presenting symptom, 6/526-528
prevalence, 49/520
progressive external ophthalmoplegia, 22/179, 62/290
progressive supranuclear palsy, 49/524
pseudoautosomal dominant, 42/215
putaminous cell change, 6/535
reproductive fitness, 6/520
retrocollis, 49/521
secondary pigmentary retinal degeneration, 13/335, 60/730
serotonin, 49/526
spasmodic torticollis, 6/529, 42/215-218, 49/521-523
status marmoratus, 49/524
striatal degeneration, 2/502, 6/666
striatopallidodentate calcification, 49/524
surgery, 1/289
symptomatic dystonia, 49/523
tetrahydrobiopterin, 49/524
treatment, 1/289, 6/537-540, 49/525
tremor, 42/218
type, 6/526
variable expression, 49/520
Wilson disease, 42/215, 49/524
writers cramp, 6/528, 42/215, 49/521-523
writing deterioration, 42/216
X-linked, *see* X-linked dystonia musculorum deformans
Dystonia musculorum deformans (Ziehen-Oppenheim), *see* Dystonia musculorum deformans
Dystonia musculorum progressiva
hereditary diurnal fluctuating juvenile dystonia, 49/536
Dystonic choreoathetosis
dystonia musculorum deformans, 49/525
hereditary exertional, *see* Hereditary exertional paroxysmal dystonic choreoathetosis
hereditary paroxysmal, *see* Hereditary paroxysmal dystonic choreoathetotis

hereditary paroxysmal kinesigenic, *see* Hereditary paroxysmal kinesigenic choreoathetosis
Dystonic lipidosis
juvenile, *see* Juvenile dystonic lipidosis
Dystonic paraplegia
amyotrophic, *see* Amyotrophic dystonic paraplegia
autosomal dominant, 49/525
dystonia musculorum deformans, 49/525
familial amyotrophic, *see* Familial amyotrophic dystonic paraplegia
hereditary, *see* Hereditary dystonic paraplegia
Dystopia canthi medialis lateroversa, *see* Klein-Waardenburg syndrome
Dystopia canthorum
acrocephalosyndactyly type II, 43/324
Klein-Waardenburg syndrome, 43/51
Saethre-Chotzen syndrome, 43/322
W index, 43/51
Dystopic cortical myelinogenesis
brain cortex heterotopia, 50/206
cerebellar vermis, 50/206
clinical features, 30/362, 50/207
definition, 30/360, 50/206
EEG, 30/362
etiology, 30/361
mental deficiency, 30/362, 50/207
microgyria, 50/206
pathology, 30/360, 50/206
Dystopic gallbladder
holoprosencephaly, 50/237
Dystrophia dermochondrocornealis
epilepsy, 14/792
Dystrophia myotonica, *see* Myotonic dystrophy
Dystrophia retinae pigmentosa, *see* Primary pigmentary retinal degeneration *and* Secondary pigmentary retinal degeneration
Dystrophia retinae pigmentosa dysacusis syndrome, *see* Lindenov-Hallgren syndrome
Dystrophic bullous epidermolysis, *see* Epidermolysis bullosa dystrophica
Dystrophin
α-actinin, 62/126
amino acid sequence, 62/125
ankyrin, 62/126
brain, 62/125
Duchenne muscular dystrophy, 62/121, 125
limb girdle syndrome, 62/187
muscle, 62/125
muscle fiber, 62/35
spectrin, 62/126
Dystrophin deficiency
incomplete, *see* Incomplete dystrophin deficiency
Dystrophy
adiposogenital, *see* Adiposogenital dystrophy
adult neuroaxonal, *see* Adult neuroaxonal dystrophy
argyrophilic, *see* Argyrophilic dystrophy
axonal, *see* Axonal dystrophy
Becker muscular, *see* Becker muscular dystrophy
butterfly shaped pigmentary foveal, *see* Pigmentary foveal dystrophy
central pontine, *see* Central pontine dystrophy
centroperipheral tapetoretinal, *see* Centroperipheral tapetoretinal dystrophy
cerebrometacarpometatarsal, *see* Pseudopseudohypoparathyroidism
cholesterol ester, 66/3
choroid, *see* Choroid dystrophy
congenital atonic sclerotic muscular, *see* Congenital atonic sclerotic muscular dystrophy
congenital muscular, *see* Congenital muscular dystrophy
corneal, *see* Corneal dystrophy
corneal lattice, *see* Corneal lattice dystrophy
distal muscular, *see* Distal muscular dystrophy
Duchenne muscular, *see* Duchenne muscular dystrophy
Emery-Dreifuss muscular, *see* Emery-Dreifuss muscular dystrophy
enzyme, 66/51
Erb scapulohumeral muscular, *see* Scapulohumeral muscular dystrophy (Erb)
Fabry corneal, *see* Fabry corneal dystrophy
facioscapulohumeral muscular, *see* Facioscapulohumeral muscular dystrophy
familial juvenile epithelial, *see* Familial juvenile epithelial dystrophy
glial, *see* Glial dystrophy
hereditary progressive cone, *see* Hereditary progressive cone dystrophy
infantile myotonic, *see* Infantile myotonic dystrophy
infantile neuroaxonal, *see* Infantile neuroaxonal dystrophy
infantile spongy, *see* Infantile spongy dystrophy
inflammatory scapuloperoneal muscular, *see* Inflammatory scapuloperoneal muscular dystrophy
juvenile neuroaxonal, *see* Juvenile neuroaxonal dystrophy
Landouzy-Dejerine, *see* Facioscapulohumeral muscular dystrophy
late infantile neuroaxonal, *see* Late infantile neuroaxonal dystrophy
limb girdle, *see* Limb girdle syndrome

lipopigment, *see* Lipopigment dystrophy
macular corneal, *see* Macular corneal dystrophy
molecular biology, 66/69
muscular, *see* Muscular dystrophy
myelin storage, *see* Myelin storage dystrophy
myotonic, *see* Myotonic dystrophy
neuroaxonal, *see* Neuroaxonal dystrophy
neuronal, *see* Neuronal dystrophy
neuronal argyrophilic, *see* Neuronal argyrophilic dystrophy
neuronal lipopigment, *see* Neuronal lipopigment dystrophy
neuronal storage, *see* Neuronal storage dystrophy
neuropil, *see* Neuropil dystrophy
oculopharyngeal muscular, *see* Oculopharyngeal muscular dystrophy
presenile astroglial, *see* Presenile astroglial dystrophy
presenile glial, *see* Presenile glial dystrophy
primary neuroaxonal, *see* Primary neuroaxonal dystrophy
progressive muscular, *see* Progressive muscular dystrophy
reflex sympathetic, *see* Reflex sympathetic dystrophy
retinal pigmentary, *see* Retinal pigmentary dystrophy
scapulohumeral distal muscular, *see* Scapulohumeral distal muscular dystrophy
scapulohumeral muscular, *see* Scapulohumeral muscular dystrophy
scapuloperoneal muscular, *see* Scapuloperoneal muscular dystrophy
secondary neuroaxonal, *see* Secondary neuroaxonal dystrophy
senile glial, *see* Senile glial dystrophy
senile neuroaxonal, *see* Senile neuroaxonal dystrophy
spinocerebellar neuroaxonal, *see* Spinocerebellar neuroaxonal dystrophy
spinocerebellofoveal, *see* Spinocerebellofoveal dystrophy
spongy, *see* Spongy dystrophy
spongy glial, *see* Spongy glial dystrophy
subcortical argyrophilic, *see* Progressive supranuclear palsy
Sudeck, *see* Reflex sympathetic dystrophy
vitelliform foveal, *see* Vitelliform foveal dystrophy
vitreotapetoretinal, *see* Vitreotapetoretinal dystrophy
X-linked muscular, *see* X-linked muscular dystrophy
X-linked neuromuscular, *see* X-linked neuromuscular dystrophy

E. coli, *see* Escherichia coli
E3 deficiency, *see* Dihydrolipoamide dehydrogenase deficiency
EAAS, *see* Electrothermal atomic absorption spectrometry
EAE, *see* Experimental allergic encephalomyelitis
Eagle ray intoxication
elasmobranch, 37/70
Eagle spirits
characteristics, 64/95
methanol, 64/95
use, 64/95
Eagle syndrome, *see* Stylohyoid syndrome
Eales disease
acanthocyte, 49/332
spastic paraplegia, 59/439
Eales vitreous hemorrhage disease
acanthocytosis, 63/290
Ear
acrocephalosyndactyly type II, 31/236
headache, 48/6
histopathology, 42/366
inner, *see* Inner ear
lop, *see* Lop ear
low set, *see* Low set ears
middle, *see* Middle ear
neuralgia, 48/495
Niemann-Pick disease type A, 10/496
pain, *see* Otalgia
Refsum disease, 21/244
satyr, *see* Satyr ear
Ear abnormality
anencephaly, 42/12
diastrophic dwarfism, 43/382
Dubowitz syndrome, 43/248
Mondini, *see* Mondini ear abnormality
8p partial monosomy, 43/508
10p partial trisomy, 43/519
4q partial monosomy, 43/496
r(13) syndrome, 50/585
r(22) syndrome, 43/547
Scheibe, *see* Scheibe ear abnormality
Seckel syndrome, 43/379
Smith-Lemli-Opitz syndrome, 43/308
trisomy 9, 43/514
Wildervanck syndrome, 43/344
Ear nose and throat disease
headache, 5/35, 208-210, 212, 48/6
sinusitis, 5/218
Ear pain, *see* Otalgia

Ear tumor
skull base tumor, 17/190
Early childhood myoclonic epilepsy
intermediate, 73/263
severe, 73/261
Early infantile encephalopathy
epilepsy, 73/235
Early infantile epileptic encephalopathy
differential diagnosis, 73/242
EEG, 73/237
etiology, 73/238
suppression burst pattern, 73/240
symptom, 73/237
Early myoclonic encephalopathy
definition, 72/8
differential diagnosis, 73/242
EEG, 73/239
epilepsy, 72/8, 73/235
survey, 73/235
symptom, 73/239
Early onset recessive stationary ataxia
cataract, 60/653
computer assisted tomography, 60/671
pes planus, 60/669
spasticity, 60/661
Eastern encephalitis
*see also* Eastern equine encephalitis
acute viral encephalitis, 56/134, 136
Aedes sollicitans, 56/136
age, 56/136
Culiseta melanura, 56/136
epilepsy, 56/136
geography, 56/136
mental deficiency, 56/136
mortality rate, 56/136
pheasant, 56/136
season, 56/136
Togavirus, 56/136
Eastern equine encephalitis
*see also* Eastern encephalitis
alphavirus, 56/12
epidemiology, 34/72
features, 34/72
thalamus, 2/486
Eastern equine encephalomyelitis
acute viral encephalitis, 34/72
Eastern hemisphere tick-borne typhus
rickettsial infection, 34/657
Eating epilepsy
clinical features, 73/186
EEG, 73/187
incidence, 73/186
pathogenic mechanism, 73/187
treatment, 73/187
Eaton-Lambert myasthenic syndrome
*see also* Myasthenic syndrome
acquired autoimmune myasthenia gravis, 62/425
4-aminopyridine, 41/360, 62/426
anticholinesterase agent, 62/426
atypical case, 62/79
autoimmune etiology, 62/422
autonomic nervous system, 75/16, 536
azathioprine, 62/426
botulism, 62/425
calcium channel, 62/392, 421, 423
cancer, 41/349-360
carcinomatous myopathy, 22/39
celiac disease, 62/421
clinical features, 41/101, 111, 129, 319, 349-352, 62/421
congenital myasthenic syndrome, 62/425
decremental EMG, 41/130
definition, 62/421
diagnosis, 62/423
3,4-diaminopyridine, 62/426
differential diagnosis, 41/290, 356-358, 62/425
edrophonium test, 62/413
electrophysiologic study, 41/352-355
EMG, 62/78, 422
evoked muscle action potential, 41/353
germine acetate, 41/360
guanidine, 41/358, 62/426
Guillain-Barré syndrome, 62/425
Hodgkin disease, 39/56
hypermagnesemia, 63/564
hyperthyroidism, 62/421
hyporeflexia, 62/426
hypothyroidism, 62/421
immunosuppression, 62/426
jitter phenomenon, 41/354
juvenile diabetes mellitus, 62/421
lung carcinoma, 62/425
magnesium intoxication, 62/425
motoneuron disease, 22/39
motor end plate, 41/335
myasthenia gravis, 62/391, 402, 415, 421
neoplasm, 62/421
neuromuscular junction disease, 62/392
neuromuscular transmission, 7/112, 41/352
paraneoplastic syndrome, 69/374, 383, 71/673, 678
paresthesia, 62/426
pathogenesis, 41/361
pernicious anemia, 62/421
plasmapheresis, 62/426
polymyositis, 62/425

polyneuropathy, 62/413, 425
postactivation exhaustion, 41/356
postactivation facilitation, 41/356
posttetanic, 41/110
presynaptic membrane, 62/425
progressive dysautonomia, 59/161
progressive spinal muscular atrophy, 59/406
proximal muscle weakness, 43/159
repetitive nerve stimulation, 62/79
serologic test, 62/425
single fiber electromyography, 62/79
Sjögren syndrome, 62/421
subacute cerebellar degeneration, 62/421
synaptic vesicle, 41/355
treatment, 41/358-361, 62/426, 71/680
vitiligo, 62/421
Eaton-Lambert syndrome, *see* Eaton-Lambert myasthenic syndrome
EB virus, *see* Epstein-Barr virus
Ebullience
migraine, 48/162
Ecchordosis physaliphora
chordoma, 18/152, 19/287
Ecchymosis
cutaneous, *see* Cutaneous ecchymosis
Eccles, J.C., 1/51, 55
ECG
acid maltase deficiency, 62/480
ataxia telangiectasia, 14/296
autonomic nervous system, 74/621
Becker muscular dystrophy, 43/86
brain aneurysm neurosurgery, 55/54
brain arteriovenous malformation, 12/246
brain embolism, 53/230, 55/151
brain hemorrhage, 55/151
brain infarction, 53/230, 55/151
carnitine deficiency, 62/493
cerebrovascular disease, 53/230
childhood myoglobinuria, 62/561
cluster headache, 75/289
debrancher deficiency, 62/482
Duchenne muscular dystrophy, 43/106
Emery-Dreifuss muscular dystrophy, 62/153
Friedreich ataxia, 21/326, 343-346, 356
glossopharyngeal neuralgia, 48/466
GM2 gangliosidosis, 10/392
head injury, 57/226
headache, 75/282, 286
hereditary amyloid polyneuropathy, 21/125
hyperkalemic periodic paralysis, 43/152
hypertensive encephalopathy, 54/213
hyperventilation syndrome, 38/319, 63/436
infectious endocarditis, 63/112
intracranial hemorrhage, 53/230
Kearns-Sayre-Daroff-Shy syndrome, 62/311
lead polyneuropathy, 64/437
LEOPARD syndrome, 43/28
leprosy, 51/218
myophosphorylase deficiency, 27/233, 41/187
myotonic dystrophy, 62/226
obstructive sleep apnea, 63/457
olivopontocerebellar atrophy (Dejerine-Thomas), 21/427
progressive external ophthalmoplegia, 62/311
rabies, 34/254, 56/389
Refsum disease, 21/9, 200-202, 66/488, 491
sleep, 3/82
spinal cord injury, 61/286, 439
subarachnoid hemorrhage, 55/10, 15, 24, 54
Sydenham chorea, 49/361
tetraplegia, 61/439
transient ischemic attack, 53/230
transverse spinal cord lesion, 61/286, 439
trichinosis, 52/568
vasomotor disorder, 24/579
ECG abnormality
encephalomyopathy, 60/670
familial spastic paraplegia, 60/670
Friedreich ataxia, 60/316, 670
Kearns-Sayre-Daroff-Shy syndrome, 60/670
Marinesco-Sjögren syndrome, 60/670
Refsum disease, 8/22, 13/314, 21/200, 60/670
sensorineural deafness, 42/388
stroke, 75/370
ECG pathology
brain hemorrhage, 63/233
cardiac nerve stimulation, 63/235
CNS lesion, 63/234
hypothalamus stimulation, 63/235
orbital frontal cortex stimulation, 63/235
stellate ganglion stimulation, 63/235
subarachnoid hemorrhage, 63/233
ECG QT interval
brain hemorrhage, 63/233
hereditary long QT syndrome, 63/238
subarachnoid hemorrhage, 63/233
ECG ST depression
brain hemorrhage, 63/233
subarachnoid hemorrhage, 63/233
ECG T wave inversion
brain hemorrhage, 63/233
subarachnoid hemorrhage, 63/233
Echinococcosis, 35/175-207, 343-357
*see also* Hydatid cyst disease
albendazole, 52/526
brain, 16/227

brain cyst, 52/524
brain stem, 52/524
cerebral hydatid cyst, 35/183-193, 347
chronic meningitis, 56/645
clinical features, 35/347-350
CNS, 52/523
computer assisted tomography, 52/525
control, 35/356
cranial hydatid disease, 35/202-207
CSF, 52/525
diagnosis, 35/350, 52/525
differential diagnosis, 52/526
distribution, 35/176
Echinococcus granulosus, 52/523
Echinococcus multilocularis, 52/523
Echinococcus vogeli, 52/523
EEG, 35/188
epidemiology, 35/12, 175, 345-347, 52/523
epilepsy, 52/524
experimental, 35/181
headache, 52/524
intracranial hypertension, 52/524
laboratory investigation, 35/188, 350-353
life cycle, 35/343-345
mebendazole, 52/526
meningeal cyst, 52/524
parasitic disease, 35/12
parasitology, 35/343-345
pathology, 35/177-179, 183-186, 353, 52/524
pathophysiology, 35/180
spinal cord, 52/524
spinal cord compression, 35/193-202, 349
spinal epidural cyst, 20/164
spinal site, 20/293
surgery, 52/526
treatment, 35/181, 190-193, 201, 206, 354-356, 52/526
vertebral hydatid disease, 35/194-198
Echinococcus
cestode, 52/523
eosinophilia, 63/371
helminthiasis, 35/217, 52/523
hydatid disease, 52/523
Echinococcus granulosus
brain abscess, 52/151
echinococcosis, 52/523
helminthiasis, 52/506
spinal epidural empyema, 52/187
spinal muscular atrophy, 59/370
Echinocytes
Duchenne muscular dystrophy, 41/410
Echinodermata intoxication
Acanthaster planci, 37/64
crinoidea, 37/64
holothurian, 37/64
sea slug, 37/64
sea urchin, 37/64
starfish, 37/64
Echinoidea intoxication
animal experiment, 37/68
araeosoma, 37/67
asthenosoma, 37/67
diadematidae, 37/67
diarrhea, 37/67
dizziness, 37/68
dyspnea, 37/68
echinothuridae, 37/67
facial edema, 37/67
facial expression, 37/68
facial paralysis, 37/68
migraine, 37/67
nausea, 37/67
pain, 37/67
sea urchin, 37/66
speech disorder, 37/68
spinal injury, 37/67
swelling, 37/67
Toxopneustes pileolus, 37/68
Tripneustes gratilla, 37/68
vomiting, 37/67
wound infection, 37/67
Echinothrix intoxication
spinal injury, 37/67
Echinothuridae intoxication
spinal injury, 37/67
ECHO virus
acquired myoglobinuria, 62/577
acute cerebellar ataxia, 34/621, 56/329
Bornholm disease, 56/198
choroid plexus, 56/318
chronic meningitis, 56/645
chronic myositis, 56/201
classification, 34/4
discovery, 56/309
enterovirus infection, 34/139
epidemic, 56/315
growth, 34/139
hemichorea, 56/330
immunocompromise, 56/475
immunocompromised host, 56/475
myopathy, 56/332
paralytic disease, 56/329
picornavirus, 34/5, 133, 139, 56/195, 307, 309
Reye syndrome, 9/549, 29/332, 49/217
transverse myelitis, 56/330
viral meningitis, 56/11, 128, 307

ECHO virus infection
*see also* Enterovirus infection
CSF, 34/140
encephalomyelitis, 34/140
epidemiology, 34/139
fetal infection, 31/220
ECHO virus meningitidis
meningitis rash, 52/25
Echoacousia
migraine, 48/162
Echocardiography, *see* EchoCG
EchoCG
bacterial endocarditis, 52/302
brain embolism, 53/230
brain infarction, 53/230
cerebrovascular disease, 53/230
debrancher deficiency, 62/483
Friedreich ataxia, 60/316
infectious endocarditis, 63/117
intracranial hemorrhage, 53/230
metastatic cardiac tumor, 63/104
mitral valve prolapse, 63/20
myotonic dystrophy, 62/227
nonbacterial thrombotic endocarditis, 63/120
primary cardiac tumor, 63/104
transient ischemic attack, 53/230
EchoEG
A scan, 23/265
acute subdural hematoma, 24/286
B scan, 23/266
brain arteriovenous malformation, 12/243, 31/191
brain atrophy, 23/279
brain contusion, 23/269
brain death, 24/771
brain edema, 23/270
brain embolism, 53/233
brain infarction, 53/233
brain tumor, 16/455
cerebellar hemorrhage, 12/58
cerebrovascular disease, 11/254-257, 53/233
chronic subdural hematoma, 24/307
commotio cerebri, 23/269
congenital toxoplasmosis, 52/359
corpus callosum agenesis, 30/289, 292
corpus callosum tumor, 17/512
EEG, 23/280
ependymoma, 18/130
epidural hematoma, 23/275, 24/272
epilepsy, 15/554
frontal lobe tumor, 17/263
hematoma echo, 23/275
hydrocephalus, 30/548
intracerebral hematoma, 11/685, 24/361
intracranial hematoma, 23/271, 274, 277
intracranial hemorrhage, 53/233
midline, 23/268, 275
normal, 23/267
North American blastomycosis, 35/406, 52/389
parietal lobe tumor, 17/306
penetrating head injury, 23/279
procedure, 23/265
pulsatile, 23/266
subarachnoid hemorrhage, 12/157
subarachnoid space, 23/268
subdural hygroma, 23/278
supratentorial brain tumor, 18/333
Sylvian fissure, 23/268
third ventricle tumor, 17/459
transient ischemic attack, 53/233
traumatic hydrocephalus, 24/236
ventricle, 23/268
Echoencephalography, *see* EchoEG
Echographia
agraphia, 4/151, 45/468
catatonia, 45/468
Echolalia
catatonia, 45/468
Gilles de la Tourette syndrome, 6/788, 42/221, 49/627, 629
jumpers of the Maine, 42/231, 49/629
Pick disease, 46/234, 240
Echopraxia
catatonia, 45/468
Gilles de la Tourette syndrome, 6/788, 49/627
Pick disease, 46/234
Eclampsia, 39/537-551
antiphospholipid antibody, 63/330
arterial hypertension, 63/72
brain hemorrhage, 39/542
clinical course, 39/541
computer assisted tomography, 54/60
convulsion, 39/547
definition, 39/537
differential diagnosis, 39/548
EEG, 39/547
epilepsy, 39/547
hypertensive encephalopathy, 54/212
incidence, 39/538
management, 39/549-551
migraine, 39/538
neurologic complication, 39/545-548
pathogenesis, 39/538-541
pathology, 39/542-545
prognosis, 39/548
psychiatric disorder, 39/547
risk factor, 39/538

striatal necrosis, 49/500
stroke, 39/545
subarachnoid hemorrhage, 55/3
systemic brain infarction, 11/459
terminology, 39/537
vascular disorder, 39/545
visual disorder, 39/546
Economo, C. Von, *see* Encephalitis lethargica
Ecothiopate iodide intoxication
chemical classification, 37/550
organophosphorus compound, 37/550
ECT, *see* Electroconvulsive treatment
Ectasia
vascular, *see* Vascular ectasia
vertebral artery, 53/380
Ectodermal dysplasia
anhidrotic, *see* Anhidrotic ectodermal dysplasia
Christ-Siemens-Touraine syndrome, 14/112
congenital, *see* Rothmund-Thomson syndrome
dentofacial syndrome (Weyers-Fülling), 14/112
differential diagnosis, 14/112
EEC syndrome, 43/395
familial, *see* Familial ectodermal dysplasia
hidrotic, *see* Hidrotic ectodermal dysplasia
hypohidrotic, *see* Hypohidrotic ectodermal dysplasia
hypotrichosis, 14/112
keratitis dystrophica, 14/112
mental deficiency, 43/269
Peutz-Jeghers syndrome, 43/41
Rothmund-Thomson syndrome, 14/112
satyr ear, 14/112
Ullrich-Fremerey-Dohna syndrome, 14/112
vertebral artery, 53/380
Werner syndrome, 14/112
Ectopallium
behavior, 2/720
temporal lobe lesion, 2/720
Ectopia, *see* Heterotopia
Ectopia cordis
anencephaly, 50/84
Ectopia lentis
homocystinuria, 39/392, 42/555, 55/327
Marfan syndrome, 39/389
Ectopia testis, *see* Testicular heterotopia
Ectopic lesion
schistosomiasis, 52/537
Ectrodactyly
EEC syndrome, 43/395
Ectropion
Möbius syndrome, 50/215
Edathamil, *see* Edetic acid
Edema
angioneurotic, *see* Angioneurotic edema
anterior tibial syndrome, 8/153, 41/385
bee sting intoxication, 37/107
beriberi, 28/6
brain, *see* Brain edema
brain tumor, *see* Brain tumor edema
central serous retinopathy, 13/142
cerebral, *see* Brain edema
cervical spinal cord, 25/262
cluster headache, 48/222
congenital ichthyosis, 43/406
conidae intoxication, 37/60
cytotoxic brain, *see* Cytotoxic brain edema
demyelination, 47/43
endoneurial, *see* Endoneurial edema
eosinophilia myalgia syndrome, 64/252
episodic pulmonary, *see* Episodic pulmonary edema
eyelid, 8/212
facial, *see* Facial edema
Galloway-Mowat syndrome, 43/431
heat, *see* Heat edema
high altitude brain, *see* High altitude brain edema
hypoxia, 16/180
keratitis bullosa, 8/350
meduso-congestin intoxication, 37/37
Melkersson-Rosenthal syndrome, 2/72, 14/790, 42/322
Milroy disease, 1/483, 8/208-210, 20/142
myelin, *see* Myelin edema
nerve paralysis, 8/382
Neu syndrome, 43/430
neurogenic pulmonary, *see* Neurogenic pulmonary edema
obstructive sleep apnea, 63/454
pertussis encephalopathy, 9/551
pickwickian syndrome, 38/300
plaque, 9/220
pseudoinflammatory foveal dystrophy (Sorsby), 13/138
pulmonary, *see* Pulmonary edema
Quincke, *see* Angioneurotic edema
spinal cord, *see* Spinal cord edema
spinal cord contusion, 25/47, 76
spinal cord experimental injury, 25/22
spinal cord injury, 25/54, 79, 262
spinal injury, 12/557
spongy, *see* Spongy edema
vasogenic brain, *see* Vasogenic brain edema
Edentulous orodyskinesia
chorea, 49/558
Edetate calcium disodium
lead intoxication, 64/439

manganese intoxication, 64/316
Edetic acid
bismuth intoxication, 64/344
calcium, *see* Calcium edetic acid
lead encephalopathy, 64/443
lead intoxication, 36/25-28
lead polyneuropathy, 51/270
manganese intoxication, 36/231
mercury intoxication, 36/181-183
organolead intoxication, 64/134
thallium intoxication, 36/272
Wilson disease, 27/404
Edetic acid intoxication
headache, 48/420
Edinburgh Index
autonomic nervous system, 74/621
Edinburgh, United Kingdom
neurology, 1/7
Edinger comb system
anatomy, 6/403
Edinger, L., 1/9, 27
Edinger tract
anterolateral funiculus, 45/232
Edinger-Westphal nucleus
oculomotor nerve injury, 24/59
oculomotor nuclear complex, 2/301
parasympathetic fiber origin, 2/107, 279
pretectal region, 2/533
Edrophonium test
acetylcholinesterase, 41/126
amyotrophic lateral sclerosis, 62/413
congenital myasthenic syndrome, 62/413
Eaton-Lambert myasthenic syndrome, 62/413
myasthenia gravis, 40/298, 303, 321, 41/126, 240, 62/413
poliomyelitis, 62/413
progressive external ophthalmoplegia, 62/303
EDTA, *see* Edetic acid
Edwards syndrome
*see also* Russell-Silver syndrome
abortion, 50/542
anencephaly, 43/537, 50/562
Arnold-Chiari malformation type I, 50/562
arthrogryposis, 50/561
associated malformation, 31/520-522
bilateral ptosis, 50/561
birth incidence, 43/537
birth weight, 50/561
blepharophimosis, 50/561
brain cortex disorganization, 50/249
brain cortex heterotopia, 50/248, 275
cebocephaly, 50/562
cerebellar agenesis, 50/177, 183
cerebellar atrophy, 43/537
cerebellar heterotopia, 50/251
cerebellar hypoplasia, 50/178, 562
cervical pterygium, 50/561
clinical features, 31/517-521
clinical symptomatology, 50/560
CNS, 31/521
coloboma, 50/562
congenital heart disease, 50/275
convulsion, 31/521, 43/535
corneal opacity, 50/562
corpus callosum agenesis, 31/522, 43/536, 50/152, 163, 167, 562
corpus callosum hypoplasia, 43/536, 50/275
cryptorchidism, 43/535, 50/561
cyclopia, 50/562
cytogenetics, 31/515
definition, 31/515, 50/560
dermatoglyphics, 31/507, 43/536
diaphragm hernia, 50/275
digital abnormality, 43/535
dolichocephalic skull, 43/535, 50/561
epidemiology, 31/516, 50/560
facial paralysis, 50/561
family, 50/562
growth failure, 50/560
gyral abnormality, 43/536
heart disease, 43/536
hippocampus heterotopia, 50/275
history, 31/516, 50/560
holoprosencephaly, 31/521, 43/537, 50/557, 562
horseshoe kidney, 43/536, 50/275
hypospadias, 50/561
inferior olivary nucleus, 50/250
inferior olivary nucleus heterotopia, 50/250, 275
inner ear, 50/562
intelligence, 50/562
life expectancy, 50/560
lissencephaly, 50/250
low set ears, 43/535, 50/275
meningomyelocele, 43/537, 50/562
mental deficiency, 31/521, 43/535, 46/70, 50/542
metabolic abnormality, 31/522
microcephaly, 30/508, 31/517, 521, 50/274, 557
micrognathia, 50/275
microphthalmia, 50/562
microretrognathia, 50/561
microstomia, 50/275
middle ear, 50/562
mosaicism, 31/523, 46/70, 50/561
muscular hypotonia, 50/561
nail hypoplasia, 43/535
neurologic finding, 50/561

olfactory nerve aplasia, 50/562
optic nerve aplasia, 50/562
pachymicrogyria, 50/562
partial trisomy 18, 50/562
Patau syndrome, 14/121
polyhydramnios, 43/535
prognosis, 31/523
psychomotor development, 50/560, 562
pterygium colli, 50/275
radiologic finding, 31/523
renal abnormality, 43/536
rockerbottom foot, 50/275, 561
Russell-Silver syndrome, 46/21
seizure, 31/517
sex ratio, 43/537
short neck, 43/535
spasticity, 50/561
spina bifida, 43/537
VACTERL syndrome, 50/514
vaginal atresia, 50/561
widely spaced nipples, 43/535, 50/275
Edwin Smyth papyrus
historic review, 2/4
EEC syndrome
cheilopalatoschisis, 43/395
dental abnormality, 43/395
ectodermal dysplasia, 43/395
ectrodactyly, 43/395
mental deficiency, 43/396, 46/91
microcephaly, 43/396
renal abnormality, 43/396
sensorineural deafness, 43/395
EEG
*see also* Magnetoencephalography
abnormality, 54/157
absence status, 15/595
acrylamide intoxication, 64/70
acute subdural hematoma, 24/286
acute viral encephalitis, 56/131
Addison disease, 39/480
adenovirus meningoencephalitis, 56/284
adult neuronal ceroid lipofuscinosis, 42/466
affective psychosis, 46/460
African trypanosomiasis, 35/76-79, 52/341
age, 15/516
Aicardi syndrome, 42/696, 50/163
akinetic mutism, 23/337
aldrin intoxication, 36/414-416
Alpers disease, 42/486
alpha coma, 57/187, 189
alpha rhythm slowing, 57/184
altitude illness, 48/396
aluminum intoxication, 63/526, 64/279
Alzheimer disease, 42/277, 46/264
amnesic shellfish poisoning, 65/167
amplitude, 57/184
amyotrophic dystonic paraplegia, 42/199
angiostrongyliasis, 35/327, 52/558
anorexia nervosa, 46/586
anoxic encephalopathy, 63/219
anticonvulsant withdrawal, 72/309
anxiety, 45/267
aphasia, 45/298, 46/150
arousal disorder, 45/108-110
astrocytoma, 18/28
ataxia telangiectasia, 14/296, 60/376
athetosis, 6/446
athetotic hemiplegia, 42/201
background activity, 54/158
background activity suppression, 57/184
ballismus, 6/484
Bardet-Biedl syndrome, 13/393, 43/233
Bartter syndrome, 42/529
basilar impression, 32/78
Batten disease, 10/617, 619
Bean syndrome, 43/11
Behçet syndrome, 34/487, 56/600
benign neonatal familial convulsion, 73/133
benign partial epilepsy with centrotemporal spike, 73/13
benign partial epilepsy with occipital paroxysm, 73/24
benign partial epilepsy with somatosensory evoked potential, 73/26
benzodiazepine intoxication, 65/338
Binswanger disease, 54/223
bismuth intoxication, 64/338, 343
bone membranous lipodystrophy, 42/281
boxing injury, 23/546
brain abscess, 23/354, 33/123, 52/155
brain anoxia, 27/42, 54/159
brain arteriovenous malformation, 12/108, 242, 31/189
brain blood flow, 23/10
brain blood flow arrest, 57/461
brain concussion, 23/76, 421, 434, 57/182
brain contusion, 23/421, 434
brain cortex, 72/290
brain death, 24/729-732, 765-769, 781, 55/263
brain edema, 16/189, 23/353
brain embolism, 11/409, 53/232
brain hemorrhage, 11/280
brain hypoxia, 23/353
brain infarction, 53/122, 232
brain injury, 23/317, 335, 350, 24/620
brain ischemia, 11/272-276, 54/159, 55/219,

63/219
brain lacunar infarction, 54/254
brain metastasis, 18/221
brain phlebothrombosis, 11/278
brain stem auditory evoked potential, 54/160
brain stem death, 57/451, 463
brain stem hemorrhage, 11/280
brain thromboangiitis obliterans, 12/390, 39/203
brain tumor, 16/418
bromide intoxication, 36/302
burst suppression, 57/189
carbamazepine intoxication, 65/506
carbon disulfide intoxication, 64/25
carbon monoxide intoxication, 64/33
cat scratch disease, 52/130
cellular explanation, 72/289
central pontine myelinolysis, 28/293, 63/553
centrencephalic epilepsy, 42/682
cerebellar ataxia, 42/124
cerebellar tumor, 17/715
cerebral amebiasis, 52/321
cerebral dominance, 4/307, 466
cerebral gigantism, 43/337
cerebral palsy, 46/493
cerebral paragonimiasis, 35/257
cerebrovascular disease, 11/267-291, 53/232, 54/157
cerveau isolé, 2/280, 3/81
Charlevoix-Saguenay spastic ataxia, 60/455
child, 57/183
childhood absence epilepsy, 73/147
chorea, 49/500
chorea-acanthocytosis, 49/329, 63/281
chronic subdural hematoma, 24/308
ciclosporin intoxication, 63/537, 65/552
ciguatoxin intoxication, 65/163
clonic seizure, 15/122
clumsiness, 46/164
cluster headache, 48/238
cocaine intoxication, 65/253
coccidioidomycosis, 52/417
Coffin-Lowry syndrome, 43/239
cognitive function, 72/289
coma, 23/337, 340
commotio cerebri, 23/421
complicated migraine, 5/64, 66
compressed spectral analysis, 57/382
compressed spectral array, 57/190
congenital cerebellar atrophy, 21/463
congenital heart disease, 38/131
congenital retinal blindness, 13/98, 272
congenital toxoplasmosis, 35/137
corpus callosum agenesis, 30/289, 292, 42/10, 50/159
corpus callosum tumor, 17/512
cortical blindness, 24/49, 45/50
corticocortical, 72/293
craniopharyngioma, 18/547
Creutzfeldt-Jakob disease, 6/732, 42/657, 46/292, 56/545
cri du chat syndrome, 43/501
CSF fistula, 23/352
cyclic oculomotor paralysis, 42/332
daytime hypersomnia, 45/131
delirium, 46/524, 550
delta wave genesis, 54/159
dementia, 46/127, 202
depolarization, 57/182
developmental dysphasia, 46/141
diagnosis, 11/286
dialysis encephalopathy, 63/526, 64/277
diamorphine intoxication, 65/352
dieldrin intoxication, 36/414-416
diencephalic epilepsy, 2/452
diuretic agent, 65/439
dominant infantile optic atrophy, 13/118
DOOR syndrome, 42/385
dopamine β-mono-oxygenase deficiency, 59/163
Down syndrome, 31/401, 405, 408, 50/529
dwarfism, 43/380
dyslexia, 46/131
dyssynergia cerebellaris myoclonica, 21/514, 60/600
dystonia, 42/219
dystopic cortical myelinogenesis, 30/362
early infantile epileptic encephalopathy, 73/237
early myoclonic encephalopathy, 73/239
eating epilepsy, 73/187
echinococcosis, 35/188
echoEG, 23/280
eclampsia, 39/547
electrocerebral silence, 23/320, 57/184, 466
encéphale isolé, 1/73
encephalitis, 46/493
endrin intoxication, 36/414-416
eosinophilic meningitis, 35/325-327
ependymoma, 18/129
epidural hematoma, 23/348, 24/272, 57/186
epilepsy, 15/498-500, 46/493, 72/288, 293
epilepsy with continuous spike wave, 72/9
epilepsy with continuous spike wave during slow sleep, 73/270
epilepsy of late onset, 57/193
epilepsy with myoclonic absence, 73/232
epilepsy with myoclonic astatic seizure, 73/227
epileptic focus, 15/513

epileptic syndrome, 72/301
epileptiform, 72/293
epileptiform activity, 72/287
erythropoietin, 63/535
ethylene glycol intoxication, 64/123
evolution, 57/185
experimental, 23/79
experimental head injury, 23/76
extra pontine myelinolysis, 63/553
Fabry disease, 42/427
febrile convulsion, 15/254-259, 72/310
filariasis, 35/168, 52/517
focal change, 57/184
formiminotransferase deficiency, 42/546
frequency analysis, 23/79
Friedman-Roy syndrome, 43/251
Friedreich ataxia, 21/349, 60/320
frontal intermittent rhythmic delta activity, 54/157
frontal lobe epilepsy, 73/45
frontal lobe tumor, 17/263
Fukuyama syndrome, 43/92
functional psychosis, 46/460
gelolepsy, 45/509
generalized change, 57/183
generalized seizure, 72/301
generalized slow wave activity, 57/184
genetics, 15/429, 522
Gilles de la Tourette syndrome, 6/790, 49/631
glioblastoma multiforme, 18/58
GM2 gangliosidosis, 10/334, 391
grand mal on awakening, 73/179
grand mal epilepsy, 15/468
grand mal status, 15/150
granulomatous CNS vasculitis, 55/391
handedness, 4/259, 260
Harvey syndrome, 42/223
head injury, 23/79, 324, 333, 344, 57/181-193, 225
head trauma chronic phase, 57/185
headache, 5/33, 43, 48/39
hepatic coma, 27/349, 370, 49/214
hepatic encephalopathy, 27/349, 370
hereditary exertional paroxysmal dystonic choreoathetosis, 49/350
hereditary olivopontocerebellar atrophy (Menzel), 21/445
hereditary paroxysmal dystonic choreoathetotis, 49/350, 352
hereditary paroxysmal kinesigenic choreoathetosis, 42/206, 49/350, 353
hereditary periodic ataxia, 60/438-441
hereditary striatal necrosis, 49/496
herpes simplex virus encephalitis, 56/215
hexachlorophene intoxication, 65/478
high-pressure nervous syndrome, 48/402
histoplasmosis, 52/439
holoprosencephaly, 30/452, 456, 459, 50/238
homocystinuria, 55/328
Huntington chorea, 6/314, 49/288
hydrocephalic dementia, 46/328
hydrocephalus, 30/558
hyperactivity, 46/183
hyperaldosteronism, 39/496
hypereosinophilic syndrome, 63/372
hyperkinetic child syndrome, 3/196, 4/363, 43/201
hypernatremia, 28/452, 63/554
hyperparathyroidism, 27/290
hyperprolinemia type I, 29/133
hyperprolinemia type II, 29/138
hypertensive encephalopathy, 11/559, 54/213
hypervalinemia, 42/574
hyperventilation, 23/356, 359
hyperventilation syndrome, 38/318, 63/435
hypnotic agent, 37/351
hypocalcemia, 28/534
hypomagnesemia, 28/549
hyponatremia, 63/546
hypoparathyroidism, 27/304, 42/577
hypoxia, 11/271
hypsarrhythmia, 42/686, 691-693, 72/303
incontinentia pigmenti, 14/216, 43/23
infantile autism, 46/191
infantile neuroaxonal dystrophy, 49/403
infantile spasm, 15/223
interhemispheric asymmetry, 57/184
intracerebellar hematoma, 11/280
intracerebral hematoma, 11/280, 686, 24/362
intracranial, 72/297
intracranial hematoma, 23/344, 350
intracranial hemorrhage, 53/232
intracranial hypertension, 55/219
intracranial infection, 57/187
isobenz intoxication, 36/414-416
juvenile absence epilepsy, 73/151
juvenile Gaucher disease type A, 66/125
juvenile myoclonic epilepsy, 73/164
juvenile parasomnia, 45/141
Kii Peninsula amyotrophic lateral sclerosis, 22/371
kinesigenic choreoathetosis, 42/208
Kjellin syndrome, 42/174
Klinefelter syndrome, 31/482, 50/549
Klinefelter variant XXYY, 43/563
Koshevnikoff epilepsy, 73/120
Lafora progressive myoclonus epilepsy, 15/383, 387-389, 27/172, 174
Landau-Kleffner syndrome, 73/281, 283, 286

language disability, 73/281
lead encephalopathy, 64/436
lead intoxication, 64/444
legionellosis, 52/256
Lennox-Gastaut syndrome, 72/305, 73/214
Lenz syndrome, 43/419
lidocaine intoxication, 37/451
lightning injury, 23/696
lindane intoxication, 65/479
linear nevus sebaceous syndrome, 43/39
lissencephaly, 30/486
lissencephaly syndrome, 42/40
lithium intoxication, 64/296
localization value, 54/157
locked-in syndrome, 53/210
Lyme disease, 52/262
lymphocytic choriomeningitis virus, 56/358
magnesium imbalance, 28/549
manic depressive psychosis, 43/208
maple syrup urine disease, 29/64
mental deficiency, 42/689, 43/268, 46/24
mesencephalon, 54/157
metastatic cardiac tumor, 63/96
methanol intoxication, 36/354, 64/100
1-methyl-4-phenyl-1,2,3,6-tetrahydropyridine intoxication, 65/372
midbrain, 72/291
midbrain tumor, 17/635
migraine, 5/42, 64, 66, 48/29, 38, 144, 149
migrainous transient global amnesia, 48/164
monophasic, 57/188
monorhythmic, 57/188
moyamoya disease, 12/366, 55/298
multiple mitochondrial DNA deletion, 62/510
multiple sclerosis, 47/65
mumps encephalitis, 56/430
mumps meningitis, 56/430
musicogenic epilepsy, 73/192
myelinoclastic diffuse sclerosis, 9/475, 47/423
myoclonic epilepsy, 15/513, 27/174, 42/701, 703, 705-707
myoclonus, 49/610
myophosphorylase deficiency, 41/187
narcolepsy, 2/448, 3/94, 45/149
neglect syndrome, 45/155, 157
neonatal seizure, 73/256
neuroamebiasis, 52/321
neurocysticercosis, 35/307
neuroleptic syndrome, 6/259
neurologic intensive care, 55/219
neuromyelitis optica, 47/402
neuronal ceroid lipofuscinosis, 42/461-468
neurosyphilis, 52/277
newborn epilepsy, 15/195, 206, 216
nicotine intoxication, 36/429, 65/262
nocturnal myoclonus, 45/139
normal, 72/285
normal paroxysmal, 72/295
North American blastomycosis, 35/406, 52/389
occipital lobe epilepsy, 73/112
occipital lobe syndrome, 45/57
oligodendroglioma, 18/91
olivopontocerebellar atrophy (Dejerine-Thomas), 21/427
opticocochleodentate degeneration, 21/542, 60/754
organic psychosis, 46/493
organic solvent intoxication, 64/40
organochlorine insecticide intoxication, 64/199
organophosphate intoxication, 37/554, 64/172, 227
organotin intoxication, 64/142
orthochromatic leukodystrophy, 42/501
Paine syndrome, 43/433, 60/293
paracoccidioidomycosis, 35/536
paranoid schizophrenia, 46/494
parietal lobe, 45/173, 72/300
parietal lobe epilepsy, 73/103
parietal lobe tumor, 17/306
Parkinson disease, 6/196
Parry disease, 42/467
partial seizure, 72/299
pattern, 72/287
pediatric head injury, 23/453, 57/183
penetrating head injury, 23/350, 57/185, 187
peritoneal dialysis, 27/341
persistent unconsciousness, 57/189
persistent vegetative state, 57/189
pertussis, 52/234
pertussis encephalopathy, 33/292
pertussis vaccine, 52/245
petit mal epilepsy, 15/470, 511, 42/681
phenylketonuria, 29/30
photosensitive epilepsy, 73/184
photosensitivity, 72/304
Pick disease, 46/241
pinealoma, 18/362, 42/767
pituitary adenoma, 17/405
porencephaly, 30/690
Porot-Filiu syndrome, 42/688
posttraumatic coma, 57/187
posttraumatic epilepsy, 57/192, 72/201
prazosin intoxication, 65/447
primary cardiac tumor, 63/96
primary pigmentary retinal degeneration, 13/172-174, 197, 313

prognostic value, 54/158
progressive multifocal leukoencephalopathy, 9/495, 47/505, 510
progressive myoclonus epilepsy, 27/174
progressive supranuclear palsy, 22/221, 46/301, 49/244
psychogenic epilepsy, 72/271
psychomotor epilepsy, 15/478, 513
psychosis, 15/601, 46/495
pyrethroid intoxication, 64/219
quantitation, 54/159
rabies, 34/254, 56/389
reactivity, 57/188
reading epilepsy, 2/621, 42/717, 73/188
recording time, 54/158
recurrent meningitis, 52/52
regional brain blood flow, 11/270
REM, 57/190
REM sleep, 46/495
renal insufficiency, 27/326, 332-334, 340
respiratory encephalopathy, 63/414
Rett syndrome, 29/316, 60/637, 63/437
Reye syndrome, 29/335, 31/168, 34/169, 56/153
Roberts syndrome, 46/80
Rothmund-Thomson syndrome, 14/777
Rubinstein-Taybi syndrome, 31/284
Salla disease, 43/306
schizencephaly, 50/201
schizophrenia, 46/421, 460, 492-496, 505
seizure origin, 72/297
senile dementia, 46/286
sensory deprivation, 46/534
septum pellucidum tumor, 17/478
Sjögren-Larsson syndrome, 22/477
Sjögren syndrome, 71/65
skull fracture, 23/318, 327, 350, 57/186
sleep, 3/80, 82, 46/495, 72/286, 294
slow wave, 72/294
slow wave foci, 57/185
Smith-Lemli-Opitz syndrome, 43/308
somatosensory evoked potential, 54/160
spasmodic torticollis, 6/592-596
spectral analysis, 57/182
spike activity, 72/295
spike and wave, 72/287
spindle coma, 57/187-189
status epilepticus, 15/513, 72/308, 73/270
striatopallidodentate calcification, 6/708, 49/422
styrene intoxication, 64/54
subacute combined spinal cord degeneration, 28/163
subacute necrotizing encephalomyelopathy, 28/353
subacute sclerosing panencephalitis, 34/355, 56/419
subarachnoid hemorrhage, 11/280, 12/157
subdural empyema, 23/252
subdural hematoma, 11/280, 23/324, 333, 344, 57/186
subdural hygroma, 23/352
subgaleal hematoma, 23/318
subsarcolemmal myofibril, 42/219
superior longitudinal sinus thrombosis, 52/178
suppression burst pattern, 73/235
supratentorial brain abscess, 33/122
supratentorial brain tumor, 18/333
survey, 72/283, 286
Sydenham chorea, 6/421-426, 49/361
syncope, 75/209
syringomyelia, 32/278
systemic lupus erythematosus, 39/287, 55/380, 71/46
technical problem, 57/181
temporal lobe, 72/300
temporal lobe agenesis syndrome, 30/358
temporal lobe epilepsy, 73/73
temporal lobe lesion, 2/718
temporal lobe tumor, 17/287-289
teratoma, 31/57
tetany, 28/534
thalamic tumor, 17/617
thalamus, 54/157, 72/290, 293
thalassemia major, 42/629
thallium intoxication, 36/253, 64/325
third ventricle tumor, 17/457
thrombotic thrombocytopenic purpura, 55/473
tonic clonic seizure, 15/115
toxic shock syndrome, 52/259
toxoplasmosis, 35/133
transient cortical blindness, 24/49
transient global amnesia, 45/206
transient ischemic attack, 45/212, 53/232, 55/299
transtentorial herniation, 1/561
traumatic epilepsy, 23/357
traumatic hydrocephalus, 24/234
traumatic intracerebral hematoma, 57/186
traumatic psychosis, 24/523
traumatic psychosyndrome, 24/540
traumatic vegetative syndrome, 24/579
trichinosis, 35/277, 52/569
tricyclic antidepressant, 65/318
triethyltin intoxication, 36/282
triphasic wave, 57/189
true porencephaly, 30/690
tuberculous meningitis, 33/225, 52/210, 55/429
tuberous sclerosis, 14/347

Turner syndrome, 50/545
uremic encephalopathy, 27/332-334, 63/505, 507, 515
uremic polyneuropathy, 63/515
valproic acid intoxication, 65/504
vascular dementia, 46/360
venous sinus occlusion, 54/441
venous sinus thrombosis, 54/399
verbal auditory agnosia, 43/229
vertebrobasilar insufficiency, 54/158
visual evoked response, 54/160, 57/197
Vogt-Koyanagi-Harada syndrome, 34/526, 56/621
Waldenström macroglobulinemia, 63/398
water intoxication, 64/241
Wernicke-Korsakoff syndrome, 28/249
xeroderma pigmentosum, 60/337
XYY syndrome, 31/489
EEG classification
brain injury, 23/227, 229
epilepsy, 15/8, 72/307
EEG frequency analysis
head injury, 23/79
Effort syndrome, *see* Hyperventilation syndrome
Egerton-Stoke-Mandeville frame
cervical vertebral column injury, 25/293
Eggshell vertebra
aneurysmal bone cyst, 20/30
Ehlers-Danlos syndrome
angioid streak, 13/32, 141
autosomal dominant inheritant pattern, 55/454
brachial plexus neuritis, 55/456
brachial plexus paralysis, 55/456
brain aneurysm, 12/102, 55/454
brain angiography, 55/455
brain embolism, 55/455
cardiac valve, 55/455
carotid cavernous fistula, 24/407, 55/455
carotid system syndrome, 53/309
cerebrovascular disease, 53/43, 55/454
chalacoderma, 14/111
chorioretinal degeneration, 13/41
congenital hip dislocation, 14/111
congenital retinal blindness, 22/528
cutaneous dysplasia, 14/101
diaphragm hernia, 43/14
epilepsy, 43/14
features, 14/778
hereditary connective tissue disease, 39/381-389
hereditary multiple recurrent mononeuropathy, 55/456
hyperextensible joint, 43/14
hyperextensible skin, 43/14
hypertelorism, 14/111
hypotonia, 40/336
intestinal pseudo-obstruction, 51/493
lumbosacral plexus neuritis, 55/456
lumbosacral plexus palsy, 55/456
mental deficiency, 46/61
nerve pressure palsy, 55/456
neuralgic amyotrophy, 51/175
neurocutaneous syndrome, 14/101
neurofibromatosis, 14/568
neurologic symptom, 55/454
optic atrophy, 13/41
Parrot syndrome, 14/111
phakomatosis, 14/492
polycystic renal disease, 55/455
prevalence, 43/16
recessive transmission, 55/454
retinal detachment, 43/15
scoliosis, 43/14, 50/418
strabismus, 43/14
subarachnoid hemorrhage, 55/29, 454
11 subtypes, 55/454
symptom, 14/111
syndactyly, 14/111
type III collagen, 55/454
Ehrhardt, H., 1/21
Eikenella
bacterial endocarditis, 52/302
Einarson classification
diffuse sclerosis, 9/472
Einarson-Neel classification
diffuse sclerosis, 10/15
Ejaculation
electroejaculation, 61/323
intrathecal neostigmine, 61/322
paraplegia, 61/315, 322
physiology, 61/315
sexual function, 26/440, 442, 448
spinal cord injury, 61/315, 322
subcutaneous physostigmine, 61/322
tetraplegia, 61/315, 322
transverse spinal cord lesion, 61/315, 322
vibration, 61/322
Ekbom other syndrome
arterial hypertension, 63/85
dementia, 63/85
hydrocephalus, 63/85
megadolichobasilar artery, 63/85
presenile dementia, 63/85
transient ischemic attack, 63/85
vascular ectasia, 63/85
Ekbom syndrome, *see* Restless legs syndrome
Ekbom syndrome type II
areflexia, 60/660

dementia, 60/665
hypoesthesia, 60/661
kyphoscoliosis, 60/669
lipoma, 60/667
myoclonus, 60/662
pes cavus, 60/669
Ekman-Lobstein disease
blue sclerae, 22/495
bone fragility, 22/494
hearing loss, 22/494
hereditary progressive cochleovestibular atrophy, 60/774
Elapid intoxication
envenomation, 37/5-8, 12-16
venom, 37/12-16
Elapid snake venom
myasthenia gravis, 62/391
myasthenic syndrome, 62/391, 64/16
myoglobinuria, 64/16
Elapidae
autonomic nervous system, 75/512
venomous snake, 65/177
Elasmobranche intoxication
bat ray, 37/70
cow-nose ray, 37/70
dasyatidae, 37/70
devil ray, 37/70
eagle ray, 37/70
gymnuridae, 37/70
heterodontus francisci, 37/70
heterodontus portusjacksoni, 37/70
manta ray, 37/70
mobulidae, 37/70
myliobatidae, 37/70
potamotrygonidae, 37/70
rhinopteridae, 37/70
river ray, 37/70
round sting ray, 37/70
Somniosus microcephalus, 37/85
squaliformes, 37/70
squalus acanthias, 37/70
stingray, 37/70
urilophidae, 37/70
whip ray, 37/70
Elastase
wallerian degeneration, 51/33
Elastorrhexis generalisata, *see* Grönblad-Strandberg syndrome
Elastosis
senile cutaneous, *see* Senile cutaneous elastosis
Elastosis performans serpiginosa
penicillamine intoxication, 49/235
Wilson disease, 49/235
Elation
migraine, 48/162
Elbow splint
heavy duty below, *see* Heavy duty below elbow splint
Elbow trauma
nerve damage, 70/32
Eldridge-Berlin-Money-McKusick syndrome
Flynn-Aird syndrome, 14/113
Elective mutism
childhood psychosis, 46/193
Electric chorea (Berland-Bergeron), *see* Chorea electrica
Electric flea
synapse function, 1/49
Electric shock
myoclonus, 49/618
Electrical injury
*see also* Lightning injury
alpha coma, 61/192
alternating current, 61/193
amyotrophic lateral sclerosis, 22/329, 23/708, 710, 61/194, 196
anosmia, 61/192
autonomic dysfunction, 23/717
autonomic nervous system, 7/378, 23/717
autonomic polyneuropathy, 61/192
basilar artery thrombosis, 61/192
bladder dysfunction, 74/321
blood clotting necrosis, 61/191
blood pressure, 74/321
brachial plexus injury, 61/192
brain edema, 23/699, 718
brain hemorrhage, 61/192
brain infection, 23/717
brain injury, 11/464, 23/683
cardiac arrest, 23/706, 61/192
cataract, 61/192
clinical features, 23/697, 704, 707, 720
encephalomalacia, 61/192
epidemiology, 61/193
epilepsy, 23/713, 716, 61/192
facial paralysis, 61/195
hearing loss, 61/195
heart, 74/321
high voltage injury, 23/684, 713, 720
history, 23/686
intracranial epidural hematoma, 61/192
intraventricular hemorrhage, 61/192
keraunoparalysis, 61/192
Lichtenberg figure, 61/192
lightning parameter, 61/191
low voltage injury, 23/684, 708, 720

motoneuron disease, 61/194, 196
multiple lightning accident, 61/193
myoglobinuria, 61/192
nerve lesion, 7/344, 349, 351, 358, 365, 375, 51/134
nervous system, 61/191
neuropathy, 51/139
otologic symptom, 61/195
paralytic ileus, 74/321
paraplegia, 61/195
pathophysiology, 7/354, 23/689, 706
physical parameter, 61/191
pregnancy, 61/194
progressive muscular atrophy, 59/18
progressive spinal muscular atrophy, 22/29
skull, 7/374, 376, 23/713
spastic paraplegia, 61/195
spinal cord, *see* Spinal cord electrotrauma
spinal cord atrophy, 23/711
spinal cord injury, 23/711
spinal muscular atrophy, 51/139
step voltage, 61/194
stride potential, 7/345
subarachnoid hemorrhage, 12/121, 23/692
systemic brain infarction, 11/464
technical electricity, 23/684, 688
telephone mediation, 61/194
tetraplegia, 61/195
thermoregulation, 74/321
transverse spinal cord lesion, 61/195
tympanic membrane rupture, 61/195
vasoconstriction, 74/321
Electrical muscle activity
spontaneous, *see* Spontaneous electrical muscle activity
Electrical trauma, *see* Electrical injury
Electricity
brain edema, 7/360
sensation, 1/105
stress, 74/320
striatal lesion, 7/363
Electrocardiography, *see* ECG
Electrocerebral silence
brain stem death, 57/464
cardiac arrest, 57/466
EEG, 23/320, 57/184, 466
reversible isoelectric EEG, 57/466
survival, 57/466
Electroconvulsive treatment
learning, 27/466
memory, 27/466
memory disorder, 3/272
phantom limb pain, 45/401
Electrocution, *see* Electrical injury
Electrode
epilepsy, 15/760
nerve conduction, 7/120, 129-131, 139
stereotaxic, 15/760
trigeminal neuralgia, 48/454
Electrodermal asymmetry
schizophrenia, 46/504
Electrodermal habituation
schizophrenia, 46/497
Electrodiagnosis
carpal tunnel syndrome, 51/94
facial nerve injury, 24/111
facial paralysis, 2/64, 8/265
muscle, 1/224
myopathy, 1/224
Electroencephalography, *see* EEG
Electrogenesis
Ranvier node, 51/41
Electrogustometry
ageusia, 24/17
facial paralysis, 8/265
Electrolyte
brain hypoxia, 15/62
CSF, 19/127
Down syndrome, 31/433
epileptic focus, 15/66-68
Reye syndrome, 56/152
Electrolyte balance
autonomic nervous system, 1/499-501
brain injury, 23/109
central pontine myelinolysis, 47/588
delirium, 46/543
epilepsy, 15/62, 311, 313
hypothalamus, 1/500
radiation injury, 23/655
sweating, 1/501
Electrolyte depletion
autonomic nervous system, 75/492
Electrolyte disorder
brain aneurysm, 55/53
brain aneurysm neurosurgery, 55/53
EMG, 63/557
heat stroke, 23/676, 679
hepatic coma, 27/354
hypercalcemia, 63/561
hyperkalemia, 63/558
hypermagnesemia, 63/563
hypernatremia, 63/554
hypocalcemia, 63/558
hypokalemia, 63/557
hyponatremia, 63/545
neurologic syndrome, 63/545

normalization risk, 63/545
serum creatine kinase, 63/557
subarachnoid hemorrhage, 55/53
tetany, 63/558
Electromagnetic wave
nerve injury, 51/139
Electromyography, *see* EMG
Electron transfer flavoprotein
flavine adenine dinucleotide, 66/409
mitochondrial matrix system, 66/409
Electron transport chain
component, 27/18
Electroneurography
compression neuropathy, 51/86
Electronystagmography, *see* ENG
Electro-oculography, *see* EOG
Electrophoresis
brain tumor, 16/365
CSF examination, 16/364, 19/127-130, 133
glycosaminoglycanosis type VII, 66/313
multiple sclerosis, 47/87, 95, 99, 103
polyacrylamide gel, *see* Polyacrylamide gel electrophoresis
Electroretinography, *see* ERG
Electrostimulation
ageusia, 24/16
akinesia, 49/67
cerebellar afferent, 2/409
nerve, 7/62
parietal lobe, 2/682
spinal cord injury, 61/491
Electrothermal atomic absorption spectrometry
lead intoxication, 64/435
Electrotrauma, *see* Electrical injury
Electrovibrator
penis, 26/458
sexual function, 26/458
Elementary hallucination
*see also* Phosphene
occipital lobe, 45/367
Elephantiasis
fibrous dysplasia, 14/185
filariasis, 35/169, 52/513
neurofibromatosis, 14/185
Elevated scapula
congenital, 32/144
Sprengel deformity, 43/483
Elfin face
leprechaunism, 42/589
Williams syndrome, 43/259
Elfin face syndrome
appetite, 31/317
ataxia, 43/259
autopsy finding, 31/319
constipation, 31/317
craniosynostosis, 31/318
dental abnormality, 31/318, 43/259
enamel defect, 31/318
epicanthus, 31/318
familial leukodystrophy, 10/54
growth retardation, 31/318, 43/259
heart disease, 43/259
hypercalcemia, 31/317
hyperreflexia, 31/318
hypertelorism, 31/318
increased intracranial pressure, 31/318
laboratory finding, 31/318
mental deficiency, 31/318, 43/259
microcephaly, 43/259
muscular hypotonia, 31/318
pulmonary stenosis, 31/318
strabismus, 31/318
treatment, 31/319
vomiting, 31/317
Elliot Smit visuosensory band, *see* Visuosensory band (Elliot Smit)
Elliptocytosis
chorioretinal degeneration, 13/40
optic atrophy, 13/40
Stargardt disease, 13/133
Ellis-Van Creveld syndrome
birth incidence, 43/348
Coffin-Siris syndrome, 31/240
deafness, 14/788
dental abnormality, 43/347
Down syndrome, 31/427
founder effect, 43/348
growth retardation, 43/347
heart disease, 43/347
hydrocephalus, 43/348
mental deficiency, 43/348
nail hypoplasia, 43/347
polydactyly, 43/343
prevalence, 43/348
Ellsworth-Howard test
multiple nevoid basal cell carcinoma syndrome, 14/584
Elmustine
antineoplastic agent, 65/528
neurotoxin, 65/528
Elongated styloid process syndrome
*see also* Stylohyoid syndrome
clinical type, 48/503
history, 48/501
Elsberg, C.A., 1/5
Elsberg sign

neurofibromatosis, 20/260-263
spinal ependymoma, 20/370
spinal epidural tumor, 20/107
spinal neurinoma, 20/260
spinal tumor, 19/59
vertebral canal, 20/210
vertebral canal stenosis, 50/466
Elzholz body
Schwann cell, 51/10
Emaciation
Huntington chorea, 6/305
hydrozoa intoxication, 37/38
hypothalamus, 2/449
Embolic cerebrovascular disease
headache, 48/279
Embolic disease
headache, 48/280
Embolic striatum apoplexy
inferior type, 11/585
superior type, 11/585
Embolism
air, *see* Air embolism
anticoagulant, 38/146
atheromatous plaque, 11/437
basilar artery, 12/22
brain, *see* Brain embolism
brain air, *see* Brain air embolism
brain fat, *see* Brain fat embolism
brain hemorrhage, 11/585
brain infarction, 53/30
brain lacunar infarction, 54/239
brain thromboangiitis obliterans, 12/393, 55/313
cardiac valvular disease, 38/143-146
cerebrovascular disease, 12/450
chronic sinoatrial disorder, 39/264
fat, *see* Fat embolism
middle cerebral artery syndrome, 53/353
neurologic complication, 38/143, 146
ophthalmoscopy, 11/214
paradoxical, *see* Paradoxical embolism
pulmonary, *see* Pulmonary embolism
respiratory system, 26/315, 345, 394
retinal artery, *see* Retinal artery embolism
spinal cord, *see* Spinal cord embolism
spinal cord injury, 26/68, 207
subarachnoid hemorrhage, 12/170
Takayasu disease, 12/415
transient ischemic attack, 53/26
vertebral artery stenosis, 12/11
Embryo development
CSF, 32/529-531
Embryogenesis
choroid plexus, *see* Choroid plexus embryogenesis
congenital retinal blindness, 22/537
dysraphia, 32/187
spina bifida, 50/488
Embryonal brain tumor
World Health Organization classification, 67/3
Embryonal cell carcinoma
CNS, 68/235
imaging, 67/179
pathology, 68/235
Embryonic lipoma, *see* Hibernoma
Embryopathy syndrome
diabetes mellitus, 31/298
hyperphenylalaninemic, *see* Hyperphenylalaninemic embryopathy syndrome
Emerita analoga intoxication
paralytic shellfish poisoning, 37/56
saxitoxin, 37/56
Emery-Dreifuss muscular dystrophy
Adams-Stokes syndrome, 40/389
areflexia, 62/149
atrial flutter, 62/171
autosomal dominant, 62/155
Bethlem-Van Wijngaarden syndrome, 62/155
blood clotting factor VIII, 62/157
cardiomyopathy, 62/145, 152
carrier, 62/154
centronuclear, 62/156
chromosome Xq28, 62/156
clinical features, 40/390
contracture, 40/390, 43/88
diagnosis, 40/391
differential diagnosis, 62/154
Duchenne muscular dystrophy, 40/389
ECG, 62/153
EMG, 62/150
facioscapulohumeral muscular dystrophy, 62/155
genetic linkage, 43/88
genetics, 62/156
heart disease, 40/389, 43/88
heredofamilial myosclerosis, 62/173
heterozygote detection, 43/88
histopathology, 62/151
history, 62/146
incidence, 40/390
joint fixation, 40/390
limb girdle syndrome, 62/189
mild form, 62/149
muscle atrophy distribution, 62/148
muscle contraction, 62/145
muscle fiber type I atrophy, 62/151
myosclerosis, 40/392
neutropenia, 62/156
preclinical detection, 43/88

prognosis, 62/154
proximal muscle weakness, 43/88
rigid spine syndrome, 40/390, 62/155
scapuloperoneal syndrome, 40/392, 426, 62/171
scapuloperoneal weakness, 62/171
severe form, 62/150
single fiber electromyography, 62/150
symptom, 62/146
syncope, 62/153
terminology, 40/389
transient ischemic attack, 62/154
treatment, 40/391
winged scapula, 62/146
X-linked muscular dystrophy, 40/389, 62/117
X-linked recessive, 62/154
Emery-Dreifuss syndrome, *see* Emery-Dreifuss muscular dystrophy
Emetine
cerebral amebiasis, 52/328
experimental myopathy, 40/157
nonprimary inflammatory myopathy, 62/372
toxic myopathy, 62/600
Emetine intoxication
cytoplasmic inclusion body myopathy, 62/345
EMG
abnormal acetylcholine-acetylcholine receptor interaction syndrome, 62/437
acetonylacetone intoxication, 64/87
acid maltase deficiency, 62/480
acromegalic myopathy, 62/538
ACTH induced myopathy, 62/537
acute hemorrhagic conjunctivitis, 56/352
acute polymyositis, 62/71
adrenomyeloneuropathy, 60/171
alcoholic polyneuropathy, 51/315, 318
almitrine dimesilate polyneuropathy, 51/308
amiodarone polyneuropathy, 51/303
amnesic shellfish poisoning, 65/167
amyloidosis, 71/505
amyotrophic lateral sclerosis, 1/639, 22/297-305, 59/185
arsenic intoxication, 36/204, 64/287
arsenic polyneuropathy, 51/267
athetosis, 1/647
audiogenic startle syndrome, 49/620
auditory hallucination, 45/356
ballismus, 6/484, 49/76
Bassen-Kornzweig syndrome, 51/394, 63/275
beta adrenergic receptor blocking agent intoxication, 65/437
bismuth intoxication, 64/338
botulism, 51/189, 65/212
brachial plexus, 51/144, 148
brain death, 24/776, 778
carbon disulfide intoxication, 64/25
carbon disulfide polyneuropathy, 51/279
cauda equina tumor, 19/80
central sleep apnea, 63/460
Charlevoix-Saguenay spastic ataxia, 60/454
childhood myoglobinuria, 62/561
chloroquine myopathy, 62/608
chorea, 1/647, 49/500
chorea-acanthocytosis, 49/329, 60/141, 63/281
chronic inflammatory demyelinating polyradiculoneuropathy, 51/536
chronic progressive external ophthalmoplegia, 62/70
ciclosporin intoxication, 65/554
ciguatoxin intoxication, 65/163
cisplatin polyneuropathy, 51/300
compression neuropathy, 51/86
congenital acetylcholine receptor deficiency/short channel opentime syndrome, 62/437
congenital ε-acetylcholine receptor subunit mutation syndrome, 62/436
congenital end plate acetylcholinesterase deficiency, 62/430
congenital familial limb girdle myasthenia, 62/439
congenital high conductance fast channel syndrome, 62/438
congenital myopathy, 62/69, 333, 350
congenital slow channel syndrome, 62/432
congenital synaptic vesical paucity syndrome, 62/429
conus medullaris tumor, 19/80
critical illness polyneuropathy, 51/575, 581
cryoglobulinemia, 63/404
cryoglobulinemic polyneuropathy, 51/434
debrancher deficiency, 62/482
decremental, *see* Decremental EMG
deficiency neuropathy, 51/326
dermatomyositis, 1/642, 62/375, 71/139
diabetic neuropathy, 8/43
diphtheria, 65/240
discomplete spinal cord injury, 61/45
distal myopathy, 40/472
disulfiram polyneuropathy, 51/309
Duchenne muscular dystrophy, 40/373
dystonia, 49/72
dystonia musculorum deformans, 6/532, 49/523
Eaton-Lambert myasthenic syndrome, 62/78, 422
electrolyte disorder, 63/557
Emery-Dreifuss muscular dystrophy, 62/150
endocrine myopathy, 62/69
eosinophilia myalgia syndrome, 63/377, 64/252
eosinophilic polymyositis, 63/385

ethambutol polyneuropathy, 51/295
ethylene oxide polyneuropathy, 51/275
facial paralysis, 8/268, 281
facioscapulohumeral muscular dystrophy, 40/420, 463, 62/166
familial hypokalemic periodic paralysis, 62/69
familial infantile myasthenia, 62/427
fibrillation, 19/276, 40/143
flaccid paraplegia, 1/224
freezing gait, 49/69
Friedreich ataxia, 21/350, 51/388, 60/320
Gerstmann-Sträussler-Scheinker disease, 56/554
gold intoxication, 64/356
Guillain-Barré syndrome, 1/638
headache, 48/349
hereditary amyloid polyneuropathy, 21/126
hereditary diurnal fluctuating juvenile dystonia, 49/533
hereditary essential myoclonus, 49/619
hereditary motor and sensory neuropathy type I, 1/641, 21/287
hereditary motor and sensory neuropathy type II, 1/641, 21/287
hereditary multiple recurrent mononeuropathy, 51/556
hereditary periodic ataxia, 60/438-441
hereditary sensory and autonomic neuropathy type III, 21/110, 60/28
hereditary tremor, 49/568, 572, 586
herpes zoster, 51/181
Hirayama disease, 59/110
Huntington chorea, 49/73
hydrocephalus, 32/548-550
hyperaldosteronism myopathy, 39/494, 62/536
hypereosinophilic syndrome, 63/373
hyperkalemic periodic paralysis, 28/595
hyperparathyroid myopathy, 62/540
hyperthyroidism, 1/642
hypokalemic periodic paralysis, 28/583, 62/459
hypomagnesemia, 28/547
hypoparathyroidism, 1/642, 27/300
hypophosphatemia, 63/565
hypothyroid myopathy, 62/533
hypothyroid polyneuropathy, 51/515
IgM paraproteinemic polyneuropathy, 51/437
inclusion body myositis, 62/375
infantile spinal muscular atrophy, 22/74, 93-95, 59/56
intention tremor, 49/590
intervertebral disc prolapse, 20/580, 25/500
ionizing radiation neuropathy, 51/138
Isaacs syndrome, 40/330
Kawasaki syndrome, 56/639
Kii Peninsula amyotrophic lateral sclerosis, 22/371
Klippel-Feil syndrome, 32/119
lead polyneuropathy, 51/270
legionellosis, 52/256
leprous neuritis, 51/221
limb girdle syndrome, 40/453, 62/186
lithium intoxication, 64/296
lower motoneuron lesion, 1/636-638
lower urinary tract, 61/298
lumbar intervertebral disc prolapse, 20/580
lumbar vertebral canal stenosis, 20/723
Marin Amat syndrome, 42/320
marine toxin intoxication, 65/153
Marinesco-Sjögren syndrome, 21/556
McLeod phenotype, 63/287
mercury polyneuropathy, 51/273
metabolic myopathy, 62/69
methanol intoxication, 64/99
1-methyl-4-phenyl-1,2,3,6-tetrahydropyridine intoxication, 65/372
Minamata disease, 64/415
motor unit, 19/275, 40/138, 62/49
multiple myeloma polyneuropathy, 51/431, 63/393
muscle biopsy, 62/124
muscle contraction, 62/52
muscle contraction headache, 48/78
muscle fibrillation, 1/636
muscular dystrophy, 1/641
myasthenia gravis, 1/639, 644, 41/96, 127, 62/398, 413
myasthenic syndrome, 62/78
myoclonus, 1/646, 49/610
myokinesia, 1/646
myokymia, 40/330, 546
myopathy, 62/49
myophosphorylase deficiency, 27/233, 41/187, 62/485
myotonia, 1/641-643, 40/327, 533, 535, 41/298
myotonia congenita, 40/539, 62/70
myotonic dystrophy, 40/519, 62/70
myotubular myopathy, 62/69
Nelson syndrome, 62/537
nerve conduction, 1/640
neuralgic amyotrophy, 51/172
neuroleptic parkinsonism, 65/293
neuromuscular transmission, 1/632, 644
nitrofurantoin polyneuropathy, 51/296
nocturnal myoclonus, 45/139
organic solvent intoxication, 64/41
organophosphate induced delayed polyneuropathy, 64/176
paralytic shellfish poisoning, 65/153

paraneoplastic polyneuropathy, 51/467
Parkinson disease, 6/195, 49/102
pellagra, 7/573
penis erection, 61/313
perhexiline maleate polyneuropathy, 51/301
peripheral neurolymphomatosis, 56/182
phosphofructokinase deficiency, 40/327, 41/191, 62/487
physiologic tremor, 49/584
pituitary gigantism myopathy, 41/249, 62/538
POEMS syndrome, 51/431, 63/394
poliomyelitis, 1/638
polyarteritis nodosa, 39/307
polymyositis, 1/642, 19/87, 22/25, 38/485, 41/60, 62/187, 375, 71/139
polyneuritis, 1/639
positive wave, 40/143
primary lateral sclerosis, 59/172
progressive bulbar palsy, 22/113
progressive external ophthalmoplegia, 22/181, 62/295
progressive muscular atrophy, 59/16
progressive spinal muscular atrophy, 22/33-38
progressive systemic sclerosis, 51/448
quadriceps myopathy, 40/440, 43/128, 62/174
reflex myoclonus, 49/610
Refsum disease, 8/23, 21/198
renal insufficiency polyneuropathy, 51/358
Rett syndrome, 29/317
rheumatoid arthritis, 38/485-487, 51/447
rigid spine syndrome, 62/173
rigidity, 1/646
rimmed vacuole distal myopathy, 62/204
Roussy-Lévy syndrome, 21/176, 42/108
rubral tremor, 49/587
scapulohumeral spinal muscular atrophy, 62/171
scapuloperoneal muscular dystrophy, 42/99, 62/169
scapuloperoneal spinal muscular atrophy, 59/45, 62/170
Sjögren syndrome, 71/65
skeletal fluorosis, 36/476
snake envenomation, 65/179
spasmodic torticollis, 1/648, 6/569-577
spasticity, 1/646
spina bifida, 32/548
spinal nerve root injury, 25/417
spinal tumor, 19/66, 267-286
steroid myopathy, 62/535
Sydenham chorea, 6/418-421, 49/73
syringomyelia, 1/639, 32/277, 50/445
tetanus, 1/643, 65/219
tetany, 1/643, 646, 63/560
tetrodotoxin intoxication, 65/156
thallium intoxication, 36/253
thoracic outlet syndrome, 51/126
thyrotoxic myopathy, 62/529
Tolosa-Hunt syndrome, 48/299
toxic oil syndrome, 63/381
tremor, 1/647, 49/571
trichinosis, 35/277, 52/568
trigemino-oculomotor synkinesis, 42/320
undetermined significant monoclonal gammopathy, 63/399
uremic encephalopathy, 63/505
uremic neuropathy, 7/166, 27/340
uremic polyneuropathy, 51/356, 361
urinary tract, 61/298
vertebral metastasis, 20/420
vincristine intoxication, 65/535
vitamin $B_1$ deficiency, 51/332
vitamin $B_{12}$ deficiency, 51/340
vitamin E deficiency, 70/445
Waldenström macroglobulinemia, 63/397
Waldenström macroglobulinemia polyneuropathy, 51/432
Whipple disease, 51/185
Wohlfart-Kugelberg-Welander disease, 22/34, 71
xeroderma pigmentosum, 51/399
EMI, *see* Computer assisted tomography
Eminentia saccularis
blood-brain barrier, 56/80
Emissary foramen
condyloid, *see* Condyloid emissary foramen
mastoid, *see* Mastoid emissary foramen
parietal, *see* Parietal emissary foramen
Emissary vein
foramina parietalia permagna, 42/28
Emotion
*see also* Emotional disorder
activation theory, 3/321
affect, 3/318, 333, 335-337
aggression, 3/337, 344
amusia, 4/197
amygdala, 2/440, 719, 3/324, 329
amygdalectomy, 45/276
amygdaloid nucleus, 45/273, 275
anger, 3/336
anxiety, 3/336, 348
arousal reaction, 3/321
autonomic nervous system, 3/320, 74/79, 369, 75/439
basic, 74/369
behavior, 2/449
Binswanger disease, 46/320
body scheme disorder, 45/380

brain amine, 3/323
brain command system, 45/273
brain edema, 3/323
brain substrate, 45/277
Cannon-Bard theory, 3/319
cataplexy, 42/711
cerebral activation, 3/321-323
change, 45/271-280
clonidine intoxication, 37/443
communication, 3/332
corticotropin, 45/279
cybernetics, 3/325-332
delirium, 46/531
depression, 3/336, 348
development, 3/332-335, 4/356-360
drive, 3/329, 332
ergotropic reaction, 3/320
ergotropic system, 3/320
experience, 3/332
expression, 3/320, 337, 356, 358
feedback, 3/317, 322, 325
feeling, 3/320
frontal cortex, 3/324, 329
frontal lobe, 45/273
grief, 3/327
hypothalamus, 2/440, 715, 719, 3/319, 45/273-275, 277
input control, 3/325, 327
James-Lange theory, 3/319
levodopa intoxication, 46/593
limbic system, 3/320, 329, 45/272, 276-278
manganese intoxication, 64/308
mania, 3/338
medial thalamic nucleus, 2/485
memory, 3/328, 332
migraine, 48/60, 156, 161
motivation, 3/329
musicogenic epilepsy, 15/446
myotonia congenita, 62/264
neural activation, 3/321-323
neurochemistry, 45/278
neurologic model, 3/323, 339
neuronal configuration, 3/322
neuropsychology, 45/516
odorant, 74/370, 372
olfaction, 74/363
organization, 3/317
orienting reaction, 3/322
pain system, 45/275
panic, 3/337
Papez-MacLean theory, 3/320
participatory processes, 3/318, 325-327, 330
prefrontal syndrome, 2/751
preparatory processes, 3/318, 325-327, 330
psychiatric headache, 5/20
psychodynamics, 3/332, 339
psychophysiological aspect, 3/316-339
psychosurgery, 45/276
Raynaud disease, 8/321, 22/261, 43/70
reinforcement, 3/327-329
repression, 3/339
reticular formation, 3/322
sensory deprivation, 3/334
septal area, 2/770
strain, 45/250
syncope, 15/824, 22/263, 75/210
temporal lobe epilepsy, 2/706, 45/274
temporal lobectomy, 45/276
test-operate test-exit unit, 3/317, 329, 381
thalamus, 3/319
theory, 3/316, 317, 322, 338
time sense, 3/330
trophotropic system, 3/320
visceral brain, 2/715, 719, 3/320
visceral glandular reaction, 3/319
visceral theory, 3/319, 322
Emotional aura
temporal lobe epilepsy, 73/72
Emotional disorder
*see also* Emotion
aggression, 3/344-346
amygdala lesion, 3/347
anatomic localization, 3/345-347, 350, 361-363
anxiety, 3/348
biological rhythm, 74/505
brain embolism, 53/206
brain infarction, 53/206
centrofacial lentiginosis, 14/782
cerebrovascular disease, 53/206
congenital heart disease, 63/2
corpus callosum syndrome, 2/761
depression, 3/348
depression etiology, 3/348
developmental dysphasia, 46/141
dysphrenia hemicranica, 5/79
emotional lability, 3/357
epilepsy, 3/344, 348, 358, 363
ethiology, 3/348
extrapyramidal disorder, 3/358
facial expression, 3/356, 358
faciorespiratory mechanism, 3/362
fou rire prodromique, 3/358, 362
gyrus cinguli lesion, 3/346
hereditary sensory and autonomic neuropathy type III, 1/475, 43/58, 60/28
hippocampal lesion, 3/347

hyperkinetic child syndrome, 43/201
hypothalamus, 74/505
intracranial hemorrhage, 53/206
Klüver-Bucy syndrome, 3/346, 349-356
kuru, 42/657
limbic system, 3/343-363
Mast syndrome, 42/282
metachromatic leukodystrophy, 42/493
multiple sclerosis, 47/178
pathologic laughing and crying, 3/356-363, 55/142
pathologic rage, 3/344
pathophysiology, 3/343-363
Pick disease, 46/235
placidity, 3/346, 352
pseudobulbar paralysis, 3/358
psychomotor epilepsy, 3/344
psychosurgery, 3/347
rabies, 3/348
risus sardonicus, 3/357
sham rage, 3/345
temporal lobe electrical stimulation, 3/344
temporal lobe epilepsy, 2/716
temporal lobe lesion, 3/346
temporal lobe tumor, 3/344, 348
temporal lobectomy, 3/344, 346
transient ischemic attack, 53/206
Witzelsucht, 3/357
Emotional stress
migraine, 48/85, 355
multiple sclerosis, 9/120, 47/153
muscle contraction headache, 48/355
transient global amnesia, 45/209, 245
Emphysema
headache, 48/6
orbital injury, 24/90
Empty delta sign
asparaginase, 69/495
superior longitudinal sinus thrombosis, 52/179
Empty sella
radiation injury, 23/654
Empty sella syndrome
CSF, 17/409, 431
CSF fistula, 17/431
endocrine abnormality, 17/409
epidural arachnoid cyst, 32/397
headache, 48/432
hypopituitarism, 17/431
PEG, 17/409, 431
visual impairment, 17/431
Empyema
cranial epidural, *see* Cranial epidural empyema
cranial subdural, *see* Cranial subdural empyema
epidural, *see* Epidural empyema
nocardiosis, 35/524
spinal epidural, *see* Spinal epidural empyema
spinal subdural, *see* Spinal subdural empyema
subdural, *see* Subdural empyema
En chapelet malacia
bilateral pontine syndrome, 2/241
paramedian pontine artery, 2/241
paraplegia, 2/241
Enalapril
dipeptidyl carboxypeptidase I inhibitor, 65/444
toxic myopathy, 62/597
Enalapril intoxication
deafness, 65/446
dysgeusia, 65/446
Huntington chorea, 65/446
parkinsonism, 65/446
Enamel defect
dental abnormality, 42/691
elfin face syndrome, 31/318
epilepsy, 42/691
hypoparathyroidism, 42/577
kernicterus, 42/583
pseudohypoparathyroidism, 42/621
yellow teeth, 42/691
Enamel hypoplasia
alaninuria, 42/515
hyperphenylalaninemia, 42/611
phenylketonuria, 42/611
Encainide
flecainide intoxication, 65/455
iatrogenic neurological disease, 65/456
Encainide intoxication
headache, 65/456
insomnia, 65/456
vertigo, 65/456
Encéphale isolé
*see also* Cerveau isolé
animal experiment, 2/280, 3/49, 81
cat, 2/280, 3/49, 81
EEG, 1/73
reticular formation, 2/280
transection, 2/280
wakefulness, 2/280
Encephalitic syndrome
post, *see* Postencephalitic syndrome
Encephalitis
*see also* Encephalomyelitis, Encephalopathy *and* Meningoencephalitis
abacterial, *see* Viral encephalitis
Abt-Letterer-Siwe disease, 42/441
acquired toxoplasmosis, 35/122, 52/352, 354, 357
acute viral, *see* Acute viral encephalitis

adenovirus, *see* Adenovirus encephalitis
African trypanosomiasis, 52/342
allergic, *see* Allergic encephalitis
amebic, *see* Granulomatous amebic encephalitis
angiostrongyliasis, 52/548
animal viral disease, 56/588
Argentinian hemorrhagic fever, 56/365
Aspergillus fumigatus, 63/532
ataxic respiration, 1/498
bacterial, *see* Bacterial encephalitis
Behçet syndrome, 16/219
Bickerstaff, *see* Brain stem encephalitis
brain hemorrhage, 31/166
brain stem, *see* Brain stem encephalitis
bulbar, *see* Bulbar encephalitis
Burkitt lymphoma, 63/352
California, *see* California encephalitis
California virus, *see* California encephalitis virus
canine distemper, 34/293
caprine arthritis, *see* Caprine arthritis encephalitis
cat scratch disease, 52/128
catatonia, 46/423
central eastern tick-borne, *see* Central European tick-borne encephalitis
central European tick-borne, *see* Central European tick-borne encephalitis
central neurogenic hyperventilation, 63/441
central sleep apnea, 63/462
cerebral amebiasis, 52/315
Cervós-Navarro, *see* Cervós-Navarro encephalitis
chorea, 6/438
chronic progressive subcortical, *see* Binswanger disease
classification, 34/63
consciousness, 45/119
corticostriate, *see* Corticostriate encephalitis
Cryptococcus neoformans, 63/532
cytomegalovirus, *see* Cytomegalovirus encephalitis
dementia, 46/128, 130, 388
demyelinating, *see* Demyelinating encephalitis
differential diagnosis, 53/237
diphtheria, 51/182, 65/240
drechsleriasis, 52/486
dystonia, 6/559
eastern, *see* Eastern encephalitis
eastern equine, *see* Eastern equine encephalitis
EEG, 46/493
enterovirus, 56/327
epidemic, *see* Encephalitis lethargica
Epstein-Barr virus, *see* Epstein-Barr virus encephalitis
Epstein-Barr virus infection, 56/252
equine, *see* Equine encephalitis
Far Eastern tick-borne, *see* Far Eastern tick-borne encephalitis
febrile status epilepticus, 15/179
granulomatous, *see* Granulomatous encephalitis
granulomatous amebic, *see* Granulomatous amebic encephalitis
headache, 48/6
hemorrhagic, *see* Hemorrhagic encephalitis
herpes B, 16/219
herpes simplex virus, *see* Herpes simplex virus encephalitis
herpes zoster, 34/173-175, 56/230, 71/268
human immunodeficiency virus infection, 71/268
immunocompromise, 56/477, 65/548
immunocompromised host, 56/474, 477
inclusion body, *see* Inclusion body encephalitis
infectious mononucleosis, 34/188, 56/252
infectious striatopallidodentate calcification, 49/417
intracranial hypertension, 55/215
intractable hiccup, 63/490
Japanese, *see* Japanese encephalitis
Japanese B, *see* Japanese B encephalitis
La Cross virus, *see* La Cross virus encephalitis
Lassa fever, 56/16
leptospirosis, 33/403
lethargy, *see* Encephalitis lethargica
leukemia, 63/352
limbic, *see* Limbic encephalitis
Listeria monocytogenes, *see* Listeria monocytogenes encephalitis
listeriosis, 52/90
louping ill, *see* Louping ill encephalitis
Lyme disease, 51/204, 208, 52/262
lymphocytic leukemia, 63/352
macrophage defense defect, 65/548
malignancy, 8/134
measles, *see* Measles encephalitis
meningococcal meningitis, 33/27, 52/24
microcephaly, 50/280
microglia, *see* Acute disseminated encephalomyelitis
moose, *see* Moose encephalitis
mouse, *see* Murine encephalitis
mucormycosis, 52/471
multifocal demyelination, 52/342, 357
mumps, *see* Mumps encephalitis
Murray Valley, *see* Murray Valley encephalitis
myeloid leukemia, 63/352
myoclonia, 6/764
myxovirus, 34/79
narcolepsy, 2/448

necrotizing, *see* Necrotizing encephalitis
neuroborreliosis, 51/204, 208
neurobrucellosis, 33/312-314, 52/587
neurologic intensive care, 55/215
neutropenia, 63/339
noncongenital rubella, *see* Noncongenital rubella encephalitis
opsoclonus, 34/613, 60/657
palatal myoclonus, 38/584
paramyxovirus, 56/14
peduncular hallucination, 45/354
perivenous, *see* Acute disseminated encephalomyelitis
phagocyte defense defect, 65/548
poliomyelitis, 56/327
postvaccinal, *see* Postvaccinal encephalitis
progressive multifocal, *see* Progressive multifocal encephalitis
progressive subacute, *see* Progressive subacute encephalitis
rabies, *see* Rabies encephalitis
Rasmussen chronic, *see* Rasmussen chronic encephalitis
rehabilitation, 46/611
renal transplantation, 63/532
Reye syndrome, 55/215, 56/159
Rickettsia, *see* Rickettsia encephalitis
rubella, *see* Rubella encephalitis
Russian spring-summer, *see* Central European tick-borne encephalitis
schistosomiasis, 52/538
segmental dystonia, 49/541
spheroid body, 6/625
spongiform, *see* Spongiform encephalitis
St. Louis, *see* St. Louis encephalitis
subacute, *see* Subacute encephalitis
subarachnoid hemorrhage, 12/170
symptomatic dystonia, 49/541
T-lymphocyte, 65/548
tick-borne, *see* Tick-borne encephalitis
toxic, *see* Toxic encephalitis
toxoplasma, *see* Toxoplasma encephalitis
toxoplasmosis, 52/352
trichinosis, 52/565, 569
trichophytosis, 35/571, 52/497
varicella zoster virus, *see* Varicella zoster virus encephalitis
Venezuelan equine, *see* Venezuelan equine encephalitis
verruga peruana, 34/664
viral, *see* Viral encephalitis
West Nile, *see* West Nile encephalitis
western, *see* Western encephalitis
western equine, *see* Western equine encephalitis
Encephalitis hemorrhagica superior
pseudo, *see* Wernicke encephalopathy
Encephalitis lethargica, 34/451-457
*see also* Epidemic encephalitis
action myoclonus, 6/767
acute cerebellar ataxia, 34/623
alteration in consciousness, 34/455
arousal disorder, 45/108
behavior disorder, 34/455
brain death, 3/67
cholestanol, 66/16
clinical features, 34/454-456
consciousness, 3/49
cranial nerve disturbance, 34/455
CSF, 34/456
dementia, 46/388
dystonia, 6/558
etiology, 34/452
extrapyramidal disorder, 34/455
history, 34/451
influenza, 34/452
mode of onset, 34/454
myoclonia, 6/763
ophthalmoplegia, 22/180
Parkinson disease, 49/89, 124
parkinsonism, 34/455
pathology, 34/453
postencephalitic parkinsonism, 22/223, 34/455
progressive supranuclear palsy, 22/225
schizophrenia, 46/455
schizophreniform psychosis, 46/420
sleep disorder, 3/80
symptomatic dystonia, 6/558
thalamus, 2/486
treatment, 34/456
Encephalitis periaxialis concentrica, *see* Concentric sclerosis
Encephalitis periaxialis diffusa, *see* Subacute sclerosing panencephalitis
Encephalitis periaxialis diffusa (Schilder), *see* Subacute sclerosing panencephalitis
Encephalitis postvaccination, *see* Acute disseminated encephalomyelitis
Encephalitogenic factor, *see* Encephalitogenic myelin basic protein
Encephalitogenic myelin basic protein
acute disseminated encephalomyelitis, 9/501
amino acid, 9/507-512
component, 47/446
concept, 9/511
difference from other CNS antigen, 9/514
microheterogenicity, 9/511

molecular weight, 9/511
multiple sclerosis, 9/133
oligodendrocyte, 9/511
peptide, 9/509-511
Encephalitogenicity
collagen like protein, 9/516
Encephalocele
*see also* Cephalocele, Encephalomeningocele *and* Iniencephaly
amniotic band syndrome, 30/116
basal, *see* Basal encephalocele
brain cortex disorganization, 50/249
brain stem, *see* Brain stem encephalocele
cause, 50/28
cerebellar agenesis, 50/177
CNS malformation, 50/57
corpus callosum agenesis, 50/150
craniolacunia, 30/272, 50/137, 139
definition, 50/97
epilepsy, 72/109
exencephaly, 50/71, 73
familial hydrocephalus, 50/289
holoprosencephaly, 50/235-237
hydrocephalus, 50/286
hydromyelia, 50/426
hyperencephalus, 50/73
incidence, 50/58
iniencephalus apertus, 50/129
iniencephaly, *see* Iniencephalus clausus
median facial cleft syndrome, 46/90
notencephalus, 50/73
occipital, *see* Occipital encephalocele
optic nerve agenesis, 50/211
orbital, *see* Orbital encephalocele
orbital tumor, 17/202
parietal, *see* Parietal encephalocele
periventricular calcification, 50/258
podencephaly, 50/73
porencephaly, 30/682, 50/356
proencephalus, 50/73
sincipital, *see* Sincipital encephalocele
triploidy, 43/566
Encephaloclastic cavity
corpus callosum agenesis, 50/155
Encephaloclastic porencephaly
definition, 50/356
Encephalocraniocutaneous lipomatosis, *see* Krabbe-Bartels disease
Encephalofacial angiomatosis, *see* Sturge-Weber syndrome
Encephalomalacia
adenovirus encephalitis, 56/286
aorta aneurysm, 63/50
concentric sclerosis, 9/440
cystic, *see* Cystic encephalomalacia
electrical injury, 61/192
epilepsy, 72/363
multicystic, *see* Multicystic encephalomalacia
multiple cystic, *see* Multiple cystic encephalopathy
pallidonigral, *see* Pallidonigral encephalomalacia
Pena-Shokeir syndrome type I, 43/438
perivenous, *see* Perivenous encephalomalacia
spheroid body, 6/626
symmetrical, *see* Symmetrical encephalomalacia
Encephalomeningitis, *see* Meningoencephalitis
Encephalomeningocele, 42/26-28
*see also* Encephalocele *and* Meningocele
anencephaly, 30/187
aqueduct stenosis, 42/26
Arnold-Chiari malformation type I, 42/26
birth incidence, 42/27
blindness, 42/26
cerebellar vermis agenesis, 42/4
corpus callosum agenesis, 42/7, 26
cryptophthalmos syndrome, 43/374
Dandy-Walker syndrome, 42/26
α-fetoprotein, 42/28
Gruber syndrome, 31/242
hydrocephalus, 42/27
hydromyelia, 42/26
hypertelorism, 42/27
hypothalamic dysfunction, 42/27
Klippel-Feil syndrome, 42/26
mental deficiency, 42/26
neuronal heterotopia, 42/26
optic nerve coloboma, 42/27
18p syndrome, 31/582
prenatal diagnosis, 42/28
recurrence risk, 42/27
Encephalomeningomyelocele
thanatophoric dwarfism, 43/388
Encephalomyelitis
*see also* Encephalitis *and* Encephalopathy
acute disseminated, *see* Acute disseminated encephalomyelitis
acute experimental allergic, *see* Acute experimental allergic encephalomyelitis
acute postinfectious disseminated, *see* Acute postinfectious disseminated encephalomyelitis
allergic, *see* Allergic encephalomyelitis
Behçet syndrome, 63/55
brain hemorrhage, 31/167
canine distemper, *see* Canine distemper encephalomyelitis
carcinoma, 38/631-647

chronic relapsing experimental allergic, *see* Chronic relapsing experimental allergic encephalomyelitis
Coxsackie virus infection, 34/136
dementia, 46/389
demyelinating, *see* Demyelinating encephalomyelitis
disseminated, *see* Disseminated encephalomyelitis
eastern equine, *see* Eastern equine encephalomyelitis
ECHO virus infection, 34/140
enterovirus infection, 34/140
eosinophilic, *see* Eosinophilic encephalomyelitis
exanthematous, *see* Exanthematous encephalomyelitis
experimental allergic, *see* Experimental allergic encephalomyelitis
experimental autoimmune, *see* Experimental autoimmune encephalomyelitis
hyperacute disseminated, *see* Hyperacute disseminated encephalomyelitis
immune mediated, *see* Immune mediated encephalomyelitis
influenza virus, 56/168
JHM virus, *see* JHM virus encephalomyelitis
leptospirosis, 33/402
Listeria monocytogenes, *see* Listeria monocytogenes encephalomyelitis
lymphocytic choriomeningitis, 9/194, 56/360
malignancy, 8/134, 38/631-644
meningococcal meningitis, 52/26
neuromyelitis optica, 9/426, 431
parainfectious, *see* Acute disseminated encephalomyelitis
paraneoplastic, *see* Paraneoplastic encephalomyelitis
periodic ataxia, 21/570
postinfectious, *see* Acute disseminated encephalomyelitis *and* Postinfectious encephalomyelitis
postvaccinal measles, *see* Postvaccinal measles encephalomyelitis
postvaccinial, *see* Postvaccinial encephalomyelitis
postviral, *see* Postviral encephalomyelitis
rabies postvaccinial, *see* Rabies postvaccinial encephalomyelitis
rubella, *see* Rubella encephalomyelitis
schistosomiasis, 52/538
sulfanilamide, 63/55
Theiler virus, *see* Theiler virus encephalomyelitis
Encephalomyelitis periaxialis scleroticans
Marburg description, 9/469, 47/419
Encephalomyelitis virus
pig, 56/33
Encephalomyelopathy
arsanilic acid, 9/648
arsenic intoxication, 9/648
arsphenamine, 9/648
cytarabine intoxication, 65/533
methotrexate, 47/572
necrotizing, *see* Necrotizing encephalomyelopathy
spongiform, *see* Spongiform encephalomyelopathy
subacute necrotizing, *see* Subacute necrotizing encephalomyelopathy
tryparsamide, 9/648
Encephalomyocarditis virus
deafness, 56/107
experimental acute viral myositis, 56/195
picornavirus, 34/5, 140, 56/195, 62/381
Reye syndrome, 56/239
vertigo, 56/107
Encephalomyopathy
childhood, *see* Childhood encephalomyopathy
combined complex I-V deficiency, 62/506
complex II deficiency, 62/504, 66/422
complex III deficiency, 62/504
complex IV deficiency, 66/423
ECG abnormality, 60/670
fatal infantile, *see* Fatal infantile encephalomyopathy
mitochondrial, *see* Mitochondrial encephalomyopathy
mitochondrial DNA point mutation, 66/427
oxoglutarate dehydrogenase deficiency, 62/503
progressive, *see* Progressive encephalomyopathy
ubidecarenone deficiency, 66/423
Encephalopathia pugilistica, *see* Boxer dementia *and* Boxer encephalopathy
Encephalopathia saturnina, *see* Lead encephalopathy
Encephalopathy
*see also* Encephalitis, Encephalomyelitis *and* Leukoencephalopathy
acivicin intoxication, 65/538
acquired immune deficiency syndrome, *see* Acquired immune deficiency syndrome dementia
acrylamide polyneuropathy, 51/281
acute, *see* Acute encephalopathy
acute childhood, *see* Reye syndrome
acute hypertensive, *see* Acute hypertensive encephalopathy
acute intermittent porphyria, 27/456
acute metabolic, *see* Acute metabolic encephalopathy

acute toxic, *see* Reye syndrome
adrenoleukodystrophy, 45/119
African trypanosomiasis, 52/343
alcoholic, *see* Alcoholic encephalopathy
altretamine, 65/528
altretamine intoxication, 65/534
aluminum, *see* Aluminum encephalopathy
aluminum intoxication, 21/57
Alzheimer neurofibrillary change, 21/57
amanitin intoxication, 65/38
aminoaciduria, 29/233
aminoglycoside, 46/602
animal viral disease, 34/297
anoxic, *see* Anoxic encephalopathy
antiepileptic agent, 65/498
antineoplastic agent, 46/600
arsenic, *see* Arsenic encephalopathy
arsenic polyneuropathy, 51/266
arsphenamine, *see* Arsphenamine encephalopathy
asparaginase, 63/359, 65/529
asparaginase intoxication, 65/538
azacitidine, 65/528
Behçet syndrome, 45/119
bilirubin, *see* Bilirubin encephalopathy
Binswanger disease, 42/278, 49/471
bismuth, *see* Bismuth encephalopathy
bismuth intoxication, 36/327-329, 64/331
bluetongue virus, 34/297
boxer, *see* Boxer encephalopathy
brain abscess, 52/400
brain death, 24/702
brain infarction, 11/314, 63/345
brain metastasis, 45/119
Burkitt lymphoma, 63/352
cancer, 69/396
candidiasis, 52/400
carbon disulfide intoxication, 64/25
carbon monoxide, *see* Carbon monoxide encephalopathy
carcinomatous, *see* Carcinomatous encephalopathy
cardiac arrest, *see* Cardiac arrest encephalopathy
carmustine, 65/528
carnitine deficiency, 43/175
cat scratch disease, 52/129
cerebellar, *see* Cerebellar encephalopathy
Chagas disease, 52/347
chemotherapy, 67/355
childhood myoglobinuria, 62/561
chloramphenicol, 46/602
chlormethine, 46/601, 65/528
ciclosporin, 63/537
ciclosporin intoxication, 65/551
cisplatin, 65/529
cisplatin polyneuropathy, 51/300
classification, 69/396
coccidioidomycosis, 52/413
collagen disease, 45/118
complex IV deficiency, 62/505
congenital tactile agnosia, 4/34
consciousness, 45/117
coronary bypass surgery, 63/187
corticosteroid, 63/359, 65/529
Crome syndrome, 43/242
cryoglobulinemia, 63/404
cyanide, *see* Cyanide encephalopathy
cytarabine, 63/356, 65/528, 67/362
cytidine analog, 67/362
deanimation, *see* Brain death
deferoxamine, 63/534
deficiency, 7/558, 561, 578, 633
diabetes mellitus, 27/123-136
diabetic ketoacidosis, 45/122
dialysis, *see* Dialysis encephalopathy
diamorphine, *see* Diamorphine encephalopathy
diphtheria, 33/485
diphtheria pertussis tetanus vaccine, 52/243, 245
doxorubicin, 65/528
drug induced, 69/482
dyssynergia cerebellaris myoclonica, 60/600
early infantile, *see* Early infantile encephalopathy
early infantile epileptic, *see* Early infantile epileptic encephalopathy
early myoclonic, *see* Early myoclonic encephalopathy
epileptic, *see* Epileptic encephalopathy
epileptogenic, *see* Epileptogenic encephalopathy
episodic, *see* Episodic encephalopathy
ethylene oxide, *see* Ethylene oxide encephalopathy
ethylene oxide polyneuropathy, 51/275
etoposide, 65/528
filariasis, 35/165, 167, 170, 52/514, 517
fludarabine, 65/528
fluorouracil, 46/600, 65/528, 67/363
gold, *see* Gold encephalopathy
granulomatous amebic encephalitis, 52/317
Hashimoto, *see* Hashimoto encephalopathy
heart transplantation, 63/180
hemochromatosis, 42/553
hemorrhagic, *see* Hemorrhagic encephalopathy
hepatic, *see* Hepatic encephalopathy
hepatogenic gliodystrophic, *see* Hepatogenic gliodystrophic encephalopathy
hexachlorophene intoxication, 65/478
hypereosinophilia, 63/138

hypereosinophilic syndrome, 63/372
hypernatremia, 63/554
hypertensive, *see* Hypertensive encephalopathy
hypoglycemic, *see* Hypoglycemic encephalopathy
hyponatremia, 63/547
hypophosphatemia, 63/564
hypothyroidism, 70/93
hypoxia, 9/472
hypoxic ischemic, *see* Hypoxic ischemic encephalopathy
idiopathic hypereosinophilic syndrome, 63/138
ifosfamide intoxication, 65/534
industrial toxin, 46/392
infantile epileptogenic, *see* Infantile epileptogenic encephalopathy
infantile myoclonic, *see* Infantile myoclonic encephalopathy
infantile spasm, 73/200
infectious endocarditis, 63/23
insect bite, *see* Insect bite encephalopathy
α-interferon, 63/359
α-interferon intoxication, 63/359, 65/529, 538, 562
β-interferon intoxication, 65/529
γ-interferon intoxication, 65/562
interleukin-2, 65/529
intrathecal chemotherapy, 61/151
irradiation, *see* Irradiation encephalopathy
ischemic, *see* Ischemic encephalopathy
lactic acidosis, 45/122
lead, *see* Lead encephalopathy
legionellosis, 52/254
Leigh, *see* Subacute necrotizing encephalomyelopathy
leukemia, 63/345, 354
lidocaine, 63/192
liver disease, *see* Hepatic encephalopathy
loiasis, 52/515, 517
lymphoma, 63/350, 69/274
lymphostatic, *see* Lymphostatic encephalopathy
macroglobulinemia, 39/191
malignant lymphoma, 69/274
manganese, *see* Manganese encephalopathy
meningitis, 45/119
metabolic, *see* Metabolic encephalopathy
metabolic acidosis, 45/122
metachromatic leukodystrophy, 8/16, 45/119
methotrexate, 46/601, 63/355, 357, 65/528
methotrexate intoxication, 39/101, 65/530
methylmercury, 51/265
metodiclorofen, 65/528
metronidazole, 46/602
mexiletine, 63/192
mitomycin C, 65/528
mitotane, 65/529
mitotane intoxication, 65/538
mitoxantrone, 65/528
multifocal ischemic, *see* Multifocal ischemic encephalopathy
multifocal necrotizing, *see* Multifocal necrotizing encephalopathy
multiple cystic, *see* Multiple cystic encephalopathy
multiple organ failure, 71/527
mumps, 31/220
murine, *see* Murine encephalopathy
myoclonic, *see* Myoclonic encephalopathy
myopathy, 43/132
neonatal, *see* Neonatal encephalopathy
neurotoxicology, 64/2
nifedipine, 63/194
nimustine, 65/528
nonbacterial thrombotic endocarditis, 55/165, 489, 63/120
nonprogressive, *see* Nonprogressive encephalopathy
OKT3, *see* OKT3 encephalopathy
open heart surgery, 63/187
organic solvent intoxication, 64/41
organophosphate intoxication, 51/284
pancreatic, *see* Pancreatic encephalopathy
pellagra, 28/59, 46/336, 70/410
perinatal, *see* Perinatal encephalopathy
pertussis, *see* Pertussis encephalopathy
polyarteritis nodosa, 39/301, 55/355
polymyxin, 46/602
portosystemic anastomosis, *see* Portosystemic anastomosis encephalopathy
postanesthetic, *see* Postanesthetic encephalopathy
postanoxic, *see* Postanoxic encephalopathy
postanoxic myoclonic, *see* Posthypoxic myoclonus
posticteric, *see* Posticteric encephalopathy
postinfectious, *see* Postinfectious encephalopathy
postnarcotic, *see* Postanesthetic encephalopathy
posttraumatic, *see* Posttraumatic encephalopathy
primary carnitine deficiency, 66/402
procarbazine, 63/359, 65/528
procarbazine intoxication, 65/534
radiotherapy, 39/51
rejection, *see* Rejection encephalopathy
renal insufficiency, 27/320-330, 341
respiratory, *see* Respiratory encephalopathy
respiratory alkalosis, 45/121
rifampicin, 46/602
rubella, *see* Rubella encephalopathy

sarcoidosis, 71/478
self-mutilation, *see* Self-mutilation encephalopathy
septic, *see* Septic encephalopathy
sparfosic acid, 65/529
sparfosic acid intoxication, 65/538
spirogermanium, 65/529
spiromustine, 65/528
spongiform, *see* Spongiform encephalopathy
spongiform cortical, *see* Spongiform cortical encephalopathy
spongy, *see* Spongiform encephalopathy
status epilepticus, 73/267
strangulation, *see* Strangulation encephalopathy
subacute myelo-optic neuropathy, 51/296
subacute spongiform, *see* Subacute spongiform encephalopathy
subcortical arteriosclerotic, *see* Binswanger disease
tamoxifen, 65/529
tetraethyllead, 51/265
thallium intoxication, 36/252
thiotepa, 63/359
thrombotic thrombocytopenic purpura, 45/119
thyroid disease, 70/91
tocainide, 63/192
toxic, *see* Toxic encephalopathy
toxic oil syndrome, 63/381
toxic shock syndrome, 52/259
toxoplasmosis, 42/662
traumatic, *see* Traumatic encephalopathy
trichinosis, 52/511, 569
tsukubaenolide intoxication, 65/558
tuberculosis, 33/207-209, 216
tuberculous meningitis, 33/207-209, 216, 52/200
uremic, *see* Uremic encephalopathy
vasculitis, 45/118
verapamil, 63/194
vincristine, *see* Vincristine encephalopathy
vincristine intoxication, 65/528, 534
viral, *see* Viral encephalopathy
vitamin B3 deficiency, *see* Vitamin B3 deficiency encephalopathy
Waldenström macroglobulinemia, 55/486, 63/396
Waldenström macroglobulinemia polyneuropathy, 39/191, 532, 51/432
Wernicke, *see* Wernicke encephalopathy
Wernicke-Korsakoff, *see* Wernicke-Korsakoff encephalopathy
Whipple disease, 7/627, 46/389
Encephaloretinofacial angiomatosis, *see* Bonnet-Dechaume-Blanc syndrome
Encephaloschisis, *see* Schizencephaly
Encephalosis
pontine infarction, 12/28
Encephalotrigeminal angiomatosis, *see* Sturge-Weber syndrome
Enchondroma
bone, 19/303
vertebral, *see* Vertebral enchondroma
End plate, *see* Motor end plate
Endarterectomy
carotid, *see* Carotid endarterectomy
vertebral artery, 12/11
Endarteritis obliterans, *see* Thromboangiitis obliterans
Endemic cretinism
CNS, 31/297
deafness, 46/84
mental deficiency, 46/84
myxedema, 31/297
nervous, 31/297
pathology, 31/298
Endemic paresis
animal, 9/671, 684
ataxia, 9/684
cause, 9/671, 684
experimental data, 9/684
progressive paraparesis, 9/684
Endemic pellagra
vitamin B3 deficiency, 51/333
Endemic periodic paralysis
cotton seed intoxication, 63/557
hypokalemia, 63/557
Endemic typhus
rickettsial infection, 34/646
Endocardial fibroelastosis
systemic brain infarction, 11/441
Endocarditis
acute bacterial, 63/111
antiphospholipid antibody, 63/122
aspergillosis, 52/379
bacterial, *see* Bacterial endocarditis
brain abscess, 33/115, 473
brain embolism, 53/156
brain infarction, 53/156
brain ischemia, 53/156
Brucella, *see* Brucella endocarditis
Candida krusei, 35/416
candidiasis, 52/401
carotid murmur, 63/111
classification, 63/111
diamorphine, 37/391
diamorphine intoxication, 65/355
endomyocardial fibrosis, 63/111
fever, 63/111

fungal, *see* Fungal endocarditis
hemiplegia, 63/111
hypereosinophilia, 63/111
infective, *see* Infectious endocarditis
Libman-Sacks, *see* Libman-Sacks endocarditis
Listeria monocytogenes meningitis, 33/83
Loeffler eosinophilic fibroblastic, *see* Idiopathic hypereosinophilic syndrome
Loeffler fibroblastic, *see* Idiopathic hypereosinophilic syndrome
marantic, *see* Marantic endocarditis
neurologic complication, 63/111
neurotrophic parietal, *see* Neurotrophic parietal endocarditis
nonbacterial thrombotic, *see* Nonbacterial thrombotic endocarditis
nonseptic thrombotic, *see* Nonseptic thrombotic endocarditis
opiate intoxication, 65/355
postoperative bacterial, *see* Postoperative bacterial endocarditis
postoperative fungal, *see* Postoperative fungal endocarditis
rheumatic, 63/111
rheumatic fever, 63/111
staphylococcal meningitis, 52/71
subacute bacterial, *see* Subacute bacterial endocarditis
syphilitic, 63/111
systemic brain infarction, 11/433
systemic lupus erythematosus, *see* Systemic lupus erythematosus endocarditis
thrombotic, *see* Thrombotic endocarditis
verrucous, 63/111
Endocrine adenoma
multiple, *see* Multiple endocrine adenoma
Endocrine adenomatosis
multiple, *see* Multiple endocrine adenomatosis
Endocrine adenomatosis syndrome
multiple, *see* Multiple endocrine adenomatosis syndrome
Endocrine candidiasis syndrome
Addison disease, 42/642
alopecia, 42/642
hypoparathyroidism, 42/642
intracerebral calcification, 42/642
keratoconjunctivitis, 42/642
mental deficiency, 42/642
papilledema, 42/642
pernicious anemia, 42/462
seizure, 42/642
thymoma, 42/642
Endocrine disorder
adrenoleukodystrophy, 42/483
Albright syndrome, 31/312
anencephaly, 50/85
anxiety, 46/431
aqueduct stenosis, 50/311
basal encephalocele, 50/106
basal meningoencephalocele, 50/106
benign intracranial hypertension, 16/151, 157
brain injury, 23/119, 123
brain stem death, 57/471
brain tumor, 16/341, 345
brain tumor localization, 16/354
craniocervical region, 32/42
craniopharyngioma, 16/347
delirium, 46/544
depression, 46/427
Flynn-Aird syndrome, 42/328
hereditary motor and sensory neuropathy type I, 21/286
hereditary motor and sensory neuropathy type II, 21/286
holoprosencephaly, 30/453, 456, 50/237
Huntington chorea, 42/225
hyperthermia, 75/62
hypoglycemia, 27/70
Kearns-Sayre-Daroff-Shy syndrome, 62/309
late cerebellar atrophy, 60/587
late cortical cerebellar atrophy, 21/477, 495
Laurence-Moon-Bardet-Biedl syndrome, 13/133
leprechaunism, 42/589-591
myotonic dystrophy, 40/510
orthochromatic leukodystrophy, 10/114
pernicious anemia, 42/609
pineal tumor, 17/658
pinealoma, 16/352, 18/362, 42/767
progressive external ophthalmoplegia, 62/309
progressive muscular atrophy, 59/19
psychosis, 46/462
sarcoidosis, 71/477
xeroderma pigmentosum, 43/13
Endocrine exophthalmos, *see* Thyroid associated ophthalmopathy
Endocrine myopathy, 41/235-253
acromegaly, 41/248, 250
aldosteronism, 40/317
EMG, 62/69
fatigue, 62/527
functional, 40/282
hyperparathyroidism, 41/246, 324, 338
hyperthyroidism, 62/528
hypoparathyroidism, 41/215, 247
hypothyroid myopathy, 41/243-246, 298, 62/528
muscle cramp, 62/527

muscle fiber type II atrophy, 62/527
myasthenia gravis, 41/240, 62/528
steroid myopathy, 41/250-253
thyroid associated ophthalmopathy, 62/528
thyroid disease, 41/240
thyrotoxic exophthalmus, 41/240
thyrotoxic myopathy, 41/235-240, 62/528
thyrotoxic periodic paralysis, 41/165, 241-243, 62/528
Endocrine neoplasia syndrome
multiple, *see* Multiple endocrine neoplasia syndrome
Endocrine polyneuropathy
*see also* Acromegalic polyneuropathy, Diabetic polyneuropathy, Hyperthyroid polyneuropathy *and* Hypothyroid polyneuropathy
review, 51/497-519
Endocrinopathic motoneuron disease
WFN classification, 59/6
Endocrinopathy
multiple myeloma polyneuropathy, 51/430
Wilson disease, 49/223
Endodermal sinus tumor
pathology, 68/235
Endogenous opioid
dynorphin, 57/83, 65/349
endorphin, 39/453, 46/268, 48/100, 151, 57/83
β-endorphin, 48/100, 51/566, 57/83, 65/349
enkephalin, 57/83
leucine enkephalin, 57/83, 65/349
metenkephalin, 49/633, 65/349
precursor molecule, 65/350
schematic structure, 65/351
Endometriosis
lumbar spinal column, 20/443
Endometrium
heterotopic, *see* Heterotopic endometrium
Endomorphin, *see* Endorphin
Endomyelography
posttraumatic syringomyelia, 61/385
syringomyelia, 50/447
Endoneural sinus
anatomy, 11/40
Endoneurial edema
lead polyneuropathy, 51/271, 64/436
Endoneurial fibrosis
Bassen-Kornzweig syndrome, 51/394
Endoneurial hypoxia
diabetic polyneuropathy, 51/506
Endoneurial mononuclear cell
Guillain-Barré syndrome, 51/243
Endoneurium
nerve, 51/2
Endophthalmitis
candidiasis, 52/401
poikilomycosis, 35/567, 52/489
Pseudoallescheria, 52/491
Endoplasmic reticulum
glycosylation, 66/82
GM2 gangliosidosis, 10/417
methionine adenosyltransferase deficiency, 29/122
nerve, 51/6
Endorphin
Alzheimer disease, 46/268
brain metabolism, 57/83
endogenous opioid, 39/453, 46/268, 48/100, 151, 57/83
gastrointestinal disease, 39/453
migraine, 48/100, 151
β-Endorphin
brain metabolism, 57/83
cluster headache, 48/233
congenital pain insensitivity, 51/566
endogenous opioid, 48/100, 51/566, 57/83, 65/349
head injury, 57/83
hypothalamus, 45/238
migraine, 48/100
pituitary gland, 45/239
pro-opiomelanocortin, 65/350
Endorphin precursor
lipotropin, 65/351
Endosulfan
chlorinated cyclodiene, 64/200
neurotoxin, 36/418, 64/201
organochlorine insecticide intoxication, 36/418
toxic encephalopathy, 64/201
Endosulfan intoxication
laboratory finding, 36/418
pathology, 36/418
treatment, 36/418
Endothelial system
reticular, *see* Reticuloendothelial system
Endothelium
arterial hypertension, 63/78
brain arteriosclerosis, 63/78
brain blood flow, 53/64
brain microcirculation, 53/80
choroid plexus, *see* Choroid plexus endothelium
critical illness polyneuropathy, 51/584
hemostasis, 63/301, 305
thrombosis, 11/130
Endotoxemia
Reye syndrome, 56/165
Endotoxin
intravascular consumption coagulopathy, 55/494
Legionella pneumophila, 52/257

meningococcal meningitis, 33/23, 25, 28, 52/27
Neisseria meningitidis, 52/23
Reye syndrome, 56/153
Endotracheal intubation
complication, 71/539
head injury, 57/216
Endrin intoxication
clinical features, 36/411-414
EEG, 36/414-416
laboratory finding, 36/414-416
pathology, 36/417
prognosis, 36/416
treatment, 36/417
Endurance exercise training
orthostatic intolerance, 74/325
stress, 74/325
Energy metabolism
brain hypoxia, 27/23
brain infarction, 27/29-31
brain ischemia, 27/29-31
CNS, 27/1-23
epilepsy, 15/61
malnutrition, 29/6
Enflurane intoxication
convulsion, 37/412
epilepsy, 37/412
hypocapnia, 37/412
inhalation anesthetics, 37/412
muscular twitching, 37/412
myoclonus, 37/412
seizure, 37/412
suppression, 37/412
ENG
*see also* Neuro-otology
acoustic nerve tumor, 17/677
Alexander law, 2/378
apparatus, 2/378
bithermal test, 24/132
caloric test, 24/133, 135
Charlevoix-Saguenay spastic ataxia, 60/455
diagnostic recording, 16/316
eye closure, 2/380-382
head injury, 24/127
hereditary periódic ataxia, 60/441
hypomagnesemia, 28/548
labyrinthine ischemia, 55/131
latent nystagmus, 24/130
magnesium imbalance, 28/548
multiple sclerosis, 47/132
Pelizaeus-Merzbacher disease, 66/569
positional test, 24/130
posttraumatic syndrome, 24/537
progressive muscular atrophy, 59/16
recording, 2/379
retinovestibular hereditary spastic ataxia, 21/385
rotation test, 24/133
technique, 2/376-380, 24/128
topographical diagnosis, 2/380-383
visual fixation, 2/378, 380
Engel syndrome
Maffucci syndrome, 14/119
Engelmann disease
course, 38/400
diagnosis, 38/400
general features, 38/396
genetics, 38/400
history, 38/396
laboratory finding, 38/399
mental deficiency, 46/46
neurologic manifestation, 38/397-399
treatment, 38/400
Enhydrina schistosa intoxication
antivenene, 37/94
Enkephalin
Alzheimer disease, 46/268
basal ganglion, 49/28
basal ganglion tier III, 49/22
endogenous opioid, 57/83
gastrointestinal disease, 39/453
migraine, 48/99
pedunculopontine tegmental nucleus, 49/12
striatonigral fiber, 49/8
striatopallidal fiber, 49/6
substantia gelatinosa, 45/230
substantia nigra, 49/8
Enkephalin intoxication
neuronal activity, 37/372
Enlarged foramen magnum
Arnold-Chiari malformation type II, 50/406
Enolase
brain glycolysis, 27/9
Enolic acid
headache, 48/174
migraine, 48/174
Enophthalmos
Horner syndrome, 43/64
ocular imbalance, 24/91
orbital injury, 24/91
9p partial trisomy, 43/515
progressive hemifacial atrophy, 31/253, 43/409
21q partial monosomy, 43/541
trisomy 9, 50/564
Enoplosidae intoxication
family, 37/78
venomous spine, 37/78
Enoplosus intoxication

family, 37/78
venomous spine, 37/78
Enoxacin intoxication
epilepsy, 65/486
ENT disease, *see* Ear nose and throat disease
Entactin
muscle fiber, 62/31
Entamoeba histolytica
*see also* Neuroamebiasis
amebiasis, 35/25, 47
Amoebae, 35/28, 30
brain abscess, 52/149, 151
brain metastasis, 69/115
calcium, 52/322
cerebral amebiasis, 35/25
concanavalin A, 52/322
culture, 52/318
cyst, 52/309
cytotoxin, 52/322
direct observation, 52/318
erythrophagocytosis, 52/322
form, 52/309
isoenzyme pattern, 52/322
isolation, 52/318
microbiology, 52/309
microscopy, 52/310
tissue cytolysis, 52/322
Enteramine, *see* Serotonin
Enteric bacillary meningitis
infantile, *see* Infantile enteric bacillary meningitis
Enteric nervous system
*see also* Gastrointestinal disorder
amino acid, 75/617
anatomy, 75/614
autonomic nervous system, 74/214
development, 74/214, 75/613
genetics, 75/614
mummy, 74/191
neurotransmitter, 75/617
nitric oxide, 75/617
peptide, 75/617
purine, 75/617
survey, 75/613
Enteric neuron
gastrointestinal disorder, 75/621
Enteric neuronopathy
autonomic nervous system, 75/535
Enterobacter
adult meningitis, *see* Gram-negative bacillary meningitis
brain abscess, 52/149
gram-negative bacillary meningitis, 52/104
infantile enteric bacillary meningitis, 33/62
Enterobacter sakazakii
gram-negative bacillary meningitis, 52/118
Enteroenteric inhibitory reflex
sympathetic pathway, 75/625
Enterogenous cyst, *see* Neurenteric cyst
Enteroglucagon, *see* Glucagon
Enteropathica
acrodermatitis, *see* Danbolt disease
Enteropathy
exudative, *see* Exudative enteropathy
gluten induced, *see* Celiac disease
Enterotome
Head's zone, 2/160
terminology, 2/160
Enterovirus
acute hemorrhagic conjunctivitis, 56/307
acute viral encephalitis, 56/126, 132
chronic meningitis, 56/646
classification, 56/309
encephalitis, 56/327
facial paralysis, 56/351
features, 56/309
genome, 56/313
Guillain-Barré syndrome, 56/330
human pathogen, 56/307
immunocompromised host, 56/475
immunodeficiency, 56/330
laboratory diagnosis, 56/333
meningitis, 56/326
myalgia, 56/11
nonneurologic syndrome, 56/331
paralytic disease, 56/327
persistent infection, 56/330
picornavirus, 56/307, 309
poliomyelitis, 34/95, 56/307
property, 56/310, 312
viral infection, 56/307
viral meningitis, 56/126, 128, 307
Enterovirus 70
acute cerebellar ataxia, 56/329
acute hemorrhagic conjunctivitis, 56/307, 349
biology, 56/349
brain stem encephalitis, 56/351
cranial nerve palsy, 56/326, 329, 351
facial paralysis, 56/351
multiple cranial neuropathy, 56/329, 351
neurotropic virus, 56/349
picorna family, 56/349
proteinocytologic dissociation, 56/351
viral meningitis, 56/307
Enterovirus 71
acute cerebellar ataxia, 56/326
bulbar paralysis, 56/329

viral encephalitis, 56/326
viral meningitis, 56/307, 326
Enterovirus 72
transverse myelitis, 56/330
viral encephalitis, 56/326
Enterovirus infection, 34/133-141
*see also* Coxsackie virus infection, ECHO virus infection, Picornavirus *and* Poliomyelitis
acute cerebellar ataxia, 34/621
animal viral disease, 34/292
animal virus, 34/140
chronic, *see* Chronic enterovirus infection
Coxsackie virus, 34/134-139
ECHO virus, 34/139
encephalomyelitis, 34/140
meningitis serosa, 34/84
paralytic, 34/292
postviral fatigue syndrome, 56/332
tropical ataxic neuropathy, 51/323
Entopallium
behavior, 2/720
function, 2/713
temporal lobe lesion, 2/720
Entrapment neuropathy, *see* Compression neuropathy
ENU, *see* Ethylnitrosourea
Enuresis
epilepsy, 45/142
etiology, 3/100, 45/142
juvenile parasomnia, 45/141
nocturnal, *see* Nocturnal enuresis
sleep disorder, 3/100
sleep phase, 3/101
spinal ependymoma, 20/367
treatment, 3/101
4-En-valproic acid, *see* 2-Propyl-4-pentenoic acid
Envelope glycoprotein
herpes simplex virus, 56/208
Envenomation
arthropod, *see* Arthropod envenomation
bee sting, 37/107-111
cone intoxication, 37/60-62
elapid intoxication, 37/5-8, 12-16
hymenoptera intoxication, 37/107-111
intravascular consumption coagulopathy, 63/315
neuropharmacology, 37/13-16
octopus intoxication, 37/62
sea snake, *see* Sea snake intoxication
snake, *see* Snake envenomation
spider, *see* Spider envenomation
venom, 37/12-16
viper, 37/9
Environment
brain tumor origin, 16/33, 58, 77
carbamate intoxication, 64/184
developmental dysphasia, 46/141
intelligence, 3/303
Kii Peninsula amyotrophic lateral sclerosis, 59/274
lead, 64/432
mental deficiency, 46/76-81
multiple sclerosis, 9/86, 111
rehabilitation, 46/612-614
Environmental deprivation
mental deficiency, 46/5
Environmental duplication
migraine, 48/162
Environmental Status Scale
multiple sclerosis, 47/149
Environmental stress
sympathetic nervous system, 74/653
Enzyme
abnormality, 66/57
active site, 66/52
adverse effect, 69/494
alkaline phosphatase, 26/403
amount, 66/55
angiotensin converting, *see* Angiotensin converting enzyme
assay, 66/54
catalyst, 66/52
ceramide lactoside lipidosis, 29/350
change, 66/55
citric acid cycle, *see* Citric acid cycle enzyme
copper, *see* Copper enzyme
CSF examination, 11/237
diagnostic, 66/54
dialkylphosphoryl, *see* Dialkylphosphoryl enzyme
dystrophy, 66/51
erythrocyte, *see* Erythrocyte enzyme
feedback inhibition, 66/51
free radical, 64/18
fucosidosis, 29/350, 66/335
Gaucher disease, 66/61
genetically engineered, 66/118
globoid cell leukodystrophy, 10/146, 29/350
glycolytic, *see* Glycolytic enzyme
glycosaminoglycanosis type III, 66/301
glycosaminoglycanosis type IVB, 66/57
glycosaminoglycanosis type VII, 66/312
$GM_2$ gangliosidosis, 66/61, 256
$GM_3$ hematoside lipodystrophy, 29/350
lysosome, 66/57
lytic, *see* Lytic enzyme
methionine activating, *see* Methionine activating enzyme

migraine, 48/85, 93
mitochondrial malic, *see* Mitochondrial malic enzyme
monoalkylphosphoryl, *see* Monoalkylphosphoryl enzyme
mucolipidosis type II, 66/381
myelin, 10/138
myelin figure, 9/33
myelin synthesis, 10/138
neurolipidosis, 66/51
nicotinamide adenine dinucleotide, 66/51
nomenclature, 66/54
oxidative, *see* Oxidative enzyme
peroxisomal β-oxidation, 66/506
porphyria, 70/307
prostatic cancer, 66/56
protein, 66/51
pyruvate dehydrogenase complex deficiency, 66/415
sialic acid, 66/355
spinal cord injury, 26/383-385, 400, 405
structure, 66/51
wallerian degeneration, 7/214
white matter, 9/14-18

Enzyme catalyzed reaction
chemistry, 66/52

Enzyme defect
α-*N*-acetylgalactosaminidase deficiency, 66/345
adrenoleukodystrophy, 47/598
alpha fucosidosis, 66/330
alpha mannosidosis, 66/330
aspartylglucosaminuria, 29/228, 66/330
beta mannosidosis, 66/330
cholesterosis, 21/50
citrullinemia, 29/97
enzyme deficiency, 66/58
Fabry disease, 29/350
fumarate hydratase deficiency, 66/420
Gaucher disease, 29/350, 352
glycoproteinosis, 66/63
glycosaminoglycanosis, 60/176
glycosaminoglycanosis type II, 66/296
glycosaminoglycanosis type VI, 66/308
GM1 gangliosidosis, 10/475, 29/367
hereditary tyrosinemia, 29/215, 42/634
histidinemia, 29/196
homocystinuria, 29/111, 115
lipid storage, 66/140
lysinuric protein intolerance, 29/210
metachromatic leukodystrophy, 29/358, 47/589
multiple respiratory, *see* Multiple respiratory enzyme defect
Niemann-Pick disease, 29/355
peroxisomal β-oxidation, 66/518
phenylketonuria, 28/128, 29/31
phosphofructokinase deficiency, 41/191, 43/183
propionic acidemia, 29/208
pyridoxine dependency, 72/134
Sjögren-Larsson syndrome, 66/618
tyrosyluria, 29/215

Enzyme deficiency
autosomal recessive, 66/56
bifunctional, *see* Bifunctional enzyme deficiency
carrier, 66/56
enzyme defect, 66/58
fumarate hydratase deficiency, 66/420
inherited, *see* Inherited enzyme deficiency
metabolism, 66/58
mitochondrial, *see* Mitochondrial enzyme deficiency
multiple, 66/58
neonatal detection, 66/56
parkinsonism, 21/53
X-linked adrenoleukodystrophy, 66/462

Enzyme deficiency polyneuropathy
β-*N*-acetylhexosaminidase A deficiency, 51/369, 375
β-*N*-acetylhexosaminidase B deficiency, 51/369, 375
*N*-acetylneuraminidase deficiency, 51/380
acid ceramidase deficiency, 51/369, 379
acid lipase deficiency, 51/369, 379
activator substance, 51/368
adrenoleukodystrophy, 51/383
*N*-aspartyl-β-glucosaminidase, 51/370
aspartylglucosaminuria, 51/370
cerebroside sulfatase, 51/368
diagnosis, 51/368
Fabry disease, 51/369, 376
Farber disease, 51/369, 379
Friedreich ataxia, 51/388
α-L-fucosidase, 51/370, 381
fucosidosis, 51/370, 381
galactosialidosis, 51/370
β-galactosidase, 51/369
α-galactosidase A, 51/369, 376
galactosylceramide β-galactosidase, 51/369, 373
galactosylsulfatide, 51/367
Gaucher disease, 51/369, 379
globoid cell leukodystrophy, 51/369
α-1,4-glucosidase, 51/371
glucosylceramidase, 51/369
glycosaminoglycanosis, 51/370, 379
GM1 gangliosidosis, 51/369, 374
GM2 gangliosidosis, 51/369, 375
infantile acid maltase deficiency, 51/371

lactosylsulfatide, 51/367
lipoproteinosis, 51/367
lysosomal storage disease, 51/367
α-mannosidase, 51/370
mannosidosis, 51/370
metachromatic leukodystrophy, 51/367, 369
mitochondrial enzyme deficiency, 51/387
mucolipidosis type I, 51/370, 380
mucolipidosis type II, 51/370, 380
mucolipidosis type III, 51/370, 381
mucolipidosis type IV, 51/380
multiple sulfatase deficiency, 51/369
neuronal ceroid lipofuscinosis, 51/371
neuropathology, 51/368
Niemann-Pick disease type A, 51/369, 373
olivopontocerebellar atrophy, 51/388
peroxisomal disease, 51/367, 383
porphyria, 51/367
Refsum disease, 51/384
Salla disease, 51/370
sialidase, 51/370
sphingomyelinase, 51/369, 373
spinal muscular atrophy, 51/375
sulfogalactoglycerolipid, 51/367
Wolman disease, 51/369, 379
Enzyme histochemistry
globoid cell, 10/79, 303
GM2 gangliosidosis, 10/402-404
multicore myopathy, 62/339
muscle tissue culture, 62/94
myofibrillar lysis myopathy, 62/339
tubular aggregate myopathy, 62/352
Enzymopathy
amino acid metabolism, 21/49
glycogen storage disease, 21/52
GM1 gangliosidosis, 21/51
α-lipoprotein, 21/52
β-lipoprotein, 21/52
Lowe syndrome, 21/49
Parkinson disease, 21/53
sphingolipidosis, 21/50
EOG
Behr disease, 13/90
brain stem ischemia, 54/164
cerebrovascular disease, 54/157
hereditary adult foveal dystrophy (Behr), 13/90
Huntington chorea, 49/289
hyaloidoretinal degeneration (Wagner), 13/274
internuclear ophthalmoplegia, 54/164
middle cerebral artery infarction, 54/162
multiple sclerosis, 47/132
neuroretinal degeneration, 13/20-22
pontine infarction, 54/163
posterior cerebral artery infarction, 54/161
primary pigmentary retinal degeneration, 13/165, 174-178, 179
Stargardt disease, 13/130
vertebrobasilar transient ischemic attack, 54/164
vitelliform foveal dystrophy, 13/267
Eosinophil
anaphylaxis, 63/370
basophil, 63/370
cell surface receptor, 63/370
Charcot-Leyden crystal, 63/370
cytokine, 63/369
eosinophil cationic protein, 63/370
eosinophil chemoattractant factor, 63/370
eosinophil derived neurotoxin, 63/370
eosinophilopoietin, 63/369
function, 63/369
granule, 63/370
granulocyte-macrophage colony stimulating factor, 63/369
helminthosporium toxin, 63/370
histamine, 63/370
interleukin-3, 63/369
interleukin-5, 63/369
leukotriene, 63/369
leukotriene C4, 63/370
lysophospholipase, 63/370
macrophage chemoattractant, 63/370
major basic protein, 63/370
parasite killing, 63/370
peroxidase, 63/370
platelet activating factor, 63/370
progenitor cell, 63/369
slow reacting substance, 63/370
Eosinophil count
ascariasis, 21/569
hereditary periodic ataxia, 21/568
primary pigmentary retinal degeneration, 13/196
Eosinophilia
allergic granulomatous angiitis, 51/453, 63/371
angioneurotic edema, 63/371
angiostrongyliasis, 35/328, 52/557
Angiostrongylus cantonensis, 63/371
aspergillosis, 63/371
classification, 63/371
coccidioidomycosis, 52/415
Echinococcus, 63/371
eosinophilia myalgia syndrome, 63/371, 64/249
eosinophilic fasciitis, 63/371
eosinophilic polymyositis, 63/371
filariasis, 35/163, 52/516
helminthiasis, 52/509, 511
hypereosinophilic syndrome, 63/371

idiopathic hypereosinophilic syndrome, 63/371
mycosis fungoides, 63/371
neurologic involvement, 63/369
polyarteritis nodosa, 8/125, 39/308, 63/371
primary eosinophilia, 63/371
sarcoidosis, 63/371
schistosomiasis, 52/540, 63/371
secondary eosinophilia, 63/371
toxic oil syndrome, 63/371
trichinosis, 52/565, 569, 63/371
tryptophan, 63/371
Wegener granulomatosis, 63/371
Eosinophilia myalgia syndrome
*see also* Carpal tunnel syndrome, Mononeuritis multiplex *and* Tryptophan
acute encephalopathy, 64/257
allergic granulomatous angiitis, 64/261
alopecia, 63/375, 64/252
areflexia, 63/376
arthralgia, 63/375, 64/252
axonal degeneration, 64/254
axonal neuropathy, 63/377
Bacillus amyloliquefaciens, 64/251
body posture distortion, 64/256
cardiomegaly, 63/376
cardiovascular involvement, 64/258
carpal tunnel syndrome, 64/255
carpopedal spasm, 64/256
cause, 63/378
CD8+ cell, 64/253
cognitive impairment, 63/376
computer assisted tomography, 64/257
confusion, 64/256
connective tissue fibrosis, 63/379
cough, 64/257
cranial neuropathy, 64/254
creatine kinase, 63/377
criteria, 64/250
cutaneous symptom, 64/258
dementia, 63/376, 64/256
demyelinating neuropathy, 63/377
dermatomyositis, 62/371
diagnostic criteria, 63/375
differential diagnosis, 63/379, 64/261
dyspnea, 63/375, 64/257
edema, 64/252
EMG, 63/377, 64/252
eosinophilia, 63/371, 64/249
eosinophilic fasciitis, 63/379
epidemiology, 64/249
1,1'-ethylidenebis(tryptophan), 64/262
fasciitis, 64/259
fatigue, 63/377
fever, 63/375, 64/252
fructose-bisphosphate aldolase, 63/377
gastrointestinal symptom, 64/260
groove sign, 63/375
HLA-DR antigen, 64/253
hypereosinophilic syndrome, 64/261
hyperesthesia, 63/375, 64/254
idiopathic hypereosinophilic syndrome, 64/257
indoleamine-pyrrole-2,3-dioxygenase, 63/378
insomnia, 63/375
interleukin-1, 64/265
interleukin-2, 64/265
laboratory finding, 64/260
laboratory test, 63/377
lactate dehydrogenase, 63/377
late muscle spasm, 64/255
macrophage, 64/253
metabolic encephalopathy, 64/257
mononeuritis multiplex, 64/255
multisystem disease, 64/252
muscle biopsy, 63/377
muscle cramp, 63/377, 64/252
muscle fiber type II atrophy, 63/378
muscle spasm, 64/252
muscle weakness, 63/376
muscular atrophy, 64/253
myalgia, 63/375, 64/249, 252
myokymia, 64/256
myopathy, 63/377, 64/252
myositis, 64/252
nerve biopsy, 63/377, 64/254
nerve conduction, 63/377
neurologic complication, 63/376
neuromuscular symptom, 64/252
neuropathology, 64/253
neuropathy, 64/252
nuclear magnetic resonance, 64/257
obstructive pulmonary disease, 63/376
occlusive microangiopathy, 63/378
oxitriptan, 64/250
painful sensory neuropathy, 63/376
paranodal demyelination, 64/255
paresthesia, 63/376, 64/252, 254
pathogenesis, 63/378
peau d'orange, 63/375, 64/259
perimyositis, 63/378
perimysial fibrosis, 63/377
3-phenylaminoalanine, 64/263
3-phenylamino-1,2-propanediol, 64/263
polyneuropathy, 64/254
polyradiculoneuropathy, 64/254
pruritus, 63/375
pulmonary hypertension, 63/376, 64/258

Satoyoshi syndrome, 64/256
scleroderma, 63/375, 71/115
scleroderma-like lesion, 64/259
segmental demyelination, 64/255
sensorimotor neuropathy, 63/376
sensorimotor polyneuropathy, 63/376, 64/254
serotonin, 63/378
serum kynurenine, 63/378
sex ratio, 64/251
Shulman syndrome, 64/261
Sjögren syndrome, 64/260
sural nerve, 64/254
toxic encephalopathy, 64/257
toxic myopathy, 62/604
toxic oil syndrome, 63/379
transforming growth factor β, 63/379
treatment, 63/379, 64/265
trichinosis, 64/261
tryptophan, 63/375
tryptophan biosynthesis, 64/251
tryptophan contaminant, 64/250
USA geography, 64/250
Eosinophilic body
Nageotte nodule, 22/238
pure autonomic failure, 22/238
Eosinophilic cytoplasmic inclusion
dermatomyositis, 62/377
inclusion body myositis, 62/377
polymyositis, 62/377
Eosinophilic encephalomyelitis
*see also* Gnathostomiasis
differential diagnosis, 35/329
gnathostomiasis, 35/329
Eosinophilic fasciitis
age, 63/384
antinuclear antibody, 63/385
carpal tunnel syndrome, 63/385
cimetidine, 63/385
corticosteroid, 63/385
eosinophilia, 63/371
eosinophilia myalgia syndrome, 63/379
exertion effect, 63/384
muscle contracture, 63/385
myalgia, 63/384
neurologic complication, 63/385
paresthesia, 63/385
pathogenesis, 63/385
peau d'orange, 63/384
penicillamine, 63/385
scleroderma, 71/115
sex ratio, 63/384
treatment, 63/385
Eosinophilic fibroblastic endocarditis
Loeffler, *see* Idiopathic hypereosinophilic syndrome
Eosinophilic granuloma
*see also* Histiocytosis X
brain, 16/231
coin-shaped vertebra, 20/31
cranial vault tumor, 17/116
CSF pleocytosis, 63/422
CSF protein, 63/422
diabetes insipidus, 63/422
headache, 5/229
histiocytosis X, 38/97
incidence, 68/530
Langerhans cell histiocytosis, 70/56
Langerhans histiocyte, 63/422
sex ratio, 63/422
skull base tumor, 17/175
spinal, *see* Spinal eosinophilic granuloma
spinal cord, 67/196
spinal epidural tumor, 20/111
spinal tumor, 20/111, 68/517
vertebra plana, 20/30
vertebral column tumor, 20/30
Eosinophilic inclusion body
intestinal pseudo-obstruction, 51/493
measles virus, 56/480
1-methyl-4-phenyl-1,2,3,6-tetrahydropyridine intoxication, 65/371
poliomyelitis, 56/319
primate 1-methyl-4-phenyl-1,2,3,6-tetrahydropyridine intoxication, 65/376
progressive bulbar palsy, 59/224
subacute sclerosing panencephalitis, 56/422
type b, 56/319
Eosinophilic meningitis, 35/321-339
angiostrongyliasis, 35/321, 327, 52/510, 553, 556
Angiostrongylus cantonensis, 52/548
clinical features, 35/322, 327
course, 35/327
CSF, 35/323
diagnosis, 35/328
differential diagnosis, 35/329
EEG, 35/325-327
helminthiasis, 35/213
incidence, 35/337
infection source, 35/334
laboratory examination, 35/323-327
life cycle, 35/331
ocular fundus, 35/327
parasitology, 35/331
pathogenesis, 35/332-334
pathology, 35/330
symptom, 52/510

transmission, 35/335-337
transmission mode, 35/335-337
treatment, 35/337
Eosinophilic polymyositis
congestive heart failure, 63/385
EMG, 63/385
eosinophilia, 63/371
hypereosinophilic syndrome, 41/63, 65
muscle biopsy, 63/385
myalgia, 63/385
pathogenesis, 63/385
polyneuropathy, 63/385
proximal muscle weakness, 63/385
Raynaud phenomenon, 63/385
symptom, 63/385
systemic symptom, 63/385
Ependyma
Niemann-Pick disease type A, 10/496
Refsum disease, 8/23, 21/234, 42/148
Ependymal cyst
*see also* Colloid cyst *and* Third ventricle tumor
false localization sign, 18/135
features, 31/88
septum pellucidum, 17/538
spinal, *see* Spinal ependymal cyst
spinal epidural cyst, 20/167
sudden death, 18/135
symptom, 18/490
third ventricle, 18/133, 359
Ependymal diverticulum
third ventricle, 18/133
Ependymal rupture
syringobulbia, 50/427
syringomyelia, 50/427
Ependymitis
acute amebic meningoencephalitis, 52/326
African trypanosomiasis, 52/340
hemorrhagic, *see* Hemorrhagic ependymitis
Listeria monocytogenes meningitis, 52/93
tuberculous meningitis, 52/198
verruga peruana, 34/665
Whipple disease, 52/139
Ependymitis granularis
nuclear magnetic resonance, 54/97
Ependymoblastoma, *see* Ependymoma
Ependymocytoma, *see* Ependymoma
Ependymoepithelioma, *see* Ependymoma
Ependymoglioma, *see* Ependymoma
Ependymoma, 18/105-135, 42/731-733
age, 18/111, 19/12
animal, 18/131
architecture, 16/11, 18/116-119
awakening epilepsy, 18/126
Bailey-Cushing classification, 18/110
biochemistry, 18/111
biopsy, 67/226
blepharoplast, 18/105
brain biopsy, 16/718
brain scintigraphy, 18/130
brain tumor, 18/502, 42/731
brain tumor carcinogenesis, 18/132
cauda equina, 18/120
cell type, 18/110, 118
cerebellar syndrome, 18/128
cerebellar tumor, 17/716
cerebellopontine angle tumor, 18/120
chemotherapy, 18/130, 68/171
child, 68/171
chromosome, 67/53
chromosome abnormality, 67/45
chronic meningitis, 56/645
classification, 16/4, 7, 22, 68/16
clinical features, 31/47
coccygomedullary organ, 20/364
congenital brain tumor, 31/46
course, 18/125
CSF, 18/129
CSF cytology, 16/397
cyst, 18/119
definition, 16/22
diagnostic error, 18/128
differential diagnosis, 18/123, 128, 133, 68/168
differentiation type, 18/121, 123
echoEG, 18/130
EEG, 18/129
ependymal tubule, 18/117
epidemiology, 16/65, 31/46
epilepsy, 18/20, 126, 72/181
epithelial type, 18/110
fatty degeneration, 18/120
filum terminale, *see* Filum terminale ependymoma
genetic factor, 42/731
glioma polyposis syndrome, 42/735
grade, 68/16-19
grading system, 20/363
headache, 18/126
hemorrhage, 68/95
histologic differentiation, 18/120
histology, 18/116-123
history, 18/105, 20/353
Homer Wright rosette, 18/116, 177
hyaline papillary type, 18/110
imaging, 67/176, 181, 68/168
infratentorial, 68/168
intramedullary, 67/204
Kernohan classification, 18/105, 110

lateral ventricle tumor, 17/604, 18/332
location, 31/46
macroscopic aspect, 18/114
malignancy, 18/120, 20/363
metastasis, 18/125, 71/661
misdiagnosis, 18/129
multiple, *see* Multiple ependymoma
multiple neurofibromatosis, 43/35
multiple spinal, *see* Multiple spinal ependymoma
myopapillary type, 18/120
nerve tumor, 8/454
neurofibromatosis, 14/35
neurofibromatosis type I, 14/140
neuroradiology, 18/129
onion peeling, 18/117
optic, *see* Optic nerve glioma
origin, 18/110
ossification, 18/120
palisading, 18/117
papillary type, 18/110
perivascular halo, 18/116
posterior fossa tumor, 18/395
preference site, 18/113
prevalence, 18/111
prognosis, 18/130
psychiatric disorder, 18/127
radiotherapy, 18/502, 68/170
recurrence, 18/123
research, 16/36
rosette formation, 16/12, 18/110, 116
Roussy-Oberling classification, 18/110
sex ratio, 18/112, 19/12
site preference, 19/12
spinal, *see* Spinal ependymoma
spinal cord compression, 19/359
spinal cord tumor, 68/507
spinal glioma, 19/359
supratentorial, 68/168
supratentorial brain tumor, 18/313
surgery, 68/170
survey, 68/165
survival, 18/121, 131
symptom, 68/167
symptomatology, 18/125
third ventricle tumor, 17/442, 18/127
tissue culture, 16/38, 17/65, 68, 18/110
trabecular papillary form, 16/11
treatment, 18/130, 31/47
type frequency, 19/12
ultrastructure, 18/106-110
Ephedrine
brain vasculitis, 55/521
drug abuse, 55/521
drug addiction, 65/259
myasthenia gravis, 41/132
neurotoxin, 55/521, 65/259
orthostatic hypotension, 75/723
reflex, 1/62
subarachnoid hemorrhage, 55/521
vasoconstriction, 75/723
Ephedrine intoxication
adrenergic receptor stimulating agent, 65/260
epilepsy, 65/260
half-life, 65/260
hypertension, 65/260
psychosis, 65/260
tachycardia, 65/260
Ephelis
neurofibromatosis type I, 14/102
Peutz-Jeghers syndrome, 14/121, 603
Epibulbar dermoid
Goldenhar syndrome, 32/127, 38/410, 43/442
Epicanthus
acrocephalosyndactyly type II, 38/422
Armendares syndrome, 43/366
brachial neuritis, 42/304
cerebrohepatorenal syndrome, 43/339
Coffin-Lowry syndrome, 43/238
Down syndrome, 43/544
Dubowitz syndrome, 43/247
elfin face syndrome, 31/318
hypertelorism, 42/459
Klinefelter syndrome, 43/560
Klinefelter variant XXXY, 43/559
Minkowitz disease, 10/552
monosomy 21, 43/540
8p partial monosomy, 43/508
4p partial trisomy, 43/500
Patau syndrome, 14/120
11q partial monosomy, 43/523
4q partial trisomy, 43/498
r(9) syndrome, 50/580
r(14) syndrome, 50/588
r(22) syndrome, 43/547
Rubinstein-Taybi syndrome, 43/234
Summitt syndrome, 43/325, 372
tetraploidy, 43/568
trisomy 22, 43/548
Turner syndrome, 43/550, 50/543
XXXX syndrome, 43/554
XXXXX syndrome, 43/554
Epiconus medullaris syndrome
differential diagnosis, 59/432
Epidemic benign myalgia
*see also* Bornholm disease
acute viral myositis, 56/199

Epidemic cerebrospinal fever, *see* Meningococcal meningitis
Epidemic encephalitis
*see also* Encephalitis lethargica
Togavirus, 56/30
Epidemic hepatitis
Down syndrome, 31/398
Epidemic meningitis, *see* Meningococcal meningitis
Epidemic myalgia, *see* Bornholm disease
Epidemic neuromyasthenia, *see* Postviral fatigue syndrome
Epidemic pleurodynia, *see* Bornholm disease
Epidemic typhus, *see* Louse-borne typhus
Epidemic vertigo, *see* Vestibular neuronitis
Epidemiology
achondroplasia, 31/269
acquired immune deficiency syndrome dementia complex, 71/248
actinomycosis, 35/16
acute hemorrhagic conjunctivitis, 38/598
acute hemorrhagic leukoencephalitis, 34/587
acute immune mediated polyneuritis, 34/393
acute viral encephalitis, 34/64, 56/132
adrenomyeloneuropathy, 60/169
African trypanosomiasis, 35/8-10
aggression, 46/177
Alzheimer disease, 46/272-274
amebiasis, 35/11, 34-37
amnesic shellfish poisoning, 65/166
amyotrophic lateral sclerosis, 21/32, 22/287, 290, 339, 341, 349, 354, 59/179-181
anencephaly, 30/150, 153, 188-191, 50/77
angiosarcoma, 68/278
angiostrongyliasis, 35/322, 334, 52/547
anterior sacral meningocele, 32/206
aqueduct stenosis, 30/157, 50/305
arachnoid cyst, 31/85, 89
Aran-Duchenne disease, 21/30
Argentinian hemorrhagic fever, 56/367
Arnold-Chiari malformation type I, 30/162, 32/103
arterial hypertension, 63/71
arthrogryposis multiplex congenita, 30/165
ascariasis, 35/358
ataxia telangiectasia, 21/21, 31/24
atlantoaxial dislocation, 32/83
atlas, 32/28, 82
autism, 46/191
bacterial meningitis, 33/2-4, 52/3, 55/416
Bannwarth syndrome, 51/200
basilar impression, 30/161, 32/67-70
Behçet syndrome, 34/476-478, 71/210
benign intracranial hypertension, 67/104
beriberi, 7/636
birth incidence, 25/169
bismuth intoxication, 64/338
Bolivian hemorrhagic fever, 56/369
botulism, 71/557
brain arteriovenous malformation, 30/160, 31/177, 54/370
brain injury, 23/23
brain metastasis, 16/66, 71/611
brain stem tumor, 68/366
brain thromboangiitis obliterans, 55/312
brain tumor, 16/50, 60, 63, 31/39, 67/129
brucellosis, 52/581
Burkitt lymphoma, 63/350
butyrylcholinesterase deficiency, 42/654
Candida, 35/413-415
candidiasis, 35/414
carbon monoxide intoxication, 64/35
carcinoma, 68/173
cardiac arrest, 63/205
case fatality rate, 30/141
cat scratch disease, 34/460, 52/126
caudal aplasia, 30/164, 32/348
cavum septi pellucidi, 30/156, 304, 322-324
central pontine myelinolysis, 28/293
central sleep apnea, 63/461
cerebellar medulloblastoma, 31/50
cerebral amebiasis, 35/34
cerebrovascular disease, 11/183-203, 53/1
cervical spinal canal, 32/340
cervical vertebral column injury, 23/31
cervico-oculoacusticus syndrome, 30/162, 32/125
cestode, 35/12
Chagas disease, 35/10, 75/385
Charlevoix-Saguenay spastic ataxia, 60/451
childhood absence epilepsy, 73/145
childhood psychosis, 46/193
cholesteatoma, 31/67
chorea-acanthocytosis, 49/327
choroid plexus papilloma, 31/44, 68/173
ciguatoxin, 65/159
classic migraine, 48/18
cloverleaf skull syndrome, 30/154
CNS angiosarcoma, 68/278
CNS malformation, 30/139-171
cocaine intoxication, 65/253
coccidioidomycosis, 35/16, 443-445
congenital aneurysm, 31/138
congenital brain tumor, 30/150, 160
congenital retinal blindness, 21/3
congenital spinal cord tumor, 30/164, 32/355-357, 359, 362-364
Conradi-Hünermann syndrome, 31/275

corpus callosum agenesis, 30/150, 155, 286
Coxsackie virus infection, 34/135
cranial cephalocele, 30/154, 209
craniolacunia, 30/272, 50/138
craniosynostosis, 30/154, 221, 225
Creutzfeldt-Jakob disease, 56/546
cri du chat syndrome, 30/161, 31/569-571
cryptococcosis, 35/16
cycad intoxication, 65/21
cytomegalic inclusion body disease, 56/264
cytomegalovirus infection, 31/215, 34/212-214, 56/264
Dandy-Walker syndrome, 30/634, 50/328
De Lange syndrome, 31/279
decompression myelopathy, 61/221
dermal sinus and dermoid, 30/165, 32/455
dermatomyositis, 71/129
dermoid cyst, 31/61-63
description, 30/139
diabetes mellitus, 70/131
diastematomyelia, 50/439
diphtheria, 52/227, 65/237
diplomyelia, 50/439
Down syndrome, 30/149-151, 160
dystonia musculorum deformans, 6/519, 49/520
eastern equine encephalitis, 34/72
echinococcosis, 35/12, 175, 345-347, 52/523
ECHO virus infection, 34/139
Edwards syndrome, 31/516, 50/560
electrical injury, 61/193
eosinophilia myalgia syndrome, 64/249
ependymoma, 16/65, 31/46
epidermoid cyst, 31/67
epilepsy, 15/491, 21/3, 72/15
ethylene glycol intoxication, 64/122
familial spastic paraplegia, 21/19, 59/302-304
febrile seizure, 73/309
fetal alcohol syndrome, 30/121, 43/197
filariasis, 35/11, 161-163, 212
focal dystonia musculorum deformans, 49/520
foramina parietalia permagna, 30/155, 276, 50/141
Friedreich ataxia, 21/3, 15, 60/307
Gaucher disease, 10/566
genetics, 30/142
germ cell tumor, 68/231
Gerstmann-Sträussler-Scheinker disease, 56/554
giant axonal neuropathy, 60/75
Gilles de la Tourette syndrome, 49/632
GM2 gangliosidosis, 10/297, 558-562
gonococcal meningitis, 33/103
Guam amyotrophic lateral sclerosis, 21/34, 59/259-262
Guillain-Barré syndrome, 34/393, 51/239, 71/551
Hallervorden-Spatz syndrome, 66/713
head injury, *see* Head injury epidemiology
helminthiasis, 35/211-218, 52/507
hemangioendothelioma, 68/281
hemangiopericytoma, 68/275
Hemophilus influenzae meningitis, 33/54, 52/59
hepatitis B virus, 56/295
hereditary amyloid polyneuropathy, 21/8
hereditary amyloid polyneuropathy type 1, 21/130, 60/96
hereditary amyloid polyneuropathy type 2, 21/130, 60/98
hereditary amyloid polyneuropathy type 3, 21/8, 60/99
hereditary motor and sensory neuropathy type I, 21/3, 5, 9, 60/187
hereditary motor and sensory neuropathy type II, 21/3, 5, 9, 60/187
hereditary sensory and autonomic neuropathy type III, 10/562, 21/5, 11, 60/24
hereditary spinocerebellar degeneration, 21/14-18
hereditary tremor, 49/569
heredodegenerative disease, 21/3-42
herpes simplex virus encephalitis, 34/149
herpes simplex virus meningitis, 56/128
herpes zoster, 56/230
herpes zoster oticus, 34/171-173
history, 30/139
holoprosencephaly, 30/156, 457
hookworm disease, 35/211, 362
human immunodeficiency virus infection, 71/353
human immunodeficiency virus neuropathy, 71/353
human T-lymphotropic virus type I associated myelopathy, 56/533
Huntington chorea, 6/303, 49/268
hydranencephaly, 30/661
hydrocephalus, 30/149-151, 157, 163, 534, 661
hydromyelia, 30/150
hypertelorism, 30/154
hypertrophic interstitial neuropathy, 21/3
hypoglycemia, 70/174
idiopathic scoliosis, 42/52
incidence rate, 30/140, 47/259
infantile autism, 46/191
infantile spasm, 73/201
infantile spinal muscular atrophy, 59/53
infectious endocarditis, 63/111
iniencephaly, 30/155, 265, 50/135
interrelation of measures, 30/142
intracranial aneurysm, 30/160
intraspinal neurenteric cyst, 32/437, 442
investigation, 30/139

JC virus, 71/413
juvenile absence epilepsy, 73/150
juvenile hereditary benign chorea, 49/339, 345
Kaposi sarcoma, 71/340
Kawasaki syndrome, 52/264, 56/637
Kii Peninsula amyotrophic lateral sclerosis, 22/354, 359, 367, 59/273
Klinefelter syndrome, 31/477-479, 50/547
Klippel-Feil syndrome, 30/150, 162, 32/115
Lafora progressive myoclonus epilepsy, 42/706
Landau-Kleffner syndrome, 73/282
Lassa fever, 34/194
late infantile neuronal ceroid lipofuscinosis, 66/672
lathyrism, 36/506, 65/5
lead intoxication, 36/21, 64/432, 448
legionellosis, 52/254
leprosy, 51/216
Leptospira interrogans, 55/415
leptospirosis, 33/398
leukemia, 69/234
levodopa induced dyskinesia, 49/196
Listeria monocytogenes infection, 31/217
listeriosis, 33/79
low grade glioma, 68/35
lumbar vertebral canal stenosis, 32/332
Lyme disease, 52/261
lymphocytic choriomeningitis, 56/362
lymphoma, 67/132
Machado-Joseph disease, 60/469
macrencephaly, 30/158, 649
malaria, 35/6, 52/365
malignant hyperthermia, 62/567
manganese intoxication, 36/217, 232, 64/312
marine toxin intoxication, 65/143, 151
medulloblastoma, 68/182
meningioma, 68/401
meningocele, 32/206, 211, 223
meningococcal meningitis, 33/22, 52/21
mental deficiency, 46/2
metachromatic leukodystrophy, 10/568
metopic suture, 30/227
microcephaly, 30/156, 515
migraine, 48/13-20, 35, 118
mitral valve prolapse, 63/21
mixed connective tissue disease, 71/122
Möbius syndrome, 30/156
Mollaret meningitis, 34/547, 56/627
mortality rate, 30/141, 47/259
motoneuron disease, 21/23
moyamoya disease, 55/294
mucormycosis, 35/543, 52/468
multiple sclerosis, 9/63-84, 93-96, 47/259-283
mumps, 56/429
mushroom toxin, 65/35
myasthenia gravis, 41/101, 62/399
Mycobacterium tuberculosis, 52/195
neurinoma, 16/65
neuroamebiasis, 52/313
neuroborreliosis, 51/200
neurocysticercosis, 35/13, 293-295, 52/529
neurodegenerative disease, 21/3-42
neurofibromatosis, 30/150, 158, 31/1
neurofibromatosis type I, 30/158, 31/1
neuroleptic akathisia, 65/286
neuroleptic dystonia, 65/282
neuroleptic parkinsonism, 65/294
neuronal ceroid lipofuscinosis, 10/576
neurosyphilis, 33/376-382
neurotoxic shellfish poisoning, 65/157
Niemann-Pick disease, 10/568-572
nocardiosis, 35/517, 52/446
North American blastomycosis, 35/16, 380, 402, 52/386
olfactory bulb agenesis, 30/150
oligodendroglioma, 68/123
olivopontocerebellar atrophy (Dejerine-Thomas), 60/521
olivopontocerebellar atrophy (Wadia-Swami), 60/494
opiate addiction, 65/351
optic nerve aplasia, 30/400
optic pathway tumor, 68/74
opticocochleodentate degeneration, 60/751
orbital hypertelorism, 30/154
osteitis deformans (Paget), 43/451
pallidal degeneration, 49/446
pallidoluysiodentate degeneration, 49/446
pallidoluysionigral degeneration, 49/446
pallidonigral degeneration, 49/446
paracoccidioidomycosis, 52/455
paragonimiasis, 35/215, 243
paraneoplastic syndrome, 69/349
Parkinson dementia, 49/169
Parkinson dementia complex, 59/274
Parkinson disease, 49/124
Patau syndrome, 31/505, 50/556
pediatric head injury, 23/446, 57/327
pellagra, 28/64-76
penetrating head injury, 57/299
pertussis encephalopathy, 33/279-282
phakomatosis, 30/158-160, 31/1-3
Pick disease, 46/233
pilonidal sinus, 30/146, 150
pineal tumor, 68/231
pinealoma, 16/64

pituitary adenoma, 67/133
pituitary gland aplasia, 30/150
platybasia, 30/150, 161
poliomyelitis, 34/93, 99-102, 42/653, 56/314
polyarteritis nodosa, 55/353
polymyositis, 71/129
pontine infarction, 12/14
porencephaly, 30/158, 681, 50/355
porphyria, 60/119
postpoliomyelitic amyotrophy, 59/36
posttraumatic brain abscess, 57/321
posttraumatic epilepsy, 72/193
posttraumatic meningitis, 57/317
pressure sore, 61/347
prevalence rate, 30/140, 47/259
primary amebic meningoencephalitis, 35/34-37, 52/313
primary CNS lymphoma, 68/257
primary lateral sclerosis, 21/30
progressive bulbar palsy, 59/220
progressive external ophthalmoplegia, 62/291
progressive multifocal leukoencephalopathy, 34/313-321, 71/413
progressive rubella panencephalitis, 56/405
prolonged transient ischemic attack, 53/258
proportional rate, 30/141
rabies, 34/250-252, 56/384
rachischisis, 30/150
ratio, 30/142
Refsum disease, 21/12
Reye syndrome, 29/336, 34/80, 168, 169, 56/161, 65/121
rheumatoid arthritis, 71/21
rickettsial infection, 34/643
rubella syndrome, 31/213
Rubinstein-Taybi syndrome, 31/282, 43/235
Ryukyan spinal muscular atrophy, 21/25
sacral arachnoid cyst, 32/410
sagittal suture, 30/224
sarcoidosis, 71/470
scoliosis, 30/162, 50/417
septo-optic dysplasia, 30/321
shellfish poisoning, 65/165
Sjögren syndrome, 39/420
skeletal fluorosis, 36/466-468
snoring, 63/450
sphenoid wing aplasia, 30/278, 50/143
sphingolipidosis, 10/556-587
spina bifida, 30/149-151, 163, 32/546-548, 50/477
spinal arachnoid cyst, 32/398, 404
spinal cord electrotrauma, 61/193
spinal cord injury, 25/141-143, 61/499
spinal cord tumor, 68/497
spinal tumor, 68/511
spinocerebellar degeneration, 21/14
spongy cerebral degeneration, 10/575
Sturge-Weber syndrome, 30/159, 31/1
subacute sclerosing panencephalitis, 34/356-358, 56/418
subarachnoid hemorrhage, 12/70, 55/1, 43
subdural empyema, 57/323
syncope, 75/209
syphilis, 33/376-382
syringomyelia, 30/164, 32/256-258, 50/451
systemic lupus erythematosus, 55/369, 71/35
systemic malignancy, 71/611
Takayasu disease, 55/335
tardive dyskinesia, 49/188
temporal arteritis, 55/341
temporal lobe epilepsy, 73/68
teratoma, 31/42, 51
tetanus, 33/495-500, 52/229, 65/216, 71/559
tetrodotoxin intoxication, 65/155
thalamic tumor, 68/63
toxic shock syndrome, 52/258
toxoplasmosis, 35/117
transient ischemic attack, 53/258
traumatic intracranial hematoma, 57/250
trichinosis, 35/213, 267, 52/563
trichloroethylene intoxication, 36/457
tropical spastic paraplegia, 56/533
tropical sprue, 7/628
tuberculoma, 16/65
tuberculous meningitis, 52/195
tuberous sclerosis, 30/159
Turner syndrome, 31/497
valproic acid induced hepatic necrosis, 65/127
varicella zoster, 34/170
venous sinus thrombosis, 54/395
vertebral canal stenosis, 50/474
vertebral column, 32/131
vestibular neurinoma, 68/423
viral meningitis, 56/127
Von Hippel-Lindau disease, 30/159, 31/1, 68/274
Whipple disease, 52/136
Wolf-Hirschhorn syndrome, 30/161, 31/559
XYY syndrome, 31/486
Epidermal growth factor
muscle tissue culture, 62/88
Epidermal growth factor receptor
astrocytoma, 67/297
Epidermal nevus syndrome, *see* Linear nevus sebaceous syndrome *and* Nevus unius lateris
Epidermoid
acquired, *see* Acquired epidermoid
arachnoid cyst, 31/87, 89

chemical meningitis, 56/130
congenital spinal cord tumor, 32/358-386
hypothalamic tumor, 68/71
Mollaret meningitis, 34/548, 56/630
myelomeningocele, 20/76
optic chiasm compression, 68/75
pathology, 68/236
spinal cord compression, 19/363
Epidermoid cyst
*see also* Cholesteatoma
acoustic nerve tumor, 17/684
age, 20/77
associated defect, 20/77
cerebellopontine angle tumor, 5/273
characteristics, 20/72, 31/87
chronic meningitis, 56/645
classification, 68/310
clinical features, 20/78, 31/67
clinical presentation, 68/331
congenital, 32/355-386
congenital brain tumor, 31/61-68
cranial vault tumor, 17/113
CSF keratin, 20/79
diagnosis, 68/331
epidemiology, 31/67
epilepsy, 72/359
frequency, 20/77
lateral ventricle tumor, 17/606
location, 20/77, 31/67
lumbar puncture, 61/159
origin, 20/75
pathogenesis, 31/61, 68/329
pathology, 20/73, 31/67
pineal tumor, 17/654
radiology, 20/78
retrospinal, 20/79
sequestration, *see* Acquired epidermoid
sex ratio, 20/78
spinal, *see* Spinal epidermoid
spinal cord, 19/16
supratentorial brain tumor, 18/317
treatment, 31/68, 68/333
vertebral canal, 31/89
Epidermoid tumor
epilepsy, 72/181
Epidermolysis bullosa dystrophica
epilepsy, 14/13
mental deficiency, 14/13
neurofibromatosis type I, 14/568
Epidermolysis bullosa dystrophica hereditaria
autosomal recessive transmission, 14/791
lethal neonatal variant, 14/791
mental deficiency, 14/791
mortality, 14/791
severe form, 14/791
Epididymal cystadenoma
Von Hippel-Lindau disease, 50/375
Epididymitis
Behçet syndrome, 56/594
Epidural anesthesia
brain hemorrhage, 61/164
central cord necrosis, 61/138
complication, 61/149
complication determinant, 61/138
complication frequency, 61/137, 65/420
complication rate, 65/422
cranial nerve palsy, 61/164
epidural hematoma, 26/18, 61/137
Horner syndrome, 61/150
ischemic myelopathy, 61/137
local anesthetic agent, 65/420
management guideline, 61/141
neurologic complication, 61/137
paraplegia, 61/137, 142
spinal epidural empyema, 52/186
spinal nerve root lesion, 61/137
toxic myelopathy, 61/137
Epidural angioma
spinal, *see* Spinal epidural angioma
Epidural arachnoid cyst
congenital, 32/394-400
diagnostic features, 31/89-96
empty sella syndrome, 32/397
microscopic features, 20/609
traumatic, *see* Postoperative meningeal diverticulum
Epidural cyst
Milroy disease, 32/395
spinal, *see* Spinal epidural cyst
Epidural disease
Hodgkin disease, 69/276
non-Hodgkin lymphoma, 69/276
Epidural empyema
clinical features, 26/39
cranial, *see* Cranial epidural empyema
filariasis, 52/516
history, 26/39
nocardiosis, 52/447, 449
operative finding, 26/40
spinal, *see* Spinal epidural empyema
spinal cord injury, 25/60
staphylococcal meningitis, 52/72
traumatic, 26/40
treatment, 26/40
Epidural ganglion cyst, 20/605-609
anatomy, 20/605

cause, 20/605
clinical features, 20/606
diagnosis, 20/607
histology, 20/609
intervertebral facet joint, 20/605
myelography, 20/607
spinal nerve root syndrome, 20/605
treatment, 20/609
Epidural hematoma
*see also* Intracranial hematoma
age, 24/265, 26/14
alcoholism, 63/313
anatomy, 26/5, 34
angiography, 24/271
anterior fossa, 24/263, 266, 639
anterior meningeal artery, 57/252
anticoagulant, 26/18
blood dyscrasia, 26/19
boxing injury, 23/547
brain herniation, 24/264
brain injury, 24/264, 268
brain injury mechanism, 57/32
brain scintigraphy, 24/272
case survey report, 26/13
cause, 54/346, 57/337
chronic, *see* Chronic epidural hematoma
clinical course, 54/346
clinical features, 24/264-269, 26/4, 20
complete cord lesion, 26/21
computer assisted tomography, 23/259, 262, 54/347, 57/103, 152, 154, 256, 274
contrecoup, 57/255
CSF, 26/21
definition, 24/261
development, 57/255
diagnosis, 24/269, 54/346
differential diagnosis, 24/270, 26/23
dura stripping, 57/255
echoEG, 23/275, 24/272
EEG, 23/348, 24/272, 57/186
epidural anesthesia, 26/18, 61/137
epilepsy, 72/216
evolution, 57/255
football injury, 23/572, 574
Glanzmann thrombasthenia, 63/325
head injury, 24/264, 57/47, 103, 335
hemophilia, 26/4, 19, 63/310
history, 24/261, 26/1
incidence, 24/261, 57/251
injury mechanism, 24/264
intracerebral calcification, 49/417
intracranial, *see* Intracranial epidural hematoma
intracranial pressure, 57/260
intradural lesion, 24/264, 268
local effect, 57/259
localization, 57/251
location, 54/346
lucid interval, 24/267, 55/182, 185
lumbar puncture, 24/272, 26/18
lymphoma, 63/350
macroscopic pathology, 25/29
management, 24/272, 57/335
midbrain, 57/256
middle fossa, 24/263, 266
middle meningeal artery, 57/252, 256
midline shift, 16/102
mode of onset, 26/16
mortality, 26/24, 57/103
mummy, 74/187
myelography, 26/22, 25
natural history, 26/12
nuclear magnetic resonance, 57/174
oculomotor nerve injury, 24/63, 268
operative treatment, 24/273
outcome, 57/283
pain, 19/15
papilledema, 24/266
parasagittal, 24/263, 266
pathogenesis, 24/262, 26/3, 17
pathology, 23/49, 24/264, 57/47
pathophysiologic cascade, 57/259
pediatric head injury, 57/335, 337
perinatal, 26/18
posterior fossa, *see* Posterior fossa epidural hematoma
posterior fossa hematoma, 24/343, 57/159, 254
posterior meningeal artery, 57/252
postoperative management, 57/278
primary hemorrhagic thrombocythemia, 38/72
prognosis, 54/349
pupil, 24/268
Queckenstedt test, 26/21
radiodiagnosis, 24/271, 57/152
result, 24/273
sagittal sinus, 57/252, 256
segmental lesion distribution, 26/14, 22
sex ratio, 24/262, 26/14
site, 24/263, 266, 270
skull fracture, 57/255
source, 24/262
space occupying lesion, 16/240
spastic paraplegia, 59/438
spinal cord vascular disease, 12/558
spinal injury, 12/558, 25/29, 47, 26/1-30
spinal site, *see* Spinal epidural hematoma
spontaneous, 26/2, 3, 17

subarachnoid hemorrhage, 12/162
subdural hematoma, 24/279
surgery, 24/267, 639, 655
surgery indication, 57/273
surgical technique, 24/273, 57/277
symptom, 54/346
tramline appearance, 57/256
transverse sinus, 57/252
trauma, 54/345
traumatic, 26/2, 17
traumatic aneurysm, 24/392
traumatic intracerebral hematoma, 24/352
traumatic intracranial hematoma, 57/152
treatment, 54/349
trivial trauma, 26/19
vascular abnormality, 26/19
vascular striatopallidodentate calcification, 49/417

Epidural lesion
back pain, 69/65
spinal, *see* Spinal epidural lesion

Epidural lymphoma
epilepsy, 63/349
headache, 63/349
Hodgkin lymphoma, 63/349
non-Hodgkin lymphoma, 63/349
spinal, *see* Spinal epidural lymphoma
spinal cord compression, 63/349

Epidural metastasis
breast cancer, 69/205, 206
glucocorticoid, 69/208
radiotherapy, 69/208
surgery, 69/209
systemic therapy, 69/209
treatment, 69/206

Epidural space
cancer, 69/63

Epidural spinal abscess, *see* Spinal epidural empyema

Epidural tumor
angiomyolipoma, 20/110
chloroma, 63/343
differential diagnosis, 26/23
leukemia, 69/239
spinal cord, 67/196
spinal cord compression, 19/356
treatment, 26/24

Epidural venous plexus
anatomy, 26/9-12

Epiduritis
acute, *see* Spinal epidural empyema
differential diagnosis, 26/23

Epigastric reflex
diagnostic value, 1/252

Epilepsia partialis continua, *see* Koshevnikoff epilepsy

Epilepsie Bravais-Jacksonienne, *see* Hereditary paroxysmal kinesigenic choreoathetosis

Epilepsy
*see also* Convulsion, Convulsive epilepsy, Epileptogenesis, Nonconvulsive seizure, Seizure *and* Status epilepticus
absence status, 73/474
acetazolamide, 15/648, 73/350
acetazolamide intoxication, 37/418
acetylcholine, 15/64, 29/451, 72/89
α-*N*-acetylgalactosaminidase deficiency, 72/222
acetylsalicylic acid intoxication, 65/460
acquired immune deficiency syndrome dementia, 56/491
acrylamide intoxication, 64/63
acute amebic meningoencephalitis, 52/317
acute bacterial meningitis, 72/158
acute infantile hemiplegia, 12/342, 53/31
acute intermittent porphyria, 72/222
acute subdural hematoma, 24/280
adenosine triphosphatase 6 mutation syndrome, 62/510
adenovirus encephalitis, 56/283
adenylosuccinate deficiency, 72/222
adolescent stigma scale, 72/434
adrenoleukodystrophy, 60/169, 72/222
adult neuronal ceroid lipofuscinosis, 60/665
Adverse Drug Events Profile, 72/431
adverse drug reaction, 73/373
adverse effect scale, 72/430
affective symptom, 73/474
age, 72/19, 22
age group, 72/176
age at onset, 15/265
aggression, 3/344, 73/479
agyria, 72/109
Aicardi syndrome, 72/129
air travel, 73/379
alcohol, 72/234
alcohol abuse, 73/455
alcoholic blackout, 72/250
alcoholic polyneuropathy, 51/317
alcoholism, 43/197
alfadolone acetate with alfaxalone intoxication, 37/409
allergic granulomatous angiitis, 63/383
allergic rhinitis, 5/235
allesthesia, 45/171
alopecia, 42/684
Alpers disease, 42/486
aluminum, 72/53

aluminum intoxication, 36/323, 63/525
aluminum neurotoxicity, 63/525
amino acid, 15/315, 72/56, 84
Ammon horn, 15/612
amnesia, 72/251
amnesic shellfish poisoning, 65/167
amoxicillin, 65/472
amphetamine, 72/234
amphetamine intoxication, 65/51, 258
amphotericin, 65/472
ampicillin, 65/472
amsacrine, 65/529
amsacrine intoxication, 65/538
amygdala, 15/99
amygdalohippocampectomy, 73/390
analgesic agent, 72/229
anatomic classification, 72/10
anesthetic agent, 72/228
angiotonic, 15/2
anhidrotic ectodermal dysplasia, 14/788
animal model, 72/2, 52, 62
animal study, 72/85
anoxia, 72/362
anoxic, *see* Syncope
antecedent oriented method, 73/444
antenatal infection, 72/347
anterior temporal lobectomy, 73/390
antiarrhythmic agent, 72/229
antidepressant, 72/229
antiepileptic agent, 72/229
antihistaminic agent, 72/229
antihypertensive agent, 63/87
anti-infective agent, 72/229
anti-inflammatory agent, 72/232
anxiety, 3/348, 15/95, 73/430, 454, 478
aphasia, 4/86, 15/80, 45/322, 46/150
apnea, 63/481
aprindine intoxication, 65/455
aqueduct stenosis, 50/312
ARG syndrome, 50/582
Argentinian hemorrhagic fever, 56/365
arginase, 72/222
argininosuccinase, 72/222
argininosuccinate synthetase, 72/222
arithmic calculation, 73/189
arteriosclerosis, 15/304
arteriovenous malformation, 72/213
arteritis, 72/216
arthropod envenomation, 65/197
arylhexylamine, 72/234
asparaginase, 65/529
aspartic acid, 72/84
Aspergillus fumigatus, 63/532
asphyxia, 72/344
assessment, 73/455
astrocytoma, 15/299, 18/19-22, 72/355
ataxia, 42/114
ataxia telangiectasia, 15/325
atonic, 15/138
atrophoderma idiopathica (Pasini-Pierini), 14/791
atrophy, 73/399
attention disorder, 3/150
atypical benign partial, *see* Atypical benign partial epilepsy
auditory, 15/83
auditory precipitation, 15/445
auditory stimulation, 15/446
automatism, 73/474, 480
autonomic dysfunction, 15/140
autonomic nervous system, 75/349, 353
autonomic symptom, 15/140
autoscopia, 4/231-233, 45/387, 509
autosomal dominant gene defect, 72/130
autosomal dominant inheritance, 50/272
awakening, *see* Awakening epilepsy
bacterial endocarditis, 52/291, 298
bacterial meningitis, 52/14, 55/420
Baltic myoclonus, *see* Baltic myoclonus epilepsy
barbiturate, 15/635, 670, 72/234
Bardet-Biedl syndrome, 13/392
basilar artery migraine, 48/137
Batten disease, 10/575, 579, 608, 42/461, 463, 466, 60/666
behavior, 72/267
behavior approach, 73/442
behavior change, 46/416
behavior technic, 73/441
behavior therapy, 73/442
Behr disease, 15/329, 331
benign frontal lobe, *see* Benign frontal lobe epilepsy
benign neonatal familial convulsion, 72/129, 131, 73/130, 253
benign neonatal idiopathic convulsion, 73/253
benign partial, *see* Benign partial epilepsy
benign rolandic, *see* Benign rolandic epilepsy
benzodiazepine, 15/645, 669, 72/234, 73/360
bilateral synchronous, 15/44
biofeedback, 73/445
biotinidase, 72/222
biplex wave, 15/9
birth incidence, 15/201, 544
bismuth intoxication, 64/338
bladder function, 15/95, 140
blepharospasm, 1/618
blood hyperviscosity, 55/484

body scheme disorder, 4/214
Bolivian hemorrhagic fever, 56/369
Bonnet-Dechaume-Blanc syndrome, 14/109
boric acid intoxication, 64/360
Börjeson-Forssman-Lehmann syndrome, 31/311, 42/530
boxing injury, 23/530
brain abscess, 15/307, 33/117, 139, 52/152, 400, 72/145
brain air embolism, 55/193
brain aneurysm, 55/53
brain aneurysm neurosurgery, 55/53
brain arteriosclerosis, 15/304
brain arteriovenous malformation, 12/241, 246, 15/303, 31/181, 54/377
brain atrophy, 15/327, 543, 547
brain chordoma, 72/181
brain cortex dysplasia, 30/364
brain cortex focal dysplasia, 50/207
brain defect, 72/249
brain edema, 42/690, 63/417
brain fat embolism, 55/183
brain hemorrhage, 72/215
brain hypoxia, 15/62, 612, 63/417, 72/214
brain infarction, 15/304, 63/345, 72/214
brain injury, 15/561, 23/234, 24/445, 57/130, 134, 73/470
brain lipoma, 72/181
brain metastasis, 18/216, 72/182
brain stab wound, 23/484
brain surgery, 45/9
brain tissue, 72/66
brain tumor, 15/295, 16/254, 259, 733, 18/19-22, 326, 72/175
brain tumor surgery, 16/265
brain vasculitis, 55/420
Bravais classification, 15/4
Brégeat syndrome, 14/475
Bridge classification, 15/13
Bright classification, 15/4
bronchodilating agent, 72/228
Burkitt lymphoma, 63/351, 72/180
calcium, 15/312, 72/54
calcium current, 72/45
California encephalitis, 56/141
callosal section, 73/408
Calmeil classification, 15/3
cancer, 69/9
Candida meningitis, 52/400
cannabinoid, 72/234
cannabis, 72/238
captopril intoxication, 65/445
carbamazepine, 15/653, 669, 73/358
carbamoylphosphate synthetase, 72/222
carbon disulfide intoxication, 64/24
carbon monoxide intoxication, 63/418, 64/31
cardiac arrest, 63/217
cardiac dysrhythmia, 63/236
cardiovascular response, 75/349
care measurement, 72/417
carmustine, 65/528
carnosinase deficiency, 72/222
carotid sinus reflex, 11/546
carotid sinus syndrome, 11/546
catecholamine, 72/91
cavernous angioma, 42/724, 72/129, 213
cavernous sinus thrombosis, 52/174
cavum veli interpositi, 30/328
cefazolin, 65/472
central motor region, 15/44
central neurocytoma, 72/181
centrencephalic, *see* Centrencephalic epilepsy
centrofacial lentiginosis, 14/782, 43/26
cerebellar agenesis, 30/383
cerebellar ataxia, 42/125
cerebellum, 15/616
cerebral amebiasis, 52/315
cerebral hemiatrophy, 50/206
cerebral heterotopia, 30/493
cerebral malaria, 35/146, 52/369
cerebrohepatorenal syndrome, 72/222
cerebrotendinous xanthomatosis, 60/167, 66/602
cerebrovascular disease, 53/36
Chagas disease, 52/347
Chalfont National Hospital Scale, 72/418
Chalfont Scale, 72/430
Chalfont Seizure Severity Scale, 72/430
Charlevoix-Saguenay spastic ataxia, 60/666
chess and card, *see* Chess and card epilepsy
child, 72/27, 86, 73/459, 465
Child Attitude Toward Illness, 72/434
childhood, *see* Childhood epilepsy
Childhood Illness Scale, 72/434
childhood myoglobinuria, 62/564
chlorambucil, 65/528
chlordecone intoxication, 64/204
chlorinated cyclodiene, 64/200
chlorinated cyclodiene intoxication, 64/200, 202
chlormethine, 65/528
chlormethine intoxication, 64/233, 65/534
chloroquine intoxication, 65/483
chlorphenotane, 64/199
chorea-acanthocytosis, 49/329, 51/398, 63/281
choreoathetosis, 42/699
choroid plexus neoplasm, 72/181
chromosomal abnormality, 72/128

chronic, 72/419
chronic enterovirus infection, 56/330
chronic model, 72/52
chronic subdural hematoma, 24/305
ciclosporin, 63/10, 189, 537
ciclosporin intoxication, 63/537, 65/551
ciguatoxin intoxication, 65/163
cilastatin, 65/472
ciprofloxacin intoxication, 65/486
cisplatin, 65/529
cisplatin intoxication, 64/358, 65/537
classic migraine, 48/159
classification, 15/1-8, 10-13, 23, 25, 90, 492, 42/682, 72/1-12, 18, 340
clonic, 15/122
CNS stimulant, 72/228
cobalt, 72/53
cocaine, 65/251, 72/236
cocaine intoxication, 65/251, 255
cognitive behavior therapy, 73/445
cognitive features, 73/474
cognitive function, 72/388
colloid cyst, 72/181
complex partial status, 73/474
complex I deficiency, 62/504
compound depressed fracture, 57/135
compromised, 73/454
computer assisted tomography, 72/337
conditioning, 15/550
confusion, 73/480
congenital heart disease, 38/119-123, 63/3, 72/212
congenital muscular dystrophy, 43/91
congenital pain insensitivity, 51/564
congenital Pelizaeus-Merzbacher disease, 15/376, 42/504
congenital retinal blindness, 13/272, 15/329, 22/529, 533
congestive heart failure, 63/132
consciousness, 3/123, 15/54, 130, 135, 594, 72/250
contrast media, 72/233
contraversive, 15/78
convulsant, 72/85
convulsive, *see* Convulsive epilepsy
convulsive syncope, 72/259
corpus callosum, 15/43
corpus callosum agenesis, 42/7, 72/109
cortical excision, 15/76
cortical hemiatrophy, 30/356
cortical myoclonus, 49/613
course, 15/785
cranial subdural empyema, 52/170, 172
craniofacial dysostosis (Crouzon), 31/234
Creutzfeldt-Jakob disease, 15/346, 46/291, 72/222
criminality, 15/810
Crome syndrome, 43/242
cross-cultural adaptation, 72/429
cryoglobulinemia, 63/404
cryptococcal meningoencephalitis, 52/432
Cryptococcus neoformans, 63/532
cutis verticis gyrata, 43/244
cyanide intoxication, 65/28
cycloserine, 28/116, 65/481
cycloserine intoxication, 65/481
cysticercotic cyst, 72/149
cysticercotic granuloma, 72/168
cytarabine intoxication, 65/533
cytotoxic agent, 72/232
dacrystic, *see* Dacrystic epilepsy
DC activity, 15/40
deafness, 42/688
debrancher deficiency, 62/482
decaborane intoxication, 64/360
decompression illness, 63/420
definition, 72/16
degenerative disease, 15/325, 72/125
Delasiauve classification, 15/4
delirium, 15/596, 46/544
dementia, 15/344, 42/114, 46/128
dental abnormality, 42/691
dentate, *see* Dentate epilepsy
depersonalization, 15/585
depressed skull fracture, 57/334
depression, 3/348, 15/597, 46/461, 73/430, 454, 478
dermoid tumor, 72/181
desensitization, 73/444
designer drug, 72/236
destructive lesion, 73/399
deterioration, 15/586
devastated, 73/454
developing animal, 72/85
developing child, 73/463
developing country, 72/26
developmental disorder, 72/340
dextropropoxyphene intoxication, 65/358
diagnosis, 15/297
diagnostic test, 73/388
dialysis, 63/525
dialysis disequilibrium syndrome, 63/523
dialysis encephalopathy, 63/526, 64/277
diazepam, 15/645
diencephalic, *see* Diencephalic epilepsy
diffuse dysrhythmia, 15/9
diffuse simultaneous corticosubcortical, *see* Diffuse simultaneous corticosubcortical

epilepsy
dihydropteridine reductase deficiency, 72/222
dihydropyrimidine dehydrogenase, 72/222
dipeptidyl carboxypeptidase I inhibitor, 65/445
dipeptidyl carboxypeptidase I inhibitor intoxication, 65/445
dipropylacetate, 15/669
disconnection surgery, 73/391
disinhibition, 72/108
disopyramide intoxication, 65/453
diurnal distribution, 15/459
Divry-Van Bogaert syndrome, 42/725, 55/318, 321
dopamine, 72/92
Down syndrome, 31/407
drivers licence, 15/808
driving, 72/421
drop attack, 72/250
drug, 72/227
drug induced, 28/114-117
drug interaction, 73/362
drug survey, 15/668
drug treatment, 15/664
Dumon-Radermecker syndrome, 42/477-479
duration, 15/298
Dutch Quality of Care, 72/433
dynamic disease, 72/41
dysembryoplastic neuroepithelial tumor, 72/181
dysphoria, 5/79, 15/596
dyspraxia, 72/133
dyspropterin synthase deficiency, 72/222
dysraphia, 72/110
dyssynergia cerebellaris myoclonica, 15/332, 60/599
dystrophia dermochondrocornealis, 14/792
early, 57/130, 134
early childhood myoclonic, *see* Early childhood myoclonic epilepsy
early infantile encephalopathy, 73/235
early myoclonic encephalopathy, 72/8, 73/235
eastern encephalitis, 56/136
eating, *see* Eating epilepsy
echinococcosis, 52/524
echoEG, 15/554
eclampsia, 39/547
education, 15/806
EEG, 15/498-500, 46/493, 72/288, 293
EEG classification, 15/8, 72/307
Ehlers-Danlos syndrome, 43/14
elderly patient, 73/376
electrical injury, 23/713, 716, 61/192
electrocorticography, 15/32
electrode, 15/760
electrolyte balance, 15/62, 311, 313
emotional disorder, 3/344, 348, 358, 363
employment, 72/421, 73/434
enamel defect, 42/691
encephalocele, 72/109
encephalomalacia, 72/363
endocrine factor, 15/311
energy metabolism, 15/61
enflurane intoxication, 37/412
enhancement, 72/337
enoxacin intoxication, 65/486
enuresis, 45/142
environmental factor, 73/460
environmental toxin, 72/239
ependymoma, 18/20, 126, 72/181
ephedrine intoxication, 65/260
epidemiology, 15/491, 21/3, 72/15
epidermoid cyst, 72/359
epidermoid tumor, 72/181
epidermolysis bullosa dystrophica, 14/13
epidural hematoma, 72/216
epidural lymphoma, 63/349
epilepsy prone, 72/70
Epilepsy Surgery Inventory 55, 72/424
Epstein-Barr virus, 56/252
Epstein-Barr virus encephalitis, 56/252
Epstein-Barr virus infection, 56/252
Epstein-Barr virus meningoencephalitis, 56/252
ergot intoxication, 65/61, 68
erythropoietin, 63/535
Esquirol classification, 15/2
essential, 15/2
essential thrombocythemia, 63/331
estrogen, 15/731
ethionamide, 65/481
ethosuximide, 15/644, 670, 37/200-202, 73/360
ethylene glycol intoxication, 64/122
etiology, 15/271, 274, 785, 72/28, 340
excitation, 72/47
experimental, 15/32, 34
extrapyramidal myoclonus, 6/766
extratemporal reaction, 73/402
eye movement, 1/582
false porencephaly, 72/109
familial amaurotic idiocy, 15/423
familial myoclonus, *see* Myoclonic epilepsy
familial progressive myoclonus, *see* Unverricht-Lundborg progressive myoclonus epilepsy
familial spastic paraplegia, 15/327, 22/426, 428
family, 72/127, 73/433, 471
family therapy, 73/448
Farber disease, 72/222

fatigue, 15/480
febrile, *see* Febrile convulsion
febrile convulsion, 15/259
febrile seizure, 75/351
felbamate, 73/360
fetal Minamata disease, 64/418, 421
filariasis, 35/168, 52/514, 517
financial status, 73/434
first seizure, 72/21
flecainide, 63/192
flexor spasm, 14/347
fluorouracil, 65/528
fluorouracil intoxication, 65/532
Flynn-Aird syndrome, 14/112
focal, *see* Epileptic focus
focal cortical, 15/8
focal dysplasia, 72/109
fonsecaeasis, 35/566, 52/487
foramen magnum tumor, 20/199
Förster syndrome, 42/202
frequency, 72/175
Friedreich ataxia, 15/327, 21/347
frontal lobe, *see* Frontal lobe epilepsy
frontal lobe lesion, 2/752
frontal lobe tumor, 17/247, 270
frontal pole, 15/45
fructose-1,6-diphosphatase deficiency, 72/222
fucosidosis, 72/222
Fukuyama syndrome, 43/91, 72/129
functional, 15/7
functional connectivity, 72/47
fungal CNS disease, 72/160
GABA, 15/62, 29/503-505, 72/54, 58, 65, 84, 87, 89, 73/348
gabapentin, 73/360
Galen classification, 15/1
Gastaut classification, 15/26
Gaucher disease, 15/426, 42/438, 72/222
gelastic, *see* Gelolepsy
gene locus, 72/129
general health, 72/426
general with local onset, 15/107
generalized, *see* Generalized epilepsy
generalized nonconvulsive, 15/130
generalized tonic clonic, *see* Generalized tonic clonic seizure
genetic, *see* Genetic epilepsy
genetics, 15/429, 72/70, 125, 340, 75/351
genuine, 15/6
Gibbs classification, 15/7
glioblastoma multiforme, 18/20
globoid cell leukodystrophy, 15/374, 425, 72/222
globus pallidus, *see* Infantile spasm
glutamate formiminotransferase deficiency, 72/222
glutamic acid, 72/84
glutaric aciduria type II, 72/222
glutaryl-coenzyme A dehydrogenase deficiency, 72/222
δ-glyceric aciduria, 72/222
glycine, 72/95
glycosaminoglycanosis type I, 15/426
glycosaminoglycanosis type IIIA, 72/222
GM1 gangliosidosis, 72/222
GM2 gangliosidosis, 60/666, 72/222
GM3 gangliosidosis, 72/222
gold encephalopathy, 51/305
Gowers classification, 15/7
gram-negative bacillary meningitis, 52/122
grand mal, *see* Grand mal epilepsy
granule cell, 72/108
guanosine triphosphate cyclohydrolase I, 72/222
gunshot injury, 57/311
gustatory, 15/83
gustatory precipitation, 15/447
gyromitrine, 64/3
gyromitrine intoxication, 65/36
Hallervorden-Spatz syndrome, 15/337
hallucination, 4/330, 15/95, 45/508
hallucinogenic agent, 72/234
Halstead Category Test, 15/565
head trauma, 72/191
headache, 5/28
heart disease, 73/378
heart transplantation, 63/178, 180
heat stroke, 23/674, 679
Heemstede Neurotoxicity Scale, 72/419
helminthic infection, 72/165
hemangioblastoma, 72/180
hemiasomatognosia, 45/509
hemimegalencephaly, 50/263, 72/342, 73/393
hemisphere asymmetry, 72/398
hemispherectomy, 73/390, 406
hemispheric atrophy, 72/109
hemoglobin SC disease, 55/465
hemolytic uremic syndrome, 63/325
hemorrhage, 72/345, 73/254
hepatic coma, 27/362
hereditary ataxia, 15/326
hereditary cerebello-olivary atrophy (Holmes), 21/410
hereditary deafness, 22/510-512
hereditary fructose intolerance, 72/222
hereditary hearing loss, 22/510-513
hereditary long QT syndrome, 63/238
hereditary motor and sensory neuropathy type I,

15/325
hereditary motor and sensory neuropathy type II, 15/325
hereditary progressive myoclonus, *see* Unverricht-Lundborg progressive myoclonus epilepsy
hereditary sensory and autonomic neuropathy type I, 60/9
hereditary sensory and autonomic neuropathy type III, 15/350, 21/110, 60/27
hereditary spinocerebellar degeneration, 15/327
hereditary striatal necrosis, 49/495
herpes simplex virus encephalitis, 56/214
herpes virus, 72/155
heterochromia iridis, 42/688
heterocyclic antidepressant, 65/316, 319
heterocyclic antidepressant intoxication, 65/323
heterotopia, 72/109, 340, 342
hexachlorophene intoxication, 65/478
higher brain function, 73/187
hippocampal sclerosis, 72/112, 114
hippocampus, 15/99
histamine, 72/94
histidinemia, 42/554
histoplasmosis, 52/438
historic aspect, 72/2
history, 15/1
Hodgkin lymphoma, 63/349
holocarboxylase, 72/222
holoprosencephaly, 30/459, 72/109, 344
homocitrullinemia, 72/222
homocystinuria, 55/328, 72/222
hormonal response, 75/351
hormone, 72/232
hot water, *see* Hot water epilepsy
human immunodeficiency virus infection, 72/145, 155
Huntington chorea, 15/339, 72/129, 131
hydantoin, 15/627, 670
hydatidosis, 72/168
hydrocephalus, 30/600, 42/691
hydroxylysinemia, 29/223
4-hydroxyphenylpyruvic acid oxidase deficiency, 72/222
hyperammonemia, 72/222
hyperbaric dioxide intoxication, 61/217
hypercitrullinemia, 72/222
hypereosinophilic syndrome, 63/372
hyperexplexia, 42/228, 72/255
hyperkinetic child syndrome, 3/196, 4/363, 46/177
hyperlysinemia, 72/222
hypernatremia, 28/451, 63/554
hyperphenylalaninemia, 42/611
hypertensive encephalopathy, 11/553, 54/212
hypertrophic interstitial neuropathy, 42/317
hyperviscosity syndrome, 55/484
hypocalcemia, 63/558
hypoglycemia, 70/185, 72/222
hypoglycin intoxication, 65/79
hypomagnesemia, 63/562
hyponatremia, 63/545
hypoparathyroidism, 27/303-305, 70/123
hypophosphatemia, 63/564
hypothalamic hamartoma, 75/354
hypothyroidism, 27/260
hypoxia, 72/61
hypoxic ischemic encephalopathy, 73/254
hysterical, *see* Hysterical seizure
hysterical seizure, 46/575
iatrogenic agent, 72/227, 229
iatrogenic neurological disease, 61/165
ICD classification, 15/21
ictal aggression, 73/479
ictus, 73/473
idiopathic, 15/2, 5
idiopathic generalized myoclonic astatic, *see* Idiopathic generalized myoclonic astatic epilepsy
idiopathic hypereosinophilic syndrome, 63/124
idiopathic partial, *see* Idiopathic partial epilepsy
ifosfamide, 65/528
ifosfamide intoxication, 65/534
illusion, 45/508
imaging technique, 72/337
imipenem, 65/472
immunosuppressive agent, 72/232
inappropriate antidiuretic hormone secretion, 28/499
inborn error of metabolism, 72/223
incidence, 15/492, 494, 69/9, 72/18, 20, 23
incontinentia pigmenti, 14/12, 85, 214, 46/68
indifference to pain, 8/194
infantile benign myoclonic, *see* Infantile benign myoclonic epilepsy
infantile neuroaxonal dystrophy, 59/75
infantile Refsum disease, 72/222
infantile spasm, 42/682, 686, 692
infarction, 72/212
infection, 72/145, 349
infectious disease, 15/306
infectious endocarditis, 63/23, 112, 116
infectious lesion, 72/115
infectious mononucleosis, 56/252
infertility, 73/376
inhibition, 72/47
inhibitory, 15/138

insulin, 15/319
insurance, 15/800, 811
intelligence, 15/589
intensive care, 73/378
α-interferon intoxication, 65/562
γ-interferon intoxication, 65/562
interictal aggression, 73/481
interictal disorder, 73/475
interictal personality, 73/163
interleukin-2 intoxication, 65/539
intermittent light stimulation, 15/442
International Classification, 72/6
interruption technic, 73/443
intoxication, 72/227
intracerebral hematoma, 24/360
intracranial abscess, 72/158
intracranial calcification, 15/537
intracranial hematoma, 72/211
intracranial hypertension, 16/103
intractable hiccup, 63/490
intrinsic membrane stability, 72/45
invasive study, 73/389
ionic current, 72/47
iron, 72/53
ischemia, 72/108, 210
ischemic encephalopathy, 72/210
isoniazid, 28/114-116
isoniazid intoxication, 65/480
isotope study, 15/553
Jackson classification, 15/5
Japanese encephalitis, 72/157
Jasper-Kershman classification, 15/8
Joubert syndrome, 50/191, 63/437
juvenile absence, *see* Juvenile absence epilepsy
juvenile Gaucher disease, 10/523
juvenile metachromatic leukodystrophy, 60/666
juvenile myoclonic, *see* Juvenile myoclonic epilepsy
kainic acid, 72/56
Kawasaki syndrome, 52/265
Kearns-Sayre-Daroff-Shy syndrome, 62/308, 72/222
keratosis pilaris, 14/791
kindling, 72/57
kinesigenic choreoathetosis, 42/208
kinesthetic reflex, *see* Hereditary paroxysmal kinesigenic choreoathetosis
Klinefelter syndrome, 31/482
Klinefelter variant XXYY, 31/484
Klippel-Trénaunay syndrome, 14/395, 43/24
Koshevnikoff, *see* Koshevnikoff epilepsy
Krabbe-Bartels disease, 14/411
Kurland-Hauser classification, 15/23
Lafora progressive myoclonus, *see* Lafora progressive myoclonus epilepsy
Lafora progressive myoclonus epilepsy, 72/129
lamotrigine, 73/360
Langelüddeke classification, 15/13
Lassa fever, 56/370
late onset, *see* Epilepsy of late onset
lead encephalopathy, 64/436, 443
lead intoxication, 64/435
learning disability, 73/461
legal issue, 72/421
legionellosis, 52/254
Lennox classification, 15/8
Lennox-Gastaut syndrome, 72/118
Lesch-Nyhan syndrome, 63/256, 72/222
lesion, 73/470
lesion zone, 72/107
lesionectomy, 73/390
leukemia, 63/345
leukodystrophy, 15/373, 376, 425
lidocaine, 63/192
lidocaine intoxication, 37/449, 451, 65/454
life expectancy, 15/794
light stimulation, 15/441
lindane, 64/202
lindane intoxication, 65/479
lipid inclusion body, 42/692
lipidosis, 15/423
lissencephaly, 72/342
Listeria monocytogenes, 63/531
Listeria monocytogenes meningitis, 52/93
lithium intoxication, 64/296
liver disease, 27/362, 73/377
Liverpool Adverse Events Profile, 72/419
The Liverpool Assessment Battery, 72/422
Liverpool Quality of Life Battery, 72/432
Liverpool Seizure Severity Scale, 72/419, 430
lobectomy, 73/390
lobular sclerosis, 72/109, 112
local anesthetic agent, 65/420
local onset, 15/74
localization, 2/5, 72/2, 339
lumbar puncture, 61/166
Lyme disease, 52/262
lymphocytic choriomeningitis virus, 56/358
lymphoma, 63/350
lymphosarcoma, 72/180
lysergide, 65/45
lysinuric protein intolerance, 72/222
lysosomal disorder, 72/222
macrencephaly, 72/109, 343
macrosomatognosia, 4/230, 45/385, 509
madurelliasis, 35/567, 52/488

magnesium, 15/313, 72/85
major resection, 73/403
malaria, 72/146, 162
malformation, 72/108, 125, 73/254
manganese deficiency, 64/305
manganese intoxication, 64/305
mania, 15/596
manic depressive psychosis, 15/600
mannosidosis, 72/222
maple syrup urine disease, 72/222
maprotiline intoxication, 65/319
Marinesco-Sjögren syndrome, 15/329, 42/184
mastery sense, 73/432
McNaughton classification, 15/10
measles, 72/154
measles encephalitis, 72/149
mechanism, 72/291
medication trial, 72/428
mefloquine intoxication, 65/485
MELAS syndrome, 60/666, 62/509, 72/222
melphalan, 65/528
memory, 73/454
memory disorder, 15/96
meningeal leukemia, 16/234, 63/341
meningeal lymphoma, 63/347
meningioma, 15/299, 18/20, 72/181, 359
meningitis, 15/306
meningococcal meningitis, 52/24, 26, 28
menstruation, 15/321
mental deficiency, 15/586, 42/684, 691, 43/284, 46/24
mental deterioration, 15/586
mental disorder, 15/593, 599, 601, 796
mental prognosis, 15/175
MERRF syndrome, 62/509, 72/222
mescaline, 72/237
mesial temporal sclerosis, 72/362
metabolic disorder, 72/221
metabolic disturbance, 72/347
metabolic factor, 15/311, 72/366
metachromatic leukodystrophy, 10/568, 15/373-418, 72/222
metastatic cardiac tumor, 63/94
methamphetamine, 72/234
methamphetamine intoxication, 65/366
methanol intoxication, 64/99
methotrexate, 63/357
3,4-methylenedioxyamphetamine intoxication, 65/366
3,4-methylenedioxymethamphetamine intoxication, 65/51, 367
methylenetetrahydrofolate reductase deficiency, 72/222
methylmercury, 64/3
metrizamide, 61/166
metronidazole intoxication, 65/487
mexiletine, 63/192
micrencephaly, 72/109
microdysgenesis, 72/109, 111
microsomatognosia, 4/230, 45/509
micturition, 15/95, 140
migraine, 5/42, 47, 265, 48/38, 143, 148, 161, 72/216, 255
migrainous headache, 72/5
Miller-Dieker syndrome, 50/592, 72/130
Minamata disease, 64/419, 421
misonidazole polyneuropathy, 51/301
mitochondrial disorder, 72/222
mitochondrial inheritance, 72/135
mitral valve prolapse, 63/21
model, 15/30, 72/83
Moro reflex, 42/690
mortality, 15/495, 812, 72/28
mortality rate, 72/29
motor sequence stimulation, 15/82
motor symptom, 15/76
mountain sickness, 63/417
movement induced, *see* Hereditary paroxysmal kinesigenic choreoathetosis
movement precipitation, 15/447
moyamoya disease, 72/212
mucolipidosis type I, 72/222
mucormycosis, 52/471
multifocal cortical, *see* Multifocal cortical epilepsy
multilobar reaction, 73/406
multiple myeloma, 18/255
multiple sclerosis, 9/178
multiple subpial transection, 73/409
multiple sulfatase deficiency, 72/222
multiple trichoepithelioma, 14/791
multiplex wave, 15/8
muscimol intoxication, 65/40
muscular dystrophy, 43/92
musicogenic, *see* Musicogenic epilepsy
myelography, 61/166
myoclonic, *see* Myoclonic epilepsy
myoclonic akinetic, *see* Myoclonic akinetic epilepsy
myoclonic astatic, *see* Myoclonic astatic epilepsy
myoclonus, 15/121, 38/579, 73/235
nalidixic acid intoxication, 65/486
narcolepsy, 3/96, 72/253
NARP syndrome, 72/222
neglect syndrome, 45/175
neonatal adrenoleukodystrophy, 47/595

neonatal convulsion, 73/235
neonatal myoclonus, 73/242
neonatal seizure, 73/251
neoplasm, 72/352
nerve cell, 15/36
network organization, 72/48
network parameter, 72/44
neuroacanthocytosis, 42/511, 63/290
neurobrucellosis, 52/587
neurochemistry, 15/60, 72/83
neurocutaneous melanosis, 14/422
neurocutaneous syndrome, 72/181
neurocysticercosis, 16/225, 52/530, 72/146, 165
neurofibromatosis, 14/155, 72/129
neurofibromatosis type I, 72/131
neurogenic pulmonary edema, 63/495
neuroimaging, 72/373
neuroleptic agent, 72/233
neurolipidosis, 15/423
neuronal network, 72/40, 42, 70
neuronal tumor, 72/180
neuropathology, 72/107
neuropeptide, 72/84
neurophysiology, 15/30
neuropsychology, 45/9, 11
neurosarcoidosis, 63/422
neurosis, 73/478
neurosurgery, 45/9
Neurotoxicity Scale, 72/419, 431
neurotransmission, 72/83
neurotransmitter, 72/83, 97
nevus unius lateris, 14/781, 46/95
newborn, *see* Newborn epilepsy
nicotine intoxication, 65/261
Niemann-Pick disease type C, 72/222
nocturnal enuresis, 72/254
nodular heterotopia, 72/109
nonconvulsive, *see* Nonconvulsive seizure
nonepileptic, 72/259
non-Hodgkin lymphoma, 63/349
nonketotic hyperglycinemia, 72/222
noradrenalin, 72/91
norfloxacin inoxication, 65/486
norpethidine intoxication, 65/357
Nottingham Health Profile, 72/422
nuclear magnetic resonance, 72/337
nystagmus, 42/690
occipital lobe, *see* Occipital lobe epilepsy
occipital lobe syndrome, 45/57
occipital lobe tumor, 17/322
OKT3, 63/536
OKT3 intoxication, 65/560
olfactory, 15/83
olfactory neuroblastoma, 72/180
olfactory stimulation, 15/447
oligodendroglioma, 18/20, 91, 72/180, 357
operative complication, 73/422
operative outcome, 73/422
opiate, 72/234
optic stimulation, 15/442
opticocochleodentate degeneration, 42/241, 60/754
oral contraception, 73/372
organic factor, 73/460
organochlorine insecticide intoxication, 64/199, 202, 204
organolead intoxication, 64/131, 144
organophosphate intoxication, 37/555, 64/172, 174, 226
organotin intoxication, 64/138, 141
ornithine carbamoyltransferase, 72/222
orthochromatic leukodystrophy, 10/123
other disease, 73/362
outcome, 72/417
outcome measure, 72/429
overt reward, 73/443
oxazolidinedione, 15/641
oxcarbazepine, 73/361
9p partial monosomy, 50/579
pachygyria, 72/109
pain insensitivity, 8/194
Paine syndrome, 60/293
pallidoluysiodentate degeneration, 49/456
paragonimiasis, 72/169
paranoid syndrome, 15/598
parasitic disease, 72/169
parathyroid gland, 15/319
parietal lobe, *see* Parietal lobe epilepsy
Parkinson disease, 15/336
paroxysmal disorder, 45/507-509
paroxysmal dysarthria, 45/508
paroxysmal dysphasia, 45/508
paroxysmal dystonia, 72/255
paroxysmal kinesigenic choreoathetosis, 72/255
paroxysmal movement disorder, 72/255
paroxysmal pain, 45/509
paroxysmal speech arrest, 45/508
Parry disease, 42/467
partial, 15/3, 51, 90, 107
partial seizure, 72/182
pathologic laughing, 3/358, 14/347
pathologic laughing and crying, 3/358, 45/221
pathologic rage, 3/344
pathology, 15/611, 72/184
pathophysiology, 72/39, 64
peak, 15/460

pediatric head injury, 23/453, 57/341
pefloxacin intoxication, 65/486
Pelger-Huët abnormality, 43/458
Pelizaeus-Merzbacher disease, 15/376, 42/503-505
pellagra, 6/748
penicillamine, 28/116
penicillin, 65/472
pentaborane intoxication, 64/360
pentazocine intoxication, 65/357
peptide, 72/96
perception, 45/508
perinatal asphyxia, 72/344
periodic ataxia, 21/577
periodical psychosis, 15/596
peroxisomal disorder, 72/222
personality, 73/472
personality disorder, 15/576
pertussis, 52/233
pertussis vaccine, 52/242, 245
pesticide, 72/238
pethidine intoxication, 65/357
petit mal, *see* Petit mal epilepsy
phakomatosis, 15/325, 72/347
phantom limb, 4/224, 45/396
phantom third limb, 45/509
pharmacokinetics, 73/357
pharmacology, 73/369
phencyclidine intoxication, 65/54, 368
phenethylamine intoxication, 65/51
phenobarbital, 15/635, 670, 73/359
phenothiazine, 15/653
phenylalanine, 72/222
phenylketonuria, 72/222
phenytoin, 15/627, 670, 73/358
phenytoin intoxication, 65/500
phonatory, 15/80
phosphene, 45/367
phosphoglycerate kinase deficiency, 62/489
photosensitieve baboon, 72/83
photogenic, *see* Photogenic epilepsy
photosensitive, *see* Photosensitive epilepsy
pilocarpine, 72/55
pineal tumor, 72/181
pituitary adenoma, 17/390
pituitary gland, 15/317, 42/11
pituitary tumor, 72/181
plasma level, 73/361
pneumocephalus, 24/207
polymicrogyria, 72/109, 344
pontine hemorrhage, 2/253, 12/39
pontine infarction, 12/20, 30
porencephaly, 30/686, 42/49, 50/359
positron emission tomography, 72/184, 371
postictal aggression, 73/480
postictal apnea, 63/481
postictal blindness, 2/623
postictal psychosis, 72/132, 73/481
postlumbar puncture syndrome, 61/166
postoperative anxiety, 73/423
postoperative care, 73/419
postoperative failure, 73/425
postoperative period, 73/421
poststatus epilepticus, 72/61
posttraumatic, *see* Posttraumatic epilepsy
postural, 15/79
postvaccinal encephalitis, 15/309
potassium current, 72/46
potassium imbalance, 28/476-482
precipitation, 15/446
prediction, 73/311
pregnancy, 73/362, 372, 375
pregnancy counselling, 73/374
prenatal onset, 42/690
presenile dementia, 15/343, 21/67
presurgical evaluation, 73/386
prevalence, 72/22-24
Prichard classification, 15/4
primary cardiac tumor, 63/94
primidone, 15/640, 670, 73/359
procainamide intoxication, 37/452
prodromal aggression, 73/479
prodrome, 73/473
prognosis, 15/783
progressive, 72/70
progressive external ophthalmoplegia, 62/308
progressive hemifacial atrophy, 14/777, 31/253, 43/408, 59/479
progressive multifocal leukoencephalopathy, 9/495
progressive myoclonus, *see* Progressive myoclonus epilepsy
progressive rubella panencephalitis, 56/409
projected subcortical, 15/8
propionic acidemia, 72/222
prosopagnosia, 45/509
provoked, 72/16
pseudoepileptic seizure, 73/448
pseudoporencephaly, 30/691
psychiatric disorder, 15/796, 804, 73/469
psychogenic, *see* Psychogenic epilepsy
psychological management, 73/441
psychological testing, 15/559
psychology, 72/387
psychomotor, *see* Psychomotor epilepsy
psychomotor retardation, 42/691

psychopathology, 73/446
psychosis, 15/593, 46/461, 73/423, 475, 477
psychosocial assessment, 72/407, 431
psychosocial handicap, 73/429
psychosocial problem, 72/409
psychotherapy, 73/441, 446
punishment, 73/443
pupil, 1/622, 15/140
Purkinje cell, 72/108
pyrethroid intoxication, 64/219
pyridoxal-5-phosphate, 28/109, 112-117, 42/716
pyridoxic acid, 28/109
pyridoxine dependent convulsion, 72/222
pyruvate decarboxylase deficiency, 62/502
pyruvate dehydrogenase deficiency, 72/222
quality of life, 72/408, 421, 423, 426, 432
quinolone intoxication, 65/486
r(14) syndrome, 50/588, 590
radiation injury, 23/652
radiology, 15/533, 72/183
random, *see* Random epilepsy
Rasmussen chronic encephalitis, 72/350
reading, *see* Reading epilepsy
recreation, 73/457
recurrence risk, 72/126
reflex, *see* Reflex epilepsy
Refsum disease, 15/325
rehabilitation, 15/805, 73/424, 453, 463
rejection encephalopathy, 63/530
relaxation, 73/444
religious state, 15/95
REM behavior disorder, 72/254
REM sleep, 15/458, 476, 482
renal insufficiency, 73/377
renal transplantation, 63/535
reoperation, 73/410
respiration, 15/140, 63/481
respiratory response, 75/350
response oriented method, 73/444
resurgical evaluation, 73/420
reticulosarcoma, 72/180
Rett syndrome, 60/666, 63/437
reward, 73/443
Reye syndrome, 65/116
Reynolds classification, 15/5
rhinitis, 5/235
right drug choice, 15/666
rolandic spike, 73/15
Rosenberg Self-Esteem Scale, 72/422
Roussy-Lévy syndrome, 21/175
rubella virus, 56/407
Rud syndrome, 13/321, 479, 14/13, 38/9, 43/284, 46/59, 51/398, 60/721
sarcoid neuropathy, 51/196
scanning, 72/184
schistosomiasis, 52/511, 538, 72/169
schizencephaly, 30/346, 72/109, 342
schizophrenia, 15/599, 46/461, 493, 507, 73/477
scorpion venom, 65/196
sea blue histiocyte, 60/674
secondary pigmentary retinal degeneration, 13/337-340
secondary subcortical, *see* Secondary subcortical epilepsy
seizure focus, 72/308
seizure frequency, 72/412
seizure precipitation, 73/187
seizure rating scale, 72/430
seizure severity, 72/418
selective serotonin reuptake inhibitor, 65/316
self-control, 73/445
self-esteem, 73/431
self-induced, *see* Self-induced epilepsy
sensory, 15/81
sensory seizure, 4/331
septo-optic dysplasia, 30/320
septum pellucidum, 72/109
septum pellucidum agenesis, 30/314-321
Sepulveda Epilepsy Battery, 72/431
serum level, 15/673
severity rating, 72/418
sex epidemiology, 15/495
sex hormone, 15/320
sex ratio, 72/22
sexual disorder, 15/584
sexual dysfunction, 75/98, 352
Shapiro syndrome, 50/167
single photon emission computer assisted tomography, 72/184, 372
site, 72/183
Sjögren-Larsson syndrome, 13/471
Sjögren syndrome, 71/74
skull defect, 24/663
skull fracture, 15/545
sleep, 3/86, 15/457, 770, 74/547, 75/353
sleep deprivation, 15/480
sleep disorder, 72/253
sleep-wake cycle, 15/457
slow virus, 72/145
Smith-Lemli-Opitz syndrome, 31/254
Sneddon syndrome, 55/402, 406
social effect, 72/431
social function, 73/433
social isolation, 73/433
social issue, 72/420
social prognosis, 15/800

sodium current, 72/44
somatic inhibitory, 15/80
somatosensory precipitation, 15/447
somnambulism, 72/254
sparfosic acid, 65/529
sparganosis, 72/169
spastic cerebral palsy, 42/167
special education, 73/459
spina bifida, 50/494
spinal meningioma, 20/199
spinal neurinoma, 20/250
spirogermanium, 65/529
split brain syndrome, 45/101
sport injury, 23/530
spread, 15/41
St. Louis encephalitis, 56/138
startle, *see* Startle epilepsy
state, *see* Status epilepticus
status epilepticus, 72/61
status marmoratus dysmyelinisatus, 72/109
status verrucosus, 50/204
stereoencephalotomy, 15/758
stereotactic lesion, 73/407
stereotaxic exploration, 15/758
stimulation, 72/83, 73/409
streptococcal meningitis, 52/81
stress, 73/454
striatal lesion, 6/676
striatal necrosis, 6/676, 49/500
striatopallidodentate calcification, 6/707, 49/419, 422
stroke, 72/116
Sturge-Weber syndrome, 14/227, 232-234, 31/3, 46/95, 55/443, 72/210, 347
Suarez classification, 15/10
subacute necrotizing encephalomyelopathy, 42/625
subacute sclerosing panencephalitis, 15/351, 56/419
subarachnoid hemorrhage, 55/9, 53, 72/215
subdural effusion, 24/339
subdural empyema, 52/73
subdural hematoma, 72/216
subdural lymphoma, 63/349
subependymal astrocytoma, 72/181
succinimide, 15/643, 670
sudden death, 63/496
sudden drop, 73/213
sudden infant death, 63/466
sulfite oxidase deficiency, 72/222
sultiame, 15/651, 670
superficial focal, 15/8
superior longitudinal sinus thrombosis, 52/178
surgery, 72/396, 401, 73/385, 419
surgery inventory, 72/424, 432
surgical candidate, 73/391
surgical procedure, 73/401
surgical section cerebral commissures, 4/274
surgical study, 72/428
surgical treatment, 15/739, 73/399
symmetrical spastic cerebral palsy, 42/167
Symonds classification, 15/11
sympathetic, 15/2
symptomatic, 15/3, 13, 495
synchronization, 72/49, 51
syncopal attack, 72/250
systemic lupus erythematosus, 55/373, 73/378
Taenia solium, 72/147, 167
Takayasu disease, 55/337
tardive amaurotic idiocy, *see* Dumon-Radermecker syndrome
taurine, 29/515
taxol, 65/529
television, 2/620
temporal arteritis, 55/344
temporal horn, 15/539
temporal lobe, *see* Temporal lobe epilepsy
temporal lobe pattern, 15/95
temporal lobectomy, 15/750
temporal reaction, 73/403-406
teratogenicity, 73/374
terminology, 15/1
test, 72/398
tetanus, 52/230
tetanus toxin, 72/54
tetany, 63/560
thalamic syndrome (Dejerine-Roussy), 2/485
thalamus, 15/615
thallium intoxication, 64/325
therapy, 69/9
thermoregulation, 15/140
thrombocythemia, 55/473
thrombotic thrombocytopenic purpura, 55/472, 63/325
thyroid gland, 15/318
thyroxine, 15/731
tiagabine, 73/348, 361
time sense, 45/509
Tissot classification, 15/2
tocainide, 63/192
tocainide intoxication, 65/455
tonic, 15/116, 119
tonic clonic, *see* Tonic clonic epilepsy
topiramate, 73/361
toxic, 72/366
toxic shock syndrome, 52/259

toxicity rating, 72/419
toxoplasmosis, 72/164, 351
transcobalamin II deficiency, 63/255
transient global amnesia, 45/206, 55/141
traumatic, *see* Traumatic epilepsy
treatment algorithm, 73/371
treatment resistance, 15/670
treatment success, 72/417
tricarcillin, 65/472
trichinosis, 52/511, 566
trichopoliodystrophy, 72/222
tricyclic antidepressant, 65/316-318
tricyclic antidepressant intoxication, 65/321
trimethadione, 15/641
triplex wave, 15/8
trisomy 13, 72/128
trisomy 18, 72/128
tropical disease, 72/146
true porencephaly, 30/686
Trypanosoma cruzi, 72/170
tryptophan nicotinic acid metabolism, 42/716
tsukubaenolide intoxication, 65/558
tuberculoma, 15/307
tuberculosis, 72/160
tuberculous meningitis, 15/307, 33/213, 232, 52/202, 216, 55/426, 72/349
tuberous sclerosis, 14/108, 346, 31/3, 6-10, 43/49, 72/129-131, 349
tumor, 69/9, 72/113
tumor like lesion, 73/400
tumor site, 16/256, 258, 260
tumor type, 16/257
twilight state, 15/595
ubidecarenone, 62/504
ulegyria, 72/109, 112
ultrasound study, 15/554
uncinate, 15/7
unconsciousness, 15/54, 130, 135
unilateral, 15/6
unilateral epileptic, *see* Unilateral epileptic seizure
unprovoked, 72/17
Unverricht-Lundborg progressive myoclonus, *see* Unverricht-Lundborg progressive myoclonus epilepsy
Urbach-Wiethe disease, 14/777
uremic encephalopathy, 63/504
US Veterans Administration Classification, 15/13
uterine, *see* Uterine epilepsy
VA composite rating, 72/419
VA Neurotoxicity Scale, 72/430
VA Systemic Toxicity Scale, 72/430
vaccine, 72/233
vagus nerve, 75/355
valproic acid, 73/357
valproic acid induced hepatic necrosis, 65/125
varicella zoster virus meningoencephalitis, 56/235
vascular, *see* Vascular epilepsy
vascular lesion, 73/400
vascular malformation, 72/115, 359
vasculitis, 46/359
vasopressin, 15/731
venous angioma, 72/213
venous sinus occlusion, 54/429
venous sinus thrombosis, 54/395, 72/212, 215
versive, 15/78
vertiginous, 15/83
vertigo, 2/359
vestibular precipitation, 15/446
vestibulogenic, 2/707
vigabatrin, 73/360
vincristine, 63/359
viral encephalitis, 15/308
virus, 72/152
visceral precipitation, 15/449
visual, 15/83
visual aura, 2/613, 622
visual field, 17/322
visual hallucination, 2/646, 17/322, 323, 55/146
visual perseveration, 45/509
visual precipitation, 15/441
visual stimulation, 15/442
visuospatial agnosia, 45/171
vitamin $B_1$ intoxication, 65/572
vitamin $B_6$, 15/63, 316, 28/109-116, 114-117
vitamin $B_w$, 72/222
vitamin D, 15/730
vitamin K, 15/730
vocational rehabilitation, 73/456
Vogt-Koyanagi-Harada syndrome, 56/620
Von Hippel-Lindau disease, 50/374
Wada test, 72/399
waking, 15/457
Washington Psychosocial Seizure Inventory, 72/409
water intoxication, 64/241
wave and spike, *see* Wave and spike seizure
Wechsler test, 15/563
well-being, 72/432
well-being scale, 72/422
Whipple disease, 52/137
Willis classification, 15/87
Wilson disease, 15/340, 27/384
Wolf-Hirschhorn syndrome, 43/497
World Health Organization classification, 15/25
X-linked lissencephaly, 72/136
XXX syndrome, 31/494

XYY syndrome, 31/489
yawning, 63/493
yellow teeth, 42/691
zone, 72/51
zonisamide, 73/361
Epilepsy with centrotemporal spike
benign partial, *see* Benign partial epilepsy with centrotemporal spike
Epilepsy with continuous spike wave
definition, 72/9
EEG, 72/9
slow sleep, 72/9
Epilepsy with continuous spike wave during slow sleep, 72/9
atypical benign partial epilepsy, 73/275
benign rolandic epilepsy, 73/274
clonazepam, 73/276
definition, 73/267
EEG, 73/270
Landau-Kleffner syndrome, 73/274
Lennox-Gastaut syndrome, 73/274
mechanism, 73/273
neuropsychology, 73/269
treatment, 73/277
Epilepsy with generalized tonic clonic seizure
awakening, 72/7
definition, 72/7
Epilepsy of late onset
EEG, 57/193
etiology, 15/265-267
head injury, 57/130, 134
head injury outcome, 57/193
posttraumatic epilepsy, 23/357, 57/193
prognosis, 15/269
seizure type, 15/268
Epilepsy with myoclonic absence
associated disorder, 73/232
cause, 73/233
cognitive impairment, 73/232
definition, 72/8
EEG, 73/232
neurologic deficit, 73/233
prognosis, 73/233
survey, 73/231
Epilepsy with myoclonic astatic seizure
definition, 72/8
differential diagnosis, 73/227
EEG, 73/227
etiology, 73/225
prognosis, 73/227
survey, 73/223
treatment, 73/227
Epilepsy with occipital paroxysm
benign childhood, *see* Benign childhood epilepsy with occipital paroxysm
benign partial, *see* Benign partial epilepsy with occipital paroxysm
childhood, *see* Childhood epilepsy with occipital paroxysm
Epilepsy with somatosensory evoked potential
benign partial, *see* Benign partial epilepsy with somatosensory evoked potential
Epileptic aura
migraine, 5/47
Epileptic encephalopathy
child, 73/235
Epileptic focus
adenosine triphosphatase, 15/68
anatomic site, 15/77
EEG, 15/513
electrolyte, 15/66-68
etiology, 15/76
genetics, 15/435
glutamate decarboxylase, 15/68
migraine, 45/510
mirror focus, 15/46
nerve cell, 15/36
neurochemistry, 15/61, 66
neurophysiologic aspect, 15/30-41
parietal lobe syndrome, 45/64
single photon emission computer assisted tomography, 72/377
Sjögren-Larsson syndrome, 22/477
somatomotor seizure, 15/76
stereotaxic exploration, 15/769
transient ischemic attack, 53/234
Epileptic mastication
temporal lobe epilepsy, 15/97
Epileptic myoclonus, *see* Progressive myoclonus epilepsy
Epileptic seizure
Alzheimer disease, 46/252
α-amino-n-butyric acid, 13/339
arteriovenous malformation, 31/181
Binswanger disease, 54/222
bone membranous lipodystrophy, 42/280
Bonnet-Dechaume-Blanc syndrome, 14/262
carnosinemia, 42/532
cavum septi pellucidi, 30/325
cavum vergae, 30/325
cerebellar aplasia, 30/383
definition, 72/16
Hooft-Bruens syndrome, 42/508
hyperargininemia, 29/102
intracerebral calcification, 42/534
mental deficiency, 43/282

moyamoya disease, 55/294
muscular dystrophy, 15/326
pseudohypoparathyroidism, 42/621
systemic lupus erythematosus, 55/375
Epileptic state, *see* Status epilepticus
Epileptiform behavior
psychogenic, 72/267
Epileptiform discharge
*see also* Seizure
autonomic control, 75/352
sexual function, 75/352
Epileptogenesis
*see also* Epilepsy
blood flow, 72/177
bursting, 72/177
denervation hypersensitivity, 72/178
distribution, 72/177
excitatory coupling, 72/177
glial function, 72/177
hemosiderin, 72/177
neuropathology, 72/107
sprouting, 72/178
Epileptogenic encephalopathy
brain atrophy, 42/693
infantile, *see* Infantile epileptogenic encephalopathy
psychomotor retardation, 42/693
Epileptogenic lesion
definition, 73/387
Epileptogenic zone
definition, 73/386
Epiloia, *see* Tuberous sclerosis
Epimeningitis, *see* Spinal epidural empyema
Epinephrine, *see* Adrenalin
Epineurium
nerve, 51/1
Epipatellar reflex
diagnostic value, 1/246
Epiphora
Charlin neuralgia, 48/486
Möbius syndrome, 50/215
Epiphyseal dysplasia
Klinefelter syndrome, 43/560
Klinefelter variant XXXXY, 43/560
microcephaly, 43/396-398
Mietens-Weber syndrome, 43/304
nystagmus, 43/396-398
short stature, 43/396
Epiphysis
goiter, 42/375
stippled, *see* Stippled epiphysis
Episodic apnea
Arnold-Chiari malformation type II, 50/410
congenital high conductance fast channel syndrome, 62/438
Joubert syndrome, 63/437
Miller-Dieker syndrome, 50/592
pyruvate decarboxylase deficiency, 62/502
Rett syndrome, 63/437
Reye syndrome, 56/151
Episodic ataxia, *see* Periodic ataxia
Episodic dystonia
kernicterus, 6/505
Episodic encephalopathy
glutaric aciduria type II, 62/495
multiple acyl-coenzyme A dehydrogenase deficiency, 62/495
Episodic headache
semantics, 48/3
Episodic hyperpnea
clinical features, 63/436
Joubert syndrome, 50/191, 63/437
pontine glioma, 63/440
Rett syndrome, 63/437
Episodic hypothermia
Shapiro syndrome, 50/167
Episodic pulmonary edema
autonomic nervous system, 43/62
intracranial pressure, 43/62
Episodic vertigo and deafness
hereditary, *see* Hereditary Meniere disease
Episodic vomiting
MELAS syndrome, 60/668
Epistaxis
ataxia telangiectasia, 60/368
clinical examination, 57/132
cluster headache, 48/222
coagulopathy, 63/308
head injury, 57/132
mucormycosis, 35/544, 52/471
pseudoxanthoma elasticum, 55/452
rhodotorulosis, 52/493
traumatic aneurysm, 24/383
venous sinus thrombosis, 54/397
Epithalamic thalamic nucleus, *see* Habenular nucleus
Epithelial cyst, *see* Intraspinal neurenteric cyst
Epithelial dystrophy
familial juvenile, *see* Familial juvenile epithelial dystrophy
Epithelial neurenteric cyst
Arnold-Chiari malformation type I, 32/437
Klippel-Feil syndrome, 32/429
Epithelioid cell
carotid body tumor, 8/487
globoid cell leukodystrophy, 10/69, 74, 79-84

histogenesis, 10/841
uninucleated, *see* Uninucleated epithelioid cell
Epithelioma
Malherbe, *see* Pilomatricoma
malignant, *see* Malignant epithelioma
myotonic dystrophy, 62/238
orbital tumor, 17/177
Epithelioma adenoides cysticum, *see* Multiple trichoepithelioma
Epitheliomatous phakomatosis (Brooke-Jarisch), *see* Multiple trichoepithelioma
Epitheliomatous phakomatosis (Poncet-Spiegler), *see* Multiple dermal cylindroma
Epithelium
neural, *see* Neuroepithelium
EPN, *see* Phenylphosphonothioic acid O-ethyl-O-(4-nitrophenyl)ester
Epoprostenol, *see* Prostacyclin
EPP, *see* Motor end plate potential
EPSP, *see* Excitatory postsynaptic potential
Epstein-Barr virus, 56/249-258
acquired immune deficiency syndrome, 56/255
acquired myoglobinuria, 62/577
acute cerebellar ataxia, 56/253
acute polymyositis, 56/200
acute viral myositis, 56/195
athetosis, 56/252
autonomic nervous system, 74/335
autonomic polyneuropathy, 56/255
biology, 56/249
brachial plexus neuritis, 56/254
Burkitt lymphoma, 56/8, 249, 255, 63/351
cancer, 69/450
cerebellar ataxia, 56/252
childhood myoglobinuria, 62/562
chorea, 56/252
cranial mononeuropathy, 56/8
cranial neuropathy, 56/253
deafness, 56/107, 252
epilepsy, 56/252
facial paralysis, 56/253
focal encephalitis, 56/8
gammaherpesvirus, 56/8
Guillain-Barré syndrome, 47/608, 51/240, 56/8, 254
headache, 56/253
herpes virus, 56/7
herpes virus family, 56/249
immunocompromised host, 56/477
infectious mononucleosis, 34/185, 56/8, 249
latency, 34/8, 56/251
lymphoma, 63/533
Marek disease virus, 56/249
mental deficiency, 56/263
metamorphopsia, 56/257
mononeuritis, 34/188
multiple cranial neuropathy, 56/253
multiple sclerosis, 47/329, 56/256
nasopharyngeal carcinoma, 56/8, 249, 255
neuralgic amyotrophy, 56/254
neurasthenia, 56/256
neurotropic virus, 56/249
ophthalmoplegia, 56/253
optic neuritis, 56/253
primary CNS lymphoma, 56/255, 63/533
proteinocytologic dissociation, 56/253
renal transplantation, 63/533
Reye syndrome, 29/332, 49/217, 56/150
schizophrenia, 56/257
serodiagnosis, 56/251
status epilepticus, 56/252
subacute sclerosing panencephalitis, 56/256
transverse myelitis, 56/253
vertigo, 56/107
viral meningitis, 56/126
X-linked lymphoproliferative syndrome, 56/250
Epstein-Barr virus antibody
multiple sclerosis, 47/274
Epstein-Barr virus encephalitis
classification, 34/188
epilepsy, 56/252
neuropathology, 34/188
Epstein-Barr virus infection, 34/185-191
autonomic neuropathy, 56/252
Burkitt lymphoma, 34/189, 39/63-65
cerebellar ataxia, 56/252
cranial nerve palsy, 56/252
CSF, 56/257
deafness, 56/252
encephalitis, 56/252
epilepsy, 56/252
facial paralysis, 56/252
Guillain-Barré syndrome, 56/252
immunoregulation, 56/250
immunosuppressive therapy, 65/549
infectious mononucleosis, 34/185-189
meningitis, 56/252
meningitis serosa, 34/85
multiple sclerosis, 56/252
myelitis, 56/252
nasopharyngeal carcinoma, 34/190
neurology pathogenesis, 56/258
neuropathology, 56/257
organic brain syndrome, 56/252
polyneuritis, 56/252
psychiatric syndrome, 56/256

subacute sclerosing panencephalitis, 56/252
transverse myelitis, 56/252
Epstein-Barr virus meningitis
symptom, 56/252
Epstein-Barr virus meningoencephalitis
ataxia, 56/252
athetosis, 56/252
chorea, 56/252
CSF, 56/252
epilepsy, 56/252
Epstein-Barr virus myelitis
transverse myelitis, 56/8, 253
Equal loudness balance test
result, 16/308
Equilibrium
*see also* Astasia abasia *and* Posture
ataxia, 1/334
coordination, 1/306-308
corpus callosum ataxia, 1/343, 17/508
labyrinthine, 1/334
Von Romberg test, 1/344
Equine encephalitis
animal viral disease, 34/292
eastern, *see* Eastern equine encephalitis
western, *see* Western equine encephalitis
Equine encephalomyelitis
eastern, *see* Eastern equine encephalomyelitis
Equine infectious anemia, 18/246
Equinovarus foot deformity, *see* Talipes equinovarus
Erasistratus
history, 2/5
Erb disease, *see* Primary lateral sclerosis
Erb myelopathy, *see* Cervical spondylarthrotic myelopathy
Erb paralysis
birth incidence, 2/141
brachial plexus, 51/143
Erb scapulohumeral muscular dystrophy, *see* Scapulohumeral muscular dystrophy (Erb)
Erb sign
hypoparathyroidism, 27/300
Erb, W.H., 1/3, 4, 8, 12, 16, 19, 37
Erdheim tumor, *see* Craniopharyngioma
Erection
penis, *see* Penis erection
sexual function, 26/437, 442, 448
spinal cord injury, 61/313
Eretmochelys imbricata intoxication
ingestion, 37/89
ERG
A wave, 2/508
autosomal recessive optic atrophy, 13/38
Bardet-Biedl syndrome, 13/292, 383
Bassen-Kornzweig syndrome, 51/327, 394, 60/670, 63/275
Behr disease, 13/89
Bruch membrane, 2/511
component, 2/511
congenital retinal blindness, 13/98, 272, 22/528, 532
cortical blindness, 24/50
dominant infantile optic atrophy, 13/117
evoked potential, 57/198
flicker fusion frequency, 13/19
head injury, 57/198
hereditary adult foveal dystrophy (Behr), 13/89
hereditary autosomal recessive congenital optic atrophy, 13/91
hereditary cerebellar ataxia (Marie), 21/386
hyaloidoretinal degeneration (Wagner), 13/274
methanol intoxication, 36/354, 64/99
neuronal intranuclear hyaline inclusion disease, 60/670
optic atrophy, 13/49
optic nerve injury, 24/37
primary pigmentary retinal degeneration, 13/166-172
Refsum disease, 8/23, 13/315, 21/193, 60/228, 66/488
retinal blindness, 13/16
Stargardt disease, 13/128
vitamin E deficiency, 70/446
Ergometrine
orofacial dyskinesia, 49/56
Ergoreceptor
autonomic nervous system, 75/428
Ergoreflex
muscle, 75/112
Ergot derivative
demyelination, 9/652
leukoencephalopathy, 9/652, 47/573
migraine, 48/5
Ergot intoxication
*see also* Ergotism
acroparesthesia, 65/67
alpha adrenergic receptor blocking agent, 65/63
area postrema, 65/68
arterial spasm, 65/69
bradycardia, 65/63
brain cortex atrophy, 65/69
Claviceps purpurea, 65/62
clinical picture, 65/68
coma, 65/68
convulsion, 65/68
delirium, 65/62
diagnosis, 65/70

dopamine receptor, 65/68
epidemic, 65/61
epilepsy, 65/61, 68
ergotamine, 65/63
ergotamine abuse, 65/73
ergotamine headache, 65/71
ergotism, 36/553, 65/61
gangrene, 65/61
headache, 65/67
histopathology, 65/70
history, 65/61
iatrogenic neurological disease, 65/68
ignis plaga, 65/62
individual susceptibility, 65/62
muscle intermittent claudication, 65/67
myalgia, 65/68
myocardial infarction, 65/69
organophosphorus compound intoxication, 9/652
paresthesia, 65/67
prevention, 65/74
prognosis, 65/73
retroperitoneal fibrosis, 65/70
stroke like symptom, 65/69
subclinical ergotism, 65/71
sympathoinhibition, 65/63
syncope, 65/67
treatment, 65/70
tremor, 65/67
vasoconstriction, 65/63
Ergotamine
blood pressure, 65/64
brain blood flow, 65/64
brain infarction, 53/164
brain ischemia, 53/164
carotidynia, 5/375, 48/339
chemistry, 65/63
chronic migrainous neuralgia, 48/251
clinical use, 65/66
cluster headache, 5/117, 48/228, 231
common migraine, 5/53
compound, 5/98
contraindication, 65/67
cough headache, 48/371
drug addiction, 65/61
ergot intoxication, 65/63
half-life, 65/65
headache, 48/6
leg artery, 65/65
migraine, 5/56, 96, 379, 48/76, 97, 145, 183-185
neuropathy, 7/520
pharmacokinetics, 65/63
pharmacology, 65/63
progressive dysautonomia, 59/140
Raynaud phenomenon, 8/337
retroperitoneal fibrosis, 65/70
side effect, 5/380, 65/67
vasoconstriction, 48/185
Ergotamine headache
characteristics, 65/72
ergot intoxication, 65/71
induction, 65/72
pathophysiology, 65/72
withdrawal, 65/72
Ergotamine induced headache, *see* Ergotamine headache
Ergotamine intoxication
myelinoclastic diffuse sclerosis, 9/481
Ergotamine tartrate
autonomic polyneuropathy, 51/487
cluster headache, 48/229
migraine, 48/125
paroxysmal migrainous neuralgia, 5/382
Ergotism, 36/547-557
*see also* Ergot intoxication
diagnosis, 36/554
ergot intoxication, 36/553, 65/61
history, 36/547
pellagra, 28/93
pharmacology, 36/548-551
prevention, 36/556
treatment, 36/556
Ergotropic reaction
autonomic nervous system, 74/17
behavior change, 3/52
consciousness, 3/52
emotion, 3/320
terminology, 3/52
Erignathus barbatus intoxication
vitamin A, 37/96
Eriphia sebana intoxication
crustacea, 37/58
death, 37/58
Erlenmeyer flask deformity
juvenile Gaucher disease, 10/522
Erysipelas
nerve lesion, 7/487
orbital cellulitis, 33/179
Erythema chronicum migrans
Lyme disease, 51/203, 52/260
neuroborreliosis, 51/203
Erythema marginatum
Sydenham chorea, 6/411
Erythema migrans
Bannwarth syndrome, 51/200
neuroborreliosis, 51/200
Erythema nodosum

Behçet syndrome, 56/594, 71/214
cat scratch disease, 52/128
Erythema nodosum leprosum
leprosy, 51/216
Erythematosus
lupus, *see* Lupus erythematosus
systemic lupus, *see* Systemic lupus erythematosus
Erythermalgia, *see* Erythromelalgia
Erythralgia, *see* Erythromelalgia
Erythritol tetranitrate
migraine, 5/98
Erythroblastosis fetalis
bilirubin encephalopathy, 27/422
double athetosis, 49/384
Erythroceramide
biochemistry, 9/9
Erythrocuprein
erythrocyte specific copper protein, 6/119
Erythrocyte
blood lead level, 64/433
burr, *see* Burr erythrocyte
Colorado tick fever virus, 56/30
Duchenne muscular dystrophy, 40/382, 385
helmet, *see* Helmet erythrocyte
hypoxia, 75/265
lead, 64/433
linoleic acid, 13/423
myotonic dystrophy, 40/497, 523
triangular, *see* Triangular erythrocyte
Erythrocyte choline level
Gilles de la Tourette syndrome, 49/633
Erythrocyte count
brain infarction, 53/432
multiple sclerosis, 47/340
Erythrocyte enzyme
chorea-acanthocytosis, 49/331
Erythrocyte membrane protein
chorea-acanthocytosis, 49/330
Erythrocyte sedimentation rate
atrial myxoma, 63/97
bacterial endocarditis, 52/291, 300
polyarteritis nodosa, 55/354
polymyositis, 41/60
primary amyloidosis, 51/416
spinal cord injury, 26/380
spinal cord metastasis, 20/419
Sydenham chorea, 6/426
systemic lupus erythematosus, 55/379
Takayasu disease, 55/338
temporal arteritis, 39/336, 48/10, 51/452, 55/344
venous sinus occlusion, 54/441
Wolfson syndrome, 20/428
Erythrocytosis
*see also* Polycythemia vera
definition, 55/467
hyperviscosity syndrome, 55/484
Von Hippel-Lindau disease, 31/18
Erythroderma
congenital ichthyosiform, *see* Congenital ichthyosiform erythroderma
ichthyosiform, *see* Ichthyosiform erythroderma
Kawasaki syndrome, 56/638
Passwell syndrome, 21/216
Refsum disease, 21/216
Erythrohepatic protoporphyria
aminolevulinic acid, 27/433
Erythrokeratodermia
mental deficiency, 46/60
sensorineural deafness, 42/373
Erythrokeratodermia ataxia syndrome
dysarthria, 21/216
locomotor ataxia, 21/216
muscle stretch reflex, 21/216
neurocutaneous syndrome, 21/216
nystagmus, 21/216
Erythromelalgia
dysautonomia, 22/256
head, *see* Cluster headache
idiopathic, *see* Idiopathic erythromelalgia
Raynaud phenomenon, 8/339
Erythromelia
unilateral hemimelic ichthyosiform, *see* Unilateral hemimelic ichthyosiform erythromelia
Erythromycin
acquired myoglobinuria, 62/577
deafness, 65/472
gram-negative bacillary meningitis, 52/121
legionellosis, 52/257
neurosyphilis, 52/283
neurotoxin, 65/472
Erythropoiesis
anemia, 38/15
regulation, 63/249
Erythropoietic coproporphyria
congenital, *see* Congenital erythropoietic porphyria
Erythropoietic disease
subarachnoid hemorrhage, 12/110
Erythropoietic porphyria
congenital, *see* Congenital erythropoietic porphyria
Erythropoietic protoporphyria, *see* Erythrohepatic protoporphyria
Erythropoietin
autonomic failure, 75/725
EEG, 63/535

epilepsy, 63/535
neurobehavior, 63/535
neuropathology, 63/535
Pearson syndrome, 66/427
positron emission tomography, 63/535
production, 63/249
renal oxygen partial pressure, 63/249
renal transplantation, 63/534
thalassemia, 63/257
Erythroprosopalgia
intermedius neuralgia, 5/115
nomenclature, 48/217
Erythropsia
temporal lobe epilepsy, 73/71
Erythrosphingosine
biochemistry, 9/9
Escherichia coli
acquired myoglobinuria, 62/577
bacterial food poisoning, 37/87
bacterial meningitis, 52/2
brain abscess, 52/149
childhood myoglobinuria, 62/562
cranial subdural empyema, 52/171
gram-negative bacillary meningitis, 52/104, 117
infantile enteric bacillary meningitis, 33/61-63
K-1 antigen, 33/61, 64, 52/118
mackerel intoxication, 37/87
polysaccharide antigen, 33/61
scombroid intoxication, 37/85
spinal epidural empyema, 52/187
spinal meningitis, 52/190
transverse sinus thrombosis, 52/177
virulence factor, 33/61
Escherichia coli meningitis
hydranencephaly, 50/342
Eserine, *see* Physostigmine
Esophageal achalasia
cerebellar ataxia, 43/260
mental deficiency, 43/260
Esophageal manometry
intestinal pseudo-obstruction, 51/494
Esophageal motility disorder
achalasia, 75/630
diagnostic criteria, 75/630
spasm, 75/630
Esophageal pressure
achalasia, 74/634
cough, 26/339
swallowing, 74/634
Esophageal spasm
hypermotility, 75/629
Esophageal sphincter
achalasia, 75/628
relaxation, 75/629
Esophagus
achalasia, 75/627
circopharyngeal dysfunction, 75/628
diabetes mellitus, 42/543-545, 75/592
myotonic dystrophy, 40/514
Esophagus abnormality
vagus nerve agenesis, 50/219
Esophagus atresia
Down syndrome, 50/527
dysmotility, 75/630
iniencephaly, 50/131
renal agenesis syndrome, 50/513
VATER syndrome, 50/513
Esophagus dysfunction
syringobulbia, 50/445
Esophagus hypertrophy
idiopathic, *see* Idiopathic esophagus hypertrophy
Esotropia, *see* Convergent strabismus
Esquirol classification
epilepsy, 15/2
Essential familial tremor, *see* Hereditary tremor
Essential fatty acid
multiple sclerosis, 47/199
prostaglandin, 47/199
Essential hypernatremia
clinical, 28/455
experimental, 28/457
Essential hypotonia, *see* Amyotonia congenita
Essential myoclonus
hereditary, *see* Hereditary essential myoclonus
multifocal, 49/614
Essential startle disease
myoclonus, 38/577
Essential thrombocythemia
ataxia, 63/331
brain hemorrhage, 54/294
brain infarction, 53/166
brain ischemia, 53/166
coagulopathy, 63/331
epilepsy, 63/331
hemiplegia, 63/331
transient ischemic attack, 63/331
Essential tremor, *see* Hereditary tremor
Essick cell band
hereditary olivopontocerebellar atrophy (Menzel), 21/435
olivopontocerebellar atrophy (Dejerine-Thomas), 21/418
Esterase
CSF, *see* CSF esterase
neurotoxic, *see* Neurotoxic esterase
tri-*o*-cresyl phosphate intoxication, 37/475

Esterase D locus
 13q interstitial deletion, 50/587
 retinoblastoma, 50/587
Esthesioneuroblastoma
 skull base, 68/466
 skull base tumor, 68/486
Estrogen
 catamenial migraine, 5/232
 chorea, 49/76
 Coccidioides immitis, 52/410
 dopaminergic nerve cell, 49/49
 epilepsy, 15/731
 headache, 5/232, 48/433, 437
 migraine, 5/381, 48/433, 437
 oral contraception, 55/525
 Sydenham chorea, 49/364
Etacrynic acid
 brain edema, 16/204
 neurologic intensive care, 55/213
 neurotoxin, 50/219
Etacrynic acid intoxication
 hereditary deafness, 50/219
État criblé, *see* Status cribrosus
État glacé
 cerebellar cortex, 21/496
 granular layer, 21/502
État lacunaire, *see* Brain lacunar infarction
État marbré, *see* Status marmoratus
Ethacrynic acid, *see* Etacrynic acid
Ethambutol
 chromatopsia, 65/481
 drug induced polyneuropathy, 51/294
 iatrogenic neurological disease, 65/481
 neuropathy, 7/517
 neurotoxicity, 71/368
 neurotoxin, 7/517, 33/228, 51/294, 52/211, 64/8, 65/481
 optic neuropathy, 51/294, 64/8, 65/481
 retrobulbar neuritis, 65/481
 toxic neuropathy, 64/8, 65/481
 tuberculosis, 51/294
 tuberculous meningitis, 33/228, 52/211
Ethambutol intoxication
 central scotoma, 65/481
 color vision, 65/481
 mental confusion, 65/481
 neuropathology, 65/481
 paresthesia, 65/481
 retrobulbar neuritis, 52/212, 65/481
 secondary pigmentary retinal degeneration, 60/736
 visual impairment, 64/8
Ethambutol polyneuropathy
 alcohol dehydrogenase, 51/295
 axonal degeneration, 51/295
 EMG, 51/295
 lactate dehydrogenase, 51/295
 neuropathology, 51/295
 optic chiasm, 51/295
 optic neuritis, 51/294
 optic neuropathy, 51/295
 optic tract, 51/295
 pyramidal decussation, 51/295
 reticular formation, 51/295
 segmental demyelination, 51/295
 spinal cord, 51/295
 zinc, 51/295
Ethanol, *see* Alcohol
Ethchlorvynol
 structural formula, 37/354
 toxic myopathy, 62/597, 609
Ethchlorvynol intoxication
 amblyopia, 37/354
 dizziness, 37/354
 drowsiness, 37/354
 dyschromatopsia, 37/354
 muscle weakness, 37/354
 nausea, 37/354
 peripheral neuritis, 37/354
 urticaria, 37/354
 visual impairment, 37/354
 vomiting, 37/354
Ether
 anesthetic agent, 37/412
 animal experiment, 37/412
 malignant hyperthermia, 41/268
 organic solvent intoxication, 64/40
 withdrawal symptom, 37/349
Ether phospholipid biosynthesis
 pathway, 66/507
Ethics
 death, 24/800, 820
 head injury management, 57/436
 law, 24/800
 patient-doctor relationship, 24/800
 XYY syndrome, 50/552
Ethidium bromide, *see* Homidium bromide
Ethidium bromide intoxication, *see* Homidium bromide intoxication
Ethionamide
 deafness, 65/481
 epilepsy, 65/481
 headache, 65/481
 iatrogenic neurological disease, 65/481
 neuropathy, 7/516, 535
 neurotoxin, 7/516, 535, 33/228, 65/481

tremor, 65/481
tuberculous meningitis, 33/228, 52/212
Ethionamide intoxication
convulsion, 65/481
deafness, 65/481
headache, 65/481
mental depression, 65/481
polyneuropathy, 65/481
psychosis, 65/481
tremor, 65/481
visual impairment, 65/481
Ethionamide polyneuropathy
sensory neuropathy, 51/298
Ethmocephaly
*see also* Holoprosencephaly
historic term, 50/225
holoprosencephaly, 30/445, 42/32, 50/233
Patau syndrome, 31/510, 512, 50/559
proboscis, 30/445
Ethmoid bone fracture
CSF fistula, 24/184
rhinorrhea, 24/184
Ethmoid cell
headache, 5/214
Ethmoidal artery
anterior, *see* Anterior ethmoidal artery
posterior, *see* Posterior ethmoidal artery
Ethmoidal nerve
sneezing, 63/492
Ethmoidal nerve syndrome
anterior, *see* Anterior ethmoidal nerve syndrome
Ethmoidal vein
anatomy, 11/53
Ethnic group
lead intoxication, 64/447
spinal cord tumor, 19/8
Ethoglucid, *see* Etoglucid
Ethosuximide
antiepileptic agent, 65/497
epilepsy, 15/644, 670, 37/200-202, 73/360
psychosis, 73/477
serum level, 15/691
side effect, 15/709
systemic lupus erythematosus, 55/370
Ethosuximide intoxication
akathisia, 65/508
digestive system, 37/203
drug induced systemic lupus erythematosus, 65/508
dyskinesia, 65/508
headache, 65/507
higher cortical function, 37/203
hydroxylated metabolite, 37/200
metabolite, 37/200
normesuximide, 37/200
parkinsonism, 65/508
psychosis, 65/507
reticular activating system, 37/203
seizure exacerbation, 65/507
systemic lupus erythematosus, 65/508
Ethotoin
side effect, 15/710
Ethyl alcohol intoxication, *see* Alcohol intoxication
Ethyl carbamate, *see* Urethan
Ethyl-*N*-diethyl phosphoramidocyanidate intoxication
chemical classification, 37/544
organophosphorus compound, 37/544
*O*-Ethyl *S*-[2-(diisopropylamino)ethyl]methylphosphonothionate
chemical structure, 64/225
neurotoxin, 64/223
organophosphate intoxication, 64/223
*O*-Ethyl *S*-(2-dimethylaminoethyl) methylphosphonothioate intoxication
chemical classification, 37/548
organophosphorus compound, 37/548
Ethyl *N*-dimethyl phosphoramidocyanidate, *see* Tabun
Ethyl-*N*-dimethyl phosphoramidocyanidate intoxication, *see* Tabun intoxication
Ethyl-*N*-dimethyl phosphoramidofluoridate intoxication
chemical classification, 37/544
organophosphorus compound, 37/544
Ethyl-*N*-dimethyl phosphoramidothiocyanidate intoxication
chemical classification, 37/544
organophosphorus compound, 37/544
Ethylene chloride, *see* 1,2-Dichloroethylene
Ethylene diamine tetra-acetate, *see* Edetic acid
Ethylene glycol
biotransformation, 64/123
chemical property, 64/121
neurotoxin, 46/393, 62/601, 64/121
nomenclature, 64/121
production, 64/121
toxic myopathy, 62/601
use, 64/121
Ethylene glycol intoxication
acidosis, 64/125
alcohol, 64/125
anion gap, 64/122, 125
ataxia, 64/123
biochemistry, 64/122
brain edema, 64/123

calcium oxalate, 64/124
cardiopulmonary effect, 64/124
clinical features, 64/122
CNS effect, 64/123
coma, 64/123
computer assisted tomography, 64/123
convulsive seizure, 64/122
cranial nerve deficit, 64/124
CSF, 64/123
delayed neurological deficit, 64/124
diagnosis, 64/122, 124
diethylene glycol, 64/122
dysarthria, 64/123
EEG, 64/123
epidemiology, 64/122
epilepsy, 64/122
facioacoustic oxalosis, 64/124
fatality, 64/122
flaccid tetraplegia, 64/123
gaze paralysis, 64/123
glycolic acid, 64/122
glyoxylic acid, 64/122
half-life, 64/122
headache, 64/123
hemodialysis, 64/125
hemorrhagic encephalopathy, 64/124
hypocalcemia, 64/122
lactic acidemia, 64/122
lethargy, 64/123
Massengill disaster, 64/122
metabolism, 64/123
4-methylpyrazole, 64/125
multiple cranial neuropathy, 64/124
necrotic encephalopathy, 64/124
nystagmus, 64/123
obtundation, 64/123
oxalate, 64/122
oxalosis, 64/124
Purkinje cell, 64/124
putaminal necrosis, 64/124
renal effect, 64/124
serum level, 64/122
slurred speech, 64/123
striatal necrosis, 64/124
stupor, 64/123
suicide, 64/122
sulfanilamide elixir, 64/122
symptom stage, 64/122
tetany, 64/122
TLV, 64/122
treatment, 64/125
vertigo, 64/123
Ethylene oxide
ataxia, 51/275
neurotoxicity, 51/275
toxic polyneuropathy, 51/275
Ethylene oxide encephalopathy
acute exposure, 51/275
Ethylene oxide polyneuropathy
axonal degeneration, 51/275
EMG, 51/275
encephalopathy, 51/275
gait disturbance, 51/275
hemodialysis, 51/275
sensorimotor neuropathy, 51/275
sterilization equipment, 51/275
*O*-Ethyl *S*-(2-ethylthioethyl) ethylphosphonothioate intoxication
chemical classification, 37/548
organophosphorus compound, 37/548
Ethylhydrocupreine
blindness, 13/220
visual disorder, 13/220
Ethyl iodoacetate
glycolysis, 13/210
*N*-Ethylmaleimide sensitive fusion attachment protein 25
soluble, *see* Synaptosomal associated protein 25
Ethylmalonic aciduria
hypoglycemia, 37/535
Jamaican vomiting sickness, 37/533, 535
toxicity, 37/535
Ethyl-4-nitrophenyl ethylphosphonate intoxication
chemical classification, 37/545
organophosphorus compound, 37/545
Ethylnitrosourea
animal experiment, 18/171
brain tumor carcinogenesis, 17/19
cerebellar medulloblastoma, 18/171
cerebellar tumor, 18/171
spongiform leukoencephalopathy, 47/570
*N*-Ethyl-1-phenylcyclohexylamine, *see* Eticyclidine
Eticyclidine
designer drug, 65/364
Etidronic acid
osteitis deformans (Paget), 38/368
Etiocholanolone
fever, 43/68
periodic fever, 43/68
Etiologic heterogeneity
CNS malformation, 30/87-89
spina bifida, 50/479
Etoglucid
neuropathy, 7/517, 531
Etomidate
brain infarction, 53/419

status epilepticus, 73/331
Etoposide
adverse effect, 69/494
antineoplastic agent, 65/528
chemotherapy, 67/284
encephalopathy, 65/528
neuropathy, 69/460
neurotoxin, 65/528, 67/363
polyneuropathy, 65/528
Etorphine
opiate, 65/350
Euchromatin
chromosome, 30/94
Eukaryotic gene
genetic structure, 66/70
molecular biology, 66/69
Eulenberg paramyotonia congenita, *see* Paramyotonia congenita
Eunuchoidism
bulbar paralysis, 60/658
choreoathetosis, 60/663
Matthews-Rundle syndrome, 60/575
mental deficiency, 60/664
pallidocerebello-olivary degeneration, 60/658, 663
Euphoria
amphetamine intoxication, 65/257
anterior cerebral artery syndrome, 53/349
basofrontal syndrome, 53/349
Behçet syndrome, 56/598
diamorphine intoxication, 65/352
frontal lobe syndrome, 2/738, 45/24-26, 53/349
frontal lobe tumor, 46/429
general paresis, 46/429
lidocaine intoxication, 37/450
mercury polyneuropathy, 51/272
migraine, 48/162
multiple sclerosis, 9/177, 46/429, 47/58, 178
nalorphine, 37/366
organic solvent intoxication, 64/42
pathologic laughing and crying, 3/357, 45/220
procainamide intoxication, 37/452
progressive supranuclear palsy, 49/243
Russell syndrome, 16/349
sea snake intoxication, 37/92
thalamic syndrome (Dejerine-Roussy), 2/484
Wernicke-Korsakoff syndrome, 45/194
Eupnea
classification, 1/657
European typhus, *see* Louse-borne typhus
Eustachius, B., 2/9
Event related potential
*see also* Readiness potential
Huntington chorea, 49/290
juvenile hereditary benign chorea, 49/339
Evoked auditory response, *see* Auditory evoked potential
Evoked brain stem response
schizophrenia, 46/500
Evoked cortical response
*see also* Visual evoked response
occipital response, 2/544
retinal degeneration, 13/22
retinal response, 2/512
schizophrenia, 46/499-504
somatosensory, *see* Somatosensory evoked potential
Evoked muscle action potential
botulism, 41/290
butyrylcholinesterase deficiency, 41/291
decremental response, 41/113
Eaton-Lambert myasthenic syndrome, 41/353
tick paralysis, 41/291
Evoked potential
aphasia, 45/298
auditory, *see* Auditory evoked potential
Behçet syndrome, 56/601
brain concussion, 57/182
brain death, 55/218, 264
brain infarction, 53/123
brain ischemia, 53/123
brain lacunar infarction, 54/254
brain stem auditory, *see* Brain stem auditory evoked potential
brain stem death, 57/469
central conduction time, 57/196
cerebellar cortex, 2/410
cerebellar stimulation, 2/409
cerebrovascular disease, 54/157
conditioned, *see* Conditioned evoked potential
ERG, 57/198
head injury, 57/112, 193-199, 385
head injury outcome, 57/385
hereditary periodic ataxia, 60/438-441
hereditary sensory and autonomic neuropathy type III, 75/151
human T-lymphotropic virus type I associated myelopathy, 56/537
Huntington chorea, 49/289
lateralization, 46/503
1-methyl-4-phenyl-1,2,3,6-tetrahydropyridine intoxication, 65/378
migraine, 48/29
motor, *see* Motor evoked potential
multimodality, 57/198, 385
multiple sclerosis, 47/66, 131-143
myelinoclastic diffuse sclerosis, 47/424

neurologic intensive care, 55/217
schizophrenia, 46/503-505
sensory, 57/385
short latency, *see* Short latency evoked potential
somatosensory, *see* Somatosensory evoked potential
spinal cord, 1/51
spinal cord experimental injury, 25/10
spinal nerve root injury, 25/422
technical problem, 57/181
tropical spastic paraplegia, 56/537
visual, *see* Visual evoked response
Evoked visual response, *see* Visual evoked response
Ewing sarcoma
brain metastasis, 69/124
cranial vault tumor, 17/119
differential diagnosis, 8/470
dural metastasis, 69/124
incidence, 68/530
ivory vertebra, 20/35
skull base, 68/466
skull base tumor, 17/169
spinal cord, 67/197
spinal cord compression, 19/368
spinal cord tumor, 68/525
vertebral column, 20/34
Exacerbation
acute intermittent porphyria, 51/390
congenital acetylcholine receptor deficiency/synaptic cleft syndrome, 62/438
familial infantile myasthenia, 62/427
guanethidine intoxication, 37/436
multiple sclerosis, 47/149
papaverine intoxication, 37/455
periodic ataxia, 60/650
Exanthematous encephalomyelitis
mortality, 9/547
mumps meningitis, 9/546
sequela, 9/547
Excessive CSF protein syndrome
muscular atrophy, 42/545
Excessive sweating
fucosidosis, 51/381
Excitability
axon, 51/46
intrinsic membrane, 72/43
nerve cell, 72/43
Excitability disorder
malabsorption syndrome, 70/227
Excitation
cerebellar function, 2/413
dysphrenia hemicranica, 5/79
epilepsy, 72/47
migraine, 48/161
nerve cell, 72/47
Excitation contraction coupling
calcium, 40/130
cross bridge, 40/130
malignant hyperthermia, 40/546
membrane, 40/549
myotonia, 40/556
myotonic disorder, 40/556
myotonic dystrophy, 40/523
paramyotonia congenita, 40/556
potassium channel, 40/550
transverse tubular system, 40/130
troponin, 40/130
Excitatory postsynaptic potential
cerebellar stimulation, 2/412
characteristics, 1/49
organochlorine insecticide intoxication, 64/200
synapse property, 1/55
Excitatory state
central, *see* Central excitatory state
Excitotoxic transmitter
striatal necrosis, 49/500
Excitotoxin
α-amino-3-hydroxy-5-methyl-4-isoxazolepropionic acid, 64/20
2-amino-3-methylaminopropionic acid, 64/20
glutamic acid, 64/20
kainic acid, 64/20
ketamine, 64/20
*N*-methyl-D-aspartic acid, 64/20
mode of action, 64/19
neurologic intensive care, 55/211
neurotoxicology, 64/19
neurotoxin, 55/211, 64/19
phencyclidine, 64/20
quinolinic acid, 64/20
Excretion
acrylamide intoxication, 64/72
autonomic nervous system, 74/1
bilirubin, 27/417
bromide intoxication, 36/293
glycosaminoglycan, *see* Glycosaminoglycan excretion
hexachlorophene intoxication, 37/482
lead intoxication, 36/4
organotin intoxication, 64/135
platinum intoxication, 64/357
porphyrin metabolism, 27/440-442
steroid, *see* Steroid excretion
stimulant, 37/237-241
thallium intoxication, 64/324
thiocyanate, 7/642

vanilmandelic acid, 8/478
Exencephaly
 *see also* Rachischisis
 anencephaly, 30/173, 187, 32/179
 brain cortex disorganization, 50/249
 brain tissue, 50/99
 CNS malformation, 32/171, 50/57
 encephalocele, 50/71, 73
 iniencephaly, 50/71, 73
 neurulation defect, 50/97
 periventricular calcification, 50/258
 postneurulation defect, 50/97
 rat, 50/28
 type, 50/71
Exercise
 acquired myoglobinuria, 62/572
 anterior tibial syndrome, 12/538, 20/798, 41/385
 arterial baroreflex, 74/251, 256
 arterial hypertension, 54/206
 autonomic nervous system, 74/14, 81, 245
 baroreflex, 75/439
 blood flow, 75/438
 blood pressure, 74/253, 75/716
 breathlessness, 74/579
 cardiovascular response, 74/245
 carnitine palmitoyltransferase deficiency, 62/494
 central command, 74/263
 experimental model, 40/148
 flexion contracture, 20/828
 hypotension, 75/716
 intensity, 74/247
 ischemic, *see* Ischemic exercise
 isometric, *see* Isometric exercise
 J reflex, 74/577
 Kegel, *see* Kegel exercise
 muscle fiber hypertrophy, 40/148
 muscle mass, 74/251
 muscle metaboreflex, 74/255, 257
 muscular dystrophy, 41/463
 myophosphorylase deficiency, 27/233
 myotonic dystrophy, 62/249
 myotubular myopathy, 40/15, 200
 neuromuscular block, 74/251
 neuropathy, 8/393
 neuropeptide, 74/11
 oxygen, 74/247
 rehabilitation, 41/463
 respiration, 8/395
 respiratory encephalopathy, 63/415
 response, 74/232
 space, 74/279
 spinal cord, 74/257
 sympathetic nervous system, 74/246
 syncope, 75/217
 thermoregulation, 74/263
 training, 74/261
 transient global amnesia, 45/209
 vagus nerve, 75/438
 vestibular system, 74/262
Exercise intolerance
 aconitate hydratase deficiency, 62/503
 carnitine palmitoyltransferase deficiency, 43/176
 complex I deficiency, 66/422
 complex II deficiency, 62/504
 complex III deficiency, 62/504, 66/423
 congenital acetylcholine receptor deficiency/synaptic cleft syndrome, 62/438
 familial, *see* Familial exercise intolerance
 glutaric aciduria type II, 62/495
 glycogen storage disease, 43/181, 183
 lactate dehydrogenase deficiency, 62/490
 mitochondrial myopathy, 43/188
 multiple acyl-coenzyme A dehydrogenase deficiency, 62/495
 multiple mitochondrial DNA deletion, 62/510
 muscle carnitine deficiency, 66/403
 myophosphorylase deficiency, 41/185, 62/485
 oxidation-phosphorylation coupling defect, 62/506
 phosphofructokinase deficiency, 27/235, 41/190, 43/183, 62/487
 phosphoglycerate kinase deficiency, 62/489
 phosphoglycerate mutase deficiency, 62/489
 phosphorylase kinase deficiency, 62/487
 pyruvate decarboxylase deficiency, 62/502
 ragged red fiber, 62/292
 ubidecarenone, 62/504
 ubidecarenone deficiency, 66/423
Exercise myoglobinuria
 myophosphorylase deficiency, 62/485
Exertional headache
 basilar impression, 48/374
 benign, 48/375
 brain tumor, 48/374
 classification, 48/6
 coitus, 48/377
 cough headache, 48/375
 Cushing disease, 48/375
 definition, 48/10
 differential diagnosis, 48/347, 373
 etiology, 48/376, 380
 hydrocephalus, 48/374
 hyperthyroidism, 48/375
 intracranial pressure, 5/148, 48/374
 meningitis, 48/374
 orgasm, 48/380

pheochromocytoma, 48/375
pulmonary disease, 48/375
sexual activity, 5/148, 48/379
subarachnoid hemorrhage, 48/374
treatment, 48/376, 381
Valsalva maneuver, 48/376
vascular abnormality, 48/374
Exertional myalgia
incomplete dystrophin deficiency, 62/135
Exertional paroxysmal dystonic choreoathetosis
hereditary, *see* Hereditary exertional paroxysmal dystonic choreoathetosis
Exhaustion
birgus latro intoxication, 37/58
heat, *see* Heat exhaustion
postactivation, *see* Postactivation exhaustion
Exner plexus
status verrucosus, 30/349
Exocytosis
bretylium, 74/145
muscle fiber, 62/18
sympathetic nerve terminal, 74/143
Exogenous poison
delirium, 46/545
Exogenous reaction type
delirium, 46/523
Exoglycosidase
degradation, 66/61
Exomphalos
hydromyelia, 50/427
Exomphalos macroglossia gigantism syndrome, *see* Beckwith-Wiedemann syndrome
Exophthalmic ophthalmoplegia, *see* Thyroid associated ophthalmopathy
Exophthalmos
acrocephalosyndactyly type I, 38/422, 43/318
acrocephalosyndactyly type V, 43/320
Bonnet-Dechaume-Blanc syndrome, 14/261
carotid cavernous fistula, 24/414
cavernous sinus thrombosis, 52/173
chloroma, 18/252, 63/343
chordoma, 18/155
craniofacial dysostosis (Crouzon), 50/122
craniosynostosis, 50/125
diaphyseal dysplasia, 43/376
dysthyroid ophthalmoplegia, 22/184
goiter, 2/320
Gruber syndrome, 14/115
Hand-Schüller-Christian disease, 38/95
Horner syndrome, 19/59
hypophosphatasia, 42/578
Melnick-Needles syndrome, 43/452
multiple myeloma, 18/256, 63/393
multiple neurofibromatosis, 14/398
neuroblastoma, 42/758
ocular imbalance, 24/89
ophthalmoplegia, 2/320
orbital cellulitis, 33/179
9p partial monosomy, 50/578
pulsating, *see* Pulsating exophthalmos
r(9) syndrome, 43/513
self-mutilation encephalopathy, 42/141
sphenoid wing aplasia, 50/143
superior orbital fissure syndrome, 33/179
supratentorial midline tumor, 18/339
thyroid gland, 41/240
traumatic, *see* Traumatic exophthalmos
traumatic carotid cavernous sinus fistula, 57/352
trichinosis, 52/567
tuberculoma, 18/418
tumor, 68/77
unilateral, *see* Unilateral exophthalmos
vitamin A intoxication, 37/96
Exosomatesthesia
body scheme disorder, 4/233, 242
sensory extinction, 3/191
Exostosis
hereditary multiple, *see* Hereditary multiple exostosis
multiple, *see* Hereditary multiple exostosis
vertebral, *see* Vertebral exostosis
Exotropia, *see* Divergent strabismus
Expanding hemangioma, *see* Aneurysmal bone cyst
Expectancy
brain command system, 45/273
life, *see* Life expectancy
Experimental acute viral myositis
bluetongue virus, 56/195
Chikungunya virus, 56/195
Coxsackie virus A2, 56/195
Coxsackie virus A4, 56/195
Coxsackie virus A9, 56/195
Coxsackie virus B2, 56/195
ECHO virus 9, 56/195
encephalomyocarditis virus, 56/195
influenza A virus, 56/195
picornavirus, 56/195
poliovirus, 56/195
rabies virus, 56/195
reovirus type 1, 56/195
reovirus type 2, 56/195
reovirus type 3, 56/195
retrovirus, 56/196
Ross river virus, 56/195
Semliki forest virus, 56/195
simian acquired immune deficiency virus type D,

56/195
Sindbis virus, 56/195
St. Louis encephalitis virus, 56/195
Theiler virus, 56/195
tick-borne encephalitis virus, 56/195
Togavirus, 56/196
Venezuelan equine virus, 56/195
vesicular stomatitis virus, 56/195
West Nile virus, 56/195
Experimental allergic encephalomyelitis, 47/429-452
*see also* Experimental demyelination
acid mucopolysaccharide, 9/38
active sensitization, 47/433
acute, *see* Acute experimental allergic encephalomyelitis
alkaline phosphatase, 9/38
animal study, 47/434-437
antibody, 9/136, 47/445
antibody mediated process, 47/489
antigen, 9/135-137
antimyelin basic protein, 47/446
autoimmune disease, 9/506-536
B-lymphocyte, 47/434, 442
blocking factor, 47/41
blood-brain barrier, 9/523, 47/438
Bordetella pertussis, 9/526, 47/433
cell mediated process, 47/489
Chédiak-Higashi syndrome, 9/526
chronic relapsing, *see* Chronic relapsing experimental allergic encephalomyelitis
chronic silent lesion, 47/449
CNS culture, 47/446
delayed hypersensitivity, 9/504, 47/429
demyelination, 9/38, 684-687, 47/445
diagnostic criteria, 47/430
disseminated encephalomyelitis, 47/467-470
encephalitogenic proteolipid, 9/13
experimental allergic neuritis, 47/467, 471-474
experimental demyelination, 9/684-687
Freund adjuvant, 9/517, 522, 47/430, 451
guinea pig, 47/468
history, 47/429
human, 9/140
hyperacute, 47/470
hyperacute disseminated encephalomyelitis, 9/526-529
IgG, 9/520
IgM, 9/520
immune response, 47/437
induction, 47/432-434, 473
inflammatory cell, 47/230
Lewis rat, 47/468
lymphocyte, 9/503, 517, 47/441-445, 449
macrophage, 9/503, 47/438
metachromasia, 9/38
monkey, 47/468
morphology, 47/449
multiple sclerosis, 9/60, 126, 135, 47/320, 337, 429-432, 447-450, 485, 488
myelin associated glycoprotein, 47/450
myelin basic protein, 10/135, 47/234, 430-433, 442, 446, 450, 468, 474
myelinoclasis, 9/656
nerve conduction block, 47/367
opsonin, 9/520
passive sensitization, 47/434
Pasteur rabies vaccine, 47/429, 609
pathogenesis, 9/522
pathology, 47/447-450
phospholipase, 9/38
phytohemagglutinin, 9/531
plasma cell, 47/442
proteolipid, 9/11
rabbit, 47/468
resistance, 9/520, 529-531
suppression, 47/450-452, 458
symptomatology, 47/434-437
T-lymphocyte, 47/234, 434, 442, 449
transfer, 9/136
wallerian degeneration, 9/38
Experimental allergic myositis
antibody formation, 40/165
inflammatory myopathy, 40/163-167
muscle tissue culture, 40/165
polymyositis, 40/165
Experimental allergic neuritis, 47/452-458
*see also* Experimental demyelination
antibody mediated process, 47/489
autoantigen, 9/505
cell mediated process, 47/489
chronic inflammatory demyelinating polyradiculoneuropathy, 51/537
delayed hypersensitivity, 47/429
demyelinating disease, 9/531-536
demyelination, 9/505
experimental allergic encephalomyelitis, 47/467, 471-474
foreign antigen, 9/505
Freund adjuvant, 9/506, 47/454
Guillain-Barré syndrome, 47/452, 458, 489, 611, 51/243, 252, 254
history, 47/452
hypertrophic interstitial neuropathy, 21/163
induction, 47/453
inflammation, 9/505, 47/457

lymphocyte, 9/531-536, 47/457
macrophage, 47/454-458
multifocal, 9/505, 531-536
myelin basic protein, 47/474
nerve conduction, 7/178
onion bulb formation, 21/252
pathology, 47/454-456
peripheral nerve, 9/531-536
perivenous, 9/531-536
protein P2, 47/453, 471
remyelination, 47/457
Schwann cell, 9/531-536
segmental demyelination, 51/70
symptomatology, 47/454

Experimental autoimmune encephalomyelitis
immunosuppression, 47/191

Experimental autoimmune myasthenia gravis
acetylcholine receptor, 41/97, 115-120
acetylcholine receptor antibody, 41/115
animal specy, 41/115
antibody formation, 40/163
decremental EMG, 41/117
electrophysiology, 41/117
lymphorrhage, 41/110
motor end plate, 41/115
muscle cell membrane, 41/115
pathogenesis, 40/163
synaptic vesicle, 41/115

Experimental brain tumor
brain tumor carcinogenesis, 16/40-43
morphologic study, 16/35
nitrosamine, 16/42
ultrastructure, 16/35

Experimental catatonia
bulbocapnine, 21/527

Experimental demyelination
*see also* Experimental allergic encephalomyelitis *and* Experimental allergic neuritis
animal model, 47/32
anoxic ischemic leukoencephalopathy, 47/530
cause, 47/29
conduction block, 47/30
cuprizone, 47/563
experimental allergic encephalomyelitis, 9/684-687
pathomechanism, 47/489-493
Schwann cell, 47/7
triethyltin, 47/564
type, 47/29

Experimental head injury
EEG, 23/76
impact, 23/98
mechanics, 23/67
model, 23/73

Experimental injury
air embolism, 23/616
boxing injury, 23/536, 557-562
cytoplasmic inclusion body myopathy, 62/345
dysbarism, 23/616
fat embolism, 23/634
laser injury, 23/665
ligament, 25/2, 6
noradrenalin, 25/13
radiation injury, 23/639
spinal cord, *see* Spinal cord experimental injury
spinal cord conduction, 25/11
spinal nerve root, 25/394
vertebral, *see* Vertebral experimental injury
vertebral column, *see* Spine experimental injury

Experimental ischemic myopathy
capillary change, 41/71
catecholamine, 40/167-170

Experimental model
Alzheimer disease, 27/487, 46/269-272
brain ischemia, 27/27-29
congenital myopathy, 62/358
exercise, 40/148
Guam amyotrophic lateral sclerosis, 59/265
head injury, 23/70, 76
head injury mechanism, 57/86
hereditary amyloid polyneuropathy, 60/109
missile injury, 57/301
Pick disease, 46/243
trophic interaction, 40/138-143

Experimental myasthenia gravis
acetylcholine, 41/120
α-bungarotoxin, 41/115
experimental myopathy, 41/97, 110, 115-120
miniature motor end plate potential, 41/117
muscle cell membrane, 41/119
Naja naja α-toxin, 41/113
neuromuscular block, 41/120
neuromuscular transmission, 41/120

Experimental myopathy, 40/133-171
acetylcholine mediated, 40/150, 163
alcohol, 40/162
calcavin, 40/160
carbonylcyanide-meta-chlorophenylhydrazone, 40/158
carnitine deficiency, 41/200
chloramphenicol, 40/158
chloroquine, 40/151
colchicine, 40/154
denervation, 40/140
dermatomyositis, 40/164
developing muscle, 40/146

2,4-dichlorophenoxyacetic acid, 40/160
dimethyl-*p*-phenylenediamine, 40/160
2,4-dinitrophenol, 40/158
disuse, 40/143, 200
emetine, 40/157
experimental myasthenia gravis, 41/97, 110, 115-120
experimental myositis, 40/163, 41/76, 78, 377
hemicholinium, 40/150
increased usage, 40/148
ischemic, 40/167, 41/71
malignant hyperthermia, 41/269
mitochondria, 40/138, 158-160
muscle fiber type I, 40/136
myofilamentous degeneration, 40/157
myotonia, 40/154
paraoxon, 40/150
phospholipase C, 40/151
reinnervation, 40/139
serotonin induced, 41/326
tenotomy, 40/143
tri-*o*-cresyl phosphate, 40/154
triethyltin sulfate, 41/5
vincristine, 40/154
vitamin E deficiency, 40/170
Experimental myositis
experimental myopathy, 40/163, 41/76, 78, 377
Experimental myotonia
azacosterol, 40/155
2,4-dichlorophenoxyacetic acid, 40/154
toxic myopathy, 62/597
triparanol, 40/155
Experimental pigmentary retinopathy
chemical injection, 13/203
nutritional deficiency, 13/203
parenteral retinotoxic drug, 13/204
posterior ciliary artery obliteration, 13/202
quinoline, 13/150, 219
retinotoxic agent, 13/204
toxic agent, 13/204-213
vitreous body, 13/203
Experimental spongiform leukoencephalopathy
isoniazid intoxication, 47/565
Experimental viral meningoencephalitis
CSF immunoperoxidase staining, 47/83
Exposures
radiation, *see* Radiation exposure
Expression
construction, 45/524
emotion, 3/320, 337, 356, 358
facial, *see* Facial expression
facial paralysis, 1/421
language, 45/523
manual, *see* Manual expression
neuropsychologic defect, 45/523
paraneoplastic syndrome, 69/333, 337
practic ability, 45/523
speech, 45/523
writing, 45/523
Expressive aphasia, *see* Motor aphasia
Extended arms test
labyrinthine disease, 1/338
Extensibility
motor activity, 4/447
Extensibility test
André-Thomas, *see* André-Thomas extensibility test
Extensor plantar response
clinical features, 1/248
cri du chat syndrome, 50/275
infantile Gaucher disease, 10/307, 42/438, 66/124
Lafora progressive myoclonus epilepsy, 27/173
metachromatic leukodystrophy, 42/493
Minamata disease, 64/420
neuropathy, 59/404
phenylketonuria, 29/30
postpoliomyelitic amyotrophy, 59/37
Wohlfart-Kugelberg-Welander disease, 22/71, 59/372
Extensor reflex
clinical features, 1/173
crossed, *see* Crossed extensor reflex
spinal automatism, 1/251
Extensor retinaculum
anterior tarsal tunnel syndrome, 51/105
Extensor spasm
anatomic classification, 1/279
Extensor thrust reflex
paleocerebellar syndrome, 2/418
External auditory stenosis
cryptophthalmos syndrome, 43/374
External carotid artery
absence, 12/291
anomalous branch, 12/291
carotid cavernous fistula, 24/407, 420
carotid rete mirabile, 12/308-311
Doppler sonography, 54/6
lateral position, 12/291
radioanatomy, 11/65
supply area, 53/298
temporal arteritis, 48/322
traumatic carotid cavernous sinus fistula, 57/353
External choroidal artery
anatomy, 11/26
External granular layer
cerebellar, *see* Cerebellar external granular layer

External jugular vein
anatomy, 11/55
External maxillary artery
anastomosis, 12/308
External ophthalmoplegia
ataxia, 43/136
Basedow disease, 22/179
Bassen-Kornzweig syndrome, 60/656
bulbar paralysis, 22/140
chorioretinal degeneration, 13/40
chronic progressive, *see* Chronic progressive external ophthalmoplegia
congenital myopathy, 62/333
dentatorubropallidoluysian atrophy, 60/656
dialysis encephalopathy, 63/528
Fazio-Londe disease, 22/140
Ferguson-Critchley syndrome, 59/319
hypertrophic interstitial neuropathy, 21/152
hypoglossal nerve agenesis, 50/220
Kearns-Sayre-Daroff-Shy syndrome, 60/655
Machado-Joseph disease, 60/656
neuronal intranuclear hyaline inclusion disease, 60/655
nuclear aplasia, 22/179
olivopontocerebellar atrophy (Dejerine-Thomas), 60/656
olivopontocerebellar atrophy variant, 21/452
optic atrophy, 13/40
phenytoin intoxication, 65/500
progressive, *see* Progressive external ophthalmoplegia
progressive bulbar palsy, 22/140
progressive supranuclear palsy, 22/181, 221
Refsum disease, 13/40, 22/212
secondary pigmentary retinal degeneration, 13/179
Shy-Drager syndrome, 1/474
superior orbital fissure syndrome, 33/179
Wohlfart-Kugelberg-Welander disease, 22/141
External otitis, *see* Otitis externa
External pallidum
choreoathetosis, 49/457
Extinction
motor, *see* Motor extinction
sensory, *see* Sensory extinction
somatosensory, *see* Somatosensory extinction
visual, *see* Visual inattention
Extinction phenomenon, *see* Tactile inattention
Extinction syndrome
sensory, *see* Sensory extinction
Extra pontine myelinolysis
biphasic course, 63/548
brain stem auditory evoked potential, 63/553
cerebellar, 63/548
computer assisted tomography, 63/552
CSF, 63/553
EEG, 63/553
hyponatremia, 63/548
hyponatremia correction, 63/548
lateral geniculate, 63/548
management, 63/553
mesencephalic, 63/548
neostriatal, 63/548
nuclear magnetic resonance, 63/552
pathogenesis, 63/553
recovery, 63/552
symptom, 63/548
thalamocapsular, 63/548
urea infusion, 63/554
Extracampine hallucination
hallucinosis, 46/562
Extracellular potassium
demyelination, 47/36
Extracellular space
brain, 56/81
brain edema, 56/81
Ranvier node, 7/22, 47/11
Extracephalic abnormality
holoprosencephaly, 50/236
Extracranial blood flow
migraine, 48/78
Extracranial insult
head injury outcome, 57/125
hypotension, 57/125
hypoxia, 57/125
Extradural abscess, *see* Epidural empyema
Extradural arachnoid cyst, *see* Epidural arachnoid cyst
Extradural empyema, *see* Epidural empyema
Extradural ganglion cyst, *see* Epidural ganglion cyst
Extradural hematoma, *see* Epidural hematoma
Extradural Nocardia granuloma
spinal cord compression, 52/449
Extradural spinal abscess, *see* Spinal epidural empyema
Extradural spinal lipoma, *see* Spinal epidural lipoma
Extradural tumor, *see* Epidural tumor
Extradural venous plexus, *see* Epidural venous plexus
Extralamellar thalamic nucleus
nuclear group, 2/470
principal reticular nucleus, 2/473
reticularis polaris nucleus, 2/473
Extramedullary compression
posttraumatic syringomyelia, 26/154
Extramedullary spinal tumor, 19/51-72

Brown-Séquard syndrome, 19/30, 59
duration, 19/54
Horner syndrome, 19/59
myelography, 19/65, 192
radicular pain, 19/55
syringomyelia, 50/456
Extraocular motor nucleus
infantile spinal muscular atrophy, 59/62, 67
Extraocular muscle
carotid cavernous fistula, 24/415
embryology, 30/397
innervation, 30/397
Extraocular muscle paralysis
puffer fish intoxication, 37/80
tetrodotoxin intoxication, 65/156
Extraocular muscle weakness
*see also* Progressive external ophthalmoplegia
botulism, 41/289
congenital myopathy, 62/334
myasthenia gravis, 41/98, 62/399
myotubular myopathy, 41/11
Extraocular palsy
chronic progressive external ophthalmoplegia, 2/323
diphtheria, 51/183
Extrapyramidal disorder
*see also* Hyperkinesia
antihistaminic agent, 6/828
benactyzine, 6/827
benzatropine methanesulfonate, 6/828
boxing injury, 23/541, 554
bradykinesia, 2/344
caramiphen, 6/827
chlorphenoxamine, 6/828
chronic mercury intoxication, 64/370
corpus callosum tumor, 17/511
10,11-dihydro-5H-dibenz[*b*,*f*]azepine, 6/829
drug induced, 6/248-266
drug treatment, 6/826-843
emotional disorder, 3/358
encephalitis lethargica, 34/455
familial spastic paraplegia, *see* Hereditary dystonic paraplegia
frontal lobe tumor, 17/257
furor bulborum, 2/344
gaze spasm, 2/344
Hallervorden-Spatz syndrome, 66/714
hypoparathyroidism, 27/305-310
indication, 6/848
MAO inhibitor, 6/262
Minamata disease, 7/642
myoclonus, 2/344
narcolepsy, 2/344
occipital lobe tumor, 17/338
ocular bobbing, 2/345
oculogyric crisis, 2/344
opsoclonus, 2/344
pellagra, 28/90
surgical treatment, 6/844-878
thephorin, 6/828
tricyclic derivative, 6/263
trihexyphenidyl, 6/827
valproic acid intoxication, 37/202
Extrapyramidal hyperkinesia
Benedikt syndrome, 2/298
congenital Pelizaeus-Merzbacher disease, 10/158, 42/503
Down syndrome, 50/526
Extrapyramidal myoclonus
epilepsy, 6/766
Extrapyramidal sign
Cockayne-Neill-Dingwall disease, 13/433
familial spastic paraplegia, *see* Ferguson-Critchley syndrome
frontal lobe tumor, 17/257
Gerstmann-Sträussler-Scheinker disease, 60/622
hereditary spastic ataxia, 21/16
linear nevus sebaceous syndrome, 43/38
parkinsonism, 49/187
spastic paraplegia, 42/175
Extrapyramidal syndrome
agraphia, 45/464
amyotrophic lateral sclerosis, 22/315, 59/203
Bardet-Biedl syndrome, 13/401, 43/233
conditional responsive, *see* Hereditary paroxysmal kinesigenic choreoathetosis
congenital retinal blindness, 42/400-402
drug induced, *see* Drug induced extrapyramidal syndrome
Pick disease, 6/668, 46/234, 240
progressive supranuclear palsy, 22/219, 46/301, 64/5
Extrapyramidal system
carbamazepine intoxication, 37/203
cerebellar ataxia, 42/126
Hallervorden-Spatz syndrome, 22/403
motor activity, 1/299
neuroanatomy, 1/157-159
phenytoin intoxication, 37/203
primidone intoxication, 37/203
Spatz concept, 6/64
Exudative enteropathy
immunocompromised host, 56/469
Eye
*see also* Vision
acrocephalosyndactyly type II, 31/236

Aicardi syndrome, 31/227
antimongoloid, *see* Antimongoloid eyes
apraxia, *see* Oculomotor apraxia
autonomic nervous system, 74/399
cholinergic receptor blocking agent, 74/418
Cockayne-Neill-Dingwall disease, 31/238
conjugate deviation, 1/284, 2/586
fatigue, 5/35, 204, 206
glycosaminoglycanosis, 10/452
headache, 5/35, 204, 206, 48/6
hereditary amyloid polyneuropathy type 1, 60/95
hereditary amyloid polyneuropathy type 2, 60/98
leukemia, 39/13
mongoloid, *see* Mongoloid eyes
Niemann-Pick disease type A, 10/496
pain, 5/330
panda, *see* Panda eyes
pharmacology, 74/418
Refsum disease, 21/244
sympathetic innervation, 5/335
toxocariasis, 52/511
Eye apraxia, *see* Oculomotor apraxia
Eye convergence
deficiency, 1/607
disorder, 2/667
eye movement, 1/606
gaze apraxia, 45/406
ocular imbalance, 24/75
paralysis, 1/607, 2/341
pathology, 2/340-342
spasm, 1/607, 2/341
test, 2/33
weakness, 24/75
Eye dryness
acute pandysautonomia, 51/475
Eye fixation
eye movement, 1/583, 599
occipital lobe tumor, 17/324, 329
ocular bradykinesia, 1/599
progressive supranuclear palsy, 49/241
spasm, 1/583, 599
Eye-hand discoordination, *see* Ocular ataxia
Eye injury
cyanea intoxication, 37/52
hydrozoa intoxication, 37/40
spinal injury, 26/190
Eye lesion
Goldenhar syndrome, 31/247
Hallermann-Streiff syndrome, 31/248
incontinentia pigmenti, 14/214
Lowe syndrome, 31/301
multiple nevoid basal cell carcinoma syndrome, 14/465
Norrie disease, 31/291
progressive hemifacial atrophy, 31/253
Eye movement
*see also* Gaze movement *and* Nystagmus
abnormality, 1/581, 49/240
agnosia, 4/18
alexia, 4/123
anesthesia, 1/579
ataxia, 1/605
brain death, 24/720
central autonomic dysfunction, 75/169
climbing, *see* Climbing eye movement
clinical examination, 57/135
conjugate, *see* Conjugate eye movement
Creutzfeldt-Jakob disease, 46/291
developmental dyslexia, 46/114-116
dissociated, 1/604
divergence, 1/606, 608
drug induced, 1/580
dysconjugate, *see* Dysconjugate eye movement
dysmetric ataxia, 1/605
epilepsy, 1/582
examination, 1/575
eye convergence, 1/606
eye fixation, 1/583, 599
eye muscle impairment, 1/608
eyelid closure, 1/579, 613
facial paralysis, 1/613
head movement, 1/605
headache, 48/50
image fusion, 1/606
intracranial hypertension, 16/132
involuntary, *see* Involuntary eye movement
medial longitudinal fasciculus, 1/605
metabolic ataxia, 21/581
1-methyl-4-phenyl-1,2,3,6-tetrahydropyridine intoxication, 65/372
Minamata disease, 64/414
motor dissociation, 1/162
multiple neuronal system degeneration, 59/137, 75/169
ocular dyskinesia, 1/584
ocular periodicity, 1/585
oculomotor apraxia, 1/605, 14/653
oculomotor nerve, 1/576, 583
periodic movement, 1/585
pineal tumor, 17/656
posture, 1/575, 598, 3/24
progressive dysautonomia, 59/137
progressive supranuclear palsy, 22/219, 49/240, 243
reading epilepsy, 1/583
recording, 1/580

saccadic, *see* Opsoclonus
seesaw, 1/605
sensorimotor stroke, 54/253
skew deviation, 1/605
sleep, 1/579, 3/82
slow, *see* Slow eye movement
smooth pursuit, *see* Smooth pursuit eye movement
synkinesis, 1/610
terminology, 2/286
vestibular stimulation, 1/577
viscosity, 14/284
visual agnosia, 4/22, 30
visual stimulation, 1/576
Eye muscle
myopathy, 1/612
paralysis, 1/608
temporal arteritis, 48/314
thyroid dysfunction, 1/612
Eyelid
abnormal opening, 1/614
agnosia, 4/18
clonus, 2/319
closure reflex, 2/293
edema, 8/212
facial paralysis, 1/613
facial spasm, 1/618
movement impairment, 4/18
neurofibromatosis type I, 14/624
normal, 1/613
oculomotor paralysis, 1/614
palpebral reflex, 1/614
spasm, *see* Blepharospasm
twitch sign, 22/205
visual stimulation, 1/614
Eyelid apraxia
Huntington chorea, 49/279
progressive supranuclear palsy, 49/243
Eyelid coloboma
Goldenhar syndrome, 38/411, 43/442
Treacher-Collins syndrome, 43/421
Eyelid nystagmus
brain embolism, 53/213
brain infarction, 53/213
cerebrovascular disease, 53/213
intracranial hemorrhage, 53/213
lesion site, 53/213
transient ischemic attack, 53/213
Eyelid retraction
Ferguson-Critchley syndrome, 59/319
myogenic, 1/616
paradoxical, *see* Paradoxical eyelid retraction
pathologic, *see* Collier sign
spastic, 1/614
Eyelid spasm, *see* Blepharospasm